S0-DXD-659

COLLEGE MISERICORDIA LIBRARY
RECEIVED
DEC 10 1973
DALLAS, PENNSYLVANIA

RODD'S CHEMISTRY OF CARBON COMPOUNDS

ELSEVIER SCIENTIFIC PUBLISHING COMPANY
335 JAN VAN GALENSTRAAT
P.O.BOX 1270, AMSTERDAM, THE NETHERLANDS

AMERICAN ELSEVIER PUBLISHING COMPANY, INC.
52 VANDERBILT AVENUE, NEW YORK, N.Y. 10017

LIBRARY OF CONGRESS CARD NUMBER: 64-4605

ISBN 0-444-41093-7

WITH 45 TABLES

PRINTED IN THE NETHERLANDS

RODD'S CHEMISTRY OF CARBON COMPOUNDS

RODD'S CHEMISTRY OF CARBON COMPOUNDS

VOLUME I

GENERAL INTRODUCTION

ALIPHATIC COMPOUNDS

*

VOLUME II

ALICYCLIC COMPOUNDS

*

VOLUME III

AROMATIC COMPOUNDS

*

VOLUME IV

HETEROCYCLIC COMPOUNDS

*

VOLUME V

MISCELLANEOUS

GENERAL INDEX

*

RODD'S CHEMISTRY OF CARBON COMPOUNDS

A modern comprehensive treatise

SECOND EDITION

Edited by

S. COFFEY

M.Sc. (London), D.Sc. (Leyden), F.R.I.C.

formerly of

I.C.I. Dyestuffs Division, Blackley, Manchester

VOLUME IV PART A

HETEROCYCLIC COMPOUNDS

Three-, four- and five-membered heterocyclic compounds with a single hetero-atom in the ring

ELSEVIER SCIENTIFIC PUBLISHING COMPANY

AMSTERDAM LONDON NEW YORK

1973

CONTRIBUTORS TO THIS VOLUME

R. LIVINGSTONE, B.SC., PH.D., F.R.I.C.
Department of Pure and Applied Chemistry, The Polytechnic, Huddersfield.

R. E. FAIRBAIRN, B.SC., PH.D., F.R.I.C.
formerly of Research Department, Dyestuffs Division, I.C.I. Ltd.,
Manchester, 9 *(Index)*.

PREFACE TO VOLUME IV A

In Volumes II and III the chemistry of saturated and unsaturated carbocyclic compounds is discussed; Volume IV deals with *heterocyclic* compounds, *i.e.* compounds containing hetereogenous rings (heterocycles) in which there is incorporated into the ring one or more elements other than carbon, the most common ones being oxygen, sulphur and nitrogen. The majority of physiologically and biologically important compounds are heterocyclic and the investigation of natural products, the chemistry of vital processes along with the search for improved pharmaceutical products, has so stimulated the study of heterocyclic compounds that the literature on this branch of organic chemistry now exceeds that on all other branches.

The plan followed in the presentation of Volume IV is essentially that used in the first edition. Sometimes, however, for the readers' convenience the original chapters have been split up into two or more, and they are regrouped to conform more nearly to the general pattern followed in Volumes I, II and III of the present edition. The introductory chapter on nomenclature in Volume IV A of the first edition is now omitted; questions of naming and numbering the various systems are discussed as they arise in the text. In accordance with I.U.P.A.C. Rules (**B** 1–11 and **C** 14.1, the main points of which are reproduced as an appendix to this volume), the most unsaturated form of each ring system is discussed first, followed by dihydro-, tetrahydro-, etc., derivatives and by compounds with a homocyclic ring fused to the parent heterocycle.

Volume IV A deals systematically with the chemistry of heterocyclic compounds in which there are present three-, four- or five-membered rings containing a single hetero-atom. Chapter 1 is concerned with compounds of all three types having three- and four-membered heterocyclic rings; Chapters 2 and 3 with those having five-membered rings containing an element from Group VI of the Periodic Table, respectively, oxygen and one of the remaining elements (sulphur, selenium or tellurium). Chapter 4 deals with five-membered heterocycles containing nitrogen (Group V) and 5 with fused-ring compounds of the same type, while Chapter 6 records five-membered heterocyclic compounds containing an element from Groups III, IV and V (other than N).

The whole of the revised text of this first sub-volume has been prepared by Dr. R. LIVINGSTONE. The editor takes this opportunity to congratulate

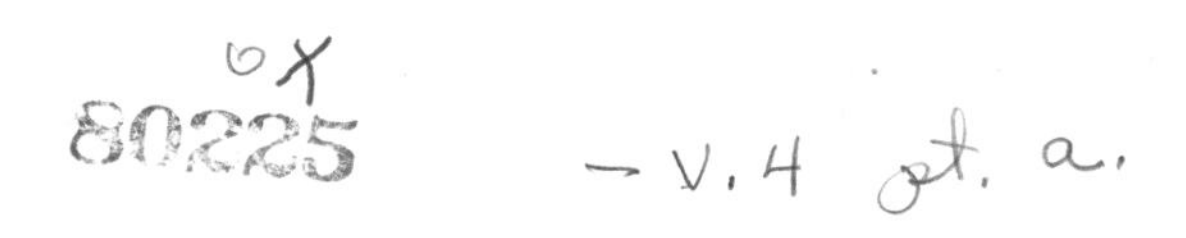

the author on the completion of so formidable a task. It is a pleasure also to express his thanks to Mr. E. B. ROBINSON for his continued help with editorial matters.

April 1973 S. COFFEY

CONTENTS

VOLUME IVA

Heterocyclic Compounds; Three-, four- and five-membered heterocyclic compounds with a single hetero-atom in the ring

PREFACE VII
OFFICIAL PUBLICATIONS; SCIENTIFIC JOURNALS AND PERIODICALS XVI
LIST OF COMMON ABBREVIATIONS AND SYMBOLS USED XVII

Chapter 1. Compounds with Three- and Four-Membered Heterocyclic Rings
by R. LIVINGSTONE

1. Introduction 1
2. Three-membered rings containing an oxygen atom 3
   a. Oxiranes (ethylene oxides) 3
   b. Oxirenes (acetylene oxides) 5
3. Three-membered rings containing one sulphur atom 5
   a. Thiiranes (ethylene sulphides) 5

   (*i*) Synthesis, 5 – (*ii*) Properties and reactions, 8 – (*iii*) Individual thiiranes, 12 –

   b. Thiiranium salts (episulphonium salts) 12
   c. Thiirenes (acetylene sulphides) 14
4. Three-membered rings containing one nitrogen atom 15
   a. Aziridines (ethyleneimines) 15

   (*i*) Synthesis, 16 – (*ii*) Reactions of aziridines, 20 – (*iii*) Properties of individual aziridines, 27 – (*iv*) Halogenoaziridines, 27 – (*v*) Cyanoaziridines and aziridine carboxylic acids, 30 –

   b. Aziridinones (α-lactams) 31
   c. Azirines 33

   (*i*) 1-Azirines: synthesis, 33 – (*ii*) Properties and reactions of 1-azirines, 35 – (*iii*) 2-Azirines, 37 –

   d. Aziridinium salts 38
5. Three-membered rings with more than one hetero-atom 40
   a. Oxaziridines (oxaziranes) 40
   b. Diaziridines and related compounds 42

   (*i*) Diaziridines and diazirines, 42 – (*ii*) Diaziridinones, 44 –

   c. Oxadiaziridine 45
6. Four-membered rings with one oxygen atom 45
   a. Oxetanes (trimethylene oxides) 45

   (*i*) Formation, 45 – (*ii*) Properties and reactions, 46 –

   b. Alkoxyoxetanes and oxetanols 48
   c. Oxetanones 49

   (*i*) Oxetan-2-ones (β-lactones), 49 – (*ii*) Oxetan-3-ones, 50 –

d. Oxetes or oxetenes . . . 52
7. Four-membered rings with one sulphur atom . . . 52
a. Thietanes (trimethylene sulphides) . . . 52

(i) Preparation, 52 – (ii) Properties and reactions, 53 –

b. Thietes or thietenes and their oxides . . . 56
c. Thietanones . . . 59
8. Four-membered rings with one selenium atom . . . 61
9. Four-membered rings containing one nitrogen atom . . . 61
a. Azetidines (trimethyleneimines) . . . 61

(i) Synthesis, 61 – (ii) Properties and reactions, 63 –

b. Azetidinones; oxoazetidines . . . 67

(i) Synthesis, 67 – (ii) Reactions, 69 –

c. Azetidine-2,4-diones (malonimides) . . . 70
10. Three- and four-membered rings containing other elements . . . 71
11. Four-membered rings with two hetero-atoms . . . 72
a. Dioxetanes . . . 72
b. Dithietanes . . . 73
c. Oxathietanes . . . 74
d. Dithietenes . . . 75
e. Diazetidines . . . 76

(i) 1,2-Diazetidinones, 76 – (ii) 1,3-Diazetidines, 78 –

f. Diazetines (dihydrodiazetes) . . . 79
g. Rings containing one nitrogen and one oxygen or sulphur atom . . . 80

(i) Oxazetidines, 80 – (ii) Thiazetidines, 81 –

Chapter 2. Compounds Containing a Five-Membered Ring with One Hetero Atom of Group VI; Oxygen
by R. Livingstone

1. The furan group . . . 83
a. Furan and its substitution products . . . 84

(i) Synthetic methods, 84 – (ii) General properties and reactions, 89 – (iii) Furan, 96 – (iv) Alkyl- and aryl-furans, 97 – (v) Halogenofurans, 102 – (vi) Furansulphonic acids, 103 – (vii) Nitrofurans, 103 – (viii) Amino- and cyano-furans, 105 – (ix) Oxodihydrofurans; butenolides (hydroxyfurans), 107 – (x) Furfuryl alcohol and related compounds, 109 – (xi) Aldehydes, 112 – (xii) Ketones, 115 – (xiii) Carboxylic acids, 117 –

b. Dihydrofurans . . . 121

(i) 2,3-Dihydrofurans, 123 – (ii) 2,5-Dihydrofurans, 125 –

c. Tetrahydrofurans . . . 128

(i) Tetrahydrofuran, 130 – (ii) Halogenotetrahydrofurans, 131 – (iii) Hydroxytetrahydrofurans, 132 – (iv) Aminotetrahydrofurans, 132 – (v) Oxotetrahydrofurans, 133 – (vi) Side-chain substituted tetrahydrofurans, 133 –

d. Compounds with two or more unfused furan nuclei . . . 136
e. Polyene derivatives of furan . . . 140

2. Benzo[*b*]furans and their hydrogenated products 141
a. Benzo[*b*]furans 141

(*i*) General synthetic methods, 141 – (*ii*) General properties and reactions, 145 – (*iii*) Alkyl- and aryl-benzofurans, 148 – (*iv*) Halogenobenzofurans, 152 – (*v*) Nitro- and amino-benzofurans, 154 – (*vi*) Hydroxybenzofurans, 156 – (*vii*) Alcohols, aldehydes and ketones, 158 – (*viii*) Benzofurancarboxylic acids, 163 –

b. Hydrobenzofurans 166

(*i*) 2,3-Dihydrobenzofurans (coumarans), 166 – (*ii*) Tetrahydrobenzofurans, 176 – (*iii*) Hexahydrobenzofurans, 178 – (*iv*) Octahydrobenzofurans, 179 –

3. Isobenzofuran (benzo[*c*]furan; 3,4-benzofuran) and its derivatives 180
a. Isobenzofurans 180

(*i*) Synthetic methods, 181 – (*ii*) Properties and reactions, 185 – (*iii*) Individual isobenzofurans, 187 –

b. Hydroisobenzofurans 188

(*i*) Dihydroisobenzofurans, 188 – (*ii*) Tetra-, hexa- and octa-hydroisobenzofurans, 190 –

4. Other bicyclic systems with a furan ring 191
5. Dibenzofuran or diphenylene oxide and its hydrogenation products . . . 194
a. Dibenzofurans 194

(*i*) General synthetic methods, 194 – (*ii*) Properties and reactions, 196 – (*iii*) Dibenzofuran and derivatives, 197 –

b. Reduced dibenzofurans 203
c. Natural products related to dibenzofuran 204
6. Other tri- and poly-cyclic systems with a furan ring 207
a. Condensed systems of benzene rings with one furan nucleus 207
b. Systems of benzene rings with more than one furan nucleus 215
c. Other tri- and poly-cyclic furans 217

Chapter 3. Compounds with Five-Membered Rings Having One Hetero Atom from Group VI; Sulphur and its Analogues
by R. LIVINGSTONE

1. Monocyclic thiophenes and hydrothiophenes 219
a. Thiophene and its substitution products 220

(*i*) Synthetic methods, 220 – (*ii*) General properties and reactions, 224 – (*iii*) Thiophene, 229 – (*iv*) Metal and metalloid derivatives of thiophenes, 231 – (*v*) Alkyl- and arylthiophenes, 233 – (*vi*) Halogenothiophenes, 238 – (*vii*) Nitro- and amino-thiophenes, 241 – (*viii*) Hydroxythiophenes, 243 – (*ix*) Sulphonic acids, thiols and related compounds, 245 – (*x*) Thiophene alcohols and related compounds, 247 – (*xi*) Aldehydes and ketones, 249 – (*xii*) Thiophenecarboxylic acids, 252 –

b. Thiolenes; dihydrothiophenes 255
c. Thiolanes, tetrahydrothiophenes 257
d. Compounds having two or more unfused thiophene rings 260

2. Benzo[*b*]thiophenes (thianaphthenes) and related compounds 265
a. Benzo[*b*]thiophenes 266

(*i*) Synthetic methods, 266 – (*ii*) General properties and reactions, 270 – (*iii*) Benzo[*b*]thiophene, 271 – (*iv*) Alkyl- and aryl-benzothiophenes, 272 – (*v*) Halogenobenzothiophenes, 273 – (*vi*) Nitrobenzothiophenes, 274 – (*vii*) Aminobenzothiophenes, 275 – (*viii*) Hydroxybenzothiophenes, 276 – (*ix*) Sulphonic acids, 277 – (*x*) Alcohols, aldehydes, ketones, 277 – (*xi*) Carboxylic acids, 279 –

b. Reduced benzothiophenes 280

(*i*) Dihydrobenzothiophenes, benzothiolenes, 280 – (*ii*) Tetrahydrobenzothiophenes, 286 – (*iii*) Hexa- and octa-hydrobenzothiophenes, 287 –

3. Benzo[*c*]thiophenes (3,4-benzothiophenes, isobenzothiophenes) and related compounds 288
a. Benzo[*c*]thiophenes 288
b. Hydrobenzo[*c*]thiophenes 291

(*i*) Dihydrobenzo[*c*]thiophenes, 291 – (*ii*) Tetrahydrobenzo[*c*]thiophenes, 293 – (*iii*) Hexa- and octa-hydrobenzo[*c*]thiophenes, 293 –

4. Cycloalkanothiophenes 294
5. Systems of two fused thiophene rings 296
6. Dibenzothiophenes 300
a. Dibenzothiophene and its derivatives 300

(*i*) Preparation and synthesis, 300 – (*ii*) Properties and reactions, 302 – (*iii*) Derivatives of dibenzothiophene, 302 –

b. Reduced dibenzothiophenes 305
7. Other tri- and poly-cyclic systems containing the thiophene ring 306
8. Compounds with five-membered heterocyclic rings having one selenium atom 313
a. Selenophenes and related compounds 313

(*i*) Preparation and properties, 314 – (*ii*) Selenophenes, 315 – (*iii*) Selenophene derivatives, 315 – (*iv*) Hydroselenophenes, 319 – (*v*) Biselenienyl (bis-selenophene), 320 – (*vi*) Metalloid derivatives, 320 –

b. Benzo[*b*]selenophenes (selenanaphthenes, 2,3-benzoselenophenes) . . . 320
c. Dibenzoselenophenes, diphenylene selenides 324
d. Systems of two fused selenophene rings 325
9. Heterocyclic five-membered ring compounds having one tellurium atom . 326
a. Tellurophenes 326
b. Hydrotellurophenes 327
c. Benzotellurophenes 327

Chapter 4. Compounds Containing a Five-Membered Ring with one Hetero Atom from Group V: Nitrogen
by R. LIVINGSTONE

1. Pyrroles 329
a. General synthetic methods 330
b. General properties and reactions 335
c. Pyrrole and its substitution products 337

(*i*) Pyrrole, 337 – (*ii*) *N*-Derivatives of pyrrole (1-substituted pyrroles), 338 – (*iii*) Individual *N*-substituted pyrroles, 340 – (*iv*) *C*-Alkyl- and *C*-aryl-pyrroles, 341 – (*v*) Halogenopyrroles, 346 – (*vi*) Nitro-, nitroso-, diazo-, amino- and cyano-pyrroles, 348 – (*vii*) Hydroxypyrroles, 351 – (*viii*) Sulphur derivatives of pyrroles, 352 – (*ix*) Pyrrolylalkanols and related compounds, 353 – (*x*) Aldehydes and ketones, 355 – (*xi*) Pyrrolecarboxylic acids and their esters, 359 –

2. Pyrrolines, dihydropyrroles 368
a. 1-Pyrrolines 368

(*i*) Synthesis, 369 – (*ii*) Properties and reactions of 1-pyrrolines, 370 –

b. 1-Pyrroline 1-oxides 371

(*i*) Synthesis, 371 – (*ii*) Properties and reactions, 372 –

c. 2-Pyrrolines 373
d. 3-Pyrrolines 374
e. Oxopyrrolines; pyrrolinones 375

(*i*) 5-Oxopyrrolines, 375 – (*ii*) 4-Oxopyrrolines, 376 – (*iii*) 3-Oxopyrrolines, 377 – (*iv*) Dioxopyrrolines, 378 –

3. Pyrrolidines; tetrahydropyrroles 379
a. Pyrrolidines 379

(*i*) Synthesis, 379 – (*ii*) Properties and reactions, 380 – (*iii*) Individual pyrrolidines, 382 –

b. Substituted pyrrolidines 383

(*i*) Hydroxypyrrolidines, 383 – (*ii*) Oxopyrrolidines, 384 – (*iii*) Poly-oxopyrrolidines, 387 – (*iv*) Pyrrolidinecarboxylic acids, 388 –

4. Compounds having two or more independent five-membered rings . . . 391
a. Bipyrrolyls 391
b. Compounds having two or more pyrrole rings linked through methylene or methine groups 393

(*i*) Dipyrrolylmethanes, 393 – (*ii*) Pyrromethenes, 395 –

Chapter 5. Compounds Containing Five-Membered Rings with One Hetero Atom from Group V: Nitrogen; Fused-Ring Compounds
by R. LIVINGSTONE

1. Indoles (2,3-benzopyrroles) and 3*H*-indoles 397
a. General synthetic methods 398

(*i*) The Fischer indole synthesis, 398 – (*ii*) The Bischler synthesis, 401 – (*iii*) The Madelung synthesis, 402 – (*iv*) Synthesis from nitro compounds with *ortho* side-chains or their reduction products, 402 – (*v*) Synthesis from benzo-quinones and their derivatives, 403 – (*vi*) From ketones and isocyanides, 404 – (*vii*) From pyrroles, 404 – (*viii*) From *o*-aminocarbonyl compounds, 404 – (*ix*) From δ-sultams, 405 –

b. General properties and reactions of indoles 405

(*i*) Reduction, 405 – (*ii*) Addition reactions and polymerisation, 405 – (*iii*) Substitution reactions, 407 – (*iv*) Metal compounds, 408 – (*v*) Oxidation, 408 – (*vi*) Reaction with benzyne, 410 – (*vii*) Condensation with nitroso and carbonyl compounds, 410 – (*viii*) Alkylation, 411 – (*ix*) Characterisation of indoles, 411 –

c. Indole and its substitution products 411

(*i*) *N*-Derivatives, 412 – (*ii*) *C*-Alkyl-, *C*-aryl- and related compounds, 414 – (*iii*) Halogeno-indoles, 418 – (*iv*) Sulphur compounds, 420 – (*v*) Nitro-, amino- and related indoles, 421 – (*vi*) Hydroxyindoles, 424 – (*vii*) Metal and metalloid derivatives, 427 – (*viii*) Indolylcarbinols, aminoalkylindoles and related compounds, 428 – (*ix*) Aldehydes and ketones, 431 – (*x*) Carboxylic acids, 434 –

d. 3*H*-Indoles (indolenines) 439
2. Hydroindoles 445
a. Indolines (dihydroindoles) 445

(*i*) Synthesis, 445 – (*ii*) General properties and reactions, 446 – (*iii*) Individual indolines, 448 –

b. Oxoindolines (indolinones) and hydroxyindoles 448

(*i*) 2-Oxoindolines (oxindoles), 448 – (*ii*) 3-Oxoindoline; indoxyl, 454 – (*iii*) 3-Hydroxy-2-oxoindolines; 3-hydroxyoxindoles; dioxindoles, 456 – (*iv*) 2,3-Dioxoindolines; isatins, 458 – (*v*) Indolines with side-chain carbonyl groups, 464 –

c. Tetra-, hexa-, and octa-hydroindoles 465
d. Compounds having more than one indole nucleus 467
3. Isoindoles and isoindolines 470
a. Isoindoles 470
b. Isoindolines 474
c. Other hydroisoindoles 477

(*i*) Tetrahydroisoindoles, 477 – (*ii*) Hexa- and octa-hydroisoindoles, 477 –

4. Other bicyclic pyrrole systems 477
a. Spirocyclic systems 477
b. Cycloalkenopyrroles 478
c. Fused systems of two heterocyclic rings 480
5. Benzoindoles 482
6. Compounds containing the carbazole nucleus 486
a. Carbazole and its derivatives 486

(*i*) Synthetic methods, 487 – (*ii*) General properties and reactions, 490 – (*iii*) Carbazole, 491 – (*iv*) *N*-Substituted carbazoles, 492 – (*v*) *C*-Alkylcarbazoles, 494 – (*vi*) Halogenocarbazoles, 495 – (*vii*) Nitro-, nitroso- and amino-carbazoles, 496 – (*viii*) Hydroxycarbazoles, 502 – (*ix*) Sulphonic acids and related compounds, 503 – (*x*) Organometallic derivatives, 503 – (*xi*) Alcohols, aldehydes and ketones derived from carbazole, 504 – (*xii*) Carbazolecarboxylic acids, 506 –

b. Hydrocarbazoles 509

(*i*) Tetrahydrocarbazoles, 509 – (*ii*) More fully reduced carbazoles, 514 – .

c. Compounds having more than one singly linked carbazole nucleus . . 517
d. Carbazoles with additional fused rings 518

(*i*) Benzocarbazoles, 518 – (*ii*) More complicated fused ring carbazoles, 524 –

7. Other tricyclic pyrrole systems 526

Chapter 6. Other Five-Membered Ring Compounds with One Hetero Atom in the Ring from Groups 3, 4 and 5
by R. LIVINGSTONE

1. Phosphorus compounds 531
a. Mononuclear compounds 531

(*i*) Phospholes, 531 – (*ii*) Phospholenes and phospholanes, 534 –

b. Polynuclear, fused ring compounds 538

(*i*) Dibenzophospholes, 538 –

2. Silicon compounds 539
a. Mononuclear compounds 539

(*i*) Silacyclopentadienes; siloles, 539 – (*ii*) Silacyclopentenes, 542 – (*iii*) Silacyclopentanes, 543 –

b. Fused ring compounds 543
3. Germanium compounds 545
a. Five-membered mononuclear compounds 545
b. Polycyclic compounds with germanium in a five-membered ring . . 547
4. Compounds containing aluminium, boron, antimony and tin as the hetero atom 549

Appendix

IUPAC Commission on the Nomenclature of Organic Chemistry

B. Fundamental heterocyclic systems 551
Rule **B-1.** Extension of the Hantzsch–Widman System 551
Rule **B-2.** Trivial and semi-trivial names 554
Rule **B-3.** Fused heterocyclic systems 562
Rule **B-4.** "a" Nomenclature 565
Rule **B-5.** Radicals 567
Rule **B-6.** Cationic hetero-atoms 568

Heterocyclic spiro compounds 569
Rule **B-10** (alternate to Rule **B-11**) 569
Rule **B-11** (alternate to Rule **B-10**) 569

Seniority of ring systems 570
Rule **C-14.1** 570

Index 575

OFFICIAL PUBLICATIONS

B.P.	British (United Kingdom) Patent
F.P.	French Patent
G.P.	German Patent
Sw.P.	Swiss Patent
U.S.P.	United States Patent
U.S.S.R.P.	Russian Patent
B.I.O.S.	British Intelligence Objectives Sub-Committee Reports, H.M. Stationery Office, London.
C.I.O.S.	Combined Intelligence Objectives Sub-Committee Reports
F.I.A.T.	Field Information Agency, Technical Reports of U.S. Group Control Council for Germany
B.S.	British Standards Specification
A.S.T.M.	American Society for Testing and Materials
A.P.I.	American Petroleum Institute Projects
C.I.	Colour Index Number of Dyestuffs and Pigments

SCIENTIFIC JOURNALS AND PERIODICALS

With few obvious and self-explanatory modifications the abbreviations used in references to journals and periodicals comprising the extensive literature on organic chemistry, are those used in the World List of Scientific Periodicals.

LIST OF COMMON ABBREVIATIONS AND SYMBOLS USED

A	acid
Å	Ångström units
Ac	acetyl
a	axial
as, asymm.	asymmetrical
at.	atmosphere
B	base
Bu	butyl
b.p.	boiling point
C, mC and μC	curie, millicurie and microcurie
c, C	concentration
conc.	concentrated
crit.	critical
D	Debye unit, 1×10^{-18} e.s.u.
D	dissociation energy
D	dextro-rotatory; dextro configuration
DL	optically inactive (externally compensated)
d	density
dec. or decomp.	with decomposition
deriv.	derivative
E	energy; extinction; electromeric effect
E1, E2	uni- and bi-molecular elimination mechanisms
E1cB	unimolecular elimination in conjugate base
e.s.r.	electron spin resonance
Et	ethyl
e	nuclear charge; equatorial
f	oscillator strength
f.p.	freezing point
G	free energy
g.l.c.	gas liquid chromatography
g	spectroscopic splitting factor, 2.0023
H	applied magnetic field; heat content
h	Planck's constant
Hz	hertz
I	spin quantum number; intensity; inductive effect
i.r.	infrared
J	coupling constant in n.m.r. spectra
K	dissociation constant
k	Boltzmann constant; velocity constant
kcal.	kilocalories
L	laevorotatory; laevo configuration
M	molecular weight; molar; mesomeric effect
Me	methyl

m	mass; mole; molecule; *meta-*
ml	millilitre
m.p.	melting point
Ms	mesyl (methanesulphonyl)
[M]	molecular rotation
N	Avogado number; normal
n.m.r.	nuclear magnetic resonance
n	normal; refractive index; principal quantum number
o	*ortho-*
o.r.d.	optical rotatory dispersion
P	polarisation; probability; orbital state
Pr	propyl
Ph	phenyl
p	*para-;* orbital
p.m.r.	proton magnetic resonance
R	clockwise configuration
S	counterclockwise config.; entropy; net spin of incompleted electronic shells; orbital state
S_N1, S_N2	uni- and bi-molecular nucleophilic substitution mechanisms
S_Ni	internal nucleophilic substitution mechanisms
s	symmetrical; orbital
sec	secondary
soln.	solution
symm.	symmetrical
T	absolute temperature
Tosyl	*p*-toluenesulphonyl
Trityl	triphenylmethyl
t	time
temp.	temperature (in degrees centigrade)
tert	tertiary
U	potential energy
u.v.	ultraviolet
v	velocity
α	optical rotation (in water unless otherwise stated)
$[\alpha]$	specific optical rotation
α_A	atomic susceptibility
α_E	electronic susceptibility
ε	dielectric constant; extinction coefficient
μ	microns (10^{-4} cm); dipole moment; magnetic moment
μ_B	Bohr magneton
μg	microgram (10^{-6} g)
λ	wavelength
υ	frequency; wave number
χ, χ_d, χ_μ	magnetic, diamagnetic and paramagnetic susceptibilities
$\sim$	about

(+)	dextrorotatory
(−)	laevorotatory
⊖	negative charge
⊕	positive charge

Chapter 1

Compounds with Three- and Four-Membered Heterocyclic Rings

R. LIVINGSTONE

1. Introduction

In heterocyclic chemistry a saturated three-membered ring containing one oxygen, sulphur, or nitrogen atom is known as an oxiran(e)* (ethylene oxide), thiiran(e) (ethylene sulphide), or aziridine (ethyleneimine). The related unsaturated compounds are oxiren(e), thiiren(e), and azirine, but in the latter case the double bond can occupy one of two positions giving 1-azirine and 2-azirine. The corresponding saturated four-membered rings are oxetan(e) (trimethylene oxide), thietan(e) (trimethylene sulphide), and azetidine (trimethyleneimine), while the unsaturated derivatives containing one double bond are oxet(e) or oxeten(e), thiet(e) or thieten(e), and azetine. Azete is used for the fully unsaturated, four-membered ring containing nitrogen and two double bonds. Alternatively the mono-saturated azetines may be named as dihydroazetes. Some of the unsaturated compounds have not yet been prepared.

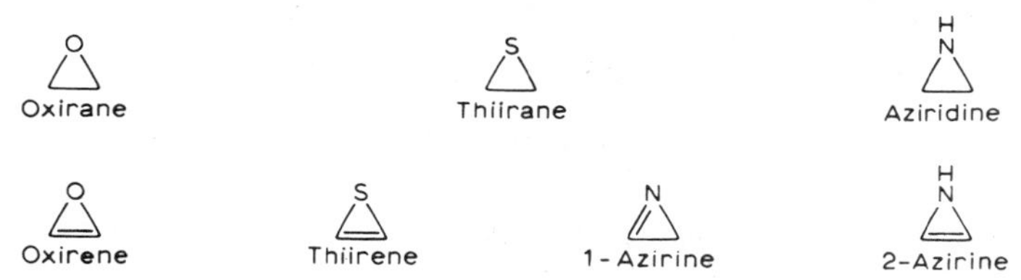

* The systematic nomenclature of monocyclic heterocyclic compounds is covered by IUPAC Rules B-1, which require the names of the fundamental ring systems to end in "e". In some systems of nomenclature, however, *e.g.* those used by the Chemical Society (London) and Chemical Abstracts, the terminal "e", shown here in parentheses, is elided.

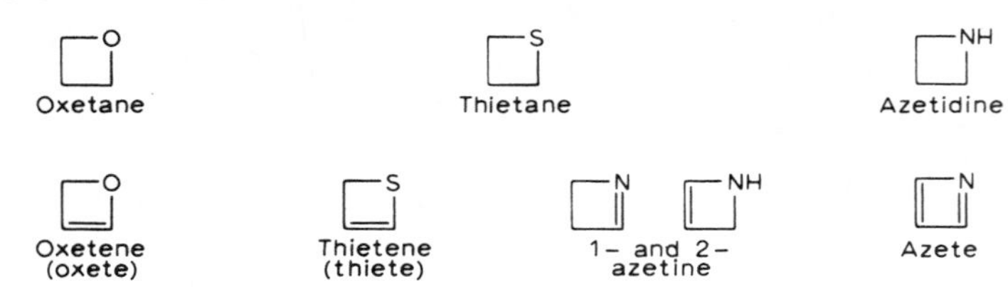

Structural data on three- and four-membered rings are given in Table 1.

TABLE 1

STRUCTURAL DATA

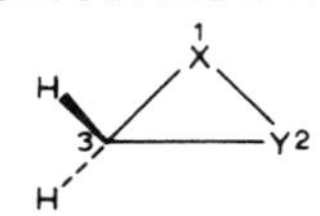

	Name	X	Y	Bond length (Å) XY	CY	CH	Bond angle (°) CXY	XYC	HCH	Ref.
I	Oxirane	O	CH_2	1.436	1.472	1.082	61.4	59.3	116.57	1
	Thiirane	S	CH_2	1.819	1.492	1.078	48.4	65.8	116.0	1
	Aziridine	NH	CH_2	1.488	1.480	—	59.6	60.2	—	2
	Diazirine	N	N	N = N 1.228 ± 0.003	1.482	1.09	—	—	117	3

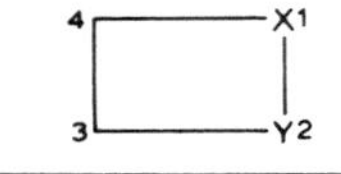

	Name	X	Y	Bond length (Å) XY	YC	CC	CX	Bond angle (°) CXY	XYC	YCC	CCX	Ref.
II	Oxetane	O	CH_2	—	—	1.54 ±0.03	1.46 ±0.03	94.5 ±3	—	88.5 ±3	—	4
	Thietane	S	CH_2	—	—	1.55	1.85	78±1	—	97±5	—	5
	Oxetan-2-one	O	CO	1.45	1.53	1.53	1.45	89	94	83	94	6
	Azetidin-2-one	NH	CO	1.39	1.49	1.51	1.38	95	92	96	87	7

References

1 *G. L. Cunningham et al.*, J. chem. Phys., 1951, **19,** 676.
2 *T. E. Turner et al., ibid.*, 1953, **21,** 564.
3 *L. Pierce* and *S. V. Dobyns*, J. Amer. chem. Soc., 1962, **84,** 2651.
4 *P. W. Allen* and *L. E. Sutton*, Acta Cryst., 1950, **3,** 46.
5 *E. Goldish*, J. chem. Educ., 1959, **36,** 408.
6 *J. Bregman* and *S. H. Bauer*, J. Amer. chem. Soc., 1955, **77,** 1955.
7 *D. Crowfoot et al.*, "The Chemistry of Penicillin" *(H. T. Clarke, J. R. Johnson* and *R. Robinson)*, Princeton University Press, N.J., 1949.

Four-membered rings in general are more difficult to prepare than three-membered, and when both types of ring closure are structurally possible, the three-membered ring is usually formed exclusively (*W. P. Evans*, Z. phys. Chem., 1891, **7**, 337; *H. Freundlich* and *H. Kroepelin*, *ibid.*, 1926, **122**, 39; *W. J. Gensler*, J. Amer. chem. Soc., 1948, **70**, 1843). Both saturated ring systems are strained but the four-membered not as highly as the corresponding three-membered, in which the normal bond angles and distances are more distorted. They are more easily opened than larger ring compounds and none shows aromatic character. Three-membered heterocyclic rings are much more easily opened than cyclopropane, since hydrolytic and similar processes are facilitated by the presence of the hetero atom. Direct ring closure is only usefully effected by intramolecular application of reactions which are essentially irreversible under the experimental conditions, *e.g.* the formation of oxirane from 2-chloroethanol by the irreversible Williamson ether synthesis.

Although four-membered rings in many instances are opened almost as easily as their three-membered analogues, in general, more vigorous conditions are required, but in particular they may break down thermally into two unsaturated molecules, sometimes in both possible directions. Thus when C is an oxygen atom and D a carbonyl group elimination of carbon dioxide usually takes place very easily giving A=B. Four-membered rings are often synthesised by a reversal of this general method of breakdown.

$$\begin{matrix} A{=}B \\ \\ C{=}D \end{matrix} \;\leftarrow\; \begin{matrix} A - B \\ | \quad\;\; | \\ C - D \end{matrix} \;\rightarrow\; \begin{matrix} A \quad B \\ \| \quad\; \| \\ C \quad D \end{matrix}$$

2. Three-membered rings containing an oxygen atom

*(a) Oxiranes (ethylene oxides)**

Oxirane (ethylene oxide) and its simpler derivatives have already been discussed (see Vol. I D, pp. 19 *et seq.*). Methods for making more complex compounds include the following:

(*1*) The Darzens condensation of an aldehyde or ketone with an α-halogenoketone or ester (*G. Darzens*, Compt. rend., 1904, **139**, 1215); benzaldehyde, for example, reacts with bromomethyl phenyl ketone in the presence of sodium ethoxide to give 2-benzoyl-3-phenyloxirane (*S. Bodforss*, Ber., 1916, **49**, 2795):

* See also *S. Winstein* and *R. B. Henderson*, in "Heterocyclic Compounds", Ed. *R. C. Elderfield*, Vol. 1, Chap. 1, Wiley, New York, 1950.

$$PhCHO + CH_2Br\cdot COPh \xrightarrow{NaOEt} PhCH\text{—}CH\cdot COPh \text{ (epoxide, O bridging)}$$

acetophenone with ethyl chloroacetate yields phenylmethylglycidic ester (*C. F. H. Allen* and *J. van Allen*, Org. Synth., Coll. Vol., **3**, 1955, p. 727):

$$Ph\cdot C(CH_3){=}O + ClCH_2\cdot CO_2Et \xrightarrow{NaNH_2} Ph\cdot C(Me)\text{—}CH\cdot CO_2Et \text{ (epoxide, O bridging)}$$

spiro compounds with one ring an epoxide have been obtained by this method (*M. S. Newman* and *B. J. Magerlein*, Org. Reactions, 1949, **5**, 413):

$$\text{cyclohexanone} + CH_2Cl\cdot CO_2Me \xrightarrow{NaOMe} \text{spiro epoxide}\ (CO_2Me)$$

and a related compound is afforded by reacting methylenecyclohexane with perbenzoic acid (*M. Tiffeneau et al.*, Compt. rend., 1937, **205**, 54).

(*2*) Oxiranes have been formed by treating an equimolecular solution of a carbonyl compound and methylene bromide in tetrahydrofuran with two gram atoms of lithium or the corresponding quantity of a diluted lithium amalgam (*F. Bertini et al.*, Chem. Comm., 1969, 1047):

$$R_2CO + CH_2Br_2 \xrightarrow{2\ Li} R_2C\text{—}CH_2\ \text{(oxirane)}$$

5-*tert*-Butyl-2,2-dimethyl-1,4-dioxaspiro[2.2]pentane, an example of the unique spiro dioxide structure, has been isolated following the buffered oxidation of 2,5,5-trimethyl-2,3-hexadiene with a peracid (*J. K. Crandall*, *W. H. Machleder* and *M. J. Thomas*, J. Amer. chem. Soc., 1968, **90**, 7346):

$$(H)(Bu^t)C{=}C{=}C(Me)_2 \longrightarrow \text{spiro dioxide}\ (Bu^t, H, Me, Me)$$

Related allene mono-oxides have also been obtained. 1,3-Di-*tert*-butylallene oxide upon heating to 100° isomerises to give 2,3-di-*tert*-butylcyclopropanone (*R. L. Camp* and *F. D. Greene*, *ibid.*, p. 7349):

$$Bu^tCH{=}C{=}CHBu^t \xrightarrow[25^\circ]{ArCO_3H} \text{oxirane}(Bu^t){=}CHBu^t \xrightarrow{100^\circ} \text{2,3-di-}Bu^t\text{-cyclopropanone}$$

The solvolysis reaction of esters of several 2-hydroxymethyloxiranes (2,3-epoxy-1-propanols) have led to the formation of esters of 3-oxetanols. Glycidol (I) and *p*-toluenesulphonyl chloride in pyridine give a good yield of 3-tosyloxyoxetane (II) (*H. G. Richey* and *D. V. Kinsman*, Tetrahedron Letters, 1969, 2505):

OH CH2 TsCl C5H5N TsO⊖ TsO (I) (II)

(b) Oxirenes (acetylene oxides)

The first authentic oxirenes are probably the *di*-n-*propyl*- and *di*-n-*butyl-oxirenes*, b.p. 75°/15 mm, and 42°/0.01 mm, obtained by oxidising the related acetylenes with peracetic acid (*H. H. Schlubach* and *V. Franzen*, Ann., 1952, **577**, 60). Oxirene intermediates, *e.g.* diphenyloxirene, are postulated in the peracid oxidation of related acetylenes (*R. N. McDonald* and *P. A. Schwab*, J. Amer, chem. Soc., 1964, **86**, 4866; *J. K. Stille* and *D. D. Whitehurst*, *ibid.*, p. 4871). Oxiren itself has not yet been described.

3. Three-membered rings containing one sulphur atom

(a) Thiiranes (ethylene sulphides)

Thiiranes (ethylene sulphides) were slow to receive more than superficial study, no doubt owing to their difficult accessibility and their strong tendency to polymerise. They have found application in rendering wool unshrinkable (*T. Barr* and *J. B. Speakman*, J. Soc. Dyers and Col., 1944, **60**, 238; *S. Blackburn* and *H. Philips*, *ibid.*, 1945, **61**, 203), and as antituberculosis agents.

(i) Synthesis

(*1*) The most generally serviceable process is to treat an oxirane with potassium thiocyanate, thiourea, or some other thioamide (*W. Davies et al.*, J. chem. Soc., 1946, 1050; 1949, 282; 1952, 4480; *F. G. Bordwell* and *H. M. Anderson*, J. Amer. chem. Soc., 1953, **75**, 4959); the reactions may be formulated:

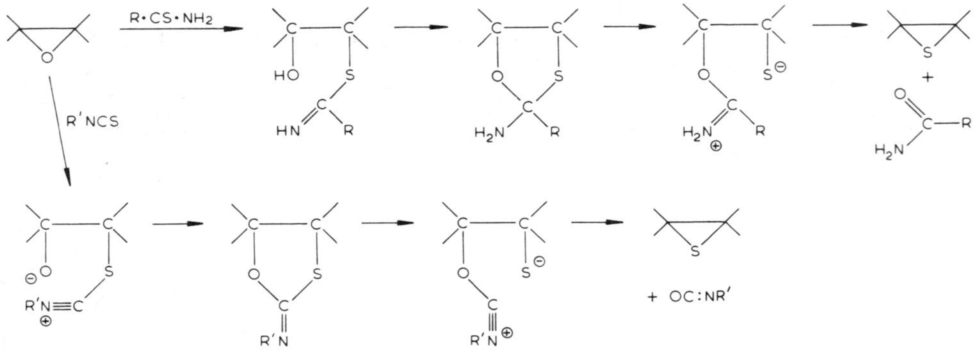

Cyclic carbonates of 1,2-diols with potassium thiocyanate yield thiiranes (*S. Searles, H. R. Hays* and *E. F. Lutz*, J. org. Chem., 1962, **27,** 2828, 2832). Methyl thiiranes (propylene sulphides) of the type

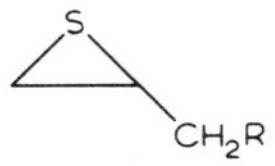

(R = SMe, SEt, OEt and NEt_2) are available from the corresponding oxiranes and thiourea (*E. P. Adams et al.*, J. chem. Soc., 1960, 2665). In the first two cases the intermediate thiouronium salts can be isolated:

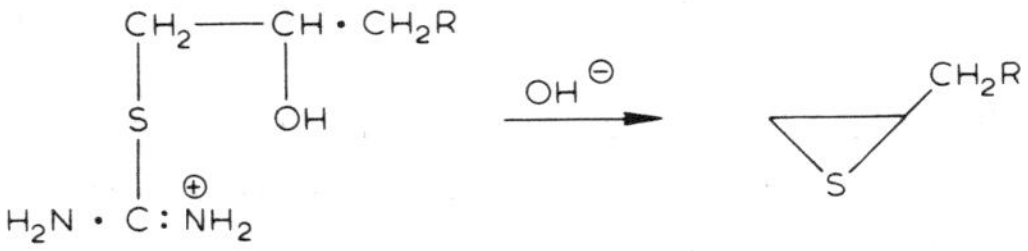

(*2*) At pH 5–11, 1,2-chlorothiols give thiiranes (*W. Coltof*, U.S.P. 2,183,860; C.A., 1940, **34,** 2395).

2-Mercaptomethylthiiranes are obtained from dimercaptoalkanols on treatment with hydrochloric acid followed by a weak base, or through their triacyl derivatives (*F. P. Doyle et al.*, J. chem. Soc., 1960, 2660).

(*3*) Treatment of oxiranes with carbonyl sulphide in the presence of lithium phosphate catalyst gives thiiranes (*J. A. Durden, H. A. Stansburg* and *W. H. Catlette*, J. org. Chem., 1961, **26,** 836). The reaction is base-catalyzed and the mechanism may be generalized as follows:

B + COS → $\overset{\oplus}{B}$—C(=O)—$\overset{\ominus}{S}$ → (epoxide) → $\overset{\oplus}{B}$—C(=O)—S—C—C—$O^{\ominus}$ → (−B) → (cyclic thiocarbonate) → (Δ,B) → thiirane + CO_2

thiirane + $\overset{\oplus}{B}$—C(=O)—$\overset{\ominus}{S}$ → $\overset{\oplus}{B}$—C(=O)—S—C—C—$S^{\ominus}$ → (−B) → (cyclic dithiocarbonate) → (Δ,B) → thiirane + COS

(*4*) Sodium sulphide converts, 1,2-dithiocyanates and 1,2-chlorothiocyanates into thiiranes (*M. Delépine* and *S. Eschenbrenner*, Bull. Soc. chim. Fr., 1923, [iv], **33,** 703).

(*5*) Diaryldiazomethanes react with compounds of the type RCSR′, in which R and R′ may be Ar, ArO, ArS, or Cl (*H. Staudinger* and *J. Siegwart*, Helv., 1920, **3,** 840; *A. Schönberg* and *L. von Vargha*, Ann., 1930, **483,** 176) giving:

(2,2-diaryl-3-R-3-R′-thiirane)

Diaryldiazomethanes and sulphur at room temperature (*N. Latif* and *I. Fathy*, J. org. Chem., 1962, **27,** 1633), yield:

(tetraarylthiirane)

(*6*) Thioketones may be reduced by Grignard reagents (*Schönberg*, Ann., 1927, **454,** 37):

Ar_2CS —(Ar′MgI)→ Ar′·Ar′ + $Ar_2C(SMgI)$—$C(SMgI)Ar_2$ → thiirane (Ar_2, Ar_2) + $S(MgI)_2$

(*7*) Diarylketoximes give tetra-arylthiiranes by successive treatment with ferricyanide and hydrogen sulphide (*Schönberg* and *M. Z. Barakat*, J. chem. Soc., 1939, 1074):

$$2\ Ph_2C:NOH \longrightarrow \text{(}Ph_2C\text{–O–}CPh_2\text{, }N{=}N\text{ ring)} \xrightarrow{H_2S} \text{(thiirane, }Ph_2\text{, }Ph_2\text{)} + N_2$$

$\beta\beta'$-Diphenyldivinyl sulphide undergoes cyclisation on irradiation to give *trans*-2,3-diphenyl-5-thiabicyclo[2.1.0]pentane, the first known example of a cyclobutene episulphide (*E. Block* and *E. J. Corey*, J. org. Chem., 1969, **34**, 896):

$$PhCH{=}CH{-}S{-}CH{=}CHPh \xrightarrow{h\nu} \text{(Ph, Ph-substituted thiabicyclopentane)}$$

(ii) Properties and reactions

See *Davies* and co-workers (J. chem. Soc., 1950, 317; 1951, 774).

(*1*) All reactions of thiiranes involve ring opening and since the electron density at the sulphur atom is lower than that at the oxygen in oxiranes, in some cases they are therefore less reactive toward electrophilic reagents. Their reactivity toward nucleophilic reagents seems similar to or a little greater than that of oxiranes. The most characteristic reaction of thiiranes is mercaptoethylation involving the addition to compounds with active hydrogen atoms. The reaction may be acid- or base-catalysed (*Delépine*, Compt. rend., 1920, **171**, 36; Bull. Soc. chim. Fr., 1920, **27**, 740; *Delépine* and *Eschenbrenner*, *loc. cit.*) and polymerisation may be regarded as a special case of mercaptoethylation:

$$\text{(thiirane)} + NaX \longrightarrow XCH_2\cdot CH_2SNa \;\xrightarrow{\text{+ thiirane}}\; XCH_2\cdot CH_2\cdot S\cdot CH_2\cdot CH_2SNa \quad etc.$$

$$\text{(thiirane)} \xrightarrow{HCl} \text{(S}^{\oplus}\text{–H thiiranium) } Cl^{\ominus} \longrightarrow ClCH_2\cdot CH_2SH$$

$$\text{(thiirane)} + ClCH_2\cdot CH_2SH \longrightarrow \text{(}^{\oplus}S\cdot CH_2\cdot CH_2SH\text{ thiiranium) } Cl^{\ominus} \longrightarrow Cl(CH_2\cdot CH_2S)_2H \quad etc.$$

Those polymers which contain terminal halogen atoms are degraded thermally to 1,4-dithiane (I) by internal sulphonium salt formation (*G. M. Bennett et al.*, J. chem. Soc., 1927, 1803):

$CH_2 \cdot S \cdot CH_2 \cdot CH_2$ / CH_2Cl / $S \cdot CH_2\cdot$ → $CH_2 \cdot S \cdot CH_2 \cdot CH_2$ / CH_2—$\overset{\oplus}{S} \cdot CH_2\cdot$ $Cl^{\ominus}$ → (I) + $ClCH_2\cdot$

(*2*) Alkaline hydrolysis of 2-chloromethylthiirane gives 3-hydroxythietane (*Adams et al., loc. cit.*) (p. 53):

CH_2Cl —($OH^{\ominus}$)→ HO

Unsymmetrical thiiranes are often opened by acids so that the sulphur remains attached to the primary carbon atom, with fission of the bond linking it to secondary or tertiary carbon, suggesting a mechanism of "unimolecular" type:

Me —($H^{\oplus}$)→ Me → $\overset{\oplus}{Me CH}$—CH_2 HS —($Cl^{\ominus}$)→ Cl SH / $MeCH$—CH_2

Acetyl chloride similarly affords 2-chloropropyl *S*-thioacetate, $CH_3 \cdot CHCl \cdot CH_2SAc$. With amines and sodio-cyanoacetic ester, reaction takes place in the opposite sense, indicating a "bimolecular" attack on primary carbon by the nucleophilic reagent:

HS NR_2 / Me_2C—CH_2 ←(R_2NH)— Me_2 —($Na^{\oplus}$ $\overset{\ominus}{C}H(CN) \cdot CO_2Et$)→ NaS $CH(CN) \cdot CO_2Et$ / Me_2C—CH_2

The thiirane ring is also opened by thiols (→ $HS \cdot CH_2 \cdot CH_2 \cdot SR$), by alcohols in presence of boron fluoride (→ $HS \cdot CH_2 \cdot CH_2OEt$; *H. R. Snyder* and *J. M. Stewart*, U.S.P. 2,497,422; C.A., 1950, **44**, 4025), by sodium bisulphite (→ $HS \cdot CH_2 \cdot CH_2 \cdot SO_3Na$), and by nitric acid (→ $HO_3S \cdot CH_2 \cdot CO_2H$).

(*3*) Highly arylated or acylated thiiranes tend to lose sulphur, giving olefins; 2-acylthiomethylthiiranes on thermal decomposition yield sulphur and 3-acylthiopropenes (*Adams et al., loc. cit.*):

$$\text{(2-thiiranyl)}CH_2\cdot S\cdot COR \xrightarrow{\Delta} CH_2{:}CH\cdot CH_2\cdot S\cdot COR$$

organolithium compounds remove sulphur from thiirane (*Bordwell et al.*, J. Amer. chem. Soc., 1954, **76,** 1082):

$$RLi + C_2H_4S \rightarrow C_2H_4 + RSLi$$

Desulphurisation of *cis*- and *trans*-2,3-dimethylthiirane with iodine in benzene proceeds stereospecifically giving *cis*- and *trans*-butene, respectively *G. K. Helmkamp* and *D. J. Pettit*, J. org. Chem., 1962, **27,** 2942).

The mechanism of the stereospecific desulphurisation of thiiranes with methyl iodide has been investigated and the formation of the following sequence of intermediates suggested; thiiranium salt, *β*-iodo sulphide, *β*-iodosulphonium iodide. None could be isolated, but with methyl bromide as the alkylating agent both the sulphide and sulphonium bromide have been obtained. These have been converted to butene by treatment with iodide ion or iodine (*Pettit* and *Helmkamp*, *ibid.*, 1964, **29,** 3258):

$$\text{2,3-dimethylthiirane} \xrightarrow{MeI} \left[\text{S-methylthiiranium } I^{\ominus}\right] \longrightarrow \left[MeCH(I)\text{—}CH(SMe)Me\right] \longrightarrow \left[MeCH(I)\text{—}CH(\overset{\oplus}{S}(Me)_2)Me \; I^{\ominus}\right]$$

$$\xrightarrow[I^{\ominus}]{E_2} MeHC{=}CHMe + (Me)_2S \xrightarrow{MeI} (Me)_3S^{\oplus} I^{\ominus}$$

$$\text{2,3-dimethylthiirane} \xrightarrow{MeBr} MeCH(Br)\text{—}CH(SMe)Me + MeCH(Br)\text{—}CH(\overset{\oplus}{S}(Me)_2)Me \; Br^{\ominus}$$

threo-2-Bromo-3-methylthiobutane

threo-2-Bromo-3-dimethylsulphonio-butane bromide

Me–CH–S–CH–Me (ring) $\xrightarrow{MeBr}$ H–C(Me)(Br)–C(Me)(H)–SMe + H–C(Me)(Br)–C(Me)(H)–$\overset{\oplus}{S}(Me)_2$ $Br^{\ominus}$

erythro-2-Bromo-3-methylthiobutane

erythro-2-Bromo-3-dimethylsulphonio-butane bromide

(*4*) Normal methods of oxidation fail to convert thiiranes into sulphones (1,1-dioxides) and sulphoxides (1-oxides); ring opening occurs giving sulphonic and sulphuric acids (*C. C. J. Culvenor*, *Davies* and *N. S. Heath*, J. chem. Soc., 1949, 282). However, oxidation of thiirane with sodium metaperiodate gives *thiirane* 1-*oxide (ethylene sulphoxide)*; it distils at 46–48° /2 mm. The thiirane 1-oxides derived from the sulphides of propylene, cyclohexene, and styrene cannot be distilled without decomposition (*G. E. Hartzell* and *J. N. Paige*, J. Amer. chem. Soc., 1966, **88,** 2616). 2,3-Dibenzoyl-2,3-diphenylthiirane is oxidised with hydrogen peroxide in acetic acid to the sulphoxide and sulphone. Two isomers of the sulphoxide have been isolated (*D. C. Dittmer* and *G. C. Levy*, J. org. Chem., 1965, **30,** 636):

PhCO(Ph)C–S–C(Ph)COPh → PhCO(Ph)C–S(O)–C(Ph)COPh + PhCO(Ph)C–SO_2–C(Ph)COPh

(*5*) Tetraphenylthiirane 1,1-dioxide and thiirane 1,1-dioxide, respectively, are isolated following the reaction between diphenyldiazomethane (*Staudinger* and *F. Pfenninger*, Ber., 1916, **49,** 1941) and diazomethane (*G. Hesse*, *E. Reichold* and *S. Mayundar*, *ibid.*, 1957, **90,** 2106), and sulphur dioxide:

RR'C : N_2 $\xrightarrow{SO_2}$ RR'C–SO_2–CRR' $\xrightarrow{\Delta}$ SO_2 + RR'C : CR'R

Thiirane 1,1-dioxides are isolated from the reaction of sulphenes ($RCH = SO_2$) with diazoalkanes (*G. Optiz* and *K. Fischer*, Angew. Chem., intern. Edn., 1965, **4,** 70). *cis*-2,3-Diphenylthiirane 1,1-dioxide is produced when phenyldiazomethane is in excess over the quantity of sulphur dioxide, *viz.*, by passing the latter into a solution of phenyldiazomethane. According to orbital symmetry predictions the cycloelimination of sulphur dioxide from thiirane 1,1-dioxide should be nonconcerted, although *cis*-thiirane 1,1-dioxides undergo a stereospecific thermal loss of sulphur dioxide to give the corresponding *cis*-olefins (*N. Tokura, T. Nagai* and *S. Matsumura*, J. org. Chem., 1966, **31,** 349; *N. P. Neureiter*, J. Amer. chem. Soc., 1966, **88,** 558). *cis*- and *trans*-Stilbene episulphone undergo stereospecific decomposition in methanol, but evidence supporting a stepwise process is now available, for it has been found that both isomers yield only *trans*-stilbene in the presence of methoxide (*Bordwell et al., ibid.*, 1968, **90,** 429, 435). Thiirane 1-oxides lose sulphur monoxide nonstereospecifically on heating, *cis*- or *trans*-but-2-ene sulphoxide giving a mixture of *cis*- and *trans*-but-2-ene (*Hartzell* and *Paige*, J. org. Chem., 1967, **32,** 459).

(*6*) Preferential C–C bond cleavage occurs giving dibenzyl sulphone on reduction of *cis*-diphenylthiirane 1,1-dioxide with lithium tetrahydridoborate in tetrahydrofuran (*Matsumara, Nagai* and *Tokura*, Tetrahedron Letters, 1966, 3939):

O_2 S, H, H, Ph, Ph ⟶ $PhCH_2 \cdot SO_2 \cdot CH_2Ph$

(*7*) 2,3-Dibenzoyl-2,3-diphenylthiirane on treatment with triphenylphosphine (*Dittmer* and *Levy*, J. org. Chem., 1965, **30,** 636), or on irradiation (*A. Padwa* and *D. Crumrine*, Chem. Comm., 1965, 506) is converted into *trans*-dibenzoylstilbene:

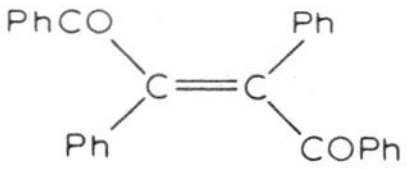

PhCO(Ph)C=C(Ph)COPh

(iii) Individual thiiranes

Thiirane (*Delépine* and *Eschenbrenner, loc. cit.*), b.p. 55°, d_4^{20} 1.0113, dipole moment 1.66 ± 0.03 D, is insoluble in water; it polymerises slowly and more quickly in presence of acids or alkali. For thermodynamic, spectroscopic data and molecular dimensions of thiirane see *H. W. Thompson* and *W. T. Cave*, Trans. Faraday Soc., 1951, **47,** 951; *G. B. Guthrie et al.*, J. Amer. chem. Soc., 1952, **74,** 2795; *Cunningham et al.*, J. chem. Phys., 1951, **19,** 676). Some homologues and derivatives are given in Table 2. For an extensive list of preparations see *M. Sander*, Chem. Reviews, 1966, **66,** 302.

(b) Thiiranium salts (episulphonium salts)

Stable thiiranium (episulphonium) (II) salts have been isolated from the reaction of alkanesulphenyl 2,4,6-trinitrobenzenesulphonates with cyclo-

TABLE 2

THIIRANES

Substituent	*b.p. (°C)*	*Ref.*
2-Methyl	75–76	1
2-Ethyl		2
2,2-Dimethyl	87	1
2-Chloromethyl	79–81/114 mm	1
7-Thiabicyclo[4.1.0]heptane (cyclohexene episulphide)	67–68/16 mm	1,3
2,3-Diphenyl	87–88/4 mm	4
Tetraphenyl	m.p. 175 (decomp.)	5

References
1 *W. Davies et al.*, J. chem. Soc., 1946, 1050.
2 *M. Delépine* and *P. Jaffeux*, Compt. Rend., 1921, **172,** 158; Bull. Soc. chim. Fr., 1921, **29,** 136.
3 *E. E. van Tamelen*, Org. Synth., 1952, **32,** 39.
4 *C. O. Guss* and *D. L. Chamberlain*, J. Amer. chem. Soc., 1952, **74,** 1342.
5 *A. Schönberg* and *M. Z. Barakat*, J. chem. Soc., 1939, 1074.

octene, and by the alkylation of cyclo-octene sulphide with oxonium salts (*e.g.* trimethyloxonium 2,4,6-trinitrobenzenesulphonate) or with a tertiary halide and silver 2,4,6-trinitrobenzenesulphonate. Related sulphonium salts from less hindered thiiranes have not been isolated (*Pettit* and *Helmkamp*, J. org. Chem., 1964, **29,** 2702, 3258):

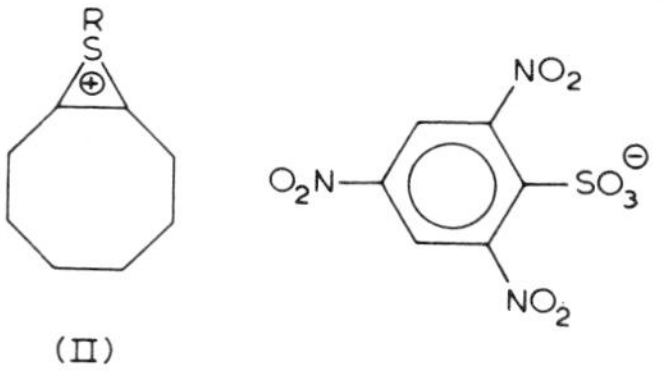

R S ⊕ NO_2 O_2N $SO_3^{\ominus}$ NO_2

(II)

The reaction of cyclo-octene-*S*-methylepisulphonium 2,4,6-trinitrobenzenesulphonate (II, R = Me) with a variety of nucleophilic reagents occurs preferentially by attack at sulphur to give cyclo-octene and a sulphenyl compound rather than at carbon to give a 1,2-disubstituted cyclo-octane (*D. C. Owsley*, *Helmkamp* and *S. N. Spurlock*, J. Amer. chem. Soc., 1969, **91,** 3606).

The intervention of a thiiranium ion (episulphonium ion) has been postulated to count for a single product (V) obtained on chlorination of the isomeric alcohols III and IV:

$$\underset{\text{(III)}}{\text{Et—S—CH}_2\text{—CH(Me)—OH}} \xrightarrow[\text{SOCl}_2]{\text{HCl}} \text{1-ethyl-2-methylthiiranium ion} \xleftarrow[\text{SOCl}_2]{\text{HCl}} \underset{\text{(IV)}}{\text{HO—CH}_2\text{—CH(Me)—S—Et}}$$

$$\downarrow$$

$$\underset{\text{(V)}}{\text{Et—S—CH}_2\text{—CH(Me)Cl}}$$

Thiiranium ion intermediates have also been proposed for the polar addition of sulphenyl halides to double bonds (*W. H. Mueller*, Angew. Chem., Intern. Edn., 1969, **8,** 482):

$$\text{R—S—X} + \text{>C=C<} \longrightarrow \text{thiiranium ion (R on S}^{\oplus}\text{)}\ X^{\ominus} \longrightarrow \text{RS—C—C—X}$$

(c) Thiirenes (acetylene sulphides)

2-Methylthiirene 1,1-dioxide is obtained by the dehydrobromination of 2-bromo-2-methylthiirane 1,1-dioxide by triethylamine at 3°. The thiirane 1,1-dioxide is prepared by treating α-bromoethanesulphonyl chloride with diazomethane in ether at 7–8° in the presence of triethylamine (*L. A. Carpine* and *R. H. Rynbrandt*, J. Amer. chem. Soc., 1966, **88,** 5682):

$$\text{MeCH(Br)SO}_2\text{Cl} \xrightarrow[\text{(Et)}_3\text{N}]{\text{CH}_2\text{N}_2} \text{2-bromo-2-methylthiirane 1,1-dioxide} \xrightarrow{\text{(Et)}_3\text{N}} \text{2-methylthiirene 1,1-dioxide}$$

Related thiirene 1,1-dioxides are generated by the action of base on αα-dichlorosulphones ($RCH_2 \cdot SO_2 \cdot CCl_2R'$) (*L. A. Paquette* and *L. S. Wittenberg*, Chem. Comm., 1966, 471).

The synthesis of 2,3-diphenylthiirene 1,1-dioxide (VI), the first example of a three-membered ring heterocycle which is potentially aromatic, because of the possible involvement of *d* orbitals at the heteroatom, is reported (*Carpine* and *L. V. McAdams*, J. Amer. chem. Soc., 1965, **87,** 5804):

$$PhCH(Br)\cdot SO_2 \cdot CHPh(Br) \xrightarrow{(Et)_3N}$$ (VI) (VII)

2,3-**Diphenylthiirene** 1,1-**dioxide** decomposes at its melting point to diphenylacetylene and sulphur dioxide, but it is very much more stable in chloroform solution at room temperature than the corresponding saturated *cis*-diphenylthiirane 1,1-dioxide which decomposes to *cis*-stilbene and sulphur dioxide.

A comparable study has been made of the rates of cleavage by alkoxide ions of diphenylthiirene 1,1-dioxide (VI) and diphenylcyclopropenone (VII). The ratio of rates was ~5000 to 1. This striking reversal of the usual much more facile attack of base on the carbonyl group than on the sulphonyl group is attributed to a marked conjugative stabilisation of VII as contrasted to a slight conjugative stabilisation of VI (*Bordwell* and *S. C. Crooks*, *ibid.*, 1969, **91**, 2084).

4. Three-membered rings containing one nitrogen atom

(a) Aziridines (ethyleneimines)

A. Ladenburg was the first to attempt the preparation of aziridine by distilling ethylenediamine hydrochloride, but this method, successful in the preparation of five-, six-, and seven-membered rings, here gave the dimer, piperazine. The simple base was prepared in the same year (1888) by *S. Gabriel*, who treated β-bromoethylamine, with alkali. *W. Marckwald* (Ber., 1900, **33,** 764), concluded that the product, which was stable to permanganate and gave an insoluble benzenesulphonyl derivative, must be aziridine and not vinylamine,, $CH_2{:}CHNH_2$, as *Gabriel* had supposed. Aziridines have attracted considerable attention, because of their easy conversion into high polymers, their use as aminoalkylating agents, and their powerful physiological action; 2,5-bisaziridinyl-*p*-benzoquinol (2,5-bisethylene-imino-*p*-benzoquinol) is strongly carcinostatic *in vivo* (*A. Marxer*, Experientia, 1955, **11,** 184).

An interesting stereochemical question relates to aziridines. Whereas the enantiomeric configurations of an unsymmetrical tertiary amine are too labile to lead to viable stereoisomers, the disposition of the nitrogen valencies in oximes or azo-compounds gives rise to isolable *syn* and *anti* modifications. Several workers pointed out simultaneously (*P. Maitland*, Annual Reports, 1939, **36,** 243) that if we regard an azomethine system C=N– as a "two-membered ring" in contrast with the open-chain saturated amines, it would be worth while to test the resolvability of such a three-membered ring compound as

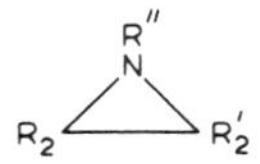

So far attempts to resolve compounds of this type have been unsuccessful, for it appears that the nitrogen atom inverts too easily at ordinary temperatures.

In fact the high rate of inversion of 1-methylsulphonyl- and 1-*p*-tolylsulphonyl-aziridine at $-37°$ is attributed to the $d\pi$–$p\pi$ interaction of the sulphur *d*-orbitals (*T. G. Traylor*, Chem. and Ind., 1963, 649).

The invertomers of 1-chloro-2-methylaziridine have been isolated (p. 29).

(i) Synthesis

(*1*) *Gabriel's* original ring-closure of β-bromoethylamine is applicable in general to primary and secondary bases (*Marckwald* and *O. Frobenius*, Ber., 1901, **34**, 3544):

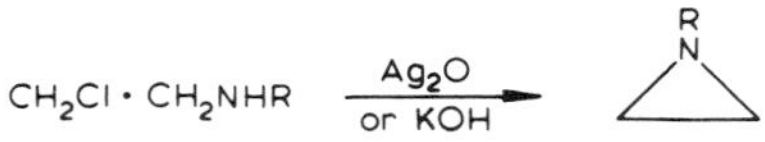

$$CH_2Cl \cdot CH_2NHR \xrightarrow[\text{or KOH}]{Ag_2O}$$

Similarly, tertiary β-chloroethylamines yield aziridinium derivatives and then piperazinium salts (*W. E. Hanby et al.*, J. chem. Soc., 1947, 513, 519; *P. D. Bartlett et al.*, J. Amer. chem. Soc., 1947, **69**, 2971):

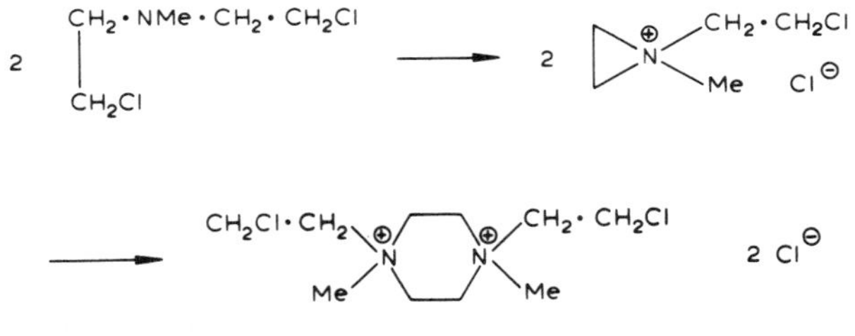

It is also possible to cyclise *N*-β-halogenoethylsulphonamides (*R. Adams* and *T. L. Cairns*, *ibid.*, 1939, **61**, 2464; *M. S. Kharasch and H. M. Priestley*, *ibid.*, p. 3425):

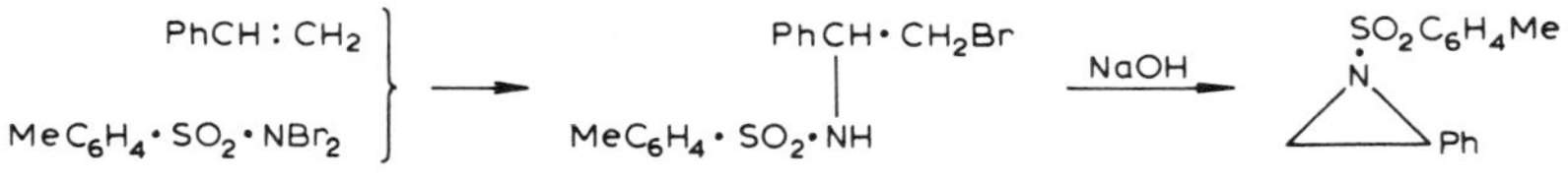

$$\left.\begin{array}{r}PhCH:CH_2\\ MeC_6H_4\cdot SO_2\cdot NBr_2\end{array}\right\} \longrightarrow \begin{array}{c}PhCH\cdot CH_2Br\\ |\\ MeC_6H_4\cdot SO_2\cdot NH\end{array} \xrightarrow{NaOH}$$

Derivatives of 2-acyl-3-arylaziridines (*cis*- and *trans*-forms) have been prepared in accordance with the following scheme (*N. H. Cromwell et al.*, *ibid.*, 1943, **65**, 312; 1949, **71**, 708):

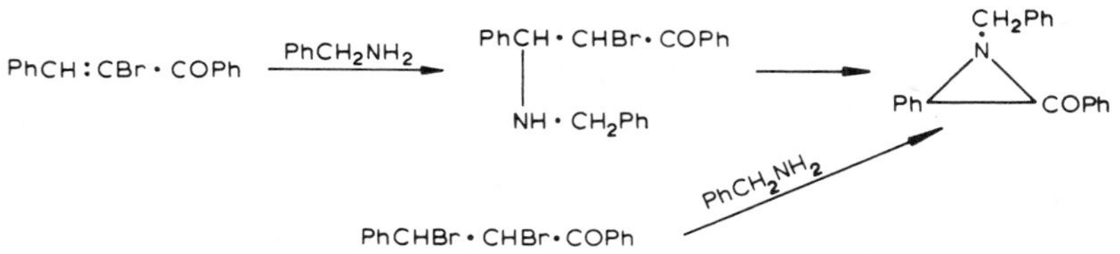

$$PhCH{:}CBr\cdot COPh \xrightarrow{PhCH_2NH_2} PhCH(NH\cdot CH_2Ph)\cdot CHBr\cdot COPh \longrightarrow \text{1-benzyl-2-Ph-3-COPh-aziridine}$$

$$PhCHBr\cdot CHBr\cdot COPh \xrightarrow{PhCH_2NH_2} \text{1-benzyl-2-Ph-3-COPh-aziridine}$$

(*2*) Some β-hydroxyethylamines yield aziridines by esterification with sulphuric acid, followed by treatment with alkali; this is a convenient route to aziridine itself (*H. Wenker*, *ibid.*, 1935, **57**, 2328) and has been successfully employed in the synthesis of 2- and 3-arylaziridines (*S. J. Brois*, J. org. Chem., 1962, **27**, 3532). Alternatively, the amino-alcohol hydrochloride may be treated with thionyl chloride and the product distilled with alkali without isolation (*G. I. Braz*, C.A., 1956, **50**, 2457):

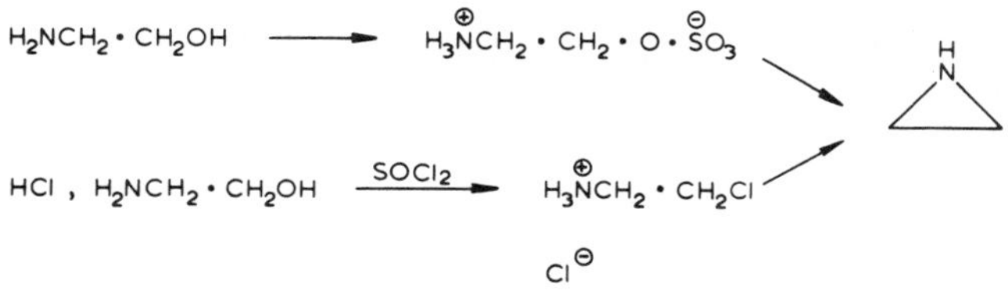

$$H_2NCH_2\cdot CH_2OH \longrightarrow H_3\overset{\oplus}{N}CH_2\cdot CH_2\cdot O\cdot \overset{\ominus}{S}O_3 \longrightarrow \text{aziridine (NH)}$$

$$HCl,\ H_2NCH_2\cdot CH_2OH \xrightarrow{SOCl_2} H_3\overset{\oplus}{N}CH_2\cdot CH_2Cl\ \ Cl^{\ominus} \longrightarrow \text{aziridine (NH)}$$

Another route to aziridines involves the nitrosochlorination of an olefin, followed by reduction and cyclisation (*G. L. Closs* and *Brois*, J. Amer. chem. Soc., 1960, **82**, 6068):

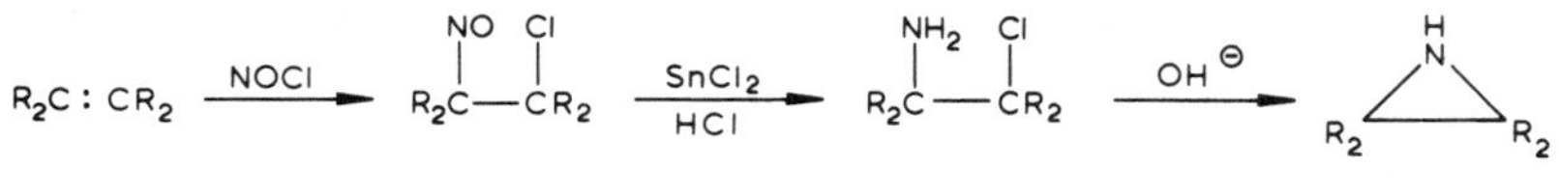

$$R_2C{:}CR_2 \xrightarrow{NOCl} R_2C(NO)\text{—}CR_2(Cl) \xrightarrow[HCl]{SnCl_2} R_2C(NH_2)\text{—}CR_2(Cl) \xrightarrow{OH^{\ominus}} \text{2,2,3,3-}R_4\text{-aziridine (NH)}$$

(*3*) Aziridines, Schiff bases or mixtures of both are formed from olefins and organic azides. At elevated temperatures the 1,2,3-triazolines, which are the products formed initially at lower temperatures, decompose with the elimination of nitrogen. Cyclopentene, cycloheptene, and *cis*-cyclooctene on heating with phenyl azide give the Schiff bases as the major products, while cyclohexene and *trans*-cyclo-octene give the aziridines (*K. R. Henery-Logan* and *R. A. Clark*, Tetrahedron Letters, 1968, 801; *K. Alder* and *G. Stein*, Ann., 1933, **501,** 1; *S. J. Davis* and *C. S. Rondestvedt*, Chem. and Ind., 1956, 845):

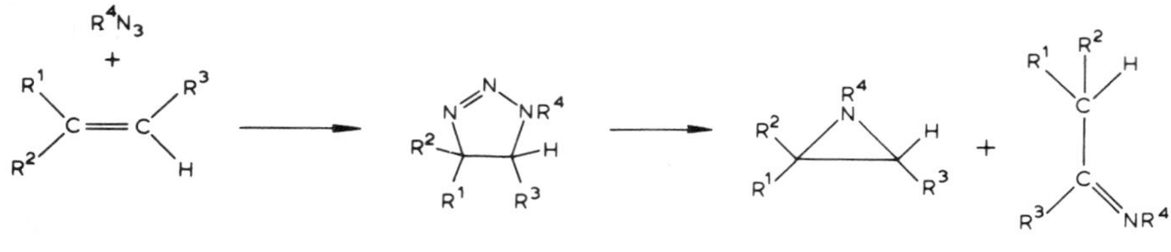

A stereospecific synthesis of aziridines is achieved by the reductive ring closure of β-iodoazides under mild conditions using lithium tetrahydrido-aluminate in ether (*A. Hassner*, *G. J. Matthews* and *F. W. Fowler*, J. Amer. chem. Soc., 1969, **91,** 5046):

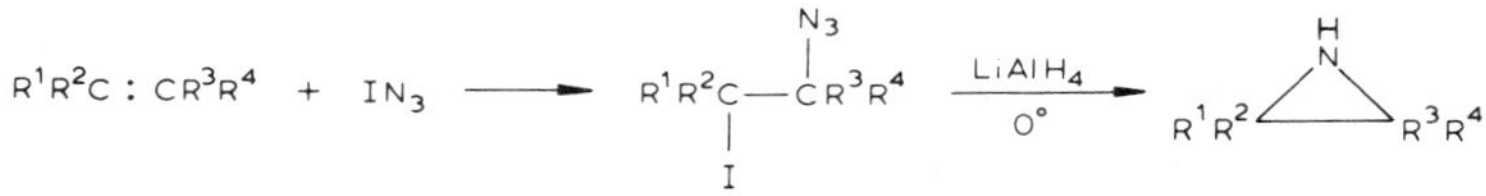

When 1,2,3-triazolines are decomposed photochemically the main products are aziridines (*P. Scheiner*, Tetrahedron, 1968, **24,** 2757).

(*4*) *J. Hoch* discovered the remarkable reaction of some ketoximes with Grignard reagents (*K. N. Campbell et al.*, J. org. Chem., 1943, **8,** 103; 1944, **9,** 194):

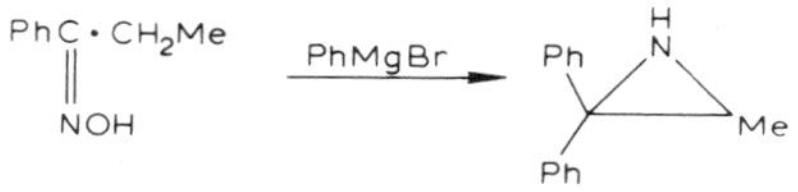

Reduction of benzyl ketoximes with lithium tetrahydridoaluminate gives aziridines as the main product (*K. Kitahonoki et al.*, Tetrahedron Letters, 1965, 1059). The reaction is stereoselective in that the aziridine obtained depends upon the configuration of the starting oxime (*K. Kotera*, *T. Okada* and *S. Miyazaki*, *ibid.*, 1967, 841):

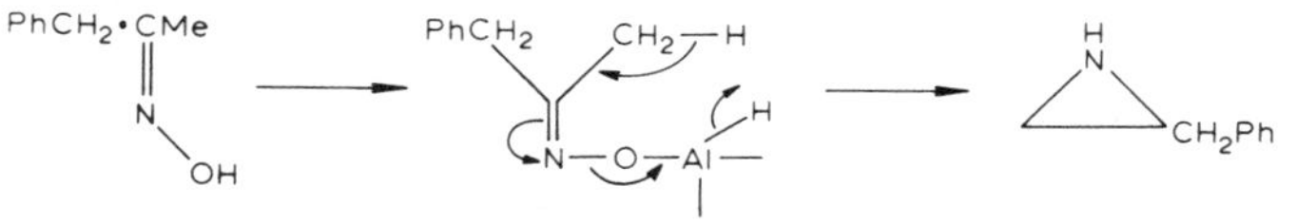

contrasting with the synthesis of α-aminoketones from oxime tosylates through the use of the Neber rearrangement (*P. W. Neber and A. Friedol-*

sheim, Ann., 1926, **449,** 109; see also p. 33), which is non-stereoselective.

Aziridines are also prepared by the reduction with lithium tetrahydridoaluminate in tetrahydrofuran of 2-isoazolines, obtained by 1,3-dipolar cycloaddition of nitrile oxides and olefins (*Kotera et al., ibid.*, 1968, 5759):

(*5*) The irradiation of ethyl azidoformate at 254 mμ in cyclohexene gives an aziridine (*W. Lwowski* and *T. W. Mattingly, ibid.*, 1962, 277):

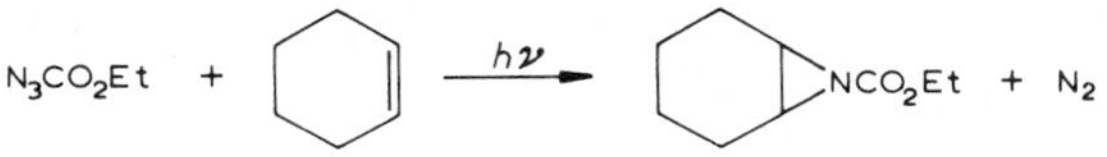

N_3CO_2Et + cyclohexene $\xrightarrow{h\nu}$ 7-azabicyclo[4.1.0]heptane-NCO_2Et + N_2

(*6*) The stereospecific addition of an aminonitrene to mono- and di-enes to give an aziridine (V) has been achieved by the oxidation of 3-aminobenzoxazolin-2-one (I) and other *N*-amino compounds (II–IV) with lead tetra-acetate in the presence of an olefin (*C. W. Rees et al.*, Chem. Comm., 1967, 1230; 1969, 146):

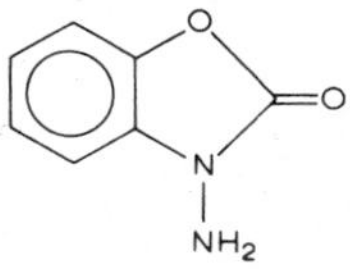

(I) 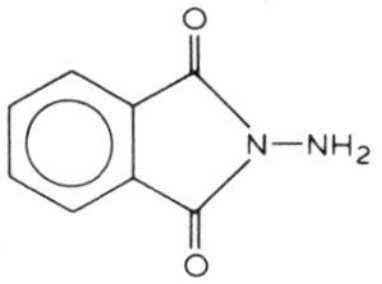 (II) 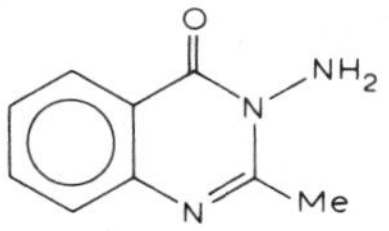 (III) 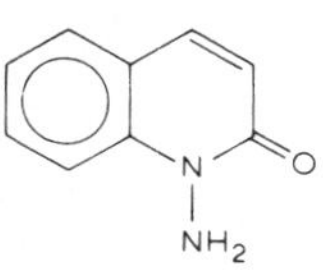 (IV)

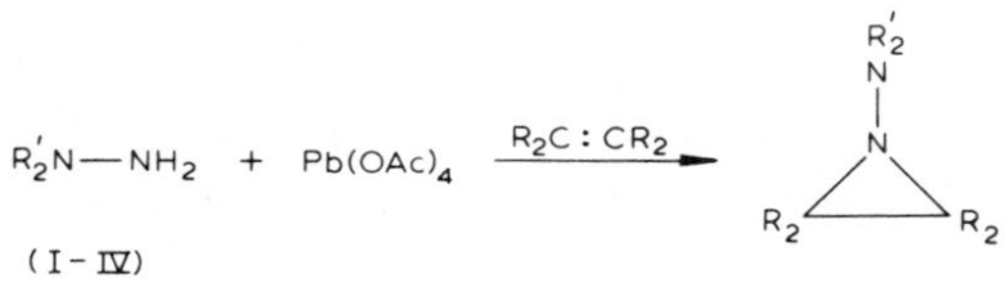

$R'_2N{-}NH_2$ + $Pb(OAc)_4$ $\xrightarrow{R_2C:CR_2}$ (V)

(I–IV)

Highly strained bridge aziridines have been synthesised by intramolecular trapping of a nitrene intermediate in the oxidation of $\delta\varepsilon$-unsaturated primary amines with lead tetra-acetate (*W. Nagata et al.*, J. Amer. chem. Soc., 1967, **89,** 5045):

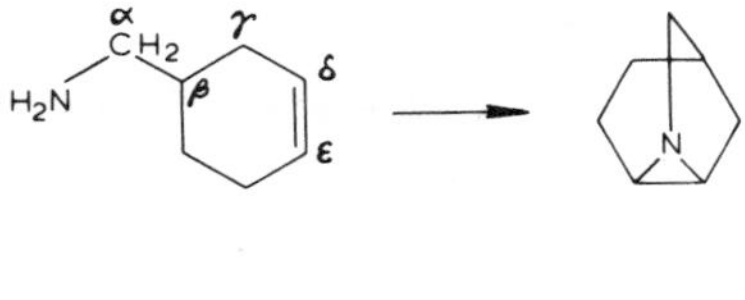

The photolysis of 3-azido-2-phenylprop-1-ene gives 3-phenyl-1-azabicyclo[1.1.0]butane (*A. G. Hortmann* and *J. E. Martinelli*, Tetrahedron Letters, 1968, 6205):

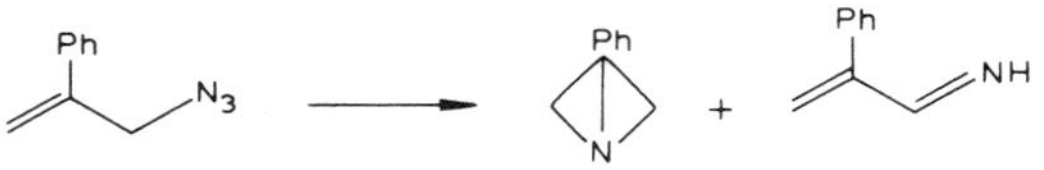

The parent compound, 1-azabicyclo[1.1.0]butane (VI) has been prepared by the action of base on the appropriate salt (R = H) (*W. Funke*, Angew. Chem., intern. Edn., 1969, **8**, 70):

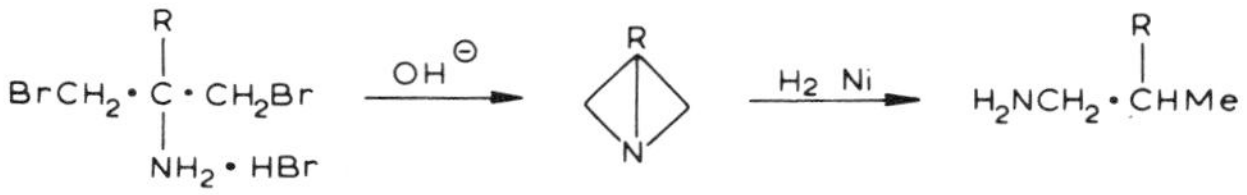

R = H, Me or Et (VI)

Catalytic hydrogenation of VI with Raney nickel and alcohol gives primary amines; and acid chlorides, acid anhydrides and thiophenol add to VI, breaking the bridge bond and forming azetidines.

(*7*) The quaternary salts formed by the action of methyl iodide on the adducts obtained from $\alpha\beta$-unsaturated carboxanilides and hydrazine undergo base-catalysed elimination of trimethylammonium iodide to give aziridine-2-carboxanilides (*G. R. Harvey*, J. org. Chem., 1968, **33**, 887):

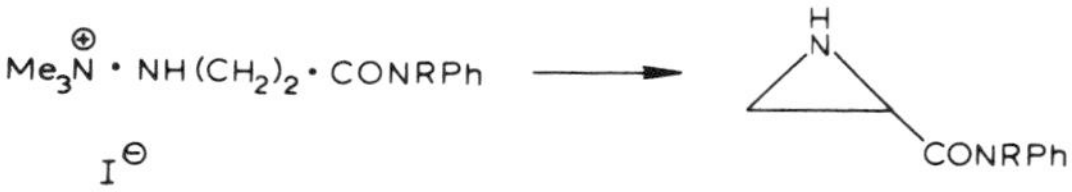

(*8*) Aziridines have been obtained by treatment of α-chloroamides with aluminium hydride (*Y. Langlois, H. P. Husson* and *P. Potier*, Tetrahedron Letters, 1969, 2085):

(ii) Reactions of aziridines

(*1*) Aziridine undergoes addition to $\alpha\beta$-unsaturated ketones, esters and sulphones (*H. Bestian et al.*, G.P. 849,407; C.A., 1958, **52**, 11945); the product from 2,2-dimethylaziridine and acrylonitrile can be selectively hydrogenated with retention of the imine ring (*D. S. Tarbell* and *D. K. Fukushima*, J. Amer. chem. Soc., 1946, **68,** 2499):

$$\text{Me}_2\text{-aziridine (NH)} \xrightarrow{CH_2{:}CH\cdot CN} \text{Me}_2\text{-aziridine-}NCH_2\cdot CH_2\cdot CN \xrightarrow[\text{or } H_2,\ \text{Raney Ni, pressure } 100^\circ]{Na/EtOH,\ PhMe} \text{Me}_2\text{-aziridine-}NCH_2\cdot CH_2\cdot CH_2NH_2$$

(*2*) *Addition to olefins.* Aziridine also adds to styrene under the influence of metallic sodium (*M. Erlenbach* and *A. Sieglitz*, B.P. 692,368; C.A., 1954, **48,** 7627).

(*3*) *Halogenation.* Aziridine reacts with aqueous sodium hypohalite at -5 to -10° to give 1-chloro- and 1-bromo-aziridine, which with 1-lithium-aziridine yields 1,1′-biaziridine (*A. F. Graefe* and *R. E. Meyer*, J. Amer. chem. Soc., 1958, **80,** 3939):

$$\text{(aziridine)}NLi + ClN\text{(aziridine)} \xrightarrow{Et_2O} \text{(aziridine)}N{-}N\text{(aziridine)}$$

(*4*) Aziridine with *nitrosyl chloride* at -60° in ether yields 1-nitrosoaziridine

$$\text{(aziridine)}N\cdot NO$$

detected by its u.v. absorption spectrum and decomposed at 20° to ethylene and nitrous oxide (*W. Rundel* and *E. Müller*, Ber., 1963, **96,** 2528).

(*5*) *Aziridinium salts.* Stable aziridinium perchlorates have been synthesised from β-chloroethylamines by treatment with silver perchlorate (*N. J. Leonard* and *J. V. Paukstelis*, J. org. Chem., 1965, **30,** 821):

$$Et_2NCH_2\cdot CH_2Cl \xrightarrow[0^\circ]{AgClO_4} \text{(aziridinium, } \overset{\oplus}{N}Et_2\text{)}\ ClO_4^{\ominus}$$

The reaction of 1,1,2,2-tetrasubstituted aziridinium salts with weak nucleophiles generally result in 1,2-bond cleavage, while the 1,3-bond is broken with strong nucleophiles (*Leonard* and *D. A. Durand*, *ibid.*, 1968, **33,** 1322). Preliminary 1,2-bond cleavage is also postulated as the initial step in the expansion of the aziridinium ring with aldehydes (*Leonard*, *E. F. Keifer* and *L. E. Brady*, *ibid.*, 1963, **28,** 2850), ketones (*Leonard*, *Paukstelis* and *Brady*, *ibid.*, 1964, **29,** 3383), and nitriles (*Leonard* and *Brady*, *ibid.*, 1965, **30,** 817).

(6) *Aminoethylation.* Processes of ring-opening, which include the normal reaction of *N*-substituted aziridines with cyanogen bromide (*R. C. Elderfield* and *H. A. Hagemann, ibid.*, 1949, **14**, 605), often display the versatility of the imine and especially of the aziridinium ion as aminoethylating agents. The ring is opened, both in aqueous acid and alkaline solutions, giving aminoethanol:

$$\text{Acid:}\quad H_2O + \overset{\oplus}{N}H_2(CH_2-CH_2) \longrightarrow H_2\overset{\oplus}{O}CH_2\cdot CH_2NH_2 \xrightarrow{-H^{\oplus}} HOCH_2\cdot CH_2NH_2$$

$$\text{Alkaline:}\quad HO^{\ominus} + \overset{\oplus}{N}H_2(CH_2-CH_2) \longrightarrow HO\cdot CH_2\cdot CH_2\overset{\ominus}{N}H \xrightarrow{H^{\oplus}} HOCH_2\cdot CH_2NH_2$$

In some cases the main reaction of an aziridine under acid conditions is dimerisation to a piperazine derivative (*L. Knorr*, Ber., 1904, **37**, 3507):

$$2\ \text{MeN}(CH_2)_2 \longrightarrow \text{MeN}(CH_2CH_2)_2\text{NMe}$$

At low pH dimerisation does not take place in the case of several aliphatic tertiary β-chloroethylamines and these react with $S_2O_3^{2\ominus}$, $HCO_3^{\ominus}$, $Cl^{\ominus}$, $EtCO_2^{\ominus}$, and $PhCO_2^{\ominus}$ *via* the aziridinium ion, the initial cyclization being over 99.5% complete in dilute aqueous solution (*E. M. Schmultz* and *J. M. Sprague*, J. Amer. chem. Soc., 1948, **70**, 48; *B. Cohen, E. R. van Artsdalen* and *J. Harris, ibid.*, 1952, **74**, 1878):

$$RR'NCH_2\cdot CH_2Cl \rightleftharpoons RR'\overset{\oplus}{N}(H_2C-CH_2) + Cl^{\ominus} \xrightarrow{X^{\ominus}} RR'NCH_2\cdot CH_2X$$

It appears, therefore, that aminoethylation by β-chloroethylamine involves the intermediate production of an aziridine or an aziridinium ion. Aziridinium ions can be determined by their aminoethylation of thiosulphate (*C. Golumbic et al.*, J. org. Chem., 1946, **11**, 518).

A detailed study of the *polymerisation of aziridine* under acid conditions indicates that it proceeds by the formation of iminium intermediates (*W. G. Barb*, J. chem. Soc., 1955, 2564, 2577; *J. E. Ewley et al.*, J. Amer. chem. Soc., 1958, **80**, 3458) thus:

$$\text{C}_2\text{H}_4\text{NH} \xrightarrow{H^{\oplus}}$$

$$\text{C}_2\text{H}_4\text{NH} + \text{C}_2\text{H}_4\overset{\oplus}{\text{N}}\text{H}_2 \longrightarrow \text{C}_2\text{H}_4\overset{\oplus}{\text{N}}\text{H}\cdot CH_2\cdot CH_2NH_2$$

$$\text{C}_2\text{H}_4\text{NH} + \text{C}_2\text{H}_4\overset{\oplus}{\text{N}}\text{HCH}_2\cdot CH_2NH_2 \longrightarrow \text{C}_2\text{H}_4\overset{\oplus}{\text{N}}\text{HCH}_2\cdot CH_2NHCH_2\cdot CH_2NH_2$$

The aminoethylation of an imino-group by the aziridinium ion is a general reaction, promoted by aluminium chloride (*G. H. Coleman* and *J. S. Callen*, *ibid.*, 1946, **68,** 2006). Aromatic hydrocarbons have been aminoethylated in the presence of aluminium chloride (*G. I. Braz*, C.A., 1954, **48,** 569).

(7) *Hydrogen sulphide*, *thiols* and *phenols* react with the free imine (*Bestian*, *loc. cit.; L. B. Clapp*, J. Amer. chem. Soc., 1951, **73,** 2584) to give bis-β-aminoethyl sulphide, $(NH_2CH_2\cdot CH_2)_2S$, a β-aminoethyl alkyl or aryl sulphide, $NH_2CH_2\cdot CH_2SAr$, and a β-aminoethyl aryl ether, $NH_2CH_2\cdot CH_2OAr$, respectively. With sulphurous acid and aziridine, *Gabriel* (Ber., 1888, **21,** 2667) observed the formation of taurine:

$$\text{C}_2\text{H}_4\text{NH} + H_2SO_3 \longrightarrow NH_2CH_2\cdot CH_2SO_3H$$

(8) *Carbonyl and thiocarbonyl compounds.* Aziridine reacting in the cold with carbonyl compounds yields aminohydrins (*W. J. Rabourn* and *W. L. Howard*, J. org. Chem., 1962, **27,** 1039), which isomerise in boiling ether either to oxazolidines (*J. B. Doughty, C. L. Lazzell* and *A. R. Collett*, J. Amer. chem. Soc., 1950, **72,** 2866):

$$\text{RCHO} + \text{C}_2\text{H}_4\text{NH} \longrightarrow \text{R—CH—OH (on N of aziridine)} \xrightarrow[\text{Et}_2\text{O}]{\text{boiling}} \text{oxazolidine (O, NH, R)}$$

$$\text{C}_2\text{H}_4\text{NH} + \text{RR}'\text{CO} \longrightarrow \text{oxazolidine (O, NH, R R}')$$

or to alkylidenebisaziridines (*A. Dornow* and *W. Schacht*, Ber., 1949, **82,** 464):

Carbon dioxide reacts with aziridine at low temperature to give aziridinium vinylcarbamate (*A. Seher*, Ann., 1952, **575,** 153):

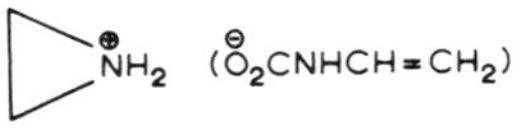

Aziridines react with carbon disulphide to give 2-thiothiazolidones in good yields when the reaction is carried out without solvent and under pressure (*Clapp* and *J. W. Watjen*, J. Amer. chem. Soc., 1953, **75,** 1490):

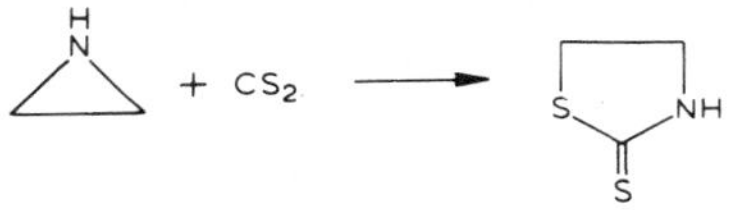

(9) *Ring expansion reactions.* On heating, *N*-aroylaziridines undergo ring-expansion to oxazolines. The same isomerisation can be effected by nucleophiles and it appears that electronic factors play a dominant role in the sodium iodide-catalysed ring-opening of certain aziridines (*H. W. Heine et al.*, J. org. Chem., 1958, **23,** 1554; 1967, **32,** 3069; J. Amer. chem. Soc., 1959, **81,** 2202; Tetrahedron Letters, 1967, 1859):

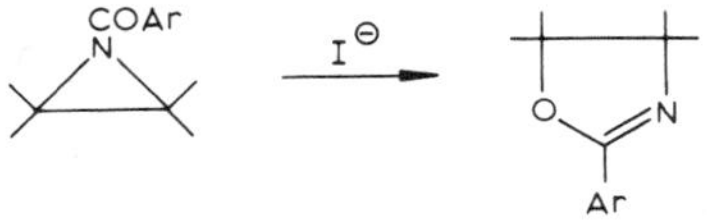

Reaction of sodium *N-tert*-butylaziridine-2-carboxylate with thionyl chloride in tetrahydrofuran dried with sodium hydride does not yield the usual acid chloride but 1-*tert*-butyl-3-chloro-azetidin-2-one (*J. A. Deyrup* and *S. C. Clough*, J. Amer. chem. Soc., 1969, **91,** 4590):

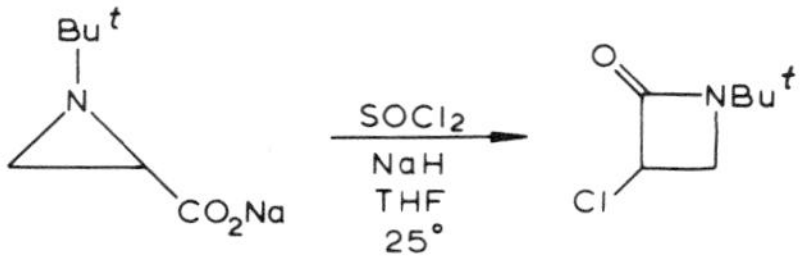

On heating simple 2-vinylaziridines rearrange to 3-pyrrolines (*R. S. Atkinson* and *Rees*, J. chem. Soc., C, 1969, 778):

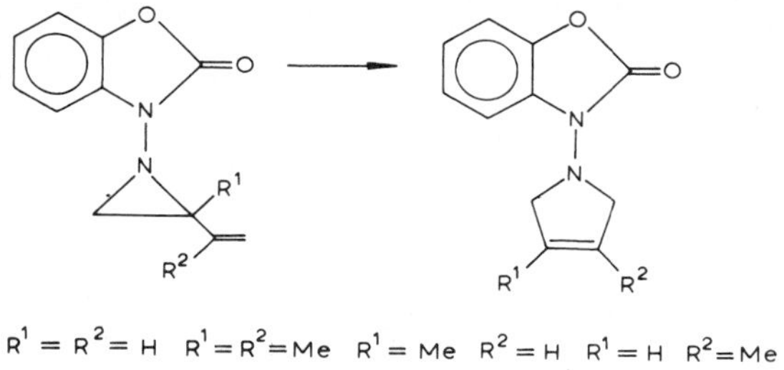

$R^1 = R^2 = H$ $R^1 = R^2 = Me$ $R^1 = Me$ $R^2 = H$ $R^1 = H$ $R^2 = Me$

(*10*) The reduction of 1-acylaziridines by lithium tetrahydridoaluminate affords a method for converting a carboxylic into an aldehyde group (*H. C. Brown* and *A. Tsukamato*, J. Amer. chem. Soc., 1961, **83**, 2016):

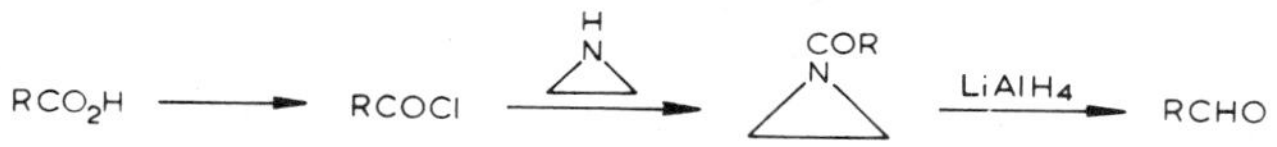

$RCO_2H \longrightarrow RCOCl \longrightarrow$ (1-acylaziridine) $\xrightarrow{LiAlH_4} RCHO$

(*11*) *Carbon–carbon fission.* The direction of fission in unsymmetrical aziridines appears to be governed by the same principles as in oxirans (*Tarbell* and *P. Noble, ibid.*, 1950, **72**, 2657); the stereochemistry of some 2,3-dimethyl- and 2,3-diaryl-aziridines and the mechanism of their ring opening have been discussed (*H. J. Lucas et al., ibid.*, 1952, **74**, 944; *N. H. Cromwell et al., ibid.*, 1953, **75**, 5384; *V. B. Schatz* and *Clapp, ibid.*, 1955, **77**, 5113). A number of examples have been reported of aziridine reactions in which the carbon–carbon bond is cleaved (*R. Huisgen, W. Scheer* and *H. Huber, ibid.*, 1967, **89**, 1753; *A. Padwa* and *L. Hamilton*, J. Heterocyclic Chem., 1967, **4**, 118; *Heine, R. Peavy* and *A. J. Durbetaki*, J. org. Chem., 1967, **32**, 3924). It may be induced by thermal or photochemical excitation and probably proceeds *via* an azomethine ylide intermediate, *e.g.*, the formation of 2,5-diphenyloxazole when *trans*-2-benzoyl-1-*tert*-butyl-3-phenylaziridine is heated at 220° (*Padwa* and *W. Eisenhardt*, Chem. Comm., 1968, 380):

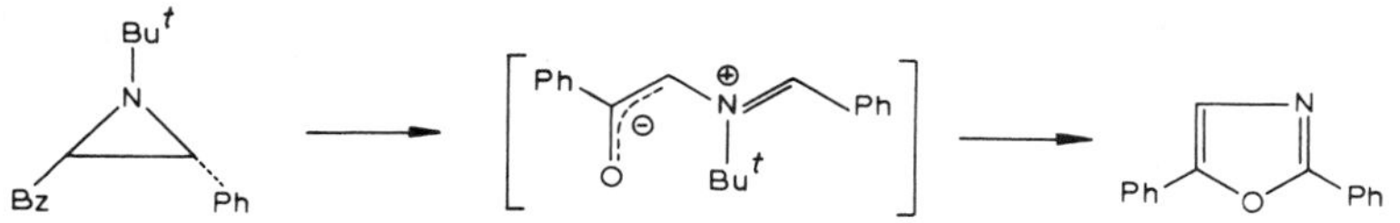

cis- or *trans*-2-Benzoyl-1-cyclohexyl-3-phenylaziridine on treatment with one equivalent of diphenylcyclopropenone in boiling benzene affords a 4-oxazoline (*J. W. Lown, R. L. Smalley* and *G. Dallas, ibid.*, 1968, 1543):

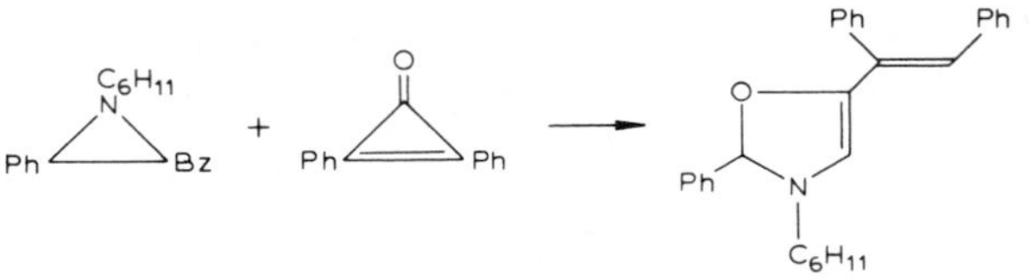

Other reactions in which the carbon–carbon bond of aziridines is cleaved have been reported (*Huisgen et al.*, Angew. Chem., intern. Edn., 1969, **8,** 602, 604). *cis*-1-(*N*-Phenylcarbamoyl)- and *cis*-1-(*N*-phenylthiocarbamoyl)-2,3-dimethylaziridine have been isomerised under acid conditions to oxazoline and thiazoline derivatives (*T. Nishiguchi et al.*, J. Amer. chem. Soc., 1969, **91,** 5835, 5841).

Irradiation of 1,2,3-triphenylaziridine in alcohol affords mainly benzaldehyde acetal, *N*-benzylaniline and *N*-benzylideneaniline together with a small amount of alkyl benzyl ether, the latter arising *via* phenylcarbene, which is trapped by the alcohol. Photolysis in cyclohexene affords a 1,3-cycloaddition product, 1,2,3-triphenyloctahydroisoindole as a mixture of two stereoisomers (*H. Nozaki, S. Fujita* and *R. Moyori*, Tetrahedron, 1968, **24,** 2193):

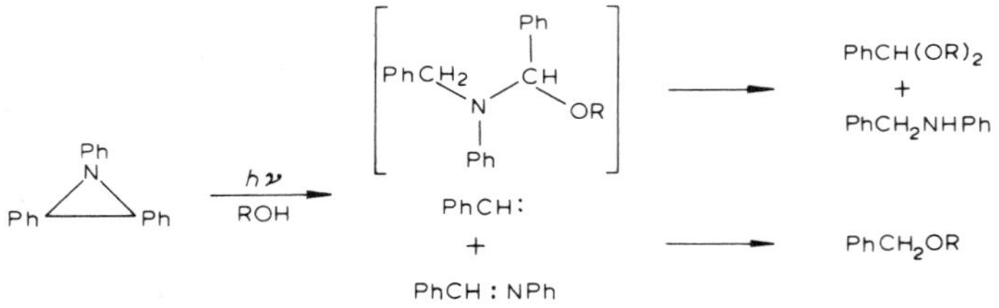

(*12*) The deamination of *cis*- and *trans*-2,3-dimethylaziridine with difluoroamine is stereospecific (*J. P. Freeman* and *W. H. Graham*, J. Amer. chem. Soc., 1967, **89,** 1761):

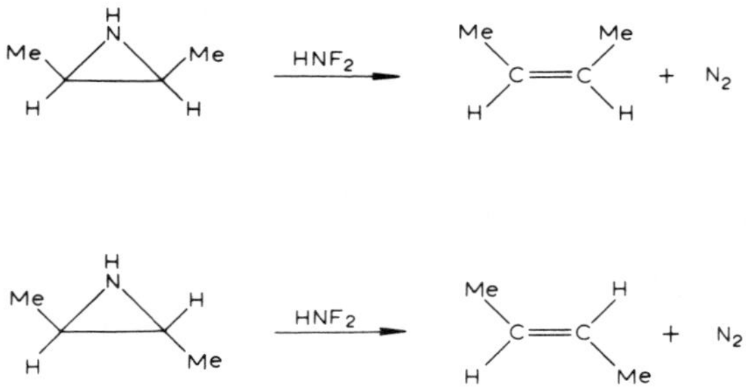

(iii) Properties of individual aziridines

Aziridine (*V. P. Wystrach et al.*, J. Amer. chem. Soc., 1955, **77**, 5915; 1956, **78**, 1263; *C. F. H. Allen et al.*, Org. Synth., 1950, **30**, 38), b.p. 56–57°, mixes with water in all proportions. Its i.r. absorption is discussed by *H. W. Thompson* and *W. T. Cave* (Trans. Faraday Soc., 1951, **47**, 951) and structural data is given by *A. B. Turner et al.*, J. chem. Phys., 1953, **21**, 564). The imine is stable in the absence of carbon dioxide, but polymerises energetically in the presence of a little hydrochloric acid. It is strongly caustic and causes acute inflammation of the mucous membrane; p*K* 7.98 (*G. J. Buist* and *H. J. Lucas*, J. Amer. chem. Soc., 1957, **79**, 6157). It behaves as a secondary amine; acylation of the imino-group proceeds normally in presence of triethylamine, giving rather unstable products while the carbamates and ureas are more stable (*H. Bestian*, Ann., 1950, **566**, 210). The *N*-lithium derivatives have been prepared and used for synthesis of 1-alkyl and 1-aryl derivatives (*H. Gilman et al.*, J. Amer. chem. Soc., 1945, **67**, 2106).

The melting points and boiling points of some representative substituted aziridines are listed in Table 3.

(+)- and (−)-Alanine have been converted into (+)- and (−)-2-methylaziridine (*Y. Minoura, M. Takebayaski* and *C. C. Price*, J. Amer. chem. Soc., 1959, **81**, 4689).

7-Azabicyclo[4.1.0]*heptane (cyclohexeneimine)*

NH

m.p. 20–21°, b.p. 149–150°, is prepared from 2-aminocyclohexanol by the Wenker process (p. 17). Acids open the heterocyclic ring with Walden inversion (*O. E. Paris* and *P. E. Fanta*, *ibid.*, 1952, **74**, 3007).

(iv) Halogenoaziridines

Dichloroaziridines are readily available from the addition of dichlorocarbene to Schiff bases [*1*] (*E. K. Fields* and *S. M. Sandri*, Chem. and Ind., 1959, 1216; *P. K. Kadale* and *J. O. Edwards*, J. org. Chem., 1960, **25**, 1431; *A. G. Cook* and *Fields*, *ibid.*, 1962, **27**, 3686). Monochloroaziridines are

$Bu^nOK + CH_2Cl_2$

Ph, H, C=N, Ph — $LiCHCl_2$ [2] → Ph, N, Ph, Cl

[1] $KOBu^t$, $CHCl_3$

Ph, N, Ph, Cl_2 — MeLi [3] → Ph, N, Ph, Me, Cl

TABLE 3

AZIRIDINES

Substituent	*b.p. (°C)*	*Ref.*
1-Methyl	24	1
1-Ethyl	49/690 mm	2
1-Phenyl	70/13 mm	3
2-Methyl	63–64	4
2-Ethyl	88–89	4
2,2-Dimethyl	69–70	5
cis-2,3-Dimethyl	83	6
trans-2,3-Dimethyl	75	6
cis-2,3-Diphenyl	m.p., 81–82	7
trans-2,3-Diphenyl	m.p., 46–47	8
1-Benzyl	84–87/8 mm	9
Methyl 1-acetate	74/24 mm	10
1-Acetyl	42/20 mm	10
1-Ethoxycarbonyl	60–63/21 mm	10
1-Carbamoyl	m.p., 106	10
1,1′-Carbonyl-bis-aziridine	m.p., 39–41	10
1-Benzoyl	m.p., 8–9	11
1-*p*-Toluenesulphonyl	m.p., 52	
Di-, tri- and tetra-substituted		12

References
1 *H. T. Hoffman et al.*, J. Amer. chem. Soc., 1951, **73,** 3028.
2 *R. C. Elderfield* and *H. A. Hageman*, J. org. Chem., 1949, **14,** 605.
3 *H. W. Heine et al.*, J. Amer. chem. Soc., 1954, **76,** 1173.
4 *G. D. Jones*, J. org. Chem., 1944, **9,** 484.
5 *T. L. Cairns*, J. Amer. chem. Soc., 1941, **63,** 871.
6 *H. J. Lucas et al.*, *ibid.*, 1952, **74,** 944.
7 *D. A. Darapsky* and *H. Spannagle*, J. pr. Chem., 1915, 272.
8 *A. Weissberger* and *H. Bach*, Ber., 1931, **64,** 1095; 1932, **65,** 631.
9 *W. S. Gump* and *N. J. Nikawitz*, J. Amer. chem. Soc., 1950, **72,** 1309.
10 *H. Bestian*, Ann., 1950, **566,** 210.
11 *A. A. Goldberg* and *W. Kelly*, J. chem. Soc., 1948, 1919.
12 *A. Hassner, G. J. Matthews* and *F. W. Fowler*, J. Amer. chem. Soc., 1969, **91,** 5046.

prepared by addition of monochlorocarbene to Schiffs bases [*2*] and also from dichloroaziridines [*3*] (*J. A. Deyrup* and *R. B. Greenwald*, Tetrahedron Letters, 1965, 321).

Monochloroaziridines undergo many smooth nucleophilic substitutions without rupture of the ring and with inversion of stereochemistry (*idem*, J. Amer. chem. Soc., 1965, **87,** 4538). 2-Chloro-2-methyl-1,3-diphenylaziridine on treatment with potassium *tert*-butoxide gives hydrocinnamanilide (*idem*, Tetrahedron Letters, 1966, 5091):

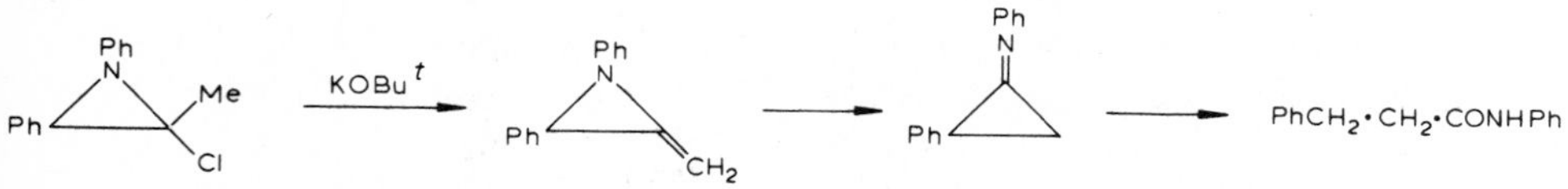

Some of the reactions of 2,2-dichloro-1,3,3-triphenylaziridine are illustrated in the following reaction scheme (*K. Ichimura* and *M. Otta*, *ibid.*, 1966, 807):

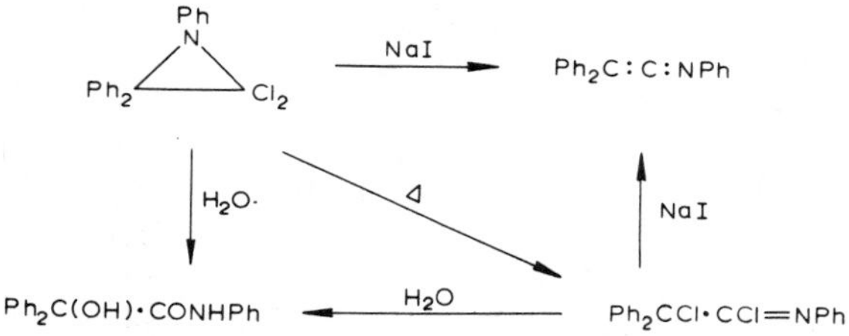

1-Halogeno-2,2-dimethyl- and 1-halogeno-2,2,3,3-tetramethyl-aziridines (p. 21) have been prepared by treating the respective aziridine with either *N*-chloro- or *N*-bromo-succinimide in ether (*S. J. Bois*, J. Amer. chem. Soc., 1968, **90**, 506):

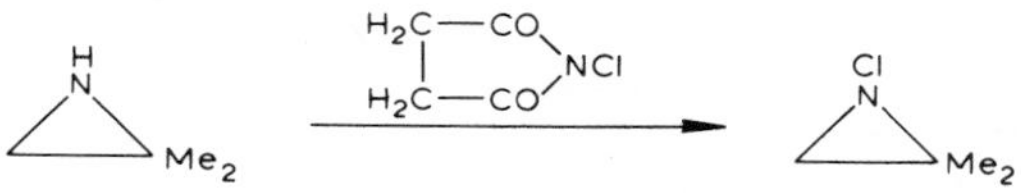

The *cis*- and *trans*-isomers of 1-chloro-2-methylaziridine have been prepared by this method and they represent the first isolable invertomers of trivalent nitrogen, thus providing evidence for the existence of a stable nitrogen pyramid in 1-chloroaziridines (*Bois*, *ibid.*, 1968, **90**, 508):

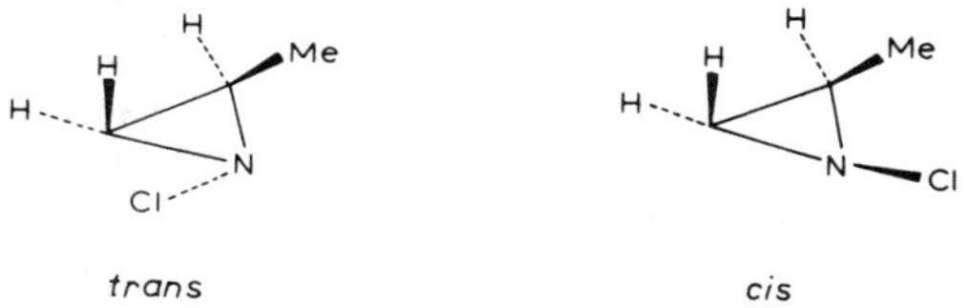

trans *cis*

Their exceptional stability demonstrates the remarkable configuration-holding power of chlorine and clearly negates the concept of *d*-orbital resonance stabilisation of pyramidal inversion in 1-chloroaziridines (*V. F.*

Bystrov et al., Opt. Spectry., U.S.S.R., 1965, **19**, 122). Related 1-halogeno-2-alkylaziridines have been prepared (*R. G. Kostyanousky, Z. E. Samojlova* and *I. I. Tchervin*, Tetrahedron Letters, 1969, 719). Though complex, the n.m.r. spectra of *N*-benzoxazolinonylaziridines (VII) show that inversion at the aziridine nitrogen is greatly retarded (*Atkinson* and *Rees*, J. chem. Soc., 1969, 772; *D. J. Anderson, D. C. Horwell* and *Atkinson*, Chem. Comm., 1969, 1189):

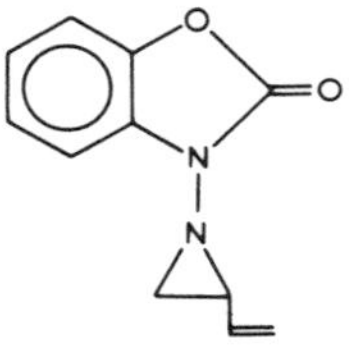

(VII)

The invertomers of 7-chloroazabicyclo[4.1.0]heptane may also be isolated (*D. Felix* and *A. Eschenmoser*, Angew. Chem., intern. Edn., 1968, **7**, 224). 1-Chloroaziridines undergo concerted carbon–carbon bond cleavage under solvolytic conditions (*P. G. Gassman* and *D. K. Dygos*, J. Amer. chem. Soc., 1969, **91**, 1543):

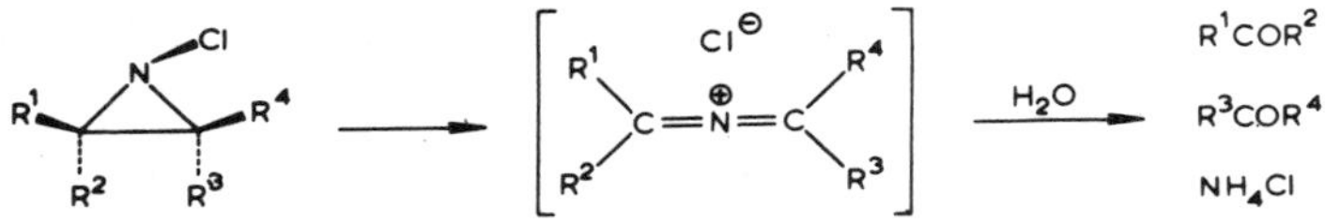

R^1COR^2 R^3COR^4 NH_4Cl

The *N*-chloro-derivative of indano [1,2-*b*]aziridine undergoes spontaneous dehydrochlorination to yield isoquinoline *via* the nitrenium ion (*Horwell* and *Rees*, Chem. Comm., 1969, 1428):

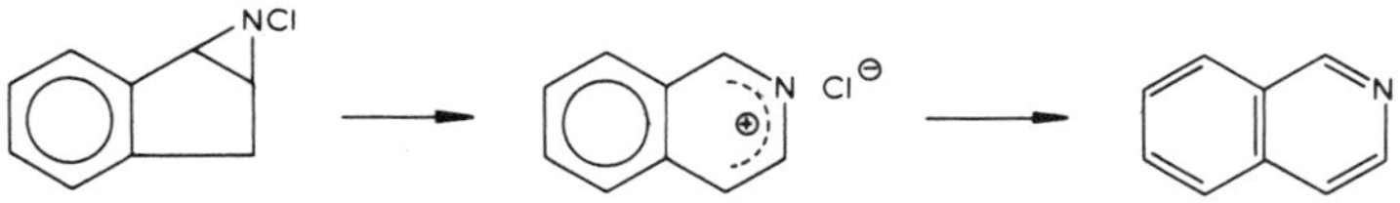

(v) Cyanoaziridines and aziridine carboxylic acids

αβ-Dibromopropionitrile reacts with primary amines in the presence of a tertiary base to give 1-alkyl-2-cyanoaziridines (*T. Wagner-Jauregg*, Helv., 1961, **44**, 1237):

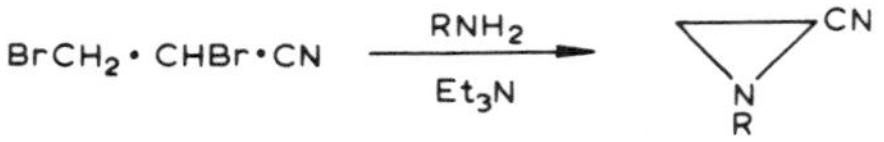

$$BrCH_2 \cdot CHBr \cdot CN \xrightarrow[Et_3N]{RNH_2}$$

When methyl 1-*tert*-butylaziridine-2-carboxylate is reacted with excess hydrazine an unusual reductive ring-opening occurs, with the evolution of gas and the formation of 3-*tert*-butylaminopropionhydrazide (*Deyrup* and *Clough*, J. Amer. chem. Soc., 1968, **90**, 3592):

$$\text{1-Bu}^t\text{-aziridine-2-}CO_2Me \xrightarrow[H_2O]{NH_2NH_2} Bu^tNH \cdot CH_2 \cdot CH_2 \cdot CONHNH_2$$

(b) Aziridinones (α-lactams)*

Phenylated aziridinones may be obtained by reacting appropriate chloroacetanilides with sodium hydride (*S. Sarel* and *H. Leader*, *ibid.*, 1960, **82**, 4752):

$$PhRCCl \cdot CONHPh \xrightarrow{NaH} \text{1-Ph-3-R-3-Ph-aziridinone} + H_2 + NaCl$$

(R = H or Ph)

1-*tert*-Butyl-3-phenylaziridinone is isolated following the treatment of *N*-*tert*-butyl-*N*-chlorophenylacetamide with potassium *tert*-butoxide (*H. E. Baumgarten et al.*, *ibid.*, 1963, **85**, 3303). Similarly 1-*tert*-butyl-3,3-dimethylaziridinone is prepared by the dehydrobromination of 2-bromo-*N*-*tert*-butyl-2-methylpropionamide:

$$Me_2C(Br) \cdot CO \cdot NHCMe_3 \xrightarrow{KOBu^t} [Me_2C(Br)-C(=O)-\overset{\ominus}{N}CMe_3] \longrightarrow \text{1-Bu}^t\text{-3,3-Me}_2\text{-aziridinone}$$

$$\xrightarrow{\text{Boil } Et_2O \text{ 1h}} CH_2{:}C(Me) \cdot CONHCMe_3 + Me_2CO + Me_3CNC$$

The α-lactam decomposes on vacuum distillation (above 30°) and on column and thin-layer chromatography. On boiling with ether conversion to *N*-*tert*-butylmethylacrylamide occurs, along with the formation of small amounts of acetone and *tert*-butyl isocyanide. Nonionic nucleophiles such as water, *tert*-butyl alcohol, benzylamine and ethyl glycinate react smoothly

* See *I. Lengyel* and *J. C. Sheehan*, Angew. Chem., intern. Edn., 1968, **7**, 25.

with 1-*tert*-butyl-3,3-dimethylaziridinone to produce α-substituted *N-tert*-butylamides (*Sheehan* and *Lengyel*, *ibid*., 1964, **86,** 1356). The effect of substituents on the ease of formation and stability of aziridinones is indicated by the reaction between 2-bromo-2-methyl-*N*-*n*-propylpropionamide and 2-chloro-*N*-butylpropionamide, and potassium *tert*-butoxide, respectively. The former produced two products, *tert*-butyl 2-methyl-2-*n*-propylaminopropionate and 2-*tert*-butoxy-2-methyl-*N*-*n*-propylpropionamide, whereas the latter only one, *tert*-butyl 2-*tert*-butylaminopropionate. In both reactions the α-lactam is detected at low temperature by infrared spectroscopy (1840 cm^{-1} band):

$$Me_2C(Br)\cdot CONHCH_2\cdot CH_2Me \xrightarrow{KOBu^t} \left[\text{1-}C_3H_7\text{-3,3-}Me_2\text{-aziridinone}\right] \longrightarrow Me_2C(NH{-}C_3H_7)\cdot CO_2\cdot CMe_3\ (41\%) + Me_2C(OCMe_3)\cdot CONHC_3H_7\ (24\%)$$

$$MeCH(Cl)\cdot CONHCMe_3 \xrightarrow{KOBu^t} \left[\text{1-}Bu^t\text{-3-Me-aziridinone}\right] \longrightarrow MeCH(NHCMe_3)\cdot CO_2\cdot CMe_3$$

The reactivity of the aziridinones so far described appears to be similar to that of β-lactones, with alcohols giving α-substituted amides and methanolic sodium methoxide amino esters. In comparison the reactivity of 1,3-di-*tert*-butylaziridinone is somewhat remarkable. The substitution of *tert*-butyl groups in the 1 and 3 positions produces an α-lactam which is relatively resistant to nucleophilic cleavage and possesses relatively high thermal stability. The resistance to nucleophilic cleavage can be attributed to the steric hindrance provided by the *tert*-butyl groups. Besides being prepared by the dehydrobromination method 1,3-*tert*-butylaziridinone was obtained by the addition of dichlorocarbene to *N*-neopentylidene-*tert*-butylamine (*Sheehan* and *J. H. Beeson*, *ibid*., 1967, **89,** 362):

$$Me_3C\cdot CH{:}NCMe_3 + {:}CCl_2 \xrightarrow{KOBu^t} \text{1,3-}Bu^t_2\text{-2,2-}Cl_2\text{-aziridine} \xrightarrow[H_2O]{NaHCO_3} \text{1,3-}Bu^t_2\text{-aziridinone}$$

Reduction of 1-*tert*-butyl-3,3-dimethylaziridinone with lithium tetrahydridoaluminate results in cleavage of the ring exclusively at the amide linkage.

2-*tert*-Butylamino-2-methyl-1-propanol is formed in good yield with only a small amount of aldehyde (*Sheehan* and *Lengyel*, J. org. Chem., 1966, **31**, 4244):

$$\text{(Me}_2\text{, N-Bu}^t\text{ }\alpha\text{-lactam)} \xrightarrow{LiAlH_4} \left[\text{Me}_2\text{, N-Bu}^t\text{, H, }\overset{\ominus}{O}-AlH_3\right] \longrightarrow Me_2C(NHBu^t)\cdot CH_2OH + Me_2C(NHBu^t)\cdot CHO$$

The formation of 2,2-diphenylindoxyl by the reaction between phenyl isocyanate and diphenyldiazomethane under the influence of ultraviolet light can be explained by assuming that the photogenerated carbene adds to the isocyanate to produce an α-lactam which then collapses to the indoxyl (*idem, ibid.*, 1963, **28**, 3252):

$$Ph_2CN_2 \xrightarrow{-N_2} [Ph_2C:] \xrightarrow{PhNCO} [\alpha\text{-lactam}, Ph_2] \longrightarrow \text{2,2-diphenylindoxyl}$$

The same α-lactam is also postulated as an intermediate in the formation of 3,3-diphenyloxindole from α-chloro-α,α-diphenylacetanilide and sodium hydride (*Sarel et al.*, Tetrahedron Letters, 1964, 1553). In this case it is assumed that there is concomitant attack of the N atom of the α-lactam on the C atom of another molecule to give a zwitterion intermediate stabilised by proton shift to give 3,3-diphenyloxindole.

(c) Azirines

(i) 1-Azirines: synthesis

(*1*) 1-Azirines are formed as the first product of the reaction of certain sulphonyl esters of ketoximes with a base (*P. W. Neber* and *G. Huh*, Ann., 1935, **515**, 283; *D. J. Cram* and *M. J. Hatch*, J. Amer. chem. Soc., 1953, **75**, 33, 38):

$$(NO_2)_2C_6H_3\cdot CH_2\cdot C(=NOTs)Ph \xrightarrow{C_5H_5N} (NO_2)_2C_6H_3\text{-azirine-}Ph \xrightarrow{H_2SO_4} (NO_2)_2C_6H_3\cdot CH(NH_2)\cdot COPh$$

(*2*) Substituted 1-azirines (*e.g.*, R = *n*-C_4H_9, C_6H_5, *o*-$CH_3\cdot C_6H_4$) are obtained by vapour-phase pyrolysis of α-alkyl- or α-aryl-α-azidoethylenes (*G. Smolinsky*, J. org. Chem., 1962, **27**, 3557):

$$RCH(Y)CH_2X \xrightarrow{NaN_3} RCH(N_3)CH_2X \xrightarrow{Bu^tO^\ominus} CH_2{=}C(R)N_3 \xrightarrow{\Delta} \text{2-R-1-azirine}$$

X = Cl

Y = Cl or TsO

(*3*) *P. F. Parcell* (Chem. Ind., 1963, 1396) during studies of the Neber rearrangement found that isobutyrophenone dimethylhydrazone methiodide with 1 mole of sodium isopropoxide at room temperature gave 3,3′-dimethyl-2-phenyl-1-azirine, which in a base-catalysed reaction added propan-2-ol reversibly, affording 3,3′-dimethyl-2-phenyl-2-propoxyaziridine:

$$Me_3\overset{\oplus}{N}{-}N{=}C(Ph)CHMe_2 \xrightarrow{PrO^\ominus} \text{3,3-}Me_2\text{-2-Ph-azirine} \underset{PrO^\ominus,\ -PrOH}{\overset{PrO^\ominus,\ +PrOH}{\rightleftharpoons}} \text{3,3-}Me_2\text{-2-Ph-2-OPr-aziridine}$$

(Me₃N⁺–N=C(Ph)CHMe₂ with excess $PrO^\ominus$ gives the aziridine directly.)

Other 3,3-disubstituted 2-phenyl-1-azirines may be prepared by the base-catalysed elimination of trimethylammonium iodide from quaternary salts (*S. Sato*, Bull. chem. Soc., Japan, 1968, **41,** 1440).

(*4*) Azirines are obtained by the photolysis of conjugated vinyl azides prepared by the addition of azide ion to conjugated allenic esters (R = H or Me) (*G. R. Harvey* and *K. W. Ratts*, J. org. Chem., 1966, **31,** 3907):

$$CH_2{:}C{:}C(R){\cdot}CO_2Et \xrightarrow[THF/H_2O]{NaN_3} MeC(N_3){:}C(R){\cdot}CO_2Et \xrightarrow[C_6H_6]{h\nu} \text{2-Me-3-R-3-}CO_2Et\text{-azirine} + MeN{:}C{:}C(R)CO_2Et$$

4 : 1

1-Azido-1-phenylethylene in benzene on irradiation gives mainly 2-phenyl-1-azirine and a small amount of 4-phenyl-3-phenylimino-1-azabicyclo-[2.1.0]pentane, which may arise by photocyclo-addition of keten phenylimine and 2-phenyl-1-azirine (*F. P. Woerner*, *H. Reimlinger* and *D. R. Arnold*, Angew. Chem., intern. Edn., 1968, **7,** 130):

$$Ph{-}C(N_3){=}CH_2 \xrightarrow{h\nu} \text{2-Ph-1-azirine} + \text{4-Ph-3-(N–Ph)-1-azabicyclo[2.1.0]pentane}$$

A fused 1-azirine (VIII) has been prepared by the photochemical transformation of the related vinyl azide, formed by the *trans*-addition of iodine azide to an alkene, followed by azide-directed *trans*-elimination of hydriodic acid (*A. Hassner* and *F. W. Fowler*, Tetrahedron Letters, 1967, 1545; J. Amer. chem. Soc., 1968, **90**, 2869):

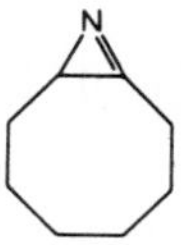

N

(VIII)

5-Alkoxyisoxazoles undergo a thermally induced ring contraction to give 1-azirines (*T. Nishiwaki*, Tetrahedron Letters, 1969, 2049):

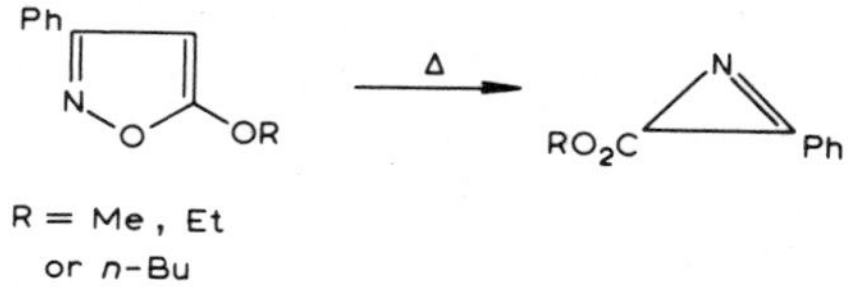

Ph N O OR Δ N RO_2C Ph

R = Me, Et or *n*-Bu

(ii) Properties and reactions of 1-azirines

The boiling points and melting points of some representative 1-azirines are listed in Table 4.

TABLE 4

1-AZIRINES *(Hassner* and *Fowler, loc. cit.)*

Substituent	*b.p. (°C)*
3-Methyl-2-phenyl-	96/15 mm
2-Butyl-	80/145 mm
2-(β-Phenylethyl)-	108–110/10 mm
2-Phenyl-	83/10 mm
2,3-Diethyl-	68/130 mm
2-Benzyl-	74/1.5 mm
2-Methoxycarbonyl-2-phenyl-	m.p. 45–45.5

(*1*) Reduction of suitable azirines with lithium tetrahydridoaluminate gives *cis*-aziridines; since the reaction is stereospecific, this method thus enables *trans*-olefins and mixtures of *cis*- and *trans*-olefins to be converted to *cis*-aziridines (*Hassner* and *Fowler*, *loc. cit.)*:

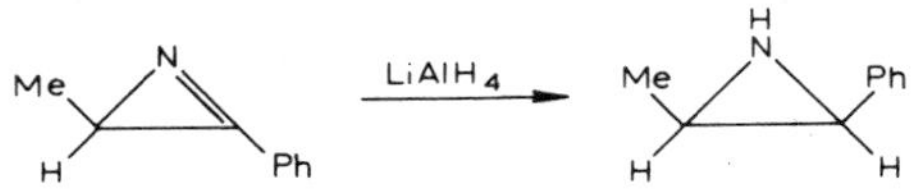

Me H N Ph —$LiAlH_4$→ Me H NH Ph H

(*2*) 2-Phenyl-1-azirine reacts with dimethylsulphonium methylide in tetrahydrofuran at $-10°$ to give 3-phenyl-1-azabicyclo[1.1.0]butane (*A. G. Hortman* and *D. A. Robertson*, J. Amer. chem. Soc., 1967, **89,** 5974):

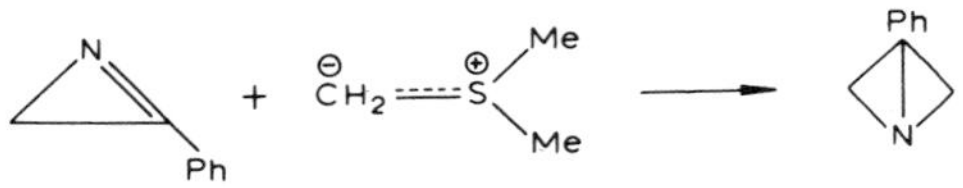

N Ph + $\overset{\ominus}{C}H_2$=$\overset{\oplus}{S}$Me Me → Ph N

(*3*) The ring expansion of an azirine has been effected, proceeding through the azirinium salt, generated *in situ*, as a probable intermediate, *e.g.*, 2,2-dimethyl-3-phenylazirine has been converted into an oxazolinium salt with perchloric acid in acetone and to an imidazolinium salt with perchloric acid in acetonitrile (*Leonard* and *B. Zwanerburg*, *ibid.*, 1967, **89,** 4456):

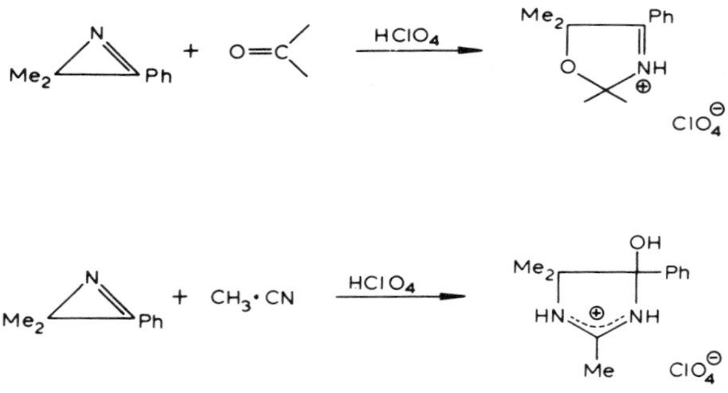

Me_2 N Ph + O=C< —$HClO_4$→ Me_2 Ph O NH ⊕ $ClO_4^{\ominus}$

Me_2 N Ph + CH_3·CN —$HClO_4$→ Me_2 OH Ph HN ⊕ NH Me $ClO_4^{\ominus}$

(*4*) Thermal rearrangement of 2-methyl-3-phenyl-1-azirine to 2-methylindole probably involves a dipolar intermediate. However, the pyrolysis of 3-phenyl-1-azirine gives equal amounts of indole and benzyl cyanide, implying that hydrogen transfer within the intermediate can compete with cyclisation (*K. Isomura, S. Kobayashi* and *H. Taniguchi*, Tetrahedron Letters, 1968, 3499):

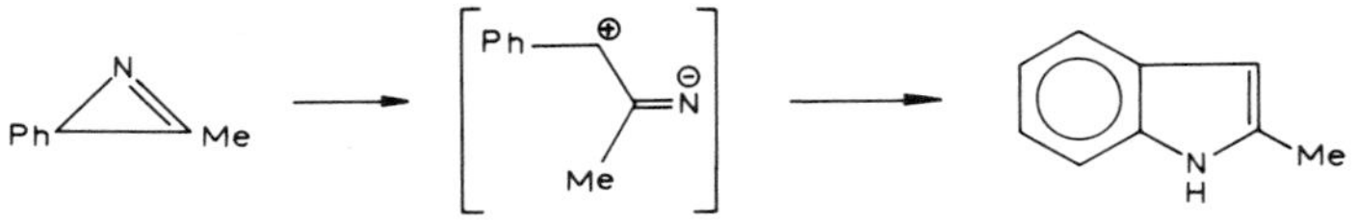

Ph N Me → [Ph ⊕ =N⊖ Me] → N H Me

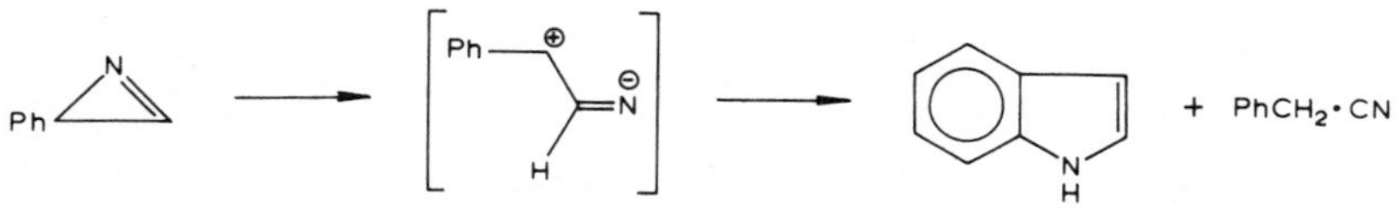

(5) Nucleophilic addition to the azirine double bond occurs readily, although the adducts generally undergo rearrangements; 2-phenyl-1-azirine and acetophenone in the presence of strong base give 2,4-diphenylpyrrole (*Sato, H. Kato* and *M. Ohta*, Bull. chem. Soc., Japan, 1967, **40,** 2936); 3-methyl-2-phenyl-1-azirine with diazomethane affords a mixture of *cis*- and *trans*-1-azido-2-phenyl-2-butene together with 3-azido-2-phenyl-1-butene (*V. Nair*, J. org. Chem., 1968, **33,** 2121). The azirine double bond adds on acid chlorides and acid anhydrides, *e.g.*, 3-methyl-2-phenyl-1-azirine with benzoyl chloride yields *N*-benzoylchloroaziridines, which are stable in benzene but readily rearrange in acetone to 4-methyl-2,5-diphenyloxazole and the dichloro-amide IX (*Sato, Kato* and *Ohta*, Bull. chem. Soc. Japan, 1967, **40,** 2938; *Fowler* and *Hassner*, J. Amer. chem. Soc., 1968, **90,** 2875):

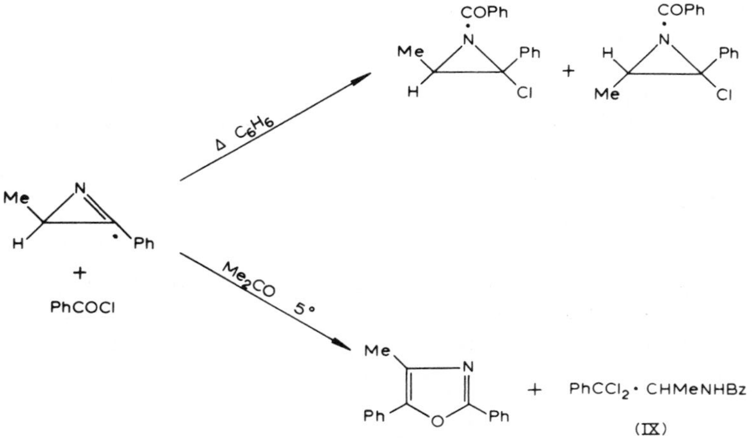

(IX)

(iii) 2-Azirines

As a cyclic planar 4π-electron system, isoelectronic with the cyclopropenyl anion, 2-azirine should be destabilised by electron delocalisation, that is, it should be anti-aromatic. In an attempt to prepare a 2-azirine XI the nitrene X was added to an acetylene; however, spontaneous rearrangement occurred to give the 1-azirine XII (*Anderson, T. L. Gilchrist* and *Rees*, Chem. Comm., 1969, 147):

$R_2N{-}NH_2 \xrightarrow{Pb(OAc)_4} R_2N{-}\ddot{N}:$ (X) $\xrightarrow{R^1C:CR^2}$ [(XI)] $\rightarrow$ (XII)

$R_2N{-}$ = phthalimido (N-)

(d) Aziridinium salts

Aziridinium salts have been prepared by the reaction between diazomethane and a tertiary iminium perchlorate, analogous to the nucleophilic attack of diazomethane on a carbonyl group.

1-*N*-Pyrrolidylcyclohexene has been converted to 2,2-pentamethylene-1,1-tetramethylenaziridinium perchlorate, and the aziridinium ring has been formed in other mono-, bi-, and tri-cyclic systems (*N. J. Leonard* and *K. Jann*, J. Amer. chem. Soc., 1960, **82**, 6418):

$\xrightarrow{HClO_4}$ $ClO_4^{\ominus}$ $\xrightarrow{CH_2N_2}$ $ClO_4^{\ominus}$

Hydrolysis and alcoholysis of aziridinium salts in these systems lead to heterocyclic ring-enlargement and provide routes to cyclic β-hydroxy- and β-alkoxy-amino compounds (*Leonard et al.*, J. org. Chem., 1963, **28**, 1499):

$\xrightarrow{CH_2N_2}$ $\xrightarrow{CH_3OH}$ $\xrightarrow{OH^{\ominus}}$

Expansion of the aziridinium ring to an oxazolidinium ring can be effected by reaction with cyclic, acyclic and aryl ketones (*idem*, Angew. Chem., 1963, **75**, 1031; J. org. Chem., 1963, **28**, 2850; 1964, **29**, 3383):

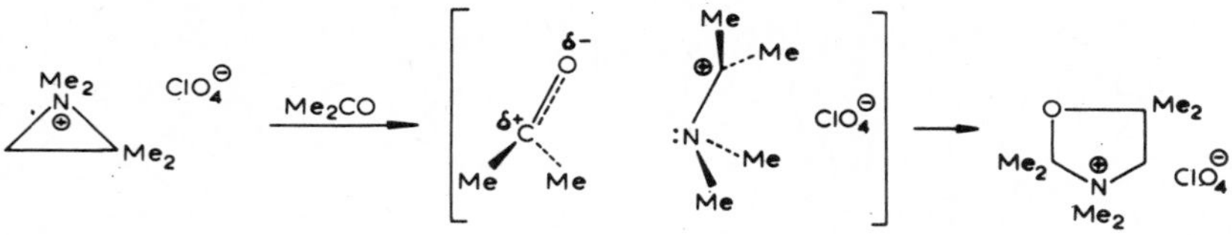

and to an imidazolinium ring with nitriles (*Leonard* and *L. E. Brady*, *ibid.*, 1965, **30,** 817):

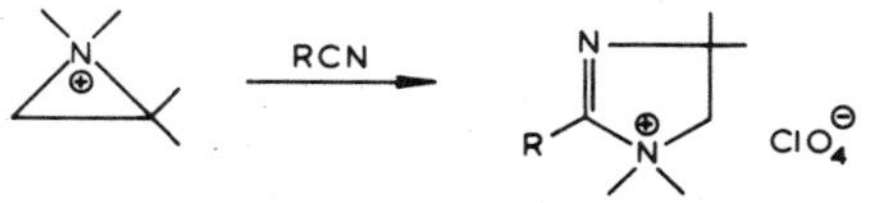

1,1,2,2-Tetramethylaziridinium perchlorate reacts with substituted 1-pyrroline 1-oxide, a three-membered charged ring, combining with a 1,3-dipolar moiety (nitrone) to form a six-membered charged ring (*Leonard*, *D. A. Durand* and *F. Uchimaru*, *ibid.*, 1967, **32,** 3607):

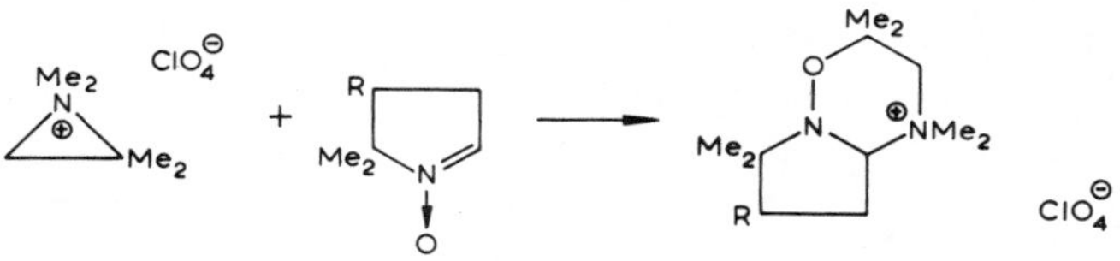

Stable aziridinium perchlorates, *e.g.*, 1,1-diethylaziridinium perchlorate and 3-azoniaspiro[2.5]octane perchlorate, have been synthesised from β-chloroethylamines by treatment with silver perchlorate (*Leonard* and *J. V. Paukstelis*, *ibid.*, 1965, **30,** 821; *A. T. Bottini* and *R. L. van Ettem*, *ibid.*, p. 575; *G. K. Helmkamp*, *R. D. Clark* and *J. R. Koskineu*, *ibid.*, p. 666):

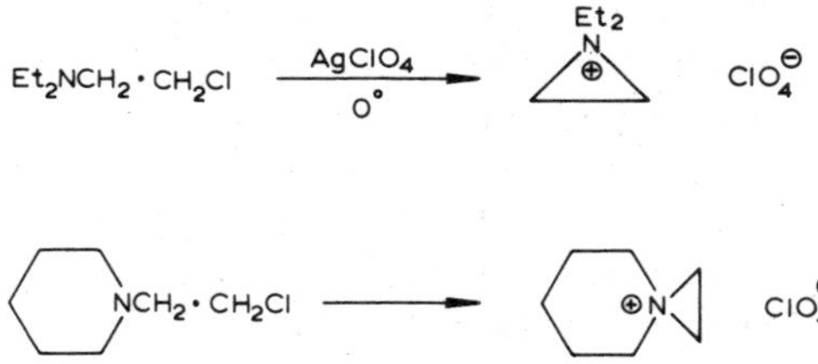

$$Et_2NCH_2 \cdot CH_2Cl \xrightarrow[0^\circ]{AgClO_4} \text{1,1-diethylaziridinium } ClO_4^\ominus$$

The reactions of small charged heterocycles have been reviewed (*D. R. Crist* and *Leonard*, Angew. Chem., intern. Edn., 1969, **8,** 962).

5. Three-membered rings with more than one hetero-atom

(a) Oxaziridines (oxaziranes)

Recently a number of three-membered rings with two hetero-atoms have been synthesised and their properties studied. Certain azomethines may be oxidised readily with anhydrous peracetic acid in methylene chloride to give oxaziridines in good yield. The latter are active oxygen compounds and may be assayed iodometrically (*W. D. Emmons*, J. Amer. chem. Soc., 1956, **78**, 6208; *L. Horner* and *E. Jürgens*, Ber., 1957, **90**, 2184):

$$RCH{=}NR' \xrightarrow{CH_3\cdot CO_3H} \text{R–}\overset{O}{\triangle}\text{–NR}'$$

$$R = H,\ Ph,\ p\text{-}O_2NC_6H_4,\ CH_3$$

$$R' = Bu^t,\ Et,\ Pr^n$$

The mechanism of oxidation of Schiff bases to oxaziridines by peroxy-acids in various solvents has been studied and it is believed that the peroxy-acid undergoes a synchronous hydrogen-exchange process with HY, where HY is the solvent, benzoic acid (a product), or another molecule of peroxy-acid. The molecule HY (if solvent) must be a protic solvent, capable of proton exchange. The transition state is composed of the peroxy-acid, the Schiff base, and a molecule of HY (*V. Madan* and *L. B. Clapp*, J. Amer. chem. Soc., 1969, **91**, 6078):

$$RC({=}O)\text{–O–O–H} + HY \rightleftharpoons [RC(O)\text{–O–O}\cdots H\text{–Y–}H] + ArCH : NR' \longrightarrow [\text{transition state}] \longrightarrow \text{Ar–}\overset{O}{\triangle}\text{–NR}'$$

Oxaziridines can be obtained by the irradiation of certain nitrones with u.v. light (*J. S. Splitter* and *M. C. Calvin*, J. org. Chem., 1958, **23**, 651), by the ozonolysis of hindered imines, thus *N*-isobutylidene-*tert*-butylamine gives 2-*tert*-butyl-3-isopropyloxaziridine (*J. S. Belew* and *J. T. Person*, Chem. and Ind., 1959, 1246), or by the reaction of an aldehyde or ketone with *N*-chloromethylamine or *N*-methylhydroxylamine-*O*-sulphonic acid in the presence of base (*E. Schmitz, R. Ohme* and *D. Murawski*, Angew. Chem., 1961, **73**, 708).

Acid hydrolysis of oxaziridines may give either aldehydes and β-alkylhydroxylamines possibly *via* the nitrones, or else aldehydes, ketones and primary amines:

$$\text{Ph-oxaziridine-NR} \xrightarrow{H^{\oplus}} \left[\overset{\oplus}{\text{PhCH}}\text{N(OH)R}\right] \underset{}{\overset{-H^{\oplus}}{\rightleftharpoons}} \text{PhCH:N(}\rightarrow\text{O)R}$$

$$\downarrow H_2O$$

$$\text{PhCH(OH)—N(OH)R} \longrightarrow \text{PhCHO} + \text{RNHOH}$$

Reduction of 2-*tert*-butyl-3-phenyloxaziridine with lithium tetrahydrido-aluminate in ether gave *N*-benzylidene-*tert*-butylamine, whereas the corresponding nitrone afforded *N*-benzyl-*N*-*tert*-butylhydroxylamine (*Emmons*, J. Amer. chem. Soc., 1957, **79**, 5739):

$$\text{Ph-oxaziridine-NBu}^t \xrightarrow{LiAlH_4} \text{PhCH:NBu}^t$$

$$\text{PhCH:N(}\rightarrow\text{O)Bu}^t \xrightarrow{LiAlH_4} \text{PhCH}_2\cdot\text{N(OH)Bu}^t$$

Oxaziridines react smoothly with one equivalent of peracetic acid to give nitrosoalkane dimers (*idem, ibid.*, 1957, **79**, 6522):

$$\text{Me}_2\text{-oxaziridine-NCHMe}_2 \xrightarrow{CH_3\cdot CO_3H} [\text{Me}_2\text{CHNO}]_2 + \text{Me}_2\text{CO}$$

Treatment of a suitable carbonyl compound with chloramine or hydroxylamine-*O*-sulphonic acid in 2 *N* sodium hydroxide, followed by reaction with acid chlorides, gives 2-acyloxaziridines (*Schmitz, Ohme* and *S. Schramm*, Tetrahedron Letters, 1965, 1857).

The stability of the oxaziridine ring and the nature of the cleavage depend on the kind of substituents present and the solvent used. Those with alkyl groups on either or both the carbon and nitrogen atoms of the ring are sufficiently stable to be isolated. With aryl groups on both of these atoms, the stability is decreased markedly so that they are observed only in solution. The 2,3-diaryloxaziridines are considerably less stable than 2-alkyl-3-aryloxaziridines (*Splitter* and *Calvin*, J. org. Chem., 1965, **30**, 3427).

2-Methyl-3,3-diphenyloxaziridine may be isolated in optically active form by the oxidation of *N*-diphenylmethylenemethylamine with (1*S*)-peroxycamphoric acid; this is the first example of asymmetric induction at trivalent nitrogen (*F. Montanari*, *I. Moretti* and *G. Torre*, Chem. Comm., 1968, 1694):

$$Ph_2C{=}NMe \longrightarrow Ph_2\overset{O}{C}{-}\overset{*}{N}Me$$

Other optically active 2-alkyl-3,3-diphenyloxaziridines have been prepared (*idem, ibid.*, 1969, 1086) and related stable oxaziridine diastereomers possessing a non-inverting nitrogen atom have been synthesised and interconverted photochemically (*D. R. Boyd*, *R. Spratt* and *D. M. Jerina*, J. chem. Soc., C. 1969, 2650).

(b) Diaziridines and related compounds

(i) Diaziridines and diazirines

A diaziridine (I) can be formed by the reaction between *N*-chloromethylamine and 3,4-dihydroisoquinoline (*Schmitz*, Angew. Chem., 1959, **71**, 127):

(I)

Diaziridines obtained by the reaction between a ketone, a primary amine and chloramine (or a Schiffs base and chloramine) are oxidised to the corresponding diazirines by mercuric oxide in neutral and by permanganate in alkaline solutions, and by silver oxide (*S. P. Paulsen*, *ibid.*, 1960, **72**, 781):

$$RR'C\langle NH, NH \rangle \underset{Red.}{\overset{Ox.}{\rightleftharpoons}} RR'C\langle N{=}N \rangle$$

Conversely diazirines on reduction yield diaziridines, and they react with Grignard reagents to give *N*-alkyldiaziridines (*Schmitz* and *Ohme*, Ber., 1961, **94**, 2166; Angew. Chem., 1961, **73**, 220).

1,2-Dialkyldiaziridines are obtained when formaldehyde is condensed with two moles of primary aliphatic amine and then treated with one mole of sodium hypochlorite (*Ohme*, *Schmitz* and *P. Dolge*, Ber., 1966, **99**, 2104).

Diaziridine formed by reaction between formaldehyde, chloramine and ammonia (*Schmitz* and *Ohme*, Tetrahedron Letters, 1961, 612) yields diazirine on oxidation *in situ* with chromic acid (*idem*, Ber., 1962, **95,** 795). The n.m.r. spectra show two resonances for the R (H or CH_3) groups on the ring-carbon atom of diaziridines (II) with different substituents on the nitrogen atoms at temperatures up to 100°. Hence it is concluded that inversion of at least one nitrogen atom is a relatively slow process (*A. Mannschreck* and *W. Seitz*, Angew. Chem., intern. Edn., 1969, **8,** 212).

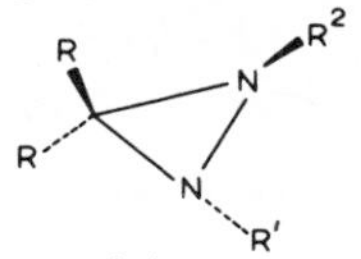

(II)

N,N'-Di-*tert*-butyl-*N''*-methylguanidine and *tert*-butyl hypochlorite give the diaziridineimine III, which undergoes slow reversible valence isomerisation (IIIa ⇋ IIIb) (*H. Quast* and *Schmidt*, *ibid.*, 1969, **8,** 448):

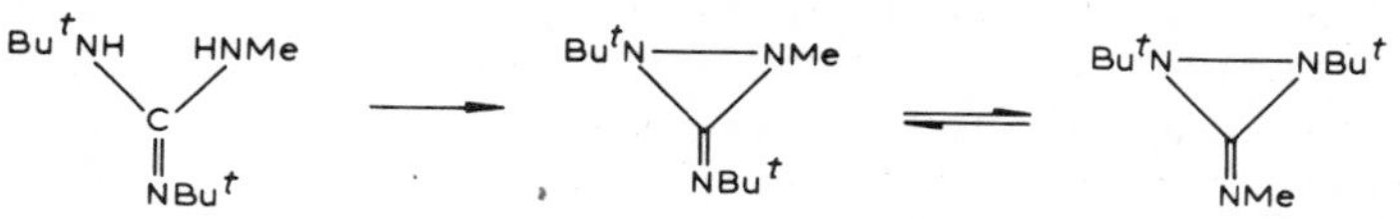

(a) (b)

(III)

Diazirine can be obtained by the reaction of difluoramine with *tert*-butylazomethine or *tert*-octylazomethine in carbon tetrachloride solution (*W. H. Graham*, J. Amer. chem. Soc., 1962, **84,** 1063), and has been prepared by the reactions between methylenediammonium sulphate and sodium hypochlorite solution (*Schmitz* and *Ohme*, Ber., 1964, **97,** 297), and *tert*-octylazomethine and dichloramine (*Graham*, J. org. Chem., 1965, **30,** 2108):

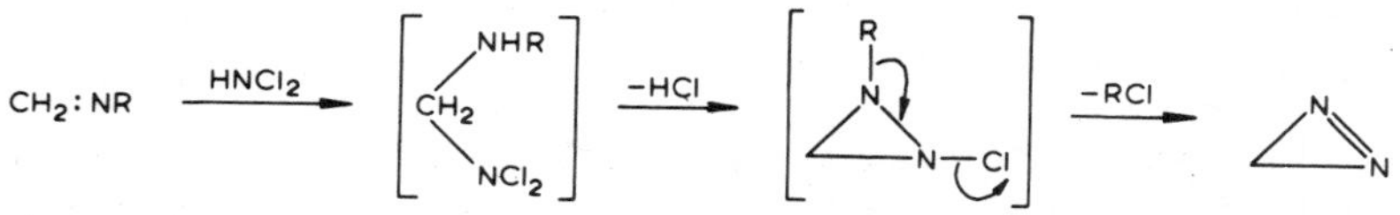

$CH_2{:}NR \xrightarrow{HNCl_2} [CH_2(NHR)(NCl_2)] \xrightarrow{-HCl} [\ldots] \xrightarrow{-RCl}$ diazirine

It is a colourless gas, b.p. −14°, μ 1.59±0.06 D (*L. Pierce* and *V. Dobyns*, J. Amer. chem. Soc., 1962, **84,** 2651), differing in chemical and physical properties from diazomethane, $CH_2{=}N^{\oplus}{=}N^{\ominus} \leftrightarrow CH_2^{\ominus}{-}N^{\oplus}{\equiv}N$, (*cf.* Vol. I B, p. 157) and isodiazomethane, $CH{\equiv}N^{\oplus}{-}NH^{\ominus} \leftrightarrow CH^{\ominus}{=}N^{\oplus}{=}NH$ (*E. Müller* and *D. Ludsteck*, Ber., 1954, **87,** 1887) with which it is isomeric; it is stable to *tert*-butoxide in *tert*-butyl alcohol, decomposed relatively slowly by sulphuric acid, can be stored in glass, and on photolysis affords nitrogen and carbene, which corresponds in energy content to that from linear diazomethane and in selectivity to the less energetic carbene from ketene (*H. M. Frey* and *I. D. R. Stevens*, Proc. chem. Soc., 1962, 79). Molecular dimensions were deduced from microwave data (*Pierce* and *Dobyns*, *loc. cit.*).

Reduction of 1-cyclohexyl-3-ethyldiaziridine with lithium tetrahydridoaluminate affords *N*-*n*-propylcyclohexylamine and ammonia, and with hydrogen and Raney nickel cyclohexylamine and *n*-propylamine. 1-Alkyldiaziridines on acid-hydrolysis yield an alkylhydrazine and a carbonyl compound (*C. Szantay* and *Schmitz*, Ber., 1962, **95,** 1759).

The halogenation of alkyl- or aryl-amidines and -isoureas in aqueous dimethyl sulphoxide solution affords the corresponding alkyl-, aryl-, or alkoxy-3-halogenodiazirine (*Graham*, J. Amer. chem. Soc., 1965, **87,** 4396):

NH
R—C—NH$_2$
NaOCl
NaCl
R
N
N
Cl

(ii) Diaziridinones

Di-*tert*-butyldiaziridinone (IV) is obtained on treating *N,N'*-di-*tert*-butyl-*N*-chlorourea with potassium in pentane or with potassium *tert*-butoxide in *tert*-butyl alcohol. Compared with the aziridinones (p. 31) it is relatively unreactive towards nucleophiles and shows high thermal stability (*F. D. Greene* and *J. C. Stowell*, J. Amer. chem. Soc., 1964, **86,** 3569; *Greene, Stowell* and *W. R. Bergmark*, J. org. Chem., 1969, **34,** 2254):

R
NCl
O=C—N—R
H
R = Bu^t
R
N
NR
O
(IV)
HCl
H
R—N—N—R
COCl
KOBu^t
R
N
NR
O
(45%)
+
CO$_2$R
RNH · NR
(25%)
H$_2$ Pd/C
NaBH$_4$
RNH · CO · NHR
..CO · NR · NHR

Diaziridinones are formed along with carbodiimides (V) and nitroalkane (VI) by the reaction between 2-methyl-2-nitrosopropane and aliphatic isocyanides at moderate temperatures. The relative yields of products are affected by substituents, concentration and temperature (*Greene* and *J. F. Pazos*, *ibid.*, 1969, **34,** 2269):

Bu^{t}NO + RNC
Bu^t
N
NR
O
+ Bu^{t}N:C:NR + Bu^{t}NO$_2$
(V) (VI)

Diaziridinones undergo oxidation–reduction and rearrangement reactions in the presence of substituted hydrazines (*Greene*, *Bergmark* and *J. G. Pacifici*, J. org. Chem., 1969, **34,** 2263):

$$\text{(diaziridinone: O=C, N-Bu}^t\text{, NBu}^t\text{)} + Bu^tNH\cdot NHBu^t \xrightarrow[2h]{100^\circ} Bu^tNH\cdot CO\cdot NHBu^t + Bu^tN{:}NBu^t$$

(c) Oxadiaziridine

Irradiation of azoxy-*tert*-butane in pentane at 10° affords an oxadiaziridine, which is thermally labile reverting quantitatively to the azoxy derivative. The oxadiaziridine is unreactive towards water and moderately stable in acidic media (*S. S. Hecht* and *Greene*, J. Amer. chem. Soc., 1967, **89,** 6762):

$$Bu^t\overset{O^{\ominus}}{\underset{\oplus}{N}}{:}NBu^t \underset{\Delta}{\overset{h\nu}{\rightleftharpoons}} Bu^tN\text{—}NBu^t \text{ (bridged by O)}$$

6. Four-membered rings with one oxygen atom

(a) Oxetanes (trimethylene oxides)

(i) Formation

(*1*) Oxetanes are produced by the action of alkalis on γ-halogeno-alcohols or sometimes preferably on their esters:

$$\begin{matrix} CH_2\cdot OAc \\ | \\ CH_2\cdot CH_2Cl \end{matrix} \xrightarrow{KOH} \text{oxetane}$$

The reaction often proceeds more slowly and less smoothly than in the corresponding synthesis of oxiranes; simple displacement of the halogen atom, and 1,2-elimination of hydrogen halide, giving an olefin, are among the competing reactions (*S. Searles* and *M. J. Gortatowski*, J. Amer. chem. Soc., 1953, **75,** 3030).

(*2*) 2,2-Dimethyloxetane has been obtained on adding sulphuric acid to neopentyl glycol and then treating the mixture with sodium hydroxide (*L. F. Schmoyer* and *L. C. Case*, Nature, 1959, **183,** 389). Some 3,3-disubstituted oxetanes can be derived from pentaerythritol di- and tri-chlorides (*A. C. Farthing*, J. chem. Soc., 1955, 3648):

$$C(CH_2OH)(CH_2Cl)_3 \xrightarrow{KOH} \text{3,3-bis(chloromethyl)oxetane } (ClCH_2,\ CH_2Cl) \longrightarrow \text{3-}RCH_2\text{-3-}CH_2R'\text{-oxetane}$$

$(R = R' = EtO,\ PhO,\ OAc,\ R = Cl,\ R' = OAc)$

and from 2,2-dialkyl-3-bromopropan-1-ols (*Searles, R. G. Nickerson* and *W. K. Witsiepe*, J. org. Chem., 1959, **24,** 1839).

(*3*) Substituted oxetanes can be obtained by the action of a Grignard reagent on a β-bromo acid chloride, excess Grignard reagent at the first stage leading directly to an oxetane (*F. Nerdal* and *P. Weyerstahl*, Angew. Chem., 1959, **71,** 339):

$$R^1R^2C(CH_2Br)COCl \xrightarrow{R^3MgBr} R^1R^2C(CH_2Br)COR^3 \xrightarrow{R^4MgBr} [R^1R^2C(CH_2Br)-C(OH)R^3R^4] \longrightarrow \text{oxetane } (R^1R^2,\ R^3R^4)$$

$$R^1R^2C(CH_2Br)COCl \xrightarrow{\text{Excess } R^3MgBr} \text{oxetane } (R^1R^2,\ R^3_2)$$

(*4*) The light-catalysed reaction of fumaronitrile and acetone provides a mixture of *cis*- and *trans*-2,3-dicyano-4,4′-dimethyloxetane (*J. J. Beereboom* and *M. Schach von Wittenau*, J. org. Chem., 1965, **30,** 1231). Carbonyl compounds react with 2-methyl-2-butene in the presence of ultraviolet light to give oxetanes (*G. Büchi, G. G. Inman* and *E. S. Lipinsky*, J. Amer. chem. Soc., 1954, **76,** 4327):

$$PhCHO + Me_2C{:}CHMe \longrightarrow \text{oxetane } (Ph,\ Me_2,\ Me)$$

(*ii*) *Properties and reactions*

Oxetane, b.p. 48°, d_4^{25} 0.8930, n^{25} 1.3897, can be prepared from either 3-chloropropanol (*E. Reboul*, Ann. Chim., 1878, **14,** 495; *A. V. Ipatov*, J. Russ. Phys. chem. Soc., 1914, **46,** 62) or 3-chloropropyl acetate and potassium hydroxide (*C. G. Derrick* and

D. W. Bissell, J. Amer. chem. Soc., 1916, **38,** 2478; *C. R. Noller*, Org. Synth., 1949, **29,** 92). For structural data see p. 2. It is miscible with water and is a hydroxypropylating agent, reacting with alcohols and phenols as readily as does oxirane, but much less readily with amines. Friedel–Crafts condensation with benzene gives γ-phenylpropyl alcohol (*Searles et al.*, J. Amer. chem. Soc., 1954, **76,** 56, 2313, 2798). The ring is also opened by hydrogen bromide or phosphorus pentachloride yielding 1,3-dihalogenopropanes, and Grignard reagents or organolithium compounds yield primary alcohols (*Searles, ibid.*, 1951, **73,** 124):

$$\text{oxetane} \xrightarrow{RMgBr} RCH_2 \cdot CH_2 \cdot CH_2OMgBr \longrightarrow RCH_2 \cdot CH_2 \cdot CH_2OH$$

In presence of boron trifluoride both oxetane and 3,3-dimethyloxetane yield a mixture of a linear polyether and cyclic tetramer, but 2-methyloxetane gives only linear polymer (*J. B. Rose*, J. chem. Soc., 1956, 542, 546). Reduction of oxetanes having two or fewer alkyl substituents with lithium tetrahydridoaluminate results in cleavage between the oxygen and the less substituted carbon atom; this holds also for the reduction of 2-phenyloxetane with sodium tetrahydrodoborate (*Searles et al.*, J. Amer. chem. Soc., 1957, **79,** 948).

2-*Methyloxetane*, b.p. 60° (*F. Sondheimer* and *R. B. Woodward*, *ibid.*, 1953, **75,** 5438); 2-*ethyloxetane*, b.p. 89° (*R. Lespieau*, Bull. Soc. chim. Fr., 1940, **7,** 254); 2,4-*dimethyloxetane*, b.p. 71°; 3,3-*dimethyloxetane*, b.p. 80° (*G. M. Bennett* and *W. G. Philip*, J. chem. Soc., 1928, 1937; *Rose, loc. cit.*); 2,3-*dimethyloxetane*, b.p. 85–86° (*Searles et al.*, J. Amer. chem. Soc., 1957, **79,** 952).

2-**Oxaspiro**[3.2]**hexane,** a highly strained oxetane, is prepared by the reaction of 3,3-bis(chloromethyl)oxetane with zinc dust in molten acetamide (*Searles* and *E. F. Lutz*, *ibid.*, 1959, **81,** 3674); enhanced reactivity is shown by both rings:

ClCH2 / CH2Cl (oxetane) —Zn→ 2-oxaspiro[3.2]hexane
—LiAlH4→ (cyclopropane: CH2OH, Cl)
—aq. HCl, 0°→ (cyclobutane: CH2OH, Cl)
—HBr, C6H6→ [(cyclopropane: CH2OH, CH2Br)] ⟶ BrCH2 · CH2 · CH(CH2Br) · CH2OH

2-Oxaspiranes $(CH_2)_nC(CH_2)_2O$, $n = 2,3,4$ or 5, have been synthesised and their properties compared (*Searles, Lutz* and *M. Tamres, ibid.*, 1960, **82,** 2932).

2,6-*Dioxaspiro*[3,3]*heptane*, m.p. 89°, b.p. 172°, containing two fused oxetane rings has been obtained by a double ring closure (*H. J. Backer* and *K. J. Keuning*, Rec. Trav. chim., 1934, **53,** 812):

$(HOCH_2)(BrCH_2)C(CH_2Br)(CH_2OH) \xrightarrow{KOH}$ 2,6-dioxaspiro[3,3]heptane

(*b*) *Alkoxyoxetanes and oxetanols*

2- and 3-Alkoxyoxetanes may be obtained by the photoaddition of carbonyl compounds to vinyl ethers; with propionaldehyde, cyclohexanone, benzaldehyde, and benzophenone the ratio of 2-:3- isomer is (25 ± 5):75 (*S. H. Schroeter* and *C. M. Orlando*, J. org. Chem., 1969, **34,** 1181; *Schroeter*, Chem. Comm., 1969, 12):

$R'COR^2 + CH_2{:}CHOR^3 \xrightarrow{h\nu}$ 2-OR^3 oxetane (R^1R^2) + 3-R^3O oxetane (R^1R^2)

In the presence of excess carbonyl compound, the relative amount of the 2-isomer decreases with irradiation time, because it is attacked by the photoexcited carbonyl compound and converted to a *β*-lactone (*Schroeter*, Tetrahedron Letters, 1969, 1591):

2-OR^3 oxetane (R^1R^2) $\xrightarrow[h\nu]{R^1COR^2}$ β-lactone (R^1R^2)

2-Alkoxyoxetanes (four-membered cyclic acetals) react quantitatively under mild conditions with water and with primary, secondary, or tertiary alcohols or phenols to give 3-hydroxyaldehydes and 3-hydroxyacetals, respectively, and with Grignard reagents and lithium tetrahydridoaluminate to afford 3-hydroxy ethers. Some 3-alkoxyoxetanes on hydrolysis yield substituted glycerol α-monoethers (*idem*, J. org. Chem., 1969, **34,** 1188).

Oxetan-3-ols have been photochemically synthesised from α-alkoxyacetophenones (*P. Yates* and *A. G. Szabo*, Tetrahedron Letters, 1965, 485);

$PhCO \cdot CH_2OCH_2R \xrightarrow[benzene]{h\nu}$ 3-Ph-3-OH-2-R oxetane $\xleftarrow{PhMgBr}$ 2-R oxetan-3-one

α-benzyloxyacetophenone gave a mixture of *cis*- and *trans*-2,3-diphenyloxetan-3-ols (*R. B. LaCourt* and *G. E. Griffin*, *ibid.*, 1965, 1549).

3-Oxetanols may be obtained also from certain esters of 2-hydroxymethyloxiranes (2,3-epoxypropanols) by solvolysis (see p. 5).

(c) Oxetanones

(i) Oxetan-2-ones (β-lactones)

These compounds, as exemplified by β-propiolactone, have already been discussed in some detail (see Vol. I D, pp. 99–109). The following information amplifies this earlier account. Since oxetan-2-ones contain a strained four-membered heterocyclic ring, they differ in the general methods available for their preparation and in their chemical reactivity from γ- and δ-lactones.

They may be prepared by the following methods:

(*1*) Treatment of β-halogeno acids with basic reagents, *e.g.*, one equivalent of sodium hydroxide, sodium carbonate, and sodium bicarbonate in aqueous solution, or moist silver oxide in ether.

(*2*) The reaction of ketenes with carbonyl compounds in the presence of sulphuric acid, except with diphenylketene when no catalyst is required (*H. E. Zaugg*, Org. Reactions, 1954, **8,** 305):

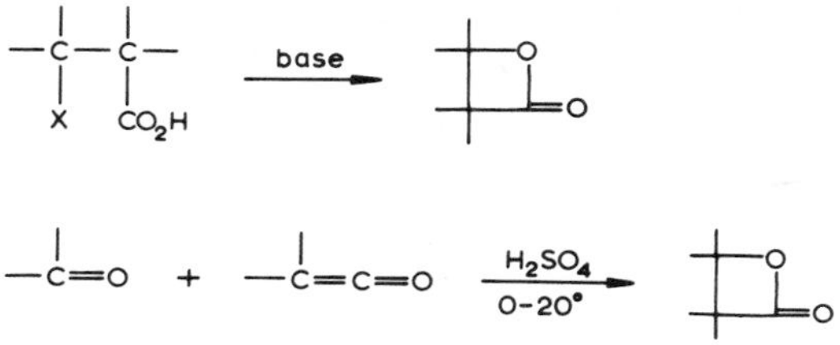
base
X
CO_2H
O
H_2SO_4
0-20°

(*3*) Treatment of α-substituted β-aminopropionic acids with nitrous acid (*E. Testa et al.*, Ann., 1961, **639,** 166):

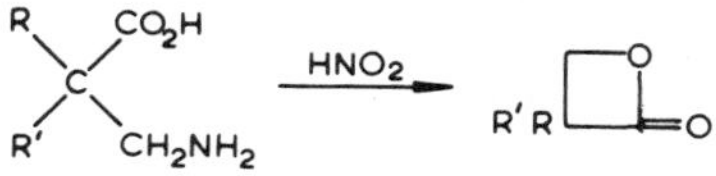
R
CO_2H
C
R'
CH_2NH_2
HNO_2
R'R
O

(*4*) Oxetan-2-ones may be obtained also by the ready thermolysis of the 4-oxo-1,3-dioxane, formed by an exceptionally facile process involving reaction of a β-hydroxy acid with an orthoester (*R. C. Blume*, Tetrahedron Letters, 1969, 1047):

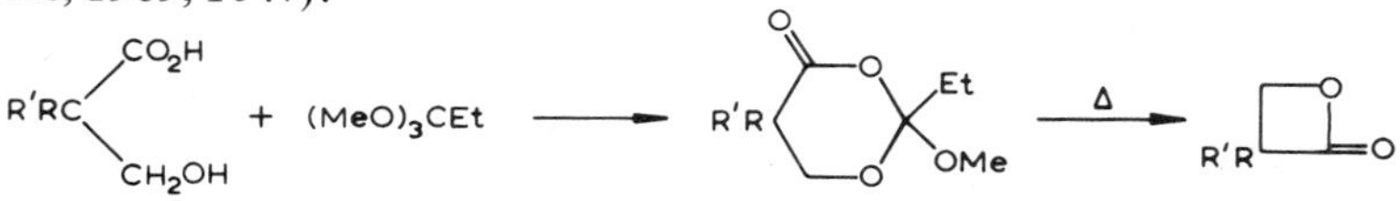
CO_2H
R'RC
CH_2OH
+ $(MeO)_3CEt$
R'R
O
Et
OMe
Δ

β-Propiolactone reacts with excess alkyl-lithium compounds to give 1,1-disubstituted propane-1,3-diols, with organocadmium compounds and benzylmagnesium chloride to give 3-substituted propionic acids, and with organomagnesium compounds to give alkyl or aryl vinyl ketones (*C. G. Stuckwisch* and *J. V. Bailey*, J. org. Chem., 1963, **28**, 2362). For structural data see p. 2.

(ii) Oxetan-3-ones

2,2,4,4-**Tetramethyloxetan-3-one** (I), m.p. 48°, may be prepared by oxidation of 2,2,4,4-tetramethyl-3-hydroxyoxetane-3-carboxylic acid, obtained from 2,5-dimethyl-3-hexyne-2,5-diol, with lead tetra-acetate (*B. L. Murr*, *G. B. Hoey* and *C. T. Lester*, J. Amer. chem. Soc., 1955, **77**, 4430; *C. Sandris* and *G. Ourisson*, Bull. Soc. chim. Fr., 1956, 958):

$Me_2C(OH){\cdot}C{\vdots}C{\cdot}CMe_2(OH)$ —($HgSO_4$, distil)→ 2,2,5,5-tetramethyltetrahydrofuran-3-one —(Br_2)→ 4,4-dibromo derivative (Br_2, Me_2, Me_2)
—(Reflux dil. aq. $NaHCO_3$ then conc. KOH)→ 2,2,4,4-tetramethyl-3-hydroxyoxetane-3-carboxylic acid (HO, CO_2H, Me_2, Me_2)
—($PbAc_4$, $CHCl_3$)→ (I) —→ Oxime and 2,4-dinitrophenylhydrazone

(I) —(Zn/Ac_2O)→ $Me_2C(OAc){\cdot}CO{\cdot}CHMe_2$ + $Me_2C(OAc){\cdot}CO{\cdot}CMe_2(OAc)$

Following the preparation of 1-oxaspiro[3.5]nonan-3-one (III), by the hydrolysis of the crude diazo-ketone (II) with methanolic potassium hydroxide and reaction of the product with acetic acid, what was believed to be the 2,4-dinitrophenylhydrazone of oxetan-3-one was obtained as a derivative of the product formed by similar treatment of chloroacetyldiazomethane (*J. R. Marshall* and *J. Walker*, J. chem. Soc., 1952, 467):

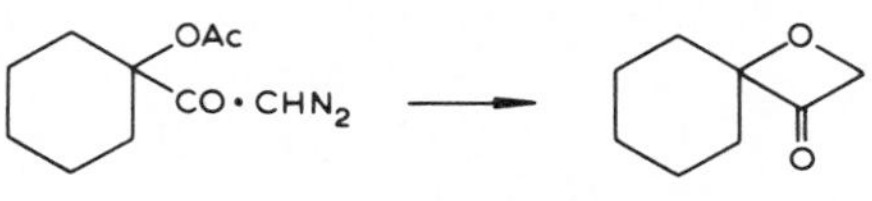

OAc
CO·CHN$_2$ ⟶

(II) (III)

ClCH$_2$·COCl $\xrightarrow{CH_2N_2}$ ClCH$_2$·CO·CHN$_2$ $\xrightarrow[(2)\ AcOH]{(1)\ MeOH/K_2CO_3}$

The autoxidation of tetraphenylacetone gives 2,2,4,4-tetraphenyloxetan-3-one and its structure has been established by the reactions indicated (*Hoey*, *D. O. Dean* and *Lester*, J. Amer. chem. Soc., 1955, **77,** 391):

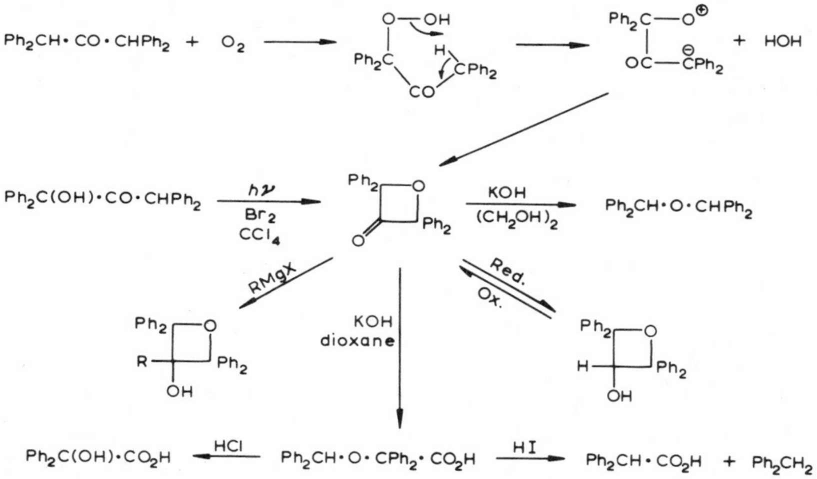

Ph$_2$CH·CO·CHPh$_2$ + O$_2$ ⟶ O—OH, Ph$_2$C, H, CPh$_2$, CO ⟶ Ph$_2$C—O$^{\oplus}$, OC—$\overset{\ominus}{C}$Ph$_2$ + HOH

Ph$_2$C(OH)·CO·CHPh$_2$ $\xrightarrow[Br_2,\ CCl_4]{h\nu}$ Ph$_2$, O, Ph$_2$, O $\xrightarrow[(CH_2OH)_2]{KOH}$ Ph$_2$CH·O·CHPh$_2$

RMgX: Ph$_2$, O, R, Ph$_2$, OH

KOH dioxane

Red. / Ox.: Ph$_2$, O, H, Ph$_2$, OH

Ph$_2$C(OH)·CO$_2$H $\xleftarrow{HCl}$ Ph$_2$CH·O·CPh$_2$·CO$_2$H $\xrightarrow{HI}$ Ph$_2$CH·CO$_2$H + Ph$_2$CH$_2$

Irradiation of tetramethyloxetan-3-one with 3130 Å light results in two competing modes of decomposition: (*1*) cleavage to acetone and dimethylketene; and (*2*) decarbonylation to yield tetramethyloxirane and other products. In polar solvents, path (*1*) occurs to the virtual exclusion of path (*2*) while in nonpolar solvents the two paths are about equally important (*P. J. Wagner et al., ibid.*, 1966, **88,** 1242):

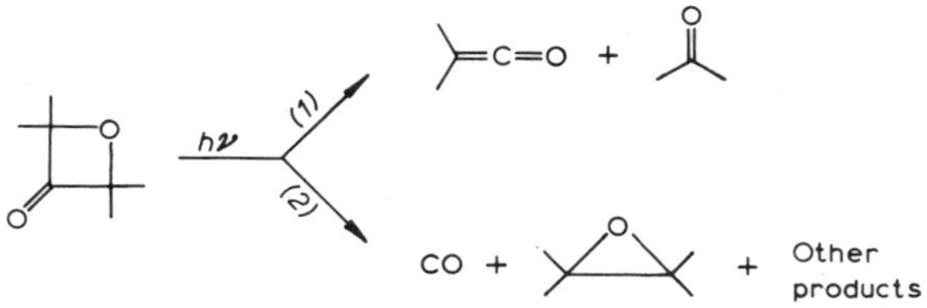

$h\nu$ (1) ⟶ C=O + O

(2) ⟶ CO + O + Other products

(d) Oxetes or oxetenes

2-Ethoxy-4-bistrifluoromethyl-oxet-2-ene (IV) has been isolated following the reaction at low temperature between hexafluoroacetone and ethoxyacetylene. If stored at room temperature for a few days the oxeten slowly isomerises to ethyl β,β-bis(trifluoromethyl)acrylate (*W. J. Middleton*, J. org. Chem., 1965, **30,** 1307):

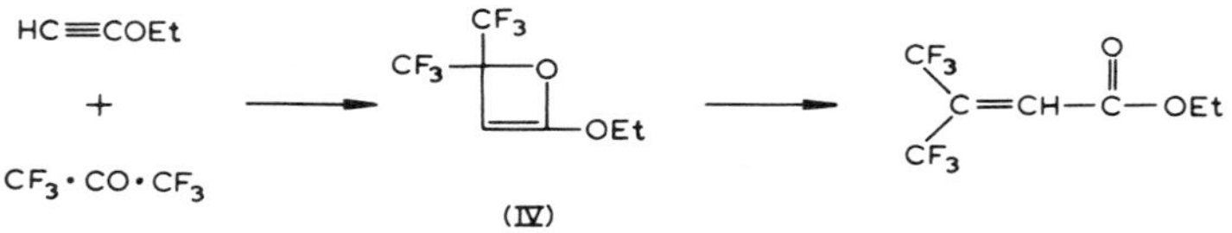

HC≡COEt + $CF_3 \cdot CO \cdot CF_3$ ⟶ (IV) ⟶ $(CF_3)_2C{=}CH{-}CO{-}OEt$

2,3,4,4-Tetramethyl-oxet-2-ene has been obtained following the irradiation of 3,4-dimethylpent-3-en-2-one in pentane (*L. E. Friedrich* and *G. B. Schuster*, J. Amer. chem. Soc., 1969, **91,** 7204):

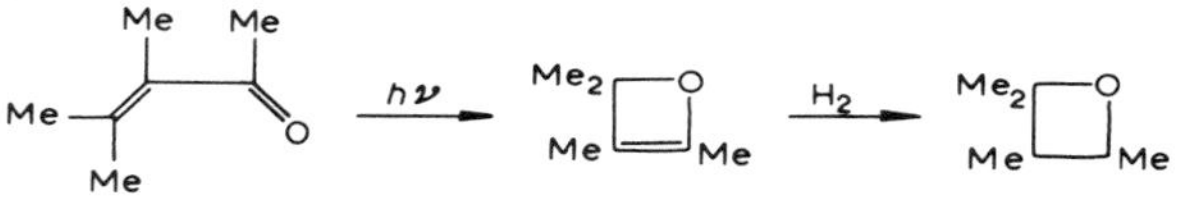

$\xrightarrow{h\nu}$ $\xrightarrow{H_2}$

7. Four-membered rings with one sulphur atom

*(a) Thietanes (trimethylene sulphides)**

(i) Preparation

Thietanes may be prepared:

(*1*) by the reaction between 1,3-dihalogenoalkanes and alkali sulphides (*E. Grishkeuich-Trochimowski*, J. Russ. Phys. chem. Soc., 1916, **48,** 880; *R. W. Bost* and *M. W. Conn*, C.A., 1955, **29,** 1350); or thiourea followed by alkaline cleavage of the resulting thiuronium salt (*W. E. Haines et al.*, J. Phys. chem., 1954, **58,** 270); by treating 3-halogenoalkanethiols with alkali (*E. P. Adams et al.*, J. chem. Soc., 1960, 2665):

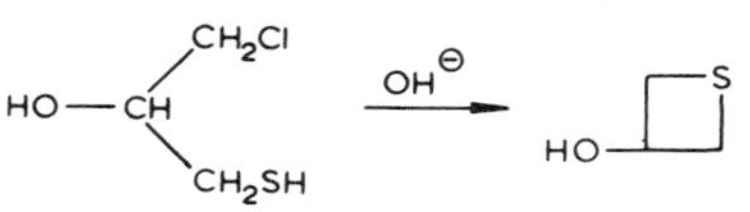

$HO{-}CH(CH_2Cl)(CH_2SH) \xrightarrow{OH^{\ominus}}$ 3-hydroxythietane

(*2*) from 3-chlorothiol esters and alkali; the 3-chlorothiol acetates may be obtained from allyl halides and thiolacetic acid (*F. G. Bordwell* and *W. A. Hewett*, J. org. Chem., 1958, **23,** 636; *D. C. Dittmer et al.*, J. Amer. chem. Soc., 1957, **79,** 4431; *R. M. Dobson* and *G. Klose*, Chem. and Ind., 1963, 450);

* *M. Sander*, Chem. Reviews, 1966, **66,** 341.

(*3*) by reacting chloromethylthiirane with either alkali or alkali phenoxides (p. 9) (*Sander*, Monatsh. Chem., 1965, **96,** 896);

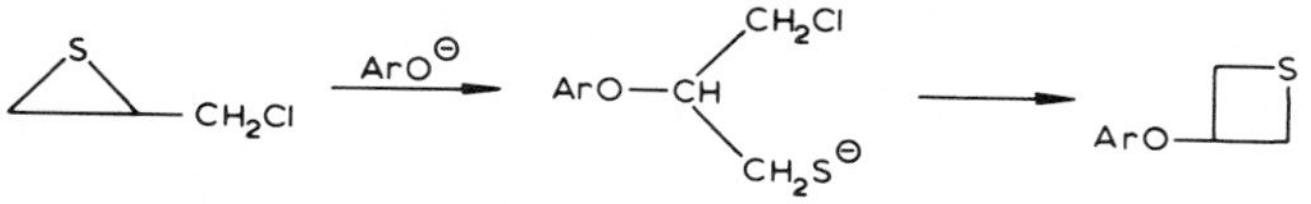

(*4*) by heating cyclic carbonates of 1,3-diols with alkali thiocyanates (*S. Searles, H. R. Hays* and *E. F. Lutz*, J. Amer. chem. Soc., 1958, **80,** 3168; J. org. Chem., 1962, **27,** 2832); the pyrolysis is thought to proceed by a mechanism similar to that proposed for the formation of thiiranes from oxiranes and thiocyanate ion (*E. E. van Tamelen*, J. Amer. chem. Soc., 1951, **73,** 3444);

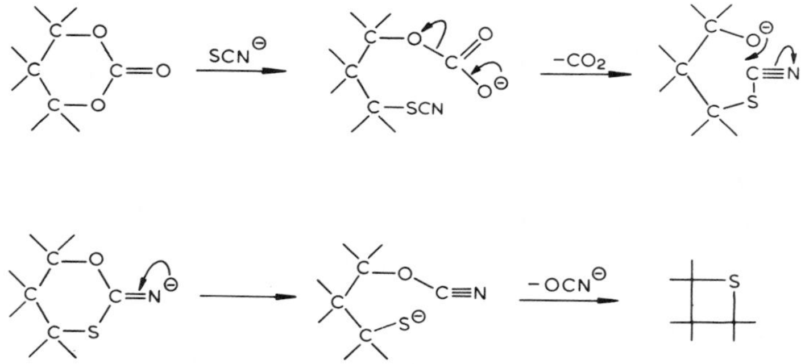

(*5*) by the reduction of thietane dioxides with lithium tetrahydridoaluminate (*Bordwell* and *W. H. McKellin*, *ibid.*, 1951, **73,** 2251):

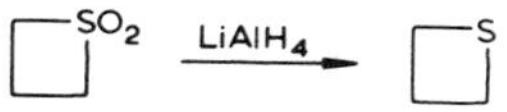

(*ii*) *Properties and reactions*

Thietane, b.p. 94°, is prepared in moderate yield from 1,3-dibromopropane and sodium sulphide (*G. M. Bennett* and *A. F. Hock*, J. chem. Soc., 1927, 2496). It is insoluble in water and can be oxidised normally to *thietane* 1,1-*dioxide*, m.p. 76°. For structural data see p. 2.

2-*Methylthietane*, b.p. 105–107°; 2,4-*dimethylthietane*, b.p. 113° (*Grischkewitsch-Trochimowski, loc. cit.*); 3,3-*dimethylthietane*, b.p. 119–121°; 1,1-*dioxide*, m.p. 54–55° (*H. J. Backer* and *K. J. Keuning*, Rec. Trav. chim., 1934, **53,** 808).

Unlike the thiiranes the reactions of thietanes do not always involve ring opening. They are less reactive than the thiiranes, have a greater resistance to heat and light, can be kept for a time without polymerisation occurring if stored in the dark at room temperature, and can be distilled at temperatures up to 150°.

(*1*) Polymerisation occurs slowly in sunlight and solid polymers are obtained by heating with water at 125° or dilute sulphuric acid at 110° (*Y. K. Yur'ev et al.*, C.A., 1955, **49,** 281).

(*2*) Thietanes form 1:1 complexes with mercuric chloride.

(*3*) They interact with methyl iodide to give 1:2 adducts (*Bennett* and *Hock, loc. cit.*);

$$\text{thietane} \xrightarrow{\text{MeI}} [\text{thietanium-SMe}]^{\oplus}\, I^{\ominus} \rightleftharpoons ICH_2\cdot CH_2\cdot CH_2SMe \xrightarrow{\text{MeI}} [ICH_2\cdot CH_2\cdot CH_2SMe_2]^{\oplus}\, I^{\ominus}$$

(*4*) They are desulphurised on treatment with Raney nickel (*Adams et al., loc. cit.*):

$$\text{3-hydroxythietane} \longrightarrow HOCH(Me)_2$$

(*5*) Thietane reacts with acetyl chloride in the presence of stannic chloride but does not react with amines (*Yur'ev et al., loc. cit.*):

$$\text{thietane} + CH_3\cdot COCl \longrightarrow CH_3\cdot COSCH_2\cdot CH_2\cdot CH_2Cl$$

(*6*) Thietane 1-oxides are obtained on oxidation with nitrous acid, chromic acid in acetic acid, or with one mole of hydrogen peroxide. They give 1:1 adducts with mercuric chloride (*Backer* and *Keuning*, Rec. Trav. chim., 1933, **52,** 499; *A. Cerniani, G. Modena* and *P. E. Todesco*, Gazz., 1960, **90,** 382; *Sander, loc. cit.*). The sulphinyl oxygen in 3-substituted thietane l-oxides (I, R = *p*-ClC_6H_4- or Me_3C) is reported to exert pseudo-equatorial preference (*C. R. Johnson* and *W. O. Siegl*, J. Amer. chem. Soc., 1969, **91,** 2796):

(I)

(*7*) Thietane 1,1-dioxides (sulphones) are obtained on treatment of the thietane with excess hydrogen peroxide, potassium permanganate, or peracetic acid (*Backer* and *Keuning*, Rev. Trav. chim., 1934, **53,** 808). Methanesulphonyl chloride and triethylamine with keten acetal give 3,3-diethoxythietane 1,1-dioxide (*W. E. Truce et al.*, J. Amer. chem. Soc., 1962, **84,** 3030):

$$CH_2{=}C(OEt)_2 + CH_3SO_2Cl \xrightarrow{Et_3N} (EtO)_2\text{-thietane-}SO_2$$

Oxidation of thietane with nitric acid gives polymers (*Bordwell et al., ibid.*, 1954, **76,** 1082).

(*8*) Thietane reacts with either chlorine or bromine to give bis-(3-halogenopropyl) disulphide (*J. M. Steward* and *C. H. Burnside, ibid.*, 1953, **75,** 243):

$$2\ \text{thietane} + Cl_2 \longrightarrow ClCH_2{\cdot}CH_2{\cdot}CH_2S{-}SCH_2{\cdot}CH_2{\cdot}CH_2Cl$$

(*9*) With butyl-lithium and phenyl-lithium at room temperature it gives a number of products including thiols, thioethers, and polymeric thioethers (*Bordwell* and *B. M. Pitt, ibid.*, 1955, **77,** 572):

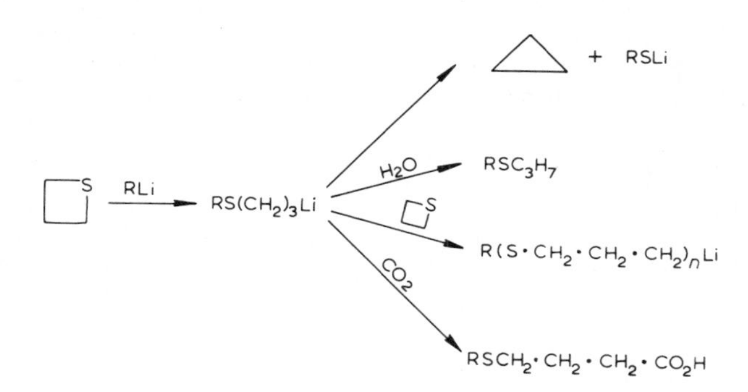

(*10*) *Cis-* and *trans-*2,4-**Diphenylthietane** have been synthesised and separated by fractional crystallisation (*Dobson* and *Klose, loc. cit.).* Oxidation with performic acid gives the 1,1-dioxides and on heating at 250° both lose sulphur dioxide with the formation of a mixture of 1,2-diphenylcyclopropanes (*Dobson* and *Klose*, Chem. and Ind., 1963, 450, 1203).

The *cis-* and *trans-*1,1-dioxides with ethylmagnesium bromide give *trans-*1,2-diphenylcyclopropanesulphinic acid, but in the presence of potassium *tert-*butoxide stereospecific ring-expansion occurs to yield *cis-* and *trans-*3,5-diphenyl-1,2-oxathiolane 2-oxide (*Dobson, P. D. Hammen* and *R. A. Davis*, Chem. Comm., 1968, 9):

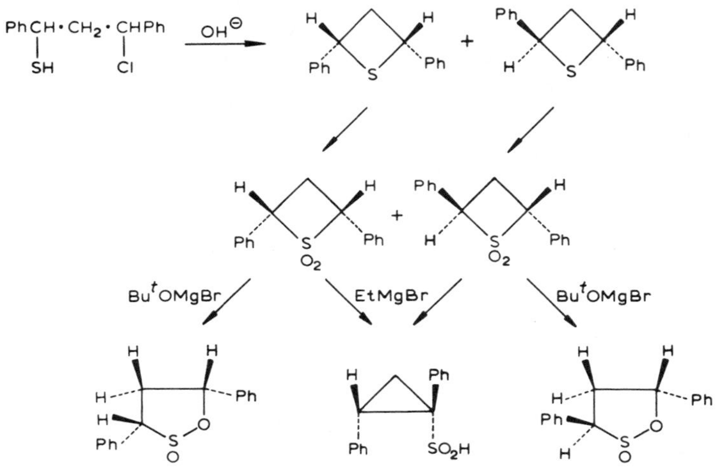

Ring contraction also results when *cis*- and *trans*-2,4-diphenylthietane 1-oxides react with potassium *tert*-butoxide to give a mixture of *cis*-1,2-diphenylcyclopropane-sulphinic acid and *cis*-1,2-diphenylcyclopropanethiol. Different products are obtained from the reaction of *cis*- and *trans*-monoxides with methylmagnesium iodide; the *cis*-isomer affords *cis*-2,4-diphenyltetrahydrothiophene and a benzothiepin, while the *trans*-isomer deoxygenates to give *trans*-2,4-diphenylthietane. In both instances *cis*- and *trans*-diphenylcyclopropanes are also formed (*Dobson* and *Hammen*, *ibid.*, 1968, 1294):

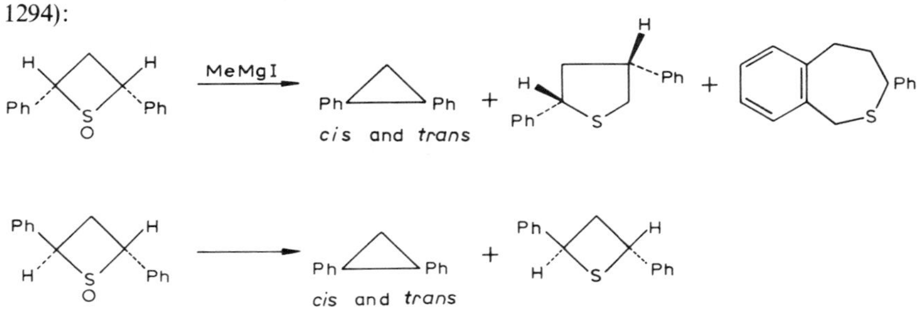

(b) Thietes* or thietenes and their oxides

Thietene 1,1-dioxide has been prepared by the following route (*D. C. Dittmer* and *M. E. Christy*, J. org. Chem., 1961, **26**, 1324):

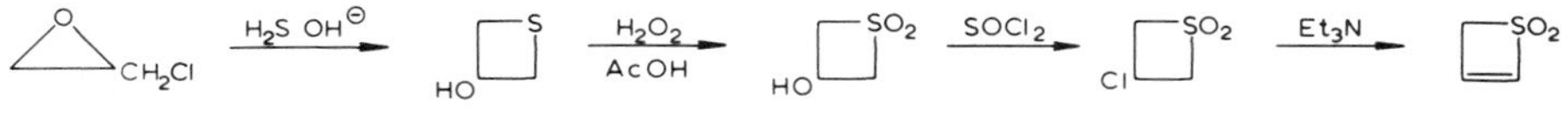

---

* See p. 1.

and also by Hofmann elimination on 3-aminothietane 1,1-dioxides obtained by the reaction of sulphonyl chlorides with appropriate enamines (*W. E. Truce et al.*, Tetrahedron Letters, 1963, 1677). Attempts to convert thietan-3-ol into thietene (thiete) gave polymers. Thietene 1,1-dioxide on reduction with lithium tetrahydridoaluminate and sodium tetrahydridoborate gives propane-1-thiol and thietane 1,1-dioxide, respectively. It forms an adduct with anthracene (*Dittmer* and *Christy*, J. Amer. chem. Soc., 1962, **84,** 399). Derivatives of thietene in which the double bond forms part of an aromatic system have been prepared by the Diels–Alder addition of isobenzofurans to thietene 1,1-dioxides, followed by dehydration and reduction (*L. A. Paquette*, J. org. Chem., 1965, **30,** 629):

Ph Ph Me_2 SO_2 reflux in C_6H_6 Ph Ph SO_2 Me_2

$H^{\oplus}$ Ph Ph SO_2 Me_2 $LiAlH_4$ Ph Ph S Me_2

An aliphatic thietene has been obtained in pentane solution following the treatment of the quaternary salt II with potassium *tert*-butoxide (*Dittmer* and *F. A. Davis*, J. Amer. chem. Soc., 1965, **87,** 2064):

$N^{\oplus}$—Me $I^{\ominus}$ S $KOBu^t$ $HCONMe_2$ $n-C_5H_{12}$ [S] H_2O_2 SO_2

(II)

The dioxide had been obtained previously from the cyclo-adduct of sulphene and cyclohexanone pyrrolidine enamine (*idem*, J. org. Chem., 1964, **29,** 3131):

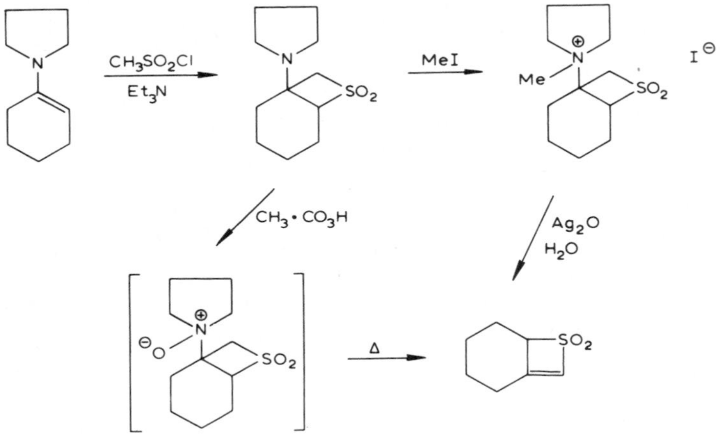

1-Diethylaminopropyne on treatment with several *para*-substituted phenylmethanesulphonyl chlorides in ether in the presence of triethylamine leads to the formation of a mixture of isomeric thietene dioxides (*Truce, R. H. Baury* and *P. S. Bailey Jr.*, Tetrahedron Letters, 1968, 5651):

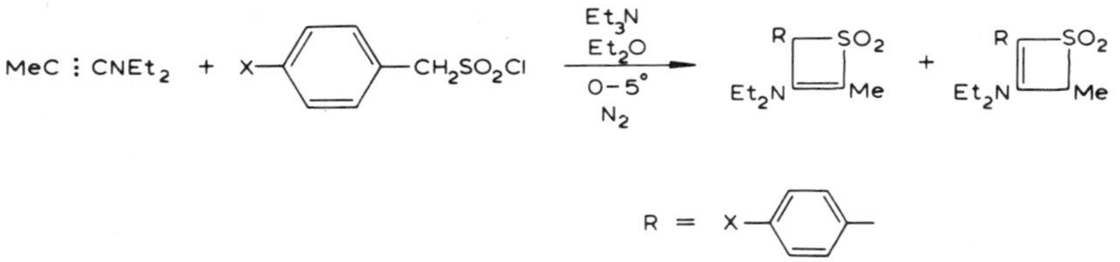

The thietene 1,1-dioxides (III, R = H or Ph) undergo a thermal rearrangement to give unsaturated cyclic sulphinic esters (*J. F. King et al.*, Chem. Comm., 1969, 31; *C. L. McIntosh* and *P. de Mayo, ibid.*, 1969, 32):

(III)

The highly strained 2-methylenethietene dioxide IV has been obtained by treating 2-dimethylaminomethyl-3-dimethylamino-4-phenylthietane 1,1-dioxide with excess 30% hydrogen peroxide in acetic acid–acetic anhydride solution at room temperature (*Paquette, M. Rosen* and *H. Stucki*, J. org. Chem., 1968, **33**, 3020). The dioxide gradually becomes coloured at room temperature and polymerises to a solid which is insoluble in all the common organic solvents. The exocyclic double bond is the more reactive, and undergoes a thermal cycloaddition with 1,3-diphenylisobenzofuran, and a photochemically induced dimerisation:

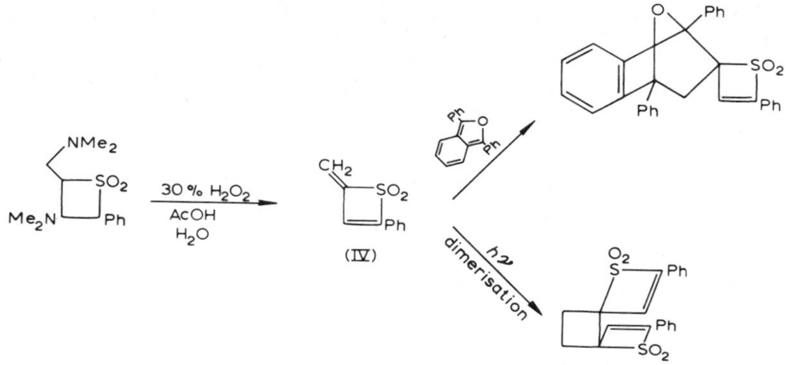

2-**Methylene**-4-**phenylthietene** 1,1-**dioxide** (IV) can be hydrogenated to a dihydro and a tetrahydro derivative, and undergoes a Michael reaction with dimethylamine (*Paquette* and *Rosen*, *ibid*., 1968, **33,** 3027):

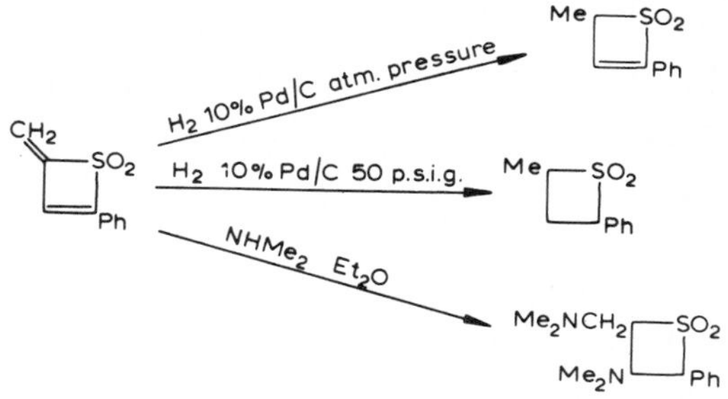

(c) Thietanones

Thietan-2-ones are β-thiolactones, *e.g.* the lactone of β-mercaptoisovaleric acid is 4,4-dimethylthietan-2-one (*I. L. Knunyant et al.*, C.A., 1956, **50,** 6312):

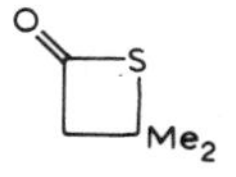

Thietan-3-ol on oxidation with benzoin and aluminium *tert*-butoxide in boiling ether or benzene gives thietan-3-one (*H. Prinzbach* and *G. von Veh*, Z. Naturforsch., 1961, **166,** 763), also obtained from 1,3-dibromoacetone dimethyl acetal and sodium sulphide with subsequent mild acid hydrolysis (*R. Mayer* and *K. F. Funk*, Angew. Chem., 1961, **73,** 578).

Thietan-3-one 1,1-**dioxide** has been obtained by the addition of methanesulphonyl chloride to ketene diethyl acetal in the presence of triethylamine and subsequent treatment with acid (*W. E. Truce* and *J. R. Norell*, Tetrahedron Letters, 1963, 1297) and from 3-morpholinothietene 1,1-dioxide by treatment with an acidic ion-exchange resin (*R. H. Hasek et al.*, J. org. Chem., 1963, **28,** 2496):

CH_3SO_2Cl + 2 CH_2 : C(N⟨morpholine⟩O)$_2$ ⟶ 3-morpholinothietene 1,1-dioxide $\xrightarrow[H_2O]{H^{\oplus}}$ thietan-3-one 1,1-dioxide

When diazobenzil and sulphur dioxide are heated together in benzene 2,2,4-triphenyl-4-benzoylthietan-3-one 1,1-dioxide and a small amount of 2-phenyl-3-diphenylmethylene-2-benzoyl-β-sultone are obtained, probably by cycloaddition of diphenylketen (formed by Wolf rearrangement) and the sulphen arising by the interaction of phenylbenzoylcarbene with sulphur dioxide (*T. Nagai, M. Tanaka* and *N. Tokura*, Tetrahedron Letters, 1968, 6293):

PhCO·C(N_2)·Ph ⟶ PhCO·C̈·Ph ⟶ $Ph_2C:C:O$ + PhCO·C̈·Ph·SO_2 ⟶ (Ph$_2$, SO$_2$, COPh, Ph thietanone) + (Ph, Ph, O—SO$_2$, COPh, Ph sultone)

Thietan-3-one 1,1-dioxide forms a 2,4-dinitrophenylhydrazone, is reduced by diborane to thietan-3-ol 1,1-dioxide, and although it dissolves in sodium bicarbonate solution with the liberation of carbon dioxide, no enol tautomer is detected by i.r. and n.m.r. spectroscopy (*Truce* and *Norell*, J. Amer. chem. Soc., 1963, **85,** 3231, 3236). With secondary amines at 20° it gives a mixture of the thioamide, MeCO·CS·NR_2, and 1-mercaptopropan-2-one (*Funk* and *Mayer*, J. prakt. Chem., 1963, **21,** 65).

Tetrabromoneopentane gives a good yield of 2,6-**dithiaspiro**[3.3]**heptane** (3,3′-**spirobithietane**) (V) (*H. J. Backer* and *K. J. Keuning*, Rec. Trav. chim., 1933, **52,** 499; 1934, **53,** 798):

C(CH_2Br)$_4$ ⟶ (V)   (VI)

The complete series of *S*-oxidation products has been prepared, including the disulphoxide VI, which was resolved by crystallising the camphorsulphonate of its cobaltous complex; the dissymmetry of VI parallels that of allene.

8. Four-membered rings with one selenium atom

Selenetane (trimethylene selenide), b.p. 118° (for i.r. spectrum see *A. B. Harvey, J. R. Durig* and *A. C. Morrissey*, J. chem. Phys., 1967, **47,** 4864) is obtained from trimethylene bromide and sodium selenide (*G. T. Morgan* and *F. H. Burstall*, J. chem. Soc., 1930, 1497); a similar process failed to give the tellurium derivative (*W. V. Farrar* and *J. M. Gulland*, *ibid*., 1945, 11). The smoother reaction between substituted trimethylene bromides and potassium selenide may be used to prepare 3,3-*dimethylselenetane*, b.p. 56° /40 mm, and the spiro-compounds I and II (*H. J. Backer* and *H. J. Winter*, Rec. Trav. chim., 1937, **56,** 492):

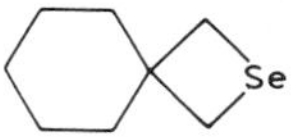

Se Se Se

(I) 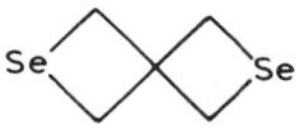 (II)

9. Four-membered rings containing one nitrogen atom

(a) Azetidines (trimethyleneimines)

Although azetidine and its derivatives have been known since 1888 (*S. Gabriel* and *J. Weiner*, Ber., 1888, **21,** 2669) comparatively little work, until recently, had been carried out on methods of preparation (*W. R. Vaughan et al.*, J. org. Chem., 1961, **26,** 138). The fact that penicillin contained a β-lactam ring, and the necessity for comparative tests in the chemotherapy of cancer stimulated interest in the chemistry of azetidine derivatives. The lower-ring homologue of proline has been isolated from natural sources.

(i) Synthesis

Azetidines may be prepared by:

(*1*) The ring-closure of γ-halogenopropylamines in alkaline solution, best results being obtained with secondary amines (*M. Kohn*, Ann., 1907, **351,** 134; *F. C. Schaefer*, J. Amer. chem. Soc., 1955, **77,** 5928); with tertiary γ-halogenopropylamines cyclisation to azetidinium salts may be accompanied by linear polymeric ammonium salts (*C. F. Gibbs* and *C. S. Marvel*, *ibid.*, 1934, **56,** 725; *C. Mannich* and *G. Baumgarten*, Ber., 1937, **70,** 210):

$$Me_2C:CH{\cdot}COMe + NH_2Me \longrightarrow Me_2C(NHMe){\cdot}CH_2{\cdot}COMe \longrightarrow Me_2C(NHMe){\cdot}CH_2{\cdot}CH(OH)Me$$

$$\longrightarrow Me_2C(NHMe){\cdot}CH_2{\cdot}CH(Br)Me \longrightarrow \text{1,2,2,4-tetramethylazetidine (Me, NMe, Me}_2)$$

(*2*) The reaction of 1,3-dihalogenoalkanes with toluenesulphonamide in presence of alkali (*W. Marckwald et al.*, Ber., 1898, **31,** 3261; 1899, **32,** 2031):

$$Br(CH_2)_3Br + K{\cdot}NH{\cdot}SO_2R \longrightarrow \text{azetidine-}NSO_2R \xrightarrow[AmOH]{Na} \text{azetidine-}NH$$

The reaction of 1,3-dibromopropane with aniline gives 1,2,3,4-tetrahydroquinoline (*A. Fischer, R. D. Topson* and *J. Vaughan*, J. org. Chem., 1960, **25,** 463) and not 1-phenylazetidine (*M. Scholtz*, Ber., 1899, **32,** 2252).

(*3*) The reaction of 3-aminoalkyl hydrogen sulphates with base (*R. C. Elderfield* and *H. A. Hageman*, J. org. Chem., 1949, **14,** 622).

(*4*) The pyrolysis of 1,3-diaminopropane is reported to give azetidine in low yield (*A. Ladenburg* and *J. Sieber*, Ber., 1890, **23,** 2727).

(*5*) The main problems in cyclisation are the competing elimination and dimerisation reactions. The former are negligible in the case of primary tosylates. The problem of azetidine synthesis has been reviewed and cyclisation to the azetidine system is considered as a conformational problem. Neither fragmentation nor E2 elimination may be expected to interfere with cyclisation if the "leaving group" on $C_{(1)}$ is primary; only dimerisation, which can be controlled by appropriate dilution need be considered. When the leaving group is secondary, conditions should be selected which favour S_N2 reaction over S_N1 and the nitrogen should be substituted by a bulk group to inhibit fragmentation. A convenient method for the preparation of certain azetidines is the reduction of the 1-tosyl derivative with sodium and amyl alcohol (*W. R. Vaughan et al., loc. cit.*):

$$CH_2(NH_2){-}CH_2{-}CH_2OH \xrightarrow[C_5H_5N,\ <5^\circ]{TosCl} \overset{③}{C}H_2(NHTos){-}\overset{②}{C}H_2{-}\overset{①}{C}H_2OTos \xrightarrow[EtOH]{Na} \text{azetidine-}NTos \xrightarrow[AmOH]{Na} \text{azetidine-}NH$$

(*6*) The reduction of 2-azetidinones to azetidines by lithium tetrahydridoaluminate is possible, providing there is no substituent on the nitrogen.

N-Alkyl derivatives undergo ring cleavage to give 3-alkylaminopropanols (*E. Testa, L. Fontanella* and *G. F. Cristiani*, Ann., 1959, **626,** 114).

(*7*) *N*-Arylazetidines may be obtained by the cyclisation of 1-arylamino-3-phenoxypropanes in benzene using aluminium chloride (*L. W. Deady et al.*, Tetrahedron Letters, 1968, 1773):

$$ArNH(CH_2)_3OPh \xrightarrow{AlCl_3} \text{azetidine-NAr}$$

(*8*) *N*-Unsubstituted azetidine-2-ones (β-lactams) react with triethoxonium tetrafluoroborate to give azetines, which may be reduced with lithium tetrahydridoaluminate to the corresponding azetidines (*G. Pifferi et al.*, J. Heterocyclic Chem., 1967, **4,** 619):

$$\text{EtO, RR'-azetine (=N)} \xrightarrow{LiAlH_4} \text{EtO, RR'-azetidine (NH)}$$

(ii) Properties and reactions

Azetidine (*Marckwald, loc. cit.*), b.p. 63°/748 mm, $d^{20.4}$ 0.8436, n_D^{24} 1.4287; *picrate*, m.p. 166–167°, as a secondary amine, gives ureas and the *N*-nitroso compound, and can

TABLE 5

AZETIDINES

Substituent	*b.p. (°)*	*Ref.*
1,3,3-Trimethyl	73–74	1
2,4,4-Trimethyl	86–88	2
1-Phenyl	242–245	3
1-Nitroso	198–199	4,5
1-Nitro	59/1.5 mm	5
1-Benzoyl	m.p. 61–62	6
1-Carbamoyl	m.p. 207	4
1-(*N*-phenylcarbamoyl)	m.p. 189–191	6
1-*p*-Toluenesulphonyl	120	4
3-Sulphonic acid	decomp.	7

References
1 *C. Mannich* and *G. Baumgarten*, Ber., 1937, **70,** 210.
2 *M. Kohn*, Ann., 1907, **351,** 143.
3 *M. Scholtz*, Ber., 1899, **32,** 2251.
4 *W. Marckwald et al.*, *ibid.*, 1898, **31,** 3251; 1899, **32,** 2031.
5 *C. L. Bumgardner*, *K. S. McCallum* and *J. P. Freeman*, J. Amer. chem. Soc., 1961, **83,** 4417.
6 *Y. Iwakura et al.*, J. org. Chem., 1965, **30,** 3410.
7 *S. Gabriel* and *J. Colman*, Ber., 1906, **39,** 2889.

be alkylated to quaternary salts (*F. G. Mann et al.*, J. chem. Soc., 1942, 163). These may polymerise on heating (*cf.* preparation) or revert to γ-halogenopropylamines; the quaternary hydroxides either undergo normal Hofmann degradation or yield γ-hydroxypropylamines (*Kohn, loc. cit.*). Strong acids open the azetidine ring, giving, *e.g.* γ-chloropropylamine, while more dilute acids can afford γ-hydroxypropylamine.

Some substituted azetidines are listed in Table 5.

1-**Nitrosoazetidine** (*C. C. Howard* and *Marckwald*, Ber., 1899, **32,** 2032; *C. L. Bumgardner, K. S. McCallum* and *J. P. Freeman*, J. Amer. chem. Soc., 1961, **83,** 4417) on oxidation with peroxytrifluoroacetic acid produces 1-**nitroazetidine:**

N·NO [O] → N·NO_2

1-**Benzoylazetidine** and related compounds are isomerised to dihydro-1,3-oxazines with either picric acid or *p*-toluenesulphonic acid in boiling toluene (*Y. Iwakura et al.*, J. org. Chem., 1965, **30,** 3410):

NCOPh → Ph, N, O

but not as readily as the 1-thioacylazetidines.

1-**Thiobenzoylazetidine** on treatment with concentrated hydrochloric acid at room temperature gives 2-phenyldihydro-1,3-thiazine (*idem, ibid.*, 1966, **31,** 3352):

N·CSPh $H^{\oplus}$ → Ph, N, S

1-Thiocarbamoylazetidines with acid give tetrahydro-2-imino-1,3-thiazines (*M. Tisler*, Tetrahedron Letters, 1959, 12).

N·CS·NHR $H^{\oplus}$ → RN, H, N, S

3-**Benzoyl**-1-*tert*-**butyl-2-phenylazetidines** have been synthesised and epimerised (*J. L. Imbach et al.*, J. org. Chem., 1967, **32,** 78). On irradiation the *cis*- and *trans*-forms undergo a ring expansion to give 1-*tert*-butyl-2,4-diphenylpyrrole which involves the migration of an aryl group from the α-position to the carbonyl carbon of an n–π^* excited state (*R. Gruber* and *L. Hamilton*, J. Amer. chem. Soc., 1967, **89,** 3077). In the case of the *trans*-isomer the major product is 1-*tert*-butyl-2,3-diphenylpyrrole (*A. Padua* and *Gruber*, *ibid.*, 1968, **90,** 4456):

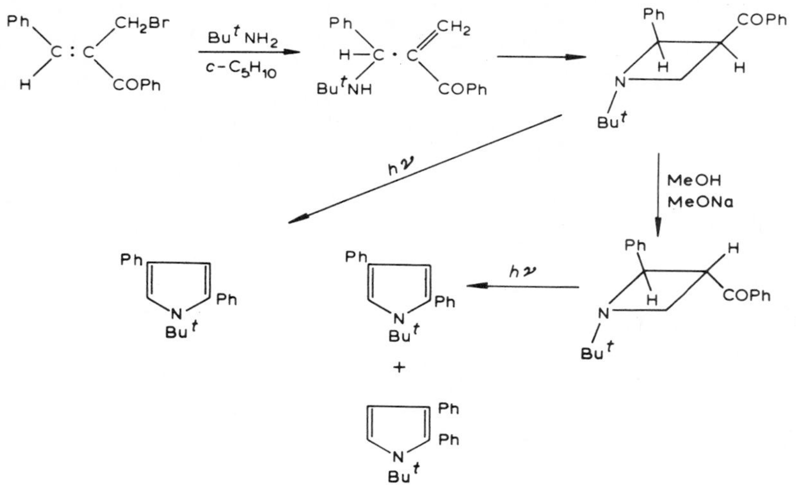

1-*tert*-**Butyl**-3-**chloroazetidine** partially isomerises on heating to give 1-*tert*-butyl-2-chloromethylaziridine. This is the first ring-contraction of an azetidine to be reported (*V. R. Gaertner*, Tetrahedron Letters, 1968, 5919):

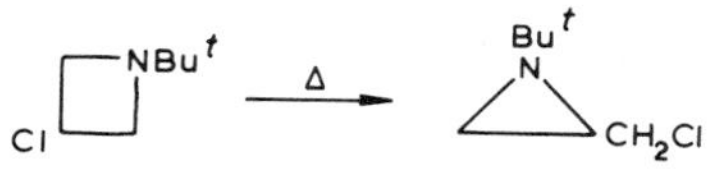

Irradiation of 3-phenyl-3,4-dihydrobenzotriazine in benzene solution with u.v. light gives 1-**phenyl**-2,3-**benzo**-2-**azetine** which reacts with aniline and acetic acid and upon thermal or photochemical excitation is rapidly converted to a reactive species as shown; the latter is conveniently trapped as a 1:1 adduct with *N*-phenylmaleimide (*E. M. Burgess* and *L. McCullagh*, J. Amer. chem. Soc., 1966, **88**, 1581):

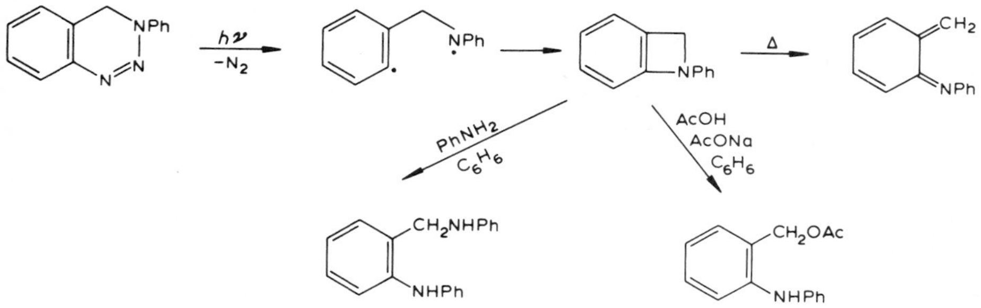

cis- and *trans*-7-**Azabicyclo**[4.2.0]**octanes** have been synthesised (*E. J. Moriconi* and *P. H. Mazzocchi*, J. Org. Chem., 1966, **31**, 1372):

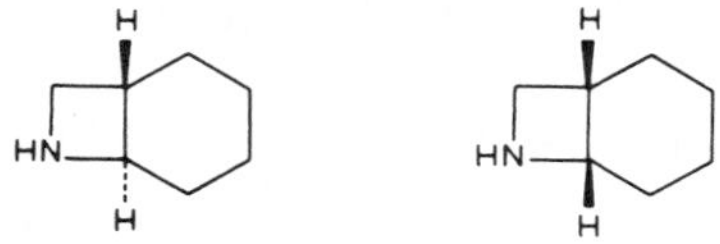

Ring-closure of 1-alkylamino-3-chloropropanes with silver perchlorate gives tertiary azetidinium perchlorates (*N. J. Leonard* and *D. A. Durand*, *ibid.*, 1968, **33**, 1322).

1,1,2,2-**Tetramethyl-** and 1-**benzyl**-1,2,2-**trimethyl-azetidinium perchlorates** have been synthesised conveniently by the following route:

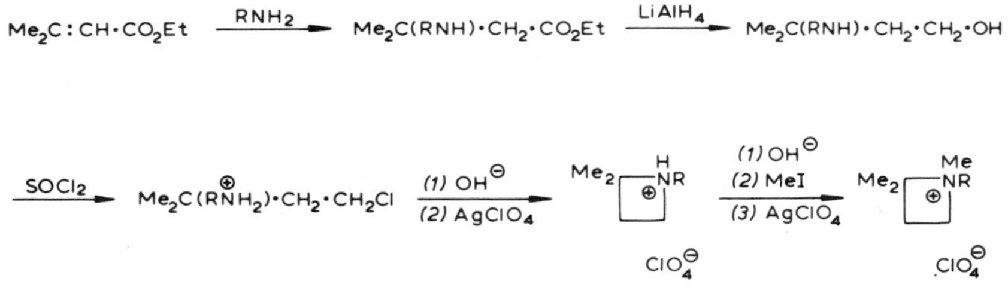

$$Me_2C:CH\cdot CO_2Et \xrightarrow{RNH_2} Me_2C(RNH)\cdot CH_2\cdot CO_2Et \xrightarrow{LiAlH_4} Me_2C(RNH)\cdot CH_2\cdot CH_2\cdot OH$$

$$\xrightarrow{SOCl_2} Me_2C(R\overset{\oplus}{N}H_2)\cdot CH_2\cdot CH_2Cl \xrightarrow[(2)\ AgClO_4]{(1)\ OH^{\ominus}}$$

The tetramethyl salt proved to be relatively unreactive, but the benzyltrimethyl salt underwent solvolytic ring-opening with alcohols to form *N*-(3-alkoxy-3-methylbutyl)-*N*-methyl-benzylamine perchlorate

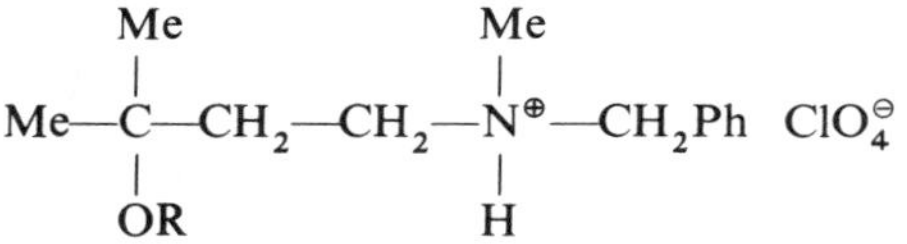

$$Me-C(Me)(OR)-CH_2-CH_2-N^{\oplus}(Me)(H)-CH_2Ph\ \ ClO_4^{\ominus}$$

With sodium methoxide, or on heating in solution, both salts show a strong tendency to undergo eliminative ring-opening to give substituted 3-methyl-3-buten-1-ylamines. 1-Benzyl-1,2,2-trimethylazetidinium perchlorate reacted with nitrones, in particular 4,5,5-trimethyl-1-pyrroline 1-oxide and 5,5-dimethyl-1-pyrroline 1-oxide to give a charged heterocycle:

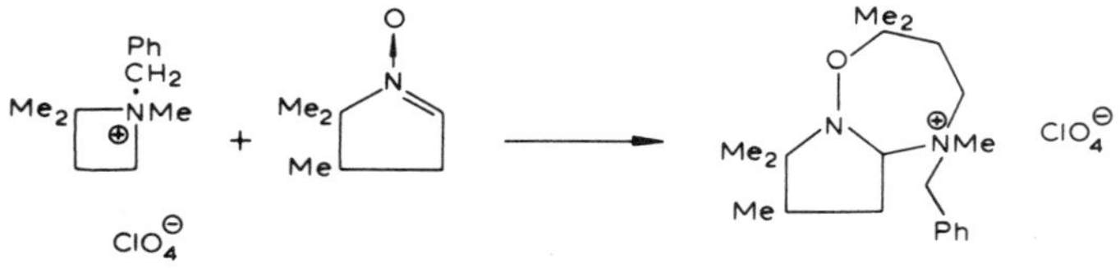

Azetidin-3-**ols** may be obtained by u.v. irradiation of α-dialkylamino-ketones (*R. A. Clasen* and *S. Searles*, Chem. Comm., 1966, 289):

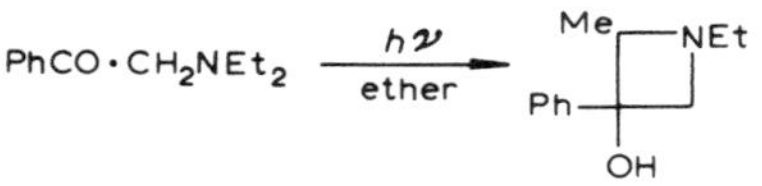

$$PhCO\cdot CH_2NEt_2 \xrightarrow[\text{ether}]{h\nu}$$

Azetidine-2-carboxylic acid, isolated from leaves of lily-of-the-valley *(Convallaria majalis)* and *Polygonatum multiflorum*, has been synthesised from γ-amino-α-bromobutyric acid (*L. Fowden*, Biochem. J., 1956, **64,** 323 and *M. Bryant*, *ibid.*, 1958, **70,** 626) and a study made of its biosynthesis (*idem*, *ibid.*, 1959, **71,** 210; *E. Leete*, J. Amer. chem. Soc., 1964, **86,** 3162).

(b) Azetidinones; oxoazetidines

(i) Synthesis

β-Lactams (Vol. I D, p. 192) are azetidin-2-ones. They may be prepared:

(*1*) by the reaction between ketenes and azomethines (*W. Kirmse* and *L. Horner*, Ber., 1956, **89,** 2759; *R. Pfleger* and *A. Jager*, *ibid.*, 1957, **90,** 2460).

(*2*) The reaction of a β-amino ester with one equivalent of a Grignard reagent is applicable to the synthesis of a variety of 1-alkylazetidin-2-ones and also to **azetidin-2-one,** m.p. 73–74° (for structural data see p. 2) (*R. W.* and *A. D. Holley*, J. Amer. chem. Soc., 1949, **71,** 2124, 2129):

$$NH_2(CH_2)_2\cdot COEt \xrightarrow{EtMgBr} \left[\begin{array}{l} NH(CH_2)_2\cdot CO_2Et \\ | \\ MgBr \end{array}\right] \longrightarrow \text{azetidin-2-one (NH)}$$

$$\begin{array}{l} R^2OCO\cdot CHR^3 \\ \quad | \\ R^1NH\cdot CHR^4 \end{array} \xrightarrow{RMgX} \text{1-}R^1\text{-3-}R^3\text{-4-}R^4\text{-azetidin-2-one} \begin{array}{l} \xrightarrow{OH^\ominus} R^3CH\cdot CO_2^\ominus / R^4CH\cdot NHR^1 \\ \xrightarrow{EtOH\ HCl} R^3CH\cdot CO_2Et / R^4CH\cdot NHR^1\ HCl \end{array}$$

3,3-Disubstituted azetidin-2-ones may be obtained by the cyclisation of β-aminopropionic esters in the presence of a Grignard reagent (*L. Fontanella* and *E. Testa*, Ann., 1959, **622,** 117; **625,** 95):

$$R^1R^2C(CO_2Et)(CH_2NH_2) \xrightarrow{RMgX} \text{3-}R^1\text{-3-}R^2\text{-azetidin-2-one (NH)}$$

and can be estimated following hydrolytic ring-scission by reaction with ninhydrin (*V. d'Amato et al.*, Ann., 1960, **635,** 127).

(*3*) *N*-Substituted aminomalonic esters *N*-acylated with α-halogeno acids and subsequently ring-closed with base give azetidin-2-ones, but not when the nitrogen is unsubstituted (*J. C. Sheehan* and *A. K. Bose*, J. Amer. chem. Soc., 1951, **73,** 1761):

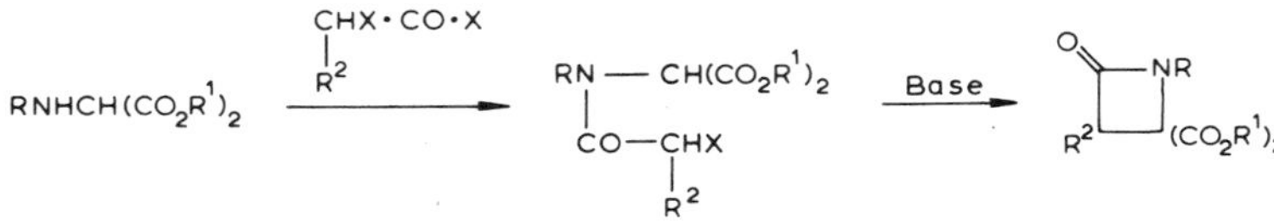

$RNHCH(CO_2R^1)_2 \xrightarrow{CHX \cdot CO \cdot X,\ R^2} RN(CO{-}CHXR^2){-}CH(CO_2R^1)_2 \xrightarrow{Base}$ (azetidinone: O, NR, R^2, $(CO_2R^1)_2$)

Sodium 1-*tert*-butylaziridine-2-carboxylate can be converted to 1-*tert*-butyl-3-chloroazetidin-2-one (see p. 24).

(*4*) The reaction between benzylideneaniline and Reformatsky reagents gives predominantly the *cis*-isomer when R = Me and the *trans*-isomer when R = Ph (*H. B. Kagan, J. J. Basselier* and *J. L. Luche*, Tetrahedron Letters, 1964, 941):

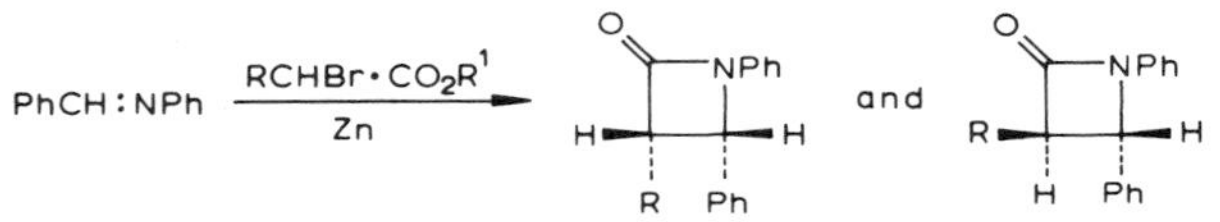

$PhCH{:}NPh \xrightarrow[Zn]{RCHBr \cdot CO_2R^1}$ (O, NPh, H, H, R, Ph) and (O, NPh, R, H, H, Ph)

(*5*) Irradiation of α,β-unsaturated amides gives *cis*- and *trans*-β-lactams (*O. L. Chapman* and *W. R. Adams*, J. Amer. chem. Soc., 1967, **89,** 4243); the *cis*-isomer predominates with aromatic amides, whereas in the previous method it was the *trans*-isomer:

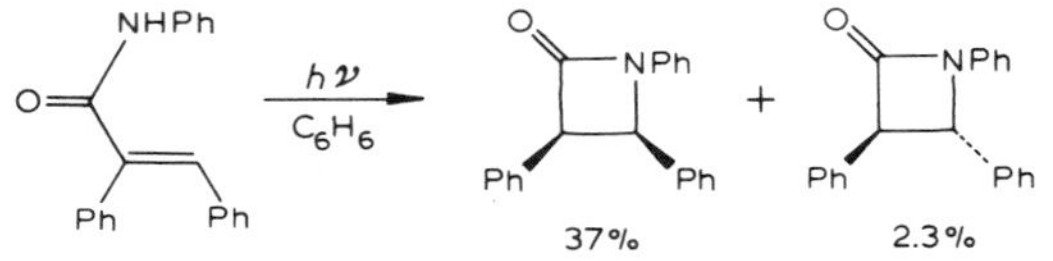

(NHPh, O, Ph, Ph) $\xrightarrow[C_6H_6]{h\nu}$ (O, NPh, Ph, Ph) 37% + (O, NPh, Ph, Ph) 2.3%

(*6*) Substituted allenes react with chlorosulphonyl isocyanate to form β-lactams containing an exocyclic double bond (*E. J. Moriconi* and *J. F. Kelly*, *ibid*., 1966, **88,** 3657):

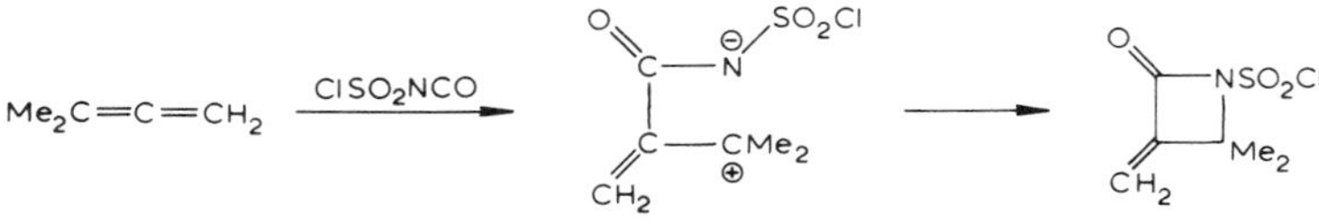

$Me_2C{=}C{=}CH_2 \xrightarrow{ClSO_2NCO}$ (O, C, $\ominus$N, SO_2Cl, C, CMe_2 $\oplus$, CH_2) $\longrightarrow$ (O, NSO_2Cl, Me_2, CH_2)

cis- and *trans*-Methylstyrene react with chlorosulphonyl isocyanate to give *cis*- and *trans*-1-chlorosulphonyl-3-methyl-4-phenylazetidin-2-one, respectively (*idem*, Tetrahedron Letters, 1968, 1435). At temperatures below $-10°$ chlorosulphonyl isocyanate reacts with 2-methylbutadiene and the 1,2-cycloaddition product is a β-lactam; on warming to 40° a 1,2→1,4 rearrangement takes place to give a δ-lactam (*Moriconi* and *W. C. Meyer*, *ibid.*, 1968, 3823):

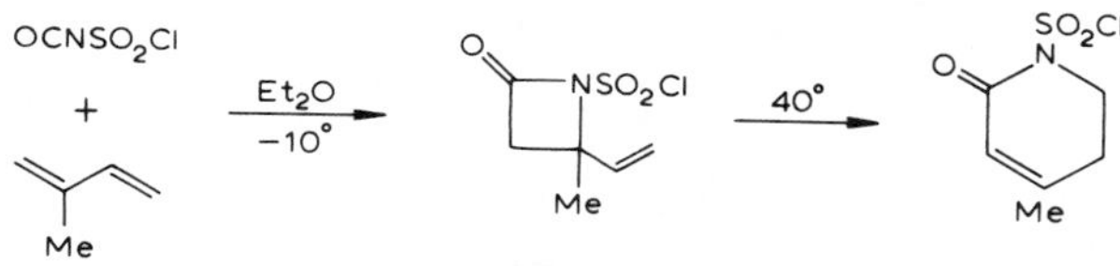

OCNSO$_2$Cl + (2-methylbutadiene, Me) $\xrightarrow[-10°]{Et_2O}$ β-lactam (NSO$_2$Cl, Me, vinyl) $\xrightarrow{40°}$ δ-lactam (SO$_2$Cl on N, Me)

Some control over β-lactam stereochemistry can be achieved by varying the sequence of addition of reagents, *e.g.*, the slow addition of azidoacetyl chloride to a solution of benzalaniline and triethylamine in methylene chloride favours the *cis*-product, while the *trans*- is favoured when triethylamine is added to a mixture of the Schiff base and azidoacetyl chloride (*Bose et al.*, Tetrahedron, 1967, **23**, 4769):

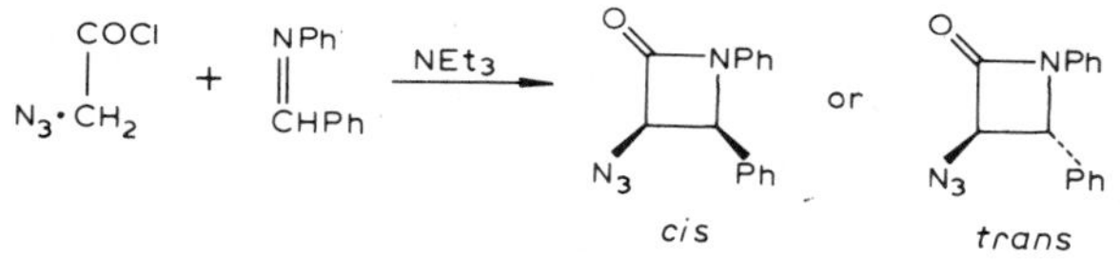

$N_3 \cdot CH_2COCl$ + PhN=CHPh $\xrightarrow{NEt_3}$ (NPh, N_3, Ph) *cis* or (NPh, N_3, Ph) *trans*

(ii) Reactions

(*1*) U.v. irradiation of 1-phenylazetidin-2-ones results in ring-opening and the course of the reaction depends on the nature of the 4-substituent. An electron-withdrawing group promotes 1–4 and 2–3 bond cleavage to afford an olefin and phenyl isocyanate, while an electron-releasing group favours 1–2 and 3–4 bond fission to yield an imine and ketene (*M. Fischer*, Ber., 1968, **101**, 2669).

(*2*) A β-thiolactam may be obtained by replacing the carbonyl oxygen of β-lactams by sulphur on treatment with phosphorus pentasulphide (*K. R. Henery-Logan, H. P. Knoepfel* and *J. V. Rodricks*, J. Heterocyclic Chem., 1968, **5**, 433):

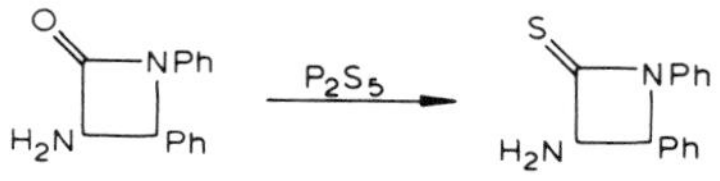

(O, NPh, H_2N, Ph) $\xrightarrow{P_2S_5}$ (S, NPh, H_2N, Ph)

(*3*) Cleavage of the triazene I with one equivalent of boron trifluoride etherate led to the evolution of the theoretical amount of nitrogen and the isolation of 1,2-diphenyl-2-azetin-4-one (*Henery-Logan* and *Rodricks*, J. Amer. chem. Soc., 1963, **85**, 3524):

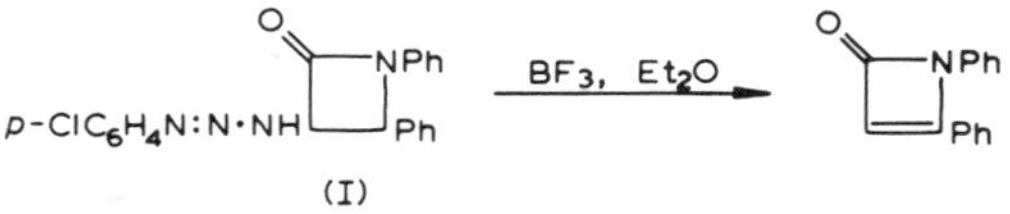

$$p\text{-}ClC_6H_4N{:}N{\cdot}NH\text{-(I)} \xrightarrow{BF_3,\ Et_2O} \text{1,2-diphenyl-2-azetin-4-one}$$

(I)

(c) Azetidine-2,4-diones (malonimides)

Azetidine-2,4-diones may be prepared from cyanoacetic esters and converted to an azetidine by treatment with lithium tetrahydridoaluminate (*Testa et al.*, Helv., 1959, **42**, 2370; Ann., 1960, **633**, 56; **635**, 119):

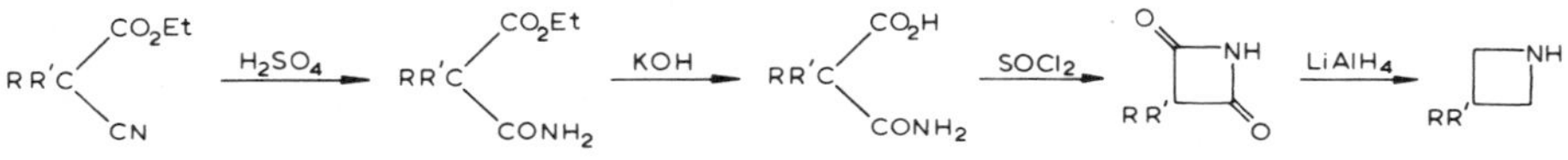

$$RR'C(CO_2Et)(CN) \xrightarrow{H_2SO_4} RR'C(CO_2Et)(CONH_2) \xrightarrow{KOH} RR'C(CO_2H)(CONH_2) \xrightarrow{SOCl_2} \text{RR'-azetidine-2,4-dione} \xrightarrow{LiAlH_4} \text{RR'-azetidine}$$

Some substituted azetidine-2,4-diones have been prepared from a ketene and an alkyl isocyanate and from a malonyl chloride and a primary amine (*A. Ebnother et al.*, Helv., 1959, **42**, 918).

2,3-Bis(*tert*-butylimino)azetidines (II) along with 3-*tert*-butylamino-2-phenylindoles are obtained from the reaction between *N*-arylimines and *tert*-butyl isocyanide with acid catalysis in non-basic solvents (*J. A. Deyrup et al.*, Tetrahedron, 1969, **25**, 1467):

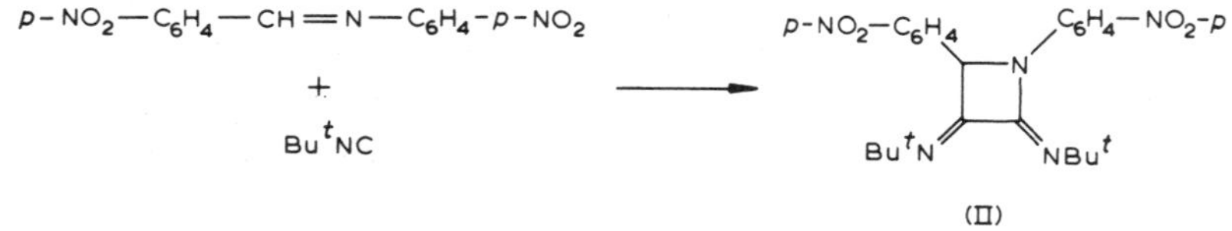

$$p\text{-}NO_2\text{—}C_6H_4\text{—}CH{=}N\text{—}C_6H_4\text{-}p\text{-}NO_2 + Bu^tNC \longrightarrow \text{(II)}$$

(II)

2,6-**Diazaspiro**[3.3]**heptane** (III) has been prepared from tetrabromoneopentane (*F. J. Govaert*, Proc. Koninkl. Nederland. Akad. Wetenschap., 1934, **37**, 156):

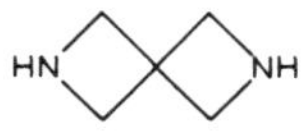

(III)

10. Three- and four-membered rings containing other elements

Three-membered rings containing silicon and germanium have been synthesised; four-membered ring compounds containing silicon and phosphorus are known. Diphenylacetylene reacts with either dichlorodimethylsilane and sodium, or polydimethylsilicone to give low yields of 1,1-**dimethyl-**2,3-**diphenylsilirene.**

1,1-**Diiodogermirene** and 1,1-**diiodo-**2,3-**diphenylgermirene** may be obtained in a similar manner from germanium diiodide and hot acetylene and diphenylacetylene, respectively. The 2,3-diphenyl compound reacts with sodium hydroxide and Grignard reagents (*M. E. Volpin et al.*, Tetrahedron, 1962, **18,** 107):

$$\text{Ph}\overline{\text{C=C(Ge}Me_2)}\text{Ph} \xleftarrow{\text{MeMgI}} \text{Ph}\overline{\text{C=C(Ge}I_2)}\text{Ph} \underset{\text{HI}}{\overset{\text{NaOH}}{\rightleftharpoons}} \text{Ph}\overline{\text{C=C(Ge(OH)}_2)}\text{Ph}$$

1,1-**Dimethylsiletane,** b.p. 81°, has been obtained by the following route (*L. H. Sommer* and *G. A. Baum*, J. Amer. chem. Soc., 1954, **76,** 5002):

$$Me_3Si(CH_2)_3Br \xrightarrow[(2)\ H_2O]{(1)\ H_2SO_4} [Br(CH_2)_3SiMe_2]_2O \xrightarrow[H_2SO_4]{NH_4Cl} Br(CH_2)_3SiMe_2Cl \xrightarrow[Et_2O]{Mg} \square\text{SiMe}_2$$

and 3,3-**bismethoxycarbonyl-**1,1-**dimethylsiletane** from bis-(iodomethyl)-dimethylsilane, dimethyl malonate and sodium methoxide (*R. West*, *ibid.*, 1955, **77,** 2339):

$$Me_2Si(CH_2I)_2 + CH_2(CO_2Me)_2 \xrightarrow{NaOMe} (MeO_2C)_2\square\text{SiMe}_2$$

1,1,2-**Triphenylsiletane** is obtained by reacting diphenyl-(3-phenylpropyl)silane with *N*-bromosuccinimide and then with magnesium.

The siletane is readily cleaved by lithium tetrahydridoaluminate (*H. Gilman* and *W. H. Atwell*, *ibid.*, 1964, **86,** 2687):

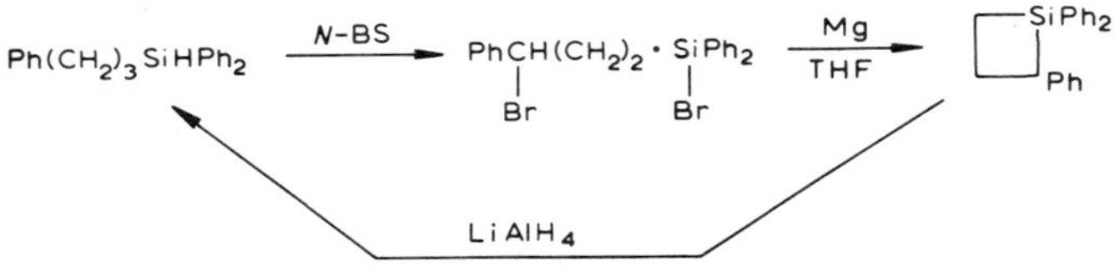

$$Ph(CH_2)_3SiHPh_2 \xrightarrow{N\text{-BS}} PhCH(Br)(CH_2)_2\cdot SiPh_2(Br) \xrightarrow[THF]{Mg} \text{siletane } (SiPh_2, Ph) \xrightarrow{LiAlH_4} Ph(CH_2)_3SiHPh_2$$

The reaction of 2,4,4-trimethyl-2-pentene with phosphorus trichloride and aluminium chloride in methylene chloride at 0–5° followed by partial hydrolysis with water yields a product which contains phosphorus and chlorine. Further hydrolysis gives a chlorine-free four-membered **phosphetane** I, which is stable to hot nitric or sulphuric acid and to aqueous sodium hydroxide (*J. J. McBride Jr. et al.*, J. org. Chem., 1962, **27,** 1833):

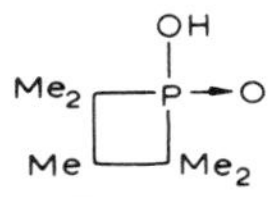

(I)

11. Four-membered rings with two hetero-atoms

At present there is little information available on four-membered rings containing two hetero-atoms, but there is every indication of a growing interest in their chemistry.

(a) Dioxetanes

A novel cannabinoid (I) possessing a four-membered cyclic peroxide ring (a dioxetane) has been isolated along with other products from the pyridine-catalysed reaction of citral with olivetol (*R. K. Razdan* and *V. V. Kane*, J. Amer. chem. Soc., 1969, **91,** 5190).

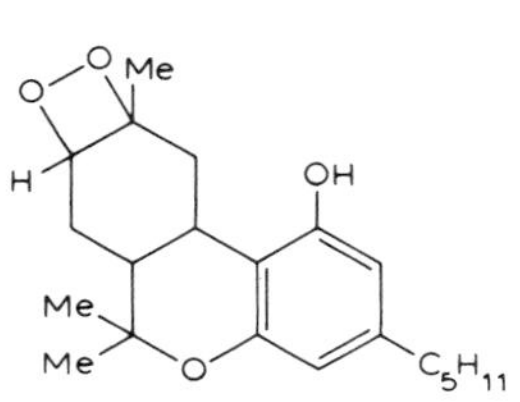

(I)

(b) Dithietanes

Tetrafluoro-1,2-**dithietane** unlike most fluorinated disulphides is not relatively stable to heat and acids, and exists only briefly at 25°. Its spontaneous polymerisation can be reserved above 250° (*W. R. Brasen et al.*, J. org. Chem., 1965, **30**, 4188):

$$(-SCF_2 \cdot CF_2S-)_n \rightleftharpoons \left[F_2C(S)S\,CF_2 + \text{dimer} \right] \xrightarrow{CF_3 \cdot CF:CF_2} \text{2-trifluoromethyl-2,4,4,5,5,6,6-heptafluoro-1,3-dithiane}$$

1,3-Dithietanes (desaurins) are obtained by the reaction of certain ketones with base and carbon disulphide (*P. Yates* and *D. R. Moore*, J. Amer. chem. Soc., 1958, **80**, 5577):

$$PhCH_2 \cdot COMe \xrightarrow{CS_2} (MeCO)(Ph)C{=}C\langle S_2 \rangle C{=}C(Ph)(COMe)$$

Dicyanomethylene-1,3-**dithietane** (II) may be prepared from dipotassio-1,1-dimercapto-2,2-dicyanoethylene and excess diiodomethane in boiling acetonitrile (*D. C. Dittmer, H. E. Simmons* and *R. D. Vest*, J. org. Chem., 1964, **29**, 497):

$$(NC)_2C:C(SK)_2 \xrightarrow[MeCN]{CH_2I_2} (NC)_2C{=}C\langle S_2 \rangle CH_2 \quad (II)$$

$$(II) \xrightarrow{C_5H_5N} (NC)_2C{=}C(S^{\ominus}){-}S{-}CH_2{-}\overset{\oplus}{N}C_5H_5$$

$$(II) \xrightarrow{(Ph)_3P} (NC)_2C{=}C(S^{\ominus}){-}S{-}CH_2{-}\overset{\oplus}{P}(Ph)_3$$

Diethyl sodiomalonate reacts with thiophosgene to give a 1,3-dithietane, which on treatment with sulphur tetrafluoride in the presence of hydrogen fluoride yields 2,4-**bis**-[2,2,2-**trifluoro**-1-(**trifluoromethyl**)**ethylidene**]-1,3-**dithietane** (*M. S. Raasch*, Chem. Comm., 1966, 577):

$$(EtO_2C)_2C{=}C\langle S_2 \rangle C{=}C(CO_2Et)_2 \xrightarrow[HF]{SF_4} (F_3C)_2C{=}C\langle S_2 \rangle C{=}C(CF_3)_2 \xrightarrow{650°} (F_3C)_2C{=}C{=}S$$

The reaction of 3-diazo-2-butanone with boiling carbon disulphide yields **bis-(1-acetylethylidene)-1,3-dithietane**(III) (*A. J. Kirby*, Tetrahedron, 1966, **22**, 3001), whereas under similar conditions 2-diazo-2-phenylacetophenone gives the five-membered ring compound IV (*Yates* and *L. L. Williams*, Tetrahedron Letters, 1968, 1205). In the latter case one of the PhCO·CPh moieties undergoes rearrangement of the Wolff type:

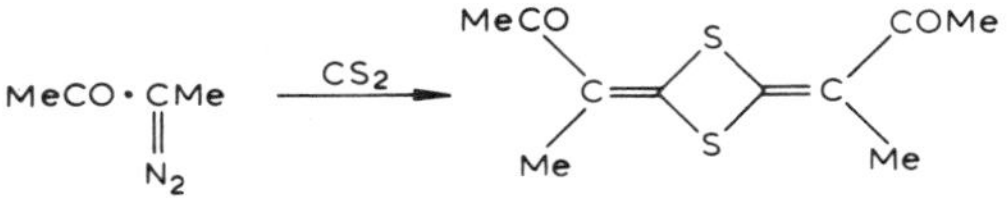

$MeCO \cdot C(=N_2)Me \xrightarrow{CS_2}$ (III)

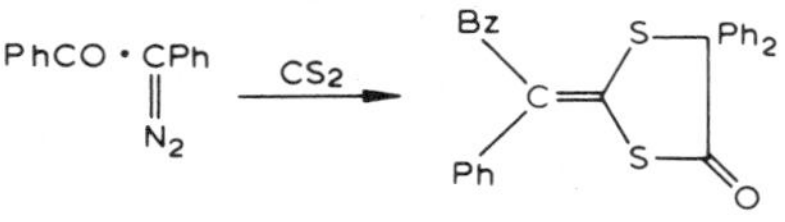

$PhCO \cdot C(=N_2)Ph \xrightarrow{CS_2}$ (IV)

Disulphene [sulphene dimer; 1,3-**dithietane tetroxide** (V)], has been obtained by treating methanesulphonyl chloride with an excess of trimethylamine in tetrahydrofuran at −20° (*G. Opitz* and *H. R. Mohl*, Angew. Chem., intern. Edn., 1969, **8**,73):

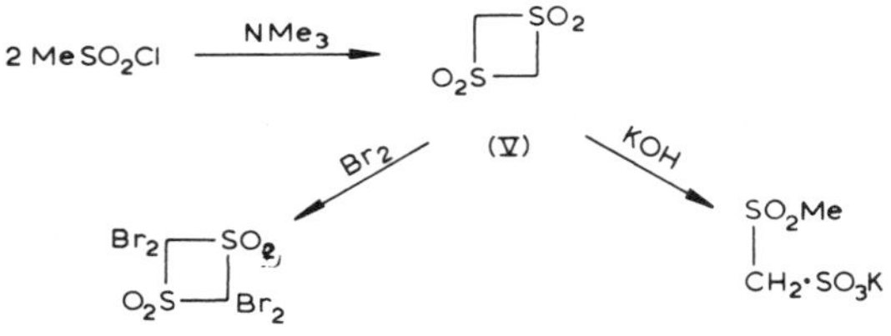

$2\,MeSO_2Cl \xrightarrow{NMe_3}$ (V); (V) $\xrightarrow{Br_2}$ tetrabromo derivative; (V) $\xrightarrow{KOH}$ $MeSO_2 \cdot CH_2 \cdot SO_3K$

(c) Oxathietanes

β-Sultones (the anhydro compounds from β-hydroxysulphonic acids) are 1,2-oxathietanes. The unstable β-sultone, **4-phenyl**-1,2-**oxathietane dioxide,** has been obtained by treating styrene with the dioxane–sulphur trioxide complex (*F. G. Bordwell et al.*, J. Amer. chem. Soc., 1954, **76**, 3945).

A number of such β-sultones have been synthesised by reaction of a variety of fluoro-olefines with sulphur trioxide. These rearrange in the presence of bases to give dihalogenosulphoacetic acid derivatives (*D. C. England, M. A. Dietrich* and *R. V. Linsey*, *ibid.*, 1960, **82**, 6181):

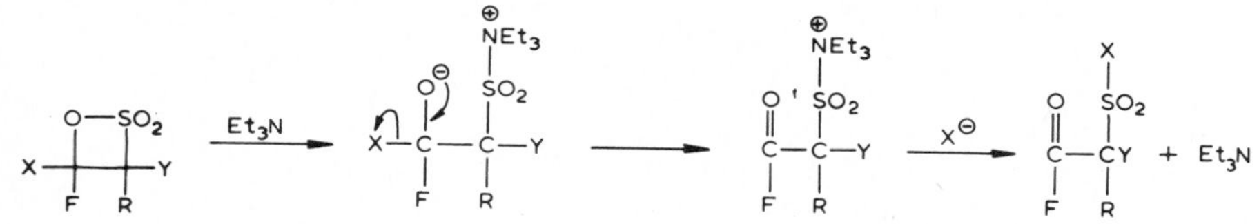

(R, X, Y = F or Cl)

A rearrangement to an oxathietane is preferred to loss of sulphur monoxide in the photochemical decomposition of a thiirane 1-oxide (*Dittmer, G. S. Levy* and *G. E. Kuhlmann, ibid.*, 1967, **89,** 2793):

2,6-**Oxathiaspiro**[3.3]**heptane** may be obtained by treating 3,3-bischloromethyloxetane with sodium sulphide (*T. W. Campbell*, J. org. Chem., 1957, **22,** 1029):

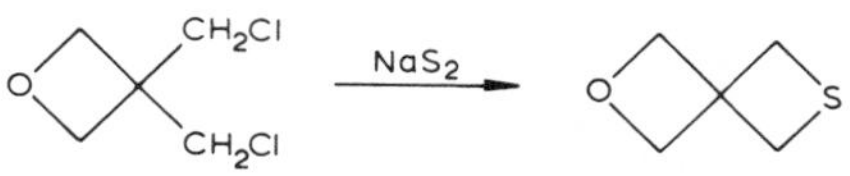

(d) Dithietenes

When hexafluoro-2-butyne was passed through vapours of boiling sulphur, **bis-(trifluoromethyl)**-1,2-**dithietene** (VI) was obtained. It adds to tetramethylethylene at 100° to give a dihydrodithiin VII and dimerises in the presence of a trace of triethylamine (*C. G. Krespan, B. C. McKusick* and *T. L. Cairns*, J. Amer. chem. Soc., 1960, **82,** 1515):

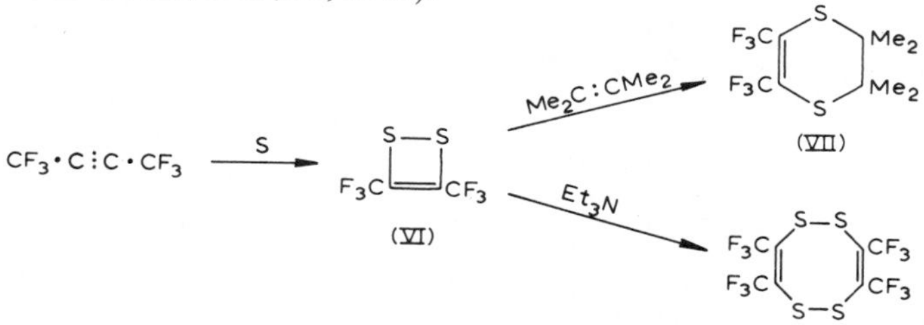

Theoretical studies suggest that 1,2-**dithietene** possesses considerable delocalisation energy and that its stability is comparable to that of 1,2-dithiolane (*G. Bergson*, Arkiv Kemi, 1962, **19,** 181, 265).

(e) Diazetidines

(i) 1,2-Diazetidinones

Ethyl benzeneazoformate combines with diphenylketene to give the fairly stable 2-**ethoxycarbonyl**-1,4,4-**triphenyl**-1,2-**diazetidinone** (VIII), which is saturated and undergoes reversible hydrolysis (*C. K. Ingold* and *S. D. Weaver*, J. chem. Soc., 1925, **127,** 378):

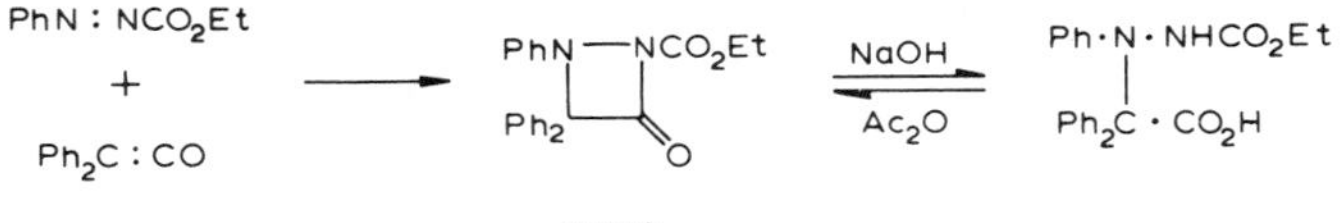

$PhN:NCO_2Et$ + $Ph_2C:CO$ → (VIII) $\underset{Ac_2O}{\overset{NaOH}{\rightleftharpoons}}$ $Ph \cdot N \cdot NHCO_2Et$ / $Ph_2C \cdot CO_2H$

(VIII)

1,2-Diazetidin-3-ones may be prepared also by the irradiation of α-diazoketones in methylene chloride in the presence of azobenzene (*W. Fischer* and *E. Fahr*, Tetrahedron Letters, 1966, 5245):

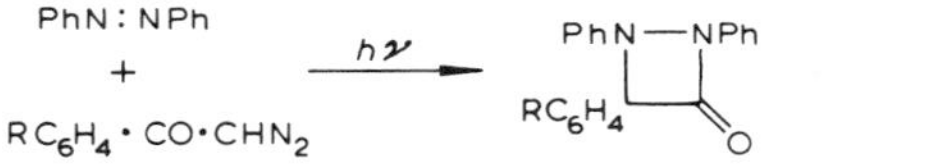

$PhN:NPh$ + $RC_6H_4 \cdot CO \cdot CHN_2$ $\xrightarrow{h\nu}$

(R = *o*-, *m*-, *p*-NO_2 ; Cl ; *o*-, *p*-Br ; *p*-Me ; *p*-OMe)

Treatment of unsymmetrical azo compounds with ketene under u.v. irradiation gives a mixture of diazetidin-3-ones:

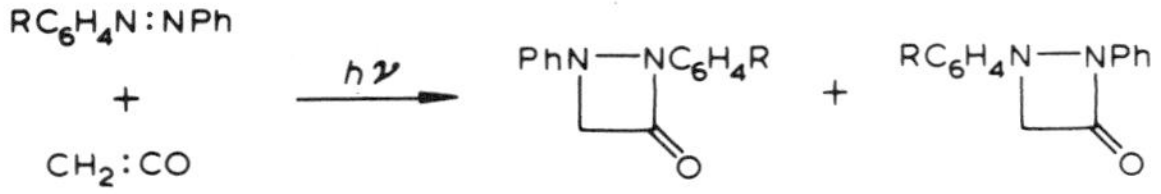

$RC_6H_4N:NPh$ + $CH_2:CO$ $\xrightarrow{h\nu}$

A study of the temperature dependence of the proton resonance spectra shows a relatively slow inversion at the N-1 atom of IX and X (*Fahr et al., ibid.*, 1967, 161):

(IX) ⇌ (X)

Ethyl 2,4,4-**triphenyldiazetidin**-3-**one**-1-**carboxylate** (XI) has been obtained from chlorodiphenylacetyl chloride and ethyl *N′*-phenylhydrazine-*N*-carboxylate in boiling toluene. The diazetidone on warming in methanol,

ethanol, or dioxane and hydrochloric acid afforded **2,4,4-triphenyldiazetidin-3-one** (XII) and 1-ethoxycarbonylamino-3,3-diphenyloxindole (XIII) (*C. W. Bird*, J. chem. Soc., 1963, 674):

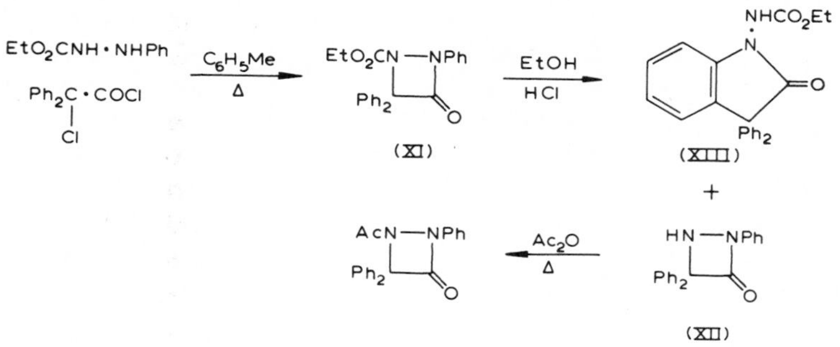

Treatment of **1,2,4,4-tetraphenyldiazetidin-3-one** with sodium ethoxide or *tert*-butoxide, followed by hydrolysis, results in cleavage of the C–N bonds to give azobenzene and diphenylacetic acid, but with methyl-lithium cleavage of the N–N and C–C bonds occurs to yield benzophenone anil and acetanilide. With lithium tetrahydridoaluminate azobenzene and diphenylacetaldehyde are obtained as the major products along with smaller amounts of 1,1-diphenylethylene glycol, 2,2-diphenylethanol, and aniline (*J. H. Hall*, J. org. Chem., 1964, **29**, 3188):

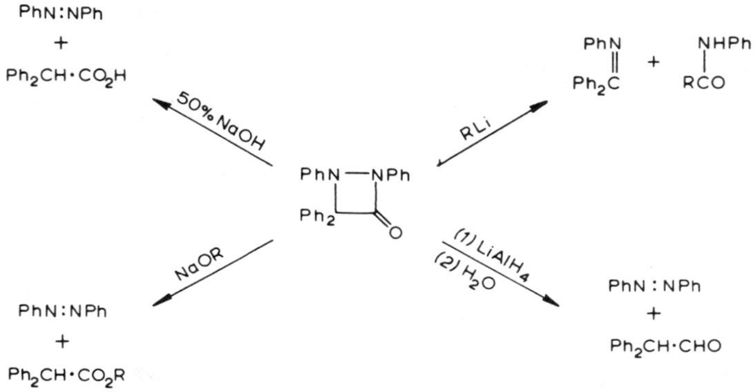

Addition of diphenylketene to *p*-chlorophenyl diazocyanide gives **2-(4-chlorophenyl)-1-cyano-4,4-diphenyldiazetidin-3-one,** which undergoes a facile isomerisation either on melting or on brief boiling in xylene (*Bird*, Chem. and Ind., 1963, 1556):

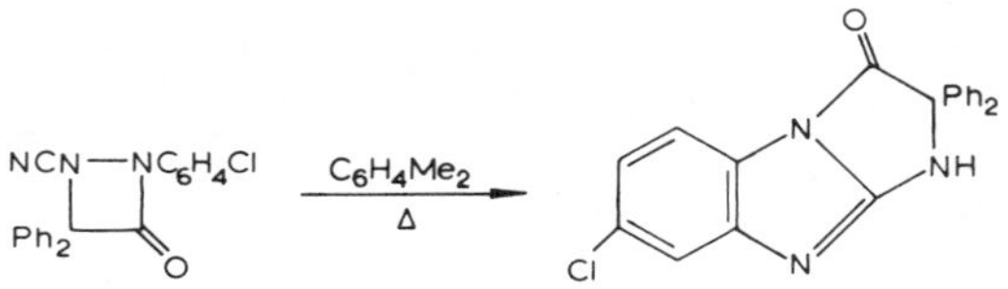

1-**Diphenylmethyldiazetidin**-3-**one** (XIV) is obtained by the reduction with sodium tetrahydridoborate of the cyclic azomethine amide prepared by the treatment of benzophenone chloroacetylhydrazone with sodium hydride or potassium *tert*-butoxide. Catalytic reduction of the cyclic azomethine amide with deactivated Raney nickel gives 2,2-diphenyl-4-imidazolidinone, suggesting the possible intermediary of the bicyclodiaziridine valence bond tautomer (*R. B. Greenwald* and *E. C. Taylor*, J. Amer. chem. Soc., 1968, **90,** 5272):

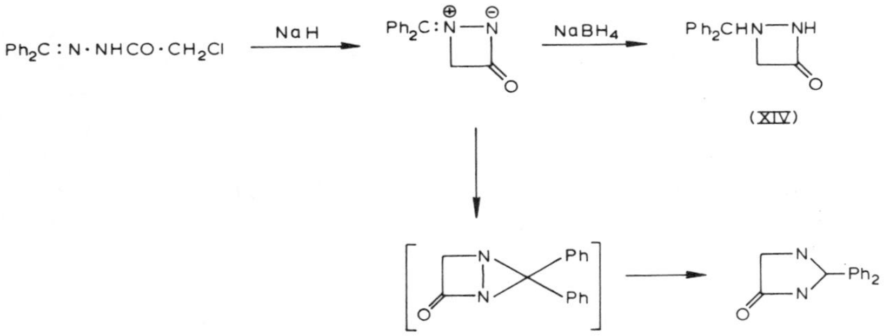

(XIV)

(ii) 1,3-Diazetidines

Formaldehyde with *p*-toluidine gives 1,3-**di**-*p*-**tolyl**-1,3-**diazetidine** (XV; R = *p*-tolyl), m.p. 127° (*Ingold* and *H. A. Piggott*, J. chem. Soc., 1923, **123,** 2745), and with benzenesulphonamide, 1,3-**dibenzenesulphonyl**-1,3-**diazetidine,** m.p. 132° (XV; R = Ph·SO_2) (*A. Magnus-Levy*, Ber., 1893, **26,** 2148):

(XV)

While a 1,3-diazetidin-2-one structure has been wrongly assigned to various aldehyde–urea condensation products, the 1,3- and 1,4-**diphenyl diazetidin**-2-**ones** (XVI and XVII) are probably correctly formulated (*A. Senier* and *F. G. Shepheard*, J. chem. Soc., 1909, **95,** 494; *W. J. Hale* and *N. A. Lange*, J. Amer. chem. Soc., 1919, **41,** 379):

PhNCO + CH_2:NPh → (XVI) HNCO + PhCH:NPh → (XVII)

1,3-**Diphenyl**-4-(**phenylimino**)-1,3-**diazetidin**-2-**one** is formed when diphenylcarbodiimide and phenyl isocyanate are heated, with or without cuprous chloride as catalyst. The diazetidone is ring-opened on treatment with sodium *n*-butoxide (*W. J. Farrisey Jr., R. J. Ricciardi* and *A. A. R. Sayigh*, J. org. Chem., 1968, **33,** 1913):

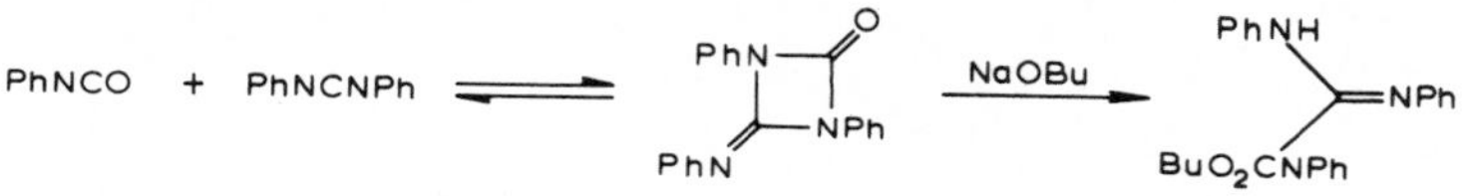

PhNCO + PhNCNPh ⇌ → NaOBu → PhNH, =NPh, BuO_2CNPh

Phenyl isocyanate dimerises, especially under the influence of triethylphosphine, giving 1,3-**diphenyl**-1,3-**diazetidine**-2,4-**dione** (XVIII), m.p. 176°; the formation is supported by X-ray diffraction data (*C. J. Brown*, J. chem. Soc., 1955, 2931):

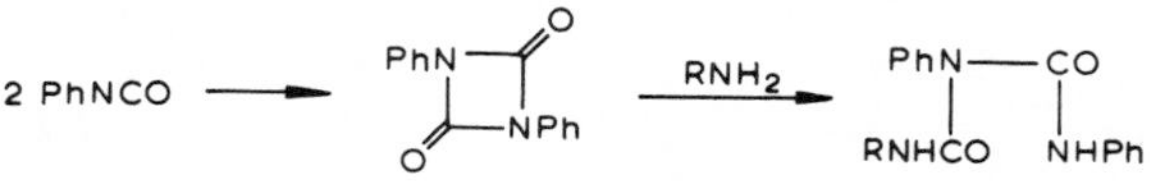

2 PhNCO → (XVIII) → RNH_2 → PhN—CO, RNHCO NHPh

The reaction is general, and the products yield trisubstituted biurets when heated with primary amines (*C. Raiford* and *H. B. Freyermuth*, J. org. Chem., 1943, **8,** 230; *O. Diels* and *H. Grube*, Ber., 1920, **53,** 854).

(f) Diazetines (dihydrodiazetes)

Tetrafluoro-Δ^1-1,2-**diazetine** (**tetrafluoro**-3,4-**dihydro**-1,2-**diazete**) (XIX) has been obtained from cyanogen and argentic fluoride (*H. J. Emeleus* and *G. L. Hurst*, J. chem. Soc., 1962, 3276) and its structure verified by its n.m.r. spectrum (*E. A. V. Ebsworth* and *Hurst*, *ibid.*, p. 4840):

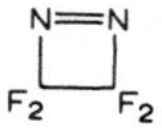

N=N, F_2 F_2

(XIX)

The synthesis of Δ^1-1,2-diazetines (3,4-dihydro-1,2-diazetes) from quadricyclane (XX) and dimethyl or diethyl azodicarboxylate has been reported (*N. Rieber et al.*, J. Amer. chem. Soc., 1969, **91,** 5668):

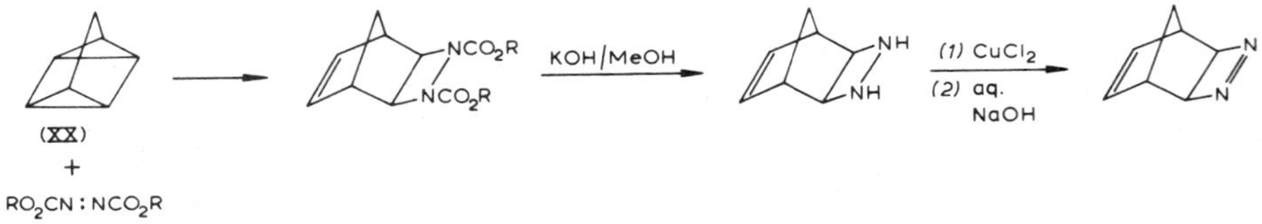

(XX) + $RO_2CN:NCO_2R$ ⟶ (bis-NCO_2R adduct) $\xrightarrow{KOH/MeOH}$ (NH—NH) $\xrightarrow{(1)\ CuCl_2;\ (2)\ aq.\ NaOH}$ (N=N)

Azosulphones react with ketene *N,N*-acetals by cycloaddition giving Δ^3-1,2-**diazetines** (1,2-**dihydro**-1,2-**diazetes**) (*F. Effenberger* and *R. Maier*, Angew. Chem., intern. Edn., 1966, **5,** 416):

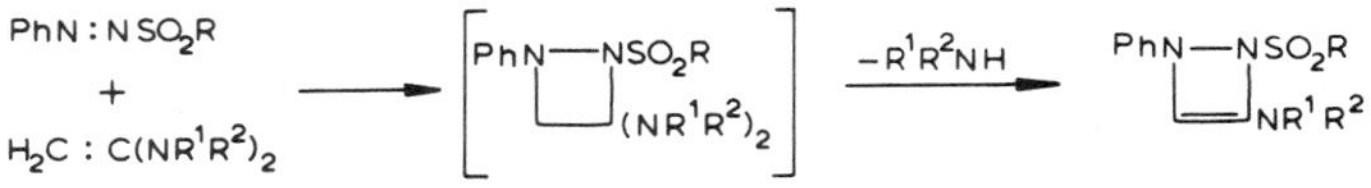

$PhN:NSO_2R$ + $H_2C:C(NR^1R^2)_2$ ⟶ [PhN—NSO_2R, $(NR^1R^2)_2$] $\xrightarrow{-R^1R^2NH}$ PhN—NSO_2R, NR^1R^2

N-Sulphonylimines undergo similar reactions.

(g) Rings containing one nitrogen and. one oxygen or sulphur atom

(i) Oxazetidines

Diphenylketene combines with nitrosobenzene to give 2,4,4-**triphenyloxazetidin**-3-**one** (XXI) the constitution of which is established as indicated; the alternative adduct 2,3,3-**triphenyloxazetidin-4-one** (XXII), apparently also formed, decomposes into carbon dioxide and benzophenone anil which combines further with diphenylketene, affording 1,2,2,3,3-**pentaphenylazetidin-4-one** (XXIII) (*H. Staudinger* and *S. Jelagin*, Ber., 1911, **44,** 365):

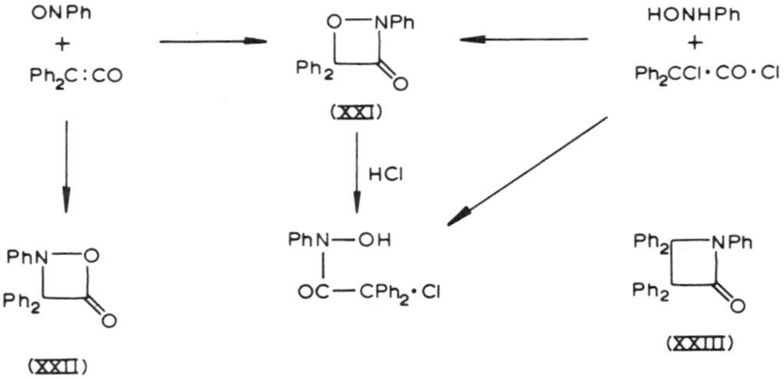

ONPh + $Ph_2C:CO$ ⟶ (XXI) [O—NPh, Ph_2, =O] ⟵ HONHPh + $Ph_2CCl\cdot CO\cdot Cl$

(XXI) $\xrightarrow{HCl}$ PhN—OH, OC—$CPh_2\cdot Cl$

(XXII) [PhN—O, Ph_2, =O]

(XXIII) [Ph_2—NPh, Ph_2, =O]

The products from nitrosobenzene and diethyl methylenemalonate, nitrosobenzene and 1,1-diphenylethylene are open chain compounds, PhN(OH)-$CH{:}C(CO_2Et)_2$ and $PhN(O){:}CPh_2$, respectively (*N. F. Hepfinger, C. E. Griffin* and *B. L. Shapiro*, Tetrahedron Letters, 1963, 1365), and not oxazetidines as originally believed (*Ingold* and *S. D. Weaver*, J. chem. Soc., 1924, 1456).

Tetrafluoroethylene and trifluoronitrosomethane at room temperature give **perfluoro-2-methyl-1,2-oxazetidine (perfluoro-2-methyl-1-oxa-2-azacyclobutane)**, b.p. −7°. The colourless gas is stable to water and dissociates in the opposite sense to its synthesis on heating (*D. A. Barr* and *R. N. Haszeldine*, J. chem. Soc., 1955, 1881):

$$ON{\cdot}CF_3 + F_2C{:}CF_2 \longrightarrow \text{(O—NCF}_3\text{, F}_2\text{C—CF}_2\text{ ring)} \xrightarrow{550°} F_2C{=}O + CF_2{=}NCF_3$$

(ii) Thiazetidines

Treatment of the *N*-methylene derivative of cyclohexylamine or benzylamine with formaldehyde and hydrogen sulphide gives 1,3-**thiazetidines** (*E. R. Braithwaite* and *J. Graymore*, *ibid.*, 1950, 208; 1953, 143) which are decomposed by acids into primary amine, formaldehyde, and trimeric thioformaldehyde:

$$H_2S \quad CH_2O + CH_2{:}NR \longrightarrow \text{(S—CH}_2\text{—NR—CH}_2\text{ ring)} + H_2O$$

Treatment of tetramethylallene with ethyl *N*-sulphonylcarbamate from fragmentation of the inner salt $EtO_2CN^-SO_2N^+Et_3$, obtained when ethoxycarbonylsulphamoyl chloride reacts rapidly with triethylamine in benzene solution yields two isomeric cyclo-adducts **2-ethoxycarbonyl-4-isopropylidene-3,3-dimethyl-1,2-thiazetidine 1,1-dioxide** (XXIV) and 6-ethoxy-3-isopropylidene-2,2-dimethyl-2,3-dihydro-1,4,5-oxathiazine 4,4-dioxide (XXV) (*G. M. Atkins* and *E. M. Burgess*, J. Amer. chem. Soc., 1968, **90**, 4744):

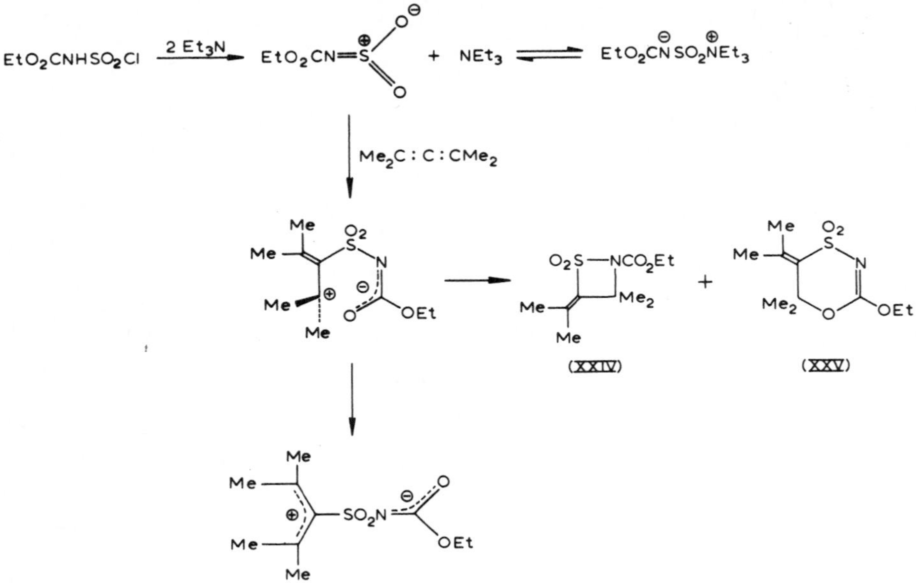

$EtO_2CNHSO_2Cl \xrightarrow{2\ Et_3N} EtO_2CN{=}\overset{\oplus}{S}(O^{\ominus}){=}O + NEt_3 \rightleftharpoons EtO_2C\overset{\ominus}{N}SO_2\overset{\oplus}{N}Et_3$

$\downarrow Me_2C:C:CMe_2$

(XXIV) (XXV)

4,4-**Dialkyl**-1,2-**thiazetidin**-3-**one** 1,1-**dioxides** (XXVI) have been prepared by the following route (*B. J. R. Nicolaus, E. Bellasio* and *E. Testa*, Helv., 1962, **45**, 717):

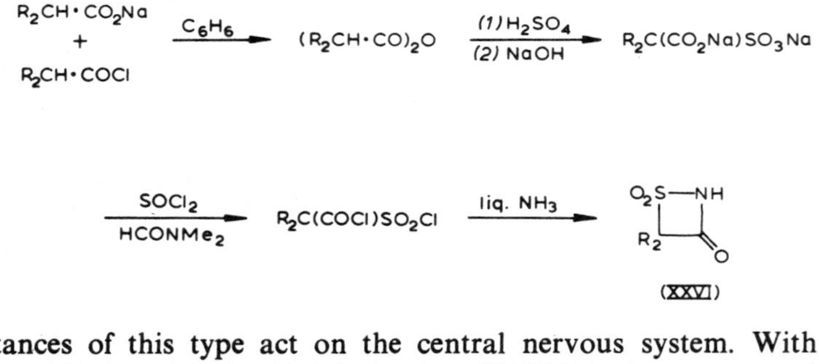

$R_2CH{\cdot}CO_2Na + R_2CH{\cdot}COCl \xrightarrow{C_6H_6} (R_2CH{\cdot}CO)_2O \xrightarrow[(2)\ NaOH]{(1)\ H_2SO_4} R_2C(CO_2Na)SO_3Na$

$\xrightarrow[HCONMe_2]{SOCl_2} R_2C(COCl)SO_2Cl \xrightarrow{liq.\ NH_3}$ (XXVI)

Substances of this type act on the central nervous system. With alkyl iodides (R′I) they yield *N*-alkyl derivatives (XXVII) in the presence of alkali, and *O*-alkyl derivatives (XXVIII) if silver oxide is used (*idem*, Helv., 1963, **46**, 450):

(XXVII) (XXVIII)

Chapter 2

Compounds Containing a Five-Membered Ring with One Hetero Atom of Group VI; Oxygen

R. LIVINGSTONE

1. The furan group

The first furan derivative to be described was 2-furoic acid (pyromucic acid, furan-2-carboxylic acid; *K. W. Scheele*, 1780); furfural (furan-2-carbaldehyde), reported by *J. W. Döbereiner* in 1831, was characterised nine years later by *J. Stenhouse*, and *H. Limpricht* prepared furan in 1870 by treating barium furoate with soda lime. The formula I is established by syntheses of furan derivatives and by the relationship of furan to tetrahydrofuran (II). The radical C_4H_3O– is designated *furyl*, and $C_4H_3OCH_2$–, *furfuryl:*

(I) 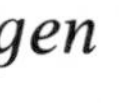 (II)

The uncondensed and unreduced furan nucleus occurs in a few constituents of essential oils. They are commonly plant products; a few are of fungal origin. Furfural, is manufactured on a large scale by the acid hydrolysis of oak hulls (husks), and its reductive and other transformations afford useful intermediates and industrial solvents.

The 2,3-benzofuran (benzo[*b*]furan) nucleus is more common in nature than that of furan, notably in the active principles of fish- and insect-poisons. Dibenzofuran (diphenylene oxide) is readily available and has been the object of much preparative work directed to the synthesis of drugs and, to some extent, dyes.

For general reviews of furan chemistry see *L. N. Owen*, "Furans", (Ann.

Reports, 1945, **42,** 157); *A. P. Dunlop* and *F. N. Peters*, "The Furans" (A.C.S. Monograph, Reinhold Publishing Corpn., New York, 1953).

(a) Furan and its substitution products

(i) Synthetic methods

(*1*) The very general Paal–Knorr method leading to substituted furans depends on the dehydration of 1,4-diketones by a wide variety of acid reagents (*C. Paal*, Ber., 1884, **17,** 2756). The requisite dicarbonyl compounds are often made by acetoacetic syntheses, and the intermediate diketonic esters themselves readily yield furancarboxylic acids:

$$MeCO{\cdot}CH_2{-}CH(CO_2Et){\cdot}COMe \longrightarrow MeCO{\cdot}CH_2{-}CH_2{\cdot}COMe \xrightarrow[150^\circ]{ZnCl_2} \text{2,5-dimethylfuran}$$

$$MeCO{\cdot}CH_2{-}CH(CO_2Et){\cdot}COMe \xrightarrow{H_2SO_4} \text{2,5-dimethylfuran-3-carboxylic acid } (CO_2H)$$

Convenient routes to serviceable dialdehydes and keto-aldehydes are described by *R. G. Jones* and *E. C. Kornfeld* (J. org. Chem., 1954, **19,** 1671; J. Amer. chem. Soc., 1955, **77,** 4069). In some cases the acetoacetic synthesis gives an enol ether (*e.g.*, III) which is dehydrated to a furan (*T. Reichstein et al.*, Helv., 1933, **16,** 276):

$$EtO_2C{\cdot}CO{\cdot}CH_2Br + Na{\cdot}CH(CO_2Et){\cdot}CO{\cdot}CO_2Et \longrightarrow \underset{\text{(III)}}{EtO_2C{\cdot}CO{\cdot}CH_2{-}O{-}C(CO_2Et){=}CH{\cdot}CO_2Et} \longrightarrow \text{furan with } EtO_2C, CO_2Et, CO_2Et$$

Although not conclusive there is evidence indicating that the Paal–Knorr synthesis probably proceeds by way of the mono-enol, *e.g.*, the behaviour of the individually stable tautomers of dimesitoylmesitylethane (IV, R = $C_6H_2Me_3$) (*R. E. Lutz* and *C. J. Kibler*, J. Amer. chem. Soc., 1940, **62,** 360):

$$\underset{\text{(IV)}}{RCO{\cdot}CHR{-}CH_2{\cdot}COR} \rightleftharpoons RC(OH){=}CR{-}CH_2{\cdot}COR \longrightarrow \text{2-hydroxy-2,3-dihydrofuran (R, R, OH)} \longrightarrow \text{2,3,5-trisubstituted furan (R, R, R)}$$

β-Chloroallyl ketones, readily available from the alkylation of enolate ions, on hydrolysis with sulphuric acid give 1,4-dicarbonyl compounds (not isolated), which undergo subsequent dehydration and cyclisation to furans (*E. J. Nienhouse, R. M. Irwin* and *G. R. Finni, ibid.*, 1967, **89,** 4557):

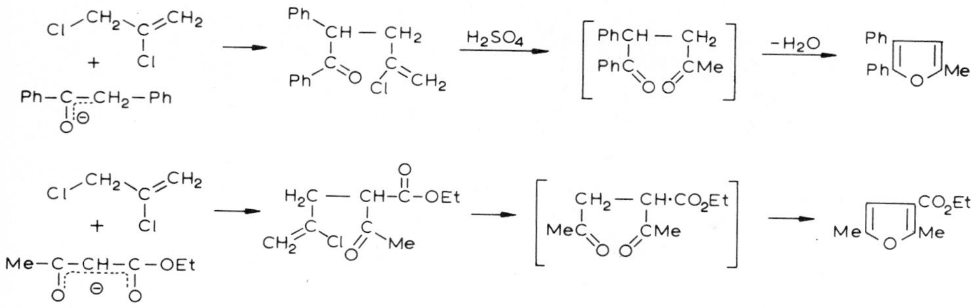

Diaroylethenes are converted (the *cis* isomers often more easily than the *trans*) into 3-substituted furans by treatment with acid reagents (*Lutz, ibid.*, 1926, **48,** 2916; *Lutz* and *C. E. McGinn, ibid.*, 1942, **64,** 2585):

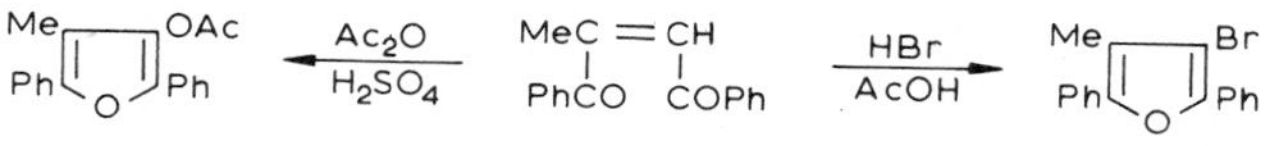

(*2*) The Feist–Benary synthesis, in which a pyrrole derivative is formed from an α-chloroketone, a β-ketonic ester and ammonia, gives furans as subsidiary or sometimes as main products; *e.g.*, ethyl acetoacetate gives predominantly pyrroles, diethyl acetonedicarboxylate chiefly furans. The pyrroles are those which would result from *C*-alkylation of the ester by the halide function of the chloro-ketone, followed by familiar conversion of a 1,4-diketone into pyrrole; the furans on the other hand, are related to the *O*-acetonyl derivatives of the ester. For the furan synthesis, pyridine or aqueous alkali can advantageously replace ammonia; and 1,2-dichloroethyl ether as a source of chloroacetaldehyde can replace the chloroketone (*F. Feist*, Ber., 1902, **35,** 1539, 1545; *I. J. Rinkes*, Rec. Trav. chim., 1931, **50,** 1127; *E. W. Scott* and *J. R. Johnson*, J. Amer. chem. Soc., 1932, **54,** 2549):

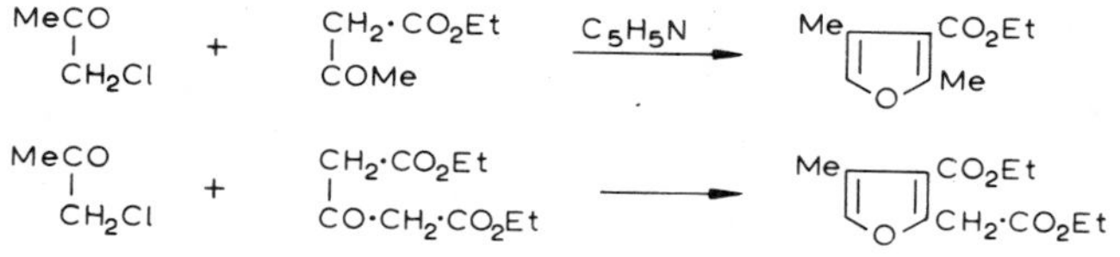

(*3*) On bromination and treatment with alkali, methyl coumalinate (α-pyrone-5-carboxylate) gives furan-2,4-dicarboxylic acid (*Feist*, Ber., 1901, **34,** 1992):

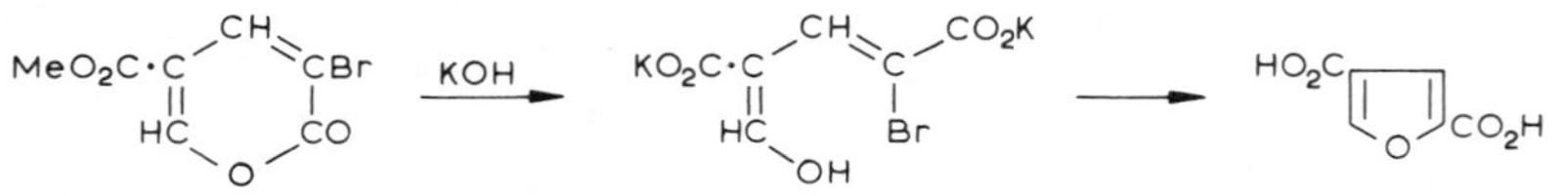

Similar treatment of 4,6-dimethyl-α-pyrone-5-carboxylic acid (isodehydracetic acid) gives, by concomitant decarboxylation, 2,4-dimethylfuran-3-carboxylic acid.

(*4*) Adducts from β-oxo-esters and α-nitro-olefins on treatment with acid, which promotes a Nef reaction and cyclisation, give furans (*F. Boberg* and *G. R. Schultze*, *ibid.*, 1957, **90,** 1215; *Boberg* and *A. Kieso*, Ann., 1959, **626,** 71):

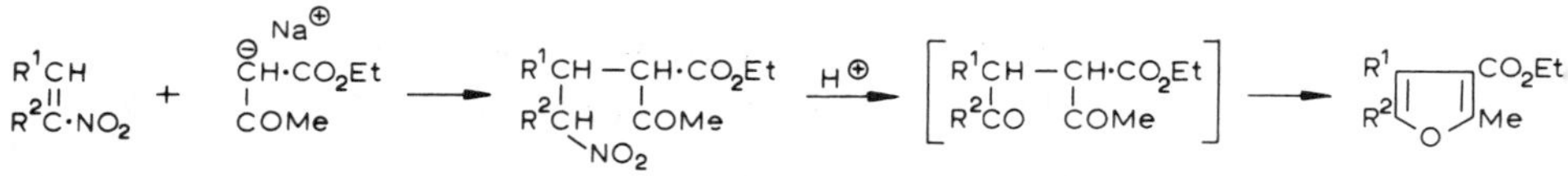

(*5*) A number of furans have been synthesised by the reduction of α,β-unsaturated-γ-lactones using di-isobutylaluminium hydride in tetrahydrofuran (*H. Minato* and *T. Nagasaki*, J. chem. Soc., C, 1966, 377):

This method enables a furan ring to be constructed at a desired position in high molecular compounds:

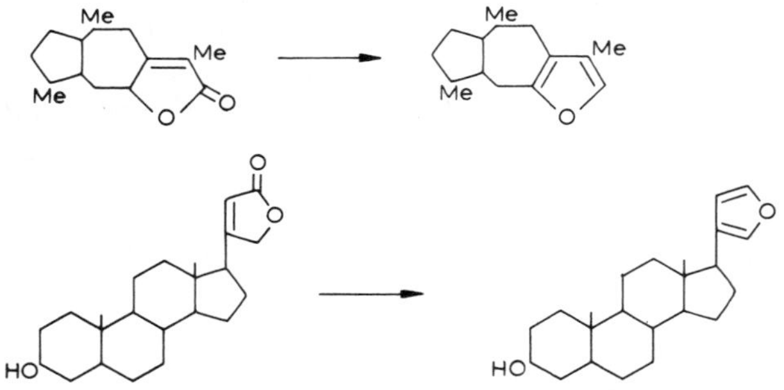

(*6*) Many carbohydrates and their derivatives are converted into furans by acids (sulphuric, hydrochloric, oxalic). Thus pentoses give furfural (furan-2-carbaldehyde) and similarly methylpentoses give 5-methylfurfural;

glucose reacts with great reluctance, but fructose gives good yields of 5-hydroxymethylfurfural (*W. N. Haworth* and *W. G. M. Jones*, J. chem. Soc., 1944, 667). By the use of the ^{14}C isotope it has been shown that the aldehyde group of the furan derivatives is derived from the 1-position *i.e.*, the potential aldehyde group of the carbohydrate (*W. A. Bonner* and *M. R. Roth*, J. Amer. chem. Soc., 1959, **81**, 5454). It seems possible that the reaction proceeds *via* the 2,5-anhydro-sugar, *e.g.*, the easy formation of 5-hydroxymethylfurfural (V) from 2,5-anhydro-D-mannose (*J. Staněk et al.*, "The Monosaccharides", Academic Press, New York and London, 1963, p. 110):

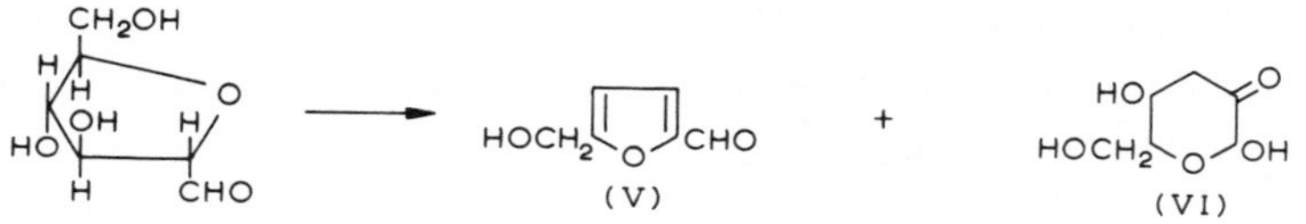

CH₂OH … CHO ⟶ HOCH₂–furan–CHO (V) + HOCH₂–pyranone–OH (VI)

3-Deoxy-D-erythrohexosone (VI) has been detected as an intermediate in the acid conversion of hexoses (*e.g.*, fructose) into 5-hydroxymethylfurfural (V) (*E. F. L. J. Anet*, Chem. and Ind., 1962, 262).

Since erythritol gives no furan, the reducing function of the sugar seems to be essential.

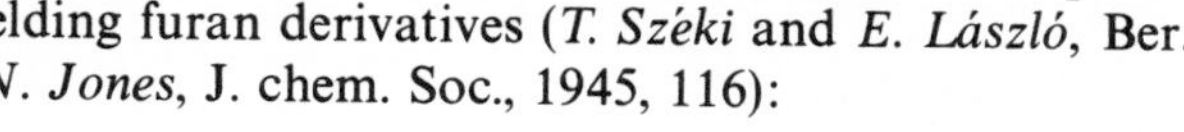

$$HO{\cdot}CH_2{\cdot}CH(OH){\cdot}CH(OH){\cdot}CH(OH){\cdot}CHO \longrightarrow \text{furan-2-CHO} + 3\,H_2O$$

Mucic acid and its isomers yield furan-2-carboxylic acid by pyrolysis and furan-2,5-dicarboxylic acid by dehydration.

(*7*) Aldoses condense with 1,3-dicarbonyl compounds in presence of zinc chloride, yielding furan derivatives (*T. Széki* and *E. László*, Ber., 1940, **73**, 924; *J. K. N. Jones*, J. chem. Soc., 1945, 116):

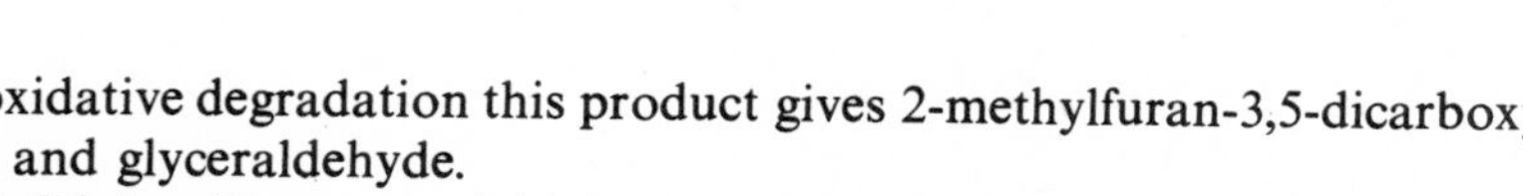

$$EtO_2C{\cdot}CH_2{\cdot}COMe + CHO{\cdot}(CHOH)_4{\cdot}CH_2OH \longrightarrow \text{3-}EtO_2C\text{-2-Me-furan-5-}(CHOH)_3{\cdot}CH_2OH$$

By oxidative degradation this product gives 2-methylfuran-3,5-dicarboxylic acid and glyceraldehyde.

(*8*) Diglycollic esters yield furans by condensation with 1,2-dicarbonyl compounds (*P. C. Guha* and *B. H. Iyer*, J. Indian Inst. Sci., 1938, **21A**, 115; *H. J. Backer* and *W. Stevens*, Rec. Trav. chim., 1940, **59**, 423):

PhCO·COPh + $RO_2C·CH_2–O–CH_2·CO_2R$ —NaOR→ 2,5-di(HO_2C)-3,4-diphenylfuran

$RO_2C·CH_2OCH_2·CO_2R$ + $EtO_2C·CO_2Et$ ⟶ 3,4-dihydroxyfuran-2,5-dicarboxylic acid

(*9*) Some furans can be obtained by exploiting the reactivity of acetylenes (*E. R. H. Jones et al.*, J. chem. Soc., 1946, 45, 54; 1947, 1586):

HC⋮CNa + $ClCH_2·CH(O)CH_2$ ⟶ HC⋮C–CH=CH–CH_2OH —$HgSO_4$, H_2SO_4→ 2-methylfuran

PrCH(OH)–C≡C–COPr —Piperidine→ 2,5-dipropyl-3-(NC_5H_{10})furan

Benzoin reacts with dimethyl acetylenedicarboxylate in boiling acetone containing potassium carbonate to yield alcohol VII, which dehydrates on boiling in methanol containing a few drops of sulphuric acid to give a furan (*J. B. Hendrickson, R. Rees* and *J. F. Templeton*, J. Amer. chem. Soc., 1964, 85, 107):

PhCO–PhCHOH + $C·CO_2Me$≡$C·CO_2Me$ ⟶ (VII) ⟶ 2,3-diphenyl-4,5-di(CO_2Me)furan

(*10*) Ethoxycarbonylcarbene generated by the copper-catalysed decomposition of diazoacetic ester, adds to α-methoxymethylene ketones in an unusual 1,4-manner to afford 2,3-disubstituted furans (*D. L. Storm* and *T. A. Spencer*, Tetrahedron Letters, 1967, 1865):

MeO·CH= (steroid ketone) + $N_2CH·CO_2Et$ —160°, $CuSO_4$→ EtO_2C-furanosteroid —(1) Saponification; (2) Decarboxylation, Cu 250°→ furanosteroid

(*11*) Furans may be obtained by the reaction between enol-ethers of β-dicarbonyl compounds in tetrahydrofuran and dimethylsulphonium methylide in dimethyl sulphoxide at 0° (*T. M. Harris, C. M. Harris* and *J. C. Cleary*, *ibid.*, 1968, 1427):

$$RCO{\cdot}CH{:}C(R')OMe \xrightarrow{Me_2\overset{\oplus}{S}\overset{\ominus}{C}H_2} \text{2-R-4-R'-furan}$$

R = R' = Ph
R = Me R' = Ph
R = Ph R' = H

Keten dimer and VIII in a closely related reaction give a 3-hydroxyfuran (*H. Takei et al.*, Bull. chem. Soc. Japan, 1968, **41,** 1738):

$$\text{CH}_2\text{=C—CH}_2\text{(O—CO)} + RCO{\cdot}\overset{\ominus}{C}H\overset{\oplus}{S}Me_2 \text{ (VIII)} \longrightarrow \text{3-HO-2-(R·CO)-5-Me-furan}$$

(*12*) Pyrylium salts are converted with ring contraction to furans by mild treatment with aqueous hydrogen peroxide (*A. T. Balaban* and *C. D. Nenitzescu*, Ber., 1960, **93,** 599):

$$\text{2,4,6-trimethylpyrylium } ClO_4^{\ominus} \xrightarrow[H_2O_2]{H_2O} \text{2-Me-3-Me-... furan (Me, Me, COMe)}$$

(ii) General properties and reactions

Properties. The furan nucleus is much less stable than that of benzene and the correct value of its resonance energy is probably 25 ± 1 kcal/mole (*D. S. Sappenfield* and *M. M. Kreevoy*, Tetrahedron, 1963, Suppl., **2,** 157). Microwave examinations show that the furan molecule is planar and gives the following molecular parameters (*B. Bak et al.*, J. mol. Spectroscopy, 1962, **9,** 124):

H 126.0°; 1.431 Å; 106.0°; 1.361 Å; 110.7°; 106.6°; H 115.9°; 1.362 Å; O; ± 0.002 Å

A theoretical treatment of the furan system has been elaborated (*F. L. Pilar* and *J. R. Morris*, J. chem. Phys., 1961, **34,** 389); the p.m.r. spectra of furan and some substituted furans have been fully worked out (*E. J. Corey et al.*, J. Amer. chem. Soc., 1958, **80,** 1204; *R. J. A. Abraham* and *H. J. Bernstein*, Canad. J. Chem., 1961, **39,** 905; *G. S. Reddy* and *J. H. Goldstein*, J. Phys. Chem., 1961, **65,** 1539; J. Amer.

chem. Soc., 1962, **84,** 583); the relative magnitude of the ring currents in benzene, thiophene, and furan have been calculated (*H. A. P. de Jongh* and *H. Wynberg*, Tetrahedron, 1965, 515); ^{13}C-magnetic resonance of furan (*T. F. Page, T. Alger* and *D. M. Grant*, J. Amer. chem. Soc., 1965, **87,** 5333); absorption in the u.v. region (*W. C. Price* and *A. D. Walsh*, Proc. roy. Soc., A, 1941, 179, 201; *D. G. Manley* and *E. D. Amstutz*, J. org. chem., 1957, **22,** 323), i.r. region (*H. W. Thompson* and *R. B. Temple*, Trans. Faraday Soc., 1945, **41,** 27; *A. R. Katritzky* and *J. M. Lagowski*, J. chem. Soc., 1959, 657; *M. Fétizon* and *P. Baranger*, Bull. Soc. chim. Fr., 1957, 1311).

Furan is best represented as a resonance hybrid; formulae IX to XIII represent the more important hypothetical limiting structures and IX is the major contributor to the resonance hybrid:

(IX) (X) (XI) (XII) (XIII)

Reactions. The "aromatic" character of the furan ring, is indicated by the benzoin condensation of furfural and by the diazotisability of some amino-furans (*H. Gilman* and *G. F. Wright*, Rec. Trav. chim., 1934, **53,** 13), but it is significant that simple hydroxy- and amino-furans behave rather as oxo- or imino-dihydrofurans. Furan is attacked more readily than benzene by electrophiles, primarily in position 2. A 2-monosubstituted furan almost always reacts further in position 5, independently of the nature of the original substituent. These substitutions take place either through addition followed by elimination, or by direct displacement, as in benzene:

The 2-position of a 3-monosubstituted furan is reactive if the substituent is *ortho, para*-directive, otherwise position 5 is the next to be attacked (*Gilman* and *Wright*, Chem. Reviews, 1932, **11,** 323; *Gilman* and *R. R. Burtner*, J. Amer. chem. Soc., 1933, **55,** 2903). Mercuration likewise takes place in α-positions, giving products useful for characterising furans and these positions are also metallated by organo-alkali metal compounds (*Gilman* and *F. Breuer*, *ibid.*, 1934, **56,** 1123):

N.m.r. studies show that there is less extensive conjugation in furans than in thiophenes, and that 3-substituents are less affected by the heteroatom than 2-substituents (*S. Gronowitz et al.*, Arkiv Kemi, 1963, **19,** 483).

(*1*) *Mineral acids* easily cause resinification, or else the furan derivative is hydrolysed, but not nearly so easily as an enol ether; the product is a 1,4-dicarbonyl compound (reversal of synthetic method *1*).

(*2*) *Oxidation.* Furans are relatively unstable to oxidation so that it is seldom possible to oxidise alkyl side-chains to carboxyl. The *ozonolysis* of simple furans at −60° is unaccompanied by the resin formation which occurs at higher temperatures; both mono- and di-ozonides are formed as intermediate products (*B. P. Jibben* and *J. P. Wibaut*, Rec. Trav. chim., 1960, **79,** 342). The ozonolysis of 2,5-diphenylfuran at −40° besides benzoic acid yielded 14% of phenylglyoxal, the result of the 1,4-addition of ozone (*P. S. Bailey* and *H. O. Colomb*, J. Amer. chem. Soc., 1957, **79,** 4238):

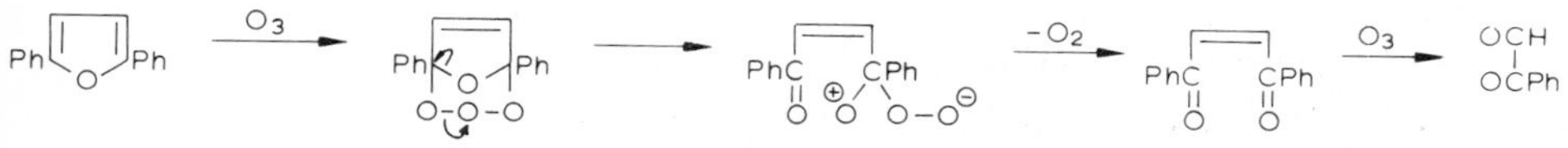

Furans are oxidised by bromine in the presence of methanol to give 2,5-dimethoxy-2,5-dihydrofurans, converted by boiling in 1% acetic acid with hydrazine to the corresponding pyridazines (*J. Levisalles*, Bull. Soc. chim. Fr., 1957, 997).

Oxidation by a strong solution of hydrogen peroxide occurs at the α-position of furans having methyl groups in the 2- and 5-positions. The product with benzoyl chloride in pyridine gives a dibenzoate and is reduced to a diketone (*R. Criegee* and *D. Seebach*, Ber., 1963, **96,** 2704; *Seebach*, *ibid.*, 1963, **96,** 2712):

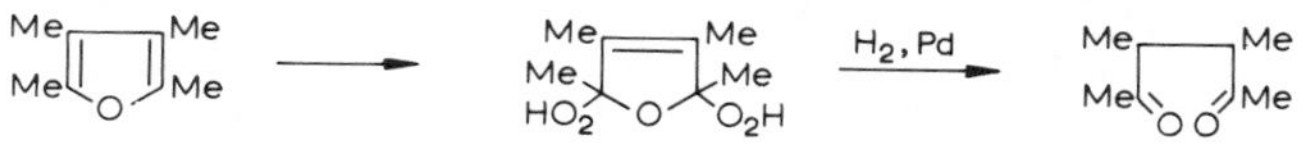

The type of product obtained by the oxidation of furans with nitric acid, permanganates, and lead tetra-acetate depends on the substituents and the conditions of the reaction. They may be derivatives of maleic or succinic acids or of related ketones.

(*3*) *Addition reactions. Halogenation* and *nitration* are preceded or accompanied by addition and the furan nucleus is more easily hydrogenated than that of benzene. The *hydrogenation* of 2-alkylfurans has been studied under various conditions with a variety of catalysts to give tetrahydrofurans, aliphatic ketones and primary alcohols (*N. I. Shuĭkin* and *I. F. Bel'skiĭ*, Doklady Akad. Nauk, S.S.S.R., 1957, 116; 1960, 131, 109; *Shuĭkin, Bel'skiĭ* and *R. A. Karakhanov*, *ibid.*, 1958, 122, 625; *Shuĭkin* and *Bel'skiĭ*, Izvest. Akad. Nauk, S.S.S.R., Otdel khim. Nauk, 1958, 309). Relative rates of the trifluoroacetylation of heterocycles including furan and some of its derivatives have been reported (*S. Clementi, F. Genel* and *G. Marino*, Chem. Comm., 1967, 498).

Diene addition. Unlike benzene or pyrrole, furan reacts as a normal diene with *maleic anhydride* (*M. G. van Campen* and *J. R. Johnson*, J. Amer. chem. Soc., 1933, **55,** 430); and the stereochemistry of the process has been thoroughly studied (*R. B. Woodward* and *H. Baer*, *ibid.*, 1948, **70,** 1161). N.m.r. studies show that, in methyl cyanide, furan and maleic anhydride afford the *exo*-adduct twice as fast as the *endo*-isomer, while maleic acid yields the *endo*-adduct four times as fast as the *exo*-isomer; the *exo*-adducts are the thermodynamically more stable (*J. A. Berson* and *R. Swidler*, *ibid.*, 1953, **75,** 1721; *F. A. L. Anet*, Tetrahedron Letters, 1962, 1219).

Phthalic acid is obtained on boiling the anhydride adduct with hydrobromic acid:

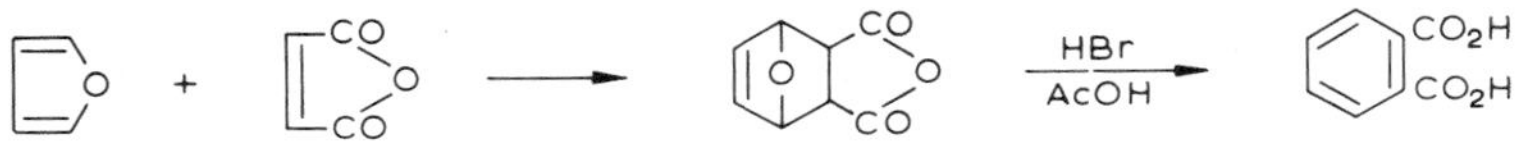

With *maleic imide* the stereochemistry of the addition may be controlled by the reaction temperature (*H. Kwart* and *I. Burchuk*, J. Amer. chem. Soc., 1952, **74,** 3094):

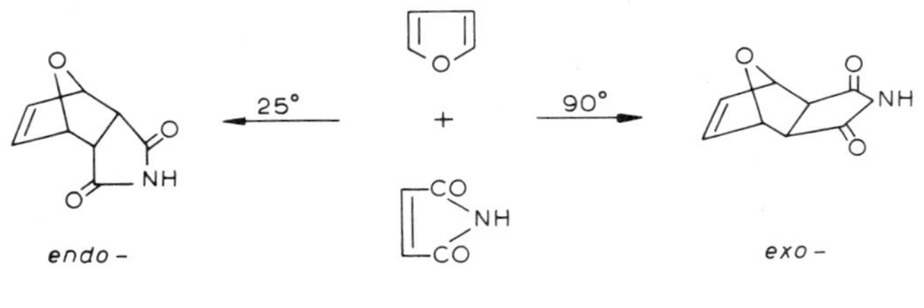

The benzophenone-photosensitised addition of furan to dimethylmaleic anhydride affords a cyclobutane (*G. O. Schenck et al.*, Ber., 1962, **95,** 1642):

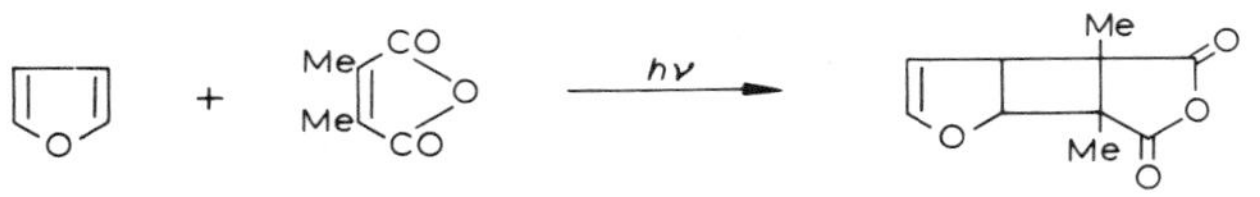

A novel type of intramolecular Diels–Alder reaction occurs with the furan XIV; on keeping at room temperature it gives compound XV, vacuum-distillation of which regenerates XIV and treatment with hydrobromic acid and acetic acid gives 2-phenylisoindoline (*D. Bilovič*, *Ž. Stojanac* and *V. Hahn*, Tetrahedron Letters, 1964, 2071):

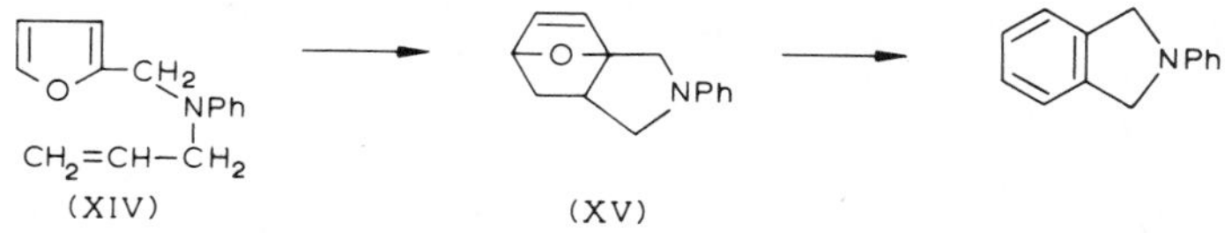

CH_2 NPh $CH_2{=}CH{-}CH_2$ NPh NPh

(XIV) (XV)

The side-chain of 2-vinylfuran takes part in the addition of maleic anhydride (*R. Paul*, Compt. rend., 1939, **208,** 1028):

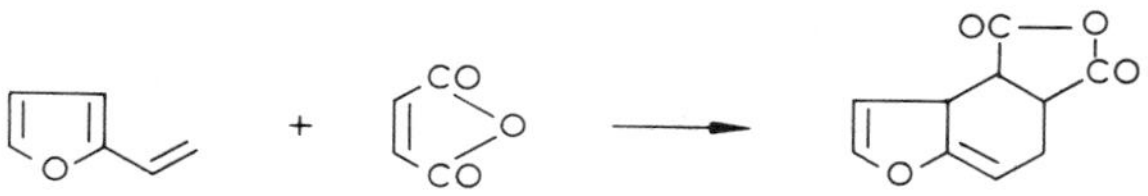

CO O CO + OC—O CO

The addition of *diethyl acetylenedicarboxylate* provides a route to furan-3,4-dicarboxylic acid (*K. Alder* and *H. F. Rickert*, Ber., 1937, **70,** 1354):

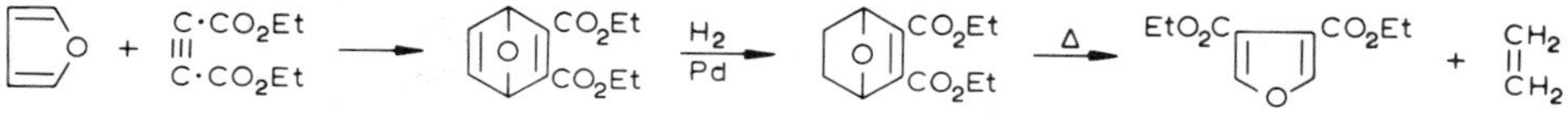

O + $C{\cdot}CO_2Et$ ≡ $C{\cdot}CO_2Et$ → CO_2Et CO_2Et $\xrightarrow[Pd]{H_2}$ CO_2Et CO_2Et $\xrightarrow{\Delta}$ EtO_2C CO_2Et + $CH_2{=}CH_2$

A number of *arynes*, generated from Grignard and organolithium reagents have been trapped with furan (*R. Harrison*, *H. Heaney* and *P. Lees*, Tetrahedron, 1968, **24,** 4589):

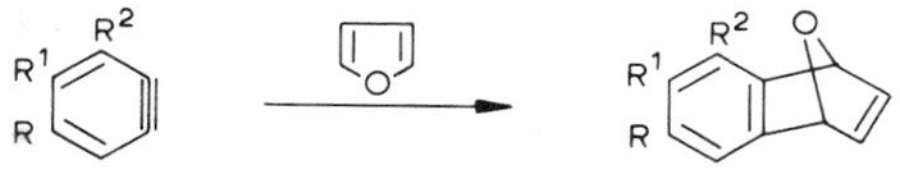

R^1 R^2 R → R^1 R^2 O R

(*4*) *Photo-chemistry.* In the presence of mercury vapour at 2537 Å, furan (and 2-alkylfurans) undergoes a photochemical *decarbonylation;* Diels–Alder adducts with cyclopropene and cyclopropenecarbaldehyde are also formed (*R. Srinivasan*, J. Amer. chem. Soc., 1967, **89,** 1758, 4812):

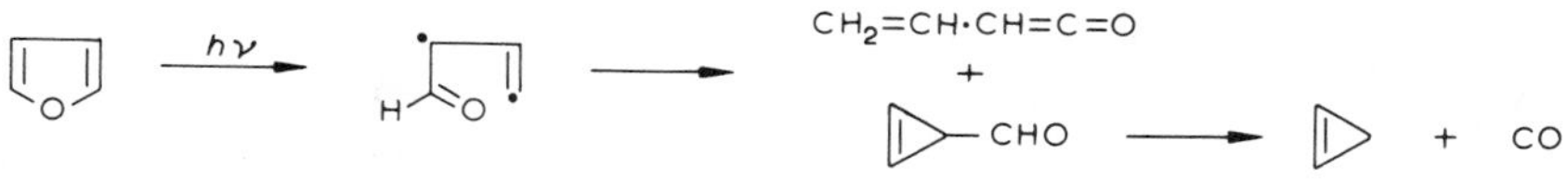

$h\nu$ → H O → $CH_2{=}CH{\cdot}CH{=}C{=}O$ + CHO → + CO

2-Methylfuran and 2,5-dimethylfuran photoisomerise to 3-methylfuran and 2,4-dimethylfuran, respectively (*H. Hiraoka* and *Srinivasan*, *ibid.*, 1968, **90,** 2720). These reactions are considered to proceed *via* cyclopropene intermediates and in the photoisomerisation of 2,5-di-*tert*-butylfuran the intermediate may be isolated (*E. E. van Tamelen* and *T. H. Whitesides*, *ibid.*, 1968, **90,** 3895):

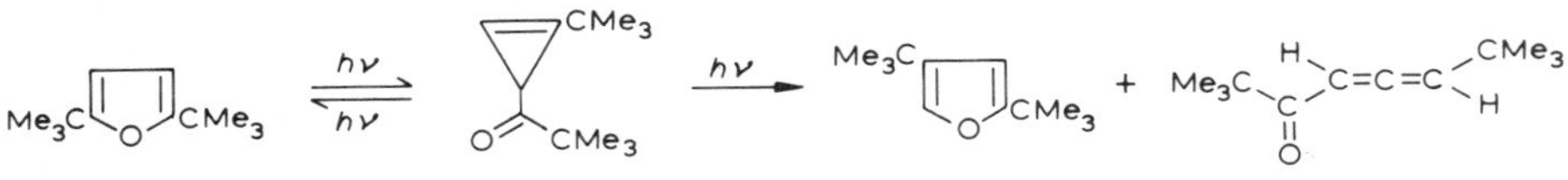

Benzophenone, often used as a photosensitiser, affords a number of adducts (XVI, XVII and XVIII) with furan in the presence of light (*G. O. Schenck*, *W. Hartmann* and *R. Steinmetz*, Ber., 1963, **96,** 498; *M. Ogata*, *H. Watanabe* and *H. Kanō*, Tetrahedron Letters, 1967, 533; *J. Leitich*, *ibid.*, 1967, 1937; *G. R. Evanega* and *E. B. Whipple*, *ibid.*, 1967, 2163):

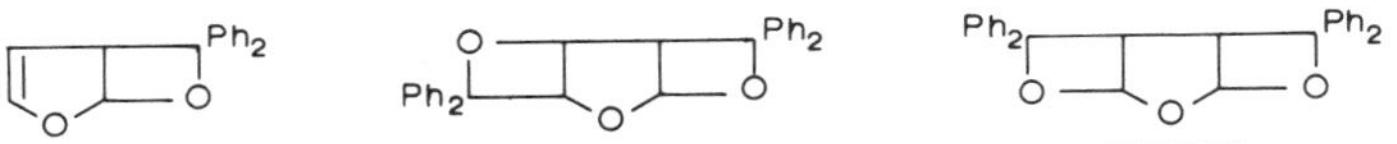

(XVI) (XVII) (XVIII)

(5) *Other addition reactions.* Furan reacts with *diazomethane* in the presence of cuprous bromide to give the bicyclo-oxin XIX and 5% of the tricyclo-oxepane XX (*E. Müller et al.*, Tetrahedron Letters, 1963, 1047). Photolysis with methyl diazoacetate affords the carboxylic ester XXI converted to methyl 4-formyl-1,3-butadiene-1-carboxylate (XXII) by methanolic hydrogen chloride (*G. O. Schenck* and *R. Steinmetz*, Ann., 1963, **668,** 19):

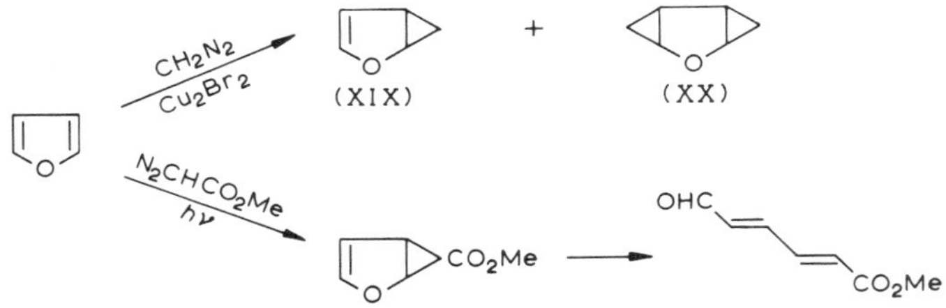

(XIX) (XX)

(XXI) (XXII)

3,4-Dimethoxyfuran gives Michael-type adducts with *ethyl 8-methoxy-1,4-naphthoquinone-2-carboxylate* in toluene (*C. H. Eugster* and *R. Good*, Chimia, 1962, **16,** 343):

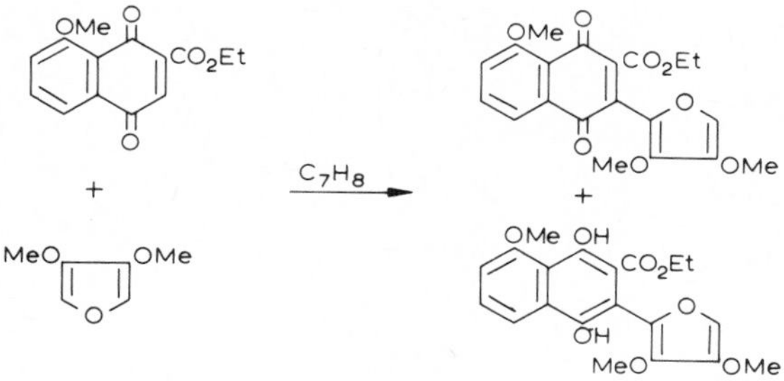

OMe O CO2Et O + MeO OMe O — C_7H_8 → OMe O CO2Et O O MeO OMe + OMe OH CO2Et OH O MeO OMe

Furan condenses with *acetone* to give octamethyltetraoxaquaterene (XXIII) (*W. H. Brown* and *W. N. French*, Canad. J. Chem., 1958, **36,** 371, 537) and annulene polyoxides XXIV, XXV and XXVI are obtained in low yield from the phosphonium salt XXVII (prepared from 5-chloromethyl-2-furfural) on treatment with lithium ethoxide in dimethylformamide (*J. A. Elix*, Chem. Comm., 1968, 343):

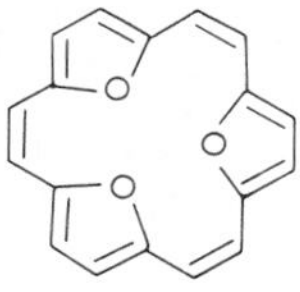

(XXIV)

+

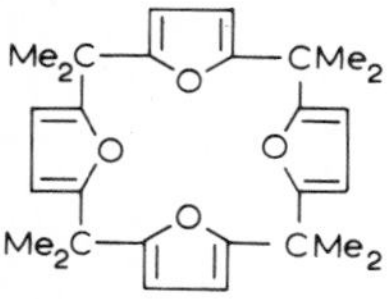

Me_2C O CMe_2 O O Me_2C O CMe_2

(XXIII)

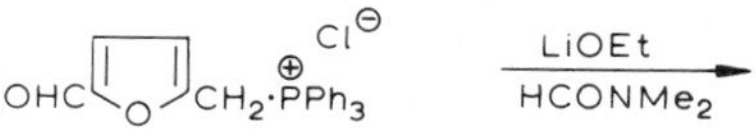

OHC O $CH_2\cdot\overset{\oplus}{P}Ph_3$ $Cl^{\ominus}$

(XXVII)

$\xrightarrow[\text{HCONMe}_2]{\text{LiOEt}}$

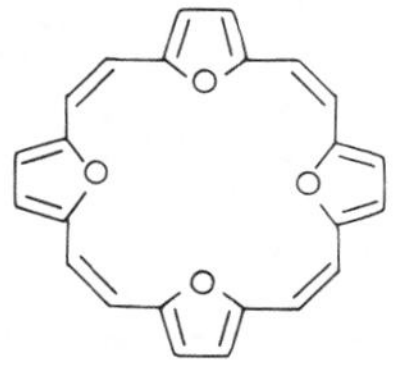

(XXV)

+

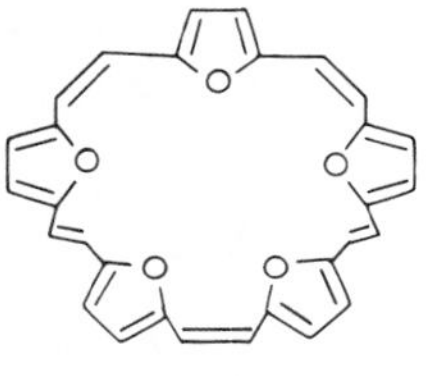

(XXVI)

(6) *Ring-fission.* The usual ring-opening by acid hydrolysis is well exemplified by the case of 2,5-dimethylfuran (*G. Benson*, Org. Synth., 1936, **16,** 26). It takes a more complex course with 2-furylcarbinols; thus furfuryl alcohol with methanolic hydrogen chloride yields $(MeO)_2CH{\cdot}CHO$-$Me{\cdot}CH_2{\cdot}COMe$, which forms α-methoxylaevulic acid, isolated as the methyl ester (*K. G. Lewis*, J. chem. Soc., 1957, 531; *L. Birkofer* and *R. Dutz*, Ann., 1957, **608,** 7). A mechanism has been proposed following the isolation of intermediates (*Birkofer* and *F. Beckmann*, *ibid.*, 1959, **620,** 21):

$$\text{(furan)}CH_2OH \xrightarrow[MeOH]{HCl} \begin{matrix} CH(OMe)-CH_2 \\ | \qquad\quad | \\ CH(OMe)_2 \quad COMe \end{matrix} \longrightarrow \begin{matrix} CH(OMe)-CH_2 \\ | \qquad\quad | \\ CO_2Me \quad COMe \end{matrix}$$

(iii) Furan

Furan, which occurs in pine wood tar, may be prepared by pyrolysis of 2-furoic acid (*W. C. Wilson*, Org. Synth., 1927, **7,** 40); by passing furfural over nickel with hydrogen at 280° (*C. L. Wilson*, J. chem. Soc., 1945, 61); by the catalytic decarbonylation of furfural with steam (*S.-T. Li et al.*, Ko Hsueh Tung Pao, 1958, **3,** 88; C.A., 1958, **52,** 20111b); by passing furfural over caustic lime in the presence of water vapour (*S. Hillers*, *ibid.*, 1959, **53,** 14076i), or over a catalyst consisting of a mixture of oxides (*A. Y. Karmilchiks* and *Hillers*, *ibid.*, 1959, **53,** 15037c). It has m.p. −85°; b.p. 31°; d_4^{20} 0.9366; μ 0.67 (*H. de Vries Robles*, Rec. Trav. chim., 1939, **58,** 111; *E. C. E. Hunter* and *J. R. Partington*, J. chem. Soc., 1931, 2062); u.v., i.r., and n.m.r. spectra (p. 90); and the Raman effect (*A. W. Reitz*, Z. Phys. Chem., 1938, **B38,** 275) have been measured.

Furan, which mixes with the common organic solvents, dissolves in about 100 parts of water, is stable to alkalis and to metallic sodium, but is resinified by acids. It colours a pinewood shaving, moistened with hydrochloric acid green, Ozonisation of furan gives glyoxal (*M. Freri*, Gazz., 1933, **63,** 281). It undergoes a diene synthesis with ethylene (150°/100 atm.), yielding 3,6-epoxycyclohexene (*W. Nudenberg* and *L. W. Butz*, J. Amer. chem. Soc., 1944, **66,** 307):

$$\text{(furan)} + \begin{matrix} CH_2 \\ \| \\ CH_2 \end{matrix} \longrightarrow \text{(3,6-epoxycyclohexene)}$$

Substitution reactions of furan have been reviewed (*M. G. Reinecke* and *H. W. Adickes*, *ibid.*, 1968, **90,** 511).

Metallic derivatives. Furans are metallated in the α-position by alkali metal alkyls (*H. Gilman* and *F. Breuer*, *ibid.*, 1934, **56,** 1123), and some halogenofurans afford Grignard reagents (*Gilman et al.*, *ibid.*, 1932, **54,** 733); both products show the usual reactions. With mercuric chloride and sodium acetate furan gives progressively the mono-, di- and tetra-chloromercuri derivatives; tetra-mercuration occurs when the reaction is carried out with mercuric acetate in boiling acetic acid; the ring in general is attacked in α-positions, often with displacement of α-carboxyl groups, which affords a useful method of decarboxylation (*Gilman* and *G. F. Wright*, *ibid.*, 1933, **55,** 3302); retention of the ester group generally occurs during the mercuration of furancarboxylic esters. The mixed salt $(C_4H_3O)\cdot CO_2HgOAc$ from 2-furoic acid gives 3-acetoxymercurifuran on pyrolysis, presumably *via* a cyclic intermediate; a way is thus opened to 3-substituted furans, although it does rearrange with some reagents (*S. Gronowitz* and *G. Sörlin*, Arkiv Kemi, 1963, **19**, 515). These mercurials give iodofurans with iodine, modest yields of acyl- or alkyl-furans with reactive acyl or alkyl halides, difurylmercury with sodium thiosulphate, and the mercury is displaced by hydrogen on treatment with acids. Mercuration of furan with mercuric cyanate gives mono- and di-mercurated derivatives and 2-methylfuran affords only the 5-mono-derivative (*E. Söderbäck*, Acta chem. Scand., 1959, **13**, 1221). The action of arsenic trichloride on chloromercurifurans leads to furan arsenicals (*W. G. Lowe* and *C. S. Hamilton*, J. Amer. chem. Soc., 1935, **57**, 1081, 2312). Furan with butyl-lithium gives 2-furyl-lithium, which affords tetrafurylgermanium (*R. W. Leeper*, C.A., 1944, **38**, 726), furylmagnesium iodide and tetrafuryl-lead (*Gilman* and *E. B. Towne*, Rec. trav. chim., 1932, **51**, 1054) with the appropriate metallic halide. Some furyl metallic compounds are assembled in Table 1.

TABLE 1

METALLIC DERIVATIVES OF FURAN

Derivative	*m.p. (°C)*	*Derivative*	*m.p. (°C)*
$(2\text{-}C_4H_3O)HgCl$	151	$(2\text{-}C_4H_3O)AsCl_2$	unstable
$(2\text{-}C_4H_3O)_2Hg$	114	$(2\text{-}C_4H_3O)_2AsCl$	unstable
$(3\text{-}C_4H_3O)HgCl$	184.5	$(2\text{-}C_4H_3O)_3As$	35
$(3\text{-}C_4H_3O)_2Hg$	73	$(2\text{-}C_4H_3O)_2AsO_2H$	138
$(2\text{-}C_4H_3O)HgOCN$	122–123	$(2\text{-}C_4H_3O)_4Ge$	99–100
$5\text{-}(2\text{-}MeC_4H_2O)HgOCN$	114–115	$(2\text{-}C_4H_3O)_4Pb$	52–53

(iv) Alkyl- and aryl-furans

(*1*) *Preparation of alkylfurans.* A few alkylfurans have been isolated from wood-tar, many have been made by direct ring-synthesis, and others by decarboxylation of the more easily synthesised acids. The Kishner–Wolff reaction has been usefully applied. The 2-derivatives are further accessible by Friedel–Crafts alkylation of methyl 2-furoate, followed by hydrolysis and decarboxylation; by dehydration and reduction of alkylfurylcarbinols obtained from furfural and Grignard reagents, or by the selective reduction of the carbinol over palladium catalyst, *e.g.*, furylmethylcarbinol and ethylfurylcarbinol over 10% Pd–C at 200–230° give 2-ethyl- and 2-propyl-furan respectively (*N. I. Shuĭkin* and *I. F. Bel'skiĭ*, Doklady Akad. Nauk, S.S.S.R., 1957, **117**, 95; Ber., 1958, **91**, 948); and from Grignard reagents and furfuryl bromide (*Gilman* and *N. O. Calloway*, J. Amer. chem. Soc., 1933, **55**, 4197;

R. Paul, Bull. Soc. chim. Fr., 1935, [v], **2**, 2220, 2227). Furan with ethylene at 20–35° in the presence of boron trifluoride-etherate yields 2-ethyl- and 2,5-diethyl-furan (*S. Hillers*, *A. Bērzina* and *L. Lauberte*, Latvijas PSR Zinālnu Akad. Vēstis, 1958, **4**, 71). Catalytic alkylation of furan may be effected with isobutylene (*Shuĭkin*, *B. L. Lebedev* and *V. G. Nikol'skii*, Doklady Akad. Nauk, S.S.S.R., 1967, 174, 621).

3,4-Epoxy-3-methylbutanal diethylacetal obtained from the reaction of 2-methylallyl chloride with magnesium and ethyl orthoformate followed by oxidation of the product by perphthalic acid, gives 3-methylfuran on boiling with 0.1 *N* sulphuric acid (*J. W. Cornforth*, J. chem. Soc., 1958, 1310):

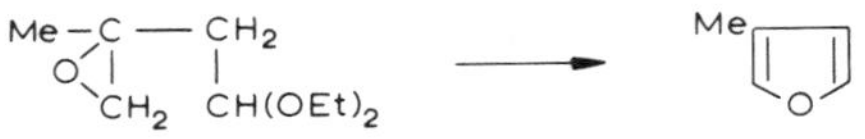

(2) *Properties and reactions of alkylfurans.* Various n.m.r. effects have been reported for methyl- and dimethylfurans (*S. Rodmar*, *B. Rodmar* and *A. Ali Khan*, Acta chem. Scand., 1966, **20**, 2515; *J. P. Morizur*, *Y. Pascal* and *F. Vernier*, Bull. Soc. chim. Fr., 1966, 2296; *M. Ohtsuru*, *K. Tori* and *H. Watanabe*, Chem. pharm. Bull., Tokyo, 1967, **15**, 1015).

The simplest members give a green, and many of the more complex, a red pine-shaving reaction (*T. Reichstein*, Helv., 1932, **15**, 1110).

The photochemistry of 2-alkylfurans has been discussed on p. 93.

Electrolysis of a solution of 2,5-dimethylfuran in methanolic sodium methoxide affords 2,5-dimethyl-2,5-dimethoxy-2,5-dihydrofuran and in acetic acid in the presence of sodium acetate it yields 2,5-bis(acetoxymethyl)furan (*A. J. Baggaley* and *R. Brettle*, Chem. Comm., 1966, 108; *K. E. Kolb* and *C. L. Wilson*, *ibid.*, p. 271):

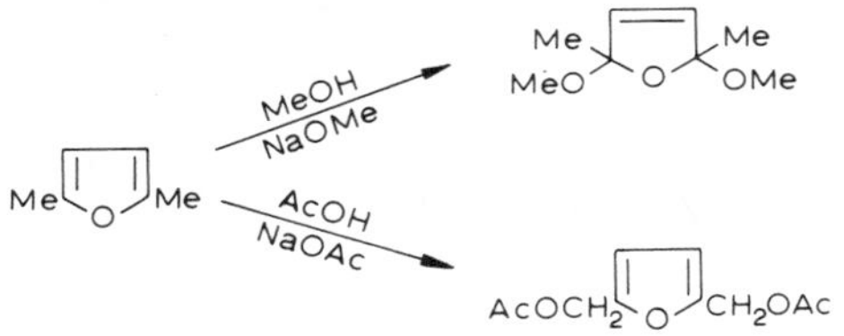

2,5-Di- and tetra-methylfuran with peracetic acid give the corresponding peroxides (*D. Seebach*, Ber., 1963, **96**, 2712):

Ozonolysis of 2-methylfuran gives a mixture of glyoxal and methylglyoxal in a ratio 1:2 approximately, and for 2,5-dimethylfuran 1:1.4. For the synthesis and reactions of alkylated furans see *W. H. Brown* and *G. F. Wright*, Canad. J. chem., 1957, **35,** 236.

Individual alkylfurans are listed in Table 2 (p. 101).

(*3*) *Arylfurans.* 2-**Arylfurans** may be prepared by a modified Gomberg synthesis (*A. W. Johnson*, J. chem. Soc., 1946, 895) in which stabilised diazonium salts derived from the corresponding amines are decomposed in a large excess of furan. The arylfurans undergo uncatalysed addition of tetracyanoethylene and of dimethyl acetylenedicarboxylate. The adducts from the latter compound rearrange in warm acetic acid to give dimethyl 3-hydroxy-4′-substituted-biphenyl-1,2-dicarboxylates (*D. C. Ayres* and *J. R. Smith*, Chem. Comm., 1967, 886):

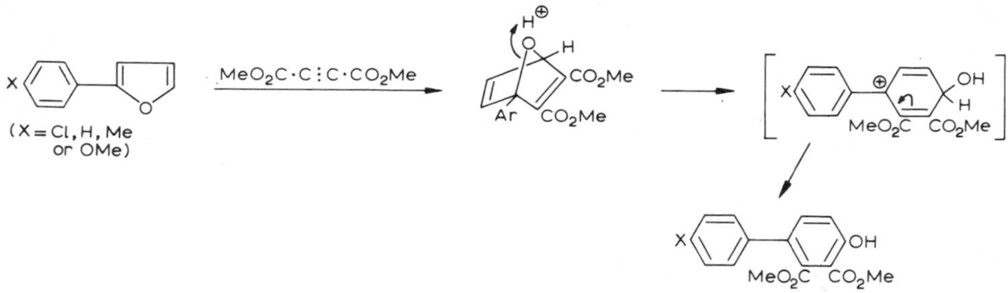

(X = Cl, H, Me or OMe)

$MeO_2C \cdot C \vdots C \cdot CO_2Me$

2,4-**Diphenyl-** and **tetraphenylfuran** are obtained respectively by pyrolysis of acetophenone and by heating benzoin with hydrochloric acid. Oxidation of 2,5-diphenylfuran with lead tetra-acetate in chloroform at 20–30° gives *cis*-benzoyl ethylene and boiling in acetic acid affords 2-acetoxy-2,5-diphenyl-2,3-dihydrofuran-3-one (*C.-K. Dien* and *R. E. Lutz*, J. org. Chem., 1957, **22,** 1355):

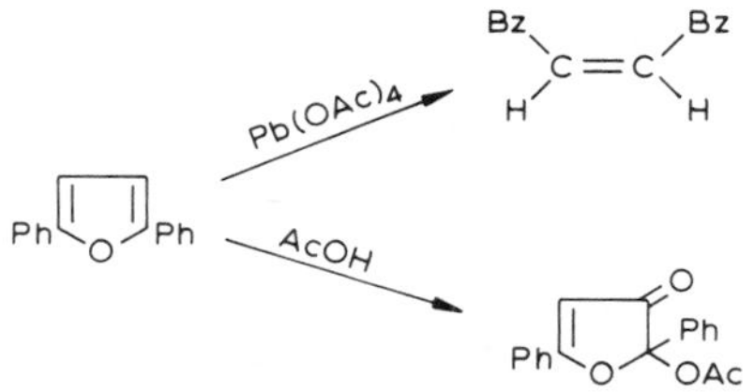

$Pb(OAc)_4$; AcOH

The synthesis of 3-**phenylfuran** has been accomplished by the following route from 1,2,4-trihydroxybutane (*H. Wynberg*, J. Amer. chem. Soc., 1958, **80,** 364):

$CH_2OH{\cdot}CHOH{\cdot}CH_2{\cdot}CH_2OH$ $\xrightarrow{p\text{-}MeC_6H_4SO_3H}$ (3-hydroxytetrahydrofuran) $\xrightarrow[H_2SO_4]{Na_2Cr_2O_7}$ (tetrahydrofuran-3-one) $\xrightarrow{PhMgBr}$ (3-phenyl-3-hydroxytetrahydrofuran) $\longrightarrow$ (3-phenyl-2,5-dihydrofuran) + [3-phenyl-4,5-dihydrofuran] ; $\xrightarrow{S\ in\ HCONMe_2}$ 3-phenylfuran

A few heavily substituted 3-phenylfurans can exist in stable optically active forms owing to restricted rotation about the bond joining the two nuclei (*A. Khawam* and *E. V. Brown*, *ibid.*, 1952, **74,** 5603).

Some arylfurans are listed in Table 2.

(*4*) *Naturally occurring furans.* A series of furan derivatives encountered in natural products includes the following:

Perilla citriodora oil affords 3-isohexenylfuran, **perillen** (XXVIII), which contains the citral skeleton; ozonolysis yields isohexylsuccinic acid and acetone (*H. Kondo* and *H. Suzuki*, Ber., 1936, **69,** 2459). **Dendrolasin** (XXIX) occurs in the mandibular glands of the ant *Dendrolasius fuliginosus Lao* and contains the farnesol skeleton; oxidation yields acetone, succinic and laevulic acid (*A. Quilico et al.*, Tetrahedron, 1957, **1,** 156, 177).

Carlina oxide, from *Carlina acaulis*, is α-benzyl-β-2-furylacetylene (XXX), since it can be reduced to 2-γ-phenylpropylfuran and oxidised to phenylacetic acid (*Gilman et al.*, J. Amer. chem. Soc., 1933, **55,** 3461). It has been synthesised from benzyl chloride and the Grignard derivative of 2-furylacetylene (*A. S. Pfau et al.*, Helv., 1935, **18,** 935).

3-furyl–$CH_2{\cdot}CH_2{\cdot}CH{:}CMe_2$ (XXVIII)

3-furyl–$CH_2{\cdot}CH_2{\cdot}CH{:}CMe{\cdot}CH_2{\cdot}CH_2{\cdot}CH{:}CMe_2$ (XXIX)

2-furyl–$C{\vdots}C{\cdot}CH_2Ph$ (XXX)

For other examples see Vol. II B, p. 170; Vol. II C, pp. 290 *et seq.*, 302 *et seq.* The biogenesis of furan rings is discussed in Vol. II E, p. 78.

TABLE 2

ALKYL-, ARYL- AND RELATED FURANS

Substituent	*b.p. (°C)*	*HgCl deriv. m.p. (°C)*	*Ref.*
2-Methyl	64	134	1
2-Ethyl	92–93	109	2
2-Isopropyl	106–109	117–118	3
2-*tert*-Butyl	119–120	136–137	3
2-Phenyl	107–108/18 mm		4
2-Benzyl	114–116/23 mm		2
2-Vinyl	99		5
2-Ethynyl	105–106	118–119	6
3-Methyl	65	142	7
3-Isopropyl	111–112	90	8
3-Phenyl	m.p. 58.5–59		9
2,3-Dimethyl	42/115 mm		10
2,4-Dimethyl	94	113	11, 12
2,5-Dimethyl	94	163–164 dec.	13
3,4-Dimethyl	49/125 mm		10
2,4-Diphenyl	m.p. 109		14, 15
2,5-Diphenyl	m.p. 91		16
3,4-Diphenyl	m.p. 110		17
4-Methyl-2-phenyl	m.p. 38–40		12
2,3,4-Trimethyl	54.5/57 mm		18
2,3,5-Trimethyl	51.5/62 mm		18
Tetramethyl	45–46/12 mm		13
Tetraphenyl	m.p. 175		19

References

1 In 92% yield from furfural by H_2/Cu chromite/250° (*L. E. Schniepp et al.*, J. Amer. chem. Soc., 1947, **69,** 672).
2 *R. Paul*, Compt. rend., 1935, **200,** 1481; 1936, **202,** 1444.
3 *H. Gilman* and *N. O. Calloway*, J. Amer. chem. Soc., 1933, **55,** 4197.
4 *A. W. Johnson*, J. chem. Soc., 1946, 895.
5 *Paul* and *S. Tchelitcheff*, Bull. Soc. chim., France, 1947, 453; *Y. Hackipama* and *M. Imoto*, J. Soc. chim. Ind., Japan, 1952, **45,** Suppl. 190; C.A., 1950, **44,** 9721.
6 *C. Moureu et al.*, Ann. Chim., 1927, **7,** 13.
7 *Gilman* and *R. R. Burtner*, J. Amer. chem. Soc., 1933, **55,** 2903.
8 *Gilman et al.*, *ibid.*, 1935, **57,** 906.
9 *H. Wynberg*, *ibid.*, 1958, **80,** 364.
10 *T. Reichstein* and *A. Grüssner*, Helv., 1933, **16,** 28.
11 *Gilman* and *Burtner*, Rec. Trav. chim., 1932, **51,** 667.

12 *T. Morel* and *P. E. Verkade*, *ibid.*, 1951, **70,** 35.
13 *R. Gaertner* and *R. G. Tonkyn*, J. Amer. chem. Soc., 1951, **73,** 5872; *J. Colonge* and *R. Gelin*, Bull. Soc. chim. Fr., 1954, 208.
14 *E. P. Kohler* and *W. N. Jones*, J. Amer. chem. Soc., 1919, **41,** 1263.
15 *T. M. Harris*, *C. M. Harris* and *J. C. Cleary*, Tetrahedron Letters, 1968, 1427.
16 *Kohler* and *H. F. Engelbrecht*, J. Amer. chem. Soc., 1919, **41,** 1382.
17 *H. J. Backer* and *W. Stevens*, Rec. Trav. chim., 1940, **59,** 423.
18 *Reichstein et al.*, Helv., 1932, **15,** 1112.
19 *Yu. Zalkind* and *V. Teterin*, J. pr. Chem., 1932, **133,** 195.

(*v*) *Halogenofurans*

(*1*) *Preparation.* (*a*) *Direct halogenation.* The direct halogenation of furan is unsatisfactory, and most compounds in this series have been prepared by halogenating 2-furoic acid or its derivatives followed by decarboxylation. In many cases the halogenation affords addition products which lose halogen acid when boiled with alkali. Halogenation of a carboxylic acid, especially in alkaline solution, may cause displacement of the 2-carboxyl group by halogen. Halogen atoms in the α-position are preferentially removed by reduction, so rendering accessible compounds halogenated in the β-position only (*A. F. Shepard et al.*, J. Amer. chem. Soc., 1930, **52,** 2083). 2-Bromofuran has been prepared directly from furan and dioxane dibromide at 0° (*A. P. Terent'ev*, *L. I. Belen'kiĭ* and *L. A. Yanovskaya*, Zhur. obshcheĭ Khim., 1954, **24,** 1265). Direct chlorination of furan with 1.6 moles of chlorine at −40° yields a mixture of 2-chloro-, 2,5-dichloro-, and 2,3,5-trichloro-furan, while increasing the amount of chlorine affords tetrachlorofuran and 2,5-dihydro-2,2,3,4,5,5-hexachlorofuran.

(*b*) *From mercurihalides.* Furylmercuric chloride and its congeners react smoothly with bromine and iodine, and the latter provides a good source of 2-iodofuran (*H. Gilman* and *G. F. Wright*, J. Amer. chem. Soc., 1933, **55,** 3302). Although 3-furylmercuric chloride rearranges with some reagents, with iodine it is converted into 3-iodofuran (*S. Gronowitz* and *G. Sörlin*, Arkiv Kemi, 1963, **19,** 515).

(*c*) *From endoxohydrophthalic acids.* 3-Halogenofurans have been prepared along with the salt of maleic acid by heating to 150° a suspension of 4,5-dihalogeno-3,6-endoxohexahydrophthalic anhydride in water, containing the reaction equivalent of alkali-metal hydroxide (*W. W. Levis Jr.*, U.S.P. 2,773,882/1956):

[X, X, O, CO, O, CO] —aq. NaOH→ [furan, X] + $CH \cdot CO_2Na$ ‖ $CH \cdot CO_2Na$

2,2,3,3-Tetrachlorobutane-1,4-diol, alumina and diphenyl oxide heated to 200–220° yield 3,4-dichlorofuran, which with maleic anhydride affords 4,5-

dichloro-3,6-endoxo-1,2,3,6-tetrahydrophthalic anhydride (*G. M. Mkryan et al.*, Arm. Khim. Zhur., 1966, **19,** 898; C.A., 1967, **67,** 64133m):

$$CCl_2{-}CCl_2(CH_2OH)_2 \longrightarrow \text{3,4-dichlorofuran} \longrightarrow \text{dichloro-endoxo-tetrahydrophthalic anhydride}$$

(*d*) *Fluoro compounds.* Tetrahydrofuran with cobalt trifluoride at 100–120° yields a mixture of $C_4H_2F_6O$ isomers, and subsequent treatment with fused potassium hydroxide gives tetrafluorofuran (*J. Burdon, J. C. Tatlow* and *D. F. Thomas*, Chem. Comm., 1966, 48). The fluorines in tetrafluorofuran are relatively inert to nucleophilic substitution.

(*2*) *Properties and reactions.* Although these halogeno compounds are often rather unstable, their halogen atoms are not easily exchanged, and few, except 2-iodofuran, readily afford Grignard reagents. Although no Grignard reagent has been obtained from 3-iodofuran, with butyl-lithium at −70° it affords 3-furyl-lithium (*Gronowitz* and *Sörlin*, Acta chem. Scand., 1961, **15,** 1419). 3-Iodofuran with sodium methoxide and cuprous cyanide in the presence of copper gives 3-methoxy- and 3-cyano-furan (*Gronowitz* and *Sörlin, loc. cit.*). The kinetics of nucleophilic displacement with piperidine of 2-halogenofurans (*D. G. Manly* and *E. D. Amstutz*, J. org. Chem., 1957, **22.** 133), and the mechanism of the reaction of 5-bromo-2-nitrofuran with amines (*V. N. Novikov* and *Z. N. Nazarova*, Zhur. organ. Khim., 1966, **2,** 1901) have been investigated. For individual compounds see Table 3.

(vi) Furansulphonic acids

Furan is destroyed by sulphuric acid, but yields **furan**-2-**sulphonic acid** (*amide*, m.p. 123°) and 2,5-**disulphonic acid** (*amide*, m.p. 199–200°) with sulphur trioxide-dioxane or better with sulphur trioxide–pyridine complex (*A. P. Terent'ev* and *L. A. Kazitsina*, C.r. Acad. Sci., U.R.S.S., 1946, **51,** 603; *J. F. Scully* and *E. V. Brown*, J. org. Chem., 1954, **19,** 894). On the other hand furans having carbonyl or carboxyl substituents can be sulphonated with sulphuric acid (*H. B. Hill* and *A. W. Palmer*, Amer. chem. J., 1888, **10,** 373, 409).

(vii) Nitrofurans

Nitric acid in acetic anhydride gives with furan 2-**nitrofuran,** m.p. 29° (*I. J. Rinkes*, Rec. Trav. chim., 1930, **49,** 1169; *B. T. Freure* and *J. R. Johnson*, J. Amer. chem. Soc., 1931, **53,** 1142), by way of an addition product which can be hydrolysed to maleic anhydride. The constitution is confirmed by the production of the same compound by nitrating 2-furoic acid, and

TABLE 3

HALOGENOFURANS

Substituent	*b.p. (°C)*	*Ref.*
2-Chloro	78	1
3-Chloro	80	1
3,4-Dichloro	123	1, 2
2,3,4-Trichloro	152/734 mm	1
2-Bromo	103	1, 3
3-Bromo	103	1
2,5-Dibromo	62–63/15 mm	
Tetrabromo	m.p. 64–65	
2-Iodo	43–45/15 mm	4
3-Iodo	133	4, 5
2,5-Di-iodo	m.p. 47	4
Tetraiodo	m.p. 165	6
Tetrafluoro	17	7

References
1 *A. F. Shepard et al.*, J. Amer. chem. Soc., 1930, **52,** 2083.
2 *G. M. Mkryan et al.*, Arm. Khim. Zhur., 1966, **19,** 898; C.A., 1967, **67,** 64133m.
3 *A. P. Terent'ev, L. I. Belen'kii* and *L. A. Yanovskaya*, Zhur. obscheĭ Khim., 1954, **24,** 1265.
4 *H. Gilman* and *G. F. Wright*, J. Amer. chem. Soc., 1933, **55,** 3302.
5 *S. Gronowitz* and *G. Sörlin*, Archiv Kemi, 1963, **19,** 515.
6 *H. Scheibler et al.*, J. pr. Chem., 1933, **136,** 232.
7 *J. Burdon, T. C. Tatlow* and *D. F. Thomas*, Chem. Comm., 1966, 48.

by its further nitration to 2,5-**dinitrofuran,** m.p. 101°, which is hydrolysed by baryta to maleate and nitrite. 2-Acetoxy-5-nitro-2,5-dihydrofuran is definitely the nitration intermediate (*N. Clauson-Kaas* and *J. Fakstorp*, Acta chem. Scand., 1947, **1,** 210). The entering nitro group always prefers a free 2- or 5-position:

AcO—(dihydrofuran)—NO_2 ; NO_2 ; O_2N … NO_2 ; CH=CH, CO_2H CO_2H ; CH=CH, CHO CHO ; CO_2H

Treatment of furan with nitronium tetrafluoroborate affords 2-nitrofuran in low yield (*G. Oláh, S. Kuhn* and *A. Mlinkó*, J. chem. Soc., 1956, 4257).

3-**Nitrofuran,** m.p. 27°, is obtained from 5-methyl-2-furoic acid (*Rinkes*, Rec. Trav. chim., 1938, **57,** 390):

HO_2C … Me → HO_2C … Me, NO_2 —Cu chromite / quinoline→ NO_2, Me —$K_3Fe(CN)_6$→ NO_2, CO_2H —Δ→ NO_2

Furans which do not contain an electron-attracting group are usually nitrated under very mild conditions in order to avoid resinification.

Treatment of 5-bromopyromucic acid in acetic anhydride with nitric acid in acetic anhydride at $-10°$ gives 5-**bromo**-2-**nitrofuran,** m.p. 48°, which on boiling with potassium iodide in acetic acid affords 5-**iodo**-2-**nitrofuran,** m.p. 76° (*Z. N. Nazarova* and *V. N. Novikov*, Zhur. obshcheĭ Khim., 1961, **31,** 263). The latter compound on heating with nitric acid yields 2,5-dinitrofuran. 5-Bromo-2-nitrofuran and sodium thiocyanate on boiling in acetone give 5,5′-**dinitro**-2,2′-**difuryl sulphide,** m.p. 98–99°, a compound with strong activity against *Staphylococcus* organisms. Nitrofurans exhibit specific activity against certain types of bacteria which do not develop resistance strains (*K. Miura* and *H. K. Reckendorf*, Progr. Med. Chem., 1967, **5,** 320).

O_2N–(furan)–Br $\xrightarrow{KSCN}$ O_2N–(furan)–S–(furan)–NO_2

For a review on nitrating agents and on mechanisms of nitration of furan and its derivatives see *P. Krkoska*, *A. Jurasek* and *J. Kovac*, Chem. Listy, 1968, **62,** 182; C.A., 1968, **69,** 18939x).

Furan derivatives containing a nitro group in a side chain are prepared by reaction between nitromethane and an aldehyde group in the side chain, separated from the nucleus by at least two carbon atoms, at $-5°$ to 0° with the exclusion of water and in the presence of alcoholic sodium hydroxide (*O. Moldenhauer et al.*, G.P. 1,006,430/1957):

(furan)–$CH_2 \cdot CH_2 \cdot CHO$ + $MeNO_2$ $\xrightarrow[NaOH,\ 0°]{MeOH}$ (furan)–$CH_2 \cdot CH_2 \cdot CH{:}CHNO_2$

(viii) Amino- and cyano-furans

Attempts to obtain 2-aminofuran by the reduction of 2-nitrofuran, results in the breakdown of the latter and the liberation of ammonia. In contrast ethyl 5-amino-2-furoate has been prepared by reduction of the related nitro compound catalytically or by means of aluminium.

Furancarboxylic acid azides give on decomposition isocyanates which are not smoothly hydrolysed to amines. By heating the azides with formic acid, formamido compounds are obtained, decomposable by alkali to the amines (*H. B. Stevenson* and *J. R. Johnson*, J. Amer. chem. Soc., 1937. **59,** 2525):

The products are easily hydrolysed with loss of ammonia, those having the amino-group in the 3-position are stable enough to be isolated. The amines do not form normal Schiff bases with benzaldehyde, and with nitrous acid give solutions which couple with 2-naphthol but do not show other diazo-reactions: the alternative ketimine formula must be considered. It has been observed that a variety of 2-amino-furans do not behave as typical aromatic amines towards an electrophilic reagent, but their reactions are characterstic of the polarised form of an enamine. Thus 2-amino-3-cyano-4,5-dimethylfuran reacts with bromoacetyl bromide to provide a route to the furopyrroles (*C. T. Wie, S. Sunder* and *C. D. Blanton*, Tetrahedron Letters, 1968, 4605):

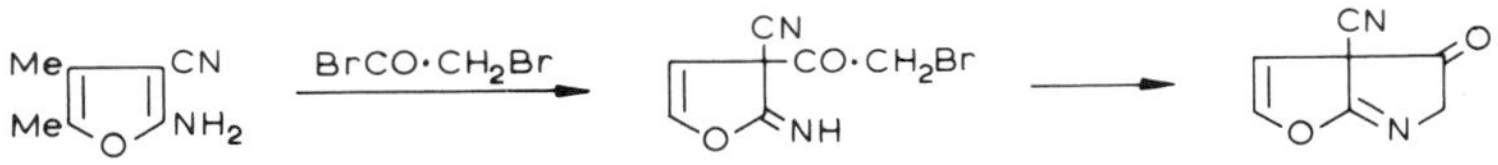

A few tertiary 3-furylamines have been obtained by ring-synthesis (*R. E. Lutz et al.*, J. Amer. chem. Soc., 1946, **68,** 2224):

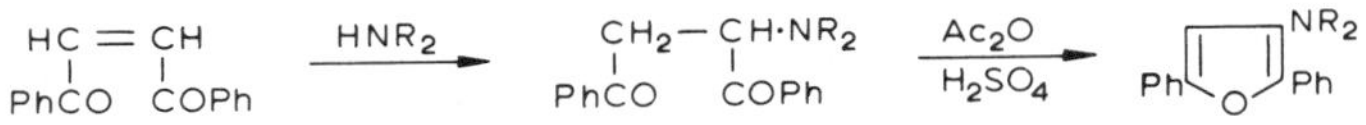

Tetracyano- and 3,4-dicyano-furan have been synthesised by the following route (*C. D. Weis*, J. org. Chem., 1962, **27,** 3520, 3693, 3514):

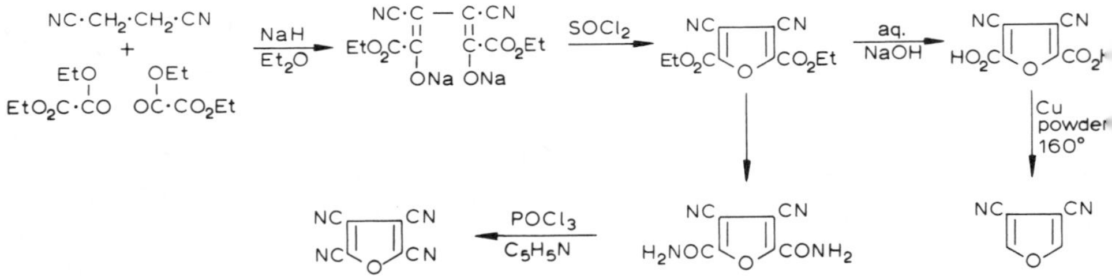

Hydrogenation of 3-amino-1-furylbutane and 3-amino-1-furylpentane over Pt–C at 200–210° affords 2-methyl-5-propylpyrrolidine and 2-ethyl-5-propylpyrrolidine, respectively (*I. F. Bel'skiĭ* and *N. I. Shuĭkin*, Doklady Akad. Nauk, S.S.S.R., 1961, **137,** 331).

3-**Amino-2-methylfuran,** b.p. 51–52°/4 mm; 3-**amino-2,5-dimethylfuran,** b.p.55–56°/4 mm (*Stevenson* and *Johnson*, *loc. cit.*).

Furoyl azide affords 2-*furyl isocyanate*, b.p. 111–112°, *methyl* N-2-*furylcarbamate*, b.p. 120°/20 mm, and N-2-*furylbenzamide*, m.p. 124.5° (*H. M. Singleton* and *W. R. Edwards*, J. Amer. chem. Soc., 1938, **60,** 540). Furan-3,4-dicarboxylic acid gives, *via* the diazide, *furyl*-3,4-*dicarbamate*, m.p. 166–167°, of interest in relation to oxybiotin (*G. Stork*, *ibid.*, 1945, **67,** 884).

(ix) Oxodihydrofurans; butenolides (hydroxyfurans)

Preparation of 2- and 3-hydroxyfurans have been reported (*H. H. Hodgson* and *R. R. Davies*, J. chem. Soc., 1939, 806; 1013) but reactions which could lead to these compounds invariably yield the tautomeric carbonyl derivatives or their decomposition products. So far true 2-hydroxyfurans are unknown (*M. P. Cave et al.*, J. Amer. chem. Soc., 1956, **78.** 2303) and little is known about 3-hydroxyfurans.

(*1*) *2-Oxodihydrofurans.* The 2-oxo-compounds are butenolides (Vol. 1 D, p. 116; Vol. II D, p. 371 *et seq.*), which behave as lactones and have no phenolic properties.

2-**Acetoxyfuran,** b.p. 49–50°/9 mm, is obtained in good yield by pyrolysis of 2,5-diacetoxy-2,5-dihydrofuran (p. 126) (*Cave et al.*, *loc. cit.*). 2-**Methoxyfuran,** b.p. 110°, can be prepared in a similar way (*idem*, Chem. and Ind., 1955, 17), or by pyrolysis of its 5-carboxylic acid obtained by heating methyl 5-bromo-2-furoate with sodium methoxide (*D. G. Mann* and *E. D. Amstutz*, J. org. Chem., 1956, **21,** 516). Both ester and ether undergo the Diels–Alder reaction, and are very easily hydrolysed to derivatives of β-formylpropionic acid. Reaction with bromine or chlorine results in elimination of acetyl halide and the formation of 5-halogeno-2,5-dihydro-2-oxofuran. Treatment with lead tetra-acetate gives 5-acetoxy-2,5-dihydro-2-oxofuran. α- and β-**Angelica lactone** (5-methyl-2-oxo-dihydrofurans) are both converted by alkali into an equilibrium mixture in which the α-compound predominates; the amount of hydroxy form XXXI present at equilibrium is exceedingly small:

Acetylangelicalactone (XXXII), m.p. 177–180°, easily accessible from ethyl diacetosuccinate, shows no enol reactions and dissolves only slowly in sodium hydroxide; on distillation it yields the *enol*, m.p. 63°, and reverts to the oxo-form when boiled with water (*L. Knorr*, Ann., 1898, **303,** 133).

The reaction between diazoketones and ketens affords 3,3-disubstituted-2-oxo-2,3-dihydrofurans ($\beta\gamma$-butenolides) (*W. Reid* and *H. Mengler*, Angew. Chem., 1961, **73,** 218):

$R{\cdot}CO{\cdot}CHN_2 + R^1R^2C{:}CO \longrightarrow$

ω-Diazoacetophenone and diphenylketen give 2-oxo-3,3,5-triphenyl-2,3-dihydrofuran (*idem*, Ann., 1962, **651,** 54).

Esters of type XXXIII have been oxidised by selenium dioxide to 2-oxo-2,5-dihydrofurans (*N. Danielli, Y. Mazur* and *F. Sondheimer*, J. Amer. chem. Soc., 1962, **84,** 875):

(XXXIII)

1,2-Diphenylcyclopropene-3,3-dicarboxylic acid on heating is decomposed to give 4,5-diphenyl-2-oxo-2,5-dihydrofuran (*R. Breslow, R. Winter* and *M. Battiste*, J. org. Chem., 1959, **24,** 415; *cf. S. F. Darling* and *E. W. Spanagel*, J. Amer. chem. Soc., 1931, **53,** 1117):

The compound obtained from pyruvic acid and benzylideneaniline and previously believed to be a dioxopyrrolidine has been reformulated as a lactone, 5-phenyl-3-phenylamino-2,5-dihydrofuran (*H. H. Wasserman* and *R. C. Koch*, Chem. and Ind., 1957, 428):

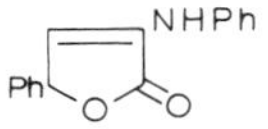

(*2*) 3-*Oxodihydrofurans.* The oxo and hydroxy forms of 2,5-diphenyl-3-hydroxyfuran (XXXIV and XXXV) have been isolated, but the enol changes rapidly and apparently completely into the oxo tautomer (*E. P. Kohler, F. H. Westheimer* and *M. Tishler*, J. Amer. chem. Soc., 1936, **58,** 264; *Kohler* and *D. W. Woodward*, *ibid.*, p. 1933):

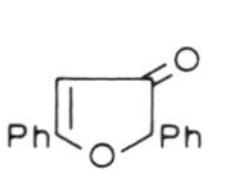

(XXXIV) 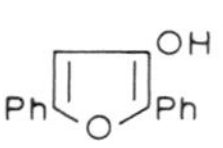 (XXXV)

3-Oxo-compounds are predominantly ketonic in character unless carbonyl substituents are also present. Some 3-oxodihydrofurans are stable towards acids and being vinylogous esters do not form normal semicarbazones (*C. H. Eugster, K. Allner* and *R. E. Rosenkranz*, Chimia, 1961, **15,** 516). 3-Oxo-2-methyl-2,3-dihydrofuran (3-hydroxy-2-methylfuran), produced from 5-oxorhamnolactone by the stages indicated, gives only 1/3 mole of methane in a Zerevitinov determination (*E. Votoček* and *S. Malachta*, Coll. Czech. chem. Comm., 1932, **4,** 87). The intermediate acid, evidently more decidedly enolic, gives a cherry-red ferric chloride reaction.

5-Oxorhamnolactone $\xrightarrow[HCl]{MeOH}$ (MeO, Me, CO_2H furan) $\xrightarrow[H_2O]{HCl}$ (HO, Me, CO_2H furan) $\xrightarrow{\Delta}$ (HO, Me furan) ⇌ (O=, Me dihydrofuran)

The enol forms give a positive reaction with ferric chloride, react rapidly with bromine, and form a peroxide with oxygen. 3-Hydroxyfuran prepared by acid hydrolysis of 3-methoxyfuran exists almost entirely in the oxo form (*A. Hofman, W. von Philipsborn* and *Eugster*, Helv., 1965, **48,** 1322). *Dihydrofuranolones* are produced by a lactolisation of 1,2,4-triketones (*R. E. Lutz* and *A. H. Stewart*, J. Amer. chem. Soc., 1936, **58,** 1885; 1941, **63,** 1143):

PhCO–CO–CH_2–COPh ⇌ PhCO–CO–CH=C(OH)Ph ⇌ (O=, Ph, HO, Ph dihydrofuranone) $\xrightarrow[HCl]{MeOH}$ (O=, Ph, MeO, Ph dihydrofuranone)

Benzylidenecrotonolactone obtained from 5-bromo- or 5-chloro-furfural and phenylmagnesium bromide is an analogue in the furan series of the methylenequinones (*H. Gilman et al., ibid.*, 1950, **72,** 3). Protoanemonin (Vol. II A, p. 102) is the simplest possible case of this kind.

PhCH= (furanone) =O

Benzylidenecrotonolactone

Maleic anhydride can be regarded as furanquinone, and resembles *p*-benzoquinone in forming coloured molecular complexes with aromatic nuclei (*P. Pfeiffer* and *T. Bottler*, Ber., 1918, **51,** 1819).

(x) Furfuryl alcohol and related compounds

Halogen atoms in the side-chain in the furan series show the usual aliphatic reactivity, with the exception that 2-**chloromethylfuran** *(furfuryl chloride)*, b.p. 37°/15 mm, does not form Grignard compounds effectively and reacts abnormally with potassium cyanide (*T. Reichstein*, Ber., 1930, **63,** 749) to give 2-cyano-5-methylfuran. Since the hydrolysis of 2-chloromethylfuran in aqueous dioxane is unimolecular (*J. Egyed* and *A. Gerecs*, Acta Chim. Acad. Sci. Hung., 1962, **29,** 91; C.A., 1962, **57,** 16524) the following mechanism seems possible:

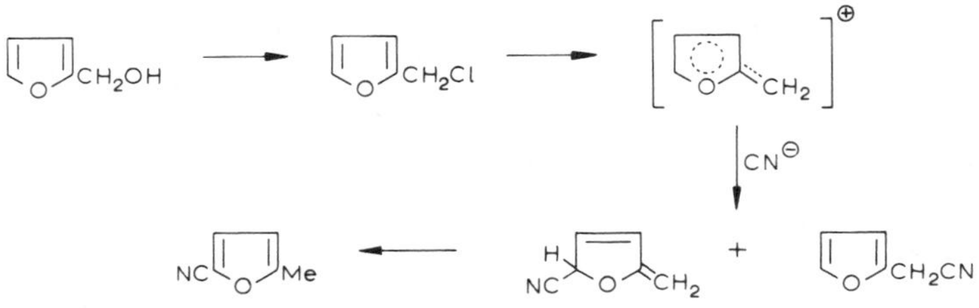

2-Chloromethyl-5-methylfuran does not undergo a similar rearrangement, but gives 2-cyanomethyl-5-methylfuran.

2-Chloromethylfuran and sodium cyanide in dimethyl sulphoxide give 2-cyanomethylfuran, whereas in formamide they yield a mixture of 2-cyanomethylfuran and 2-cyano-5-methylfuran (*K. Yu. Novitskii, Kh. Gresl* and *Yu. K. Yur'ev*, Khim. Geterotsiki. Soedin., 1966, 829):

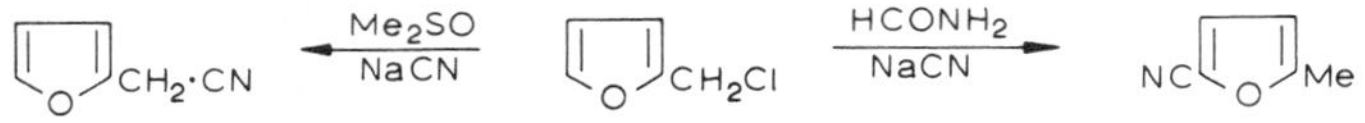

The very unstable, reactive 2-**bromomethylfuran,** b.p. 32.5–34.5/2 mm, can be obtained by treating the alcohol with phosphorus tribromide (*J. E. Zanetti* and *J. T. Bashour*, J. Amer. chem. Soc., 1939, **61,** 2249; *R. B. Woodward, ibid.*, 1940, **62,** 1478) or by treating methylfuran with *N*-bromosuccinimide (*N. P. Buu-Hoï* and *J. Lecocq*, Compt. rend., 1946, **222,** 1441).

Unlike the 2-isomer, 3-**chloromethylfuran,** b.p. 42–43°/17 mm, prepared from the alcohol, forms a Grignard reagent (*E. Sherman* and *E. D. Amstutz*, J. Amer. chem. Soc., 1950, **72,** 2195). 3-Chloromethylfurans with secondary amines give 3-dialkylaminomethylfurans (*Novitskii et al.*, Khim. Geterotsikl. Soedin., 1966, 835).

2-**Furylcarbinol** (*furfuryl alcohol; 2-hydroxymethylfuran)*, b.p. 170°, which occurs in roasted coffee oil, is obtained from furfural by hydrogenation over copper chromite (*G. Calingaert* and *G. Edgar*, Ind. Eng. Chem., 1934, **26,** 878), by Cannizzaro reaction (*W. C. Wilson*, Org. Synth., Coll. Vol. 1, 1932, p. 270) or by "crossed" Cannizzaro reaction with formaldehyde (*A. M. Berkenheim* and *T. F. Dankova*, J. gen. Chem., U.S.S.R., 1939, **9,** 924). The alcohol is miscible with water and with organic solvents except paraffins, and with acids is easily resinified or undergoes ring opening to laevulic acid (p. 96). The fluid resin passes into a hard polymer and furfuryl alcohol condensates find application in corrosion-resisting cements. *Furfuryl methyl ether*, b.p. 134–136°; *difurfuryl ether*, b.p. 88–89°/1 mm (*W. R. Kirner*, J. Amer. chem. Soc., 1928, **50,** 1955). *Furfuryl acetate*, b.p. 69–70°/7 mm (*The Miner Laboratories*, Org. Synth., 1927, **7,** 44); for other esters of furfuryl alcohol see *Zanetti*, J. Amer. chem. Soc., 1925, **47,** 535.

Sodium hydrogen sulphide converts furfural into **difurfuryl disulphide,** which can be reduced further to **furfuranethiol,** b.p. 155° (*H. Gilman* and *A. P. Hewlett*, *ibid.*, 1930, **52,** 2141). Epichlorohydrin reacts with lithium or sodium salts of ethylthioethyne in liquid ammonia to give ethyl furfuryl sulphide (*G. Vollema* and *J. F. Arens*, Rec. Trav. chim., 1959, **78,** 140):

CH_2Cl (epoxide) + LiC:CSEt ⟶ $HOCH_2 \cdot CH:CH \cdot C:CSEt$ ⟶ (furan)CH_2SEt

Furyl carbinols have been prepared by the Grignard reaction from furfural or ethyl furoate.

2-**Furylmethylcarbinol,** b.p. 162–163°, $[\alpha]_{5893}$ −17.0° (EtOH) (*D. I. Duveen* and *J. Kenyon*, J. chem. Soc., 1936, 621); 2-**furyldimethylcarbinol,** b.p. 62°/11 mm (*T. Reichstein et al.*, Helv., 1932, **15,** 1118); 2-**furylphenylcarbinol,** b.p. 138–142°/10 mm (*R. Paul*, Compt. rend., 1936, **202,** 1444); 2-**furyldiphenylcarbinol,** m.p. 91° (*S. H. Mahood* and *H. F. Aldrich*, J. Amer. chem. Soc., 1930, **52,** 4477).

Bouveault–Blanc reduction of ethyl 2-furylacetate gives *β*-2-**furylethanol,** $C_4H_3O \cdot CH_2 \cdot CH_2OH$, b.p. 86–88°/21 mm (*Amstutz* and *J. Plucker*, *ibid.*, 1941, **63,** 206).

3-**Furylcarbinol,** b.p. 79–80°/17 mm, is obtained by the reduction of the 3-carboxylic acid with lithium tetrahydridoaluminate (*Sherman* and *Amstutz*, *loc. cit.*).

3-**Furyldimethylcarbinol,** b.p. 53–58°/8 mm, n_D^{25} 1.4718, is prepared either by the Grignard reaction between methyl 3-furoate and methylmagnesium iodide, or from 3-furyl-lithium and acetone (*J. T. Wrobel* and *K. Galuszko*, Rocz. Chem., 1966, **40,** 1005; C.A., 1967, **66,** 18630c).

The following formula has been assigned to **aucubin,** the glycoside from *Aucuba japonica* and other plants (*P. Karrer* and *H.Schmid*, Helv., 1946, **29,** 525):

HO
HO
Aucubin

Furfurylamine (*2-aminomethylfuran*), b.p. 145°, is soluble in water and is obtained by reducing furfuraldoxime over Raney nickel at ordinary pressure (*Paul*, Bull. Soc. chim. Fr., 1937, **4,** 1121), or by similar reduction of furfural in presence of ammonia under 150 atm. (*E. J. Schwoegler* and *H. Adkins*, J. Amer. chem. Soc., 1939, **61,** 3499).

Secondary amines of the type $(C_4H_3O)CHRNHR'$ are produced from furfurylideneamines and Grignard reagents (*B. L Emling et al.*, *ibid.*, 1949, **71,** 703).

Di-2-furfurylamine, b.p. 135–142°/15 mm. **Dimethylfurfurylamine,** b.p. 142–145° (*methiodide*, m.p. 117–118°) is prepared by heating furfural with dimethylformamide (*E. A. Weilmuenster* and *C. N. Jordan*, *ibid.*, 1945, **67,** 415). Secondary and tertiary furfurylamines can also be made by direct aminomethylation of furans (*R. F. Holdren* and *R. M. Hixon*, *ibid.*, 1946, **68,** 1198; *E. L. Eliel* and *P. E. Peckham*, *ibid.*, 1950, **72,** 1209).

$$\text{Me(furan)} + CH_2O + HNMe_2 \longrightarrow \text{Me(furan)}CH_2NMe_2$$

5-Aminomethylfurfuryl alcohol on treatment with acid is converted into 5-hydroxy-2-methylpyridine (*N. Elming* and *N. Clauson-Kaas*, Acta chem. Scand., 1956, **10,** 1603; 1664):

$$NH_2CH_2\text{(furan)}CH_2OH \xrightarrow{H^{\oplus}} \text{HO(pyridine)Me}$$

(xi) Aldehydes

While furfural and some others are readily accessible from carbohydrates, it is possible to insert the aldehyde group directly into the furan ring by the Gatterman synthesis (*T. Reichstein*, Helv., 1930, **13,** 345) or to reduce furancarboxylic acids by the Sonn–Müller or the Rosenmund method. Furans with a free α-position are readily formylated by dimethylformamide and phosphoryl chloride (*V. J. Traynelis, J. J. Miskel* and *J. R. Sowa*, J. org. Chem., 1957, **22,** 1269).

Furan-2-carbaldehyde, furfural, was first obtained by Döbereiner in 1832 when he treated starch or sugar with sulphuric acid and manganese dioxide. Pentoses afford it in substantial quantities on treatment with acids, and it is manufactured on the large scale from oat hulls and corn cobs, which are rich in pentosans. Furfural is formed during the thermolysis of pentosans (*A. N. Kislitsyn, A. V. Sokolov* and *E. V. Fomina*, Zh. Prikl. Khim., 1968, **4,** 222; C.A., 1968, **69,** 11555k). Furfural is a colourless liquid, m.p. −36°, b.p. 161.7°, d_4^{20} 1.1598, which turns brown in air; it dissolves in 12 parts of cold water and mixes with organic solvents except paraffins. It shows the normal properties of an aromatic aldehyde: condenses in the presence of potassium cyanide to yield furoin $(C_4H_3O)CHOH{\cdot}CO(C_4H_3O)$, gives Cannizzaro's reaction, reacts with ammonia to give furfuramide (hydrofurfuramide), $(C_4H_3O)CH(N{:}CH{\cdot}C_4H_3O)_2$, gives Perkin's reaction yielding furylacrylic acid (XXXVI), and with dimethylaniline affords an analogue of malachite green. Furylacrylic acid (XXXVI) is oxidised by bromine water to "furonaldehyde" (XXXVII), or hydrolysed by hydrochloric acid to ketopimelic acid (XXXVIII), both of which have been converted into *n*-pimelic acid (XXXIX), showing that furylacrylic acid has a straight chain of seven carbon atoms and that furfural is a 2-substituted furan:

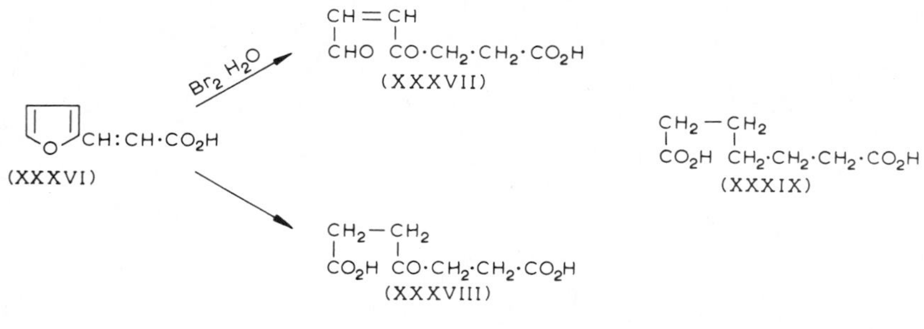

Br_2, H_2O → CH=CH(CHO)·CO·CH_2·CH_2·CO_2H (XXXVII)

(XXXVI) CH:CH·CO_2H

→ CH_2—CH_2(CO_2H)·CO·CH_2·CH_2·CO_2H (XXXVIII)

CH_2—CH_2(CO_2H)·CH_2·CH_2·CH_2·CO_2H (XXXIX)

The autoxidation of furfural yields β-formylacrylic acid and γ-formyl-α-oxobutenoic acid (*D. A. Isacescu et al.*, Bull. Soc. chim. Fr., 1967, 2171).

When furfural reacts with a primary arylamine in acid solution, the ring is opened, giving a reddish-violet polyene salt, which is transformed on warming with acid into a 3-hydroxypyridinium salt (XL) (*T. Zincke* and *G. Mühlhausen*, Ber., 1905, **38,** 3824; *W. M. Foley et al.*, J. Amer. chem. Soc., 1952, **74,** 5489). Treatment of the polyene salts with alkali gives colourless bases formulated as **XLI**, which are also obtained from furfural and arylamines in the absence of acids (*J. C. McGowan*, J. chem. Soc., 1954, 4032). The higher vinylogues, furylacraldehyde and furylpentadienal, react analogously (*W. König*, Ber., 1934, **67,** 1274). The absorption spectra of the polyene salts have been carefully studied.

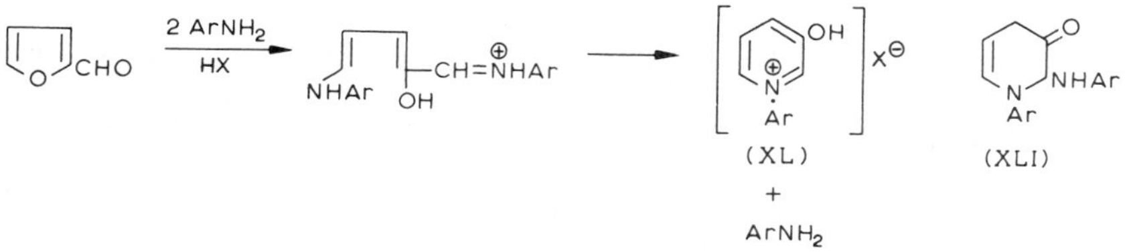

CHO —(2 $ArNH_2$, HX)→ NHAr, OH, CH=$\overset{\oplus}{N}$HAr → [OH, $\overset{\oplus}{N}$, Ar] $X^{\ominus}$ (XL) + $ArNH_2$; O, N, NHAr, Ar (XLI)

The resin from the reaction between aniline and furfural in the presence of a small amount of hydrogen chloride yields a 1:1 adduct, a 1:3 adduct considered to be 5-anilinomethyl-1-phenyl-3-phenylimino-2*H*-pyrroline (XLII), and a 1:2 adduct of aniline:furfural, similar to the product XLIII obtained by condensation of furoin with aniline (*M. S. Barvinck et al.*, Zh. org. Khim., 1967, **3,** 1107):

PhNH·CH_2, NPh, N, Ph (XLII) CH(OH)·C(:NPh) (XLIII)

Furfural does not react with maleic anhydride, but the diacetate does and furfural acetals condense with maleic anhydride and dimethyl acetylenedicarboxylates (*L. M. Gomes*, Bull. Soc. chim. Fr., 1967, 1753). Furfural reacts as a dienophile *e.g.*, with two moles of butadiene:

$$\text{C}_4\text{H}_3\text{O}\cdot\text{CHO} + 2\ \text{CH}_2{:}\text{CH}\cdot\text{CH}{:}\text{CH}_2 \longrightarrow \text{(CHO adduct)}$$

On treatment with isopropyl chloride and aluminium chloride furfural unexpectedly gives 4-isopropylfurfural and none of the 5-isomer (*H. Gilman, N. O. Calloway* and *R. R. Burtner*, J. Amer. chem. Soc., 1935, **57,** 906).

The mechanism of the oxidation of furfural and furfuryl alcohol with bromine and bromine chloride in aqueous solution has been studied (*M. Szakacs-Pinter* and *L. Maros*, Acta Chim. Acad. Sci. Hung., 1968, **56,** 199; C.A., 1968, **69,** 51274z).

Furfural finds application as a solvent, notably in the selective extraction of unsaturates from petroleum fractions or from vegetable oils, or of aromatics from lubricants and Diesel oils. It is a source of resins when condensed with formaldehyde either alone or with phenol in the manufacture of bakelite.

The quantitative *determination of furfural* has attracted attention since it permits in turn that of pentoses and/or uronic acids which yield it on treatment with acid. The aldehyde has been weighed as its insoluble compounds with phloroglucinol or thiobarbituric acid, or less specifically with dinitrophenylhydrazine, and determined colorimetrically by its reaction with aniline acetate (*R. A. Stillings* and *B. L. Browning*, Ind. Eng. Chem., anal. Ed., 1940, **12,** 499). Methyl- and hydroxymethyl-furfuraldehyde do not interfere with the last method.

The *furfural oximes, anti-*(α-), m.p. 75–76°, *syn-*(β), 91–92°; *phenylhydrazone*, m.p. 97–98°; 2,4-*dinitrophenylhydrazone*, m.p. 212–214° (yellow form), and 230° (red form) (229°, decomp. corr.) (*H. Bredereck*, Ber., 1932, **65,** 1833); *azine*, m.p. 113°; *semicarbazone* recorded m.p. 190–192°, 202–203°, and 214–215°; *diacetate*, m.p. 52–53° (*R. T. Bertz*, Org. Synth., 1953, **33,** 39); *diethyl acetal*, b.p. 189–191°. *Furfurylidenenitromethane* (2-*furylnitroethene*), m.p. 78°; *furfuramide ("hydrofurfuramide")*, m.p. 117° (*W. N. Hartley* and *J. J. Dobbie*, J. chem. Soc., 1898, **73,** 598).

2,3-Dibromofuran on treatment with butyl-lithium at −30° and subsequent boiling in dimethylformamide gives 3-**bromofurfural,** b.p. 60°/0.7 mm (*M. Robba, M. C. Zaluski* and *B. Roques*, C.r. Acad. Sci., Paris, Ser. C, 1967, **264,** 413; C.A., 1967, **67,** 32662s). 5-**Nitrofurfural,** m.p. 35–36° (*Gilman* and *G. F. Wright*, J. Amer. chem. Soc., 1930, **52,** 2550; *D. Nardi et al.*, J. med. Chem., 1967, **10,** 530), and its derivatives, notably the *semicarbazone*, have powerful antibacterial properties.

Successive treatment of sucrose with hydrogen chloride and stannous chloride gives 5-**methylfurfural,** b.p. 83–84°/15 mm, *phenylhydrazone*, m.p. 147–148° (*I. J. Rinkes*, Org. Synth., 1934, **14,** 62); it is also formed by distilling methylpentoses with acid. When sucrose is heated under pressure with aqueous oxalic acid, it affords, in good yield, 5-**hydroxymethylfurfural,** m.p. 31.5°, b.p. 110°/0.02 mm, *semicarbazone*, m.p. 192° (194–195°), also obtained in high yield when sucrose is heated at 100° in dimethylformamide in the presence of catalytic amounts of iodine (*T. G. Bonner, E. J. Bourne* and *M. Ruskiewicz*, J. chem. Soc., 1960, 787). This is derived only from the fructose moiety; but glucose previously treated with alkali affords some hydroxymethylfurfural. Hydroxymethylfurfural can be dehydrated to the *ether*, $O(CH_2{\cdot}C_4H_2O{\cdot}CHO)_2$, m.p. 112° (*J. Kiermayer*, Chem. Ztg., 1895, **19,** 1003).

With dry hydrogen chloride sucrose gives 5-**chloromethylfurfural,** m.p. 37–38° (*W. H. Haworth* and *W. G. M. Jones*, J. chem. Soc., 1944, 667).

Furan-3-**carbaldehyde,** (3-**furaldehyde**), b.p. 144°/732 mm, is obtained from 3-furoic acid by the Rosenmund method; *phenylhydrazone*, m.p. 149.5°, *diacetate*, m.p. 50° (*Gilman* and *Burtner*, J. Amer. chem. Soc., 1933, **55,** 2903); and 2,4-**dimethylfuran**-3-**carbaldehyde,** b.p. 73°/11 mm, is obtained from the corresponding acid by the Sonn–Müller process (*T. Reichstein* and *H. Zschokke*, Helv., 1932, **15,** 1105). 5-**Nitrofuran**-3-**carbaldehyde,** m.p. 76°, *diacetate*, m.p. 87° (*Gilman* and *Burtner*, *loc. cit.*).

2-**Furylacetaldehyde,** b.p. 58°/10 mm, results from the decomposition of the glycidic ester,

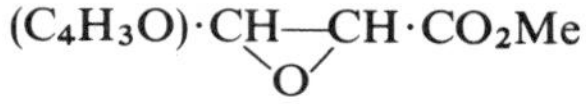

from furfural and methyl chloroacetate (*Reichstein*, Ber., 1930, **63,** 749); and 2-**furylacraldehyde,** m.p. 54°, is prepared from furfural and acetaldehyde (*A. Hinz et al., ibid.*, 1943, **76,** 676). For polyene aldehydes see p. 140.

(*xii*) *Ketones*

(*1*) *Preparation.* 2-Furyl ketones can be obtained by the Friedel–Crafts reaction from furans and acyl chlorides (*Gilman* and *Calloway*, J. Amer. chem. Soc., 1933, **55,** 4197), from 2-furoyl chloride and aromatic compounds and by treating 2-furonitrile with a Grignard reagent (*W. Borsche et al.*, Ber., 1938, **71,** 957) or from 2-furodimethylamide with Grignard reagents (*N. N. Maxim*, C.A., 1931, **25,** 513). In the catalysed acetylation of furan and thiophene with acetic anhydride in dichloromethane, it has been shown that furan is more reactive than thiophene by a factor 9.3 or 11.9 according to the catalyst used (iodine or stannic chloride) (*P. Linda* and *G. Marino*, Tetrahedron, 1967, 1739).

3-Furyl ketones can be prepared from 3-furoyl chloride and Grignard reagents (*P. Grunanger* and *A. Mantegani*, Gazz., 1959, **89,** 913). The reaction between chloroacetaldehyde and the sodium salt of hydroxymethyleneacetone affords a mixture containing 2-methylfuran-3-carbaldehyde and 3-acetylfuran (*M. Valenta*, Coll. Czech. chem. Comm., 1967, **32,** 897).

(*2*) *Reactions.* Furyl ketones are converted by arylamines into 3-hydroxy-1-arylpyridinium salts in the same way as furfural (*Borsche et al., loc. cit.*):

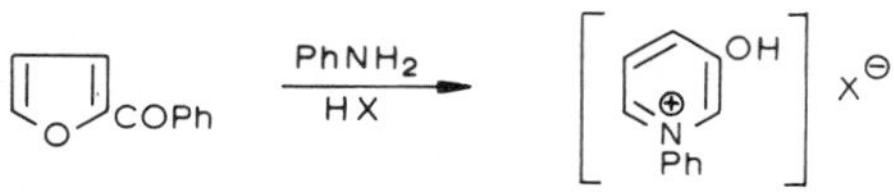

With ammonia and ammonium chloride the ketones yield tertiary 3-hydroxypyridines (*H. Leditschke*, Ber., 1952, **85,** 202). *N. Elming* and *N. Clauson-Kaas* have applied this reaction to the synthesis of pyridoxine (see formula on p. 127). The enamines of 2-acetylfuran on distillation are transformed into *o*-aminophenols (*L. Birkofer* and *G. Daum*, Angew. Chem., 1960, **72,** 707):

$$\text{2-furyl-COMe} \xrightarrow[\text{AcOH}]{\text{NHR}_2} \text{2-furyl-C(NR)=CH}_2 \xrightarrow{\Delta} \text{C}_6\text{H}_4(\text{OH})\text{NR}_2$$

Mixtures of 2-acylfurans, secondary amines, benzene and acetic acid on azeotropic distillation yield the crude enamines (*idem*, Ber., 1962, **95,** 183).

Hydrogenation of 5-acyl-2-alkyl furans over 10% platinum/carbon catalyst at atmospheric pressure and 300–310° gives a mixture of mainly phenols and ketones, *e.g.*, 5-acetyl-2-ethylfuran affords 60% of a mixture of 3-ethyl- and 2,3-dimethyl-phenol along with 30% of 2- and 3-octanones, and traces of ethylbenzene and *o*-xylene (*N. I. Shuĭkin, I. F. Bel'skiĭ* and *G. K. Vasilevskaya*, Doklady Akad. Nauk, S.S.S.R., 1960, **132,** 861). Hydrogenation of 5-acyl-2-alkylfurans over platinum gives exclusively the products of ring-opening at the C–O bond proximate to the carboxyl group; the resulting 1,5-diketones cyclise directly to mono- and di-substituted homologues of cyclohexanone or phenols. At 200–230°, phenols are almost totally absent and only the cyclohexanone and cyclohexanol derivatives are obtained. Under favourable temperature conditions (200–220°) the 1,5-diketones can be converted into diols, which cyclise to tetrahydropyrans (*idem, ibid.*, 1959, **127,** 359; 1961, **136,** 591).

2-**Acetylfuran,** m.p. 30–32° (*Reichstein*, Helv., 1930, **13,** 356; *H. D. Hartough* and *A. I. Kosak*, J. Amer. chem. Soc., 1947, **69,** 1012); *oxime*, m.p. 104°; *semicarbazone*, m.p. 150°; p-*nitrophenylhydrazone*, m.p. 185–186°. 2-*Chloroacetylfuran*, m.p. 30.5° (*E. B. Knott*, J. chem. Soc., 1947, 1656). 2-**Benzoylfuran,** b.p. 284° (*Hartough* and *Kosak*, J. Amer. chem. Soc., 1947, **69,** 3095, 3098); 2-**cinnamoylfuran,** m.p. 89°, 45°, 41° (trimorphic) (*C. Weygand* and *F. Strobelt*, Ber., 1935, **68,** 1839). 2-**Furylideneacetone,** m.p. 39–40° (*G. L. Leuck* and *L. Cejka*, Org. Synth., 1927, **7,** 42). 3-**Acetylfuran,** b.p. 84°/21 mm, m.p. 53–54°; *semicarbazone*, m.p. 161–162°; p-*nitrophenylhydrazone*, m.p. 215° (*Gilman, B. L. Woolley* and *Wright*, J. Amer. chem. Soc., 1933, **55,** 2609; *Grunanger* and *Mantegani, loc. cit.*).

Ethyl 3-**furyl ketone,** b.p. 48–49°/2.5 mm (*Grunanger* and *Mantegani, loc. cit.*); 3-**acetyl**-2,5-**dimethylfuran,** b.p. 95–96°/18 mm (*Gilman* and *Calloway, loc. cit.*).

Oxidation of 2-acetylfuran with selenium dioxide gives 2-**furylglyoxal hydrate,** m.p. 69° (*F. Kipnis* and *J. Ornfelt*, J. Amer. chem. Soc., 1948, **70,** 3948). When sucrose is treated with oxalic acid, the resulting hydroxymethylfurfural is accompanied by a little of the isomeric 2-*hydroxyacetylfuran*, m.p. 83–85° (*R. E. Miller* and *S. M. Cantor*, *ibid.*, 1952, **74,** 5236).

Perilla ketone (see Vol. II B, p. 171) has been obtained from furan-2-carbonyl chloride and diisopentylcadmium (*T. Matsuura*, Bull. chem. Soc. Japan, 1957, **30,** 430).

The ethereal oil of *Elsholtzia cristata* affords **elsholtzione,** *isobutyl 3-methyl-2-furyl ketone*, b.p. 210°, (Vol. II B, pp. 170–171), which has been degraded successively to 3-methylfuran-2-carboxylic acid (elsholtzic acid) and to furan-2,3-dicarboxylic acid, and synthesised (*Reichstein*, Helv., 1931, **14,** 1277):

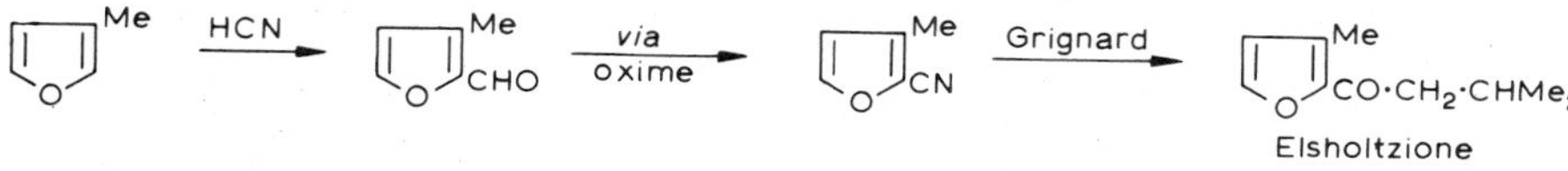

Me HCN → Me CHO *via* oxime → Me CN Grignard → Me CO·CH_2·$CHMe_2$

Elsholtzione

Black-rotted sweet potato affords a volatile diketone, **ipomeanine,** b.p. 70–75°/0.003 mm, identified by synthesis as 3-laevuloylfuran, $(C_4H_3O)CO{\cdot}CH_2{\cdot}CH_2{\cdot}COMe$, and **ipomeamarone,** b.p. 131°/3 mm, which appears to be stereoisomeric with **ngaione** from *Eremophila latrobei* (see Vol. II C, p. 290). Degradation of the two substances gives 3-furoic acid with pyruvic and isovaleric acids, and they are formulated as XLIV (*T. Kubota et al.*, Chem. and Ind., 1954, 902).

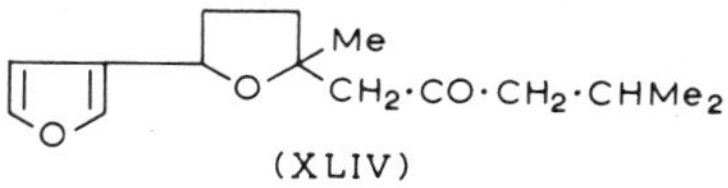

Me CH_2·CO·CH_2·$CHMe_2$

(XLIV)

(xiii) Carboxylic acids

(*1*) *Preparation.* Furancarboxylic acids, which are listed in Table 4 (p. 122) are occasionally prepared by oxidising other furan derivatives or by carbonation of furan organo-metallic compounds, *e.g.*, 2,3-dibromofuran on treatment with butyl-lithium at −20° followed by carbonation yields 3-bromofuran-2-carboxylic acid (*Robba*, *Zaluski* and *Roques*, *loc. cit.*):

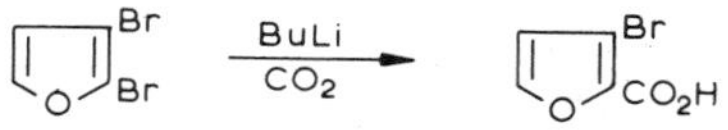

Br Br BuLi CO_2 → Br CO_2H

but more often by ring-synthesis, using the Paal–Knorr (p. 84) or the Feist–Benary method (p. 85). The 2-mono- and the 2,5-di-carboxylic acids, pyromucic and dehydromucic acid, respectively, are readily obtained by thermal or acid dehydration of mucic acid or its stereoisomers:

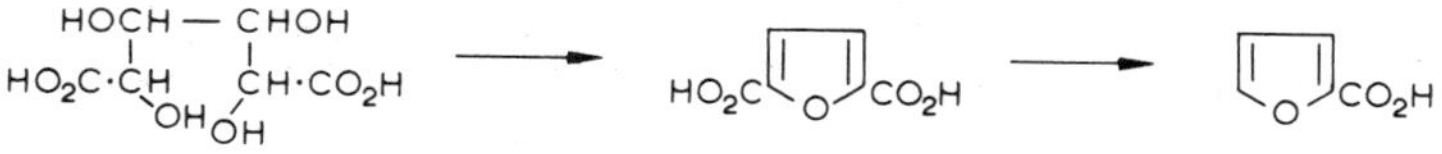

HOCH — CHOH / HO_2C·CH CH·CO_2H / OH OH → HO_2C CO_2H → CO_2H

A curious reaction, not limited to a single case, leads to *methronic acid* (XLVII; R = H), m.p. 204°, and its methyl derivative (XLVII; R = Me) (*T. Reichstein* and *A. Grüssner*, Helv., 1933, **16**, 6). In the former preparation succinic anhydride is condensed with ethyl acetoacetate to give the enol lactone (XLV; R = H), converted by alkali into XLVI (R = H), *via* shift of the double bond followed by internal Michael addition, and thence, by cleavage of the lactone ring, dehydration and hydrolysis into methronic acid. The mechanism accounts for the fact that αα-dimethylsuccinic acid does not undergo the reaction, no double-bond shift being possible in the structure analogous to XLVI (*O. Červinka*, Chem. and Ind., 1961, 472):

$EtO_2C{\cdot}CH_2$ / MeCO + $CH_2{\cdot}CO$ / $RCH{\cdot}CO$ >O $\xrightarrow{Ac_2O}$ (XLV)

$CH_2{\cdot}CO_2Na$ / $RCH{\cdot}CO_2Na$ $\xrightarrow{Ac_2O}$ succinic anhydride

(XLV) → (XLVI) → (XLVII)

Methronic acid (XLVII; R = H) gives on heating the long-known *pyrotritaric acid* (*uvic acid*) (XLVIII), which *J. Wislicenus* first obtained in 1868 by distilling tartaric acid and which is produced in serviceable yield by heating pyruvic acid with acetic anhydride and sodium acetate.

Its constitution is established by synthesis from acetonylacetoacetic ester (*H. B. Stevenson* and *J. R. Johnson*, J. Amer. chem. Soc., 1937, **59**, 2525); diethyl diacetosuccinate similarly gives *carbopyrotritaric acid* (XLIX), which is easily decarboxylated to uvic acid:

$CH_2{-}CH{\cdot}CO_2Et$ / MeCO COMe → (XLVIII) ← XLVII

$EtO_2C{\cdot}CH{-}CH{\cdot}CO_2Et$ / MeCO COMe → (XLIX)

(2) *General properties and reactions.* Furan-2-carboxylic acid is stronger than its isomer, and in general the α-carboxylic acids are more easily decarboxylated than the β (*Reichstein* and *H. Zschokke*, Helv., 1932, **15,** 268). Likewise in mercuration, only α-carboxyls are replaced by the chloromercuri group (*H. Gilman* and *G. F. Wright*, J. Amer. chem. Soc., 1933, **55,** 3302). The kinetics of bromination of methyl furan-2-carboxylate in acetic acid shows exactly the same characteristics as the bromination of benzene derivatives. The reaction is first order in bromine and in substrate when it is carried out in the presence of a large excess of lithium bromide (*P. Linda* and *G. Marino*, Chem. Comm., 1967, 499).

Methyl 5-methylfuran-2-carboxylate on treatment with nitric acid and acetic anhydride unexpectedly yields methyl furan-5-carbaldehyde-2-carboxylate (*S. Šrogyl* and *J. Peterek*, Coll. Czech. chem. Comm., 1966, **31,** 1578):

Me–C₄H₂O–CO_2Me ⟶ OHC–C₄H₂O–CO_2Me

also obtained by the hydrolysis with dilute sulphuric acid of the nitrone obtained by the condensation of methyl 5-chloromethylfuran-2-carboxylate with 4-nitrosodimethylaniline in alkaline medium (*M. Valenta*, Sb. Vys. Sk. Chem. Technol. Praze, Org. Technol., 1966, **8,** 103; C.A., 1967, **67,** 43607d).

Furan-2-carboxylic acid (*2-furoic acid, pyromucic acid*), first obtained from mucic acid by *Houtou–Labillardière* in 1818, is prepared from furfural by the Cannizzaro reaction or by oxidation (*C. D. Hurd et al.*, J. Amer. chem. Soc., 1933, **55,** 1082). The acid, m.p. 133–134°, b.p. 230–232°, dissolves in 26 parts of cold water, and has K_c $6.8 \cdot 10^{-4}$. With halogens it undergoes addition, substitution, and fission to derivatives of maleic semialdehyde (*cf.* p. 102). *Methyl ester*, b.p. 181°; *ethyl ester*, m.p. 34°; p-*phenylphenacyl ester*, m.p. 110–111°; *anhydride*, m.p. 73°; *chloride*, b.p. 173°; *amide*, m.p. 142°; *anilide*, m.p. 123°. *Furoylglycine*, m.p. 165°, is excreted by dogs fed on furfural (*M. Jaffe* and *R. Cohn*, Ber., 1887, **20,** 2311).

Thiofuroic acid, b.p. 101–103°/16 mm, is obtained from furoyl chloride and sodium hydrogen sulphide (*S. Patton*, J. Amer. chem. Soc., 1949, **71,** 3571).

By a surprising reaction methyl furoate reacts with benzene in presence of aluminium chloride to give a substantial yield of α-naphthoic acid (*C. C. Price* and *C. F. Huber*, *ibid.*, 1942, **64,** 2137):

C_6H_6 + furan-2-CO_2Me ⟶ naphthalene-1-CO_2H

The sensitiveness of the furan ring to oxidation is used in a synthesis of tri-substituted acetic acids (*Reichstein et al.*, Helv., 1935, **18,** 721):

Under controlled conditions and keeping the temperature $\langle 10°$ the main products from the reaction of methyl and ethyl 2-furoates with acetic anhydride and stannic chloride are methyl and ethyl 5-acetoacetyl-2-furoates (*R. Ercoli et al.*, J. org. Chem., 1967, **32,** 2917).

Furantetracarboxylic acid is a surprisingly strong acid ($pk_1 = 1.37$) forming a series of monosalts (*W. Cocker et al.*, Tetrahedron, 1959, **7,** 299). It is prepared by the successive action of bromine and sulphuric acid on diethyl sodio-oxaloacetate and is decarboxylated in turn to the 2,3,4-tri-, the 3,4-di-, and the 3-mono-carboxylic acids (*Reichstein et al.*, Helv., 1933, **16,** 276):

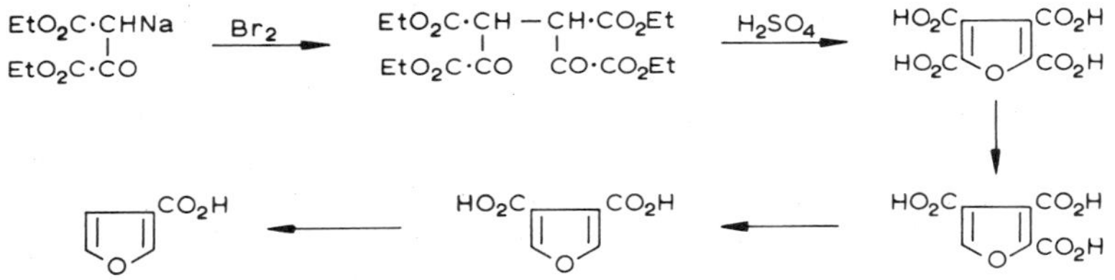

Furan 3,4-**dicarboxylic acid** can also be prepared by a striking application of the diene synthesis (*K. Adler* and *H. F. Rickert*, Ber., 1937, **70,** 1354):

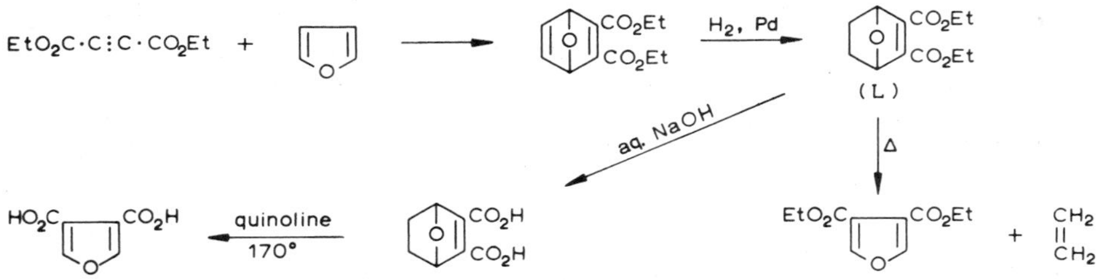

However dimethyl 3,6-expoxy-3,4,5,6-tetrahydrophthalate (dimethyl ester of L) on pyrolysis at 10 mm gives 3,6-epoxy-hexahydrobenzoic acid and 3,6-epoxy-hexahydrophthalic acid (*E. J. Forbes* and *M. K. A. Khan*, Pakistan J. Sci. Res., 1957, **10,** 223). The 3,4-dicarboxylic acid is probably best obtained by ring-synthesis (*1*) (p. 84) (*Jones* and *Kornfeld, loc. cit.*).

Oxidation of 7-hydroxybenzo[*b*]furan-2-carboxylic acid gives by destruction of the benzene nucleus **furan**-2,3,5-**tricarboxylic acid** (*Reichstein* and *A. Grüssner*, Helv., 1933,

16, 555), which when heated yields the 2,4-dicarboxylic acid, best prepared by synthetic method (*3*) (p. 85).

The last acid is in turn decarboxylated to **furan-3-carboxylic acid** (3-**furoic acid**) which occurs naturally in *Evonymus atropurpureus* (*Reichstein* and *Zschokke*, *loc. cit.*). The earliest known furanpolycarboxylic acid is **furan-2,5-dicarboxylic acid** (p. 87), *dehydromucic acid;* the remaining isomeride, the 2,3-compound, is decarboxylated to furan-3-carboxylic acid (*Reichstein* and *Zschokke*, *loc. cit*).

Bromination of methyl 4,5-dihydrofuran-3-carboxylate with *N*-bromosuccinimide and heating of the resulting product with 50% aqueous potassium hydroxide solution on acidification gives a mixture of methyl furan-3-carboxylate and furan-3-carboxylic acid in the ratio 2:1 (*J. T. Wrobel* and *K. Galuszko*, Roczniki Chem., 1966, **40,** 1005; C.A., 1967, **66,** 18630c).

The melting points of some furancarboxylic acids and their esters and amides are listed in Table 4.

2-**Furylacetic acid,** m.p. 62°, and β-2-furylalanine are obtained from furfural by the rhodanine method (*C. Gränacher*, Helv., 1922, **5,** 610; *E. D. Amstutz et al.*, J. Amer. chem. Soc., 1940, **62,** 1512; 1950, **72,** 2195). β-2-**Furylpropionic acid,** m.p. 58° (*A. W. Ingersoll*, Org. Synth., 1929, **9,** 45).The Perkin or the Knoevenagel reaction yields *trans-β*-**furylacrylic acid,** m.p. 141° (*S. Rajagopalan* and *P. V. A. Raman*, *ibid.*, 1945, **25,** 51) (*methyl ester*, m.p. 27°); *cis-β*-2-**furylacrylic acid,** m.p. 104°. Synthetic method (*2*) affords 4-**methylfuran-3-carboxylic-2-acetic acid,** m.p. 196° (*Reichstein* and *Zschokke*, Helv., 1932, **15,** 268). 3,4-**Dihydroxyfuran-2,5-dicarboxylic acid** (p. 88), m.p. 220°, and 5-**hydroxymethyl-2-furoic acid,** m.p. 166° (decomp.) (*Reichstein*, *ibid.*, 1926, **9,** 1066). 2-**Furylglyoxylic acid,** m.p. 98°, yields with sodium amalgam 2-**furylglycollic acid,** m.p. 117°, the analogue of mandelic acid (*E. Fischer* and *F. Brauns*, Ber., 1913, **46,** 892; *F. Nerdel et al.*, *ibid.*, 1954, **87,** 276). *Ethyl 2-furoylacetate*, b.p. 170°/32 mm (*G. Barger et al.*, J. chem. Soc., 1937, 718).

(*b*) *Dihydrofurans**

Two types of dihydrofuran are possible 2,3-(≡ 4,5)dihydrofurans (LI) and the 2,5-isomers (LII):

* *H. Normant* has provided a useful summary of the chemistry of reduced furans (Bull. Soc. chim. Fr., 1951, 115).

TABLE 4

FURANCARBOXYLIC ACIDS

Acid	*m.p.(°C)*	*Derivative*	*Ref.*
2-Carboxylic acid (2-furoic acid)	133–134	*Me ester*, b.p. 181°; *amide*, m.p. 143°	1
3-Bromo-2-carboxylic acid	163		2
4-Bromo-2-carboxylic acid	129	*Et ester*, m.p. 29°	3,4
5-Bromo-2-carboxylic acid	184	*Me ester*, m.p. 81°	1,3
3,5-Dibromo-2-carboxylic acid	168		5
4,5-Dibromo-2-carboxylic acid	171		6
3-Chloro-2-carboxylic acid	149	*Et ester*, m.p. 30°	7
4-Chloro-2-carboxylic acid	148–149		8
5-Chloro-2-carboxylic acid	179–180	*Et ester*, b.p. 216–218°	7,8
4,5-Dichloro-2-carboxylic acid	—	*Et ester*, b.p. 123–129°/ 20 mm	8
3-Nitro-2-carboxylic acid	125		9
5-Nitro-2-carboxylic acid	185		10
3-Methyl-2-carboxylic acid (elsholtzic acid)	134		see p. 117
4-Methyl-2-carboxylic acid	131–132		11
5-Methyl-2-carboxylic acid	109	*Me ester*, b.p. 98°/15 mm	12
3,5-Dimethyl-2-carboxylic acid	147	*Me ester*, b.p. 160–161	10
3-Carboxylic acid (3-furoic acid)	122	*Amide*, m.p.169°	13,14
5-Bromo-3-carboxylic acid	130		15
5-Nitro-3-carboxylic acid	138	*Et ester*, m.p. 56°	15
2-Methyl-3-carboxylic acid	102	*Et ester*, b.p. 85–87°/ 20 mm	16
4-Methyl-3-carboxylic acid	139		11
5-Methyl-3-carboxylic acid	119	*Et ester*, b.p. 69–71°/ 6 mm	17
2,4-Dimethyl-3-carboxylic acid	122	*Et ester*, b.p. 97°/10mm	18
2,5-Dimethyl-3-carboxylic acid (pyrotritaric acid)	135	*Me ester*, b.p. 198°	19
2,3-Dicarboxylic acid	225	*diMe ester*, m.p. 37°	20
2,4-Dicarboxylic acid	266	*diMe ester*, m.p. 109°	15
2,5-Dicarboxylic acid	320	*diMe ester*, m.p. 112°	
3,4-Dicarboxylic acid	217	*diMe ester*, m.p. 46°	21
2,5-Dimethyl-3,4-dicarboxylic acid (carbopyrotritaric acid)	230	*diMe* ester, m.p. 63–64°	19
2,3,4-Tricarboxylic acid	273 dec.	*triMe ester*, m.p. 108°	22
2,3,5-Tricarboxylic acid	259	*triMe ester*, m.p. 80°	23
2,3,4,5-Tetracarboxylic acid	247	*tetraMe ester*, m.p. 107°	22

References
1 *R. M. Whittaker*, Rec. Trav. chim., 1933, **52,** 352.
2 *M. Robba, M. C. Zaluski* and *B. Roques*, Compt. rend. Acad. Sci., Paris, Ser. C, 1967, **264,** 413.
3 *H. B. Hill* and *C. R. Sanger*, Ann., 1886, **232,** 58.
4 *R. J. van der Wal*, C.A., 1937, **31,** 2207.
5 *H. Gilman, A. M. Jannery* and *C. W. Bradley*, C.A., 1934, **28,** 763.
6 *L. M. Nazarova* and *Ya. K. Syrkin*, Zhur. obshcheĭ Khim. (J. Gen. Chem.), 1949, **19,** 777; C.A., 1950, **44,** 1092.
7 *A. F. Shepard, N. R. Winslow* and *J. R. Johnson*, J. Amer. chem. Soc., 1930, **52,** 2083.
8 *Merck*, Neth. P. Appl., 6,606,229/1966; C.A., 1967, **67,** P11490u.
9 *I. J. Rinkes*, Rec. Trav. chim., 1938, **57,** 392.
10 *B. T. Freuve* and *J. R. Jackson*, J. Amer. chem. Soc., 1931, **53,** 1142.
11 *T. Reichstein* and *H. Zschokke*, Helv., 1931, **14,** 1275.
12 *Rinkes*, Rec. Trav. chim., 1930, **49,** 1118.
13 *E. Sherman* and *E. D. Amstutz*, J. Amer. chem. Soc., 1950, **72,** 2195.
14 *A. Gosh* and *C. R. Raha*, J. Indian chem. Soc., 1954, **31,** 461.
15 *Gilman* and *R. R. Burtner*, J. Amer. chem. Soc., 1933, **55,** 2903.
16 *Gilman et al.*, Rec. Trav. chim., 1932, **51,** 407.
17 *Gilman, Burtner* and *E. W. Smith*, J. Amer. chem. Soc., 1933, **55,** 403.
18 *E. R. Alexander* and *S. Baldwin*, *ibid.*, 1951, **73,** 356.
19 *Gilman* and *Burtner*, Rec. Trav. chim., 1932, **51,** 667.
20 *R. G. Jones*, J. Amer. chem. Soc., 1955, **77,** 4069.
21 *O. Moldenhauer*, Ann., 1953, **580,** 169.
22 *Reichstein et al.*, Helv., 1933, **16,** 276.
23 *Reichstein* and *A. Grüssner*, *ibid.*, 1933, **16,** 555.

(i) 2,3-Dihydrofurans

2,3-**Dihydrofuran** (LI), b.p. 55°, is the internal enol ether of γ-hydroxybutaldehyde, $HOCH_2{\cdot}CH_2{\cdot}CH_2{\cdot}CHO$, into which it is quickly converted by acid hydrolysis. Conversely, γ-hydroxy-aldehydes or -ketones (tautomeric with 2-hydroxy-tetrahydrofurans), as well as 2-alkoxy- or 2-acyloxy-tetrahydrofurans, easily afford 2,3-dihydrofurans by loss of water, alcohol, or acid:

The same dihydrofuran is produced by pyrolysis of tetrahydrofurfuryl alcohol, tetrahydrofuronitrile, or methyl tetrahydrofuroate (*C. L. Wilson*, J. chem. Soc., 1945, **52,** 58), and also less directly from tetrahydrofuran (*H. Normant*, Compt. rend., 1949, **228,** 102; **229,** 1348). A very convenient route to 2,3-dihydrofurans is by the alkaline decomposition of the toluene-*p*-sulphonylhydrazones of tetrahydrofuran-3-ones (*M. A. Gianturco, P. Friedel* and *V. Flanagan*, Tetrahedron Letters, 1965, 1847).

Reactions. The adduct from 2,3-dihydrofuran and dichlorocarbene on heating alone affords 2,3-dichloro-5,6-dihydro-2*H*-pyran, but 2-chloropenta-2,4-diene-1-al if heated with an excess of quinoline (*J. C. Anderson, D. G. Linday* and *C. B. Reese*, Tetrahedron, 1964, **20,** 2091):

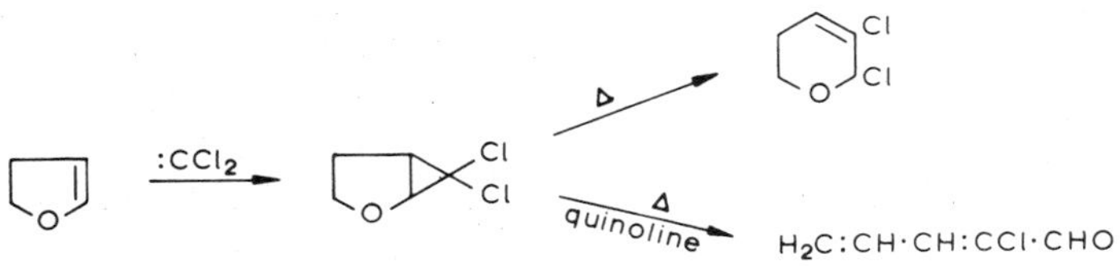

:CCl_2 → Δ; Δ quinoline → $H_2C{:}CH{\cdot}CH{:}CCl{\cdot}CHO$

2,3-Dihydrofuran suffers a curious thermal isomerisation to cyclopropanecarbaldehyde (*Wilson*, J. Amer. chem. Soc., 1947, **69,** 3002):

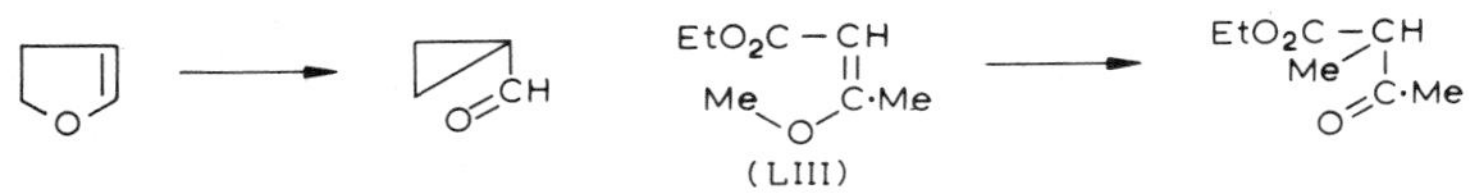

(LIII)

The change corresponds formally to the rearrangement of an enol ether LIII. The stereochemistry of the substituted cyclopropanes obtained following the irradiation of methyl-substituted 2-phenyl-2,3-dihydrofurans suggests that the ring-closure of the intermediate diradicals is sterically controlled; a mixture of *cis*- and *trans*-3-methyl-2-phenyl-2,3-dihydrofurans is isomerised to the cyclopropane carbaldehyde (LIV) (*P. Scribe, M. R. Monot* and *J. Wiemann*, Tetrahedron Letters, 1967, 5157):

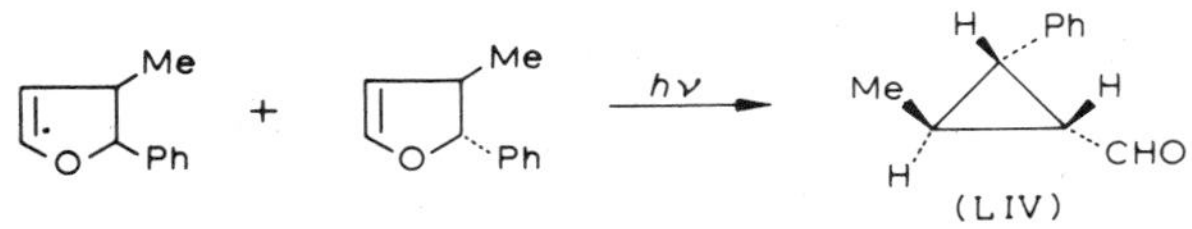

(LIV)

2,2,4,5-Tetraphenyl-2,3-dihydrofuran on photolysis isomerises to 1-benzoyl-1,2,2-triphenylcyclopropane, but with the 2,5-dihydro derivative LV phenyl migration occurs to give ketone LVI (*D. W. Boykin* and *R. E. Lutz*, J. Amer. chem. Soc., 1946, **86,** 5046):

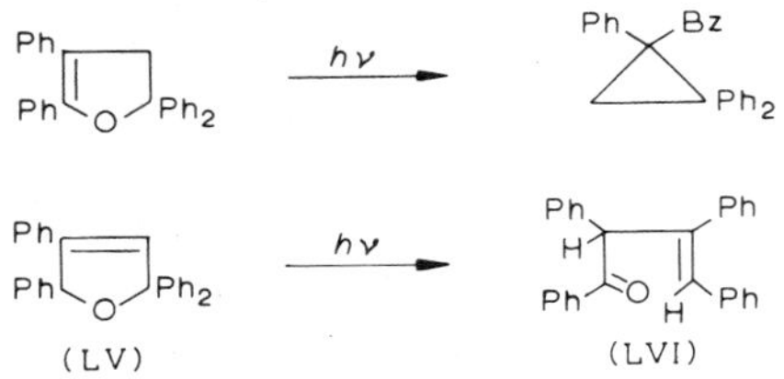

(LV) (LVI)

The thermal rearrangement of 2,5-dimethyl-2-vinyl-2,3-dihydrofuran gives 4-methyl-4-cycloheptenone (*S. J. Rhoads* and *C. F. Brandenburg*, *ibid.*, 1966, **88**, 4294):

2,2-**Dimethyl**-2,3-**dihydrofuran,** b.p. 78°, like other 2,3-dihydrofurans, is polymerised by boron fluoride (*D. A. Barr* and *J. B. Rose*, J. chem. Soc., 1954, 3766).

Butenolides (p. 107) are 2,3- and 2,5-dihydrofuran-2-ones.

When 2,3-dihydrofuran is treated with pentylsodium and carbonated, it affords 4,5-**dihydro**-2-**furoic acid,** m.p. 117°, which on ring-opening and re-closure gives lactone LVII (*R. Paul* and *S. Tchelitcheff*, Compt. rend., 1952, **235,** 1226):

(LVII)

The acid-catalysed hydrolysis of 2-methylenetetrahydrofuran, 2-methyl-4,5-dihydrofuran and 2,3-dihydrofuran has been kinetically studied in aqueous solutions containing different acid catalysts (*A. Kankaanpera*, *E. Taskinen* and *P. Salomaa*, Acta chem. Scand., 1967, **21,** 2487).

2,3-Dihydrofurans of the general formula (LVIII, R, R^1, and R^2=H or Me) have been obtained by the isomerisation of the corresponding 2,5-dihydrofurans at 60–170° in the presence of nickel, palladium or platinum (*W. Hofmann* and *H. Pasedach*, G.P. 1,248,669/1967):

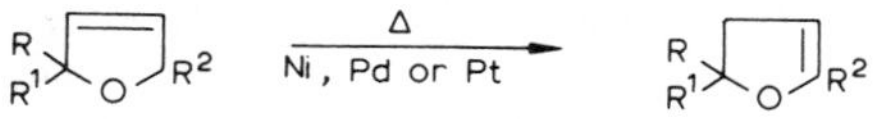

(LVIII)

Cross-linked polymers have been prepared from 2,3-dihydrofuran and an olefin in the presence of a cationic and/or Ziegler–Natta catalyst (*H. Naarmann* and *K. Schneider*, F.P. 1,443,778/1966).

(ii) 2,5-Dihydrofurans

Partial reduction of 2-butyne-1,4-diol, $HOCH_2{\cdot}C{\equiv}C{\cdot}CH_2OH$, readily accessible from formaldehyde and acetylene (Vol. I A, p. 453), gives 2-butene-1,4-diol, which is dehydrated by acids to 2,5-**dihydrofuran** (LII), b.p. 67° (*W. Reppe*, Ann., 1955, **596,** 80; *N. O. Brace*, J. Amer. chem. Soc., 1955, **77,** 4157).

A novel and general synthesis of 2,5-dihydrofurans, substituted in the 2 and/or 3 position is available from α-hydroxyketones using vinyltriphenylphosphonium bromide as the cyclisation reagent (*E. E. Schweiger* and *J. G. Liehr*, J. org. Chem., 1968, **33**, 583):

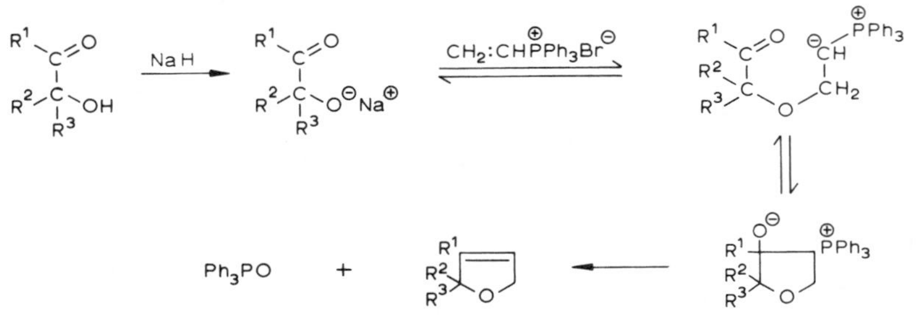

N. Clauson-Kaas and his associates have effected 2,5-addition to furans, leading to reactive products shown in Table 5. Thus by treatment with bromine in methanol in presence of alkali, or by electrolysis in methanolic solution, furan gives 2,5-dimethoxy-2,5-dihydrofuran (LIX); lead tetra-acetate, or bromine and potassium acetate in acetic acid, give 2,5-diacetoxy-2,5-dihydrofuran (LX). Both products are hydrolysed to malealdehyde, which is also produced from furan by oxidation with hydrogen peroxide and osmium tetraoxide. These reactions are generally applicable to furans not bearing "negative" substituents, and the substances can be hydrogenated smoothly to tetrahydrofurans:

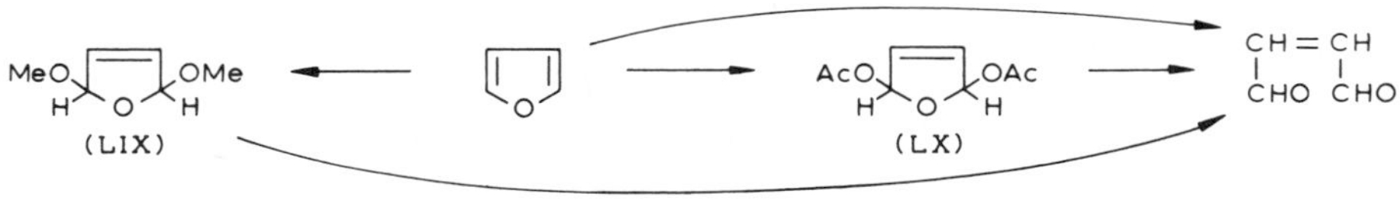

Both the di- and tetra-hydrofurans so obtained appear to be mixtures of *cis* and *trans* stereoisomerides, separated only in a few cases.

The products offer attractive synthetic possibilities. Thus the tetrahydro derivatives give pyrroles with ammonia or amines (*Clauson-Kaas* and *Elming*, Acta chem. Scand., 1952, **6**, 867):

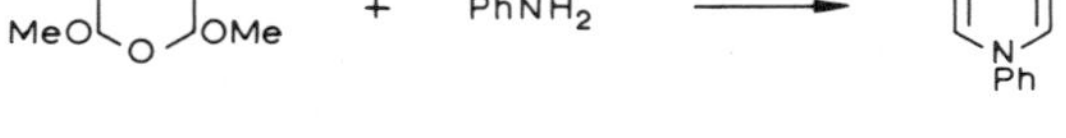

The formation of 3-hydroxypyridines from dihydro-derivatives having a side-chain amino-group has been exploited in an interesting synthesis of pyridoxine (*idem, ibid.*, 1955, **9**, 23):

Hydroxybenzoic acids, difficultly accessible otherwise, have also been produced (*Clauson-Kaas* and *P. Nedenskov*, *ibid.*, 1955, **9,** 27):

When furfural is submitted to a modified Kishner–Wolff reduction, the normal product, 2-methylfuran, is accompanied by an isomer, b.p. 77–79°, which appears to be 2-**methylene**-2,5-**dihydrofuran,** and is converted by acid

TABLE 5

SUBSTITUTED DI- AND TETRA-HYDROFURANS

Substituents	*B.p. (°C) of 2,5-dihydro*	*Ref.*	*B.p. (°C) of tetrahydro*	*Ref.*
2-OEt	35/13 mm	1		
2,5-$(OMe)_2$	47/8 mm	2	147–150	3
2,5-$(OEt)_2$	50–53/1 mm	4	30–31/1 mm	4
2,5-$(OAc)_2$	129–132/9 mm (m.p. 51–52)	3	125–128/10 mm	3
2,5-$(OMe)_2$:2-$CH(OMe)_2$	107–110/13 mm	5		
2,5-$(OMe)_2$:2-$CH(OAc)_2$	122–126/1 mm	6		
2,5-$(OMe)_2$:2-CO_2Me	121/13–14 mm	7	105–108/13 mm	7

References

1 *F. Quennehen* and *H. Normant*, Compt. rend., 1949, **228,** 1301.
2 *N. Clauson-Kaas et al.*, Acta chem. Scand., 1952, **6,** 531.
3 *Clauson-Kaas* and *N. Elming*, *ibid.*, 1950, **4,** 1233.
4 *J. Fakstorp et al.*, J. Amer. chem. Soc., 1950, **72,** 869.
5 *Clauson-Kaas et al.*, Acta chem. Scand., 1952, **6,** 545.
6 *Clauson-Kaas* and *J. Fakstorp*, *ibid.*, 1947, **1,** 415.
7 *Clauson-Kaas* and *F. Limborg*, *ibid.*, 1952, **6,** 551.

into 2-methylfuran (*N. Kishner*, J. gen. Chem., U.S.S.R., 1931, **1,** 1212; *H. L. Rice*, J. Amer. chem. Soc., 1952, **74,** 3193):

$$\text{2-furaldehyde (CHO)} \longrightarrow \text{2-methylene-2,3-dihydrofuran (}=CH_2\text{)} \longrightarrow \text{2-methylfuran (Me)}$$

The existence of the methylene compound as a distinct individual illustrates the limited "aromaticity" of the unreduced furan nucleus.

2,5-Dihydrofuran is less sensitive to acids than its isomeride, into which it is converted by treatment with *tert*-butyl alcoholic potash (*Paul et al.*, Bull. Soc. chim. Fr., 1950, 668), and it combines slowly with 1,3-dienes. Dehydration of the appropriate glycol similarly affords 2,5-**dimethyl**-2,5-**dihydrofuran,** b.p. 90–93° (*Reppe, loc. cit.*) and 2,2,5,5-**tetramethyl**-2,5-**dihydrofuran,** b.p. 102° (*J. Salkind*, Ber., 1923, **56,** 187) while 2,2,5,5-**tetraphenyldihydrofuran,** m.p. 190–191°, is produced directly from phenylmagnesium bromide and diethyl maleate (*T. Purdie* and *P. S. Arup*, J. chem. Soc., 1910, **97,** 1537). The unsymmetrical 2,2-**dimethyl**-2,5-**dihydrofuran** boils at 85° (*Barr* and *Rose, loc. cit.*).

(c) Tetrahydrofurans

Tetrahydrofurans are produced from 1,4-halogeno-alcohols by treatment with alkali, and from 3,4- or 4,5-olefinic alcohols with sulphuric acid. In the latter case and in the ring-closure of 4,5-dihalogeno-alcohols or of 1,2,5-triols, tetrahydrofurans are formed in preference to alternative rings. (*R. Paul*, Ann. Chim., 1932, **18,** 303; *Paul* and *H. Normant*, Bull. Soc. chim. Fr., 1944, **11,** 365). Dehydration of tetramethylene glycols, especially with phosphoric acid at a high temperature, affords tetrahydrofurans in good yield; many 1,4-glycols are easily accessible from acetylene (*G. F. Hennion* and *G. M. Wolf*, J. Amer. chem. Soc., 1940, **62,** 1368; *W. Reppe*, Ann., 1955, **596,** 25, 38, 80):

$$C_2H_2 + 2\,CH_2O \rightarrow HOCH_2{\cdot}C{:}C{\cdot}CH_2OH \rightarrow HOCH_2{\cdot}CH_2{\cdot}CH_2{\cdot}CH_2OH$$

Tetrahydrofurans may be synthesised by the following route (*W. B. Renfrow et al.*, J. org. Chem., 1961, **26,** 935):

$$Br(CH_2)_3OH + Me{-}\overset{Me}{\overset{|}{C}}{=}CH{\cdot}Me \xrightarrow{H_2SO_4} Br(CH_2)_3O{-}\underset{Me}{\overset{Me}{C}}{-}Et \xrightarrow[Et_2O]{Mg} BrMg(CH_2)_3OCMe_2{\cdot}Et$$

$$\xrightarrow{RCOR'} R{-}\underset{OH}{\overset{R'}{C}}{\cdot}(CH_2)_3OCMe_2{\cdot}Et \xrightarrow[p\text{-}Me{\cdot}C_6H_4{\cdot}SO_3H]{\Delta} \text{2-R-2-R'-tetrahydrofuran} + C_5H_{10}$$

Tetrahydrofurans have been alkylated by alkenes or unsaturated ethers in the presence of *tert*-butyl peroxide, the reaction taking place at the α-atom and the unsaturated C atom (*N. I. Shuĭkin* and *B. L. Lebedev*, Izv. Akad. Nauk, S.S.S.R., Ser. Khim., 1967, 639; *Shuĭkin*, *Lebedev* and *I. P. Yakoulev*, *ibid.*, 1967, 644). The acetone, acetophenone and benzophenone-initiated photochemical alkylation of tetrahydrofuran with olefins has been described (*T. Rosenthal* and *D. Elad*, Tetrahedron, 1967, 3193), as has also the photochemical decomposition of diazomethane in tetrahydrofuran, giving 2- and 3-**methyltetrahydrofuran;** no tetrahydropyran could be detected in the product (*W. von E. Doering*, *L. H. Knox* and *M. Jones*, J. org. Chem., 1959, **24,** 136).

For a review of the polymerisation of tetrahydrofurans see *H. Meerwein*, *D. Delfs* and *H. Morschel*, Angew. Chem., 1960, **72,** 927.

The furan nucleus can be reduced catalytically, and important solvents and intermediates are so produced on the large scale from furfural (*B. H. Wojcik*, Ind. Eng. Chem., 1948, **40,** 210; *O. W. Cass*, *ibid.*, p. 216; *Paul*, Bull. Soc. chim. Fr., 1956, 838). A copper–chromium catalyst often affects only the side-chain; the ring is effectively hydrogenated in presence of Raney nickel (*H. E. Burdick* and *H. Adkins*, J. Amer. chem. Soc., 1934, **56,** 438) or of iron-activated platinum oxide (*W. E. Kaufmann* and *R. Adams*, *ibid.*, 1923, **45,** 3029). Under some conditions the side-chain of furfural is lost. Some important transformations are shown below:

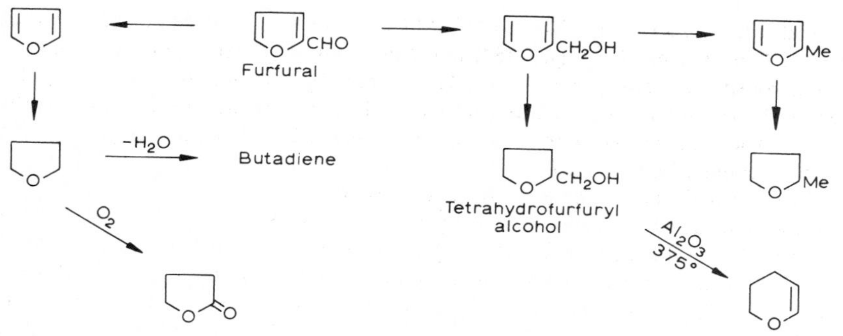

Furfural

$-H_2O$ Butadiene

O_2

Tetrahydrofurfuryl alcohol

Al_2O_3 375°

By selective use of catalysts and conditions, the following products, among others, have been obtained from β-2-furfurylacraldehyde:

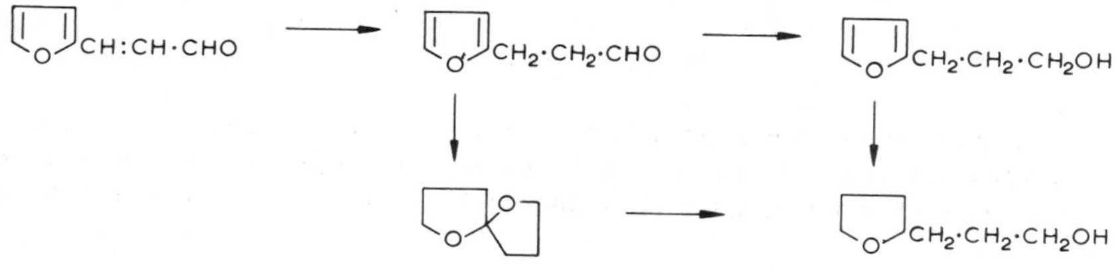

By reduction, often with magnesium and acetic acid, $\alpha\beta$-unsaturated aldehydes and ketones give derivatives of di- and tetra-hydrofuran. Thus mesityl oxide affords LXI and LXII (*M. Kolobielski*, Ann. Chim., 1955, **10,** 271):

Me$_2$, Me, Me, O, CH:CMe$_2$ (LXI) Me$_2$, Me, Me, HO, O, CH:CMe$_2$ (LXII)

(i) Tetrahydrofuran

Tetrahydrofuran, b.p. 66°, m.p. −108 to −107°, d_4^{20} 0.888 (*D. Starr* and *R. M. Hixon*, Org. Synth., 1936, **16,** 77; *W. Reppe*, Ann., 1955, **596,** 80) is a valuable solvent miscible with water, for example, aqueous tetrahydrofuran greatly facilitates the formation and subsequent reactions of certain diazonium salts in cases where amine salts and/or diazonium salts are difficulty soluble (*J. A. Cade* and *A. Pilbeam*, Chem. and Ind., 1959, 1578). γ-Butyrolactone and tetrahydrofuran have been prepared by a two-step catalytic hydrogenation of maleic anhydride in the presence of a nickel catalyst (*Mitsubishi Petrochemical Co.*, Neth. P. Appl. 6,611,853/1965; C.A., 1968, **68,** 49113g). The interchange

$$[Et_3\overset{\oplus}{O}][\overset{\ominus}{S}bCl_6] + C_4H_8O \rightarrow Et_2O + [C_4H_8\overset{\oplus}{O}Et][\overset{\ominus}{S}bCl_6]$$

affords the *oxonium salt* $[C_4H_8OEt]^{\oplus}[SbCl_6]^{\ominus}$, m.p. 132° (decomp.) (*H. Meerwein et al.*, J. pr. Chem., 1939, [ii], **154,** 83). Autoxidation of tetrahydrofuran gives the dangerously explosive *peroxide*, b.p. 62°/0.2 mm (*R. Criegee*, Angew. Chem., 1950, **62,** 120). Tetrahydrofuran yields succinic acid nearly quantitatively when oxidised with nitric acid; passed with ammonia over alumina at 300–320°, it gives 80–85% of pyrrolidine (*Reppe*, *loc. cit.*).

The tetrahydrofuran ring is easily opened by acids or by acid chlorides in presence of zinc chloride (*O. Grummitt et al.*, Org. Synth., 1949, **29,** 89), thus making accessible 1,4-glycols, 1,4-dihalides, and 1,4-halogeno-alcohols as important materials for aliphatic synthesis. Cleavage by methanesulphonyl bromide in the presence of zinc chloride gives 4-bromobutylmethanesulphonate, 4,4′-dibromodibutyl ether, and 1,4-bis-methanesulphonyloxybutane (*G. Sieber*, Ann., 1960, **631,** 180); lithium tetrahydridoaluminate–aluminium chloride yields butan-1-ol (*W. J. Bailey* and *F. Marktscheffel*, J. Org. Chem., 1960, **25,** 1787); aqueous bromine gives 4-hydroxybutyraldehyde (*N. C. Deno* and *N. H. Potter*, J. Amer. chem. Soc., 1967, **89,** 3550). The nucleus is also susceptible to reductive cleavage; thus hydrogenation of furfuryl alcohol can yield as by-products pentane-1,2- and -1,5-diols (*R. Connor* and *Adkins*, *ibid.*, 1932, **54,** 4678).

Although benzyltriphenoxyphosphonium bromide has been used as a reagent in tetrahydrofuran, this solvent reacts exothermically with triphenoxyphosphine dibromide giving triphenylphosphate and 1,4-dibromobutane (*A. Zamojski*, Chem. and Ind., 1963, 117).

3,4-**Diphenyltetrahydrofuran** is dehydrogenated by sulphur giving 3,4-diphenylfuran (*D. G. Farnum* and *M. Burr*, J. org. Chem., 1963, **28,** 1387):

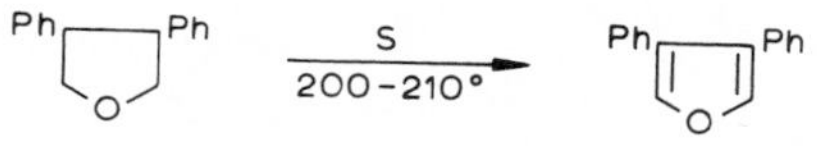

(ii) Halogenotetrahydrofurans

Chlorination of tetrahydrofuran gives, in good yield, 2,3-*dichlorotetrahydrofuran* in which the 2-chlorine atom is very readily replaced by alkoxyl, acetoxyl, or the group R of the Grignard reagent. The products of the last interchange yield unsaturated alcohols when treated with sodium, by a reaction analogous to the production of ethylene from β-halogenoethers by the action of metals:

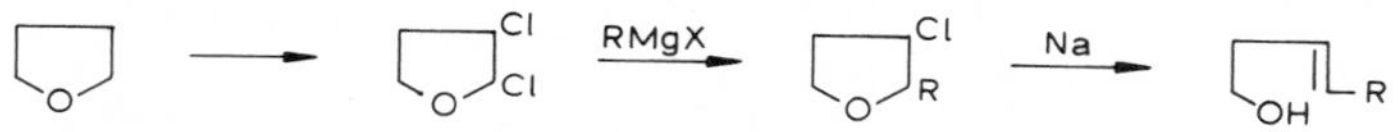

A related process involves side-chain halogen:

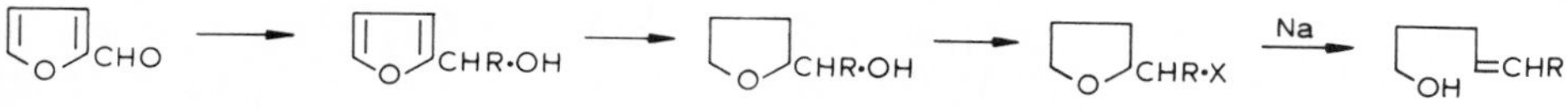

and the two afford a simple means of lengthening the carbon chain in R by four and five atoms (*Normant*, Compt. rend., 1948, **226,** 185, 733; *L. Crombie et al.*, J. chem. Soc., 1950, 1707,1714; 1956, 136). By the action of sodamide, these halogeno-compounds give the acetylenes corresponding to the olefins just formulated (*E. R. H. Jones et al.*, Org. Synth., 1953, **33,** 68).

Chlorination of tetrahydrofuran at $-30°$ to $-40°$ with irradiation affords 2-chloro- and 2,5-dichloro-tetrahydrofuran, which is unstable and reactive, the chlorine atoms undergoing nucleophilic displacement readily. Acid hydrolysis gives succindialdehyde, and with primary and tertiary amines the dihalide yields pyrroles and furan respectively (*M. Kratochvil* and *J. Hort*, Coll. Czech. chem. Comm., 1962, **27,** 52; *H. Gross*, Angew. Chem., 1960, **72,** 268; Ber., 1962, **95,** 83). At room temperature the 2-chloro-derivative loses hydrogen chloride and adds chlorine, forming 2,3-dichloro-tetrahydrofuran.

2,3-Dichlorotetrahydrofuran on chlorination at 80° gives 2,3,5-**trichloro-tetrahydrofuran,** and is converted to 2,3,3-**trichlorotetrahydrofuran** *via* 3-*chloro*-4,5-*dihydrofuran*, b.p. 56°/100 mm (*L. M. Bolotina*, *N. I. Kutsenko* and *P. A. Moshkin*, Doklady Acad. Nauk, S.S.S.R., 1967, **175,** 85):

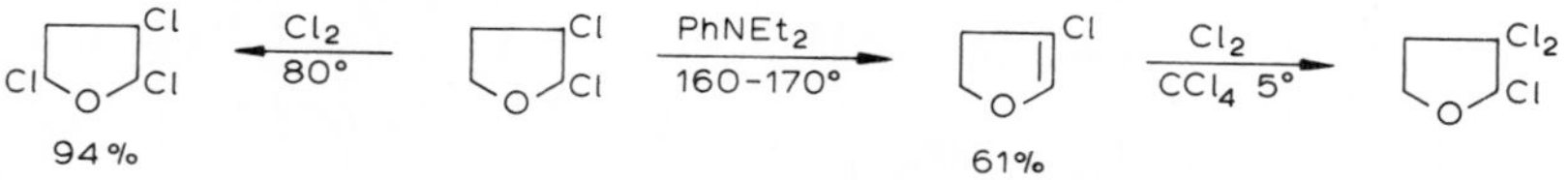

On passing a stream of butadiene through a solution of 2,3-dichlorotetrahydrofuran in ether containing zinc chloride at -12 to $-10°$ both the 1,2- and 1,4- addition products, 3-chloro-4-(3-chlorotetrahydrofur-2-yl)but-1-ene and 1-chloro-4-(3-chlorotetrahydrofur-2-yl)but-2-ene are obtained and may be separated (*S. A. Vartanyan, R. A. Kuroyan* and *A. O. Tosunyan*, Zhur. org. Khim., 1968, **4,** 51):

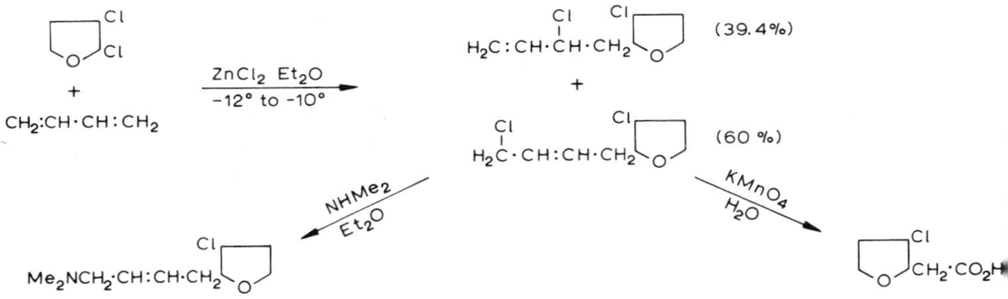
Cl
Cl
+
CH2:CH·CH:CH2
ZnCl2 Et2O
-12° to -10°
Cl Cl
H2C:CH·CH·CH2
(39.4%)
+
Cl Cl
H2C·CH:CH·CH2
(60 %)
NHMe2
Et2O
KMnO4
H2O
Cl
Me2NCH2·CH:CH·CH2
Cl
CH2·CO2H

(iii) Hydroxytetrahydrofurans

2-**Hydroxytetrahydrofurans** are the "lactols"*, *i.e.*, the dehydration products of the hydrated carbonyl forms of γ-hydroxyaldehydes and ketones, the furanose sugars being a most important special case (see Vol. I F). Dehydration of butane-1,2,4-triol yields 3-**hydroxytetrahydrofuran** *(Reppe, loc. cit.)*, also obtained as acetate by reaction of formaldehyde with allyl acetate (*S. Olsen*, Acta chem. Scand., 1950, **4,** 473):

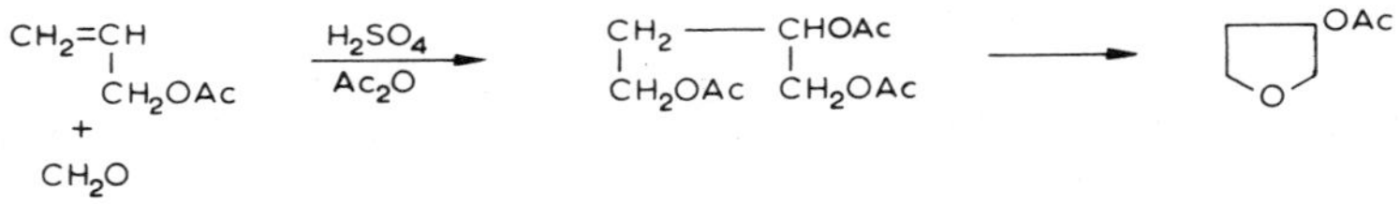
CH2=CH
CH2OAc
+
CH2O
H2SO4
Ac2O
CH2 — CHOAc
CH2OAc CH2OAc
OAc

Heating tetrahydrofuran with *tert*-butyl perbenzoate and an alcohol in the presence of cuprous chloride introduces alkoxy groups into the α-position (*S. O. Lawesson* and *C. Berglund*, *ibid.*, 1960, **14,** 1854).

By successive treatment with hypochlorous acid and alkali, 2,5-dihydrofuran gives 3,4-oxiranotetrahydrofuran (LXIII) from which 3,4-**dihydroxytetrahydrofuran** is produced by hydrolysis *(Reppe, loc. cit.)*:

(LXIII)

(iv) Aminotetrahydrofurans

Amines of the type LXIV, which are easily hydrolysed and otherwise very reactive, are prepared by treating 2-hydroxytetrahydrofurans or dihydrofurans with ammonia or amines (*C. Glacet*, Bull. Soc. chim. Fr., 1954, 575):

* This term, suggesting the reduction product of a lactone, has no official standing.

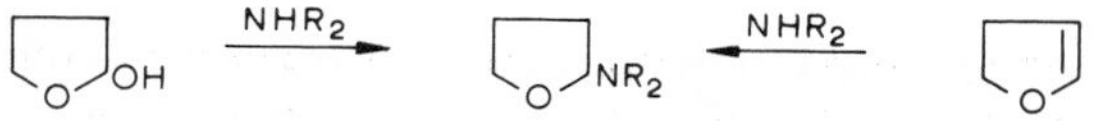

(LXIV)

Derivatives of 3,4-diaminotetrahydrofuran have been used in the synthesis of biotin.

(v) Oxotetrahydrofurans

Two processes leading by direct ring-synthesis to 3-oxotetrahydrofurans are reported by *J. Salkind* (J. gen. Chem., U.S.S.R., 1940, **10,** 1432) and by *M.* and *W. Bradley* (B.P. 568,402/1943):

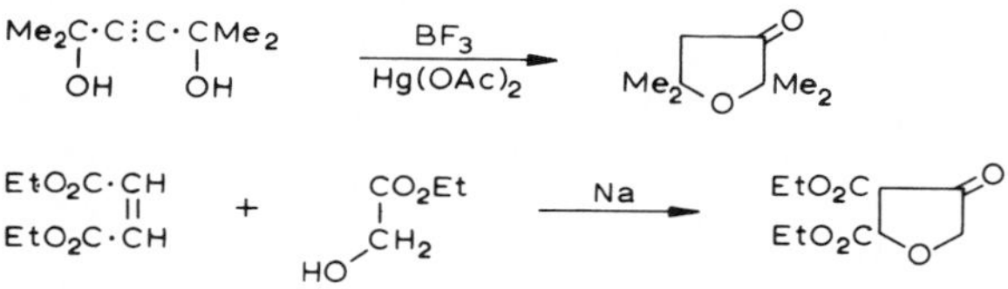

Tetronic acid, the lactone of γ-hydroxyacetoacetic acid, is an enolic modification of 2,4-**dioxotetrahydrofuran,**

3,4-**Dioxotetrahydrofuran** (LXV) has been synthesised starting with 1,4-dibromobiacetyl. It forms monohemiacetals (LXVI) with primary and secondary alcohols, and with water the tetrol LXVII (*E. C. Kendall* and *Z. G. Hajos*, J. Amer. chem. Soc., 1960, **82,** 3219, 3220):

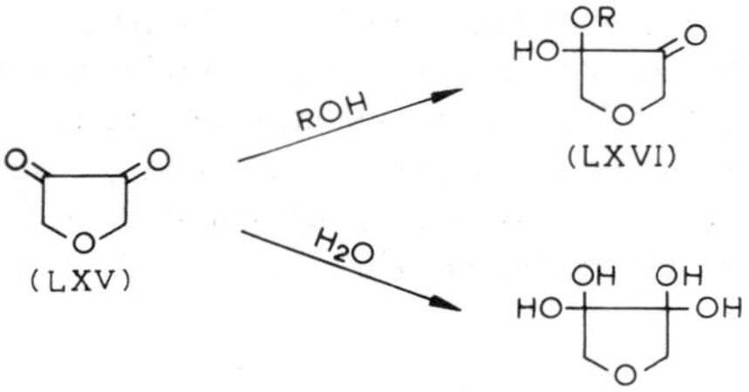

(LXV) (LXVI) (LXVII)

(vi) Side-chain substituted tetrahydrofurans

Tetrahydrofurfuryl alcohol, b.p. 177–178°/750 mm (p. 129) undergoes a curious and technically important ring-enlargement to dihydropyran (*L. E. Schniepp* and *H. H. Geller*, J. Amer. chem. Soc., 1946, **68,** 1646). Tetrahydrofurfuryl acetate on heating with zinc chloride is isomerised to 3-acetoxytetrahydropyran, with stereospecific inversion at the asymmetric centre (*D. Gagnaire* and *A. Butt*, Bull. Soc. chim. Fr., 1961, 309).

Tetrahydrofurfuryl esters when pyrolysed at 450–480° afford methyl propenyl ketone in fair yield (*G. T. Baumgartner* and *C. L. Wilson*, J. Amer. chem. Soc., 1959, **81,** 2440).

2-**Methylenetetrahydrofuran** has been prepared by the dehydrobromination of tetrahydrofurfuryl bromide, and it isomerises easily by heat or acid to 2-methyl-4,5-dihydrofuran. At higher temperatures both these compounds rearrange first to cyclopropyl methyl ketone and then to methyl propenyl ketone (*D. M. A. Armitage* and *Wilson*, *ibid.*, 1959, **81,** 2437). 2-Methylenetetrahydrofuran with pentylsodium affords 4-pentynol (*Paul* and *S. Tchelitcheff*, Compt. rend., 1951, **232,** 2330; Bull. Soc. chim. Fr., 1950, 520) and pyrolysis proceeds as shown:

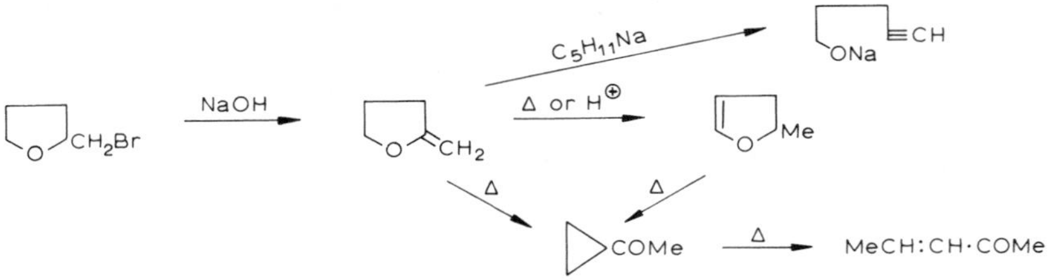

3,4-**Dimethylenetetrahydrofuran** (LXVIII) is obtained by the pyrolysis of 3,4-di-(acetoxymethyl)tetrahydrofuran (*Bailey* and *S. S. Miller*, J. org. Chem., 1963, **28,** 802):

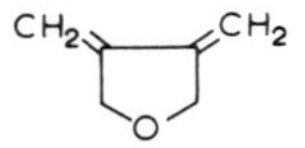

(LXVIII)

When tetrahydrofurfuryl alcohol, labelled with ^{14}C in the exocyclic methylene group, is passed over hot alumina it is rearranged to 2,3-dihydropyran, which is equally radioactive at C-2 and C-6. The rearrangement is not the result of double-bond isomerisations and the oxonium ion LXIX may account for these observations *via* a 1–5 hydride transfer (*W. J. Gensler*, *J. E. Stouffer* and *R. G. McInnis*, J. org. Chem., 1967, **32,** 200; *G. Descotes*, *B. Giroud-Abel* and *J. C. Martin*, Bull. Soc. chim. Fr., 1967, 2472):

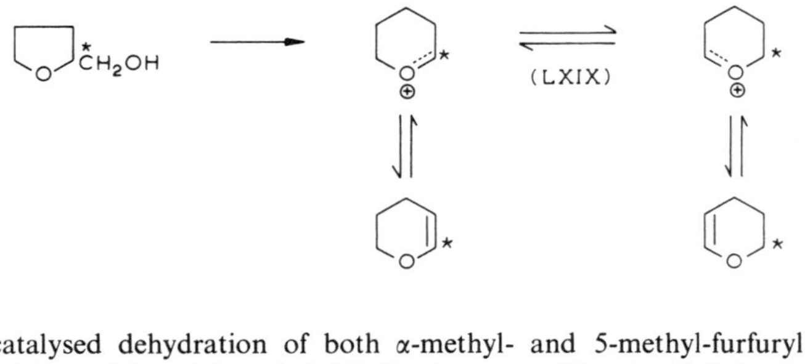

(LXIX)

Alumina-catalysed dehydration of both α-methyl- and 5-methyl-furfuryl alcohols affords a mixture of 6- and 2-methyl-2,3-dihydropyrans (*Gensler*, *I. Ruks* and *S. Marburg*, Chem. Comm., 1966, 782):

TABLE 6

SUBSTITUTED TETRAHYDROFURANS

Substituents	*b.p. (°C)*	*Ref.*
2-Methyl	78	1
2-Ethyl	105–107/740 mm	2
3-Methyl	83	3
2,2-Dimethyl	93	4
2,5-Dimethyl	92	1
2,2,5,5-Tetramethyl	119	1
3,4-Diphenyl	m.p. 86	5
2-Methylene	98–99	6
3-Chloro	52–54/13 mm	1, 7
2,3-Dichloro, *trans*	55/14 mm	7, 8
2,3,3-Trichloro	78/10 mm	9
2,3,5-Trichloro	84/10 mm	9
Perfluoro	0.8–1.2/733 mm	10
3-Chloro-2-methyl	42/14 mm	7
3-Hydroxy	181	1
3-Oxo	34–35/9 mm	11
3-Amino-2,2,5,5-tetramethyl	66–67/23 mm	12
2-Chloromethyl	41–42/11 mm	13
2-Bromomethyl	61–62/13 mm	14
2-Hydroxymethyl	178	2, 15
2-α-Hydroxyethyl	68/11 mm	16
2-Aminoethyl	146	17
2-Carbaldehyde	145	18
2-Carboxy	128–129/14 mm	19
2-Ethoxycarbonyl	80/11 mm	19
2-Carboxamide	m.p. 80	20
2,5-Dicarboxy	(*cis*, m.p. 126–127) *trans*, m.p. 94	21

References
1 *W. Reppe*, Ann., 1955, **596,** 25, 38, 80.
2 *H. E. Burdick* and *H. Adkins*, J. Amer. chem. Soc., 1934, **56,** 438.
3 *G. M. Bennett* and *W. G. Philip*, J. chem. Soc., 1928, 1937.
4 *J. Colonge* and *P. Garnier*, Compt. rend., 1946, **222,** 803.
5 *D. G. Farnum* and *M. Burr*, J. org. Chem., 1963, **28,** 1387.
6 *R. Paul* and *S. Tchelitcheff*, Bull. Soc. chim. Fr., 1950, 520.
7 *L. Crombie et al.*, J. chem. Soc., 1956, 136.
8 *H. Normant*, Compt. rend., 1948, **226,** 185.
9 *L. M. Bolotina, N. I. Kutsenko* and *P. A. Moshkin*, Doklady Acad. Nauk, S.S.S.R., 1967, **175,** 85.
10 *A. L. Henne* and *S. B. Richter*, J. Amer. chem. Soc., 1952, **74,** 5420.
11 *Reppe*, Ann., 1955, **596,** 183.
12 *S. Kanao* and *S. Inagawa*, J. pharm. Soc., Japan, 1928, **48,** 40.

13 *L. A. Brooks* and *H. R. Snyder*, Org. Synth., 1945, **25,** 84.
14 *L. H. Smith, ibid.*, 1943, **23,** 88.
15 *H. Wienhaus*, Ber., 1920, **53,** 1656.
16 *D. I. Duveen* and *J. Kenyon*, Bull. Soc. chim. Fr., 1940, **7,** 165.
17 *Adkins et al.*, J. Amer. chem. Soc., 1933, **55,** 2051; 1934, **56,** 2419.
18 *J. G. M. Bremmer et al.*, J. chem. Soc., 1949, S25, 527.
19 *Paul* and *G. Hilly*, Compt. rend., 1939, **208,** 359.
20 *Wienhaus* and *H. Sorge*, Ber., 1913, **46,** 1927.
21 *D. Lean*, J. chem. Soc., 1900, **77,** 103; *W. N. Haworth, ibid.*, 1945, 1.

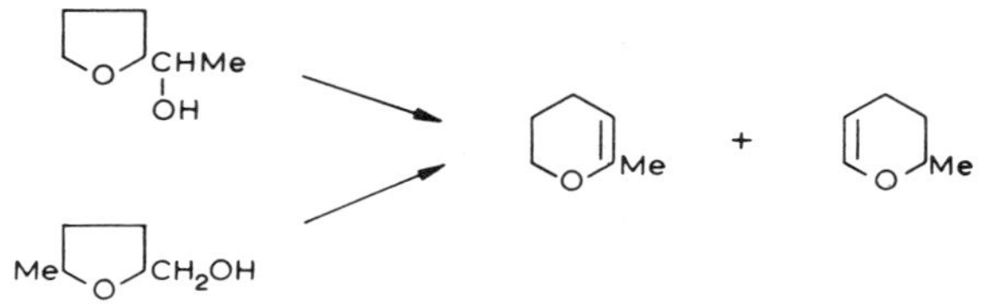

Tetrahydrofurfuryl alcohol has been reported as a product of the destructive hydrogenation of lignin (*A. Bailey*, J. Amer. chem. Soc., 1943, **65,** 1165).

Tetrahydrofurfurylamine is obtained by hydrogenating furfural in liquid ammonia, and along with bis(tetrahydrofurfuryl)amine, by hydrogenating tetrahydrofurfuramide.

Furoic acid yields **tetrahydrofuroic (tetrahydropyromucic) acid,** m.p. 21°, on hydrogenation (*H. Wienhaus* and *H. Sorge*, Ber., 1913, **46,** 1927; *Kaufmann* and *Adams, loc. cit.*). **Tetrahydrofuran-2,5-dicarboxylic acid** has been isolated in *cis* and *trans* forms, of which the latter alone gives a monomeric anhydride (*D. Lean*, J. chem. Soc., 1900, **77,** 103; *W. N. Haworth et al., ibid.*, 1945, 1). **Ascaridic acid,** derived from ascaridole (Vol. II B, p. 207), is 2-methyl-5-isopropyltetrahydrofuran-2,5-dicarboxylic acid. Some tetrahydrofuran derivatives are listed in Table 6 (see also Table 5, p. 127).

(d) Compounds with two or more unfused furan nuclei

Symmetrically substituted 2,2′-bifurans (2,2′-bifuryls) have been prepared by the Ullmann reaction and the unsymmetrically substituted 2,2′-bifurans by the construction of the second ring utilising the β-oxo-ester substituent of the alkyl furoylacetate (*R. Grigg, J. A. Knight* and *M. V. Sarjent*, J. chem. Soc., C, 1966, 976):

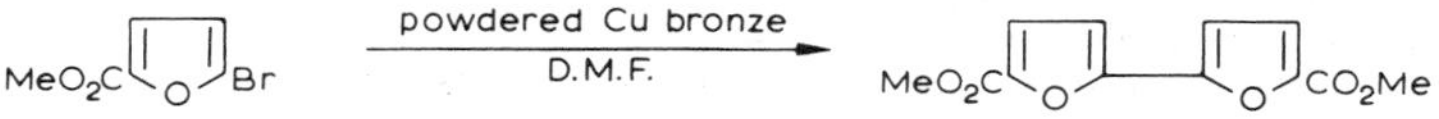

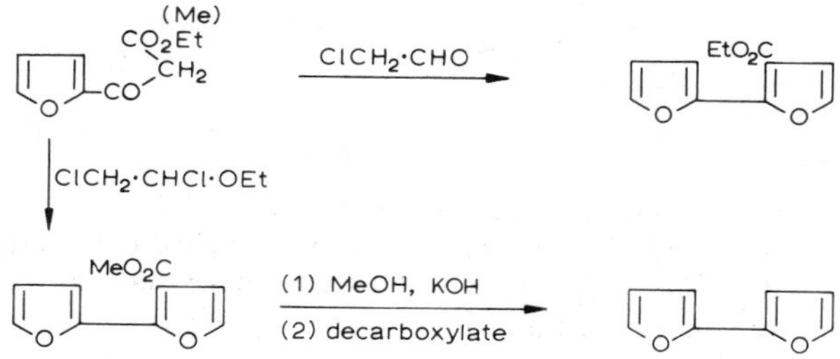

2,2′-**Bifuran,** b.p. 63–64°/11 mm, may be prepared by the latter method (*T. Reichstein, A. Grüssner* and *H. Zschokke*, Helv., 1932, **15,** 1066) and by the addition of cobaltous chloride to 2-furylmagnesium iodide in tetrahydrofuran (*R. E. Atkinson, R. F. Curtis* and *G. T. Phillips*, J. chem. Soc., C, 1967, 2011). Ethyl acetoacetate reacts with butynedial tetraethyl acetal to give diethyl 5,5′-dimethyl-2,2′-bifuran-4,4′-dicarboxylate, converted to 5,5′-dimethyl-2,2′-bifuran on heating in quinoline with copper chromite (*Zh. A. Krasnaya, S. L. Portnova* and *V. F. Kucherov*, Khim. Geterotsiki, 1967, 585; *Krasnaya et al.*, Khim. Atsetilena, 1968, 246; C.A., 1969, **70,** 114533b).

The Diels–Alder reaction between 2,2′-bifuran and dimethyl acetylenedicarboxylate gives compounds LXX and LXXI (*Grigg, P. Roffey* and *Sarjent*, J. chem. Soc., C, 1967, 2327):

2,2′-Bifuran is easily formylated by the Vilsmeyer–Haack method to give 2,2′-bifuran-5-carbaldehyde (*Atkinson, Curtis* and *Phillips, loc. cit.*), previously prepared in low yield by a Gatterman reaction (*Reichstein, Grüssner* and *Zschokke, loc. cit.*). Reduction of the aldehyde yields 5-methyl-2,2′-bifuran and acid–base condensation with malonic acid affords the acrylic acid derivative:

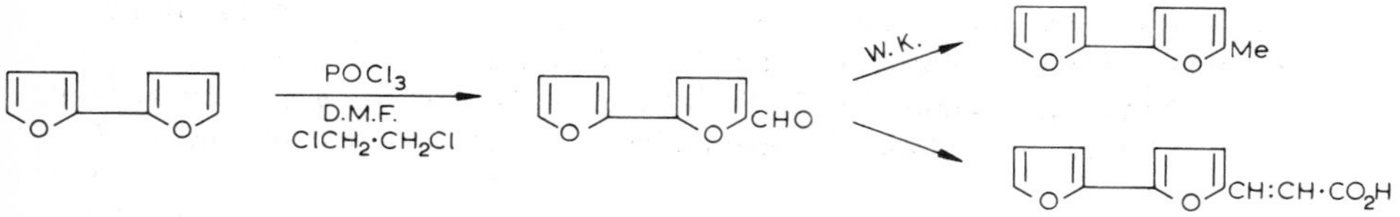

3,3′-**Bifuran,** m.p. 45.5–46°, has been prepared from 3-furyl-lithium and 3-oxotetrahydrofuran by the following route (*H. Wynberg* and *J. W. van Reijendam*, Rec. Trav. chim., 1967, **86,** 381):

(1) Add at −60° allow to warm to 0° (2) Aq. NH_4Cl (3) Distil; S, D.M.F.

Ngaione and ipomeamarone (p. 117) are derived from 2,3′-bifuran.

U.v. irradiation of 1,4-bis(diazomethyl)benzene gives **octahydro**-2,2′-**bifuran** (*I. Moritani, T. Nagai* and *Y. Shirota*, Kogyo Kagaku Zasshi, 1965, **68,** 296), also produced by the photodimerisation of tetrahydrofuran in the presence of mercuric chloride (*Y. Odaire et al.*, Nippon Kagaku Zasshi, 1967, **88,** 39).

Octahydro-3,3′-**bifuran** may be obtained from butane-1,2,3,4-tetracarboxylic acid (*E. Buchta* and *K. Greiner*, Ber., 1961, **94,** 1311):

Butanetetracarboxylic acid $\xrightarrow[H_2SO_4]{EtOH}$ $EtO_2C{\cdot}CH_2{\cdot}CH(CO_2Et){\cdot}CH(CO_2Et){\cdot}CH_2CO_2Et$ $\xrightarrow[T.H.F.]{LiAlH_4}$

$HOCH_2{\cdot}CH_2{\cdot}CH(CH_2OH){\cdot}CH(CH_2OH){\cdot}CH_2{\cdot}CH_2OH$ $\xrightarrow[180-190°]{KHSO_4}$

The absolute configuration of 5-**methyl**-2,3,4,5-**tetrahydro**-2,3′-**bifuran** (LXXII) from sweet potato fusel oil has been studied by n.m.r. spectroscopy (*M. Ogawa* and *Y. Hirose*, Nippon Kagaku Zasshi, 1965, **86,** 1335):

Me

(LXXII)

5-**Methyl**-2,2′-**bifuran,** b.p. 99–100°/10 mm; 2,2′-**bifuran**-5-**carbaldehyde,** m.p. 59–60°; *trans-β*-5-(2,2′-**bifuranyl)acrylic acid,** m.p. 170–171° (*Atkinson, Curtis* and *Phillips, loc. cit.*); 5-**methyl**-2,2′-**bifuran**-3-**carboxylic acid,** m.p. 194°, *methyl ester*, b.p. 115–120°/0.3 mm; 3-**methyl**-2,2′-**bifuran**-3′-**carboxylic acid,** m.p. 109–110°; **dimethyl** 2,2′-**bifuran**-5,5′-**dicarboxylate,** m.p. 226–228°; **dimethyl** 4,4′-**dimethyl**-2,2′-**bifuran**-5,5′-**dicarboxylate,** m.p. 202–202.5°. 5,5′-**Dinitro**-2,2′-**bifuran,** m.p. 215–217° (*Grigg, Knight* and *Sarjent, loc. cit.*); 5,5′-**dimethyl**-2,2′-**bifuran,** m.p. 37–39°; 5,5′-**dimethyl**-2,2′-**bifuran**-4,4′-**dicarboxylic acid,** m.p. 300°, *diethyl ester*, m.p. 95.5–97°; 5,5′-**dimethyl**-4,4′-**di(hydroxymethyl)**-2,2′-**bifuran,** m.p. 136–138° (*Krasnaya, Portnova* and *Kucherov, loc. cit.*). **Octahydro**-2,2′-**bifuran,** b.p. 78–80°/14 mm (*V. M. Micovic et al.*, Tetrahedron Letters, 1965, 1559); **octahydro**-3,3′-**bifuran,** b.p. 92°/11 mm (*Buchta* and *Greiner, loc. cit.*).

Cyanofuran and furylmagnesium iodide give **di**-2-**furyl ketone,** m.p. 33°, which by reduction affords **di**-2-**furylmethane,** b.p. 80°/18 mm (*H. Gilman* and *G. F. Wright*, J. Amer. chem. Soc., 1933, **55,** 3302).

5,5′-**Bis**(2-**carboxyethyl**)-2,2′-**difurylmethanes** and their diesters (see Table 7), have been synthesised by the following method (*V. G. Glukhovtsev* and *S. V. Zakharova*,

Khim. Geterotsikl. Soedin., 1968, 947):

$$\text{(2-furyl)}CH_2\cdot CH_2\cdot CO_2Et + CH_3CHO \xrightarrow[\text{and } H_2SO_4,\ 25°]{\text{trace of } 1,4\text{-di(OH)}C_6H_4} EtO_2C\cdot CH_2\cdot CH_2\text{-(furan)-}CH(Me)\text{-(furan)-}CH_2\cdot CH_2\cdot CO_2Et$$

$$\xrightarrow{8\%\ NaOH} \overset{(R)}{HO_2C}\cdot CH_2\cdot CH_2\text{-(furan)-}\overset{(R^1)(R^2)}{CH(Me)}\text{-(furan)-}CH_2\cdot CH_2\cdot \overset{(R)}{CO_2H}$$

TABLE 7

ACIDS AND ESTERS OF THE TYPE

$$RO_2C\cdot CH_2\cdot CH_2\text{-(furan)-}C(R^1)(R^2)\text{-(furan)-}CH_2\cdot CH_2\cdot CO_2R$$

R	*R*^{1}	*R*^{2}	*b.p. (°C)*
Me	H	Me	170–171/2 mm
Me	Me	Me	170–172/3 mm
Et	Me	Me	171–172/2 mm
Me	H	H	175–176/4 mm; m.p. 37
H	H	Me	m.p. 128
H	H	H	m.p. 141–141.5
H	Me	Me	m.p. 117.5–118

As an aromatic aldehyde, furfural undergoes the benzoin condensation, yielding **furoin,** m.p. 135° and thence **furil,** m.p. 165° (*W. W. Hartman* and *J. B. Dickey*, J. Amer. chem. Soc., 1933, **55,** 1228). Furil, by the normal benzilic changes, gives the unstable **furilic acid.** *Furil dioxime*, m.p. 166–168°, has been used as an analytical reagent for palladium and nickel (*S. A. Reed et al.*, J. org. Chem., 1947, **12,** 792).

$$2\ (C_4H_3O)\cdot CHO \xrightarrow{KCN} \underset{\text{Furoin}}{(C_4H_3O)\cdot CHOH\text{–}CO\cdot(C_4H_3O)} \xrightarrow[C_5H_5N]{CuSO_4} \underset{\text{Furil}}{(C_4H_3O)\cdot CO\text{–}CO\cdot(C_4H_3O)} \longrightarrow \underset{\text{Furilic acid}}{(C_4H_3O)_2C(OH)CO_2H}$$

Thermal decomposition of polymeric thiofurfural or of furfuraldazine (*P. Pascal* and *L. Normand*, Bull. Soc. chim. Fr., 1911, **9,** 1029) yields 1,2-**bis(2-furyl)ethene,** m.p. 100°. The hexaphenylethane analogue 1,2-bis(2-furyl)-1,1,2,2-tetraphenylethane, does not exhibit radical dissociation (*H. E. French* and *D. R. Smith*, J. Amer. chem. Soc., 1945, **67,** 1949).

Furans condense with ketones R_2CO in presence of hydrochloric acid to give macrocyclic products reminiscent of reduced porphyrins (p. 95), through linear intermediates such as LXXIII and LXXIV (*Wright et al.*, J. org. Chem., 1955, **20,** 1147):

CR_2 (LXXIII) CR_2 CR_2 (LXXIV)

(e) Polyene derivatives of furan

These have been studied, on the one hand as objects of spectroscopic investigation as such and as halochromic materials, and on the other as sources of difficulty accessible aliphatic compounds by hydrogenation, followed by opening of the furan ring (*J. Schmitt*, Ann., 1941, **547,** 270; *A. Hinz et al.*, Ber., 1943, **76,** 676; *P. Karrer et al.*, Helv., 1946, **29,** 1836; *J. C. Smith*, J. chem. Soc., 1953, 618). The polyene aldehydes result from the condensation of furfural with successive molecules of acetaldehyde. They yield the ketones by condensation with acetone or diacetyl, and the acids by Claisen condensation with ethyl acetate. The properties of the furan polyenes are set out in Table 8.

TABLE 8

FURAN POLYENES (F = 2-FURYL)

Substance	*M.p. (°C)*	*Colour*
$F{\cdot}CH{=}CH{\cdot}CHO$	54	colourless
$F(CH{=}CH)_2CHO$	66	pale yellow
$F(CH{=}CH)_3CHO$	111	golden yellow
$F(CH{=}CH)_4CHO$	155	orange red
$F(CH{=}CH)_5CHO$	194	brown-red
$F(CH{=}CH)_6CHO$	218	brown-violet
$F(CH{=}CH)_7CHO$	230 decomp.	violet-black
$F(CH{=}CH)_2CO_2H$	153–154	pale yellow
$F(CH{=}CH)_3CO_2H$	200	brown yellow
$(F{\cdot}CH{=}CH)_2CO$	60	lemon-yellow
$[F{\cdot}(CH{=}CH)_2]_2CO$	121	golden-yellow
$[F(CH{=}CH)_3]_2CO$	175–176	deep orange
$(F{\cdot}CH{=}CH{\cdot}CO)_2$	156–157	light yellow
$[F(CH{=}CH)_2CO]_2$	182–184	red
$[F(CH{=}CH)_3CO]_2$		red-violet

2. Benzo[b]furans and their hydrogenated products*

(a) Benzo[b]furans

The first compound of this series to be described was coumarilic acid (benzofuran-2-carboxylic acid) which Perkin obtained from coumarin in 1870:

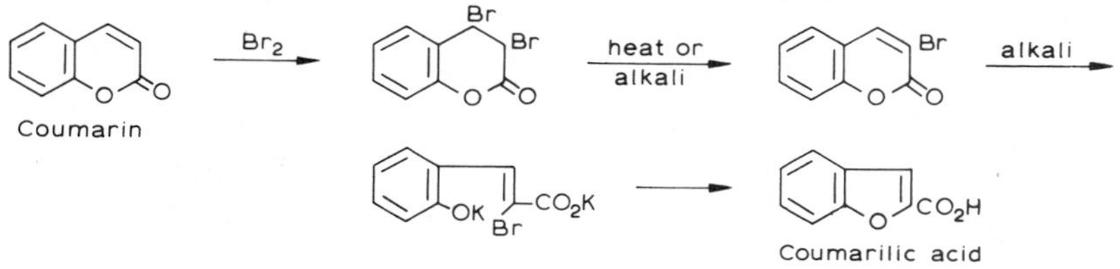

By distilling coumarilic acid with lime, in 1883, Fittig and Ebert prepared the parent substance and called it "coumarone", a term easily confused with "coumarin". As a result of this genetic relationship this name was accepted over the years for benzofuran and its derivatives; similarly 2,3-dihydrobenzofurans were named "coumarans" and the related ketones as "coumaranones". In accordance with I.U.P.A.C. rules these trivial names are now regarded as obsolete, the official systematic name for the parent compound being *benzo*[b]*furan* or alternatively 2,3- (or 4,5)-*benzofuran*, numbered as shown:

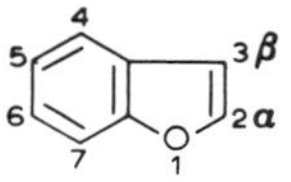

A method of numbering was long used in which the α-position was numbered 1 as starting point and the oxygen disregarded. The constitution of benzofuran follows from its conversion, by alkali, into *o*-hydroxyphenylacetic acid, and from various syntheses of the parent substance and its derivatives. Benzofuran occurs in coal-tar, and is an important source of the so-called coumarone resins. The benzofuran nucleus is fairly often encountered in nature, especially fused with a pyran ring in a series of fish poisons.

(i) General synthetic methods

The important routes involve closure of the heterocyclic ring in a sub-

* The extensive early work is summarised by *R. Stoermer* (Ann., 1900, **312**, 237).

stance having the benzene nucleus ready formed. Firstly, closure can take place between the oxygen atom and $C_{(2)}$.

(*1*) The intramolecular alkylation exemplified by the synthesis of coumarilic acid is widely used in preparing benzofurans and 2,3-dihydrobenzofurans (*G. Komppa*, Ber., 1893, **26,** 2968):

(*2*) 2-Acyl- and 2-carboxy-benzofurans (or their products of decarboxylation) are accessible by a very general method of ring-closure in derivatives of *o*-hydroxycarbonyl compounds (*R. Stoermer* and *M. Schäffer*, Ber., 1903, **36,** 2863; *A. Robertson et al.*, J. chem. Soc., 1948, 2254) (R = H, alkyl, aryl):

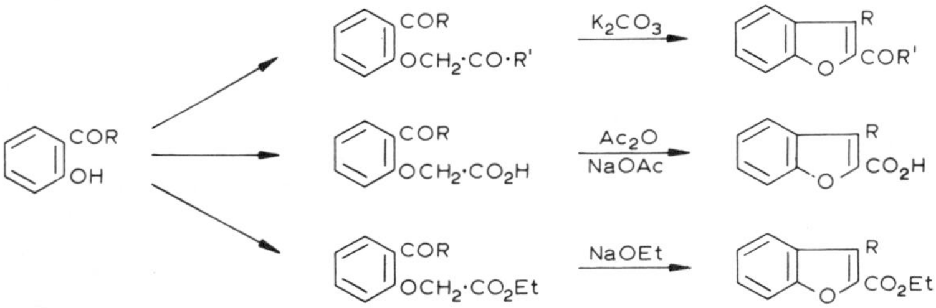

Several methods involve ring-closure between $C_{(3)}$ and the benzene nucleus with or without simultaneous or closely associated linking of oxygen to $C_{(2)}$.

(*3*) Benzofurans are produced in acceptable yield by dehydration of aryloxyacetones and in poor yield from aryloxyacetals (*Stoermer*, Ann., 1900, **312,** 237):

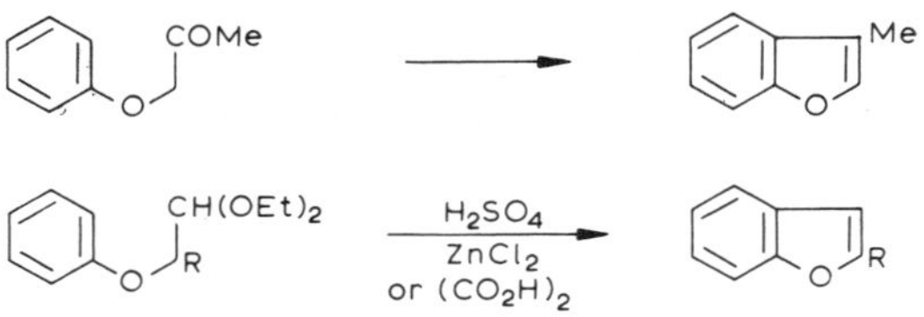

(*4*) Dihydric phenols react with benzoin in presence of 73% sulphuric acid (*F. R. Japp* and *A. N. Meldrum*, J. chem. Soc., 1899, **75,** 1035; *O. Dischendorfer*, in a series of papers in Monatsh., 1933–1948):

HO OH + OCPh HOCHPh → HO Ph Ph O

Condensation of 2-hydroxy-4,6,4′-trimethoxybenzil with ethyl bromoacetate in acetone containing potassium carbonate and hydrolysis of the crude ester with aqueous potassium hydroxide gives 2-carboxymethoxy-4,6,4′-trimethoxybenzil (I); cyclised to 3-p-*anisoyl-4,6-dimethoxybenzofuran* (II), m.p. 130°, rather than to the isomeric 5,7,4′-trimethoxyisoflavone (*S. A. N. N. Bokhari* and *W. B. Whalley*, J. chem. Soc., 1963, 5322):

eO OH CO·OC OMe MeO + $BrCH_2·CO_2Et$ —(1) Me_2CO, K_2CO_3 (2) KOH, H_2O→ MeO $OCH_2·CO_2H$ CO·CO OMe MeO (I) —Ac_2O, NaOAc→ OMe CO OM MeO O (II)

(5) Quinones having one free nuclear position react with the sodio-derivatives of β-dicarbonyl compounds, giving products easily convertible into benzofurans (*L. I. Smith et al.*, J. Amer. chem. Soc., 1936, **58,** 629; 1940, **62,** 133):

O Me Me Me O —$NaCHAc_2$→ Me HO $CHAc_2$ Me OH Me —HCl→ Me HO Me O Me Me

—$NaCH(CO_2Et)_2$→ [Me HO $CH(CO_2Et)_2$ Me OH Me] → Me HO Me O O Me

Keten acetal and *p*-benzoquinone afford 2-ethoxy-5-hydroxybenzofuran (*S. M. McElvain* and *E. L. Engelhardt*, *ibid.*, 1944, **66,** 1077):

O O + CH_2 ‖ $C(OEt)_2$ → HO O OEt —acid→ HO $CH_2·CO_2H$ OH

(6) By an extension of the Fischer indole synthesis benzofurans may be obtained from *O*-phenyloximes (*T. Sheradsky*, Tetrahedron Letters, 1966, 5225):

$$\text{PhO–N=CR'CH}_2\text{R} \xrightarrow[\text{BF}_3\cdot\text{Et}_2\text{O, AcOH}]{100^\circ} \text{2-R'-3-R-benzofuran}$$

The oximes are prepared from the appropriate carbonyl compound and *O*-phenylhydroxylamine, $PhONH_2$, in alcohol containing a trace of hydrogen chloride. *o*-Aryloximes have been obtained by treating the metal salt of an oxime with an activated halide, such as a *p*-halogenonitrobenzene, in a polar solvent (*A. Mooradian, ibid.*, 1967, 407).

(7) Benzofuran has been prepared catalytically from phenol and ethanol and a mechanism derived (*B. Sila* and *T. Lesiak*, Rocz. Chem., 1961, **35,** 1519; *A. Krause, ibid.*, 1963, **37,** 827); from phenol, acetylene and carbon dioxide at about 650° over alumina (*Lesiak, ibid.*, 1962, **36,** 1533); by the vapour-phase dehydrocyclisation of 2-ethylphenol in the presence of carbonyl sulphide or hydrogen sulphide and a catalyst such as magnesia or alumina (*D. E. Boswell et al.*, Ind. Eng. Chem., Process Des. Develop., 1968, **7,** 215); by passing ethylene through phenol at 170–180° and then over an alumina/iron oxide catalyst at 650° (*Lesiak*, Rocz. Chem., 1965, **39,** 757).

The reaction between cuprous acetylides and *o*-halogenophenols is a general and convenient route to 2-substituted benzofurans (*C. E. Castro, E. J. Gaughan* and *D. C. Owsley*, J. org. Chem., 1966, **31,** 4071):

$$o\text{-BrC}_6\text{H}_4\text{OH} + \text{CuC:CR} \xrightarrow[\text{N}_2]{\text{C}_5\text{H}_5\text{N}} \text{2-R-benzofuran} + \text{CuBr}$$

(*8*) When an equimolar mixture of a salicylaldehyde and bis-(ethylsulphonyl)methane is heated in boiling toluene in the presence of a slight excess of an equimolar quantity of piperidine, which neutralises the phenolic hydroxyl group of the aldehyde and acts as catalyst, 2-ethylsulphonylbenzofuran is formed (*M. L. Oftedahl, J. W. Baker* and *M. W. Dietrich, ibid.*, 1965, **30,** 296):

$$o\text{-OHC–C}_6\text{H}_4\text{OH} \longrightarrow o\text{-CH(OH)·CH(SO}_2\text{Et)}_2\text{–C}_6\text{H}_4\text{O}^\ominus \longrightarrow \text{[}o\text{-}^\ominus\text{O–C}_6\text{H}_4\text{–CH=C(SO}_2\text{Et)–SO}_2\text{Et]} \longrightarrow \text{2-SO}_2\text{Et-benzofuran} + \text{EtSO}_2^\ominus$$

(*9*) Benzofuran may be obtained by the catalytic dehydrogenation of

2,3-dihydrobenzofuran at 600–630° over an alumina/iron oxide catalyst, activated with moist air at 800° (*Lesiak*, Rocz., Chem., 1964, **38**, 1015).

(*10*) The closure of a benzene nucleus on to an already formed furan ring has been effected with one of the stereoisomeric products of the reaction of 2-benzoylfuran with diethyl succinate (*E. B. Knott*, J. chem. Soc., 1945, 189):

(ii) General properties and reactions

The benzofuran nucleus is a little less sensitive to acids than the unfused furan ring, but such reagents as sulphuric acid or aluminium chloride cause polymerisation. Nitration takes place energetically, with primary attack on the furan moiety, the defined product of mono-nitration being the 2-compound. This contrasts with the behaviour of indole which is substituted preferentially in position 3. Halogens combine additively, and the 2,3-dihalogeno-2,3-dihydrobenzofurans lose hydrogen halide under varying conditions to give varying proportions of 2- and 3-halogenobenzofurans. Attempted Friedel–Crafts synthesis usually leads to resinification, but the Gatterman–Hoesch procedure can give good yields of aldehydes (*R. T. Foster* and *A. Robertson*, *ibid.*, 1939, 921).

Oxidising agents readily attack the furan nucleus, giving derivatives of salicylic acid; ozonolysis likewise gives salicylic acid and salicylaldehyde, with some catechol (*A. v. Wacek et al.*, Ber., 1940, **73**, 521). The heterocyclic ring is reduced catalytically or by sodium and alcohol, with some reluctance, giving dihydrobenzofurans (coumarans), and then octahydrobenzofurans, often with much concurrent hydrogenolysis (*J. Entel et al.*, J. Amer. chem. Soc., 1951, **73**, 4152).

Hydrogenation of benzofurans under pressure in the presence of tungsten disulphide catalyst results in ring cleavage at the O site (*S. Landa* and *J. Chuckla*, Sb. Vysoke Skoly Chem.-Technol. Praze, Technol. Paliv, 1961, **5**, 35; C.A., 1966, **65**, 10555f). Furan ring opening is observed in the reduction of III and IV using lithium in liquid ammonia, when a limited amount of alcohol is present. The corresponding 5-methoxy-2,3,4,7-tetrahydrobenzofurans (V and VI) are formed in the presence of excess alcohol. A side product from the reduction of V is 2-methyl-2,3,4,5,6,7-hexahydrobenzofuran (VII) (*S. D. Darling* and *K. D. Wills*, J. org. Chem., 1967, **32**, 2794):

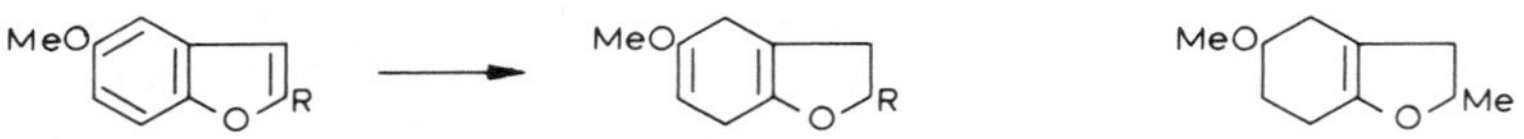

(III) R = Me
(IV) R = H
(V) R = Me
(VI) R = H
(VII)

The 1,2-bond of 3-formyl- or 3-acyl-benzofuran (VIII) is broken with ammonia to give (2-hydroxyphenyl)-β-enamino ketones (IX), which regenerate the initial VIII by losing ammonia either spontaneously, or on heating, or on treatment with dilute hydrochloric acid. 3-Cyano-, 2-ethyl-3-cyano-, and 3-ethoxycarbonyl-2-ethyl-benzofuran do not undergo this degradation (*M. Hubert-Habart et al.*, Bull. Soc. chim. Fr., 1966, 1587):

(VIII) (IX)

Halogenobenzofurans do not usually give normal Grignard reagents, but they react with butyl-lithium:

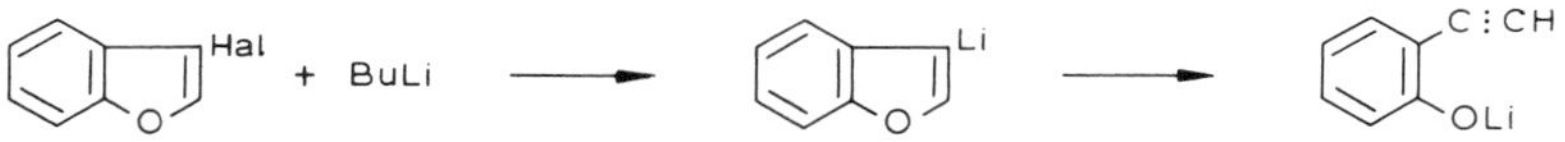

The 3-lithio-compound readily changes into the lithium compound of *o*-hydroxyphenylacetylene, which is also obtained by attempted Grignardisation of 3-bromobenzofuran (*H. Gilman* and *D. S. Melstrom*, J. Amer. chem. Soc., 1948, **70**, 1655). This type of fission is common with β-halogenoethers (*cf.* p. 131).

Hydroxybenzofurans having the substituent in the furan ring behave predominantly as the tautomeric oxodihydrobenzofurans (coumaranones) and attempts to prepare 2- and 3-aminobenzofurans apparently lead to isomeric iminodihydrobenzofurans (iminocoumarans) which are quickly hydrolysed to oxodihydrobenzofurans (coumaranones) (*I. J. Rinkes*, Rec. Trav. chim., 1932, **51**, 349):

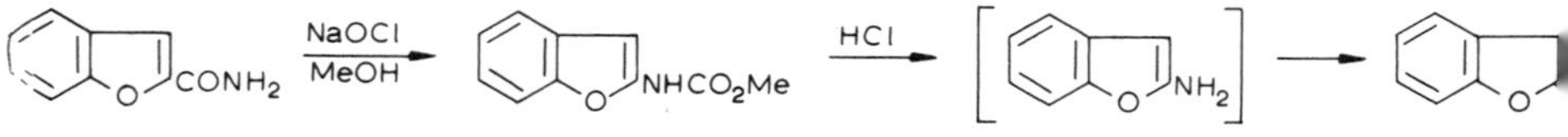

Benzofuran in benzene irradiated in the presence of acetophenone, propiophenone, acetone and xanthone yields X and XI (3:1), but in the presence of benzophenone, benzaldehyde or thioxanthone affords good yields of the 1:1 adducts XII and XIII. Irradiation of equimolar amounts of benzofuran and duroquinone in benzene gives XIV; in the absence of benzene, two isomeric 2:1 adducts XV are formed (*C. H. Krauch, W. Metzner* and *G. O. Schenck*, Ber., 1966, **99**, 1723):

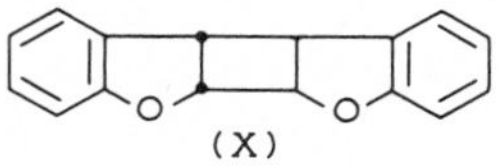

(X) 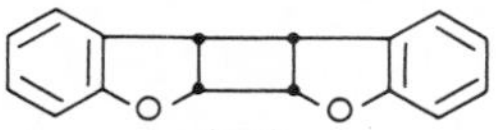 (XI)

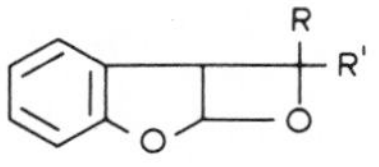

(XII) 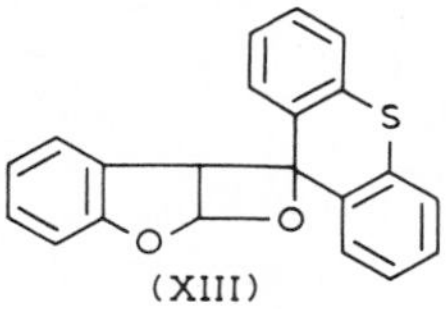 (XIII)

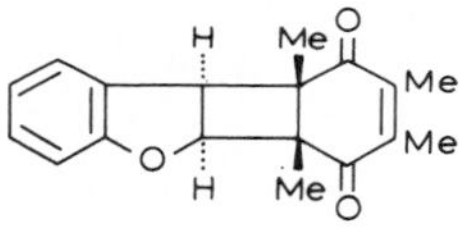

(XIV) 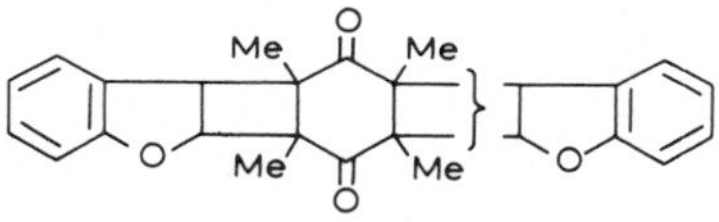 (XV)

On reaction with dichlorocarbene in hexane, benzofuran affords an adduct, which on hydrolysis with water is converted into *bis*[3-*chloro*-2-(3-*chromenyl*)]*ether*, m.p. 181–182°, formed apparently as shown (*W. E. Parham et al.*, J. org. Chem., 1963, **28**, 577):

+ $:CCl_2$ → → $\xrightarrow{H_2O}$

Benzofuran with hydrogen sulphide or ammonia under pressure over an alumina catalyst at a high temperature gives benzothiophene and indole, respectively (*T. Lesiak*, Rocz. Chem., 1965, **39**, 589). Similarly benzothiophene may be obtained in the presence of chromium oxide–copper chromate and molybdenum sulphide (*H. M. Foster*, U.S.P. 3,381,018/1965). The pyrolysis of an equimolar mixture of benzene and benzofuran at 750° gives 2-phenylbenzofuran as the main product (*J. Lore* and *J. Chopin*, Bull. Soc. chim. Fr., 1965, 1652).

Benzofuran is present in the fraction of coal-tar, b.p. 155–185°, along with indene, in which it can be enriched by selective extraction with ethylene glycol, and it has been isolated through the picrate. It is obtained synthetically by the general methods (*2*) and (*3*) above and by decomposition of the 2-carboxylic acid, coumarilic acid (*J. Entel et al.*, J. Amer. chem. Soc., 1951, **73**, 4152). Benzofuran is produced in 10% yield along with furan-2,5-dicarboxylic acid by distilling saccharic acid with hydrobromic acid (*A. C. Cope* and *R. T. Keller*, J. org. Chem., 1956, **21**, 141); succindialdehyde is thought to be an intermediate:

Saccharic acid ⟶ $CH_2(CHO)\cdot CH_2(CHO)$ + $OHC\cdot CH_2\cdot CH_2\cdot CHO$ ⟶ benzofuran

The colourless liquid has b.p. 171°, $d_4^{22.7}$ 1.0913, $n_D^{22.2}$ 1.565, μ 0.79 (*E. A. Shott-Lvova* and *Y. K. Syrkin*, J. phys. Chem., U.S.S.R., 1938, **12,** 479). For spectroscopic data, see *P. Lambert* and *J. Lecomte* (Compt. rend., 1939, **208,** 1148) and *H. Berton (ibid.*, 1948, **227,** 342). It is very stable towards alkali, but with sulphuric acid yields a series of polymers which on heating regenerate benzofuran with charring and production of much phenol. The polymers, often involving indene as well as benzofuran, find application as neutral, chemically inert resins in the manufacture of varnishes and other materials. Optically active polybenzofuran is obtained by the polymerisation of benzofuran using aluminium chloride as catalyst complexed with (+)- or (−)-phenylalanine (*G. Natta* and *M. Farina*, Tetrahedron Letters, 1963, 703). *Benzofuran picrate*, m.p. 102–103°, and the *trinitrobenzene adduct*, m.p. 104°, serve, with the dibromide, to identify the substance.

Spectra: u.v. (*G. M. Badger* and *B. J. Christie*, J. chem. Soc., 1956, 3438; *J. M. Hollas*, Spectrochim. Acta, 1963, **19,** 753); i.r. (*A. Cheutin, M. L. Desvoye* and *R. Roger*, Compt. rend., 1964, **258,** 2559); n.m.r. (*P. J. Black* and *M. L. Heffernan*, Australian J. Chem., 1965, **18,** 353: *Black, R. D. Brown* and *Hefferman, ibid.*, 1967, **20,** 1325; *J. A. Elvidge* and *R. G. Foster*, J. chem. Soc., 1963, 590; 1964, 981); ^{13}C-n.m.r. (*K. Tori* and *T. Nakagawa*, J. phys. Chem., 1964, **68,** 3163; *E. Lippmaa, A. Olivson* and *J. Past*, C.A., 1966, **64,** 7552g); m.s. (*N. S. Vul'fson, V. I. Zaretskii* and *V. G. Zaikin*, Izvest. Akad. Nauk S.S.S.R., Ser. Khim., 1963, 2215; *B. Willhalm, A. F. Thomas* and *F. Gautschi*, Tetrahedron, 1964, 1185; *C. S. Barnes* and *J. L. Occolowitz*, Australian J. Chem., 1964, **17,** 975).

(iii) Alkyl- and aryl-benzofurans

These compounds (Table 9, p. 151) are produced by the usual methods of ring-synthesis, and may often be characterised by their crystalline picrates. Methylbenzofurans are obtained by boiling with alkali the dibromides of *o*-allylphenols, and treatment of 3-oxo-2,3-dihydrobenzofurans with Grignard reagents give the 3-alkyl compounds. A general process for 2-arylbenzofurans depends on the condensation of salicylaldehyde with arylbromoacetic esters:

salicylaldehyde (CHO, OH) + $BrCH(Ph)CO_2Et$ ⟶ 3-hydroxy-2-phenyl-2,3-dihydrobenzofuran-2-carboxylate (OH, CO_2Et, Ph) $\xrightarrow[\text{distil}]{\text{saponify}}$ 2-phenylbenzofuran + H_2O + CO_2

2-Alkylbenzofurans may be obtained from 2-alkyl-3-oxo-2,3-dihydrobenzofurans following reduction with either lithium tetrahydridoaluminate or sodium tetrahydridoborate (*Y. Kawase* and *S. Nakamoto*, Bull. chem. Soc., Japan, 1962, **35,** 1624):

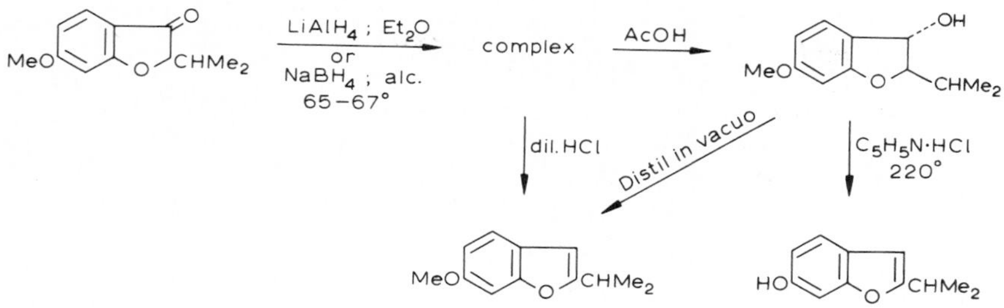

Dehydrocyclisation of *o*-allylphenol occurs to give 2-methylbenzofuran when it is passed over a catalyst composed of alumina spheres containing 0.75% of platinum and 0.274% of lithium, at 550° (*G. E. Illingworth* and *J. J. Louvar*, U.S.P. 3,285,932/1964).

When 2′-hydroxychalcones (XVI) are kept for a week at room temperature in aqueous ethanol containing sodium hydroxide (up to 16%) and hydrogen peroxide (1–15%), a mixture of the expected flavanol (yield up to 30%) and the acid XVII (yield up to 30%) is obtained. The 2-arylbenzofuran-3-carboxylic acids (XVII) are decarboxylated on treatment with copper–quinoline to the parent 2-arylbenzofuran (*D. M. X. Donnelly et al.*, Chem. and Ind., 1961, 1453):

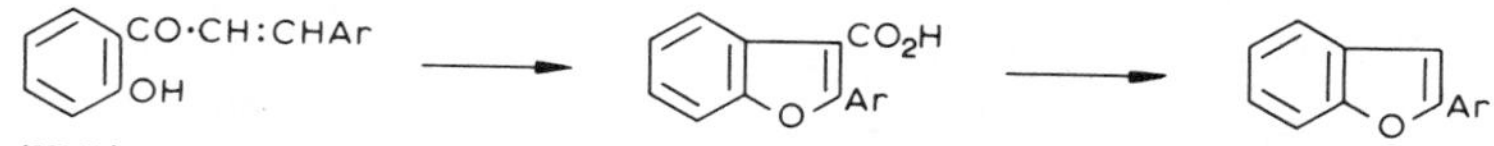

Aryloxyacetophenones (XVIII) in acetic acid are cyclised to 3-arylbenzofurans (XIX) by the addition of concentrated sulphuric acid; compound XIX ($R^1=R^3=R^4=H$; $R^2=Me$; $R^5=Me_2CH$) is the only one obtained by heat alone. The aryloxyacetophenones are prepared by boiling ω-halogenoacetophenones and sodium phenolates in benzene (*R. Royer* and *E. Bisagni*, Bull. Soc. chim. Fr., 1959, 1468):

2,3-Diphenylbenzofurans may be prepared by the cyclodehydration of α-aryloxydeoxybenzoins in 15–20 times their weight of polyphosphoric acid at 80–120° (*C. Perrot* and *E. Cerutti*, Compt. rend., Ser. C, 1967, **264,** 1301).

The benzoins are formed by the condensation of various phenols and α-bromodeoxybenzoin in absolute alcohol in the presence of sodium ethoxide. Benzoin following reaction with *o*- and *p*-cresol in the presence of boric acid is treated with dilute aqueous sodium hydroxide solution to yield 7-methyl- and 5-methyl-2,3-diphenylbenzofuran, respectively (*B. Arventiev* and *H. Offenberg*, Acad. Rep. Populare Romine, Filiala Iasi, Studii Cercetari Stiint., Chim., 1960, **11,** 305; C.A., 1962, **56,** 11554c).

Benzofuran readily undergoes acylation in the 2-position by aliphatic acid anhydrides in the presence of phosphoric acid, the solvent being the acid corresponding to the anhydride. The resultant ketones may be reduced by the Huang–Minlon method to the 2-alkylbenzofurans (*N. P. Buu-Hoï, N. D. Xuong* and *N. V. Bac*, J. chem. Soc., 1964, 173). Similarly 3-methylbenzofuran-2-carbaldehyde affords 2,3-dimethylbenzofuran (*Bisagni* and *Royer*, Bull. Soc. chim. Fr.,, 1962, 925). The effect of the position of substituents on the Friedel–Crafts diacetylation of alkylbenzofurans in carbon disulphide, with excess acetyl chloride and aluminium trichloride, has been studied (*Royer et al., ibid.*, 1966, 211). For acylation and formylation of 2,3-dimethylbenzofuran derivatives see *Royer et al., ibid.*, 1965, 2607. For the nitration and chloromethylation see pp. 154 and 158.

Benzofurans substituted in the 5 and 5,7 positions may be prepared by condensing a substituted salicylaldehyde with dimethyl bromomalonate or ethyl 2-bromophenylacetate in ethyl methyl ketone in the presence of anhydrous potassium carbonate, followed by hydrolysis and decarboxylation (*A. A. Rao* and *N. V. S. Rao*, Symp. Syn. Heterocycl. Compounds Physiol. Interest, Hyderabad, India, 1964, **26,** (Pub. 1966); C.A., 1968, **69,** 18955z).

The photolysis of a 1 % solution of thymoquinone in methanol, followed by methylation using dimethyl sulphate and alkali gives 5-methoxy-2,6-dimethylbenzofuran as one of the products (*C. M. Orlando et al.*, Chem. Comm., 1966, 714); *tert*-butyl-*p*-benzoquinone behaves similarly:

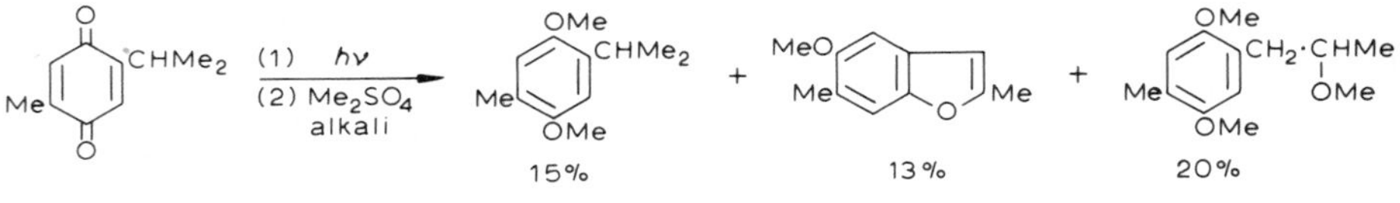

(1) hν; (2) Me_2SO_4, alkali → 15% + 13% + 20%

For u.v. spectra of some 2-arylbenzofurans see *A. S. Angeloni, F. Delmoro* and *M. Tramontini*, Ann. Chim. Rome, 1963, **60,** 1751; i.r. spectra of a number of 2-alkylbenzofurans, *F. Binon* and *C. Goldenberg*, Bull. Soc. chim. Belges, 1965, **74,** 306; n.m.r. of monomethylbenzofurans, their long-range and other proton–proton couplings, *J. A. Elvidge* and *R. G. Foster*, J. chem. Soc., 1964, 981.

TABLE 9

ALKYL- AND ARYL-BENZOFURANS

Substituent	b.p. (°C)	Ref.
2-Methyl	197	1,2
3-Methyl	196	3
5-Methyl	197–199	3
6-Methyl	192–193	3
7-Methyl	190–191	3
2,3-Dimethyl	56–57/15 mm	4
2,5-Dimethyl	220	5
2,6-Dimethyl	217–218	3
2,7-Dimethyl	215/749 mm	3,6
3,5-Dimethyl	222	3
3,6-Dimethyl	222	3
3,7-Dimethyl	217–217	3
2,3,4-Trimethyl	45–47/2 mm	4
2,3,5-Trimethyl	110/10 mm	4,6
2,3,6-Trimethyl	112/13 mm	7
2,3,7-Trimethyl	45–49/3 mm	4
3-Ethyl	217	8
2-*n*-Propyl	61–62/0.75 mm	9
2-Benzyl	180–185/15 mm	10
2-Vinyl	74/1.8 mm	11
2-Phenyl	m.p. 120	9,12
3-Phenyl	m.p. 42	13
2,3-Diphenyl	m.p. 123	14
5-Methyl-2,3-diphenyl	m.p. 114	15
7-Methyl-2,3-diphenyl	m.p. 64–65	15

References

1 *L. Claisen*, Ann., 1919, **418,** 69; Ber., 1920, **53,** 322.
2 *T. Sheradsky*, Tetrahedron Letters, 1966, 5225.
3 *R. Stoermer*, Ann., 1900, **312,** 237; *W. R. Boehme*, Org. Synth., 1953, **33,** 43.
4 *B. Sila*, Rocz. Chem., 1968, **42,** 1773.
5 *K. von Auwers* and *L. Anschütz*, Ber., 1921, **54,** 1559.
6 *Von Auwers*, Ann., 1921, **422,** 133.
7 *R. Royer et al.*, Compt. rend., Ser. C, 1966, **262,** 1286.
8 *Stormer* and *E. Barthelmes*, Ber., 1915, **48,** 62.
9 *C. E. Castro, E. J. Gaughan* and *D. C. Owsley*, J. Org. Chem., 1966, **31,** 4071.
10 *Stoermer et al.*, Ber., 1924, **57,** 72.
11 *E. D. Elliot*, J. Amer. chem. Soc., 1951, **73,** 754.
12 *S. Kawai et al.*, Ber., 1939, **72,** 1146.
13 *Stoermer* and *O. Kippe*, *ibid.*, 1903, **36,** 3992.
14 *B. I. Arventi*, Bull. Soc. chim. Fr., 1936, **3,** 598.
15 *Arventiev* and *H. Offenberg*, C.A., 1962, **56,** 11554c.

(iv) Halogenobenzofurans

Chloro- and bromo-benzofurans substituted in the furan ring are usually prepared by direct halogenation of the parent substance, chlorination at low temperatures affording a mixture of 2- and 3-chlorobenzofurans in which the former predominates; addition products are often intermediates which may lose hydrogen halide only on heating or treatment with alkali. Thus benzofuran forms a dibromide converted by heat into 2-bromobenzofuran and by alcoholic potash into the 3-bromo compound. Further bromination of either isomeride gives 2,3-dibromobenzofuran.

3-Chloro-2-phenylbenzofuran may be obtained by the addition of thionyl chloride to a solution of 2′-hydroxybenzoin in pyridine (*D. J. Cooper* and *L. N. Owen*, J. chem. Soc., C, 1966, 533):

6-Chloro-2-phenylbenzofuran may be prepared by the Sandmeyer reaction following the diazotisation of 6-amino-2-phenylbenzofuran (*A. S. Angeloni* and *M. Tramontini*, Ann. Chim., Rome, 1965, **55,** 1029).

Bromination of 2-phenylbenzofuran using molar quantities in dry carbon disulphide at $-10°$ affords 3-bromo-2-phenylbenzofuran; with 2 moles of bromine, 3,6-dibromo-2-phenylbenzofuran is formed (*idem*, Boll. Sci. Fac. Chim. Ind., Bologna, 1963, **21,** 243; C.A., 1964, **60,** 15808e).

Some halogeno derivatives of benzofurans are obtained by indirect methods, *e.g.*, 7-chloro-2,3-dimethylbenzofuran from *o*-chlorophenol (*C. Goldenberg, F. Binon* and *E. Gillyns*, Chim. Therap., 1966, 221; C.A., 1966, **65,** 16925g):

2,4-Dibromophenol reacts with cuprous phenylacetylide and cuprous *n*-propylacetylide in pyridine to give 5-bromo-2-phenylbenzofuran and 5-bromo-2-*n*-propylbenzofuran, respectively (*C. E. Castro, E. J. Gaughan* and *D. C. Owsley*, J. org. Chem., 1966, **31,** 4071).

Some bromo-derivatives of benzofurans have been prepared using compounds in which the 2-position is blocked by an easily removable group (*V. S. Salvi* and *S. Sethna*, J. Indian chem. Soc., 1967, **44,** 135):

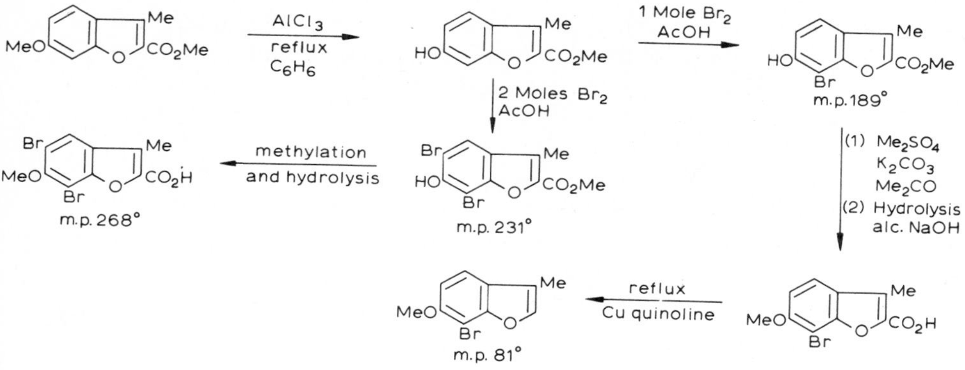

4,5,6,7-Tetrafluorobenzofuran is obtained by the reduction and dehydration of the related 3-oxo-2,3-dihydrobenzofuran, prepared by the cyclisation and decarboxylation of ethyl 6-ethoxycarbonyl-2,3,4,5-tetrafluorophenoxyacetate or by the reaction of 2,3,4,5-tetrafluorophenoxyacetic acid with butyl-lithium followed by carbonation (*G. M. Brooke* and *B. S. Furniss*, J. chem. Soc., C, 1967, 869). Tetrafluorobenzofuran reacts with sodium methoxide to afford a mixture of 4-, 6-, and 7-methoxytrifluorobenzofurans in the ratio 27:57:16, acetylates to give a mixture of the 2- and 3-acetyl derivatives in the ratio 85:15, and with butyl-lithium yields the 2-lithio-compound, which may be converted into the 2-carboxylic acid and the 2-carbaldehyde (*Brooke, Furniss* and *W. K. R. Musgrave, ibid.*, 1968, 580).

TABLE 10

HALOGENOBENZOFURANS

Substituent	*b.p. (°C)*	*Ref.*	*Substituent*	*m.p. (°C)*	*Ref.*
2-Bromo	221–223	1	2,3-Dichloro	25–26	1
3-Bromo	m.p. 39	2	2,3-Dibromo	27	1
2-Chloro	203	1	3-Chloro-2-phenyl	47	3
3-Chloro	199–201	1	5-Bromo-2-phenyl	158–159	4
5-Chloro	215–217	1	5-Bromo-2-*n*-propyl	b.p. 102–103	4
7-Chloro	210–212	1	7-Bromo-2,3-dimethyl	34.5	5

References
1 *R. Stoermer*, Ann., 1900, **312,** 237; **313,** 78.
2 *Stoermer* and *B. Kahlert*, Ber., 1902, **35,** 1633.
3 *D. J. Cooper* and *L. N. Owen*, J. chem. Soc., C, 1966, 533.
4 *C. E. Castro, E. J. Gaughan* and *D. C. Owsley*, J. org. Chem., 1966, **31,** 4071.
5 *C. Pene et al.*, Bull. Soc. chim. Fr., 1966, 586.

Halogen atoms, whether in the furan or the benzene ring, are firmly held. In most cases the orientation has been confirmed by preparing the 2- or 3-halogenobenzofurans from the corresponding oxodihydrobenzofuran and phosphorus halide.

Some halogenobenzofurans are listed in Table 10.

(v) Nitro- and amino-benzofurans

Direct nitration of benzofuran in acetic acid gives 2-nitrobenzofuran, which is transformed by sodium ethoxide into 2-oximino-3-oxo-2,3-dihydrobenzofuran:

$\xrightarrow{HNO_3}$ NO_2 $\xrightarrow{NaOEt}$ O NOH

5,7-Dinitrobenzofurans may be obtained by the Fischer cyclisation of *O*-(2,4-dinitrophenyl)oximes, prepared by condensing *tert*-butyl *N*-hydroxycarbamate with chloro-2,4-dinitrobenzene, treating the resulting carbamate with trifluoroacetic acid and condensing the product *O*-(2,4-dinitrophenyl)hydroxylamine with either aldehydes or ketones (*T. Sheradsky*, J. heterocycl. Chem., 1967, **4,** 413). 2-Methyl-5,7-dinitrobenzofuran and some related nitro derivatives of benzofurans are formed by refluxing the necessary *O*-aryloxime, obtained from the aryl halide and the alkali metal salt of the oxime, with alcoholic hydrogen chloride (*A. Mooradian* and *P. E. Dupont*, *ibid.*, 1967, **4,** 411; *Mooradian*, Tetrahedron Letters, 1967, 407).

A double ring-synthesis affords 5-nitro-2-methylbenzofuran (*W. J. Hale*, Ber., 1912, **45.** 1596):

$O_2N{-}CH(CHO)_2$ + $CH_2{\cdot}CH_2{\cdot}COMe$ / CO / CH_3 $\xrightarrow{NaOH}$ O_2N COMe $\xrightarrow{ZnCl_2}$ O_2N Me

The nitration of 2,3-diphenylbenzofuran yields the 6-nitro derivative (*B. Arventiev* and *H. Offenberg*, C.A., 1962, **56,** 11554c). Treatment of 2-phenylbenzofuran with nitric acid in anhydrous acetic acid at 50–60° yields a mixture of 3-nitro- and 6-nitro-2-phenylbenzofuran, whereas at 80–90° with acetic anhydride as solvent the 3,6-dinitro derivative is obtained (*A. S. Angeloni* and *M. Tramontini*, Ann. Chim., Rome, 1965, **55,** 1028).

2-**Nitrobenzofuran,** m.p. 134°, 3-*bromo-2-nitrobenzofuran*, m.p. 132° (*R. Stoermer*, Ann., 1900, **312,** 237; *Stoermer* and *B. Kahlert*, Ber., 1902, **35,** 1633). The potassium salt of 2-nitrobenzofuran-3-ol is prepared from 3-bromo-2-nitrobenzofuran by treatment with a secondary amine followed by alkali (*Stoermer*, Ber., 1909, **42,** 199; *Stoermer* and *K. Brachmann*, Ber., 1911, **44,** 316). On decomposition the salt yields oxindigotin:

Br, NO_2 —R_2NH→ NR_2, NO_2 —KOH→ OK, NO_2 —H_2O→ [ ]$_2$

Oxindigotin

5-*Nitrobenzofuran*, m.p. 114–115° (*H. Erlenmeyer et al.*, Helv., 1948, **31,** 75); 6-*nitrobenzofuran*, m.p. 109–110° (*P. Rumpf* and *C. Gausser*, *ibid.*, 1954, **37,** 435); 7-*nitrobenzofuran*, m.p. 96–97° (*S. Tanaka*, J. chem. Soc., Japan, 1952, **73,** 282).

3-**Nitro-2-phenylbenzofuran,** m.p. 110–111°; 6-*nitro-2-phenylbenzofuran*, m.p. 150–151°; 3,6-**dinitro-2-phenylbenzofuran,** m.p. 192–193° (*Angeloni* and *Tramontini*, *loc. cit.*); 2,3-**diphenyl-6-nitrobenzofuran,** m.p. 165–166° (*Arventiev* and *Offenberg*, *loc. cit*); 2,3-**diphenyl-5-methyl-6-nitrobenzofuran,** m.p. 142–143°; 2,3-*diphenyl-7-methyl-6-nitrobenzofuran*, m.p. 158° (*idem*, C.A. 1963, **59,** 7464h); 5-**chloro-2,3-diphenyl-6-nitrobenzofuran,** m.p. 148–149°; 4,7-**dimethyl-2,3-diphenyl-6-nitrobenzofuran,** m.p. 189° (*idem*, C.A., 1964, **60,** 493d).

When a mixture of benzofuran-2-carbaldehyde, nitromethane, methylamine hydrochloride, sodium carbonate and ethanol is allowed to stand at 25–30° for two weeks 2-(2′-*nitrovinyl*)*benzofuran* (XX; R = H), m.p. 106°, is formed; 2-(2′-*methyl-2′-nitrovinyl*)*benzofuran* (XX; R = Me), m.p. 103°, may be prepared similarly (*K. Kobayashi* and *H. Suzuki*, Jap. P. 1429/1964; C.A., 1964, **60,** 11987A):

CHO + $MeNO_2$ ⟶ $CH:CRNO_2$

(XX)

Reduction of 2-nitrobenzofuran gives, not 2-aminobenzofuran but the product of its hydrolysis, 2-oxo-2,3-dihydrobenzofuran (XXI), which is also formed by hydrolysis of methyl 2-benzofurylcarbamate obtained from coumarilamide (*I. J. Rinkes*, Rec. Trav. chim., 1932, **51,** 349):

NO_2 ⟶ NH_2 ⟶ NH ⟶ (XXI)

$CONH_2$ —NaOCl, MeOH→ $NHCO_2Me$ —HCl→ (XXI)

Coumarilamide

Methyl 2-benzofuryl carbamate, m.p. 139°; 5-**aminobenzofuran,** b.p. 133–134°/14 mm (*Erlenmeyer et al.*, *loc. cit.*); 6-*dimethylamino-3-methylbenzofuran*, m.p. 58° (*H. von Pechmann* and *M. Schaal*, Ber. 1899, **32,** 3690).

Benzofurans with an amino group in a side chain may be obtained in the following way (*Byk-Gulden Lomberg*, G.P. 1,203,277/1963):

3-**Acetylamino**-2-**phenylbenzofuran,** m.p. 202–203°; 6-**amino**-2-**phenylbenzofuran,** m.p. 200–201° (*Angeloni* and *Tramontini, loc. cit.*).

(vi) Hydroxybenzofurans

Benzofurans hydroxylated in the 2- or the 3-position behave rather as the tautomeric 2- or 3-oxo-2,3-dihydrobenzofurans (2- or 3-coumaranones), unless there is also an acyl substituent in the 3- or 2-position. Successive methylation, hydrolysis, and decarboxylation of ethyl 3-hydroxybenzofuran-2-carboxylate gives 3-**methoxybenzofuran,** b.p. 109–110°/17 mm (*K. von Auwers*, Ann., 1912, **393,** 357). Keten diethyl acetal and *p*-benzoquinone give 2-ethoxy-5-hydroxybenzofuran (p. 143). 5-Hydroxybenzofurans may also be synthesised *via* the reaction between benzoquinones and enamines (*G. Domschke*, J. pr. Chem., 1966, **32,** 144). 4-, 5-, 6- and 7-Hydroxybenzofurans react as normal phenols; several have been prepared from hydroxy- or alkoxy-salicylaldehydes by synthetic method (*2*) (p. 142) (*T. Reichstein et al.*, Helv., 1935, **18,** 816), and others have been made by method (*5*).

cis-4-(2-Furyl)-3-methoxycarbonylbut-3-enoic acid is cyclised by acetic anhydride and sodium acetate to methyl 4-acetoxybenzofuran-6-carboxylate, hydrolysed by alcoholic potash to 4-hydroxybenzofuran-6-carboxylic acid. The ester may also be transformed to 4-methoxybenzofuran (*S. M. Abel-Wahhab* and *L. S. El-Assal*, J. chem. Soc., C, 1968, 867):

7-Hydroxybenzofuran is present in coal tar (*O. Kruber* and *W. Schmieden*,

Ber., 1939, **72,** 658); and other hydroxybenzofurans have been encountered in the degradation of rotenone, usnic acid, and karanjin. 5-Methoxybenzofuran is the simplest of the naturally occurring benzofurans (*J. H. Birkinshaw, P. Chaplen* and *W. P. K. Findlay,* Biochem. J., 1957, **66,** 188). For the biosynthesis of 5-methoxybenzofuran see *J. D. Bu'lock, A. T. Hudson* and *B. Kaye,* Chem. Comm., 1967, 814.

TABLE 11

HYDROXYBENZOFURANS

Substituents	*m.p.(°C)*	*Ref.*
4-Hydroxy	57.5	1
4-Methoxy	b.p. 220–222	2, 3
4-Hydroxy-6-carboxylic acid	243–244	3
5-Methoxy	31.5	4
5-Hydroxy-2,3-dimethyl	79–80	5, 6
5-Hydroxy-2,3-diphenyl	158–160	7
6-Hydroxy	57–58	8
6-Hydroxy-3-methyl	103	9
6-Hydroxy-2,3-dimethyl	108	6
6-Hydroxy-2,3-diphenyl	117.5	7
7-Hydroxy	43	
7-Hydroxy-2,3-dimethyl	100–101	6
4,6-Dimethoxy	b.p. 109/0.15 mm	10
6,7-Dihydroxy	72.5–74	11
4-Methoxy-2,3,7-trimethyl	71	12
5-Hydroxy-2,3,6-trimethyl	124	12
6-Methoxy-2,3,5-trimethyl	60	12
7-Hydroxy-2,3,4-trimethyl	92	12
7-Hydroxy-2,3,6-trimethyl	90.5	12
4,7-Dihydroxy-2,3,6-trimethyl	127	12

References

1 *T. Reichstein* and *R. Hirt,* Helv., 1933, **16,** 121.
2 *D. B. Limaye,* C.A., 1937, **31,** 2206.
3 *S. M. Abdel-Wahhab* and *S. L. El-Assal,* J. chem. Soc., C, 1968, 867.
4 *G. R. Ramage* and *C. V. Stead,* J. chem. Soc., 1953, 3602.
5 *A. Robertson et al., ibid.,* 1953, 1262.
6 *H. Kipper* and *H. Randsepp,* Tr. Tallinsk. Politekhn. Inst., Ser. A., 1965, **230,** 67; C.A., 1967, **66,** 10800b.
7 *F. R. Japp* and *A. N. Meldrun,* J. chem. Soc., 1899, **75,** 1035.
8 *A. Sonn* and *E. Patschke,* Ber., 1925, **58,** 96.
9 *K. Fries* and *M. Nöhren,* Ber., 1925, **58,** 1027.
10 *R. T. Foster* and *Robertson,* J. chem. Soc., 1939, 921.
11 *H. Bickel* and *H. Schmid,* Helv., 1953, **36,** 664.
12 *R. Royer et al.,* Bull. Soc. chim. Fr., 1965, 2607.

Acylation of 2,3-dimethyl-4-hydoxybenzofuran with acetic acid, phenylacetic acid, 2-methoxyphenylacetic acid, or 2,4,5-trimethoxyphenylacetic acid and polyphosphoric acid, affords mixtures of the 5- and 7-acyl derivatives. The isomeric 2,3-dimethyl-6-hydroxybenzofuran yields the 5-acyl compounds (*Y. Kawase, M. Nanbu* and *F. Miyoshi*, Bull. chem. Soc., Japan, 1968, **41,** 2672). Compounds of type (XXII, R = H and Me, R′ = NMe_2, NEt_2, piperazino) have been prepared by boiling 1 mole of β-chloroethylamine or β-chloropropylamine hydrochloride in ethanol with 2,3-dimethyl-4-hydroxybenzofuran (*M. Descamps, J. Vander Elst* and *F. Binon*, Ing. Chim. Fr., 1967, **49,** 34; C.A., 1968, **69,** 59013j).

OCHR·CH_2R′ Me Me

(XXII)

Individual hydroxybenzofurans are collected in Table 11.

(vii) Alcohols, aldehydes and ketones

By reduction with lithium tetrahydridoaluminate coumarilic acid affords 2-**benzofurylcarbinol** (*2-hydroxymethylbenzofuran*), m.p. 26°, also obtained by the Cannizzaro reaction from the related aldehyde. It yields *2-chloromethylbenzofuran*, m.p. 33–36°, which reacts abnormally with magnesium, giving by ring-fission *o*-hydroxyphenylallene, in which the ring is easily reconstituted (*R. Gaertner*, J. Amer. chem. Soc., 1951, **73,** 4400):

CH_2MgCl → CH:CH:CH_2 OMgCl → CH:CH:CH_2 OH —alkali→ Me

2-Methylbenzofuran can be chloromethylated directly in position 3 giving 3-chloromethyl-2-methylbenzofuran; by treatment with magnesium and carbonation, the product gives an acid (XXIII) distinct from the expected 2-methylbenzofuran-3-acetic acid (*Gaertner, ibid.*), 1952, **74,** 5319):

CH_2MgCl Me —CO_2→ CH_2 CO_2H Me (XXIII) —Δ→ Me Me + CO_2

Chloromethylation of 2-phenylbenzofuran affords *3-chloromethyl-2-phenylbenzofuran*, m.p. 74–75°, and *bis(2-phenyl-3-benzofuranyl)methane*, m.p. 177–178° (*A. S. Angeloni* and *M. Tramontini*, Ann. Chim., Rome, 1965, **55,** 1028):

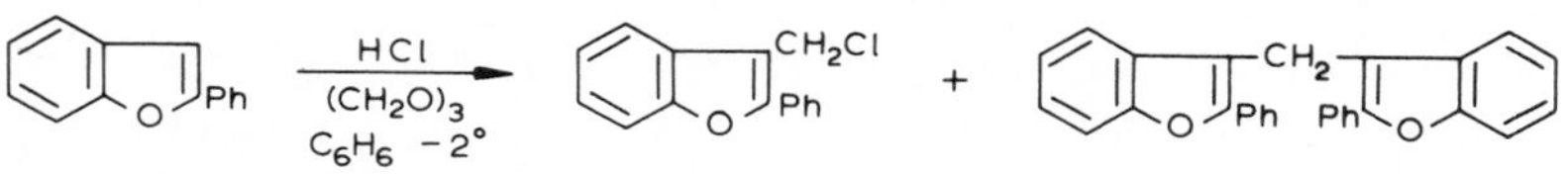

3-**Benzofurylcarbinol** (3-*hydroxymethylbenzofuran*), m.p. 47°, b.p. 175°/0.01 mm, has been prepared by the reduction of methyl benzofuran-3-carboxylate with lithium tetrahydridoaluminate (*L. Capuano*, Ber., 1965, **98,** 3659).

Benzofuran-2-**carbaldehyde,** b.p. 130–131°/13 mm, 135–136°/18 mm, has been prepared from coumariloyl cyanide (2-cyanobenzofuran) (*T. and I. Reichstein*, Helv., 1930, **13,** 1275), and directly from benzofuran by treatment with dimethylformamide and phosphorus oxychloride, a method of general application (*N. P. Buu-Hoï et al.*, J. chem. Soc., 1955, 3688; Bull. Soc. chim. Fr., 1962, 1875); benzofuran-2-carbaldehydes have also been made by the Gattermann process (*H. F. Birch* and *A. Robertson*, J. chem. Soc., 1938, 306). In contrast with furfural (p. 112), the aldehyde gives only a yellow colour with aniline acetate. In absolute alcohol it reacts with acetone on treatment with 10% sodium hydroxide at −5° to give *bis(benzofurylidene)acetone*, m.p. 120° (*M. Descamps, F. Binon* and *J. van der Elst*, Bull. Soc. chim. Belges, 1963, **72,** 513). The Wittig reaction of benzofuran-2-carbaldehyde with methoxymethylenetriphenylphosphorane yields an equimolar mixture of *cis*- and *trans*-2-(β-methoxyvinyl)benzofuran (XXIV and XXV) (*W. J. Davidson* and *J. A. Elix*, Tetrahedron Letters, 1968, 4589):

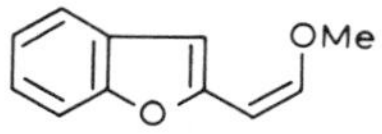

(XXIV)

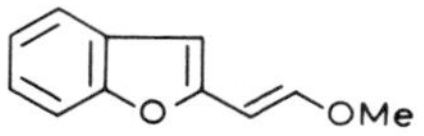

(XXV)

3-*Methylbenzofuran*-2-*carbaldehyde*, m.p. 66°, has been obtained from 3-methylcoumariloyl chloride by the Rosenmund reaction (*B. Sila*, Rocz. Chem., 1964, **38,** 1387). 3-*Chlorobenzofuran*-2-*carbaldehyde*, m.p. 75°, is given by the reaction between 3-oxo-2,3-dihydrobenzofuran, phosphorus oxychloride and dimethylformamide; similarly 3,5-*dichlorobenzofuran*-2-*carbaldehyde*, m.p. 147–148° (*Y. Anmo et al.*, Yakugaku Zasshi, 1963, **83,** 807; C.A., 1963, **59,** 15239b). **Benzofuran**-3-**carbaldehyde,** m.p. 39°, b.p. 145°/35 mm, is prepared by oxidation of 3-**benzofurylcarbinol** in carbon tetrachloride using manganese dioxide (*Capuano, loc. cit.*).

2,3-**Dimethylbenzofuran**-7-**carbaldehyde,** b.p. 93°/0.3 mm, may be obtained from salicylaldehyde and 3-chlorobutan-2-one (*C. Goldenberg, Binon* and *E. Gillyns*, Chim. Therap., 1966, 221; C.A., 1966, **65,** 16926b):

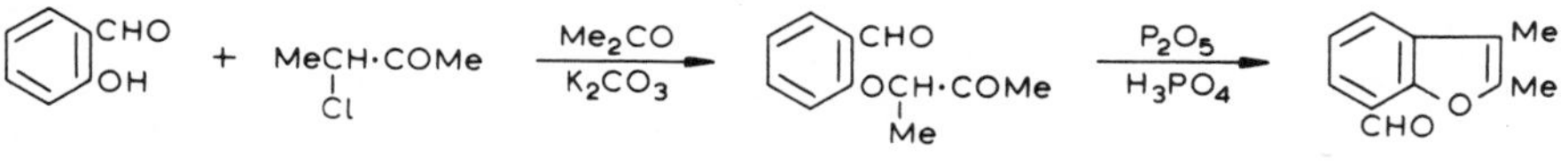

By synthetic method (*2*) with the appropriate halogenoketone, salicylaldehyde gives 2-**acetylbenzofuran,** m.p. 76° (*E. D. Elliott*, J. Amer. chem. Soc., 1951, **73,** 754), and 2-*benzoylbenzofuran*, m.p. 90–91° from which 2-*benzofuryldiphenylcarbinol*, m.p. 133–

134°, is derived (*R. C. Fuson et al.*, J. org. Chem., 1941, **6,** 845). The acetyl compound gives, with benzaldehyde, 2-*cinnamoylbenzofuran*, m.p. 114–115° (*M. Polonowski*, Bull. Soc. chim. Fr., 1953, 200); by reduction, 2-*benzofurylmethylcarbinol*, m.p. 41° (*R. L. Shriner* and *J. Anderson*, J. Amer. chem. Soc., 1939, **61,** 2705); and by reduction of the oxime, α-2-*benzofurylethylamine*, b.p. 140°/20 mm (*R. Stoermer* and *M. Schäffer*, Ber., 1903, **36,** 2863). The *p*-toluenesulphonyl derivative of the oxime of 2-acetylbenzofuran decomposes in methanol into ammonium toluenesulphonate, 2-oxo-2,3-dihydrobenzofuran (XXVI), and 3-hydroxy-2-methyl-1,4-benzopyrone (XXVII) (*T. A. Geissman* and *A. Armer*, J. Amer. chem. Soc., 1955, **77,** 1623):

(XXVI) ← $C(Me)=N\cdot SO_2\cdot C_7H_7$ → (XXVII)

A solution of 6-methoxy-3-methylbenzofuran and isobutenyl methyl ketone, Me_2C-:CHAc, in ethanol on saturation with hydrogen chloride gives 2,2-*bis*(6-*methoxy*-3-*methyl*-2-*benzofuryl*)*propane*, m.p. 140–141° (56%), and 4-(6-*methoxy*-3-*methyl*-2-*benzofuryl*)-4-*methyl*-3-*pentanone*, b.p. 123–125°/0.5 mm (23%), (*T. Abe* and *T. Shimizu*, Nippon Kagaku Zasshi, 1966, **87,** 870):

MeO, Me + Me_2C:CHAc —HCl, EtOH, 0°→ MeO, Me, CMe_2, MeO, Me + MeO, Me, Me, C—CO—CH_2Me, Me

Similarly treated 3-methylbenzofuran affords 4-(3-*methyl*-2-*benzofuryl*)-4-*methyl*-2-*pentanone* (47%), b.p. 124–125°/0.5 mm, and 2,2-*bis*(3-*methyl*-2-*benzofuryl*)*propane* (26%), b.p.170–174°/0.5 mm, m.p. 91–92°.

3-**Acetylbenzofuran,** m.p. 37°, is produced from 3-cyanobenzofuran with methylmagnesium iodide (*M. Martynoff*, Bull. Soc. chim. Fr., 1952, 1056); 2-alkylbenzofurans undergo Friedel–Crafts acylation in position 3 in presence of stannic chloride (*Buu-Hoï, loc. cit.*).

3-Acylbenzofurans react with hydroxylamine yielding either the expected oxime or the corresponding isoxazoles (XXVIII). The ratio of the two products depends on the structure of the benzofuran used and the reaction conditions, particularly the pH (*R. Royer* and *E. Bisagni*, Bull. Soc. chim. Fr., 1963, 1746).

With hydrazine, pyrazoles (XXIX) are obtained (*Descamps*, *Binon* and *Van der Elst*, Bull. Soc. chim. Belges, 1964, **73,** 459):

CHO, Et —(1) $NH_2OH\cdot HCl$, aq. EtOH; (2) HCl or AcOH→ OH, Et, N, O (XXVIII)

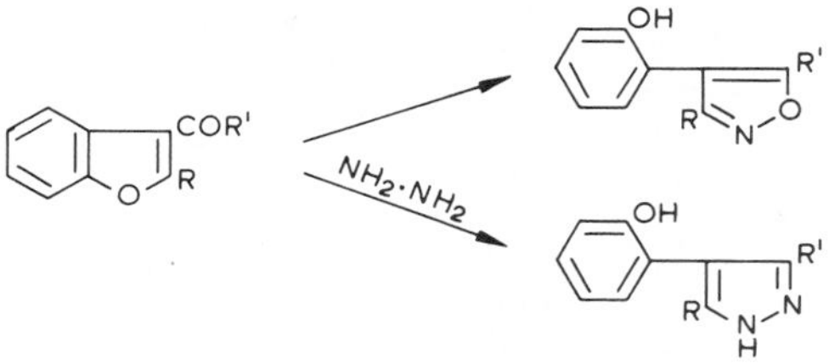

(XXIX)

2-Ethylbenzofuran-3-carbaldehyde on boiling in absolute ethanol with an appropriate amine affords a Schiff base, XXX, which on reduction with lithium tetrahydridoaluminate yields 2-ethyl-3-benzofurfurylamine (XXXI) (*R. Landi-Vittory et al.*, Farmaco, [Pavia], Ed. Sci., 1963, **18,** 465; C.A., 1963, **59,** 9941d):

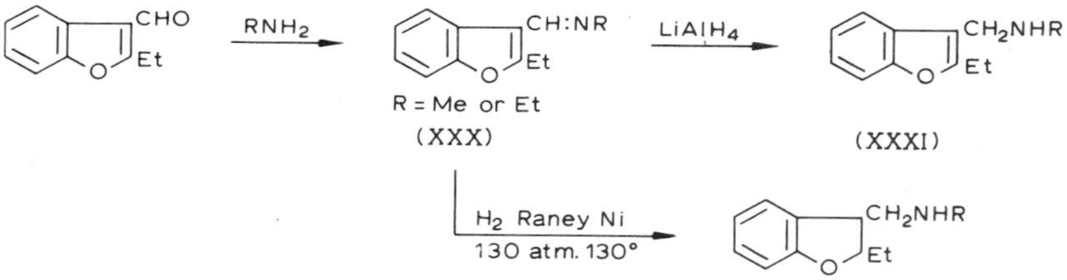

R = Me or Et
(XXX) (XXXI)

Acetylation of 2,3-dimethylbenzofuran with acetyl chloride (2 moles) in the presence of excess aluminium chloride affords *2,3-dimethyl-4,6-diacetylbenzofuran*, m.p. 101° (*Royer et al.*, Compt. rend., 1964, **258,** 5007). *7-Acetyl-2,3-dimethylbenzofuran*, m.p. 72°, is obtained from *o*-hydroxyacetophenone and 3-chlorobutan-2-one (p. 159) (*Goldenberg, Binon* and *Gillyns, loc. cit.*).

With aqueous acid, *2-benzofurancarbonyldiazomethane*, $(C_8H_5O)CO{\cdot}CH{=}N_2$, m.p. 117–118°, gives *2-hydroxyacetylbenzofuran*, m.p. 128–129° (*R. B. Wagner* and *J. M. Tome*, J. Amer. chem. Soc., 1950, **72,** 3477).

Mono- and di-alkylaminoacetylbenzofurans have been prepared by treating 2-bromoacetylbenzofuran with amines (*A. Burger* and *A. J. Deinet, ibid.*, 1945, **67,** 566).

The primary alcohol, **egonol,** $C_{19}H_{18}O_5$, m.p. 116°, occurs in Japanese egonoki fruits (*Styrax* spp.). *S. Kawai* and his associates degraded it to piperonylic acid and an acid $C_{11}H_{14}O_5$ which on monomethylation and oxidation gave isohemipinic acid (XXXII). Thermal decarboxylation of the C_{11}-acid gave dihydroconiferyl alcohol and egonol was formulated as shown:

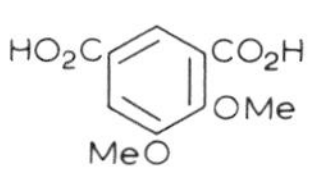

(XXXII) 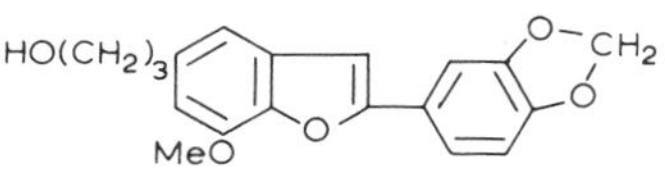 Egonol

Egonol was synthesised from XXXIII (*Kawai, T. Nakamura* and *N. Sugiyama*, Ber., 1939, **72,** 1146):

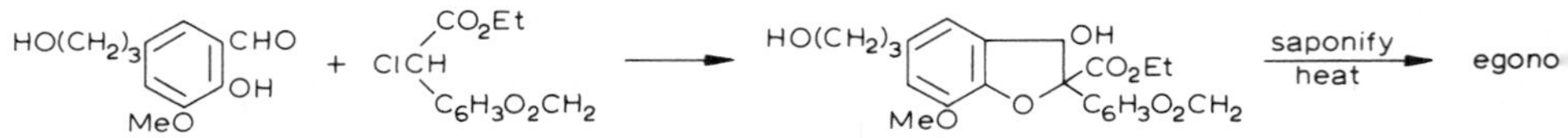
HO(CH2)3
CHO
OH
MeO
+
CO2Et
ClCH
C6H3O2CH2
HO(CH2)3
OH
CO2Et
MeO
O
C6H3O2CH2
saponify
heat
egono

(XXXIII)

The glycoside fraction isolated from the seeds of *S. officinalis* yields egonol on acid hydrolysis and a new benzofuran derivative, which has been formulated as 2-(3,4-dimethoxyphenyl)-5-(3-hydroxypropyl)-7-methoxybenzofuran (XXXIV) (*R. Segal et al.*, J. chem. Soc., C, 1967, 2402).

2-(6-Hydroxy-2-methoxy-3,4-methylenedioxyphenyl)benzofuran (XXXV) has been isolated from yeast (*M. A. P. Meisinger et al.*, J. Amer. chem. Soc., 1959, **81,** 4979; *A. F. Wagner et al.*, *ibid.*, p. 4983):

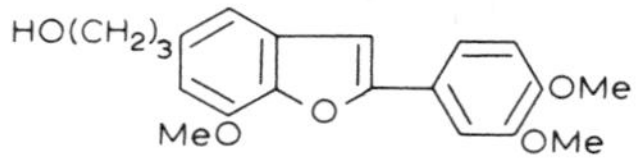
HO(CH2)3
MeO
O
OMe
OMe

(XXXIV)

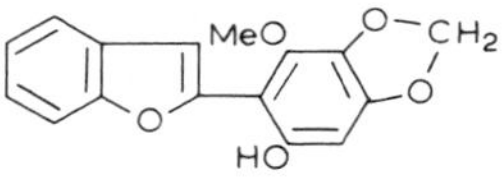
MeO
O
CH2
O
O
HO

(XXXV)

The roots of *Eupatorium purpureum* afford the yellow phenolic ketone, **euparin,** $C_{13}H_{12}O_3$, which is a conjugated diene, and by Beckmann change of its oxime gives the acetyl derivative of an aromatic amine. It has the properties of an *o*-hydroxy-ketone, and oxidation of its methyl ether gives the monomethyl ether of 5-acetyl-2,4-dihydroxybenzoic acid (XXXVI). Of the possible alternative formulae, the one shown was proved to be correct by synthesis of tetrahydroeuparin (*B. Kamthong* and *Robertson*, J. chem. Soc., 1939, 925, 933):

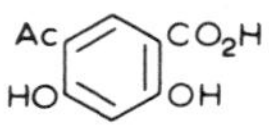
Ac
CO2H
HO
OH

(XXXVI)

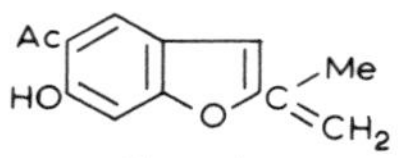
Ac
HO
O
C
Me
CH2

Euparin

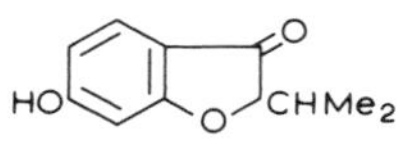
O
HO
O
CHMe2

(XXXVII)

The oxime of XXXVII was reduced to the corresponding amine, which lost ammonia when heated, giving the 6-hydroxy-2-isopropyl benzofuran (XXXVIII); this was reduced and acetylated (Hoesch) to tetrahydroeuparin (5-acetyl-6-hydroxy-2-isopropylcoumaran).

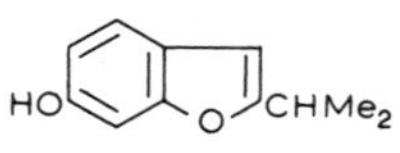
HO
O
CHMe2

(XXXVIII)

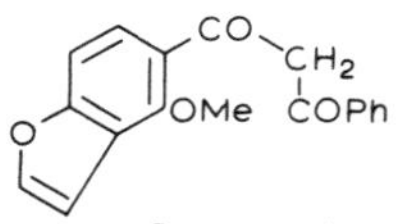
CO
CH2
OMe
COPh
O

Pongamol

Pongamol, m.p. 128–129°, from *Pongamia glabra*, is a β-diktone with the carbon skeleton of a furoflavone. The constitution is disclosed by the isolation of all four normal products of alkaline fission, and by its synthesis by condensing ethyl 4-methoxybenzofuran-5-carboxylate with acetophenone (*T. R. Seshadri et al.*, *ibid.*, 1954, 1871; 1955, 2048).

The structure of the fungal pigment **thelephoric acid** a bisbenzofuranquinone is confirmed by a synthesis involving the condensation of chloranil with two moles or 3,4-dimethoxyphenol, followed by demethylation (*J. Gripenberg*, Tetrahedron, 1960, 135):

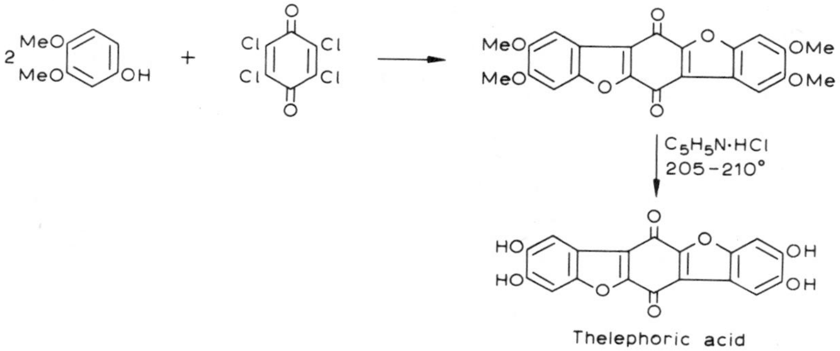

2 MeO/MeO-phenol (OH) + chloranil (Cl, Cl, Cl, Cl) ⟶ MeO, MeO … OMe, OMe bisbenzofuranquinone

$C_5H_5N{\cdot}HCl$, 205–210°

HO, HO … OH, OH

Thelephoric acid

(viii) Benzofuran carboxylic acids

(*1*) *Monocarboxylic acids.* Benzofuran-2-carboxylic acids are obtained by Perkins original method from coumarins, by synthetic route (*2*), or from sodium aryloxides and ethyl α-chloroacetoacetate:

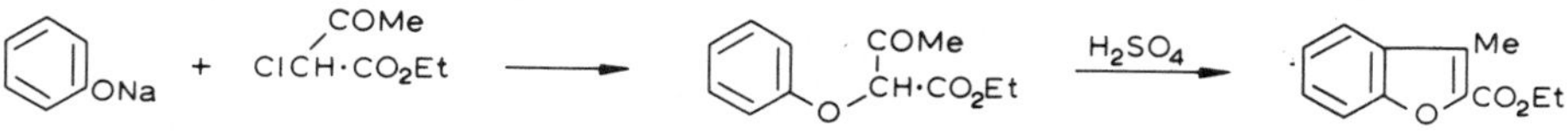

PhONa + $ClCH(COMe){\cdot}CO_2Et$ ⟶ $PhO{\cdot}CH(COMe){\cdot}CO_2Et$ $\xrightarrow{H_2SO_4}$ 3-Me, 2-CO_2Et benzofuran

Methyl 3,5-dimethylbenzofuran-2-carboxylate on treatment with anhydrous aluminium chloride in benzene affords 3-methylbenzofuran-2-carboxylic acid (*V. S. Salvi* and *S. Sethna*, J. Indian chem. Soc., 1968, **45,** 433). A number of esters of 3-methylbenzofuran-2-carboxylic acid have been prepared (*B. Sila*, Rocz. Chem., 1964, **38,** 1387).

The reaction of 3-chlorocoumarin (XXXIX) with sodium methoxide to produce methyl coumarilate (benzofuran-2-carboxylate) proceeds only if methanol is present and a mechanism has been proposed. Several 4-substituted 3,6-dichlorocoumarins have been converted into alkyl 3-substituted 5-chlorobenzofuran-2-carboxylates; previously these had been thought to be other coumarin derivatives (*M. S. Newman* and *C. K. Dalton*, J. org. Chem., 1965, **30,** 4126):

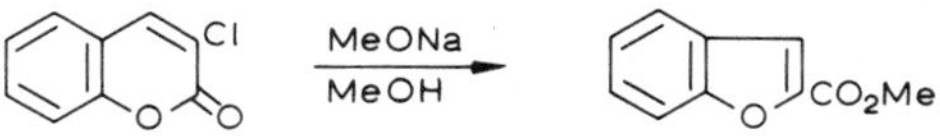

(XXXIX) $\xrightarrow[MeOH]{MeONa}$ benzofuran-2-CO_2Me

Benzofuran-3-carboxylic acid is produced in small quantity by carbonation of the Grignard reagent from 3-bromobenzofuran, and some derivatives by the following process due to *K. Fries et al.* (Ann., 1914, **402,** 327):

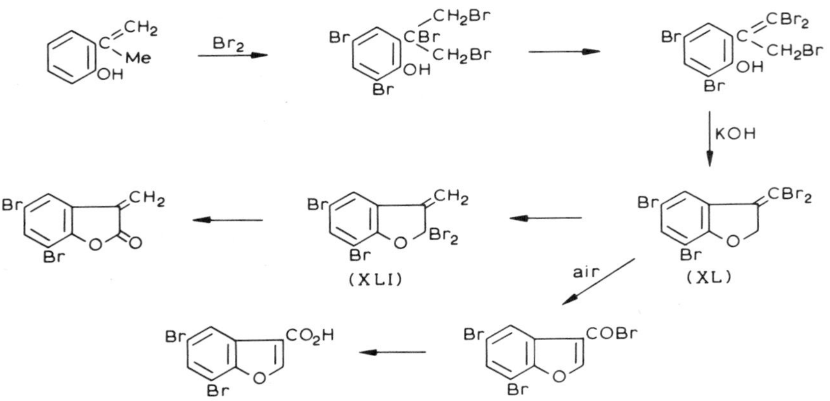

The transformation XL → XLI can be formulated as an intramolecular anionotropic migration of the aryloxy system from the nuclear to the side-chain carbon atom.

Condensation of 1,4-benzoquinone with excess ethyl 3-aminocrotonate in boiling dichloroethane gives hydroquinone and a number of products, including ethyl 3-amino-2-(2,5-dihydroxyphenyl)crotonate, which on acid-catalysed hydrolysis yields ethyl 2-(2,5-dihydroxyphenyl)acetoacetate, cyclised with zinc chloride to ethyl 5-hydroxy-2-methylbenzofuran-3-carboxylate (*S. A. Monti*, J. org. Chem., 1966, **31,** 2669):

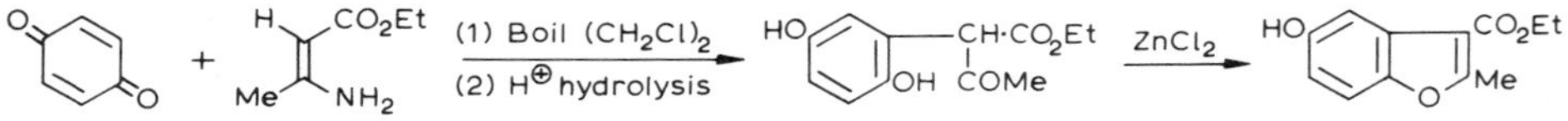

The action of *o*-tolylmagnesium bromide on ethyl coumarilate affords the alcohol XLII, whereas 2-biphenylmagnesium iodide gives the ketone XLIII (*G. A. Holmberg, F. Malmstron* and *S. O. Eriksson*, Acta Acad. Abo., Math. Phys., 1968, **28,** 7; C.A., 1969, **70,** 106314r):

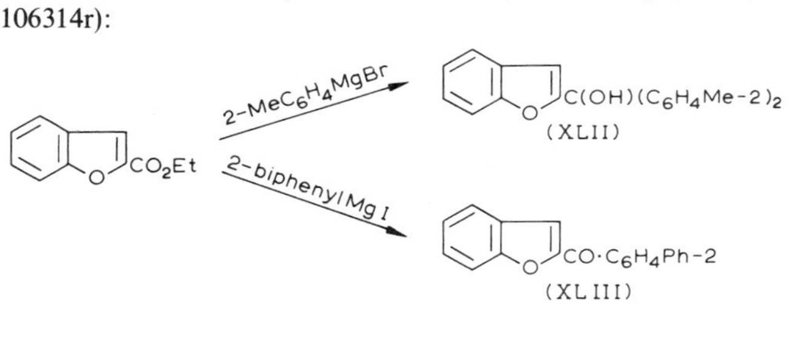

The rates of alkaline hydrolysis of carboxylic acid methyl esters in "70% dioxane" solution follow the sequence 3-benzo[*b*]furyl > 2-benzo[*b*]furyl > 2-benzo[*b*]thienyl > indol-2-yl > indol-3-yl > phenyl > 3-benzo[*b*]thienyl (*A. Feinstein, P. H. Gore* and *G. L. Read*, J. chem. Soc., B, 1968, 205).

cis-4-(2-Furyl)-3-methoxycarbonylbut-3-enoic acid is cyclised by acetic anhydride and sodium acetate to methyl 4-acetoxybenzofuran-6-carboxylate (*S. M. Adel-Wahhab* and *L. S. El-Assal*, J. chem. Soc., C, 1968, 867).

(2) *Dicarboxylic acids.* Benzofuran-2,3-dicarboxylic acid, best obtained from 2,3-dioxo-2,3-dihydrobenzofuran (coumarandione), yields the 3-monocarboxylic acid when heated:

$CO \cdot CO_2H$ / OH → $CO \cdot CO_2H$ / $O \cdot CH_2 \cdot CO_2H$ —Ac_2O→ CO_2H, CO_2H → CO_2H

The addition of 2-vinylfuran to dimethyl acetylenedicarboxylate at room temperature yields a mixture of *dimethyl benzofuran-4,5-dicarboxylate*, m.p. 64–66°, and the ester XLIV (*W. J. Davidson* and *J. A. Elix*, Tetrahedron Letters, 1968, 4589):

$CH{:}CH_2$ —$MeO_2C \cdot C{:}C \cdot CO_2Me$→ MeO_2C, CO_2Me + $CH{:}CH_2$, CO_2Me, CO_2Me

(XLIV)

Benzofuran-2-carboxylic acid (coumarilic acid), m.p. 192–193°; *ethyl ester*, m.p. 30°; *amide*, m.p. 159°; *nitrile*, m.p. 36° (*R. C. Fuson et al.*, Org. Synth., 1944, **24**, 33; *S. Tanaka*, J. Amer. chem. Soc., 1951, **73**, 872). The method of *Fuson et al.*, has been modified and a number of esters including phenyl esters and amides have been prepared (*Foo Pan* and *Tsan-Ching Wang*, J. Chinese chem. Soc., Taiwan, Ser. II, 1961, **8**, 220; C.A., 1963, **59**, 535d). The mono- and di-*N*-substituted amides of benzofuran-2-carboxylic acid have been reduced with lithium tetrahydridoaluminate to the corresponding 2-benzofurfurylamines. Ethyl benzofuran-2-carboxylate with phenylmagnesium bromide gives *diphenyl-2-benzofurylcarbinol*, m.p. 132–133°; similarly, from ethyl 3-methylbenzofuran-2-carboxylate, diphenyl-(3-methyl-2-benzofuryl)-carbinol, m.p. 82°, may be obtained (*Holmberg* and *C. E. Avellan*, C.A., 1965, **53**, 6944d). 3-*Methylbenzofuran-2-carboxylic acid*, m.p. 188° (*A. Hantzsch*, Ber., 1886, **19**, 1290). The amide when heated with phosphorus pentoxide yields 2-*cyano-3-methylbenzofuran* (3-*methylcoumarilonitrile*), m.p. 64–65°, b.p. 137–140°/4 mm (*Sila, loc. cit.*). I.r. spectra of coumarilic acid esters (*C. Goldenberg* and *F. Binon*, Bull. Soc. chim. Belges, 1966, **73**, 129); i.r. spectra of some substituted benzofuran-2-carboxylic acids (*W. W. Epstein, W. J. Horton* and *C. T. Lin*, J. org. Chem., 1965, **30**, 1246); high resolution n.m.r. spectrum of methyl 5,6-dimethylbenzofuran-2-carboxylate (*A. D. Cohen* and *K. A. McLaughlan*, Mol. Phys., 1965, **9,** 49).

Benzofuran-3-carboxylic acid, m.p. 162°, (*E. H. Huntress* and *W. H. Hearon*, J. Amer. chem. Soc., 1941, **63,** 2762), *methyl ester*, b.p. 142–143°/20 mm (*L. Capuano*, Ber., 1965, **98,** 3659), *nitrile* (*3-cyanobenzofuran*), m.p. 93° (*M. Martynoff*, Bull. Soc. chim. Fr., 1952, 1056). *Ethyl 5-hydroxy-2-propylbenzofuran-3-carboxylate*, m.p. 105–106°, obtained by adding benzoquinone to a solution of ethyl propionoacetate in ethanol containing zinc chloride at 80–85° (*W. Schoetensack*, *G. Hallmann* and *K. Haegele*, U.S.P. 3,235,566/1963).

Benzofuran-2,3-dicarboxylic acid, m.p. 259–260° (*Huntress* and *Hearon*, *loc. cit.*). A number of benzofuran-dicarboxylic acids have been prepared and a study made of the acetylation of their esters (*Y. Kawase* and *M. Takashima*, Bull. chem. Soc., Japan, 1967, **40,** 1224).

Benzofuran-2- and -3-acetic acids, the first obtained from the lower homologue *via* the acyl cyanide and glyoxylic acid, and the second by Arndt–Eistert chain-lengthening, show slight activity as plant growth hormones. Substituted benzofuran-3-acetic-2-carboxylic acids are accessible by condensation of phenols with acetonedicarboxylic acid, followed by application of the Perkin coumarilic acid synthesis to the resulting coumarin-4-acetic acids.

Benzofuran-2-acetic acid, m.p. 98° (*R. B. Wagner* and *J. M. Tome*, J. Amer. chem. Soc., 1950, **72,** 3477); **benzofuran-3-acetic acid,** m.p. 89–90° (*T. Reichstein et al.*, Helv., 1937, **20,** 883); *5-methylbenzofuran-3-acetic-2-carboxylic acid*, m.p. 224° (*B. B. Dey*, J. chem. Soc., 1915, **107,** 1606); *β-3-benzofurylalanine* decomposes at 240° (*H. Erlenmeyer* and *W. Grubenmann*, Helv., 1947, **30,** 297).

Benzofurylalkanoic acids may be prepared by the application of the Willgerodt–Kindler reaction to 2- or 3-acylbenzofurans; **benzofuran-2-propionic acid,** m.p. 108–109°; *2-ethylbenzofuran-3-acetic acid*, m.p. 84–85°; *2-ethylbenzofuran-3-propionic acid*, m.p. 70–71°; *2-ethylbenzofuran-3-butyric acid*, m.p. 77–78° (*E. Bisagni* and *R. Royer*, Bull. Soc. chim. Fr., 1962, 86).

Dieckmann cyclisation of diethyl *o*-carboxyphenoxyacetate affords ethyl 3-hydroxybenzofuran-2-carboxylate, a keto–enolic substance giving *C*- and *O*-alkyl derivatives (*K. von Auwers*, Ann., 1912, **393,** 352; Ber., 1928, **61,** 408). 4-Hydroxy-3-isopropylbenzofuran-5-carboxylic acid is produced from elliptone and from karanjin.

(b) Hydrobenzofurans

(i) 2,3-Dihydrobenzofurans (coumarans)

(*1*) 2,3-Dihydrobenzofurans are obtainable by hydrogenation of benzofurans and may be prepared by the action of hydrogen bromide on *o*-hydroxy-*β*-phenylethyl alcohols (*L. I. Smith et al.*, J. org. Chem., 1941, **6,** 229) or on *o*-allylphenols, when an *o*-hydroxy-*β*-phenylethyl bromide may be an intermediate:

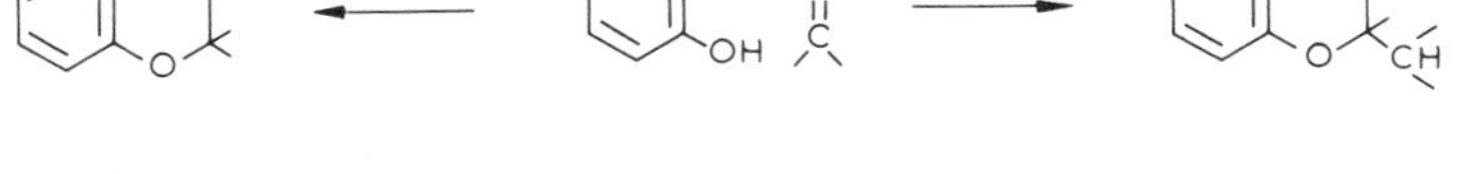

The product of the latter reaction may be a 2,3-dihydrobenzofuran or a chroman; the direction of addition conforms to Markownikow's rule of addition to an olefinic system, subject to the influence of peroxides (*C. D. Hurd* and *W. A. Hoffman*, *ibid*., 1940, **4,** 212).

(*2*) *o*-Allylphenol in the presence of diphenyl hydrogen phosphate cyclises rapidly to give 2-methyl-2,3-dihydrobenzofuran (*J. A. Miller* and *H. C. S. Wood*, J. chem. Soc., C, 1968, 1837).

(*3*) 2,3-Dihydrobenzofurans may be obtained by the photolysis of some *o*-allylphenols in benzene at 20° (*G. Frater* and *H. Schmid*, Helv., 1967, **50,** 255), and as the result of the pyrolysis of *tert*-butyl-1,4-benzoquinones (*C. M. Orlando et al.*, Tetrahedron Letters, 1966, 3003) and *o*-benzyloxybenzaldehydes (*S. P. Pappas* and *J. E. Blackwell*, *ibid*., 1966, 1171). The irradiation of 4-(2,3-dibromopropyl)-2,6-di-*tert*-butyl-4-methylcyclohexa-2,5-dienone gives 2-bromomethyl-4,7-di-*tert*-butyl-2,3-dihydrobenzofuran (*B. Miller*, Chem. Comm., 1966, 327):

Me_3C CMe_3 Me $CH_2{\cdot}CHBr{\cdot}CH_2Br$ $\xrightarrow[45°]{h\nu}$ CMe_3 CH_2Br CMe_3

(*4*) Treatment of salicylaldehyde with dimethylsulphoxonium methylide yields 2,3-dihydro-3-hydroxybenzofuran (*B. Holt* and *P. A. Lowe*, Tetrahedron Letters, 1966, 683):

CHO OH $\xrightarrow{Me_2\overset{\oplus}{S}O\overset{\ominus}{C}H_2}$ OH

o-Hydroxyacetophenone under similar conditions gives 3-methylbenzofuran; no intermediate hydroxy compound has yet been isolated.

(*5*) 2,3-Dihydrobenzofurans hydroxylated in the benzene nucleus may be prepared by condensation of quinols with allyl alcohols, allyl halides, or 1,3-dienes in presence of acid reagents (*Smith et al.*, J. Amer. chem. Soc., 1939, **61,** 2615; 1941, **63,** 1887); when the γ-carbon atom carries two hydrogen atoms, the dihydrobenzofuran predominates; when it bears none, the chroman is the sole product.

HO OH + {$HO{\cdot}CH_2{-}\dot{C}{=}\dot{C}{-}$ or; $Br{\cdot}CH_2{-}\dot{C}{=}\dot{C}{-}$ or; $CH_2{=}\dot{C}{-}\dot{C}{=}$} ⟶ HO $CH_2{\cdot}\dot{C}{=}\dot{C}{-}$ OH ⟶ HO $CH_2{-}$; HO

(*6*) The reaction between phenol and *tert*-butylacetylene at 60–75° yields some 3-*tert*-butyl-2,3-dihydrobenzofuran (*A. K. Sopkina* and *V. D. Ryabov*, Zhur. organ. Khim., 1965, **1,** 2164):

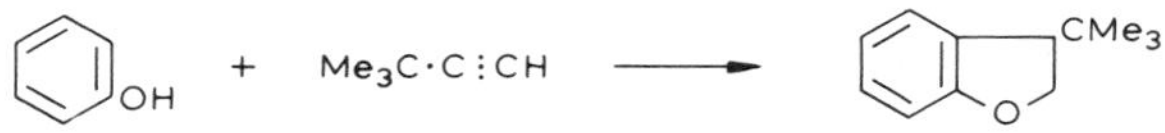

$$C_6H_5OH + Me_3C{\cdot}C{:}CH \longrightarrow \text{3-}CMe_3\text{-2,3-dihydrobenzofuran}$$

(*7*) Crotyl 2,4-dichloro-5-methylphenyl ether gives a mixture of the expected product, 2,4-dichloro-5-methyl-6-(α-methylallyl)phenol (XLV) and 2,4-dichloro-3-methyl-6-(α-methylallyl)phenol when boiled in dimethylaniline. The phenol XLV and its methyl ether are cyclised by concentrated sulphuric acid to *cis*- and *trans*-5,7-dichloro-2,3,4-trimethyl-2,3-dihydrobenzofurans, which have been separated (*D. P. Brust et al.*, J. org. Chem., 1966, **31,** 2192):

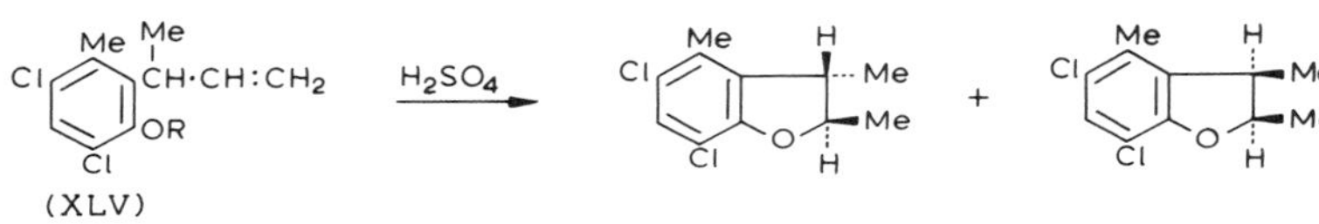

(XLV)
(R = H or Me)

$$\text{(XLV)} \xrightarrow{H_2SO_4} \textit{cis} + \textit{trans}$$

(*8*) 2,2,6-Trimethyl-2,3-dihydrobenzofuran has been obtained *via* the following abnormal Grignard reaction (*F. Bohlmann* and *C. Zdero*, Tetrahedron Letters, 1968, 3683):

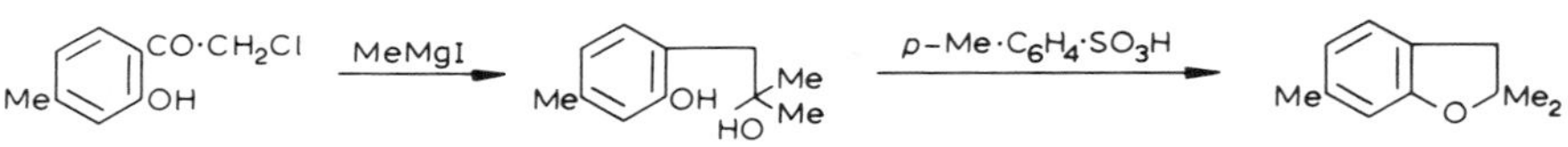

$$Me{-}C_6H_3(OH){\cdot}CO{\cdot}CH_2Cl \xrightarrow{MeMgI} Me{-}C_6H_3(OH){\cdot}CH_2{\cdot}C(OH)Me_2 \xrightarrow{p\text{-}Me{\cdot}C_6H_4{\cdot}SO_3H} \text{6-Me-2,2-}Me_2\text{-2,3-dihydrobenzofuran}$$

2,3-**Dihydrobenzofuran** *(coumaran)*, b.p. 188°, d_4^{20} 1.083, is obtained from phenol and ethylene dibromide by successive treatment with alkali and zinc chloride, from *o*-amino-β-phenylethyl alcohol by diazotisation (*G. M. Bennett* and *M. M. Hafez*, J. chem. Soc., 1941, 287), or from *o*-methoxy-β-phenylethyl alcohol with hydrogen bromide (*G. Chatelus* and *P. Cagniant*, Compt. rend., 1947, **224,** 1777). 2,3-Dihydrobenzofurans can be dehydrogenated to benzofurans by bromination with *N*-bromosuccinimide and treatment with a tertiary base (*T. A. Geissman*, U.S.P. 2,659,734/1951; C.A., 1954, **48,** 13722). Mononitration of 2,3-dihydrobenzofuran has been achieved by the use of acetyl chloride and silver nitrate in acetonitrile; the products are the 5- and 7-nitrobenzofurans. Attempted nitration with nitronium tetrafluoroborate resulted in extensive ring opening (*N. R. Raulins et al.*, J. heterocycl. Chem., 1968, **5,** 1). Treatment of 2,3-dihydrobenzofurans with sodium in pyridine results in ring cleavage and the formation of *o*-vinylphenols (*J. Gripenberg* and *T. Hase*, Acta chem. Scand., 1966, **20,** 1561):

Addition of bromine to benzofuran gives 2,3-*dibromo*-2,3-*dihydrobenzofuran*, m.p. 88°, which easily loses hydrogen bromide; phosphorus pentachloride converts 3-oxo-2,3-dihydrobenzofuran into 3,3-*dichloro*-2,3-*dihydrobenzofuran*, b.p. 115–120°/30 mm (*R. Stoermer* and *F. Bartsch*, Ber., 1900, **33**, 3175).

Although it is commonly assumed that the coupling constants of protons at C-2 and C-3 in 2,3-disubstituted-2,3-dihydrobenzofurans are larger for *cis*- than for *trans*-isomers, examples have been noticed in which this situation is reversed (*L. H. Zalkow* and *M. Ghosal*, Chem. Comm., 1967, 922).

2-**Methyl**-2,3-**dihydrobenzofuran,** b.p. 199° (*R. Adams* and *R. E. Rindfusz*, J. Amer. chem. Soc., 1919, **41,** 648; *Miller* and *Wood, loc. cit.*). A number of substituted 2,3-dihydrobenzofurans have been synthesised including the following: 2,5-*dimethyl*-, b.p. 69°/3 mm, 2,6-*dimethyl*-, b.p. 62–65°/3 mm, 5-tert-*butyl*-2-*methyl*-, b.p. 95–96°/3.5 mm, 7-*isopropyl*-2,4-*dimethyl*-, b.p. 93°/3 mm, 2,4,6-*trimethyl*-2,3-*dihydrofuran*, b.p. 74°/2 mm (*T. A. Rudol'fi et al.*, Zh. obshcheĭ Khim., 1965, **35,** 886).

White snakeroot *(Eupatorium urticaefolium)* contains several toxic constituents derived from benzofuran, **tremetone** (the major toxic component), **hydroxytremetone** and **dehydrotremetone** (*W. A. Bonner et al.*, Tetrahedron Letters, 1961, 417):

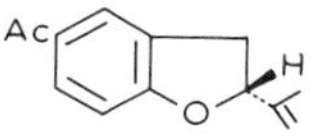

Tremetone 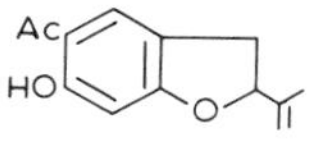 Hydroxytremetone 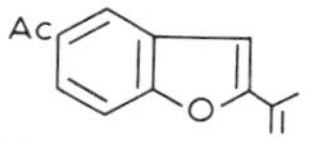 Dehydrotremetone 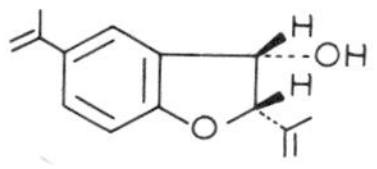 Toxol

The racemic form of tremetone has been synthesised (*J. I. DeGraw, D. M. Brown* and *Bonner*, Tetrahedron, 1963, 19) and the absolute configurations of tremetone and **toxol** have been determined (*Bonner et al.*, Tetrahedron, 1964, 1419).

The structure of **cyanomaclurin** has been derived from n.m.r. data (*G. Chakravarty* and *T. R. Deshadri*, Tetrahedron Letters, 1962, 787):

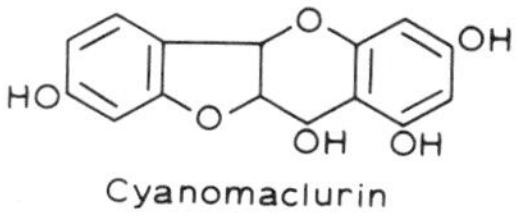

Cyanomaclurin

Numerous alkylated 5-hydroxy-2,3-dihydrobenzofurans have been prepared in connection with the study of Vitamin E and a few have shown biological activity (*Smith*, Chem. Reviews, 1940, **27**, 320; *H. M. Evans et al.*, J. org. Chem., 1939, **4,** 376); their most characteristic reaction is easy oxidation to quinones with fission of the heterocyclic ring (*P. Karrer et al.*, Helv., 1938, **21,** 939):

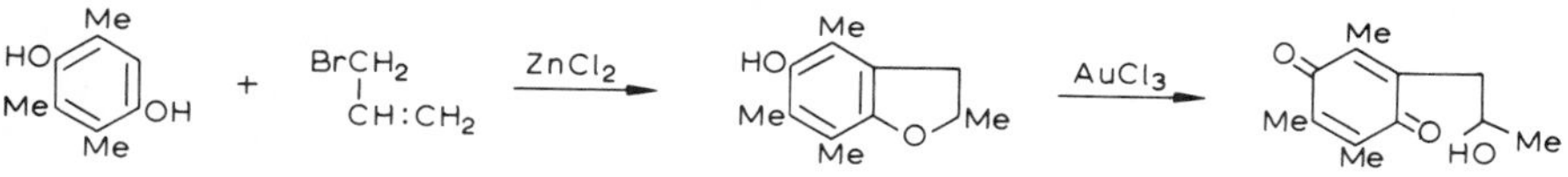

Application of synthetic method (*2*) to resorcylaldehyde 4-benzyl ether gives 6-**hydroxy**-2,3-**dihydrobenzofuran,** m.p. 60°.

The *ethyl* 2-(2,3-*dihydrobenzofuryl)carbamate*, m.p. 105°, obtained from 2,3-dihydrobenzofuran-2-carboxylic acid by the Curtius reaction, gives on hydrolysis only benzofuran and no 2-amino-2,3-dihydrobenzofuran. Reduction of 3-oxo-2,3-dihydrobenzofuran oxime affords the water-soluble, strongly basic 3-**amino**-2,3-**dihydrobenzofuran,** b.p. 122°/18 mm (*R. Stoermer* and *W. König*, Ber., 1906, **39,** 492), which yields 3-**hydroxy**-2,3-**dihydrobenzofuran** with nitrous acid.

The lactone of *o*-hydroxyphenylacetic acid, 2-**oxo**-2,3-**dihydrobenzofuran,** m.p. 49° and 28.5° (dimorphous), does not react like a hydroxybenzofuran (*S. Czaplicki et al.*, Ber., 1909, **42,** 827). By condensation with benzaldehyde the lactone or the acid gives 3-*benzylidene*-2-*oxo*-2,3-*dihydrobenzofuran*, m.p. 76°. Alkaline degradation of some naturally occurring 3-hydroxyflavanones gives hydroxylated benzylidene-2-oxo-2,3-dihydrobenzofurans, which can also be prepared by condensing phenylpyruvic acid with phenols (*Gripenberg* and *B. Juselius*, Acta chem. Scand., 1953, **7,** 1323; 1954, **8,** 734; *D. Molho et al.*, Bull. Soc. chim. Fr., 1954, 1397):

HO–C6H4–OH + CO·CH2Ph / CO2H —AlCl3→ [HO-benzofuranone with =CHPh]

The 4-, 5-, 6-, and 7-methyl-2-oxo-2,3-dihydrobenzofurans have been prepared by reducing the cyanohydrins of the methylsalicylaldehydes with hydriodic acid (*O. Aubert et al.*, Acta chem. Scand., 1952, **6,** 433):

Me–C6H3(OH)–C(H)(OH)CN —HI→ [Me-2-oxo-2,3-dihydrobenzofuran]   [3-OH-2-oxo-2,3-dihydrobenzofuran] (XLVI)

Indirect hydrolysis of salicylaldehyde cyanohydrin through the imino-ether gives 3-**hydroxy**-2-**oxo**-2,3-**dihydrobenzofuran** (XLVI), m.p. 107–108° (*K. Ladenburg et al.*, J. Amer. chem. Soc., 1936, **58,** 1292).

2-Oxo-2,3-dihydrobenzofuran reacts with pyridine and aromatic acid halides to give dark-red polyene colouring matters of type XLVII (*P. Pfeiffer* and *E. Enders*, Ber., 1951, **84,** 313):

[2-oxo-2,3-dihydrobenzofuran] —Ar·COCl / C5H5N→ [benzofuranone]=CAr·N:CH·CH=CH·CH=CH–[2-hydroxybenzofuran] (XLVII)

With pyridine and acetic anhydride it gives 3-acetyl-2-oxo-2,3-dihydrobenzofuran (*T. A. Geissman* and *A. Armen*, J. Amer. chem. Soc., 1955, **77,** 1623).

3-**Oxo**-2,3-**dihydrobenzofuran[benzofuran**-3(2*H*)-**one]**, steam-volatile needles, m.p. 102°, is most effectively prepared by treating with alkali *o*-hydroxy-*ω*-chloroacetophenone, the product of Fries rearrangement of phenyl chloroacetate (*K. Fries* and *W.*

Pfaffendorf, Ber., 1910, **43,** 212). It is also produced smoothly from *o*-methoxybenzoyldiazomethane with acid (*A. K. Bose* and *P. Yates*, J. Amer. chem. Soc., 1952, **74,** 4703):

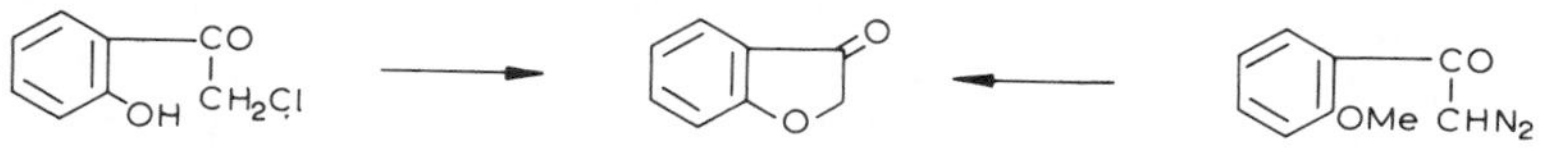

3-Oxo-2,3-dihydrobenzofuran dissolves in aqueous sodium hydroxide, but otherwise reacts as an oxodihydrobenzofuran rather than as a hydroxybenzofuran: *oxime*, m.p. 159°. On the other hand, 2-acyl-3-oxo-2,3-dihydrobenzofurans behave as enols (*K. von Auwers* and *E. Auffenberg*, Ber., 1919, **52,** 92). Most characteristic is the reactivity of the 2-methylene group of 3-oxo-2,3-dihydrobenzofuran: nitrous acid gives the 2-oxime of 2,3-dihydrobenzofuran-2,3-dione; nitrosodimethylaniline the 2-*p*-dimethylaminoanil (*P. Chovin*, Bull. Soc. chim. Fr., 1944, **11,** 82); and benzenediazohydroxide the 2-phenylhydrazone.

Benzaldehyde gives 2-benzylidene-3-oxo-2,3-dihydrobenzofuran (*vide infra*), while alipathic aldehydes and ketones may also yield alkylidenebis(3-oxo-2,3-dihydrobenzofurans) (*R. L. Shriner* and *J. Anderson*, J. Amer. chem. Soc., 1938, **60,** 1415). Chlorine and bromine likewise attack the 2-position, giving products in which the halogen is highly mobile. Acylation of 3-oxo-2,3-dihydrobenzofuran gives the enol ester, 3-acyloxybenzofuran, while alkylation may be effected either on oxygen or on carbon (*von Auwers et al.*, Ber., 1917, **50,** 1149, 1585; 1919, **52,** 77, 92).

Catalytic reduction of a 3-oxo-2,3-dihydrobenzofuran may afford successively the benzofuran and the 2,3-dihydro derivative (*G. R. Ramage* and *C. V. Stead*, J. chem. Soc., 1953, 3602). The Grignard reagent usually gives a 3-substituted benzofuran by dehydration of the normal carbinol (*Stoermer* and *E. Barthelmes*, Ber., 1915, **48,** 62):

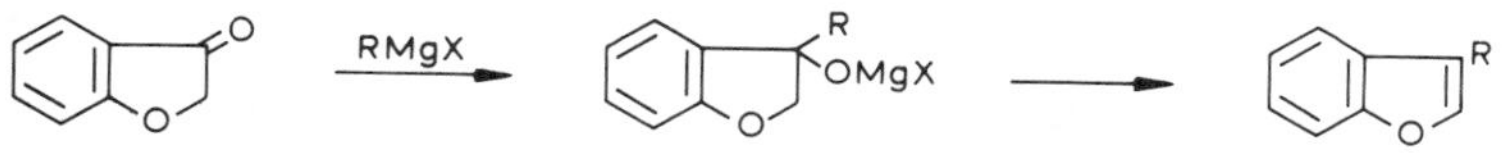

In the reactions of 4-chloro-3-methyl- and 3,4-dimethyl-phenoxyacetyl chlorides with aluminium chloride in benzene the major cyclic product is the 5,6-disubstituted 3-oxo-2,3-dihydrobenzofuran (*M. H. Palmer* and *N. M. Scollick*, J. chem. Soc., C, 1968, 2833):

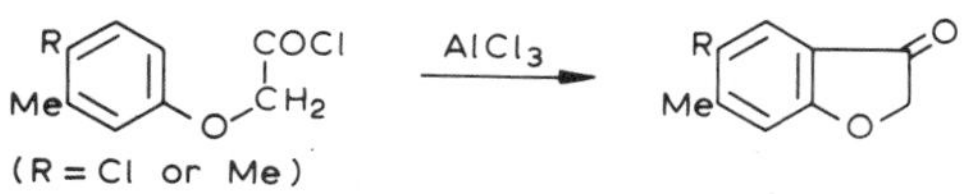

(R = Cl or Me)

It appears that electron-releasing substituents assist decarbonylation and cyclisation to the 3-oxo-2,3-dihydrobenzofuran [benzofuran-3(2*H*)-one], while electron-attracting substituents lead to a high proportion of the 2-aryloxyacetophenone (*Palmer* and *G. J. McVie*, J. chem. Soc., B, 1968, 745).

Benzofuran-3(2*H*)-ones may also be prepared from *o*-anisoyldiazomethane and methyl orthoformate in the presence of catalytic amounts of boron trifluoride dietherate. Similarly ethyl orthothioformates afford the dithioacetal XLVIII, which on pyrolysis with potassium hydrogen sulphate gives XLIX (*A. Schoenberg, K. Praefcke* and *J. Kohtz*, Ber., 1966, **99**, 3076):

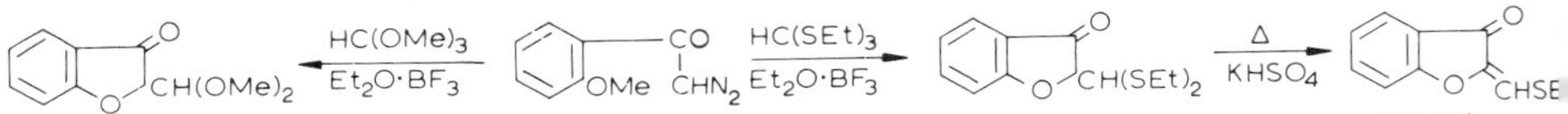

$CH(OMe)_2$ ← $HC(OMe)_3$ / $Et_2O{\cdot}BF_3$ — CO, OMe, CHN_2 — $HC(SEt)_3$ / $Et_2O{\cdot}BF_3$ → $CH(SEt)_2$ (XLVIII) — Δ / $KHSO_4$ → CHSEt (XLIX)

2-**Methyl**-3-**oxo**-2,3-**dihydrobenzofuran,** b.p. 119°/15–16 mm (*von Auwers*, Ber., 1919, **52,** 113); 5-, 6-, and 7-*methyl*-3-*oxo*-2,3-*dihydrobenzofurans*, m.p. 54°, 86° and 89°, respectively (*L. Higginbotham* and *H. Stephen*, J. chem. Soc., 1920, **117,** 1534); 2,5-*dimethyl*-, b.p. 123–126°/13 mm and 2,2,5-*trimethyl*-3-*oxo*-2,3-*dihydrobenzofuran*, b.p. 127°/21 mm (*von Auwers* and *H. Schutte*, Ber., 1919, **52,** 77); 2-*bromo*- and 2,2-*dibromo*-3-*oxo*-2,3-*dihydrobenzofurans*, m.p. 87° and 142° (*Fries* and *Pfaffendorff*, *ibid.*, 1912, **45,** 154).

The acid-catalysed condensation of 6-methyl-3-oxo-2,3-dihydrobenzofuran with acetic anhydride gives compound L (*Bohlmann* and *Zdero*, *ibid.*, 1968, **101,** 3941).

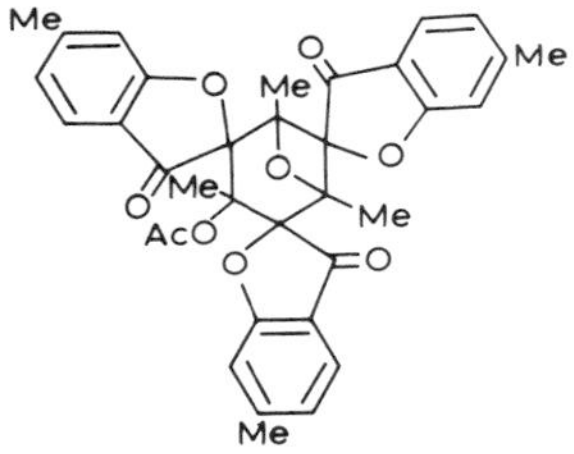

(L)

Side-chain bromination of diacetoxyacetophenone, followed by hydrolysis and ring closure, give 4-, 6-, and 7-**hydroxy**-3-**oxo**-2,3-**dihydrobenzofurans,** m.p. 121°, 243° and 186–188° (*Shriner* and *M. Witte*, J. Amer. chem. Soc., 1939, **61,** 2328); 2,3-dihydroxy-α-bromoacetophenone is formed as an intermediate and u.v. spectral data have been reported for the above (*G. Schenck* and *M. Huke*, Tetrahedron Letters, 1968, 2375). Related methods afford the 4,6- and 6,7-*dihydroxy*-3-*oxo*-2,3-*dihydrobenzofurans*, m.p. 250° and 229°.

α-Chloroacetoxy-3,4-dihydroxybenzene and 2,4,5-trihydroxy-α-chloroacetophenone have been isolated as intermediates in the preparation of 5,6-dihydroxy-3-oxo-2,3-dihydrobenzofuran by the Hoesch synthesis from hydroxyhydroquinone (*Schenck* and *Huke*, *loc. cit.*), *i.e.*, the reaction between polyhydric phenols and chloroacetonitrile (*A. Sonn*, Ber., 1917, **50,** 1262):

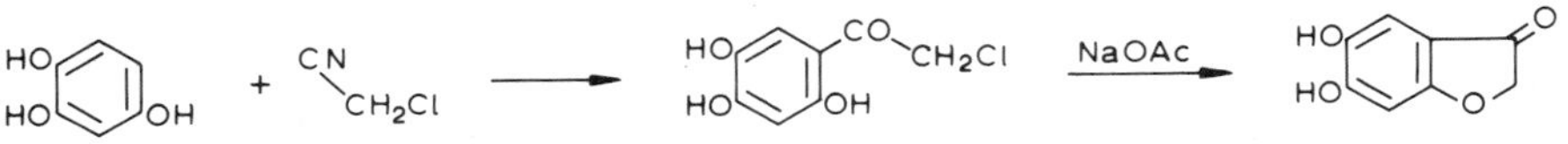

+ CN–CH_2Cl → $COCH_2Cl$ —NaOAc→

A number of 2-hydroxy-α-chloroacetophenones have been converted by the Hoesch ketone synthesis or the Fries rearrangement to 2,3-dihydrobenzofuran-3-ones (*Schenck, Huke* and *K. Goerlitzer*, Tetrahedron Letters, 1967, 2059; *Fries* and *G. Finck*, Ber., 1908, **41,** 4271). Suitable initial materials arise from dihydric phenols with chloroacetyl chloride and aluminium chloride (*K. Horváth*, Monatsh., 1951, **82,** 901). Its reaction with 6,7-dihydroxy-3-oxo-2,3-dihydrobenzofuran has been used for the spectrophotometric determination of germanium (*V. A. Nazakenko* and *E. N. Poluektova*, Zhur. anal. Khim., 1967, **22,** 895).

A compound erroneously described as a 2,3-dihydro-3-hydroxy-2-oxobenzo[*b*]-furan, the lactone of 2-hydroxymandelic acid has proved to be 2-**hydroxy**-3-**oxo**-2,3-**dihydrobenzo**[*b*]**furan,** m.p.108°, the cyclic hemiacetal of 2-hydroxyphenylglyoxal (*R. Howe, B. S. Rao* and *H. Heyneker*, J. chem. Soc., C, 1967, 2510):

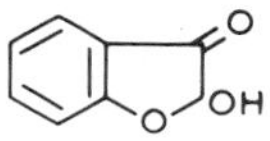

Under acidic conditions 6-methoxy-3-oxo-2-(2,4,6-trimethoxybenzoyl)-2,3-dihydrobenzofuran does not demethylate to give the expected 6-methoxy-2-(2-hydroxy-4,6-dimethoxybenzoyl)-3-oxo-2,3-dihydrobenzofuran but yields 1,3,5-trimethoxybenzene and 6-methoxy-3-oxo-2,3-dihydrobenzofuran (*R. Bryant* and *D. L. Haslam*, *ibid.*, 1967, 1345):

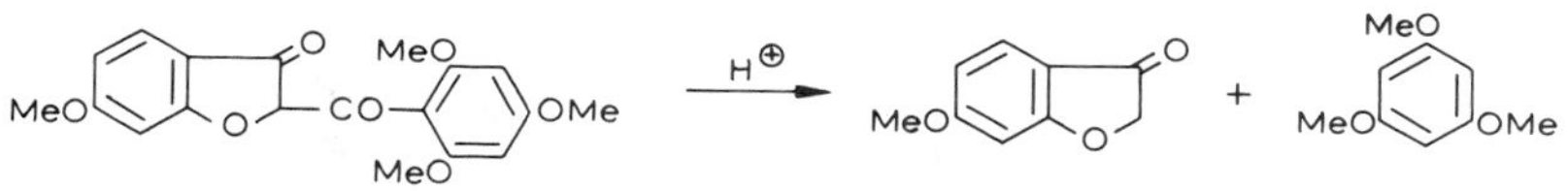

2-**Benzylidene**-3-**oxo**-2,3-**dihydrobenzofuran,** which was first mistaken for flavone, forms yellowish needles, m.p. 111° (*P. Friedländer* and *J. Neudörfer*, Ber., 1897, **30,** 1077); the dibromide is converted by alcoholic potash into flavanol (*von Auwers et al.*, *ibid.*, 1908, **41,** 4233; 1916, **49,** 809; Ann., 1914, **405,** 243):

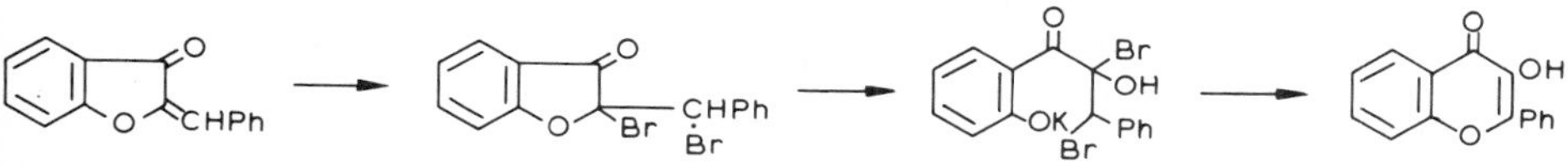

Benzylidene-3-oxo-2,3-dihydrobenzofurans are also converted into flavones or flavanols by treatment with potassium cyanide or hydrogen peroxide (*T. S. Wheeler et al.*, Chem. and Ind., 1955, 652). Some *o*-hydroxyphenyl styryl ketones (chalkones) give benzylidene-3-oxo-2,3-dihydrobenzofurans on oxidation (*Wheeler et al.*, *ibid.*, 1956, 1018).

A few hydroxylated benzylidene-3-oxo-2,3-dihydrobenzofurans — "*aurones*" — occur in nature. **Areusin,** from yellow *Antirrhinum majus* (*M. K. Seikel* and *Geissman*, J. Amer. chem. Soc., 1950, **72,** 5725), and **cernuoside,** from *Oxalis cernua* (*G. B. Marini-Bettólo et al.*, Gazz., 1953, **83,** 224) are isomeric glucosides of 2-(3′,4′-dihydroxybenzylidene-4,6-dihydroxy-3-oxo-2,3-dihydrobenzofuran:

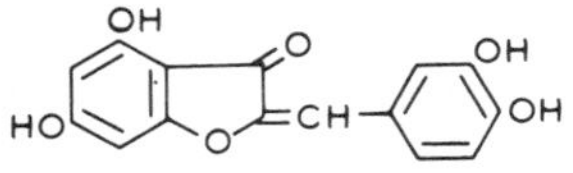

Coreopsis grandiflora affords the glucoside, **leptosin** which has been synthesised as follows (*Geissman et al.*, J. Amer. chem. Soc., 1943, **65,** 677; 1944. **66,** 486; 1951, **73,** 5765):

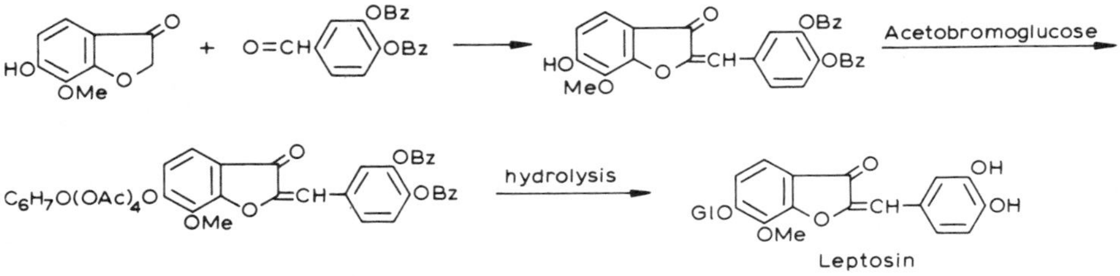

Leptosin

Sulphurein, from *Cosmos sulphureus*, differs from leptosin in lacking the methoxyl group (*M. Shimokoriyama* and *S. Hattori*, *ibid.*, 1953, **75,** 1900; *Geissman* and *L. Jurd*, *ibid.*, 1954, **76,** 4475).

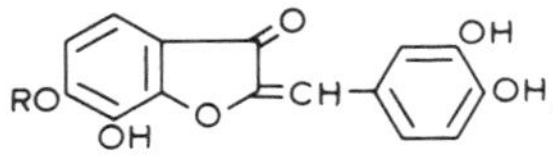

(LI) R = $C_6H_{11}O_5$
(LII) R = H

Maritimein (LI) which has been synthesised gives on hydrolysis the aglucon LII (*L. Farkas*, *L. Pallos* and *M. Nogradi*, Magy. Kem. Folyoirat, 1965, **71,** 270; C.A., 1965, **63,** 9901a). **Bractein** a glycoside from *Helichrysum bracteatum* has also been synthesised (*Farkas* and *Pallos*, Ber., 1965, **98,** 2930).

Oxidation of aurones by alkaline hydrogen peroxide gives aurone epoxides and flavanols; with sodium peroxide the products are aroyl-3-oxo-2,3-dihydrobenzofurans and with *tert*-butyl hydroperoxide in the presence of Triton-B the products are flavanols (*M. Geoghegan et al.*, Tetrahedron, 1966, 3203).

Griseofulvin, $C_{17}H_{17}ClO_6$, m.p. 218–219°, $[\alpha]_{5461}$ +417°, a colourless antibiotic from the mould *Penicillium griseo-fulvum*, is one of the few natural products which contains chlorine. It is the methyl enol ether of the *β*-diketone, griseofulvic "acid". Hydrogenation of the enol double bond gives dihydrogriseofulvin which contains the system $\cdot CH_2 \cdot CO \cdot CH_2 \cdot$ Oxidation of griseofulvin affords the acid LIII with (+)-methylsuccinic acid, and under other conditions methoxytoluquinone (LIV). The mould also elaborates the chlorine-free analogue and, if grown in a medium containing potassium bromide, the corresponding bromo-compound:

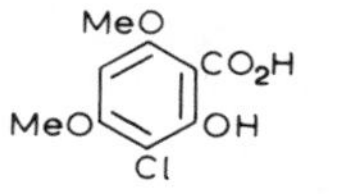

(LIII) 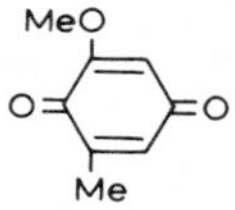 (LIV)

The oral antibiotic griseofulvin is formulated as shown (*A. E. Oxford et al.*, Biochem. J., 1939, **33,** 240; *J. MacMillan, T. P. C. Mulholland et al.*, J. chem. Soc., 1952, 3949–4002; 1953, 1697; 1954, 2585).

It has been synthesised by a number of routes (*A. C. Day, J. Nabney* and *A. I. Scott*, Proc. chem. Soc., 1960, 284; *A. Brossi et al.*, Helv., 1960, **43,** 1444; *C. H. Kuo et al.*, Chem. and Ind., 1960, 1627), and the racemic form has been obtained in a remarkable manner from 7-chloro-4,6-dimethoxy-2,3-dihydrobenzofuran-3-one and methoxyethynyl prop-1-enyl ketone in the presence of potassium *tert*-butoxide (*G. Stork* and *M. Tomasz*, J. Amer. chem. Soc., 1962, **84,** 310):

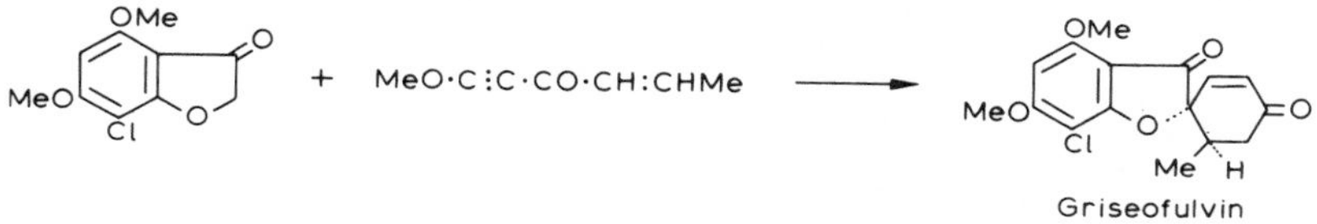

+ MeO·C⋮C·CO·CH:CHMe ⟶

Griseofulvin

2,3-**Dioxo**-2,3-**dihydrobenzofuran,** best obtained from isatin by the diazo reaction (*E. H. Huntress* and *W. M. Hearon*, *ibid.*, 1941, **63,** 2762), forms yellow plates, m.p. 134°, and is easily hydrolysed to *o*-hydroxyphenylglyoxylic acid. It reacts with carbonyl reagents at the 3-position; 3-*phenylhydrazone*, m.p. 185° (*Fries* and *Pfaffendorf*, Ber., 1921, **45,** 157); *oxime*, m.p. 175° (decomp.), formed by nitrosation of 2,3-dihydrobenzofuran-3-one (*E. Mameli*, Gazz., 1926, **56,** 768), and yields on further oximation the *dioxime*, m.p. 203°.

5- and 6-*Methyl*-2,3-*dioxo*-2,3-*dihydrobenzofuran*, m.p. 149° and 112°, are prepared from their 2-oximes, obtained by nitrosation of the related 3-oxo compounds (*Fries*, Ber., 1909, **42,** 234), the 6-compound also from the aryl chloroglyoxylate (*R. Stollé* and *E. Knebel*, *ibid.*, 1921, **54,** 1213):

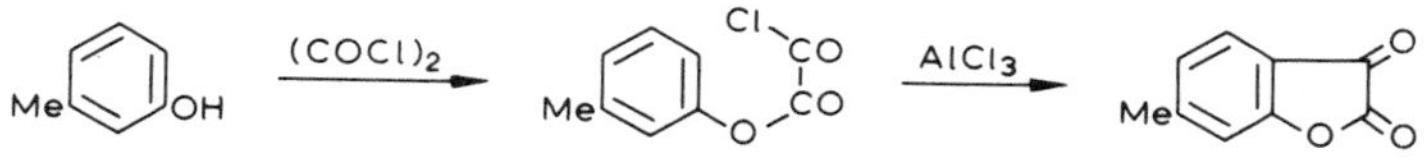

$(COCl)_2$ ⟶ $AlCl_3$ ⟶

5-*Methyl*-2,3-*dihydrobenzofuran*-2,3-*dione* 3-*phenylhydrazone*, obtained directly, melts at 148°; the 2-*phenylhydrazone*, m.p. 224°, is produced by coupling 5-methyl-2,3-dihydrobenzofuran-3-one with diazotised aniline (*von Auwers*, Ann., 1911, **381,** 265).

By successive treatment with halogen and alkali, *o*-allylphenyl acetate gives 2-*chloromethyl*-, b.p. 118–119°/11 mm, and 2-*bromomethyl*-2,3-*dihydrobenzofuran;* the chloro compound reverts to allylphenol when treated with sodium (*cf.* p. 131) and yields on hydrolysis 2-**hydroxymethyl**-2,3-**dihydrobenzofuran,** b.p. 138°/9.5 mm (*H. Normant*, Bull. Soc. chim. Fr., 1945, **12,** 609):

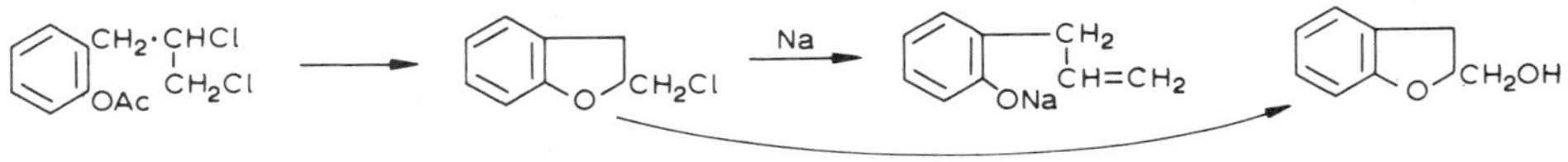

With mercuric salts the allylphenol yields compounds of the type 2-(chloromercurimethyl)-2,3-dihydrobenzofuran, which with halogens give 2-halogenomethyl-2,3-dihydrobenzofuran and with hydrogen sulphide or sodium stannite regenerate the allylphenol (*R. Adams et al.*, J. Amer. chem. Soc., 1922, **44**, 1781; 1923, **45**, 1842).

Reduction of 2-acetylbenzofuran with sodium amalgam yields 2-**acetyl**-2,3-**dihydrobenzofuran,** b.p. 132–133°/12 mm, while hydrogenation over nickel gives 2-α-**hydroxyethyl**-2,3-**dihydrobenzofuran,** b.p. 145°/20 mm (*Shriner* and *Anderson*, *ibid.*, 1939, **61,** 2705). Numerous 2-acyl-2,3-dihydrobenzofuran-3-ones have been prepared by treating *o*-acyloxy-ω-chloroacetophenones with a basic reagent, notably sodium hydride; the reaction involves migration of the acyl group to carbon (*Wheeler et al.*, J. chem. Soc., 1954, 4174):

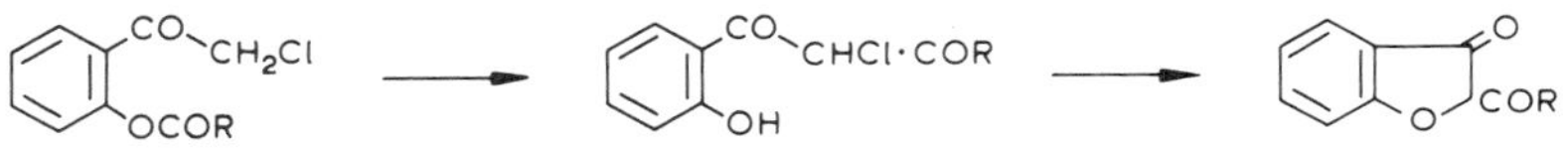

By Friedel–Crafts acetylation, 2-methyl-2,3-dihydrobenzofuran gives the expected 5-*acetyl-2-methyl-2,3-dihydrobenzofuran*, b.p. 145–146°/6 mm (*R. T. Arnold* and *J. C. McCool*, J. Amer. chem. Soc., 1942, **64,** 1315).

2,3-**Dihydrobenzofuran**-2-**carboxylic acid,** m.p. 116.5°, is obtained by reducing benzofuran-2-carboxylic acid (coumarilic acid) with sodium amalgam. By a Dieckmann ring-closure (compare synthetic method *2*, p. 142), the ester LV affords *ethyl 2,3-dihydrobenzofur-3-one-2-carboxylate*, m.p. 66°, which is probably better represented as an enol, giving a blue ferric chloride reaction; acylation takes place on oxygen, but alkylation mainly on carbon (*von Auwers*, Ann., 1912, **393,** 338):

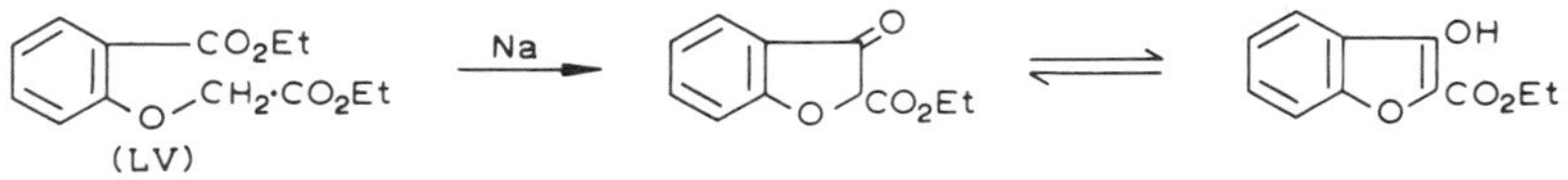

(LV)

An internal Michael addition affords diethyl 2,3-dihydrobenzofuran-3-acetate-2-carboxylate (*C. F. Koelsch*, J. Amer. chem. Soc., 1945, **67,** 569):

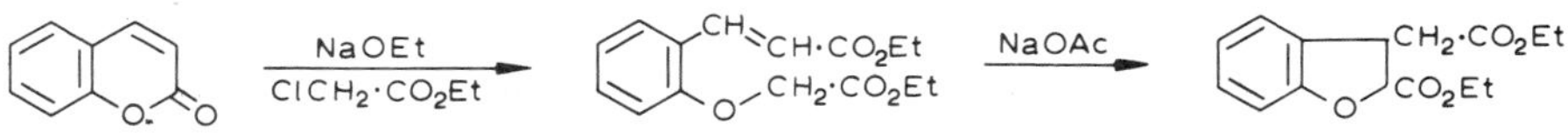

Tubaic acid, an important product of the degradation of rotenone, is 4-hydroxy-3-isopropenyl-2,3-dihydrobenzofuran-5-carboxylic acid.

(ii) Tetrahydrobenzofurans

The adduct LVI of benzil and cyclohexanone is converted by acid dehydrating agents into 2,3-*diphenyl-5,6-dihydrobenzofuran*, m.p. 102–105°, which affords 2,3-**diphenyl**-4,5,6,7-**tetrahydrobenzofuran,** m.p. 120° (*C. F. H. Allen* and *J. A. van Allen*, J. org. Chem., 1951, **16,** 716):

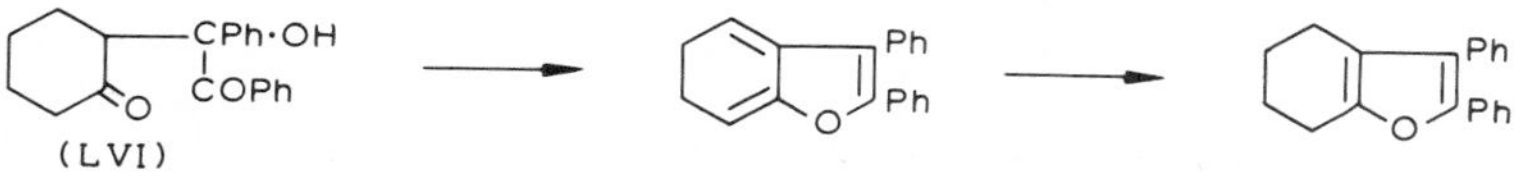

CPh·OH
COPh
(LVI)
Ph
Ph
Ph
Ph

Cyclohexane-1,3-diones condense with ethyl α-chloroacetoacetate to give 3-methyl-4-oxo-4,5,6,7-tetrahydrobenzofurans (*H. Stetter* and *R. Lauterbach*, Ber., 1960, **93,** 603):

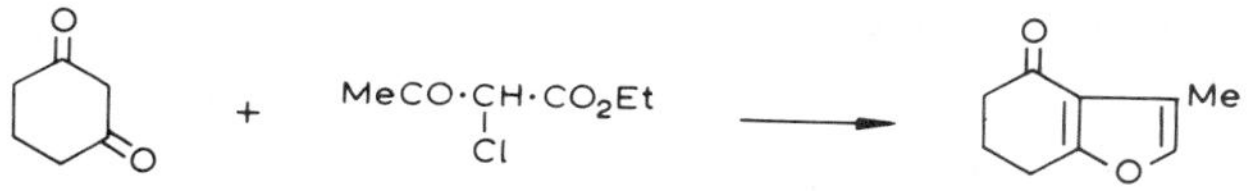

+ MeCO·CH·CO_2Et
Cl
Me

1-Methylcyclohexane-3,5-dione affords **evodone,** a ketone occurring in *Evodia hortensis* Forst (*Stetter* and *Lauterbach*, Angew. Chem., 1959, **71,** 673):

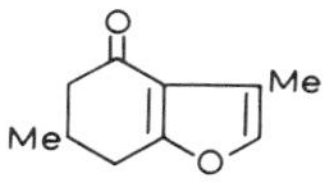

Me
Me
Evodone

On reduction evodone yields **menthofuran,** b.p. 80°/18 mm, which is present in peppermint oil, and is produced by treating pulegone with sulphuric acid and heating the resulting unsaturated sultone (*W. Treibs*, Ber., 1937, **70B,** 85):

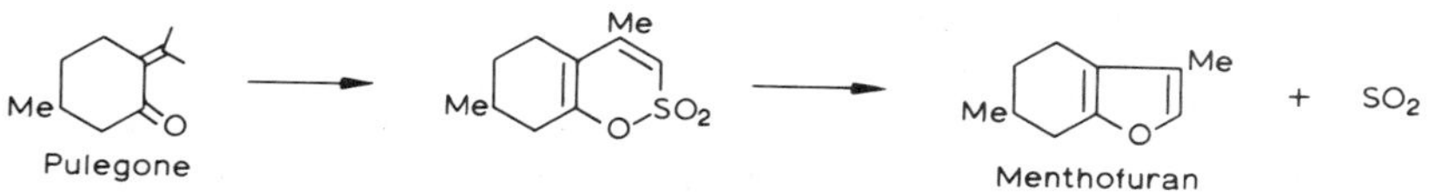

Me
Pulegone
Me
Me
SO_2
Me
Me
Menthofuran
+ SO_2

Menthofuran is also a product of the reaction of isopulegone with mercuric acetate (*Treibs et al.*, Ann., 1953, **581,** 59).

The alkylation of enolate ions followed by hydrolysis–dehydration of the resulting β-chloroallyl ketone with sulphuric acid gives substituted 4,5,6,7-tetrahydrobenzofurans, *e.g.*, 2-methyl-4,5,6,7-tetrahydrobenzofuran (LVII) and 2,7-dimethyl-4,5,6,7-tetrahydrobenzofuran (LVIII). 2-(2-Chloroallyl)cyclopentanones give 1,4-diketones instead of the desired furan (*E. J. Nienhouse, R. M. Irwin* and *G. R. Finni*, J. Amer. chem. Soc., 1967, **89,** 4557):

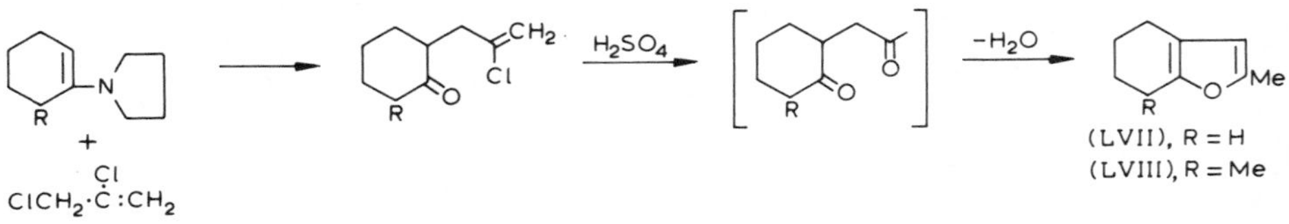

(LVII), R = H
(LVIII), R = Me

3-Methyl-2,4,5,6,7,7a-hexahydrobenzofuran-2-one yields 3-*methyl*-4,5,6,7-*tetrahydrobenzofuran*, b.p. 55°/8 mm on treatment with diisobutylaluminium hydride in tetrahydrofuran at −20° to −25° under nitrogen, followed by 2 *N* sulphuric acid (*Shinogi and Co. Ltd.*, B.P. 1,081,647/1965):

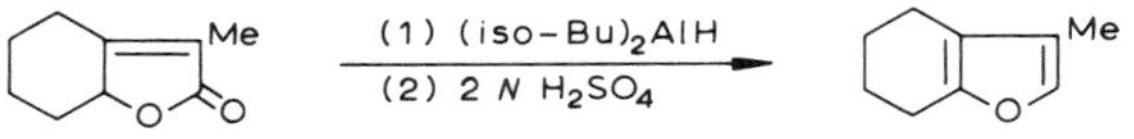

A number of 4,5,6,7-tetrahydrobenzofurans have been synthesised in good yield by this method, which is simpler and more efficient than the earlier method (*H. Minato* and *T. Nagasaki*, J. chem. Soc., C, 1966, 377).

2-Methyl-4,5,6,7-tetrahydrobenzofuran-3-carboxylic acid is obtained from chlorocyclohexanone and ethyl acetoacetate (*F. Ebel et al.*, Helv., 1929, **12,** 16):

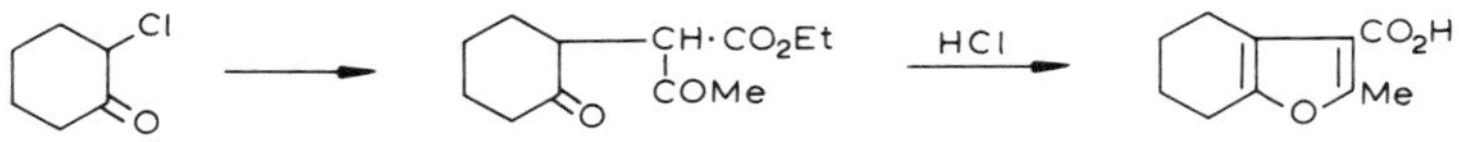

5-**Methoxy**-2,3,4,7-**tetrahydrobenzofurans** may be obtained from the corresponding benzofurans (p. 145).

(iii) Hexahydrobenzofurans

The unsaturated γ-hydroxy acids, 2-hydroxy-6-oxocyclohex-1-enylacetic acid (LIX) and 2-hydroxy-4,4-dimethyl-6-oxocyclohex-1-enylacetic acid (LX) on treatment with acetic anhydride give the corresponding lactones 2,3,4,5,6,7-hexahydrobenzofuran-2,4-dione and 6,6-dimethyl-2,3,4,5,6,7-hexahydrobenzofuran-2,4-dione respectively. The unsaturated lactones undergo acylation on the carbon adjacent to the lactone carbonyl when acetic anhydride is used in the presence of sodium acetate (*B. M. Goldschmidt, B. L. van Duuren* and *C. Mercado*, J. chem. Soc., C, 1966, 2100):

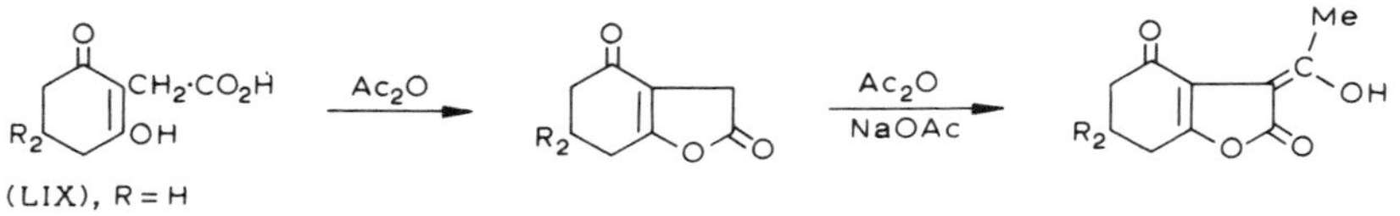

(LIX), R = H
(LX), R = Me

Diethyl malonate and cyclohexene oxide give, after hydrolysis and decarboxylation, trans-*hexahydrobenzofuran-2-one*, b.p. 118–119°/6 mm, which has been converted, indirectly, into the cis-*isomer*, b.p. 112°/6 mm (*M. S. Newman* and *C. A. van der Werf*, J. Amer. chem. Soc., 1945, **67,** 233).

2,4,5,6,7,7a-**Hexahydrobenzofuran** may be prepared from 2-hydroxycyclohexanone and vinyltriphenylphosphonium bromide (*E. E. Schweizer* and *J. G. Liehr*, J. org. Chem., 1968, **33,** 583):

2-(2-Hydroxyethyl)cyclohexanone on distillation *in vacuo* cyclises to give 2,3,4,5,6,7-*hexahydrobenzofuran*, b.p. 56°/10 mm (*W. E. Harvey* and *D. S. Tarbell*, *ibid.*, 1967, **32,** 1679); 2-*methyl*-2,3,4,5,6,7-*hexahydrobenzofuran*, b.p. 85–86°/30 mm.

2-Phenyl-7a-(*N*-pyrrolidino)perhydrobenzofuran (LXII) prepared from 1-pyrrolidino-1-cyclohexene (LXI) affords 2-phenyl-2,3,4,5,6,7-hexahydrobenzofuran (LXIII) on treatment with oxalic acid (*P. Jakobsen* and *S.-O. Lawesson*, Tetrahedron, 1968, 3671):

(LXI) (LXII) (LXIII)

(iv) Octahydrobenzofurans

Octahydrobenzofuran, b.p. 172°, is obtained by hydrogenating benzofuran over platinum (*N. I. Shuikin et al.*, C.A., 1941, **35,** 2508). cis-*Octahydrobenzofuran*, b.p. 66°/15 mm, is obtained by reducing 2,3,4,5,6,7-hexahydrobenzofuran (*Harvey* and *Tarbell, loc. cit.*); cis-2-*methyl-octahydrobenzofuran*, b.p. 63°/16 mm, is obtained similarly. For conformations of octahydrobenzofurans and n.m.r. spectra see *N. Belorizky* and *D. Gagnaire*, Compt. rend., 1969, **268,** 688. Ethyl 2-oxocyclohexaneglyoxylate when hydrogenated at 70 atm. in the presence of Raney nickel affords 3-*hydroxy-2-oxo-octahydrobenzofuran*, m.p. 133–134° (*A. Barco, M. Anastasia* and *G. P. Pollini*, Chem. Ind., Milan, 1969, **51,** 165; C.A., 1969, **70,** 106283e).

2-Oxo-octahydrobenzofurans (lactones) of types LXIV (R = H, Me, Et or Pr) are prepared by treatment of diethyl sodiomalonate with the epoxide of the corresponding cyclohexene and subsequent hydrolysis, decarboxylation, and cyclisation of the condensation product. On heating in the presence of acid, LXIV isomerises to LXV (*J. Ficini* and *A. Manjean*, Compt. rend., 1966, **263**, 425):

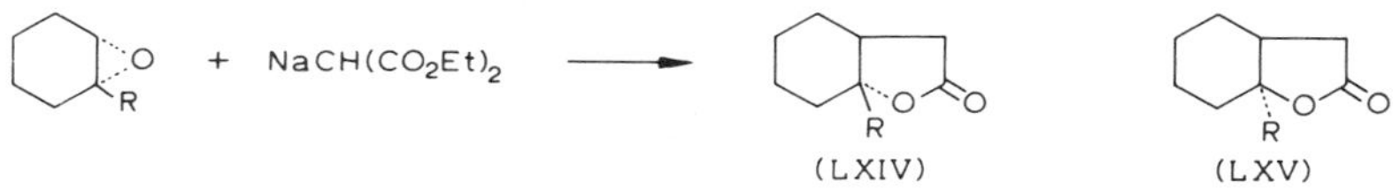

+ $NaCH(CO_2Et)_2$ → (LXIV) (LXV)

2,3-**Dioxo-octahydrobenzofuran** (LXVI), 99–101°, is prepared from 2-oxocyclohexylglyoxylic acid; being the derivative at once of a β-hydroxyketone and of an α-oxo acid, it suffers pyrolysis to tetrahydrobenzaldehyde (*P. A. Plattner* and *L. M. Jampolsky*, Helv., 1943, **26**, 687):

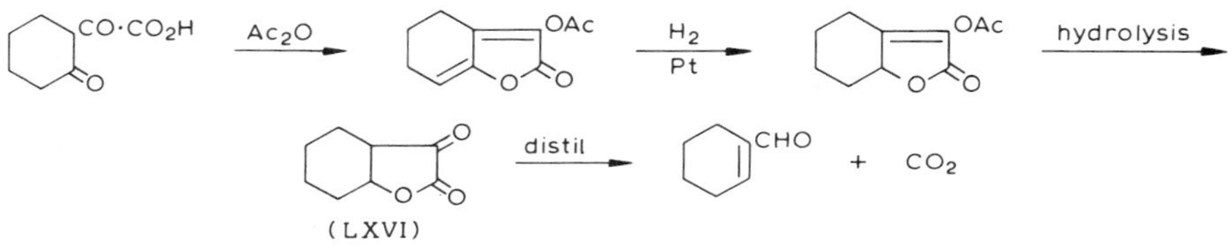

$CO \cdot CO_2H$ —Ac_2O→ OAc —H_2, Pt→ OAc —hydrolysis→

(LXVI) —distil→ CHO + CO_2

3. Isobenzofuran (benzo[*c*]furan; 3,4-benzofuran) and its derivatives

(*a*) *Isobenzofurans*

The compounds of this series are much less stable and far less numerous than those of the 2,3-benzofuran (benzo[*b*]furan) group. Without using formal charges, only one structure (I) can be written for the molecule, which may be compared with that of isoindole (p. 470). 1,3-Dihydroisobenzofuran (II)* is a normal cyclic ether, and among its derivatives are such familiar compounds as phthalic anhydride and phthalide:

(I) (II)

Nearly all the known unreduced isobenzofurans have aryl substituents in positions 1 and 3.

* In the earlier nomenclature 1,3-dihydroisobenzofuran is called "phthalan"; this trivial name is not now accepted officially. It is also referred to as "*o*-xylylene oxide".

(i) Synthetic methods

(*1*) With Grignard reagents, phthalides give hydroxyketones, tautomeric with hydroxy-1,3-dihydroisobenzofurans and then glycols which are easily dehydrated to 1,3-dihydroisobenzofurans (R = H or aryl):

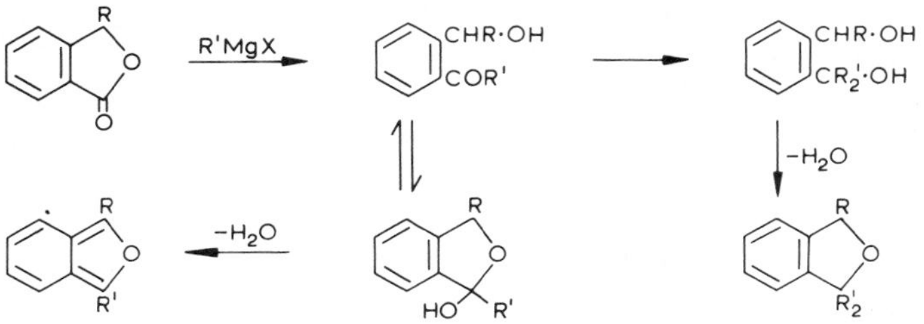

When R′ is aryl, the first product is dehydrated to an unreduced isobenzofuran (*A. Guyot* and *J. Catel*, Bull. Soc. chim. Fr., 1906, [iii], **35,** 551, 1124, 1135); when R′ is alkyl, a methylene-1,3-dihydroisobenzofuran may be produced (*S. Natelson* and *A. Pearl*, J. Amer. chem. Soc., 1936, **58,** 2448):

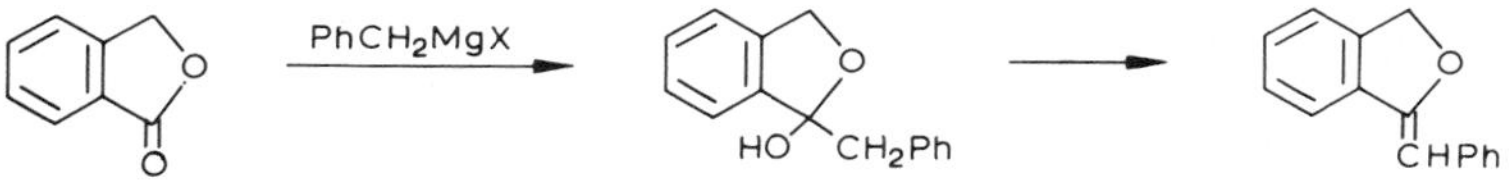

Phthalaldehyde yields 1,3-dialkyl-1,3-dihydroisobenzofurans (*F. Nelken* and *H. Simonis*, Ber., 1908, **41,**986):

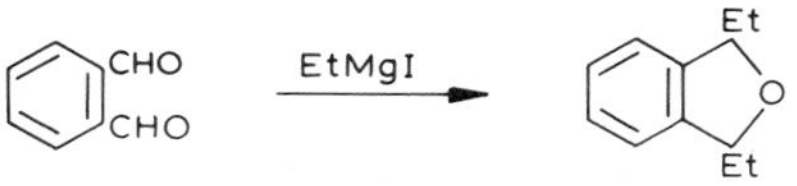

and 2-formylbenzoyl chloride with phenylmagnesium bromide affords 1,3-diphenylisobenzofuran (*M. Renson*, Bull. Soc. chim. Belg. 1961, **70,** 77):

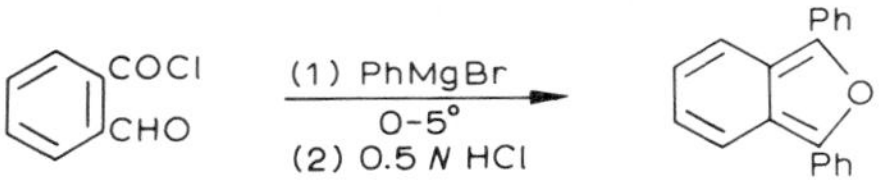

In a similar way, organometallic compounds react with phthalic esters, giving either hydroxy-ketones or glycols (*Guyot* and *Catel*, *loc. cit.*; *G. Wittig* and *M. Leo*, Ber., 1931, **64,** 2395):

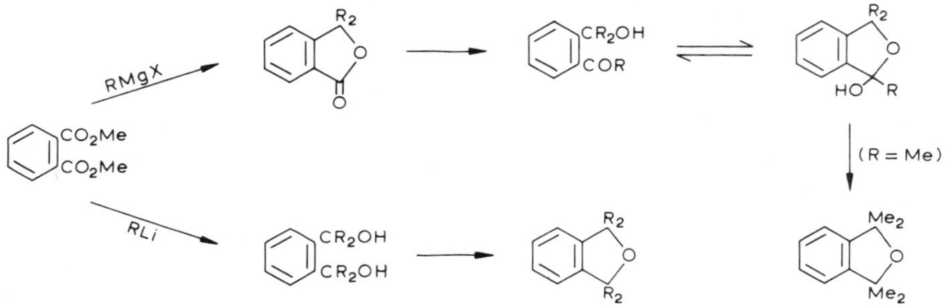

When R is alkyl, the hydroxyketone is easily dehydrated to a methylene derivative (*Y. Shibata*, J. chem. Soc., 1909, **95,** 1449).

(*2*) When a triarylmethane-*o*-carboxylic acid (*e.g., III*) is treated with sulphuric acid (*F. F. Blicke* and *R. A. Patelski*, J. Amer. chem. Soc., 1936, **58,** 273, 559), the product may be a diarylisobenzofuran, probably produced as follows:

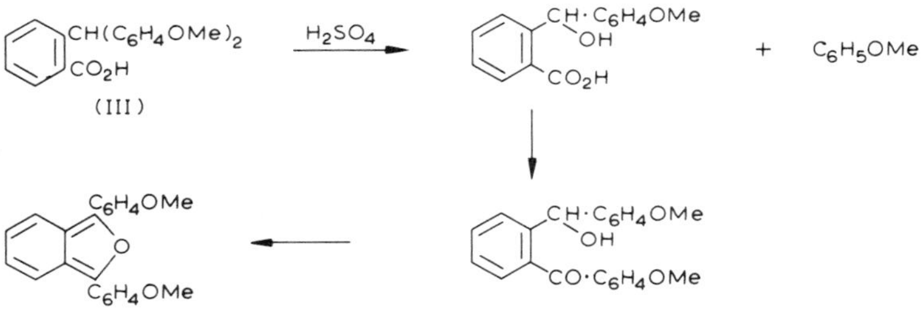

(*3*) Treatment of diphenylindenone with alkali in presence of an oxidising agent or of diphenylindenone epoxide with alkali alone gives diphenyl- or benzoyl(phenyl)-isobenzofuran according to conditions (*C. F. H. Allen* and *J. A. van Allen*, *ibid.*, 1948, **70,** 2069):

(*4*) Butadienes combine with 1,2-diacylethenes, giving 1,2-diacylcyclohexenes which can be dehydrated to 4,7-dihydroisobenzofurans; these in turn can be dehydrogenated most generally by the indirect route shown (*R. Adams* and *M. H. Gold*, *ibid.*, 1940, **62,** 56):

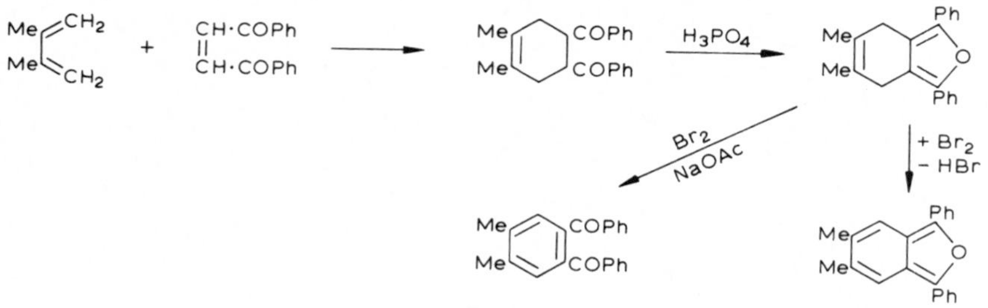

(5) 1,3-Diphenylisobenzofurans may be obtained by the dehydration of 1,2-dibenzoylcyclohexa-1,4-dienes (IV) prepared by the Diels–Alder addition of butadienes to dibenzoylacetylene (*M. E. Mann* and *J. D. White*, Chem. Comm., 1969, 420):

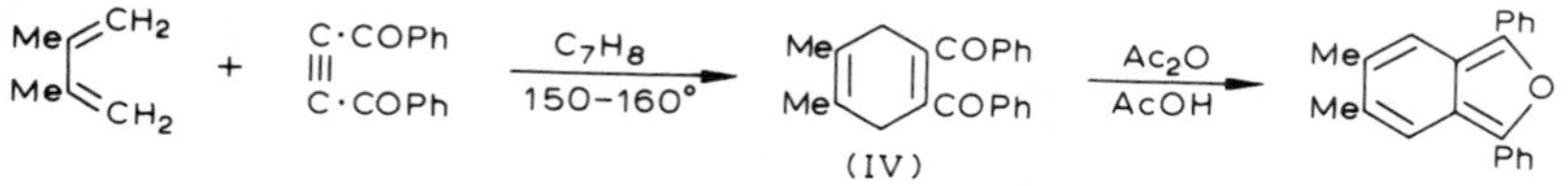

(IV)

(6) On treatment with 80% sulphuric acid, 2,3-dibenzoylquinol gives 1,3-diphenylisobenzofuran-4,7-quinone (4,7-dioxo-1,3-diphenyl-4,7-dihydroisobenzofuran), which with acetic anhydride and sulphuric acid reverts to dibenzoylquinol diacetate (*R. Pummerer* and *G. Marondel*, Ber., 1956, **89**, 1454):

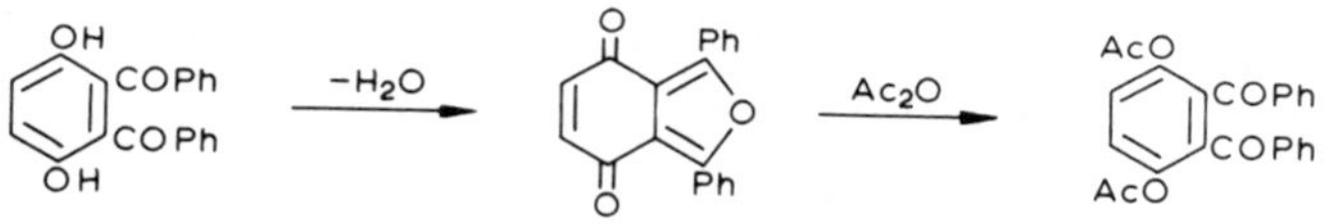

(7) (*o*-Duroylphenyl)phenylcarbinol on treatment with strong mineral acid gives 1-duryl-3-phenylisobenzofuran (*C. Fuson* and *W. C. Rife*, J. org. Chem., 1960, **25**, 2226):

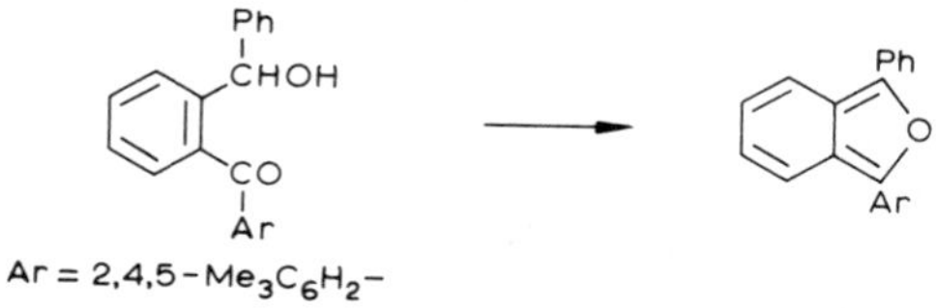

Ar = 2,4,5-$Me_3C_6H_2$–

(8) The photolysis of 1,4-diphenylphthalazine *N*-oxide results in the formation of 1,3-diphenylisobenzofuran (*O. Buchardt*, Tetrahedron Letters, 1968, 1911):

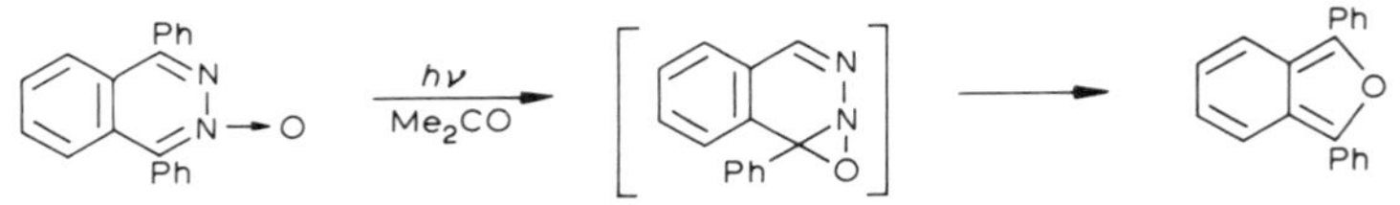

(9) A series of unsymmetrical substituted isobenzofurans has been prepared from the appropriate 2-benzoylbenzophenones by treatment with potassium tetrahydridoborate (*W. W. Zajec* and *D. E. Pichler*, Canad. J. Chem., 1966, **44,** 833). Hydrolysis of the product from the reaction between *o*-dibenzoylbenzene and an equivalent of sodium affords 3,3′-*dihydroxy*-1,1′,3,3′-*tetraphenyl*-1,1′-*bis*-1,3-*dihydroisobenzofuranyl*, m.p. 170–180° (decomp.), which with acid decomposes instantly with loss of water to give an equimolar mixture of *o*-dibenzoylbenzenes and 1,3-diphenylisobenzofuran (*B. J. Herold*, Tetrahedron Letters, 1962, 75). 1-Hydroxy-1,3-diphenyl-1,3-dihydroisobenzofuran with acid gives 1,3-diphenylisobenzofuran (*A. Rieche* and *M. Schulz*, Ann., 1962, **653,** 32). 5,6-Dimethyl-1,3-diphenylisobenzofuran is obtained when 2,3-dibenzoyl-6,7-dimethyl-1,4-endoxo-1,4-diphenyl-1,4-dihydronaphthalene is treated with sodium hydroxide (*W. Ried* and *K. H. Boenninghausen*, *ibid.*, 1961. **639,** 61).

The adduct, which is prepared from 1,4-dihydronaphthalene-1,4-endooxide (V) and tetraphenylcyclopentadienone, is decomposed in refluxing "diglyme" in the presence of V to give adducts VI and VII, which shows that isobenzofuran is present in the first reaction (*L. F. Fieser* and *M. J. Haddadin*, J. Amer. chem. Soc., 1964, **86,** 2081):

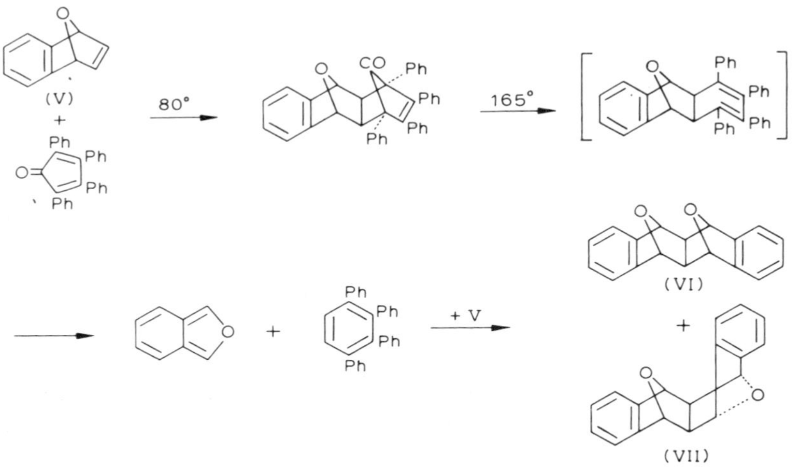

(V) + 80° 165° + + V (VI) + (VII)

(ii) Properties and reactions

The deep yellow diarylisobenzofurans show an intense blue-green fluorescence in solution; they dimerise when exposed to light; and they autoxidise in air to *o*-diaroylbenzenes, with which they are thus interconvertible (*Guyot* and *Catel*, *loc. cit.*). *Adams* and *Gold* (J. Amer. chem. Soc., 1940, **62,** 2038) suggested that all these phenomena were related to a ready transition to a di-radical form (VIII):

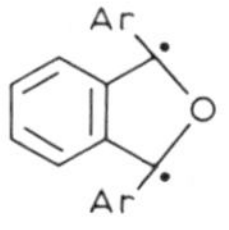

(VIII)

The most characteristic reaction of isobenzofurans is their combination with philodienes, which makes available otherwise inaccessible polycyclic compounds (*R. Weiss et al.*, Monatsh., 1932, **61,** 143, 162; *C. Dufraisse* and *P. Compagnon*, Compt. rend., 1938, **207,** 585; *A. Etienne et al.*, Bull. Soc. chim. Fr., 1952, 750; *J. A. Berson*, J. Amer. chem. Soc., 1953, **75,** 1240) *e.g.*:

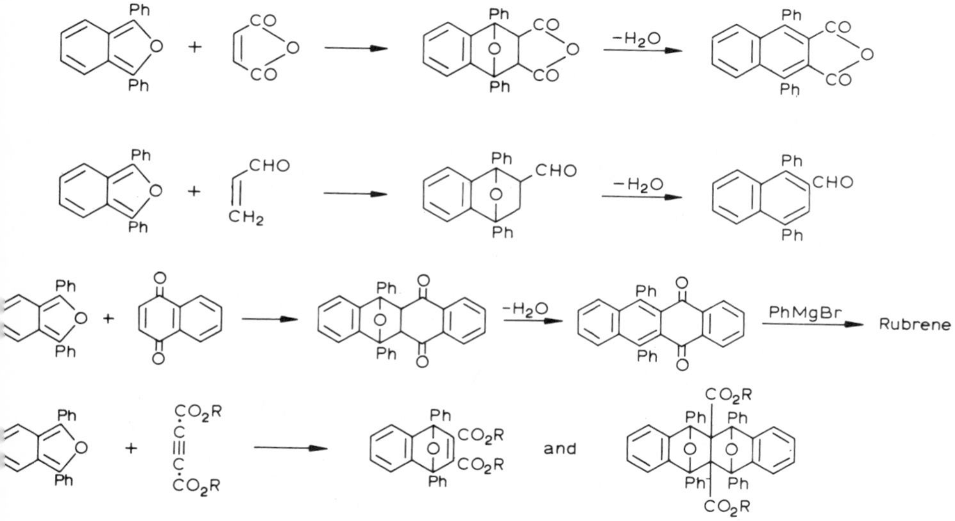

$\xrightarrow{-H_2O}$

$\xrightarrow{-H_2O}$

$\xrightarrow{-H_2O}$ $\xrightarrow{PhMgBr}$ Rubrene

and

Norbornadiene reacts with 1,3-diphenylisobenzofuran to give a single 1:1 adduct assigned the *exo,exo* structure IX. It also reacts to give a mixture of two isomeric 1:2 adducts; the principal bis-adduct is assigned the *exo,exo,exo* structure X (*M. P. Cava* and *F. M. Scheel*, J. org. Chem., 1967, **32,** 1304):

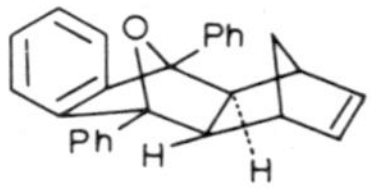

(IX) 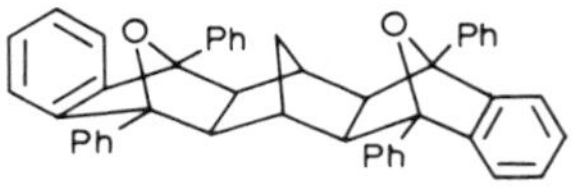 (X)

The kinetics of the Diels–Alder reaction with acrylonitrile in different solvents have been studied and the results show that in the presence of a large excess of acrylonitrile the reaction is first order, but the rate is proportional to the concentration of acrylonitrile (*J. Gillois-Doucet* and *P. Rumpf*, Bull. Soc. chim. Fr., 1959, 1823). 1,3-Diphenylisobenzofuran has been employed for the trapping of the nonisolable intermediates benzyne and cyclopentyne as adducts (*G. Wittig et al.*, Angew. Chem., 1958, **70,** 166; 1960, **72,** 324); a number of arynes generated from Grignard and organolithium reagents have been trapped with furan and 1,3-diphenylisobenzofuran (*R. Harrison, H. Heaney* and *P. Lees*, Tetrahedron, 1968, 4589). It is an excellent trapping agent for benzocyclobutadiene and for halogenated benzocyclobutadiene (*Cava* and *R. Pohlke*, J. org. Chem., 1962, **27,** 1564):

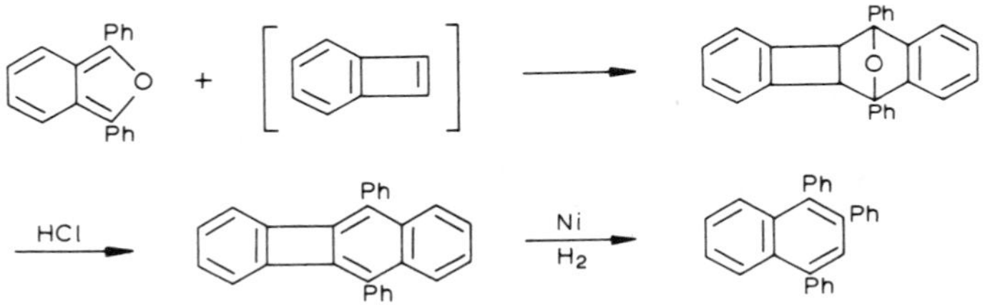

o,o'-Dibenzoyl tetraphenylethene (XI) is obtained *via* the dimerisation of 1,3-diphenylisobenzofuran through intermediate complexes when it is treated with alkanesulphonic acids (*A. le Berre* and *G. Lonchambon*, Bull. Soc. chim. Fr., 1967, 4328):

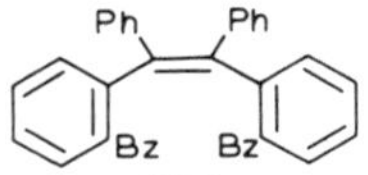

(XI)

Treatment of 1,3-diphenylisobenzofuran in benzene at room temperature in the dark with oxygen gives a white amorphous *peroxide* which explodes when heated to 150–170°. Reduction of the polymeric peroxide gives *o*-dibenzoylbenzene (*Le Berre* and *R. Ratzimbazafy*, *ibid.*, 1963, 229).

The electrochemiluminescence emission, polarographic half-wave oxidation and reduction potentials, anion- and cation-radical stabilities, fluorescence spectra, and fluorescence efficiencies of a number of aryl-substituted isobenzofurans have been examined in dimethylformamide solution (*A. Zweig et al.*, J. Amer. chem. Soc., 1967, **89,** 4091; U.S.P. 3,399,328/1968). The oxidation of 1,3-diphenylisobenzofuran by photochemically produced singlet states of oxygen has been studied (*T. Wilson*, J. Amer. chem. Soc., 1966, **88,** 2898; *D. F. Evans*, Chem. Comm., 1969, 367).

1-Substituted 1-hydroxy-3,3-dimethyl-1,3-dihydroisobenzofurans (XII; Ar=*o*-Me-, *o*-MeO-, *o*-EtO-, *o*-Cl-, *o*-Br- or *p*-EtOC$_6$H$_4$-) are converted to the isobenzofurylium (phthalylium) salt with perchloric acid, ferric chloride and antimony pentachloride (*L. A. Pavlova* and *V. S. Sorokina*, Zhur. org. Khim., 1968, **4,** 2228):

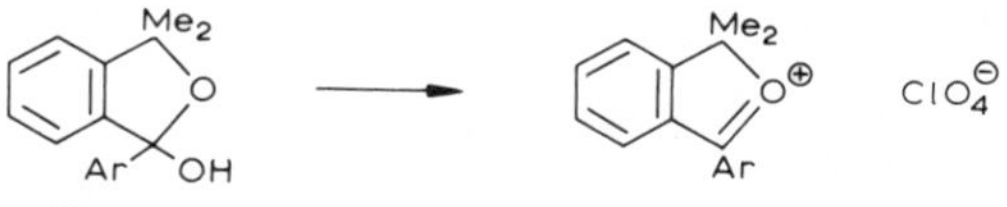

(XII)

Isobenzofurylium tetrachloroferrates (XIII, X = FeCl$_4^{\ominus}$) have been reduced using zinc and hydrochloric acid to give the corresponding 1,1′-bis(1,3-dihydroisobenzofuranyls) (biphthalyls) (*T. G. Melent'eva* and *Pavlova*, *ibid.*, 1967, **3,** 743):

(XIII)

1-Hydroxy-1,3,3-trimethyl-1,3-dihydroisobenzofuran may be prepared from diethyl phthalate and methylmagnesium iodide and then converted to the corresponding isobenzofurylium perchlorate (*A. Fabrycy*, Rocz. Chem., 1962, **36,** 243):

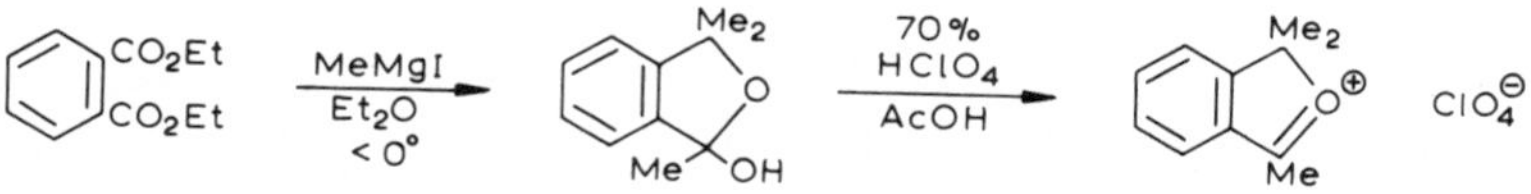

(iii) Individual isobenzofurans

Isobenzofurans: 1,3-*diphenyl-*, m.p. 131°; 1,3-*di-*p-*tolyl-*, m.p. 125°; 1,3-*di-*p-*anisyl-*, m.p. 126°; 1,3-*di-α-naphthyl-*, m.p. 166° (*C. Seer* and *O. Dischendorfer*, Monatsh., 1913, **34,** 1493); 4,7-*dimethyl-*1,3-*diphenyl-*, m.p. 129–131°; 5,6-*dimethyl-*1,3-*diphenyl-*, m.p. 194–195°; 1,3,4,7-*tetraphenyl-*, m.p. 265–266°; 1,3,4,5,6,7-*hexaphenyl-*, m.p. 254–257° (*W. Ried* and *K. H. Boenninghausen*, Ann., 1961, **639,** 61).

1,3,3-*Trimethylisobenzofurylium perchlorate*, m.p. 160–161°. N.m.r. of some isobenzofurans see *P. J. Black, R. D. Brown* and *M. L. Heffernan*, Australian J. Chem., 1967, **20,** 1305.

(b) Hydroisobenzofurans

(i) Dihydroisobenzofurans

(*1*) 1,3-*Dihydroisobenzofuran* is obtained in poor yield from *o*-xylylene dibromide with alkali, and in good yield by reducing phthaloyl chloride with lithium tetrahydridoaluminate to *o*-xylylene glycol and dehydrating this over aluminia (*J. Entel et al.*, J. Amer. chem. Soc., 1952, **74,** 441). The homogeneous partial oxidation of *o*-xylene in air at 1 atmosphere and 455–525° produces 1,3-dihydroisobenzofuran as a major product (*J. Loftus* and *C. N. Satterfield*, J. Phys. Chem., 1965, **69,** 909). The methiodides of two carbinolamines (XIV and XV) are cyclised thermally with the elimination of trimethylamine and hydriodic acid to give 76 and 56% yields of 1,3-dihydroisobenzofurans XVI and XVII, respectively (*R. L. Vaulx, F. N. Jones* and *C. R. Hauser*, J. org. Chem., 1964, **29,** 505).

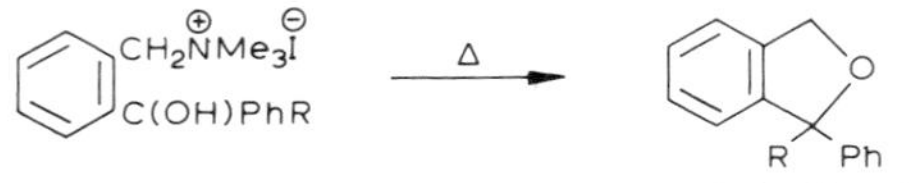

(XIV), R = H
(XV), R = Ph

(XVI), R = H
(XVII), R = Ph

1,3-Dihydroisobenzofuran undergoes autoxidation in air, and hydrogen-abstraction by *tert*-butoxy radicals gives the radical XVIII which does not disproportionate, but dimerises to give 1,1′-bis(1,3-dihydroisobenzofuranyl) (XIX) in the normal and iso-forms (*R. L. Huang* and *H. H. Lee*, J. chem. Soc., 1964, 2500):

(XVIII) (XIX)

The reaction of *o*-phthalaldehyde with alkyl hydroperoxides affords the corresponding 1,3-diperoxy-1,3-dihydroisobenzofuran (*A. Rieche* and *M. Schulz*, Ber., 1964, **97,** 190):

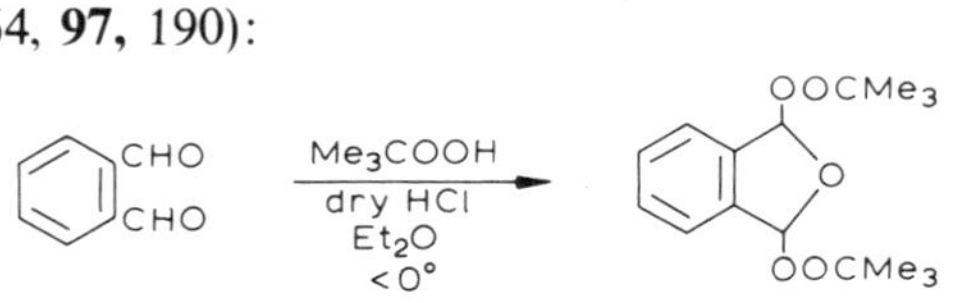

1,1,3,3-Tetrachloro-1,3-dihydroisobenzofuran is one of the "phthalylene tetrachlorides" obtained by the action of phosphorus pentachloride on phthaloyl chloride (*E.Ott*, *ibid.*, 1922, **55,** 2108), while phthalide and phthalic anhydride are mono- and di-oxo-dihydroisobenzofurans, and the 1-hydroxy-derivative, "hydrophthalide", is tautomeric with *o*-formylbenzyl alcohol.

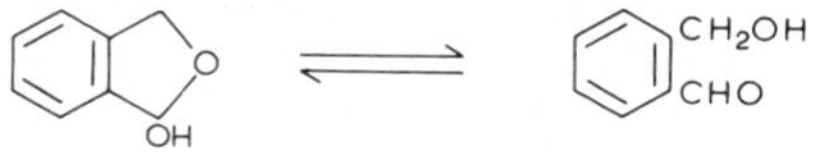

1,3-Dihydroisobenzofurans substituted in position 1 and 3 are made by dehydrating glycols produced from phthalic esters, *o*-benzoylbenzoic esters, or phthalides by treatment with organometallic compounds and/or lithium tetrahydridoaluminate. Alkylene-1,3-dihydroisobenzofurans have been mentioned as products of synthesis (*1*) (p. 181).

The reaction of tetrachlorofuran with 2,5-dihydrofuran gives a mixture of the *endo* and *exo* isomers of 4,5,6,7-tetrachloro-4,7-endoxo-1,3,3a,4,7,7a-hexahydroisobenzofuran, which on treatment with 98% sulphuric acid at 20–30° affords 5,6-dichloro-4,7-dihydroxy-1,3-dihydroisobenzofuran (XX) and at −5°, the related dichloro-1,3-dihydroisobenzofuran-4,7-quinone (XXI) and 4,5,6-trichloro-7-hydroxy-,13-dihydroisobenzofuran (XXII) (*Ruhr Chemie A.G.*, G.P. 1,170,963/1961; *H. Feichtinger* and *H. Linden*, U.S.P. 3,176,024/1962); all are effective fungicides:

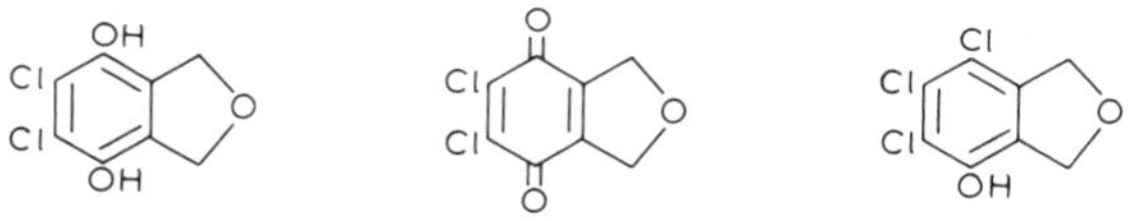

(XX) (XXI) (XXII)

Pyrolysis of the diazonium fluoroborate (decomp. 99°) obtained from 5-aminophthalide yields 5-fluorophthalide, affording 5-fluoro-1,1,3,3-tetrachloro-1,3-dihydroisobenzofuran on reacting with phosphorus pentachloride in phosphorus oxychloride. The tetrachloro derivative may be converted into 1,1,3,3,5-pentafluoro-1,3-dihydroisobenzofuran (*L. M. Yagupol'skii* and *R. V. Belinskaya*, Zhur. obshcheĭ Khim., 1963, **33**, 2358):

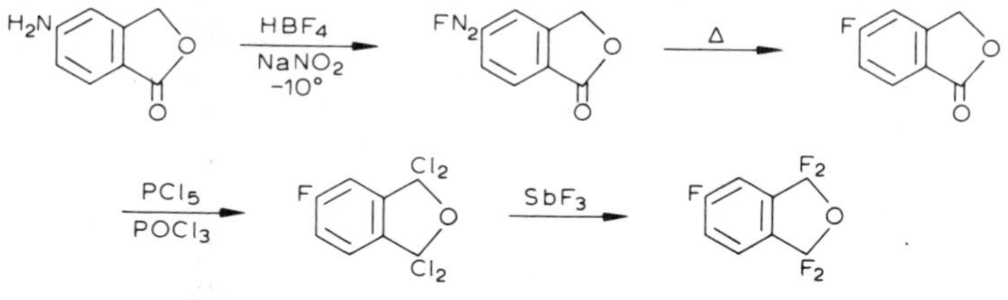

$\xrightarrow[NaNO_2,\ -10°]{HBF_4}$ $\xrightarrow{\Delta}$ $\xrightarrow[POCl_3]{PCl_5}$ $\xrightarrow{SbF_3}$

1,3-**Dihydroisobenzofurans** (*Rieche* and *Schulz*, Ann., 1962, **653**, 32): 1,3-*dihydroisobenzofuran*, f.p. 6°, b.p. 192°, d_4^{25} 1.0873, n_D^{25} 1.5440; 1-*phenyl*-, m.p. 35° (*A. Pernot* and *A. Willemart*, Bull. Soc. chim. Fr., 1953, 321); 1,1-*dimethyl*-, b.p. 76°/11 mm (*A. Ludwig*, Ber., 1907, **40,** 3060); 1,3-*dimethyl*-, b.p. 137°/50 mm; 1,1-*diphenyl*-, m.p. 95°, and 1,3-*diphenyl*-, m.p. 95° (*Guyot* and *Catel*, *loc. cit.*; *F. Seidel*, Ber., 1928, **61,** 2267); 1,1,3-*triphenyl*-, m.p. 120° (*Seidel* and *O. Bezner*, *ibid.*, 1932, **65,** 1566); 4,5,6-*trichloro-7-hydroxy*-, m.p. 228°; 5,6-*dichloro-4,7-dihydroxy*-, m.p. 243–244° (*Ruhr Chemie A.G.*, *loc. cit.*); 1,1,3,3-*tetrachloro-5-fluoro*-, b.p. 82°/2 mm; 1,1,3,3,5-*pentafluoro*-, b.p. 141–142°; 1,1,3,3,5-*pentachloro*-, m.p. 66–67°; 5-*chloro*-1,1,3,3-*tetrafluoro*-, b.p. 169–170°; 5-*bromo*-1,1,3,3-*tetrachloro*-, m.p. 96°; 5-*bromo*-1,1,3,3-*tetrafluoro*-, b.p. 86°/25 mm; 1,1,3,3-*tetrachloro-5-iodo*-, m.p. 118–119°; 1,1,3,3-*tetrafluoro-5-iodo*-1,3-*dihydroisobenzofuran*, b.p. 90°/23 mm (*Yagupol'skii* and *Belinskaya*, *loc. cit.*); 1,3-*dihydroisobenzofuran-4-carboxylic acid*, m.p. 196–199° (*J. Blair*, *W. R. Logan* and *G. T. Newbold*, J. chem. Soc., 1958, 304). 1-*Benzylidene*-1,3-*dihydroisobenzofuran* forms golden crystals, m.p. 94°; 3,3-*dimethyl*-1-*methylene*-1,3-*dihydroisobenzofuran*, b.p. 145–146°/5 mm, can be oxidised to dimethylphthalide. For u.v. and i.r. spectra of 1,3-dihydroisobenzofurans see *R. Allison et al.*, J. chem. Soc., 1958, 4311.

Shihunine, an alkaloid isolated from the Chinese drug Tsung Huan-tsao is a spiro derivative of 1,3-dihydroisobenzofuran (*Y. Inubushi et al.*, Chem. Pharm. Bull., Tokyo, 1964, **12,** 749):

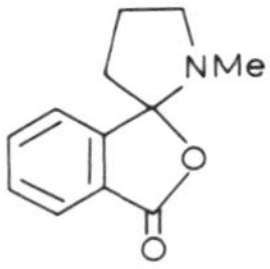

NMe
O
O
Shihunine

(*2*) 4,7-*Dihydroisobenzofurans.* The 4,7-dihydroisobenzofurans which result from synthesis (*4*) (p. 182) are pale yellow and fluoresce strongly, like the unreduced isobenzofurans.

4,7-**Dihydroisobenzofurans:** 1,3-*dimethyl*-, b.p. 198–203° (*G. O. Schenck*, Ber., 1947, **80,** 226); 1,3-*diphenyl*-, m.p. 120°–121°; 5,6-*dimethyl*-1,3-*diphenyl*-4,7-*dihydroisobenzofuran*, m.p. 225–226°.

(*3*) 4,5-*Dihydroisobenzofurans.* A solution of 1,2-dibenzoyl-1-cyclohexene in chloroform on treatment with concentrated hydrochloric acid affords 1,3-*diphenyl*-4,5-*dihydroisobenzofuran*, m.p. 71–72° (*R. C. Fuson* and *J. A. Haefner*, J. org. Chem., 1960, **25**, 2226):

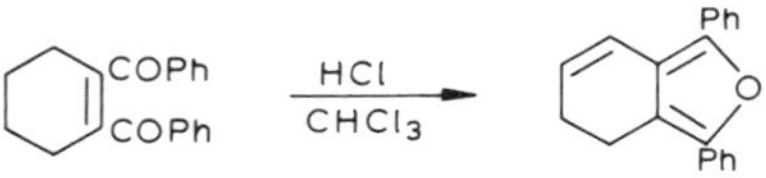

COPh
COPh
$\xrightarrow[CHCl_3]{HCl}$
Ph
O
Ph

(ii) Tetra-, hexa- and octa-hydroisobenzofurans

(*1*) *Tetrahydroisobenzofurans.* Hexahydrophthaloyl chloride and benzene give 1,2-dibenzoylcyclohexane and thence by dehydration 1,3-*diphenyl-4,5,6,7-tetrahydroisobenzofuran*, m.p. 97.5° (*Fuson et al., ibid.*, 1945, **10,** 55).

(*2*) *Hexahydroisobenzofurans.* In the presence of acids, cyclohexene and formaldehyde yield the *alcohol* XXIII, m.p. 205–208°, which is dehydrated to 1,3,3a,4,5,6-*hexahydroisobenzofuran*, b.p. 61°/10 mm (*S. Olsen et al.*, Z. Naturforsch., 1946, **I,** 448):

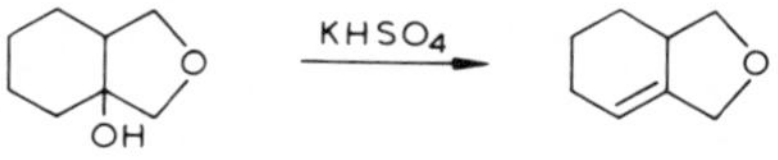

$KHSO_4$

(XXIII)

cis-Tetrahydrophthalic anhydride yields a cis-1,3,3a,4,7,7a-*hexahydroisobenzofuran*, b.p. 178–179°, n_D^{25} 1.4888 (*J. E. Ladbury* and *E. E. Turner*, J. chem. Soc., 1954, 3885):

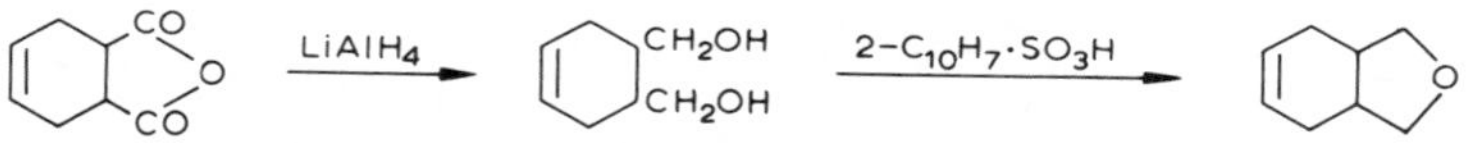

$LiAlH_4$ CH_2OH CH_2OH $2\text{-}C_{10}H_7\cdot SO_3H$

(*3*) *Octahydroisobenzofurans. Cis-* and *trans*-ethyl hexahydrophthalates yield the glycols, which are dehydrated to cis-, b.p. 175°, n_D^{21} 1.4712, and trans-*octahydroisobenzofuran*, b.p. 173°, n_D^{17} 1.4695, $[\alpha]_D^{22}$ + 110° (benzene) (*G. A. Haggis* and *L. N. Owen, ibid.*, 1953, 389; *B. T. Gillis* and *P. E. Beck*, J. org. Chem., 1963, **28,** 1388). Octahydroisobenzofuran is also produced by hydrogenation of the dihydro compound over nickel (*Entel et al., loc. cit.*).

4. Other bicyclic systems with a furan ring

The *cis* and *trans* forms of cyclopentane-1,2-dicarboxylic acid have been converted into the related 1,2-bishydroxymethylcyclopentanes, and then into cis-, b.p. 149°, $n_D^{20°}$ 1.4590, and trans-3-*oxabicyclo*[3.3.0]*octanes* (I), b.p. 158°, $n_D^{20°}$ 1.4631, (*L. N. Owen* and *A. G. Peto*, J. chem. Soc., 1955, 2383):

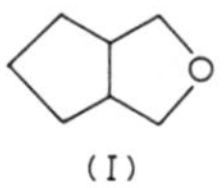

$Me_2C\cdot CH\cdot CH_2\cdot H_2C$ OH OH HO $Me_2C=$

(I) 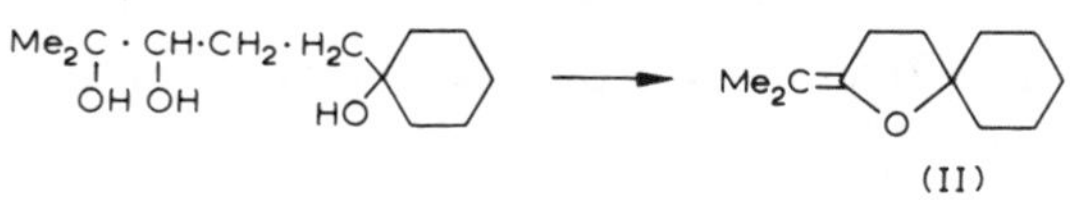 (II)

Dehydration of 1-(3′,4′-dihydroxy-4′-methylpentyl)cyclohexanol affords the spiro-compound 2-isopropylidene-1-oxaspiro[4.5]decane (II) (*A. Mondon*, Ann., 1954, **585,** 43).

Oxetones, the "dianhydrides" of γ,γ'-dihydroxyketones contain a spirocyclic system of two tetrahydrofuran rings with an α-carbon atom in common. Two molecules of γ-lactone under the influence of sodium or sodium alkoxides condense with loss of water to form compounds, which readily lose carbon dioxide forming the oxetone; *e.g.* γ-valerolactone gives *dimethyloxetone*, b.p. 209°, which is formed also by the hydrolysis of 2,8-dibromononan-5-one (*J. Volhard, ibid.*, 1892, **267**, 90):

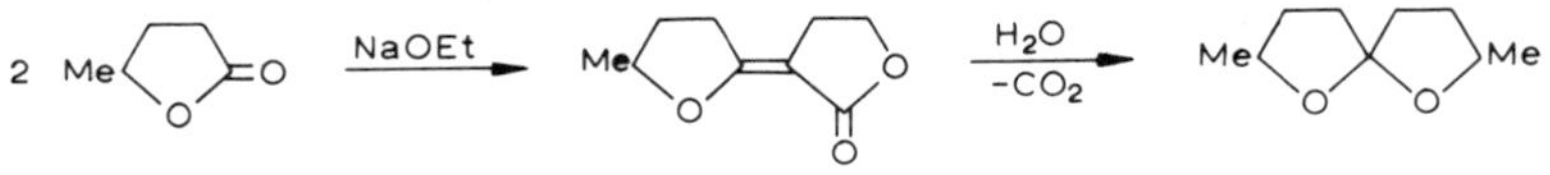

2 Me–(lactone) →(NaOEt)→ … →(H_2O, $-CO_2$)→ …

A system in which a tetrahydrofuran and a cyclohexane ring share three carbon atoms, 2-oxabicyclo[3.2.1]octane, is produced by dehydration of 3-hydroxymethylcyclohexanol (*M. F. Clarke* and *Owen*, J. chem. Soc., 1950, 2108):

By ring-closure of δ-2-furylvaleric acid a derivative of 2,3-cyclohepteno-furan is produced, which with selenium dioxide gives, probably, a yellow *dihydroxyfuranotropolone* (III), m.p. 194° (*W. Treibs* and *W. Heyer*, Ber., 1954, **87**, 1197; Ann., 1955, **595**, 203).

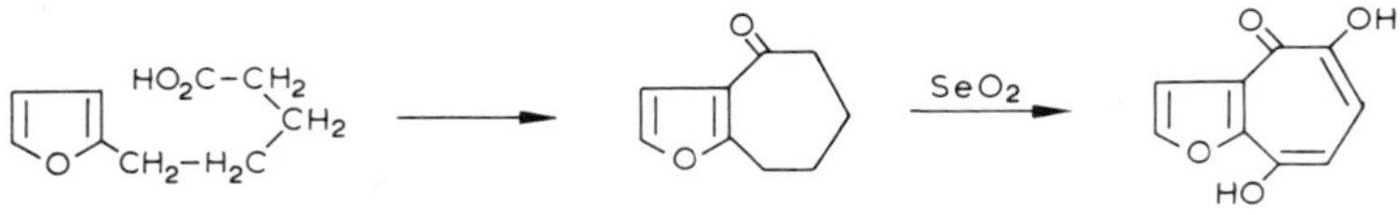

(III)

The reduction of 1,2-bis(methoxycarbonyl)cyclo-octatetraene with aluminium hydride at −75° gives in addition to the mono- and di-hydroxymethyl derivatives, the presumed aldehydic derivative IV, converted in ether or by passage over silica gel to cyclo-octa[*c*]furan (V), which is a $4n\pi$-electron system and readily reacts with oxygen and forms an adduct VI with tetracyanoethene (*E. le Goff* and *R. B. LaCount*, Tetrahedron Letters, 1965, 2787):

(IV) (V) (VI)

The furan and ethylene oxide rings are fused in 3,4-*epoxy*-2,2,5,5-*tetra*-

methyltetrahydrofuran, b.p. 90–91°/100 mm (*O. Heuberger* and *Owen*, J. chem. Soc., 1952, 910).

Two furan rings can be fused in various ways, *e.g.* as in isomannide (VII) produced from mannitol by hydrochloric acid (*L. F. Wiggins*, J. chem. Soc., 1945, 4), and in the case where mercuric sulphate converts the diacetylenic glycol VIII into a substance IX containing the same ring-system (*G. Dupont et al.*, Bull. Soc. chim. Fr., 1955, 1078):

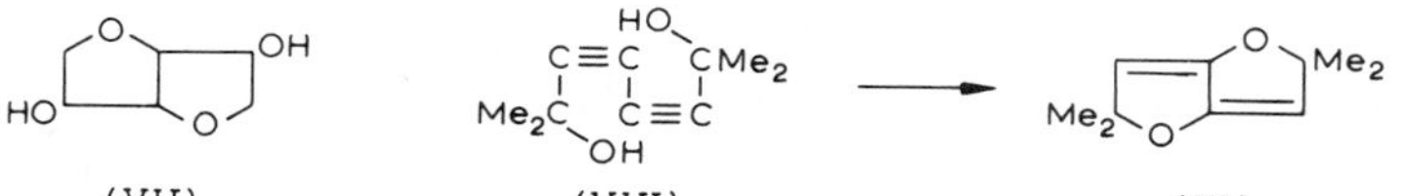

(VII) (VIII) (IX)

The lignan **gmelinol,** isolated from *Gmelina leichhardtii*, on treatment with acid gives, initially **isogmelinol.** Reduction of gmelinol and isogmelinol with sodium in liquid ammonia gives dihydro-derivatives, *e.g.* X, the reactions, rotations, and p.m.r. spectra of which lead to the absolute configurations shown for the respective parent compounds (*A. J. Birch* and *M. Smith*, J. chem. Soc., 1964, 2705):

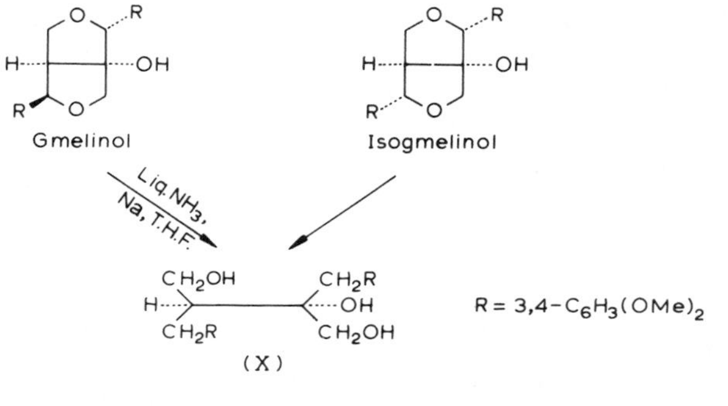

Gmelinol Isogmelinol

Liq. NH_3, Na, T.H.F.

(X) $R = 3,4\text{-}C_6H_3(OMe)_2$

Compounds having the bridged 1,4-oxidocyclohexane system, in which two tetrahydrofuran rings share the oxygen and two carbon atoms, are relatively common.

Some are natural products or closely related to them (see Vol. II B, pp. 31, 73, 150; cantharidin, Vol. II E, p. 70; 1,4-cineole, ascaridole, Vol. II B, p. 206), and many are obtained by Diels–Alder diene syntheses with furan derivatives. Thus addition of furan to dimethyl acetylenedicarboxylate gives *dimethyl* 1,4-*epoxycyclohexa*-2,5-*diene*-2,3-*dicarboxylate*, b.p. 130-133°/2 mm (*G. Stork et al.*, J. Amer. chem. Soc., 1953, **75**, 384):

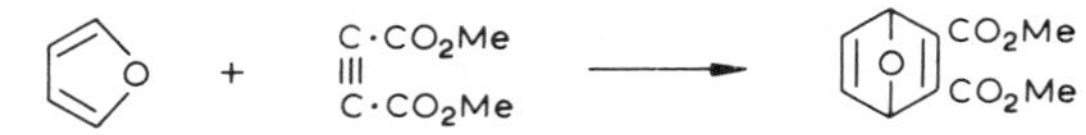

5. Dibenzofuran or diphenylene oxide and its hydrogenation products

(a) Dibenzofurans

Dibenzofuran (diphenylene oxide) was first prepared by Lesimple in 1866, by decomposing phenyl phosphate over lime; eight years later Graebe produced it more effectively by heating phenol with lead oxide. The constitution (I) follows from the rational syntheses (*1*), (*2*), and (*3*) below. The older system of numbering (I) has been generally replaced by scheme (II), used here.

(I) (II)

The chemistry of dibenzofuran has been studied in much detail since the 1930's, partly in the search for dye intermediates, and partly because the occurrence of the nucleus in alkaloids of the morphine group instigated chemotherapeutic work. The ring system also occurs in such lichen products as usnic acid, substances apparently produced by the oxidation of phenols. Dibenzofurans are also present in coal tar (*O. Kruber et al.*, Ber., 1932, **65,** 1382; 1936, **69,** 107; 1938, **71,** 2478; 1941, **74,** 1693).

(i) General synthetic methods

In contrast with the chemistry of most heterocyclic systems, derivatives are often made by substitution in the easily available dibenzofuran rather than by ring-synthesis from appropriately substituted intermediates. Ring closure may be effected in *o*-substituted biphenyls (methods *1*, *2*), or in biphenyl ethers (method *3*).

(*1*) Dibenzofurans are produced in good yield by dehydrating 2,2′-dihydroxybiphenyls in presence of acid reagents (*S. Rajagopalan* and *K. Ganapati*, Proc. Indian Acad. Sci., 1942, **15A**, 432; C.A., 1943, **37,** 1124), or by fusing 2-halogeno-2′-hydroxybiphenyls with potash (*K. Schimmelschmidt*, Ann., 1950, **566,** 184):

Heat / $ZnCl_2$

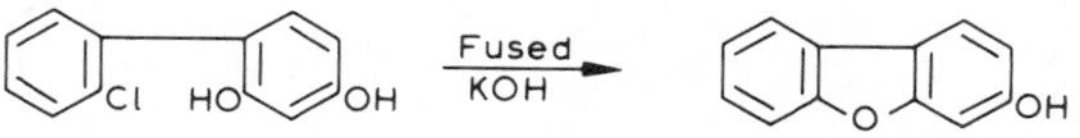

Heating 2,2′-dihydroxybiphenyl with 48% hydrobromic acid and acetic acid gives dibenzofuran, but 2,3,2′-trihydroxybiphenyl and 2,3,2′,3′-tetrahydroxybiphenyl, prepared by heating their tri- and tetra-methyl ethers with hydrobromic acid do not give the ring-closure products on similar treatment. Fusion of trihydroxybiphenyl with zinc chloride affords 4-hydroxydibenzofuran (*B. G. Pring* and *N. E. Stjernstrom*, Acta chem. Scand., 1968, **22,** 681). Demethylation of 2,2′,3,3′,6,6′-hexamethoxybiphenyl with a mixture of hydrobromic and acetic acids under reflux gives 1,2,8,9-, and 1,4,8,9-tetrahydroxydibenzofuran (*S. Forsen* and *Stjernstrom*, Arkiv Kemi, 1963, **21,** 65).

Both the above processes take place much more easily, under alkaline conditions, when nitro groups are present *o*- or *p*- to hydroxyl or halogen (*J. van Alphen*, Rec. Trav. chim., 1932, **51,** 715; *F. H. Case* and *R. U. Schock*, J. Amer. chem. Soc., 1943, **65,** 2086):

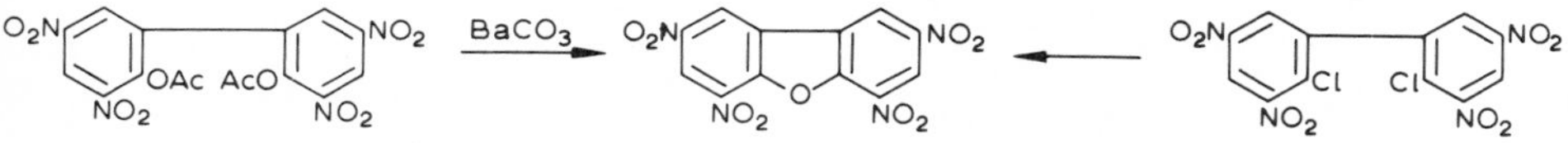

(*2*) Bisdiazonium salts from 2,2′-diaminobiphenyl on decomposition give dibenzofurans with, in some cases, phenazones (*E. Täuber* and *E. Halberstadt*, Ber., 1892, **25,** 2745):

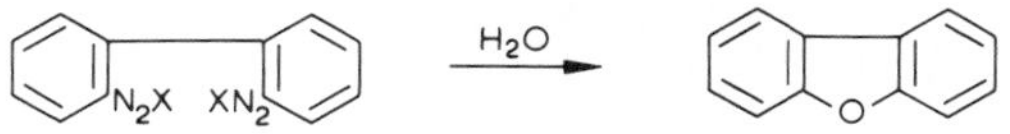

(*3*) *o*-Aminobiphenyl ethers undergo ring closure on diazotisation (*H. McCombie et al.*, J. chem. Soc., 1931, 529):

The diazonium salt derived from 2-aminophenyl 2,4-xylyl ether with hot dilute sulphuric acid (1:1) affords a mixture, from which 2,4-dimethyldibenzofuran (III) (8.3%) and 2-hydroxyphenyl 2,4-xylyl ether (24.6%) are isolated. Treatment of the isomeric 6-amino-2,4-xylyl phenyl ether, in a similar manner yields compound III (46%) and 6-hydroxy-2,4-xylyl phenyl ether (8.5%) (*W. E. Parham* and *R. W. Strassburg*, J. org. Chem., 1961, **26,** 4749):

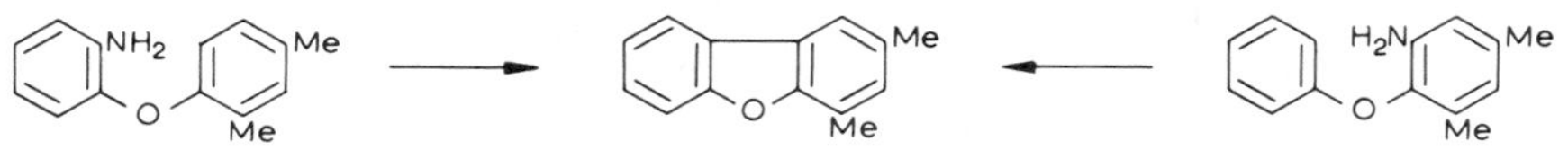

(III)

Dibenzofuran has also been prepared by pyrolysis of phenol (*C. Delaunois*, Ann. Mines Belg., 1968, 9; C.A., 1968, **69,** 60469u) or diphenyl ether, and by the dehydrogenation of *o*-phenylphenol using a palladium/carbon catalyst at 400° (*M. Suzumura* and *H. Yasin*, Jap. P. 28,092/1965; C.A., 1966, **64,** P11177f).

(ii) Properties and reactions

The furan ring is in general very stable, but dibenzofuran yields 2,2′-dihydroxybiphenyl on fusion with alkali (*H. Gilman et al.*, J. Amer. chem. Soc., 1940, **62,** 1963), a method which has been used to determine the orientation of substituted derivatives. Nitro groups in positions 2 and 4 greatly facilitate the ring-fission. The ring is also opened by boiling solutions of dibenzofurans in dioxan or ether with lithium (*Gilman* and *D. J. Esmay*, *ibid.*, 1953, **75,** 2947).

Halogenation, sulphonation, Friedel–Crafts acylation, and arsonation with arsenic acid, all take place in the 2-position. The bromo compound has been converted through the Grignard reaction into the 2-carboxylic acid, also obtained from 2-methyldibenzofuran, the structure of which is established by synthesis (*F. Mayer* and *W. Krieger*, Ber., 1922, **55,** 1659). The product of sulphonation (*I. Lielbriedis* and *E. Gudriniece*, C.A., 1963, **59,** 8692h) has been converted into 2-bromodibenzofuran (*Gilman et al.*, J. Amer. chem. Soc., 1934, **56,** 1412) by the stages:

$$RSO_3H \rightarrow RSO_2H \xrightarrow{HgCl_2} RHgCl \xrightarrow{Br_2} RBr$$

On the other hand, nitration takes place successively in positions 3 and 8 (*N. M. Cullinane*, J. chem. Soc., 1930, 2267; 1932, 2365); the 3-compound has been orientated by conversion into 3-chlorodibenzofuran (*cf.* synthetic method *3*).

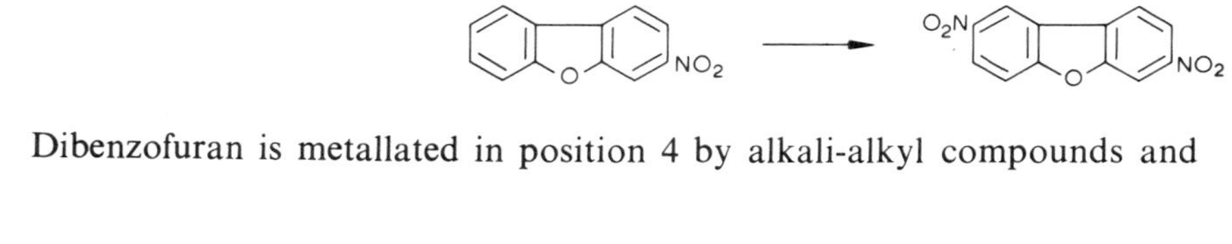

10 % 70 %

Dibenzofuran is metallated in position 4 by alkali-alkyl compounds and by mercuric salts (*Gilman* and *R. V. Young*, J. Amer. chem. Soc., 1934, **56,** 1415; 1935, **57,** 1121) as is shown by conversion of the products into the 4-carboxylic acid, obtained from 4-methyldibenzofuran of established constitution (*Kruber*, Ber., 1932. **65,** 1382).

These methods make available a wide range of dibenzofurans substituted in 2, 3 or 4 positions but not in position 1; moreover, the usual synthetic procedures fail in the case of the 1-derivatives. The 4-amino-compound can be made by several methods from 4-lithiodibenzofuran, and its acetyl derivative can be brominated in position 1 and deaminated. Since bromodibenzofurans yield the lithio-compounds by "interchange" (RBr + BuLi → RLi + BuBr) all four lithiodibenzofurans are accessible, and the lithium atoms can be replaced by carboxyl, methyl (by methyl sulphate), halogen, or amino-groups (by *O*-methylhydroxylamine).

As would be expected, when one benzene nucleus of dibenzofuran is nitro-substituted, electrophilic reagents substitute in the other nucleus; whereas if an *ortho, para*-directive group is present, further substitution takes place in the same ring. A hydroxyl or amino group directs substitution *ortho, para* to itself, taking control in competition with the ring-oxygen atom.

As a reactive aromatic system dibenzofuran undergoes direct chloromethylation, and the Gattermann process yields the 2-aldehyde, from which the analogues of benzoin, benzil, benzilic acid, and cinnamic acid are obtained (*L. E. Hinkel et al.*, J. chem. Soc., 1937, 778).

The technique of the production of ketones by the Friedel–Crafts reaction has been studied by *N. P. Buu-Hoï* and *R. Royer* (Rec. Trav. chim., 1950, **69,** 861).

For u.v. spectra of 2-substituted dibenzofurans (*M. Kuroki*, Nippon Kagaku Zasshi, 1968, **89,** 681); u.v. and i.r. spectra of dimethylbenzofurans (*C. Karr et al.*, U.S. Bur. Mines Bull., 1967, **637,** 198; C.A., 1967, **67,** 66435); n.m.r. (*P. J. Black, R. D. Brown* and *M. L. Heffernan*, Australian J. Chem., 1967, **20,** 1325; *Black* and *Heffernan*, *ibid.*, 1965, **18,** 353) and m.s. of some dibenzofuran derivatives (*Pring* and *Stjernstrom*, Acta chem. Scand., 1968, **22,** 549; *J. H. D. Eland* and *C. J. Danby*, J. chem. Soc., 1965, 5935; *C. S. Barnes* and *J. L. Occolowitz*, Australian J. Chem., 1964, **17,** 975).

(iii) Dibenzofuran and derivatives

Dibenzofuran, m.p. 83° (*A. F. Williams*, Nature, 1948, **162,** 925), b.p. 287°, is obtainable from coal tar, and in 17–20% yield by distilling phenol with lead oxide (*Cullinane*, J. chem. Soc., 1930, 2268; *V. M. Krauchenko* and *V. I. Mil'skii*, Tr. Donetsk. Ind. Inst., 1960, **23,** 137; C.A., 1962, **57,** 16525f), or nearly quantitatively by dehydration of 2,2′-dihydroxybiphenyl (*Cullinane*, Rec. Trav. chim., 1936, **55,** 881). It resists the action of hydriodic acid even under extreme conditions, and is very stable towards alkalis; contrast the sensitiveness of furan and monobenzofuran.

The pyrolysis of nitrobenzene with benzene gives a mixture containing, amongst other products, dibenzofuran, phenyl-, diphenyl-, and triphenyl-dibenzofuran (*E. K. Fields* and *S. Meyerson*, J. Amer. chem. Soc., 1967, **89**, 3224). The short lived existence of **dehydrobenzofuran** (IV) has been demonstrated by formation of 1,2,3,4-tetraphenyldibenzofuran (V) in 70% yield from fusion of tetracyclone with bis(bromobenzofuran)-mercury (VI) derived from 2-lithio-3-bromobenzofuran (*G. Wittig*, Rev. Chim., Acad. Rep. Populaire Roumaine, 1962, **7**, 1393; C.A., 1964, **61**, 4297d):

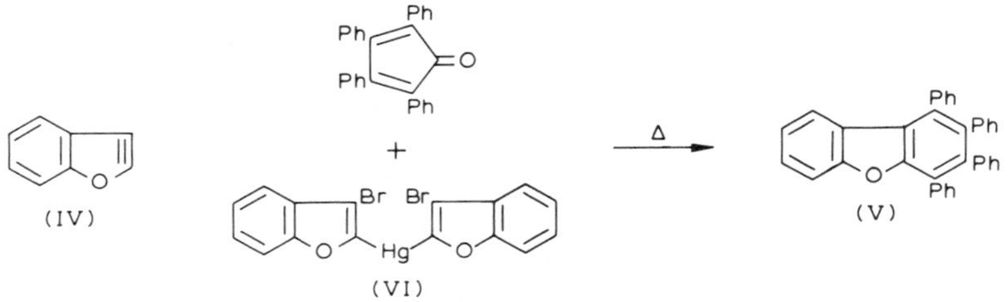

Pyrolysis of octafluorodibenzothiophene 5,5-dioxide affords *perfluorodibenzofuran*, which reacts with sodium methoxide in methanol to give the mono- and di-ethers VII and VIII (*R. D. Chambers, J. A. Cunningham* and *D. J. Spring*, J. chem. Soc., C, 1968, 1560):

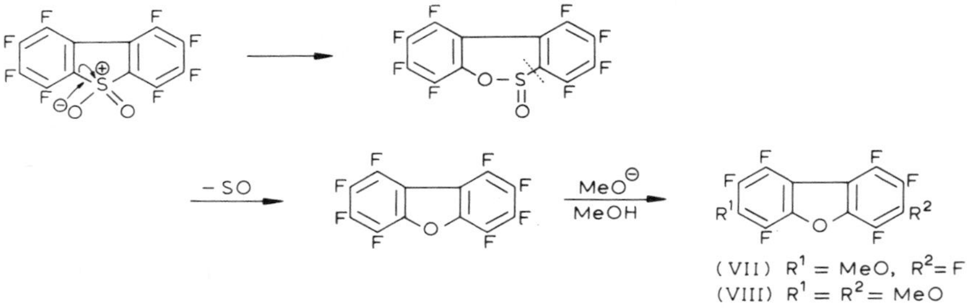

(VII) R^1 = MeO, R^2 = F
(VIII) R^1 = R^2 = MeO

1,2,3,4-**Tetrafluorodibenzofuran** has been prepared from hexafluorobenzene and *o*-lithioanisole. It undergoes nucleophilic substitution reactions with sodium hydrogen sulphide, sodium methiolate and lithium tetrahydridoaluminate, involving replacement of the 3-fluorine atom (*P. J. N. Brown, R. Stephens* and *J. C. Tatlow*, Tetrahedron, 1967, 4041):

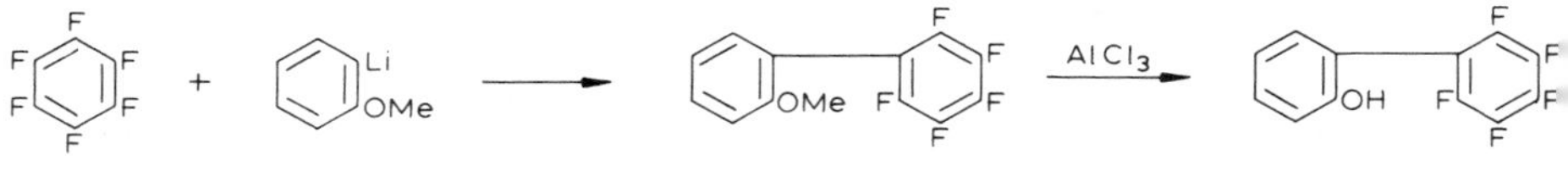

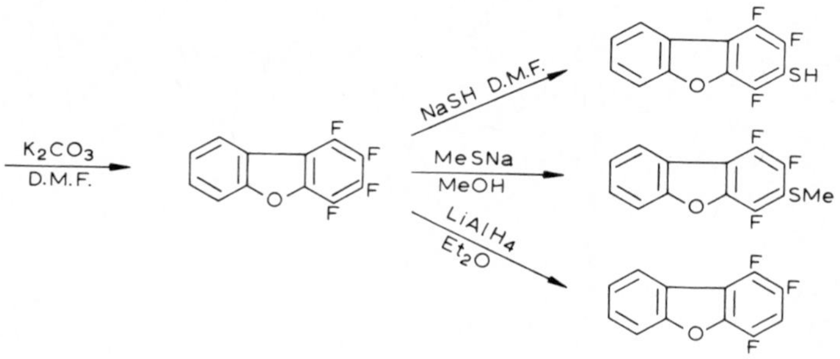

trans-2-(β-Methoxyvinyl)benzofuran (p. 159) undergoes a smooth cycloaddition with dimethyl acetylenedicarboxylate in boiling toluene to yield compound IX, which oxidises slowly in air to *dimethyl 3-methoxydibenzofuran*-1,2-*dicarboxylate* (X), m.p. 104–106°, also obtained on a preparative scale by boiling IX with *N*-bromosuccinimide in carbon tetrachloride. The major by-product from the addition reaction is dimethyl dibenzofuran-1,2-dicarboxylate (XI) (p. 165) (*W. J. Davidson* and *J. A. Elix*, Tetrahedron Letters, 1968, 4589):

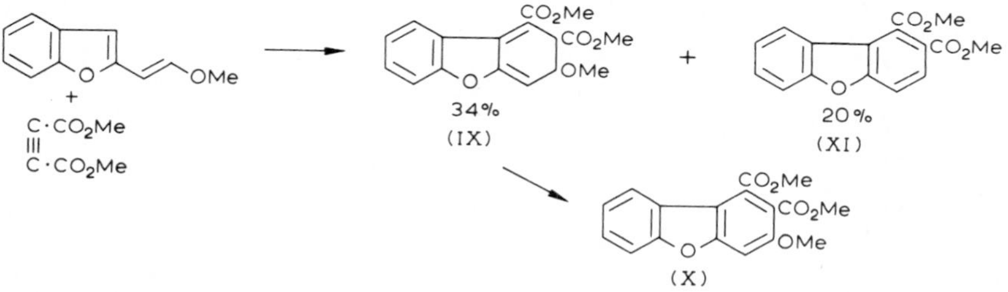

Dibenzofuran-2,8-**dicarboxylic acid** is obtained by oxidation of 2,8-diacetodibenzofuran (*J. D. Behun*, U.S.P. 3,190,853/1959; C.A., 1965, **63,** P5814c):

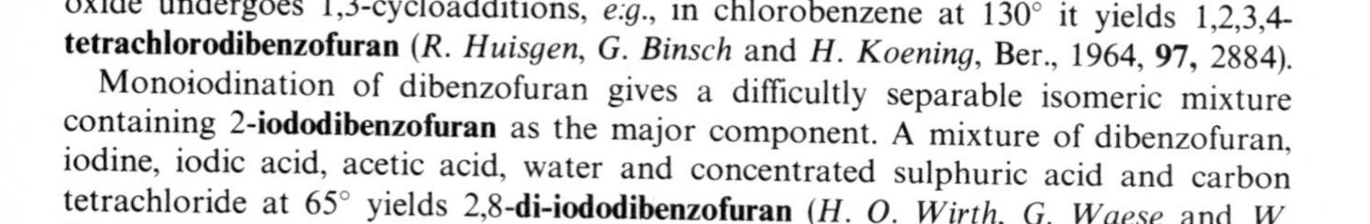

The oxo-carbene from the decomposition of 3,4,5,6-tetrachlorobenzene-2-diazo-1-oxide undergoes 1,3-cycloadditions, *e.g.*, in chlorobenzene at 130° it yields 1,2,3,4-**tetrachlorodibenzofuran** (*R. Huisgen, G. Binsch* and *H. Koening*, Ber., 1964, **97,** 2884).

Monoiodination of dibenzofuran gives a difficultly separable isomeric mixture containing 2-**iododibenzofuran** as the major component. A mixture of dibenzofuran, iodine, iodic acid, acetic acid, water and concentrated sulphuric acid and carbon tetrachloride at 65° yields 2,8-**di-iododibenzofuran** (*H. O. Wirth, G. Waese* and *W. Kern*, Macromol. Chem., 1965, **86,** 139; C.A., 1966, **64,** 19529h).

TABLE 12

DIBENZOFURANS

Substituent	*m.p.(°C)*	*Ref.*
1-Methyl *picrate*	58–59	1
2-Methyl	45	2, 3
3-Methyl	66	4
2-Ethyl	liquid	6, 7
1-Phenyl	63–64	8
2-Phenyl	99–100	8
3-Phenyl	130–131	8
4-Phenyl	b.p. 157–159/0.2 mm	8
3-Butenyl	b.p. 145/0.06 mm	9
4-Allyl	b.p. 130/0.05 mm	9
4-Methallyl	b.p. 135/0.05 mm	9
4-Pentenyl	b.p. 137/0.04 mm	9
4-Vinyl	b.p. 122/0.45 mm	9, 10
2,4-Dimethyl	42–43	11
1,2-Diphenyl	188–189	12
1,4-Diphenyl	110–111	12
1-$SiMe_3$	b.p. 163–164/7 mm	13
2-$SiMe_3$	44	13
3-$SiMe_3$	49	13
4-$SiMe_3$	41	13
2-$SiPh_3$	138	14
4-$SiPh_3$	153–154	14
2-Fluoro	89	15
3-Fluoro	88,5	15
1,2,4-Trifluoro	81.5–83	16
1,2,3,4-Tetrafluoro	100–101	16
Octafluoro	100	17
2-Chloro	106	18
3-Chloro	101	19
1,2,3,4-Tetrachloro	169–170	20
1-Bromo	67	20
2-Bromo	110	21
3-Bromo	120	18
4-Bromo	72	22
2,8-Dibromo	195	18
2-Iodo	112	23
3-Iodo	142	24
2,8-Diiodo	77	25
1-Hydroxy	140	20
2-Hydroxy	134	20, 26
2-Methoxy	46	20, 26

TABLE 12 *(continued)*

Substituent	*m.p. (°C)*	*Ref.*
3-Hydroxy	138	27
4-Hydroxy	101–102	28
4-Methoxy	52	29
2-Chlorosulphonyl	140	30
1-Nitro	120–121	31
2-Nitro	151	32
3-Nitro	186	18, 24
3,8-Dinitro	245	33
1-Amino	74	21
2-Amino	129–130	27
3-Amino	94	25, 34
4-Amino	85	21, 32
3-Methylamino	48–49	34
3-Dimethylamino	98	34
3-Acetamide	178	7
3-Hydrazino	174–175	35, 36
2-AsO_3H_2	250	37
3-AsO_3H_2	275	37
2-Hydroxymethyl	124	38
2-CH(OH)Me	63–64	39
2-Carbaldehyde	68	40
2-Acetyl	82	39
2-Bromoacetyl	105–106	39
2,8-Diacetyl	162–163	41
2-Benzoyl	136–138	42
1-Carboxy	232–233	21
1-Methoxycarbonyl	63	21
2-Carboxy	248–249	38, 43
2-Methoxycarbonyl	82–83	43
2-Cyano	138	3
3-Carboxy	272	21, 25
3-Methoxycarbonyl	138.5	21
3-Cyano	120	25
4-Carboxy	209	5, 44
4-Methoxycarbonyl	92–94	45
1,2-Bis(methoxycarbonyl)	121–123	46
1,2-$(CO_2Me)_2$-3-OMe	104–106	46
2,8-Bis(chlorocarbonyl)	232	41
2,8-Bis(methoxycarbonyl)	168	41, 47
1-$CH_2 \cdot CO_2H$	171–172	48
2-$CH_2 \cdot CO_2H$	162–163	49
4-$CH_2 \cdot CO_2H$	217–218	48

References

1 *J. N. Chatterjee* and *R. R. Ray*, Ber., 1959, **92,** 998.
2 *F. Mayer* and *W. Krieger*, *ibid.*, 1922, **55,** 1659.
3 *M. Kuroki*, Nippon Kagaku Zasshi, 1968, **89,** 527.
4 *O. Kruber et al.*, Ber., 1938, **71,** 2478.
5 *Kruber*, *ibid.*, 1932, **65,** 1382.
6 *N. P. Buu-Hoï* and *R. Royer*, Rec. Trav. chim., 1950, **69,** 861.
7 *W. Borsche* and *R. Schacke*, Ber., 1923, **56,** 2498.
8 *E. B. McCall, A. J. Neale and T. J. Rawlings*, J. chem. Soc., 1962, 4900.
9 *International Business Machines Corpn.*, B.P. 1,077,086/1963; C.A., 1968, **68,** 78123r.
10 *W. A. Hewett* and *E. Gipstein*, J. Polym. Sci., B, 1968, **6,** 565.
11 *W. E. Parham* and *R. W. Strassburg*, J. org. Chem., 1961, **26,** 4749.
12 *McCall, Neale* and *Rawlings*, J. chem. Soc., 1962, 5291.
13 *C. Eaborn* and *J. A. Sperry*, *ibid.*, 1961, 4921.
14 *R. H. Meen* and *H. Gilman*, J. org. Chem., 1955, **20,** 73.
15 *Gilman et al.*, *ibid.*, 1956, **21,** 457.
16 *P. J. N. Brown, R. Stephens* and *J. C. Tatlow*, Tetrahedron, 1967, 4041.
17 *R. D. Chambers, J. A. Cunningham* and *D. J. Spring*, J. chem. Soc., C, 1968, 1560.
18 *H. McCombie et al.*, J. chem. Soc., 1931, 529.
19 *N. M. Cullinane*, *ibid.*, 1930, 2267.
20 *R. Huisgen, G. Binsch* and *H. Koenig*, Ber., 1964, **97,** 2884.
21 *Gilman* and *P. R. van Ess*, J. Amer. chem. Soc., 1939, **61,** 1365.
22 *Buu-Hoï* and *Royer*, Rec. Trav. chim., 1948, **67,** 175.
23 *Gilman et al.*, J. Amer. chem. Soc., 1939, **61,** 2836.
24 *Gilman et al.*, *ibid.*, 1934, **56,** 2473.
25 *Borsche* and *W. Bothe*, Ber., 1908, **41,** 1940.
26 *H. O. Wirth, G. Waese* and *W. Kern*, Macromol. Chem., 1965, **86,** 139.
27 *K. Schimmelschmidt*, Ann., 1950, **566,** 184.
28 *K. Tatematsu* and *B. Kubota*, Brit. Abstr. A, 1935, 220.
29 *Gilman et al.*, J. Amer. chem. Soc., 1939, **61,** 951.
30 *Gilman* and *R. V. Young*, *ibid.*, 1935, **57,** 1121.
31 *Gilman et al.*, *ibid.*, 1934, **56,** 1412.
32 *Gilman* and *J. Swiss*, *ibid.*, 1944, **66,** 1884.
33 *Cullinane*, J. chem. Soc., 1932, 2365.
34 *W. H. Kirkpatrick* and *P. T. Parker*, J. Amer. chem. Soc., 1935, **57,** 1123.
35 *Gilman* and *L. C. Cheney*, *ibid.*, 1939, **61,** 3149.
36 *V. Grinsteins* and *J. Uldrikis*, C.A., 1964, **60,** 5432a.
37 *B. F. Skiles* and *C. S. Hamilton*, J. Amer. chem. Soc., 1937, **59,** 1006.
38 *F. Weygand* and *R. Mitgau*, Ber., 1955, **88,** 301.
39 *E. Mosettig* and *R. A. Robinson*, J. Amer. chem. Soc., 1935, **57,** 2186.
40 *L. E. Hinkel et al.*, J. Chem. Soc., 1937, 778.
41 *J. D. Behun*, U.S.P., 3,190,853/1965; C.A., 1965, **63,** P5814c.
42 *Gilman et al.*, J. Amer. chem. Soc., 1954, **76,** 6407.
43 *Gilman et al.*, *ibid.*, 1940, **62,** 346.
44 *Gilman* and *Young*, *ibid.*, 1934, **56,** 1415.
45 *Gilman et al.*, *ibid.*, 1939, **61,** 643.
46 *W. J. Davidson* and *J. A. Elix*, Tetrahedron Letters, 1968, 4589.
47 *Y. Sugii* and *T. Sengoku*, Brit. Abstr. A, 1936, 611.
48 *P. L. Soutwick, M. W. Munsell* and *E. A. Bartkus*, J. Amer. chem. Soc., 1961, **83,** 1358.
49 *W. Wenner*, J. org. Chem., 1950, **15,** 548.

4-**Vinyldibenzofuran** may be obtained from 4-lithiodibenzofuran and acetaldehyde (*W. A. Hewett* and *E. Gipstein*, J. Polym. Sci., B, 1968, **6**, 565; C.A., 1968, **69**, 77826f):

(1) Li, BuCl; (2) MeCHO → CH(OH)Me; $KHSO_4$, 1% 1,4-$C_6H_4(OH)_2$ → $CH{=}CH_2$

Some derivatives of dibenzofuran are listed in Table 12.

(b) Reduced dibenzofurans

Reduction of dibenzofuran with sodium and alcohols, or better with hydrogen over platinum black, yields 1,2,3,4-**tetrahydrodibenzofuran,** b.p. 154–156°/20 mm, n_D^{20} 1.5795, d_{20}^{20} 1.0938 (*Cullinane* and *H. J. H. Padfield*, J. chem. Soc., 1935, 1131), *picrate*, m.p. 91°. It is also produced by heating 2-chlorocyclohexanone with sodium phenoxide (*F. Ebel*, Helv., 1929, **12,** 3). On ozonisation it gives δ-*o*-hydroxybenzoylvaleric acid:

ONa + O, Cl → ; O_3 → [CO, O, CO] → $CO(CH_2)_4CO_2H$, OH

Tetrahydrodibenzofuran undergoes nitration, sulphonation, and Friedel–Crafts acetylation in position 7 (*meta* to oxygen), and metallation in position 6 (*Gilman et al.*, J. Amer. chem. Soc., 1935, **57,** 2095).

When dibenzofuran is hydrogenated over Raney nickel, it gives 1,2,3,4,4a,9b-**hexahydrodibenzofuran** (XII), b.p. 140–142°/20 mm, 125–127°/14 mm, n_D^{20} 1.5515, and then the **perhydrodibenzofuran,** b.p. 134–137°/20 mm, with some ring-fission to cyclohexylcyclohexanol and bicyclohexyl (*J. I. Jones* and *A. S. Lindsey*, J. chem. Soc., 1950, 1836).
The ring in 1,2,3,4,4a,9b-hexahydrodibenzofuran on treatment with sodium in pyridine opens to give *o*-(1-cyclohexenyl)phenol (*J. Gripenberg* and *T. Hase*, Acta Chem. Scand., 1966, **20,** 1561):

(XII) Na, C_5H_5N → OH

Some phenols react with cyclohexanones to give 1,2,3,4,4a,9b-hexahydrodibenzofurans (*J. B.* and *V. Niederl*, J. Amer. chem. Soc., 1939, **61,** 1785):

Me, OH + O, Me; HCl → Me, O, Me

1,2,3,4,6,7,8,9-**Octahydrodibenzofuran** (dicyclohexanofuran) (XIII), b.p. 139–140°/12 mm, is obtained *via* the photo-oxidation of 1,1-dicyclohexenyl and may be converted into the methoxyhydroperoxide XIV by photo-oxidation in methanol sensitised by Rose Bengal (*G. O. Schenck et al.*, Tetrahedron, 1967, 2583):

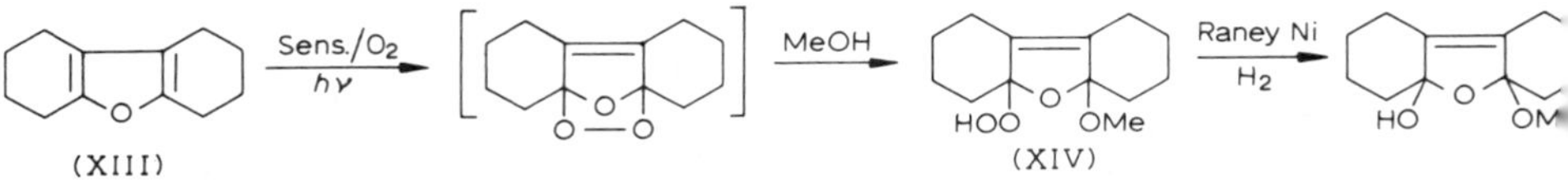

(XIII) —Sens./O_2, hν→ [ ] —MeOH→ (XIV) —Raney Ni, H_2→

Addition of concentrated sulphuric acid to cyclohexanone in benzene at 20–30° yields XIII and 2-methyl-2,3-cyclopentano-4,5-cyclohex-5-eno-2,3,4,5-tetrahydrofuran; both are also formed by the Beckmann rearrangement of cyclohexanone oxime (*K. L. Feller et al.*, Izvest. Akad. Nauk S.S.S.R., Ser. Khim., 1966, 1958; Tr. po Khim: Khim. Tecknol., 1965, 500; C.A., 1966, **65,** 673h):

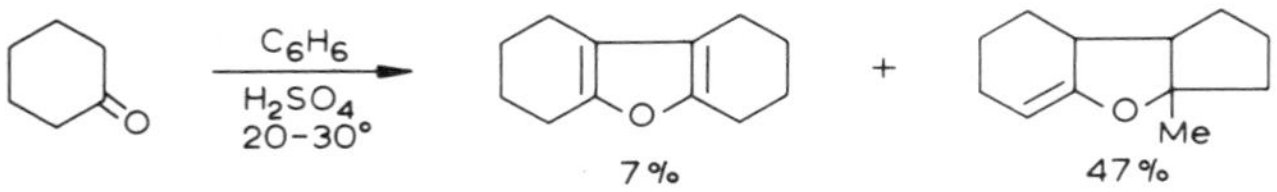

—C_6H_6, H_2SO_4, 20–30°→ 7% (XIII) + 47%

The condensation of cyclohexane-1,3-dione with 2-chlorocyclohexanone, an elaboration of the Feist–Benary furan synthesis is a useful route to 1-**oxo**-1,2,3,4,6,7,8,9-**octahydrodibenzofuran** (*J. N. Chatterjee* and *R. R. Ray*, Ber., 1959, **92,** 998; *E. B. McCall, A. J. Neale* and *T. J. Rawlings*, J. chem. Soc., 1962, 5291):

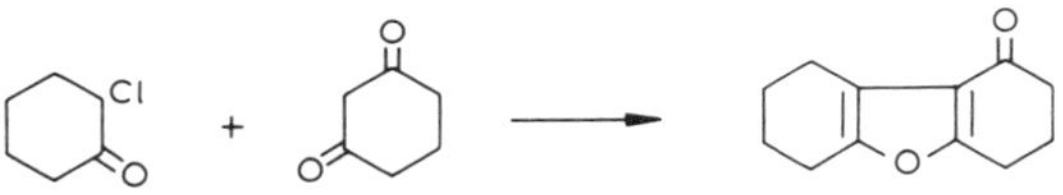

(c) Natural products related to dibenzofuran

Porphyrilic acid (XV), $C_{16}H_{10}O_7$, m.p. 285° (decomp.), from the lichens *Haematomma porphyrium* and *H. coccineum*, contains two phenolic hydroxyl groups, one carboxyl group, a lactone ring, one *C*-methyl substituent, and one indifferent (ethereal?)oxygen atom (*C. A. Wachmeister*, Acta chem. Scand., 1954, **8,** 1433; 1956, **10,** 1404; *H. Erdtman* and *Wachtmeister*, Chem. and Ind., 1956, 960). Gibb's reagent gives a blue colour with porphyrilic acid but not with methyl porphyrilate obtained from methyl *O,O*-diacetylporphyrilate by methanolysis. This indicates that porphyrilic acid contains a carboxyl group in the *para*-position to a hydroxyl group. Since, in the formation of compound XVI no carbon atom is lost and as the *C*-methyl group and lactonic properties disappear, porphyrilic acid must contain both a phthalide ring and an *o*-toluic acid group. Compound XVI affords 1,7-dihydroxybenzofuran and controlled permanganate oxidation of the diacetate of porphyrilic acid, so as only to affect the lactonic system, yields a tricarboxylic acid, subsequently converted to 1,7-dihydroxy-3-methyldibenzofuran:

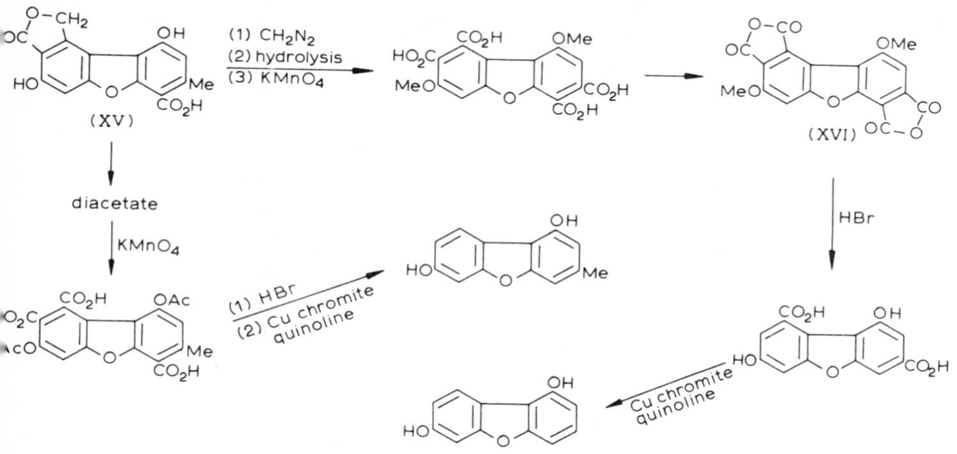

(XV) → (1) CH_2N_2 (2) hydrolysis (3) $KMnO_4$ → → (XVI) → HBr → Cu chromite quinoline →

(XV) → diacetate → $KMnO_4$ → (1) HBr (2) Cu chromite quinoline →

Usnic acid, $C_{18}H_{16}O_7$, m.p. 204°, $[\alpha]_D^{25°}$ + 503° ($CHCl_3$), which occurs in *Usnea barbata* and many other lichens, was first isolated by Rochleder and Heldt in 1843. It is not a carboxylic acid, owing its acidity to a β-diketone system, and also contains two phenolic hydroxyl groups. By alkaline hydrolysis it gives acetoacetic acid and usnetic acid, $C_{14}H_{14}O_6$, thermal decarboxylation of which gives usnetol, $C_{13}H_{14}O_4$, while vigorous treatment with alkali affords pyrousnic acid, $C_{12}H_{12}O_5$, and acetic acid. Decarboxylation of pyrousnic acid or alkaline decomposition of usnetol gives usneol, $C_{11}H_{12}O_3$:

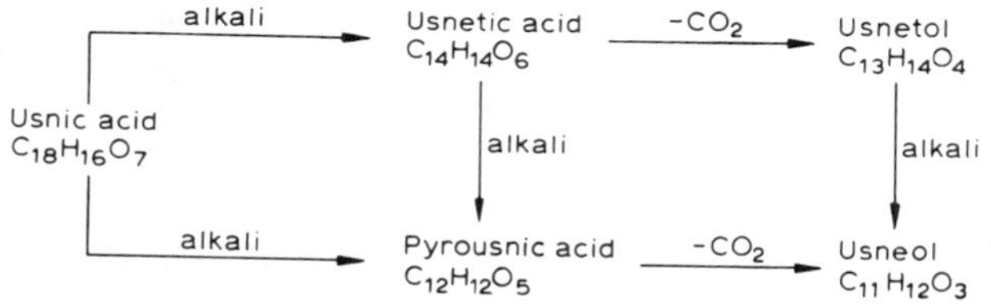

Usnic acid $C_{18}H_{16}O_7$ —alkali→ Usnetic acid $C_{14}H_{14}O_6$ —$-CO_2$→ Usnetol $C_{13}H_{14}O_4$

Usnic acid $C_{18}H_{16}O_7$ —alkali→ Pyrousnic acid $C_{12}H_{12}O_5$ —$-CO_2$→ Usneol $C_{11}H_{12}O_3$

Usnetic acid —alkali→ Pyrousnic acid; Usnetol —alkali→ Usneol

Usneol yields on ozonisation a monoacetyl derivative of 2,4,6-trihydroxy-3-methylacetophenone (*C. Schöpf* and *K. Heuck*, Ann., 1927, **459,** 233); and usnetic acid has been degraded to 3-methylfuran-2-acetic-4,5-dicarboxylic acid (*Y. Asahina* and *M. Yanagita*, Ber., 1937, **70,** 1500). This indicates that the compounds contain a benzofuran skeleton, and the following structures for usneol and usnetol have been established by synthesis of the methyl ethers (*F. H. Curd* and *A. Robertson*, J. chem. Soc., 1933, 714, 1173):

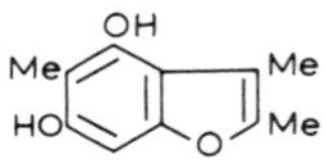

Usneol

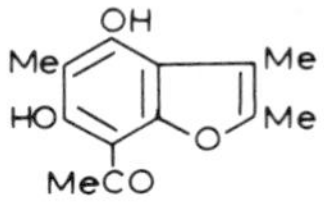

Usnetol

Usnetic acid is therefore as shown below, and the attachment of the aceto-acetic acid fragment remains to be determined. When usnic acid is heated with 95% alcohol at 150°, it affords decarbousnic acid, $C_{17}H_{18}O_6$ (addition of H_2O and loss of CO_2). Decarbousnic acid, a β-diketone, is hydrolysed by alkali to usnetic acid and acetone, and is therefore formulated as shown:

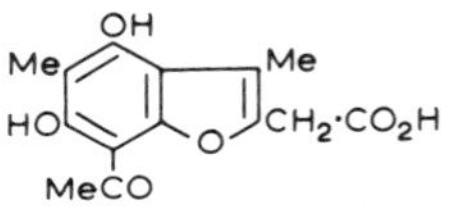

Usnetic acid

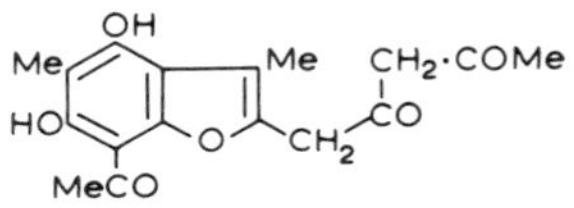

Decarbousnic acid

Usnic acid would then be, formally, the product of a carboxylation of decarbousnic acid, followed by dehydration.

The possibilities are limited, since usnic acid does not behave as a lactone, and is optically active. *Curd* and *Robertson* (*ibid.*, 1937, 894) proposed the following formula, which has been confirmed by a simple synthesis:

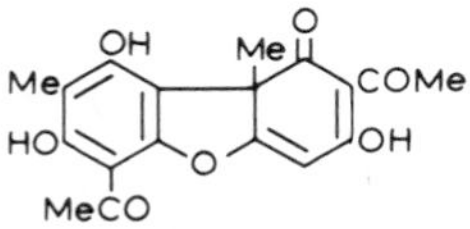

Usnic acid

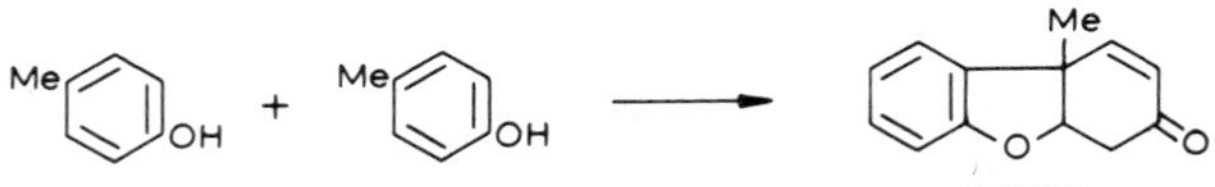

(XVII)

The dimeric product of oxidation of *p*-cresol with ferricyanide is XVII (*D. H. R. Barton et al., ibid.*, 1956, 530; *Robertson et al., ibid.*, p. 2322).

Similar oxidation of methylphloracetophenone gave XVIII, which was dehydrated to usnic acid:

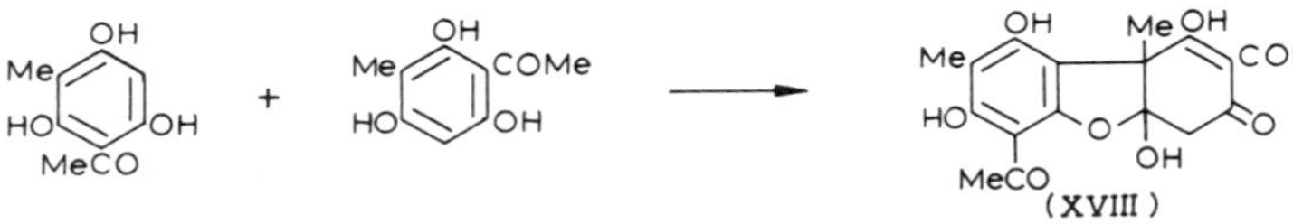

(XVIII)

6. Other tri- and poly-cyclic systems with a furan ring

(a) Condensed systems of benzene rings with one furan nucleus

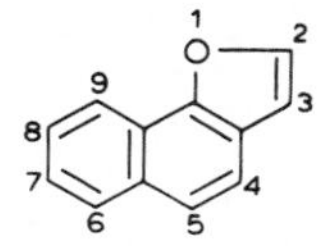

(I) Naphtho[1,2-*b*]furan [naphtho(1',2':2,3)furan, 6,7-benzocoumarone*]

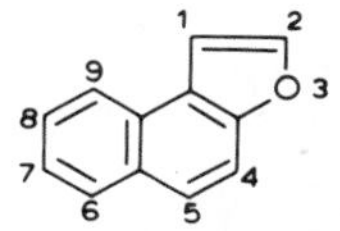

(II) Naphtho[2,1-*b*]furan [naphtho(2',1':2,3)furan, 4,5-benzocoumarone]

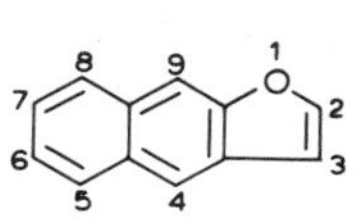

(III) Naphtho[2,3-*b*]furan [naphtho(2',3':2,3)furan, 5,6-benzocoumarone]

Coal tar contains **naphtho**[1,2-*b*]**furan** (*6,7-benzocoumarone*) (I), b.p. 284°. *R. Stoermer* (Ann., 1900, **312,** 308, 311) prepared both it and the 3-*methyl* derivative, m.p. 38°, by method (*3*) for benzofurans (p. 142). *2-Phenylnaphtho*[1,2-b]*furan*, m.p. 88–90° (*J. N. Chatterjea, V. N. Mehrotra* and *S. K. Roy*, Ber., 1963, **96,** 1167); *3-phenylnaphtho*[1,2-b]*furan*, m.p. 111–112°, is prepared from 1-naphthol and chloroacetophenone (*idem, ibid.*, 1963, **96,** 1156; *K. K. Thomas* and *M. M. Bokadia*, J. Indian chem. Soc., 1966, **43,** 713):

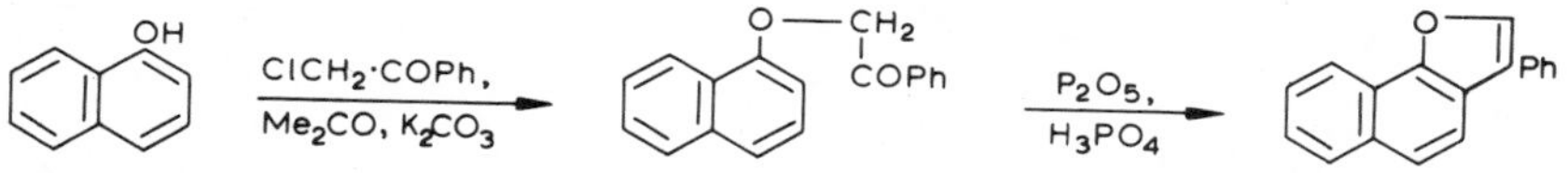

$ClCH_2{\cdot}COPh$, Me_2CO, K_2CO_3 → P_2O_5, H_3PO_4 →

2-Methylnaphtho[1,2-b]*furan*, m.p. 234° (*Chatterjea, ibid.*, 1959, **36,** 76); *2-methyl-3-phenylnaphtho*[1,2-b]*furan*, b.p. 237–238°/13 mm, m.p. 80°; *3-methyl-2-phenylnaphtho*[1,2-b]*furan*, m.p. 118° (*Royer, Bisagni* and *C. Hudry*, Bull. Soc. chim. Fr., 1960, 1178). Several methylated benzofurans have been converted *via* their succinoylation products to substituted furotetralones and hydroxynaphthofurans (*Royer, Bisagni* and *M. Hubert-Habart, ibid.*, 1965, 1794) and a number of substituted naphtho[1,2-*b*]furans have been synthesised (*A. N. Grinev, G. K. Prokof'eva* and *A. P. Terent'ev*, Zhur. obshcheĭ Khim., 1957, **27,** 1688; J. gen. Chem. U.S.S.R., 1957, **27,** 1757). *3-Acetyl-5-hydroxy-2-methylnaphtho*[1,2-b]*furan*, m.p. 262°; *3-benzoyl-5-hydroxy-2-methylnaphtho*[1,2-b]*furan*, m.p. 240°; *ethyl 5-hydroxy-2-phenylnaphtho-*[1,2-b]*furan-3-carboxylate*, m.p. 194–195° (*E. Bernatek*, Acta chem. Scand., 1956, **10,** 273).

Stoermer (*loc. cit.*) prepared **naphtho**[2,1-*b*]**furan,** (*4,5-benzocoumarone*) (II), m.p. 65°, present in coal tar, and its 1-*methyl* homologue, m.p. 59°. A 96% yield of the 1-methyl derivative may be obtained by cyclisation of 2-naphthoxyacetone with anhydrous hydrogen fluoride (*J. S. Moffatt*, J. chem. Soc., C, 1966, 725). Naphtho[2,1-*b*]furan may also be obtained from 3-chlorobenzo[*f*]doumarin, first converted to *naphtho*[2,1-b]*furan-2-carboxylic acid*, m.p. 197–199° (*M. le Corre* and *E. Levas*, Compt. rend., 1963, **257,** 1622; *Le Corre*, Ann. Chim., Paris, 1968, [14], **3,** 193):

* This trivial nomenclature is obsolete (*cf.* p. 141).

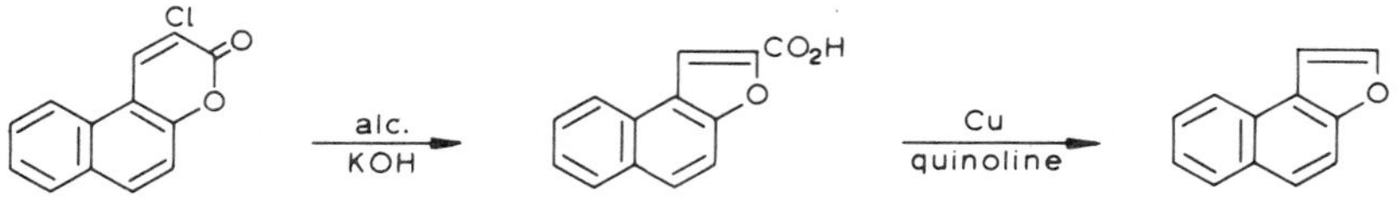

The photocyclodehydrogenation of 2-styrylfuran yields naphtho[2,1-*b*]furan (*C. E. Loader* and *C. J. Timmons*, J. chem. Soc., C, 1967, 1677):

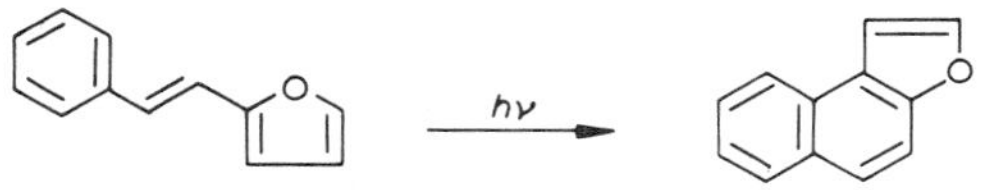

N.m.r. spectra of naphtho[2,1-*b*]furan (*T. J. Batherham* and *J. A. Lamberton*, Australian J. Chem., 1964, **17,** 1305).

2-*Methylnaphtho*[2,1-b]*furan*, m.p. 54° (*Chatterjea* and *Roy*, J. Indian chem. Soc., 1968, **45,** 45; *Distillers Co.*, B.P., 858,470/1958); 1-*phenylnaphtho*[2,1-b]*furan*, b.p. 191–193°/4,5 mm, is obtained from 2-naphthoxyacetophenone (*Thomas* and *Bokadia*, *loc. cit.*); 2-*phenylnaphtho*[2,1-b]*furan*, m.p. 140°; 1-*methyl-2-phenylnaphtho*[2,1-b]*furan*, m.p. 125° (*Chatterjea, Mehrotra* and *Roy*, Ber., 1963, **96,** 1156); 1-*methyl-2-phenylnaphtho*[2,1-b]*furan*, m.p. 104°; 2-*methyl-1-phenylnaphtho*[2,1-b]*furan*, b.p. 238°/17 mm, m.p. 55° (*Royer*, *Bisagni* and *Hudry*, *loc. cit.)*; 1,2-*diphenylnaphtho*[2,1-b]*furan*, m.p. 106°, is prepared by condensing benzoin and 2-naphthol in presence of 73% sulphuric acid (*O. Dischendorfer* and *E. Ofenheimer*, Monatsh., 1943, **74,** 135; *B. Arventiev*, *H. Wexler* and *M. Strul*, Acad. rep. populare Romîne, Filiala Iasi Studii cercetări stiint., Chim., 1960, **11,** 63; C.A., 1961, **55,** 15453a); 1-(p-*methoxyphenyl*)-2-*phenyl*-, m.p. 133–135°, 1,2-*bis*(p-*methoxyphenyl)*-, m.p. 116–117°, and 1,2-*bis*(o-*methoxyphenyl*)-*naphtho*[2,1-b]*furan*, m.p. 125–126° (*B. R. Brown*, *G. A. Somerfield* and *P. D. J. Weitzman*, J. chem. Soc., 1958, 4305; *S. S. Tiwari* and *S. C. Srivastava*, J. med. Chem., 1967, **10,** 983); 7-*ethylnaphtho*[2,1-b]*furan*, m.p. 70° (*P. Cagniant*, *P. Faller* and Mme. *P. Cagniant*, Bull. Soc. chim. Fr., 1961, 1938); *naphtho*[2,1-b]*furan-2-carbaldehyde*, m.p. 76°; 3-*acetyl-2-methylnaphtho*[2,1-b]*furan*, m.p. 82.5°; 3-*benzoyl-2-methylnaphtho*[2,1-b]*furan*, m.p. 112° (*Chatterjea* and *Roy*, *loc. cit.*); 1,5,8-*trimethylnaphtho*[2,1-b]*furan*, m.p. 98–100°, is the major product formed by the pyrolysis of curzerenone, n.m.r. and u.v. data are given (*H. Hikina et al.*, Tetrahedron Letters, 1968, 4417); 2-*diphenylmethylnaphtho*[2,1-b]*furan*, m.p. 177° (*R. Livingstone*, *D. Miller* and *S. Morris*, J. chem. Soc., 1960, 5148).

Naphtho[2,3-*b*]**furan,** (5,6-*benzocoumarone*) (III), m.p. 120°, is prepared by heating 3-formyl-2-naphthyloxy-acetic acid, acetic anhydride and fused sodium acetate (*P. Emmott* and *R. Livingstone*, J. chem. Soc., 1957, 3144).

2-*Methylnaphtho*[2,3-b]*furan*, m.p. 71°, may be obtained by the following route (*N. S. Narasimhan* and *M. V. Paradkar*, Chem. and Ind., 1963, 1529):

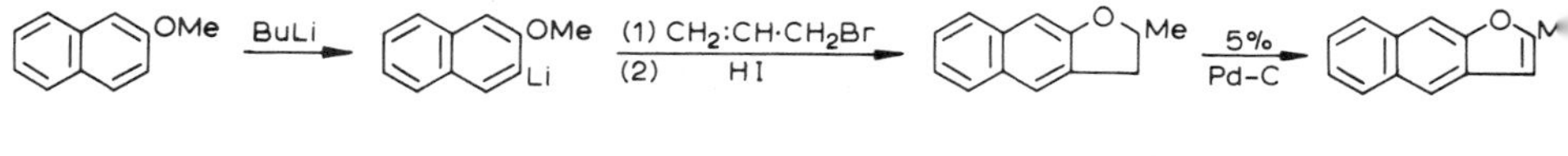

3-*Methylnaphtho*[2,3-b]*furan*, m.p. 63°, is prepared *via* the Grignard reaction between 3-oxo-2,3-dihydronaphtho[2,3-*b*]-furan and methylmagnesium iodide (*Emmott* and *Livingstone*, J. chem. Soc., 1958, 4629):

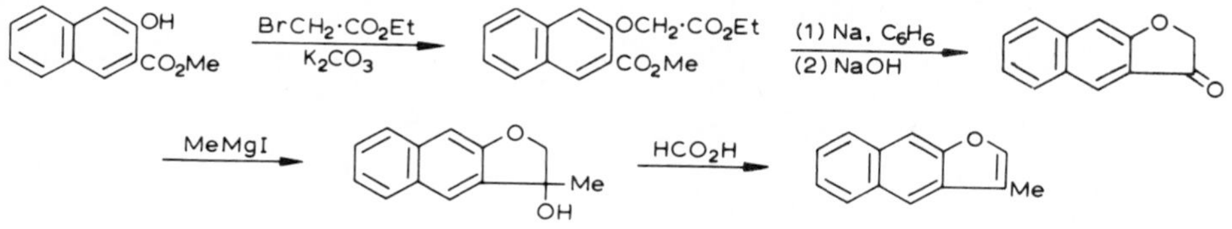

3,5-*Dimethylnaphtho*[2,3-b]*furan*, m.p. 72–74°, is afforded by the dehydrogenation of lindestrene, isolated from the root of *L. strychnifolia* (*K. Takeda et al.*, Tetrahedron, 1964, 2655); 2,3,9-*trimethylnaphtho*[2,3-b]*furan*, b.p. 174°/10 mm, m.p. 20°, is prepared either from the related benzofuran or naphthalene derivative (*Royer et al.*, Bull. Soc. chim. Fr., 1964, 1259; 1965, 1794):

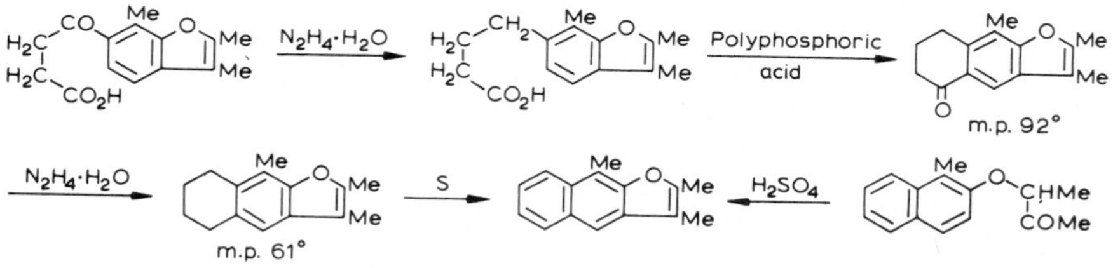

Naphtho[2,1-*b*]-, naphtho[2,3-*b*]-, and 5-methoxynaphtho[1,2-*b*]-furan, and their 2-carboxylic acids have been prepared by a general method from hydroxynaphthaldehydes (*Emmott* and *Livingstone*, J. chem. Soc., 1957, 3144):

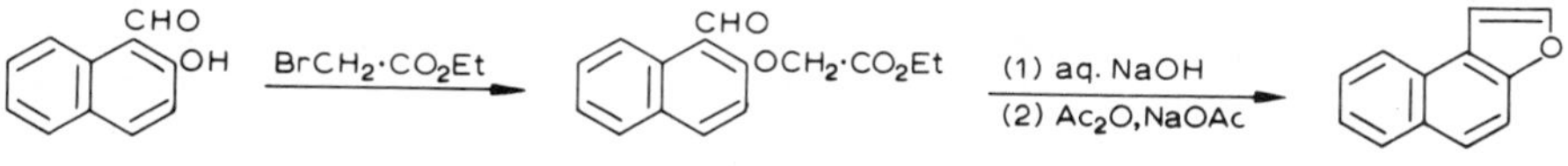

2-Methylnaphtho[1,2-*b*]- and [2,1-*b*]-furans have been obtained along with the corresponding naphthol and acetone following the irradiation of the related naphthoxy ketone. They are formed by cyclisation during work up of products resulting from an *ortho*-rearrangement (*J. R. Collier, M. K. M. Dirania* and *J. Hill*, *ibid.*, C, 1970, 155):

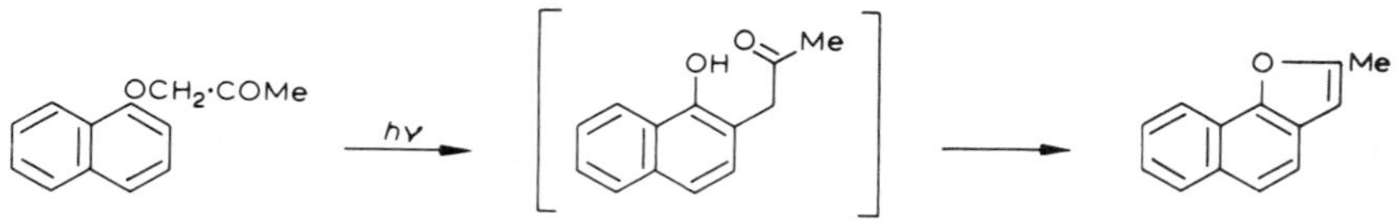

Hydrogenation over palladium–charcoal in ethanol of the appropriate naphthofuran affords 3-**methyl**-2,3-**dihydronaphtho**[1,2-*b*]**furan,** b.p. 92–94°/0.5 mm, 1-*methyl*-1,2-*dihydronaphtho*[2,1-b]*furan*, b.p. 76–82°/0.05 mm, and 3,9-*dimethyl*-2,3-*dihydronaphtho*[2,3-b]*furan*, b.p. 91–95°/0.05 mm (*J. S. Moffatt, ibid.*, 1966, 724).

The reaction of the 3-methoxymethylene ketone (IV) with ethoxycarbonylcarbene (from ethoxycarbonyldiazomethane) yields *ethyl* 4,5-*dihydronaphtho*[1,2-b]*furan*-2-*carboxylate*, m.p. 68–69° (*S. T. Murayama* and *T. A. Spencer*, Tetrahedron Letters, 1969, 4479):

(IV) $\xrightarrow[\text{boiling } Cl_2CH\cdot CHCl_2,\ CuSO_4]{EtO_2C\cdot CHN_2}$

1,2-*Dihydronaphtho*[2,1-b]*furan*, b.p. 154°/10 mm, is produced by treating the acid phthalate of β-(2-hydroxy-1-naphthyl)ethanol, $HOC_{10}H_6\cdot CH_2\cdot CH_2OH$, with alkali (*C. O. Guss*, J. Amer. chem. Soc., 1951, **73,** 608). 2-Naphthol and vinyl carbinols with phosphoric acid yield 1-alkyl derivatives, which cyclise to 1,2-dihydronaphtho[2,1-*b*]furans (*A. I. Kakhniashvili* and *G. Sh. Glonti*, Zhur. obshcheĭ Khim., 1964, **34,** 3135). 1,5,8-*Trimethyl*-6,7-*dihydronaphtho*[2,1-b]*furan (pyrocurzerenone)* (V), m.p. 76.5–77.5°, isolated from the rhizome of zedoary, *Curcuma zedoaria*, is obtained also by the pyrolysis of curzerenone (*Hikino, loc. cit.*):

(V)

2-*Methyl*-2,3-*dihydronaphtho*[2,3-b]*furan*, m.p. 35°, and its 4- and 9-methoxy derivatives have been synthesised. The hydrogenolysis was carried out in ethyl acetate over solid palladium chloride (*D. C. C. Smith* and *D. E. Steere*, J. chem. Soc., 1965, 1545):

(1) $MeCHBr\cdot CO_2Et$, Me_2CO, K_2CO_3 (2) KOH, aq. EtOH → Ac_2O, NaOAc → m.p. 134 – 136°

(1) MeOH, KOH (2) KBH_4 → Two isomers m.p. 128–130° and 83– 84° → hydrogenolysis →

Ring closure of 1- and 2-naphthyloxyacetic acids give 3-**oxo**-2,3-**dihydronaphtho**[1,2-*b*]**furan,** m.p. 119°, and 1-**oxo**-1,2-**dihydronaphtho**[2,1-*b*]**furan,** m.p. 133°, respectively

(*B. H. Ingham et al.*, J. chem. Soc., 1931, 895; *M. Darnault, G. Fontaine* and *P. Maitte*, Compt. rend., 1968, **266,** 1712; C.A., 1968, **69,** 106372n); while the orange-yellow 1,2-**dioxo**-1,2-**dihydronaphtho**[2,1-*b*]**furan,** m.p. 183° (decomp.) is obtained by Friedel–Crafts condensation between oxalyl chloride and methyl 2-naphthyl ether or between oxalbis(phenylimidyl) dichloride, Ph·N:CCl·CCl:NPh, and the free naphthol (*H. Staudinger et al.*, Helv., 1921, **4,** 334, 342); 3-**hydroxy**-2,3-**dihydronaphtho**[2,1-*b*]**furan,** m.p. 103° (*Darnault, Fontaine* and *Maitte, loc. cit.*).

3-**Oxo**-2,3-**dihydronaphtho**[2,3-*b*]**furan,** m.p. 148°, is obtained on treating ethyl 3-oxo-2,3-dihydronaphtho[2,3-*b*]furan-2-carboxylate in ethanol with a 30% solution of sodium hydroxide; 2,4-*dinitrophenylhydrazone*, m.p. 262° (*Emmott* and *Livingstone*, J. chem. Soc., 1958, 4629; *G. Haberland* and *G. Kleinert*, Ber., 1938, **71,** 470). The furanone gives no colour with neutral ferric chloride solution and does not decolourise bromine water; treatment with phosphorus pentachloride yields the dichloro derivative which readily loses hydrogen chloride to form 3-**chloronaphtho**[2,3-*b*]**furan,** m.p. 81–82° (*Emmott* and *Livingstone, loc. cit.*). Derivatives of the [1,2-*b*] and [2,3-*b*] systems are encoutered in the tranformation of lapachol (*S. C. Hooker*, J. Amer. chem. Soc., 1936, **58,** 1168).

3a,9b-**Dimethyl**-1,2,3a,9b-**tetrahydronaphtho**[2,1-*b*]**furan,** b.p. 88–92°/0.8 mm; 2,3,9-*trimethyl-5,6,7,8-tetrahydronaphtho*[2,3-b]*furan*, m.p. 61° (*Royer et al.*, Bull. Soc. chim.

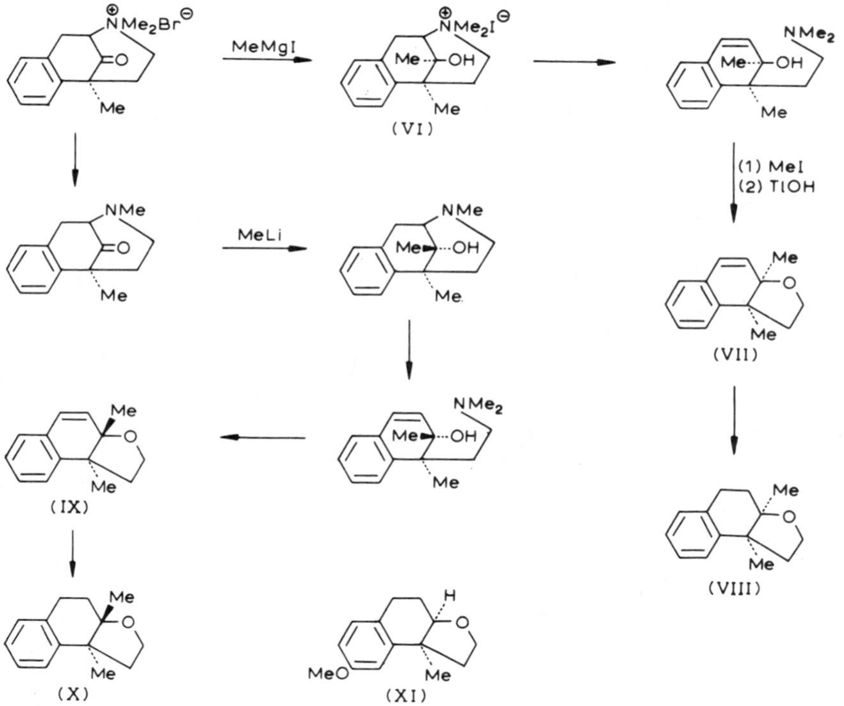

MeMgI

(VI)

(1) MeI
(2) TlOH

MeLi

(VII)

(IX)

(VIII)

(X)

(XI)

Fr., 1964, 1259; 1965, 1794); 3a,9b-**dimethyl**-1,2,3a,4,5,9b-**hexahydronaphtho**[2,1-*b*]**furan,** b.p. 85–88°/0.2 mm; *4-bromo*-3a,9b-*dimethyl*-1,2,3a,4,5,9b-*hexahydronaphtho*[2,1-b]*furan*, b.p. 130°/0.3 mm (*E. M. Fry*, J. org. Chem., 1957, **22,** 1710); 1-*methyl*-1,2,6,7,8,9-*hexahydronaphtho*[2,1-b]*furan*, b.p. 72–75°/0.05 mm *(Moffatt, loc. cit.)*. A number of naphthofuran derivatives, *e.g.*, 3a,5a-**dimethyl-dodecahydronaphtho**[2,1-*b*]**furan,** b.p. 105–110°/0.2 mm, have been obtained as intermediates during studies in the synthesis of terpenes (*T. G. Halsall* and *M. Moyle*, J. chem. Soc., 1960, 4931), 3a,6,6,9a-*Tetramethyldodecahydronaphtho*[2,1-b]*furan*, m.p. 75–76°, is a flavouring material for tobacco (*J. N. Schumacher*, U.S.P. 2,905,576/1958; C.A., 1960, **54,** P2679a).

Degradation of 9-hydroxy-2,5,9-trimethyl-6,7-benzomorphan methiodide (VI) by two Hofmann elimination reactions gives *cis*-3a,9b-dimethyl-1,2,3a,9b-tetrahydronaphtho[2,1-*b*]furan (VII), hydrogenated catalytically to the corresponding *cis*-3a,9b-dimethyl-1,2,3a,4,5,9b-hexahydro derivative (VIII). If methyl-lithium is used instead of methylmagnesium iodide and the free base in place of the methobromide, then the *trans*-isomers IX and X may be obtained (*E. L. May* and *H. Kugita*, J. org. Chem. 1961, **26,** 188; *May, Kugita* and *J. Harrison, ibid.*, p. 1621); *cis*-9b-methyl-8-methoxy-1,2,3a,4,5,9b-hexahydronaphtho[2,1-*b*]furan (XI) (*Kugita* and *May*, *ibid.*, p. 1954).

The pyrrolidine enamine (XII) of *trans*-9-methyl-2-decalone is alkylated with ethyl α-bromopropionate and the product hydrolysed with 5% potassium hydroxide in methanol to give the oxo-carboxylic acid XIII. The αβ-unsaturated γ-lactone XIV, obtained on heating the acid with acetic anhydride and sodium acetate, is reduced with di-isobutylaluminium hydride or diethylaluminium hydride in tetrahydrofuran to yield 3,8a-**dimethyl**-4,4a,5,6,7,8,8a,9-**octahydronaphtho**[2,3-*b*]**furan** (XV), m.p. 35–37.5° (*H. Minato* and *T. Nagasaki*, J. chem. Soc., C, 1966, 377):

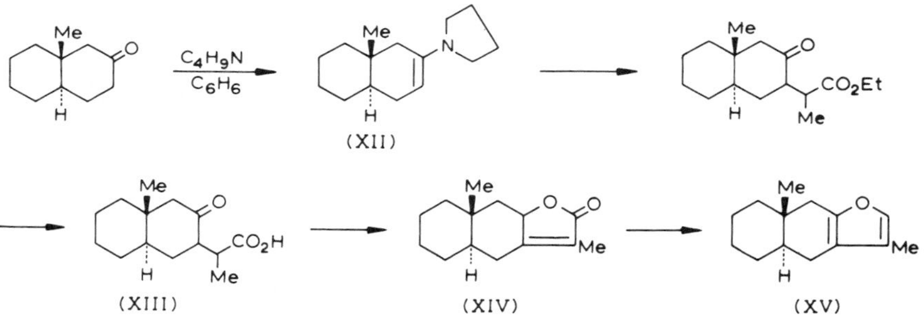

Some sesquiterpene derivatives possess a naphthofuran ring system; **erivanin,** isolated from *Artemisia fragrans*, being a naphtho[1,2-*b*]furan derivative (*R. I. Evstratowa et al.*, Khim. Prir. Soedin., 1969, **5,** 239; C.A., 1970, **72,** 32051v).

Erivanin

The structure of **atractylon,** m.p. 38°, [α]D 40.0°, a furanosequiterpene isolated from *Atractylis ovata* Thunb, and *Actractylodes japonica* Kizumi has been established (*H. Hikino, Y. Hikino* and *I. Yosioka*, Chem. Pharm. Bull., Tokyo, 1962, **10,** 641; 1964, **12,** 755), shown to be identical with the racemic compound, 3,8a-dimethyl-5-methylene -4,4a,5,6,7,8,8a,9-octahydronaphtho[2,3-*b*]furan obtained from 2-oxo-2,3,4,6,7,8-hexahydronaphthalene (*Minato* and *Nagasaki*, J. chem. Soc., C, 1966, 1866):

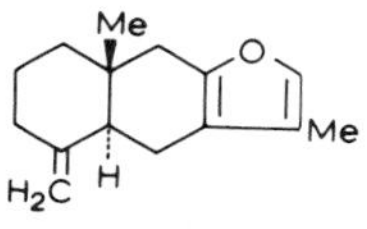

Atractylon

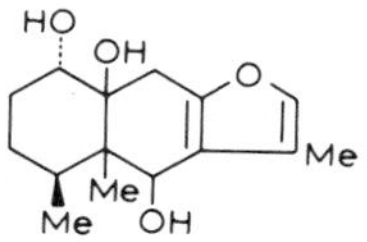

(XVI)

Euryopsol, m.p. 173–174°, isolated from the resin of *Euryops* spp, is 1α,6β,10β-trihydroxyfuranoeremophilane (XVI), 4β,8α,8aβ-trihydroxy-3,4a,5-trimethyl-4,4a,5,6,7,8,8a,9-octahydronaphtho[2,3-*b*]furan (*G. A. Eagle et al.*, Tetrahedron, 1969, 5277).

The 1-ethynyl-1-hydroxy-8a-methyl-6-oxo-1,2,3,4,6,7,8,8a-octahydronaphthalenes (XVII) and (XVIII) undergo rearrangement when boiled with *p*-toluenesulphonic acid in chloroform to give 2,6-dimethyl-4,5-dihydro-3*H*-naphtho[1,8-*bc*]furan (XIX) (*S. Swaminathan et al.*, J. org. Chem., 1966, **31,** 656):

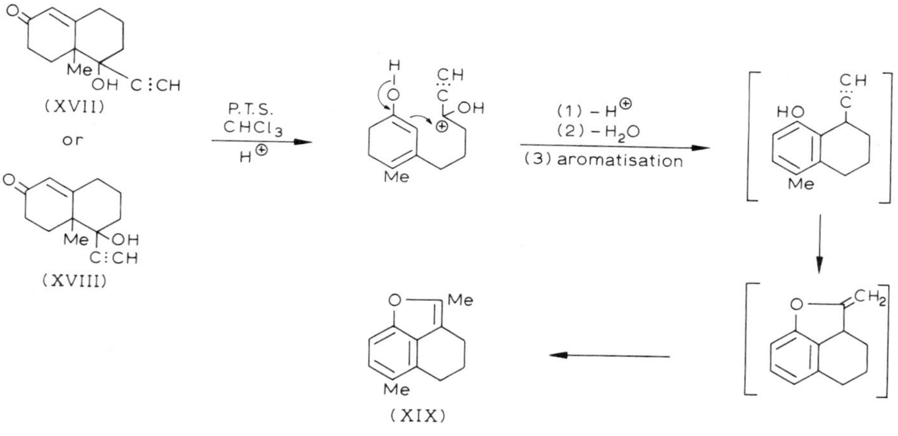

(XVII) or (XVIII) —P.T.S., $CHCl_3$, $H^{\oplus}$→ … —(1) $-H^{\oplus}$ (2) $-H_2O$ (3) aromatisation→ … → … → (XIX)

When the Grignard complex from the lactone of 8-hydroxy-1-naphthoic acid and methylmagnesium iodide is decomposed with acid an intractable product is formed, but if a neutral solution of sodium ethylenediaminetetra-acetate is used 2,2-**dimethyl-**2*H*-**naphtho**[1,8-*bc*[**furan** (XX), b.p. 145–150°/15 mm, *picrate*, m.p. 106–107°, and 8-acetyl-1-naphthol are obtained (*R. J. Packer* and *D. C. C. Smith*, J. chem. Soc., C, 1967, 2194):

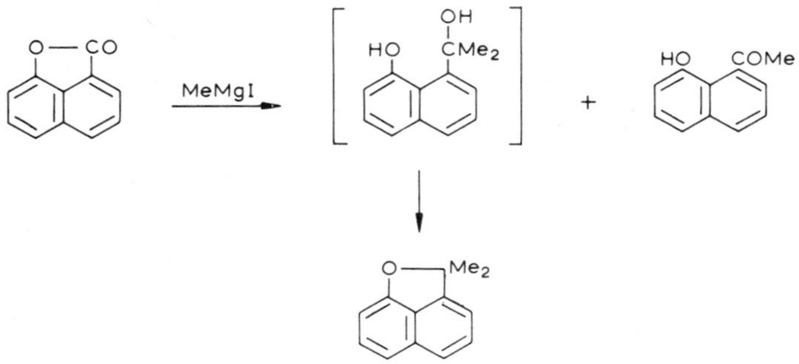

MeMgI

(XX)

A few substances are known in which the furan ring shares two carbon atoms with anthracene (*C. Marschalk*, Bull. Soc. chim. Fr., 1942, **9,** 801) and phenanthrene (*A. L. Wilds*, J. Amer. chem. Soc., 1942, **64,** 1421). The tanshinones, from *Salvia miltiorrhizae*, appear to be *o*-quinones derived from a phenanthrafuran (*F. v. Wessely et al.*, Ber., 1942, **75,** 617, 958; *Y. Okumura et al.*, Bull. chem. Soc., Japan, 1961, **34,** 895; *K. Takiura* and *K. Koizumi*, Chem. Pharm. Bull., Tokyo, 1962, **10,** 112); and vinhaticoic acid (Vol. II C, pp. 378, 391; *F. E.* and *T. J. King*, J. chem. Soc., 1953, 4158) and cafestol (Vol. II C, pp. 380, 396; *C. Djerassi et al.*, J. org. Chem., 1955, **20,** 1046; J. Amer. chem. Soc., 1960, **82,** 4342; *A. I. Scott et al.*, Tetrahedron, 1964, 1339) are also phenanthrafuran derivatives. Anthracene and furan are fused in another manner in the violet phenylbenzoyleneisobenzofuran (XXI); and in yet another in benzoin yellow, a mordant dye produced from benzoin and gallic acid (*C. Graebe*, Ber., 1898, **31,** 2975):

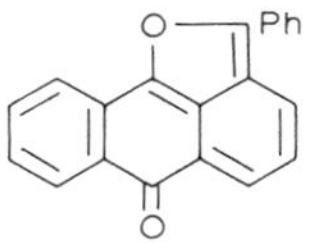

(XXI) 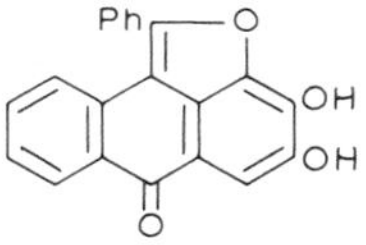 Benzoin yellow

Of tetracyclic systems in which the furan ring has all its carbon atoms in common with benzene nuclei, the most significant are the linear benzonaphthofuran, β-brazan, and 3-hydroxy-4,5-oxidophenanthrene, morphenol.

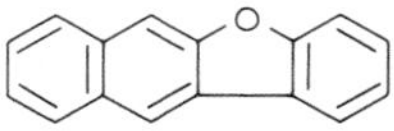

β – Brazan 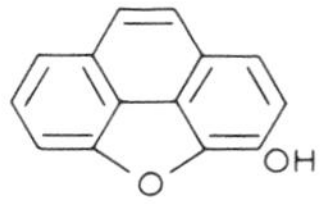 Morphenol

β-**Brazan** (*benzo*[b]*naphtho*[2,3-d]*furan*), m.p. 209°, present in coal tar (*O. Kruber*, Ber., 1937, **70,** 1556; *Chatterjea*, J. Indian chem. Soc., 1954, **31,** 101), is a degradation product of brazilin.

On oxidation it yields the yellow quinone 2,3-*phthaloylbenzofuran*, m.p. 245–246° (*R. A. Robinson* and *E. Mosettig*, J. Amer. chem. Soc., 1939, **61,** 1148). *β*-Brazan has been succinoylated and some derivatives prepared (*Chatterjea* and *Mehrotra*, J. Indian chem. Soc., 1963, **40,** 203); derivatives of α-**brazan,** benzo[*b*]naphtho[2,1-*d*]furan and γ-**brazan,** benzo[*b*]naphtho[1,2-*d*]furan have also been reported (*Chatterjea et al.*, Ber., 1963, **96,** 1156; J. Indian chem. Soc., 1963, **40,** 144), and the latter has been synthesised (*Chatterjea* and *K. D. Banerji*, Ber., 1965, **98,** 2738):

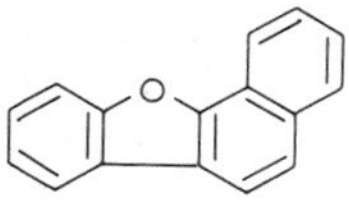

α-Brazan

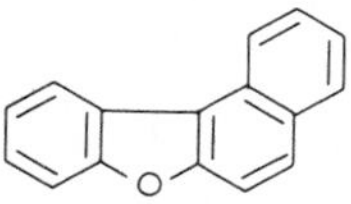

γ-Brazan

Three **dinaphthofurans** are accessible by dehydration of dinaphthols: XXII, m.p. 179°, from α-dinaphthol; XXIII, m.p. 156°, and XXIV, m.p. 159°, from *β*-dinaphthol (*G. R. Clemo* and *R. Spence*, J. chem. Soc., 1928, 2811; *N. P. Buu-Hoï*, *ibid.*, 1952, 489). A number of angular dinaphthofurans have been synthesised (*Chatterjea et al.*, *loc. cit.*):

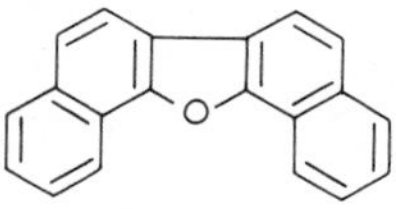

(XXII)

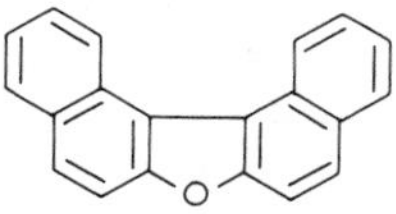

(XXIII)

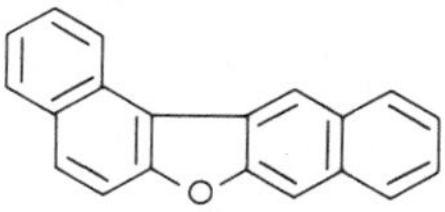

(XXIV)

(b) Systems of benzene rings with more than one furan nucleus

The photochemical cyclodehydrogenation of *trans*-1,2-di(2-furyl)ethylene yields benzo[1,2-*b*:4,3-*b*′]difuran (XXV) in low yield (*C. E. Loader* and *C. J. Timmons*, J. chem. Soc., C, 1967, 1677):

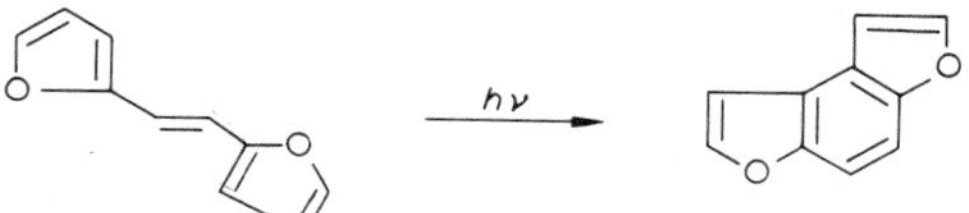

hν

(XXV)

G. J. Gie (Arkiv Kemi, 1945, **19,** [2], No. II; Brit. Abstr. AII, 1950, 563) dehydrated the pinacols of *o*-hydroxyacetophenones to derivatives of 3a,6a-dimethyl-2,3:5,6-dibenzo-3a,6a-dihydrofuro[3,2-*b*]furan (XXVI).

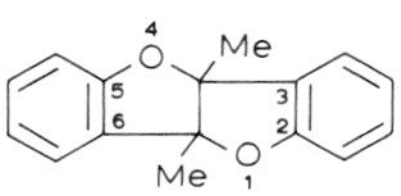

(XXVI) 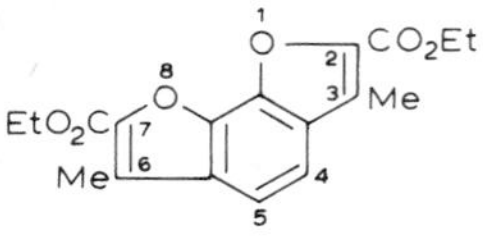 (XXVII) 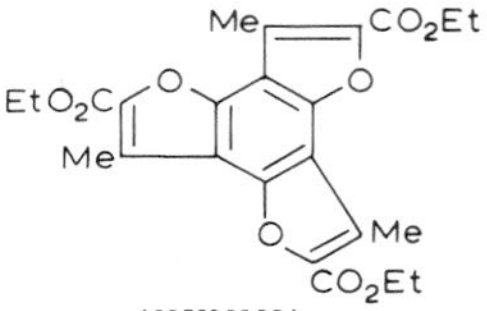 (XXVIII)

Disodio-catechol with ethyl α-chloroacetoacetate gives the 3,6-dimethylbenzo[1,2-*b*:6,5-*b'*]difurandicarboxylic ester (XXVII) (*G. Nuth*, Ber., 1887, **20,** 1332). Resorcinol and quinol give analogous products of ambiguous constitution, and phloroglucinol the trimethylbenzotrifurantricarboxylic ester XXVIII (*E. Lang*, Ber., 1886, **19,** 2935). 3,8-*Dimethylbenzo*[1,2-b:3,4-b']*difuran* (XXIX), m.p. 27°, has been synthesised unambiguously (*D. B. Limaye* and *T. B. Panse*, C.A., 1942, **36,** 1037), as has its linear isomer 3,5-*dimethylbenzo*[1,2-b:5,4-b']*difuran* (XXX), m.p. 107–108° (*J. Algar et al.*, Proc. roy. Irish Acad., 1932, **41B,** 8; Brit. Abstr. A, 1932, 860):

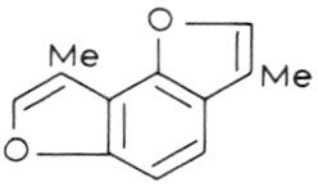

(XXIX) 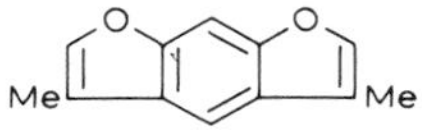 (XXX) 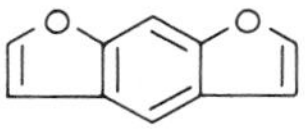 (XXXI)

Benzo[1,2-*b*:5,4-*b'*]**difuran** (XXXI) has been obtained from 6-hydroxybenzofuran-5-carbaldehyde by reacting with ethyl bromoacetate and cyclising the product (*L. R. Worden et al.*, J. heterocycl. Chem., 1969, **6,** 191), and a number of benzodifuran derivatives have been prepared (*R. Royer et al.*, Bull. Soc. chim. Fr., 1963, 1003; 1965, 2607; Compt. rend., 1966, **262,** 1286).

Condensation of benzoin with the three dihydroxybenzenes gives three isomeric tetraphenylbenzodifurans XXXII, m.p. 237°; XXXIII, m.p. 223°; and XXXIV, m.p. 281° (*F. R. Japp* and *A. N. Meldrum*, J. chem. Soc., 1899, **75,** 1035; *R. J. W. LeFèvre et al., ibid.*, 1948, 1992):

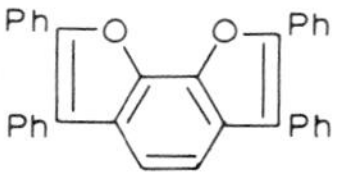

(XXXII) 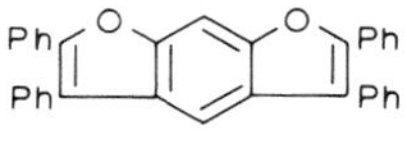 (XXXIII) 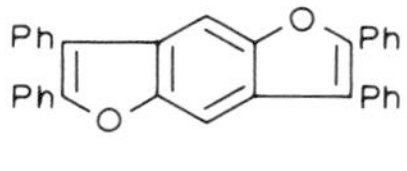 (XXXIV)

3-Oxo-2,3-dihydrobenzofurans can yield by autocondensation products of type XXXV (*W. Baker* and *R. Banks*, *ibid.*, 1939, 279):

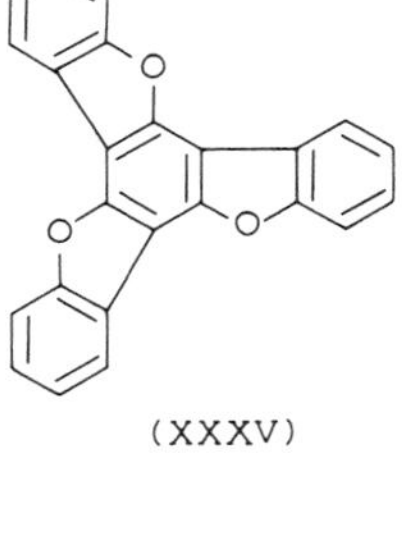

(XXXV)

(c) Other tri- and poly-cyclic furans

Systems involving cyclohexane or bridged cyclohexane rings include XXXVI, b.p. 85.5–88°/25 mm, from cyclopentadiene and 2,5-dihydrofuran (*N. O. Brace*, J. Amer. chem. Soc., 1955, **77,** 4157) and its dihydro-derivative (*K. Alder* and *W. Roth*, Ber., 1954, **87,** 161); XXXVII, m.p. 144–146°, obtained by dehydration of the Diels–Alder adduct from cyclohexadiene and dibenzoylethylene (*G. O. Schenck*, Ber., 1949, **82,** 123); and XXXVIII, b.p. 123–125°/16 mm and the related saturated compound, derived ultimately from cyclohexanone and acetylene by way of the glycol, C_5H_{10} > C(OH)·C≡C·C(OH) < C_5H_{10}(*W. Reppe et al.*, Ann., 1955, **596,** 110, 112):

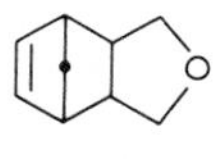

(XXXVI) 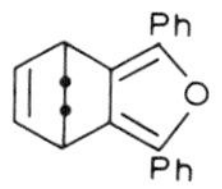 (XXXVII) 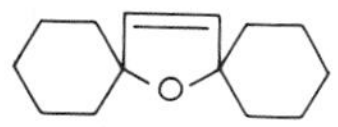 (XXXVIII)

Chapter 3

Compounds with Five-Membered Rings Having One Hetero Atom from Group VI; Sulphur and its Analogues

R. LIVINGSTONE

1. Monocyclic thiophenes and hydrothiophenes*

In 1883 *V. Meyer* observed that the indophenin colour reaction (p. 228), then recognised as diagnostic for benzene, was not given by the pure hydrocarbon, but was due to a sulphur-containing contaminant of the coal-tar product. He separated this contaminant, making use of its easy sulphonation, established the formula C_4H_4S, and named it *thiophen* from its close resemblance to benzene (phene). In coal tar, benzene and naphthalene with their homologues are accompanied by small quantities respectively of thiophene, thionaphthene (benzo[*b*]thiophene), and their homologues. Reduced and unreduced thiophenes and thionaphthene are present in shale oil (*F. Challenger*, J. Soc. chem. Ind., 1929, **48,** 622; *W. Steinkopf* and *W. Nitschke*, Arch. Pharm., 1940, **278,** 360), and reduced thiophenes in petroleum. Commercial outlets for thiophene are comparatively few, but it is used in the synthesis of some drugs and as a constituent of some copolymers. The important nutritional factor, biotin, is a derivative of tetrahydrothiophene.

The constitution follows from the numerous syntheses of thiophenes and

```
 4     3            β'     β
HC —— CH           HC —— CH
‖5   2‖            ‖α'   α‖
HC     CH          HC     CH
  \   /              \   /
    S                  S
    1
   (I)                (II)
```

* *W. Steinkopf*, "Die Chemie des Thiophens", Steinkopf, Dresden and Leipzig, 1941; *H. D. Hartough*, "Thiophen and its Derivatives", Interscience, New York, 1952; *S. Gronowitz*, Adv. in Heterocyclic Chem., Vol. 1, Academic Press, New York, 1963, p. 1.

tetrahydrothiophenes. The molecular fine structure is discussed by *H. C. Longuet-Higgins* (Trans. Faraday Soc., 1949, **45**, 173). Numbering is as shown in I, in the older literature often as II. The radical C_4H_3S- is designated "thienyl".

(a) Thiophene and its substitution products

*(i) Synthetic methods**

(*1*) *C. Paal* and his pupils (Ber., 1885, **18,** 367, 2251; 1886, **19,** 551, 555; 1887, **20,** 2557; 1890, **23,** 1495) showed that 1,4-dicarbonyl compounds gave thiophenes when heated with phosphorus "trisulphide" or pentasulphide. One or both carbonyl groups may be part of a carboxyl or ester function; in this case the product may be a hydroxy-, alkoxy-, or alkylthio-thiophene; or alternatively, reduction takes place, giving a thiophene with the α-position free:

$H_2C—CH_2$ / MeCO COMe —P_2S_3→ 2,5-dimethylthiophene

$H_2C—CH_2$ / MeCO CO_2H —P_2S_3→ [2-methyl-5-hydroxythiophene] —P_2S_3→ 2-methylthiophene; also directly —P_2S_3→ 2-methylthiophene

(see p. 243)

For an advantageous route to suitable intermediates, see *R. G. Jones* (J. Amer. chem. Soc., 1955, **77,** 4069).

Thiophenes may be prepared by the reaction between 1,4-diketones and hydrogen sulphide in the presence of anhydrous hydrogen chloride at 0–40°, *e.g.* acetonylacetone gives 2,5-dimethylthiophene (91%) (*P. D. May*, U.S.P. 3,014,923/1958; C.A., 1962, **56,** 8692a).

(*2*) In *Hinsberg*'s method, ethyl thiodiglycollate reacts with a 1,2-diketone and, with subsequent decarboxylation, the method provides a route to 3,4-disubstituted thiophenes:

PhCO·COPh + $EtO_2C·CH_2SCH_2·CO_2Et$ —NaOMe→ 3,4-diphenylthiophene-2,5-dicarboxylic acid (HO_2C, CO_2H, Ph, Ph) → 3,4-diphenylthiophene

Experiments show that the half-ester acid (isolated in high yield) must be formed by a nonhydrolytic route, and a δ-lactone intermediate suggests itself at once. The ring synthesis may then follow the following course

* For a general treatment *cf. D. E. Wolf* and *K. Folkers*, Org. Reactions, 1951, **6,** 410.

(*H. Wynberg* and *H. J. Kooreman*, J. Amer. chem. Soc., 1965, **87**, 1739):

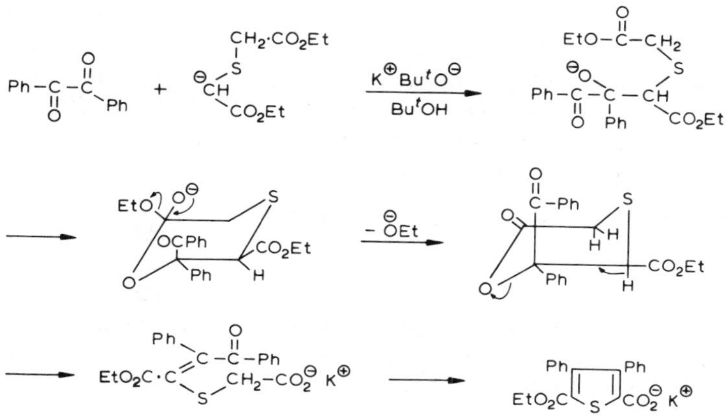

The process can be applied to ethyl oxalate or pyruvate as dicarbonyl compounds, giving hydroxythiophenes (*O. Hinsberg*, Ber., 1910, **43**, 901; *H. J. Backer* and *W. Stevens*, Rec. Trav. chim., 1940, **59**, 423, 899):

(*3*) Condensation of ethyl thioglycollate with methyl vinyl ketone or the related Mannich base $Me_2NCH_2 \cdot CH_2Ac$ in the presence of piperidine gives ethyl 3-hydroxy-3-methyltetrahydrothiophene-2-carboxylate (III); dehydration of this with polyphosphoric acid yields the dihydrothiophene IV, which disproportionates in the presence of polyphosphoric acid to a small extent to thiophene and tetrahydrothiophene derivatives. Dehydrogenation of IV affords ethyl 3-methylthiophene-2-carboxylate (*B. D. Tilak, H. S. Desai* and *S. S. Gupta*, Tetrahedron Letters, 1964, 1609):

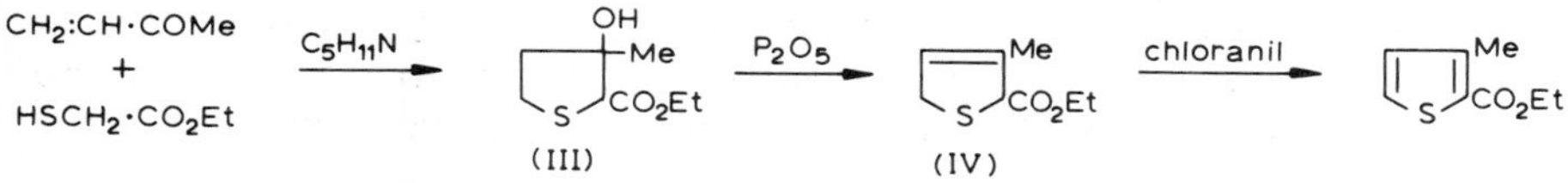

2-Chlorovinyl aldehydes condense with ethyl thioglycollate to give thiophenes (*S. Hauptmann et al.*, *ibid.*, 1968, 1317):

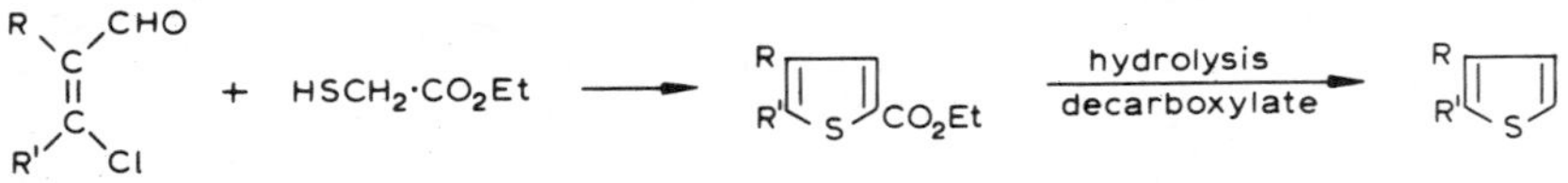

(*4*) Thiophenes may be produced by the reactions between diacetylenes and hydrogen sulphide under weakly basic conditions at 20–80°, yields being good except in the case of thiophene itself (*K. E. Schulte, J. Reisch* and *L. Hoerner*, Ber., 1962, **95**, 1943):

Diacetylenes may also be transformed into the corresponding thiophenes in the presence of sodium ethoxide and hydrogen, alkali-metal hydrogen or ammonium hydrogen, sulphide (*Schulte et al.*, Arch. Pharm., 1963, **296,** 456).

(*5*) The 1:1 adducts of β-iminocarbonyl derivatives and alkyl, aryl or acyl isothiocyanates react with phenacyl bromide to produce tetra-substituted thiophenes (*S. Rajappa* and *B. G. Advani*, Tetrahedron Letters, 1969, 5067):

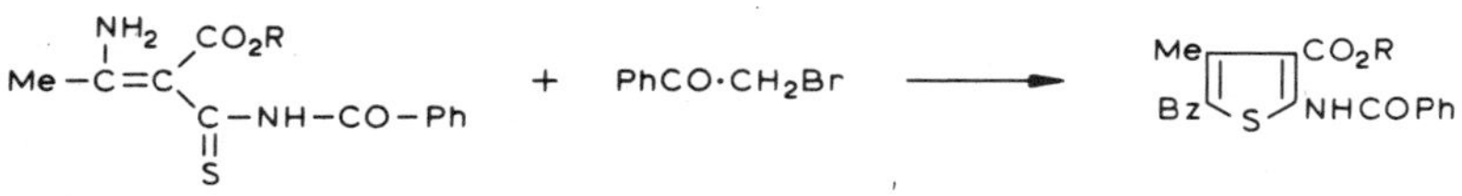

Thiophenes have also been obtained from the condensation of α-bromocarbonyl compounds with (α-thiobenzoyl)-acetophenone under basic conditions (*M. Takaku, Y. Hayasi* and *H. Nozaki*, Bull. chem. Soc., Japan, 1970, **43,** 1917).

(*6*) A hydroxythiophene may be obtained by the following route (*N. K. Chakrabarty* and *S. K. Mitra*, J. chem. Soc., 1940, 1385):

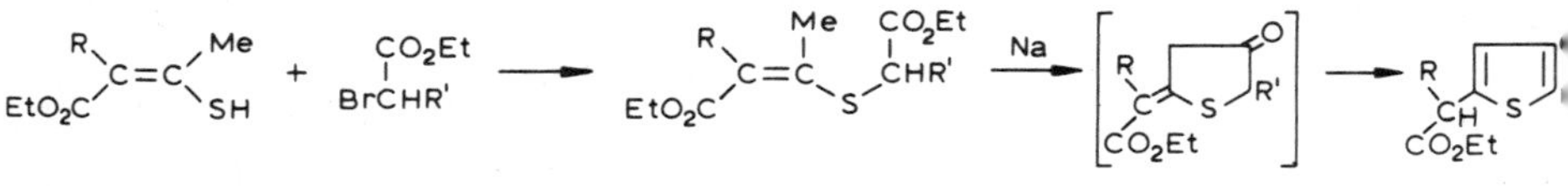

(*7*) Another route to 3-hydroxythiophenes involves the chloracetylation of ethyl β-aminocrotonate and treatment of the product with potassium hydrogen sulphide (*E. Benary* and *A. Baravian*, Ber., 1915, **48,** 593):

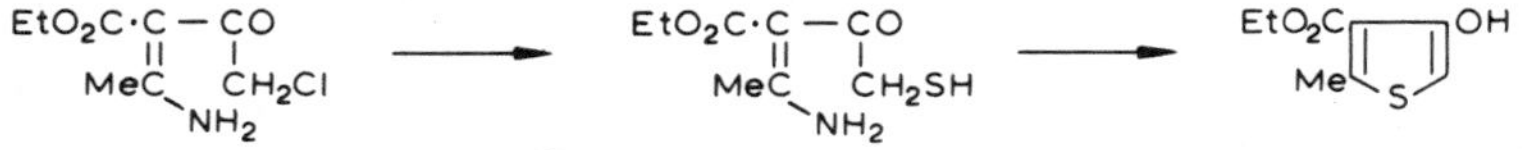

(*8*) Mucic acid, heated at 210° with barium sulphide, gives some thiophene-2-carboxylic acid.

(*9*) The monosulphoxides (VI) of 2,5-diphenyl-1,4-dithiin (V) and of the 3-nitro-, 3-bromo-, and the 3-bromo-6-nitro-derivatives have been prepared. They are rather unstable and decompose to give substituted thiophenes (*H. H. Szmant* and *L. M. Alfonso*, J. Amer. chem. Soc., 1957, **79**, 205):

(V) $\xrightarrow{MeCO_3H}$ (VI) $\xrightarrow{\Delta}$ 2,5-diphenylthiophene

(*10*) Methyl 3-morpholinodithioacrylate in acetone reacts with an α-halogenocarbonyl compound in the presence of excess triethylamine to yield ethyl 5-methylthiothiophene-2-carboxylate (*E. J. Smutny*, *ibid.*, 1969, **91**, 208):

$C_4H_8ON{-}CH{:}CH{\cdot}C(=S)SMe + BrCH_2{\cdot}CO_2Et \xrightarrow[Me_2CO]{Et_3N}$ MeS–thiophene–CO_2Et

(*11*) 2-Thiabicyclo[2.2.1]heptane (VII) has been prepared by the following route from thiophosgene and cyclopentadiene (*C. R. Johnson*, *J. E. Keiser* and *J. C. Sharp*, J. org. Chem., 1969, **34**, 860):

cyclopentadiene + $Cl_2C{=}S$ → dichloro-thiabicycloheptene $\xrightarrow[Et_2O]{3\text{-}ClC_6H_4{\cdot}CO_3H}$ dichloro-sulphone $\xrightarrow[NaOAc]{H_2/Pd(OH)_2}$ chloro-sulphone $\xrightarrow{LiAlH_4}$ (VII)

The oxidation of 2-thiabicyclo[2.2.1]heptane by a variety of reagents has been studied, and the ratio of *endo* and *exo* sulphoxide isomers obtained is related to their relative stabilities and the mechanism of oxidation (*Johnson et al.*, Tetrahedron, 1969, 5649).

(*12*) Less important methods of preparation include the following: Cyclic acid anhydrides may be reduced under pressure in the presence of hydrogen sulphide and cobalt sulphide at 180–325°; thus succinic anhydride affords thiophene (11%) along with other products (*R. W. Campbell*, U.S.P. 3,345,381/1965; C.A., 1968, **68**,

29587m). Thiophenes are prepared by the passage of an aliphatic hydrocarbon and hydrogen sulphide over a transition metal oxide (*B.A.S.F.*, B.P. 887,426/1959; C.A., 1962, **57,** 11169d), and furan and 2-methylfuran vapour and hydrogen sulphide over alumina containing potassium phosphotungstate (*T. E. Deger, R. H. Goshorn* and *B. Buchholz*, G.P. 1,228,273/1962; C.A., 1967, **66,** 18665t; Belg. P. 623,801/1962; C.A., 1963, **59,** 8705g). *S*-Labelled thiophene may be obtained from radioactive thiophene-2-carbaldehyde prepared by passing hydrogen [^{35}S]sulphide into an aqueous solution of 3-chloro-1,2-dioxocyclopentane (*N. P. Buu-Hoï*, Bull. Soc. chim. Fr., 1958, 1407). Methods are available for the preparation of 2-[^{13}C]thiophene and 3-[^{13}C]thiophene (*B. Bak, J. Christiansen* and *J. T. Nielsen*, Acta Chem. Scand., 1960, **14,** 1865).

(ii) General properties and reactions

(*1*) The thiophene ring shows typically aromatic behaviour, though its resemblance to benzene can be overstressed. The unreduced nucleus is stable to alkalis, moderately stable to acids and fairly resistant to oxidation whereby thiophene homologues give carboxylic acids in modest yield. Substitution takes place much more easily in thiophene than in the benzene ring, the ratio of speeds of nitration being reported as 850:1 (*E. Imoto* and *R. Motoyama*, C.A., 1954, **48,** 9997). The products of the easy nuclear mercuration, which are valuable intermediates in synthetic work, serve also to characterise the parent thiophenes. The point of attack is determined primarily by the principle that α-positions (2 and 5) are more easily substituted than β (3 and 4), and to a minor degree by the usual *ortho*, *para-* or *meta*-directive effect of an existing substituent (*H. D. Hartough*, J. chem. Educ., 1950, **27,** 500). Thus nitration gives first mainly 2-nitrothiophene, then a mixture of the 2,4- and 2,5-dinitro compounds, whereas dibromination gives almost exclusively 2,5-dibromothiophene. An *ortho,para*-directive group in position 3 directs towards 2, and a *meta*-directive one to 5.

(*2*) Addition reactions are few; thiophene does not react with maleic anhydride (*R. Delaby*, Bull. Soc. chim. Fr., 1937, [v]. **4** 765), but in the presence of light it combines with dimethylmaleic anhydride sensitised by benzophenone (*G. O. Schenck, W. Hartmann* and *R. Steinmetz*, Ber., 1963, **96,** 498), and with ethyl diazoacetate at 130° gives in small yield ethyl 2-thiabicyclo[3.1.0]-3-hexene-6-carboxylate (VIII) (*W. Steinkopf* and *H. Augestad-Jensen*, Ann., 1922, **428,** 154):

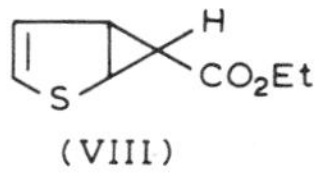

H
S
CO_2Et

(VIII)

Thiophene reacts with ethoxycarbonylnitrene formed from ethyl azidoformate to give ethyl pyrrole-1-carboxylate (*K. Hafner* and *W. Kaiser*, Tetrahedron Letters, 1964, 2185):

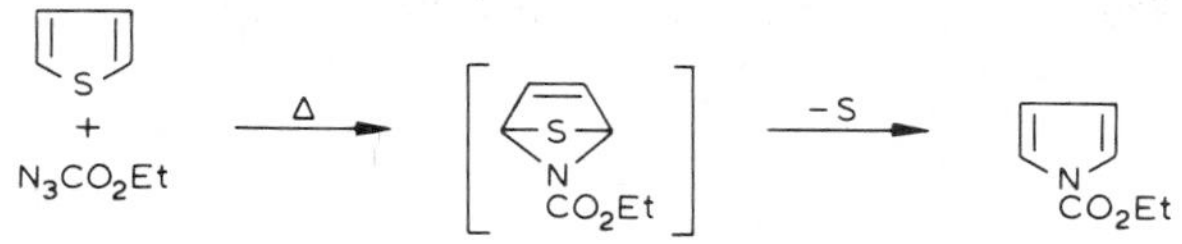

$$\text{thiophene} + N_3CO_2Et \xrightarrow{\Delta} [\text{S-bridged N-}CO_2Et\text{ adduct}] \xrightarrow{-S} \text{pyrrole-N-}CO_2Et$$

Tetrafluorobenzyne forms an unstable adduct with thiophene which decomposes to give 1,2,3,4-tetrafluoronaphthalene. This is believed to be the first recorded example of the participation of thiophene in a Diels–Alder reaction (*D. D. Callander*, *P. L. Coe* and *J. C. Tatlow*, Chem. Comm., 1966, 143):

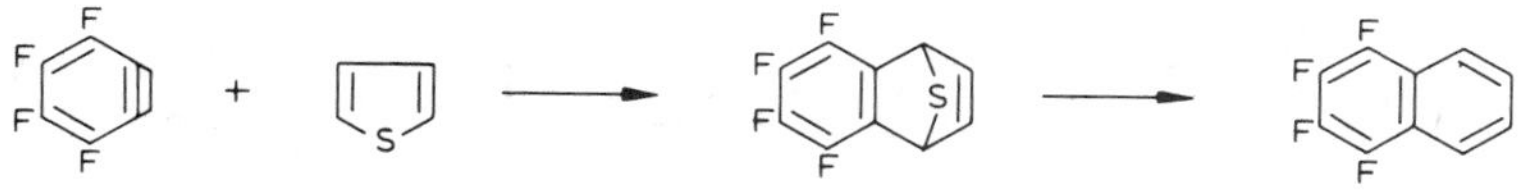

Arynes react with thiophene (benzothiophene) at high temperatures to give products of insertion and 1,2- and 1,4-addition. Hydrogen transfer also occurs, evidently with the formation of thiophyne (*E. K. Fields* and *S. Meyerson*, *ibid*., 1966, 708; C.A., 1967, **67**, 53183b).

(*3*) The polymerisation of thiophene by 100% orthophosphoric acid gives a *trimer* (IX), m.p. 37°, and a supposed "*pentamer*" (X), m.p. 112°. The structure of IX was deduced from spectral data and destructive hydrogenation–desulphurisation (*S. L. Meisel*, *G. C. Johnson* and *Hartough*, J. Amer. chem. Soc., 1950, **72**, 1910), and confirmed from its dehydrogenation with chloranil to 2,2′:4′,2″-terthienyl. The so-called "pentamer" has been unambiguously established by X-ray analysis and dehydrogenation to 4,7-di(2-thienyl)benzo[*b*]thiophene to have structure X with the thienyl groups *cis*, derived from four molecules of thiophene (*R. F. Curtis et al*., Chem. Comm., 1969, 165):

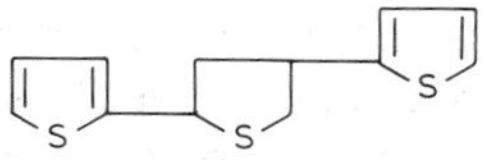

(IX)

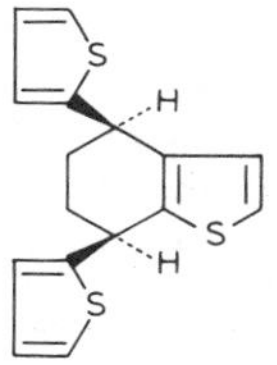

(X)

(*4*) Although thiophene does not combine with alkyl halides it does react either with a cold suspension of trimethyloxonium tetrafluoroborate in methylene chloride or with methyl iodide and silver perchlorate to give, on the addition of a saturated solution of sodium hexafluorophosphate, methylthiophenium hexafluorophosphate (XI) (*G. C. Brumlik, A. I. Kosak* and *R. Pitcher*, J. Amer. chem. Soc., 1964, **86**, 5360):

(XI)

(5) Hydrogen peroxide oxidises thiophene to a product having the composition of a "sesquioxide", $(C_4H_4S)_2O_3$, which is formulated as XII, and may arise by Diels–Alder combination of the sulphoxide and sulphone (*W. Davis* and *F. C. James*, J. chem. Soc., 1953, 15):

(XII)

Solutions containing the unstable sulphone, thiophene 1,1-dioxide (XIV) have been prepared from the butadiene–sulphur dioxide adduct (XIII) (*W. J. Bailey* and *E. W. Cummings*, J. Amer. chem. Soc., 1954, **76**, 1932, 1936, 1940):

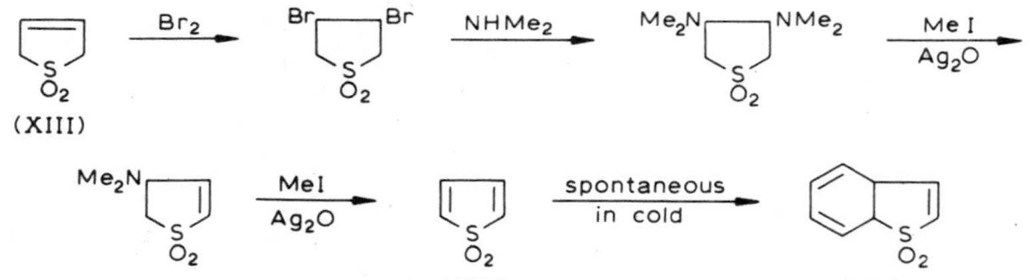

(XIII) (XIV) (XV)

In cold, concentrated solutions, thiophene 1,1-dioxide gives 3a,7a-dihydrobenzo[*b*]thiophene 1,1-dioxide (dihydrothianaphthene 1,1-dioxide) (XV). The aromatic character of thiophenes is destroyed by their conversion into the 1,1-dioxides, which behave like cyclic dienes, and also as dienophiles owing to the electron-attracting character of the sulphone groups (*J. M. Whelan*, Diss. Abs., 1959, **20**, 1180; C.A., 1960, **54**, 4535e). 3,4-Diphenylthiophene, in contrast to thiophene, is oxidised by peroxyacetic acid to an isolable stable 1,1-dioxide, which on boiling in phenol dimerises and immediately loses one molecule of sulphur dioxide to form 3,3a,5,6-tetraphenyl-3a,7a-dihydrobenzo[*b*]thiophene 1,1-dioxide (XVI) (*C. G. Overberger* and *Whelan*, J. org. Chem., 1961, **26,** 4328).

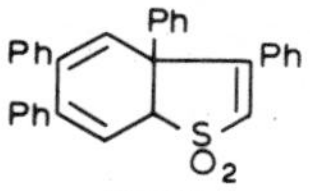
Ph
Ph
Ph
Ph
S
O2

(XVI)

Thiophene 1-oxide has been obtained from *trans*-3,4-dihydroxytetrahydrothiophene; it is unstable and polymerises readily (*M. Procházka*, Coll. Czech. chem. Comm., 1965, **30,** 1158):

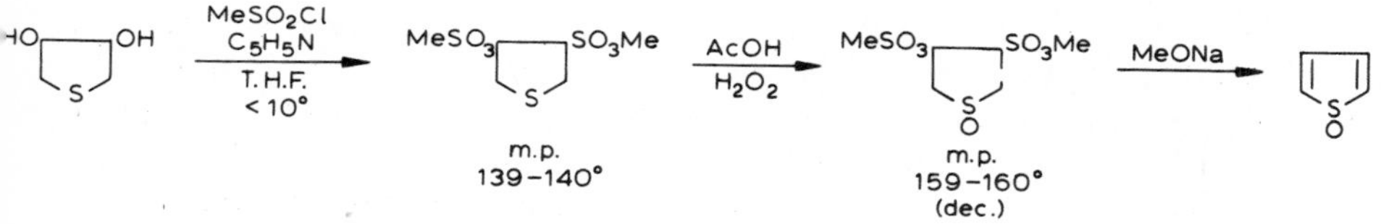
HO
OH
S
MeSO2Cl
C5H5N
T.H.F.
< 10°
MeSO3
SO3Me
S
m.p.
139–140°
AcOH
H2O2
MeSO3
SO3Me
S
O
m.p.
159–160°
(dec.)
MeONa
S
O

(6) The usually stable thiophene ring can sometimes be opened smoothly by reduction: thiophene in ether added to sodium in liquid ammonia affords some *n*-butanethiol, also obtained by adding sodium to a solution of thiophene in ethanol (*W. Hückel* and *I. Nabih*, Ber., 1956, **89,** 2115); tetraphenylthiophene with sodium and amyl alcohol affords tetraphenylbutane (*E. Bergmann*, J. chem. Soc., 1936, 505), and the ring in tetrahydrothiophenes may be opened to give the corresponding butanes by (*hydrogen-containing*) Raney nickel (*V. du Vigneaud et al.*, J. biol. Chem., 1942, **146,** 475). Since thiophenes are accessible by flexible syntheses and are easily substituted by alkyl groups, this reductive desulphurisation affords a useful route to complex aliphatic compounds (*G. M. Badger et al.*, J. chem. Soc., 1954, 4162). The hydrodesulphurisation of thiophene to butane and hydrogen sulphide over chromia has been studied (*P. J. Owens* and *C. H. Amberg*, Canad. J. chem., 1962, **40,** 941) and also the hydrodesulphurisation using other catalysts (*R. S. Mann*, J. chem. Soc., 1964, 1531; *S. Kolboe* and *Amberg*, Canad. J. Chem., 1966, **44,** 2623; *Kolboe*, *ibid.*, 1969, **47,** 352). Compounds of the type XVII, formed by high-dilution cyclisation can be desulphurised to cyclic hydrocarbons or ketones (*Ya. L. Gol'dfarb*, *S. Z. Taĭts* and *L. I. Belen'kii*, Izvest. Akad. Nauk, S.S.S.R., Otdel. Khim. Nauk, 1957, 1262):

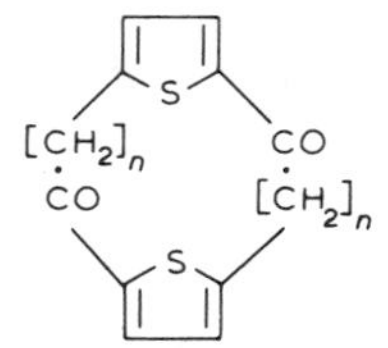

(XVII)

(7) Ozonisation of thiophene gives glyoxal and oxalic acid; and thallous hydroxide opens the ring, yielding thallous sulphide with formation of some succinate (*G. T. Morgan* and *W. Ledbury*, J. chem. Soc., 1922, **121,** 2893). Thiophene, 3-methylthiophene and 3,4-dimethylthiophene on treatment with oxygen at elevated temperatures in the presence of a Group Va oxide catalyst, *i.e.* of vanadium, antimony or phosphorus, give thiomaleic anhydride, methylthiomaleic anhydride and dimethylthiomaleic anhydride, respectively (*T. J. Jennings*, U.S.P. 3,265,712/1964; C.A., 1966, **65,** 12113a).

(*8*) Thiophenes give the *indophenin reaction*, intense colourations with isatin or with other 1,2-dicarbonyl compounds such as benzil, phenanthraquinone, phenylglyoxylic acid or alloxan, in presence of sulphuric acid. The true "α-indophenins", formed from thiophenes unsubstituted in the 2- and 5-positions, can be degraded to derivatives of 2,2′-bithienyl, leading to the formulation XVIII for the deep-blue isatin compound (*Steinkopf* and *W. Hanske*, Ann., 1939, **541,** 238). 2,3-Unsubstituted thiophenes can afford "β-indophenins", formulated as XIX:

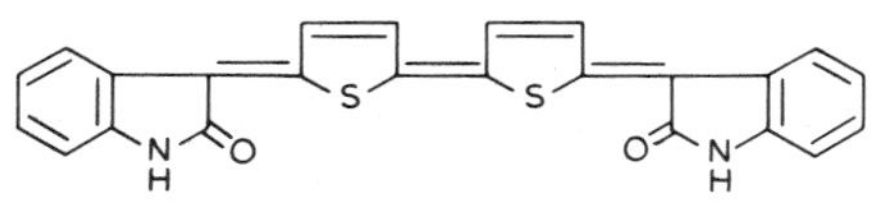

(XVIII) 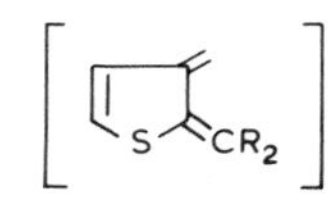 (XIX)

(*9*) Corresponding derivatives of benzene (b.p. 80°), and of thiophene (b.p. 84°) have, as a rule, nearly identical b.p.s., but often differ substantially in m.p. (Table 1).

TABLE 1

CORRESPONDING BENZENE AND THIOPHENE DERIVATIVES

X =	*Et*	*Cl*	NO_2	*NHAc*
2-(C_4H_3S)X	b.p. 133°	b.p. 128°	m.p. 46.5°	m.p. 160–161°
3-(C_4H_3S)X	b.p. 136°	b.p. 137°	m.p. 78–79°	m.p. 145–148°
C_6H_5X	b.p. 136°	b.p. 132°	m.p. 5°	m.p. 113°

(*10*) Thiophene derivatives are often characterised as the crystalline thienylmercuric salts (p. 231). Substitution reactions of thiophene derivatives (*M. G. Reinecke* and *H. W. Adickes*, J. Amer. chem. Soc., 1968, **90,** 511), macrocycles involving the thiophene ring system (*O. Meth-Cohn*, Quart. Reports Sulphur Chem., 1970, **5,** 129), and developments in thiophene chemistry during the period 1962–1968 (*C. D. Hurd*, *ibid.*, 1969, **4,** 75, 159) have all been reviewed.

(iii) Thiophene

Thiophene is best made on the small scale by heating sodium succinate with phosphorus "trisulphide" (*R. Phillips*, Org. Synth., 1932, **12,** 72) or in larger quantities by passing acetylene over pyrites at 300° (*W. Steinkopf*, Ann., 1922, **428,** 123). It can also be obtained by passing hydrogen sulphide with acetylene over alumina at 425°, or from acetylene and carbon disulphide at 700°; on the commercial scale, *n*-butane and sulphur are heated at 600° (*H. E. Rasmussen et al.*, Ind. Eng. Chem., 1946, **38,** 376). It is a colourless liquid, m.p. −38.3°, b.p. 84.12°, d_4^{20} 1.0644, n_D^{20} 1.5287 (*F. S. Fawcett* and *Rasmussen*, J. Amer. chem. Soc., 1945, **67,** 1705), μ 0.54D (*H. de V. Robles*, Rec. Trav. chim., 1939, **58,** 111). For a general discussion of the thermodynamic properties, see *G. Waddington et al.*, J. Amer. chem. Soc., 1949, **71,** 797, and for electron diffraction data see *V. Schomaker* and *L. Pauling*, *ibid.*, 1939, **61,** 1769). The dimensions of the thiophene molecule have been determined from data obtained from the microwave spectrum (*B. Bak et al.*, J. Mol. Spectr., 1961, **7,** 58),

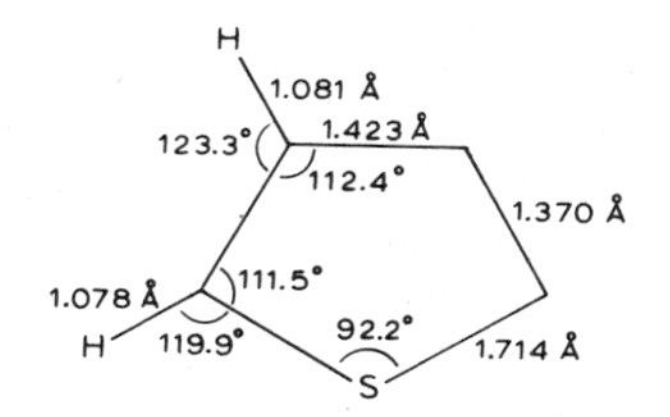

and an empirical method has been described whereby they are calculated from rotational spectra measurements (*C. W. N. Cumper*, Trans. Faraday Soc., 1958, **54,** 1266).

From the above information it is suggested that thiophene is not well represented by the classical structure Ia, but that it is in a state of resonance and that the major contributing resonance structures are Ia to Ie. If account is taken of the *s* and *d* contribution to the *p* bonding, then additional structures If to Ij contribute to the resonance form of thiophene (*D. S. Sappenfield* and *M. M. Kreevoy*, Tetrahedron, 1963, **19,** Suppl. 2, 157). The fact that thiophene can be alkylated on the sulphur atom (p. 226) supports the idea that the resonance structures with a negative charge on the sulphur have significance:

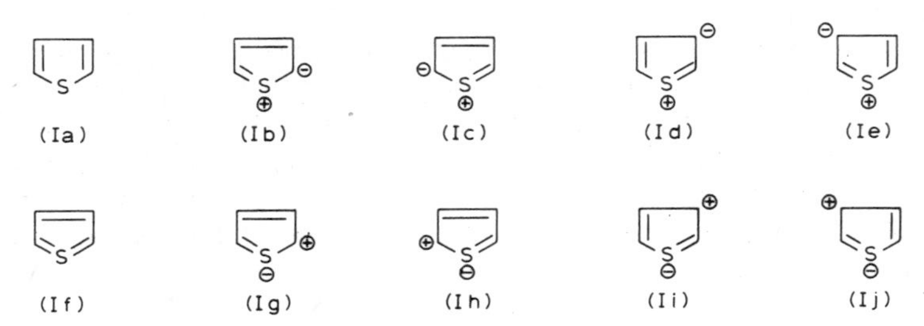

Bond angles and bond lengths for the thiophene ring have been determined on the basis of ^{13}C–H coupling constants and m.o. calculations and confirmed by dipole moment values (*E. S. Vincent et al.*, Bull. Soc. chim. Fr., 1966, 3530). The ultraviolet absorption has been discussed and compared with that of furan, pyrrole and cyclopentadiene (*W. C. Price* and *A. S. Walsh*, Proc. roy. Soc., 1941, **A179,** 201). For infrared absorption, see *H. W. Thompson* and *R. B. Temple*, Trans Faraday Soc., 1945, **41,** 27. N.m.r.: the ring currents in thiophene and furan have been measured by comparison of their proton shifts with those of molecules without ring currents (*R. I. Abraham et al.*, Chem. Comm., 1965, 43); along with that in pyrrole they are less than in benzene (*J. A. Elvidge*, *ibid.*, 1965, 160; *H. A. P. de Jongh* and *H. Wynberg*, Tetrahedron, 1965, **21,** 515); calculations based on chemical shift data support the accepted order of aromaticity thiophene > pyrrole > furan (*Elvidge*, *loc. cit.*; *D. W. Davies*, Chem. Comm., 1965, 258).

Tetradeuterothiophene, prepared by treating tetra(chloromercuri)-thiophene with deuterium chloride, m.p. −38.6, b.p. 83.3, n_D^{20} 1.52660, d_4^{20} 1.11382 (*Steinkopf* and *M. Boëtius*, Ann., 1941, **546,** 208). The best method for the preparation of [^{35}S]-thiophene is by exchange between thiophene vapour and Ni^{35}S at 600° (*Ya. D. Zebrenskii et al.*, Zhur. obshcheĭ Khim., 1965, **35,** 1369). The heterogeneous exchange of deuterium and tritium in the 2- and 3-positions of thiophene have been studied in aqueous sulphuric acid (*B. Östman* and *S. Olssen*, Arkiv Kemi, 1960, **15,** 275).

Thiophene is thermally very stable, more so than 2-methylthiophene; both give hydrogen sulphide, hydrogen and methane as the only gaseous products on pyrolysis at 800–825°; at 825° thiophene affords some benzene (*C. D. Hurd*, *R. V. Levetan* and *A. R. Macon*, J. Amer. chem. Soc., 1962, **84,** 4515). The pyrolysis of thiophene at 690° is reported to give bithienyl as the main product and three minor products, benzothiophene, thiophthene, and phenylthiophene, which are thought to arise *via* the formation of thiophyne (p. 225). Thiophene is slowly decomposed by concentrated sulphuric acid so that sulphonation requires caution, and boiling hydriodic acid removes the sulphur atom. It can be dried by distillation over sodium, whereas potassium abstracts sulphur. Thiophene in benzene can be determined by mercuration (*G. Claxton* and *W. H. Hoffert*, J. Soc. chem. Ind., 1946, **65,** 333) or colorimetrically by the indophenin reaction (*C. Schwalbe*, Chem. Ztg., 1905, **29,** 895).

Thiophene is used as a chain extender in the synthesis of branched-chain alkanoic acids (*J. F. McGhie et al.*, J. chem. Soc., 1962, 350) and along with its homologues as starting compounds for the preparation of aliphatic amino acids (*L. Goldfarb et al.*, Tetrahedron, 1962, 21). Decaborane reacts with thiolane to give $B_{10}H_{12}{\cdot}2C_4H_8S$

(*E. L. Muetties*, U.S.P. 3,154,561/1958; C.A., 1965, **62**, 11685b), and thiophene reacts with iron pentacarbonyl to afford dicarbonyl(thiophene)iron, $[(C_4H_4S)Fe(CO)_2]$ (*R. Burton et al.*, Chem. and Ind., 1958, 1592). The decomposition of dibenzoyl peroxide in thiophene at 80° led to the isolation of a mixture of 2-phenylthiophene, 2,2'-bithienyl and 2,3'-bithienyl (*C. E. Griffin* and *K. K. R. Martin*, Chem. Comm., 1965, 154).

(iv) Metal and metalloid derivatives of thiophenes

With metallic sodium, 2-chlorothiophene gives 2-**thienylsodium,** also produced in good yield from thiophene, 63% sodium amalgam and ethyl chloride or bromobenzene. In 3-methylthiophene, which usually undergoes substitution in position 2, the sodium atom enters position 5, so opening up synthetic possibilities (*J. W. Schick* and *H. D. Hartough*, J. Amer. chem. Soc., 1948, **70,** 286). The metallation of thiophene with *n-butyl-lithium* yields 2-**thienyl-lithium** (*H. Gilman* and *D. A. Shirley*, J. Amer. chem. Soc., 1949, **71,** 1870), but 3-thienyl-lithium could only be obtained through halogen–metal interconversion between 3-bromothiophene and *n*-butyl-lithium at −70°, the only temperature at which it is stable (*S. Gronowitz*, Arkiv. Kemi, 1954, **7,** 361; *P. Moses* and *Gronowitz*, *ibid.*, 1962. **18,** 119). Butyl-lithium displaces the SeR group in 2- and 5-substituted thiophenes; thus treatment of methyl 2-thienyl selenide (R = Me) with butyl-lithium followed by solid carbon dioxide affords thiophene-2-carboxylic acid (*Ya. L. Gol'dfarb, V. P. Litvinov* and *A. N. Sukiasyan*, Izvest. Akad. Nauk, S.S.S.R., Ser. Khim., 1967, 2585).

Thienylmagnesium halides are formed normally from 2-bromo- and 2-iodo-thiophene and from 2,5-dibromothiophene; the 3-compounds and polyhalogeno-compounds usually react only by the "entrainment" method in which a substantial quantity of, say, methyl iodide is simultaneously converted into Grignard reagent, but the Grignard reagent may be obtained by reacting 3-**thienyl-lithium** with magnesium bromide (*Gronowitz*, Arkiv Kemi, 1958, **12,** 533). The formation of a Grignard reagent, followed by hydrolysis, can be used to effect partial dehalogenations (*cf.* p. 238). With ethylmagnesium iodide at 160–170°, thiophene evidently gives some 2-thienylmagnesium iodide, since subsequent carbonation yields thiophene-2-carboxylic acid (*F. Challenger* and *G. M. Gibson*, J. chem. Soc., 1940, 305). The Grignard compounds show the usual reactions, and in particular afford other metallic derivatives by double decomposition (*E. Krause* and *G. Renwanz*, Ber., 1927, **60,** 1582; 1929, **62,** 1710; 1932, **65,** 777); in the following list, R = 2-thienyl, C_4H_3S:

$R{\cdot}B(OH)_2$	m.p. 134°	R_4Sn	m.p. 156°	R_3Bi	m.p. 137.5°
R_2TlBr	dec. 270°	R_4Pb	m.p. 154.5°	R_2Te	m.p. 50.5°
R_4Si	m.p. 135.5°	R_3Sb	m.p. 49°	R_3TeBr	m.p. 253°
R_4Ge	m.p. 150°	R_3SbCl_2	m.p. 229°		

Cuprous iodide reacts with 2-thienylmagnesium iodide to give 2-**thienyl-copper,** which on treatment with quinoline and iodobenzene is converted into 2-phenylthiophene (*M. Nilsson,* Tetrahedron Letters, 1966, 679).

2-Iodothiophene reacts readily with activated **calcium** in tetrahydrofuran at −35°. Carboxylation gives thiophene-2,5-dicarboxylic acid and 2-iodo-thiophene-5-carboxylic acid, but none of the expected thiophene-2-carboxylic acid. The reaction sequence shown involves metallation and calcium–halogen exchange reactions (*D. Bryce-Smith* and *A. C. Skinner,* J. chem. Soc., 1963, 577):

2-iodothiophene $\xrightarrow{Ca}$ 2-thienyl-CaI $\xrightarrow{(C_4H_3S)I}$ ICa-thiophene-I $\xrightarrow[\text{or Ca}]{(C_4H_3S)CaI}$ ICa-thiophene-CaI

ICa-thiophene-I $\xrightarrow{CO_2,\ H^{\oplus}}$ HO_2C-thiophene-I

ICa-thiophene-CaI $\xrightarrow{CO_2,\ H^{\oplus}}$ HO_2C-thiophene-CO_2H

H. Sachs (Ber., 1892, **25,** 1514) has prepared 2(?)-substituted **phosphorus compounds:**

$$C_4H_4S \xrightarrow[\text{red. heat}]{PCl_3} (C_4H_3S)PCl_2 \longrightarrow (C_4H_3S)P(OH)_2 \longrightarrow (C_4H_3S)PO(OH)_2$$

$$(C_4H_3S)PCl_2 \xrightarrow{Et_2Zn} (C_4H_3S)PEt_2 \quad \text{(b.p. 225°)}$$

Tri-2-thienylarsine, m.p. 25–26°, is obtained from 2-bromothiophene, arsenic trichloride and sodium. Arsenic chloride and dithienylmercury give the same compound along with chlorodithienylarsine, $(C_4H_3S)_2AsCl$, and dichlorothienylarsine, $(C_4H_3S)AsCl_2$. Hydrolysis and oxidation of the chloro-compounds affords the thienylarsenic acids $(C_4H_3S)_2AsO{\cdot}OH$ and $(C_4H_3S)AsO(OH)_2$, the latter of which can be reduced to 2-thienylarsine oxide and 2,2′-arsenothiophene (*C. Finzi,* Gazz., 1915, **45,** II, 280; 1925, **55,** 824; 1932, **62,** 244).

The **mercury derivatives** are important, both as means of isolating and characterising thiophenes, and as reactive intermediates in synthetic work. Mercuric chloride buffered with sodium acetate replaces α-hydrogen atoms at moderate temperatures; β-positions are attacked in special cases. The monochloromercuri-compounds are soluble and crystalline, the di-substituted amorphous and insoluble. Hot dilute acids remove the mercury with quantitative regeneration of the parent thiophene; halogens replace the mercury by halogen, and acid chlorides by acyl groups; in many cases sodium iodide yields a dithienylmercury (*Steinkopf*, Ann., 1921, **424,** 23; *E. Cherbuliez* and *C. Giddey*, Helv., 1952, **35,** 160):

$$2\,RHgCl \;+\; 2\,NaI \;\rightarrow\; R_2Hg \;+\; HgI_2 \;+\; 2\,NaCl$$

Di-2-thienylmercury, m.p. 198–199°, 2-**chloromercurithiophene,** m.p. 182–183°. Mercuric acetate in hot acetic acid progressively replaces all nuclear hydrogen atoms by HgOAc, the α-position alone being usually attacked at 50°. The products yield the corresponding chloromercuri compounds on warming with aqueous sodium chloride (*Steinkopf* and *A. Killingstad*, Ann., 1937, **532,** 288). The mercuration of thiophene with mercuric cyanate affords mono- and di-mercurated derivatives and 2-methylthiophene only a mono-derivative (*E. Söderbäck*, Acta Chem. Scand., 1959, **13,** 1221).

(v) Alkyl- and aryl-thiophenes

Synthesis. (*1*) *S*-**Alkylthiophenium** salts may be obtained from thiophenes, silver tetrafluoroborate and alkyl halides in 1,2-dichloroethane (see p. 226). They are powerful alkylating agents and their n.m.r. spectra are consistent with a pyramidal arrangement for the sulphur–carbon bonds (*R. M. Acheson* and *D. R. Harrison*, Chem. Comm., 1969, 724).

(*2*) Many *C*-alkyl- and -aryl-thiophenes are best prepared by ring-synthesis (*1*), (p. 220). Normal Friedel–Crafts alkylation of thiophene fails in simple cases, but olefins and alcohols give fair yields of a mixture of 2- and 3-alkylthiophenes in the presence of phosphoric acid or other acid catalysts (*W. G. Appleby et al.*, J. Amer. chem. Soc., 1948, **70,** 1552; *P. D. Caesar*, *ibid.*, p. 3623); heating thiophene and isobutylene with phosphoric acid on kieselguhr in an autoclave at 150° gives a mixture of 2- and 3-*tert*-butylthiophene (*N. I. Shuikin* and *B. L. Lebedev*, Izvest. Akad. Nauk, S.S.S.R., Ser. Khim., 1967, 1154). Alkyl-substituted thiophenes are easily prepared from the more accessible acylated compounds by the Clemmensen, or better by the Huang–Minlon modification of the Wolff–Kishner reduction (*W. J. King* and *F. F. Nord*, J. org. Chem., 1949, **14,** 638; *N. P. Buu-Hoï et al.*, J. chem. Soc., 1953, 547; 1954, 1975; 1958, 4202; *P.* and *D. Cagniant*, Bull. Soc. chim. Fr., 1953, 62; 1953, 713; *Gol'dfarb*, *M. A. Kalik* and *M. L. Kirmalova*, Zhur. obshcheĭ Khim., 1960, 1012). Early workers used the Wurtz–Fittig synthesis with advantage. Long-chain alkylthiophenes are described by *Buu-Hoï et al.* (J. chem. Soc., 1955, 1581) and by *Cagniant* (Bull. Soc. chim. Fr., 1955, 359). The vacuum-pyrolysis of 3,4-bisacetoxymethyltetrahydrothiophene gives 3,4-dimethylthiophene with small amounts of 3,4-dimethylenethiolane (*E. J. Fetter*, C.A., 1962, **57,** 758e).

Several thienylacetylenes which occur naturally have been synthesised (*F. Bohlmann et al.*, Ber., 1965, **98,** 155, 369; Tetrahedron Letters, 1965, 1385; *T. S. Sorensen et al.*, Acta chem. Scand., 1964, **18,** 2182; *R. E. Atkinson, R. F. Curtis* and *G. T. Phillips*, J. chem. Soc., 1965, 7109; *Atkinson* and *Curtis*, Tetrahedron Letters, 1965, 297). Thiophene can be arylated in position 2 in modest yield by arenediazohydroxides (*M. Gomberg* and *W. E. Bachmann*, J. Amer. chem. Soc., 1924, **46,** 2339). Phenyl radicals from *N*-nitrosoacetanilide attack thiophene mainly in the 2-position (*J. Degani et al.*, Ann. Chim., 1961, **51,** 434; *N. R. Watt*, Diss. Abstr. B, 1968, **28,** 3234; C.A., 1968, **69,** 27145n). Heating a mixture of thiophene and nitrobenzene (1:5 mole ratio) at 400–600° yields a mixture of 2-phenylthiophene (75%) and the 3-isomer (25%) (*E. K. Fields* and *S. Meyerson*, J. Amer. chem. Soc., 1967, **89,** 724).

(*3*) 2-Arylthiophenes can be produced by the decarboxylative coupling of thiophene-2-carboxylic acid with iodobenzene or its *p*-methoxy, -methyl or -nitro derivatives using cuprous oxide and boiling quinoline (*Nilsson* and *C. Ullenius*, Acta chem. Scand., 1968, **22,** 1998).

(*4*) On irradiation 2-phenylthiophene is irreversibly rearranged to the 3-phenyl isomer.

A phenyl migration is not involved in this reaction since 2-phenyl-[2-^{14}C]thiophene affords 3-phenyl[3-^{14}C]thiophene. The simplest mechanism for this rearrangement deduced from a study of the isomerisation of 5-deuterio-2-pentadeuteriophenylthiophene and methyl-substituted 2-phenylthiophenes involves the formation of a cyclopropenethiocarbaldehyde XX (*H. Wynberg et al.*, J. Amer. chem. Soc., 1965, **87,** 3998; 1967, **89,** 3487, 3495, 3501):

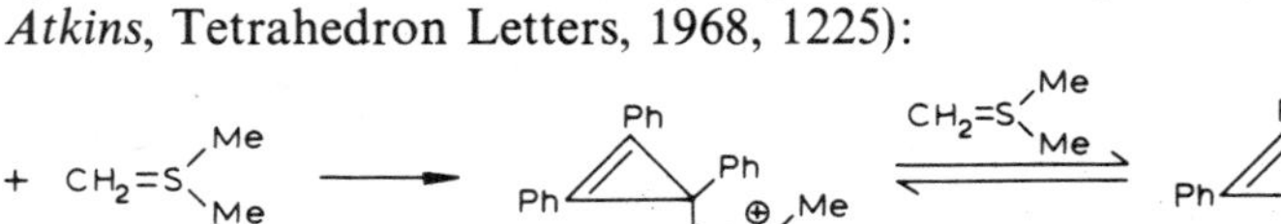

(XX)

(*5*) 2,3,4-Triphenylthiophene is obtained by the treatment of triphenylcyclopropenium bromide with dimethylsulphonium methylide (*B. M. Trost* and *R. Atkins*, Tetrahedron Letters, 1968, 1225):

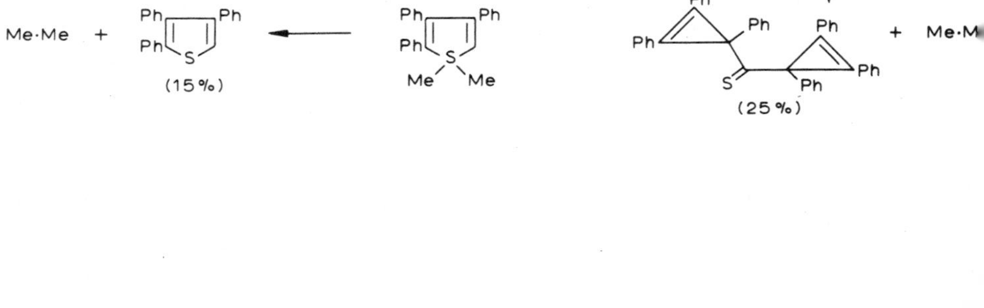

(15%)

(25%)

(*6*) *A. Laurent* (Ann., 1844, **52,** 354) prepared the first thiophene derivative to be characterised, tetraphenylthiophene, "thionessal", by heating polymeric thiobenzaldehyde. It is also produced by heating with sulphur a variety of aromatic compounds, including tetraphenylcyclopentadienone (*W. Dilthey et al.*, Ber., 1935, **68,** 1159).

Reactions. Oxidation of alkylthiophenes gives thiophenecarboxylic acids in low yield. The base-catalysed oxidation of several alkylthiophenes with molecular oxygen has been studied in the polar solvent, hexamethylphosphoramide; 2-methyl- and 2,5-dimethylthiophene give thiophene-2-carboxylic acid, and 3-methylthiophene yields thiophene-3-carboxylic acid (*T. J. Wallace* and *F. A. Baron*, J. org. Chem., 1965, **30,** 3520).

Isomerisation of 2- and 3-*tert*-butylthiophene in the presence of aluminium chloride gives polymeric products plus traces of dialkyl products; 2,5-di-*tert*-butylthiophene is converted to the 2,4-isomer. Transalkylation of 2,5-dimethylthiophene by 2,5-di-*tert*-butylthiophene affords 3-*tert*-butyl-2,5-dimethylthiophene (*Wynberg* and *U. E. Wiersum*, J. org. Chem., 1965, **30,** 1058).

Hydrodesulphurisation of 2- and 3-methylthiophenes over a sulphided, alumina-supported cobalt molybdate catalyst, conform to the reaction sequence (previously proposed for thiophene) C–S bond fission followed by diene and mono-olefin formation, and complete saturation (*P. Desikan* and *C. H. Amberg*, Canad. J. Chem., 1963, **41,** 1966).

2-Vinylthiophene undergoes *diene addition* with maleic anhydride and benzoquinone (*W. Davies* and *Q. N. Porter*, J. chem. Soc., 1957, 4958) and is chloromethylated in the side-chain as well as in the nucleus to give 2-chloromethyl-5-(3′-chloropropen-1-yl)thiophene (*R. Lukeš, M. Janda* and *K. Kefurt*, Coll. Czech. chem. Comm., 1960, **25,** 1058).

Halogenation: under all conditions free halogens attack the nucleus of thiophene homologues, but side-chain bromination is effected by *N*-bromosuccinimide (*K. Dittmer et al.*, J. Amer. chem. Soc., 1949, **71,** 1201).

Effect of substitution: Aryl substitution weakens the aromatic character of the thiophene ring. Thus tetraphenylthiophene is reduced by tin and hydrochloric acid to tetraphenylbutane and oxidised by hydrochloric acid and potassium chlorate to *cis*-dibenzoylstilbene or by hydrogen peroxide to sulphone which reverts to tetraphenylthiophene on reduction (*O. Hinsberg*, Ber., 1915, **48,** 1611).

Thiophenes bearing such electron-donating substituents as alkyl or aryl are oxidised by perbenzoic acid to fairly stable 1,1-dioxides (*H. J. Backer* and *J. L. Melles*, Rec. Trav. chim., 1952, **71,** 869; 1953, **72,** 314). Unlike simple alkyl or aryl sulphones, these can be reduced to the parent thiophenes by zinc and hydrochloric acid. As olefinic sulphones they combine with piperidine and with phenylmethanethiol and they react as dienes with maleic anhydride at 150–200°:

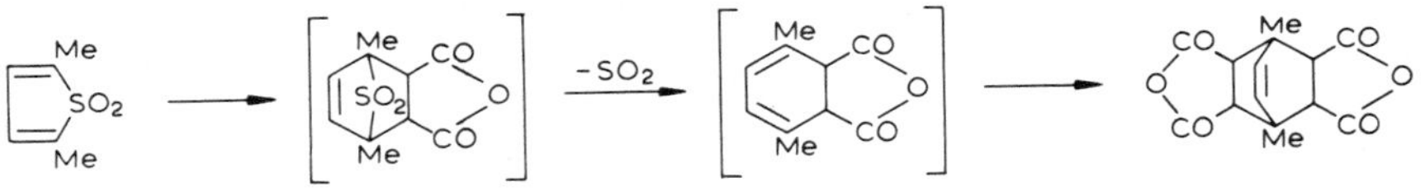

Restricted rotation about a phenyl–thienyl bond is shown by the production of 2-(3′-nitrophenyl)thienyl-3-carboxylic acid (XXI), in optically active but rather unstable form (*L. J. Owen* and *Nord*, J. org. Chem., 1951, **16,** 1864):

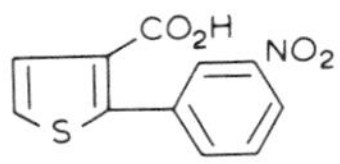

(XXI)

In boiling deuterioacetic acid 3-alkyl-, 3-phenyl-, and 3-phenylthio-thiophenes undergo deuterium–hydrogen exchange at the 2-position while the corresponding 2-substituted derivatives are exchanged at the 5 position (*Wynberg et al.*, J. org. Chem., 1968, **33,** 2902).

The properties of a number of substituted thiophenes have been reported (*K. E. Miller*, J. chem. Eng. Data, 1965, **10,** 305; C.A., 1965, **63,** 11474b) and the aromaticity of monosubstituted thiophenes estimated by n.m.r. dilution shifts (*M. M. Dhingra et al.*, Proc. Indian Acad. Sci., A, 1967, **65,** 203; C.A., 1967, **67,** 63605e). Data on alkyl- and aryl-thiophenes are collected in Table 2.

TABLE 2

ALKYL- AND ARYL-THIOPHENES, *etc.*

Thiophene	*B.p. (°C)*	*Source*	*Ref.*
2-Methyl-	113	sodium laevulate – P_4S_7;	1
		thiophene, Na liquid NH_3, alkyl halide (MeBr)	2
3-Methyl-	115	sodium methylsuccinate – P_4S_7	3
2,3-Dimethyl-	141	β-methyllaevulic acid	4
2,4-Dimethyl-	139	3-methylthiophene	5

TABLE 2 *(continued)*

Thiophene	*B.p. (°C)*	*Source*	*Ref.*
2,5-Dimethyl-	136	acetonylacetone – P_4S_7	6
3,4-Dimethyl-	144–146	dimethylbutadiene – S	7
2,3,4-Trimethyl-	160–163	αβ-dimethyllaevulic acid	8
2,3,5-Trimethyl-	163	2,4-dimethylthiophene	5
Tetramethyl-	182–184	2-iodo-3,4,5-trimethylthiophene	9
2-Ethyl-	132–134	2-acetylthiophene	10
3-Ethyl-	135–136	sodium ethyl succinate – P_4S_7	
2-Isopropyl-	152	thiophene }	11
3-Isopropyl-	155.5	propylene – H_3PO_4 }	11
2-*tert*-Butyl-	164	thiophene }	11,12
3-*tert*-Butyl-	169	isobutylene – H_3PO_4 }	11,12
2-Allyl-	159	$(C_4H_3S)MgI + C_3H_5Br$	
2-Benzyl-	257–262	thiophene–benzyl alcohol	13
2-Vinyl-	62–63/50 mm	2-α-chloroethylthiophene	14
3-Vinyl-	156–158	methyl 3-thienylcarbinol	15
2-Acetylenyl-	32/3 mm	β-2-thienylvinyl bromide	16
3-Acetylenyl-	65–67/60 mm	3-vinylthiophene dibromide	15
2-Styryl-	m.p. 111	$(C_4H_3S)CHO + CH_2Ph{\cdot}MgCl$	
2-Phenyl-	m.p. 37	sodium β-benzoylpropionate	17
3-Phenyl-	m.p. 92	sodium phenylsuccinate	17
2,4-Diphenyl-	m.p. 121	acetophenone – H_2S	18
2,5-Diphenyl-	m.p. 153	diphenacyl	17
3,4-Diphenyl-	m.p. 114	related dicarboxylic acid	19
2,3,4-Triphenyl-	m.p. 215–216		20
Tetraphenyl-	m.p. 184	stilbene – S	

References

1 *A. Chrzaszczewska*, Rocz. Chim., 1952, **5,** 53.
2 *W. J. Zimmerschied* and *R. C. Arnold*, U.S.P. 2,585,292/1952.
3 *R. F. Feldkamp* and *B. F. Tullar*, Org. Synth., 1954, **34,** 73.
4 *A. F. Shepard*, J. Amer. chem. Soc., 1932, **54,** 2951.
5 *J. Sicé*, J. org. Chem., 1954, **19,** 70.
6 *G. N. Jean* and *F. F. Nord*, *ibid.*, 1955, **20,** 1363.
7 *Shepard*, *A. L. Henne* and *T. Midgley*, J. Amer. chem. Soc., 1934, **56,** 1355.
8 *N. Zelinsky*, Ber., 1887, **20,** 2025.
9 *Idem*, *ibid.*, 1888, **21,** 1835.
10 *W. J. King* and *Nord*, J. org. Chem., 1949, **14,** 638.
11 *W. G. Appleby et al.*, J. Amer. chem. Soc., 1948, **70,** 1552.
12 *N. P. Buu-Hoï*, J. chem. Soc., 1954, 1975.
13 *W. Steinkopf* and *W. Hanske*, Ann., 1939, **541,** 257.
14 *W. S. Emerson* and *T. M. Patrick*, J. org. Chem., 1948, **13,** 729.
15 *C. Troyanowsky*, Compt. rend., 1951, **32,** 2236; Bull. Soc. chim. Fr., 1955, 424.
16 *A. Vaitiekunas* and *Nord*, J. org. Chem., 1954, **19,** 902.
17 *H. J. Backer* and *J. L. Melles*, Rec. Trav. chim., 1953, **72,** 314, 491.
18 *E. Campaigne*, J. Amer. chem. Soc., 1944, **66,** 684.
19 *Backer* and *W. Stevens*, Rec. Trav. chim., 1940, **59,** 423.
20 *B. M. Trost* and *R. Atkins*, Tetrahedron Letters, 1968, 1225.

(vi) Halogenothiophenes

(*1*) *Chlorothiophenes.* Thiophene is readily brominated and still more readily chlorinated. Addition accompanies substitution, and the elements of hydrogen halide can be removed from the addition products by pyrolysis or treatment with alkali (see Scheme 1).

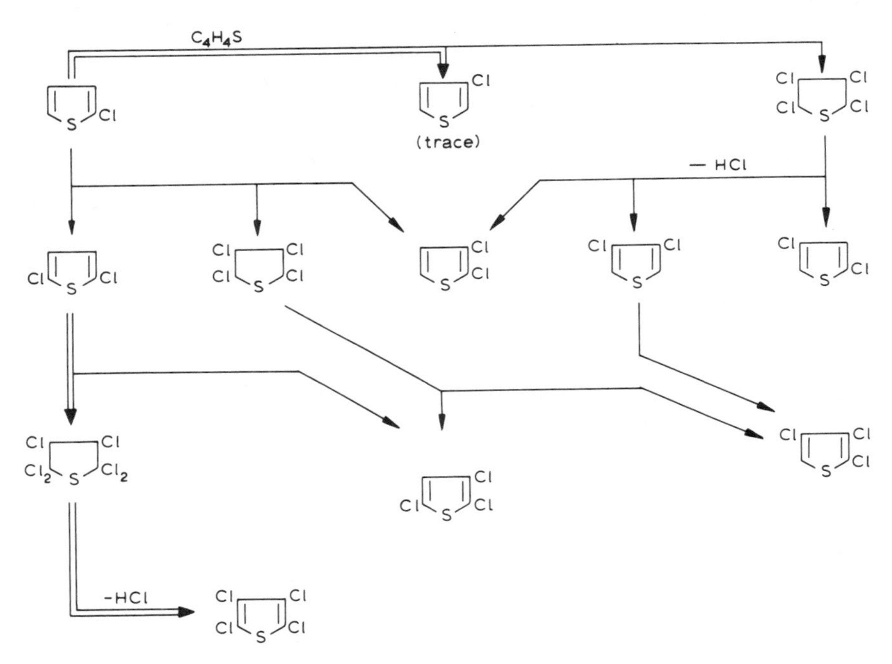

Scheme 1.

The substitution products of the major route shown by a double line can be separated on the small scale through the mercury complexes (*W. Steinkopf* and *W. Köhler*, Ann., 1937, **532,** 250). Others may be obtained as follows (see Scheme 2).

The Grignard reaction (entrainment method) on tetrachlorothiophene followed by hydrolysis, gives 2,3,4-trichlorothiophene. A study has been made of the vapour-phase bromination and chlorination of thiophene and 2-methylthiophene (*C. D. Hurd* and *H. J. Anderson*, J. Amer. chem. Soc., 1953, **75,** 3517). Compared with benzene the uncatalysed chlorination and bromination of thiophene in acetic acid proceed 10^7 and 10^9 times faster

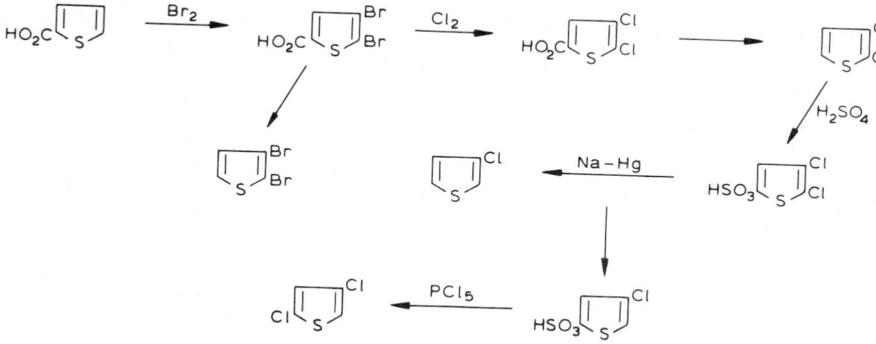

Scheme 2.

(*G. Marino*, Tetrahedron, 1965, 843). Tetrachlorothiophene may be obtained by the reaction between hexachlorobuta-1,3-diene and sulphur (*E. J. Geering*, J. org. Chem., 1959, **24,** 1128):

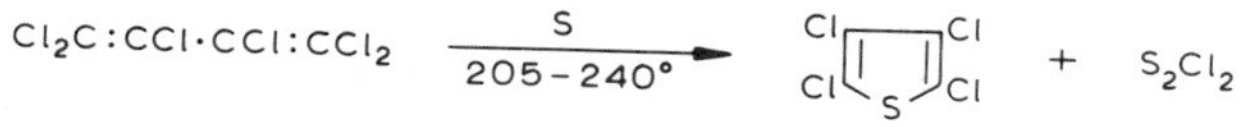

$$Cl_2C{:}CCl{\cdot}CCl{:}CCl_2 \xrightarrow[205-240^\circ]{S} C_4Cl_4S + S_2Cl_2$$

(2) *Bromothiophenes.* Bromination of thiophene gives the 2-, 2,5-di-, 2,3,5-tri- and 2,3,4,5-tetra-bromo derivatives, and no addition products are isolated. The mechanism is stated to be different to that of chlorination (*S.-O. Lawesson*, Arkiv Kemi, 1957, **11,** 373). The bromothiophenes are easily separated by distillation and by using cyanogen bromide, substitution can be stopped at the 2-bromothiophene stage (*Steinkopf*, Ann., 1923, **430,** 78), Bromination of some substituted thiophenes has been accomplished under remarkably mild conditions. Upon treatment with *N*-bromosuccinimide in acetic acid–chloroform solution at room or slightly elevated temperatures, 3-methyl-, 3-phenyl-, 3-phenylthio-, and 3-bromothiophenes are rapidly brominated in the 2 position in nearly quantitative yields. Bromination of 2-methyl-, 2-phenyl-, 2-phenylthio-, and 2-bromothiophene under the same conditions yields the 5-bromo derivatives (*H. Wynberg et al.*, J. org. Chem., 1968, **33,** 2902). In solutions of hydrogen bromide in acetic acid some 2-bromothiophenes undergo rearrangements, *e.g.*, 2-bromo-3-phenylthiophene gives 2-bromo-4-phenylthiophene, 3-phenylthiophene and 2,5-dibromo-3-phenylthiophene. 2-Thienylmagnesium chloride in tetrahydrofuran on reacting with bromine gives 2-bromothiophene (*Metal* and *Thermit Corpn.*, B.P. 826,619/1955; C.A., 1960, **54,** 15218h), also obtained on treating thiophene in methanol containing potassium carbonate with bromine at below −7° (*M. Janda*,

Coll. Czech. chem. Comm., 1963, **28,** 2524). 3-Bromothiophene may be prepared by boiling 2,3,5-tribromothiophene with zinc dust in acetic acid (*S. Gronowitz*, Acta chem. Scand., 1959, **13,** 1045). 2-Bromothiophene reacts with potassium amide in liquid ammonia, under typical benzyne-producing conditions to give 3-aminothiophene and a small amount of 3-bromothiophene (*M. G. Reinecke* and *H. W. Adickes*, J. Amer. chem. Soc., 1968, **90,** 511). 3-Bromothiophene with excess butyl-lithium in ether at low temperatures affords 3-, and 2,3-di-lithiothiophene (*B. Östman*, Arkiv Kemi, 1964, **22,** 551).

(*3*) *Iodothiophenes.* Direct iodination, best with iodine and mercuric oxide, ceases at the 2,5-diiodo-compound, but tetraacetoxymercurithiophene is iodinated to tetraiodothiophene. 2-Iodothiophene has been prepared from thiophene, iodine and iodic acid in aqueous acetic acid containing sulphuric acid (*H. O. Wirth et al.*, Ann., 1960, **634,** 84); it is an important source of thiophene derivatives through its Grignard compound, *e.g.* with ethylene oxide it affords 2-vinylthiophene (*L. V. Andreeva* and *M. M. Koton*, Zhur. obshcheĭ Khim., 1957, **27,** 997), and with trialkyl borates (*e.g.*, triethyl borate), 2-thienyl boronic acid, $(C_4H_3S)B(OH)_2$ (*H. J. Roth* and *B. Miller*, Arch. Pharm., 1964, **297,** 513). 3-Thienyl radicals have been generated by the photolysis of 3-iodothiophene and their reactions studied (*G. Martelli, P. Spagnolo* and *M. Tiecco*, J. chem., Soc., B, 1968, 901).

(*4*) *Fluorothiophenes.* Treatment of 2-thienyl-lithium with perchloryl fluoride gives 2-fluorothiophene (*R. D. Schuetz et al.*, J. org. Chem., 1963, **28,** 1420), also obtained in poor yields from the reaction between arsenic trifluoride and 2-iodothiophene (*R. T. van Vleck*, J. Amer. chem. Soc., 1949, **71,** 3256). 2-Fluoro-5-methylthiophene has been prepared by the former method.

Thiophene has been fluorinated with the supposed potassium tetrafluorocobaltate(III) prepared from potassium trifluorocobaltate(II) and fluorine to give XXII and XXIII as major products. The former on bubbling through molten potassium hydroxide at ~250° affords *tetrafluorothiophene*, b.p. 60° (68°), in low yield (*J. Burdon et al.*, Chem. Comm., 1969, 27):

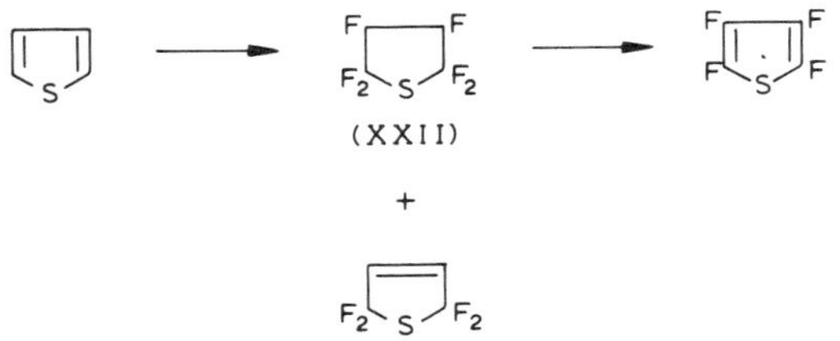

(XXII)

+

(XXIII)

Perfluorothiophene is obtained by treating hexafluoro-1,4-diiodobutane with sulphur at 250° (*G. van D. Tiers*, J. org. Chem., 1961, **26,** 2538).

A halogen atom directly attached to the unreduced nucleus is in general firmly held, but chlorine in position 2 is labilised by nitro-groups in positions 3 or 5 (*Hurd* and *K. L. Kreuz*, J. Amer. chem. Soc., 1952, **74,** 2965). It is, however, possible to effect the Grignard reaction; in presence of an alkyl halide – entrainment method – even polyhalogenothiophenes react, with successive replacement of α-halogen atoms.

Individual halogenothiophenes are assembled in Table 3.

TABLE 3

HALOGENOTHIOPHENES

Substitution	*Chloro-*	*Ref.*	*Bromo-*	*Ref.*	*Iodo-*	*Ref.*
2-Mono	b.p. 130°	3	b.p. 149–151°	1	b.p. 73°/15 mm	2
3-Mono	b.p. 136–137°	4	b.p. 157–158°	5	b.p. 68°/12 mm	6
2,3-Di	b.p. 173–174°	4	b.p. 219°	4		
2,4-Di	b.p. 174–175°	4	b.p. 210°	7		
2,5-Di	b.p. 162°	4	b.p. 210°	4	m.p. 41°	
3,4-Di	b.p. 185°	4	b.p. 221–222°	7		
Tetra	m.p. 36°	8	m.p. 117–118°	7	m.p. 199°	6

References

1 *N. G. Buu-Hoï*, Ann., 1944, **556,** 1.
2 *W. Minnis*, Org. Synth., 1932, **12,** 44; *H. U. Lew* and *C. R. Noller*, *ibid.*, 1950, **30,** 53.
3 *C. D. Hurd* and *H. J. Anderson*, J. Amer. chem. Soc., 1953, **75,** 3517.
4 *W. Steinkopf* and *W. Köhler*, Ann., 1937, **532,** 250.
5 *C. Troyanowski*, Bull. Soc. chim. Fr., 1955, 424.
6 *Steinkopf*, *H. F. Schmitt* and *H. Fiedler*, Ann., 1937, **527,** 237.
7 *Steinkopf*, *H. Jacob* and *H. Penz*, *ibid.*, 1934, **512,** 136.
8 *H. L. Coonradt* and *H. D. Hartough*, J. Amer. chem. Soc., 1948, **70,** 1158.

(vii) Nitro- and amino-thiophenes

The mono-nitration of thiophene, best conducted in acetic anhydride (*V. S. Babasinian*, Org. Synth., 1934, **14,** 76; *A. H. Blatt*, *S. B. Rosenberg* and *L. W. Kresch*, J. org. Chem., 1957, **22,** 1693) gives mainly 2-**nitrothiophene,** m.p. 46.5°, b.p. 224°, which is almost colourless and gives a deep brown-red solution in boiling alkali. 2-Nitrothiophene of high purity can be obtained also by chlorosulphonation of the isomeric mixture of nitrothiophenes obtained by the nitration of thiophene (*B. Östman*, Acta chem. Scand., 1968, **22,** 1687). Nitration of thiophene with benzoyl nitrate in methyl cyanide gives 2- and 3-nitrothiophenes in the ratio 6.2:1 and with [2-^{3}H]-thiophene a weak secondary α-isotope effect is observed (*idem*, Arkiv Kemi, 1963, **19,** 499). Thiophene may be converted into 2-nitrothiophene with cupric nitrate in acetic anhydride (*N. I. Puthokin*, C.A., 1957, **51,** 16419) and this reagent has been used successfully with thiophene derivatives which decompose on nitration by conventional methods (*V. N. Ivanova*, J. gen. Chem., U.S.S.R., 1958, **28,** 1288). *Steinkopf* (Ann., 1914, **403,** 17) established the orientations of a series of thiophene derivatives by relating them to 2-methylthiophene (synthesis, p. 220):

$$2\text{-}(C_4H_3S)Me \longrightarrow 2\text{-}(C_4H_3S)CO_2H \longleftarrow 2\text{-}(C_4H_3S)COMe \longrightarrow 2\text{-}(C_4H_3S)CMe{:}NOH$$

$$2\text{-}(C_4H_3S)NO_2 \longrightarrow 2\text{-}(C_4H_3S)NH_2 \longrightarrow 2\text{-}(C_4H_3S)NHAc \longleftarrow$$

3-**Nitrothiophene,** m.p. 78–79°, is prepared by hydrolysing the product of nitration of thiophene-2-sulphonyl chloride (*Steinkopf* and *T. Höpner*, Ann., 1933, **501,** 174; *Blatt, Rosenberg* and *Kresch, loc. cit.*) or of 2-cyano-thiophene (*P. Reynaud* and *R. Delaby*, Bull. Soc. chim. Fr., 1955, 1614). It may be obtained by simultaneous debromination and decarboxylation of 5-bromo-4-nitrothiophene-2-carboxylic acid prepared by nitrating methyl 5-bromothiophene-2-carboxylate (*R. Motoyama et al.*, Nippon Kagaku Zasshi, 1957, **78,** 950). Further nitration of 3-nitrothiophene gives only 2,4-*dinitrothiophene*, m.p. 56°, while 2-nitrothiophene gives a mixture of this with 2,5-*dinitrothiophene*, m.p. 80–82° (*Blatt, Rosenberg* and *Kresch, loc. cit.*), *e.g.*, fuming nitric acid affords 56% of the 2,4- and 44% of the 2,5-isomer (*Östman*, Arkiv Kemi, 1963, **19,** 499); both dinitrothiophenes form coloured addition complexes with aromatic hydrocarbons. Nitration of substituted thiophenes occasionally leads to displacement of halogen atoms or carboxyl or acyl groups.

Thiophene-2-carbaldehyde, thiophene-3-carbaldehyde, 2-cyano-, 2-nitro-, 3-cyano-, and 3-nitro-thiophene have been nitrated at different temperatures in trifluoroacetic acid and the reactivity orders for the ring positions are $5 \gg 4 \gg 3$ and $5 > 2 > 4$ for 2- and 3-substituted thiophenes, respectively (*idem*, Acta chem. Scand., 1968, **22,** 2754).

The adduct formed between methoxide ion and 2-methoxy-3,5-dinitro-thiophene has a Meisenheimer structure (XXIV) (*G. Doddi, G. Illuminati* and *F. Stegel*, Chem. Comm., 1969, 953):

NO2 O2N S OMe —MeO⊖→ NO2 OMe OMe O2N S⊖

(XXIV)

Aminothiophenes can be prepared by reducing nitrothiophenes with tin and hydrochloric acid at 40–45°; the free bases are oils which resinify quickly in the air; 2-**aminothiophene,** b.p. 77–79°/11 mm, rapidly oxidises and polymerises, *acetyl* derivative, m.p. 60°; 3-**aminothiophene** is unstable, *acetyl* derivative, m.p. 145–148° (*Steinkopf*, Ann., 1914, **403,** 17; *Steinkopf* and *Höpner*, *loc. cit.*). 3-Aminothiophene may be obtained from 2-bromothiophene (p. 241). The n.m.r. spectra of 2- and 3-aminothiophenes show that they are true amines (*R. A. Hoffman* and *Gronowitz*, Arkiv Kemi, 1960, **16,** 515). Diazotisation has been effected in a few cases only (*Steinkopf* and *P. J. Müller*, Ann., 1926, **448,** 210; *Putokhin* and *V. I. Yakovlev*, C.A., 1955, **49,** 12431; *Putokhin* and *A. N. Sorokin*, C.A., 1957, **51,** 16419. Hydrogenation of 2-nitrothiophene in acetic anhydride at 100° with rhenium heptasulphide as catalyst yields 2-acetylaminothiophene; at 200° the thiophenic nucleus is reduced (*C. Aretos* and *J. Vialle*, C.A., 1963, **59,** 5145). 2-*Dimethylaminothiophene*, m.p. 165–169° (*methiodide*, m.p. 153°) has been prepared by boiling together thiophene, bis-(dimethylamino)methane [$(Me_2N)_2CH_2$], and acetic acid (*M. Muehlstaedt* and *W. Rauner*, J. pr. Chem., 1965, **29,** 319). While 2-*thienylcarbamate*, m.p. 52°, is made by the Curtis reaction from thiophene-2-carboxylic acid, and 2-*acetamidothiophene*, m.p. 160°, by Beckmann rearrangement of methyl thienyl ketoxime (*J. Cymerman-Craig* and *D. Willis*, J. chem. Soc., 1955, 1071), the Hofmann reaction fails with the amide of thiophene-2-carboxylic acid. But the same reaction applied to the 3-isomer gives, after acetylation of the product 3-*acetamidothiophene*, m.p. 145–148° (*E. Campaigne* and *P. A. Monroe*, J. Amer. chem. Soc., 1954, **76,** 2447). 2-Acetamidothiophene affords 2-methylaminothiophene by methylation and hydrolysis. Carboxylated 3-aminothiophenes are obtained in good yield by treating oximinotetrahydrothiophenes with dry hydrogen chloride (*L. C. Cheney* and *J. R. Piening*, *ibid.*, 1945, **67,** 729):

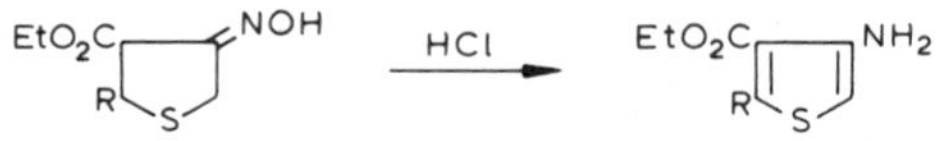

$$EtO_2C, NOH, R, S \xrightarrow{HCl} EtO_2C, NH_2, R, S$$

(viii) Hydroxythiophenes

Treatment of γ-ketonic acids with phosphorus pentasulphide gives the socalled 2-hydroxythiophenes, more correctly formulated in most cases as 5-oxothiolenes (2-oxo-dihydrothiophenes) *e.g.* XXV and XXVI:

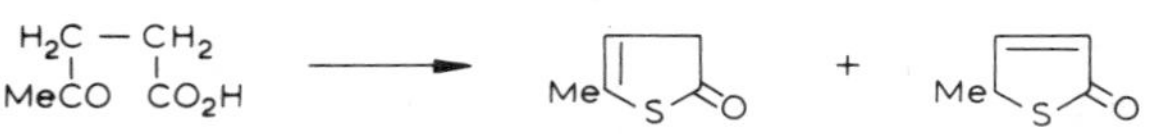

$$H_2C-CH_2, MeCO\ CO_2H \longrightarrow Me, S, O\ (XXV) + Me, S, O\ (XXVI)$$

2-**Hydroxythiophene,** m.p. 7–9°, b.p. 217–219°, has been prepared in 22% yield by oxidising 2-thienylmagnesium bromide with gaseous oxygen in presence of isopropylmagnesium bromide (*C. D. Hurd* and *K. L. Kreuz*, J. Amer. chem. Soc., 1950, **72,** 5543); by heating 2-*tert*-butoxythiophene with *p*-toluenesulphonic acid at 150° (*H. J. Jakobsen, E. H. Larsen* and *S. O. Lawesson*, Rec. Trav. chim., 1963, **82,** 791); along with its alkyl derivatives by the action of hydrogen peroxide on thien-2-ylboronic acids (*A. B. Hörnfeldt* and *S. Gronowitz*, Arkiv Kemi, 1963, **21,** 239); by the reaction of 2-thienyl-lithium with 1,2,3,4-tetrahydro-1-naphthyl hydrogen peroxide (*Hurd* and *H. J. Anderson*, J. Amer. chem. Soc., 1953, **75,** 5124). The so-called 2-hydroxythiophene, which according to the n.m.r. spectrum of the liquid compound exists as 2-oxo-3-thiolene (2-oxo-2,5-dihydrothiophene) (XXVII) (*Hörnfeldt* and *Gronowitz*, *loc. cit.*), yet also gives the reactions characteristic of the enol form.

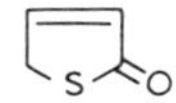

(XXVII)

It gives a red colour with ferric chloride and couples with diazo compounds, but has no phenolic smell and is brominated far less readily than phenol. Infrared absorption indicates O – H, both aliphatic and aromatic C – H, and C = O bonds. The *methyl ether*, b.p. 153–155°, has the odour of anisole (*Hurd* and *Kreuz*, *loc. cit.*; *J. Sicé*, J. Amer. chem. Soc., 1953, **75,** 3697). 2-Cyanomethyl-5-methoxythiophene in attempted Hoesch synthesis yielded 2-cyanomethyl-5-(2′,4′,6′-trihydroxyphenyl)thiophene (XXVIII) (*Gronowitz* and *B. Jägersten*, Arkiv Kemi, 1961, **18,** 213):

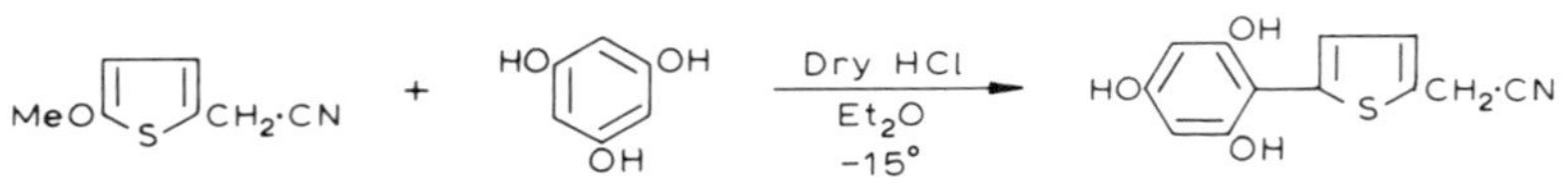

Dry HCl, Et_2O, −15°

(XXVIII)

The reaction between 2-methylthiophene, butyl-lithium and hydrogen proxide yields *2-methyl-5-oxo-2-thiolene* (*5-methyl-2-oxo-2,3-dihydrothiophene*) (XXV), b.p. 65–66°/10 mm, and a mixture of this compound and the 2,5-dihydro compound XXVI, b.p. 66–85°/10 mm; addition of hydrochloric acid to XXV gives *2-methyl-5-oxo-3-thiolene* (XXVI), b.p. 79–82°/8 mm, almost completely (*Hörnfeldt* and *Gronowitz*, *loc. cit.*); the mixture of isomers XXV and XXVI is alkali-soluble and can be *O*-acetylated, *O*-benzoylated, and with benzaldehyde yields 4-benzylidene-2-methyl-5-oxo-2-thiolene (XXIX) (*Steinkopf* and *F. Thormann*, Ann., 1939, **540,** 1):

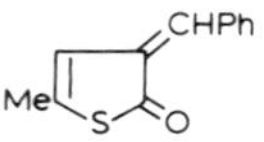

(XXIX)

As an analogue of thio-oxindole (see p. 281), the mixture of isomers reacts with suitable dicarbonyl compounds to give indigoid type pigments. 5-Aryl-2-hydroxythiophenes exist in methanol solution partly as enols and partly as ketones, whereas the 5-alkyl compounds show no enolic form (*Hörnfeldt*, Arkiv Kemi, 1964, **22,** 211).

Oxidation of 3-thienylmagnesium bromide affords a small yield of the unstable 3-**hydroxythiophene** (*M. C. Ford* and *D. Mackay*, J. chem. Soc., 1956, 4985), also obtained by decarboxylation of 3-hydroxythiophene-2-carboxylic acid (*H. Fiesselmann, P. Schipprack* and *L. Seitler*, Ber., 1954, **87,** 841), and by treating 1,4-dibromo-2-hydroxybutane with sodium hydrogen sulphide in methanol in a sealed tube at room temperature (*E. P. Adams et al.*, J. chem. Soc., 1960, 2649). Unlike the 2-isomer it is appreciably phenolic and its infrared spectrum indicates that it exists as a tautomeric mixture consisting mainly of the enolic form (*Ford* and *Mackay, loc. cit.*). 3-Hydroxy-5-phenylthiophene (XXXI), accessible from XXX, obtained by heating ethyl cinnamate with sulphur, behaves as a tautomer (*P. Friedländer* and *S. Kielbasinski*, Ber., 1912, **45,** 3389):

(XXX) $\xrightarrow[ClCH_2 \cdot CO_2Na]{Na_2S}$ $Ph \cdot C(=CH \cdot COSCH_2 \cdot CO_2H) - S - CH_2 \cdot CO_2H$ $\xrightarrow[NaOAc]{Ac_2O}$ 3-acetoxy-5-phenylthiophene ⟶ (XXXI)

Thiotetronic acid (p. 257) may be regarded as 2,4-**dihydroxythiophene.** The *dibenzoyl* derivative, m.p. 110°, of 3,4-**dihydroxythiophene** is accessible by a Hinsberg synthesis (*E. W. Fager*, J. Amer. chem. Soc., 1945, **67,** 2217). For the preparations and tautomeric structures of some methyl-, bromo-, and methoxy-substituted 2-oxodihydrothiophenes see *Hörnfeldt* and *Gronowitz* (*loc. cit.*) and for 5-substituted 2-oxodihydrothiophenes see *Hörnfeldt* (*loc. cit.*). 3,5-Dinitro- and 5-acetyl-3-nitro-2-hydroxythiophene are prepared by treatment of the corresponding chlorides with sodium formate in methanol; they are unstable solids giving coloured, stable sodium salts (*Hurd* and *Kreuz*, J. Amer. chem. Soc., 1952, **74,** 2965).

(ix) Sulphonic acids, thiols and related compounds

Sulphonation of thiophene by sulphuric acid, often conducted in ligroin as solvent to minimise resinification, takes place predominantly in position 2; the smooth reaction with chlorosulphonic acid gives the free acid with some sulphonyl chloride; the latter product is obtained also when the reaction is carried out in chloroform at −10° (*A. Buzas* and *J. Teste*, Bull. Soc. chim. Fr., 1960, 793). Thiophenes are also sulphonated by mild reagents such as the compounds of pyridine and of dioxane with sulphur trioxide (*G. M. Kadatski* and *A. P. Terentiev*, J. gen. Chem., U.S.S.R., 1952, **22,** 153). Sulphonic acids of thiophene homologues are sometimes best made by sulphonating the acetyl derivative of the thiophene; thus 5-acetyl-2-ethylthiophene affords 2-ethylthiophene-5-sulphonic acid. With oleum 2,5-dibromothiophene gives the cyclic 3,4-disulphonic anhydride XXXII, or, with a limited quantity of oleum, the monosulphonic acid from which thiophene-3-sulphonic acid is obtained by reductive debromination (*J. Langer*, Ber., 1884, **17,** 1566):

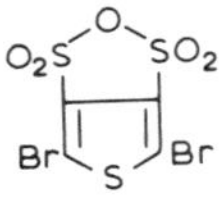

(XXXII)

Thiophene-2- and -3-**sulphonyl chlorides,** m.p. 32–33° and 47–48°; *sulphonamides*, m.p. 147° and 155–157°, respectively. Thiophene-2,4-disulphonic acid is obtained from acetamidothiophene by disulphonation and deamination (*H. Scheibler et al.*, Ber., 1954, **87,** 1184).

Reduction of thiophene-2-sulphonyl chloride affords the **thiophene-2-sulphinic acid,** m.p. 72–73° (*H. Burton* and *W. A. Davy*, J. chem. Soc., 1948, 525), the 2,2′-**disulphide,** m.p. 56° (*F. Challenger et al., ibid.*, 1948, 769), and **thiophene-2-thiol,** b.p. 166° (*W. H. Houff* and *R. D. Schuetze*, J. Amer. chem. Soc., 1953, **75,** 6316), which is also a by-product in the preparation of thiophene from sodium succinate. Thiophenethiols may be prepared also by reaction of sulphur with thienylmagnesium halides (*P. D. Caesar* and *P. D. Branton*, Ind. Eng. Chem., 1952, **441,** 122) or thienyl-lithium compounds (*S. Gronowitz et al.*, Arkiv Kemi, 1960, **16,** 309; 1961, **17,** 237); by the cleavage of alkyl thienyl sulphides with sodium in liquid ammonia (*Ya. L. Gol'dfarb, M. A. Kalik* and *M. L. Kirmalova*, Izvest. Nauk, S.S.S.R., Otdel. Khim. Nauk, 1960, 1696; C.A., 1961, **55,** 8377d); and by a simple synthesis involving the Friedel–Crafts reaction of 2,4-dinitrobenzenesulphenyl chloride (XXXIII) with the thiophene nucleus followed by basic cleavage of the resulting sulphide (*Schuetze* and *W. L. Fredericks*, J. org. Chem., 1962, **27,** 1301):

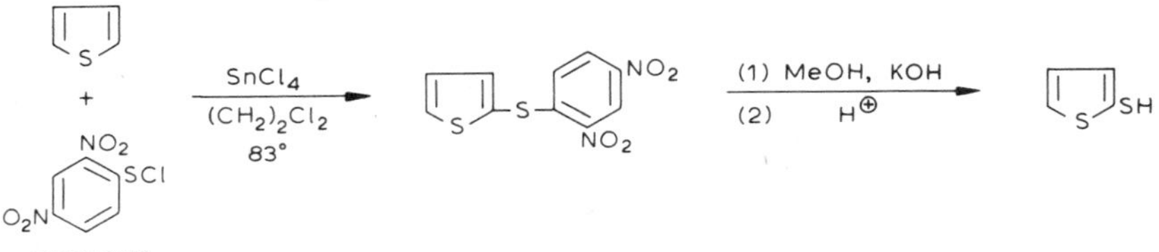

$SnCl_4$, $(CH_2)_2Cl_2$, 83°; (1) MeOH, KOH (2) $H^{\oplus}$

(XXXIII)

The most effective method for the preparation of alkyl 2-thienyl sulphides [2-(1-thia-alkyl)thiophenes, 2-thienyl thio-ethers] is by the alkylation of thiophene-2-thiols (*H. Aft* and *B. E. Christensen, ibid.*, 1962, **27,** 2168):

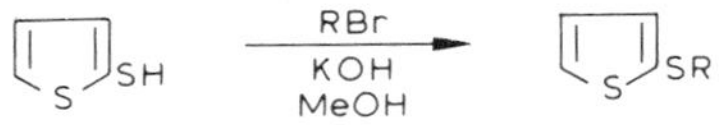

RBr, KOH, MeOH

2-Iodothiophene with thiosalicyclic acid in the presence of copper acetate and potassium carbonate affords *o*-carboxyphenyl thienyl sulphide, $SC_4H_3 \cdot S \cdot C_6H_4 \cdot CO_2H$ (*W. Steinkopf et al.*, Ann., 1937, **527**, 237).

Phenyl 2-thienyl sulphoxide, phenyl 2-thienyl sulphone, and *methyl 2-thienyl sulphone*, m.p. 69–70°, 124° and 47° (*Burton* and *Davy, loc. cit.*). Treated with sulphur, thienylmagnesium bromide yields 2,2′-**diethienyl sulphide,** b.p. 155°/14.5 mm, and thence **dithienyl sulphone,** m.p. 130–131° (*Challenger* and *J. B. Harrison*, J. Inst. Pet., 1935, **21,** 135). Thiocyanation with thiocyanogen and aluminium chloride yields 2-**thiocyanothiophene** (2-*thienyl thiocyanate*), b.p. 104°/9 mm and 2,5-**dithiocyanothiophene,** m.p. 87–88° (*E. Söderback*, Acta chem. Scand., 1954, **8,** 1851); the former compound may be prepared by the reaction between 2-chloromercurithiophene and thiocyanogen (*E. Cherbuliez* and *C. Giddey*, Helv., 1952, **35,** 160).

Thiophene-3-thiol, b.p. 171°, obtainable from the 3-sulphonic acid, is a by-product in the commercial synthesis of thiophene from butane (*Caesar* and *Branton, loc. cit.*); it affords 3-*thienylthioacetic acid*, m.p. 51° (*Challenger et al., loc. cit.*). Infrared (*Gronowitz, P. Moses* and *A. B. Hörnfeldt*, Arkiv Kemi, 1961, **17,** 237) and n.m.r. spectra of thiophene-2- and -3-thiols indicate that they exist in the thiol forms (*R. A. Hoffman* and *Gronowitz, ibid.*, 1960, **16,** 563).

Thiophene-3,4-dithiol, b.p. 88–93°/1.5 mm, may be prepared by reacting 3-bromothiophene-4-thiol with *n*-butyl-lithium in ether at −70° and then adding sulphur. Spectroscopic data showed that the dithiol form was the most stable tautomer (*Gronowitz* and *Moses*, Acta chem. Scand., 1962, **16,** 105).

(x) Thiophene alcohols and related compounds

2-**Hydroxymethylthiophene** (2-**thienylcarbinol, thenyl alcohol**)*, b.p. 207°, is prepared from 2-thienylmagnesium iodide and formaldehyde (*W. Steinkopf* and *G. Bokor*, Ann., 1939, **540,** 23) or by Cannizzaro dismutation of thiophene-2-carbaldehyde (*F. W. Dunn* and *K. Dittmer*, J. Amer. chem. Soc., 1946, **68,** 2561). Grignard processes also lead to *methyl-2-thienylcarbinol* (α-2-*thienylethanol*), $MeCH(C_4H_3S)OH$, $[\alpha]_D^{20}$ −23.7° (*L. G. Anderson, M. P. Balfe* and *J. Kenyon*, J. chem. Soc., 1950, 1866), *dimethyl-2-thienylcarbinol*, m.p. 33°, and *diphenyl-2-thienylcarbinol*, m.p. 131° (*W. Minnis*, J. Amer. chem. Soc., 1929, **51,** 2143). The bis-*p*-dimethylamino derivative of the last-named carbinol is similar in its preparation and properties to the analogous colour-base of malachite green (*C. D. Mason* and *F. F. Nord*, J. org. Chem., 1951, **16,** 722). Thienylsodium and ethylene oxide give *β*-2-**thienylethanol,** $(C_4H_3S)CH_2 \cdot CH_2OH$, b.p. 99–100°/7 mm (*J. W. Schick* and *H. D. Hartough*, J. Amer. chem. Soc., 1948, **70,** 1646). *Di-2-thienylcarbinol*, m.p. 56–57°, is obtained by the reaction between the Grignard reagent from 2-bromothiophene and thiophene-2-carbaldehyde (*Ya. L. Gol'dfarb* and *M. L. Kirmalova*, Izvest. Akad. Nauk, S.S.S.R., Otdel. Khim. Nauk, 1956, 746; C.A., 1957, **51,** 1935g). **Methyl-3-thienylcarbinol,** 1-(3-**thienylethanol**), b.p. 85°/4 mm, is obtained by reduction of 3-acetylthiophene with lithium tetrahydridoaluminate (*M. Janda* and *L. Paviensky*, Coll. Czech. chem. Comm., 1967, **32,** 2675).

**Thenyl* = $(SC_4H_3)CH_2$–

With hydrogen halides, 2-thienylcarbinol affords lachrymatory 2-**chloromethylthiophene** (2-**thenyl chloride**), b.p. 53–55°/6 mm and 2-**bromomethylthiophene** (2-**thenyl bromide**), b.p. 55°/1.5 mm; 3-*bromomethylthiophene*, b.p. 76°/1 mm, can be prepared, like its isomeride, by treating 3-methylthiophene with *N*-bromosuccinimide in the presence of dibenzoyl peroxide (*E. Campaigne* and *B. F. Tullar*, Org. Synth., 1953, **33**, 96; *Dittmer et al.*, J. Amer. chem. Soc., 1949, **71**, 1201; *S. Gronowitz*, Arkiv Kemi, 1955, **8**, 441). 2-Chloromethylthiophene may be obtained also by chloromethylation of thiophene (*N. K. Kochetkov* and *N. V. Dudykina*, Zhur. obshcheĭ Khim., 1957, **27**, 1399). Three different methods have been used for the preparation of 3-substituted halogenoalkylthiophenes; side-chain bromination of 3-methylthiophene, 3-chloromethylation of thiophene and halogen-lithium interchange (*Gronowitz*, *loc. cit.*).

2-Thenyl chlorides appear to be more active than the benzyl analogues in nucleophilic aliphatic substitution (*T. L. Cairns* and *B. C. McKusick*, J. org. Chem., 1950, **15**, 790; *K. Petterson*, Acta chem. Scand., 1950, **4**, 395). The Grignard compound obtained from 2-thenyl chloride under special conditions reacts normally with carbon dioxide, while other reagents attack the nuclear position 3 (*R. Gaertner*, J. Amer. chem. Soc., 1951, **73**, 3934). 2-Thenyl chloride by treatment with thiourea, yields 2-**thienyl-methanethiol** (2-**mercaptomethylthiophene**), b.p. 86°/15 mm (*P. Cagniant*, Compt. rend., 1949, **229**, 1342), and with liquid ammonia a mixture of primary, secondary, and a small amount of tertiary amine; the yield of primary amine is increased with a large excess of liquid ammonia (*Y. Matsuki* and *T. Sano*, Kogyo Kagaku Zasshi, 1960, **63**, 834). The chloromethylation of 2-vinylthiophene gives 2-(3-chloro-1-propenyl)-5-chloromethylthiophene (*R. Lukeš*, *Janda* and *K. Kefurt*, Coll. Czech. Chem. Comm., 1960, **25**, 1058).

Depending on the conditions a Mannich-type reaction of general application between thiophene, formaldehyde and ammonium chloride affords 2-**aminomethylthiophene** (2-**thenylamine**), b.p. 63–65°/4 mm, together with the related secondary amine, di-2-thenylamine; like the benzylamines, these amines are strong bases (*H. D. Hartough* and *S. L. Meisel*, J. Amer. chem. Soc., 1948, **70**, 4018). A similar reaction with hydroxylamine and formaldehyde gave 2-*thenylhydroxylamine*, m.p. 58–60° (*Hartough*, *ibid.*, 1947, **69**, 1355). *N,N'*-Di-(2-thenyl)-1,3-diazacyclobutane can be hydrolysed in dilute acid solution to give thiophene-2-carbaldehyde (48 %) and N-*methyl-2-thenylamine* (47 %), b.p. 184–185° (*Hartough* and *J. J. Dickert*, *ibid.*, 1949, **71**, 3922):

$$C_4H_4S + NH_4Cl + CH_2O \longrightarrow (C_4H_3S)-CH_2-N\langle {}^{CH_2}_{CH_2} \rangle N-CH_2-(C_4H_3S)$$

$$\xrightarrow{H^{\oplus}} NH_4^{\oplus} + CH_2O + (C_4H_3S)CHO + (C_4H_3S)CH_2NHMe$$

Thiophene-2-carbaldehyde affords by a Leuckart–Wallack reaction *thenyldimethylamine*, $(C_4H_3S)CH_2NMe_2$, b.p. 165–169° (*E. A. Weilmuenster*, J. org. Chem., 1952, **17**, 404). The same aldehyde gives *β-2-thienylethylamine*, $(C_4H_3S)CH_2 \cdot CH_2NH_2$, b.p.

201°, either *via* β-thienylpropionic acid (*G. Barger* and *A. P. T. Easson*, J. chem. Soc., 1938, 2100) or *via* thienylnitroethene (*R. T. Gilsdorf* and *Nord*, J. org. Chem., 1950, **15**, 807). It is preferable to reduce thienylacetonitrile with lithium tetrahydridoaluminate (*W. Herz* and *L. Tsari*, J. Amer. chem. Soc., 1955, **77**, 3529). 2-Thienylacetonitrile is best prepared by boiling 2-thenyl chloride and sodium cyanide in acetone (*K. Pettersson*, Acta chem. Scand., 1950, **4**, 395).

(xi) Aldehydes and ketones

Thiophene carbaldehydes are made by treating thiophene or its derivatives with methylformanilide and phosphorus oxychloride (*N. P. Buu-Hoï et al.*, J. chem. Soc., 1950, 2130), from thienylmagnesium halides and ethyl orthoformate (*V. du Vigneaud et al.*, J. biol. Chem., 1945, **159**, 385), or from thiophene and an alkyl dichloromethyl ether in the presence of a Friedel–Crafts catalyst (*A. Rieche, H. Gross* and *E. Höft*, Ber., 1960, **93**, 88):

Cl_2CHOMe; $SnCl_4$; CH_2Cl_2 or CS_2 → CHO 90 %

The initial product of the aminomethylation of thiophene is thiophene-2-carbaldimine, from which the aldehyde is easily obtained (p. 248).

Thiophene-2-carbaldehyde, b.p. 198° (*A. W. Weston and R. J. Michaels*, Org. Synth., 1951, **31**, 108; *Campaigne* and *W. W. L. Archer*, J. Amer. chem. Soc., 1953, **75**, 989), is also obtainable by a remarkable transformation of 1-chlorocyclopentane-2,3-dione, derived ultimately from the chlorination of phenol (*A. Hantzsch*, Ber., 1889, **22**, 2827):

H_2S → CHO

and by boiling a suspension of thiopyrylium iodide in chloroform with manganese dioxide (*I. Degani et al.*, Gazz., 1967, **97**, 397). It yields an *oxime*, m.p. 111°, *semicarbazone*, m.p. 224°, decomp., *phenylhydrazone*, m.p. 134–135°, and a hydrobenzamide analogue, and undergoes the Cannizzaro reaction. The benzoin condensation gives 2,2′-**thenoin,** $(C_4H_3S)CHOH{\cdot}CO(C_4H_3S)$, m.p. 107–108° and thence **thenil,** $[(C_4H_3S)CO]_2$, m.p. 82–84°, and thenilic acid, $(C_4H_3S)_2C(OH){\cdot}CO_2H$ (*S. Z. Cardon* and *H. P. Lankelma*, J. Amer. chem. Soc., 1948, **70**, 4248; *F. F. Nord et al.*, J. org. Chem., 1949, **14**, 184). The aldehyde condenses with nitromethane, giving *β-nitro-α-2-thienylethene*, m.p. 83° (*W. Ried* and *M. Wilk*, Ann., 1954, **590**, 111) and with acylglycines, *e.g.* cinnamoylglycine, to yield substituted (2-thenal)-5-oxazolones (*O. P. Singhal* and *P. I. Ittyerah*, J. Indian chem. Soc., 1965, **42**, 579).

Thiophene-2-carbaldehydes: 5-methyl-, b.p. 112–113°/21 mm; *5-chloro-*, b.p. 91–92°/13 mm; *5-bromo-*, b.p. 114–115°/14 mm; *3-methyl-*, b.p. 113–114°/25 mm; and 5-tert-*butyl-thiophene-2-carbaldehyde*, b.p. 134–136/25 mm, are prepared by the reaction of the corresponding thiophene with formamide, $HCONR^1R^2$ ($R^1 = H$ or alkyl and R^2 = Me or Et), and phosphorus oxychloride (*A. W. Weston*, G.P. 953.082/1956; C.A., 1959, **53,** 8163; U.S.P. 2,853,493/1958; C.A., 1959, **53,** 10251). Formylation of 3-methylthiophene with dimethylformamide and phosphorus oxychloride gives 3-methylthiophene-2-aldehyde (*Campaigne* and *Archer*, *loc. cit.*), but metallation to the sodio-compound followed by reaction with dimethylformamide yields 3-methylthiophene-5-carbaldehyde (*J. Sicé*, J. org. Chem., 1954, **19,** 70). In its chemical properties, 2,4-dichlorothiophene-2-carbaldehyde behaves as a normal aromatic aldehyde (*E. Profft* and *H. Wolf*, Ann., 1959, **628,** 96). **Junipol,** a metabolic product of the wood-rotting fungus *Daedalea junisperina*, has been synthesised (*K. E. Schulte* and *N. Jantos*, Arch. Pharm., 1959, **292,** 536):

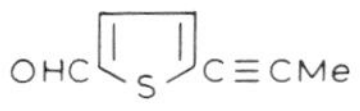

OHC–(C₄H₂S)–C≡CMe

Junipol

Side-chain bromination of 3-methylthiophene, followed by combination with hexamine and hydrolysis, gives **thiophene-3-carbaldehyde,** b.p. 195–199°; *oxime*, m.p. 113–114°; *phenylhydrazone*, m.p. 138° (*Campaigne et al.*, Org. Synth., 1953, **33,** 93). With potassium cyanide it yields 3,3′-**thenoin,** m.p. 116–117° (*Campaigne* and *W. M. Le Suer*, J. Amer. chem. Soc., 1948, **70,** 1555). By hydrolysing the hexabromo derivative of 2,5-dimethylthiophene, 3,4-dibromothiophene-2,5-dicarbaldehyde is obtained (*W. Steinkopf* and *W. Köhler*, Ann., 1937, **532,** 250). A number of thiosemicarbazones of thiophenecarbaldehydes have been prepared in connection with the study of tuberculostatic compounds.

A twofold Curtis reaction with thenylmalonic azide gives a dicarbamate hydrolysable to 2-**thienylacetaldehyde,** b.p. 52–54°/4 mm; *semicarbazone*, m.p. 131° (*C. D. Mason* and *Nord*, J. org. Chem., 1951, **16,** 1869):

$$(C_4H_3S)CH_2{\cdot}CH(CON_3)_2 \rightarrow (C_4H_3S)CH_2{\cdot}CH(NHCO_2Et)_2 \rightarrow (C_4H_3S)CH_2{\cdot}CHO$$

Acetaldehyde condenses with thiophene-2-carbaldehyde to give *β-2-thienylacraldehyde*, b.p. 55°/0.05 mm (*E. A. Braude et al.*, J. chem. Soc., 1952, 4155).

Acid chlorides or anhydrides react with thiophenes to give ketones in the presence of phosphoric acid (*H. D. Hartough* and *A. I. Kosak*, J. Amer. chem. Soc., 1947, **69,** 3093) or of Friedel–Crafts catalysts such as stannic chloride (*G. Stadnikov* and *I. Gol'dfarb*, Ber., 1928, **61,** 2341) and boron trifluoride (*M. W. Farrar* and *R. Levine*, J. Amer. chem. Soc., 1950, **72,** 3697); ketones are also obtainable from thiophenes and carboxylic acids with phosphorus pentoxide (*Hartough* and *Kosak*, *loc. cit.*). Thiophene has been acetylated with acetic anhydride in the presence of perchloric acid at 54° to yield 2-acetylthiophene (*M. Sy* and *B. de Malleray*, Bull. Soc. chim. Fr., 1963, 1276).

The above methods afford 2-**acetylthiophene,** m.p. 10–11°, b.p. 214° (*J. R. Johnson* and *G. E. May*, Org. Synth., 1938, **18,** 1); 3-*acetyl-2,5-dimethylthiophene*, b.p. 224°; 2-*benzoylthiophene*, m.p. 57° (*W. Minnis, ibid.*, 1932, **12,** 62). In the catalysed acetylation of furan and thiophene with acetic anhydride furan reacted faster by a factor of 9.3 or 11.9, the catalyst being iodine or stannous chloride, respectively (*P. Linda* and *G. Marino*, Tetrahedron, 1967, 1739). Thiophene is acylated in the α- and α,α'-positions by the Gomberg reaction and by the *N*-nitrosoacetanilide method (*A. Tundo*, Boll. Sci, Facolta Chim. Ind. Bologna, 1960, **18,** 102).

Various 2-alkylthiophenes have been obtained by the Huang–Minlon reduction of alkyl thienyl ketones prepared from, for example, thiophene, *n*-butyryl chloride and stannic chloride in benzene (*T. F. Grey, J. F. McGhie* and *W. A. Ross*, J. chem. Soc., 1960, 1502), and by reduction of the ketones over a tungsten–nickel sulphide catalyst (*P. Truitt et al.*, J. org. Chem., 1957, **22,** 1107). 2-Acetylthiophene has been converted to a mixture of 2-ethylthiophene (52%) and 2-ethylthiolane (p. 257) (26%) by reduction with synthesis gas in the presence of a cobalt catalyst (*H. Greenfield et al.*, J. org. Chem., 1958, **23,** 1054). Condensation of 2-acetylthiophene with 5-nitrofurfural in methanol in the presence of sulphuric acid gives compound XXXIV which has antiseptic and fungicidal properties (*Ciba Ltd.*, B.P. 929,450/1961; C.A., 1964, **60,** 1704g):

$$(C_4H_3S)CO\cdot CH:CH(C_4H_2O)NO_2$$

(XXXIV)

2-Benzoylthiophene is reduced to 2-benzylthiophene by lithium tetrahydridoaluminate/aluminium chloride in ether at 35° (*R. F. Nystrom et al.*, J. Amer. chem. Soc., 1958, **80,** 2896).

The chloride and nitrile of thiophene-3-carboxylic acid give with dimethylcadmium and phenylmagnesium bromide respectively 3-**acetylthiophene,** m.p. 57° (*Campaigne et al.*, J. Amer. chem. Soc., 1948, **70,** 1555; 1955, **77,** 5365) and 3-**benzoylthiophene,** m.p. 63–64°. 3-Acetylthiophene may be obtained also by reacting 3-cyanothiophene with methylmagnesium iodide (*M. Janda* and *L. Paviensky*, Coll. Czech. chem. Comm., 1967, **32,** 2675), and by converting 3-thienyl-lithium to a Grignard compound with magnesium bromide, which is then treated with acetic anhydride (*S. Gronowitz*, Arkiv Kemi, 1958, **12,** 533). By condensation with benzaldehyde, 2-acetylthiophene gives 2-*cinnamoylthiophene*, m.p. 82° (*C. Weygand* and *F. Strobelt*, Ber., 1935, **68,** 1839), and by the Mannich reaction dimethylaminoethyl 2-thienyl ketone (*G. A. Levvy* and *H. B. Nisbet*, J. chem. Soc., 1938, 1053). Acid hydrolysis (HCl) of the compound of hexamine with 2-bromoacetylthiophene yields 2-aminoacetylthiophene hydrochloride (*C. F. Huebner et al.*, J. org. Chem., 1953, **18,** 21).

The 1,2-diketone, 1,2-dioxo-1-phenyl-2-(2'-thienyl)ethane (*Steinkopf* and *G. Bokor*, Ann., 1939, **540,** 14) and 2-*thienylglyoxal* (*hydrate*, m.p. 94°) can be prepared from related oximinomonoketones, the latter also by oxidising acetylthiophene with selenium dioxide (*F. Kipnis* and *J. Ornfelt*, J. Amer. chem. Soc., 1946, **68,** 2734). The complexing agent, thenoyltrifluoroacetone, $(C_4H_3S)CO\cdot CH_2\cdot CO\cdot CF_3$, has been applied to the separation of zirconium and hafnium (*B. G. Schulz* and *E. M. Larsen*, *ibid.*, 1950, **72,** 3610).

(xii) Thiophenecarboxylic acids

(*1*) *Nuclear carboxylic acids.* The most general route to thiophene-carboxylic acids is the carbonation of thienyl Grignard compounds which affords among others **thiophene-3-carboxylic acid** (**3-thenoic acid**) (*Steinkopf* and *H. F. Schmitt*, Ann., 1938, **533,** 264), also obtainable from the 3-aldehyde (*E. Campaigne* and *W. M. Le Suer*, Org. Synth., 1953, **33,** 94) and in good yield by the oxidation of 3-methylthiophene with aqueous sodium dichromate at 225–250° (*L. Friedman, D. L. Fishel* and *H. Shechter*, J. org. Chem., 1965, **30,** 1453; see also p. 220).

Carbonation of 2-thienylsodium or the more generally applicable method of oxidising 2-acetylthiophene (*H. D. Hartough* and *L. G. Conley*, J. Amer. chem. Soc., 1947, **69,** 3096; *M. Sy* and *B. de Malleray, loc. cit*) affords **thiophene-2-carboxylic acid** (**2-thenoic acid**); its production by heating mucic acid with barium sulphide is indicative of the orientation. It may also be obtained by hydrolysis of the ester prepared by reacting thiophene with 1,2-dichloromethylenedioxybenzene (readily obtained from catechol carbonate and phosphorus pentachloride) in the presence of either aluminium or stannic chlorides (*H. Gross, J. Rusche* and *M. Mirsch*, Ber., 1963, **96,** 1382); by the carboxylation of thiophene with phosgene in the absence of a catalyst (*G. Kühnhanss, H. Reinhardt* and *J. Teubel*, J. pr. Chem., 1956, **3,** 137); by oxidation of 2-methyl- and 2,5-dimethyl-thiophene (*cf.* p. 235).

Three-dimensional refinements of the structure of thiophene-2-carboxylic acid have been reported (*P. Hudson*, Acta Cryst., 1962, **15,** 919; *M. Nardelli, G. Fava* and *G. Giraldi, ibid.*, p. 737; *Hudson* and *J. H. Roberston, ibid.*, p. 913). Properties of these thiophenecarboxylic acids and some of their functional derivatives are as follows:

Acid	*m.p.*	10^5K_a	*Et ester*	*Anhydride*	*Chloride*	*Amide*	*Nitrile*
Thiophene-2-carboxylic	129°	31.6	b.p. 218°	m.p. 62°	b.p. 208°	m.p. 180°	b.p. 196°
Thiophene-3-carboxylic	138°	7.8	b.p. 208°	m.p. 55°	b.p. 204°	m.p. 178°	b.p. 59°/3 mm

Thiophene-2-[^{14}C]carboxylic acid is prepared by treatment of 2-thienylmagnesium bromide in tetrahydrofuran with [^{14}C]carbon dioxide (*L. Pichat* and *P. Carbonnier*, C.A., 1960, **54,** 15262b).

Thiophenecarboxylic acids can be desulphurised by Raney nickel–aluminium alloy in tritiated water to give good yields of labelled acids, *e.g.*, valeric, caproic and stearic acids, for use in the study of biological processes (*N. P. Buu-Hoï*, Nature, 1957, **180,** 385). In the esterification of thiophene-2-carboxylic acid with alcohol in the presence of an inorganic acid catalyst, the addition of hydrogen peroxide reduces the reaction time and the amount of alcohol required; examples include methyl, *tert*-butyl, and 2-hydroxyethyl (*S. Togashi*, Jap. P. 24,619/1965; C.A., 1966, **64,** 5050h).

On passing the toluene-*p*-sulphonates of several 2-thienyl ketoximes down an alumina column the Beckmann rearrangement occurs to produce thiophene-2-carboxamides (*J. C. Craig* and *A. R. Naik*, J. Amer. chem. Soc., 1962, **84,** 3410). **Thiophene-2-carboxamide** may also be obtained by the following route (*R. Graf*, Ann., 1963, **661,** 111):

thiophene $\xrightarrow[Et_2O,\ C_6H_6]{OCNSO_2Cl,}$ 2-$CONHSO_2Cl$-thiophene $\xrightarrow[70°]{H_2O}$ 2-$CONH_2$-thiophene

or by reacting thiophene with carbamoyl chloride (*Kühnhanss, Reinhardt* and *Teubel*, East Ger. P. 13,495/1957; C.A., 1959, **53,** 7196), a reagent previously used by *L. Gatterman* (Ann., 1888, **224,** 29) in an application of his synthesis to the preparation of carboxylic acids from thiophene homologues (see also *Kühnhanss, Reinhardt* and *Teubel*, J. pr. Chem., 1956, **3,** 137). Saponification of the carboxamide yields the acid.

With cyanogen chloride thienylmagnesium bromide gives 2-**cyanothiophene** (*Steinkopf* and *W. Ohse*, Ann., 1925, **437,** 14), and 3-**cyanothiophene** is prepared by dehydration of thiophene-3-aldoxime (*Campaigne* and *H. L. Thomas*, J. Amer. chem. Soc., 1955, **77,** 5365). Tetrabromothiophene, by Grignard reaction and carbonation, gives 3,4,5-tribromothiophene-2-carboxylic acid, not obtainable by direct substitution (*Steinkopf et al.*, Ann., 1934, **510,** 160). By the use of mercuric acetate α-carboxyl groups can be replaced by acetoxymercuri, replaceable in its turn by hydrogen (*Steinkopf* and *W. Hanske*, *ibid.*, 1937, **532,** 236); a similar displacement occurs frequently in chlorination and sometimes in nitration.

Thiophenedicarboxylic acids may be made by oxidising dialkylthiophenes or acetylalkylthiophenes (*B. R. Baker et al.*, J. org. Chem., 1953, **18,** 138) or by the Hinsberg synthesis (p. 220). The **tricarboxylic acids** are prepared by ring-synthesis (*R. G. Jones*, J. Amer. chem. Soc., 1955, **77,** 4163), and **thiophenetetracarboxylic acid** results from the action of sulphur on ethyl acetylenedicarboxylate at 150° (*A. Michael*, Ber., 1895, **28,** 1635), or the hydrolysis of tetracyanothiophene, obtained by heating tetracyano-1,4-dithiin at its melting point (*H. E. Simmons et al.*, J. Amer. chem. Soc., 1962, **84,** 4746):

(NC)(NaS)C=C(SNa)(CN) + 2 $SOCl_2$ $\xrightarrow[0°]{(MeOCH_2)_2}$ tetracyano-1,4-dithiin $\xrightarrow{207–208°}$ tetracyanothiophene

tetracyanothiophene $\xrightarrow{HCl,\ 100°}$ thiophene-2,3,4,5-tetra(CO_2H)

tetracyanothiophene $\xrightarrow{HCl,\ 25°}$ 3,4-di(NC)-2,5-di($CONH_2$)thiophene (H_2NOC, NC / CN, $CONH_2$)

tetracyanothiophene $\xrightarrow{0.1\ N\ NaOH}$ thiophene-2,3,4,5-tetra($CONH_2$)

Table 4 gives the melting points of the methyl esters.

TABLE 4

THIOPHENE-DI-, TRI-, AND TETRA-CARBOXYLIC ACIDS

Orientation	2,3-[1]	2,4-	2,5-[2]	3,4-[3]	2,3,4-	2,3,5-	2,3,4,5-
Me ester m.p.	60°	125°	169°	60°	87°	83°	167°

References
1 *B. R. Baker et al.*, J. org. Chem., 1953, **18,** 138.
2 *J. M. Griffing* and *L. F. Salisbury*, J. Amer. chem. Soc., 1948, **70,** 3416.
3 *E. C. Kornfeld* and *R. G. Jones*, J. org. Chem., 1954, **19,** 1671.

(*2*) *Acids having the carboxyl group in the side chain.* 2-**Thienylacetic acid,** m.p. 76°, is prepared *via* the nitrile from 2-thenyl chloride (*F. F. Blicke* and *F. Leonard*, J. Amer. chem. Soc., 1946, **68,** 1934); the *amide*, m.p. 146–147°, is produced in 68% yield from 2-acetylthiophene by the Willgerodt reaction (*O. Dann* and *H. Distler*, Ber., 1951, **84,** 423). Oxidation of 2-acetylthiophene gives 2-**thienyl-glyoxylic acid**, m.p. 91.5°, best prepared by Friedel–Crafts condensation of thiophene with ethoxalyl chloride (*Blicke* and *M. U. Tsao*, J. Amer. chem. Soc., 1944, **66,** 1645). By reduction it yields 2-**thienylglycollic acid,** m.p. 115°, and 2-thienylacetic acid; 3-**thienylacetic acid,** m.p. 79–80°, is similarly prepared (*E. Campaigne* and *W. M. Le Suer*, *ibid.*, 1948, **70,** 1555). 2-*Thienylacrylic acid*, m.p. 144°, is obtained from thiophene-2-carbaldehyde by the Perkin reaction or through condensation with malonic acid (*C. Barger* and *A. P. T. Easson*, J. chem. Soc., 1938, 2100; *P. Fournari* and *J. P. Chane*, Bull. Soc. chim. Fr., 1963, 479); 2-**thienylpropiolic acid,** m.p. 130–133° (*A. J. Osbahr et al.*, J. Amer. chem. Soc., 1955, **77,** 1911). *D.* and *P. Cagniant* (Bull. Soc. chim. Fr., 1954, 1349) describe long-chain ω-2-thienyl fatty acids. The ester and nitrile of thienylacetic acid can be ethoxycarbonylated to esters of 2-thienyl-malonic and -cyanoacetic acids (*Blicke* and *Leonard*, *loc. cit.*). The Reformatsky reaction as a source of β-thienyl-β-hydroxypropionic acids has been examined by *R. D. Schuetz* and *W. H. Houff* (J. Amer. chem. Soc., 1955, **77,** 1836). *P. Cagniant* (Ann. Chim., 1952, **7,** 476) prepared 2-**thienylpyruvic acid,** m.p. 245° (decomp.):

$$(C_4H_3S)CH_2{\cdot}CN \xrightarrow{(CO_2Et)_2} (C_4H_3S)\underset{\displaystyle CO{\cdot}CO_2Et}{\underset{|}{CH}}{\cdot}CN \xrightarrow{HCl} (C_4H_3S)CH_2{\cdot}CO{\cdot}CO_2H$$

The isomeric thiophenecarbaldehydes have been converted by standard methods into 2-thienylalanine, $(C_4H_3S)CH_2{\cdot}CH(NH_2){\cdot}CO_2H$ (*V. du Vigneaud et al.*, J. biol. Chem., 1945, **159,** 385) and 3-thienylalanine (*R. Garst et al. ibid.*, 1949, **180,** 1013). Both show "anti-phenylalanine" properties in inhibiting growth of micro-organisms.

Hydroxy-acids with nuclear hydroxyl in β- and carboxyl in α-positions are obtained by Hinsberg's synthesis (p. 220); the reverse orientation is produced (*S. Mitra et al.*, J. chem. Soc., 1939, 1116) thus:

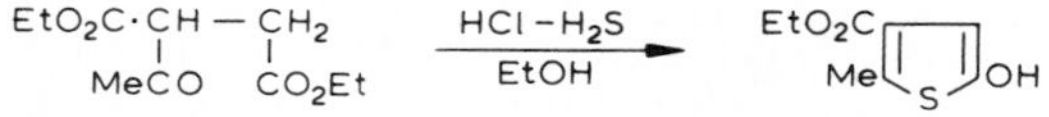

$$EtO_2C\cdot CH(COMe)-CH_2CO_2Et \xrightarrow[EtOH]{HCl-H_2S} \text{EtO}_2\text{C-, Me-, OH-thiophene}$$

Ethyl 4-hydroxy-2-methylthiophene-3-carboxylate undergoes a curious transformation into a pyrazole derivative (*E. Benary* and *L. Silberstrom*, Ber., 1919, **52**, 1605):

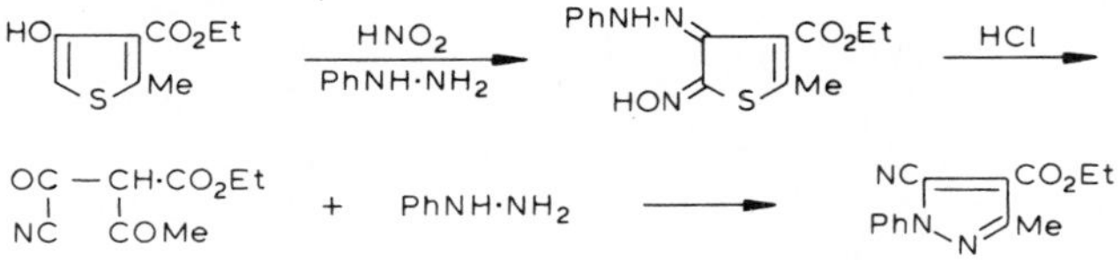

$$\xrightarrow[PhNH\cdot NH_2]{HNO_2} \quad \xrightarrow{HCl}$$

$$OC(NC)-CH(COMe)\cdot CO_2Et + PhNH\cdot NH_2 \longrightarrow$$

(b) Thiolenes; dihydrothiophenes*

Two isomeric dihydrothiophenes are possible; 2-**thiolene,** 2,3-**dihydrothiophene** (XXXV), b.p. 112°, d^{20} 1.0361, n_D^{20} 1.5312, and 3-**thiolene,** 2,5-**dihydrothiophene** (XXXVI), b.p. 122°, d^{20} 1.0591, n_D^{20} 1.5306. They are obtained, with products of ring-fission, by reducing thiophene with sodium –liquid ammonia–methanol, and each gives the corresponding *S*-dioxide (1,1-dioxide) on oxidation (*S. F. Birch* and *D. T. McAllan*, J. chem. Soc., 1951, 2556):

(XXXV) (XXXVI)

Application of the Bamford–Stevens reaction to the *p*-tolylsulphonylhydrazone of thiolan-3-one gives a mixture of 2- and 3-thiolenes, separated by vapour-phase chromatography; m.s., i.r., and n.m.r. data are reported (*M. A. Gianturco*, *P. Friedel* and *V. Flanagan*, Tetrahedron Letters, 1965, 1847):

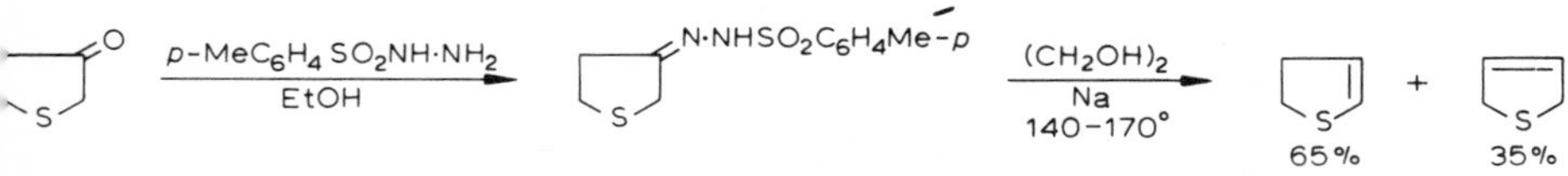

$$\xrightarrow[EtOH]{p\text{-}MeC_6H_4SO_2NH\cdot NH_2} N\cdot NHSO_2C_6H_4Me\text{-}p \xrightarrow[140-170°]{(CH_2OH)_2,\ Na} 65\% + 35\%$$

Butadiene and its derivatives combine with sulphur dioxide, commonly in ether at 100° in presence of pyrogallol, to give 3-**thiolene** 1,1-**dioxide,** m.p. 65.5°, and its derivatives, which decompose into their generators above

**Cf.* I.U.P.A.C. Rule B-1.2.

100°. 3-Thiolene 1,1-dioxide is converted by alkali, best under ultra-violet irradiation, into the more stable 2-**thiolene** 1,1-**dioxide,** m.p. 49°; the structures assigned to the isomers are confirmed by ozonisation:

HO_2C CO_2H / $H_2C{\cdot}SO_2{\cdot}CH_2$ HO_2C-CH_2 / HO_3S-CH_2

The double bond of 2-thiolene adds in alkaline solution the elements of water, alcohols, thiols, and amines, while that of the 3-isomer shows normal olefinic reactions; thus many types of substituted thiolane 1,1-dioxides are available. Both thiolene 1,1-dioxides can be hydrogenated to thiolane 1,1-dioxide (*H. J. Backer* and *J. Strating*, Rec. Trav. chim., 1934, **53,** 525; 1943, **62,** 815; *E. de Roy van Zuydewijn, ibid.*, 1937, **56,** 1047; *H. Staudinger* and *B. Ritzenthaler*, Ber., 1935, **68,** 455).

2- and 3-thiolene 1,1-dioxides undergo addition reactions with hydrazine and substituted hydrazines to give either *N,N*- or *N,N'*-substituted hydrazines, depending on the nature of the original substituent and on the reaction conditions (*C. S. Argyle et al.*, J. chem. Soc., C, 1967, 2156).

Oxidation of 2- and 3-thiolene with hydrogen peroxide give the corresponding **thiolene** 1-**oxides;** 3-thiolene 1-oxide with hydrogen bromide in carbon tetrachloride affords 3,4-dibromothiolane and with hydrogen iodide it yields 3-thiolene (*R. C. Krug* and *D. E. Boswell*, J. heterocycl. Chem., 1967, **4,** 309). The oxidation of 3,4-disubstituted 2- and 3-thiolenes with 30% hydrogen peroxide in acetic acid gives the corresponding thiophene derivatives. However, the oxidation of the same compounds with perbenzoic acid in chloroform yields the corresponding 1,1-dioxides (*T. Takaya et al.*, Bull. chem. Soc., Japan, 1968, **41,** 2086).

3,4-Disubstituted thiolenes react with iodosobenzene to give the related thiophenes (*Takaya* and *S. Hijikata*, *ibid.*, 1968, **41,** 2532).

5-Isopropyl-2-methyl-2-thiolene has been obtained from 2-methyl-2-hepten-6-one by the following route (*L. Bateman* and *R. W. Glazebrook*, J. chem. Soc., 1958, 2834):

$Me_2C{:}CH{\cdot}(CH_2)_2COMe \xrightarrow{MeCOSH} Me_2CH{\cdot}CH(SCOMe){\cdot}(CH_2)_2{\cdot}COMe \xrightarrow{EtOH,\ KOH}$

$Me_2CH{\cdot}CH(SH){\cdot}(CH_2)_2{\cdot}COMe \longrightarrow$ [ring: Me_2CH, S, Me, OH] $\longrightarrow$ [ring: Me_2CH, S, Me]

The reaction of 2,6-dimethylocta-2,6-diene with sulphur at 140° yields a number of cyclic monosulphides. Certain of the cyclic sulphide fractions which are subjected to considerable heating during distillation probably contain some **5-ethyl-2-isopropyl-5-methyl-2-thiolene** (XXXVII). However, this compound does not appear to be a primary reaction product but results from the sulphur- or polysulphide-catalysed isomerisation of XXX-VIII (*Batemen, Glazebrook* and *C. G. Moore, ibid.*, 1958, 2838, 2846):

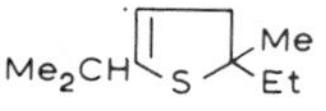

(XXXVII) 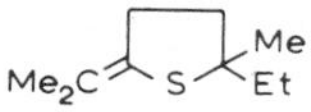 (XXXVIII)

Thiophene on treatment with hypochlorous acid affords 2,5-*di-chloro*-3-*thiolene* 1-*oxide*, m.p. 112°, along with mono- and di-chloro-thiophene, whereas alkylthiophenes give alkylchlorothiophenes (*A. Kergomard* and *S. Vincent*, Bull. Soc. chim. Fr., 1967, 2197):

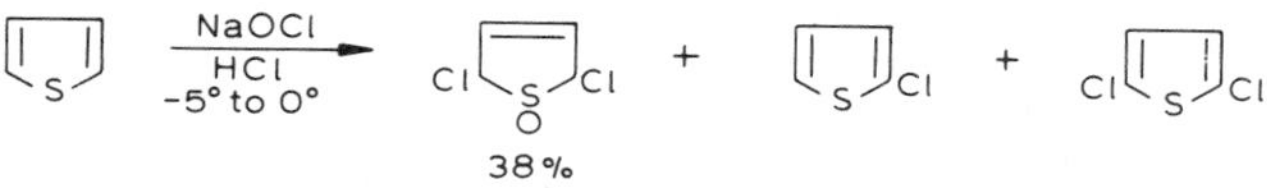

NaOCl, HCl, −5° to 0°; 38%

By chlorination in presence of iodine, thiophene gives, nearly quantitatively, hexachloro-2,5-dihydrothiophene (XXXIX) (*H. D. Hartough et al.*, J. Amer. chem Soc., 1952, **74,** 163):

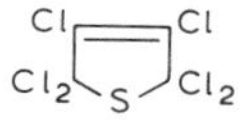

(XXXIX)

Treatment of acetylthioglycollyl chloride with diethyl sodiomalonate, followed by hydrolysis with sodium carbonate, gives ethyl 4-hydroxy-2-oxo-3-thiolene-3-carboxylate (XL), which is further decomposed by boiling water to "thiotetronic acid" (XLI) (p. 245) or a tautomeric variant, a strong monobasic acid giving an intense red ferric reaction (*Benary*, Ber., 1913, **46,** 2103):

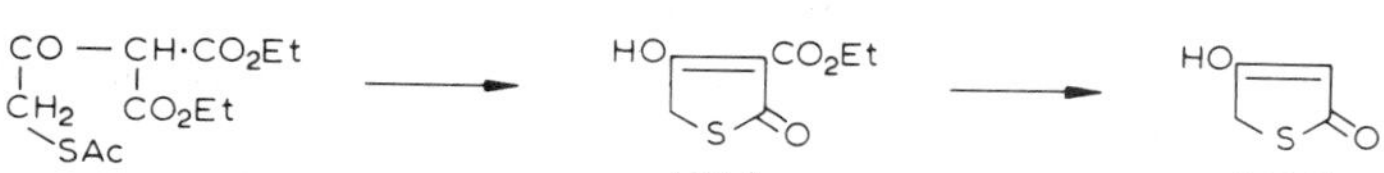

(XL) (XLI)

(c) *Thiolanes, tetrahydrothiophenes*

Though the direct hydrogenation of thiophene has been effected over palladium–charcoal (*K. Folkers et al.*, J. Amer. chem. Soc., 1945, **67,** 2092), and in iso-octane over a $CoO_3/MoO_3/Al_2O_3$ catalyst (*R. Papadopoulos* and *M. J. G. Wilson*, Chem. and Ind., 1965, 427), and thiophene and substituted thiophenes have been reduced at 180–190° with excess synthesis gas using a cobalt catalyst (*H. Greenfield et al.*, J. org. Chem., 1958, **23,** 1054), most compounds of this series have been made by ring-synthesis, the simple ones from 1,4-dihalides and alkali sulphides (*J. von Braun* and *A. Trümpler*, Ber., 1910, **43,** 545). The passage of tetramethylene glycol with hydrogen sulphide over alumina at 400° affords thiolane (*J. K. Juriev* and *N. G. Medovschtschikov*, J. gen. Chem., U.S.S.R., 1939, **9,** 628).

In appropriate cases thiolanes have been obtained by the Dieckmann condensation, most frequently in synthetic approaches to β-biotin (*S. A. Harris et al.*, J. Amer. chem. Soc., 1944, **66,** 1757; 1945, **67,** 2096):

Cysteine → $H_2NCH-CO_2H$, CH_2, $CH_2 \cdot CO_2H$, S → $BzNHCH-CO_2Me$, CH_2, $CH_2 \cdot CO_2Me$, S →(NaOMe) BzNH, O, S, CO_2Me

→($OCH \cdot (CH_2)_3 \cdot CO_2Me$) BzNH, O, S, $CH \cdot (CH_2)_3 \cdot CO_2Me$ →(reduce oxime) BzNH, NHAc, S, $(CH_2)_4 \cdot CO_2Me$

→((1) $Ba(OH)_2$ (2) H_2SO_4) H_2N, NH_2, S, $(CH_2)_4 \cdot CO_2Me$ →($COCl_2$, aq. $NaHCO_3$) O, HN, NH, S, $(CH_2)_4 \cdot CO_2H$

β- Biotin

In many cases cyclisation takes place mainly in the opposite sense (*B. R. Baker et al.*, J. org. Chem., 1947, **12,** 138, 155; *H. Fiesselmann et al.*, Ber., 1954, **87,** 835, 841):

$EtO_2C \cdot CH$ = RCH, CO_2Et, CH_2, HS → $EtO_2C \cdot CH_2$, RCH, CO_2Et, CH_2, S → EtO_2C, O, R, S

Homologues of thiolane occur in petroleum (*C. F. Mabery* and *W. O. Quayle*, Amer. chem. J., 1906, **35,** 404).

Thiolane*, tetrahydrothiophene, b.p. 121.2°, d^{20} 0.9998, n_D^{20} 1.5047, from 1,4-dibromobutane or 1,4-dichlorobutane (*J. K. Lawson, W. K. Easley* and *W. S. Wagner*, Org. Synth., 1956, **36,** 89), gives a methiodide; *mercuric chloride adduct*, m.p. 130°; 1-*oxide*, b.p. 105–107°/12 mm; 1,1-*dioxide*, m.p. 10° (*D. S. Tarbell* and *C. Weaver*, J. Amer. chem. Soc., 1941, **63,** 2939; *S. F. Birch* and *D. T. McAllan*, J. chem. Soc., 1951, 2556).

Tetrachlorothiolane is produced in stereoisomeric forms, m.p. 113° and 46°, in the

*Sometimes referred to as "thiophane".

chlorination of thiophene (p. 238), and that of 2-iodothiophene leads ultimately to **octachlorothiolane,** m.p. 215°. **Perfluorothiolane,** b.p. 40.7°, d_4^{25} 1.6339, n_D^{25} 1.3052, has been obtained by reacting hexafluoro-1,4-diiodobutane with sulphur at 250° (*G. van D. Tiers*, J. org. Chem., 1961, **26,** 2538).

Hofmann decomposition of 3-dimethylaminomethyl-4-methylenethiolane gives 3,4-dimethylthiophene rather than 3,4-dimethylenethiolane (XLII) (*C. S. Marvel, R. M. Nowak* and *J. Economy*, J. Amer. chem. Soc., 1956, **78,** 6171):

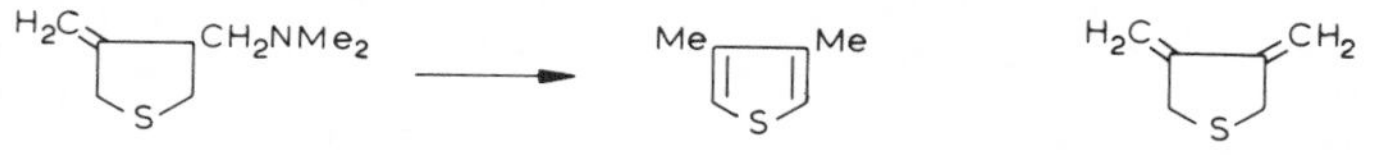

(XLII)

Treatment of *cis*- or *trans*-2,5-bishydroxymethylthiolane with hydrochloric acid results in ring expansion to afford *cis*- or *trans*-5-chloro-2-chloromethylthiane (*J. V. Černý* and *J. Hora*, Coll. Czech. chem. Comm., 1960, **25,** 711):

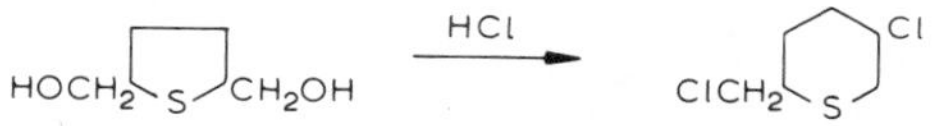

Water-soluble 3-methylnitrosoaminothiolane 1,1-dioxide (XLIII), a useful source of diazomethane, is prepared by reaction of divinyl sulphone with methylamine, followed by nitrosation (*K. T. Potts*, Chem. Reviews, 1961, **61,** 87):

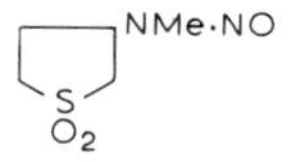

(XLIII)

Oxidation of thiolane with ammonia and sulphur in water gives succinic acid (*W. G. Toland et al.*, J. Amer. chem. Soc., 1958, **80,** 5423). Mono- or di-carboxylate ions catalyse its oxidation to thiolane 1-oxide by aqueous iodine at pH 8 and 25° (*K.-H. Gensch, I. H. Pitman* and *T. Higuchi, ibid.*, 1968, **90,** 2096). Hydrogenolysis of thiolane over sulphided cobalt molybdate–alumina gives butanethiol, which is converted to butenes and butane; some dehydrogenation occurs to give thiophene, which is also desulphurised (*P. Desikan* and *C. H. Amberg*, Canad. J. Chem., 1964, **42,** 843).

P. Karrer and *H. Schmid* (Helv., 1944, **27,** 116) have prepared 3-**oxothiolane,** b.p. 84–85°/24 mm, by a Dieckmann synthesis, and the simple ring-closure shown below leads to a 3-oxothiolane 1,1-dioxide (*W. E. Truce* and *R. H. Knospe*, J. Amer. chem. Soc., 1955, **77,** 5063):

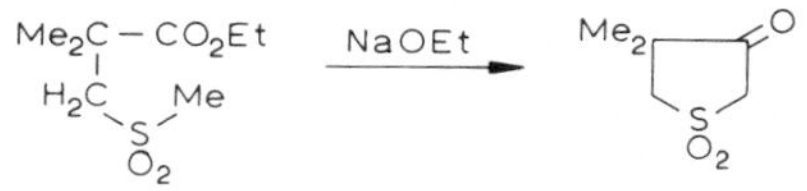

2-**Hydroxythiolane,** one of the first examples of a simple cyclic hemithioacetal with sulphur in the ring, has been synthesised as shown (*J. M. Cox* and *L. N. Owen*, J. chem. Soc., C, 1967, 1130):

$$\text{thiolane} \xrightarrow[\text{Cu}_2\text{Cl}_2,\ \text{C}_6\text{H}_6]{\text{Bu}^t\text{O}_2\text{Ac}} \text{2-acetoxythiolane} \xrightarrow[(2)\ \text{CO}_2]{(1)\ \text{Na, MeOH}} \text{2-hydroxythiolane}$$

By reduction with sodium amalgam, thiophene-2-carbaldoxime gives 2-**aminomethylthiolane,** b.p. 69.5°/10 mm (*N. I. Putokhin* and *V. S. Egorova*, J. gen. Chem., U.S.S.R., 1948, **18,** 1866), and thiophene-2-carboxylic acid gives **thiolane-2-carboxylic acid,** m.p. 53°. Racemic and *meso*-$\alpha\alpha'$-dibromoadipic acid give with sodium sulphide respectively racemic and *meso*-**thiolane-2,5-dicarboxylic acids,** m.p. 165–166° and 144–145°; the active forms melt at 179–180° (*A. Fredga*, J. pr. Chem., 1938, ii, **150,** 124). 2,5-**bis(Ethoxycarbonylamino)thiolane** (XLIV), m.p. 152–154°, is readily hydrolysed to give ammonia, hydrogen sulphide and succinaldehyde (*G. B. Brown* and *G. W. Kilmer*, J. Amer. chem. Soc., 1943, **65,** 1674):

$$\underset{\text{(XLIV)}}{\text{EtO}_2\text{CHN-C}_4\text{H}_6\text{S-NHCO}_2\text{Et}} \xrightarrow[\text{or HCl, aq. EtOH}]{\text{Ba(OH)}_2\ \text{or NaOH}} \text{OHC-CH}_2\text{-CH}_2\text{-CHO} + \text{H}_2\text{S} + \text{NH}_3$$

3,4-Disubstituted thiolanes are of interest in relation to biotin. The *cis*(?)-3,4-dicarboxylic ester is recorded by *Marvel* and *E. E. Ryder* (*ibid.*, 1955, **77,** 66). A 3,4-**dihydroxythiolane,** m.p. 54–58°, was obtained by ring-synthesis from dichlorodihydroxybutane; the corresponding stereosiomeric 3,4-**diaminothiolanes** (*benzoyl* derivatives, m.p. 295–300° and 238–239°) were prepared as follows (*V. du Vigneaud et al.*, J. biol. Chem., 1942, **145,** 495; *Kilmer* and *H. M. Kennis*, *ibid.*, 1944, **152,** 103):

$$(\text{EtO}_2\text{C})_2\text{CH}-\text{CH}(\text{CO}_2\text{Et})_2 + \text{ClCH}_2\text{SCH}_2\text{Cl} \longrightarrow \text{3,4-(HO}_2\text{C)}_2\text{-thiolane} \xrightarrow{\text{Curtius}} \text{3,4-(H}_2\text{N)}_2\text{-thiolane}$$

The early stages of the synthesis of β-biotin itself from cysteine have already been mentioned (p. 258).

(d) Compounds having two or more unfused thiophene rings

The action of cupric chloride (*W. Steinkopf* and *J. Roch*, Ann., 1930, **482,** 251) or of a small amount of cobaltous chloride in the presence of bromobenzene or butyl bromide (*J. P. Morizur*, Bull. Soc. chim. Fr., 1964, 1331) on 2-thienylmagnesium bromide gives 2,2′-**bithienyl** (**bithiophene**), m.p. 33°, also obtained from 2-iodothiophene and copper or silver, or by pyrolysis of thiophene itself (*H. Wynberg* and *A. Bantjes*, J. org. Chem., 1959, **24,** 1421). This last process yields in addition 3,3′-**bithienyl,** m.p. 132°, rationally synthesised by heating sodium butanetetracarboxylate with "phosphorus trisulphide" (*K. von Auwers* and *T. V. Bredt*, Ber., 1894, **27,** 1741):

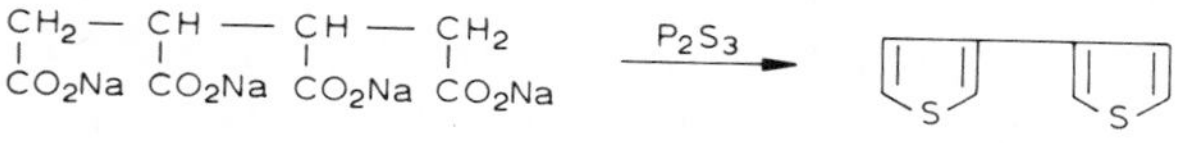

$$\underset{CO_2Na}{CH_2}-\underset{CO_2Na}{CH}-\underset{CO_2Na}{CH}-\underset{CO_2Na}{CH_2} \xrightarrow{P_2S_3}$$

The Grignard reagent from 2-iodothiophene reacts with cuprous iodide in ether to give 2-thienylcopper, which decomposes in air to 2,2′-bithienyl (*M. Nilsson*, Tetrahedron Letters, 1966, 679).

2,2′- and 3,3′-Bithienyl may be prepared by reacting 2- or 3-thienyl-lithium with an equimolar quantity of cupric chloride in ether. 3,3′-Bithienyl has been obtained from 3-thienyl-lithium and 3-oxothiolane (*S. Gronowitz* and *H. O. Karlsson*, Arkiv Kemi, 1960, **8,** 89):

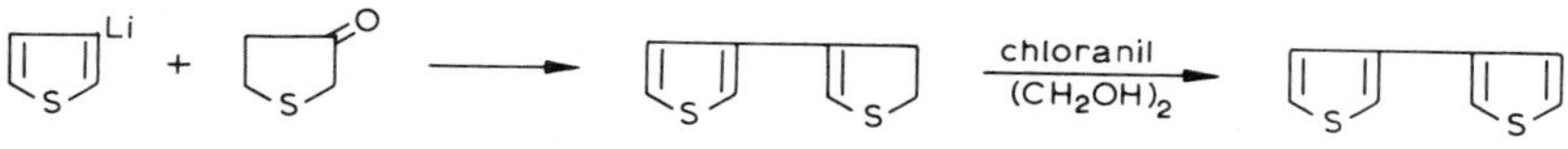

chloranil $(CH_2OH)_2$

The former method offers a route to sterically hindered 3,3′-bithienyls, thus 4-bromo-2-methylthiophene on treatment with butyl-lithium and cupric chloride at −70° yields *5,5′-dimethyl-3,3′-bithienyl*, m.p. 133–136°; boiling with excess butyl-lithium and carbonation gives *5,5′-dimethyl-3,3′-bithienyl-2,2′-dicarboxylic acid*, m.p. 254–257° (decomp.). *4,4′-Dibromo-5,5′-dimethyl-3,3′-bithienyl-2,2′-dicarboxylic acid*, m.p. 312–313° (decomp.) has been resolved (*Gronowitz* and *H. Frostling*, Tetrahedron Letters, 1961, 604; Acta Chem. Scand., 1962, **16,** 1127).

2,3′-**Bithienyl,** m.p. 65°, is produced by heating 2-(2′-thienyl)but-2-ene, $CH_3 \cdot CH{:}C(C_4H_3S) \cdot CH_3$, with sulphur (*J. Teste* and *N. Lozac'h*, Bull. Soc. chim. Fr., 1954, 492), or from 3-oxothiolane and 2-thienylmagnesium iodide (*Wynberg*, *A. Logothetis* and *D. VerPloeg*, J. Amer. chem. Soc., 1957, **79,** 1972):

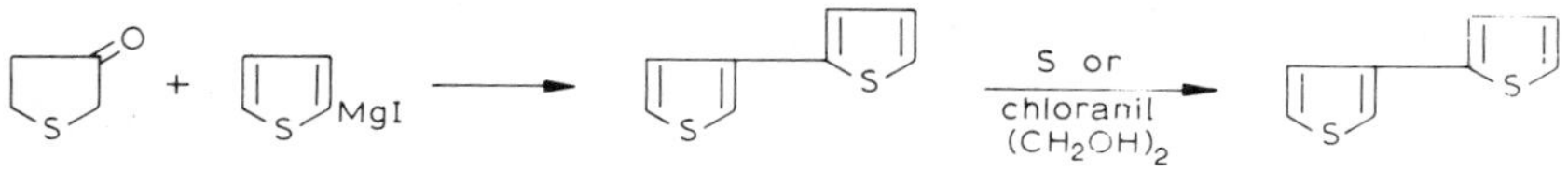

S or chloranil $(CH_2OH)_2$

When 2,2′-bithienyl is irradiated with u.v. light in benzene at 60°, 2,3′-bithienyl and 3,3′-bithienyl are formed; 2,3′-bithienyl on irradiation affords 3,3′-bithienyl (*Wynberg* and *H. van Driel*, *ibid.*, 1965, **87,** 3998):

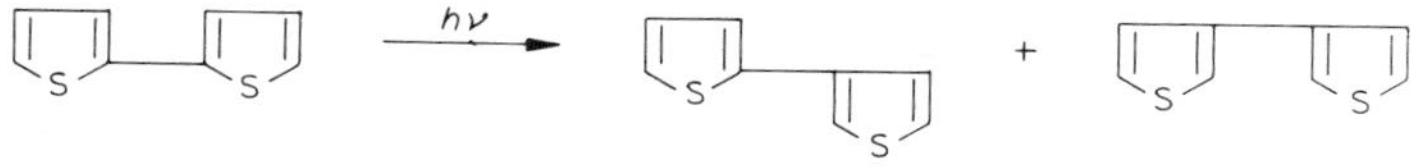

$h\nu$

3,3′-**Bithiolane,** [3,3′-bi(tetrahydrothiophene)], may be obtained from 2,2′-bis(1,4-diiodobutyl), $ICH_2 \cdot CH_2 \cdot CH(CH_2I) \cdot CH(CH_2I) \cdot CH_2 \cdot CH_2I$, and sodium sulphide in ethanol (*E. Buchta* and *K. Greiner*, Naturwiss., 1959, **46,** 532).

2,2′Bithienyl on treatment with phosphorus oxychloride and dimethylformamide in toluene gives 2,2′-**bithienyl-5-carbaldehyde,** m.p. 59°, *thiosemicarbazone*, m.p. 197°, Wolff–Kishner reduction affords 5-methyl-2,2′-bithienyl; *5′-methyl-2,2′-bithienyl-5-carbaldehyde*, m.p. 98°, 5,5′-**dimethyl**-2,2′-**bithienyl,** m.p. 68°; 5,5′-**dimethyl**-2,2′-**bithienyl-3-carbaldehyde,** m.p. 65° (*E. Lescot, N. P. Buu-Hoï* and *N. D. Xuong*, J. chem. Soc., 1959, 3243).

Methyl 2,2′-bithienyl-5-carboxylate on treatment with iodine and mercuric oxide gives **methyl** 5′-**iodo**-2,2′-**bithienyl-5-carboxylate,** m.p. 161–162°. Alkaline hydrolysis yields *5′-iodo-2,2′-bithienyl-5-carboxylic acid*, m.p. 256–258°. *Methyl 5′-bromo-2,2′-bithienyl-5-carboxylate*, m.p. 125–126°, *5′-bromo-2,2′-thienyl-5-carboxylic acid*, m.p. 261–262°, 5-**bromo**-2,2′-**bithienyl,** m.p. 30–33°; *methyl 5′-chloro-2,2′-bithienyl-5-carboxylate*, m.p. 95°, *5′-chloro-2,2′-bithienyl-5-carboxylic acid*, m.p. 270–271°, 5-**chloro**-2,2′-**bithienyl,** b.p. 55°/0.05 mm. Direct iodination of 2,2′-bithienyl with iodine and yellow mercuric oxide yields 5-**iodo**-2,2′-**bithienyl,** b.p. 150°/3.5 mm and 5,5′-**diiodo**-2,2′-**bithienyl,** m.p. 164° (*R. F. Curtis* and *G. T. Phillips*, J. chem. Soc., 1965, 5134). Attempts to resolve *3,3′-diiodo-2,2′-bithienyl-5-carboxylic acid*, m.p. 185–187°, have failed (*Gronowitz* and *L. Karlsson*, Arkiv Kemi, 1964, **22,** 119). 5-Iodo-2,2′-bithienyl reacts with cuprous cyanide or thiocyanate in pyridine to give 5-**cyano**-2,2′-**bithienyl,** m.p. 76–77°, converted on hydrolysis to *2,2′-bithienyl-5-carboxylic acid*, m.p. 183–184° (*R. E. Atkinson, Curtis* and *Phillips*, J. chem. Soc., C, 1967, 2011).

Nitration of 2,2′-bithienyl using nitric acid in acetic anhydride gives 40% 5-**nitro**-2,2′-**bithienyl,** m.p. 105–107°, and 9.5% 3-**nitro**-2,2′-**bithienyl,** m.p. 38–39°; nitration of 5-nitro-2,2′-bithienyl yields 52% 5,5′-**dinitro**-2,2′-**dithienyl,** m.p. 260°, and 24% 3,5-**dinitro**-2,2′-**dithienyl,** m.p. 185–186°; 3,5,5′-**trinitro**-2,2′-**dithienyl,** m.p. 160–161°, 3,5,3′-**trinitro**-2,2′-**dithienyl,** m.p. 157–158°, 3,5,3′,5′-**tetranitro**-2,2′-**bithienyl,** m.p. 193–194° (*C. Carpanelli* and *G. Leandri*, Ann. Chim. (Rome), 1961, **51,** 181).

Alkanoyl and alkyl 2,2′-bithienyls are prepared by reacting bithienyls with an acyl chloride in presence of stannic chloride and reduction of the ketone obtained by hydrazine hydrate (*K. E. Miller*, J. chem. Eng. Data, 1963, **8,** 605; C.A., 1964, **60,** 487). 2,2′-Bithienyl-5-carbaldehyde condenses with an aromatic or heterocyclic methyl ketone in alcohol in the presence of a few drops of 20% aqueous sodium hydroxide giving 2(5,2′-bithienyl)vinyl ketones (*A. E. Lipkin* and *N. I. Putokhin*, Khim. Geterolsikl. Soedin., 1967, 243):

S S CHO —RCOMe→ S 4 3 5 1 2 S CH:CH·COR

R = 4-biphenyl, 2-thienyl, 5-phenyl-2-thienyl, 2-naphthyl

The reaction of 2,2′-, 2,3′-, and 3,3′-dithienyl with 2 equivalents of *N*-bromosuccinimide in chloroform–acetic acid leads to the formation in virtually quantitative yield of 5,5′-dibromo-2,2′-, 2′,5-dibromo-2,3′-, and 2,2′-dibromo-3,3′-dithienyl, respectively. The bromothienyls are reduced by zinc and deuterioacetic acid to the respective deuterio compounds:

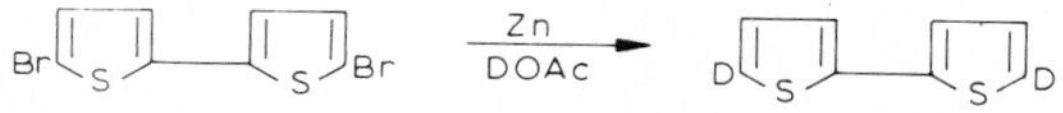

Deuterium–hydrogen exchange in boiling deuterioacetic acid occurs exclusively at the positions shown in XLV, XLVI and XLVII (*R. M. Kellog*, *A. P. Schaap* and *Wynberg*, J. org. Chem., 1969, **34,** 343):

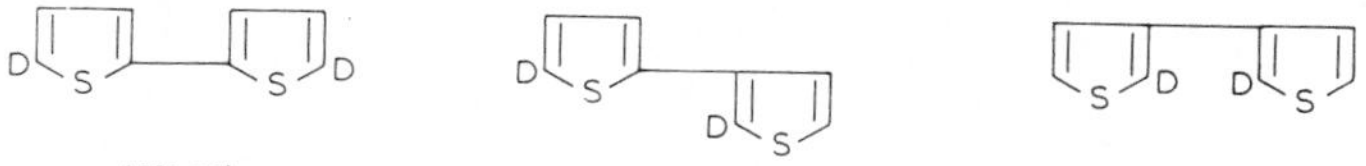

(XLV) (XLVI) (XLVII)

The acylation and metal–hydrogen interchange of 2,2′-bithienyl, 2,3′-bithienyl, and 2,2′,5′,2″-terthienyl have been studied and the structures of the resulting products determined by reductive desulphurisation to aliphatic compounds (*Wynberg* and *Bantjes*, J. Amer. chem. Soc., 1960, **82,** 1447). The results are summarised in Scheme 3.

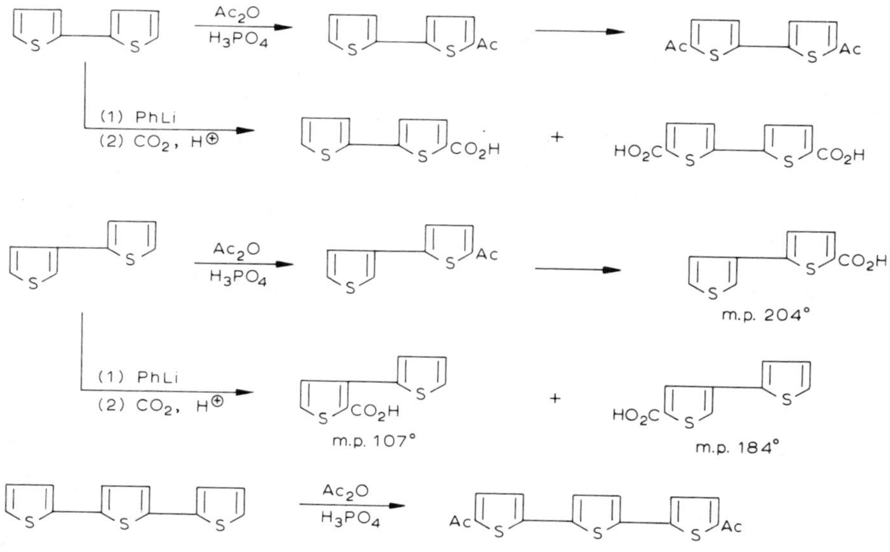

m.p. 204°

m.p. 107° m.p. 184°

Scheme 3

2,2′-Bithienyl forms a derivative, m.p. 133–134°, of undetermined constitution, with dodecacarbonyltriiron (*T. A. Manuel* and *T. J. Meyer*, Inorg. Chem., 1964, **3,** 1049; C.A., 1964, **61,** 6614f). The crystal structures of the three isomeric dithienyls have been determined by 3-dimensional X-ray methods and in all cases the molecules in the solid state are planar (*G. J.*

Vissers et al., Acta Cryst., Sect. B, 1968, **24**, 467). Electron-diffraction studies on 2,2′-bithienyl vapour show a non-planar molecule with an angle of twist of 34° (*A. Almenningen, O. Bastiansen* and *P. Svendsås*, Acta chem. Scand., 1958, **12**, 1671).

2-Thienyl-lithium affords with thenyl chloride 2,2′-**dithienylmethane,** m.p. 44–45°, and by carbonation 2,2′-**dithienyl ketone,** m.p. 90° (*N. M. Löfgren* and *G. Tegner*, *ibid.*, 1952, **6,** 1021; *R. M. Acheson et al.*, J. chem. Soc., 1956, 698).

3,3′-**Dithienyl ketone,** m.p. 72–73° (*Steinkopf* and *H. F. Schmitt*, Ann., 1938, **533,** 264), 2,3′-**dithienyl ketone,** m.p. 63° (*E. Campaigne* and *H. L. Thomas*, J. Amer. chem. Soc., 1955, **77,** 5365) and 2-**furyl** 2-**thienyl ketone** (*H. Gilman et al.*, Rec. Trav. chim., 1933, **52,** 395) are made by standard methods.

Tri-2-**thienylmethane,** m.p. 49–50°, is obtained from thiophene-2-carbaldehyde and thiophene with phosphorus pentoxide; **tri-**2-**thienylcarbinol,** from thiophene-2-carboxylic ester and thienyl-2-magnesium iodide, is unstable and gives a coloured crystalline perchlorate resembling triphenylmethyl perchlorate (*W. Schlenk* and *K. Ochs*, Ber., 1915, **48,** 676). The thienyl analogues of hexaphenylethane appear not to undergo radical dissociation (*T. L. Chu* and *T. J. Weismann*, J. Amer. chem. Soc., 1955, **77,** 2189).

Members of the series of polythienyls are produced by heating 2-iodothiophene with copper; brominated derivatives are formed by the action of sulphuric acid on bromothiophene or on bromodithienyl. 2,2′:5′,2″-**Terthienyl** (XLVIII; $n = 1$) occurs in the petals of the yellow marigold, *Tagetes erecta*, L. The members from terthienyl onwards are coloured (see Table 5) (*Steinkopf et al.*, Ann., 1941, **546,** 180; *J. W. Sease* and *L. Zechmeister*, J. Amer. chem. Soc., 1947, **69,** 270, 273):

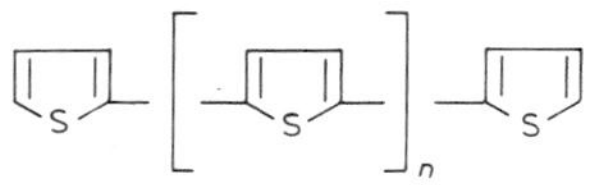

(XLVIII)

The reaction of 5-iodo-2,2′-bithienyl with cuprous acetate in pyridine unexpectedly gives 2,2′:5′,2″:5″,2‴-**quaterthienyl,** m.p. 215–216° (XLVIII; $n = 2$) in good yield (*Atkinson, Curtis* and *Phillips, loc. cit.*).

5,5″-**Diphenyl**-2,2′:5′,2″-**terthienyl,** m.p. 276° (*K. E. Schulte et al.*, Arch. Pharm., 1963, **296,** 456); 5-**methyl**-5′-(2-**thienylmethyl**)-2,2′-**bithienyl,** m.p. 42° (*Lescot, Buu-Hoï* and *Xuong, loc. cit.*).

TABLE 5

POLYTHIENYLS OF TYPE XLVIII

	$n = 0$	$n = 1$	$n = 2$	$n = 3$	$n = 4$	$n = 5$
m.p.	33°	95°	215°	253°	304°	328°
$\lambda max^{(Benzene)}(m\mu)$	305	355	391	418		
Visual colour	Colour-less	Greenish yellow	Chrome yellow	Orange	Light red	Wine red

2. Benzo[b]thiophenes (thianaphthenes)[1] and related compounds

Benzo[b]thiophene, 2,3-benzothiophene, thionaphthene, thianaphthene*, (I), present in coal tar, in various crude petroleum oils, and in shale oils, was first obtained by *L. Gattermann* and *A. Lockhart* (Ber., 1893, **26,** 2808) by treating diazotised *o*-amino-ω-chlorostyrene successively with potassium xanthate and hydroxide:

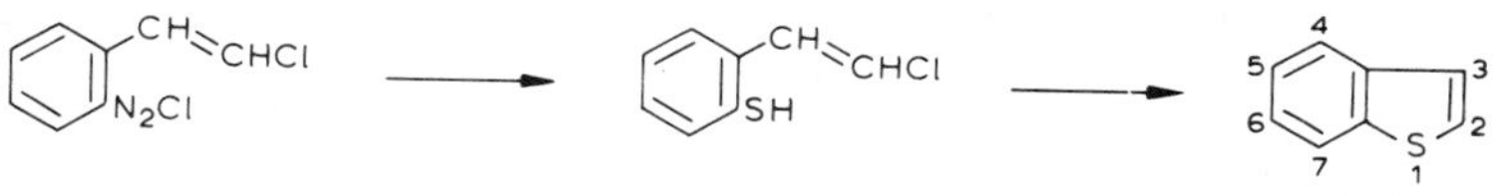

(I)

The chemical properties of benzo[b]thiophene indicate that it is best represented as a resonance hybrid of I and the charged structures II, III and IV; I being the most important and IV the least:

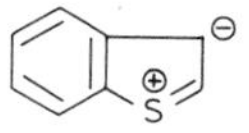

(II) 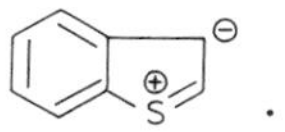 (III) 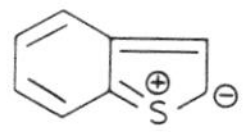 (IV)

Vat dyes of the thioindigo series are the most significant derivatives of benzothiophene, the constitution of which follows from numerous syntheses, but interest in them is declining.

* The names "thianaphthene" and "thionaphthene" are no longer accepted in I.U.P.A.C. Rules.

[1] *H. D. Hartough* and *S. L. Meisel*, "Compounds with Condensed Thiophen Rings", Wiley (Interscience), New York, 1954, p. 17; *B. Iddon* and *R. M. Scrowston*, Advances in Heterocyclic Chemistry, 1970, **11,** 177.

(a) Benzo[b]*thiophenes*

(i) Synthetic methods

Most methods involve closure of the thiophene ring in materials having the benzene nucleus already formed.

(*1*) By an unusual ring-closure between the sulphur atom and the benzene ring, α-mercaptocinnamic acid is dehydrogenated to benzothiophene-2-carboxylic acid (*E. Campaigne* and *R. S. Cine*, J. org. Chem., 1956, **21,** 39):

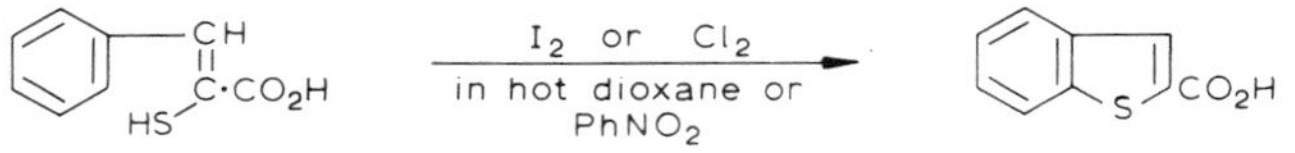

I_2 or Cl_2 in hot dioxane or $PhNO_2$

(*2*) The most important methods depend on ring-closure between $C_{(2)}$ and $C_{(3)}$.

(*A*) Reaction can be effected between the carbonyl function of an aromatic aldehyde or ketone and a reactive methylene group attached to an *ortho* sulphur atom (*F. Mayer*, Ann., 1931, **488,** 259) (R = Me, Ph, OH):

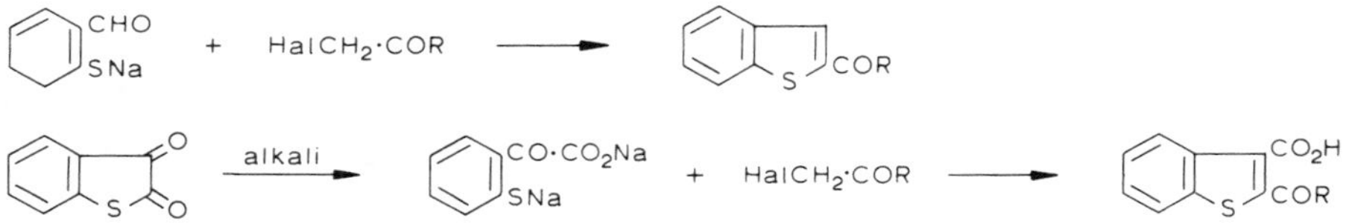

$HalCH_2{\cdot}COR$

alkali; $HalCH_2{\cdot}COR$

F. Krollpfeiffer et al. (Ann., 1950, **566,** 139) record a case in which the reactive methyl group is attached to sulphonium sulphur:

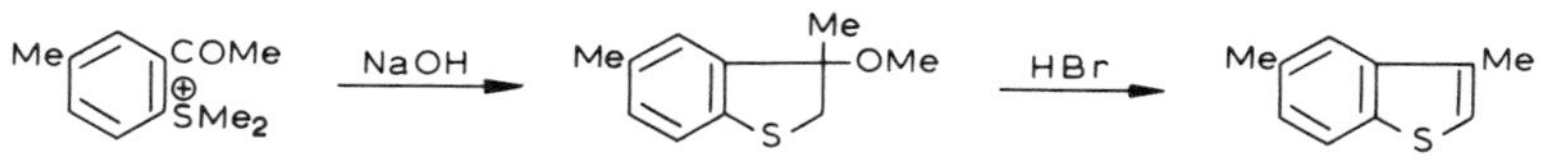

NaOH; HBr

with concomitant migration of the other *S*-methyl group. They have also used methylthio derivatives as initial materials:

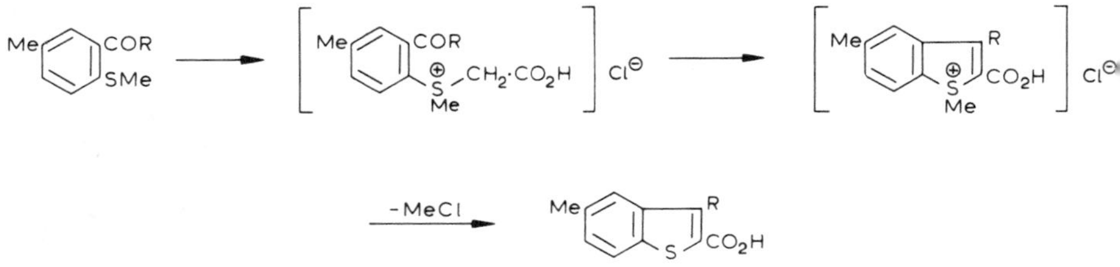

−MeCl

An increased yield of 3,5-dimethylbenzothiophene is obtained by a modification of Krollpfeiffer's method, and use of hydrogen bromide in the final stage (*R. A. Guerra*, C.A., 1965, **63,** 5581g).

(*B*) Similar ring-closures through a carboxyl function lead to technically important derivatives of 3-hydroxybenzothiophene (more correctly called 3-oxo-2,3-dihydrobenzothiophene, p. 281). Diazotised anthranilic acid affords *o*-carboxyphenylthioglycollic acid (V), either directly with thioglycollic acid or by way of *o*-mercaptobenzoic acid and its reaction with chloroacetic acid. By fusion with alkali at 150°, V yields 3-hydroxybenzothiophene-2-carboxylic acid, while heating with acetic anhydride and sodium acetate gives 3-acetoxybenzothiophene (*P. Friedländer*, Ann., 1907, **351,** 390; *C. Hansch* and *H. G. Lindwall*, J. org. Chem., 1945, **10,** 381):

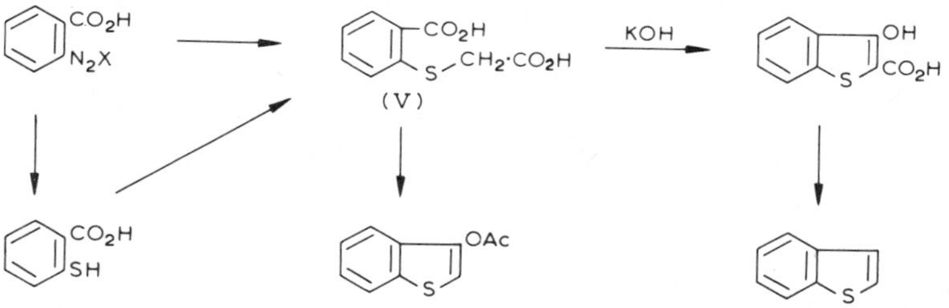

CO_2H N_2X → CO_2H $S{-}CH_2{\cdot}CO_2H$ (V) $\xrightarrow{KOH}$ OH CO_2H S

CO_2H SH   OAc S   S

(*C*) From an arylamine salt *via* the *o*-aminoarenethiol by the Herz reaction (*cf. L. Cassella and Co.*, B.P. 17,417/1914; G.P. 360,690/1919; Friedl., 1926, **14,** 908; *R. Herz*, U.S.P. 1,243,171/1917; J. Soc. chem. Ind., 1921, **40,** 619A). Arylamine salts (hydrochlorides) react with sulphur chloride to give benzo-1,3-thiaza-2-thionium chlorides (VI) (accompanied by nuclear chlorination when the *para* position to the amino group is unsubstituted), which produce *o*-aminoarenethiols on careful hydrolysis. Treatment of the latter with sodium chloroacetate followed by diazotisation and replacement of the diazo group by cyano, yield nitriles, which on reductive alkaline hydrolysis afford benzo[*b*]thiophene-(thioindoxyl)-2-carboxylic acids, the penultimate intermediates in the manufacture of thioindigoid dyes as shown below for Vat Red I (C.I. 73,360):

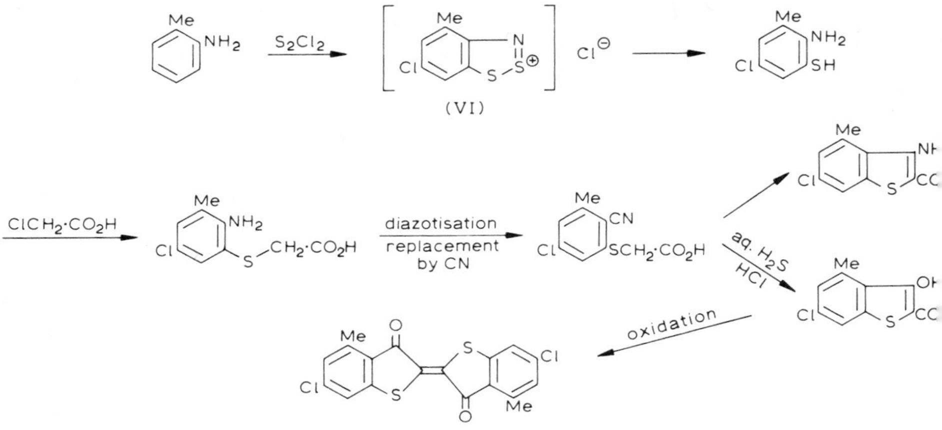

(*D*) From thiosalicylic acid and 1,3-diketones; *e.g.* acetylacetone and benzoylacetone react with thiosalicylic acid in concentrated sulphuric acid to form 3-hydroxybenzothiophene (thioindoxyl) (*S. Smiles* and *B. N. Ghosh*, J. chem. Soc., 1915, **107,** 1377) and 2-benzoyl-3-hydroxybenzothiophene (2-benzoylthioindoxyl) (*Hassle, Apotekare P. Nordstrom Fabrik A.B.*, B.P. 1,101,946/1965; C.A., 1969, **70,** 87301r), respectively:

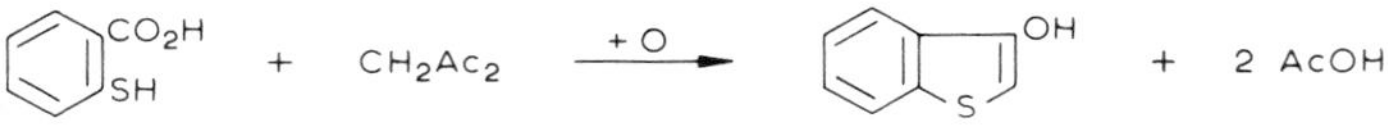

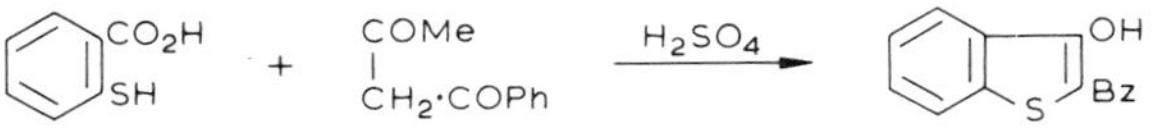

(*3*) The ring can also be closed between $C_{(3)}$ and the benzene nucleus. Thus acidic dehydrating agents convert arylthioacetaldehyde dialkylacetals, $ArSCH_2{\cdot}CH(OR)_2$, into benzothiophenes (*A. V. Sunthankar et al.*, C.A., 1952, **46,** 4524; 1954, **48,** 10725; *W. Davies et al.*, J. chem. Soc., 1956, 2603; *B. D. Tilak*, Tetrahedron, 1960, 76):

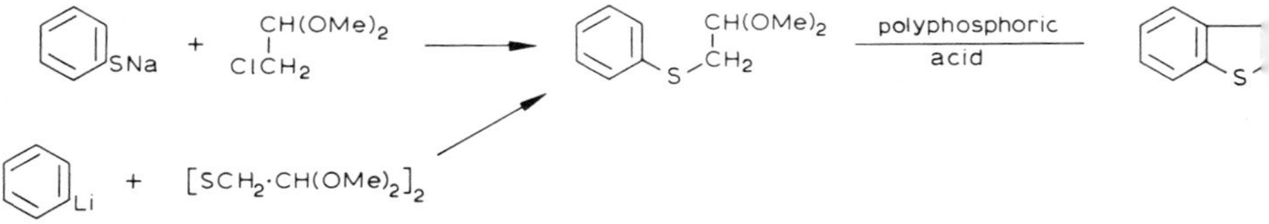

In some cases arylthioalkyl ketones can be cyclised in good yield (*E. G. G. Werner*, Rec. Trav. chim., 1949, **68,** 509):

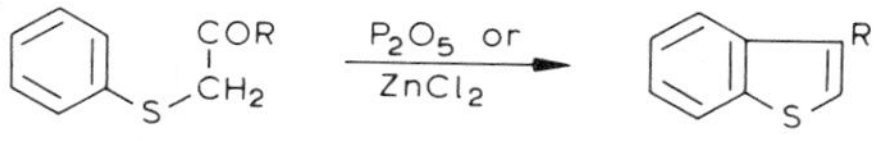

Aryl phenacyl sulphides (and 3-phenylbenzothiophene) in hydrogen fluoride at room temperature yield, with phenyl migration, 2-phenylbenzothiophenes (*O. Dann* and *M. Kokorudz*, Ber., 1958, **91,** 172).

(*4*) Benzothiophenes may be obtained by treating the product VII, from dimethyl acetylenedicarboxylate and methyl thiosalicylate, with methylamine; ammonia causes cyclisation only in the case of the *cis* isomers. The original product VII is the *trans* isomer, which is thermally isomerised to the corresponding *cis* isomer (*N. D. Heindel et al.*, J. org. Chem., 1967, **32,** 2678):

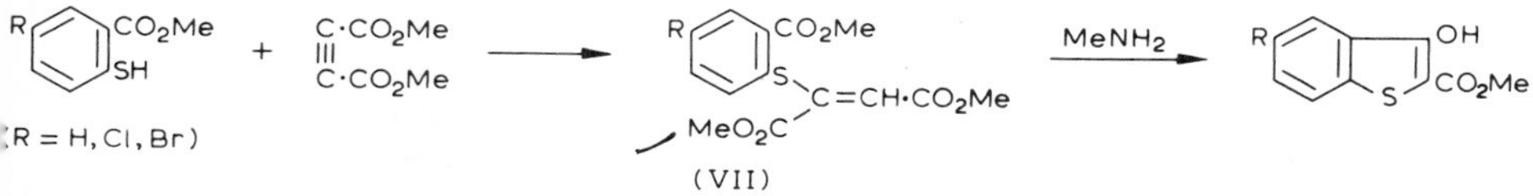

(R = H, Cl, Br)

(VII)

(5) In exceptional cases the benzene ring has been built on to an existent thiophene nucleus. 2-(Butenyl)thiophene yields 95% of benzothiophene when passed over phosphorous pentoxide on kieselguhr at 600° (*K. L. Kreutz*, U.S.P. 2,686,185/1954; C.A., 1955, **49,** 11019c); butadiene combines with 2,3-dihydrothiophene 1,1-dioxide (*K. Alder et al.*, Ber., 1938, **71,** 2451):

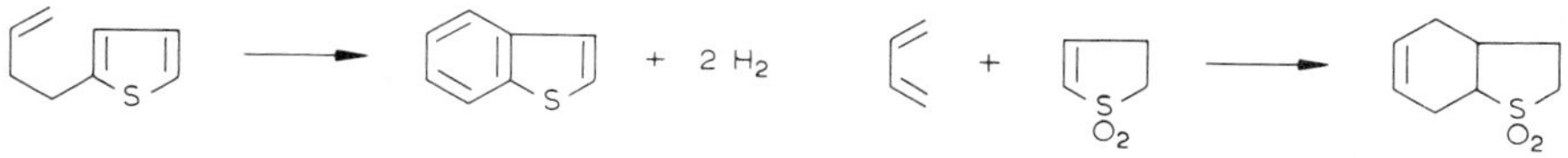

The adduct of 2-vinylthiophene with maleic anhydride on hydrolysis and dehydrogenation gives benzothiophene-4,5-dicarboxylic acid anhydride, whereas dehydrogenation, followed by hydrolysis and decarboxylation yields benzothiophene (*J. F. Scully* and *E. V. Brown*, J. Amer. chem. Soc., 1953, **75,** 6329; *Davies* and *Q. N. Porter*, J. chem. Soc., 1957, 4958):

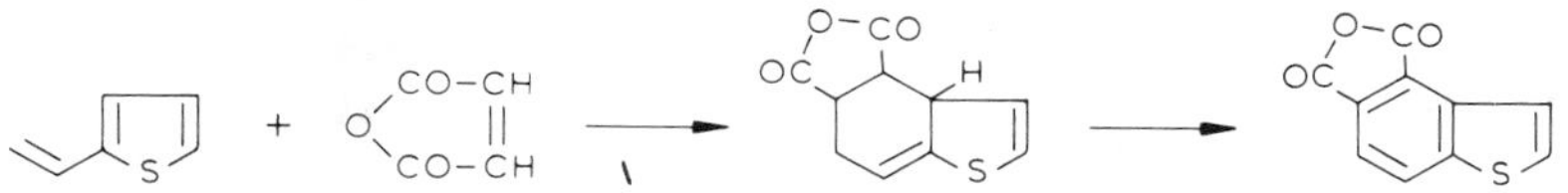

(ii) General properties and reactions

(*1*) *Pyrolysis.* The stability of the benzothiophene system is shown by its formation in quantity by pyrolytic processes. Benzothiophene is much more stable and less reactive than thiophene.

(*2*) *Oxidation* with hydrogen peroxide gives the 1,1-dioxide, in which the thiophene nucleus is no longer "aromatic"; hydrogen bromide, alcohols in the presence of alkali, thiophenols, and malonic ester all add to the 2,3 double bond (*F. Challenger* and *P. H. Clapham*, J. chem. Soc., 1948, 1615; *F. G. Bordwell et al.*, J. Amer. chem. Soc., 1950, **72,** 1985; 1954, **76,** 3637). The primary products of the degradation of benzothiophene with ozone are *o*-mercapto-benzaldehyde and -benzoic acid (*A. v. Wacek et al.*, Ber., 1940, **73,** 521).

(*3*) The *reduction* of benzothiophene with sodium and alcohol affords 2,3-dihydrobenzthiophene (*S. F. Birch*, J. Inst. Pet., 1954, **40,** 76) and with sodium in liquid ammonia it yields 2-ethylbenzenethiol (*W. Hückel* and *I. Nabih*, Ber., 1956, **89,** 2115; *M. Nakazaki*, Nippon Kagaku Zasshi, 1959, **80,** 687). Hydrogenation first affects the thiophene ring; and treatment with Raney nickel leads to desulphurisation with production of ethylbenzene (*F. F. Blicke* and *D. G. Sheets*, J. Amer. chem. Soc., 1949, **71,** 4010):

Na, EtOH; Raney Ni; Na, liq. NH_3

(*4*) *Substitution.* Direct nitration, bromination, Friedel–Crafts acylation, and mercuration (*G. Komppa*, J. pr. Chem., [ii], 1929, **122,** 319; *W. König* and *G. Hamprecht*, Ber., 1930, **63,** 1546; *Challenger* and *S. A. Miller*, J. chem. Soc., 1939, 1005) take place predominantly in position 3, as shown by reference to 3-hydroxybenzothiophene, oriented by synthesis:

$(C_8H_5S)NO_2 \rightarrow (C_8H_5S)NH_2 \rightleftharpoons (C_8H_5S)OH(3)$

$C_8H_6S \rightarrow (C_8H_5S)Br \rightarrow (C_8H_5S)MgBr \rightarrow (C_8H_5S)CHO$

$C_8H_6S \rightarrow (C_8H_5S)COMe \rightarrow (C_8H_5S)CO_2H$

$C_8H_6S \rightarrow (C_8H_5S)HgOAc \rightarrow (C_8H_5S)CO_2Et$

With sulphonation the position is uncertain and the sole product is either 2- or 3-sulphonic acid (p. 277). Although acylation gives mainly the 3-isomer, in the particular case of Friedel–Crafts acetylation, the products contain 12 to 33% of the 2-isomer, depending on conditions (*M. W. Farrar* and *R. Levine*, J. Amer. chem. Soc., 1950, **72,** 4433). The isopropylation of benzothiophene with several reagents using various methods and catalysts gives a mixture of the 2- and 3-substituted products in which the 2-isomer predominates (*S. F. Bedell, E. C. Spaeth* and *J. M. Bobbitt*, J. org. Chem., 1962, **27,** 2026). Sodamide, metallic sodium and butyl-lithium metallate benzothiophene in position 2 (*R. Weissgerber* and *O. Kruber*, Ber., 1920, **53,** 1551; *D. A. Shirley* and *M. D. Cameron*, J. Amer. chem. Soc., 1952, **74,** 664).

Benzothiophenes substituted in position 5 by hydroxyl or amino functions are further substituted first in position 4, showing the "naphthoid" character of the benzothiophene system (*Bordwell* and *H. Stange*, *ibid.*, 1955, **77,** 5939). Benzothiophene 1,1-dioxide is nitrated in position 6, *para* in the benzene nucleus to the vinyl substituent constituted by $C_{(3)}$:$C_{(2)}$ (*Challenger* and *Clapham*, *loc. cit.*).

(iii) Benzo[b]*thiophene*

Benzo[*b*]thiophene, m.p. 32°, b.p. 221–222°, d_4^{36} 1.1486; *picrate*, m.p. 149°; *trinitrobenzene adduct*, m.p. 148–150° (*P. Cagniant*, Bull. Soc. chim. Fr., 1949, 382), smells like naphthalene. It is prepared by reduction of thioindoxyl (p. 281) or thioindoxylcarboxylic acid (*Friedländer et al.*, Ber., 1908, **41,** 227; *Hansch* and *Lindwall*, J. org. Chem., 1945, **10,** 381), and by oxidation of *o*-mercaptocinnamic acid with ferricyanide. It is produced in good yield by passing styrene with hydrogen sulphide over ferrous sulphide–alumina at 600° (*R. J. Moore* and *B. S. Greensfelder*, J. Amer. chem. Soc., 1947, **69,** 2008; *V. M. Kravchenko* and *I. S. Pastukhova*, Dokl. Akad. Nauk, S.S.S.R., 1958, **119,** 285) or from benzenethiol and acetylene at 400–800° (*T. Me*, C.A., 1954, **48,** 731). Benzothiophene is also obtained from acetophenone and hydrogen sulphide by passage over chromium–aluminium oxides at 550°; diphenylthiophene is a major by-product at 400° (*P. B. Venuto, P. S. Landis* and *D. E. Boswell*, Ind. Eng. Chem. Prod. Res. Develop., 1968, **7,** 44); from benzenethiol and ethylene (*T. Lesiak*, Rocz. Chem., 1965, **39,** 757) or acetylene (*idem, ibid.*, 1964, **38,** 1923; *S. Horie*, Nippon Kagaku Zasshi, 1957, **78,** 1171) in the presence of a catalyst; and by the cyclodehydrogenation of 2-ethylbenzenethiol over molybdenum sulphide–alumina, ferrous sulphide–alumina or copper chromite on carbon at 380–480° (*Lesiak*, Rocz. Chem., 1965, **39,** 681). To isolate benzothiophene from crude coal-tar naphthalene, the latter is heated with a little sulphuric acid; the enriched benzothiophene naphthalene mixture recovered from the sulphonic acids is then heated at 100–120° with sodium or sodamide, giving sodio-benzothiophene which is decomposed by water (*Weissgerber* and *Kruber*, *loc. cit.*).

Benzothiophene on treatment with hydrogen peroxide affords **benzo[*b*]thiophene**

1,1-**dioxide,** m.p. 142–143°, which gives on irradiation a *dimer*, m.p. 330–331°. When heated in high-boiling solvents, the monomer or the dimer gives the tetracyclic product VIII dehydrogenated to IX; the alternative system X results by heating the dry dioxide (*Davies et al.*, J. chem. Soc., 1952, 4678; 1955, 315, 1565). The kinetics of the above self-condensation in various solvents have been studied (*W. F. Taylor* and *T. J. Wallace*, Tetrahedron, 1968, 5081):

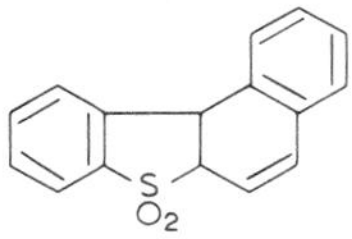

S O2 (VIII) S O2 (IX) S O2 (X)

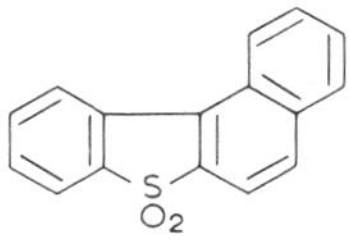

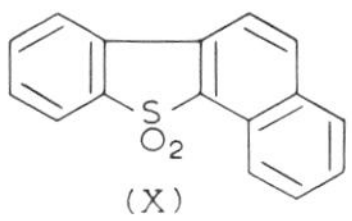

Benzothiophene 1,1-dioxide forms 2,3-adducts with butadiene (*O. C. Elmer*, U.S.P. 2,664,426/1953; C.A., 1955, **49,** 1106), dimethyl acetylenedicarboxylate (*E. W. Duck*, C.A., 1956, **50,** 9376) and 1-vinylnaphthalene (*Davies* and *Porter*, J. chem. Soc., 1957, 459), but not with maleic anhydride and diethyl acetylenedicarboxylate.

It is interesting that in the base-catalysed addition of arenethiols (thiophenols) to the 1,1-dioxide, the new sulphur atom attaches itself to $C_{(3)}$, whereas the peroxide-promoted addition takes place in the opposite sense (*Bordwell et al.*, J. Amer. chem. Soc., 1954, **76,** 3637).

(iv) Alkyl- and aryl-benzothiophenes

Homologues of benzothiophene are usually prepared by ring-synthesis. Isopropylation of benzothiophene gives a mixture of 2- and 3-isopropylbenzothiophenes (p. 271) whereas it is believed that alkylation with isobutene in the presence of either sulphuric acid or phosphoric acid yields only 3-*tert*-butylbenzothiophene (*R. E. Conary* and *R. F. McCleary*, U.S.P. 2,652,405/1953; C.A., 1954, **48,** 121786; *B. B. Corson et al.*, J. org. Chem., 1956, **21,** 584), and with *tert*-amyl alcohol in the presence or stannic chloride only 3-*tert*-amylbenzothiophene (*A. C. Cope* and *W. D. Burrows*, *ibid.*, 1966, **31,** 3093).

2-**Methylbenzothiophene,** m.p. 51–52°, is prepared from 2-benzothienyl-lithium (*Shirley* and *Cameron*, *loc. cit.)* and by ring-synthesis (*M. Pailer* and *E. Romberger*, Monatsh., 1960, **91,** 1070); 3-*methylbenzothiophene*, b.p. 112–113°/10 mm (*A. Ricci*, Ann. chim., Rome, 1953, **43,** 323; C.A., 1954, **48,** 10725d; *N. B. Chapman*, *K. Clarke* and *B. Iddon*, J. chem. Soc., 1965, 774); 5-*methylbenzothiophene*, 37–38° (*A. V. Sunthankar* and *B. D. Tilak*, Proc. Indian Acad. Sci., 1950, **32A,** 396; *Chapman et al.*, J. chem. Soc., C, 1968, 514); 6-*methylbenzothiophene*, m.p. 42–42.5° (*D. S. Tarbell* and *D. K. Fukushima*, J. Amer. chem. Soc., 1946, **68,** 1456; *Tilak*, Proc. Indian Acad. Sci., 1950, **32A,** 390); 2,3-**dimethylbenzothiophene,** b.p. 96–98°/1.5 mm (*Bordwell* and *T. W. Cutshall*, J. org. Chem., 1964, **29,** 2020); 2,5-*dimethylbenzothiophene*, m.p. 52°; 2,6-*dimethylbenzothiophene*, m.p. 62°; 2,7-*dimethylbenzothiophene*, b.p. 132–136°/11 mm; 2-**ethylbenzothiophene,** b.p. 118–125°/10 mm *(Pailer* and *Romberger*, *loc. cit.);* 3-*ethylbenzothio-*

phene, b.p. 143°/24 mm (*G. D. Gal'pern, I. U. Numanov* and *I. M. Nasyrov*, Dokl. Akad. Nauk. Tadzh. SSR, 1964, **7,** 34; C.A., 1965, **62,** 6450h; *N. P. Buu-Hoï* and *Cagniant*, Ber., 1943, **76,** 1269); 2-**isopropylbenzothiophene,** b.p. 72°/0.5 mm; 3-*isopropylbenzothiophene*, b.p. 69–70°/0.1 mm (*Bedell, Spaeth* and *Bobbitt, loc. cit.;* 3-*tert*-**butylbenzothiophene,** b.p. 149°/20 mm (*Corson et al., loc. cit.*). 2-**Phenylbenzothiophene,** m.p. 176°, results from pyrolysis of dibenzyl sulphide or disulphide (*A. W. Horton*, J. org. Chem., 1949, **14,** 761) and from the reaction between 2-benzothienyl-lithium and fluorobenzene (*S. Middleton et al.*, J. chem. Soc., 1956, 4791; Australian J. Chem., 1959, **12,** 218); 3-*phenylbenzothiophene*, b.p. 148–152°/1.75 mm (*O. Dann* and *M. Kokurudz*, Ber., 1958, **91,** 172; *D. S. Rao* and *Tilak*, J. Sci. Ind. Res., India, 1959, **18B,** 77) may be prepared by the reaction between thioindoxyl and phenylmagnesium bromide (*T. S. Murthy* and *Tilak*, *ibid.*, 1960, **19B,** 395). Treatment with hydrofluoric acid converts 3-phenylbenzothiophene into the 2-isomer (*Dann* and *Kokorudz, loc. cit.*). 2,3-Diarylbenzothiophenes may be obtained by cyclisation in the presence of boron trifluoride of the adducts of arylsulphenyl 2,4,6-trinitrobenzenesulphonates and tolane or 4,4'-dimethyltolane (*G. Capozzi et al.*, Tetrahedron Letters, 1968, 4039).

2-*Benzothienyltriphenylsilicane*, $C_8H_5S{\cdot}SiPh_3$, m.p. 148–149° (*R. H. Meen* and *H. Gilman*, J. org. Chem., 1955, **20,** 73).

(v) *Halogenobenzothiophenes*

Direct bromination of benzothiophene under carefully controlled conditions affords an excellent yield of 3-**bromobenzothiophene** (92%), b.p. 98°/0.3 mm (*W. H. Cherry et al.*, Australian J. Chem., 1967, **20,** 313); excess bromine gives 2,3-*dibromobenzothiophene*, m.p. 59° (*W. Reid* and *H. Bender*, Ber., 1955, **88,** 34; *Y. Matsuki* and *K. Fujieda*, Nippon Kagaku Zasshi, 1967, **88,** 1193); 2-*bromobenzothiophene*, m.p. 46° (*Matsuki* and *Y. Adachi, ibid.*, 1968, **89,** 192). Chlorination yields 3-**chlorobenzothiophene** (69%), b.p. 111–113°/10 mm, 2,3-*dichlorobenzothiophene*, m.p. 56° (*A. H. Schlesinger* and *D. T. Mowry*, J. Amer. chem. Soc., 1951, **73,** 2614) and a small amount of 2-*chlorobenzothiophene*, m.p. 34° (*G. van Zyl et al.*, Canad. J. chem., 1966, **44,** 2283); excess chlorine gives 2,3-dichlorobenzothiophene. 2-**Iodobenzothiophene,** m.p. 63.5–64°, is prepared from 2-benzothienyl-lithium and iodine, and 3-*iodobenzothiophene*, b.p. 120–121°/1.6 mm, by iodination of benzothiophene in benzene in the presence of mercuric oxide (*R. Gaertner*, J. Amer. chem. Soc., 1952, **74,** 4950). 2-**Fluorobenzothiophene,** b.p. 93–94°/25 mm, is prepared by the reaction of perchloryl fluoride on benzothienyllithium (*R. D. Schuetz et al.*, J. org. Chem., 1963, **28,** 1420). Halogenobenzothiophenes may be obtained from the appropriate diazonium salts (*Chapman, Clarke* and *S. D. Saraf*, J. chem. Soc., C, 1967, 731).

3-**Bromobenzothiophene** 1,1-**dioxide,** m.p. 184°, exchanges its bromine readily with secondary amines, and reluctantly with silver nitrate; 2-*bromobenzothiophene* 1,1-*dioxide*, m.p. 150–151°, obtained from benzothiophene 1,1-dioxide by successive addition of bromine and removal or hydrogen bromide, behaves similarly (*Bordwell et al.*, J. Amer. chem. Soc., 1948, **70,** 1558; 1949, **71,** 1702).

Bromo- and iodo-benzothiophenes readily form Grignard reagents; chlorobenzothiophenes are converted on heating with cuprous cyanide into the corresponding nitriles (*P. Cagniant, P. Faller* and *D. Cagniant*, Bull. Soc. chim. Fr., 1966, 3055);

2-bromobenzothiophene with sodium methoxide in the presence of cupric oxide and potassium iodide affords 2-methoxybenzothiophene *(Matsuki* and *Adachi, loc. cit.);* and the halogen atom in halogenobenzothiophenes can be replaced by deuterium by reacting the Grignard reagent with deuterium oxide, or by reducing the bromobenzothiophene with zinc and deuterated acetic acid (*B. Caddy et al.*, Australian J. Chem., 1968, **21,** 1853; *K. Takahashi, I. Ito* and *Matsuki*, Bull. chem. Soc., Japan, 1966, **39,** 2316).

Dehalogenation of 2-halogenobenzothiophenes may be effected by hydrogen in the presence of palladium, or sodium amalgam in methanol (*Matsuki* and *Ito*, Nippon Kagaku Zasshi, 1967, **88,** 751). The former method does not affect the 3-halogeno compounds, but the halogen atom may be removed by the latter method, by Raney nickel and hydrazine hydrate (*M. Martin-Smith* and *S. T. Reid*, J. chem. Soc., 1960, 938), and by hydriodic acid (*A. H. Lamberton* and *J. E. Thorpe, ibid.*, C, 1967, 2571).

(vi) Nitrobenzothiophenes

Nitration of benzothiophene gives predominantly 3-**nitrobenzothiophene,** m.p. 81° (*R. Zahradník et al.*, Coll. Czech. chem. Comm., 1963, **28,** 776; *D. E. Boswell et al.*, J. heterocycl. Chem., 1968, **5,** 69); with nitric acid in acetic anhydride at 25° the only product isolated was the 3-isomer, but with nitric acid in acetic acid at 10° besides the 3-isomer, the 4- and 6-isomers are also obtained. 2-Methylbenzothiophene on nitration yields *2-methyl-3-nitrobenzothiophene*, m.p. 98–98.5° (*D. A. Shirley, M. J. Danzig* and *F. C. Canter*, J. Amer. chem. Soc., 1953, **75,** 3278). If an electron-withdrawing group is present in the 2-position then nitration occurs mainly in the benzene ring (*V. P. Mamaev* and *O. P. Shkurko*, Khim. Geterotsikl. Soedin., Akad. Nauk Latv., S.S.S.R., 1965, 516; C.A., 1966, **64,** 675; *Van Zyl et al.*, Canad. J. Chem., 1966, **44,** 2283); an electron-donating group in the 3-position gives the 2-nitro compound *(Shirley, Danzig* and *Canter, loc. cit.; Ricci, loc. cit.)*, and an electron-withdrawing group in the 3-position affords a mixture of products possessing the nitro group in the benzene ring.

2-**Nitrobenzothiophene** may be obtained either by the debromination of 3-bromo-2-nitrobenzothiophene or by decarboxylation of 2-nitrobenzothiophene-3-carboxylic acid *(Van Zyl, loc. cit.; Boswell et al., loc. cit.)*. 2,3-Dimethylbenzothiophene reacts with acetyl nitrate to give *3-methyl-2-nitromethylbenzothiophene*, m.p. 103–104°, some *2,3-dimethyl-6-nitrobenzothiophene*, m.p. 124–125° and a small amount of 3-methylbenzothiophene-2-carbaldehyde (*Bordwell* and *T. W. Cutshall*, J. org. Chem., 1964, **29,** 2020):

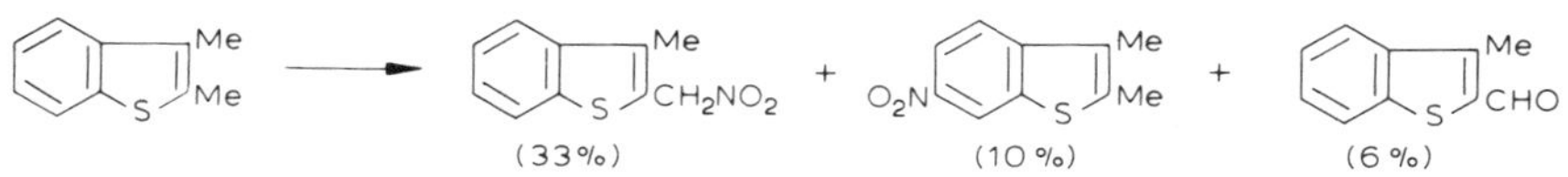

4-**Nitrobenzothiophene,** m.p. 84°, is prepared by reduction and deamination with hydrogen sulphide in alcoholic ammonia of *3,4-dinitrobenzothiophene*, m.p. 199.5–200°; 5-**nitrobenzothiophene,** m.p. 148–149°, is obtained by decarboxylation of 5-nitrobenzothiophene-2-carboxylic acid (*W. Davies* and *Q. N. Porter*, J. chem. Soc., 1957, 826), and also obtained from (*p*-nitrophenylthio)-acetaldehyde dimethyl acetal (*K. Rabindran, A. V. Sunthankar* and *B. D. Tilak*, Proc. Indian Acad. Sci., 1952, **A36,** 405):

O_2N-C6H4-S-CH_2-CH(OMe)$_2$ —polyphosphoric acid→ O_2N-benzothiophene

6-**Nitrobenzothiophene** 1,1-**dioxide,** m.p. 188°, is prepared by nitrating benzothiophene 1,1-dioxide with fuming nitric acid *(Davies* and *Porter, loc. cit.)*.

3-**Methyl**-2-**nitrobenzothiophene,** m.p. 148–149° *(Shirley, Danzig* and *Canter, loc. cit.; Ricci, loc. cit.)*; *7-methyl-3-nitrobenzothiophene*, m.p. 117–118° (*Matsuki* and *T.-C. Li*, Nippon Kagaku Zasshi, 1965, **86,** 853). 3,4-**Dinitrobenzothiophene,** m.p. 169–170°; 3,6-*dinitrobenzothiophene*, m.p. 172–173° 3,7-*dinitrobenzothiophene*, m.p. 183–184°; 4,6-*dinitrobenzothiophene*, m.p. 206–207° (*K. J. Armstrong et al.*, Quart. Repts. Sulphur. Chem., 1968, **3,** 357; J. chem. Soc., C, 1969, 1766).

5-**Hydroxy**-4-**nitrobenzothiophene,** m.p. 119°, and related derivatives (*Martin-Smith* and *Reid, loc. cit.)*.

(vii) Aminobenzothiophenes

2-Nitrobenzothiophene on reduction with ammonium sulphide in ethanol yields thio-oxindole (*Boswell et al., loc. cit.)*:

2-nitrobenzothiophene (NO_2) → thio-oxindole

but the rather unstable 2-**aminobenzothiophene,** m.p. 115–117°, may be obtained either by the mild hydrolysis and decarboxylation of ethyl 2-acetamidobenzothiophene-3-carboxylate (*K. Gewald, G. Neumann* and *H. Böttcher*, Z. Chem., 1966, **6,** 261; Ber., 1968, **101,** 1933) or by treating nitrile XI in benzene with aluminium bromide or in ether with hydrogen chloride (*G. W. Stacy, F. W. Villaescusa* and *T. E. Wollner*, J. org. Chem., 1965, **30,** 4074):

(XI) C6H4(CH_2·CN)(SCH_2Ph) → 2-aminobenzothiophene (NH_2)
- Ac_2O → 2-acetamidobenzothiophene (NHAc), m.p. 224-226°
- (1) HNO_2 (2) Heat → thio-oxindole
- (1) HNO_2 (2) CuBr → 2-bromobenzothiophene (Br)

3-*Aminobenzothiophene*, m.p. 37–38°, is less stable than its 2-isomer and is obtained by reduction of 3-nitrobenzothiophene with stannous chloride; *acetyl deriv.*, m.p. 169–171° *(M. S. El Shanta, R. M. Scrowston* and *M. V. Twigg*, J. chem. Soc., C, 1967,

2364; *Boswell et al., loc. cit.*). 3-Aminobenzothiophene is easily hydrolysed to thioindoxyl. When 3-nitrobenzothiophene (*K. Fries* and *E. Hemmecke*, Ann., 1929, **470,** 1) is reduced with sodium sulphide, it gives the bimolecular product XII, reduced by stannous chloride to the diamine XIII which can be dehydrogenated to the di-imine of thioindigotin; XII has also been converted into 2,2′-*bibenzothienyl*, m.p. 262°.

(XII) (XIII)

4-*Aminobenzothiophene*, m.p. 52° (*Boswell et al., loc. cit.*); 5-*aminobenzothiophene*, m.p. 71° (*Rabindran, Sunthankar* and *Tilak, loc. cit.*); 6-*aminobenzothiophene*, m.p. 114–115° (*C. Hansch et al.*, J. org. Chem., 1955, **20,** 1056; 1956, **21,** 265); 7-aminobenzothiophene (*Boswell et al., loc. cit.*).

5-Acetamido-3-bromobenzothiophene reacts with excess of nitric acid in hot acetic acid to give 3-bromo-2,7-dinitrobenzothiophene-5-diazo-4-oxide (*I. Brown et al.*, Chem. and Ind., 1962, 982):

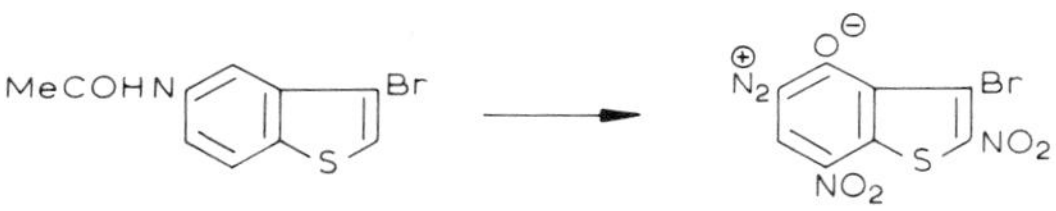

(*viii*) *Hydroxybenzothiophenes*

Of hydroxybenzothiophenes, the 2- and 3-compounds, "thio-oxindole" (p. 281) and "thioindoxyl" (p. 281), are treated later as oxodihydrobenzothiophenes. 3-**Methoxybenzothiophene,** b.p. 260°, 107–108.5°/3 mm (*P. Friedländer*, Ann., 1907, **351,** 409; *Matsuki* and *Adachi, loc. cit.*); 4-**hydroxybenzothiophene,** m.p. 80–81° (*Bordwell* and *H. Stange*, J. Amer. chem. Soc., 1955, **77,** 5939); 5-*hydroxybenzothiophene*, m.p. 103–105° (*M. Martin-Smith* and *M. Gates, ibid.*, 1956, **78,** 5351; *Martin-Smith et al.*, J. chem. Soc., C, 1967, 1899); 6-*hydroxybenzothiophene*, m.p. 102–102.5° (*Hansch* and *B. Schmidhalter*, J. org. Chem., 1955, **20,** 1056); 7-hydroxybenzothiophene is reported to be unstable (*Sunthankar* and *Tilak*, Proc. Indian Acad. Sci., 1951, **33A,** 35). The halogenoacetal synthesis (*3*) (p. 268) has afforded 5-, 6- and 7-methoxybenzothiophenes (*Sunthankar* and *Tilak, loc. cit.*), and 5-*methoxybenzothiophene*, m.p. 215–216°, and 6-*methoxybenzothiophene*, m.p. 251°, may be obtained by decarboxylation of the 2-carboxylic acids, prepared by cyclisation of the corresponding α-mercaptocinnamic acid (synthetic method *1*, p. 266) (*E. Campaigne* and *W. E. Kreighbaum*, J. org. Chem., 1961, **26,** 1326); this method also leads to 4,5-**dimethoxybenzothiophene,** m.p. 240–241°; 5,6-*dimethoxybenzothiophene*, m.p. 260–261°; and 5,6-*diethoxybenzothiophene*, m.p. 245–246° (*idem, ibid.*, 1961, **26,** 359; *Campaigne* and *R. E. Cline, ibid.*, 1956, **21,** 39). 4,7-**Dihydroxybenzothiophene,** m.p. 171–172° (decomp.) (*A. Blackhall* and *R. H. Thomson*, J. chem. Soc., 1954, 3916); 5,6-*dihydroxybenzothiophene*, m.p. 134–137° (*D. G. Bew* and *G. R. Clemo, ibid.*, 1953, 1314); yellow **benzothiophene**-4,7-**quinone,** m.p. 130–131° (*L. F. Fieser* and *R. G. Kennelly*, J. Amer. chem. Soc., 1935, **57,** 1611); 5-*hydroxy*-6-

methylbenzothiophene-4,7-quinone (*5-hydroxy-6-methyl-4,7-thianaphthenequinone*), m.p. 155°, an analogue of phthiocol (2-hydroxy-3-methyl-1,4-naphthoquinone) (*P. Cagniant*, Compt. rend., 1951, **232,** 734).

Benzothiophene-2-thiol (*2-mercaptobenzothiophene*), b.p. 106–110°/2 mm, is prepared from 2-benzothienyl-lithium and sulphur (*R. B. Mitra, L. J. Pandya* and *Tilak*, J. Sci. Ind. Research, India, 1957, **16B,** 348; C.A., 1958, **52,** 5371h).

(*ix*) *Sulphonic acids*

There is little reliable information on the sulphonation of benzothiophene; reaction with sulphuric acid gives a mono-, or a mixture of mono-, di-, and tri-sulphonic acids, depending on the concentration of acid and the reaction conditions; the orientations of the products have not yet been determined (*P. N. Gorelov*, C.A., 1963, **59,** 11398; *P. S. Landis, J. A. Brennan* and *P. B. Venuto*, J. chem. Eng. Data, 1967, **12,** 610), but it has been reported that sulphonation gives the 2- and 3-isomers (*Boswell et al., loc. cit.*). Benzothiophene-3-sulphonic acid is believed to be formed along with 3-acetylbenzothiophene on treating benzothiophene with concentrated sulphuric acid in acetic anhydride at 20° (*M. Pailer* and *E. Romberger*, Monatsh., 1961, **92,** 677).

(*x*) *Alcohols, aldehydes, ketones*

2-**Hydroxymethylbenzothiophene,** m.p. 99–100°, is prepared by reducing benzothiophene-2-carboxylic acid with lithium tetrahydridoaluminate (*F. F. Blicke* and *D. G. Sheets*, J. Amer. chem. Soc., 1949, **71,** 2856); also obtained from 2-benzothienyl-sodium and formaldehyde. Substituted 2-hydroxymethylbenzothiophenes may be obtained by the above methods and by either reacting the 2-lithium derivative with formaldehyde (*R. Gaertner, ibid.*, 1952, **74,** 766; *Y. Matsuki* and *T.-C. Li*, Nippon Kagaku Zasshi, 1965, **86,** 102) or by reducing the corresponding aldehyde or acid chloride with sodium tetrahydridoborate (*F. G. Bordwell* and *T. W. Cutshall*, J. org. Chem., 1964, **29,** 2020; *O. P. Shkurka* and *V. P. Mamaev*, Izvest. Sibirsk. Otdel Akad. Nauk S.S.S.R., Ser. Khim. Nauk, 1965, 81; C.A., 1966, **64,** 2040b; *E. Campaigne* and *E. S. Neiss*, J. heterocycl. Chem., 1966, **3,** 46). The methods involving reduction are also applicable to the preparation of 3-hydroxymethylbenzothiophenes. 3-**Hydroxymethylbenzothiophene,** b.p. 124–125°/1.5 mm, is prepared from the 3-carboxylic acid (*Blicke* and *Sheets, loc. cit.*) and by a crossed Cannizzaro reaction between benzothiophene-3-carbaldehyde and formaldehyde (*Campaigne* and *Neiss, loc. cit.*).

trans-3,4-Dihalogenothiochromans on boiling in aqueous dioxane lose halogen halide and rearrange to give 2-hydroxymethylbenzothiophene (*H. Hofmann* and *G. Salbeck*, Angew. Chem., intern. Edn., 1969, 456):

Br Br S ⟶ S CH_2OH

Chloromethylation of benzothiophene gives 3-**chloromethylbenzothiophene,** m.p. 44–45° (*S. Avakian et al.*, J. Amer. chem. Soc., 1948, **70,** 3075); the following derivatives are obtained by similar means: 3-chloromethyl-2-

methyl- (*Merck A. G.*, G.P. 1,121,054/1960; C.A., 1962, **57,** 835d); -5-*methyl-*, m.p. 78–79°; -6-*methyl-*, b.p. 122–126°/1 mm (*N. B. Chapman et al.*, J. chem. Soc., C, 1968, 514); -7-*methyl-benzothiophene*, m.p. 59–60° (*Matsuki* and *Li*, Nippon Kagaku Zasshi, 1965, **86,** 853); 2-*chloromethyl-3-methylbenzothiophene*, b.p. 123–125°/0.8 mm, m.p. 40.5–41.5° (*Gaertner*, J. Amer. chem. Soc., 1952, **74,** 2991). 2-**Chloromethylbenzothiophene,** m.p. 55–56°, is obtained from the related alcohol on treatment with thionyl chloride (*Blicke* and *Sheets*, *loc. cit.*). The Grignard reagent from 2-chloromethylbenzothiophene gives "abnormal" products of 3-substitution with carbon dioxide, acetyl chloride, ethyl benzoate, or formaldehyde, while the 3-chloromethyl compound gives mixtures (*Gaertner*, J. Amer. chem. Soc., 1952, **74,** 766, 2186, 2991); the Grignard reagent from 2-chloromethyl-3-methylbenzothiophene undergoes a curious transformation, also observed in the benzofuran series (p. 158):

Me, CH_2Cl —Mg; ClCOR→ CMe:C:CH_2, COR —alkali→ Me, Me

3-Chloromethylbenzothiophene may be oxidised to the 3-carbaldehyde with cupric nitrate in nitric acid (*M. Martin-Smith* and *W. E. Sneader*, J. chem. Soc., C, 1967, 1899) or reduced to the 3-methylbenzothiophene with lithium tetrahydridoaluminate–lithium hydride (*Gaertner*, J. Amer. chem. Soc., 1952, **74,** 2991). Reduction of 3-chloromethylbenzothiophene with stannous chloride gives a lower yield of 3-methylbenzothiophene because of the formation of another product (*idem*, *ibid.*, 1952, **74,** 2185).

Treatment of methylbenzothiophenes with *N*-bromosuccinimide in the presence of benzoyl peroxide affords **bromomethylbenzothiophenes** (*Chapman et al.*, J. chem. Soc., 1965, 774; C, 1968, 518).

Benzothiophen-2-carboxylamide affords by reduction 2-**aminomethylbenzothiophene,** m.p. 58–59° (*D. A. Shirley* and *M. D. Cameron*. J. Amer. chem. Soc., 1952, **74,** 664). 3-Dialkylaminomethylbenzothiophenes have been prepared by the reaction of amines with 3-bromomethylbenzothiophenes (*Chapman et al.*, *loc. cit.*).

β-3-**Benzothienylethyl alcohol,** $(C_8H_5S)CH_2{\cdot}CH_2OH$, b.p. 188–190°/21 mm (*P. Cagniant*, Bull. Soc. chim. Fr., 1949, 382); 2-*isomeride*, m.p. 80°, derived *β*-2-**benzothienylethylamine,** b.p. 164–167.5/11.5 mm (*D. B. Capps* and *C. S. Hamilton*, J. Amer. chem. Soc., 1953, **75,** 697).

Benzothiophene-2-**carbaldehyde,** m.p. 42°, is obtained by the Sommelet method, *i.e.* by reacting 2-bromomethylbenzothiophene with hexamethylenetetramine and treating the product with aqueous acetic acid (*Matsuki* and *Li*, Nippon Kagaku Zasshi, 1966, **87,** 186), or form benzothienyl-lithium and methylformanilide (*Shirley* and *M. J. Danzig*, J. Amer. chem. Soc., 1952, **74,** 2935); *benzothiophene-3-carbaldehyde*, m.p. 58°, from 3-benzothienylmagnesium bromide and ethyl orthoformate (*Cagniant, loc. cit.*), from benzothiophene-3-carboxylic acid (*D. F. Elliot* and *C. Harington*, J. chem. Soc., 1949, 1374), or by the Sommelet method (*Matsuki* and *Li, loc. cit.*; *Campaigne* and *Neiss*, J. heterocycl. Chem., 1966, **3,** 46); *benzothiophene-4-carbaldehyde*, m.p. 34°; *benzothiophene-5-carbaldehyde*, m.p. 57°; *benzothiophene-6-carbaldehyde*, m.p. 43°; *benzothiophene-7-carbaldehyde*, m.p. 42° (*Matsuki* and *Li, loc. cit.*); *3-methylbenzothiophene-2-carbaldehyde*, m.p. 88–88.5° (*V. V. Ghaisas, B. J. Kane* and *F. F. Nord*, J. org. Chem., 1958, **23,** 560); *7-methylbenzothiophene-2-carbaldehyde*, m.p. 40° (*Matsuki* and *Li*, Nippon Kagaku Zasshi, 1965, **86,** 102); *7-methylbenzothiophene-3-carbaldehyde*, m.p. 64° (*idem, ibid.*, 1965, **86,** 853).

2-**Acetylbenzothiophene,** m.p. 88–89° (*M. Martnoff*, Compt. rend., 1952, **234,** 736; *R. Royer et al.*, J. org. Chem., 1962, **27,** 3808) is prepared by synthesis (*2A*) (p. 266); *3-acetylbenzothiophene*, m.p. 64° (*N. P. Buu-Hoï* and *Cagniant*, Ber., 1943, **76,** 1269; *Shirley, B. H. Gross* and *Danzig*, J. org. Chem., 1958, **23,** 1024) and *3-benzoylbenzothiophene*, b.p. 220°/12 mm (*idem*, Rec. Trav. chim., 1948, **67,** 64), by Friedel–Crafts reactions. Acetylation of benzothiophene using acetyl chloride in the presence of stannic chloride gives a mixture of 2- and mainly 3-acetyl derivative (*Royer, P. Demerseman* and *A. Cheutin*, Bull. Soc. chim. Fr., 1961, 1534). *2-Acetyl-3-methylbenzothiophene*, m.p. 77°; *3-acetyl-2-methylbenzothiophene*, m p. 69° (*H. E. Baumgarten* and *A. L. Krieger*, J. Amer. chem. Soc., 1955, **77,** 2438; *F. Sauter* and *L. Golser*, Monatsh., 1967, **98,** 2039).

(xi) *Carboxylic acids*

Benzothiophene-2-**carboxylic acid,** m.p. 240–241°, is produced by carbonation of 2-benzothienyl-sodium (*A. Schönberg et al.*, Ber., 1933, **66,** 233) or -lithium (*R. W. Goettsch* and *G. A. Wiese*, J. Amer. pharm. Assoc., 1958, **47,** 319), by oxidation of the aldehyde (*Matsuki* and *Li*, Nippon Kagaku Zasshi, 1966, **87,** 186), or by dehydrogenation of α-mercaptocinnamic acid (*Campaigne* and *R. E. Cline*, J. org. Chem., 1956, **21,** 29), *methyl ester*, m.p. 72–73°. *Benzothiophene-3-carboxylic acid*, m.p. 176–177°, is obtained by carbonation of 3-benzothienylmagnesium bromide (*E. M. Crook* and *W. Davies*, J. chem. Soc., 1937, 1697; *G. M. Badger, P. Cheuychit* and *W. H. F. Sasse*, Australian J. Chem., 1964, **17,** 371), by oxidation of the aldehyde (*Campaigne* and *Neiss*, J. heterocycl. Chem., 1966, **3,** 46), or by hydrolysis of the *nitrile*, m.p. 74° (*R. B. Mitra, K. Rabindran* and *B. D. Tilak*, J. Sci. Ind. Res., India, 1956, **15B,** 627); *methyl ester*, b.p. 165–166°/17 mm.

All the isomeric benzothiophenecarboxylic acids, except the 4-acid have been synthesised, mostly by methods involving ring-closure (*Badger et al.*, J. chem. Soc., 1957, 2624). Ring-synthesis (*2A*) (p. 266) affords *benzothiophene-2,3-dicarboxylic acid*, m.p.

250–251°, *anhydride*, m.p. 171°, *dimethyl ester*, m.p. 91° (*P. Friedländer et al.*, Ber., 1908, **41,** 237).

Benzothiophene-4-carboxylic acid, m.p. 190–191° (*R. L. Titus, M. Choi* and *P. M. Hutt*, J. heterocycl. Chem., 1967, **4,** 651); *benzothiophene-5-carboxylic acid*, m.p. 211–212°; *benzothiophene-6-carboxylic acid*, m.p. 216–217° (*Matsuki* and *Li*, Nippon Kagaku Zasshi, 1966, **87,** 186; *C. Hansch* and *B. Schmidhalter*, J. org. Chem., 1955, **20,** 1056); *3-methylbenzothiophene-2-carboxylic acid*, m.p. 244–246° (*Campaigne* and *Neiss*, J. heterocycl. Chem., 1965, **2,** 231); *2-methylbenzothiophene-3-carboxylic acid*, m.p. 194–195° (*Sauter, Golser* and *P. Stütz*, Monatsh., 1967, **98,** 2089).

Benzothiophene-2-acetic acid, m.p. 141–142°, is produced by Arndt–Eistert chain-lengthening of the lower homologue (*N. P. Kefford* and *J. M. Kelso*, Australian J. biol. Sci., 1957, **10,** 80); *benzothiophene-3-acetic acid*, m.p. 109°, which has growth-promoting activity, is prepared from 3-chloromethylbenzothiophene (*Blicke* and *Sheets*, J. Amer. chem. Soc., 1948, **70,** 3768; *Crook* and *Davies, lic. cit.*) and by the Arndt–Eistert reaction (*Kefford* and *Kelso, loc. cit.*) and by ring-synthesis method (*3*) (p. 268) (*O. Dann* and *M. Kokorudz*, Ber., 1958, **91,** 172; *Sauter* and *A. Barakat*, Monatsh., 1967, **98,** 2393).

Benzothiophene-4-acetic, m.p. 146–146.5°, also shows auxin activity (*M. C. Kloetzel, J. E. Little* and *D. M. Frisch*, J. org. Chem., 1953, **18,** 1511; *Kloetzel* and *Frisch*, U.S.P. 2,930,800/1960; C.A., 1960, **54,** 16465g); benzothiophene-7-acetic acid (*D. W. H. MacDowell* and *T. D. Greenwood*, J. heterocycl. Chem., 1965, **2,** 44).

β-3-**Benzothienylalanine** (*β-3-thianaphthenylalanine*), m.p. 248–250° or 279–280°, is an effectual tryptophan antagonist (*Avakian et al.*, J. Amer. chem. Soc., 1948, **70,** 3075; *Elliott* and *Harington*, J. chem. Soc., 1949, 1374).

(b) Reduced benzothiophenes

(i) Dihydrobenzothiophenes, benzothiolenes*

The most important compounds in this group are the 2- and 3-oxo-2,3-dihydrobenzothiophenes, thio-oxindole and thioindoxyl, and 2,3-dioxodihydrobenzothiophene, benzothiophenequinone.

(*1*) 2,3-**Dihydrobenzothiophene,** 2,3-**benzothiolene,** b.p. 235.6°, $d^{25°}$ 1.1358, n_D^{25} 1.6170, is made by reducing benzothiophene with sodium and alcohol (*S. F. Birch et al.*, J. Inst. Pet., 1954, **40,** 76), from *o*-aminoethylbenzene by diazotisation and successive treatment with xanthate and acid (*G. M. Bennett* and *M. M. Hafez*, J. chem. Soc., 1941, 287), or like some other 2,3-dihydrobenzothiophenes by the reduction of the corresponding 2,3-dihydrobenzothiophene 1,1-dioxide with lithium tetrahydridoaluminate (*G. van Zyl* and *R. A. Koster*, J. org. Chem., 1964, **29,** 3558); *2,3-dihydrobenzothiophene 1,1-dioxide*, m.p. 92° (*F. Challenger* and *P. H. Clapham*, J. chem.

* See I.U.P.A.C. Rule B 1.2.

Soc., 1948, 1615). Pyrolysis of *o*-crotylthiobenzoic acid gives 2-*ethyl*-2,3-*dihydrobenzothiophene*, b.p. 69–72°/0.7–0.8 mm (*J. C. Petropoulos, M. A. McCall* and *D. S. Tarbell*, J. Amer. chem. Soc., 1953, **75,** 1130) and 3-methyl-2,3-dihydrobenzothiophene is obtained by catalytic reduction of 3-methyl-benzothiophene (*H. Kwart* and *C. M. Hackett*, *ibid*., 1962, **84,** 1754).

Reduction of benzothiophene 1,1-dioxides with zinc and sodium hydroxide gives the corresponding 2,3-dihydrobenzothiophene 1,1-dioxides (*W. E. Truce* and *J. P. Milionis*, *ibid*., 1952, **74,** 974), also obtained by oxidation of the dihydro compound with hydrogen peroxide (*Bennett* and *Hafez*, *loc. cit.*) or peracetic acid. 2,3-Dihydrobenzothiophenes are dehydrogenated with sulphur to yield the related benzothiophenes (*B. R. Muth* and *A. I. Kiss*, J. org. Chem., 1956, **21,** 576; *W. Carruthers, A. G. Douglas* and *J. Hill*, J. chem. Soc., 1962, 704).

(*2*) **Thio-oxindole,** 2-**oxo**-2,3-**dihydrobenzothiophene,** exists in two forms, m.p. 33–34° and 44–45°; it is the lactone of *o*-mercaptophenylacetic acid with which it is easily interconvertible, and is best prepared by Wolff–Kishner reduction of 2,3-dioxo-2,3-dihydrobenzothiophene (thianaphthenequinone) (*R. H. Glauert* and *F. G. Mann*, J. chem. Soc., 1952, 2127). Thio-oxindole may also be prepared by the oxidation of 2-benzothienyl-lithium in the presence of an alkyl Grignard reagent (*Van Zyl et al*., J. org. Chem., 1961, **26,** 4946) or by heating an acidic solution of the diazonium salt obtained from 2-aminobenzothiophene (*G. W. Stacy, F. W. Villaescusa* and *T. E. Wollner*, *ibid*., 1965, **30,** 4074).

Thio-oxindole exists solely in the oxo form, in solution and in the solid phase (*Stacy* and *Wollner*, *ibid*., 1967, **32,** 3028; *D. E. Boswell et al*., J. heterocycl. Chem., 1968, **5,** 69). The 3-methylene group is reactive; nitrous acid and *p*-dimethylaminobenzaldehyde give the 3-oxime and the 3-dimethylaminoanil of 2,3-dioxo-2,3-dihydrobenzothiophene; indigoid colouring matters (*q.v.*) are produced readily from thioxindole with isatin-3-anil and similar partners (*C. Marschalk*, Ber., 1912, **45,** 1481; J. pr. Chem., 1913, [ii], **88,** 227), and other condensation products have been obtained (*J. N. Chatterjea* and *A. K. Mitra*, J. Indian Chem. Soc., 1959, **36,** 315). Thio-oxindole reacts with diazomethane to yield a spiro-oxadiazole XIV (*D. G. Hawthorne* and *Q. N. Porter*, Australian J. Chem., 1966, **19,** 1751):

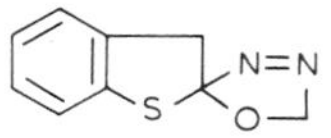

(XIV)

3-*Acetylthio-oxindole*, m.p. 105° (*Glauert* and *Mann*, *loc. cit.*).

(*3*) **Thioindoxyl,** 3-**oxo**-2,3-**dihydrobenzothiophene,** m.p. 71°, exists predominantly in the oxo form (*R. H. Thomson*, Quart. Reviews, 1956, **10,** 34;

G. M. Oksengendler and *M. A. Mostoslavskiĭ*, Ukr. Khim. Zhur., 1960, **26,** 69; C.A., 1960, **54,** 14934b) and smells like 1-naphthol. Thioindoxyl can be produced by cyclodehydration of phenylthioacetic acid (phenylthioglycollic acid), either directly with hydrogen fluoride (*O. Dann* and *M. Kokorudz*, Ber., 1953, **86,** 1449; *Y. Matsuki* and *F. Shoji*, Nippon Kagaku Zasshi, 1967, **88,** 755), or more generally *via* the acid chloride (*C. E. Dalgliesh* and *Mann*, J. chem. Soc., 1945, 893); yields are generally low with phosphorus pentoxide (*G. M. Badger et al., ibid.*, 1957, 2624). Other arylthioacetic acids give corresponding thioindoxyls.

$$PhSCH_2CO_2H \xrightarrow{PCl_3} PhSCH_2COCl \xrightarrow{AlCl_3} \text{thioindoxyl}; \quad PhSCH_2CO_2H \xrightarrow{HF} \text{thioindoxyl}$$

Thioindoxyl is more readily obtained from phenylthioglycollic-*o*-carboxylic acid *via* thioindoxyl-2-carboxylic acid (*cf.* synthesis *C*, p. 267).

Thioindoxyl is easily oxidised to thioindigo. It reacts in the oxo-form as a reactive methylene compound at position 2; aldehydes condense readily; (*benzylidene deriv.*, m.p. 131.5°); isatin chloride, isatin-2-anil, and many related compounds give indigoid dyestuffs. Nitrous acid affords 2,3-dioxo-2,3-dihydrobenzothiophene-2-oxime. Thioindoxyl is methylated on oxygen giving 3-methoxybenzothiophene (p. 276); it likewise affords 3-*acetoxybenzothiophene*, b.p. 165°/18 mm. Hydrogen peroxide oxidises thioindoxyl to **thioindoxyl** 1,1-**dioxide,** m.p. 136°, also obtained by cyclisation of *o*-methanesulphonylbenzonitrile (*Truce, W. W. Bannister* and *R. H. Knospe*, J. org. Chem., 1962, **27,** 2821):

$$o\text{-}C_6H_4(CN)(SO_2Me) \xrightarrow[(2)\ H^{\oplus}]{(1)\ NaNH_2} \text{thioindoxyl 1,1-dioxide}$$

or by acidic hydrolysis of 3-diethylaminobenzothiophene 1,1-dioxide (*W. Davies et al.*, J. chem. Soc., 1955, 1565).

Thioindoxyl 1,1-dioxide is monobrominated (*M. A. Matskanova* and *G. Ya. Vanags*, Doklady Akad. Nauk, S.S.S.R., 1960, **132,** 615) and mononitrated (*M. Mackanova*, C.A., 1961, **55,** 27231a; *Mackanova* and *G. Vanags*, C.A., 1963, **59,** 6341f) in the 2-position; with phosphorus pentachloride it gives 3-*chlorobenzothiophene* 1,1-*dioxide*, m.p. 166–167° (*Davies et al., loc. cit.*), and on treatment with alkali it gives *o*-methanesulphonylbenzoic acid (*R. G. Pearson, D. H. Anderson* and *L. L. Alt*, J. Amer. chem. Soc., 1955, **77,** 527):

$$\text{thioindoxyl 1,1-dioxide} \longrightarrow o\text{-}C_6H_4(CO_2H)(SO_2Me)$$

Alkaline decomposition of the reaction mixture obtained from the reduction of thioindoxyl with sodium tetrahydridoborate, affords 3-**hydroxy**-2,3-**dihydrobenzothiophene,** m.p. 55–56° (*N. Kucharczyk* and *V. Horák*, Chem. and Ind., 1964, 976; Coll. Czech. chem. Comm., 1968, **33,** 92):

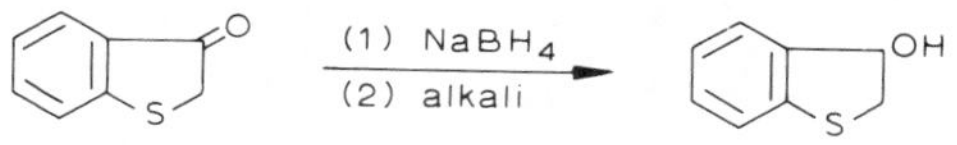

(1) $NaBH_4$ (2) alkali

The Reformatsky reaction between thioindoxyl and ethyl bromoacetate affords 3-benzothienylacetic acid (*M. Martin-Smith* and *S. T. Reid*, J. chem. Soc., C, 1967, 1897).

(4) *Substituted thioindoxyls.* In the preparation of some thioindoxyls linkage of the sulphur atom is often effected prior to ring closure (*K. von Auwers* and *F. Arndt*, Ber., 1909, **42,** 537; *F. Krollpfeiffer et al.*, Ann., 1949, **563,** 15):

$ClCH_2{\cdot}COCl$ / $AlCl_3$; + MeCl

Dalgliesch and *Mann (loc. cit.)* prepared substituted thioindoxyls as follows (*cf.* the Herz process p. 267):

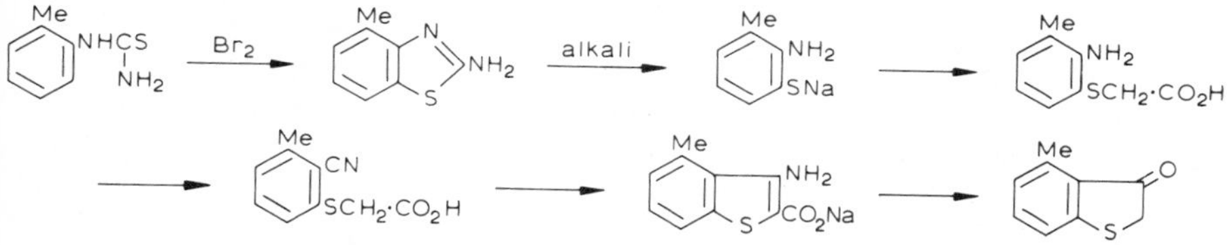

Br_2; alkali

The transformation, by sodium acetate and acetic anhydride, of 2,2′-bisthiobenzoic acid and its derivatives into 3-acetoxybenzothiophenes probably involves ring formation between $C_{(2)}$ and sulphur (*J. L. D'Silva* and *E. W. McClelland*, J. chem. Soc., 1932, 2883):

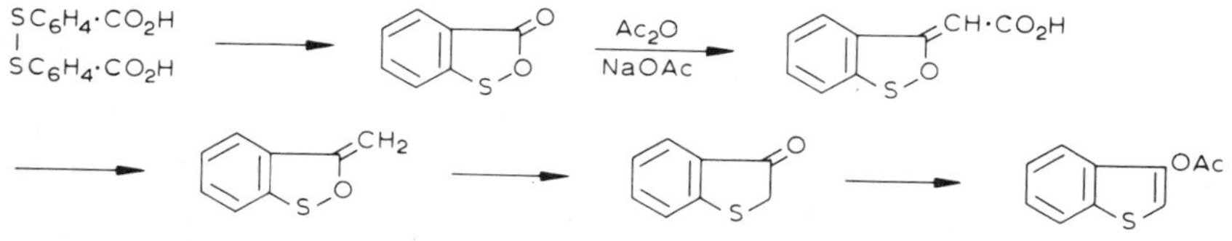

Ac_2O / NaOAc

6-**Ethoxythioindoxyl-2-carbaldehyde,** m.p. 147–149°, 2-**acetylthioindoxyl,** m.p. 79.5–80° (*Glauert* and *Mann*, *loc. cit.*). 2-Arylthioindoxyls undergo alkylation in the presence of potassium *tert*-butoxide both at the oxygen and at $C_{(2)}$ (*Merck A. G.*, Neth. P. Appl. 6,413,199/1963; C.A., 1965, **63,** 18038g).

(5) 2,3-**Dioxo**-2,3-**dihydrobenzothiophene, benzothiophene**-2,3-**quinone,** *thianaphthenequinone* (XV), is prepared from thioindoxyl by methods of general application (*F. Mayer*, Ann., 1931, **488,** 259; *Dalgliesh* and *Mann*, *loc. cit.*):

Br_2 hydrolysis (XV) NH_2OH ArNO $ArNH_2$ hydrolysis NAr NOH

The yellow quinone, m.p. 128°, is easily hydrolysed to *o*-mercaptophenylglyoxylic acid. Hydroxylamine and phenylhydrazine attack the truly ketonic 3-carbonyl group; the 2-oxime and 2-arylhydraxones are prepared by treating thioindoxyl with nitrous acid and diazonium salts, respectively (*P. Friedländer et al.*, Ber., 1908, **41,** 227): 2-*oxime*, m.p. 172° decomp., 3-*oxime*, m.p. 186°. Alkaline hydrolysis of thioindigo 1,1-dioxide gives 2,3-**dioxo**-2,3-**dihydrobenzothiophene** 1,1-**dioxide** (*thianaphthenequinone S-dioxide*), m.p. 265° (*T. Posner* and *E. Wallis*, Ber., 1924, **57,** 1680). With chloramine-T in alcohol, 2,3-dioxo-2,3-dihydrobenzothiophene gives the ester XVI (*Dalgliesch* and *Mann*, J. chem. Soc., 1945, 913):

$CO{\cdot}CO_2Et$ $NHSO_2C_7H_7$ $NSO_2C_7H_7$

(XVI)

2,3-Dioxo-2,3-dihydrobenzothiophenes undergo ring-expansion reactions with diazomethane and related compounds to give derivatives of 3-hydroxy-1-thiochromone (XVII) and in some cases small amounts of 4-substituted 3-hydroxy-1-thiacoumarin (XVIII) (*B. Eistert* and *H. Selzer*, Ber., 1963, **96,** 1234; *A. Schönberg, K. Junghans* and *E. Singer*, Tetrahedron Letters, 1966, 4667; *Schönberg* and *Junghans*, Ber., 1966, **99,** 1241):

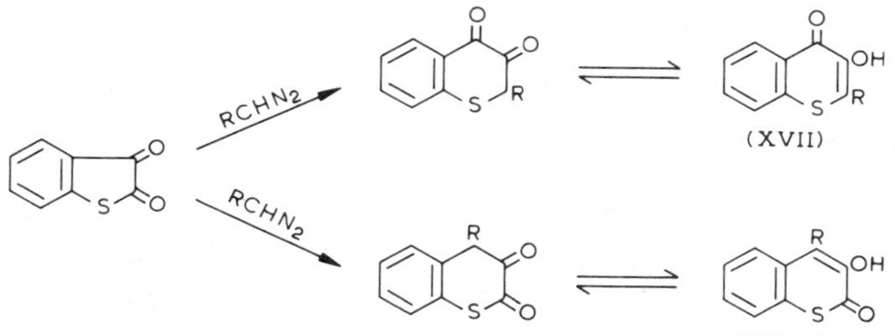

(XVII)

(XVIII)

The 3-*p*-tosylhydrazone of 2,3-dioxo-2,3-dihydrobenzothiophene on treatment in methylene dichloride with sodium hydroxide yields 2-*oxo*-3-*diazo*-2,3-*dihydrobenzothiophene* (3-*diazothianaphthen*-2-*one*) (XIX), m.p. 68° (*W. Ried* and *R. Dietrich*, Ber., 1961, **94,** 387):

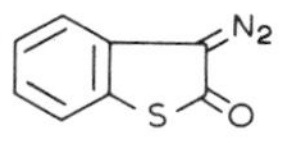

(XIX) 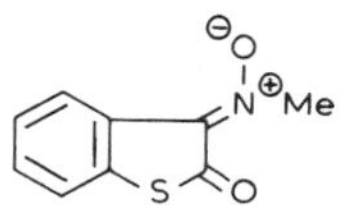 (XX)

Treatment of 2,3-dioxo-2,3-dihydrobenzothiophene 3-oxime with diazomethane or of the quinone with *N*-methylhydroxylamine yields the *N*-methyl nitrone XX (*Eistert et al.*, Ber., 1964, **97,** 2469).

2,3-**Dioxo-4-methyl**-2,3-**dihydrobenzothiophene** (4-*methylthianaphthenequinone*), m.p. 120–121°; *C. E. Dalgliesh* and *F. G. Mann*, J. chem. Soc., 1945, 893); 2,3-*dioxo*-5-*methyl*-2,3-*dihydrobenzothiophene*, m.p. 146–147° (*D. Walker* and *J. Leib*, J. org. Chem., 1963, **28,** 3077); 2,3-*dioxo*-7-*methyl*-2,3-*dihydrobenzothiophene*, m.p. 126–127° (*Dalgliesh* and *Mann, loc. cit.*); 5-*chloro*-2,3-*dioxo*-2,3-*dihydrobenzothiophene*, m.p. 100°, on oxidation with hydrogen peroxide in the presence of ammonia affords 5-chloro-1,2-benzisothiazole-3-carboxamide (XXI) (*M. S. E. Shanta, R. M. Scrowston* and *M. V. Twigg*, J. chem. Soc., C, 1967, 2364):

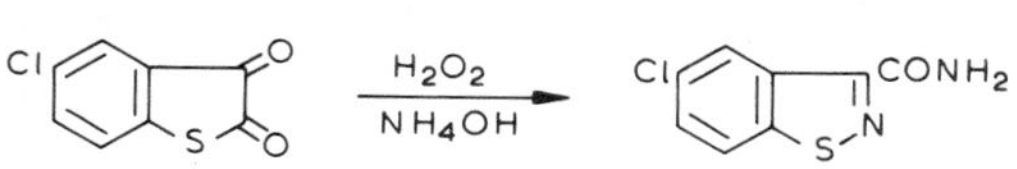

(XXI)

Degradation by alkali of thioindigoid pigments gives 3-**oxo**-2,3-**dihydrobenzothiophene-2-carbaldehyde (thioindoxyl-2-carbaldehyde),** m.p. 107°, and 2-**oxo**-2,3-**dihydrobenzothiophene**-3-**carbaldehyde,** m.p. 130°, both synthesised from the related oxodihydrobenzothiophene by the Gatterman method (*Mann et al.*, *ibid.*, 1952, 2127; 1955, 30). They condense with thio-oxindole or thioindoxyl, giving red methine colouring matters of type XXII:

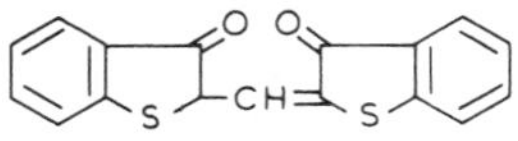

(XXII)

(*6*) **Thioindoxyl-2-carboxylic acid** (p. 267), is the rather unstable intermediate in the production of thioindigo. The dimethyl ester of phenylthioglycollic-2-carboxylic acid gives by Dieckmann ring-closure *methyl thioindoxyl-2-carboxylate*, m.p. 108–109°, which undergoes simultaneous *C*- and *O*-methylation. The O-*methyl deriv.*, m.p. 68°, yields 3-*methoxybenzothiophene*-2-*carboxylic acid*, m.p. 173° (*K. von Auwers*, Ann., 1912, **393,** 372). The diethyl ester of *o*-carboxybenzenesulphonylacetic acid gives, on ring-closure with sodium ethoxide, *ethyl* 3-*hydroxybenzothiophene*-2-*carboxylate* 1,1-*dioxide*, m.p. 140°, which is fully enolic (*W. B. Price* and *S. Smiles*, J. chem. Soc., 1028, 2858).

(ii) Tetrahydrobenzothiophenes

A compound believed to be a derivative of 3a,4,5,6-tetrahydrobenzothiophene (p. 235) is obtained from 2-vinylthiophene and maleic anhydride. Thiophene, condensed with the methyl ester-chloride of succinic acid, gives β,2-thenoylpropionic acid $(C_4H_3S)CO{\cdot}CH_2{\cdot}CH_2{\cdot}CO_2H$, whence the usual methods yield **4-oxo-4,5,6,7-tetrahydrobenzothiophene** (XXIII), m.p. 35.5–37°, and 4,5,6,7-**tetrahydrobenzothiophene,** b.p. 88°/13.3 mm, d_4^{20} 1.090, n_D^{20} 1.5572 (*D.* and *P. Cagniant*, Bull. Soc. chim. Fr., 1953, 62; *M. C. Kloetzel, J. E. Little* and *D. M. Frisch*, J. org. Chem., 1953, **18,** 1511; *M. Maillet* and *M. Sy*, Compt. rend., 1967, **C264,** 1193), also obtained by the reduction of 7-oxo-4,5,6,7-tetrahydrobenzothiophene (*D. W. H. MacDowell* and *T. D. Greenwood*, J. heterocycl. Chem., 1965, **2,** 44):

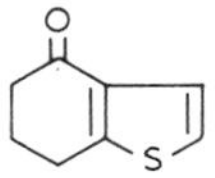

(XXIII)

Some 4,5,6,7-tetrahydrobenzothiophenes may be synthesised from cyclohexanone and its derivatives (*K. Gewald*, Z. Chem., 1962, **2,** 305; 1967, **7,** 186; Ber., 1965, **98,** 3571; *Gewald, E. Schinke* and *H. Böttcher*, Ber., 1966, **99,** 94):

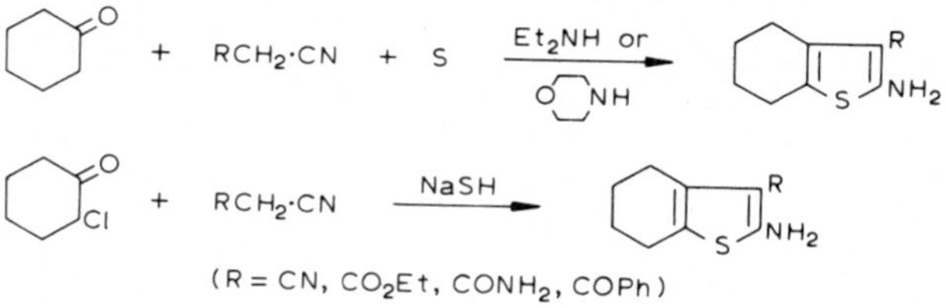

Et_2NH or (morpholine)

+ $RCH_2\cdot CN$ + S → 2-amino-3-R-4,5,6,7-tetrahydrobenzothiophene

+ $RCH_2\cdot CN$ (NaSH) →

(R = CN, CO_2Et, $CONH_2$, COPh)

2-Mercaptocyclohexanone reacts with methyl acetylenecarboxylate in the presence of potassium *tert*-butoxide to yield methyl 4,5,6,7-tetrahydrobenzothiophene-3-carboxylate together with other products (*F. Bohlmann* and *E. Bresinsky*, Ber., 1964, **97**, 2109):

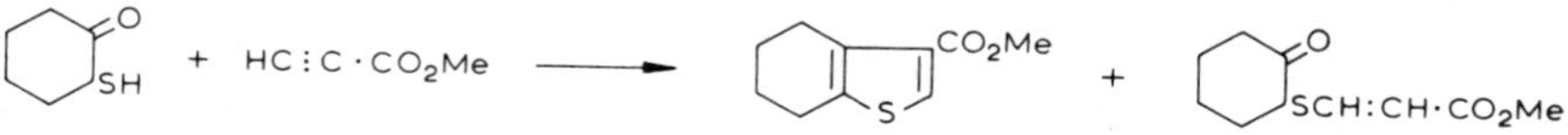

+ $HC\!:\!C\cdot CO_2Me$ → (CO_2Me) + ($SCH\!:\!CH\cdot CO_2Me$)

4,5,6,7-Tetrahydrobenzothiophene behaves like thiophene in electrophilic substitution reactions; it is formylated with *N*-methylformanilide and phosphorus oxychloride (*N. P. Buu-Hoï* and *M. Khenissi*, Bull. Soc. chim. Fr., 1958, 359), brominated by *N*-bromosuccinimide, iodinated in the presence of mercuric oxide (*P.* and *D. Cagniant*, *ibid.*, 1955, 1252), and undergoes Friedel–Crafts reactions (*Buu-Hoï* and *Khenissi*, *loc. cit.*), all in the 2-position. Depending on reaction conditions the bromination of 4-oxo-4,5,6,7-tetrahydrobenzothiophene gives either a 2-bromo or a 5-bromo derivative (*J. Sam* and *G. G. Advani*, J. Pharm. Sci., 1965, **54**, 753; *S. Nishimura et al.*, Nippon Kagaku Zasshi, 1962, **83**, 343). The oximes of 4-oxo- and 7-oxo-4,5,6,7-tetrahydrobenzothiophene undergo the Beckmann rearrangement to yield compounds **XXIV** and XXV (*Nishimura et al., loc. cit.; B. P. Fabrichnyĭ et al.*, Zhur. obshcheĭ Khim., 1961, **31**, 1244; Zhur. organ. Khim., 1968, **4**, 680):

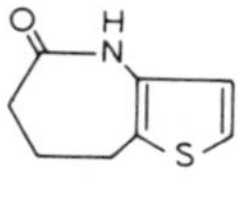

(XXIV) 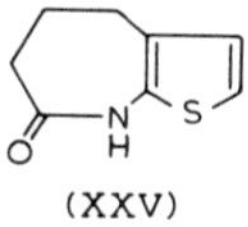 (XXV)

2-Propynylcyclohexanone reacts with a mixture of hydrogen sulphide and hydrogen chloride to afford 2-methyl-4,5,6,7-tetrahydrobenzothiophene (*K. E. Schulte*, *J. Reisch* and *D. Bergenthal*, Ber., 1968, **101**, 1540):

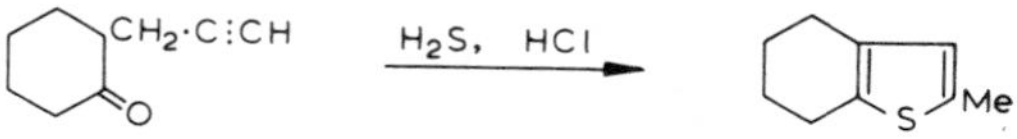

$CH_2\cdot C\!:\!CH$ —(H_2S, HCl)→ Me

(iii) Hexa- and octa-hydrobenzothiophenes

3-Mercaptocyclohexanone on treatment with chloroacetaldehyde in the presence of a catalytic amount of *p*-toluenesulphonic acid gives **4-oxo-**

2,4,5,6,7,7a-**hexahydrobenzothiophene** (XXVI) (*H. M. Foster, R. P. Napier* and *C. C. Chu*, U.S.P. 3,346,591/1965; C.A., 1968, **68,** 78126), also obtained by acidic hydrolysis of the product from 3-mercaptocyclohexanone ethylene acetal and chloroacetaldehyde (*Foster* and *Napier*, U.S.P. 3,357,997/1965; C.A., 1968, **69,** 2863):

$$\text{3-mercaptocyclohexanone} + ClCH_2{\cdot}CHO \xrightarrow{p\text{-}MeC_6H_4SO_3H} \text{(XXVI)}$$

The **octahydrobenzothiophenes** are prepared from cyclohexanone as follows:

$$\text{cyclohexanone} \longrightarrow \text{1-ethynylcyclohexanol (OH, C:CH)} \xrightarrow[-H_2O]{H_2;} \text{1-vinylcyclohexene (CH=CH}_2) \xrightarrow{SO_2} \text{sulphone (S, O}_2)$$

$$\xrightarrow{H_2-Ni} \text{cis-sulphone} \xrightarrow{LiAlH_4} \text{cis-octahydrobenzothiophene}$$

Alkali converts the *cis*-sulphone into the *trans* form, reduced to the *trans*-octahydrobenzothiophene (*Birch et al.*, J. org. Chem., 1955, **20,** 1178); cis-*octahydrobenzothiophene*, b.p. 102°/20 mm, d^{25} 1.0341; 1,1-*dioxide*, m.p. 38°; trans-*form*, b.p. 103°/21 mm, d^{25} 1.0167; 1,1-*dioxide*, m.p. 93°.

Treatment of 3-β-chloroethylcyclohex-1-ene with sodium hydrogen sulphide affords a mixture of *cis*-octahydrobenzothiophene and *cis*-2-thiabicyclo[3.3.1]nonane (XXVII) (*N. P. Volynskii, G. D. Gal'pern* and *A. B. Urin*, Khim. Geterotsikl. Soedin., Akad. Nauk. Latv. S.S.R., 1967, 1031; C.A., 1968, **69,** 77035):

$$\text{3-}(CH_2{\cdot}CH_2Cl)\text{cyclohexene} \xrightarrow{NaSH} \text{octahydrobenzothiophene} + \text{(XXVII)}$$

3. Benzo[*c*]thiophenes (3,4-benzothiophenes, isobenzothiophenes) and related compounds

(*a*) *Benzo*[c]*thiophenes*

Benzo[*c*]thiophene derivatives are much less common, and less stable than benzo[b]thiophenes. Acetonylacetone condenses with 2,5-dimethylthiophene in presence of hydrogen fluoride to give 1,3,4,7-**tetramethylbenzo[c]-thiophene** (I), m.p. 150–152°, which combines with maleic anhydride to form the adduct II (*O. Dann et al.*, Ber., 1954, **87**, 140):

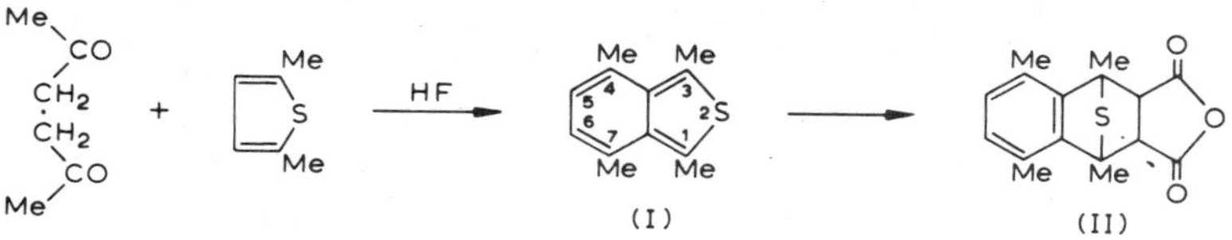

(I) (II)

Catalytic dehydrogenation of 1,3-dihydrobenzo[c]thiophene over palladium on carbon at 330° in the absence of oxygen gives **benzo[c]thiophene** (III), m.p. 50–51°.

Benzo[c]thiophene is stable in solution in the presence of stabilisers, but only for a few days as a solid at −30° under nitrogen. In odour it resembles naphthalene and it forms a *maleic anhydride adduct*, m.p. 153–154°, which with sodium hydroxide affords naphthalene-2,3-dicarboxylic acid (*R. Mayer et al.*, Angew. Chem., 1962, **74**, 118; J. pr. Chem., 1963, **20**, 244):

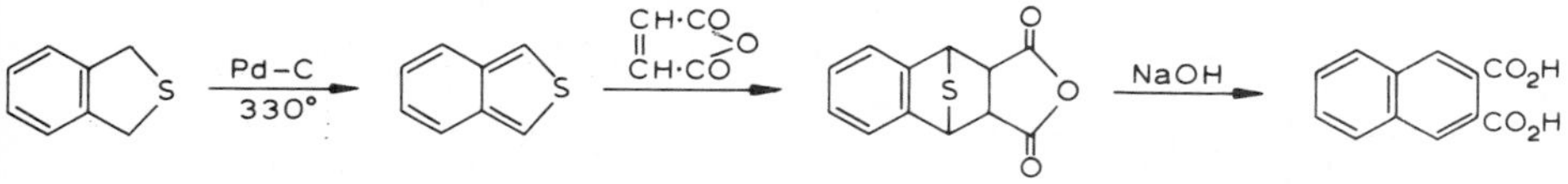

(III)

The thermal decomposition of 1,3-dihydrobenzo[c]thiophene 2-oxide gives no evidence of the formation of sulphur monoxide and *o*-benzoquinodimethane (*o*-benzoquinonedimethide), but dehydration occurs to give benzo[c]thiophene (*M. P. Cava* and *M. M. Pollack*, J. Amer. chem. Soc., 1966, **88**, 4112).

A convenient and simple method for the generation of the unstable *o*-benzoquinodimethane (IV) is by the thermal decomposition of 1,3-dihydrobenzo[c]thiophene 2,2-dioxide (*Cava* and *A. A. Deana*, *ibid.*, 1959, **81**, 4266):

(IV)

The condensation of 2-ethoxymethylenecyclohexanone with ethyl thioglycollate in the presence of sodium ethoxide gives 4-ethoxy-

2-ethoxycarbonyl-1-hydroxybicyclo[4.3.0]-3-thianonane (V), which on dehydration affords 1-ethoxycarbonyltetrahydrobenzo[*c*]thiophene, b.p. 118°/2.5 mm. Treatment of this compound with *N*-bromosuccinimide followed by sodium methoxide yields **benzo[*c*]thiophene-1-carboxylic acid** (VI), m.p. 198°, *methyl ester*, b.p. 110°/0.01 mm (*B. D. Tilak*, *H. S. Desai* and *S. S. Gupte*, Tetrahedron Letters, 1966, 1953):

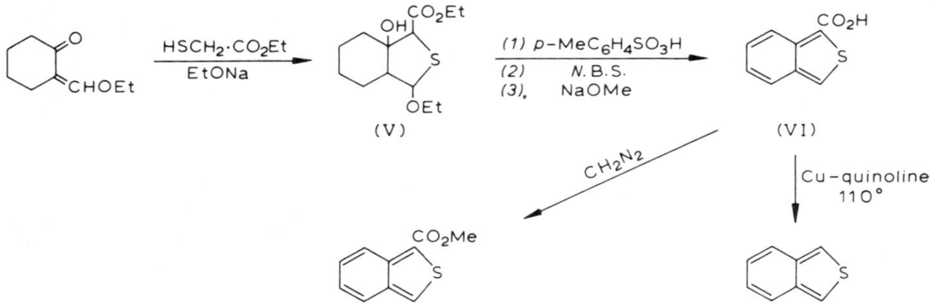

The synthesis of the reasonably stable methyl benzo[*c*]thiophene-5-carboxylate has been described (*H. Wynberg*, *J. Feijen* and *D. J. Zwanenburg*, Rec. Trav. chim., 1968, **87**, 1006). 5,6-**Dimethoxy**-1-**methylbenzo**[*c*]**thiophene**, m.p. 110–113°, results from the pyrolysis of the isothiachromanone enamine VII (*F. H. M. Deckers*, *W. N. Speckamp* and *H. O. Huisman*, Chem. Comm., 1970, 1521):

1,3-**Diphenylbenzo**[*c*]**thiophene** (VIII), m.p. 118–119°, obtained from benzaldehyde and thiobenzilic acid, does not react readily with maleic anhydride (*C. Dufraisse* and *D. Daniel*, Bull. Soc. chim. Fr., 1937, **4**, 2063):

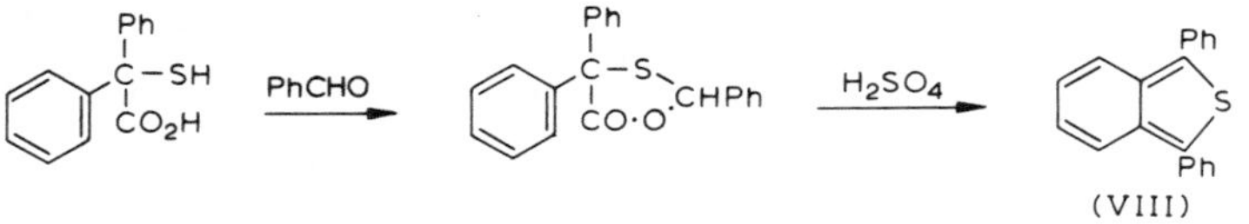

Condensation of 1,2-dibenzoyl-4,5-dimethylcyclohexa-1,4-diene (IX) with phosphorus pentasulphide in boiling toluene gives 5,6-**dimethyl**-1,3-

diphenylbenzo[c]thiophene, m.p. 182.5–183° (*M. E. Mann* and *J. D. White*, Chem. Comm., 1969, 420):

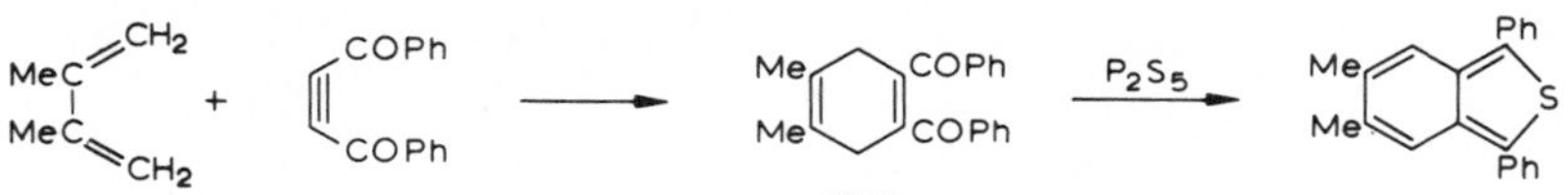

(IX)

1,3,4,7-**Tetraphenylbenzo[c]thiophene,** m.p. 283°, is obtained by heating compound X with sulphur at 220°. It may also be prepared by heating 4,7-dihydro-1,3,4,7-tetraphenylbenzo[c]furan with sulphur at 160° (*E. D. Bergmann et al.*, Tetrahedron, 1964, 195):

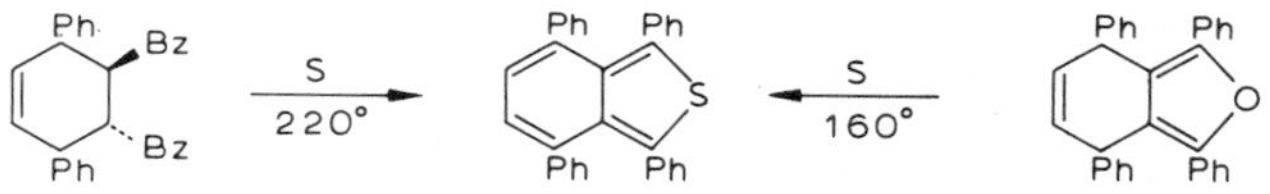

(X)

A Wittig reaction between 1,2-benzocyclobutadiene quinone (XI) and the bis-ylide XII gives *benzo*[3,4]*cyclobuta*[1,2-c]*thiophene* (XIII), m.p. 98–98.5°, which on treatment with 6% hydrogen peroxide in acetic acid affords *benzo*[3,4]*cyclobuta*[1,2-c]*thiophene* 2,2-*dioxide* (XIV), m.p. 213–215°. Comparison of the n.m.r. spectra of XIII and XIV dramatically demonstrates the presence of a paramagnetic ring current shielding effect of the 4π-cyclobutadienyl ring in XIII (*P. J. Garratt* and *K. P. C. Vollhardt*, Chem. Comm., 1970, 109):

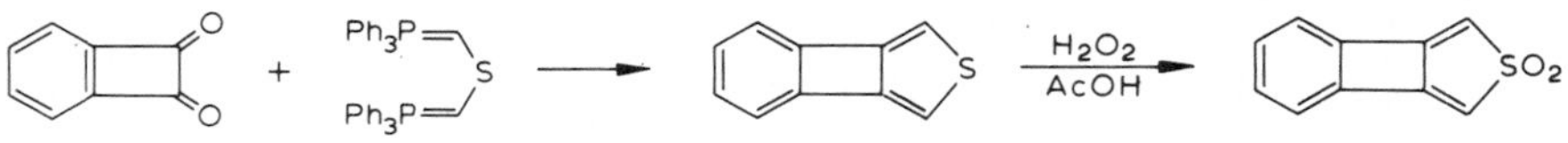

(XI) (XII) (XIII) (XIV)

(b) *Hydrobenzo*[c]*thiophenes*

(i) *Dihydrobenzo*[c]*thiophenes*

1,3-**Dihydrobenzo[c]thiophene,** m.p. 23°, b.p. 241.5°, 94.7°/5 mm, d^{25} 1.1446, n_D^{25} 1.6038, is readily obtained from α,α'-dibromo-*o*-xylene and sodium sulphide (*J. A. Oliver* and *P. A. Ongley*, Chem. and Ind., 1965, 1024); 1,3-*dihydrobenzo*[c]*thiophene* 2,2-*dioxide* (S-*dioxide*), m.p. 150–152°; *methiodide*, m.p. 175° (*J. von Braun*, Ber., 1925, **58**, 2165).

The pyrolysis of 1,3-dihydrobenzo[c]thiophene 2,2-dioxide provides probably the best route to benzocyclobutene:

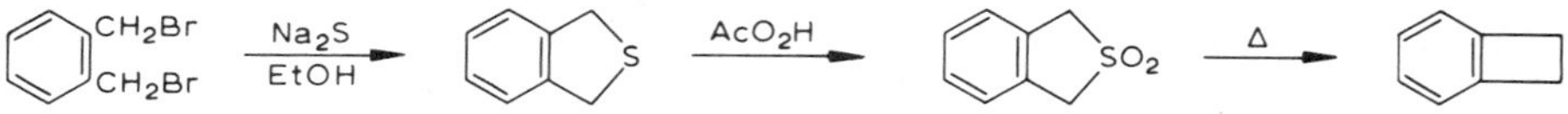

The synthesis of benzocyclobutene by photochemical desulphonation of 1,3-dihydrobenzo[*c*]thiophene 2,2-dioxide has also been studied (*Y. Odaira, K. Yamaji* and *S. Tsutsumi*, Bull. chem. Soc., Japan, 1964, **37,** 1410).

1,3-**Dimethyl**-1,3-**dihydrobenzo**[*c*]**thiophene,** m.p. 43–44°, 2,2-*dioxide*, m.p. 112–113°; on pyrolysis the latter did not give the expected 1,3-dimethylbenzocyclobutene but rather *o*-ethylstyrene (*Cava* and *M. J. Mitchell*, Rev. Chim. Acad. Rep. Populaire Roumaine, 1962, **7,** 737; C.A., 1964, **61,** 4281b):

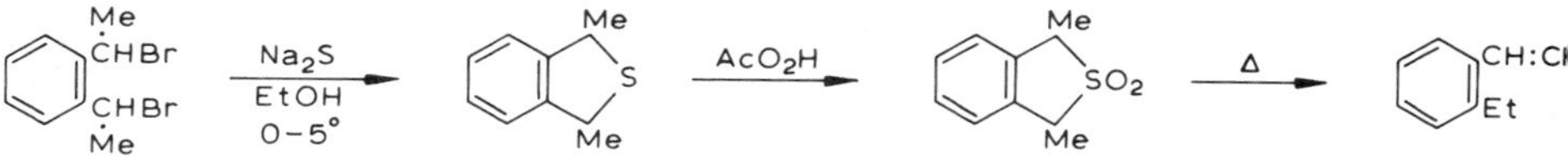

The photolytic decomposition of 1,3-diphenyl-1,3-dihydrobenzo[*c*]thiophene 2,2-dioxide yields 13% *trans*-1,2-diphenylbenzocyclobutene (*Cava, R. H. Schlèssinger* and *J. P. van Meter*, J. Amer. chem. Soc., 1964, **86,** 3173):

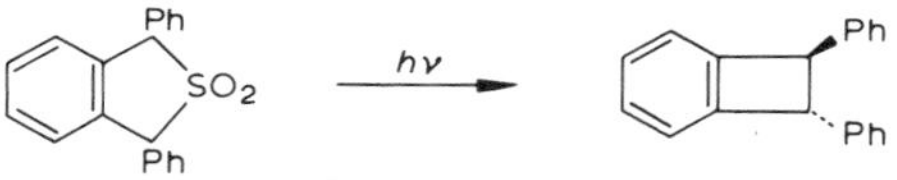

The light-catalysed bromination of 1,3-diphenyl-1,3-dihydrobenzo[*c*]thiophene 2,2-dioxide in carbon tetrachloride gives 1-*bromo*-1,3-*diphenyl*-1,3-*dihydrobenzo*[c]*thiophene* 2,2-*dioxide*, m.p. 198–200° or 1,3-*dibromo*-1,3-*diphenyl*-1,3-*dihydrobenzo*[c]*thiophene* 2,2-*dioxide*, m.p. 225–227°, depending on the amount of bromine used (*Cava* and *J. McGrady*, Chem. Comm., 1968, 1648).

The n.m.r. spectrum of 1,3-dihydrobenzo[*c*]thiophene 2-oxide has been studied under varying conditions of temperature and acidity, and the data obtained suggests that it is dimeric and the dimer bonding increases as the temperature decreases (*R. F. Watson* and *J. F. Eastham*, J. Amer. chem. Soc., 1965, **87,** 664).

1-**Phenyl**-1,3-**dihydrobenzo**[*c*]**thiophene,** b.p. 135–140°/0.1 mm, m.p. 60–62°; 1-*methyl*-3-*phenyl*-1,3-*dihydrobenzo*[c]*thiophene*, b.p. 138–148°/0.1 mm; 3,3-*dimethyl*-1-*phenyl*-1,3-*dihydrobenzo*[c]*thiophene*, m.p. 70–72° (*P. V. Petersen, N. Lassen* and *T. Ammitzboell*, S. African P 6,800,199/1969; C.A., 1970, **72,** 90271c).

The initial product of the action of sulphur dioxide on diphenyldiazomethane gives on heating a yellow substance, converted by pyridine into colourless 1,1,3-**triphenyl**-1,3-**dihydrobenzo**[*c*]**thiophene** 2,2-**dioxide** (XV) (*H. Kloosterziel* and *H. J. Backer*, Rec. Trav. chim., 1952, **71,** 1235):

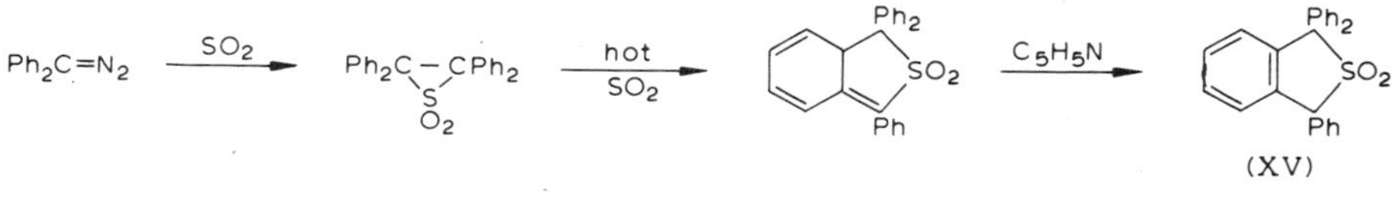

(XV)

The final product loses sulphur on pyrolysis, giving derivatives of dihydroanthracene.

Phthalic anhydride (or acid) may be reduced by hydrogen sulphide to **1-oxo-1,3-dihydrobenzo[*c*]thiophene** (2-*thiophthalide*), m.p. 59–60°, in high yield, produced also by rearrangement of 1-thiophthalide (XVI) on heating with aniline in a sealed tube (*V. Prey et al.*, Monatsh., 1958, **89,** 5051; 1960, **91,** 319):

(XVI) —190°→

L. M. Yagupol'skii and *R. V. Belinskaya* (Zhur. obshcheĭ Khim., 1966, **36,** 1414) describe 1,1,3,3-**tetrachloro**-1,3-**dihydrobenzo**[*c*]**thiophene,** m.p. 112–113°, some tetrafluoro-1,3-dihydrobenzo[*c*]thiophenes and related derivatives; 4,5,6,7-**tetrafluoro**-1,3-**dihydrobenzo**[*c*]**thiophene,** m.p. 33–34° (*P. L. Coe, B. T. Croll* and *C. R. Patrick*, Tetrahedron, 1967, 505).

(ii) Tetrahydrobenzo[c]*thiophenes*

By condensation with succinic anhydride, followed by reduction and ring-closure, 2,5-dimethylthiophene gives 4-*oxo*-1,3-*dimethyl*-4,5,6,7-*tetrahydrobenzo*[c]*thiophene* (XVII), m.p. 46°, and thence 1,3-**dimethyl-4,5,6,7-tetrahydrobenzo**[*c*]**thiophene,** b.p. 245° (*W. Steinkopf et al.*, Ann., 1938, **536,** 128; *N. P. Buu-Hoï et al.*, Rec. Trav. chim., 1950, **69,** 1053):

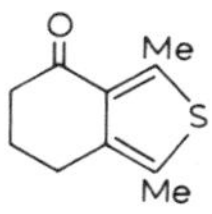

(XVII)

(iii) Hexa- and octa-hydrobenzo[c]*thiophenes*

Reduction of cyclohexene- or cyclohexane-1,2-dicarboxylic anhydride with hydrogen and hydrogen sulphide at 145° gives 3,4,5,6-tetrahydrothiophthalide (1-oxo-1,3,4,5,6,7-hexahydrobenzo[*c*]thiophene), expecially when carried out in a steel autoclave (*W. G. Toland* and *R. W. Campbell*, J. org. Chem., 1963, **28,** 3124):

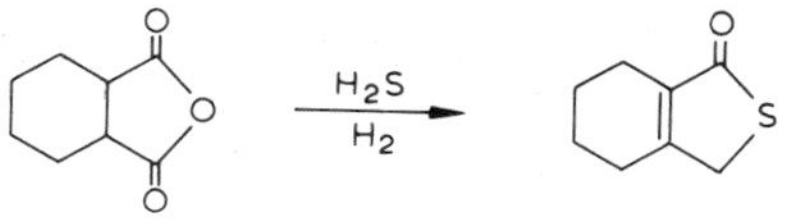

—H_2S, H_2→

Stereoselectivity has been observed in the oxidation of 1,3,3a,4,7,7a-hexahydrobenzo[*c*]thiophene to the 2-oxide and 2,2-dioxide in the presence of *Aspergillus niger*. The absolute configurations of these compounds are known because the (−)-enantiomers have been prepared from the (−)-di-acid (XVIII) (*B. J. Auret, D. R. Boyd* and *H. B. Henbest*, J. chem. Soc., C, 1968, 2374):

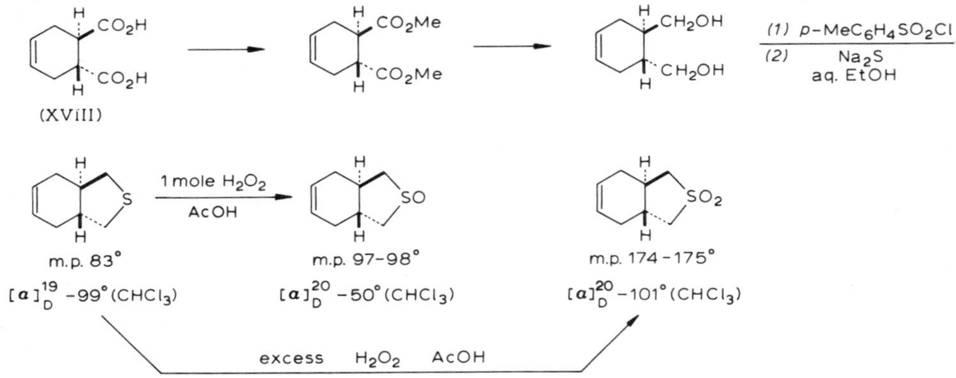

(XVIII)

(1) p-MeC$_6$H$_4$SO$_2$Cl
(2) Na$_2$S aq. EtOH

1 mole H$_2$O$_2$ AcOH

m.p. 83° $[\alpha]_D^{19}$ −99° (CHCl$_3$)

m.p. 97–98° $[\alpha]_D^{20}$ −50° (CHCl$_3$)

m.p. 174–175° $[\alpha]_D^{20}$ −101° (CHCl$_3$)

excess H$_2$O$_2$ AcOH

By successive treatment with hydrogen bromide and sodium sulphide, the 1,2-bishydroxymethylcyclohexanes give the stereoisomeric **octahydrobenzo[*c*]thiophenes,** *cis*, b.p. 108.5°/21 mm, d^{25} 1.0388, 2,2-*dioxide*, m.p. 39.5–41°; *trans*, b.p. 105°/20 mm, d^{25} 1.0162, 2,2-*dioxide*, m.p. 105–105.5° (*S. F. Birch et al.*, J. org. Chem., 1954, **19,** 1449). A study has been made of the optical properties of octahydrobenzo[*c*]thiophene (XIX) [(−)-(8*R*,9*R*)-*trans*-2-thiahydrindan] and ORD curves obtained (*P. Law et al.*, *ibid.*, 1967, **32,** 498):

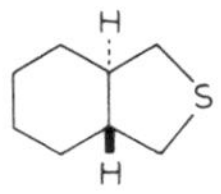

(XIX)

Octahydrobenzo[*c*]thiophene may be prepared by the reduction of 1,2,3,6-tetrahydrophthalic anhydride in the presence of hydrogen sulphide and cobalt sulphide at 180–325° (*R. W. Campbell*, U.S.P. 3,345,381/1965; C.A., 1968, **68,** 29587m).

4. Cycloalkanothiophenes

The combination of sulphur dioxide with 1-vinylcycloalkenes of type I can be applied to the preparation of the sulphones (II, $n = 3$–6), which are isomerised by alkali, and are converted into fully-reduced 2,3-cycloalkanotetrahydrothiophenes (III) by the following route (*H. J. Backer* and *J. R. van der Bij*, Rec. Trav. chim., 1943, **62,** 561; *S. F. Birch et al.*, J. org. Chem., 1955, **20,** 1178):

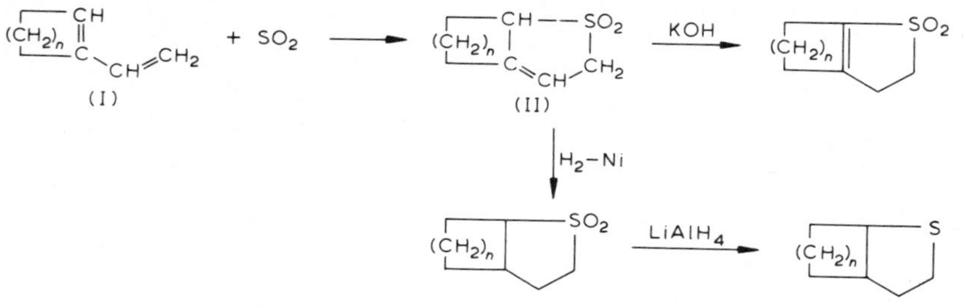

(I) (II) (III)

Isopropenyl 2-thienyl ketone, CH_2:CHMe·CO·C_4H_3S, is cyclised by sulphuric acid to 5-methyl-6-oxocyclopenteno[*b*]thiophene (IV) (*J. H. Burckhalter* and *J. Sam*, J. Amer. chem. Soc., 1951, **73,** 4460). The fully reduced *cis*-**cyclopentano[*b*]thiophene,** b.p. 88°/25 mm, d^{25}1.0379, 1,1-*dioxide*, m.p. 36°, is prepared as shown above for III. The isomeric 3,4-**cyclopentanotetrahydrothiophene, (tetrahydrocyclopentano[*c*]thiophene)** (V), is obtained in *cis* and *trans* forms, *cis*, b.p. 88°/25 mm, d^{25} 1.0386, n_D^{31} 1.5218, 2,2-*dioxide* (S-*dioxide*), m.p. 72°; *trans*, b.p. 96°/28 mm, m.p. 28°, n_D^{31} 1.5233, 2,2-*dioxide*, m.p. 120°, from the corresponding cyclopentanedicarboxylic acids by reduction to the glycols, followed by reaction of their toluenesulphonic esters with sodium sulphide (*Birch et al., loc. cit.; L. N. Owen* and *A. G. Peto*, J. chem. Soc., 1955, 2383):

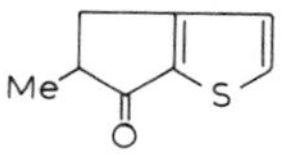

(IV) 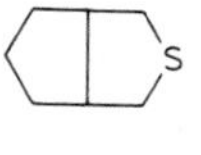 (V) 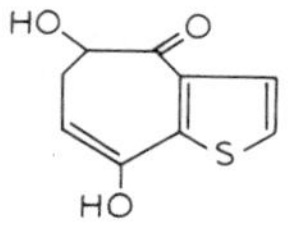 (VI)

Condensation of thiophene with the mono-ethyl ester chloride of glutaric acid, followed by the usual processes of reduction, ring-closure, and reduction, gives 2,3-**cycloheptanothiophene (cycloheptano[*b*]thiophene),** b.p. 99°/12.5 mm (*D.* and *P. Cagniant*, Bull. Soc. chim. Fr., 1955, 680); the intermediate cyclic ketone is oxidised by selenium dioxide to the **hydroxythiophenotropolone** (VI), m.p. 184° (*W. Heyer* and *W. Treibs*, Ann., 1955, **595,** 203). Heating 1,3-dimethyl-6*H*-cyclohepta[*c*]thiophen-6-one (VII) with 1,2,3,4-tetrachlorocyclopentadiene in acetic anhydride gives the potentially dipolar 6π–10π system VIII (*G. Seitz* and *H. Mönnighoff*, Angew. Chem., intern. Edn., 1970, **9,** 907):

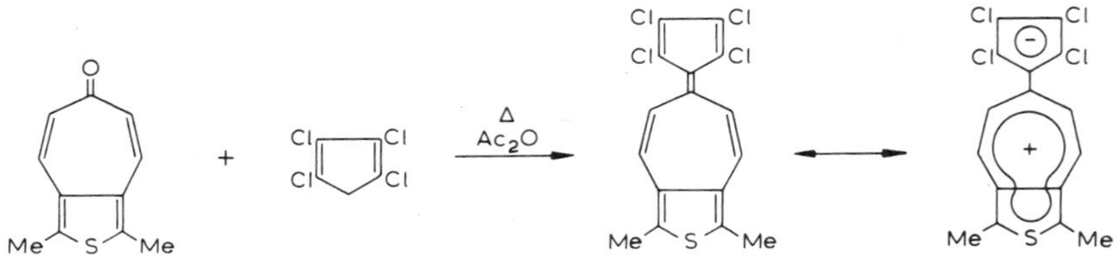

(VII) (VIII)

Thienotropylium perchlorate (IX) has been synthesised and in contrast to the oxygen analogue can be recrystallised from water without decomposition (*D. Sullivan* and *R. Pettit*, Tetrahedron Letters, 1963, 401):

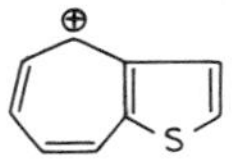

(IX) 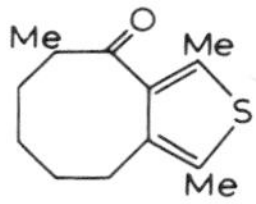 (X)

Ketones of type X have been obtained by cyclisation of ω-(3-thienyl)alkanoyl chlorides with stannic chloride–carbon disulphide or of the corresponding acids with polyphosphoric acid (*P. Cagniant*, *G. Merle* and *D. Cagniant*, Bull. Soc. chim. Fr., 1970, **1**, 322).

5. Systems of two fused thiophene rings

In most compounds of this group, the thienothiophenes (thiophthenes), two thiophene rings have two carbon atoms in common; with the nuclei unreduced three isomerides are possible:

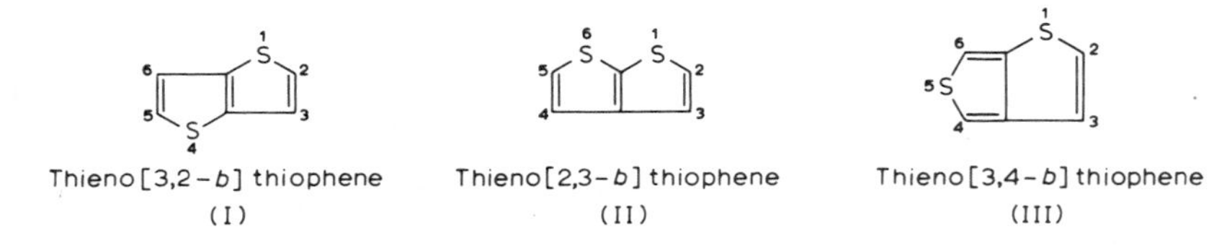

Thieno[3,2-*b*]thiophene (I) Thieno[2,3-*b*]thiophene (II) Thieno[3,4-*b*]thiophene (III)

Acetylene and sulphur at 440° yield, among other products, two thienothiophenes, of which one, m.p. 56° (*picrate*, m.p. 146°), having zero dipole moment, must be **thieno[3,2-*b*]thiophene** (I); the other, m.p. 6°, μ 1.03 D (*picrate*, m.p. 137–138°), regarded as **thieno[2,3-*b*]thiophene** (II), is produced also by distilling citric acid with "P_2S_3" (*F. Challenger* and *J. B. Harrison*, J. Inst. Pet., 1935, **21**, 135). Both substances have been synthesised by ring-closure of the 2- and 3-(2′,2′-dialkoxyethylthio)thiophenes, $(C_4H_3S)\cdot S\cdot$-$CH_2\cdot CH(OR)_2$, obtained by reaction of thiophene-2- and -3-thiols with

β-bromoacetals (*V. V. Ghaisas* and *B. D. Tilak*, C.A., 1954, **48,** 12731). The same synthesis, applied to 2,5-dimethylthiophene-3-thiol gives a dimethyl derivative of III, 4,6-**dimethylthieno**[3,4-*b*]**thiophene** (*picrate*, m.p. 116°) (*O. Dann* and *W. Dimmling*, Ber., 1954, **87,** 373).

Compound II undergoes Friedel–Crafts acetylation α to the sulphur atom (*Challenger* and *B. Fishwick*, J. Inst. Pet., 1953, **39,** 220). For the preparation of 2-methylthieno-[3,2-*b*]thiophene see *S. Gronowitz, U. Ruden* and *B. Gestblom*, Arkiv Kemi, 1963, **20,** 297.

Thieno[2,3-*b*]thiophene and thieno[3,2-*b*]thiophene react with butyl-lithium in ether to give metallation at the α-position. Both compounds are about six times as reactive as thiophene itself (*A. Bugge*, Acta Chem. Scand., 1968, **22,** 63).

Methyl thiophene-2-carboxylate may be chloromethylated and the product subjected to ring closure, hydrolysis, and decarboxylation to give 4,6-**dihydrothieno**[3,4-*b*]**thiophene** (IV), b.p. 78–79°/1.2 mm (*D. J. Zwanenburg, H. de Haan* and *H. Wynberg*, J. org. Chem., 1966, **31,** 3363):

MeO2C–thiophene →(ClCH2OMe, ZnCl2)→ MeO2C–thiophene(CH2Cl)(CH2Cl) →(Na2S, MeOH)→ MeO2C–dihydrothienothiophene →(Cu, quinoline)→ (IV)

which may be synthesised also by the following route (*D. W. H. MacDowell* and *T. B. Patrick*, *ibid.*, p. 3592):

3-(1,3-dioxolan-2-yl)thiophene →((1) BuLi, (2) D.M.F.)→ 2-CHO derivative →(H⊕)→ thiophene-2,3-dicarbaldehyde (CHO, CHO) →(LiAlH4)→ CH2OH, CH2OH →(PBr3)→ CH2Br, CH2Br →(Na2S)→ dihydrothieno[3,4-b]thiophene

Its sulphoxide (V) on boiling in acetic anhydride gives acetic thieno[3,4-*b*]-thiophene-2-carboxylic anhydride which on hydrolysis with aqueous methanol and acidification affords **thieno**[3,4-*b*]**thiophene-2-carboxylic acid,** m.p. 210° decomp., *methyl ester*, m.p. 65–66.5°. The acid on decarboxylation yields **thieno**[3,4-*b*]**thiophene** (VI), m.p. 7–7.5°, n_D^{20} 1.6905, *picrate*, m.p. 113° (*Wynberg* and *Zwanenburg*, Tetrahedron Letters, 1967, 761):

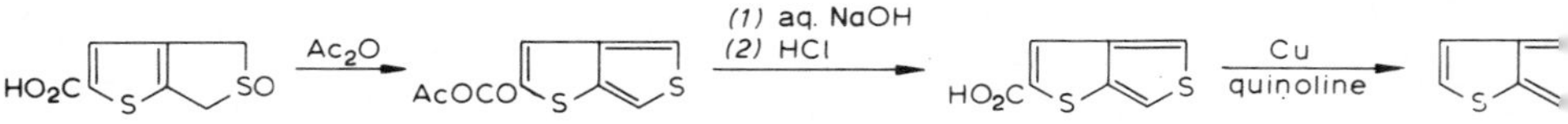

HO$_2$C (V) SO —Ac$_2$O→ AcOCO S S —(1) aq. NaOH (2) HCl→ HO$_2$C S S —Cu, quinoline→ (VI)

Evidence for the transient existence of a derivative of thieno[3,4-*c*]thiophene (VIII), a reactive tetravalent sulphur intermediate, is indicated by heating the sulphoxide VII in acetic anhydride in the presence of *N*-phenylmaleimide, when the *exo*-adduct IX and its *endo*-isomer are isolated (*M. P. Cava* and *N. M. Pollack*, J. Amer. chem. Soc., 1967, **89**, 3639):

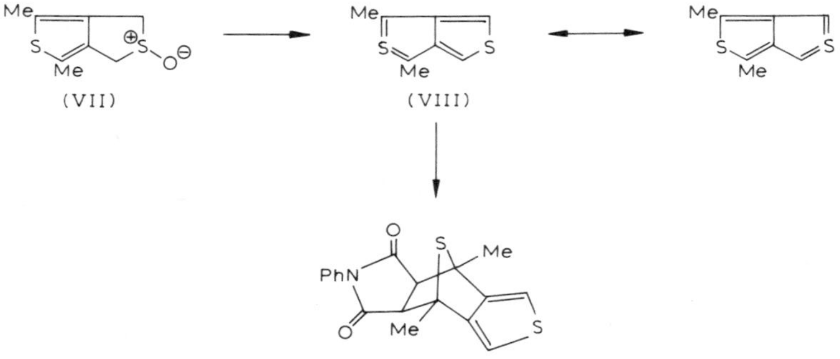

(VII) → (VIII) ↔

↓

(IX)

In order to study the effect of ring strain on the properties of five-membered heteroaromatics **1,3-dihydrothieno**[3,4-*c*]**thiophene** (1*H*,3*H*-**thieno**[3,4-*c*]-**thiophene**) (X), m.p. 61–62°, has been synthesised using a greatly improved thiophene ring synthesis (*Wynberg* and *Zwanenburg*, J. org. Chem., 1964, **29**, 1919):

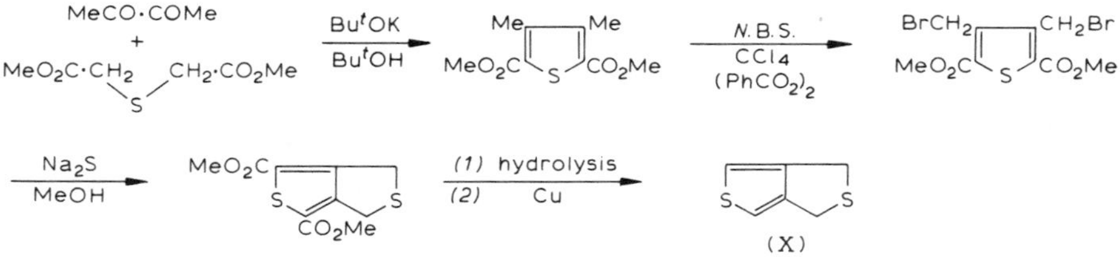

MeCO·COMe + MeO$_2$C·CH$_2$–S–CH$_2$·CO$_2$Me —Bu^{t}OK / Bu^{t}OH→ —N.B.S., CCl$_4$, (PhCO$_2$)$_2$→ BrCH$_2$, CH$_2$Br, MeO$_2$C, CO$_2$Me —Na$_2$S / MeOH→ MeO$_2$C, CO$_2$Me —(1) hydrolysis (2) Cu→ (X)

3-Phenylbenzo[*b*]thiophene on heating with sulphur is converted into **benzothieno**[2,3-*b*]**benzothiophene** (XI), but the same reaction in the presence of aluminium chloride affords **benzothieno**[3,2-*b*]**benzothiophene** (XII) (*T. S. Murthy*, *L. J. Pandya* and *B. D. Tilak*, J. Sci. Ind. Res., India, 1961, **20B**, 169):

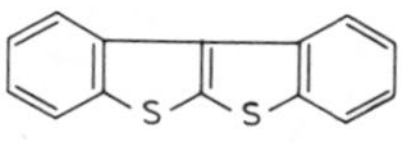

(XI) 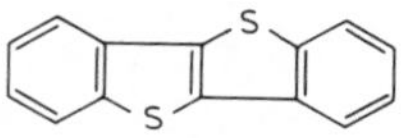 (XII)

Cyclisation of 1,4-bis-(ω-dimethoxyethylthio)benzene gives **benzo**[1,2-*b*:4,3-*b*′]**dithiophene** (XIII), m.p. 118°, *picrate*, m.p. 148–149°; **benzo-**[1,2-*b*:4,5-*b*′]**dithiophene** (XIV), m.p. 198°, *picrate*, m.p. 159° (*D. S. Rao* and *Tilak*, *ibid.*, 1957, **16B,** 65):

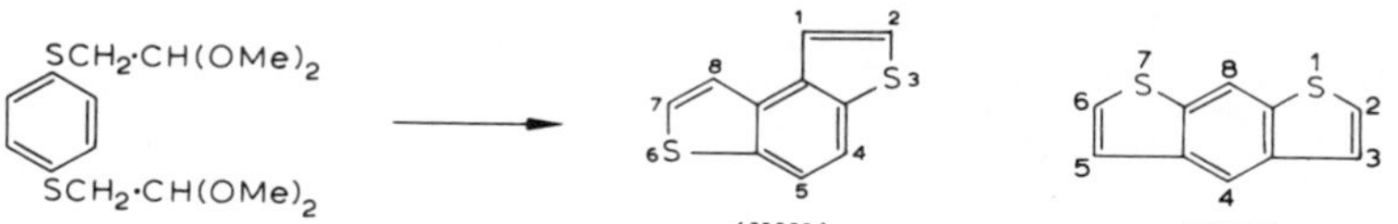

(XIII) (XIV)

Benzo[1,2-*b*:4,3-*b*′]dithiophene (XIII) may be obtained also from *trans*-1,2-di-(2-thienyl)ethene (p. 215) and the same method affords thieno-[3,2-*e*]benzofuran from *trans*-1-(2-furyl)-2-(2-thienyl)ethene.

After several routes had been tried in vain to synthesise **cyclopenta-**[1,2-*b*:4,3-*b*′]**dithiophene,** m.p. 66–67°, a thiophene analogue of fluorene, the following synthetic scheme proved successful (*Wynberg* and *A. Kraak*, J. org. Chem., 1964, **29,** 2455):

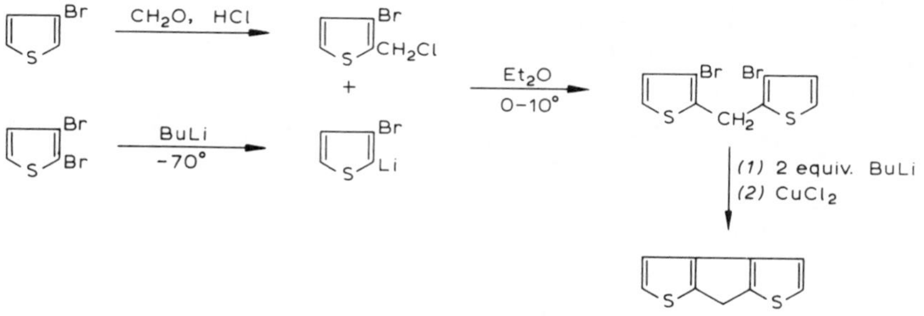

Two hexaheterohelicenes, benzo[*d*]naphtho[1,2-*d*′]benzo[1,2-*b*:4,3-*b*′]dithiophene (XV) and thieno[3,2-*e*]benzo[1,2-*b*:4,3-*b*′]bisbenzothiophene (XVI) and a heptaheterohelicene have been prepared and resolved (*Wynberg* and *M. B. Groen*, J. Amer. chem. Soc., 1968, **90,** 5339; Chem. Comm., 1969, 964) and the crystal structure of XV has been determined (*G. Stulen* and *G. J. Visser*, *ibid.*, 1969, 965). An optically active undecaheterohelicene has been synthesised starting from a partially resolved heptaheterohelicene (*Wynberg* and *Groen*, J. Amer. chem. Soc., 1970, **92,** 6664):

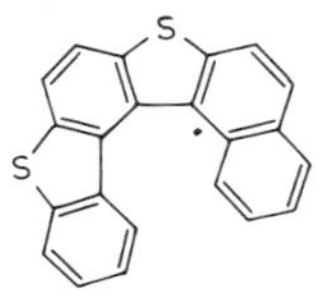

(XV) 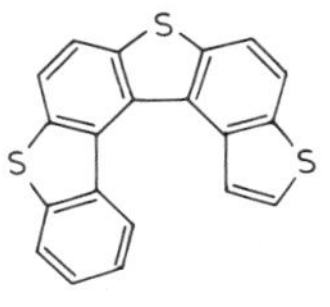 (XVI)

A highly toxic sulphonium salt, 1-**thioniabicyclo**[2.2.1]**heptane chloride** (XVII), has been prepared, in which two hydrogenated thiophene nuclei have the sulphur and two carbon atoms in common (*V. Prelog* and *E. Cerkovnikov*, Ann., 1939, **537,** 214; *cf. W. F. Cockburn* and *A. F. Mackay*, J. Amer. chem. Soc., 1954, **76,** 5703):

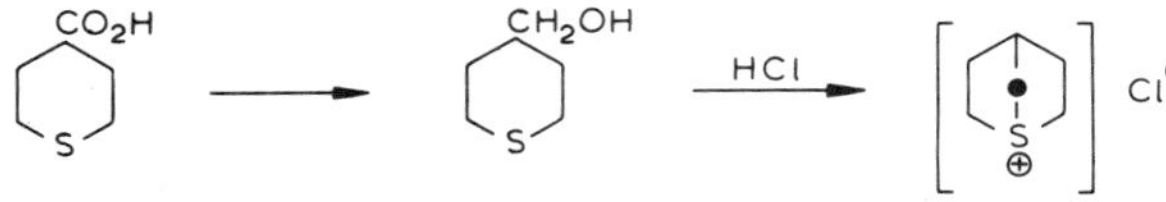

CO_2H → CH_2OH —HCl→ $Cl^{\ominus}$

(XVII)

6. Dibenzothiophenes

(a) Dibenzothiophene and its derivatives

Dibenzothiophene (diphenylene sulphide, I), present in coal tar (*O. Kruber*, Ber., 1920, **53,** 1566) along with the 4,6-dimethyl derivative which also occurs in Kuwait petroleum (*Kruber* and *A. Raeithel*, Ber., 1954, **87,** 1469; *W. Carruthers*, Nature, 1955, **176,** 790), was first obtained by pyrolysis of diphenyl sulphide, and more effectively by dehydrating diphenyl sulphoxide with sodamide or phenyl-sodium (*A. Schönberg*, Ber., 1923, **56,** 2275; *F. Fuchs*, Monatsh., 1929,**53/54,** 438). The older system of numbering (II) has been gradually replaced by scheme (I) used here.

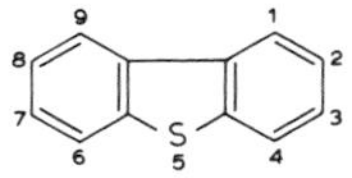

(I) 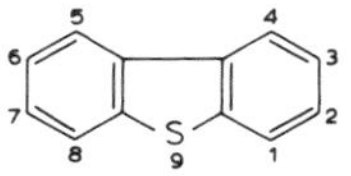 (II)

(i) Preparation and synthesis

(*1*) Dibenzothiophene is best prepared by fusing biphenyl with sulphur and aluminium chloride (*H. Gilman* and *A. L. Jacoby*, J. org. Chem., 1938, **3,** 108), and is also produced by appropriate diazo-reactions from 2,2′-diaminobiphenyl (*H. W. Schwechten*, Ber., 1932, **65,** 1608) or from 2-aminodiphenyl sulphide (*N. M. Cullinane et al.*, J. chem. Soc., 1939, 151).

(*2*) Dibenzothiophene has been synthesised from thiophene, by succinoylation and subsequent reduction to ethyl γ-(2-thienyl)butyrate, followed by further succinoylation to an oxo acid, reduction and cyclisation to give III, which is then submitted to reduction of the carbonyl group, cyclisation, further reduction, and dehydrogenation as shown in the following reaction scheme (*A. Rahman* and *O. L. Tombesi*, Tetrahedron Letters, 1968, 3925):

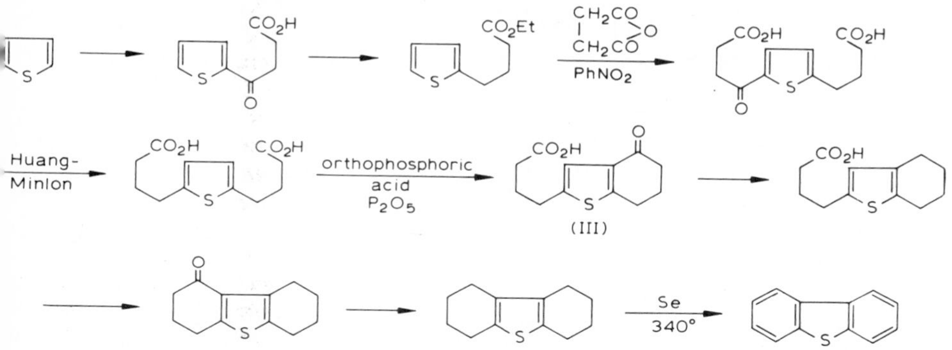

CO_2H CO_2Et CH_2CO CH_2CO $PhNO_2$ CO_2H CO_2H

Huang-Minlon CO_2H CO_2H orthophosphoric acid P_2O_5 CO_2H (III) CO_2H

Se 340°

(*3*) Cyclisation of the dilithium derivative of diphenyl sulphone with cupric chloride gives a 20% yield of dibenzothiophene and 50% of its 5,5-dioxide, which may be reduced to the former by lithium tetrahydridoaluminate (*V. N. Gogte et al., ibid.*, 1960, 30):

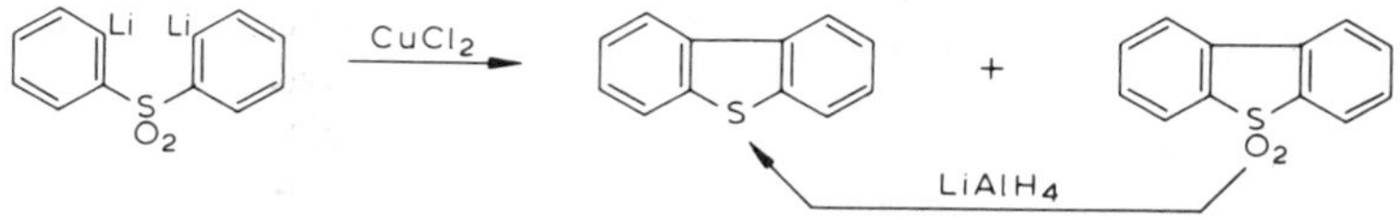

Li Li $CuCl_2$ + $LiAlH_4$

(*4*) Dibenzothiophene is formed along with other products from the reaction between benzene and sulphur dioxide at 500° (*A. Levy* and *C. J. Ambrose*, J. Amer. chem. Soc., 1959, **81**, 249).

(5) Thianthrene 5-oxide on addition to butyl-lithium in ether gives dibenzothiophene and some thianthrene (*Gilman* and *D. R. Swayanpati*, J. Amer. chem. Soc., 1955, **77**, 3387; J. org. Chem., 1956, **21**, 1278):

BuLi

(*6*) Various dibenzothiophene derivatives have been prepared by use of a halogen–metal interconversion, *e.g.*, 1-bromodibenzothiophene on treatment with butyl-lithium and dimethyl sulphate gives 1-methyldibenzothiophene (*Gilman* and *G. R. Wilder*, J. org. Chem., 1957, **22**, 523).

(7) All the isomeric monophenyl derivatives of dibenzothiophene have

been synthesised and the free radical phenylation of dibenzothiophene has been discussed (*E. B. McCall et al.*, J. chem. Soc., 1962, 4900, 5288).

(*8*) Bis(2-biphenyl) disulphide obtained from 2-aminobiphenyl, reacts with iodine to give dibenzothiophene and a small amount of 2,3-dihydro-7-phenylbenzo[*b*]thiophene when ethylene glycol is employed as solvent (*E. Campaigne et al.*, J. org. Chem., 1962, **27,** 4111):

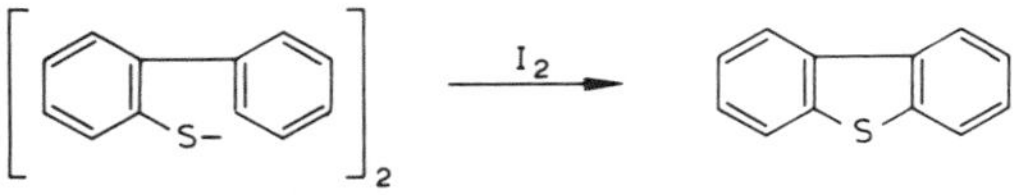

I_2

(ii) Properties and reactions

Dibenzothiophene, m.p. 99°, is successively oxidised by hydrogen peroxide to *dibenzothiophene 5-oxide* (S-*oxide*), m.p. 188°, and *dibenzothiophene 5,5-dioxide*, m.p. 232–233° (*Gilman* and *D. L. Esmay*, J. Amer. chem. Soc., 1952, **74,** 2021); oxidation of dibenzothiophene by hydrogen peroxide in acetic acid has been studied kinetically (*B. N. Heimlich* and *T. J. Wallace*, Tetrahedron, 1966, 3571), and the rates of the two stages of oxidation with perbenzoic acid in 1:1 aqueous dioxane at 25° have been measured (*A. Greco, G. Modena* and *P. E. Todesco*, Gazz., 1960, **90,** 6711). A kinetic study has been made of the formation of dibenzothiophene 5-oxide from dibenzothiophene and chlorine in aqueous acetic acid (*E. Baciocchi* and *L. Mandolini*, Ric. Sci., 1967, **37,** 863). Raney cobalt has been used to desulphurise dibenzothiophene, but it is less effective than nickel (*G. M. Badger, N. Kowando* and *W. H. F. Sasse*, J. chem. Soc., 1959, 440).

(iii) Derivatives of dibenzothiophene

Substitution of dibenzothiophene usually takes place in position 2, numbering as in I (*Cullinane et al.*, J. chem. Soc., 1936, 1435; *Gilman* and *Jacoby*, *loc, cit.*); thus the products of *nitration* have been oriented by synthesis:

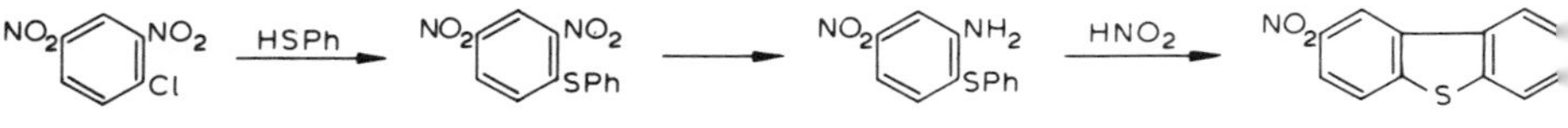

NO_2 NO_2 Cl HSPh NO_2 NO_2 SPh → NO_2 NH_2 SPh HNO_2 → NO_2 S

The 5,5-dioxide, on the other hand, is nitrated in positions 3 and 7, since reduction of the product gives "benzidine sulphone", also obtained by heating benzidine with sulphuric acid *(Cullinane, loc. cit.)*.

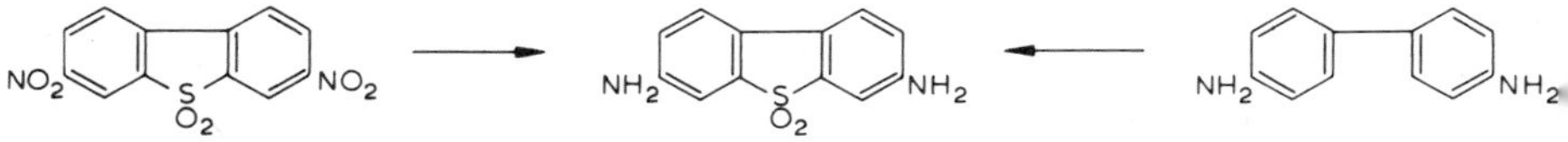

NO_2 S O_2 NO_2 → NH_2 S O_2 NH_2 ← NH_2 NH_2

The 5-oxide also is nitrated in position 3; the product can be oxidised to the known 3-nitrodibenzothiophene 5,5-dioxide and reduced to 3-**amino-**

dibenzothiophene (*R. K. Brown et al.*, J. Amer. chem. Soc., 1948, **70,** 1748). This amine, which is a source of 3-derivatives of dibenzothiophene, is obtained, surprisingly, by treating 4-iodobenzothiophene with sodamide (*Gilman* and *J. F. Nobis*, *ibid.*, 1945, **67,** 1479).

The difference in reactivity between positions *ortho* and *para* to the bridging group in dibenzothiophene is explained by a theory recently proposed to account for the Mills–Nixon effect (*R. Taylor*, J. chem. Soc., B, 1968, 1559). Homolytic phenylation of dibenzothiophene takes place at the 1, 2, 3 and 4 positions in the ratio 3:1:2:3 (*McCall*, *A. J. Neale* and *T. J. Rawlings*, J. chem. Soc., 1962, 5288).

An ethereal solution of dibenzothiophene and butyl-lithium after boiling for a long time and then carbonating gives **dibenzothiophene-4-carboxylic acid** (*Gilman* and *S. H. Eidt*, J. Amer. chem. Soc., 1956, **78,** 2633); comparisons have been made using methyl-, butyl-, and phenyl-lithium in ether and/or tetrahydrofuran (*Gilman* and *S. Gray*, J. org. Chem., 1958, **23,** 1476). Dibenzothiophene is readily cleaved by lithium in tetrahydrofuran and if the product is carbonated, then 3,4-benzothiocoumarin and some 2-mercaptobiphenyl-2′-carboxylic acid are obtained (*Gilman* and *J. J. Dietrich*, *ibid.*, 1957, **22,** 851):

S —Li, T.H.F., CO_2→ S, O + SH CO_2H

Benzyl chloride reacts with dibenzothiophene in the absence of catalyst to give the four *monobenzyl derivatives*, but in the presence of aluminium chloride, the 4-isomer is obtained mainly (*Monsanto Chemicals Ltd.*, F.P. 1,351,472/1964; C.A., 1964, **61,** 9468a). **Aryl-** and **cycloalkyl-dibenzothiophenes** have been prepared from the necessary bromo derivatives *via* the Grignard reaction with a cyclohexanone, *e.g.*, the 1-, 2-, 3-, and 4-(1-cyclohexenyl) derivatives of dibenzothiophene have been obtained and dehydrogenated to the corresponding phenyl derivatives using a palladium–carbon catalyst (*idem*, Fr. Addn. 80,790/1963; C.A., 1964, **60,** 1704e). 2-Bromodibenzothiophene on treatment with phenol and potassium phenoxide in the presence of copper bronze at 185–195° gives 2-**phenoxydibenzothiophene.** These types of materials are thermally stable and useful as hydraulic fluids and heat-transfer media (*idem*, B.P. 932,813/1960; C.A., 1964, **60,** 504c). **Octafluorodibenzothiophene** (IV) has been prepared by nucleophilic attack of pentafluorothiophenolate anion on tetrafluorobenzyne followed by ring closure (*R. D. Chambers* and *D. J. Spring*, Tetrahedron Letters, 1969, 2481);

it may also be obtained by Ullmann coupling of bis(*o*-bromotetrafluorophenyl) sulphide, and undergoes nucleophilic substitution in the 2-position by methoxide ion (*Chambers, J. A. Cunningham* and *Spring*, Tetrahedron, 1968, 3997):

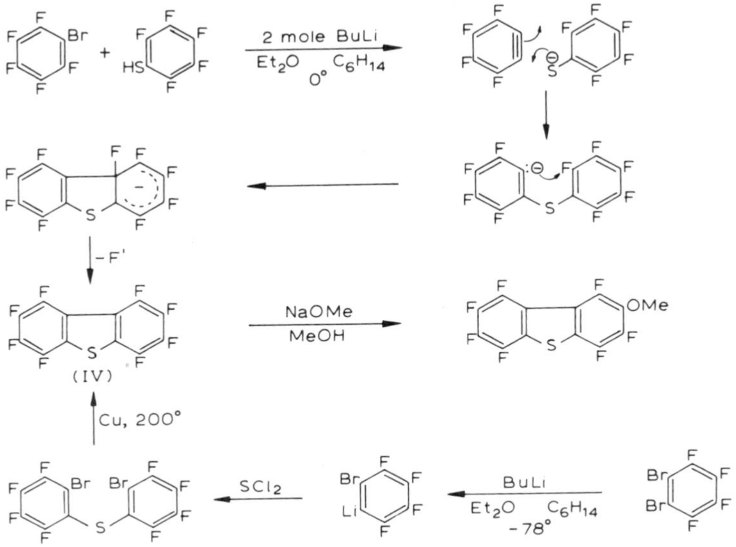

The pyrolysis of **dibenzothiophene** 5,5-**dioxide** gives dibenzofuran instead of biphenylene, the expected product of sulphur dioxide extrusion. This decomposition closely parallels the behaviour of the dioxide under electron impact (*E. K. Fields* and *S. Meyerson*, J. Amer. chem. Soc., 166, **88,** 2836). The heterocyclic ring of dibenzothiophene 5,5-dioxide is opened by fusion with alkali, giving biphenyl-2-sulphonic acid, a process which can be reversed by the Friedel–Crafts reaction; the method has been used in orientating 2-bromodibenzothiophene (*M. Chaix* and *F. de Rocheboüet*, Bull. Soc. chim. Fr., 1935, [v], **2,** 273). Decomposition of dibenzothiophene 5,5-dioxide in excess of sodium or potassium hydroxide at 300° gives a high yield of dibenzofuran (*Wallace* and *Heimlich*, Tetrahedron, 1968, 1311; Chem. and Ind., 1966, 1885). The well-characterised **dibenzothiophenecarboxylic acids** serve as reference compounds in orientation. Hydroxydibenzothiophenecarboxylic acids are recorded as dyestuff intermediates (*F. Muth*, G.P. 593,506/1934; C.A., 1934, **28,** 3422).

Substituted dibenzothiophenes are listed in Table 6.

TABLE 6

SUBSTITUTED DIBENZOTHIOPHENES

Substituents	*m.p. (°C)*	*Ref.*
1-Methyl	67–68	1
2-Methyl	88–89	1
3-Methyl	78–79	1
4-Methyl	65	2
1-Phenyl	45.5–48	3
2-Phenyl	73–74	3
3-Phenyl	178.5–179.5	3
4-Phenyl	67.5–69	3
2,8-Dimethyl	122–123	1
1-Chloro	88	1
3-Chloro	80–81	1
4-Chloro	84–85	1
1-Bromo	84	4
2-Bromo	127	5
4-Bromo	83–84	6
2,8-Dibromo	229	7
1-Iodo	78–79	1
2-Iodo	87–88	7
3-Iodo	112–113	1
2-Hydroxy	159	5
4-Hydroxy	167	
1-Nitro	97	4
2-Nitro	186	5
2,8-Dinitro	324–325 dec.	8
2-Amino	133	5
3-Amino*	122	
4-Amino	110	9
2-$SiPh_3$	153–154	10
2-Acetyl	111	2
1-CO_2Me	72	2
2-CO_2Me	75	2
3-CO_2Me	130	11
4-CO_2Me	95	2
Octafluoro	99–100	12
Heptafluoro-2-methoxy	105–107	12

* *Acetyl derivative* carcinogenic (*E. C. Miller et al.*, C.A., 1949, **43,** 9238).

References
1 *H. Gilman* and *G. R. Wilder*, J. org. Chem., 1957, **22,** 523.
2 *Gilman* and *A. L. Jacoby*, *ibid.*, 1938, **3,** 108.
3 *Monsanto Chemicals Ltd.*, Fr. Addn. 80,790/1963; C.A., 1964, **60,** 1704e.
4 *Gilman* and *Wilder*, J. Amer. chem. Soc., 1954, **76,** 2906.
5 *N. M. Cullinane et al.*, J. chem. Soc., 1939, 151.
6 *Gilman* and *D. L. Esmay*, J. Amer. chem. Soc., 1954, **76,** 5786.
7 *C. Courtot et al.*, Compt. rend., 1928, **186,** 1624.
8 *Courtot* and *C. Pomonis*, *ibid.*, 1926, **182,** 893.
9 *Gilman* and *S. Avakian*, J. Amer. chem. Soc., 1946, **68,** 1514.
10 *R. H. Meen* and *Gilman*, J. org. Chem., 1955, **20,** 73.
11 *Gilman et al.*, *ibid.*, 1938, **3,** 120.
12 *R. D. Chambers, J. A. Cunningham* and *D. J. Spring*, Tetrahedron, 1968, 3997.

(b) Reduced dibenzothiophenes

Dibenzothiophene is reduced by sodium in liquid ammonia to 1,4-**dihydrodibenzothiophene,** m.p. 76° *(Gilman* and *Jacoby, loc. cit.)*. 1,2,3,4-**Tetrahydrodibenzothiophene,** b.p. 180°/3.3 mm, is obtained in good yield by dehydration of 2-oxocyclohexyl phenyl sulphide (*K. Rabindran* and *A. V. Sunthankar*, C.A., 1954, **48,** 10725), or from benzothiophene by the succinic anhydride process (*N. P. Buu-Hoï* and *P. Cagniant*, Ber., 1943, **76,** 1269). Benzothiophene 1,1-dioxide with butadiene yields the *tetrahydrodibenzothiophene* 5,5-*dioxide* (V), m.p. 91° (*O. C. Elmer*, U.S.P. 2,664,426/1953; C.A., 1955, **49,** 1106). When cyclohexanone or cyclohexenylcyclohexanone is boiled with sulphur, 1,2,3,4,6,7,8,9-**octahydrodibenzothiophene** (VI), m.p. 32°, is produced (*W. Cooper*, J. chem. Soc., 1955, 1386):

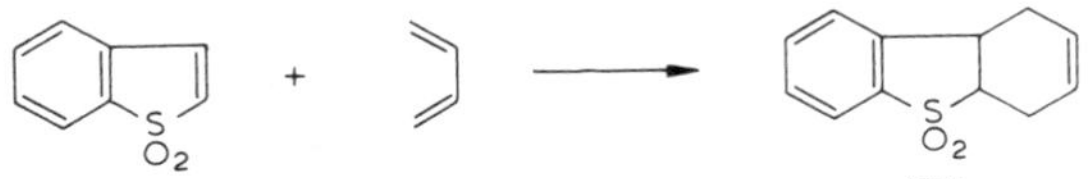

(V) 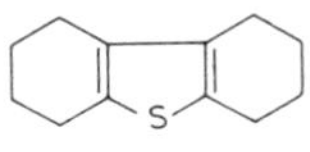 (VI)

Sulphur dioxide and dicyclohexenyl combine to form *decahydrodibenzothiophene* 5,5-*dioxide*, m.p. 78° (*H. J. Backer et al.*, Rec. Trav. chim., 1941, **60,** 381):

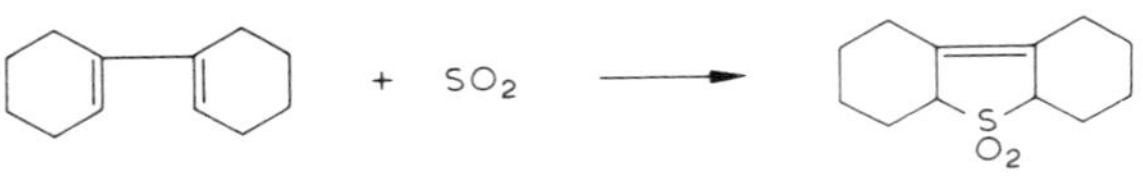

7. Other tri- and poly-cyclic systems containing the thiophene ring

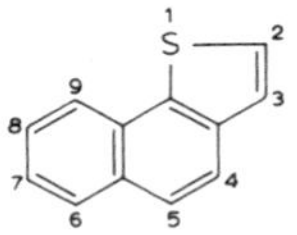

(I)
Naphtho[1,2-*b*]thiophene (6,7-benzothianaphthene)

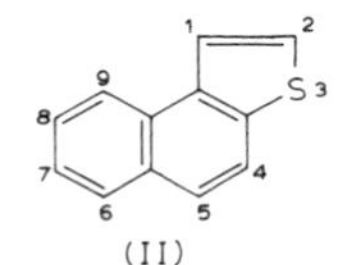

(II)
Naphtho[2,1-*b*]thiophene (4,5-benzothianaphthene)

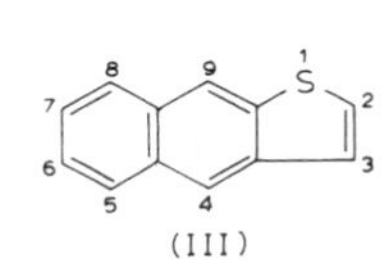

(III)
Naphtho[2,3-*b*]thiophene (5,6-benzothianaphthene)

Naphtho[1,2-*b*]**thiophene** (I), b.p. 120°/0.2 mm, m.p. 27–28°, *picrate* m.p. 146–147.5°, 1,1-*dioxide*, m.p. 179° (*J. E. Banfield et al.*, J. chem. Soc., 1956, 2603), isolated from a coal-tar fraction (*O. Kruber* and *A. Raeithel*, Ber., 1953, **86,** 366), may be obtained by the reduction of 6,7-benzothioindoxyl with zinc dust and acetic acid (*W. Carruthers*, J. chem. Soc., 1953, 4186) and by the cyclisation of 2,2-dimethoxyethyl 1-naphthyl sulphide, which affords also a small amount of naphtho[1,8-*bc*]thiopyran (IV) (*H. S. Desai, D. S. Rao* and *B. D. Tilak*, Chem. and Ind., 1957, 464); the cyclisation of related arylthioacetaldehyde diethyl acetals has been examined and the most reliable, although not infallible, reagent is a mixture of phosphorus pentoxide and phosphoric acid (*Banfield et al., loc. cit.*):

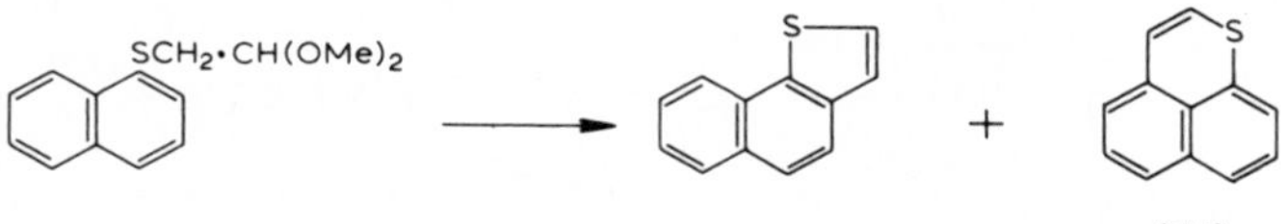

$SCH_2 \cdot CH(OMe)_2$ → [S] + [S]

(IV)

Bis(1-carboxy-2-β-naphthylvinyl) disulphide (V) on treatment with iodine in dioxane at 50° yields naphtho[1,2-*b*]thiophene-2-carboxylic acid; similarly bis(β-2-naphthylethyl) disulphide (VI) with slight excess of iodine in boiling ethylene glycol affords 2,3-**dihydronaphtho**[1,2-*b*]**thiophene** (VII), *picrate*, m.p. 129–131°, whereas naphtho[1,2-*b*]thiophene is obtained when an excess of iodine is used (*E. Campaigne* and *B. G. Heaton*, Chem. and Ind., 1962, 96; J. org. Chem., 1964, **29**, 2372):

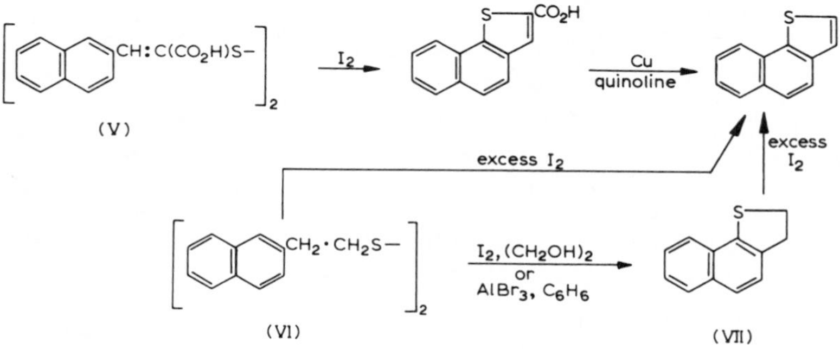

$[CH{:}C(CO_2H)S-]_2$ (V) —I_2→ S, CO_2H —Cu, quinoline→ S

excess I_2

$[CH_2 \cdot CH_2S-]_2$ (VI) —$I_2, (CH_2OH)_2$ or $AlBr_3, C_6H_6$→ S (VII) —excess I_2→

A compound regarded as 2-**phenylnaphtho**[1,2-*b*]**thiophene** (VIII), m.p. 56–57°, is obtained by the cyclisation of 1-naphthyl phenacyl sulphide using sulphuric acid; the appropriate sulphides on heating with zinc chloride afford compounds regarded as 2-(4-**methoxyphenyl)naphtho**[1,2-*b*]**thiophene,** m.p. 164–165°, and 2-(4-**methoxyphenyl)naphtho**[2,1-*b*]**thiophene,** m.p. 157–158° (*Banfield et al.*, J. chem. Soc., 1956, 4791):

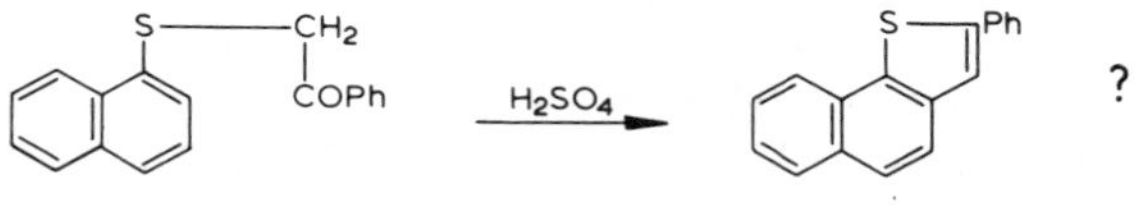

S—CH_2—COPh —H_2SO_4→ S, Ph ?

(VIII)

Naphtho[2,1-*b*]**thiophene** (II), m.p. 111–112°, is prepared from 2-naphthylthioacetaldehyde diethyl acetal in chloroform on treatment with anhydrous stannic chloride (*idem, ibid.*, 1956, 2603); the cyclisation may also be effected with polyphosphoric acid. 4,5,6,7-Tetrahydro-4-oxonaphtho[2,1-*b*]thiophene may also be converted to naphtho[2,1-*b*]thiophene (*Carruthers, A. G. Douglas* and *J. Hill, ibid.*, 1962, 704). *Naphtho*[2,1-*b*]*thiophene* 3,3-*dioxide*, melts at 136–137° or 141–142° with evolution of sulphur dioxide (*W. Davies* and *Q. N. Porter, ibid.*, 1956, 2609). 2-α-Styrylthiophene (IX) in hexane on irradiation with u.v. light in the presence of iodine gives naphtho-[2,1-*b*]thiophene in almost quantitative yield (*Carruthers* and *H. N. M. Stewart*, Tetrahedron Letters, 1965, 301):

hν

(IX)

Methyl derivatives of naphtho[2,1-*b*]thiophene are conveniently prepared by irradiation of appropriately substituted styrylthiophenes (*idem*, J. chem. Soc., 1965, 6221).

2-Vinylthiophene reacts slowly with benzoquinone to give *naphtho*-[2,1-b]*thiophene*-6,9-*quinone* (X), m.p. 167–168°, and so far it has not proved possible to isolate an adduct, which has not undergone dehydrogenation by excess quinone (*Davies* and *Porter, ibid.*, 1957, 4958):

AcOH

$LiAlH_4$ / T. H. F.

(X)

1-**Methylnaphtho**[2,1-*b*]**thiophene,** m.p. 59°, *picrate*, m.p. 152–153°; 1-**phenylnaphtho**[2,1-*b*]**thiophene,** m.p. 88°, 3,3-*dioxide*, m.p. 182°; 2-**phenylnaphtho**[2,1-*b*]**thiophene,** m.p. 108° (*Banfield et al., ibid.*, 1956, 4791; *O. Dann* and *M. Kokorudz*, Ber., 1958, **91,** 172; *T. S. Murthy* and *Tilak*, J. Sci. Ind. Res., 1960, **19B,** 395; C.A., 1961, **55,** 11388c); 6,7,8,9-**tetrahydronaphtho**[2,1-*b*]**thiophene,** b.p. 119–121°/0.3 mm (*Davies* and *Porter*, J. chem. Soc., 1956, 2609).

The nitration of naphtho[2,1-*b*]thiophene gives the 2-nitro derivative as the sole product and formylation, acylation, lithiation, and mercuration also take place in the 2-position. Bromination gives a mixture of the 2-bromo- and 2,5-dibromo derivatives (*K. Clarke, G. Rawson* and *R. M. Scrowston, ibid.*, C, 1969, 537), melting points are recorded in Table 7.

TABLE 7

NAPHTHO[2,1-*b*]THIOPHENE DERIVATIVES

Substituent	*m.p.(°C)*	*Substituent*	*m.p. (°C)*
2-Acetyl	110–112	2,5-Dibromo	119–121
2-Formyl	119–121	2-Carboxylic acid	277–279
2-Bromo	90–92	2-Nitro	90–92
5-Bromo	76–78	2-Propionyl	105–107

Spectral data of naphtho[2,1-*b*]thiophene and 1- and 2-alkyl and -aryl derivatives: i.r. (*F. R. McDonald* and *C. L. Cook*, C.A., 1967, **67**, 58846e); m.s. (*S. Meyerson* and *E. K. Fields*, J. org. Chem., 1968, **33**, 847); u.v. of naphtho-[1,2-*b*]-, -[2,1-*b*]- and -[2,3-*b*]-thiophenes (*Carruthers* and *J. R. Crowder*, J. chem. Soc., 1957, 1932; *N. Trinajstic* and *A. Hinchliffe*, Z. phys. Chem., 1968, **59**, 271).

The substitution of naphtho[1,2-*b*]- and naphtho[2,1-*b*]-thiophenes alkylated in the 2- and 3-, and 1- and 2-positions, respectively, has been studied (*P.* and *D. Cagniant* and *P. Faller*, Bull. Soc. chim. Fr., 1964, 1756). The self-condensation of naphtho-[1,2-*b*]thiophene 1,1-dioxide and of naphtho[2,1-*b*]thiophene 3,3-dioxide (see p. 272) afford a route to 4,2′- (XI) and 1,1′-naphthylphenanthrene (XII), respectively (*Davies* and *Porter*, J. chem. Soc., 1956, 2609):

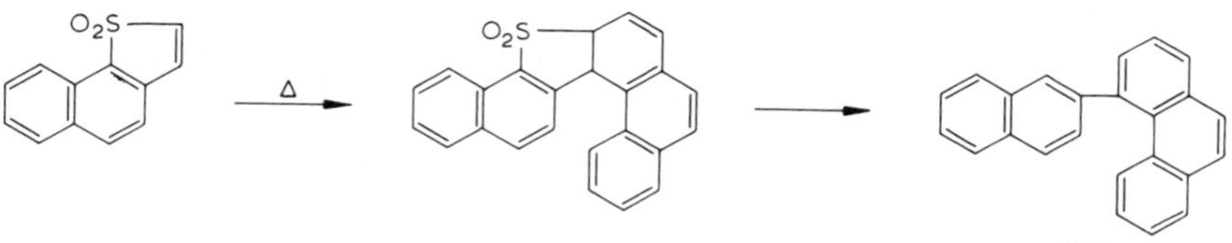

O_2S Δ O_2S

(XI)

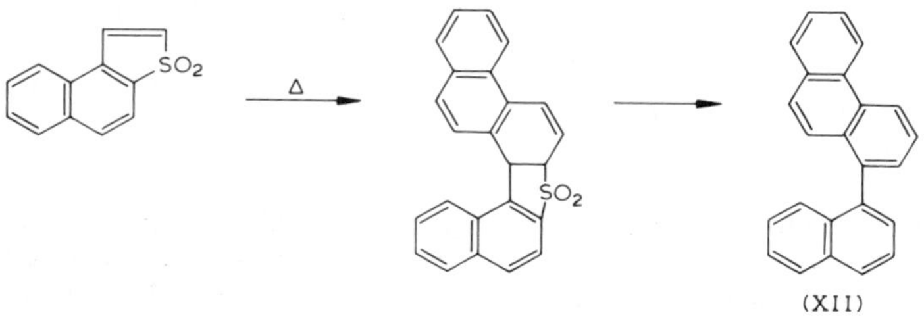

SO_2 Δ SO_2

(XII)

The hydrolysis of *trans*-3,4-dibromo-3,4-dihydro-2*H*-naphtho[1,2-*b*]thiopyran (XIII) abd *trans*-1,2-dibromo-2,3-dihydro-1*H*-naphtho[2,1-*b*]thiopyran (XIV) yields 2-hydroxymethylnaphtho[1,2-*b*]thiophene, and 2-hydroxymethylnaphtho[2,1-*b*]thiophene, respectively (*R. Livingstone et al.*, J. chem. Soc., C, 1972, 787):

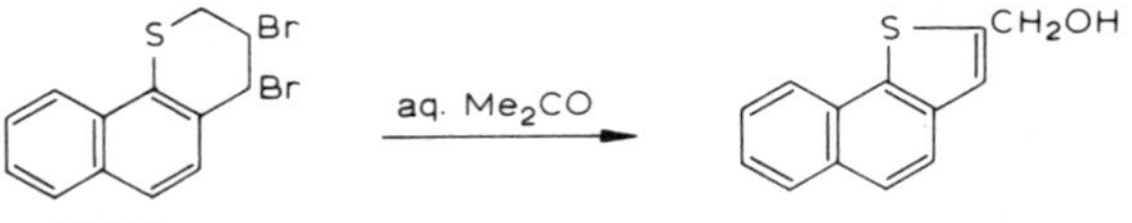

(XIII) aq. Me_2CO

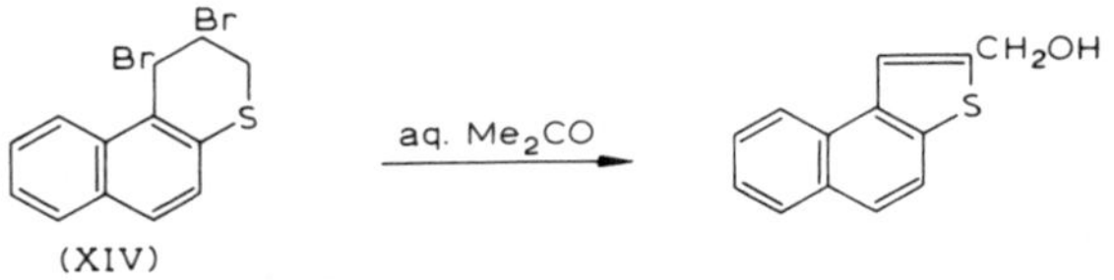

(XIV) aq. Me_2CO

Naphtho[2,3-*b*]**thiophene** (III), m.p. 192–193°, has been prepared by reduction of the 4-acetoxy derivative, with zinc dust and sodium hydroxide or with tricyclohexyloxyaluminium in cyclohexanol. In a second method, 5-(3′-carboxypropyl)-2,3-dihydrobenzo[*b*]thiophene (XV) is cyclised to the ketone XVI from which naphtho[2,3-*b*]thiophene is obtained by reduction of the carbonyl group and dehydrogenation (*Carruthers*, *Douglas* and *Hill*, *loc. cit.*):

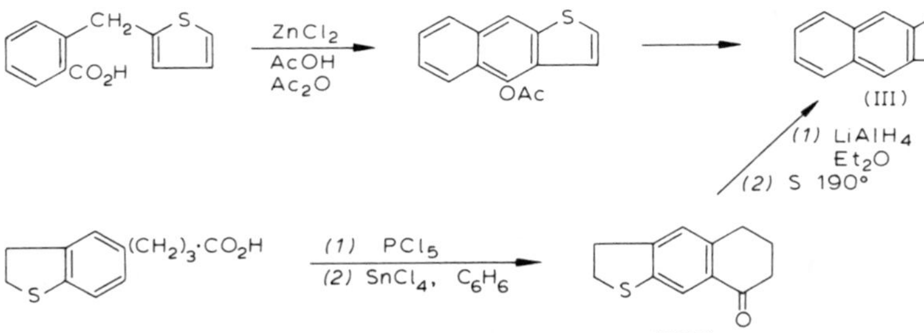

$ZnCl_2$, AcOH, Ac_2O (III) (1) $LiAlH_4$, Et_2O (2) S 190° (XV) (1) PCl_5 (2) $SnCl_4$, C_6H_6 (XVI)

Naphtho[2,3-*b*]thiophene forms *meso*-addition compounds much less readily than does anthracene, but a maleic anhydride adduct can be made under forcing conditions. Reduction with sodium and pentyl alcohol yields 4,9- and 2,3-dihydronaphtho-[2,3-*b*]thiophene. Friedel–Crafts acetylation in nitrobenzene leads to 3-*acetylnaphtho-*[2,3-b]*thiophene*, m.p. 126–127°; in methylene chloride a mixture containing the 3- and the 4-*acetyl* derivative, b.p. 160–165°/0.8 mm is obtained (*Carruthers*, J. chem. Soc., 1963, 4477). The asymmetric triptycene, 5,12-dihydro-5,12-[2′,3′-*b*]thienonaphthacene (XVII) has been synthesised (*H. Wynberg*, *J. de Wit* and *H. J. M. Sinnige*, J. org. Chem., 1970, **35,** 711):

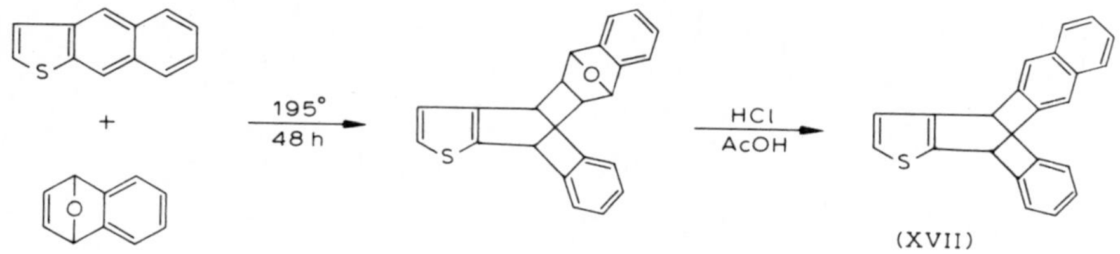

(XVII)

Polycyclic analogues of "thioindoxyl" (XVIII–XXIII) are produced by standard methods (XVIII, XIX, *P. Friedländer* and *N. Woroshzow*, Ann., 1912, **388**, 1; XX, XII, *CIBA*, B.P. 210,413/1923; 249,489/1925; XXI, *P. Ruggli* and *W. Heitz*, Helv., 1931, **14**, 257; XXIII, *E. Havas*, U.S.P. 2,094,595/1937):

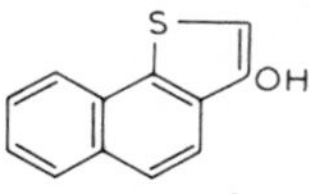

3-Hydroxynaphtho[1,2-*b*]-thiophene (XVIII)

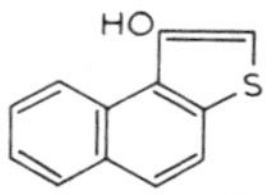

1-Hydroxynaphtho[2,1-*b*]-thiophene (XIX)

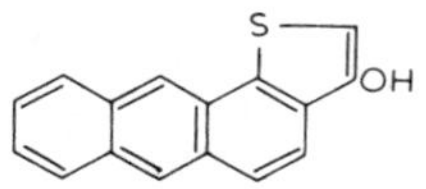

3-Hydroxyanthra[1,2-*b*]-thiophene (XX)

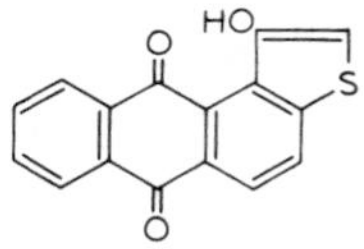

1-Hydroxyanthra[2,1-*b*]-thiophene-6,11-quinone (XXI)

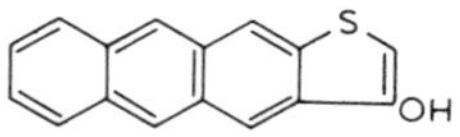

3-Hydroxyanthra[2,3-*b*]-thiophene (XXII)

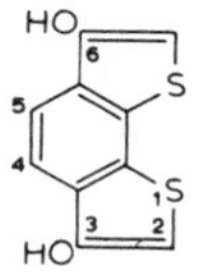

3,6-Dihydroxybenzo-[2,1-*b*:3,4-*b*′]dithiophene (XXIII)

Naphtho[1,8-*bc*]**thiolane** (XXIV) has been synthesised (*D. G. Hawthorne* and *Porter*, Australian J. Chem., 1966, **19**, 1909) and several derivatives described (*Campaigne* and *D. R. Knapp*, J. heterocycl. Chem., 1970, **7**, 107):

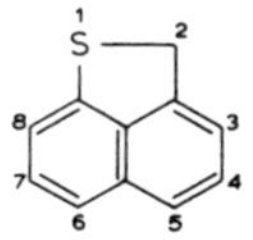

(XXIV)

Benzo[*b*]**naphtho**[2,3-*d*]**thiophene** (XXV), m.p. 160°, occurs in coal-tar chrysene (*Kruber* and *L. Rappen*, Ber., 1940, **73**, 1184; *E. G. G. Werner*, Rec. Trav. chim., 1949, **68**, 520); **benzo**[*b*]**naphtho**[1,2-*d*]**thiophene** (XXVI); **benzo**[*b*]**naphtho**[2,1-*d*]**thiophene** (XXVII), m.p. 185–186°, 11,11-*dioxide*, m.p. 236° (*K. Rabindran* and *B. D. Tilak*, C.A., 1954, **48**, 10724; 1955, **49**, 993; *Davies*, *Porter* and *J. R. Wilmshurst*, J. chem. Soc., 1957, 3366); m.s. (*S.*

Meyerson and *R. W. Vander Haar*, J. chem. Phys., 1962, **37,** 2458). Several methyl derivatives of benzo[*b*]naphtho[2,1-*d*]thiophene have been synthesised (*Carruthers* and *Stewart*, J. chem. Soc., 1965, 6221):

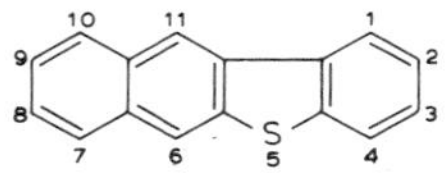

(XXV) 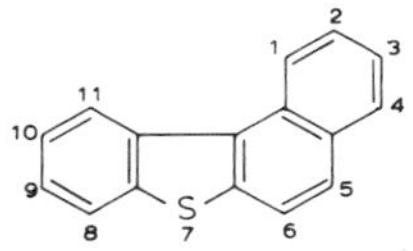 (XXVI) 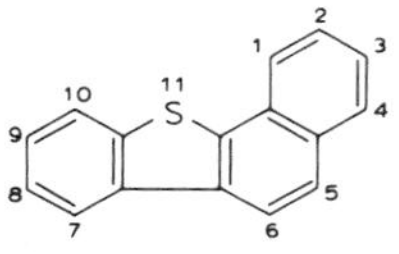 (XXVII)

Anthraquinone analogues (XXVIII–XXXI) — "thiophanthrenequinones" — have been made, usually by condensing phthalic anhydride with thiophenes or benzo[*b*]thiophenes (XXVIII, *W. Steinkopf*, Ann., 1915, **407,** 94; XXIX, *Steinkopf et al., ibid.*, 1939, **540,** 7; XXX, *F. Mayer, ibid.*, 1931, **488,** 259; XXXI, *K. Brass* and *L. Köhler*, Ber., 1922, **55,** 2543):

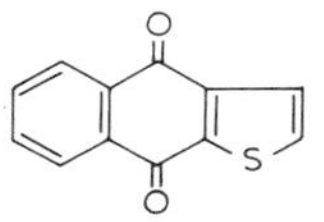

(XXVIII) 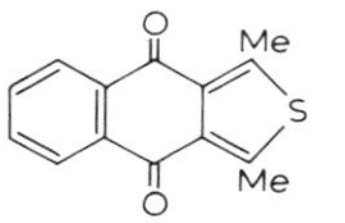 (XXIX) 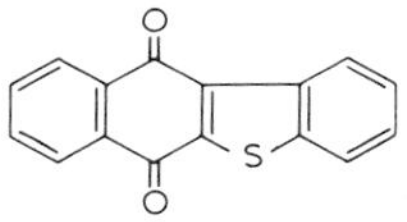 (XXX)

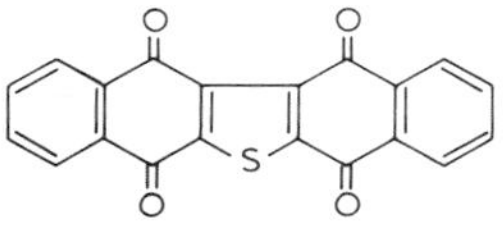

(XXXI) 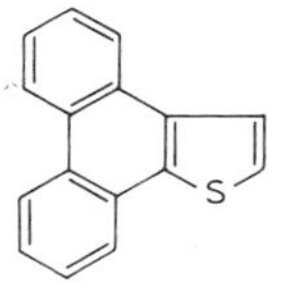 (XXXII) 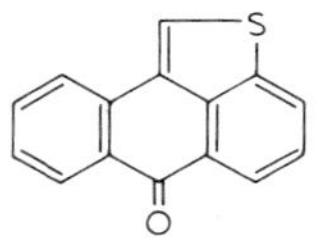 (XXXIII)

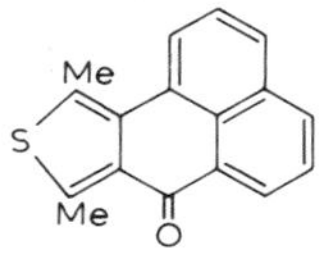

(XXXIV) 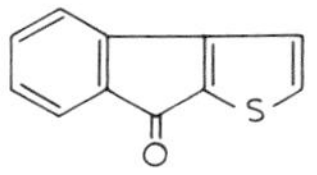 (XXXV)

Other compounds, to which references are given below, are illustrated by formulae XXXII to XXXV:

XXXII, *O. Hinsberg, ibid.*, 1910, **43,** 901; XXXIII, *L. Gatterman*, Ann., 1912, **393,** 113; XXXIV, *N. P. Buu-Hoï* and *N. Hoán*, Rec. Trav. chim., 1948, **67,** 309; XXXV, *Steinkopf* and *E. Günther*, Ann., 1936, **522,** 28.

Tetraphenylthiophene is dehydrogenated by aluminium chloride to XXXVI (*Steinkopf, ibid.*, 1935, **519,** 297). By heating acenaphthylene with sulphur, a red compound XXXVII is obtained, which combines with maleic anhydride (*D. B. Clapp*, J. Amer. chem. Soc., 1939, **61,** 2733):

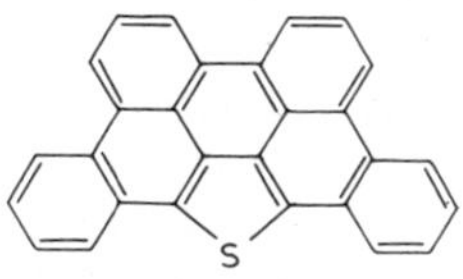

(XXXVI) 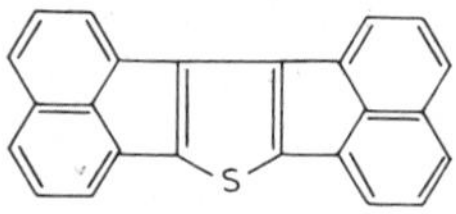 (XXXVII)

Polycyclic thiahydrocarbons containing one thiophene and two to four benzene rings are prepared from the adducts of 3-vinylbenzo[*b*]thiophene with maleic anhydride, *p*-benzoquinone, and 1,4-naphthoquinone, respectively (*Davies* and *Porter*, J. chem. Soc., 1957, 4961).

P. Cagniant describes benzothiopheno[2,3-*b*]cyclopentenone (XXXVIII) (Bull. Soc. chim. Fr., 1949, 382) and the benzothiopheno[2,3]cycloheptene XXXIX (*ibid.*, 1952, 629):

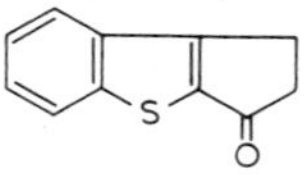

(XXXVIII) 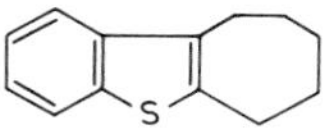 (XXXIX)

8. Compounds with five-membered heterocyclic rings having one selenium atom

(a) Selenophenes* and related compounds

The first selenophene to be described was 2,5-dimethyl-selenophene (I) obtained on heating acetonylacetone with phosphorus pentaselenide (*C. Paal*, Ber., 1885, **18,** 2255):

$$\mathrm{MeCO{-}CH_2{-}CH_2{-}COMe} \xrightarrow{P_2Se_5} \text{(I)} \qquad \text{(II)}$$

The syntheses of selenophene, its chemical properties and structure are, in general, similar to those of thiophene. It is stable and has been known for about forty years, but the study of its chemistry has been slow, probably because there was no convenient method of obtaining selenophenes and their derivatives.

* *N. N. Magdesieva*, Adv. Heterocycl. Chem., Academic Press, 1970, Vol. 12, p. 1.

(i) Preparation and properties

(*1*) Selenophene derivatives have been made by methods analogous to those used in the thiophene series: 2,5-**dimethylselenophene,** b.p. 155°, from acetonylacetone, and 2,4-**diphenylselenophene,** m.p. 112°, by heating acetophenone with selenium (*M. T. Bogert* and *C. N. Andersen*, J. Amer. chem. Soc., 1926, **48,** 223).

(*2*) The Hinsberg synthesis (*cf. p.* 220) can be applied to the production of selenophenes (*H. J. Backer*, Rec. Trav. chim., 1940, **59,** 423):

(*3*) Certain dienes and selenium on heating yield selenophenes (*B. A. Arbuzov* and *E. G. Kataev*, Doklady Akad. Nauk S.S.S.R., 1954, **96,** 983).

(*4*) Some conjugated dienes react with selenium dioxide to give dihydroselenophene dioxides (*Backer* and *J. Strating*, Rec. Trav. chim., 1934, **53,** 1113):

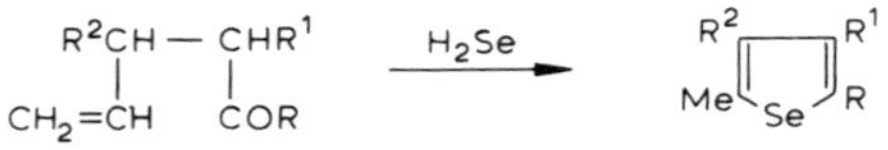

and some selenophenes may be obtained from the same reagent and paraffins or olefins in the presence of chromic oxide on alumina (*Yur'ev* and *L. I. Khmel'nitskii*, Diklady Akad. Nauk, S.S.S.R., 1954, **94,**265).

(*5*) Selenophenes have been prepared by the reaction of hydrogen selenide and hydrogen chloride in ethanol below 0° with unsaturated ketones of the following type (*K. E. Schulte et al.*, Ber., 1968, **101,** 1540):

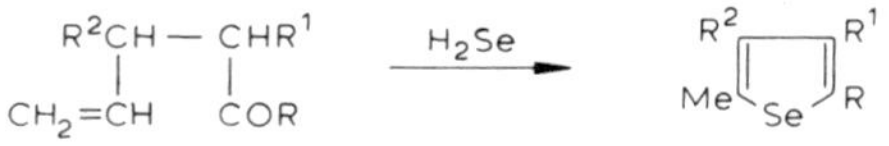

and from hydrogen selenide and acetylenic α-epoxides, vinylacetylenic α-epoxides, or hydroxy-α-epoxides in the presence of barium hydroxide (*F. Ya. Perveev*, *N. I. Kudryashova* and *D. N. Glebovskii*, Zhur. obshcheĭ Khim, 1956, **26,** 3331):

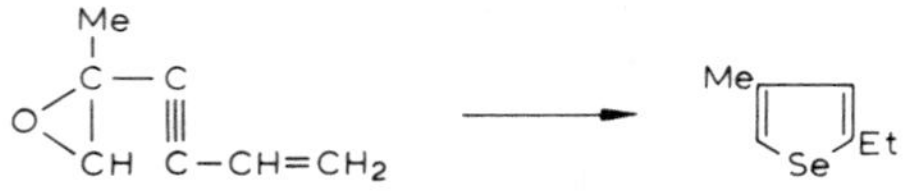

(6) Furan with hydrogen selenide in the presence of magnesium oxide at 450° yields selenophene (*Yur'ev*, *ibid.*, 1946, **16,** 851).

(ii) Selenophenes

Selenophene (II), m.p. about −38°, b.p. 110°, n_D^{25} 1.568, d_4^{15} 1.5301, is prepared by passing acetylene over selenium at 400° (*H. V. A. Briscoe et al.*, J. chem. Soc., 1928, 1741, 2628; *H. Suginome* and *S. Umezawa*, Bull. chem. Soc., Japan, 1936, **11,** 157). It is little affected under moderate conditions by acids, alkali, or oxidising or reducing agents, and gives a greenish-blue indophenin reaction. The bond lengths and angles of the selenophene molecule have been determined (*J. Kraitchman*, Amer. J. Phys., 1953, **21,** 17; *C. C. Costain*, J. chem. Phys., 1958, **29,** 864):

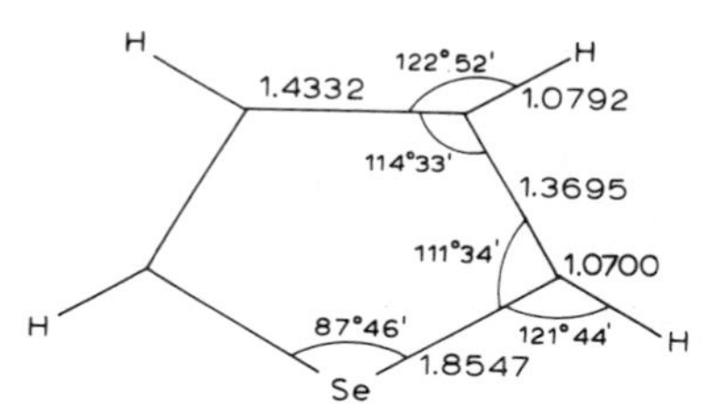

Spectral data of selenophene and derivatives: u.v. (*N. A. Chumayevskii*, *V. M. Tatevskii* and *Yu. K. Yur'ev*, Opt. Spektrosk., 1959, **6,** 45; C.A., 1959, **53,** 7764; *E. G. Treshchova*, *D. Ekkhardt* and *Yur'ev*, Zhur. fiz. Khim., 1964, **38,** 295; *L. Chierici et al.*, Gazz., 1958, **88,** 453; 1959, **89,** 560; 1960, **90,** 1125); i.r. (*Yur'ev et al.*, Opt. Spektrosk., 1967, **22,** 216; C.A., 1967, **67,** 48641v; *N. M. Pozdeev et al.*, Doklady Akad. Nauk, S.S.S.R., 1969, **185,** 384); n.m.r. (*J. Read*, *C. Mathis* and *J. Goldstein*, Spectrochim. Acta, 1965, **21,** 85).

The m.s. of five-membered heterocycles containing O, S, Se, Te, N, P, Si and Ge have been measured and the differing modes of fragmentation upon electron impact have been determined from deuterium labelling studies (*A. M. Duffield*, *H. Budzikiewicz* and *C. Djerassi*, J. Amer. chem. Soc., 1965, **87,** 2920).

Tetraphenylselenophene, m.p. 183–184° is prepared from 1,4-dilithiotetraphenylbutadiene and diselenium dibromide or from 1,4-diiodotetraphenylbutadiene and lithium selenide (*E. H. Braye*, *W. Hübel* and *J. Caplier*, J. Amer. chem. Soc., 1961, **83,** 4406).

Mono-, di- and **tetra-deuterioselenophenes** have been prepared by heating the corresponding iodoselenophenes with deuterium oxide, zinc filings and acetic anhydride (*Yur'ev*, *Magdesieva* and *L. Ya. Petrova*, Khim. Geterotsikl. Soedin., 1966, 6, 910).

(iii) Selenophene derivatives

Mercuration of the selenophene nucleus with mercuric salts occurs readily at the 2-position. If the 2-position is occupied by a bromo, nitro, acetyl, or ethoxycarbonyl group, then mercuration takes place at the 5-position (*Yu.K. Yur'ev et al.*, Zhur. obshcheĭ Khim., 1959, **29,** 1970; Khim. Geterotsikl. Soedin., 1966, 897; *M. T. Bogert* and *C. N. Andersen*, J. Amer. chem. Soc., 1926, **48,** 223).

Direct *halogenation* occurs at the 2-position and affords 2-**bromo-** and 2-**chloro-selenophene,** but the **iodo-selenophenes** are obtained either by reaction with iodine in the presence of yellow mercuric oxide or from the mercurated selenophene. The 3-monohalogeno derivatives cannot be obtained by direct methods; 3-**bromoselenophene** is prepared by the debromination of tribromoselenophene using zinc and acetic acid (*Yur'ev, N. K. Sadovaya* and *E. A. Grekova*, Zhur. obshcheĭ Khim., 1964, **34,** 847). 3-**Iodoselenophene** is prepared by treating the tetraiodo derivative with aluminium amalgam (*C. Paulmier* and *P. Pastour*, Compt. rend., Ser. C, 1967, **265,** 926). Fluorinated tetrahydroselenophenes (selenolanes) are prepared by reaction of selenium, a fluoro-olefin and an olefin at 150° (*C. G. Krespan*, J. org. Chem., 1962, **27,** 3588).

On *nitration* with nitric acid in acetic anhydride selenophene affords mainly 2-**nitroselenophene** and a small amount of the 3-isomer (*Yur'ev, E. L. Zaĭtseva* and *G. G. Rozantsev, ibid.*, 1960, **30,** 2207). 3-**Nitroselenophene** may be obtained by the decarboxylation of 4-nitroselenophene-2-carboxylic acid (*Yur'ev* and *Zaĭtseva, ibid.*, 1960, **30,** 859). Nitration occurs preferentially at the 2-position, but if this position is blocked by an electron-accepting group, *e.g.* nitro, formyl, or acetyl, then nitration takes place mainly at the 4-position. Selenophene is *sulphonated* at the 2-position by sulphuric acid (*S. Umezawa*, Bull. chem. Soc. Japan, 1936, **11,** 157), or by pyridine–sulphur trioxide; selenophene-2-carbaldehyde with dioxane–sulphur trioxide affords selenophene-2-carbaldehyde-5-sulphonic acid.

Selenophenes are readily *formylated* in the 2-position by dimethylformamide in the presence of phosphorus oxychloride (*Yur'ev* and *N. N. Mezentsova*, Zhur. obshcheĭ Khim., 1957, **27,** 179); 2-methyl- and 3-methyl-selenophene yield the 5- and 2-carbaldehyde, respectively (*Yur'ev, N. K. Sadovaya* and *M. A. Gal'bershtam, ibid.*, 1958, **28,** 620).

Selenophene-2-carbaldehyde reacts readily with ammonia, aromatic amines and diamines (*Yur'ev* and *Mezentsova, ibid.*, p. 3041), gives β-(2-selenienyl)acrylic acid *via* either the Perkin reaction or the Knoevenagel condensation with malonic acid (*Yur'ev* and *Mezentsova, ibid.*, 1957, **27,** 179), and reacts with methylmagnesium iodide to yield 2-**selenienylcarbinol,** the ethyl and propyl compounds have also been prepared (*Yur'ev, Mezentsova* and *V. E. Vas'kovskiĭ, ibid.*, 1958, **28,** 3262).

Selenophene-3-carbaldehyde is obtained from 3-bromomethylselenophene by the Sommelet reaction, from 3-selenienyl-lithium by treatment with dimethylformamide (*Paulmier* and *Pastour, loc. cit.*) and from selenophene-3-carbonitrile (*Yur'ev, N. N. Magdesieva* and *A. T. Monakhova*, Khim. Geterotsikl. Soedin., 1968, 649).

Selenophene is *acylated* in the 2-position using acyl chlorides under Friedel–Crafts conditions (*Umezawa*, Bull. chem. Soc. Japan, 1936, **11,** 775; *N. P. Buu-Hoï, P. Demerseman* and *R. Royer*, Compt. rend., 1953, **237,** 397), by tetra-acyloxysilanes in the presence of stannic chloride (*Yur'ev et al.*, Doklady Akad. Nauk S.S.S.R., 1955, **102,** 763; Zhur obshcheĭ Khim., 1956, **26,** 930; 1958, **28,** 3036):

Se (selenophene) —[$EtCO_2H$, $SiCl_4$ / $SnCl_4$]→ 2-COEt-selenophene

and by acid anhydrides in the presence of phosphoric acid (*E. G. Kataev* and *M. V. Palkina*, C.A., 1956, **50,** 937; 1958, **52,** 3762). 3-**Acetylselenophene** is obtained by treating the corresponding acyl chloride with diethyl ethoxymagnesiummalonate followed by acidification (*Yur'ev*, *Sadovaya* and *Grekova*, Zhur. obshcheĭ Khim., 1964, **34,** 847):

3-COCl-selenophene —[$EtOMgCH(CO_2Et)_2$]→ 3-$CO{\cdot}CH(CO_2Et)_2$-selenophene → 3-COMe-selenophene

Selenienyl ketones may be reduced by the Huang-Minlon method to give alkylselenophenes (*Yur'ev* and *Sadovaya*, *ibid.*, 1961, **31,** 3535).

Selenophene-2-carboxylic acid may be obtained by the oxidation of 2-acetylselenophene with a 1% solution of potassium permanganate in the presence of potassium hydroxide (*Kataev* and *Palkina*, C.A., 1956, **50,** 938; 1958, **52,** 3762); by carbonation of 2-selenienylmagnesium iodide, which requires initiation to prepare the Grignard reagent (*A. N. Nesmeyanov et al.*, Izvest. Akad. Nauk, S.S.S.R., Otdel. khim. Nauk, 1957, 1389; *Yur'ev* and *Sadovaya*, Zhur. obshcheĭ Khim., 1958, **28,** 2162); by oxidation of selenophene-2-carbaldehyde with a 3% solution of hydrogen peroxide in the presence of sodium hydroxide (*Yur'ev*, *Mezentsova* and *Vas'kovskiĭ*, *ibid.*, 1958, **28,** 3262), and by reacting selenophene with butyl-lithium followed by carbonation (*Yur'ev* and *Sadovaya*, *ibid.*, 1964, **34,** 1803). The salts of selenophene-2-carboxylic acid undergo thermal disproportionation to give **selenophene-2,5-dicarboxylic acid** and selenophene (*A. S. Angeloni* and *M. Tramontini*, Boll. sci. Fac. Chim. ind. Bologna, 1965, **23,** 301; C.A., 1966, **64,** 17520c). **Selenophene-3-carboxylic acid** has been prepared by the carbonation of 3-selenienyl-lithium and by hydrolysis of the related nitrile (*Yur'ev*, *Sadovaya* and *Grekova*, Zhur. obshcheĭ Khim., 1964, **34,** 847):

3-Br-selenophene —[PhLi / −50°]→ 3-Li-selenophene —[CO_2]→ 3-CO_2H-selenophene

3-Br-selenophene —[CuCN / quinoline]→ 3-CN-selenophene —[HCl]→ 3-CO_2H-selenophene

3-CN-selenophene —[$LiAlH_4$ / Et_2O]→ 3-CH_2NH_2-selenophene

Methyl selenophene-3-carboxylate is obtained on keeping the acid for a day in methanol saturated with hydrogen chloride (*Yur'ev*, *Magdesieva* and *Monakhova*, Zhur. org. Khim., 1965, **1,** 1094; C.A., 1965, **63,** 11474c).

Passing hydrogen chloride into selenophene and formalin in ethylene dichloride affords 2-**chloromethylselenophene** and a small amount of 2,5-bischloromethylselenophene (*Yur'ev*, *Sadovaya* and *Gal'bershtam*, Zhur. obshcheĭ Khim., 1962, **32,** 259):

$$\text{selenophene} \longrightarrow \text{2-}CH_2Cl\text{-selenophene}$$

2-Chloromethylselenophene may be converted into β-2-**selenienyl-propionic acid** (*Yur'ev* and *Gal'bershtam*, *ibid.*, p. 3249):

$$\text{Se-ring-}CH_2Cl \xrightarrow{NaCH(CO_2Et)_2} \text{Se-ring-}CH_2{\cdot}CH(CO_2Et)_2 \xrightarrow{\text{aq. KOH}} \text{Se-ring-}CH_2{\cdot}CH_2{\cdot}CO_2H$$

Some derivatives of selenophene are collected in Table 8.

TABLE 8

SOME DERIVATIVES OF SELENOPHENE

Substituent	*b.p.(°C)/m.p.(°C)*	*Ref.*
2-Chloro	b.p. 152–153/740 mm b.p. 42/12.5 mm	1
2,5-Dichloro	b.p. 67/12 mm	2
2,3,4,5-Tetrachloro	m.p. 35	3
2,3,4,5-Tetrachlorotetrahydro	m.p. 97	2
2,2,5,5-Tetrachlorotetrahydro	m.p. 96–98	2
2,2,3,4,5,5-Hexachlorotetrahydro	m.p. 55	2
2-Bromo	b.p. 59/13 mm	2
3-Bromo	b.p. 71–73/12 mm	4
2,3-Dibromo	b.p. 98–110/10 mm	4
2,5-Dibromo	b.p. 42/0.02 mm	2
2,3,5-Tribromo	m.p. 38	2
Tetrabromo	m.p. 102	2
2,2,5,5-Tetrabromotetrahydro	m.p. 97	2
2-Iodo	b.p. 84–84.5/11 mm	5
3-Iodo	b.p. 56/1 mm	6
3,4-Diiodo	b.p. 123/1 mm	6
2,3,4-Triiodo	m.p. 104	6
Tetraiodo	m.p. 208	6
2-Nitro		7
3-Nitro	m.p. 77.5–78	7
2-SO_3H		8
2-SO_2Cl	m.p. 31–32°	8
2-SO_2NH_2	m.p. 157–158°	8,9
2-CHO	b.p. 86–87/7 mm	10
3-CHO	b.p. 54/0.5 mm	6
5-Cl-2-CHO	b.p. 93/5 mm	1
5-Br-2-CHO	b.p. 124–125/12 mm	1
5-NO_2-2-CHO	m.p. 88	1
2-Me-5-CHO	b.p. 96–97/7 mm	11
2-COMe	b.p. 105–106/12 mm	12
3-COMe	m.p. 72–72.5	4
2-Me-5-COMe	b.p. 114–116/12 mm	13

TABLE 8 *(continued)*

Substituent	*b.p.(°C)/m.p.(°C)*	*Ref.*
2,3-Di-Me-5-COMe	b.p. 133–134/15 mm	13
2-COPh	m.p. 57–58	2
2-CO_2H	m.p. 120	14
3-CO_2H	m.p. 145–146	4
2-Me-5-CO_2H	m.p. 134–136	12
3-Br-2-CO_2H	m.p. 184–185	4
2,4-Di-CO_2H	m.p. 286–290 (decomp.)	13
2-CH = CH·CO_2H	m.p. 144	9
2-Me	b.p. 136/769 mm	11
2,5-Di-Me	b.p. 154–154.5/755 mm	11
2-Ph	m.p. 38	15
3-Ph	m.p. 97	15

References
1 *Yu.K. Yur'ev, N. N. Mezentsova* and *A. T. Monakhova*, Zhur. obshcheĭ Khim., 1960, **30,** 2726.
2 *H. Suginome* and *S. Umezawa*, Bull. chem. Soc., Japan, 1936, **11,** 157.
3 *W. Mack*, Angew. Chem., 1965, **77,** 260.
4 *Yur'ev, N. K. Sadovaya* and *E. A. Grekova*, Zhur. obshcheĭ Khim., 1964, **34,** 847.
5 *Yur'ev* and *Sadovaya, ibid.*, 1956, **26,** 3154.
6 *C. Paulmier* and *P. Pastour*, Compt. rend., Ser. C, 1967, **265,** 926.
7 *Yur'ev* and *E. L. Zaĭtseva*, Zhur. obshcheĭ Khim., 1960, **30,** 859, 2207.
8 *Yur'ev* and *Sadovaya, ibid.*, 1964, **34,** 1803.
9 *Idem, ibid.*, p. 2190.
10 *Yur'ev* and *Mezentsova, ibid.*, 1957, **27,** 179.
11 *Yur'ev, Mezentsova* and *V. E. Vas'kovskiĭ, ibid.*, p. 3155.
12 *E. G. Kataev* and *M. V. Palkina*, C.A., 1956, **50,** 938.
13 *Yur'ev, Sadovaya* and *E. N. Lyubimova*, Zhur. obshcheĭ Khim., 1960, **30,** 2732.
14 *Kataev* and *Palkina*, C.A., 1958, **52,** 3762.
15 *Yur'ev et al.*, Zhur. obshcheĭ Khim., 1957, **27,** 2260.

(iv) Hydroselenophenes

Tetramethylene dibromide and sodium selenide afford **tetrahydroselenophene,** b.p. 135°, (*methiodide*, m.p. 174°), which shows the additive reactions of a dialkyl selenide (*G. T. Morgan* and *F. H. Burstall*, J. chem. Soc., 1929, 1096):

$$Br(CH_2)_4Br \xrightarrow{Na_2Se} \text{(tetrahydroselenophene ring, Se)}$$

A. Fredga (J. pr. Chem., 1930, [ii], **127,** 103; 1931, [ii], **130,** 180) described the stereoisomeric **tetrahydroselenophene**-2,5-**dicarboxylic acids.**

5-**tert**-Butoxy-2-selenienyl-lithium (III) may be converted to the corresponding derivatives (IV, R = Me, CHO, Ac, SMe and OBu^t). Compound IV (R = Me or SMe) on distillation in the presence of *p*-toluenesulphonic acid gives a mixture of 2-**oxo**-2,3-**dihydro**- and 2-**oxo**-2,5-**dihydro-selenophenes** (V and VI); IV (R = H) yields only the 2,5-dihydro derivative (VI, R = H). Derivative (VIII, R = Br or OBu^t) is prepared from 2,3,5-tribromoselenophene (VII) and on treatment with acid affords 2,5-**dioxo**-2,5-**dihydroselenophene** (IX) (*J. Morel et al.*, Compt. rend., Ser. C, 1970, **270,** 825).

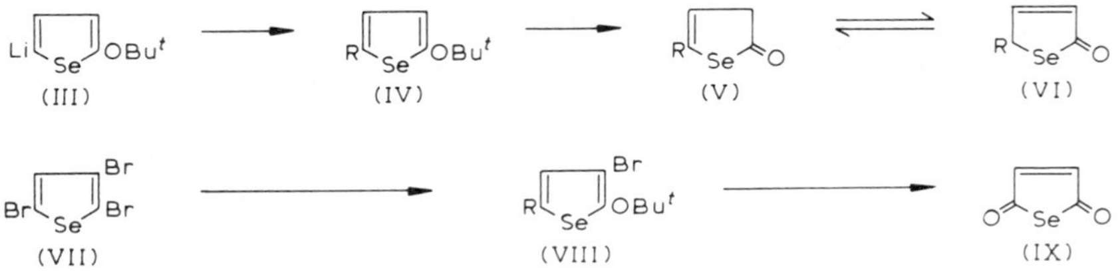

(v) Biselenienyl (bis-selenophene)

2-Iodoselenophene on treatment with activated copper affords 2,2′-**biselenienyl** (*L. Chierici et al.*, Ric. Sci., 1958, **8A**, 1537):

Octahydro-3,3′-biselenophene (**octahydro-3,3′-biselenienyl**), m.p. 105–106°, is obtained by boiling 1,6-diiodo-3,4-bisiodomethylhexane, $ICH_2 \cdot CH_2 \cdot CH(CH_2I) \cdot CH(CH_2I) \cdot CH_2 \cdot CH_2I$, and sodium selenide in absolute alcohol under nitrogen (*E. Buchta* and *K. Greiner*, Ber., 1961, **94**, 1311).

(vi) Metalloid derivatives

Reaction between 2-selenienylmagnesium iodide and potassium tetrafluoroborate yields **potassium tetra(2-selenienyl)borate** (X) (*Nesmeyanov, V. A. Sazonova* and *V. N. Drozd*, Izvest. Akad. Nauk, S.S.S.R., Otdel. khim. Nauk, 1957, 1389; C.A., 1958, **52**, 7269).

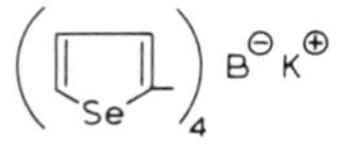

(b) Benzo[b]selenophenes (selenanaphthenes, 2,3-benzoselenophenes)

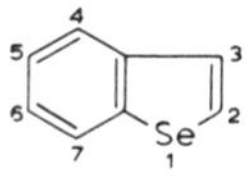

Benzo[*b*]selenophene (XI), m.p. 50–51°, b.p. 239°, *picrate*, m.p. 156–157°, has been prepared by ring-closure of the product from benzeneselenol (selenophenol) and bromoacetal (*R. B. Mitra et al.*, C.A., 1955, **49**, 6222), by reduction of 3-hydroxybenzo-[*b*]selenophene (*G. Komppa* and *G. A. Nyman*, J. pr. Chem., 1934, [ii], **139**, 229), by the passage of styrene over a chromium–aluminium oxide catalyst at 450° in the presence of selenium dioxide vapour (*Yu.K. Yur'ev et al.*, Zhur. obshcheĭ Khim., 1957, **27,** 2260), and along with ethylbenzene by dehydrocyclisation of *o*-ethylbenzene-selenol at 400° over a platinum/carbon or a copper/chromium/carbon catalyst (*C. Hansch* and *C. F. Geiger*, J. org. Chem., 1959, **24,** 1025):

$$C_6H_5CH_2Me\ (80\%) + H_2Se \xleftarrow{H_2} o\text{-}(CH_2Me)C_6H_4SeH \longrightarrow \text{benzo[b]selenophene}\ (18\%) + 2\,H_2$$

For u.v. spectral data on benzoselenophene in solution and in the vapour phase see *M. R. Padhye* and *J. C. Patel*, J. Sci. Ind. Res. India, 1956, **15B,** 171; C.A., 1956, **51,** 854c; *A. I. Kiss* and *B. R. Muth*, Acta Chim. Acad. Sci. Hung., 1957, **11,** 57, 365.

2-**Methylbenzo[*b*]selenophene,** m.p. 63°, *picrate*, m.p. 118.5° (*Muth* and *Kiss*, J. org. Chem.; 1956, **21,** 576); 3-**methylbenzo[*b*]selenophene,** b.p. 135°/15 mm, *picrate*, m.p. 130° (*L. Christiaens* and *M. Renson*, Bull. Soc. chim. Belg., 1968, **77,** 153). The reactivity of 2-methyl- and 3-methyl-benzo[*b*]selenophene towards electrophilic agents has been examined (*Christiaens, R. Dufour* and *Renson, ibid.*, 1970, **79,** 143) and the i.r. behaviour of heterocycles, including benzoselenophenes according to the nature of the hetero-atom and the substituents has been discussed (*P. Bassignana, C. Cogrossi* and *M. Gandino*, Chim. Ind., Paris, 1963, **90,** 370; C.A., 1964, **60,** 4957e).

Allyl phenyl selenide on boiling in excess of quinoline under dry nitrogen gives 2-**methyl**-2,3-**dihydrobenzo[*b*]selenophene** (XII), b.p. 103–104°/6 mm (*E. G. Kataev et al.*, Zhur. org. Khim., 1967, **3,** 597), no *o*-allylselenophenol or selenochroman (XIII) were obtained. No cyclisation occurs when the selenide is heated without solvent at a temperature > 200°, so the cyclisation mechanism is different from the classical Claisen rearrangement. Boiling the selenide with an equimolar amount of quinoline affords a mixture of 2-methyl-2,3-dihydrobenzo[*b*]selenophene and a compound, m.p. 47–48°, probably *selenochroman* (XIII) (*G. A. Chmutova*, C.A., 1969, **70,** 87444q):

$$C_6H_5Se\cdot CH_2\cdot CH{:}CH_2 \xrightarrow[\text{quinoline}]{\text{excess}} \text{2-methyl-2,3-dihydrobenzo[b]selenophene (XII)}$$

$$C_6H_5Se\cdot CH_2\cdot CH{:}CH_2 \xrightarrow[\text{amount quinoline}]{\text{equimolar}} \text{(XII)}\ 15\% + \text{selenochroman (XIII)}\ 25\%$$

Benzo[*b*]selenophene-2-carbaldehyde, m.p. 83°, may be prepared by adding selenochromene in pyridine to a mixture of selenium dioxide in pyridine and then boiling the mixture (*A. Ruwet, J. Meessen* and *Renson*, Bull. Soc. chim. Belg., 1969, **78,** 459).

Other 2-carbaldehydes may be obtained by this method, and also by the oxidation of the hydrolysis products of selenochromylium perchlorates with manganese dioxide; 3-*methylbenzo*[b]*selenophene-2-carbaldehyde*, m.p. 88–89° (2,4-*dinitrophenylhydrazone*, m.p. 283–284°); 3-*phenylbenzo*[b]*selenophene-2-carbaldehyde*, m.p. 85°; 5-*methyl-3-phenylbenzo*[b]*selenophene-2-carbaldehyde*, m.p. 133° (2,4-*dinitrophenylhydrazone*, m.p. 283°). **Benzo[*b*]selenophene-3-carbaldehyde,** m.p. 85° (2,4-*dinitrophenylhydrazone*, m.p. 285–286°) prepared by hydrolysis of the complex, obtained by treating 3-methylbenzo[*b*]selenophene with *N*-bromosuccinimide in the presence of benzoyl peroxide, and then with hexamethylenetetramine *(Christiaens* and *Renson, loc. cit.).*

Acetic acid and silicon tetrachloride in benzene add to benzo[*b*]selenophene and stannic chloride in benzene to give 2-**acetylbenzo[*b*]selenophene,** m.p. 81° (2,4-*dinitrophenylhydrazone*, m.p. 261°) (*Yur'ev et al.*, Zhur. obshcheĭ Khim., 1957, **27,** 2260):

$$\text{benzo[}b\text{]selenophene} \xrightarrow[\text{SnCl}_4,\ C_6H_6]{\text{AcOH/SiCl}_4,} \text{2-acetylbenzo[}b\text{]selenophene}$$

Ethyl bromoacetate reacts abnormally with *o*-methylselenoacetophenone in the presence of zinc through the Reformatsky reaction; the following compounds (XIV) have been obtained by this method:

(XIV) [3-R^1-benzo[*b*]selenophene-2-COR^2]

benzo[*b*]selenophene-2-carboxylic acid ($R^1 = H$, $R^2 = OH$), m.p. 240°, *chloride* ($R^1 = H$, $R^2 = Cl$), m.p. 71°, *methyl ester* ($R^1 = H$, $R^2 = OMe$), m.p. 66°, *ethyl ester* ($R^1 = H$, $R^2 = OEt$), m.p. 55°; 3-**methylbenzo[*b*]selenophene-2-carboxylic acid** ($R^1 = Me$, $R^2 = OH$), m.p. 245°, *methyl ester* ($R^1 = Me$, $R^2 = OMe$), m.p. 106°, *ethyl ester* ($R^1 = Me$, $R^2 = OEt$), m.p. 46° (*Ruwet* and *Renson*, Bull. Soc. chim. Belg., 1970, **79,** 75).

o-Selenocyanatoacetophenone after boiling with sodium dithionite and 10% sodium hydroxide solution followed by further boiling with an aqueous mixture of sodium chloroacetate, affords *o*-acetylphenylselenoacetic acid (XV). This compound on boiling with 30% aqueous potassium hydroxide solution gives 3-methylbenzo[*b*]selenophene-2-carboxylic acid (*Christiaens* and *Renson, loc. cit.*):

$$o\text{-}(COMe)C_6H_4SeCH_2{\cdot}CO_2H\ \text{(XV)} \longrightarrow \text{3-Me-benzo[}b\text{]selenophene-2-}CO_2H$$

3-**Hydroxybenzo[*b*]selenophene** (XVI), m.p. 76–77°, obtained from anthranilic acid *via* "selenosalicylic acid" in the same way as thioindoxyl, affords the red, rather unstable 2,3-**dioxo-2,3-dihydrobenzo[*b*]selenophene (selenonaphthenequinone)**, m.p. 102–103° (*R. Lesser et al.*, Ber., 1912, **45,** 1835; 1914, **47,** 2292). In neutral solution 3-hydroxybenzo[b]selenophene is present predominantly in the oxo form, 3-oxo-2,3-dihydrobenzo[b]selenophene (XVII) (*Kiss* and *Muth*, C. A., 1959, **53**, 3876g):

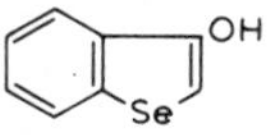

(XVI) 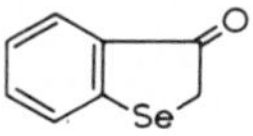 (XVII)

For the u.v. spectra of 3-hydroxybenzo[b]selenophene and selenoindigo see *Kiss* and *Muth*, Acta Chim. Acad. Sci. Hung., 1957, **11**, 57.

6-**Chloro**-3-**oxo**-2,3-**dihydrobenzo**[b]**selenophene** (XVIII), m.p. 140°, is obtained from 2-amino-4-chloroacetophenone (*E. Giesbrecht* and *I. Mori*, Anais Acad. Brasil. Cienc., 1958, **30**, 521; C.A., 1963, **58**, 5611c):

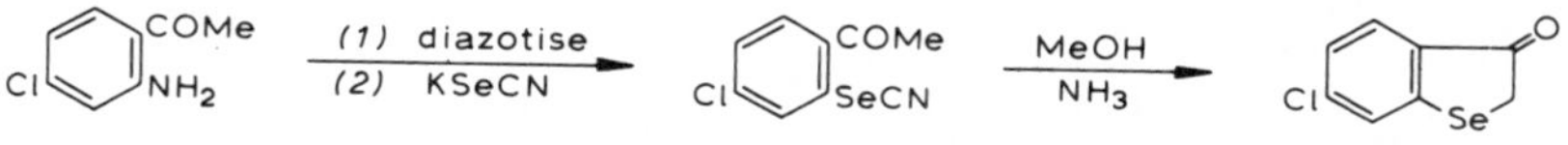

(1) diazotise (2) KSeCN; MeOH, NH_3 → (XVIII)

similarly 2-amino-4-methoxyacetophenone affords 6-**methoxy**-3-**oxo**-2,3-**dihydrobenzo**[b]**selenophene,** m.p. 123°. 2-Methyl-3-oxo-2,3-dihydrobenzo[b]selenophene may be prepared by the following route from *o,o'*-diselenodibenzoic acid (*Muth* and *Kiss*, J. org. Chem., 1956, **21**, 576):

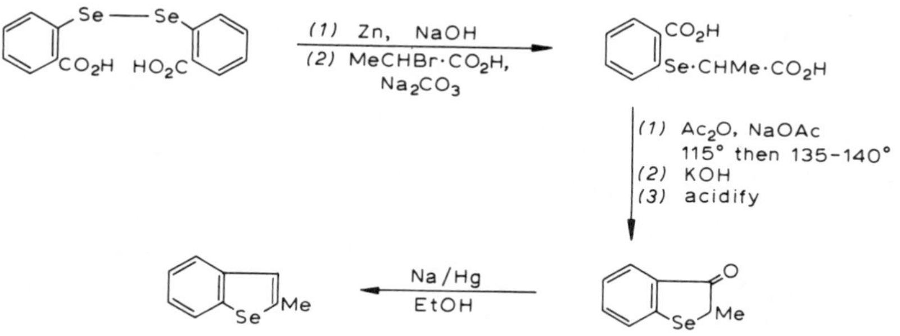

(1) Zn, NaOH (2) $MeCHBr\cdot CO_2H$, Na_2CO_3 → $Se\cdot CHMe\cdot CO_2H$; (1) Ac_2O, NaOAc 115° then 135–140° (2) KOH (3) acidify; Na/Hg, EtOH

Selenoflavanone on treatment with bromine in carbon disulphide gives a dibromide, which on boiling in a little acetic acid and cooling yields 2-*benzylidene*-2,3-*dihydrobenzo*[b]*selenophene* (2-*benzylidene*-2,3-*dihydroselenonaphthenone*), m.p. 117–118° (*J. Gosselck* and *E. Wolters*, Ber., 1962, **95**, 1237).

Some derivatives of benzo[b]selenophene are collected in Table 9.

TABLE 9

SOME DERIVATIVES OF BENZO[*b*]SELENOPHENE

Substituents	*m.p. (°C)*	*Ref.*	*Substituents*	*m.p. (°C)*	*Ref.*
2-Br, 3-Me	b.p. 135°	1	2-CHO, 3-CO_2H	205–210	1
2-Br, 3-CH_2Br	100	1	3-CHO, 2-CO_2H	199–200	1
2-Br, 3-CHO	142	1	2-COPh	71	2
2-Br, 3-$CH(OEt)_2$	38	1	2-COPh, 3-Me	87	2
3-Br, 2-Me	b.p. 125–127°	1	2,3-Di-Me	b.p. 130–132°	1
3-Br, 2-CHO	107	1	2-CO_2H, 3-Ph	216	2
3-Br, 2-CH_2OH	105	1	3-CO_2H, 2-Me	208	1
3-Br, 2-CO_2H	285	1	3-CO_2Me, 2-Me	41	1
3-Br, 2-CO_2Me	82	1	2,3-Di-CO_2H	230	1
3-Br, 2-$CH(OEt)_2$	b.p. 146–148°	1	2,3-Di-CO_2H-anhydride	185	1
2-CHO, 3-Me	91	1	2-CH_2OH	104	2
2-Ac, 3-Me	93–95	1, 2	2-CH:$CHCO_2H$	230	2
3-CHO, 2-Me	139–140	1	2-CH(Ph)OH	100	2
3-Ac, 2-Me	75–76	1			

References
1 *L. Christiaens, R. Dufour* and *M. Renson*, Bull. Soc. chim. Belg., 1970, **79**, 143.
2 *Christiaens* and *Renson*, *ibid.*, 1968, **77**, 153.

(*c*) *Dibenzoselenophenes, diphenylene selenides*

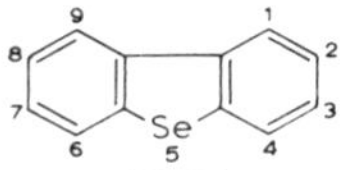

(XIX)

Dibenzoselenophene (XIX), m.p. 78°, is obtained by heating selenanthrene with copper (*N. M. Cullinane et al.*, J. chem. Soc., 1939, 151) or from bis-2-diphenylyl diselenide with bromine (*J. D. McCullough et al.*, J. Amer. chem. Soc., 1950, **72**, 5753). The electron spin resonance spectra of the radical anions of dibenzoselenophene, dibenzofuran, and dibenzothiophene have been measured and the hyperfine splitting constants assigned to the various protons by measurements of the splitting constants of various methylated and deuterated derivatives (*R. Gerdil* and *E. A. C. Lucken*, *ibid.*, 1965, **87**, 213). Dibenzoselenophene gives on nitration mainly 2-**nitrodibenzoselenophene,** m.p. 184–185°, reducible to 2-**aminodibenzoselenophene,** m.p. 99° (*E. Sawicki* and *F. E. Ray*, *ibid.*, 1952, **74**, 4120; 1955, **77**, 957; *G. E. Wiseman* and *E. S. Gould*, *ibid.*, p. 1061). For u.v. spectra of dibenzoselenophene and 2-nitrodibenzoselenophene see *R. Passerini* and *G. Purrello*, Ann. chim., Rome, 1958, **48**, 738. Nitration of **dibenzoselenophene 5-oxide (dibenzoselenophene** *Se***-oxide),** m.p. 229–230° (decomp.), yields 3-**nitrodibenzoselenophene,** m.p. 154–155°, reduced by stannous chloride to 3-**aminodi-**

benzoselenophene, m.p. 132–133° (*Sawicki* and *Ray*, J. org. Chem., 1953, **18,** 946). 2,7-**Dinitrodibenzoselenophene,** m.p. 275°; 3,7-**dinitrodibenzoselenophene),** m.p. 369°; 2,7-*bis(acetylamino)dibenzoselenophene,* m.p. 284–285°; 3,7-**diaminodibenzoselenophene,** m.p. 185°, 3,7-*bis(acetylamino)dibenzoselenophene,* m.p. 318° (*A. Magelli* and *Passerini*, Boll. sedute accad. Gioenia sci. nat. catania, 1956, **3,** 185; C.A., 1958, **52,** 10049f). 2-**Iododibenzoselenophene,** m.p. 96–97°; 2-**iodo-8-nitrodibenzoselenophene,** m.p. 263–264° (*N. Marziano*, Ricerca sci., 1960, **30,** 743; C.A., 1961, **55,** 1581e).

Heating 1,2-diiodotetrafluorobenzene with selenium at 375° gives some **octafluorodibenzoselenophene,** m.p. 118.5–120.5° (*S. C. Cohen, M. L. N. Reddy* and *A. G. Massey*, Chem. Comm., 1967, 451):

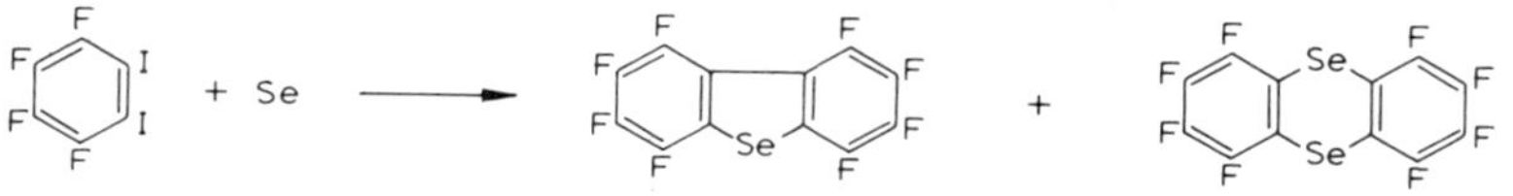

Treatment of 5-(*p*-toluenesulphonylimino)dibenzoselenophene (XX) with 2,2′-dilithiobiphenyl gives the spiro compound **bis(2,2′-biphenylylene)selenium** (XXI), which may be ring opened on treatment with aqueous potassium iodide solution to yield 2-biphenylyl-2,2′-biphenylyleneselenonium iodide (XXII). Bis(4,4′-dimethyl-2,2′-biphenylylene)selenium has also been obtained, but is much less stable than XXI (*D. Hellwinkel* and *G. Fahrbach*, Ann., 1968, **715,** 68):

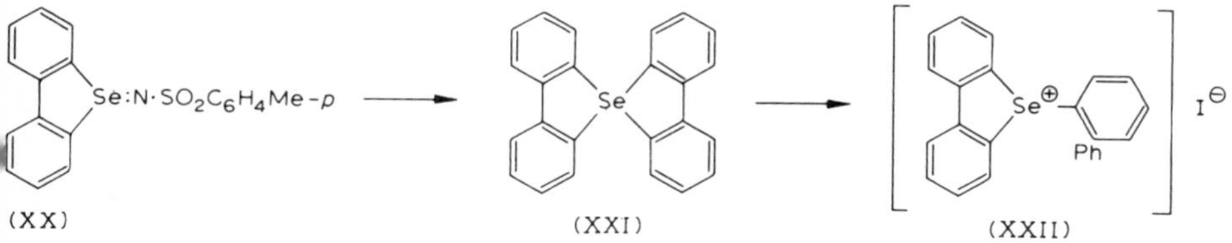

(XX) (XXI) (XXII)

Dibenzoselenophene is metallated by butyl-lithium probably in position 1 (*W. J. Burlant* and *Gould*, J. Amer. chem. Soc., 1953, **76,** 5775).

(d) Systems of two fused selenophene rings

The three **selenoloselenophenes (selenophthenes)** (XXIII, liquid, μ 1.52 D), (XXIV, m.p. 124°, μ 1.07 D), and (XXV, m.p. 53°, μ zero), oriented by their dipole moments, are among the products of the action of acetylene on heated selenium (*B. Tamamusi, H. Akiyama* and *S. Umezawa*, Bull. chem. Soc., Japan, 1939, **14,** 318; *Umezawa, ibid.*, p. 363; C.A., 1940, **34,** 424, 1309):

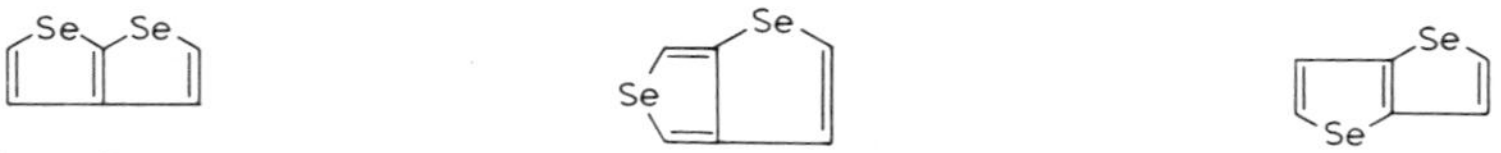

Selenolo[2,3-*b*] selenophene (XXIII)

Selenolo[3,4-*b*]selenophene (XXIV)

Selenolo[3,2-*b*]selenophene (XXV)

Benzoselenopheno[2,3-*b*]benzoselenophene **(selenanaphtheno[2,3-*b*]selenanapththene),** m.p. 190–191°, is prepared by the slow addition of excess selenium oxychloride to 1,1-diphenylethene in ether, *picrate*, m.p. 161°:

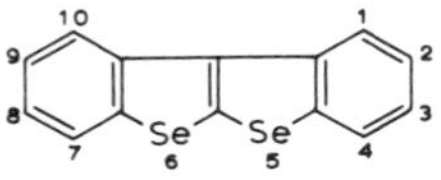

Benzoselenopheno[2,3-*b*]benzoselenophene

Similarly 1,1-bis(4-bromophenyl)ethene, 1,1-bis(4-chlorophenyl)ethene, and 1,1-bis(4-methylphenyl)ethene yield 3,8-dibromo-, 3,8-dichloro-, and 3,8-dimethyl-benzoselenopheno[2,3-*b*]benzoselenophene, respectively. Bromination or chlorination of benzoselenopheno[2,3-*b*]benzoselenophene affords the 3,8-dibromo- or 3,8-dichloro derivatives. Benzoselenopheno[2,3-*b*]benzoselenophene may also be synthesided from 1,1-diphenylethene and selenium dioxide at 230–250° (*S. Patai*, *M. Sokolovsky* and *A. Friedlander*, Proc. chem. Soc., 1960, 181). A number of derivatives of benzoselenopheno[2,3-*b*]benzoselenophene have been prepared by the action of selenium oxychloride on 1,1-diarylethenes, and the scope of the reaction, the influence of substituents in the aryl groups, and some reactions of the products have been discussed (*Patai*, *K. A. Muszkat* and *Sokolovsky*, J. chem. Soc., 1962, 734).

9. Heterocyclic five-membered ring compounds having one tellurium atom

(a) Tellurophenes

Tellurophene (I), b.p. 148°/714 mm (150°), is obtained when a solution of diacetylene in methanol is added to sodium telluride in methanol at 0° and the mixture kept overnight (*W. Mack*, Angew. Chem., 1966, **78,** 940; *Consortium für Elektrochemische Industrie, G.m.b.H.*, B.P. 1,107,698/1968; C.A., 1968, **69,** 77110t):

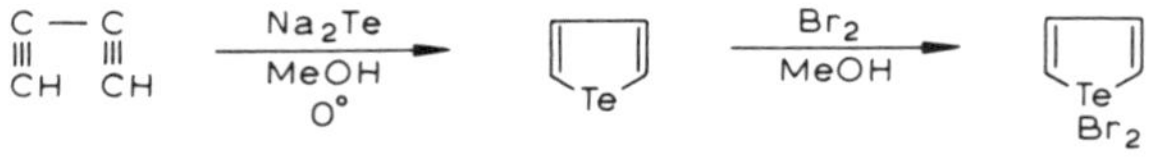

(I) (II)

Tellurophene on treatment with bromine gives 1,1-**dibromotellurophene** (II), m.p. 125° (decomp.), which is reduced to the parent compound with bisulphite solution.

2,5-Disubstituted tellurophenes may be obtained in a similar manner from substituted diacetylenes and some are collected in Table 10.

TABLE 10

DERIVATIVES OF TELLUROPHENE

Substituents 2-	5-	*m.p. (°C)*	*1,1-Dibromo- m.p. (°C)*
$C(OH)Me_2$	$C(OH)Me_2$	94	—
CH_2OH	CH_2OH	107	—
Ph	Ph	223–225	205 (decomp.)
$C(OH)Me_2$	H	40	—
Bu	Bu	—	175–180 (decomp.)

Tetraphenyltellurophene, m.p. 239.5°, is prepared from 1,4-dilithiotetraphenylbutadiene and tellurium tetrachloride, or from 1,4-diiodotetraphenylbutadiene and lithium telluride (*E. H. Braye, K. W. Huebel* and *I. H. Caplier*, J. Amer. chem. Soc., 1961, **83,** 4406; U.S.P. 3,151,140/1960; U.S.P. 3,149,101/1960.

If tellurium tetrachloride is removed by concentrated hydrochloric acid prior to distillation, then shaking powdered tellurium with hexachloro-1,3-butadiene at 250° yields **tetrachlorotellurophene,** m.p. 49°, which on chlorination affords **hexachlorotellurophene** (III), m.p. 200°:

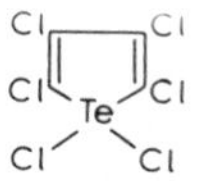

Cl Cl Cl Cl Te Cl Cl

(III)

The hexachloro compound on reduction with aqueous bisulphite solution yields tetrachlorotellurophene (*Mack*, Angew. Chem., 1965, **77,** 260).

(b) Hydrotellurophenes

Tetrahydrotellurophene (tellurolane), b.p. 166°, (1,1-*dibromide*, m.p. 130°), is obtained from sodium telluride and tetramethylene dibromide (*W. V. Farrar* and *J. M. Gulland*, J. chem. Soc., 1945, 11); m.s. (*A. M. Duffield, H. Budikiewicz* and *C. Djerassi*, J. Amer. chem. Soc., 1965, **87,** 2920). **Octahydro**-3,3′-**bitellurophene,** (3,3′-**bistellurolane**), m.p. 145°, from 1,6-diiodo-3,4-bisiodomethylhexane and sodium telluride (*cf.* p. 320).

(c) Benzotellurophenes

F. P. Mazza and *E. Melchionna* (Rend. Accad. Sci. Fis. Mat. Napoli, 1928, [iii], **34,** 54; Brit. Abstr. A, 1929, 336) describe 3-hydroxybenzo[*b*]tellurophene (3-hydroxytelluranaphthene) and the m.s. of benzo[*b*]tellurophene (IV) has been reported (*N. P. Buu-Hoï et al.*, J. heterocycl. Chem., 1970, **7,** 219):

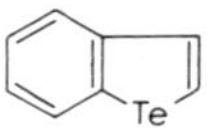

(IV) 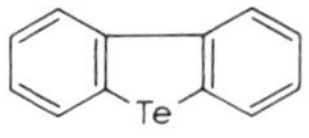 (V)

Thianthrene 5,5,10,10-tetroxide on heating almost to the m.p. with finely powdered tellurium (430–450°) in an atmosphere of carbon dioxide till evolution of sulphur dioxide ceases, gives **dibenzotellurophene** (V), m.p. 86–97° (*R. Passerini* and *G. Purrello*, Ann. chim., Rom, 1958, **48,** 738), also obtained from tellurium tetrachloride and biphenyl, *via* the dichloride derivative (*C. Courtot* and *M. G. Bastani*, Compt. rend., 1936, **203,** 197). Nitration of dibenzotellurophene gives 2-**nitrodibenzotellurophene,** m.p. 184°, and the mother liquor yields a compound, m.p. 158–161°, possibly impure 4-*nitrodibenzotellurophene (Passerini* and *Purrello, loc. cit.).*

Octafluorodibenzotellurophene, m.p. 116–119° (*S. C. Cohen, M. L. N. Reddy* and *A. G. Massey*, Chem. Comm., 1967, 451).

Chapter 4

Compounds Containing a Five-Membered Ring with one Hetero Atom from Group V: Nitrogen

R. LIVINGSTONE

1. Pyrroles

Pyrrole ("red-oil") was the name given by Runge in 1834 to certain coal-tar fractions, which turned a pine-shaving moistened with hydrochloric acid red. It was characterised by *T. Anderson* (1858) and its structure given by *A. von Baeyer* (Ber., 1870, **3,** 517). Besides occurring in coal-tar, pyrrole is found in quantity along with methyl homologues in bone oil. Compounds containing the pyrrole ring system are widely distributed in nature, *e.g.* chlorophyll, haemin, vitamin B_{12}, indigo, bile pigments and many alkaloids. The phthalocyanins are important synthetic pigments. The ring is lettered (I) or preferably numbered (II), and derivatives of 2*H*-pyrrole (III) and 3*H*-pyrrole (IV) are known. These tautomeric forms of pyrrole are described respectively also as 2*H*- and 3*H*-pyrrolenines but this nomenclature is no longer accepted in the I.U.P.A.C. rules. I.r. and Raman spectra show that the tautomers III and IV are absent from pyrrole itself (*R. C. Lord* and *F. A. Miller*, J. chem. Phys., 1942, **10,** 328), but these are the prevailing structures in several substituted pyrroles:

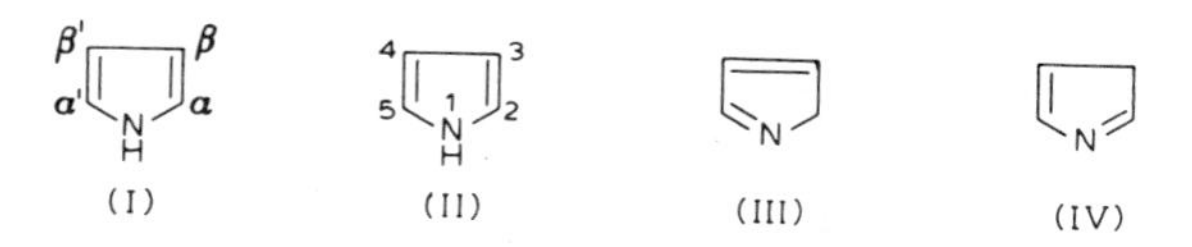

(I) (II) (III) (IV)

The dihydropyrroles are called 1-, 2-, and 3-pyrrolines (Δ^1-, Δ^2-, and Δ^3-pyrrolines) (V, VI and VII), and the tetrahydropyrroles, pyrrolidines (VIII):

The corresponding radical names are pyrrolyl, pyrrolinyl and pyrrolidinyl.

The pyrrolidine ring is present in many alkaloids and in the amino acids proline and hydroxyproline, constituents of some proteins. Another important derivative is *N*-bromosuccinimide used for free-radical bromination. For a review of the chemistry of pyrrole see *A. Treibs*, Rev. Chem. Acad. Populaire Roumaine, 1962, **7,** 1345.

The pyrrole molecule is planar and its molecular dimensions have been obtained from microwave studies (*W. S. Wilcox* and *J. H. Goldstein*, J. chem. Phys., 1952, **20,** 1656; *B. Bak et al., ibid.*, 1956, **24,** 720; *C. W. N. Cumper*, Trans. Faraday Soc., 1958, **54,** 1266), confirming the experimental results previously indicated by electron-diffraction (*V. Schomaker* and *L. Pauling*, J. Amer. chem. Soc., 1939, **61,** 1769):

1.429 Å; 107.45°; 1.371 Å; 108.08°; 108.9°; 1.383 Å

Pyrrole is not adequately represented by structure I; it is best regarded as a resonance hybrid:

π-Electron density is greater at positions 2 and 5 than at 3 and 4 (*H. C. Longuet-Higgins* and *C. A. Coulson*, Trans. Faraday Soc., 1947, **43,** 87; *R. D. Brown*, Australian J. Chem., 1955, **8,** 100). For self-consistent field molecular orbital calculations for pyrrole see *J. P. Dahl* and *A. E. Hansen*, Theoret. Chim. Acta, 1963, **1,** 199.

(a) General synthetic methods

The main methods of formation of a pyrrole are by the cyclisation of a four-carbon chain with an appropriate *N*-containing group, and by the linking of carbon to carbon to give the 3,4-bond.

(*1*) The *Knorr synthesis* (*L. Knorr* and *H. Lange*, Ber., 1902, **35,** 2998; *S. F. MacDonald*, J. chem. Soc., 1952, 4176; *Treibs* and *R. Schmidt*, Ann.,

1952, **577,** 105) depends on the reduction of an α-oximinoketone in acetic acid with zinc in the presence of a ketone possessing an active methylene group. The reduction may be carried out catalytically (*C. F. Winans* and *H. Adkins*, J. Amer. chem. Soc., 1933, **55,** 4167) or by the use of sodium dithionite. Yields are highest if an oximino-β-diketone or -ketonic ester is used, and lowest if the second component is a monoketone:

MeCO–C(=NOH)–CO$_2$Et $\xrightarrow{\text{Zn, AcOH}}$ MeCO–CH(NH$_2$)–CO$_2$Et + CH$_2$·CO$_2$Et (COMe)–COMe ⟶ 2-EtO$_2$C-3-Me-4-CO$_2$Et (COMe)-5-Me-pyrrole (NH)

Ready-made amino-ketones or amino-aldehydes, often in alkaline solution, may be condensed with the second component (*I. J. Rinkes*, Rec. Trav. chim., 1937, **56,** 1224; *O. Piloty* and *P. Hirsch*, Ann., 1913, **395,** 63; *H. Fischer* and *W. Kutscher*, *ibid.*, 1930, **481,** 199); the method is applicable to *N*-substituted pyrroles. Self-condensation of the amino ketone may give a pyrazine as a by-product:

MeCO–MeCH–NH$_2$ + Me–COMe ⟶ 2,3,5-trimethylpyrrole (NH) + tetramethylpyrazine

A related reaction to the Knorr synthesis is that between an aminoketone and an activated acetylene to give a dihydropyrrole which then forms the pyrrole by elimination of water (*J. B. Henrickson, R. Rees* and *J. F. Templeton*, J. Amer. chem. Soc., 1964, **86,** 107):

MeCO–CH(NH$_2$)–CO$_2$Et + C(CO$_2$Me)≡C(CO$_2$Me) $\xrightarrow{\text{AcONa}}$ 3-OH-3-Me-2-EtO$_2$C-4,5-di-CO$_2$Me-dihydropyrrole $\xrightarrow{\text{trace } H^{\oplus}}$ 2-EtO$_2$C-3-Me-4,5-di-CO$_2$Me-pyrrole (NH)

Another variation of the Knorr synthesis consists in the reduction of phenylazo derivatives of β-dicarbonyl compounds (*Treibs, Schmidt* and *R. Zinsmeister*, Ber., 1957, **90,** 79).

(*2*) The Paal–Knorr synthesis of pyrroles involves the treatment of a 1,4-dicarbonyl compound with ammonia or a primary amine (including hydrazine and hydroxylamine) (*L. Lederer* and *C. Paal*, *ibid.*, 1885, **18,** 2591; *Knorr*, *ibid.*, p. 299; *F. Lions et al.*, J. Proc. roy. Soc. N. S. Wales, 1938, **71,** 92; *W. S. Bishop*, J. Amer. chem. Soc., 1945, **67,** 2261; *W. Borsche* and *A. Fels*, Ber., 1906, **39,** 3877; *C. A. C. Haley* and *P. Maitland*, J. chem. Soc., 1951, 3155):

$CH_2{-}CH_2$ / MeCO COMe $\xrightarrow{NH_3}$ 2,5-dimethylpyrrole (Me, Me, N–H); $\xrightarrow{RNH_2}$ 1-R-2,5-dimethylpyrrole (Me, Me, N–R)

$CH_2{-}CH{\cdot}CO_2Et$ / PhCO COMe $\xrightarrow{MeNH_2}$ pyrrole (Ph, CO_2Et, Me, N–Me)

Appropriate initial materials are accessible by acetoacetic ester synthesis. Succindialdehyde and ammonia give pyrrole itself in poor yield.

The use of hydrazine in this method gives 1,1-bipyrrolyls (p. 391), dihydropyridazines or 1-aminopyrroles (*C. Bülow*, Ber., 1902, **35,** 4311). Thus diethyl diacetosuccinate gives diethyl 1-amino-2,5-dimethylpyrrole-3,4-dicarboxylate and thence 1-amino-2,5-dimethylpyrrole. In the same way hydroxylamine affords 1-hydroxypyrroles (*Knorr*, Ann., 1886, **236,** 290; *E. E. Blaise*, Compt. rend., 1914, **158,** 1686):

$EtCO_2{\cdot}C - C{\cdot}CO_2Et$ / MeCO COMe $\xrightarrow{NH_2{\cdot}NH_2}$ pyrrole ($EtCO_2$, CO_2Et, Me, Me, N–NH_2) $\longrightarrow$ pyrrole (Me, Me, N–NH_2)

(*3*) Pyrroles are formed by interaction in the cold of ammonia or a primary amine, a β-oxo-ester, and an α-halogenoketone or $\alpha\beta$-dichloroethyl ethyl ether (source of chloroacetaldehyde) *(Hantzch synthesis)*. Furans are by-products (*F. Feist*, Ber., 1902, **35,** 1537, 1545; *E. Benary*, Ber., 1911, **44,** 493; *G. Korschun*, Ber., 1905, **38,** 1125):

CH_2Cl / MeCO + $CH_2{\cdot}CO_2Et$ / COMe $\xrightarrow{NH_3}$ CH_2Cl / MeCO $CH{\cdot}CO_2Et$ / $\|$ CMe / H_2N $\longrightarrow$ pyrrole (CO_2Et, Me, Me, N–H)

$\downarrow$

$MeC = C{\cdot}CO_2Et$ / H_2CCl COMe $\longrightarrow$ furan (Me, CO_2Et, Me)

(*4*) A route to pyrroles is offered *via* the interaction between a nitro-olefin or a β-acyloxynitro-compound (which gives nitro-olefin on decomposition), a β-ketonic ester and a primary amine; nitro-olefin and base may be used in combination as a β-nitro-amine, or amine and ester as, *e.g.*, β-benzylaminocrotonic ester (*C. A. Grob et al.*, Helv., 1953, **36,** 49; 1955, **38,** 1121):

MeCH / $\|$ / CH / NO_2 + $CH_2{\cdot}CO_2Et$ / CMe / RN= $\longrightarrow$ MeCH——$CH{\cdot}CO_2Et$ / CH_2 CMe / NO_2 N–R $\xrightarrow[-2H]{-HNO_2}$ pyrrole (Me, CO_2Et, Me, N–R)

(5) Unsaturated aldehydes with α-arylamino-arylacetonitriles give good yields of 1,2,3-trisubstituted pyrroles (*Treibs* and *R. Derna*, Ann., 1954, **589,** 176):

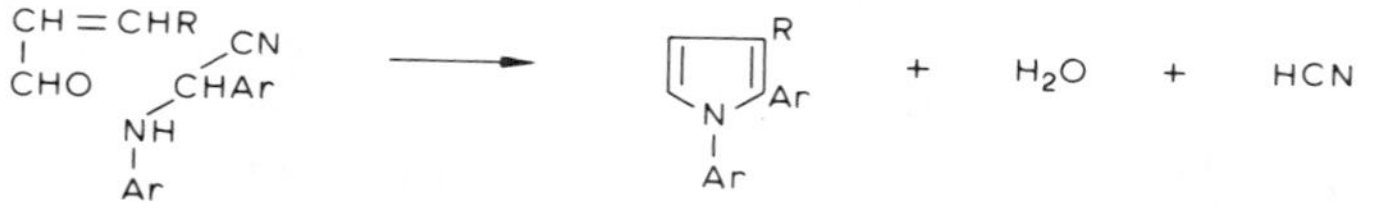

(6) Pyrrole and its *N*-alkyl or -aryl derivatives are obtained with the corresponding α-carboxylamides, by heating the mucates or saccharates of ammonia or primary amines (*A. Pictet* and *A. Steinmann*, Ber., 1902, **35,** 2530; *T. Reichstein*, Helv., 1927, **10,** 389); ammonium rhamnonate gives α-methylpyrrole (*E. Votoček* and *S. Malachta*, Coll. Czech. chem. Comm., 1936, **8,** 77):

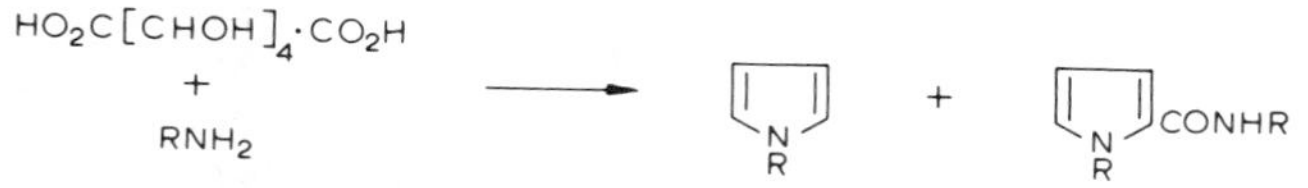

(7) Pyrrole derivatives may be obtained from tetracyanoethane which is very prone to undergo addition reactions involving two cyano groups on adjacent carbon atoms; *e.g.*, base-catalysed addition of hydrogen sulphide gives 2,5-diamino-3,4-dicyanothiophene, which can be rearranged to a pyrrolethiol (*T. L. Cairns et al.*, J. Amer. chem. Soc., 1958, **80,** 2775):

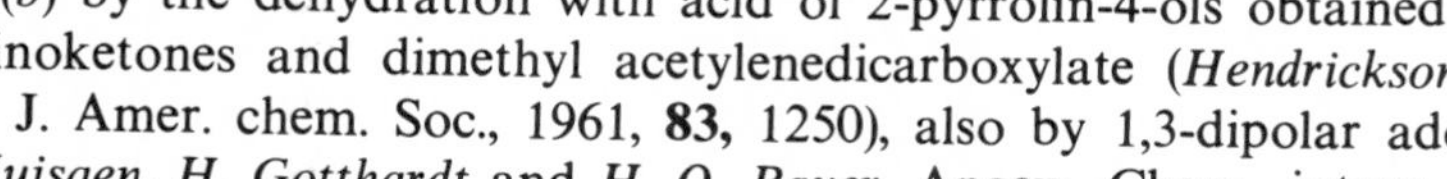

(8) Pyrroles may be prepared (*a*) by the reaction between diacetylenes and ammonia, primary aliphatic and aromatic amines in the presence of cuprous chloride (*J. Reisch* and *K. E. Schulte*, Angew. Chem., 1961, **73,** 241), (*b*) by the dehydration with acid of 2-pyrrolin-4-ols obtained from α-aminoketones and dimethyl acetylenedicarboxylate (*Hendrickson* and *Rees*, J. Amer. chem. Soc., 1961, **83,** 1250), also by 1,3-dipolar addition (*R. Huisgen, H. Gotthardt* and *H. O. Bayer*, Angew. Chem., intern. Edn., 1964, **3,** 135; *Gotthardt, Huisgen* and *F. C. Shaefer*, Tetrahedron Letters, 1964, 487):

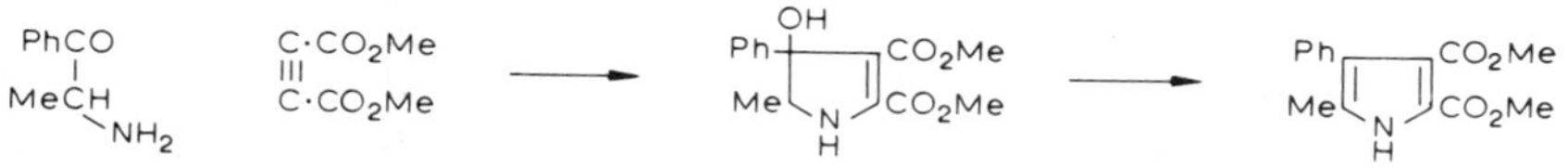

(*9*) Pyrroles are obtained by the following route from α,β-unsaturated ketones and glycine derivatives (*A. H. Jackson, G. W. Kenner* and *W. G. Terry, ibid.*, 1962, 921; *cf. R. V. Robertson, J. E. Francis* and *B. Witkop*, J. Amer. chem. Soc., 1962, **84,** 1709):

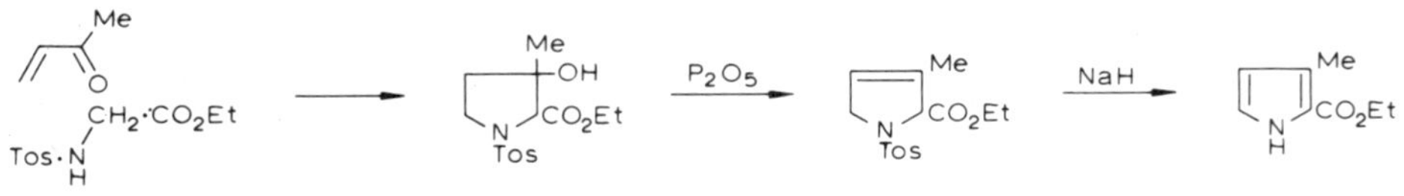

(*10*) The pyrolysis of δ-sultams gives pyrroles and sulphur dioxide. Benzo analogues afford indoles (*B. Helferich et al.*, Ann., 1961, **646,** 32, 45):

(*11*) Furanamines when treated with hydrogen over platinum at 300° are converted to pyrroles (*I. F. Bel'skii*, Zhur. obshcheĭ. Khim., 1962, **32,** 2905, 2908):

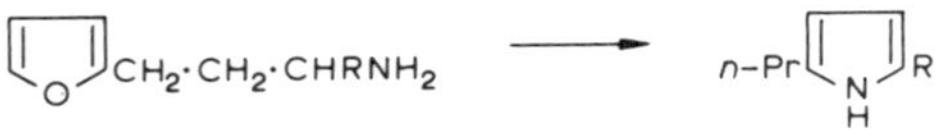

(*12*) Pyrroles have been prepared from the Diels–Alder adducts of dienecarboxylic esters and nitroso compounds (*J. Firl* and *G. Kresze*, Ber., 1966, **99,** 3695).

(*13*) *cis*-α-Formyl-β-chloro-olefins react with ethyl *N*-methylaminoacetate to give pyrroles (*S. Hauptmann et al.*, Tetrahedron Letters, 1968, 1317):

(*14*) Acetylenic amino-alcohols $RC{:}C \cdot CR'(OH) \cdot CH_2NH_2$ on heating give pyrroles and cyclopentadiene derivatives (*F. Ya. Perveer* and *V. M. Demidova*, Zhur. org. Khim., 1965, **1,** 2244).

(*15*) Pyrroles have been obtained by the dehydrogenation of pyrrolidines over heated zinc dust or palladised asbestos (*Adkins* and *L. G. Lundsted*, J. Amer. chem. Soc., 1949, **71,** 2964).

(b) General properties and reactions

(*1*) Towards electrophilic reagents pyrrole behaves mainly as a very reactive aromatic compound, comparable with phenol. Pyrrole is less stable than benzene but more stable than furan and is often formed in pyrolytic processes. Calculations based on chemical shift data support the order of aromaticity thiophene > pyrrole > furan (*J. A. Elvidge*, Chem. Comm., 1965, 160; *D. W. Davis*, *ibid.*, 1965, 258).

(*2*) Although only moderately resistant to oxidation it may occur without the ring opening, thus chromic acid gives maleimides (*P. Karrer* and *A. P. Smirnoff*, Helv., 1922, **5,** 836) and nitrous acid, maleimidoximes (*O. Piloty* and *E. Quitmann*, Ber., 1909, **42,** 4693) with loss of α-substituents. Photo-oxidation of pyrrole in the presence of eosin gives 2-hydroxy-3-pyrrolin-5-one which, on oxidation with manganese dioxide yields maleimide (*P. de Mayo* and *S. T. Reid*, Chem. and Ind., 1962, 1576):

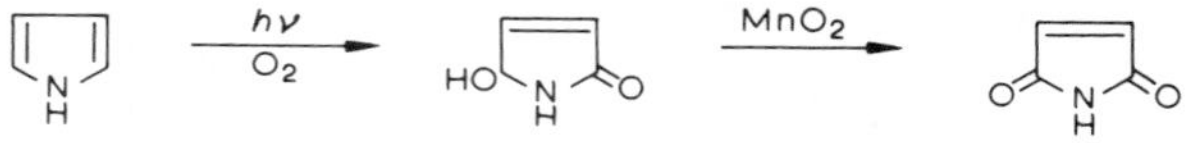

Oxidation of 2,3,4,5-tetraphenylpyrrole with nitrous acid affords dibenzoylstilbene (*R. Kuhn* and *H. Kainer*, Ann., 1952, **578,** 227).

(*3*) With acids pyrrole and its homologues, both *N*- and *C*-alkylpyrroles, like cyclic dienes, often *polymerise* or undergo hydrolytic ring-fission, but pyrroles containing an electron-withdrawing substituent are either unaffected or form simple proton salts (*A. Treibs* and *H. G. Kolm*, Ann., 1957, **606,** 166). In aqueous sulphuric acid these pyrroles accept protons at the α-positions giving stable cations (*E. B. Whipple*, *Y. Chiang* and *R. L. Hinman*, J. Amer. chem. Soc., 1963, **85,** 26), which are very reactive and readily polymerise (*G. F. Smith*, Adv. Heterocyclic Chem., 1963, **2,** 287).

Protonation of pyrrole gives a cation which has no aromatic stability due to a π-electron sextet and is therefore very reactive. The electrophile will attack a neutral pyrrole molecule with the result that the pyrrole polymerises. A homogenous trimer, **tripyrrole,** m.p. 99–100°, may be obtained by treating pyrrole with dry hydrogen chloride in ether or with a 6 *N* solution of hydrochloric acid. The structure of the trimer has been proved by degradation to 1,4-di(2′-pyrrolyl)butane (*H. A. Potts* and *Smith*, J. chem. Soc., 1957, 4018):

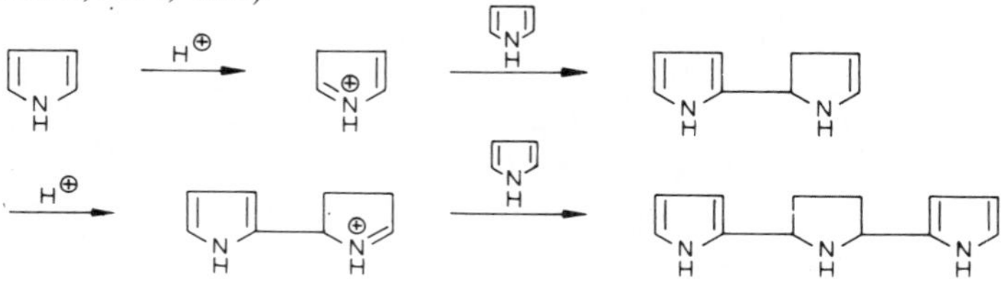

Tripyrrole, although it can be isolated, is quite a reactive compound and in acid solution undergoes further polymerisation in the presence or absence of pyrrole. The structures of higher polymers of pyrrole have not been elucidated. Pyrolysis of tripyrrole gives indole, pyrrole and ammonia. A number of 2-alkyl and 2,3-dialkyl derivatives of pyrrole afford crystalline salts of the corresponding dimers with dry hydrogen chloride or picric acid in ether, *N*-methylpyrrole polymerises but no dimer or trimer has been isolated, 3-methylpyrrole gives an amorphous polymeric material, and 2-phenylpyrrole is dimerised by hydrogen chloride in ether *(Smith, loc. cit.)*. Some 2,4-di-, tri-, and tetra-alkylpyrroles are soluble in aqueous mineral acid forming stable solutions (*H. Fischer* and *J. Klarer*, Ann., 1926, **450,** 199). Some of the dimers when heated with acid lose ammonia to give indoles also obtained from the monomers by treatment with hot aqueous sulphuric acid and almost certainly involving the formation of the dimers as intermediates; thus 2-methylpyrroles gives 2,4-dimethylindole (*C. F. H. Allen, D. M. Young* and *M. R. Gilbert*, J. org. Chem., 1937, **2,** 235, 400).

(*4*) Pyrroles seldom exhibit the additive reactions of olefins or behave as 1,3-dienes, but pyrrole is reported to undergo the Diels–Alder reaction with hexafluoro-Dewar benzene to give adduct IX and a bis-adduct (*M. G. Barlow, R. N. Haszeldine* and *R. Hubbard*, Chem. Comm., 1969, 301):

(IX)

With triphenylmethyl radicals pyrrole gives 2,5-di(triphenylmethyl)-3-pyrroline (*J. B. Conant* and *B. F. Chow*, J. Amer. chem. Soc., 1933, **55,** 3475):

$Ph_3C\cdot$

(*5*) Pyrrole is not reduced using sodium and alcohol; two atoms of hydrogen are taken up using various metal–acid combinations but hydrogenation to pyrrolidine occurs more easily using platinum in acid solution than under neutral conditions (*Treibs* and *Kolm, loc. cit.)*. A vinyl side-chain can be hydrogenated without affecting the nucleus (*Fischer* and *B. Walach*, Ber., 1925, **58,** 2821). Pyrrole is stated to react with sodium hydrogen sulphite at 100° to give sodium pyrrolidine-1,5-disulphonate (X) which, with reagents such as hydroxylamine, yields succindialdehyde derivatives (*Treibs* and *R. Zimmer-Galler*, Ann., 1963, **664,** 140):

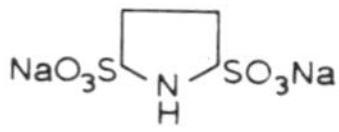

NaO_3S–(pyrrole, N–H)–SO_3Na

(X)

(6) Pyrrole is easily attacked by electrophilic reagents, substitution taking place preferentially in the 2 and 5 positions; halogenation gives tetrahalogenated pyrroles and in the case of chlorine a pentachloropyrrole; sulphonation, using 1-proto-1-pyridinium sulphonate affords the 2-sulphonic acid (p. 353), polymerisation occurring under ordinary conditions in sulphuric acid; the 2-nitro derivative is formed by treatment with acetyl nitrate at low temperatures (*K. J. Morgan* and *D. P. Morrey*, Tetrahedron, 1966, **22,** 57), but it is difficult to isolate pure products following direct nitration owing to resinification. Pyrrole undergoes diazo-coupling, Friedel–Craft, and Mannich reactions. De-activated pyrroles give the Gattermann and Hoesch reactions. The Reimer–Tiemann reaction gives pyrrole-2-carbaldehyde and 3-chloropyridine.

(c) Pyrrole and its substitution products

(i) Pyrrole

Pyrrole is prepared (*1*) from ammonium mucate (*S. M. McElvain* and *K. M. Bolliger*, Org. Synth., 1929, **9,** 78), (*2*) by passing butynediol, $HOCH_2{\cdot}C{\equiv}C{\cdot}CH_2OH$ and ammonia over thoria-alumina (*W. Reppe*, Ann., 1955, **596,** 155), (*3*) from diacetylene, ammonia, and cuprous chloride at 140–160° (*K. E. Schulte, J. Reisch* and *H. Walker*, Ber., 1965, **98,** 98), (*4*) by dehydrogenation of pyrrolidine over a rhodium–alumina catalyst at 650° (*J. M. Patterson* and *P. Drenchko*, J. org. Chem., 1959, **24,** 878), and (*5*) by catalytic dehydrogenation of diethylamine (*A. L. Liberman, O. V. Brogin* and *B. A. Kazanskii*, Izvest. Akad. Nauk, S.S.S.R., Otdel. Khim. Nauk, 1961, 525). (*6*) The passage of acetylene and ammonia, butadiene and ammonia or nitric oxide (*G. G. Schneider et al.*, Ber., 1937, **70,** 425) over aluminia, and furan and ammonia over heated aluminium silicate (*B. Bak, T. Pedersen* and *G. O. Soerensen*, Acta chem. Scand., 1964, **18,** 275) give pyrrole, which is obtained also (*7*) by passing succinimide over heated zinc dust or with hydrogen over platinum, and in the pyrolysis of diethylamine and glutamic acid. (*8*) Pyrrole has been obtained from furfural and ammonia in a one-step preparation consisting of the combination of two vapour-phase heterogeneous processes in space (*L. Meszaros, M. Bartok* and *A. S. Gilde*, Acta Univ. Szeged., Acta Phys. Chem., 1967, **13,** 121; C.A., **69,** 35846g):

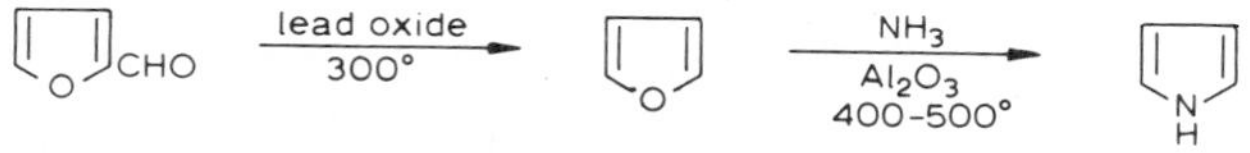

furfural (O, CHO) $\xrightarrow[\text{300°}]{\text{lead oxide}}$ furan (O) $\xrightarrow[\text{Al}_2\text{O}_3,\ 400\text{–}500°]{\text{NH}_3}$ pyrrole (N–H)

Pyrrole is a colourless liquid, b.p. 129°/760 mm; d_4^{20}. 0.9691; n_D^{20} 1.5085; μ 1.84 (gas), 1.8 (in benzene) (*R. J. W. LeFèvre et al.*, J. chem. Soc., 1953, 1626; *M. Gomel* and *H. Lumbroso*, Bull. soc. Chim. France, 1962, 2200) with an odour similar to chloroform. It is readily soluble in organic solvents, sparingly soluble in water and darkens slowly especially in the air. Spectroscopic studies include u.v. (*S. Monczel*, Z. phys. Chem., 1927, **125,** 161; *G. Scheibe* and *H. Grieneisen*, *ibid.*, 1934, **25B,** 52), i.r. (*R. Manzoni-Ansidei* and *M. Rolla*, Atti Accad. Lincei, 1938, **27,** 410; *L. R. Zumwalt* and *R. M. Badger*, J. chem. Phys., 1939, **7,** 629), Raman (*A. W. Reitz*, Z. phys. Chem., 1936, **33B,** 186; 1938, **38B,** 275), n.m.r. (*J. A. Elvidge*, Chem. Comm., 1965, 160; *H. Wynberg*, Tetrahedron, 1962, **21,** 515; *M. Freymann* and *R. Freymann*, Compt. rend., 1959, **248,** 677) and mass spectra of some derivatives of pyrrole (*H. Budzikiewicz et al.*, J. chem. Soc., 1964, 1949).

Pyrrole is a very weak base ($pK_a = -3.8$) (*Y. Chiang* and *E. B. Whipple*, J. Amer. chem. Soc., 1963, **85,** 2763), insoluble in aqueous alkali, dissolves slowly in acids with polymerisation, gradual separation of "pyrrole-red" and loss of ammonia; with hydrogen chloride in ether it gives tripyrrole hydrochloride (p. 335). Pyrrole with chromic acid gives maleimide; with hydrogen peroxide succinimide and "pyrrole-black", also produced by treatment with ozone and with hydrogen peroxide in acetic acid, 5-(2-pyrrolyl)-2-pyrrolidinone (*L. Chierici* and *G. P. Gardini*, Tetrahedron, 1966, **22,** 53).

The pyrrole ring can be expanded by the action of chlorocarbene to give pyridine (*G. L. Cross* and *G. M. Schwartz*, J. org. Chem., 1961, **26,** 2609). Benzoylation gives *C*-benzoyl derivatives. A pine-wood shaving moistened with hydrochloric acid is turned red by pyrrole vapour.

Picrate, m.p. 69° (decomp.); *trinitrobenzene adduct*, m.p. 95° (*A. Treibs* and *P. Dieter*, Ann., 1934, **513,** 65).

N-*Deuteropyrrole*, d_4^{25} 0.9725, n_D^{25} 1.5056; the Raman effect has been studied (*O. Redlich* and *W. Stricks*, Monatsh., 1936, **68,** 47; *G. B. Bonino* and *Manzoni-Ansidei*, Brit. Abstr. AI, 1937, 10). *Deuteropyrroles* (*F. A. Miller*, J. Amer. chem. Soc., 1942, **64,** 1543); n.m.r. of *N*-deuteropyrrole and other pyrrole derivatives (*R. J. Abraham* and *H. J. Bernstein*, Canad. J. Chem., 1959, 1056); i.r. (*J. Moricillo* and *J. M. Orza*, Anales real Soc. espan. fis. y quim., 1960, **56B,** 231; *R. D. Hill* and *G. D. Meakins*, J. chem. Soc., 1958, 760).

(ii) N-Derivatives of pyrrole (1-substituted pyrroles)

(*1*) *Metal salts*. The action of potassium (*G. R. Clemo* and *G. R. Ramage*, J. chem. Soc., 1931, 49) or solid potassium hydroxide at 130° gives the **potassium salt,** $C_4H_4N{\cdot}K$, completely hydrolysed by moisture to pyrrole and potassium hydroxide. Sodium does not give the corresponding derivative, but pyrroles with electron-attracting substituents afford *N*-sodium derivatives with sodium ethoxide (*A. Treibs* and *H. G. Kolm*, Ann., 1957, **606,** 166). The respective 1-substituted pyrroles are obtained by treating the potassium salt with alkyl halides, acyl chlorides and ethyl chloroformate (*C. F. Hobbs et al.*, J. Amer. chem. Soc., 1962, **84,** 43). With ferrous chloride in liquid ammonia potassium pyrrole yields the complex salt dipotassium tetrapyrroleferrate, $K_2[Fe(C_4H_4N)_4]$ (*O. Schmitz-Dumont* and *S. Pateras*, Z. anorg. Chem., 1935, **224,** 62), and with silicon tetrachloride, *tetra*-N-*pyrrolylsilicon*, m.p. 173° (*J. E. Reynolds*, J. chem. Soc., 1909, **95,** 505).

(*2*) *Pyrrolylmagnesium halides.* Pyrrole with ethylmagnesium bromide, and other Grignard reagents yields a pyrrolylmagnesium bromide, represented as XI or as an ionic hybrid XII (*B. Oddo*, Mem. Accad. Lincei, 1923, **14,** 510; Ber., 1914, **47,** 2477; Gazz., 1934, **64,** 584; *H. Gilman* and *L. L. Heck*, J. Amer. chem. Soc., 1930, **52,** 4949):

(XI) (XII)

Support for this latter suggestion is obtained from the n.m.r. spectrum of pyrrolylmagnesium chloride in ether (*M. G. Reinecke, H. M. Johnson* and *J. F. Sebastian*, *ibid.*, 1963, **85,** 2859). Alkylation of pyrrolylmagnesium bromide gives the isomeric 2- and 3-alkylpyrroles and polyalkylpyrroles, whereas acylation with either acyl halides or esters gives only the 2-acylpyrroles. Alkylation proceeds with complete inversion of configuration (*P. S. Skell* and *G. P. Bean*, *ibid.*, 1962, **84,** 4655, 4660).

(*3*) *Alkylated pyrroles* may be obtained from pyrrolylmagnesium bromide at low temperatures but on heating they isomerise to the 2-substituted pyrrole; with carbon dioxide pyrrole-2-carboxylic acid is formed.

1-*n*-Butylpyrrole at about 500° yields 2-*n*-butylpyrrole which equilibrates with the 3-isomer (*I. A. Jacobson* and *H. B. Jensen*, J. phys. Chem., 1962, **66,** 1245; 1964, **68,** 3068); other alkylpyrroles have been pyrolysed (*Patterson* and *Drenchko*, J. org. Chem., 1962, **27,** 1650; *Patterson, J. Brasch* and *Drenchko*, *ibid.*, p. 1652). The thermal isomerisation of 1-(*sec*-butyl)pyrrole to the 2-isomer occurs with a 77% retention of configuration; the 3-isomer is formed with an estimated 10% retention of configuration (*Patterson* and *L. T. Burka*, J. Amer. chem. Soc., 1966, **88,** 3671).

In the *C*-acylation of *N*-methylpyrrole by ethylmagnesium bromide and acyl chlorides, the Grignard reagent may be replaced by magnesium bromide (*W. Herz*, J. org. Chem., 1957, **22,** 1260).

The following syntheses of *N*-alkyl- and *N*-aryl-pyrroles are in addition to those described in ring-synthesis 6 (p. 333) or, where *C*-substituents are required, syntheses 1 (p. 330) and 2 (p. 331).

(*4*) In presence of alkali, pyrrole reacts with acrylonitrile and acetylene to give *β*-*N*-pyrrolylpropionitrile and *N*-vinylpyrrole, respectively (*I. G. Farbenind.*, B.P. 457,621, 470,116/1936).

(*5*) *N*-Arylpyrroles are obtained by the reaction between butadienes and thionylarylamines (*J. Roček*, C.A., 1955, **49,** 1055):

RC — CR / ‖ ‖ / CH_2 CH_2 —(ArN=SO)→ [R-substituted dihydro-1,2-thiazine N-oxide, N–Ar, SO] —(KOH)→ [3,4-R-1-Ar-pyrrole]

(*6*) Arylhydroxylamines react with two moles of dimethyl acetylenedicarboxylate giving tetramethyl *N*-arylpyrrole tetracarboxylates (*E. H. Huntress, T. E. Lesslie* and *W. H. Hearon*, J. Amer. chem. Soc., 1956, **78,** 419).

(*7*) *N*-Substituted pyrroles have been prepared by the interaction of primary amines with 1,4-bis-(dimethylamino)-1,3-butadiene in the presence of acid (*M. F. Fegley, N. M. Bortnick* and *C. H. McKeever*, *ibid.*, 1957, **79,** 4144), αα′-dibromoadiponitrile (*Treibs* and *F. Neumayr*, Ber., 1957, **90,** 76) and 1,2,3,4-tetrabromobutane (*Treibs* and *O. Hitzler*, *ibid.*, p. 787).

(*8*) *N*-Alkyl-, -aryl-, or -acyl-amines with an alkoxycarbonyl- or nitrile-activated methylene group in the α- and α′-positions condense with glyoxal in the presence of potassium *tert*-butoxide to give 1-substituted pyrroles (*K. Dimroth* and *U. Pintschovius*, Ann., 1961, **639,** 102):

R^3CH_2–N(R^1)–CH_2R^2 + HC(=O)–CH(=O) ⟶ [2-R^3-5-R^2-1-R^1-pyrrole]

R^1 = alkyl, aryl, acyl
R^2 = R^3 = CO_2Me, CO_2Et, CN

(*9*) Diels–Alder cycloaddition of 1-*p*-tolylsulphonylpyrrole and dimethylacetylenedicarboxylate gives dimethyl 7-azonorbornadiene-1,2-dicarboxylate, which undergoes photochemical ring-closure to give the 3-azaquadricyclane XIII (*H. Prinzbach, R. Fuchs* and *R. Kitzing*, Angew. Chem., intern. Edn., 1968, **7,** 67, 727):

[1-Tos-pyrrole] + C·CO_2Me ≡ C·CO_2Me ⟶ [7-Tos-7-azanorbornadiene-2,3-(CO_2Me)$_2$] —(hν)→ [Tos-N aza-quadricyclane, (CO_2Me)$_2$] (XIII)

(*iii*) *Individual* N-*substituted pyrroles*

1-**Methylpyrrole,** b.p. 114–115° (*Oddo*, Ber., 1914, **47,** 2427), 1-*ethylpyrrole*, b.p. 129–130°, and 1-*allylpyrrole*, b.p. 105°/48 mm, are prepared from potassium pyrrole, although allyl bromide and potassium pyrrole afford 2- rather than 1-allylpyrrole (*P. A. Cantor* and *C. A. van der Werf*, J. Amer. chem. Soc., 1958, **80,** 970). 1-*Methyl*, 1-*ethyl*-, 1-n-*butylpyrrole*, b.p. 53–54°/11 mm (*T. Reichstein*, Helv., 1927, **10,** 387), 1-*benzylpyrrole*, b.p. 248°, and 1-*phenylpyrrole*, m.p. 62° (*H. Adkins* and *H. L. Coonradt*, J. Amer. chem. Soc., 1941, **63,** 1563) are prepared by synthetic method 6 (p. 333).

1-Phenylpyrrole and other pyrrole derivatives (*R. Huisgen* and *E. Laschtuvka*, Ber., 1960, **93,** 81). 1-Methylpyrrole and dimethyl acetylenedicarboxylate give the corresponding indoles (*R. M. Acheson* and *J. M. Vernon*, J. chem. Soc., 1962, 1148). 1-*Vinylpyrrole*, b.p. 122°, is obtained from pyrrole and acetylene (*W. Reppe et al.*, Ann., 1956, **601,** 128; Review, *M. F. Shostakovskii, G. G. Skvortsova* and *E. S. Domnina*, Russ. Chem. Reviews, 1969, **38,** 407).

1-**Acetylpyrrole,** b.p. 180–181°, and 1-*benzoylpyrrole*, b.p. 276° are prepared from potassium pyrrole. 1-Acetylpyrrole may be obtained in excellent yield also by heating an equimolecular mixture of pyrrole and *N*-acetylimidazole (*G. S. Reddy*, Chem. and Ind., 1965, 1426). The acetyl derivative reacts in methanol with diazomethane as a catalyst to give pyrrole and methyl acetate (*H. Bredereck et al.*, Ber., 1956, **89,** 1169) and is reduced to 1-ethylpyrrole with lithium tetrahydridoaluminate/aluminium chloride (*R. F. Nystrom* and *C. R. A. Berger*, J. Amer. chem. Soc., 1958, **80,** 2896). *N*-Benzoylpyrroles are also obtained by the Schotten–Baumann method (*Treibs* and *K. H. Michl*, Ann., 1952, **577,** 115). 1,1′-*Carbonylbispyrrole*, m.p. 62–63° (*G. Ciamician* and *P. Magnaghi*, Ber., 1885, **18,** 414).

Pyrrole and formaldehyde afford the unstable 1-*hydroxymethylpyrrole*, b.p. 45.5°/1.5 mm, *phenylcarbamate*, m.p. 111–112° (*M. S. Taggart* and *G. H. Richter*, J. Amer. chem. Soc., 1934, **56,** 1385). 1-*Hydroxy*-2,5-*dimethylpyrrole*, m.p. 44–45°, yields with amyl nitrite 3-hydroxyimino-2,5-dimethyl-1-pyrroline *N*-oxide (XIV).

The oxime XV exhibits ring-chain tautomerism (*A. H. Blatt*, *ibid.*, 1934, **56,** 2774; 1936, **58,** 590), affording the "glycoside" XVI:

NOH; Me N Me; O (XIV) HC = CPh; PhC COPh; NOH ⇌ Ph; Ph; Ph N OH; O (XV) MeOH/HCl → Ph; Ph; Ph N OMe; O (XVI)

(*iv*) *C-Alkyl- and C-aryl-pyrroles*

(*1*) *Synthesis.* (*a*) *C*-Alkylpyrroles are obtained from the ethoxycarbonyl and acyl derivatives prepared by methods *1* and *2* (pp. 330, 331). The acyl groups may be converted into alkyl groups by Kishner–Wolff reduction, ethoxycarbonyl groups are lost; on hydrogenation over copper–chromium oxide under pressure, the ester group reacts less readily (*F. K. Signaigo* and *H. Adkins*, J. Amer. chem. Soc., 1936, **58,** 709); esters are often resistant to reduction using lithium tetrahydridoaluminate (*A. Treibs* and *H. Derra-Scherer*, Ann., 1954, **589,** 188).

Convenient syntheses have been devised for all the *C*-methylpyrroles with the exception of 3-methylpyrrole. Reduction of a *C*-acyl to a *C*-alkyl with lithium tetrahydridoaluminate is the key step in most of the syntheses. This method is unsuccessful for the preparation of *C*-alkyl *N*-methylpyrroles because reduction of *C*-acyl-*N*-methylpyrroles stops at the hydroxyalkyl stage, regardless of whether the acyl group is in the 2- or 3-position. 1,2-Dimethyl- and 1,2,3,5-tetramethyl-pyrroles are prepared by methylation of the potassium salt of the appropriate *C*-methyl derivatives (*R. L. Hinman* and *S. Theodoropulos*, J. org. Chem., 1963, **28,** 3052).

(*b*) Alkyl groups may be introduced by first forming an aldehyde or ketone (Gattermann or Hoesch reaction), followed by Kischner–Wolff reduction.

(*c*) The direct *C*-alkylation of pyrrole derivatives may be effected by heating at 220° with sodium alkoxide often with the replacement of acyl and ester groups (*H. Fischer* and *E. Bartholomäus*, Z. physiol. Chem., 1912, **80,** 6; *J. W. Cornforth et al.*, J. chem. Soc., 1942, 682).

Other less preparatively useful methods include (*i*) thermal rearrangement of 1-alkylpyrroles (p. 339), when sometimes a mixture of 2- and 3-substituted isomers is obtained (*J. M. Patterson* and *S. Soedigdo*, J. org. Chem., 1968, **33,** 2057; *Patterson, L. T. Burka* and *M. R. Boyd*, *ibid.*, 1968, **33,** 4033); (*ii*) the alkylation of pyrrolylmagnesium halides (*A. J. Castro et al.*, *ibid.*, 1965, **30,** 344). (*iii*) Pyrroles unsubstituted in the 2 and 5 positions may be obtained from ketones (*H. Plieninger* and *W. Bühler*, Angew. Chem., 1959, **71,** 163):

$$RCO{\cdot}CH_2R' \xrightarrow[(2)\ MeOH/H_2SO_4]{(1)\ Formylation} RCO{\cdot}CHR'{\cdot}CH(OMe)_2 \xrightarrow[(2)\ dihydropyran]{(1)\ HCN} RC(O\text{-}tetrahydropyranyl)(CN)-CHR'{\cdot}CH(OMe)_2 \xrightarrow[(2)\ H^{\oplus}]{(1)\ LiAlH_4} \text{R-pyrrole (NH)}$$

and 2,3,5-trimethylpyrrole by treating the following hydrazide derivative with sodium ethoxide and an excess of hydrazine (*E. Baltazzi*, Compt. rend., 1962, **224,** 702):

$$\text{2-CONH·NHTos-3,5-dimethylpyrrole (Me, Me, CONH·NHTos, NH)} \longrightarrow \text{2,3,5-trimethylpyrrole (Me, Me, Me, NH)}$$

(*iv*) Methylpyrroles can be obtained by the reduction of pyrrole Mannich bases (*Treibs* and *R. Zinsmeister*, Ber., 1957, **90,** 87). (*v*) 2,4-Dialkylpyrroles have been obtained by the catalytic conversion of furanamines (*I. F. Bel'skii, N. I. Shuikin* and *G. E. Skobtsova*, Izvest. Akad. Nauk S.S.S.R., Ser. Khim., 1964, 1118), and 2,5-dialkylpyrroles by the reductive amination of furfurylidenealkyl ketones using Raney nickel in methanol saturated with ammonia (*Shuikin, Bel'skii* and *G. E. Abgaforova*, *ibid.*, 1965, 163).

(*d*) *C*-Arylpyrroles are usually prepared by ring-synthesis, sometimes by isomerisation of the *N*-derivatives (p. 339).

(*2*) *Preparation from natural products.* Reduction of haemin or chlorophyll derivatives with hydriodic acid yields a mixture of the following four pyrrole homologues XVII–XX (*R. Willstätter* and *Y. Asakina*, Ann., 1911, **385,** 188; *H. Fischer et al.*, Ber., 1914, **47,** 1820; Ann., 1930, **478,** 283):

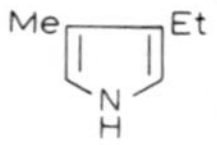

Opsopyrrole (XVII) Haemopyrrole (XVIII) Cryptopyrrole (XIX) Phyllopyrrole (XX)

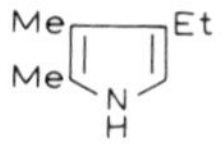

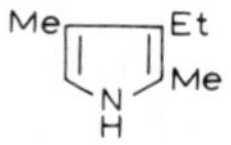

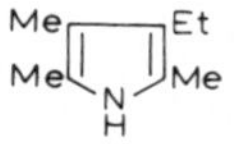

Opsopyrrole is insoluble in hydrochloric acid which dissolves the remaining three bases; these are eventually separated by fractional formation and crystallisation of the picrates. All yield ethylmethylmaleimide on oxidation, so locating the β-substituents and establishing the constitution of opsopyrrole and phyllopyrrole which can be isolated by coupling the others with diazobenzenesulphonic acid. The structures are further established by the following syntheses:

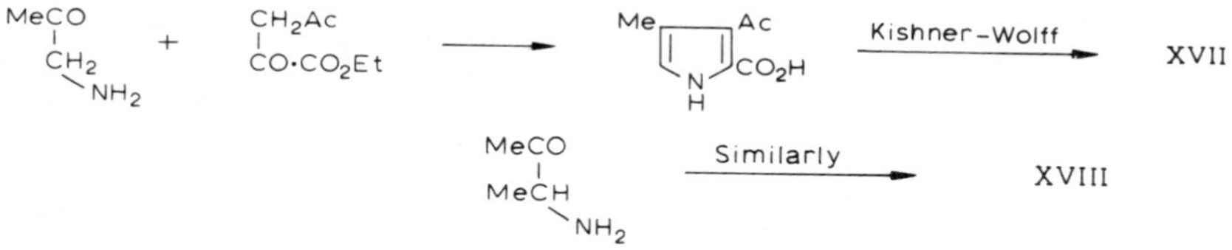

(*H. Fischer et al.*, Ann., 1928, **461,** 244; *O. Piloty* and *A. Blömer*, Ber., 1912, **45,** 3749);

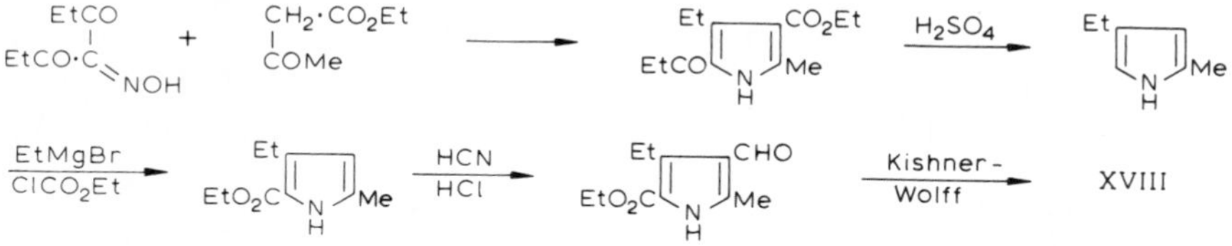

(*Fischer* and *J. Klarer*, Ann., 1926, **450,** 181);

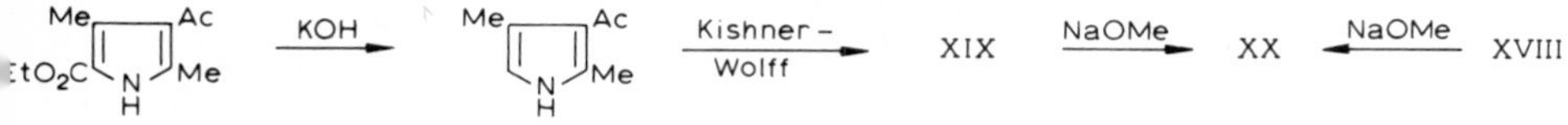

(*Fischer et al.*, Ann. 1929, **475,** 238; Z. physiol. Chem., 1938, **255,** 2; *W. Siedel*, *ibid*., 1935, **231,** 167; *S. F. MacDonald*, J. chem. Soc., 1952, 4176).

Distillation of chitin with zinc dust yields 1-*n*-hexyl-2-methyl-pyrrole (*P. Karrer* and *A. F. Smirnoff*, Helv., 1922, **5,** 856).

(3) *Reactions of* C-*alkyl- and -aryl-pyrroles*. (*a*) 2,5-Disubstituted pyrroles undergo Mannich reactions with substitution in the 3-position or in both 3- and 4-positions (*W. Herz* and *R. L. Settine*, J. org. Chem., 1959, **24,** 201); condensation with acetone, *e.g.*, 2,5-dimethylpyrrole gives *dimethylbis-*

(*2,5-dimethylpyrrol-3-yl)methane*, m.p. 174° (*J. D. White*, Chem. Comm., 1966, 711):

2 Me–pyrrole(NH)–Me $\xrightarrow{Me_2CO}$ Me_2C(2,5-dimethylpyrrol-3-yl)$_2$

They are reduced with zinc and hydrochloric acid to give a 1:3.5 mixture of *cis*- and *trans*-2,5-dimethyl-3-pyrroline in the case of 2,5-dimethylpyrrole (*D. M. Lemal* and *S. D. McGregor*, J. Amer. chem. Soc., 1966, **88,** 1335).

(*b*) N.m.r. studies show that while a number of alkylpyrroles form α-protonated salts in aqueous sulphuric acid (*R. J. Abraham, E. Bullock* and *S. S. Mitra*, Canad. J. Chem., 1959, **37,** 1859), competitive protonation at the β-position can occur, particularly in stronger acid; pK_a's of the protonation sites have been deduced (*Y. Chiang* and *E. B. Whipple*, J. Amer. chem. Soc., 1963, **85,** 2763; *Whipple, Chiang* and *R. L. Hinman, ibid.*, p. 26).

(*c*) *N*-Benzyl-2,3,4,5-tetramethylpyrrole with benzyne gives a stable 2,5-adduct (*E. Wolthuis et al.*, J. org. Chem., 1965, **30,** 190, 3225).

(*d*) Pyrrole and its alkyl derivatives react with pentacarbonylmanganese giving alkyltricarbonylpyrrolylmanganese, also accessible from pentacarbonylmanganese bromide, [$BrMn(CO)_5$], and potassium salts of pyrroles. The latter reacts similarly with dicarbonylcyclopentadienyliron iodide to give azaferrocenes (*A. R. Qazi* and *W. H. Stubbs, J. organometal. Chem.*, 1964, **1,** 471).

(*e*) 1-Benzoyl derivatives have been obtained from some 2-alkyl-, 2,4-dialkyl-, 2,5-dialkyl- and 2,3,4-trialkyl-pyrroles (*A. Jones* and *R. L. Laslett*, Australian J. chem., 1964, **17,** 1056).

(*f*) As the number of alkyl groups increases, pyrrole homologues become more basic, more sensitive to air, and less readily polymerised by acids.

(*g*) The side chains can be halogenated by sulphuryl chloride (*Fischer et al.*, Ann., 1928, **461,** 244). or by bromine in acetic acid (*Fischer* and *H. Scheyer, ibid.*, 1923, **434,** 237), simultaneously with the nucleus, and more easily in 2- than in 3-alkyls. Methyl side chains can be eliminated by successive trihalogenation, hydrolysis, and decarboxylation, or by the "pyrrolenephthalide" reaction (p. 357).

(*4*) *Characterisation of pyrrole homologues.* Pyrrole homologues have been characterised as 2-phthalidylidene-2*H*-pyrroles (pyrrolenephthalides) or as *C*-acetylpyrroles, the latter often further converted into the high-melting benzylidene derivatives (cinnamoylpyrroles). The picrates (*Treibs* and

TABLE 1

C-ALKYL- AND -ARYL-PYRROLES

Substituents	*b.p. (m.p.) °C*
2-Methyl[1]	148
3-Methyl[2]	45/11 mm
2,3-Dimethyl[3]	65/14 mm
2,4-Dimethyl[4]	162–164
2,5-Dimethyl[5]	78–80/25 mm
3,4-Dimethyl[6]	m.p. 33
2,3,4-Trimethyl[7]	m.p. 39
2,3,5-Trimethyl[8]	180–181
2,3,4,5-Tetramethyl[7]	m.p. 110
2-Ethyl[9]	68/14 mm
3-Ethyl[2]	65/14 mm
3-Ethyl-2-methyl[2]	75–77/15 mm
1-Ethyl-2,5-dimethyl[22]	102/79 mm
4-Ethyl-2-methyl[10]	86/20 mm
4-Ethyl-3-methyl[11] (opsopyrrole)	73–75/13 mm
4-Ethyl-2,3-dimethyl[10, 11] (haemopyrrole)	113/16 mm
3-Ethyl-2,4-dimethyl[12] (cryptopyrrole)	86/11 mm
3-Ethyl-2,4,5-trimethyl[7] (phyllopyrrole)	m.p. 69
2-Allyl[13]	82–83/24 mm
2-Vinyl[14]	64–67/18 mm
1-Phenyl-2,5-dimethyl[22]	m.p. 51–52
2-Phenyl[15]	m.p. 129
2,4-Diphenyl[16]	m.p. 179
2,5-Diphenyl[17]	m.p. 143–144
2,3,5-Triphenyl[18]	m.p. 140–141
Pentaphenyl[19]	m.p. 282
2-Diphenylmethyl[20]	m.p. 267
2-Triphenylmethyl[21]	m.p. 253
2,2,3,4,5-Pentamethyl-2*H*-pyrrole[23]	160–162

References

1 *H. Fischer et al.*, Ber., 1928, **61,** 1074; *N. Elming* and *N. Clauson-Kaas*, Acta chem. Scand., 1952, **6,** 867.
2 *Fischer* and *W. Rose*, Ann., 1935, **519,** 21.
3 *Fischer* and *K. Hansen*, *ibid.*, 1936, **521,** 128.
4 *A. H. Corwin* and *R. H. Krieble*, J. Amer. chem. Soc., 1941, **63,** 1829.
5 *C. F. H. Allen* and *D. M. Young*, Org. Synth., 1936, **16,** 25.
6 *Fischer* and *H. Höfelmann*, Ann., 1938, **533,** 216.
7 *A. Treibs* and *H. Derra-Scherer*, *ibid.*, 1954, **589,** 188, 196.
8 *Fischer* and *E. Bartholomäus*, Z. physiol. Chem., 1912, **80,** 6.
9 *Fischer et al.*, Ann., 1931, **486,** 69; *M. DeJong*, Rec. Trav. chim., 1929, **48,** 1029.
10 *Fischer* and *J. Klarer*, Ann., 1926, **450,** 181.
11 *Fischer et al.*, *ibid.*, 1928, **461,** 244; *O. Piloty* and *A. Blömer*, Ber., 1912, **45,** 3449.
12 *Fischer et al.*, Ann., 1929, **475,** 238; Z. physiol. Chem., 1938, **255,** 2; *W. Siedel*, *ibid.*, 1935, **231,** 167; *S. F. MacDonald*, J. chem. Soc., 1952, 4176.
13 *K. Hess*, Ber., 1913, **46,** 3125.
14 *W. Herz* and *C. F. Courtney*, J. Amer. chem. Soc., 1954, **76,** 576.
15 *H. Adkins* and *H. L. Coonradt*, *ibid.*, 1941, **63,** 1563.
16 *Allen* and *C. V. Wilson*, Org. Synth., 1947, **27,** 33.
17 *Allen et al.*, J. org. Chem., 1937, **2,** 235.
18 *F. Angelico* and *E. Calvello*, Gazz., 1901, **31,** II, 4.
19 *W. Dilthey et al.*, J. pr. Chem., 1940, **156,** 27.
20 *A. H. Cook* and *J. R. Majer*, J. chem. Soc., 1944, 482.
21 *E. Khotinsky* and *R. Patzewitch*, Ber., 1909, **42,** 3104.
22 *F. Lions et al.*, C.A., 1938, **32,** 1695.
23 *G. Plancher* and *T. Zambonini*, Atti Accad. Lincei, 1913, **22,** II, 708.

P. Dieter, Ann., 1934, **513,** 65), styphnates, and picrolonates have been employed more frequently. The more highly alkylated pyrroles tend to form yellow salt-like picrates, while the compounds formed by pyrrolecarboxylic esters are often deeply coloured molecular complexes analogous to the trinitrobenzene adducts. The 2,3-dialkylpyrroles are exceptional in giving compounds in which base:acid = 2:1 and from which only polymeric base can be recovered (*O. Piloty* and *K. Wilke*, Ber., 1912, **45,** 2590). Thermal analysis discloses unstable complexes of pyrroles with many acids and phenols (*M. Deželić*, Ann., 1935, **520,** 290).

(5) *Properties of* C-*alkyl- and -aryl-pyrroles.* Infrared spectra of 2-monosubstituted pyrroles (*R. A. Jones*, Australian J. Chem., 1963, **16,** 93); N–H stretching frequencies of di-, tri- and tetra-methylpyrrole (*Jones* and *A. G. Moritz*, Spectrochim. Acta, 1965, **21,** 295); i.r., u.v., n.m.r. spectra for pyrrole and a number of *N*- and *C*-methylpyrroles (*Hinman* and *Theodoropulos*, J. org. Chem., 1963, **28,** 3052) and mass spectra (*A. L. Jennings* and *J. E. Boggs*, *ibid.*, 1964, **29,** 2065), have been recorded.

The physical constants of some *C*-alkyl- and *C*-aryl-pyrroles are listed in Table 1.

(v) Halogenopyrroles

Halogenopyrroles are formed by direct substitution using sulphuryl chloride, bromine in alcohol or acetic acid, and iodine in potassium iodide. It is difficult to stop the halogenation of pyrrole before four hydrogen atoms have beeen replaced, and the products, especially the partially substituted ones are very unstable (*G. Ciamician*, Ber., 1904, **37,** 4200).

Chlorination of pyrrole with sulphuryl chloride gives some 2-**chloropyrrole,** an explosive oil (*K. Hess* and *F. Wissing*, *ibid.*, 1914, **47,** 1416) along with **di-, tri-,** and **tetrachloropyrrole,** which decomposes at 110° (*G. Mazzara* and *A. Borgo*, Gazz., 1905, **35,** I, 477; II, 19). Excess sulphuryl chloride gives **pentachloro-3*H*-pyrrole,** b.p. 209°, which on hydrolysis yields dichloromaleimide, converted back to the pentachloro derivative by phosphorus pentachloride (*Mazzara*, *ibid.*, 1920, **32,** II, 28). Bromination and iodination result in the formation of *tetrabromo-* (*Kalle and Co.*, G.P., 38,423/1886) and *tetraiodo-pyrrole* (*R. L. Datta* and *N. Prosad*, J. Amer. chem. Soc., 1917, **39,** 441; *A. Treibs* and *H. G. Kolm*, Ann., 1958, **614,** 176), which decompose at 154–155° and 140–150°, respectively. Tetrachloropyrrole on treatment with potassium iodide gives tetraiodopyrrole. Substituted pyrroles are also halogenated without difficulty, and although the halogen atom is usually firmly held and is stable to alkali it can be removed by reduction with zinc and alkali or by catalytic hydrogenation (*A. H. Corwin*, *W. A. Bailey* and *P. Viohl*, J. Amer. chem. Soc., 1942, **64,** 1267), and can be replaced in the case of iodo derivatives by bromo, chloro, aryl, azo, nitro and alkylidene groups (*Treibs* and *Kolm*, *loc. cit.*). A series of polymers has been formed by heating tetraiodopyrrole at temperatures between 120 and 700° in a stream of nitrogen (*R. McNeill et al.*, Australian J. Chem., 1963, **16**, 1056). Pyrryl ketones and acids or esters yield more stable bromo-compounds (*Fischer* and *Scheyer*, Ann., 1923, **434**, 237). The chlorination of methyl pyrrole-2-carboxylate and some related pyrroles with sulphuryl chloride in ether has been shown to involve both heterolytic and homolytic mechanisms, and the direct syntheses of methyl 4- and 5-chloro-, 4,5- and 3,5-dichloro, 4- and 5-bromo- and 4,5-dibromo-pyrrole-2-carboxylates have been reported together with an indirect synthesis of methyl 3,4-dichloropyrrole-2-carboxylate (*P. Hodge* and *R. W. Rickards*, J. chem. Soc., 1965, 459). The bromination of ethyl 2,4-dimethylpyrrole-3-carboxylate gives a blue dye (*Treibs* and *H. Bader*, Ann., 1959, **627**, 182):

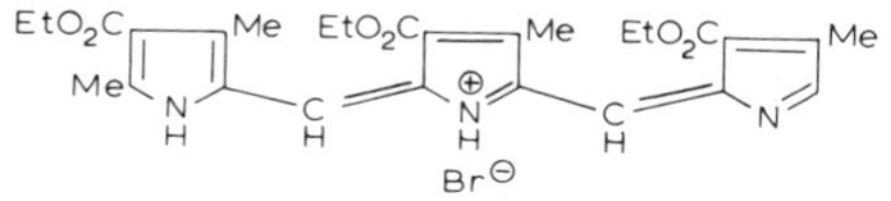

EtO_2C Me EtO_2C Me EtO_2C Me
Me N H C H $\overset{\oplus}{N}$ H C H N
$Br^{\ominus}$

3-*Chloro*-4-*phenylpyrrole*, a pale yellow oil, which is unstable at room temperature has been prepared by the following route (*N. Yoshida*, Yakugaku Zasshi Japan, 1966, **86**, 158; *Sankyo Co. Ltd.*, Neth. Patent Appl., 1967, 6,609,837):

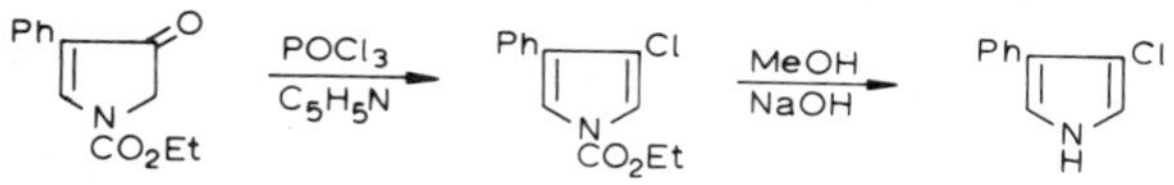

Ph O N CO_2Et $\xrightarrow[C_5H_5N]{POCl_3}$ Ph Cl N CO_2Et $\xrightarrow[NaOH]{MeOH}$ Ph Cl N H

1-*Bromo*-2,3,4,5-*tetraphenylpyrrole* is obtained when sodium 2,3,4,5-tetraphenylpyrrole is treated with bromine in ether (*K. Schilffarth* and *H. Zimmermann*, Ber., 1965, **98**, 3124).

3,4-**Dichloropyrrole,** m.p. 74°, is obtained from its dicarboxy derivative (*Fischer* and *K. Gangl*, Z. physiol. Chem., 1941, **267**, 188); 2,3-*dibromo*-4-*phenylpyrrole*, m.p. 144–146°, from 4-bromo-3-phenylpyrrole-5-carboxylic acid (*O. V. Kil'disheva, M. G. Lin'kova* and *I. L. Knunyants*, Izvest. Akad. Nauk, S.S.S.R., Otdel. Khim., Nauk, 1957, 719); 2,4-diiodo-3,5-dimethyl-, 3,4-diiodo-2,5-dimethyl-, 3,4-diiodo-1,2,5-trimethyl-pyrrole, 5-*iodo*-2,3,4-*trimethylpyrrole*, m.p. 90°, and 3-iodo-1,2,5-trimethylpyrrole, stable for 1–2 days under extreme precautions, from the corresponding methylpyrrole derivatives (*Treibs* and *Kolm, loc. cit.*).

The side chains in alkylpyrroles can be halogenated by sulphuryl chloride (*Fischer et al.*, Ann., 1928, **261**, 244), or by bromine in acetic acid (*Fischer* and *Scheyer, loc. cit.*), simultaneously with the nucleus. α-Alkyls are more readily substituted than β-alkyls.

2-Dichloromethyl-2,5-dimethyl-2*H*-pyrrole when treated with strong bases rearranges to give a pyridine (*R. L. Jones, C. W. Rees* and *C. E. Smithen*, Proc. chem. Soc., 1964, 217):

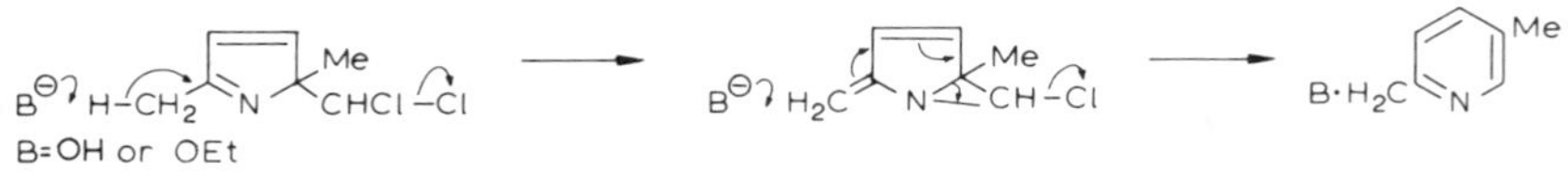

B=OH or OEt

(vi) Nitro-, nitroso-, diazo-, amino- and cyano-pyrroles

In these substituted pyrroles tautomerism in the ring-structure is much more pronounced, particularly with nitro and nitroso compounds where the nitronic acid and oximino structures derived from 2*H*- and 3*H*-pyrrole are much more in evidence (see p. 351).

(*1*) *Nitropyrroles.* Although nitropyrroles are prepared almost exclusively by replacing other substituents by the nitro group (*Treibs* and *Kolm*, *loc. cit.; H. Fischer* and *W. Zerweck*, Ber., 1922, **55,** 1949; *I. J. Rinkes*, Rec. Trav. chim., 1934, **53,** 1167; 1937, **56,** 1142), direct nitration can be effected under strictly controlled conditions.

Nitration of pyrrole with acetyl nitrate at low temperature and the treatment of sodio-pyrrole with isoamyl nitrate give 2-**nitropyrrole,** m.p. 65°, in the latter case in poor yield (*K. J. Morgan* and *D. P. Morrey*, Tetrahedron, 1966, **22,** 57). When prepared using a cold solution of nitric acid in acetic anhydride (*H. J. Anderson*, Canad. J. Chem., 1957, **35,** 21) the low m.p. usually reported is due to the presence of small amounts of 3-nitropyrrole. The best route to 3-**nitropyrrole,** m.p. 101°, is by the decarboxylation of 4-nitropyrrole-2-carboxylic acid prepared by hydrolysis of the product obtained by the condensation of nitromalondialdehyde and glycine ester (*W. J. Hale* and *W. Hoyt*, J. Amer. chem. Soc., 1915, **37,** 2551; *Morgan* and *Morrey*, *loc. cit.*):

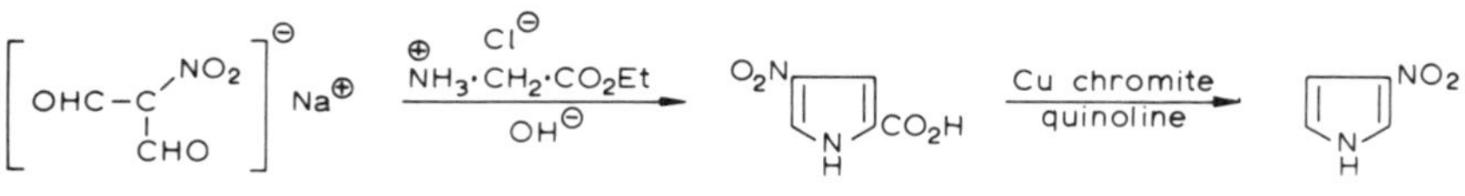

Further nitration of 2-nitropyrrole below −25° using fuming nitric acid yields mainly 2,4-**dinitro-,** and a little 2,5-**dinitro-pyrrole,** m.p. 152° and 173°, respectively (*P. Fournari*, Bull. Soc. chim. Fr., 1963, 484).

The 3-nitro derivatives of 1,2,5-triphenyl-, 2,5-diphenyl-1-*p*-tolyl-, 1-benzyl-2,5-diphenyl- and 1-methyl-2,5-diphenyl-pyrrole are obtained along with some 3,4-dinitro derivatives on warming with nitric acid and acetic acid (*V. Sprio* and *I. Fabra*, Ann. chim. (Rome), 1956, **46,** 263; *S. Giambrono* and *Fabra*, *ibid.*, 1960, **50,** 237).

2,4,5-Triphenylpyrrole reacts with amyl nitrite in ether to give 3-nitro-2,4,5-triphenylpyrrole (*J. M. Tedder* and *B. Webster*, J. chem. Soc., 1960, 3270).

The nitro derivatives of 1-methylpyrroles are prepared by the methylation of the corresponding nitropyrroles and by the nitration of the appropriate α-substituted derivatives (*Fournari*, Bull. Soc. chim. Fr., 1963, 488).

The ratios of the isomeric mono-nitro derivatives formed under various conditions have been determined polarographically (*J. Tirouflet* and *Fournari*, *ibid.*, 1963, 1651).

Nitration of pyrrole-2-carbaldehyde yields a mixture of the 4- and 5-nitropyrrolecarbaldehydes (*idem*, *ibid*., 1963, 484).

3-Acetyl- or 3-benzoyl-2,5-diphenylpyrrole on treatment with amyl nitrite in ether gives the corresponding 4-nitro, and with amyl nitrite and sodium ethoxide the 4-nitroso, derivatives (*Sprio* and *Fabra*, Ric. Sci., Rend. Sez., 1964, **B4,** 581). 2-Cyano- and 2-cyano-1-methyl-pyrrole on nitration give the corresponding 4- and 5-nitropyrroles, the cyano-group appearing to be less strongly "*meta*"-directing than when in the benzene ring (*Anderson*, Canad. J. Chem., 1959, **37,** 2053).

(*2*) *Nitrosopyrroles and diazopyrroles*. 5-**Methyl**-3-**nitroso**-2-**phenylpyrrole,** m.p. 160° (dec.), and 2,5-**dimethyl**-3-**nitrosopyrrole,** m.p. 127° (dec.), have been obtained by reacting the respective pyrroles in a solution of sodium ethoxide in ethanol with amyl nitrite. Treatment of nitrosopyrroles with nitric oxide affords 3-**diazopyrroles,** which are obtained also by the direct introduction of the diazo group with nitrous acid (*Tedder* and *G. Theaker*, J. chem. Soc., 1958, 2573), and by the normal diazotisation of 3-aminopyrroles (*F. Angelico*, Atti R. Accad. Lincei, 1905, **14,** II, 167). The first two reactions are successful only with pyrroles in which the α-positions are blocked. The 3-diazopyrroles are converted into diazonium salts by mineral acids, but these fail to couple even with resorcinol in acid media. Azo compounds can be prepared either by adding the 3-diazopyrrole to fused 2-naphthol or by refluxing a neutral solution of β-naphthol with the diazo compound (*Tedder* and *Webster*, *loc. cit.; Giambrone* and *Fabra*, *loc. cit.*). 3-Diazopyrroles are regarded as resonance hydrids:

R^2, R^3, R^1; $\overset{\oplus}{N}=\overset{\ominus}{N}$ ⟷ R^2, R^3, R^1; $\overset{\oplus}{N}\equiv N$

The first **diazonium salt** of an *N*-substituted pyrrole that behaves as expected for a normal arenediazonium salt, has been obtained by the following route (*A. Mineo*, Corriere Farm., 1966, **21,** 318; C.A., 1967, **67,** 43617g):

Bz(Ph)CH·CH_2Bz + $PhNH_2$ —(AcOH, boil)→ 1,2,5-triphenylpyrrole —(AcOH, add HNO_3/AcOH)→ 3-nitro-1,2,5-triphenylpyrrole (NO_2) —(EtOH, Al, KOH)→ 3-amino-1,2,5-triphenylpyrrole (NH_2) —(HCl, $NaNO_2$)→ 1,2,5-triphenylpyrrole-3-diazonium chloride (N_2Cl)

3,3′-**Azopyrroles** have been obtained from 2,5-diphenylpyrrole-3-diazonium chloride (*A. Kreutzberger* and *P. A. Kalter*, J. phys. Chem., 1961, **65,** 624) and pyrroles substituted in the α-positions, while the structures of 2,3′-**azopyrroles** have been assigned to those from pyrroles having free α-positions (*idem*, J. org. Chem., 1961, **20,** 3790).

Treatment of 2,4-diphenylpyrrole with buffered nitrous acid for a similar period to that used in the preparation of 3-diazopyrroles yields only the corresponding nitroso compound, but much more prolonged treatment gives 2-diazo-3,5-diphenylpyrrole and 2-diazo-4-nitro-3,5-diphenylpyrrole (*Tedder* and *Webster*, J. chem. Soc., 1962, 1638):

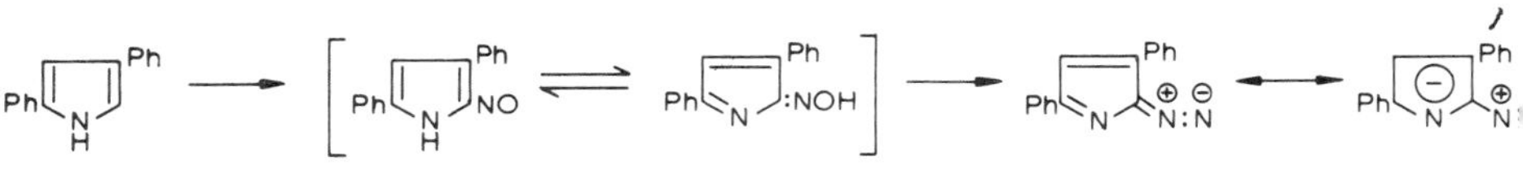

2-**Diazopyrroles** are more reactive and less stable than 3-diazopyrroles, they couple readily with 2-naphthol and are more sensitive to light.

(*3*) *Aminopyrroles.* The preparation of 1-**aminopyrroles** by the Paal–Knorr reaction, *e.g.*, 1-amino-2,5-dimethylpyrrole, has already been mentioned on p. 331. These are weak bases which yield distinctly acidic acyl derivatives and with nitrous acid lose the amino groups as nitrous oxide.

An unambiguous synthesis of a 1-aminopyrrole (*D. M. Lemal* and *T. W. Rave*, Tetrahedron, 1963, **19,** 1119) is achieved by condensing *N*-aminophthalimide with acetonylacetone in hot acetic acid to give 1-phthalimido-2,5-dimethylpyrrole, which on hydrazinolysis yields 1-*amino*-2,5-*dimethylpyrrole*, m.p. 51–52° (*R. Epton*, Chem. and Ind., 1965, 425; *C. G. Overberger et al.*, J. Amer. chem. Soc., 1955, **77,** 4100). 1-*Amino*-2,5-*diphenylpyrrole*, m.p. 219–221°, has been prepared by treating 1,4-diphenyl-1,4-butanedione with *tert*-butyl carbazate in ethanol/acetic acid followed by hydrogen chloride cleavage of the resulting *tert*-butoxycarbonyl derivative (*L. A. Carpino*, J. org. Chem., 1965, **30,** 736).

Aminopyrroles carrying the amino group on the pyrrole carbon atoms can be obtained by reducing the appropriate nitro or nitroso compounds with sodium or aluminium amalgam, or with zinc and acetic acid (*Angelico*, *loc. cit.*; *Fischer et al.*, Ber., 1923, **56,** 512; Ann., 1930, **483,** 257; 1934, **512,** 195); by catalytic reduction of sulphobenzeneazopyrroles; from α-aminoketones and malonodinitrile (*K. Gewald*, Z. Chem., 1961, **1,** 349), and from tetracyanoethene and hydrazine hydrate (*C. L. Dickinson*, *W. J. Middleton* and *V. A. Engelhart*, J. org. Chem., 1962, **27,** 2470).

Reduction of 1-benzyl-3-nitro-2,5-diphenylpyrrole with zinc in acetic acid/acetic anhydride gives 3-*amino*-1-*benzyl*-2,5-*diphenylpyrrole*, isolated as its *hydrochloride*, m.p. 256° (*Sprio*, *T. Aiello* and *Fabra*, Ann. Chim., Rome, 1966, **56,** 866).

3-*Amino*-2,5-*diphenyl*- and 3-*amino*-2,4,5-*triphenyl-pyrrole*, m.p. 183° and 190°, are formed by the hydrogenation of suitably substituted isoxazole oximes in alcohol over Raney nickel (*Sprio* and *E. Aiello*, *ibid.*, 1966, **56,** 858):

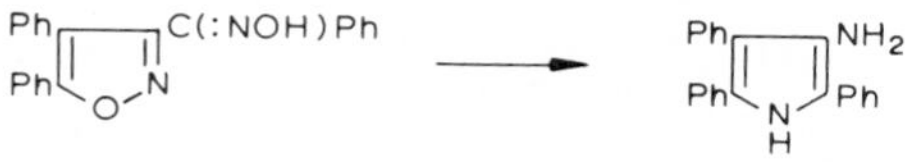

Ph C(:NOH)Ph ⟶ Ph NH$_2$ Ph Ph N H

3-Amino-2,4,5-triphenylpyrrole can be prepared also by treating the nitro compound in acetic acid with zinc dust and can be dehydrogenated to the iminotriphenyl 3*H*-pyrrole (*Aiello* and *G. Sigillò*, Gazz., 1938, **68,** 681):

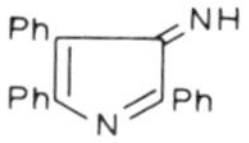

Ph NH Ph N Ph

The *C*-amines are colourless, well defined bases, moderately stable (β- more than α-derivatives), which can be alkylated or acylated. 4-*Acetyl-2-amino-3,5-dimethylpyrrole* decomposes at 235°; 3-*amino-2,4-dimethylpyrrole*, m.p. 127°; 3-*amino-5-ethoxycarbonyl-2,4-dimethylpyrrole*, m.p. 125.5°; 5-*ethoxycarbonyl-2,4-dimethyl-3-dimethylaminopyrrole*, m.p. 87°.

Acetamidoacetaldehyde in hot, slightly alkaline, aqueous solution gives 3-acetamido-1-acetylpyrrole which is hydrolysed by 0.1 *N* hydrochloric acid to **3-acetamidopyrrole** (*J. W. Cornforth*, J. chem. Soc., 1958, 1174).

3-*Acetyl-2,4-dimethylpyrrole*, m.p. 127°, reacts with ammonia to give 3-*acetamido-2,4-dimethylpyrrole*, m.p. 205°, 2-acetyl-3,5-dimethylpyrrole does not react (*S. Palazzo* and *B. Tornetta*, Ann. Chim., Rome, 1959, **49,** 842).

Condensation of dihydromuconic dinitrile with diethyl oxalate gives a hexatriene, which when heated with a primary aliphatic or aromatic amine yields a 3-cyanopyrrole (*R. Huisgen* and *E. Laschtuvka*, Ber., 1960, **93,** 65):

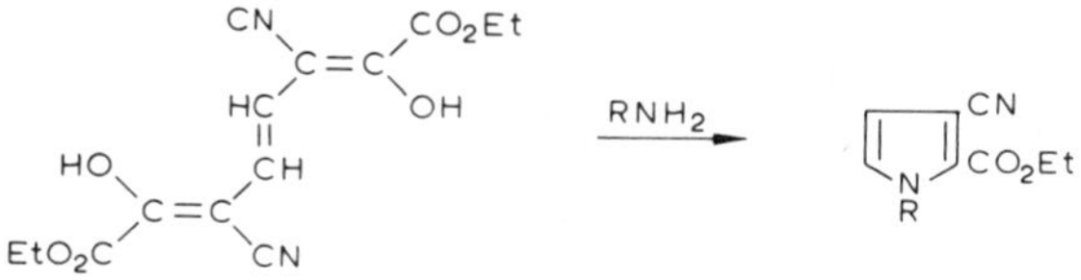

CN CO$_2$Et C=C HC OH CH HO C=C EtO$_2$C CN $\xrightarrow{RNH_2}$ CN CO$_2$Et N R

(*vii*) *Hydroxypyrroles*

1-**Hydroxypyrroles** are obtained by the Paal–Knorr reaction (p. 331) when the amine is replaced by hydroxylamine. Thus diethyl diacetosuccinate gives 1-*hydroxy-2,5-dimethylpyrrole*, m.p. 44–45°, a weakly acidic substance, which with amyl nitrite yields 3-hydroximino-2,5-dimethyl-2*H*-pyrrole *N*-oxide (*Knorr*, Ann., 1886, **236,** 290; *E. E. Blaise*, Compt. rend., 1914, **158,** 1686):

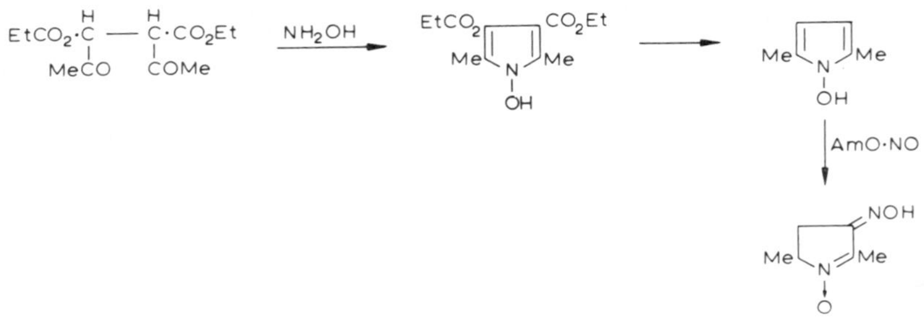

1-Hydroxypyrroles can be obtained also by the following route, *e.g.*, 3-*acetyl-1-hydroxy-2-methyl-5-phenylpyrrole*, m.p. 170° (*Sprio* and *G. C. Vaccara*, Ann. chim., Rome, 1959, **49,** 2075):

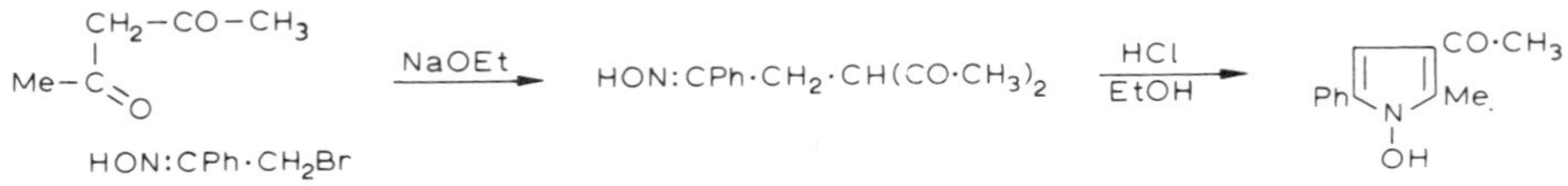

3-*Acetyl-1-hydroxy-2,5-dimethylpyrrole*, m.p. 150° (*Sprio* and *P. Madonia*, *ibid.*, 1960, **50,** 1627). Reduction of 1-hydroxypyrroles using zinc and 80% acetic acid gives the corresponding pyrrole.

C-Hydroxypyrroles, from spectral evidence particularly, are to be regarded as the tautomeric ketones, *i.e.* as pyrrolinones (pyrrolones). They are discussed therefore on pp. 375 *et seq.*

Dihydroxypyrroles are tautomeric with dioxopyrrolidines (p. 378).

(viii) Sulphur derivatives of pyrroles

Thiocyanation of pyrrole with methanolic thiocyanogen at 70°, or at 0° with cupric thiocyanate as catalyst, gives 2-**thiocyanopyrrole,** which can be converted into 2-**methylthiopyrrole** (*S. Gronowitz et al.*, J. org. Chem., 1961, **26,** 2615) on treatment with alkali and methyl iodide. 2-Methylthiopyrrole-5-carbaldehyde is obtained through the Vilsmeier formylation of 2-methylthiopyrrole.

2-Thiocyanopyrrole reacts with bromoacetic acid to afford (2-**pyrrolylthio**)**acetic acid,** which on treatment with polyphosphoric acid followed by Raney nickel gives 2-acetylpyrrole (*M. S. Matheson* and *H. R. Snyder*, *ibid.*, 1957, **22,** 1500; J. Amer. chem. Soc., 1957, **79,** 3610).

The cyclisation of 2-pyrrolylthioacetic acid by polyphosphoric acid is thought to proceed as indicated (*Gronowitz et al.*, Arkiv Kemi, 1962, **18,** 151; Acta Chem. Scand., 1962, **16,** 155):

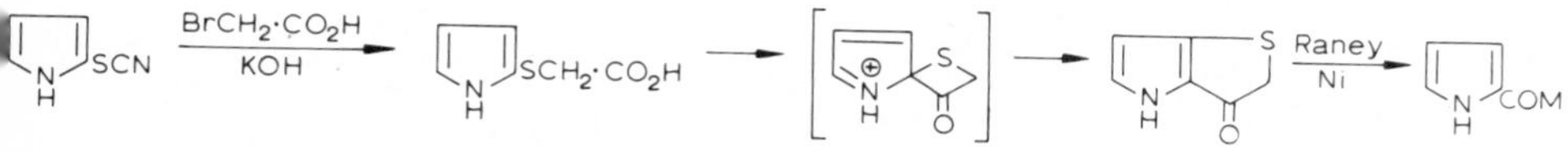

2-Methylpyrrole on thiocyanation yields 2-methyl-5-thiocyanopyrrole.

The thiocyanation of pyrrole in benzene or ether at higher temperatures yields 2,5-**dithiocyanopyrrole** (*Gronowitz et al.*, Arkiv Kemi, 1962, **18,** 151; Acta chem. Scand., 1961, **15,** 227). Thiocyanopyrroles on reduction yield mercaptans.

Aryl pyrrolyl sulphides (*Fischer et al.*, Ann., 1928, **641,** 244; *Fischer* and *M. Herrmann*, Z. physiol. Chem., 1922, **122,** 1), **dipyrrolyl sulphides** (*Fischer* and *Z. Csukàs*, Ann., 1934, **508,** 180), and **dipyrrolyl disulphides** are obtained from pyrroles and arylsulphenyl chlorides, sulphur dichloride, and sulphur monochloride, respectively.

Sulphonation of pyrrole by the sulphur trioxide complex gives **pyrrole**-2-**sulphonic acid;** *anilide*, m.p. 142° (*A. P. Terentiev et al.*, Zhur. obshcheĭ Khim., 1949, **19,** 538; 1950, **20,** 510; C.A., 1949, **43,** 7015; 1950, **44,** 7828). Sulphones are recorded (*F. Ingraffia*, Gazz., 1933, **63,** 584).

(ix) Pyrrolylalkanols and related compounds

(*1*) Formaldehyde in alkaline solution affords 2- or 3-**hydroxymethylpyrroles** with pyrrole or pyrrolecarboxylic esters (*H. Fischer* and *C. Nenitzescu*, Ann, 1925, **443,** 113); obtainable also by reducing the corresponding aldehydes with sodium or aluminium amalgam (*Fischer* and *A. Stern*, *ibid.*, 1926, **446,** 229). These hydroxymethyl compounds readily lose formaldehyde to give dipyrrolylmethanes and are important intermediates in the production of dipyrromethenes.

(*2*) Pyrrolylcarbinols have been obtained from acetylenic and vinylacetylenic oxides and ammonia and amines (*F. Ya. Perveev, E. M. Vekshina* and *L. N. Surenkova*, Zhur, obshcheĭ. Khim., 1957, **27,** 1526).

(*3*) Higher alcohols have been obtained by reducing ketones with lithium tetrahydridoaluminate (*W. Herz* and *C. F. Courtney*, J. Amer. chem. Soc., 1954, **76,** 576) and by Grignard synthesis (*Fischer* and *M. Kaan*, Z. physiol. Chem., 1922, **120,** 267). Pyrrolylmagnesium halides with ethylene oxides give derivatives of *β*-2-**pyrrolylethanol** (*A. Treibs* and *H. Scherer*, Ann., 1952, **577,** 139).

(*4*) Cryptopyrrole-5-carboxylic acid is oxidised by lead tetra-acetate to a carbinol which is converted by heat into aetioporphyrin I (*W. Siedel* and *F. Winkler*, *ibid.*, 1943, **554,** 162):

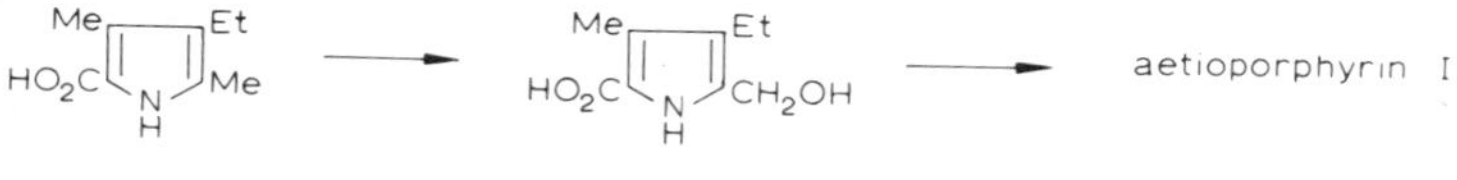

aetioporphyrin I

Glucosamine and ethyl acetoacetate give ethyl 2-methyl-5-tetrahydroxybutylpyrrole-3-carboxylate (XXI) the structure of which is indicated by its oxidation with lead tetra-acetate to ethyl 5-formyl-2-methylpyrrole-3-carboxylate (*A. Müller* and *I. Varga*, Ber., 1939, **72,** 1993):

$HO{\cdot}CH_2{\cdot}(CHOH)_3$–[pyrrole, NH]–Me, 3-$CO_2Et$

(XXI)

2-**Hydroxymethylpyrrole,** b.p. 81–83°/2 mm, from the reduction of pyrrole-2-carbaldehyde with sodium tetrahydridoborate (*R. M. Silverstein et al.*, J. Amer. chem. Soc., 1954, **76,** 4485); *acetate*, m.p. 109° (*Treibs* and *W. Ott*, Ann., 1958, **615,** 137); i.r. spectrum (*R. A. Jones*, Australian J. chem., 1963, **1,** 93); 2,5-**bis(hydroxymethyl)pyrrole,** m.p. 117° (*W. Tschelinzew* and *B. W. Maxorow*, Chem. Ztbl., 1923, I, 1505); N-*methyl-2-hydroxymethylpyrrole*, m.p. 28–30° (*E. E. Ryskiewicz* and *Silverstein*, J. Amer. chem. Soc., 1954, **76,** 5802; *F. P. Doyle et al.*, J. chem. Soc., 1958, 4458).

3-**Hydroxymethyl**-1,2,5-**trimethylpyrrole** has been obtained by reduction of 3-formyl-1,2,5-trimethylpyrrole with lithium tetrahydridoaluminate in boiling tetrahydrofuran (*R. L. Hinman* and *S. Theodoropulos*, J. org. Chem., 1963, **28,** 3052):

1,2,5-trimethyl-3-CHO-pyrrole ⟶ 1,2,5-trimethyl-3-CH_2OH-pyrrole

2-**Aminomethylpyrrole,** b.p. 96°/8 mm, from reduction of pyrrole-2-aldoxime (*N. Putochin*, Ber., 1926, **59,** 1987; C.A., 1931, **25,** 3996), with nitrous acid gives pyridine.

Reduction of ketoximes affords amines (*H. Adkins et al.*, J. Amer. chem. Soc., 1944, **66,** 1293).

Alkylated pyrrolylmethylamines, *e.g.*, *dimethyl-(2-pyrrolylmethyl)amine*, $C_4H_4N{\cdot}CH_2NMe_2$, m.p. 64°, are obtained by a Mannich reaction between pyrrole, formaldehyde, and a primary or secondary amine (*W. Herz et al.*, *ibid.*, 1947, **69,** 1698; 1948, **70,** 504; 1953, **75,** 483; *W. Kutscher* and *O. Klamerth*, Ber., 1953, **86,** 352). The methiodide is an effective alkylating agent, giving with sodium cyanide 2-pyrrolylacetonitrile; the free dimethyl(pyrrolylmethyl)amine alkylates acylamidomalonic esters:

2-(CH_2NMe_2)pyrrole + $NaC(CO_2Et)_2(NHCOR)$ ⟶ 2-[$CH_2{\cdot}C(CO_2Et)_2(NHCOR)$]pyrrole

β-2-**Pyrrolylethylamine,** b.p. 90–98°/2 mm, is obtained by the reduction of 2-pyrrolylacetonitrile or 2-pyrrolylacetamide with lithium tetrahydridoaluminate (*Kutscher* and *Klamerth*, Z. physiol. Chem., 1952, **289,** 229):

2-($CH_2{\cdot}CN$)pyrrole   2-($CH_2{\cdot}CH_2NH_2$)pyrrole

(x) Aldehydes and ketones

(*1*) *Pyrrolecarbaldehydes* are important intermediates in synthesis partly because the formyl group can be conveniently reduced to methyl, but mainly through their use in preparing dipyrromethenes and thence porphyrins.

Synthesis

The formyl group can be easily introduced by the Gattermann reaction (*H. Fischer* and *W. Zerweck*, Ber., 1922, **55,** 1942; *T. Reichstein*, Helv., 1930, **13,** 349) and by the use of phosphorus oxychloride and dimethylformamide (*E. E. Ryskiewicz* and *R. M. Silverstein*, J. Amer. chem. Soc., 1954, **76,** 4485, 5802; *T. C. Chu* and *E. J. H. Chu*, J. org. Chem., 1954, **19,** 266). Dichloromethylpyrroles have been converted into aldehydes and bromomethylpyrroles have been oxidised with chromic acid or reacted with aniline and then oxidised with potassium permanganate to yield aldehydes (*Fischer et al.*, Ann., 1926, **447,** 139; 1928, **461,** 244).

Pyrrolecarbaldehydes may be obtained by the addition of pyrroles to isocyanides followed by alkaline hydrolysis (*A. Treibs* and *A. Dietl*, Ber., 1961, **94,** 298):

EtO_2C, R, R, N, H + CNPh $\xrightarrow{HCl}$ EtO_2C, R, R, N, H, CH:NPh ⟶ EtO_2C, R, R, N, H, CHO

Pyrrole-2-carbaldehyde, m.p. 50–51°, b.p. 100–102°/12 mm, p-*nitrophenylhydrazone*, m.p. 183°, *oxime*, m.p. 164–165°, has been prepared by the Reimer–Tiemann reaction (*Fischer et al.*, Ber., 1928, **61,** 1074); from pyrrolylmagnesium iodide and ethyl, propyl or isoamyl formate (*W. Tschelinzeff* and *A. Terentjeff*, Ber., 1914, **47,** 2653; *N. Putochin*, Zhur. Russ. Fiz. Khim. Obshchestva, 1927, **59,** 809; Ber., 1926, **59B,** 1993); or better by the reaction of pyrrole with dimethylformamide in presence of phosphorus oxychloride (*Silverstein et al.*, J. org. Chem., 1955, **20,** 668; Org. Synth., 1956, **36,** 74; *G. F. Smith*, J. chem. Soc., 1954, 3842). In the Reimer–Tiemann reaction besides the aldehyde 3-chloropyridine is also obtained and an improved yield of the latter compared with that obtained in the normal liquid-phase reaction, results when the dichlorocarbene is added to the pyrrole in the vapour phase (*F. S. Baker et al.*, Chem. and Ind., 1969, 1344). Formylation of pyrrole has been effected using the 1:2 adduct of dimethylformamide and cyanuryl chloride (*R. Oda* and *K. Yamamoto*, Nippon Kagaku Zasshi, 1962, **83,** 1292).

Degradation of D-glucuronamide yields pyrrole-2-carbaldehyde (*Y. Nitta, J. Ide* and *N. Ogikubo*, Yakugatu Zasshi, 1962, **82,** 1057). Treatment with dimethyl sulphate and alkali affords 1-*methylpyrrole-2-carbaldehyde*, b.p. 73–75°/11 mm, (*E. Fischer*, Ber., 1913, **46,** 2504), also obtained from 1-methylpyrrole, dimethylformamide and phosphorus oxychloride; *semicarbazone*, m.p. 205° (*Treibs* and *Dietl*, Ann., 1958, **619,** 80;

Ryskiewicz and *Silverstein, loc. cit.*). Like the water-soluble pyrrole-2-carbaldehyde the 1-methyl derivative fails to give Schiff's, Fehling's or Tollen's tests and does not add hydrogen cyanide or undergo the Cannizzaro, benzoin, or Perkin condensation reactions.

Properties

These aldehydes may be regarded as vinylogues of an amide or α-aminoaldehyde, for the carbonyl absorption in the i.r. region is in the range of the amide rather than the aldehyde group and the molecules are highly polarised (*W. Herz* and *J. Brasch*, J. org. Chem., 1958, **23,** 1513; *M. K. A. Khan* and *K. J. Morgan*, J. chem. Soc., 1964, 2579):

Pyrrole-2-carbaldehyde reacts with triphenylvinylphosphonium bromide and sodium hydride to give an excellent yield of 3*H*-pyrrolizine (*E. E. Schweizer* and *K. K. Light*, J. Amer. chem. Soc., 1964, **86,** 2963):

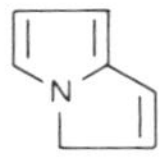

Reduction of pyrrole-2-carbaldehyde with sodium amalgam gives a pinacol (*Ryskiewicz* and *Silverstein*, J. Amer. chem. Soc., 1954, **76,** 5802; *M. S. Taggart* and *G. H. Richter*, *ibid.*, 1934, **56,** 1385) and bromination causes mostly 4-substitution (*H. J. Anderson* and *S. F. Lee*, Canad. J. chem., 1965,**43,** 409).

Decarboxylation of 3-formyl-2-methylpyrrole-5-carboxylic acid to 2-methylpyrrole-3-carbaldehyde does not occur (*Treibs* and *W. Ott*, Ann., 1958, **615,** 137).

(*2*) 2-*Pyrrolylketones* are prepared (*a*) by isomerisation of *N*-acylpyrroles; (*b*) by treating pyrroles with acid anhydrides at high temperature or in presence of strong acids (*G. Ciamician*, Ber., 1904, **37,** 4200); (*c*) by the Friedel–Crafts reaction (*Fischer* and *F. Schubert*, Z. physiol. Chem., 1926, **155,** 99); (*d*) by heating pyrroles with diazo-ketones (*O. Diels* and *H. König*, Ber., 1938, **71,** 1179):

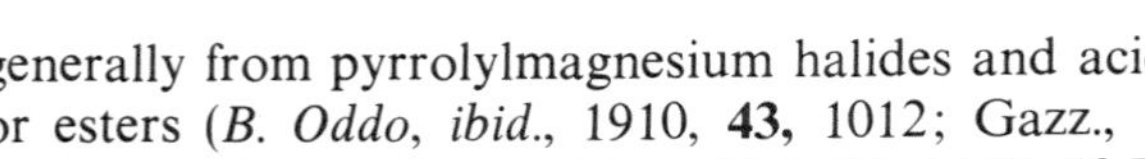

$$\text{pyrrole} + RCO\cdot CR'N_2 \xrightarrow{-N_2} \text{2-}(CO\cdot CHRR')\text{pyrrole}$$

or (*e*) more generally from pyrrolylmagnesium halides and acid chlorides, anhydrides or esters (*B. Oddo*, *ibid.*, 1910, **43,** 1012; Gazz., 1912, **42,** I, 727; *Tschelinzeff* and *Terentjeff*, Ber., 1914, **47,** 2647, 2652). (*f*) The adducts of phosphoryl chloride and 2-dimethylaminocarbonylpyrroles condense smoothly with pyrroles having a free 2-position to give, following hydrolysis, 2,2′-dipyrrolyl ketones which are readily reduced with diborane to pyrromethanes (*J. A. Ballantine et al.*, Tetrahedron Supplement 7, p. 241). (*g*) 2-Acylpyrroles can be synthesised by reductive condensation of a *β*-diketone with a hydroxyimino-*β*-diketone, a modification of the Knorr method (*G. G. Kleinspehn* and *A. H. Corwin*, J. org. Chem., 1960, **25**, 1048):

R–CH·COMe (MeCO) + HO·N=C(COMe)COMe —(Zn, AcOH)→ pyrrole (R, Me, Me, COMe; N–H)

(*h*) 3-Acylated pyrroles are obtained by the reaction between oximino ketones and *β*-diketones (p. 330), aminoacetaldehyde and acylpyruvates (*Khan*, *Morgan* and *D. P. Morrey*, Tetrahedron, 1966, **22**, 2095), and by removal of the ester group with W5 Raney nickel from 4-acyl-2-pyrrolethiolcarboxylates (*C. E. Loader* and *H. J. Anderson*, *ibid.*, 1969, **25**, 3879).

Properties and reactions

Pyrrolyl ketones show normal carbonyl reactions, except in the case of some methyl ketones the acyl group is split off by hot dilute acids. Pyrrole-2-carbaldehyde has been condensed with amino-acids, barbituric acid and barbiturates (*M. Dezelic* and *B. Bobarevic*, Glasnik Drustva Hemicara Tehnol. N. R. Bosne Hercegovine, 1961, **10**, 5; C.A., 1963, **58**, 2424a; Croat. chem. Acta, 1963, **34**, 71).

Pyrrole and phthalic anhydride in acetic acid give the yellow *2-phthalidylidene-*2H*-pyrrole* (*pyrrolenephthalide*), m.p. 240–241° (*H. Fischer* and *F. Krollpfeiffer*, Z. physiol. Chem., 1912, **82**, 267; *Oddo*, Gazz., 1925, **55**, 242), also obtained from pyrrolylmagnesium bromide and diethyl phthalate (*T. N. Godnew*, Ber., 1935, **68**, 422), while phthaloyl chlorides gives di-2-pyrrolylphthalide (pyrrolephthalein) and phenyldipyrrolylmethane-*o*-carboxylic acid (*Oddo* and *F. Tognacchini*, Gazz., 1923, **53**, 265; *F. Ingraffia*, *ibid.*, 1934, **64**, 289, 714). Heating 2-phthalidylidene-2*H*-pyrrole with alkali affords *o*-(2-pyrrolylcarbonyl)benzoic acid:

phthalic anhydride + pyrrole —(AcOH)→ 2-phthalidylidene-2H-pyrrole —(alkali)→ o-(CO_2H)C$_6$H$_4$·CO·(2-pyrrolyl)

Physical data on some pyrrolecarbaldehydes and pyrrolyl ketones are given in Table 2.

TABLE 2

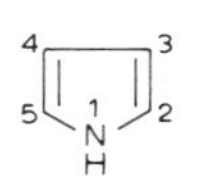

PYRROLECARBALDEHYDES AND PYRROLYL KETONES

Substituents					m.p.°	Ref.
1	*2*	*3*	*4*	*5*		
	CHO	Et	Me	Me	85	1
	CHO	Me	Et	Me	105–106	2, 3
Me	CHO				75–76	4
	CHO	Me		Me	89–90	3
	CHO	Me	Me	Me	147	3
Me	CHO	Me		Me		5
	CHO	Me	CO_2Et	Me	164–165	6
Me	CHO		NO_2		158–160	7
Me	CHO			NO_2	69–70	7
	CHO	Me	Ac	Me	166	8
	CHO	I	Me	I	192	9
	Me	CHO	Me	Me	143.5	10
	Me	CHO		Me	144	5
Me	Me	CHO		Me	96	5
		CHO	Ac	Me	132–134	11
	Me	CHO	CO_2Et	Me	151–152	6
	Me	CHO	Me	CO_2Et	143–145	6
	Ac				90	12
		Ac			115–116	13
	Me	Ac	Me		137	14
	PhCO				79	15
Me	CHO	Me	CHO	Me	152–153	5
Me	Me	CHO	CHO	Me	160	5
	Ac			Ac	161–162	16
CO_2Et	CHO				b.p. 144/14 mm	17
CO_2Et	CHO	Me		Me	b.p. 155/10 mm	17

References

1 *H. Fischer* and *G. Stangler*, Ann., 1927, **459**, 98.
2 *Fischer et al.*, Ber., 1923, **56**, 1209; Ann., 1927, **452**, 292.
3 *E. Ghigi* and *A. Drusiani*, Atti accad. sci. ist. Bologna, Classe sci. fis., 1956, II, No. 3, 1.
4 *Fischer* and *H. Orth*, "Die Chemie des Pyrrols", Vol. I, p. 175; *Treibs* and *Dietl*, Ann., 1958, **619**, 80.
5 *Ghigi* and *Drusiani*, Atti accad. sci. ist. Bologna, Classe sci. fis., 1957, II, No. 4, 14.
6 *M. Dezelic* and *K. Grom-Dursun*, Glasnik Drustva Hemicara Technol. N. R. Bosne Hercegovine, 1960, **9**, 49; C.A., 163, **58**, 2423h.
7 *P. Fournari et al.*, Compt. Rend., 1961, **253**, 1059.
8 *Y. Joh*, Seikagaku, 1961, **33**, 767.
9 *G. M. Badger et al.*, J. chem. Soc., 1962, 4339.
10 *Fischer* and *W. Zerweck*, Ber., 1923, **56**, 521.
11 *F. G. González*, *A. G. Sánchez* and *J. G. Gómez*, Anales real soc. españ. fis. y quim., 1958, **543**, 513.
12 *H. Adkins et al.*, J. Amer. chem. Soc., 1944, **66**, 1293.
13 *I. J. Rinkes*, Rec. Trav. chim., 1938, **57**, 423.
14 *L. Knorr* and *H. Lange*, Ber., 1902, **35**, 2998.
15 *B. Oddo*, *ibid.*, 1910, **43**, 1012.

16 *G. Ciamician* and *P. Silber*, Ber., 1885, **18,** 1466.
17 *Treibs* and *Dietl, loc. cit.*

(*3*) *Pyrrolepolyenecarbaldehydes.* Vinylogues of *N*-methylformanilide, NMePh(CH:CH)$_n$·CHO, condense with pyrroles in presence of acids, giving products of type XXIa hydrolysable to aldehydes XXII:

$ClO_4^{\ominus}$ $\overset{\oplus}{N}$MePh:CH·(CH:CH)$_n$–(3,5-dimethylpyrrol-2-yl) (XXIa) ⟶ OHC·(CH:CH)$_n$–(3,5-dimethylpyrrol-2-yl) (XXII)

Compound XXI, $n = 1$, is brown and XXI, $n = 2$, blue; the yellow aldehyde 3-(3′,5′-*dimethylpyrrolyl-2)propenal* (XXII, $n = 1$), has m.p. 135°, and the orange *vinylogue* XXII, $n = 2$, m.p. 178° (*M. Strell et al.*, Ber., 1954, **87,** 1011, 1019). Further condensation with pyrroles gives polymethine colours like the perchlorates shown below (red, $n = 1$; blue-black, $n = 2$), which are vinylogues of dipyrromethenes:

(3,5-dimethylpyrrol-2-yl)(CH:CH)$_n$·CH=(2,4,5-trimethyl-pyrrolium) $ClO_4^{\ominus}$

(*xi*) *Pyrrolecarboxylic acids and their esters*

The various pyrrolecarboxylic acids and their esters have been synthesised as follows:

(*1*) **Pyrrole-1-carboxylic acid.** Methyl pyrrole-1-carboxylate is prepared by (*a*) reacting potassium pyrrole with methyl chloroformate (*R. M. Acheson* and *J. M. Vernon*, J. chem. Soc., 1961, 457), or (*b*) from butyl-lithium, pyrrole, and methyl chloroformate in ether (*P. Hodge* and *R. W. Richards*, *ibid.*, 1963, 2543). Hydrolysis of the ester with water gives pyrrole-1-carboxylic acid (*D. A. Shirley, B. H. Gross* and *P. A. Roussel*, J. org. Chem., 1955, **20,** 225).

Ethyl pyrrole-1-carboxylate is obtained by route (*a*) (*A. Treibs* and *A. Dietl*, Ann., 1958, **619,** 80), or (*c*) like 1-phenylpyrrole and related compounds by reacting 2,5-dichlorotetrahydrofuran with the appropriate primary amine, ethyl carbamate (*H. Gross*, Angew. Chem., 1960, **72,** 268; *Gross* and *U. Beier*, Ber., 1962, **95,** 2270):

2,5-dichlorotetrahydrofuran ⟶ N-CO_2Et pyrrole

(*d*) Thermolysis of ethoxycarbonylnitrene in thiophene at 130° yields ethyl pyrrole-1-carboxylate and similarly 2,5-dimethylthiophene gives ethyl 2,5-

dimethylpyrrole-1-carboxylate (*K. Hafner* and *W. Kaiser*, Tetrahedron Letters, 1964, 2185). When the thermolysis is carried out in pyrrole, ethyl 2-aminopyrrole-1-carboxylate is obtained. (*e*) Methyl pyrrole-1-carboxylate reacts with dimethyl acetylenedicarboxylate to give acetylene and trimethyl pyrrole-1,3,4-tricarboxylate, which with potassium hydroxide in methanol yields dimethyl pyrrole-3,4-dicarboxylate (*Acheson* and *Vernon*, *loc. cit.*; *N. W. Gabel*, J. org. Chem., 1962, **27,** 301).

Ethyl pyrrole-1-carboxylate does not react with 2,4,6-trinitrobenzenediazonium salts (*Treibs* and *G. Fritz*, Ann., 1958, **611,** 162).

(*2*) **Pyrrole-2-carboxylic acid** is present in the alkaline hydrolysate of mucoproteins (*A. Gottschalk*, Biochem. J., 1955, **61,** 298); it can be prepared (*a*) by a modification of *Oddo's* method (Gazz., 1912, **42,** 257) by treating pyrrolylmagnesium bromide with powdered solid carbon dioxide (*F. P. Doyle et al.*, J. chem. Soc., 1958, 4458). Similarly, *N*-methylpyrrole-2-carboxylic acid has been prepared by metalation of *N*-methylpyrrole with *n*-butyl-lithium followed by carbonation (*Shirley*, *Gross* and *Roussel*, *loc. cit.*).

Alternatively pyrrole can be carbonated under pressure in potassium carbonate solution at 100° (*E. E. Smissman*, *M. B. Graber* and *R. J. Winzler*, J. Amer. pharm. Assocn., 1956, **45,** 509).

A number of reported methods for preparing pyrrole-2-carboxylic acid and its derivatives give but poor yields, but the following two are reliable, efficient routes.

(*b*) Reaction of pyrrolylmagnesium bromide with methyl chloroformate by the Signaigo and Adkin's method (*Treibs* and *Dietl*, Ann., 1958, **619,** 80) gives on distillation, methyl pyrrole-1-carboxylate, and a mixture containing mainly the 1,2-dicarboxylate with a little of the 2-carboxylate. Mild alkaline hydrolysis of this mixture, with concomitant decarboxylation of the carbamic acid, affords pure pyrrole-2-carboxylic acid. Ethyl pyrrole-2-carboxylate has been prepared from pyrrolyl-lithium and ethyl chloroformate (*idem*, *loc. cit.*).

(*c*) Alternatively, pyrrole-2-carbaldehyde, readily available by formylation of pyrrole, is oxidised with alkaline silver oxide to pyrrole-2-carboxylic acid (*Hodge* and *Richards*, *loc. cit.*). Pyrrole-2-carboxylic acid has been prepared also (*d*) from D-glucosamine and pyruvic acid (*Gottschalk*, Arch. Biochem. Biophys., 1957, **69,** 37), and (*e*) from hydroxyproline (*A. N. Radhakrishman* and *A. Meister*, J. biol. Chem., 1957, **226,** 559).

(*f*) A novel synthesis of pyrrole-2-carboxylic acid derivatives involves treating the 1,2-oxazines, obtained by reacting butadienecarboxylic esters with nitrosobenzene or 1-chloro-1-nitrosocyclohexane, with a base in ethanol (*G. Kresze* and *J. Firl*, Angew. Chem., 1964, **76,** 439):

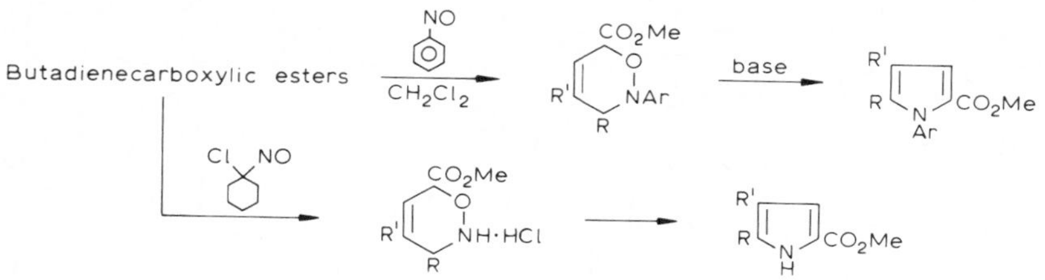

Butadienecarboxylic esters

(*g*) A convenient synthesis of pyrrole-2-carboxylate is based on the addition of toluene-*p*-sulphonylglycine to $\alpha\beta$-unsaturated ketones (*W. G. Terry et al.*, J. chem. Soc., 1965, 4389).

(*h*) α-Methyl groups can be converted to carboxyl groups by successive halogenation and hydrolysis (*H. Fischer et al.*, Ann., 1928, **461,** 244; **462,** 246).

(*i*) Irradiation of 3-diazo-2-pyridone yields pyrrole-2-carboxylic acid, and 3-diazo-4-pyridone, pyrrole-3-carboxylic acid (*O. Süs* and *K. Möller*, *ibid.*, 1955, **593,** 91):

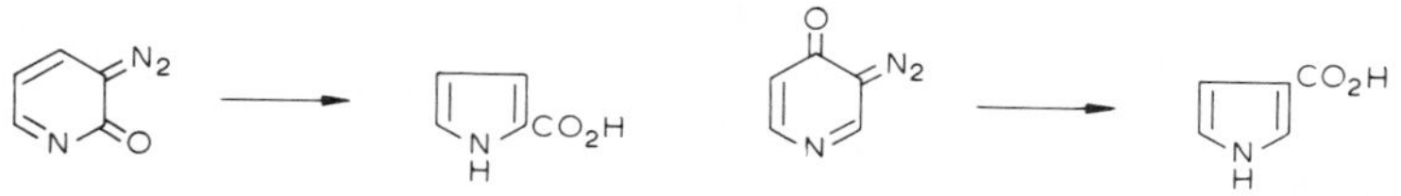

(*3*) **Pyrrole-3-carboxylic acids** are frequently obtained by ring syntheses (see pp. 331, 332).

(*a*) In one method the alkyl group of the pyrrolecarboxylic ester obtained from *tert*-butyl acetoacetate or its oximino-derivative in the Knorr synthesis is removed as isobutylene by heating with a trace of toluenesulphonic acid or with aqueous acetic acid; α-carboxyl groups are simultaneously eliminated (*Treibs* and *K. Hintermeier*, Ber., 1954, **87,** 1167).

Other methods of preparation include (*b*) boiling 1,3-bis-(ethoxycarbonyl)-4-methoxy-3-pyrroline, with alkali (*H. Rapoport* and *C. D. Willson*, J. org. Chem., 1961, **26,** 1102):

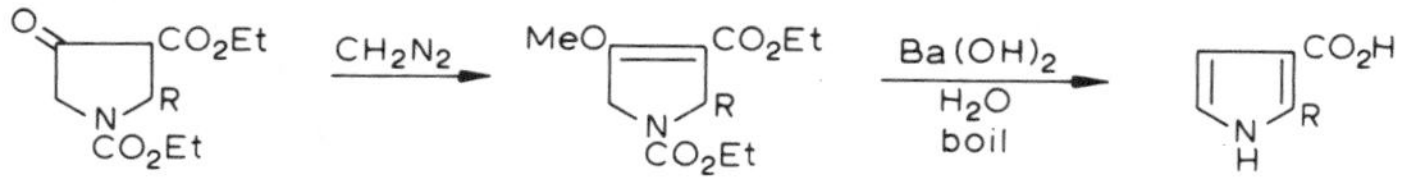

(*c*) reacting β-oxo-esters with α-halogenoketoximes in ethanol in the presence of sodium (*V. Sprio* and *J. Fabra*, Ann. chim., 1960, **50,** 1635):

$$\begin{matrix} PhCO{\cdot}CH_2{\cdot}CO_2Et \\ + \\ BrCH_2{\cdot}C(:NOH)Ph \end{matrix} \xrightarrow[EtOH]{Na} PhC(:NOH){\cdot}CH_2{\cdot}CH(COPh){\cdot}CO_2Et \xrightarrow{EtOH/HCl}$$

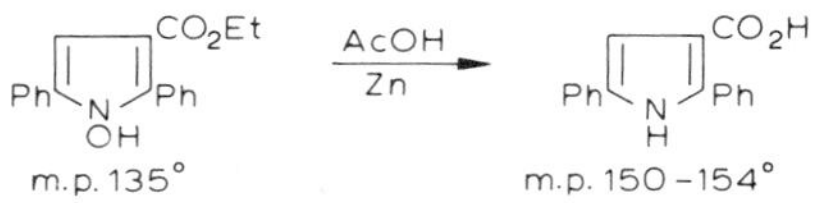

(*d*) the reaction between *β*-oxo-esters and 1,2-dibromoethyl acetate in aqueous ammonia (*H. Shinohara, S. Misaki* and *E. Imoto*, Nippon Kagaku Zasshi, 1962, **83,** 637):

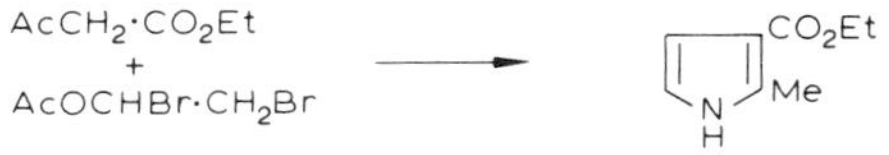

(*e*) the reaction of some derivatives of 4-*H*-azepine with hot aqueous alcohol and sodium acetate or with aqueous ammonia at room temperature (*M. Anderson* and *A. W. Johnson*, Proc. chem. Soc., 1964, 263):

(*4*) **Pyrrole-polycarboxylic acids.** (*a*) The $-CONH_2$ group can be inserted into the α- or the *β*-position of pyrroles by treatment with carbamoyl chloride in ether (*Treibs* and *R. Derra*, Ann., 1954, **589,** 174). (*b*) Oxidative degradation of cuttlefish melanin affords pyrrole-2,3,5-tricarboxylic acid, synthesised as shown (*R. Nicolaus*, Gazz., 1953, **83,** 239):

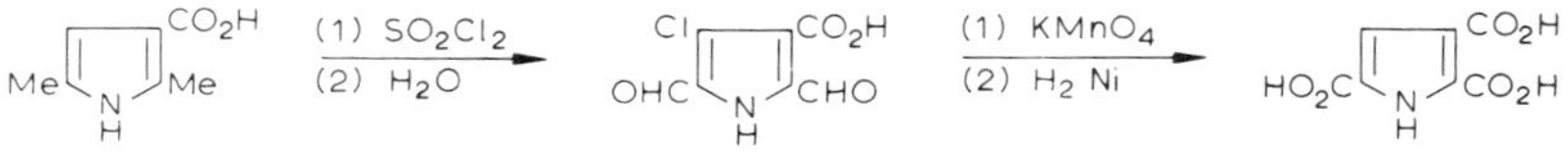

(*5*) *Properties and some reactions of pyrrolecarboxylic acids.*
In general the esters are more accessible than the free acids. The acids easily lose carbon dioxide, the α-compounds even on boiling with water. Heated with dehydrating agents, some α-derivatives yield dioxopiperazines, *e.g.* "pyrocoll" (XXIII), first obstained by distilling glue; *β*-acids yield bispyrrobenzoquinones *e.g.* XXIV:

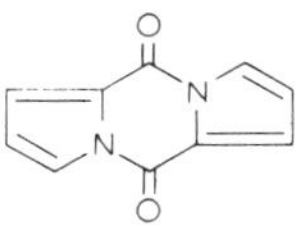

(XXIII)

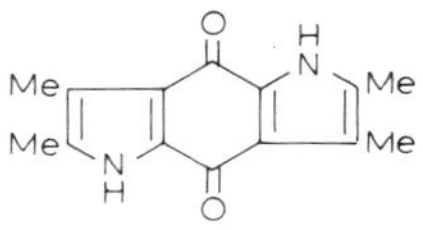

(XXIV)

Equimolar proportions of pyrrole-2-carboxylic acid and *p*-dimethylaminobenzaldehyde react together in hydrochloric acid to give 2-*p*-dimethylaminobenzylidene-2*H*-pyrrole-5-carboxylic acid (XXV), but in 2:1 ratio yield the *p*-dimethylaminophenyl-2-pyrrolyl-2-(2*H*-pyrrolylidene)methanedicarboxylic acid (XXVI), also obtained from *N*-acetylneuraminic acid and Ehrlich's reagent (*L. R. Morgan* and *R. Schunior*, J. org. Chem., 1962, **27,** 3696):

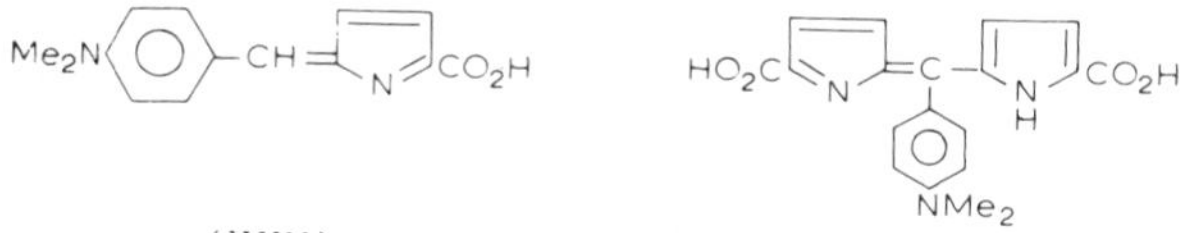

(XXV) (XXVI)

The esters of pyrrole-2-carboxylic acid are more easily hydrolysed by alkali than the 3-acids, but the 3-group is preferentially attacked in the partial hydrolysis of diesters by strong sulphuric acid (*Fischer* and *B. Walach*, Ber., 1925, **58,** 2818; *A. H. Corwin* and *J. L. Straughn*, J. Amer. chem. Soc., 1948, **70,** 2968). Dilute alkali above 160° removes carboxylic or ester groups from both positions (*idem, ibid.*, p. 1416). Esters of pyrrolecarboxylic acids react with Grignard reagents to produce carbinols in the usual way, which readily lose water to form, for example, the yellow benzhydrylidene-2*H*-pyrrole (*Fischer* and *M. Kaan*, Z. physiol. Chem., 1922, **120,** 267):

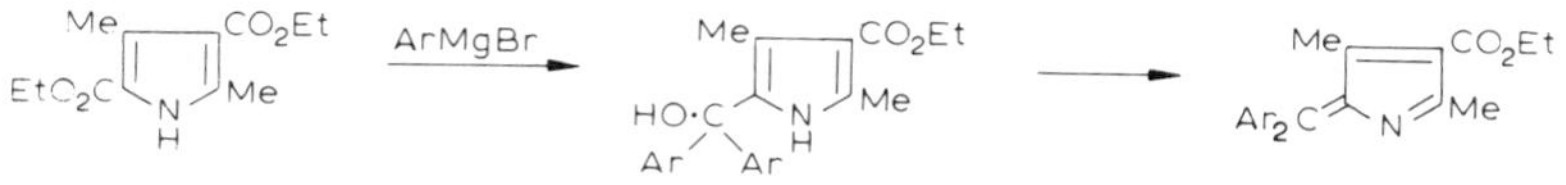

(*6*) **Side-chain carboxylic acids; pyrrolylalkanoic acids.** Ethyl diazoacetate and pyrroles give **α-pyrroleacetic esters** in good yield (*C. Nenitzescu* and *E. Solomonica*, Ber., 1931, **64,** 1924; *O. Diels* and *H. König*, *ibid.*, 1938, **71,** 1179).

β-Pyrrolylpropionic acids are important products of the reductive fission of haemin and chlorophyll (*Fischer et al. ibid.*, 1914, **47,** 791; Ann., 1926, **450,** 132; 1930, **478,** 283). They correspond to the basic products (p. 343) with a propionic acid group in place of ethyl and are separated by fractional precipitation and crystallisation of the methyl ester picrates after removal of the feebly basic opsopyrrolecarboxylic acid (XXVII)

(XXVII) Opso- (XXVIII) Haemo- (XXIX) Crypto- (XXX) Phyllo-pyrrolecarboxylic acids

All have been oxidised to haematic acid or its oxime. Since this acid yields ethylmethylmaleimide on distillation and succinic acid on oxidation; it must be methylmaleimidopropionic acid, and the related 2-pentene-2,3,5-tricarboxylic acid 2,3-anhydride (XXXII) has been synthesised from α-acetylglutaric acid cyanohydrin (XXXI) (*W. Küster*, Ann., 1906, **345,** 1; *Küster* and *J. Weller*, Ber., 1914, **47,** 532):

$$MeC(OH)(CN)-CH(CO_2Et)\cdot CH_2\cdot CH_2\cdot CO_2Et \ \text{(XXXI)} \xrightarrow{HCl} MeC(OH)(CO_2H)-CH(CO_2H)\cdot CH_2\cdot CH_2\cdot CO_2H \xrightarrow{\Delta} \text{(XXXII)}$$

The acid XXVIII has been decarboxylated to haemopyrrole and methylated by sodium methoxide to XXX. The structures are established by these observations, together with the following syntheses which exemplify general synthetic methods:

(*Fischer et al.*, Ann., 1925, **443,** 113; 1928, **462,** 246);

(*idem*, Ber., 1927, **60,** 377; Ann., 1924, **439,** 175; 1931, **491,** 186).

Acrylonitrile and ethyl acrylate add to many pyrroles, giving derivatives of β-2-pyrrolylpropionic acid (*Treibs* and *K. H. Miehl*, *ibid.*, 1954, **589,** 163).

The diene synthesis is seldom applicable to pyrroles. Maleic acid or anhydride yield derivatives of pyrrolylsuccinic acid, while pyrrolylmaleic and dipyrrolylsuccinic acids result from the addition of dimethyl acetylenedicarboxylate; with *N*-methylpyrrole and the latter reagent a true diene synthesis follows the primary addition (*O. Diels* and *K. Alder*, Ann., 1931, **486,** 211; **490,** 267; 1932, **498,** 1; *cf. Fischer* and *H. Gademann*, *ibid.*, 1942, **550,** 196):

Boiling point and melting point data on some pyrrolecarboxylic acids and their esters are listed in Table 3.

TABLE 3

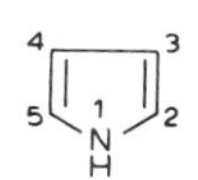

PYRROLECARBOXYLIC ACIDS AND ESTERS

Substituent groups in pyrrole nucleus					*m.p.°*	
Position: 1	2	3	4	5		*Ref.*
CO_2H	—	—	—	—	123–125	1, 2
CO_2Me	—	—	—	—	b.p. 50–50.5/12 mm	1, 3
					168–170	4
CO_2Et	—	—	—	—	b.p. 65/12 mm	5
CO_2Et	Me	—	Me	—	b.p. 118/22 mm	5
CO_2Et	CHO	—	—	—	b.p. 144/14 mm	5
CO_2Et	CHO	Me	—	Me	b.p. 155/10 mm	5
CO_2Me	Me	—	—	—	b.p. 63–65/10 mm	3
CO_2Me	Me	—	—	Me	38	3
CO_2Me	Ph	—	—	Ph	100	3
CO_2Me	—	CO_2Me	CO_2Me	—	69	3, 4
CO_2Me	Me	CO_2Me	CO_2Me	—	64	3
—	CO_2H	—	—	—	208	1, 6, 7
—	CO_2Me	—	—	—	73	1
—	CO_2Et	—	—	—	39	7
Me	CO_2Me	—	—	—	b.p. 62/1 mm	6
—	CO_2H	—	MeO	—	179–180	8
—	CO_2Me	—	MeO	—	55–58	8
—	CO_2H	—	OEt	—	158–160	9
—	CO_2H	—	CHO	—	106–107	10
—	CO_2H	—	—	CHO	75	10

TABLE 3 *(continued)*

Substituent groups in pyrrole nucleus					*m.p.°*	
Position: 1	2	3	4	5		*Ref.*
—	CO_2H	—	—	Me	101	11
Ph	CO_2Me	—	—	Me	39	11
p-ClC_6H_4	CO_2Me	—	—	—	90	11
p-ClC_6H_4	CO_2Me	—	—	Me	83	11
—	CO_2Et	Me	—	Me	125	15
—	—	CO_2Me	—	—	88	13
—	CS_2H	—	—	—	unstable	12
—	—	CO_2H	—	—	150	13, 14
—	Me	CO_2Et	—	—	78–79	16, 17
Me	Me	CO_2H	—	—	178	17
Me	Me	CO_2Et	—	—	24	17
Ph	Me	CO_2H	—	—	184	17
Ph	Me	CO_2Et	—	—	b.p. 160–161/4 mm	17
$PhCH_2$	Me	CO_2H	—	—	165–166	17
$PhCH_2$	Me	CO_2Et	—	—	b.p. 196–198/7 mm	17
—	Me	CO_2Et	Me	—	75–76	18
—	Me	CO_2Et	—	Me	117	19
—	Ph	CO_2H	—	Ph	150–154	20
—	Ph	CO_2Et	—	Ph	135	20
—	Ph	CO_2H	—	Me		20
—	Ph	CO_2Et	—	Me		20
—	—	CO_2H	MeO	—	203–204	8
—	—	CO_2Me	MeO	—	115–117	8
—	—	CO_2Et	MeO	—	107–109	8
—	CO_2H	CO_2Me	—	—	201	13
—	CO_2H	—	—	CO_2H	260 dec.	21
—	CO_2H	CO_2Me	Me	Me	201	22
—	CO_2Et	Me	CO_2Et	Me	136	23
—	—	CO_2H	CO_2H	—	290–292 dec.	4, 24
—	—	CO_2Me	CO_2Me	—	244	4
—	—	CO_2Et	CO_2Et	—	153–155	4, 24
—	Me	CO_2H	CO_2H	—	236–237 dec.	4, 24
—	Me	CO_2Me	CO_2Me	—	159	4
—	Me	CO_2Et	CO_2Et	—	124–125	4, 24
—	Me	CO_2H	CO_2H	Me	260–265 dec.	4, 25
—	Me	CO_2Me	CO_2Me	Me	118–119	4
—	Me	CO_2Et	CO_2Et	Me	99	4, 25
CO_2Me	—	CO_2Me	CO_2Me	—	69	4
—	CO_2Me	CO_2Me	CO_2Me	CO_2Me	124–125	26
Ph	CO_2Me	CO_2Me	CO_2Me	CO_2Me	117	27

TABLE 3 *(continued)*

Substituent groups in pyrrole nucleus					*m.p.°*	
Position: 1	2	3	4	5		*Ref.*
—	—	Me	p*	—	119	28
(opsopyrrolecarboxylic acid)						
—	Me	Me	p	—	130–131	29
(haemopyrrolecarboxylic acid)						
—	—	Me	p	Me	140	30
(cryptopyrrolecarboxylic acid)						
—	Me	Me	p	Me	86–88	31
(phyllopyrrolecarboxylic acid)						
—	CHO	Me	p	—	152	32
—	Me	Me	p	CHO	155	33
—	CHO	Me	p	Me	151	34
—	$COCO_2H$	—	—	—	113–115 dec.	35
—	CO_2H	Ac	Me	Me	204 dec.	36
—	CO_2Et	Me	Ac	Me	143	37

* p = 2-carboxyethyl, $HO_2C{\cdot}CH_2{\cdot}CH_2$–.

References

1 *P. Hodge* and *R. W. Richards*, J. chem. Soc., 1963, 2543.
2 *D. A. Shirley, B. H. Gross* and *P. A. Roussell*, J. org. Chem., 1955, **20,** 225.
3 *N. W. Gabel*, *ibid.*, 1962, **27,** 301.
4 *R. M. Acheson* and *J. M. Vernon*, J. chem. Soc., 1961, 457.
5 *A. Treibs* and *A. Dietl*, Ann., 1958, **619,** 80.
6 *F. P. Doyle et al.*, J. chem. Soc., 1958, 4458.
7 *N. Maxim et al.*, Bull. Soc. chim. Fr., 1938, **5,** 46.
8 *H. Rapoport* and *C. D. Willson*, J. Amer. chem. Soc., 1962, **84,** 630.
9 *R. Kuhn* and *G. Osswald*, Ber., 1956, **89,** 1423.
10 *W. A. M. Davies, A. R. Pinder* and *I. G. Morris*, Tetrahedron, 1962, 405.
11 *G. Kresze* and *J. Firl*, Angew. Chem., 1964, **76,** 439.
12 *B. Oddo* and *Q. Mingoia*, Gazz., 1926, **56,** 782.
13 *I. J. Rinkes*, Rec. Trav. chim., 1937, **56,** 1224; 1938, **57,** 423.
14 *Rapoport* and *Willson*, J. org. Chem., 1961, **26,** 1102.
15 *H. Fischer*, Org. Synth., 1937, **17,** 48; *Treibs* and *K. Hintermeier*, Ber., 1954, **87,** 1167.
16 *E. Benary*, Ber., 1911, **44,** 493.
17 *H. Shinohara, S. Misaki* and *E. Imoto*, Nippon Kagaku Zasshi, 1962, **83,** 637.
18 *L. Knorr* and *H. Lange*, Ber., 1902, **35,** 2998; *Treibs* and *Hintermeier*, *loc. cit.*
19 *W. Küster*, Z. physiol. Chem., 1922, **121,** 144.
20 *V. Sprio* and *J. Fabra*, Ann. chim., 1960, **50,** 1635.
21 *Rinkes*, Rec. Trav. chim., 1937, **56,** 1142.
22 *Fischer* and *W. Kutscher*, Ann., 1930, **481,** 199.
23 *Fischer*, Org. Synth., 1935, **15,** 17; *S. F. MacDonald*, J. chem. Soc., 1952, 4176.
24 *E. C. Kornfeld* and *R. G. Jones*, J. org. Chem., 1954, **19,** 1671.
25 *Knorr*, Ber., 1885, **18,** 1558.
26 *R. Nicolaus*, C.A., 1954, **48,** 12732.
27 *E. H. Huntress et al.*, J. Amer. chem. Soc., 1956, **78,** 419.
28 (p. 363); *Me ester picrate*, m.p. 104°.
29 (p. 363); *Mac Donald*, *loc. cit.*; *Me ester picrate*, m.p. 121–122°.

30 (p. 363); *Fischer et al.*, Ber., 1924, **57**, 602; Ann., 1926, **450**, 201; *Me ester picrate*, m.p. 107–108°.
31 (p. 363); *Me ester picrate*, m.p. 97–98°.
32 *Fischer* and *Z. Csukàs*, Ann., 1934, **508**, 167.
33 *Fischer et al.*, Z. physiol. Chem., 1929, **182**, 280.
34 *Fischer* and *M. Schubert*, Ber., 1924, **57**, 612.
35 *G. Ciamician* and *M. Dennstedt*, *ibid.*, 1883, **16**, 2348.
36 *O. Piloty* and *A. Blomer*, *ibid.*, 1912, **45**, 3749.
37 *Fischer et al.*, Ann., 1929, **475**, 238.

2. Pyrrolines, dihydropyrroles

There are three possible isomeric forms representing respectively, 1-, 2-, and 3-pyrrolines:

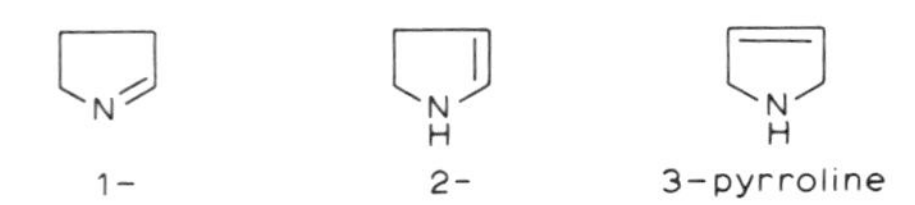

The exact assignment of the position of the double bond has been the subject of repeated misinterpretation and contradiction. This stems from the arbitrary assignment of a 2-pyrroline structure to the examples of these compounds reported in the earlier literature.

(a) 1-Pyrrolines

The application of both physical and chemical methods has provided evidence that many pyrrolines, such as the tobacco alkaloid myosmine 2,3′-pyridyl-1-pyrroline (*B. Witkop*, J. Amer. chem. Soc., 1954, **76**, 5597), those obtained from the reaction between γ-chlorobutyronitrile and a Grignard reagent [1], and those afforded by the isomerisation on heating of ketimines obtained from 1-aryl-1-cyanocyclopropanes and Grignard reagents [2] (*L. C. Craig, H. Bulbrook* and *R. M. Hixon, ibid.*, 1931, **53**, 1831; *J. B. Cloke et al., ibid.*, 1929, **51**, 1174; 1945, **67**, 1249, 2155; *P. Lipp* and *H. Seeles*, Ber., 1929, **62**, 2456):

$Cl(CH_2)_3CN$ + RMgBr ⟶ $Cl(CH_2)_3CR{:}NMgBr$ ⟶ (1-pyrroline, 2-R) [1]

Ar·C—CH₂ / Ph·C(=NH) CH₂ (cyclopropane ketimine) ⟶ (1-pyrroline, Ar, Ph) [2]

are best represented by Δ^1-structures, *i.e.* as 1-pyrrolines (*M. C. Kloetzel, J. L. Pinkus* and *R. B. Washburn*, J. Amer. chem. Soc., 1957, **79**, 4222;

J. H. Burckhalter and *J. H. Short*, J. org. Chem., 1958, **23,** 1278). Infrared measurements indicate that most of the tertiary bases hitherto regarded as 2-alkyl-1-methyl-2-pyrrolines are in fact 2-alkylidene-1-methylpyrrolidines (I), only 1,2-dimethyl-2-pyrroline has an endocyclic double bond (*R. Lukeš, V. Dědek* and *L. Novotný*, Coll. Czech. chem. Comm., 1959, **24,** 1117):

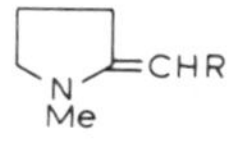

=CHR
N
Me
(I)

(i) Synthesis

γ-Aminoketones spontaneously yield pyrrolines, which are obtained also when the product resulting from the addition of a nitroparaffin to an $\alpha\beta$-unsaturated ketone or ester is reduced by zinc and acetic acid (*A. Sonn*, Ber., 1935, **68,** 148; *Sir A. Todd et al.*, J. chem. Soc., 1959, 2087); catalytic reduction gives a pyrrolidine (*Kloetzel*, J. Amer. chem. Soc., 1947, **69,** 2271):

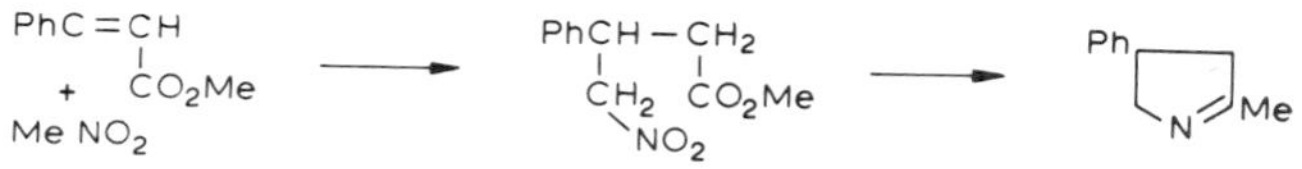

PhC=CH + CO_2Me / Me NO_2 ⟶ PhCH—CH_2 / CH_2 CO_2Me / NO_2 ⟶ Ph, N, Me

When an $\alpha\beta$-unsaturated ester is used in place of the ketone, the first product of reduction is a pyrrolidone (*R. B. Moffett* and *J. L. White*, J. org. Chem., 1952, **17,** 407).

1-Pyrroline is formed by the partial dehydrogenation of pyrrolidine over 5% palladium-on-charcoal or by boiling with mercuric acetate and acetic acid, and by the partial hydrogenation of pyrrole at low pressure and room temperature using a rhodium–alumina catalyst. It may be obtained also (*a*) by reacting pyrrolidine acetate with sodium hypochlorite to give *N*-chloropyrrolidine, which is then treated carefully with alcoholic potassium hydroxide:

N / Cl ⟶ N

(*b*) by the action of sodium periodate on 3-hydroxypiperidine (*D. W. Fuhlhage* and *C. A. van der Werf*, J. Amer. chem. Soc., 1958, **80,** 6249); (*c*) from 1-hydroxypiperidin-2-one, which with hot polyphosphoric acid loses carbon monoxide (*G. di Maio* and *P. A. Tardella*, Proc. chem. Soc., 1963, 224).

1-Pyrrolines may be obtained also by a modified Bischler–Napieralski reaction between acyl derivatives of 4-phenylbut-3-enylamine and phosphorus oxychloride (*S. Sugasawa* and *S. Ushioda*, Tetrahedron, 1959, 48):

and by the thermal rearrangement of cyclopropylimines in the presence of a trace of acid (*R. V. Stevens* and *M. C. Ellis*, Tetrahedron Letters, 1967, 5185).

Cyclopent-2-enylacetoxime on treatment with zinc and acetic acid gives a bridged pyrroline, which may be tautomeric (*R. Groit* and *T. Wagner-Jauregg*, Helv., 1959, **42,** 121, 605):

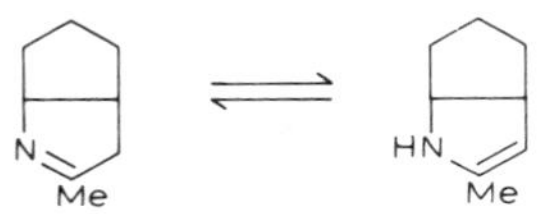

The product obtained from the action of sulphuryl chloride on Knorr's pyrrole, 2,4-bis-ethoxycarbonyl-3,5-dimethylpyrrole (see p. 331) is compound II and not III as thought previously (*H. Fischer*, *E. Sturm* and *H. Friedrick*, Ann., 1928, **461,** 244; *J. H. Mathewson*, J. org. Chem., 1963, **28,** 2153):

(II) (III)

The action of a Grignard reagent on 2-pyrrolidones gives a disubstituted pyrrolidine besides the pyrroline, *e.g.*, 1-methyl-2-pyrrolidine affords 1,2,2-trimethylpyrrolidine (*Lukeš*, Coll. Czech. chem. Comm., 1930, **2,** 531).

(ii) Properties and reactions of 1-pyrrolines

1-**Pyrroline** polymerises easily, probably to the trimer, tripyrroline (IV):

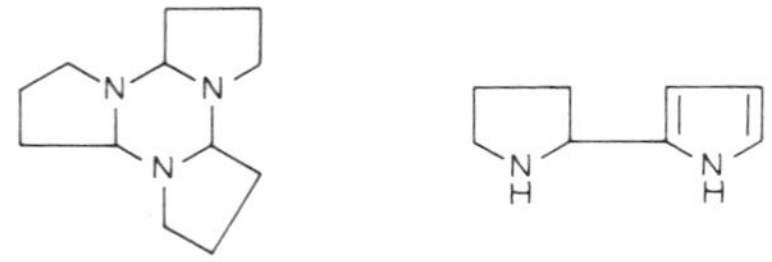

(IV) (V)

and on heating with pyrrole gives 2,2′-pyrrolidinylpyrrole (V) as the main product. The addition of 1-pyrrolines to *N*-unsubstituted pyrroles with free 2- or 3-positions is a general reaction (*Fuhlhage* and *Van der Werf*, *loc. cit.*). Addition of hydrogen cyanide to 1-pyrrolines unsubstituted in the 2-position gives 2-cyanopyrrolidines which can be hydrolysed to the corresponding proline derivatives (*Todd et al.*, *loc. cit.*):

KCN / HCl → 2-cyanopyrrolidine → 6*N* HCl → Proline (CO_2H)

The 1-pyrrolines of relatively low molecular weight are water-soluble bases which, unlike 3-pyrroline, tend to resinify in the air or even on distillation under ordinary pressure, and are reduced to pyrrolidines by tin and hydrochloric acid. Benzoyl chloride and alkali open the ring of, *e.g.* 2-methyl-1-pyrroline giving γ-benzamidopropyl methyl ketone (*S. Gabriel*, Ber., 1909, **42,** 1240).

The following pyrrolines are usually described in the literature as 2-pyrrolines, but for the reasons already given on p. 368 they are most probably 1-pyrrolines.

2-**Methylpyrroline,** b.p. 95–97°, 45°/100 mm, *picrate*, m.p. 120–121° (*P. S. Ugryumov*, Zhur. obshchei Khim., 1950, **20,** 1848); 4-*methylpyrroline*, *picrate*, m.p. 216–217° (*A. Pictet* and *G. Court*, Ber., 1907, **40,** 3777; *A.* and *R. Pictet*, Helv., 1927, **10,** 594); 2-*ethylpyrroline*, b.p. 125–126°; 1-*phenylpyrroline*, m.p. 166–167.5° (decomp.) (*G. Wittig* and *H. Sommer*, Ann., 1955, **594,** 1). Hydrogenation of β-benzoylpropionitrile gives 2-*phenylpyrroline*, m.p. 45° (*E. B. Knott*, J. chem. Soc., 1948, 186); 2-*methyl*-4-*phenylpyrroline*, b.p. 135/15 mm (*A. Sonn*, Ber., 1935, **68,** 148). Phosphorus pentachloride converts 2-pyrrolidone into 2-*chloropyrroline*, m.p. 50–51°.

(b) 1-Pyrroline 1-oxides

4,4-Dimethyl-5-nitropentan-2-one on mild reduction gives 2,4,4-**trimethyl-1-pyrroline 1-oxide.** Along with 4,5,5-**trimethyl-1-pyrroline 1-oxide** these were the first non-aromatic monomeric nitrones (*R. F. C. Brown*, *V. M. Clark* and *Sir A. Todd*, Proc. chem. Soc., 1957, 97). The 2-methyl group undergoes base-catalysed condensations with aromatic aldehydes:

Me_2 … $\overset{\oplus}{N}$ … Me, $O^{\ominus}$

(i) Synthesis

1-**Pyrroline** 1-**oxide** has been prepared from 1-ethylpyrrolidine 1-oxide by pyrolysis followed by dehydrogenation using mercuric oxide (*J. Thesing* and *W. Sirrenberg*, Ber., 1959, **92**, 1748):

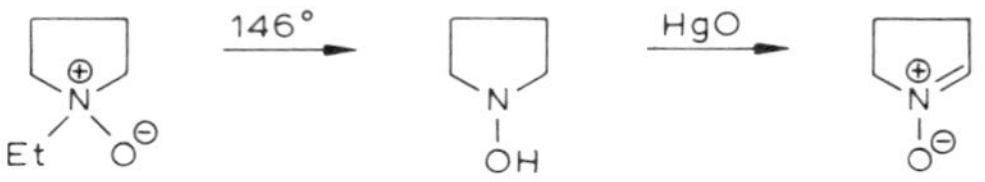

The reduction of γ-nitrocarbonyl compounds with zinc dust in aqueous ammonium chloride solution gives 1-pyrroline 1-oxides

(ii) Properties and reactions

1-Pyrroline 1-oxides add in a 1,3-manner to $\alpha\beta$-unsaturated esters to give isoxazolidines, which on thermolysis give the starting materials, and on reduction with lithium tetrahydridoaluminate yield diols, which on further reduction give pyrrolidines (*G. R. Delpierre* and *M. Lamchen*, Proc. chem. Soc., 1960, 386):

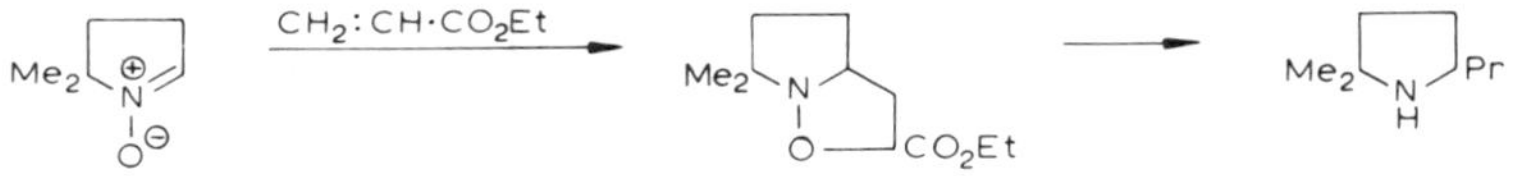

Dimethyl acetylenedicarboxylate with 1-pyrroline 1-oxides forms 1,1-adducts and those of the type VI are stable at room temperature but on heating rearrange and finally cyclodehydrate to give pyrrole derivatives (*R. Grigg*, Chem. Comm., 1966, 607):

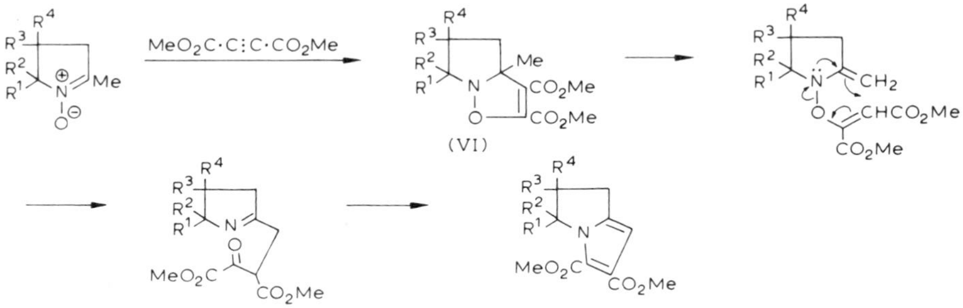

(VI)

Sodamide in liquid ammonia catalyses the dimerisation of 1-pyrroline 1-oxides to 2,2′-dipyrrolidinyl derivatives (*Todd et al.*, J. chem. Soc., 1959, 2116):

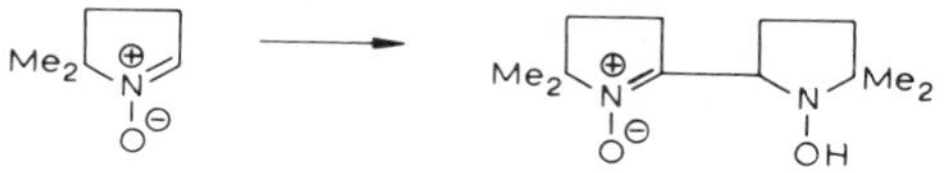

5,5-Dimethyl-1-pyrroline 1-oxide on irradiation gives the oxaziridine VII, also obtained by oxidation of 5,5-dimethylpyrroline with aqueous hydrogen peroxide, which on heating rearranges to yield 5,5-dimethylpyrrolid-2-one; acid hydrolysis affords laevulic aldehyde (*R. Bonnett, Clark* and *Todd, ibid.*, 1959, 2102):

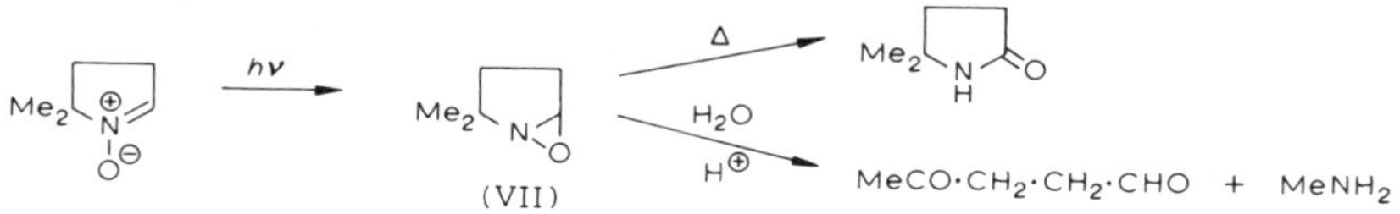

(VII)

$MeCO{\cdot}CH_2{\cdot}CH_2{\cdot}CHO + MeNH_2$

The photolysis of 5,5-dimethyl-2,4-diphenyl-1-pyrroline 1-oxide (VIII) gives the *trans*-oxaziridine IX (*J. B. Bapat* and *D. St. C. Black*, Chem. Comm., 1967, 73):

(VIII) (IX)

(c) 2-Pyrrolines

As mentioned on p. 371 very few compounds having a 2-pyrroline structure are known with certainty.

1,1-*Dimethyl-2-pyrroline*, b.p. 125–127°, obtained along with some dimeric material by the oxidation of 1,2-dimethylpyrrolidine with mercuric acetate, appears to be authentic.

Oxidation of 1-methylpyrrolidine, yields two major products, the dimer 1-methyl-3-(1′-methyl-2′-pyrrolidinyl)-2-pyrroline (X) and a trimer XI (*N. J. Leonard* and *A. G. Cook*, J. Amer. chem. Soc., 1959, **81**, 5627):

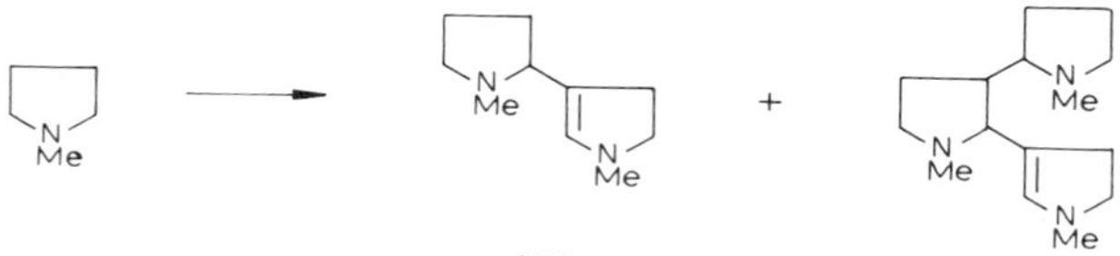

(X) (XI)

1,2,2-Trimethylpyrrolidine gives mainly the dimer, whereas 1,3,4-trimethylpyrrolidine gives as the major product the expected enamine 1,3,4-trimethylpyrroline plus some dimer:

1-Methyl-2-methylthio-2-pyrroline has been synthesised by the following route (*R. Gompper* and *W. Elser*, Tetrahedron Letters, 1964, 1971; *Org.* Synth., 1968, **48,** 97):

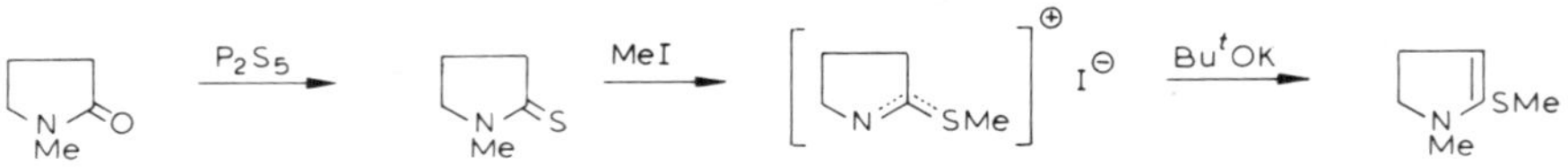

The suggested method for the preparation of 2-pyrrolines by the action of a Grignard reagent on 1-methyl-2-pyrrolidones (*R. Lukeš*, *ibid.*, 1930, **2,** 531) leads to 1-methyl-2-alkenylpyrrolidines (*Leonard* and *V. W. Gash*, J. Amer. chem., Soc., 1954, **76,** 2781).

(d) 3-Pyrrolines

The 3-pyrrolines are relatively stable, strong bases obtained by reducing many pyrroles (including *N*-methylpyrrole) with zinc and acetic or hydrochloric acid. They are not easily reduced further, but platinum causes disproportionation into pyrroles and pyrrolidines (*J. P. Wibaut* and *W. Proost*, Rec. Trav. chim., 1933, **52,** 333). Reduction of 2,5-dimethylpyrrole with zinc and hydrochloric acid gives a mixture of 2,5-dimethyl-3-pyrroline (80%) and 2,5-dimethyl-1-pyrroline (20%) (*G. G. Evans*, J. Amer. chem. Soc., 1951, **73,** 5230):

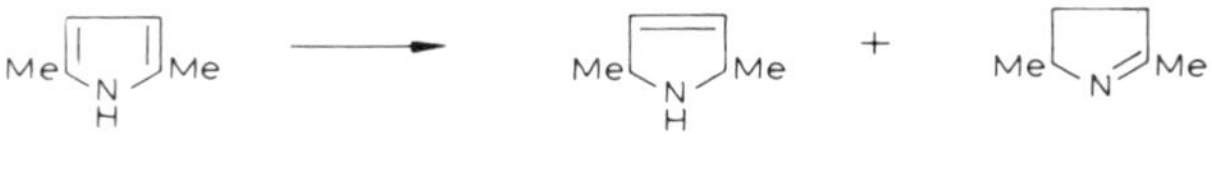

3-Pyrroline has been obtained by the reaction between hydroxylamine-*O*-sulphonic acid, sodium ethoxide and butadiene (*R. Appel* and *O. Büchner*, Angew. Chem., 1962, **65,** 430). 3-Pyrrolines may be obtained also by reacting *cis*-1,4-dichlorobut-2-ene with primary amines, the corresponding *trans*-isomers give diamines or polyamines (*J. M. Bobbitt*, *L. H. Amundsen* and *R. I. Steiner*, J. org. Chem., 1960, **25,** 2230).

Some complex pyrrolines and pyrroles have been obtained from nitrile ylides (*R. Huisgen et al.*, Angew. Chem., 1962, **74,** 31).

Water-soluble 3-**pyrroline** (*G. Ciamician* and *M. Dennstedt*, Ber., 1883, **16,** 1536; *L. H. Andrews* and *S. M. McElvain*, J. Amer. chem. Soc., 1929, **51,** 887; *A. Treibs* and *D. Dinelli*, Ann., 1935, **517,** 172) has b.p. 91°, d_{20}^{4} 0.9097, n_D^{20} 1.4664. Energetic treatment with hydriodic acid gives pyrrolidine, butylamine, and ammonia; *hydrochloride*, m.p. 173–174°, *picrate*, m.p. 156°, 1-*nitroso*-deriv., m.p. 37–38°; 1-*benzoyl-3-pyrroline*, b.p. 160–161°/2 mm; 1-*methyl-3-pyrroline*, b.p. 79–80°, occurs in tobacco (*Pictet* and *Court*, *loc. cit.*) and in belladonna leaves (*A. Goris* and *A. Larsonneau*, Chem. Ztbl., 1922, I, 757), *methiodide*, m.p. 286° (dec.); 1-*phenyl-3-pyrroline*, m.p. 100–101° (*Y. A. Arbuzor* and *A. A. Futaev*, Doklady Akad. Nauk, S.S.S.R., 1952, **85,** 1017; C.A., 1953, **47,** 5927), is obtained also from dihydrofuran and aniline (*W. Reppe*, Ann., 1955, **596,** 155); 2-*phenyl-3-pyrroline*, m.p. 45° (*Lipp* and *Seeles*, *loc. cit.*; *A. Wohl*, Ber., 1901, **34,** 1922; *F. E. King et al.*, J. chem. Soc., 1951, 239), *hydrochloride*, m.p. 240°; 2,2,5,5-*tetramethyl-3-pyrroline*, b.p. 114–116°.

The following products of reduction of pyrroles are probably correctly formulated as 3-pyrrolines: 2,4- and 2,5-*dimethyl-3-pyrroline*, b.p. 121° and 110° (*L. Knorr* and *P. Rabe*, Ber., 1901, **34,** 3491), 3-*pyrroline-2-carboxylic acid*, m.p. 235° (dec.) (*E. Fischer* and *F. Gerlach*, Ber., 1912, **45,** 2453).

(e) Oxopyrrolines; pyrrolinones

Until comparatively recently these compounds are described in the literature as hydroxypyrroles or as pyrrolones, a term no longer in use.

(i) 5-Oxopyrrolines

Synthesis. 5-Oxopyrrolines, which are of particular interest in relationship to the bile pigments, are obtained as follows:

(*a*) from the reaction of ammonia or amines with monoacyl- or hydroxymethylene-succinates (*W. O. Emery*, Ann., 1890, **260,** 137):

$EtO_2C\cdot CH(MeCO) - CH_2CO_2Et$ ⟶ (ring: EtO_2C, Me, N, O)

or acylacetates (*G. K. Almström*, *ibid.*, 1916, **411,** 350; 1918, **416,** 279):

$PhCO\cdot CH_2\cdot NHAr$ + $CH_2Ac\cdot CO_2Et$ ⟶ (ring: Ph, Ac, N–Ar, O)

(*b*) by the dehydration of the product obtained following the reactions between a succinimide and a Grignard reagent (*Lukeš*, Coll. Czech. chem. Comm., 1929, **1,** 119; *E. Walton*, J. chem. Soc., 1940, 438):

(*c*) by the reaction between pyrroles and hydrogen peroxide, when di-pyrrolyl peroxides, also prepared by atmospheric oxidation of pyrroles, are formed (*H. Fischer* and *P. Hartman*, Z. physiol. Chem., 1934, **226,** 116; *W. Metzger* and *Fischer*, Ann., 1937, **527,** 1);

(*d*) on warming 5-bromopyrrole-2-carboxylic acids with hydrochloric or hydrobromic acid:

When carried out in methanol a methoxypyrrole is obtained (*W. Siedel*, Ann., 1943, **554,** 144).

(*e*) Photo-oxidation of pyrrole in the presence of eosin gives 5-hydroxy-2-oxo-3-pyrroline, which on oxidation with manganese dioxide affords a simple synthesis of maleimide (*P. de Mayo* and *S. T. Reid*, Chem and Ind., 1962, 1576). 3,4-Dimethyl-2-oxo-3-pyrroline possesses a reactive methylene group at the 5-position and with ethyl sulphate gives 2-ethoxy-3,4-dimethylpyrrole (*H. Plieninger*, *H. Bauer* and *A. R. Katritzky*, Ann., 1962, **654,** 165).

Properties and reactions of 5-*oxopyrrolines*. Contrary to the fact that 5-oxopyrrolines (also 4-oxopyrrolines) having nuclear ester or acyl groups dissolve in aqueous alkali, couple readily with diazo-compounds, give colours with ferric chloride, and can be esterified thus indicating the properties of an enol (*Fischer* and *J. Müller*, J. physiol. Chem., 1924, **132,** 72; *Almström, loc. cit.*), other evidence, in particular spectral evidence, shows that they are better formulated as ketones (*C. A. Grob* and *P. Ankli*, Helv., 1949, **32,** 2010). Simple 5-oxopyrrolines do not behave like phenols and if *N*-substituted give no Zerewitinoff reaction (*Lukeš* and *V. Šperling*, Coll. Czech. chem. Comm., 1936, **8,** 464).

(*ii*) *4-Oxopyrrolines*

Synthesis. Although 4-oxo-2-pyrroline has not yet been prepared its derivatives can be synthesised as follows:

(*a*) from chloroacetyl chloride and ethyl *β*-aminocrotonates (*E. Benary* and *R. Konrad*, Ber., 1923, **56,** 44):

$ClCH_2{\cdot}COCl$ + $RNHCR':CH{\cdot}CO_2Et$ → $OC(CH_2Cl){-}C(CO_2Et){=}CR'(NHR)$ →(KOH) 1-R-2-R'-3-ethoxycarbonyl-4-oxo-2-pyrroline (*b*) by a route in which the first stages involve the hydrogenation of MeC(OH)(CN)·CHMe·CO_2Et (*Plieninger* and *M. Decker*, Ann., 1956, **598,** 198);

(*c*) by the cyclisation under Dieckmann conditions of Schiffs bases from β-oxo-esters and esters of amino acids (*A. Treibs* and *A. Ohorodnik*, *ibid.*, 1958, **611,** 139);

(*d*) by the cyclisation of the amine adducts of pentadiyn-3-ones in boiling xylene in which case pyridones are also formed (*T. Metler*, *A. Uchida* and *S. I. Miller*, Tetrahedron, 1968, **24,** 4285):

R, R^1NH, CO·C≡CR^2 → R, N, R^1, O, CHR^2

(*e*) The products from α,α-dialkyl-substituted *N*-phenylglycines and hot acetyl chloride on hydrolysis yield 2-methyl-4-oxo-1-phenyl-2-pyrrolines (*R. J. S. Beer*, *W. T. Gradwell* and *W. J. Oates*, J. chem. Soc., 1958, 4693):

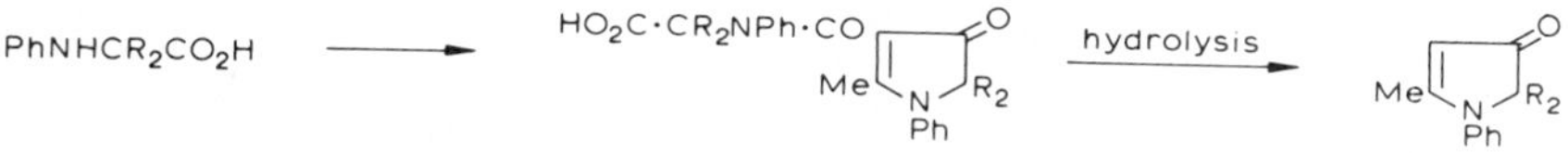

PhNHCR_2CO_2H → HO_2C·CR_2NPh·CO, O, Me, N, Ph, R_2 —hydrolysis→ O, Me, N, Ph, R_2

(*f*) 1-Anilino*cyclo*hexanecarboxylic acid after treatment with acetyl chloride and hydrolysis of the product affords cyclohexane spiro-5-(2′-methyl-4′-oxo-1′-phenyl-2′-pyrroline):

O, Me, N, Ph

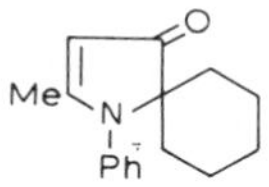

Spectral studies show that the 4-oxo-2-pyrroline (XII) does not tautomerise appreciably to the corresponding 3-hydroxypyrrole

RO_2C, O, Me, N, H (XII)   O, CO_2·$(CH_2)_7$·CH_3, N, H, Me (XIII)

and the structure of compound XIII is best represented as indicated (*R. S. Atkinson* and *E. Bullock*, Canad. J. Chem., 1963, **41,** 625).

(iii) 3-Oxopyrrolines

Oxidation of 5,5-dimethyl-1-pyrroline 1-oxide with selenium dioxide gives 5,5-dimethyl-3-oxo-1-pyrroline 1-oxide (*V. M. Clark*, *B. Sklarz* and *Sir A. Todd*, J. chem. Soc., 1959, 2123):

Although 4-oxo-2-pyrrolines are reported to show no ketonic properties and to give only some of the usual hydroxyl reactions, *C*-acylation occurs more readily than *O*-acylation (*Treibs* and *Ohorodnik*, Ann., 1958, **611,** 149; *J. Davoll*, J. chem. Soc., 1953, 3802); spectroscopic studies by i.r., u.v., and n.m.r. show that the so-called 3-hydroxypyrroles exist in the 4-oxo-2-pyrroline form and are not tautomeric substances (*Atkinson* and *Bullock*, *loc. cit.*).

In the presence of acid, benzaldehyde condenses with 4- and 5-oxopyrrolines, giving benzylidene-3*H*-pyrroles (*Fischer* and *Müller*, *loc. cit.*):

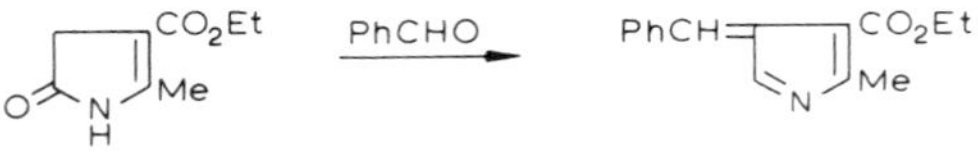

Condensation also takes place with formic and oxalic esters.

(iv) Dioxopyrrolines

By an interesting ring-opening and resynthesis, iso-oxazolium salts can afford 4,5-dioxo-2-pyrrolines, red substances, which like the analogous isatins, form carbonyl derivatives at the 4-position (*O. Mumm et al.*, Ber., 1910, **43,** 3345; 1937, **70,** 1930):

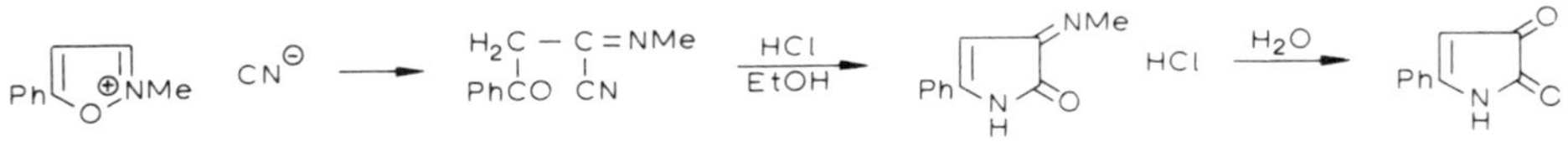

5-**Oxo**-2-**pyrroline** (*monohydrate*, m.p. 83°), is obtained from methyl γ-bromocrotonate by the phthalimide method (*W. Langenbeck* and *H. Boser*, Ber., 1951, **84,** 526). Haemopyrrole gives with hydrogen peroxide 4-*ethyl*-2,3-*dimethyl*-5-*oxo*-2-*pyrroline*, m.p. 95°, *peroxide*, m.p. 228° (dec.); cryptopyrrole yields 3-*ethyl*-2,4-*dimethyl*-5-*oxo*-2-*pyrroline*, m.p. 84°, peroxide, m.p. 219–220° (*Fischer et al.*, Z. physiol. Chem., 1932, **212,** 150); 1,2-*dimethyl*-5-*oxo*-2-*pyrroline*, m.p. 62–63° (*Lukeš* and *Walton*, *loc. cit.*); 4-*acetyl*-1,3-*diphenyl*-5-*oxo*-2-*pyrroline*, m.p. 111–112° (*Almström*, *loc. cit.*), affords 1,3-*diphenyl*-5-*oxo*-2-*pyrroline*, m.p. 169–170°; *ethyl* 2-*methyl*-5-*oxo*-2-*pyrroline*-3-*carboxylate*, m.p. 136°, *benzylidene* deriv., m.p. 184° (*Fischer* and *Müller*, *loc. cit.*); *ethyl* 2-*methyl*-4-*oxo*-2-*pyrroline*-3-*carboxylate*, m.p. 215° (*Benary* and *Konrad*, *loc. cit.*), *benzylidene* deriv., m.p. 228° (*Fischer* and *M. Herrmann*, Z. physiol. Chem., 1922, **122,** 1).

3. Pyrrolidines; tetrahydropyrroles

(a) Pyrrolidines

(i) Synthesis

(*1*) The simpler pyrrolidines are often prepared by reduction of the corresponding pyrroles and related derivatives, *e.g.*, by catalytic hydrogenation using platinum oxide (*M. de Jong* and *J. P. Wibaut*, Rec. Trav. chim., 1930, **49,** 237), or Raney nickel (200°/200–300 atm.). *C*-Acyl groups are first reduced to alkyl, then ester groups to methyl simultaneously with the nucleus (N–CO_2Et strongly promotes nuclear reduction, *H. Adkins et al.*, J. Amer. chem. Soc., 1936, **58,** 709; 1939, **61,** 1104). 3-Pyrrolines when reduced with hydriodic acid and red phosphorus, also undergo extensive ring-fission. 2-Pyrrolidones may be reduced by sodium and amyl alcohol, or by lithium tetrahydridoaluminate (*A. Pernot* and *A. Willemart*, Bull. Soc. chim. Fr., 1953, 324); succinimides by the latter reagent also (*L. M. Rice et al.*, J. org. Chem., 1954, **19,** 884). Hydrogenation of oximes of cyclopropyl ketones also gives pyrrolidines (*H. Normant*, Compt. rend., 1951, **232,** 1358).

(*2*) Pyrrolidines have been obtained by passing tetrahydrofurans or tetramethylene-1,4-glycols with ammonia or primary amines over alumina at 300° (*W. Reppe*, Ann., 1955, **596,** 143) or by passing the 1,4-glycols with primary amines over copper–chromium oxide (*Adkins et al.*, J. Amer. chem. Soc., 1936, **58,** 2491; 1938, **60,** 1033). Numerous other methods are described including the dry distillation of hydrochlorides of 1,4-diamines (*H. Oldach*, Ber., 1887, **20,** 1654; *F. C. Petersen*, *ibid.*, 1888, **21,** 290):

$$H_2NCH_2{\cdot}CH_2{\cdot}CH_2{\cdot}CH_2NH_2{\cdot}HCl \longrightarrow \text{pyrrolidine (NH)}$$

the treatment of 1,4-dihalides with primary amines (*M. Scholtz* and *P. Friemehlt*, *ibid.*, 1899, **32,** 850; *R. C. Elderfield* and *H. A. Hageman*, J. org. Chem., 1949, **14,** 605; *A. H. Sommers*, J. Amer. chem. Soc., 1956, **78,** 2439); the reaction between benzenesulphonic esters of 1,4-glycols and primary amines (*D. D. Reynolds* and *W. O. Kenyon*, *ibid.*, 1950, **72,** 1597); the liberation of hydrogen halide from 1,4-halogeno-amines (*W. Jacobi* and *G. Merling*, Ann., 1894, **278,** 1; *J. von Braun* and *E. Beschke*, Ber., 1906, **39,** 4120):

$$CH_2(NH_2{\cdot}HBr){-}CH_2{-}CH_2{-}CH_2Br \xrightarrow{\text{alkali}} \text{pyrrolidine (NH)}$$

By a curious counterpart to the above method, *N*-halogenated butylamines can furnish pyrrolidines (*K. Löffler*, *ibid.*, 1910, **43,** 2035; *G. H. Coleman* and *G. E. Goheen*, J. Amer. chem. Soc., 1938, **60,** 730):

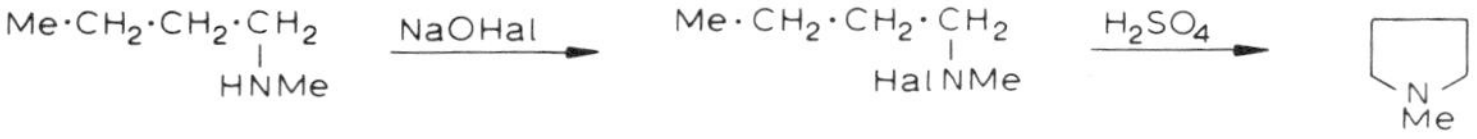

$$Me{\cdot}CH_2{\cdot}CH_2{\cdot}CH_2{\cdot}NHMe \xrightarrow{NaOHal} Me{\cdot}CH_2{\cdot}CH_2{\cdot}CH_2{\cdot}N(Hal)Me \xrightarrow{H_2SO_4} \text{1-methylpyrrolidine}$$

Tetramethylene 1,4-dibromide and toluenesulphonamide give 1-toluenesulphonylpyrrolidine (*A. Müller* and *A. Sauerwald*, Monatsh., 1927, **48,** 155). The photochemical cyclisation of aliphatic secondary *N*-chloroamines with sulphuric acid depends on the size of the alkyl groups; when they are *n*-pentyl or higher, pyrrolidines predominate, but the yield of the isomeric piperidines increase as the groups become smaller (*S. Wawzonek* and *T. P. Culbertson*, J. Amer. chem. Soc., 1959, **81,** 3367; 1960, **82,** 441).

(*3*) The piperidine ring has been converted into that of pyrrolidine *via* the product resulting from a Hofmann degradation

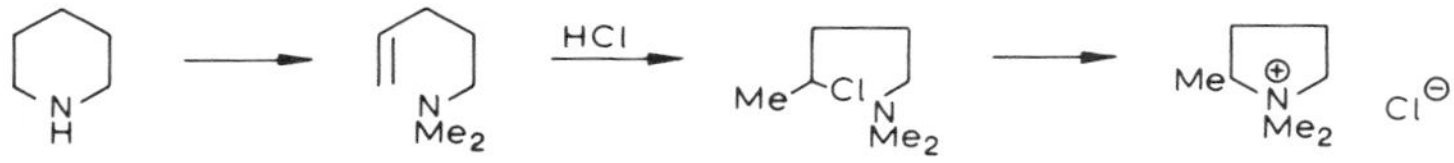

Ring-contraction occurs also during the Clemmensen reduction of 3-oxopiperidines (*N. J. Leonard* and *E. Barthel*, *ibid.*, 1950, **72,** 3632):

(*4*) The product from 5,5-dimethyl-1-pyrroline 1-oxide and ethyl acrylate on reduction with lithium tetrahydridoaluminate yields first a diol and then 2,2-dimethyl-5-propylpyrrolidine (*G. R. Delpierre* and *M. Lamchen*, Proc. chem. Soc., 1960, 386):

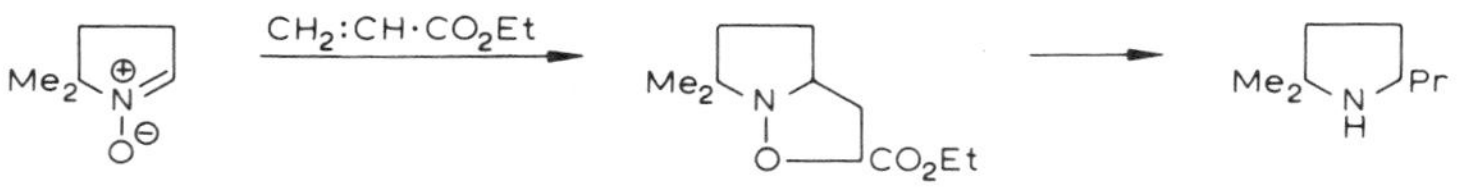

$$\xrightarrow{CH_2{:}CH{\cdot}CO_2Et}$$

(*5*) Pyrrolidine is formed from putrescine and Raney nickel in an inert solvent (*K. Kindler* and *D. Matthies*, Ber., 1962, **95,** 1992).

(ii) Properties and reactions

Pyrrolidines are stable, strong bases with all the properties of secondary aliphatic amines. The ring is opened in the normal manner by the Hofmann method or by the cyanogen bromide process *(Elderfield* and *Hageman, loc. cit.)*. 5-Dimethylaminopent-2-yne is obtained by the Hofmann degradation of 1,1-dimethyl-2-methylenepyrrolidinium hydroxide (*Lukeš, J. Pliml* and *J. Trojánek*, Coll. Czech. chem. Comm., 1959, **24,** 3109):

Pyrrolidine, unlike pyrrole or 3-pyrroline, reacts with cyanogen to give the cyanoformamidine (I) and in the absence of solvent the oxamidine (II) (*H. M. Woodburn* and *W. S. Zehrung*, J. org. Chem., 1959, **24,** 1184):

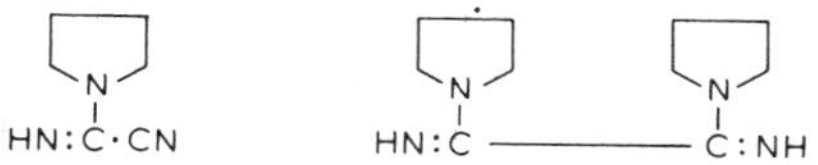

1,1-Dimethylpyrrolidinium bromide reacts with organolithium compounds to give ethene and dimethylaminoethene. Tracer evidence indicates that the decomposition proceeds *via* the α-ylid III (*F. Weygand* and *H. Daniel*, Ber., 1961, **94,** 1688) and not the ylid IV, because on exposure to butyl-lithium it affords mainly 1-methylpyrrolidine (*G. Wittig* and *W. Tochtermann, ibid.*, 1961, **94,** 1692):

The 1,1-dimethyl-2-phenylpyrrolidinium cation with sodamide undergoes ring enlargement to give a benzazocine (*G. C. Jones* and *C. R. Hauser*, J. org. Chem., 1962, **27,** 3572):

When 2-(2′-chloroethyl)-1-methylpyrrolidine reacts with nucleophiles, besides the expected products, approximately equal amounts of perhydroazepine derivatives are formed (*A. Ebnöther* and *E. Jucker*, Helv., 1964, **47,** 745):

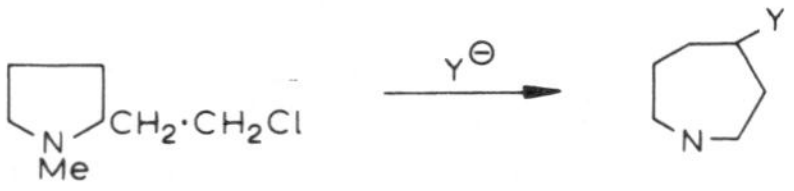

2-Ethylidene-1-methylpyrrolidine and ethyl acrylate in dioxane afford the nine-membered oxolactam perhydro-1,5-dimethylazonine-2,6-dione (V), which cyclises spontaneously to the hexahydroindole VI (*O. Červinka*, Chem. and Ind., 1959, 1129):

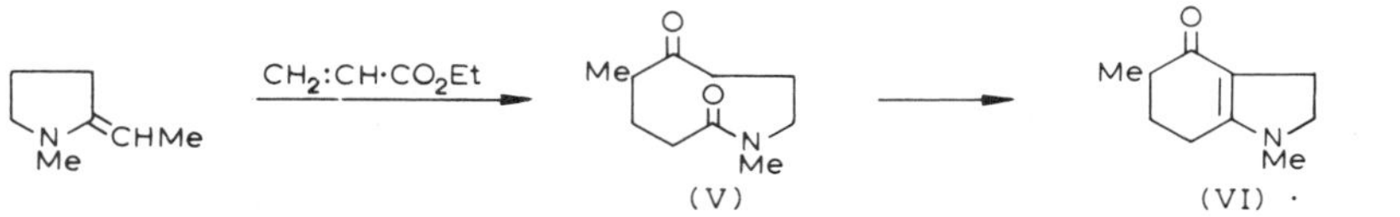

(V)

(VI) .

(iii) Individual pyrrolidines

Pyrrolidine, present in tobacco leaves, is prepared by hydrogenation of pyrrole over Adams catalyst (*L. H. Andrews* and *S. M. McElvain*, J. Amer. chem. Soc., 1929, **51,** 887) or by *Reppe's* method *(loc. cit.)*, b.p. 88.5–89°, $d^{22.5}$ 0.850, n_D^{15} 1.4270, k_b 1.3×10^{-3} (*L. C. Craig* and *R. M. Hixon*, J. Amer. chem. Soc., 1931, **53,** 4367). It is soluble in water and smells like piperidine; *picrate*, m.p. 111–112°, p-*toluenesulphonyl* deriv. (p. 380), m.p. 123°, *carbamoyl* deriv., m.p. 218° *(Reppe, loc. cit.)*; 1-*nitrosopyrrolidine*, b.p. 214° (decomp.); 1-*formylpyrrolidine (pyrrolidine*-1-*carbaldehyde)* (from pyrrolidine and chloral), b.p. 87–89°/16 mm (*F. F. Blicke* and *C.-J. Lu*, *ibid.*, 1952, **74,** 3933); 1-*benzoylpyrrolidine*, b.p. 169–170°/8 mm (*J. L. Rainey* and *Adkins*, *ibid.*, 1939, **61,** 1104).

1-**Ethyl-,** 1-**methyl-,** and 1-**phenyl-pyrrolidine,** b.p. 106°, 79°, and 124°/14 mm, respectively, are prepared by Reppe's method, *picrates* of latter two compounds, m.p. 221° and 115–116°; *dimethylpyrrolidinium iodide*, m.p. > 300°. Hydrogenation of the corresponding pyrrole gives *e.g.*, 1-*methyl-* and 1-*benzyl-pyrrolidine*, b.p. 237° (*Craig* and *Hixon*, *ibid.*, 1930, **52,** 807; 1931, **53,** 187), 1-n-*butylpyrrolidine*, b.p. 154–155°/758 mm (*Coleman*, *G. Nichols* and *T. F. Martens*, Org. Synth., 1945, **25,** 14).

The following *C*-alkyl- and -aryl-pyrrolidines are made by standard methods *(vide supra)*: 2-*methyl-*, b.p. 100°, and 2-*ethyl-*, b.p. 120–122° *(De Jong* and *Wibaut*, *loc. cit.)*; 2-*phenyl-*, b.p. 236–238°, 115°/15 mm, and 1-*methyl-2-phenyl-*, b.p. 225–227° (*F. B. la Forge*, J. Amer. chem. Soc., 1928, **50,** 2477); 3-*methyl-*, b.p. 103–105°, and 3-*phenyl-*, b.p. 120–122°/12 mm (*E. Späth* and *F. Breusch*, Monatsh., 1928, **50,** 352; *F. Bergel et al.*, J. chem. Soc., 1944, 269); 1,2,2-*trimethyl-* (p. 370), b.p. 130–135°; 3,4-*dimethyl-*[meso, b.p. 128°, (±), b.p. 122°] (*G. E. McCasland* and *S. Proskow*, J. Amer. chem. Soc., 1954, **76,** 6087); 3,3-*dimethyl-*, b.p. 114–115° (*R. F. Brown* and *N. M. van Gulick*, *ibid.*, 1955, **77,** 1083). Hydrogenation of 3-acetyl-2,4-dimethylpyrrole gives 3-*ethyl-2,4-dimethylpyrrolidine*, *cryptopyrrolidine*, b.p. 151–152° (*R. B. Moffett* and *J. L. White*, J. org. Chem., 1952, **17,** 407). 2,5-*Diphenylpyrrolidine*, b.p. 122–123.4°/0.2 mm (*C. G. Overberger*, *M. Valentine* and *J.-P. Anselme*, J. Amer. chem. Soc., 1969, **91,** 687).

(b) Substituted pyrrolidines

(i) Hydroxypyrrolidines

Hydroxypyrrolidines with the hydroxyl group in the β-position or in a side-chain are obtainable by the usual methods and show the usual reactions of alcohols, but the α-compounds undergo spontaneous dehydration to pyrrolines. 2-Ethoxy-1-benzenesulphonylpyrrolidine has been synthesised from acrolein (*A. Wohl et al.*, Ber., 1905, **38,** 4157):

$$CH_2{:}CH{\cdot}CHO \longrightarrow CH_2Cl{\cdot}CH_2{\cdot}CH(OEt)_2 \longrightarrow NC{\cdot}CH_2{\cdot}CH_2{\cdot}CH(OEt)_2 \longrightarrow NH_2{\cdot}(CH_2)_3{\cdot}CH(OEt)_2 \xrightarrow{PhSO_2Cl} \text{2-OEt-1-}SO_2Ph\text{-pyrrolidine}$$

1-**Hydroxypyrrolidine** has been obtained by the pyrolysis of 1-ethylpyrrolidine 1-oxide (p. 371).

γ-Acylaminovaleraldehydes are believed to be tautomeric with 1-acyl-2-hydroxypyrrolidines. They yield pyrrolines on gentle heating, and with methyl alcohol afford neutral compounds which give no reaction of aldehydes until treated with acids (*B. Helferich* and *W. Dommer*, Ber., 1920, **53,** 2004):

Me, N-Acyl pyrroline $\xleftarrow{-H_2O}$ $H_2C{-}CH_2$, $Me{\cdot}HC$, CHO, NHAcyl $\rightleftharpoons$ Me, OH, N-Acyl pyrrolidine $\xrightarrow[HCl]{MeOH}$ Me, OMe, N-Acyl pyrrolidine

1-**Methyl**-3-**hydroxypyrrolidine,** b.p. 77°/16 mm, is made from 1-dimethylaminobutan-3-one as follows (*C. Mannich* and *T. Gollash*, Ber., 1928, **61,** 263):

$CO{-}CH_2$, Me, CH_2, NMe_2 $\xrightarrow[(2)\ Ba(OH)_2]{(1)\ Br_2}$ 3-oxo-pyrrolidinium $\overset{\oplus}{N}Me_2\ Br^{\ominus}$ $\xrightarrow{H_2,Pt}$ 3-HO-pyrrolidinium $\overset{\oplus}{N}Me_2\ Br^{\ominus}$ $\xrightarrow{\Delta}$ 3-HO-1-Me-pyrrolidine

Primary amines with 1,4-dibromobutane-2,3-diol give *meso*-1-alkyl- or -aryl-3,4-dihydroxypyrrolidines, *e.g.*, meso-1-n-*butyl*-, b.p. 100–105°/1 mm and -1-*phenyl*-3,4-*dihydroxypyrrolidine*, m.p. 157° (*A. J. Hill* and *M. G. McKeon*, J. Amer. chem. Soc., 1954, **76,** 3548); the *trans*-form of the latter base melts at 124° (*J. J. Roberts* and *W. C. J. Ross*, J. chem. Soc., 1952, 4288). Bouveault–Blanc reduction of methyl hygrate (methyl 1-methylpyrrolidine-2-carboxylate) gives 1-*methyl*-2-*hydroxymethylpyrrol-*

idine, hygrinol, b.p. 67–68°/12 mm, *methobromide (stachydrinol),* m.p. 313–314° (dec.) (*R. R. Renshaw* and *W. E. Cass,* J. Amer. chem. Soc., 1939, **61,** 1195; *Lukeš* and *Čěrvinka,* C.A., 1955, **49,** 288, 289). Ethyl pyroglutamate is reduced by lithium tetrahydridoaluminate to 2-*hydroxymethylpyrrolidine,* b.p. 96–98°/14 mm (*Blicke* and *Lu,* J. Amer. chem. Soc., 1955, **77,** 29). Reduction of 2-aminoethylpyrrole gives 2-*aminomethylpyrrolidine,* b.p. ~50°/7 mm (*N. I. Putochin,* Chem. Ztbl., 1931, II, 442).

(ii) Oxopyrrolidines

2-Oxopyrrolidines (α- or 2-pyrrolidones) are lactams of γ-amino acids, very feebly basic, water-soluble substances which regenerate the amino acid when boiled with acid or alkali. They may be prepared by electrolytic reduction of succinimides (*J. Tafel* and *M. Stern,* Ber., 1900, **33,** 2224) or from butyrolactone with ammonia or amines *(Reppe, loc. cit.).* **2-Oxopyrrolidine** (α- or 2-**pyrrolidone**), b.p. 245°; 1-*methyl*-2-*pyrrolidone,* m.p. 206°; 1-*ethyl*-2-*pyrrolidone,* m.p. 218°. With ethanolamine, butyrolactone affords 1-hydroxyethyl-2-oxopyrrolidine, dehydrated to 1-*vinyl*-2-*pyrrolidone,* m.p. 17°, b.p. 64–66°/2 mm (*B. Puetzer et al.,* J. Amer. chem. Soc., 1952, **74,** 4959), which is also produced from 2-oxopyrrolidine and acetylene (*I. Pogany,* C.A., 1956, **50,** 952). Polymerised vinylpyrrolidone (PVP) is used as a blood-plasma extender.

Acetate VII on pyrolysis affords 2-oxo-3-vinylpyrrolidine, the double bond of which is not sufficiently activated for polymerisation. The corresponding chloride VIII on treatment with potassium *tert*-butoxide yields 1-pyrrolino[2,3-*b*]oxolane (IX) (*W. A. W. Cummings* and *A. C. Davis,* J. chem. Soc., 1964, 4591):

$CH_2{\cdot}CH_2{\cdot}OAc$ (VII) —Δ→ $CH{:}CH_2$

$CH_2{\cdot}CH_2Cl$ (VIII) —Bu^tOK→ (IX)

Hydrogenation of laevulic acid in presence of ammonia or methylamine gives 5-*methyl*-2-*oxopyrrolidine* (5-*methyl*-2-*pyrrolidone*), m.p. 41–42° (*Y. Hachihama* and *I. Hayashi,* C.A., 1955, **49,** 6226) and 1,5-*dimethyl*-2-*oxopyrrolidine,* b.p. 84–86°/13 mm (*R. L. Frank et al.,* Org. Synth., 1947, **27,** 28). Hydrogenation of the adduct from methyl acrylate and 2-nitropropane gives 5,5-*dimethyl*-2-*oxopyrrolidine,* m.p. 42–43° (*R. B. Moffett, ibid.,* 1952, **32,** 59, 86). Similarly 3-methyl-3-nitro-*n*-butyl cyanide, from 2-nitropropane and acrylonitrile, gives by reduction with iron and hydrochloric acid, 2-*amino*-5,5-*dimethyl*-1-*pyrroline* (XI), m.p. 73–74°, together with the corresponding 2-*amino*-5,5-*dimethyl*-1-*pyrroline* 1-*oxide* (X), m.p. 238° (*G. D. Buckley* and *T. J. Elliott,* J. chem. Soc., 1947, 1509):

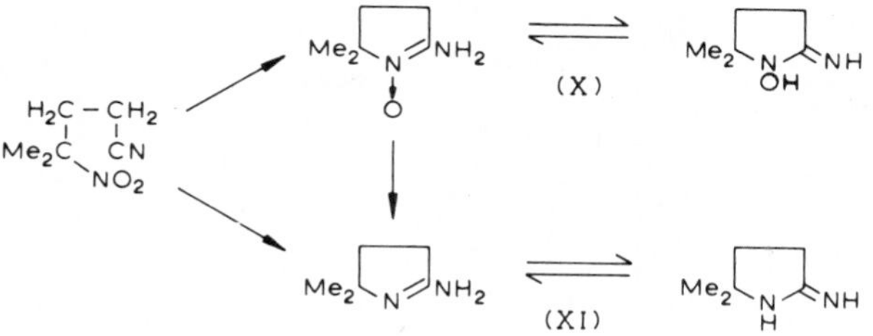

(X) (XI)

Both these compounds may be regarded as tautomeric with the corresponding 5-iminopyrrolidine.

The oxaziridine 2,2,4-trimethyl-6-oxa-1-azabicyclo[3.1.0]hexane (XII) on heating at 150° is isomerised to 2-oxo-3,5,5-trimethylpyrrolidine (XIII) and similar rearrangements occur with 3,3,5-trimethyl-6-oxa-1-azabicyclo[3.1.0]hexane (XIV) and 2,2,5-trimethyl-6-oxa-1-azabicyclo[3.1.0]hexane (XV) at 300°. Compound XIV when heated at 150° affords 1-acetyl-3,3-dimethylazetidine (XVI) and 2,4,4-trimethyl-1-pyrroline (XVII) (*L. S. Kaminsky* and *Lamchen*, J. chem. Soc., 1967, 2128):

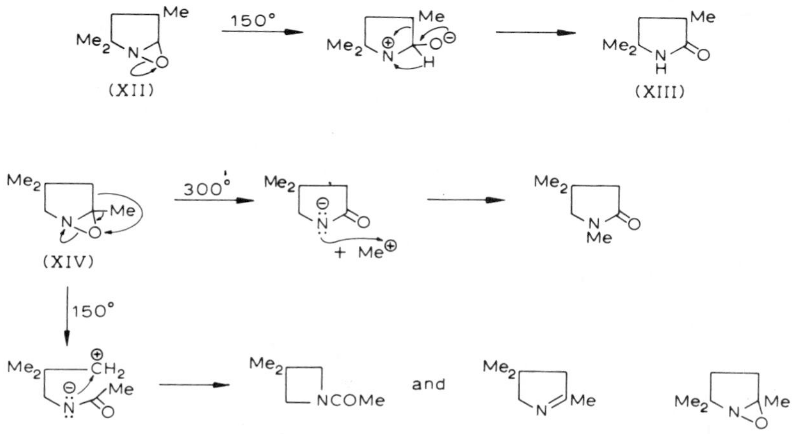

(XII) (XIII) (XIV) (XVI) (XVII) (XV)

Cyxlidation of γ-halogenobutyrohydroxamic acids gives 1-**hydroxy-2-oxopyrrolidines** (*J. Smrt, J. Beránek* and *M. Horák*, Coll. Czech. chem. Comm., 1959, **24,** 1672):

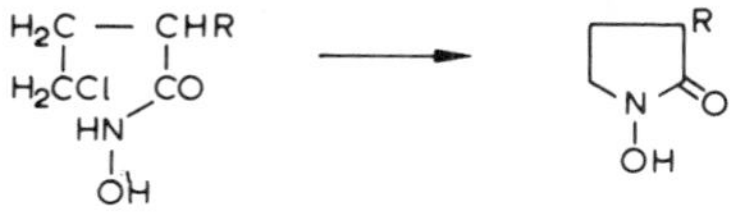

Irradiation of 2-oxopyrrolidine gives a mixture of the stereoisomers of 5,5′-dioxo-2,2′-bipyrrolidinyl (XVIII) (*M. Pesaro, I. Felner-Caboga* and *A. Eschenmoser*, Chimia, Switz., 1965, **19,** 566).

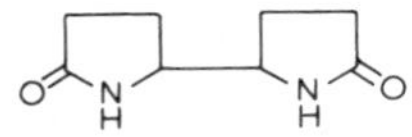

(XVIII)

Unlike 2-oxopyrrolidines, the 3-isomers (β-compounds) give the normal reactions of ketones as do substances having a carbonyl group in the side-chain, like the alkaloids hygrine and cuscohygrine.

3-**Oxopyrrolidine** *hydrochloride*, m.p. 143–144°, is obtained on treating 1-ethoxycarbonyl-3-oxopyrrolidine with 18% hydrochloric acid at 110–115°. Exhaustive reduction of the same intermediate with lithium tetrahydridoaluminate gives DL-3-hydroxy-1-methylpyrrolidine (*R. Kuhn* and *G. Osswald*, Angew., Chem., 1957, **69,** 60):

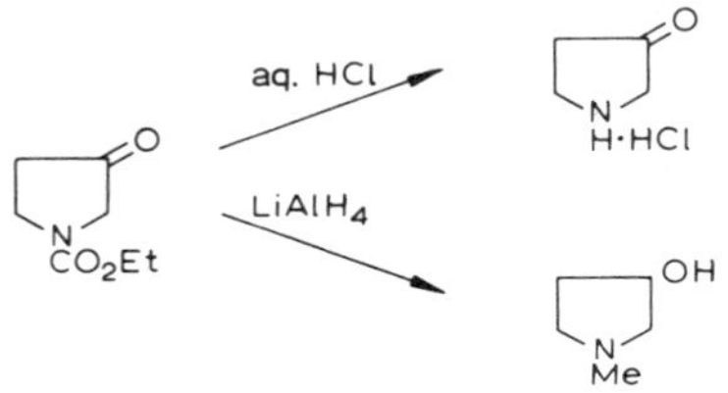

aq. HCl; $LiAlH_4$; CO_2Et; H·HCl; OH; Me

A Dieckmann ring-closure followed by decarboxylation yields 1-*methyl*-3-*oxopyrrolidine*, b.p. 46–47°/18 mm (*G. A. Prill* and *S. M. McElvain*, J. Amer. chem. Soc., 1933, **55,** 1233):

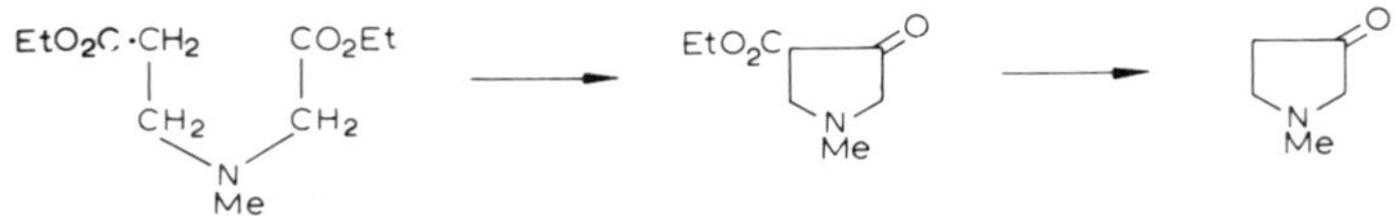

$EtO_2C·CH_2$ CO_2Et / CH_2 CH_2 / N Me ⟶ EtO_2C (N Me) ⟶ (N Me)

Similarly ethoxycarbonylmethyl-β-cyanoethylmethylamine is cyclised by sodium to give 4-cyano-1-methyl-3-oxopyrrolidine and 1-methyl-3-oxopyrrolidine-4-carboxamide (*A. H. Cook* and *K. J. Reed*, J. chem. Soc., 1945, 399):

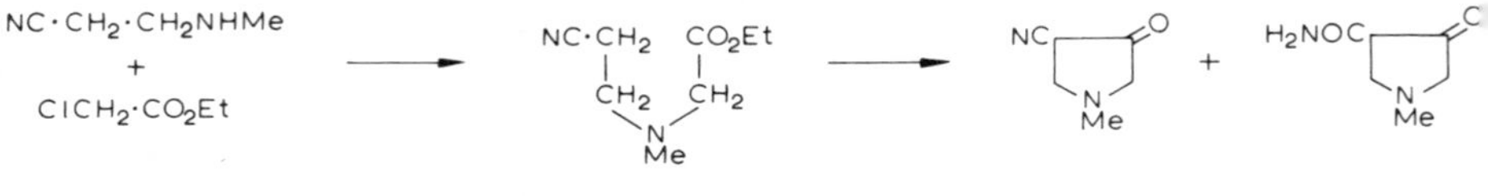

$NC·CH_2·CH_2NHMe$ + $ClCH_2·CO_2Et$ ⟶ $NC·CH_2$ CO_2Et / CH_2 CH_2 / N Me ⟶ NC (N Me) + H_2NOC (N Me)

The readily available iodomethyl styryl ketone, PhCH=CH·CO·CH_2I, with primary amines yields, *e.g.*, 3-*oxo*-1,5-*diphenylpyrrolidine*, m.p. 119–121° (*P. L. Southwick* and *H. L. Dimons*, J. Amer. chem. Soc., 1954, **76,** 5667).

(iii) Poly-oxopyrrolidines

Oxalic esters condense with *β*-alanine esters to give 2,3-**dioxopyrrolidines,** leading to 1-*benzyl-2,3-dioxopyrrolidine*, m.p. 99–100°, on hydrolysis (*Southwick* and *R. T. Crouch, ibid*., 1953, **75,** 3413):

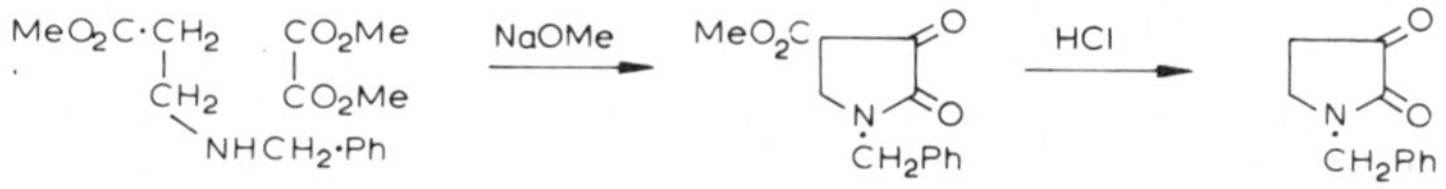

A number of 2,3-dioxopyrrolidines have been obtained by a useful one-step synthesis from ethyl acrylate, ethyl oxalate and primary amines (*Southwick et al*., J. org. Chem., 1956, **21,** 1087):

2,3-Dioxopyrrolidines become the main products when Döbner's cinchoninic acid synthesis is applied to substituted pyruvic acids or to halogeno- or nitro-anilines (*W. Borsche*, Ber., 1908, **41,** 3884; 1909, **42,** 4072):

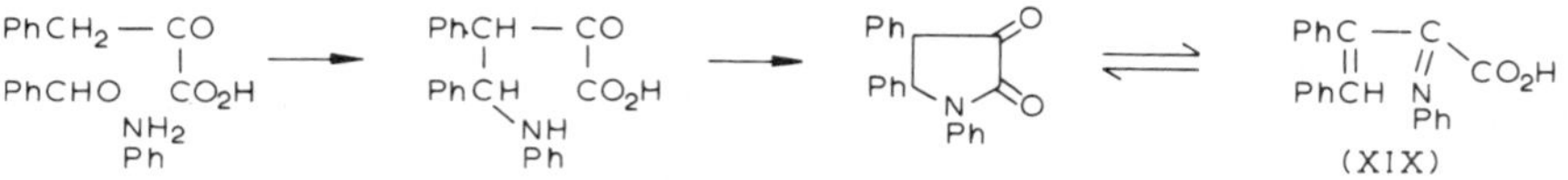

(XIX)

It is often advantageous to use a ready-made arylidene-pyruvic acid with an amine. The products are decomposed thermally into cinnamylidenearylamines and carbon dioxide, and are thought to be tautomeric with the isomeric arylimino-acids (XIX) (*W. R. Vaughan et al*., J. org. Chem., 1953, **18,** 382; 1955, **20,** 143).

2,4-**Dioxopyrrolidines** have been prepared from α-benzamidoisobutyryl chloride and methyl sodio-malonate (*S. Gabriel*, Ber., 1913, **46,** 1319), and by acetoacetylating α-amino-esters with diketene followed by ring-closure (*R. N. Lacey*, J. chem. Soc., 1954, 850):

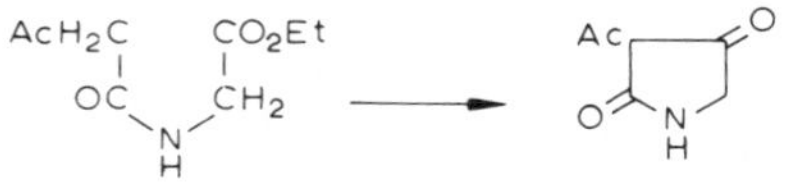

Succinimides are 2,5-**dioxopyrrolidines;** *R. P. Linstead et al*. report reactions of 2,5-diiminopyrrolidine (J. chem. Soc., 1956, 235). Ethyl oxalate condenses with ethyl ethyliminodiacetate to give (probably enolic) *ethyl 3,4-dioxo-1-ethylpyrrolidine-2,5-dicarboxylate*, m.p. 83° (*R. H. Eastman* and *R. M. Wagner*, J. Amer. chem. Soc., 1949, **71,** 4089).

Ethyl oxalate and phenylacetamide with sodium ethoxide yield 2,4,5-**trioxo-3-phenylpyrrolidine,** m.p. 217° (*G. S. Skinner* and *C. B. Miller*, *ibid.*, 1953, **75,** 977).

(iv) Pyrrolidinecarboxylic acids

Pyrrolidinecarboxylic acids are nearly neutral compounds resembling the aliphatic amino acids.

Synthesis. The α-compounds, including important products of the degradation of natural materials, have been synthesised by instructive variations on one theme:

(*a*) Diethyl phthalimidomalonate is condensed with trimethylene bromide to give XX which, by the route shown, is converted into proline (*S. P. L. Sörensen* and *A. C. Andersen*, Z. physiol. Chem., 1908, **56,** 236; *cf. E. van Heyningen*, J. Amer. chem. Soc., 1954, **76,** 3043):

$BrCH_2 \cdot CH_2 \cdot CH_2 \cdot C(CO_2Et)_2 \cdot N(CO_2)_2C_6H_4$ (XX) $\xrightarrow{\text{hydrolysis}}$ $HO \cdot CH_2 \cdot CH_2 \cdot CH_2 \cdot CH(NH_2) \cdot CO_2H$ $\xrightarrow{HCl}$ Proline

(*b*) Diethyl γ-bromopropylmalonate is brominated to diethyl 1,3′-dibromopropylmalonate, from which hygric acid is obtained by the following route (*R. Willstätter* and *F. Ettlinger*, Ann., 1903, **326,** 91):

$BrCH_2 \cdot CH_2 \cdot CH_2 \cdot CBr(CO_2Et)_2$ $\xrightarrow{MeNH_2}$ N-methylpyrrolidine-2,2-$(CONHMe)_2$ $\xrightarrow{HCl}$ Hygric acid (N-Me, 2-CO_2H)

(*c*) Malonic ester is condensed with epichlorohydrin to give XXI, which is converted into hydroxypyroline by the following route (*H. Leuchs* and *J. F. Brewster*, Ber., 1913, **46,** 986):

(XXI) (lactone: $ClCH_2 \cdot CH$—O—$CO \cdot CH(CO_2Et)$—CH_2) $\xrightarrow{Cl_2;HCl}$ (lactone: $ClCH_2 \cdot CH$—O—$CO \cdot CHCl$—CH_2) $\xrightarrow{NH_3}$ Hydroxyproline (4-HO, 2-CO_2H, NH)

The preparation of 2,2,5,5-tetramethylpyrrolidine-3-carboxylic acid and a number of hydropyrrole derivatives from dibromotriacetonamine has been described (*H. Pauly et al*, *ibid*., 1899, **32,** 2000; 1901, **34,** 2287, 2289; Ann., 1902, **322,** 77).

Proline, pyrrolidine-2-carboxylic acid is a product of the hydrolysis of many proteins, acids yielding the L-form, which is largely racemised by alkalis. For its isolation from gelatine, and purification, see *J. Kapfhammer* and *R. Eck* (Z. physiol. Chem., 1927, **170,** 294) and *P. B. Hamilton* and *P. J. Ortiz* (J. biol. Chem., 1950, **184,** 607); use has been made of the insoluble salt with rhodanilic acid, $H[Cr(SCN)_4(NH_2Ph)_2]$ (*M. Bergmann*, *ibid*., 1935, **110,** 471). The most serviceable synthesis is probably the following (*N. F. Albertson* and *J. L. Fillman*, J. Amer. chem. Soc., 1949, **71,** 2818):

$HC{=}CH_2$ (CN) + $CH_2(CO_2Et)_2$ —NaOEt→ $H_2C{-}CH_2$ (NC, $CH(CO_2Et)_2$) —H_2, Raney Ni→ ethyl 2-oxopiperidine-3-carboxylate (H_2C, H_2C, $C H_2$, $CH{\cdot}CO_2Et$, CO, NH)

—SO_2Cl_2→ ethyl 3-chloro-2-oxopiperidine-3-carboxylate ($C(Cl)CO_2Et$, CO, NH) —HCl; NaOH→ proline (pyrrolidine ring, NH, CO_2H)

DL-*m*-Nitrobenzoylproline has been resolved (*E. Fischer* and *G. Zemplen*, Ber., 1909, **42,** 2992).

The L-acid, which is soluble in alcohol, has m.p. 220°, $[\alpha]_D$ −77.4 to −80.9° in water; when it is boiled in water at pH 7 with a little isatin and a thread of acetate rayon, the latter is coloured blue (*W. Grassmann* and *K. v. Arnim*, Ann., 1935, **519,** 192). The *ethyl ester* has b.p. 78°/13 mm; the *picrate*, m.p. 152–154°; *dioxopiperazine*, m.p. 146° (*Kapfhammer* and *A. Matthes*, Z. physiol. Chem., 1934, **223,** 43). This (−)proline has the same stereochemical configuration as the natural open-chain amino acids; pyroglutamic acid yields proline when reduced with sodium and alcohol. Several of the alkaloids of ergot afford D-proline on hydrolysis.

3,4-**Dehydropyrrolidine-2-carboxylic acid** (3,4-*dehydroproline*), m.p. 236–237°, $[\alpha]_D^{20}$ −375° (H_2O) (*R. V. Robinson* and *B. Witkop*, J. Amer. chem. Soc., 1962, **84,** 1697; *A. Corbella*, *P. Gariboldi* and *G. Jommi*, Chem. and Ind., 1969, 583).

(3,4-dehydropyrrolidine ring, NH, CO_2H)

The alkaloid hygrine yields on oxidation **hygric acid,** N-*methyl*-L-*proline*, m.p. 169°, which is also obtained by degradation of nicotine. **Stachydrine,** which occurs inactive in *Stachys tuberifera* (*A. v. Planta* and *E. Schulze*, Ber., 1893, **26,** 939), and in both DL- and L-forms in other plants, is the dimethylbetaine of proline, synthesised by methylation of L-proline (*J. W. Cornforth* and *A. J. Henry*, J. chem. Soc., 1952, 601). At 235° it is rearranged to methyl hygrate:

$$Me_2\overset{\oplus}{N}\cdot CH_2\cdot CH_2\cdot CH_2\cdot CH\cdot C\overset{\ominus}{O_2} \rightarrow MeN\cdot CH_2\cdot CH_2\cdot CH_2\cdot CH\cdot CO_2Me$$

The *hydrochloride*, $[\alpha]_D^{20°}$ −28.1 in water, has m.p. 222° (dec.), and the *picrate*, m.p. 199–200°.

Protein hydrolysates also afford L-**hydroxyproline** $[\alpha]_D$, −76.3°, m.p. 270°; *picrate*, m.p. 188° (*Kapfhammer et al., Hamilton* and *Ortiz, loc. cit.*); O,N-*dibenzoyl* deriv., m.p. 121° (*H. E. Carter* and *Y. H. Loo*, J. biol. Chem., 1948, **174**, 723); N-*acetyl-O-benzoyl* deriv., m.p. 185–186° (*R. L. M. Synge*, Biochem. J., 1939, **33**, 1924). The most effective synthesis (*R. Gaudry* and *C. Godin*, J. Amer. chem. Soc., 1954, **76**, 139) proceeds:

$$CH{=}CH_2 \ldots CH_2 \quad CH(CO_2Et)_2 \xrightarrow[HCl]{SO_2Cl_2;} CH_2Cl\text{-lactone (CH—CH}_2\text{, O, CO}\cdot CHCl) \xrightarrow[100°]{NH_4OH} \text{HO-pyrrolidine-}CO_2H$$

The synthetic acid is a mixture of diastereoisomers, hydroxyproline and allohydroxyproline, isolated as copper salts; each has been resolved. All the stereoisomers can however be prepared more easily by inversion from L-hydroxyproline (*D. S. Robinson* and *J. P. Greenstein*, J. biol. Chem., 1952, **195**, 383). **Phalloidine,** the toxic polypeptide of *Amanita phalloides*, contains (−)*allohydroxyproline*, m.p. 248°, $[\alpha]_D$ −58.1°. The two naturally occuring hydroxyprolines have the absolute configurations shown (*A. Neuberger*, J. chem. Soc., 1954, 429; *C. S. Hudson* and *Neuberger*, J. org. Chem., 1950, **15**, 24; *J. Zussman*, Brit. Abstr. **AI**, 1951, 166):

Hydroxyproline *allo*-Hydroxyproline

N-**Methylhydroxyproline** (4-*hydroxyhygric acid*) occurs in the bark of *Croton gubouga* (*J. A. Goodson* and *H. W. B. Clewer*, J. chem. Soc., 1919, **115**, 923). Methylation of hydroxyproline (*A. Küng*, Z. physiol. Chem., 1913, **85**, 217) gives the diastereoisomeric dimethylbetaines, **betonicine** ($[\alpha]_D$ −36.6°) and **turicine** ($[\alpha]_D$ +40.9°), isolated from *Betonica officinalis* (*E. Schulze* and *G. Trier*, Z. physiol. Chem., 1911, **76**, 258; 1912, **79**, 236), which differ in configuration about $C_{(2)}$. The fruit of *Courbonia virgata* affords diastereoisomeric, dextrorotatory 3-hydroxystachydrines, which are structurally distinct from betonicine and turicine, and can be dehydrated and reduced to stachydrine (*Cornforth* and *Henry*, J. chem. Soc., 1952, 597).

Methylamine and αα′-dibromoadipic acid give 1-*methylpyrrolidine*-2,5-*dicarboxylic acid*, dec. 274°; analogues similarly (*J. von Braun* and *J. Seemann*, Ber., 1923, **56**, 1840). **Kainic acid,** from *Digenea simplex*, m.p. 251° (dec.), $[\alpha]_D^{29°}$ −14.8°, is regarded as having the formula below (*S. Murakami et al.*, C.A., 1956, **50**, 4123):

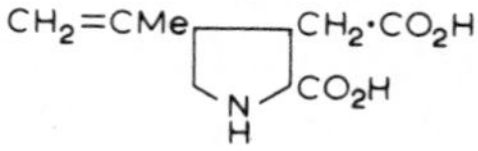

$CH_2{=}CMe$ ... $CH_2{\cdot}CO_2H$, CO_2H, NH

Kainic acid

4. Compounds having two or more independent five-membered rings

(a) Bipyrrolyls

N,N- or 1,1-**Bipyrrolyls** are synthesised from hydrazine and 1,4-diketones, and like 1-arylpyrroles, can exhibit molecular dissymmetry analogous to that of biphenyls (*R. Adams et al.*, J. Amer. chem. Soc., 1931, **53,** 374, 2353, 3519; 1934, **56,** 2089). Thus 2,2′-*dimethyl*-1,1-*bipyrrolyl*-3,3′-*dicarboxylic* acid (I), $[\alpha]_D^{20°}$ +27.5° and −25.1°, and the corresponding (2′-carboxyphenyl)-2-methylpyrrole-3-carboxylic acid (II) have been resolved and show high optical stability; analogues of II with an additional carboxyl group in the pyrrole ring, or with the carboxyl in the *meta*- or *para*-positions in the benzene ring, could not be resolved. In the case of the 1,3-dimethyl-4,6-bis(1′-pyrrolyl-3′-carboxy)-benzene (III), with two asymmetric centres, a racemic and a *meso* form have been isolated, and only the former could be resolved:

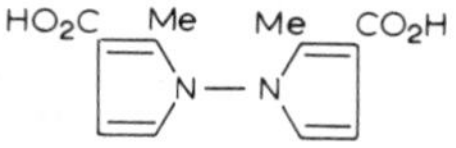

(I) 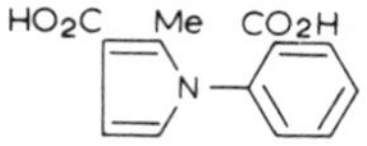 (II) 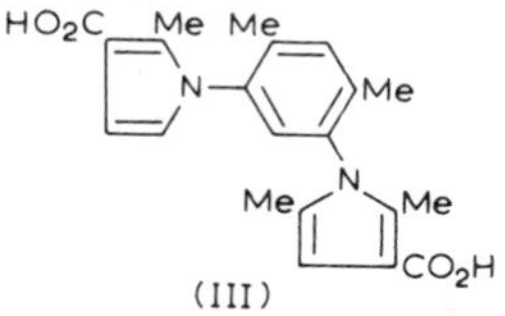 (III)

Octa-arylbipyrrolyls are known; they are obtained by treating appropriate alkali metal tetra-arylpyrroles with chlorine or bromine in an inert solvent (*e.g.* diethyl ether) under nitrogen. In solution they dissociate into deeply coloured reactive tetra-arylpyrrolyl radicals (*R. J. Abrahams* and *H. J. Bernstein*, Canad. J. Chem., 1959, 1056; *H. Zimmerman, H. Baumgartel* and *F. Bakke*, Angew. Chem., 1961, **63,** 808):

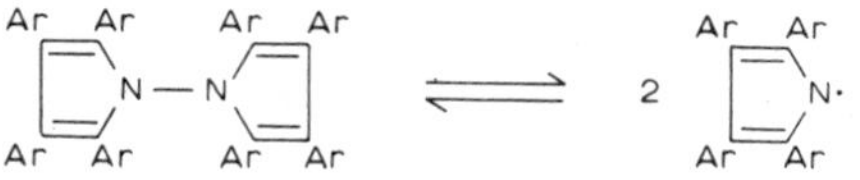

$$\text{Ar}_4\text{C}_4\text{N–NC}_4\text{Ar}_4 \rightleftharpoons 2\ \text{Ar}_4\text{C}_4\text{N}\cdot$$

The equilibrium has been measured spectroscopically.

The literature on 2,2′-bipyrrolyls has been summarised, and the synthesis and resolution of 1,3,5,1′,3′,5′-hexamethyl-2,2′-bipyrrolyl-4,4′-dicarboxylic acid described (*J. L. Webb* and *R. R. Threlkeld*, J. org. Chem., 1953, **18,** 1406, 1413).

1-Pyrroline reacts with pyrrole to afford 2,2′-pyrrolidinylpyrrole (*D. W. Fuhlhage* and *C. A. van der Werf*, J. Amer. chem. Soc., 1958, **80,** 6249), which is dehydrogenated in refluxing xylene in the presence of 5% palladium-on-carbon to give 2,2′-**bipyrrolyl,** m.p. 189–190° (*H. Rapoport* and *K. G. Holden*, *ibid.*, 1962, **84,** 635):

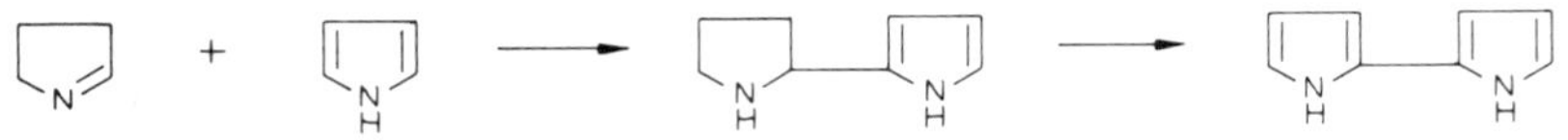

2,2′-Bipyrrolyl may also be obtained by the catalytic dehydrogenation of 2,2′-(1′-pyrrolinyl)pyrrole formed by the reaction between the complex from 2-oxopyrrolidine, dimethylformamide and phosphorus oxychloride, and pyrrole (*Rapoport* and *N. Castagnoli*, *ibid.*, 1962, **84,** 2178):

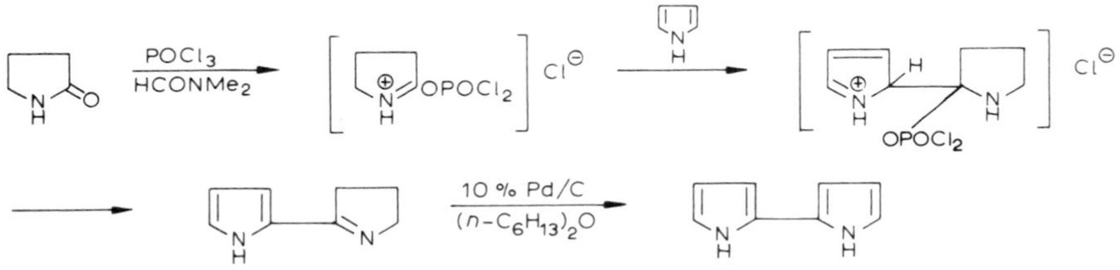

$POCl_3$ / $HCONMe_2$; $Cl^{\ominus}$; OPOCl_2; 10 % Pd/C / $(n\text{-}C_6H_{13})_2O$

Treatment of a 2,2′-bipyrrolyl with a 2-oxopyrrolidine gives a pyrrolinylbipyrrolyl, which on hydrogenation yields a 2,2′,2″-**terpyrrolyl** (*Rapoport*, *Castagnoli* and *Holden*, J. org. Chem., 1964, **29,** 883):

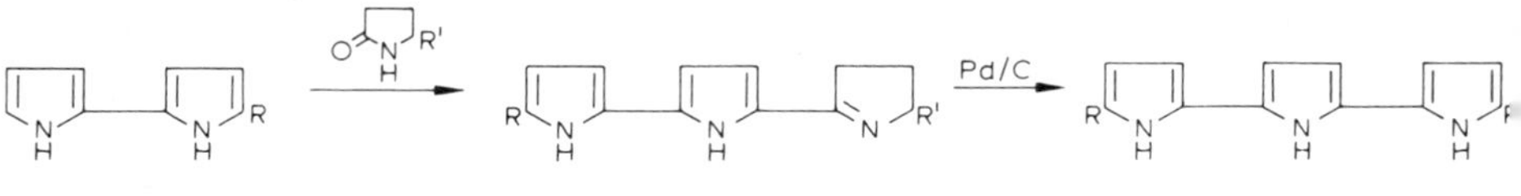

Pd/C

1,2′-(1′-**Pyrrolinyl)pyrrole** is obtained by heating together 2-methoxy-1-pyrroline and pyrrole; on heating it rearranges to 2,2′-(1′-**pyrrolinyl)pyrrole** (*R. Bonnett*, *K. S. Chan* and *I. A. D. Gale*, Canad. J. Chem., 1964, **42,** 1073):

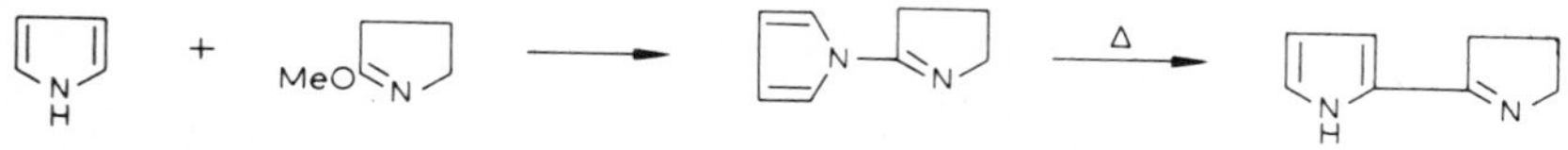

Polyoxobipyrrolidylidenes linked through a carbon–carbon double bond, having an indigoid-like structure, are known (*S. Ruhemann*, J. chem. Soc., 1906, **89,** 1236, 1847; *E. Benary* and *B. Silbermann*, Ber., 1913, **46,** 1364; *R. P. Linstead et al.*, J. chem. Soc., 1956, 244).

Pyrrole rings are connected indirectly through nitrogen in carbonylbispyrrole (p. 341), and the pyrrole and furan rings similarly in N-**furfurylpyrrole,** $C_4H_3O \cdot CH_2NC_4H_4$. b.p. 76–78°/1 mm, obtained from furfurylamine mucate (*T. Reichstein*, Helv., 1930, **13,** 353).

(b) Compounds having two or more pyrrole rings linked through methylene or methine groups

There are a number of compounds formed from four pyrrolylene residues linked *via* the 2-positions through four methine groups. The parent compound, porphin, does not occur naturally, but substituted derivatives known as porphyrins are widely distributed in nature both in the free form and as complexes with iron *e.g.*, haemoglobin which on hydrolysis gives haemin. Chlorins are dihydroporphyrins and are related to the chlorophylls, and the tetrahydroporphyrins to bacteriochlorophyll. The bile pigments all consist of a chain of four pyrrole rings linked through a carbon atom.

Porphin

These compounds will be described in a subsequent chapter dealing with pyrrole pigments.

(i) Dipyrrolylmethanes

2,2′-Dipyrrolylmethanes are important intermediates in the synthesis of porphyrins. 2,2′-**Dipyrrolylmethane,** m.p. 73°, which resinifies in air, is best obtained by Kischner–Wolff reduction of the corresponding ketone (*K. Hess* and *F. Anselm*, Ber., 1921, **54,** 2319).

A more generally used method is to boil 2-bromomethylpyrroles with water or alcohol (*Fischer et al.*, Ann., 1926, **447,** 123; **450,** 201):

$$RCH_2Br + H_2O \rightarrow RH + CH_2O + HBr$$
$$2\ RH + CH_2O \rightarrow CH_2R_2 + H_2O$$

Symmetrical 2,2′- and 3,3′-dipyrrolylmethanes are produced by condensing a pyrrole with formaldehyde, or other aldehydes or ketones, usually in the presence of acid (*Fischer* and *E. Bartholomäus*, Z. physiol. Chem., 1913, **83,** 50; **87,** 255; *Fischer* and *F. Schubert*, *ibid.*, 1926, **155,** 72).

Unsymmetrical dipyrrolylmethanes are obtained by treating 2-bromomethylpyrroles with pyrrolylmagnesium halides or with pyrroles having a free 2- or 3-position, in hydrobromic acid (*Fischer* and *P. Halbig*, Ann., 1926, **447,** 123) or by reacting 2-bromomethylpyrroles with pyrrole-2-carboxylic acids. Attempts to extend the process to include three pyrrole rings were unsuccessful however (*A. Hayes*, *G. W. Kenner* and *N. R. Williams*, J. chem. Soc., 1958, 3779):

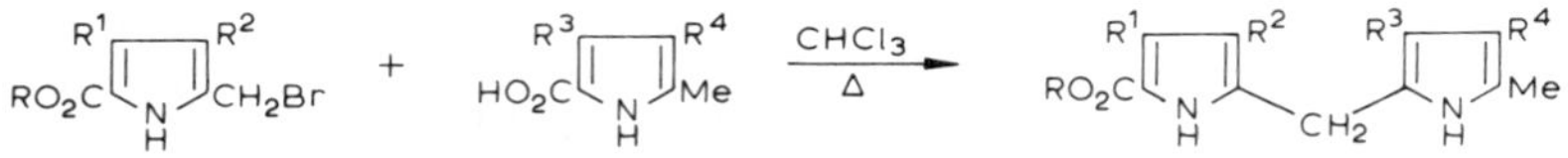

Dipyrrolylmethanes without nuclear stabilising groups, *e.g.* compound IV, have been prepared and shown to undergo characteristic pyrrole reactions without modification to the methylene group (*S. F. MacDonald*, J. Amer. chem. Soc., 1957, **79,** 2659):

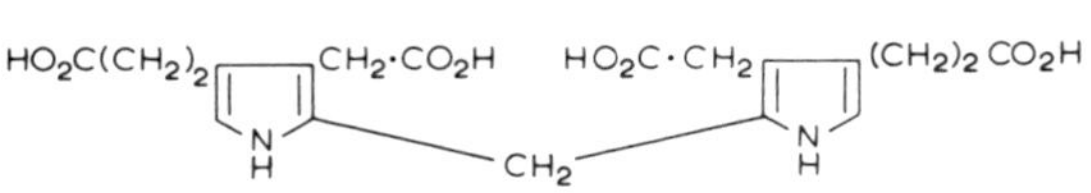

(IV)

Esters of 4-substituted 3,5-dimethylpyrrole-2-carboxylic acids have been used by the route indicated to obtain a variety of dipyrrolylmethanes, dipyrromethenes and porphyrins (*A. W. Johnson et al.*, J. chem. Soc., 1958, 4254; 1959, 3416):

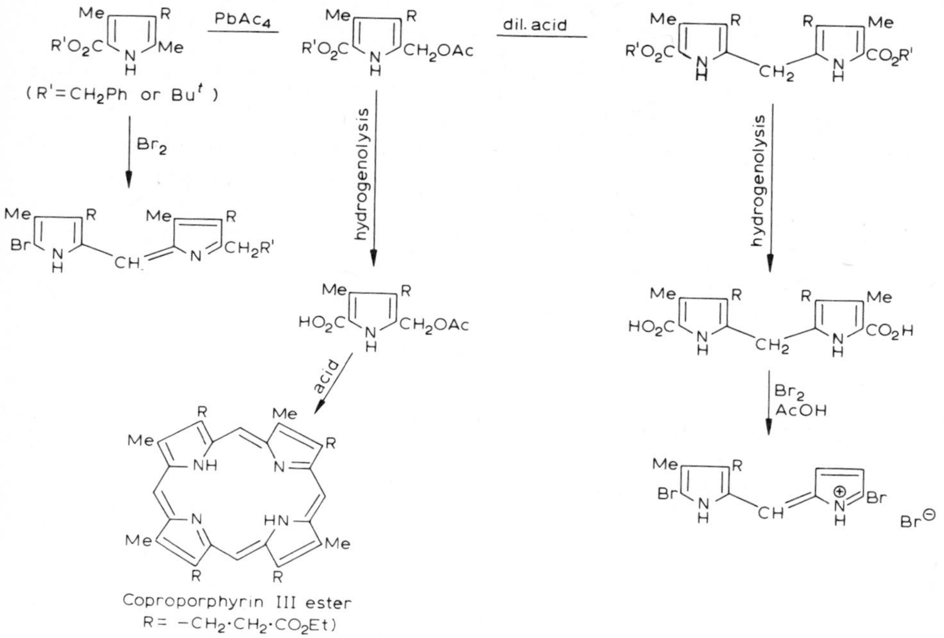

The adducts of 2-dimethylaminocarbonylpyrroles condense with pyrroles having an unsubstituted 2-position. The products on hydrolysis give 2,2′-dipyrrolyl ketones which are readily reduced with diborane to dipyrrolylmethanes (*J. A. Ballantine et al.*, Tetrahedron Suppl. 7, p. 241).

(ii) Pyrromethenes

The colourless dipyrrolylmethanes of which the most important are the bilirubic acids, yield dipyrromethenes on oxidation and with hydriodic acid fission is effected in both possible directions to a pyrrole unsubstituted in the 2-position and a 2-methylated pyrrole *(Fischer* and *Bartholomäus, loc. cit.)* thus:

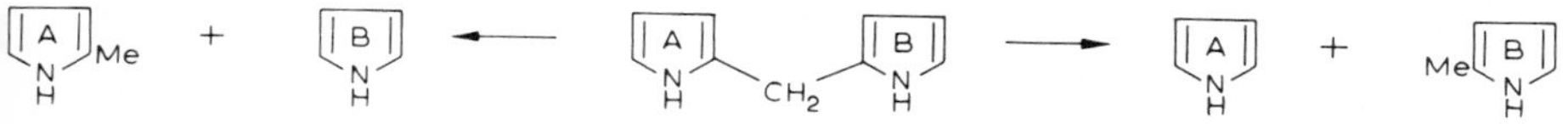

Dipyrrolylcarbinols are intermediate products in the synthesis of dipyrromethenes from pyrrolecarbaldehydes and pyrroles in the presence of acids (*Fischer* and *G. Fries*, Z. physiol. Chem., 1935, **231,** 243):

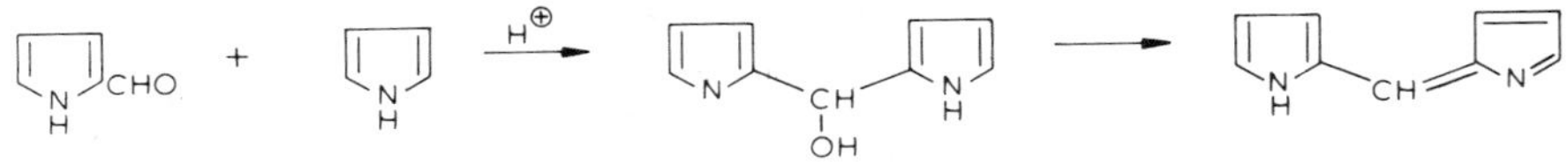

The cationoid reactivity of the methine bridge has been confirmed by the successful Michael addition of ethyl cyanoacetate to certain pyrromethenes (*A. C. Jain* and *Kenner*, J. chem. Soc., 1959, 185):

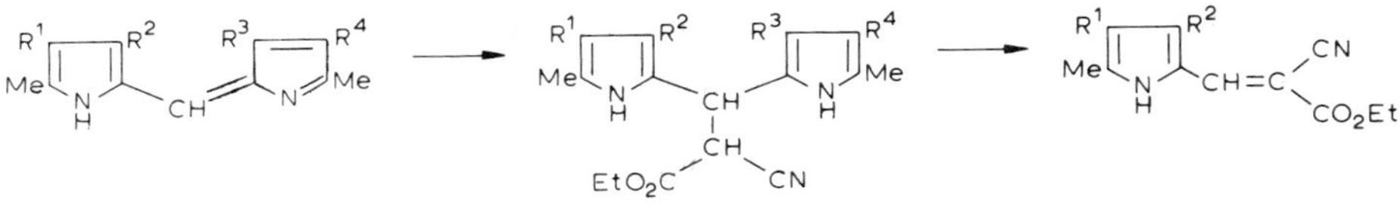

Undecylprodigiosin (V), a C-25 prodigiosin analogue isolated from a strain of *Streptomyces* has been synthesised (*H. H. Wasserman, G. C. Rodgers* and *D. D. Keith*, Chem. Comm., 1966, 825):

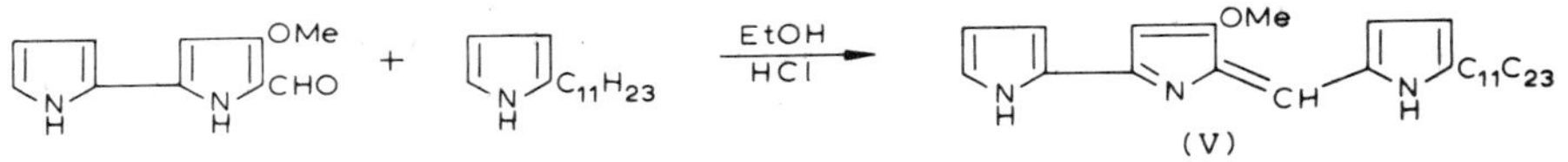

(V)

Metacycloprodigiosin, a brilliant red pigment also isolated from the same strain of *Streptomyces*, has the tripyrrole structure VI, which has been confirmed by synthesis (*idem, ibid.*, 1969, **91,** 1263; *Wasserman, Keith* and *J. Nadelson, ibid.*, p. 1264):

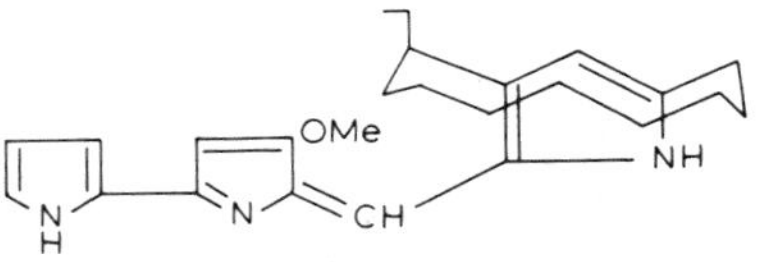

(VI)

Chapter 5

Compounds Containing Five-Membered Rings with One Hetero Atom from Group V: Nitrogen; Fused-Ring Compounds

R. LIVINGSTONE

1. Indoles (2,3-benzopyrroles)* and 3*H*-indoles

In his researches on the degradation products of indigo *Adolf Baeyer* isolated two compounds, oxindole and dioxindole. These he regarded as mono- and di-hydroxy derivatives of a parent compound C_8H_7N which he named *indole* and obtained by distilling oxindole with zinc dust (*Baeyer* and *W. Knop*, Ann., 1866, **140,** 1). Indole occurs with its homologues in coal tar, and its nucleus is present in many vat dyestuffs. The biologically indispensable amino acid tryptophan (β-3-indolylalanine), is the source of indole derivatives formed from proteins by putrefactive, digestive or violent chemical degradation. It is no doubt related biogenetically to numerous indole natural products: alkaloids, toad poisons, the plant growth hormone heteroauxin. The structure of indole as benzopyrrole was deduced by *Baeyer* (Ann. Suppl., 1870, **7,** 56; Ber., 1870, **3,** 517; 1880, **12,** 2254; 1900, **33,** S. LI–LVIII) and is shown by its oxidative degradation to anthranilic acid, and by many syntheses, notably that from *o*-amino-ω-chlorostyrene (*A. Lipp*, *ibid.*, 1884, **17,** 1067):

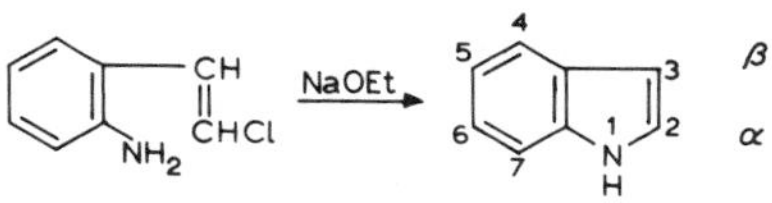

CH=CHCl, NH$_2$ $\xrightarrow{NaOEt}$ indole (positions 1–7; N–H at 1; β = 3, α = 2)

---

* General reviews: *R. B. van Order* and *H. G. Lindwall*, Chem. Reviews, 1942, **30,** 69; *J. D. White* and *M. E. Mann*, Adv. Heterocyclic Chem., 1969, **10,** 113; Early work: *G. Ciamician*, Ber., 1904, **37,** 4227; *J. Martinet*, "Matières Colorantes: l'indigo et ses dérivés", Baillière et Fils, Paris, 1926.

Nomenclature. Three fundamental structures are involved in naming 2,3-benzopyrroles and 2,3-benzopyrrolines in accordance with the I.U.P.A.C. Rules, namely:

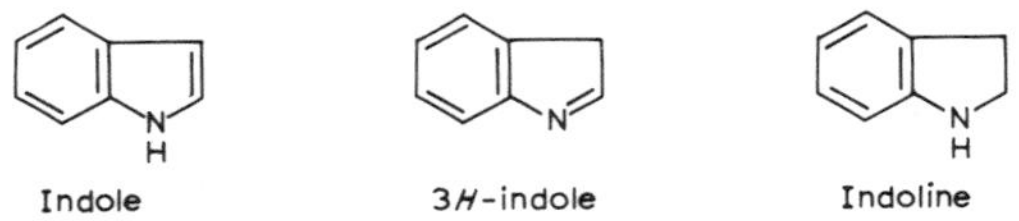

Indole 3*H*-indole Indoline

The systematic name 3*H*-indole replaces the older trivial names *indolenine*, which is still in current use although not officially acceptable, and *pseudoindole*, which is now obsolete. In many cases it is still uncertain whether particular compounds have an indole or a 3*H*-indole structure and many compounds behave as if they were hybrids. In the present account the two types are discussed under separate headings where overwhelming evidence for the 3*H*-indole exists.

(a) General synthetic methods

The following processes lead to indoles having no hydroxyl or carbonyl groups in the pyrrole nucleus. Processes leading to these latter compounds are dealt with under the appropriate derivatives.

(i) The Fischer indole synthesis

This synthesis has been the subject of much experimental work and is now the most versatile method for the preparation of indoles. The synthesis, discovered by *Emil Fischer* (Ber., 1886, **19,** 1563; Ann., 1886, **236,** 116), can be regarded as the elimination of ammonia from the arylhydrazone of an aldehyde or ketone by treatment with an acid or various metal and anhydrous metal salt catalysts. The following mechanism is well supported by numerous experimental observations (*R. B. Carlin* and *E. E. Fischer*, J. Amer. chem. Soc., 1948, **70,** 3421; *B. Robinson*, Chem. Reviews, 1963, **63,** 373):

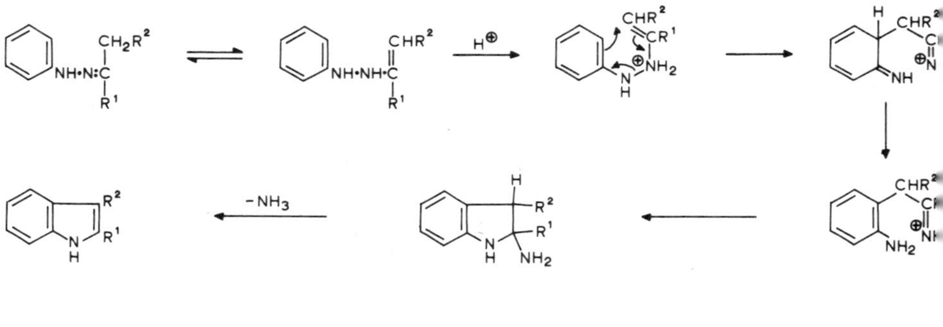

and is basically the same as the one proposed in 1918, *i.e.*, an *o*-benzidine conversion of the enehydrazine tautomer of the arylhydrazone followed by formation of the indole ring (*G. M. Robinson* and *R. Robinson*, J. chem. Soc., 1918, **113,** 639; *ibid.*, 1924, **125,** 827). Evidence that the nitrogen atom linked to the aryl group is retained and the other eliminated as ammonia has been obtained by the use of isotopically labelled compounds (*C. F. H. Allen* and *C. V. Wilson*, J. Amer. chem. Soc., 1943, **65,** 611; *K. Clusius* and *H. R. Weisser*, Helv., 1952, **35,** 400). In the preparation outlined below the postulated intermediate was isolated by passing dry hydrogen chloride into an ethanolic solution of butyrolactone phenylhydrazone, when the hydrochloride separated owing to its insolubility in the reaction medium. Subsequent heating with acid gave the indole (*H. Plieninger* and *I. Nógrádi*, Ber., 1950, **83,** 273; 1955, **88,** 1964):

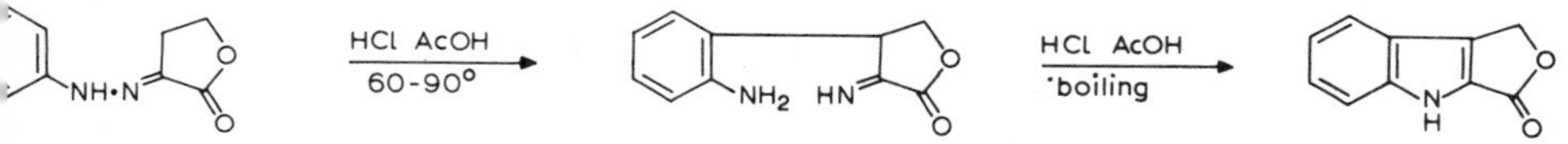

The cyclisation of arylhydrazones containing electron-releasing *meta*-substituents produces a preponderance of 6- over 4-substituted-indoles; for electron-attracting substituents the opposite is usually true (*D. W. Ockenden* and *K. Schofield*, J. chem. Soc., 1957, 3175).

Originally the arylhydrazone was fused with excess of zinc chloride, but this method can be improved by the use of a high-boiling solvent such as methylnaphthalene (*Ges. für Teerverwertung*, G.P. 238,138/1910; Chem. Ztbl., 1911, II, 1080), and it often suffices to use smaller quantities of zinc chloride or another metallic halide (*A. E. Arbusow et al.*, Ber., 1910, **43,** 2301; 1935, **68,** 1792); good results have been obtained with phosphoric acid and it has been shown to have fairly wide applicability (*H. M. Kissman, D. W. Farnsworth* and *B. Witkop*, J. Amer. chem. Soc., 1952, **74,** 3948). A number of substituted indoles can be obtained by heating the corresponding arylhydrazones in ethylene glycol, diethylene glycol or tetralin (*J. T. Fitzpatrick* and *R. D. Hiser*, J. org. Chem., 1957, **22,** 1703).

It is now common practice to subject an equimolar mixture of the arylhydrazine (or a salt) and aldehyde or ketone directly to indolization conditions, without the isolation of the arylhydrazone (*B. Robinson*, Chem. Reviews, 1969, **69,** 227). Hydrazones derived from *N,N*-disubstituted hydrazine or from easily enolisable ketones such as cyclohexanone or pyruvic acid react more easily on heating with zinc chloride or hydrogen chloride in alcohol, or even on boiling with acetic acid. If the original carbonyl group is linked to a secondary group normal indole formation is blocked and a 3*H*-indole (p. 439) is obtained:

The phenylhydrazone of a ketone of the type $RCH_2{\cdot}CO{\cdot}CHR'_2$ often yields the 3*H*-indole in preference to the alternative indole (*J. McLean et al.*, J. chem. Soc., 1955, 2519). If, however, the hydrazine is also *N,N*-disubstituted the product is an indolinol, which, when the nature of R permits, undergoes dehydration to a methyleneindoline:

With polyphosphoric acid phenylhydrazones of type I give primarily 2-substituted indoles of type II, but with zinc chloride 3*H*-indoles (III) are formed (*R. E. Lyle* and *L. Skarlos*, Chem. Comm., 1966, 644):

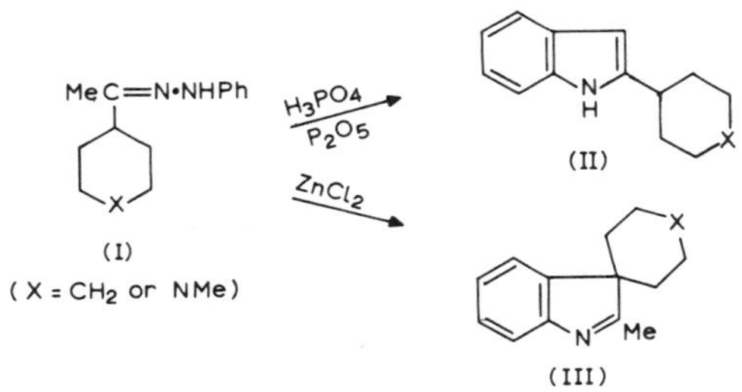

(I)
(X = CH_2 or NMe)

(II)

(III)

Hydrazone IV under Fischer conditions gives more of the quinolone than of the expected indole (*H. Rapoport* and *J. R. Tretter*, J. Amer. chem. Soc., 1958, **80,** 5574):

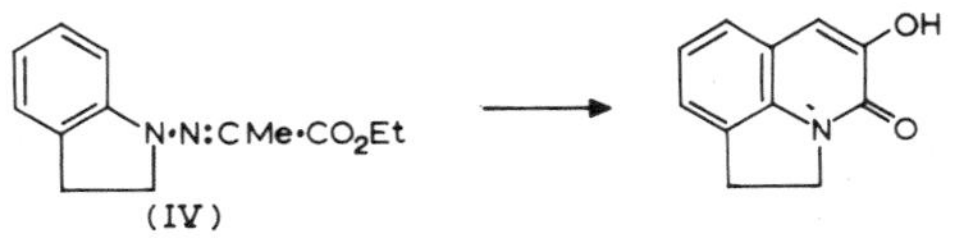

(IV)

An indole has been obtained from a dialkylaryl-substituted vinylhydrazine (*P. W. Neber et al.*, Ann., 1929, **471,** 113):

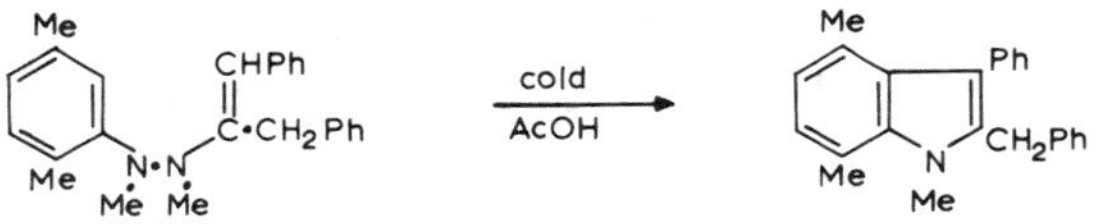

In the Japp–Klingemann reaction for making arylhydrazones an aryldiazonium cation reacts with an activated methinyl group to yield an intermediate azo compound which subsequently yields the hydrazone by loss of one of the original carbon substituents upon warming alone or in alkaline or mildly acidic media. Spontaneous indolisation occurs where possible under more vigorous acidic conditions (*B. Robinson, loc. cit.*, p. 233):

$$RC_6H_4N_2^{\oplus} + R^1R^2R^3C^{\ominus} \rightarrow RC_6H_4{\cdot}N{:}N{\cdot}CR^1R^2R^3 \rightarrow RC_6H_4{\cdot}NH{\cdot}N{:}CR^2R^3$$

In an important special case of the Fischer synthesis, the hydrazone a product of the Japp–Klingemann reaction affords an indole with a 2-carbonyl substituent (*G. K. Hughes* and *F. Lions*, J. Proc. roy. Soc. N.S.W., 1938, **71**, 475):

PhN$_2$Cl + AcCMe·CO$_2$Et (Na) → PhNH·N:CMe·CO$_2$Et → (indole, NH, 2-CO$_2$Et)

The products of the Fischer indole synthesis on the unsymmetrical ketones MeCO·CHR^{1}R^2 (R^1 = H or Me; R^2 = Me or Ar) vary with the acid catalyst, and particularly with the concentration of phosphoric oxide dissolved in orthophosphoric acid (*M. H. Palmer* and *P. S. McIntyre*, J. chem. Soc., B, 1969, 446).

(ii) The Bischler synthesis

In this synthesis (*A. Bischler* and *H. Brion*, Ber., 1892, **25**, 2860; *Bischler* and *P. Firemann*, *ibid.*, 1893, **26**, 1336) an α-arylaminoketone is heated with an acid reagent; it is equally feasible to heat an α-halogenoketone with an arylamine:

PhNH$_2$ + BrCH$_2$COPh → PhNHCH$_2$COPh + HBr → (indole, NH, 2-Ph)

Phenacylaniline on heating with aniline hydrochloride gives a 74% yield of 2-phenylindole; alternatively, benzoin is heated with an arylamine and an arylamine hydrohalide:

PhNH$_2$ + HOCHPh·COPh → (indole, NH, 2-Ph, 3-Ph)

This method encounters the complication that the arylaminoketone may undergo structural change in the presence or absence of added amine and the amine moiety may be exchanged if the added amine differs from the arylamine incorporated in the phenacyl compound:

RCO–R′CHNHAr $\xrightarrow{Ar'NH_2,\ HBr}$ [RCNHAr′ ═ R′CNHAr ?] → RCHNHAr′–RCO

These complications are less frequent when a *tert*-arylaminoketone is used, and may be excluded when R = R′ (*P. E. Verkade, E. F. J. Janetzky et al.*, Rec. Trav. chim., 1943, **62,** 763, 774; 1944, **63,** 123; 1945, **64,** 129, 139; 1946, **65,** 691, 905, 912; 1947, **66,** 317; *P. L. Julian et al.*, J. Amer. chem. Soc., 1945, **67,** 1203; *F. G. Mann et al.*, J. chem. Soc., 1943, 58; 1948, 847, 858; *R. M. Cowper* and *T. S. Stevens*, *ibid.*, 1947, 1041). A suitably substituted oxo-base may give a 3*H*-indole (*M. Garry*, Ann. Chim., 1942, **17,** 5):

COR^2 CRR^1 N H → RR^1 R^2 N

(iii) The Madelung synthesis

Here very easily accessible materials are submitted to drastic conditions; acyl-*o*-toluidines yield indoles when treated with sodamide or potassium *tert*-butoxide at high temperatures (*W. Madelung*, Ber., 1912, **25,** 1128, 3521; Ann., 1914, **404,** 1; G.P. 262,327/1911; *A. Verley*, Bull. Soc. chim. Fr., 1924, [iv], **35,** 1039; 1925, **37,** 189; *Allen* and *J. van Allan*, Org. Synth., 1942, **22,** 94; *L. Marion* and *W. R. Ashford*, Canad. J. Res., 1945, **23B,** 26; *F. Tyson*, J. Amer. chem. Soc., 1941, **63,** 2024):

Me N H COMe —($NaNH_2$, 250°)→ N H Me

In a modified Madelung synthesis, 2-methylindole is obtained readily by heating *N*-acetyl-*o*-toluidine with sodium and copper in boiling diethylaniline (*H. A. Piggott* and *E. H. Rodd*, B.P. 330,332/1929) and other 2-alkylindoles similarly.

(iv) Synthesis from nitro compounds with ortho *side-chains or their reduction products*

(*1*) Several methods of obtaining indoles are based on the interaction of a nitro group, or the product of its reduction, and the β-carbon atom of an *ortho* side-chain, *e.g.*, *o*-nitrobenzyl ketones. *o*-Nitrophenylpyruvic acid gives indole-2-carboxylic acid (*A. Reissert*, Ber., 1897, **30,** 1030; *W. O. Kermack et al.*, J. chem. Soc., 1921, **119,** 1625); occasionally *N*-hydroxyindoles are obtained, probably derived from a hydroxyamino-ketone:

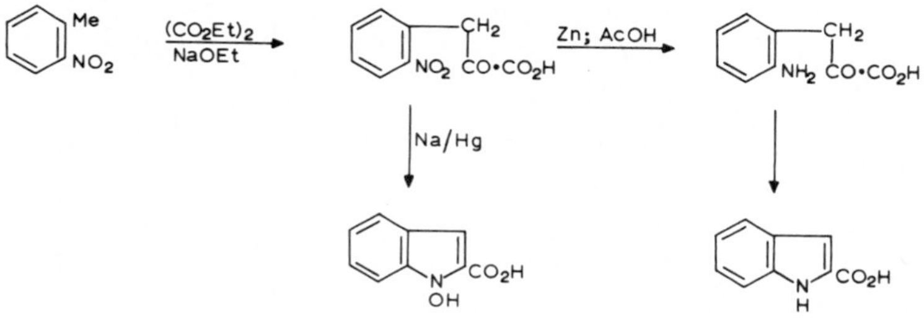

(*2*) Catalytic reduction of 2-nitrobenzyl cyanides over palladium-on-carbon leads to indoles (*H. R. Snyder et al.*, J. Amer. chem. Soc., 1958, **80,** 4622).

(*3*) *o*-Chloronitrobenzenes undergo smooth nucleophilic substitution by the potassium derivatives of ethyl cyanoacetate and diethyl malonate in *tert*-butyl alcohol, to give the *o*-nitrophenyl derivatives of these esters which are reductively cyclised to indoles and oxindoles, respectively (*C. A. Grob* and *O. Weissback*, Helv., 1961, **44,** 1748).

(v) Synthesis from benzoquinones and their derivatives

(*1*) *The Nenitzescu synthesis.* This synthesis of 5-hydroxyindoles from the reaction between *p*-benzoquinones and *β*-aminocrotonates has been developed by *A. Robertson et al.* (J. chem. Soc., 1951, 2029; 1953, 1262) and its mechanism clarified (*C. R. Allen Jr.* and *M. J. Weiss*, Chem. and Ind., 1966, 117):

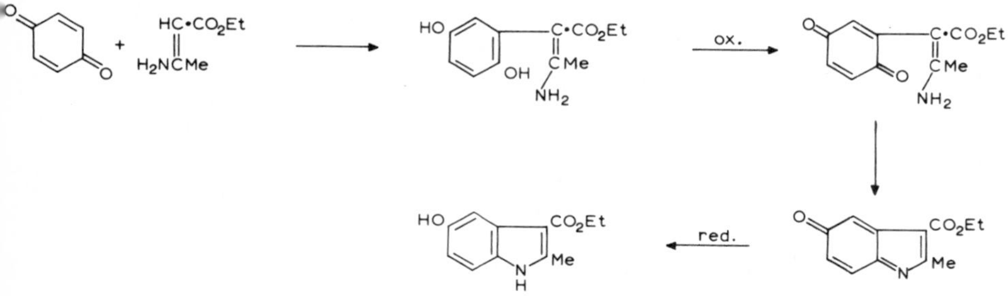

(*2*) Indoles have been obtained from tetrahalogeno-*o*-benzoquinones and *β*-aminocrotonic esters (*W. Reid* and *P. Weidermann*, Ber., 1969, **102,** 2684).

(*3*) Indole derivatives have been prepared by acid-catalysed cyclisation of the adducts obtained from 1,3-diketones and *p*-benzoquinone dibenzoylimide (*R. Adam, L. M. Werbel* and *M. D. Nair*, J. Amer. chem. Soc., 1958, **80,** 3291):

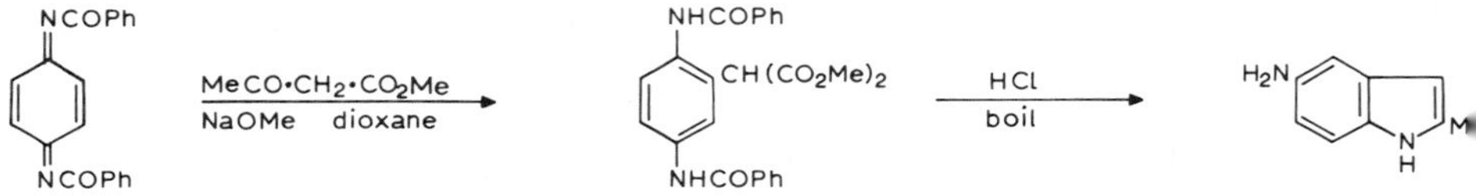

NCOPh NCOPh $\xrightarrow[\text{NaOMe dioxane}]{\text{MeCO·CH}_2\text{·CO}_2\text{Me}}$ NHCOPh CH$(CO_2Me)_2$ NHCOPh $\xrightarrow[\text{boil}]{\text{HCl}}$ H_2N N H Me

(vi) From ketones and isocyanides

Indoles may be obtained by the reaction between an aromatic ketone and an isocyanide in the presence of boron trifluoride (*B. Zeeh*, Tetrahedron Letters, 1967, 3881):

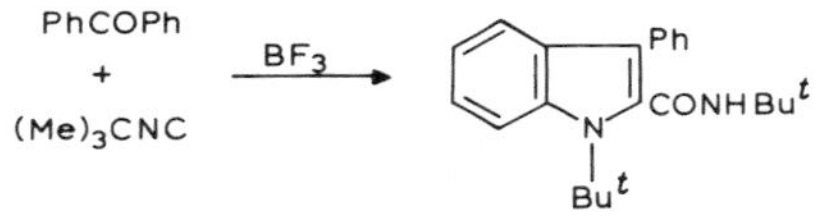

PhCOPh + $(Me)_3CNC$ $\xrightarrow{BF_3}$ Ph CONHBu^t N Bu^t

(vii) From pyrroles

Indoles may be produced by heating some pyrrole homologues with zinc acetate and acetic acid, or by heating the dimers of these homologues with sulphuric acid (*M. Dennstedt*, Ber., 1888, **24**, 3429) and from pyrroles and acetonylacetone (*Allen et al.*, J. org. Chem., 1938, **2**, 235):

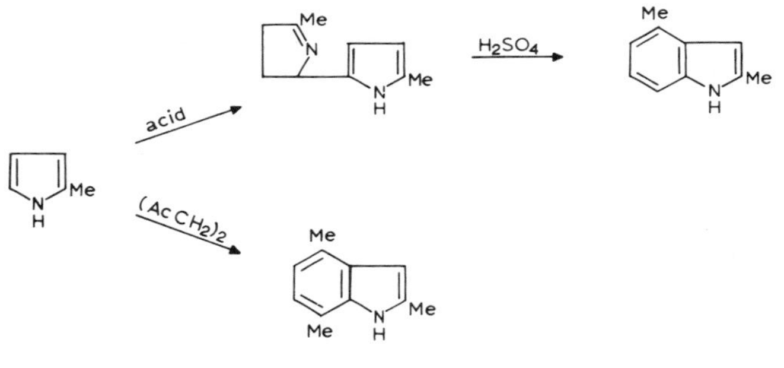

Me N H $\xrightarrow{\text{acid}}$ Me N N H Me $\xrightarrow{H_2SO_4}$ Me N H Me

$\xrightarrow{(AcCH_2)_2}$ Me N H Me Me

(viii) From o*-aminocarbonyl compounds*

Indoles may be synthesised by the reaction between dimethylsulphonium methylide and an aromatic *o*-aminocarbonyl compound (*P. Bravo, G. Gaudiano* and *A. Umani-Ranchi*, Tetrahedron Letters, 1969, 679):

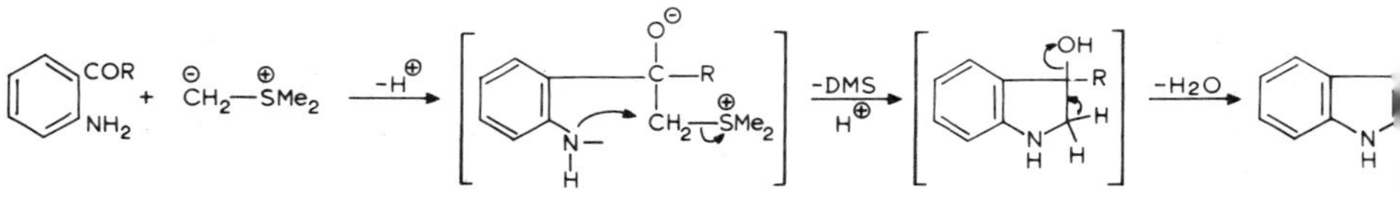

COR NH_2 + $\overset{\ominus}{C}H_2-\overset{\oplus}{S}Me_2$ $\xrightarrow{-H^{\oplus}}$ [$O^{\ominus}$ C—R $CH_2-\overset{\oplus}{S}Me_2$ N— H] $\xrightarrow[H^{\oplus}]{\text{-DMS}}$ [OH R H N H H] $\xrightarrow{-H_2O}$ N H

(ix) From δ-sultams

Benzo analogues of δ-sultams afford indoles on pyrolysis (*B. Helferich et al.*, Ann., 1961, **646,** 32, 45).

(b) General properties and reactions of indoles

Indoles closely resemble pyrroles in the feeble basicity of the nitrogen and the acidity of the imino-hydrogen atom.

(i) Reduction

Indoles are reduced to indolines catalytically, electrolytically or less generally by metal–acid combinations.

Over a nickel catalyst there may be much ring-fission to an *ortho*-substituted aniline and sometimes preferential reduction of the benzene ring (*J. von Braun et al.*, Ber., 1924, **57,** 392). Indole is reduced by dissolved metals in liquid ammonia, in the presence of alcohol or ammonium chloride to give a mixture of 4,5,6,7-tetrahydro- and 4,7-dihydro-indole. The *N*-alkyl derivatives are reduced more readily than indole, possibly because reduction of the latter is hindered by salt-formation (*S. O'Brien* and *D. C. C. Smith*, J. chem. Soc., 1960, 4609). In the case of some indoles the course of reduction with lithium and methanol in liquid ammonia depends upon when the alcohol is added to the reaction mixture; 5-methoxy-1-methylindole gives 5-methoxy-1-methylindoline if methanol is added near the end of the reaction and 5-methoxy-1-methyl-4,7-dihydroindole as the major product if methanol is present throughout the reaction (*W. A. Remers et al.*, J. Amer. chem. Soc., 1967, **89,** 5513).

(ii) Addition reactions and polymerisation

Normal addition of maleic anhydride is not observed (*O. Diels* and *K. Alder*, Ann., 1931, **490,** 277). At 0° in ethyl acetate the formation of a 2:1 molar adduct of indole to maleic anhydride is reported, while at room temperature or at boiling point a 3:1 molar adduct is obtained (*R. M. Acheson, R. S. Feinberg* and *A. R. Hands*, J. chem. Soc., 1964, 526).

Other additive reactions are rare, apart from the acid-catalysed polymerisation of indoles. Depending on experimental conditions indole yields a dimer or a trimer.

The structure of indole dimer has been deduced by systematic degradation (*H. F. Hodson* and *G. F. Smith*, *ibid*., 1957, 3544), and that of indole trimer by the fact that it was found to contain an aromatic amino group (*Smith*, Chem. and Ind., 1954, 1451), which led to a proposed structure and subsequent synthesis (*W. E. Noland* and *W. C. Kuryla*, J. org. Chem., 1960, **25,** 486):

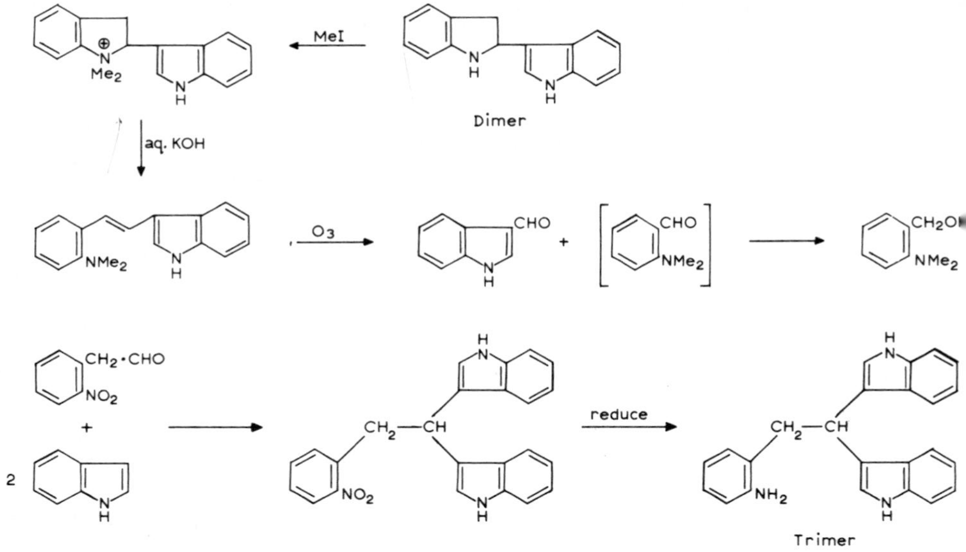

MeI

Dimer

aq. KOH

O_3

reduce

Trimer

The mechanism for the formation of the dimer can be explained by the protonation of a molecule of indole at the 3-position to give a cation, which will attack the 2-position of a neutral indole molecule. Protonation of the 3-position of the indole ring in the dimer leads to the intermediate cation, then to elimination and a reversal of the dimerisation:

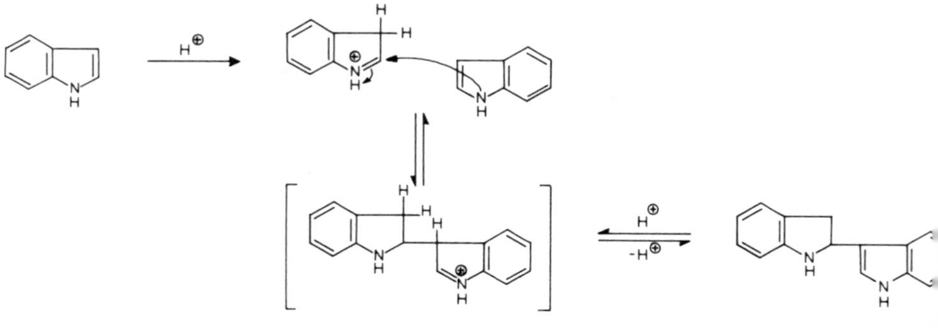

The protonation of the 3-position has been shown by deuterium exchange studies (*M. Koizumi* and *T. Titani*, Bull. chem. Soc. Japan, 1938, **13,** 307) and by nuclear magnetic resonance and ultraviolet absorption spectral studies (*R. L. Hinman* and *E. B. Whipple*, J. Amer. chem. Soc., 1962, **84,** 2534).

Compound V is however obtained by the acid-catalysed reaction between 1,3-dimethylindole and mesityl oxide (*D. A. Cockerill, Sir R. Robinson* and *J. E. Saxton*, J. chem. Soc., 1955, 4369; *B. Robinson* and *G. F. Smith, ibid.*, 1960, 4574):

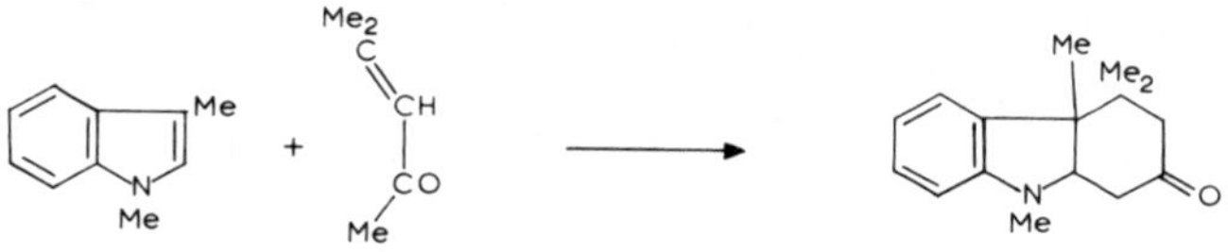

(V)

and treatment of ester VI with aqueous acid results in the cleavage of the indole ring and the formation of 7-phenacylindoline (VII). The driving force for the conversion is believed to be relief of steric strain, as simple indoles do not undergo this reaction (*J. Blake et al.*, J. Amer. chem. Soc., 1966, **88,** 4061):

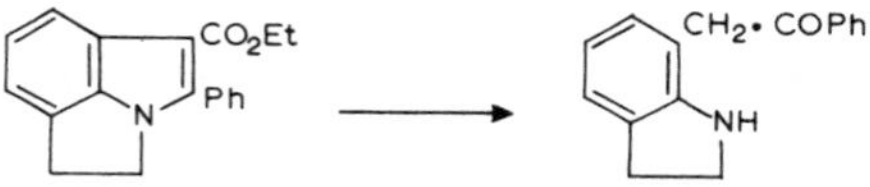

(VI) (VII)

(iii) Substitution reactions

The very easy reactions involving electrophilic substitution of pyrroles are all applicable to indoles and take place mainly at C-3, but if this position already contains a substituent then 2,3-disubstituted indoles may be formed. Calculations by molecular-orbital methods of the π-electron densities at the various atoms in the indole ring are in agreement with experimental observations and show that the 3-position is the most susceptible to electrophilic attack. Indication has been obtained that direct substitution probably does not occur at C-2, but rather a 3,3-disubstituted 3*H*-indole is first formed which rearranges to give the product (*A. H. Jackson* and *P. Smith*, Chem. Comm., 1967, 264). The reactivity of the 3-position is shown by the Mannich reaction (p. 429); the reaction with an olefinic bond to give a 3-indolyl derivative, *e.g.*, the combination with nitroethylenes (*Noland et al.*, J. Amer. chem. Soc., 1955, **77,** 456); and the first stage in the formation of highly coloured products with *o*- and *p*-quinones (*J. D. Bu'Lock* and *J. Harley-Mason*, J. chem. Soc., 1951, 703):

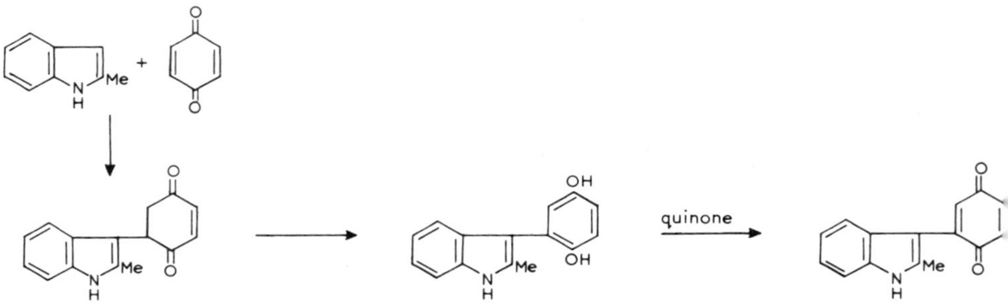

(iv) Metal compounds

Indole forms 1-indolylmagnesium halides (Grignard reagents) which are alkylated and carbonated mainly at the 3-position, while alkali metal derivatives afford predominantly 1-substituted derivatives; 1-methylindole gives mainly the 2-lithium compound (*D. A. Shirley* and *P. A. Roussel*, J. Amer. chem. Soc., 1953, **75**, 375).

The alkali metal derivatives of indole protonate predominantly on nitrogen, but the indolylmagnesium halides are the only organometallic derivatives which give substantial protonation on carbon. The extent of protonation at C-3 of the indole Grignard reagent is dependent upon the amount of water added; up to 75% exchange with a moderate excess of water, while swamping with water virtually leads to complete *N*-protonation (*J. C. Powers, W. P. Meyer* and *T. G. Parsons, ibid.*, 1967, **89**, 5812). The n.m.r. spectrum of indolylmagnesium bromide in tetrahydrofuran indicates that the compound is largely ionised (*M. G. Reinecke, H. W. Johnson* and *J. F. Sebastian*, Tetrahedron Letters, 1963, 1183).

For a review on indole Grignard reagents see *R. A. Heacock* and *S. Kasparek* (Adv. heterocyclic Chem., 1969, **10**, 43).

(v) Oxidation

Oxidative degradation, which in the pyrrole series leads by α-attack to maleimides, may transform indoles *via* the 3-hydroxy derivatives into indigos or substituted anthranilic acids.

Ozonolysis of indoles, best performed in formamide or ethyl acetate, affords *o*-acylaminoketones, $RCO{\cdot}C_6H_4NHCOR'$ (*B. Witkop*, Ann., 1944, **556**, 103; *C. Mentzer et al.*, Bull. Soc. chim. Fr., 1950, 555; *D. W. Ockenden* and *K. Schofield*, J. chem. Soc., 1953, 612), also obtained by oxidation with chromic acid (*C. M. Atkinson et al., ibid.*, 1954, 165).

Osmium tetroxide converts *N*-acetylindoles into 2,3-dihydroxyindolines (*Ockenden* and *Schofield, loc. cit.*).

Autoxidation. Some 2,3-disubstituted indoles are autoxidised to peroxides (VIII), producing on reduction 3-hydroxy-3*H*-indoles (IX), which are isomerised by alkali to 2,2-dialkylindolin-3-ones (X) (*A. Robertson et al.*, *ibid.*, 1954, 4139):

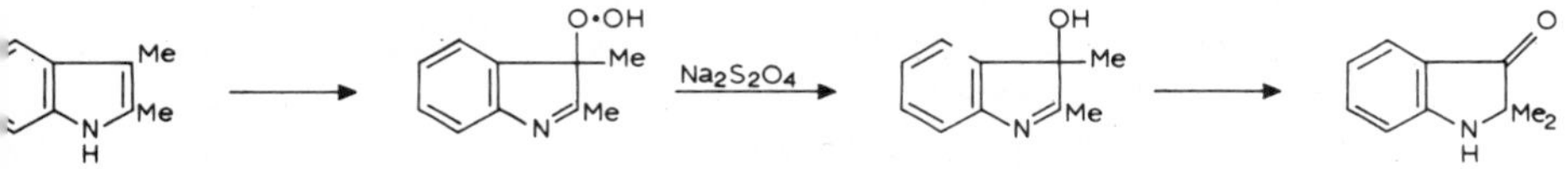

(VIII) (IX) (X)

Treatment of a solution of the 3-indolylhydroperoxide in dimethyl sulphoxide with potassium *tert*-butoxide gives a beautiful, green, chemiluminescent light; the anion XI produced intermediately is the light-emitting species (*F. McCapra* and *Y. C. Chang*, Chem. Comm., 1966, 522):

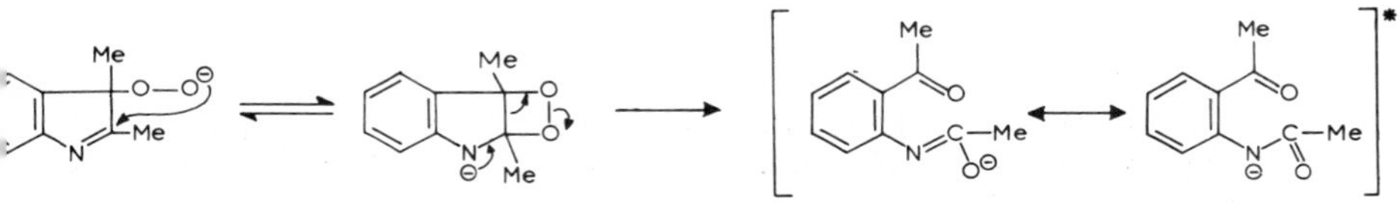

(VIII) (XI)

$$(XI)^{*} \longrightarrow (XI) + h\nu$$

The autoxidation of 2,3-diethylindole gives 2-acetyl-3-ethylindole *via* 2,3-diethyl-3-hydroperoxy-3*H*-indole (*E. Leete*, J. Amer. chem. Soc., 1961, **83**, 3645; *H. H. Wasserman* and *M. B. Floyd*, Tetrahedron Letters, 1963, 2009). This appears to be a special case (where X = Y = O·OH) of a general reaction of certain types of indole which leads to substitution in a 2- side-chain (*W. I. Taylor*, Proc. chem. Soc., 1962, 247):

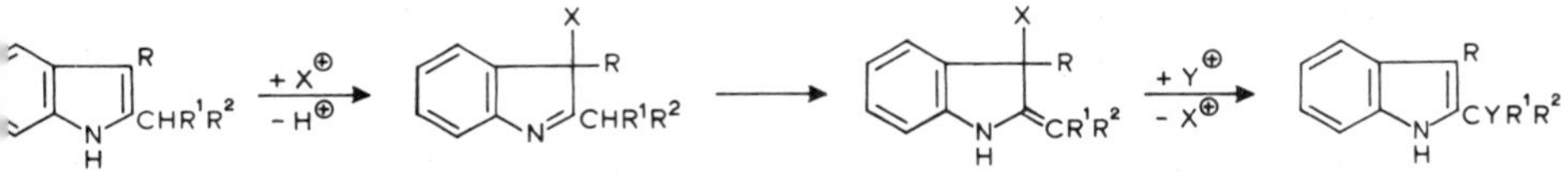

A further case, where X = Br; Y = OH, is the formation of 2-hydroxymethyl-3-methylindole from 2,3-dimethylindole by bromination and subsequent treatment with base (*S. G. P. Plant* and *M. L. Tomlinson*, J. chem. Soc., 1933, 955).

A solution of 2-benzyl-3-phenylindole in ethyl acetate on exposure to light and air gives 3-phenyldioxindole (XII), whereas shaking the solution with oxygen in the presence or absence of platinum affords 2-benzoyl-3-phenylindole (XIII); again in each case, the initial step is considered to be the formation of the 3-hydroperoxy-3*H*-indole (*F. Ying-Hsiuch Chen* and *Leete*, Tetrahedron Letters, 1963, 2013):

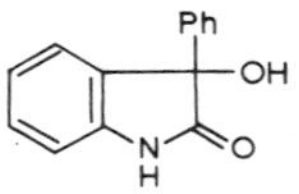

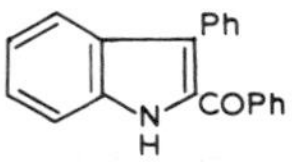

(XII) (XIII)

(vi) Reaction with benzyne

Indole reacts with benzyne, generated from bromobenzene and sodamide in liquid ammonia, to yield the *N*-phenyl- and the 3-phenyl- derivatives (5 and 15%). Treatment of the lithium salt of indole with benzyne generated from *o*-bromofluorobenzene and magnesium yields 2,3-dihydro-2,3-diphenyleneindole as the principal product (*M. E. Kuehne* and *T. Kitagawa*, J. org. Chem., 1964, **29**, 1270):

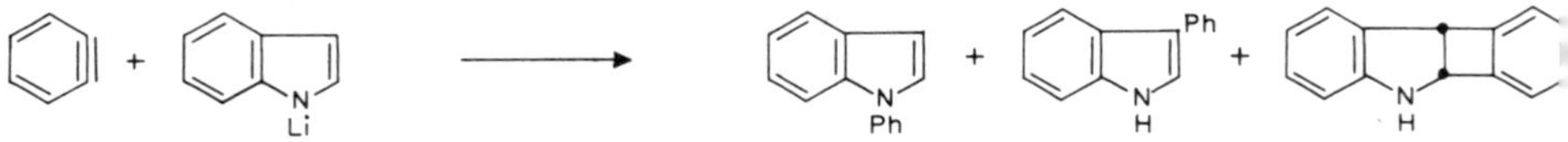

(vii) Condensation with nitroso and carbonyl compounds

The reaction between an aliphatic ketone and an aromatic isocyanide may be used to synthesise a 3*H*-indole (*B. Zeeh*, Angew. Chem., intern. Edn., 1967, **6**, 453). Keeping the homocyclic ring benzenoid, it is possible to formulate only one tautomer of indole, 3*H*-indole (XIV). Those reactions

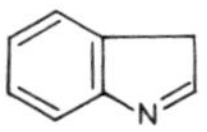

(XIV)

of pyrrole which suggest a 3*H*-pyrrole modification all yield 3-derivatives when applied to indole. Condensation with nitroso-compounds gives 3-arylimino-3*H*-indoles (*W. J. Levy* and *N. Campbell*, J. chem. Soc., 1939, 1442).

Indoles having a free 3-position condense with aldehydes (and similarly some ketones) giving alkylidene- or arylidene-bis-indoles, but in strongly acid media arylidene-3*H*-indoles (XV), unstable, coloured, weak bases, are formed (*G. O. Burr* and *R. A. Gortner*, J. Amer. chem. Soc., 1924, **46**, 1224); in alcoholic sodium ethoxide benzaldehyde yields 3-(α-ethoxybenzyl)indoles (XVI) (*M. Scholtz*, Ber., 1913, **46**, 1082, 2138):

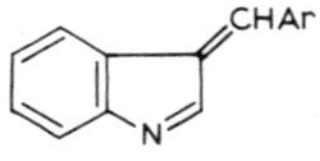

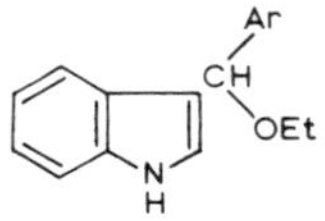

(XV) (XVI)

p-Dimethylaminobenzaldehyde attacks position 3 if available in preference to 2, but strongly acid conditions promote 2-substitution. The first products are red arylidene-

3*H*-indoles, changing into colourless aryldiindolylmethanes; these are then dehydrogenated to red 3- or blue 2-diindolomethenes (p. 468) (*H. von Dobeneck et al.*, Ann., 1952, **289,** 271; 1954, **299,** 214). Bis-(1,2,3,3-tetramethyl-3*H*-indolium) sulphate on nitration affords the 5-nitro derivative in good yield, indicating that substitution occurs *para* to the iminium group (*K. Brown* and *A. R. Katritzky*, Tetrahedron Letters, 1964, 803).

(viii) Alkylation

The direct alkylation of indoles leads by replacement of the hydrogen atoms of the pyrrole ring to fully alkylated 3*H*-indoles (see p. 442 *et seq.*) and their quaternary derivatives, and to still more complicated products (for summary and early literature: *G. Ciamician*, Ber., 1904, **37,** 4227).

(ix) Characterisation of indoles

Indoles respond to the pine-shaving and Ehrlich (*p*-dimethylaminobenzaldehyde-hydrochloric acid) colour reactions if either the 2- or the 3-position is free (*R. H. F. Manske*, Canad. J. Res., 1939, **17,** 293). They may be characterised as the picrates or as the sparingly soluble derivatives formed with xanthydrol in acetic acid (*G. Illari*, Gazz., 1938, **68,** 103).

(c) Indole and its substitution products

Indole occurs in coal tar (*O. Kruber*, Ber., 1926, **59,** 2752), in jasmine, orange blossom and other flower oils, with skatole in faeces, and is prepared by heating formyl-*o*-toluidine with potassium *tert*-butoxide (*F. T. Tyson*, Org. Synth., 1943, **23,** 42), by reduction of indoxylic acid (3-hydroxyindole-2-carboxylic acid) with sodium amalgam, or from indole-2-carboxylic acid. It is produced by dehydrogenation of methyl-*o*-toluidine, and from acetylene and aniline vapour at red heat.

Indole is steam-volatile, readily soluble in most organic solvents, crystallises from hot water as leaflets, m.p. 52.5°, b.p. 253°; heat of combustion 1020.4 kcal/mol (const. vol., *A. Stern* and *G. Klebs*, Ann., 1933, **500,** 91); μ 2.05D (*E. G. Cowley* and *J. R. Partington*, J. chem. Soc., 1936, 47); absorption spectra (*P. Lambert* and *J. Lecomte*, Compt. rend., 1939, **208,** 1148; *R. Grinbaum* and *L. Marchlewski*, Bull. Acad. Polonaise, 1937, **A,** 171); Raman effect (*F. Kohlrausch* and *R. Seka*, Ber., 1938, **71,** 1563); fluorescence spectra (*D. Bertrand*, Bull. Soc. chim. Fr., 1945, 1010, 1029).

The imino-hydrogen atom is quickly exchanged for deuterium, and in acid solution the 2- and 3-hydrogens also (*M. Koizumi* and *T. Titani*, Bull. chem. Soc., Japan, 1938, **13,** 307). Cold hydrochloric acid converts indole reversibly into the **dimer,** m.p. 110°, and irreversibly into the **trimer,** m.p. 169°; both revert to indole above 200° (p. 405).

Indole is oxidised by ozone, Caro's acid, and other reagents, to indigo. It gives a cherry-red pine-shaving reaction, a violet-red Ehrlich test which can be used for its determination (*C. B. Allsop*, Biochem. J., 1941, **35,** 965), and a blue colouration and crystalline precipitate with sodium β-naphthaquinonesulphonate: determination; distinction from skatole (*C. A. Herter* and *M. L. Foster*, J. biol. Chem., 1906, **1,** 257; **2,** 267).

(i) N-*Derivatives*

(*1*) *Metal compounds.* Heated with sodium or sodamide, indole affords the **sodium salt,** $C_8H_6N \cdot Na$ and with potassium hydroxide at 130° the **potassium salt,** which with nickel sulphate in liquid ammonia gives the complex $K_2Ni(C_8H_6N)_4$ (*O. Schmitz-Dumont* and *S. Pateras*, Z. anorg. Chem., 1935, **244,** 62). The hydrogen at the 1-position can also be replaced by Grignard reagents to give 1-**indolylmagnesium halides** which are largely ionic (see p. 408) and give with esters 1-acylindoles and some 3-indolyl ketone (*R. Majima* and *T. Shigematsu*, Ber., 1924, **67,** 1449) and mainly the 3-derivatives with other reagents (*B. Oddo* and *L. Sessa*, Gazz., 1911, **41,** I, 284; *A. R. Katritzky* and *R. Robinsom*, J. chem. Soc., 1955, 2481):

RMgX → (N–MgX) RCO₂Et → (N–COR) + (3-COR, N–H)

(*2*) *1-Alkyl-, 1-aryl- and 1-acyl-indoles.* 1-**Alkyl-** and 1-**aryl-indoles** in general are obtained by syntheses (*i*) (p. 398) and (*ii*) (p. 401). Addition of acetylene and acrylonitrile to indole gives 1-vinylindole (*W. Reppe* and *E. Keyssner*, G.P. 618,120/1935) and 1-(β-cyanoethyl)indole (*I. G. Farbenind.*, B.P. 457,621/1935), respectively.

In contrast to the behaviour of free indoles (p. 411), sodio-indole on treatment with methyl iodide gives mainly 1-methylindole.

1-Acylindoles are obtained from the metallic derivatives; acetic anhydride with the free indole gives the 3-derivative. Alkylarylamines in general, when heated with glyoxal bisulphite give products hydrolysable to oxindoles with an alkyl substituent at the 1-position (*O. Hinsberg* and *J. Rosenzweig*, Ber., 1894, **27,** 3253):

NHMe + HOCHSO₃Na / HOCHSO₃Na → (N-Me, 2-SO₃Na) → (N-Me, 2-oxo)

1-Alkylindoles can be obtained by the dehydrogenation in mesitylene with palladium on carbon of 1-alkyltetrahydro-indoles, repared by the reaction of enamines such as 1-alkyl-2-methyl-2-pyrroline with methyl vinyl ketone in tetrahydrofuran (*R. E. Ireland*, Chem. and Ind., 1958, 979):

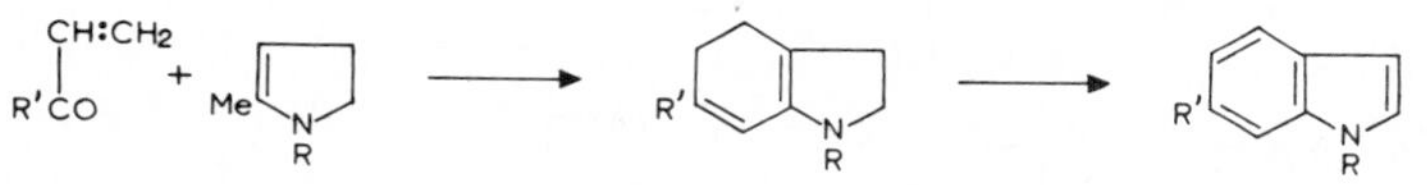

(3) *1-Amino-, 1-hydroxy- and 1-nitroso-indoles.* The dehydration of diphenylglycolaldehyde phenylhydrazone affords a product, which in acid solution isomerises to 1-**anilino**-3-**phenylindole** (*H. H. Wasserman* and *H. R. Nettleton*, Tetrahedron Letters, 1960, **7**, 33):

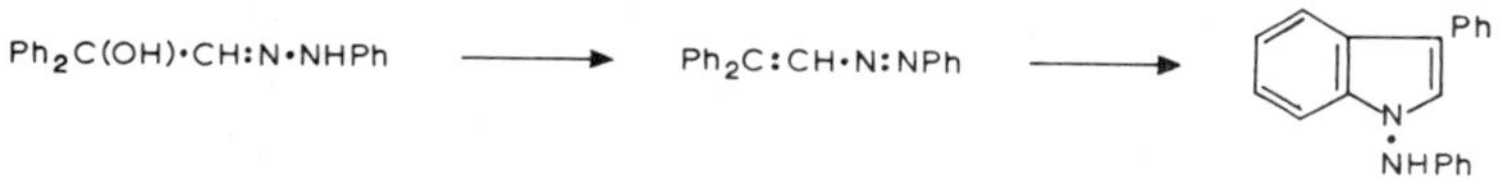

$Ph_2C(OH){\cdot}CH{:}N{\cdot}NHPh \longrightarrow Ph_2C{:}CH{\cdot}N{:}NPh \longrightarrow$

Benzoin oxime with sulphuric acid gives the weakly acidic 1-**hydroxy**-2-**phenylindole,** which is converted by amyl nitrite and sodium ethoxide into 3-oximino-2-phenyl-3*H*-indole 1-oxide:

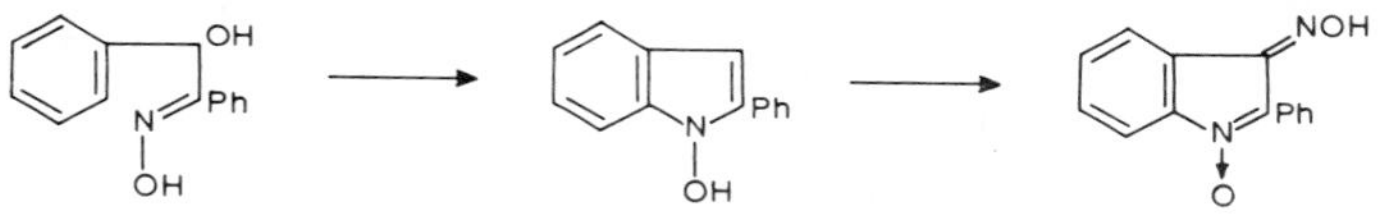

The adduct from 1,1-diphenylethene and ethyl azodicarboxylate can be degraded to 1-**amino**-3-**phenylindole** (*K. Alder* and *H. Niklas*, Ann., 1954, **585,** 97). Direct nitrosation of 3-substituted indoles having a free imino-group gives yellow 1-**nitroso-compounds** which revert to the original indole on reduction.

N-*Substituted indoles.* 1-**Methylindole,** b.p. 242–245°, *picrate*, m.p. 150° (*P. L. Julian* and *H. C. Printy*, J. Amer. chem. Soc., 1949, **71,** 3206; *D. A. Shirley* and *P. A. Roussel*, *ibid.*, 1953, **75,** 375; *K. T. Potts* and *J. E. Saxton*, J. chem. Soc., 1954, 2641); 1-*ethylindole*, b.p. 252–253°, *picrate*, m.p. 105°; 1-*isopropylindole*, b.p. 250° (*A. Michaelis* and *A. G. Robisch*, Ber., 1897, **30,** 2809). 1-*Allylindole*, b.p. 252° (*Michaelis* and *K. Luxembourg*, *ibid.*, 1893, **26,** 2174); 1-*benzylindole*, m.p. 44.5° (*O. Antrick*, Ann., 1885, **227,** 360) and 1-**phenylindole,** b.p. 326–327° (*A. Pfülf*, *ibid.*, 1887, **239,** 220), all from the 2-carboxylic acids prepared by method (*iv*) (p. 402).

1-Benzylindole is made also from indole, benzyl chloride and sodamide in liquid ammonia (*H. Plieninger*, Ber., 1954, **87,** 127). 1-**Vinylindole,** b.p. 70–75°/1 mm; 1-*β-cyanoethylindole*, m.p. 47–48°; 1-**formylindole,** b.p. 136–137°/15 mm; 1-*acetylindole*, b.p. 144–145°/10 mm (*N. Putochin*, *ibid.*, 1926, **59,** 1987); 1-*benzoylindole*, m.p. 68°, from sodioindole and benzoyl chloride (*R. Weissgerber*, *ibid.*, 1910, **43,** 3520). **Indole**-1-**carboxylic acid,** decomp. 108°, from magnesylindole and carbon dioxide (*B. Oddo* and *Q. Mingoia*, Gazz., 1927, **57,** 473).

1-**Hydroxy**-2-**phenylindole,** m.p. 171° (*E. Fischer* and *H. Hutz*, Ber., 1895, **28,** 585; *A. Angeli* and *F. Angelico*, Chem. Ztbl., 1907, I, 732). 1-**Nitroso**-2,3-**dimethylindole,** m.p. 63° (*L. Wolff*, Ber., 1888, **21,** 125). 1-**Amino**-3-**phenylindole,** m.p. 104°.

(ii) C-Alkyl-, C-aryl- and related compounds

(*1*) Alkyl- and aryl-indoles, of which the 2-, 4- and 7-methyl derivatives occur in coal tar, are commonly prepared by ring-synthesis (methods *i*, *ii* and *iii*, pp. 398, 401, 402). The Fischer indole synthesis from acetophenone 2,6-dimethylphenylhydrazone resulting in the formation of 4,7-dimethyl-2-phenylindole involves a 1,2-shift of a methyl group. The major product of the reaction is a 4-oxo-3a,4,7,7a-tetrahydro-3*H*-indole (*R. B. Carlin* and *D. P. Carlson*, J. Amer. chem. Soc., 1959, **81**, 4673):

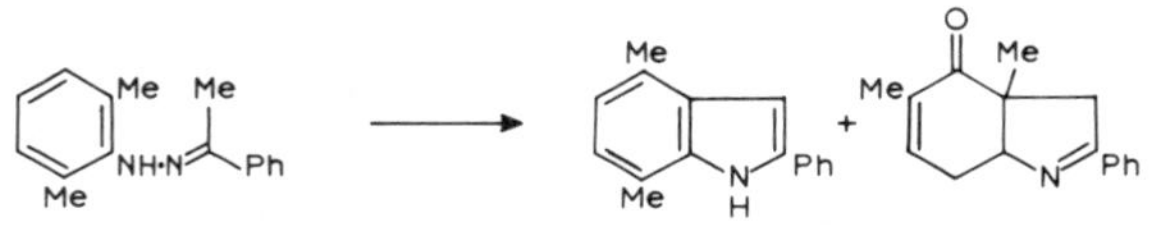

Isobutyrophenone phenylhydrazone or 3-phenyl-2-butanone phenylhydrazone on treatment with polyphosphoric acid at 150° gives an equilibrium mixture of 3,3-dimethyl-2-phenyl- and 2,3-dimethyl-3-phenyl-3*H*-indole, but in boiling acetic acid no appreciable rearrangement occurs (*F. J. Evans et al.*, J. org. Chem., 1962, **27**, 1553).

(*2*) The oxo group has been introduced into certain indoles by means of phosphorus oxychloride and the appropriate dimethylamide, a Wilsmeier–Haack type of reaction (*W. C. Anthony*, *ibid.*, 1960, **25**, 2049):

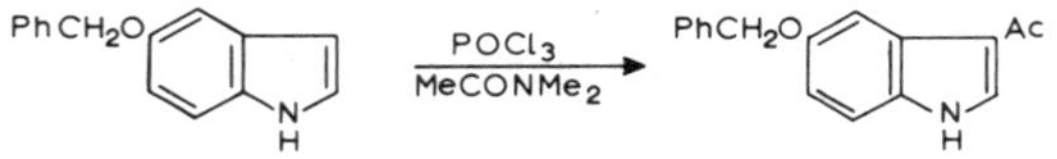

(*3*) The acid-catalysed rearrangement of *N*-alkenylanilines *via* an *o*-alkenylaniline intermediate provides a convenient and simple method for the preparation of many alkyl-substituted indoles and indolines (*A. R. Bader*, *R. J. Bridgewater* and *P. R. Freeman*, J. Amer. chem. Soc., 1961, **83**, 3319):

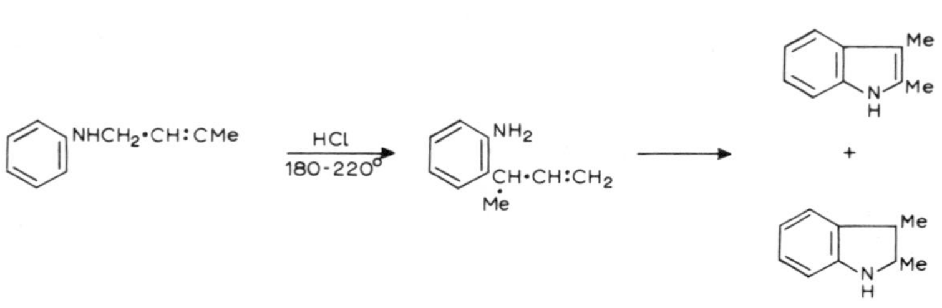

(*4*) Indoles may be alkylated directly in position 3 by the appropriate sodium alkoxide at 210–220°, in the same way as pyrroles (*R. H. Cornforth*

and *R. Robinson*, J. chem. Soc., 1942, 680). A number of alkylated indoles have been prepared by refluxing indole, an alcohol, *p*-cymene and potassium hydroxide in the presence of U.O.P. nickel^{II} (*H. F. Pratt* and *L. W. Botimer*, J. Amer. chem. Soc., 1957, **79,** 5248).

(5) When 3-arylindoles are fused with zinc chloride the aryl group migrates to position 2 (*E. Fischer* and *T. Schmidt*, Ber., 1888, **21,** 1811); the same change has been observed when 3-alkylindoles are heated with aluminium chloride–sodium chloride (*G. R. Clemo* and *J. C. Seaton*, J. chem. Soc., 1954, 2582).

2-**Methylindole** (methylketole), is obtained from acetone phenylhydrazone or from acetyl-*o*-toluidine and sodamide (method *iii*). 3-**Methylindole** (skatole), is formed by physiological or drastic chemical degradation of tryptophan, and occurs with indole in faeces, and in civet (0.1%). Skatole and 2-methylindole are oxidised by ferric chloride giving colouring matters (*H. von Dobeneck* and *W. Lehnerer*, Ber., 1957, **90,** 161), and skatole with persulphuric acid affords 3-methyloxindole (*C. E. Dalglies* and *W. Kelly*, J. chem. Soc., 1958, 3726):

and with benzaldehyde di-(3-methyl-3-indolyl)phenylmethane (*W. E. Noland* and *D. N. Robinson*, Tetrahedron, 1958, **3,** 68). Since electrophilic substitution in skatole generally occurs at the 2-position some products in the literature derived from 1,3-dimethylindole and 2,5-diones or α,β-unsaturated ketones have been reformulated (*Sir R. Robinson* and *J. E. Saxton*, J. chem. Soc., 1963, 2596; *D. A. Cockerill*, *Robinson* and *Saxton*, *ibid.*, 1955, 4369; *Noland* and *D. N. Robinson*, Tetrahedron, 1958, **3,** 68). Treatment of the product obtained by the reaction between 3-methylindole, methyl-lithium and methylene chloride with acetyl chloride gives compound XVII, which on reaction with ethanolic potassium hydroxide is converted into 4-methylquinoline obtained also by hydrolysis of the unacetylated product (*H. E. Dobbs*, Chem. Comm., 1965, 56):

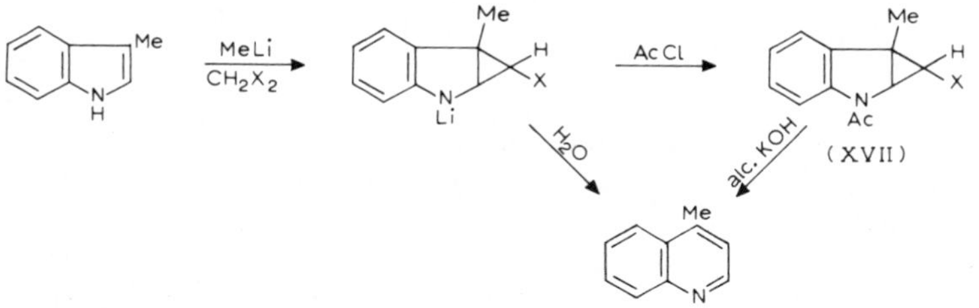

(XVII)

Indole reacts similarly (*G. L. Closs* and *G. M. Schwartz*, J. org. Chem., 1961, **26,** 2609).

3-Alkylindoles when allowed to react with two molar proportions of *N*-bromosuccinimide in *tert*-butyl alcohol yield 3-**bromo-oxindoles** (*R. L. Hinman* and *C. P. Bauman, ibid.*, 1964, **29,** 1206; *T. Kobayashi* and *N. Inokuchi*, Tetrahedron, 1964, **20,** 2055):

Skatole readily forms a *dimer* hydrochloride with hydrogen chloride in ether or with 15% aqueous hydrogen chloride. The dimerisation proceeds in a manner analogous to that of indole and the dimer possesses structure XVIII (*G. Berti, A. da Settimo* and *D. Segnini*, Tetrahedron Letters, 1960, 13; *G. F. Smith* and *A. E. Walters*, J. chem. Soc., 1961, 940; *Hinman* and *E. R. Shull*, J. org. Chem., 1961, **26,** 2339):

(XVIII)

(XIX)

It appears that the dimerisation of 3-substituted indoles is subjected to steric hindrance; 3-*n*-propylindole dimerises, but 3-isopropyl- and 3-*tert*-butyl-indole do not. 2-Methyl-indoles do not dimerise, but a mixed dimer XIX of 3-methyl- and 2-methylindole has been obtained (*Hinman* and *Shull, loc. cit.*). Adducts are formed from 3-alkylindoles and *p*-benzoquinone with the loss of hydrogen, and the compound from 3-methylindole and *p*-benzoquinone undergoes thermal or acid-catalysed isomerisation to give 2,5-bis[2-(3-methylindolyl)]hydroquinone (*W. E. Noland* and *F. J. Baude*, J. org. Chem., 1966, **31,** 3321):

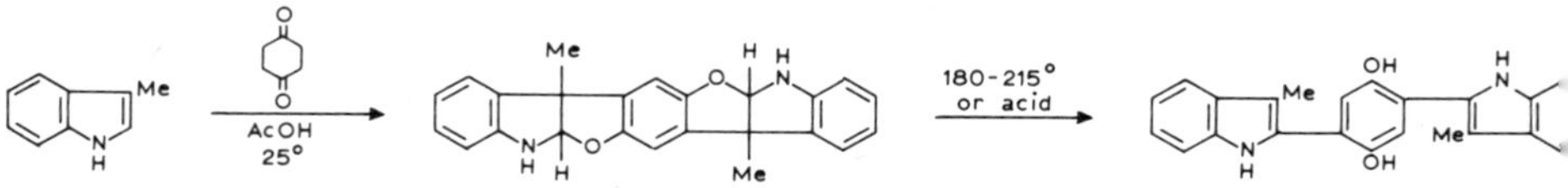

AcOH 25°

180-215° or acid

Since the chemical shifts of the H-2 and H-3 atoms differ markedly in polar and non-polar solvents the position of substituents in the indole ring can be determined (*R. V. Jardine* and *R. K. Brown*, Canad. J. Chem., 1963, **41,** 2067); long-range couplings have been reported in the proton magnetic resonance spectra of mono-methylindoles (*J. A. Elvidge* and *R. G. Foster*, J. chem. Soc., 1964, 981). 2,3-**Dimethylindole** reacts with dichlorocarbene to give a mixture of 3-dichloromethyl-2,3-dimethyl-3*H*-indole and 3-chloro-2,4-dimethylquinoline; the 3*H*-indole is not converted into the quinoline under the reaction conditions (*C. W. Rees* and *C. E. Smithen*, Chem. and Ind., 1962, 1022; J. chem. Soc., 1964, 928, 938):

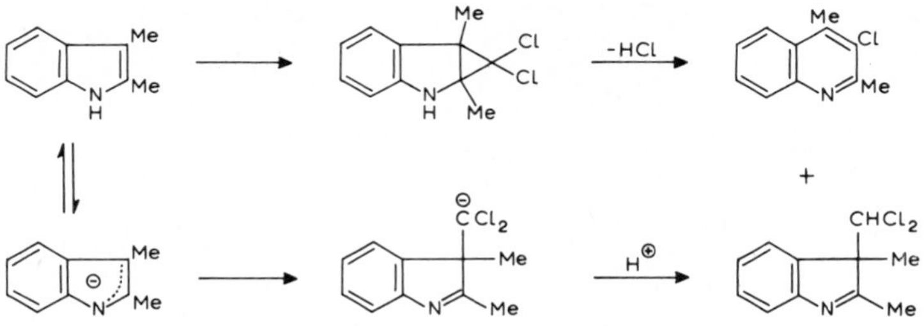

2-**Phenylindole** can be obtained by the usual methods and also by refluxing *cis*- and *trans*-2-nitrostilbenes with triethyl phosphite (*J. I. G. Cadogan* and *M. Cameron-Wood*, Proc. chem. Soc., 1962, 361).

The ultra-violet absorption spectra of 2-arylindoles in acid media are consistent with a 3*H*-indolium salt structure (*M. J. Kamlet* and *J. C. Dacons*, J. org. Chem., 1961, **26,** 220):

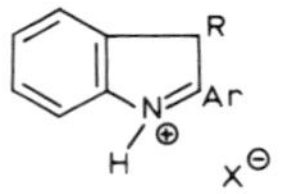

The *N*-quaternised indole, 1,1-dimethylindolium perchlorate has been prepared as shown and its ultraviolet absorption is closely similar to that of styrene and indene (*Hinman* and *J. Lang*, *ibid.*, 1964, **29,** 1449):

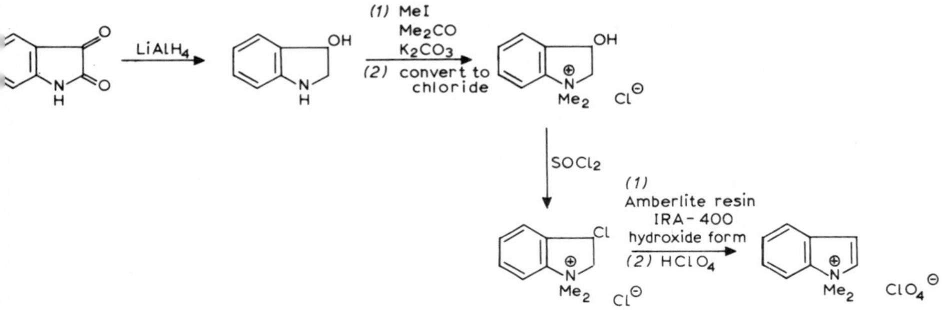

Melting points and boiling points of some representative *C*-alkyl-, -aryl- and -aralkyl-indoles are given in Table 1.

TABLE 1

C-ALKYL-, *C*-ARYL-, AND RELATED INDOLES

Substituents	*m.p. (°C)*	*Ref.*	*Substituents*	*m.p. (°C)*	*Ref.*
2-Methyl	62	1, 2	1,4-Dimethyl	b.p. 73–75/0.2 mm	1
3-Methyl (skatole)	96	1, 3	1,5-Dimethyl	b.p. 262	1
4-Methyl	b.p. 267	1, 4	1,6-Dimethyl	b.p. 68–73/0.2 mm	1
5-Methyl	59	1	1,7-Dimethyl	78	1, 12
6-Methyl	28	1	2,3-Dimethyl	108	1
7-Methyl	85	1, 5	2,4-Dimethyl	b.p. 94–96/0.5 mm	1
2-Ethyl	43	6, 7	2,5-Dimethyl	115	1
3-Ethyl	43	3	2,6-Dimethyl	88.5	1
3-Isopropyl	b.p. 288	3	2,7-Dimethyl	35–37	1
2-*tert*-Butyl	73	8	3,4-Dimethyl	117–118	13
2-Phenyl	188–189	9	Other dimethyl derivs.		1
3-Phenyl	88–89	10	1-Methyl-2-phenyl	100–101	14
3-Benzyl	103	3	2,3-Diphenyl	125	15
1,2-Dimethyl	56	1	1,2,3-Trimethyl	18	16
1,3-Dimethyl	b.p. 257–260	1, 11	Other trimethyl derivs.		17

References

1 *L. Marion* and *C. W. Oldfield*, Canad. J. Res., 1947, **25B,** 1.
2 *C. F. H. Allen* and *J. van Allan*, Org. Synth., 1942, **22,** 94.
3 *R. H. Cornforth* and *R. Robinson*, J. chem. Soc., 1942, 680.
4 *O. Kruber*, Ber., 1929, **62,** 2877.
5 *Idem, ibid.*, 1926, **59,** 2753.
6 *A. Verley* and *J. Beduwè*, Bull. Soc. chim. Fr., 1925 [iv], **37,** 189.
7 *I.C.I.*, B.P. 330,332; Chem. Ztbl., 1930, II, 2055.
8 *G. Plancher, ibid.*, 1902, II, 1322.
9 *R. L. Shriner et al.*, Org. Synth., 1942, **22,** 98; *J. I. G. Cadogan* and *M. Cameron-Wood*, Proc. chem. Soc., 1962, 361.
10 *E. Fischer* and *T. Schmidt*, Ber., 1888, **21,** 1811.
11 *E. F. J. Janetzky et al.*, Rec. Trav. chim., 1946, **65,** 193.
12 *O. Čěrvinka*, Chem. and Ind., 1959, 1129.
13 *W. A. Jacobs* and *L. C. Craig*, J. biol. Chem., 1939, **128,** 715.
14 *F. G. Mann et al.*, J. chem. Soc., 1943, 58.
15 *C. F. Koelsch*, J. Amer. chem. Soc., 1944, **66,** 1983.
16 *Janetzky* and *P. E. Verkade*, Rec. Trav. chim., 1946, **65,** 691.
17 *Idem, ibid.*, p. 700.

(iii) Halogeno-indoles

Mild treatment of oxindole with phosphorus pentachloride affords 2-chloroindole and 3-chloroindole is obtained from indole and sulphuryl chloride. On hydrolysis both chloro derivatives give oxindole:

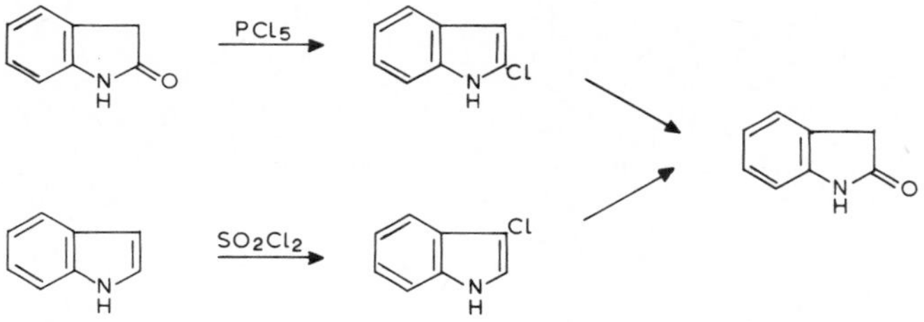

2-Methylindole is quite sensitive to chlorination giving polymeric materials with dilute chlorine or sulphuryl chloride solutions. Characterisable products 3-chloro-2-methylindole (XX) and 3-(2-methylindolyl)phosphonic dichloride (XXI) are obtained only with *N*-chlorosuccinimide and phosphorus pentachloride, respectively (*J. C. Powers*, J. org. Chem., 1966, **31,** 2627):

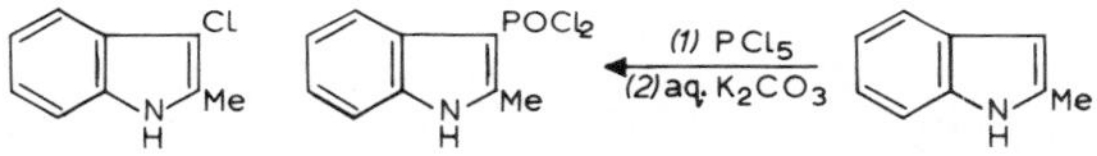

(XX) (XXI)

The action of sulphuryl chloride (*G. Mazzara* and *A. Borgo*, Gazz., 1905, **35,** II, 563) or of iodine and alkali (*H. Pauly* and *K. Gunderman*, Ber., 1908, **41,** 3999) on indoles, and the treatment of 1-benzoylindole with chlorine or bromine followed by alkali (*R. Weissgerber*, *ibid.*, 1913, **46,** 651) or of chloromercuri-indoles with iodine (*Q. Mingoia*, Gazz., 1930, **60,** 509), furnish halogenoindoles which are easily hydrolysed by acids to oxindoles. These halogenoindoles are nevertheless regarded as 3-derivatives, since they closely resemble products similarly obtained from 2-methylindole, and 3-methylindole is known to be less easily substituted. Trimethyl 1-methylindole-2,3,4-tricarboxylate on bromination in anhydrous acetic acid gives trimethyl 6-bromo-1-methylindole-2,3,4-tricarboxylate (*R. M. Acheson* and *J. M. Vernon*, J. chem. Soc., 1963, 1907), but if water is present rearrangement also takes place to yield an oxindole brominated in the 5-position and in which transfer of a carboxylate group occurs (*Acheson* and *R. W. Snaith*, Proc. chem. Soc., 1963, 344):

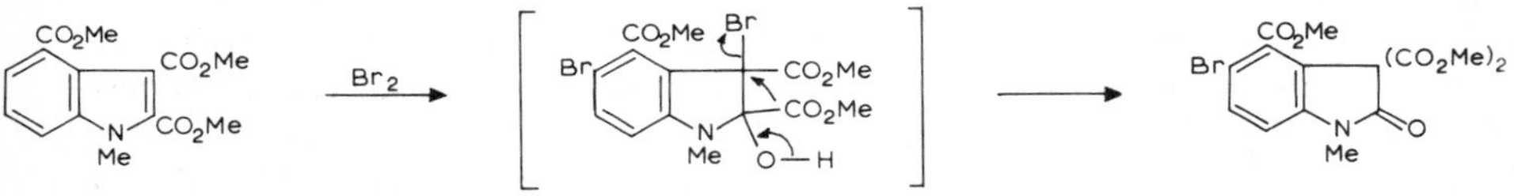

Further halogenation of indoles gives 2,3-dihalogeno compounds, which are alkali-soluble, and may be *N*-methylated directly.

N-Bromosuccinimide in *tert*-butyl alcohol converts 3-alkylindoles, indole-3-acetic acid, and indole-3-alkanoic acids to the corresponding oxindoles when a 1:1 ratio of NBS to indole is used. A 2:1 ratio of reactants affords 3-bromo-oxindoles (*R. L. Hinman* and *C. P. Bauman*, J. org. Chem., 1964, **29,** 1206):

Halogenoindoles substituted in the benzene ring are obtained by synthesis. 4-, 5-, 6-, and 7-**Fluoroindoles,** m.p. 30°, 46°, 75° and 61–62°, respectively (*F. J. Allen et al.*, J. chem. Soc., 1955, 1283); 3-, 5-, 6-, and 7-**chloroindoles,** m.p. 94°, 71–72°, 86–87° and 57–58°, respectively, 4-*chloroindole*, b.p. 106–108°/0.6 mm (*H. L. Rydon* and *J. C. Tweddle*, *ibid.*, 1955, 3499); 3-**bromoindole,** decomp. 67°; 6-*bromoindole*, m.p. 94°; 3-**iodoindole,** decomp. 72° (*H. Plieninger*, Ber., 1955, **88,** 370); 2,3-**dichloroindole,** m.p. 104° (*Mazzara* and *Borgo*, Gazz., 1906, **35,** II, 566).

(iv) Sulphur compounds

(*1*) A few indole homologues have been sulphonated directly in the benzene ring. Sulphonation of indole with the pyridine–sulphur trioxide complex gives at low temperatures **indole-1-sulphonic acid,** and at 120° probably the 2-**sulphonic acid** (*A. P. Terentiev et al.*, C.A., 1950, **44,** 1095). The products of the reaction between alkylarylamines and glyoxal bisulphite (p. 449) are probably sodium indole-2-sulphonates, although they have been alternatively regarded as oxindolesulphurous esters (*O. Hinsberg*, Ber., 1908, **41,** 1367). They are easily hydrolysed by acids, usually to the oxindole and sulphurous acid, sometimes to the indole and sulphuric acid.

(*2*) Reaction of an aqueous methanolic solution of indole and thiourea with one equivalent of iodine–potassium iodide reagent at room temperature gives *S*-(3-**indolyl)isothiuronium iodide,** m.p. 214–216° (decomp.), which on treatment with aqueous alkali followed by acidification affords 3-**indolethiol,** m.p. 100–101° (*R. L. N. Harris*, Tetrahedron Letters, 1969, 4465), obtained also on reducing 3,3′-diindolyl disulphide (*G. P. Marchese* and *E. Panzeri*, Chim. Ind. (Milan), 1969, **51,** 41; C.A., 1969, **70,** 77694g):

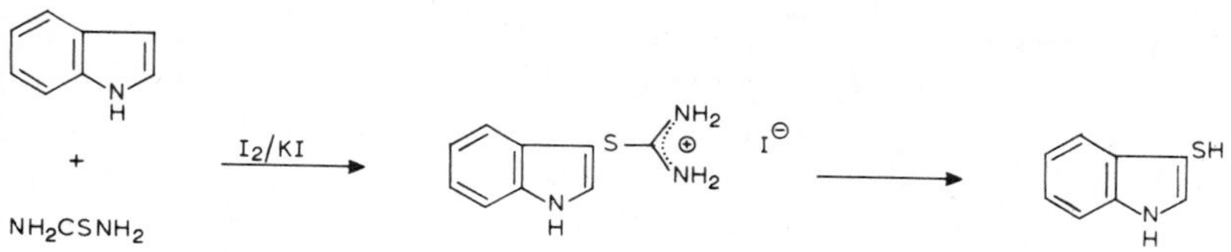

(*3*) Thiocyanation of indole gives 3-**thiocyanoindole,** m.p. 75–76°, in good yield (*M. S. Grant* and *H. R. Snyder*, J. Amer. chem. Soc., 1960, **82,** 2742). Under the influence of bases the thiocyano compound is converted to 3,3′-**diindolyl disulphide,** which yields the 2,4-dinitrophenyl 3-indolyl sulphide on reduction with lithium tetrahydridoaluminate and treatment of the reduction mixture with 2,4-dinitrochlorobenzene. The nucleus of 3-thiocyanoindole is strongly deactivated, but the indolic nitrogen is attacked by some reagents, *e.g.*, boiling with acetic anhydride affords 1-*acetyl-3-thiocyanoindole*, m.p. 123°:

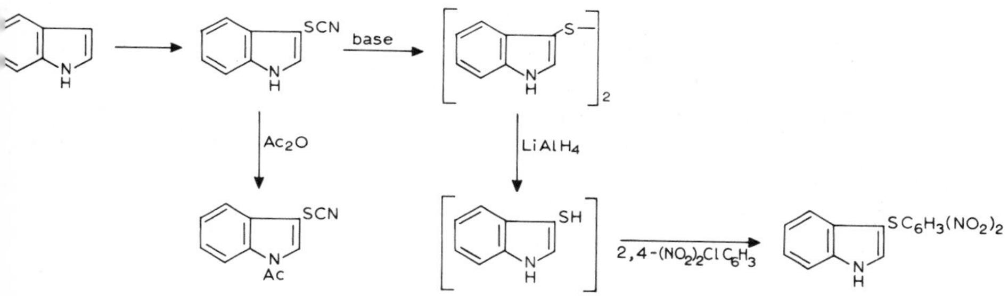

(*4*) Magnesylindole reacts with sulphur (or sulphur chloride), sulphur dioxide and sulphuryl chloride to give respectively, **diindolyl sulphide,** m.p. 232°, **sulphoxide,** m.p. 157°, and **sulphone,** m.p. 151–152°, they are probably the 3,3′-compounds (*B. Oddo* and *Q. Mingoia*, Gazz., 1932, **62,** 299).

(*5*) With sulphur chloride 3-substituted indoles give mainly 2,2-diindolyl disulphides (*3,3′-dimethyl-diindolyl-2,2′-disulphide*, m.p. 126–127°), with some mono- and tri-sulphide; methanesulphenyl chloride, MeSCl, gives with 3-phenylindole, *2-methylthio-3-phenylindole*, m.p. 85–86° (*T. Wieland et al.*, Ann., 1954, **587,** 146).

(v) Nitro-, amino- and related indoles

Nitroindoles substituted in the benzene nucleus have occasionally been obtained by direct nitration (*S. G. P. Plant* and *M. L. Tomlinson*, J. chem. Soc., 1933, 955), sometimes by ring syntheses which do not always go smoothly.

Nitration of indoles in position 3 can be effected by treatment with an alkyl nitrate and sodium alkoxide, giving yellow, acidic products also obtained by oxidising the corresponding oximino-3*H*-indoles. Elimination of the methyl group in 2-methyl-3-nitroindole confirms the orientation of the nitro group in 3-nitroindole, obtained from indole and ethyl nitrate. 3-Nitroindole can be ethylated directly in position 1:

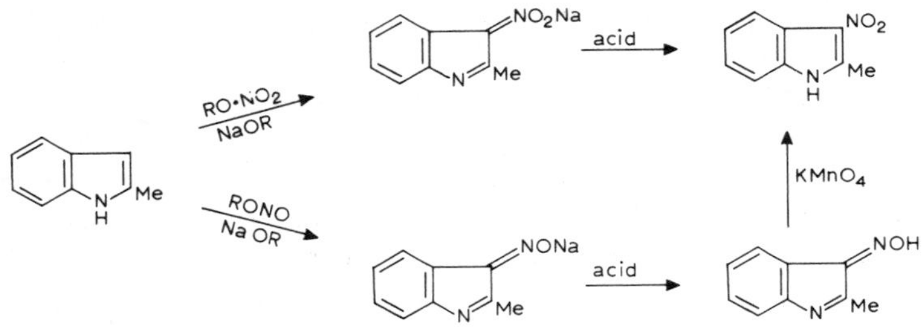

2-Alkylindoles when nitrated below room temperature in concentrated sulphuric acid give 5-nitro derivatives, but in nitric acid alone or in acetic acid, after attack at the 3-position further substitution occurs at the 4- and 6-positions (*W. E. Noland, L. R. Smith* and *K. R. Rush*, J. org. Chem., 1965, **30,** 3457).

1-Acetyl-2,3-dimethylindole on nitration affords a mixture of the 6-nitro derivative, 2-acetamidoacetophenone, and a *trans*-indoline-2,3-diol:

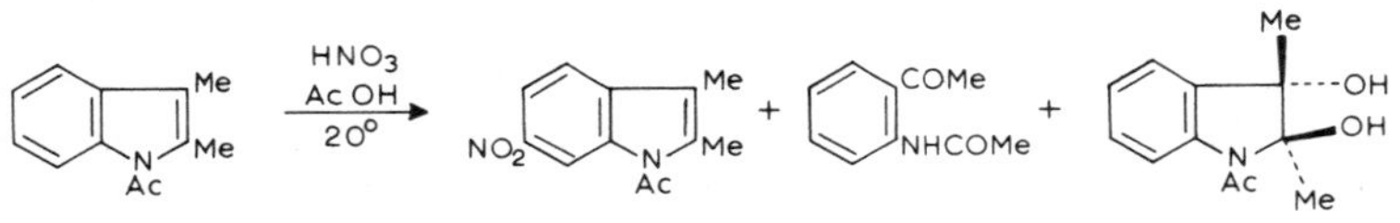

Nitrosoindoles, as synthesised by treating an indole having a free 3-position with alkyl nitrite and (usually) sodium alkoxide, are yellow amphoteric solids believed to be oximino-3*H*-indoles as shown in the penultimate reaction scheme above. On the other hand, true emerald-green nitrosoindoles result from the action of nitrous acid on 1,2-disubstituted indoles; these condense very easily through the nitroso-group, with "reactive methylene" compounds, and a 2-methyl group is so reactive that attempted preparation of 1-substituted 2-methyl-3-nitrosoindoles gives the 2-oximinomethyl-3-nitroso-compounds (*N. Campbell* and *R. C. Cooper*, J. Chem. Soc., 1935, 1208; *F. G. Mann* and *R. C. Haworth*, *ibid.*, 1944, 670).

3-**Nitroindole,** m.p. 210° (1-*ethyl* derivative, m.p. 102°) and 2-**methyl**-3-**nitroindole,** m.p. 248° (decomp.) (*F. Angelico* and *G. Velardi*, Gazz., 1904, **34,** II, 57); 6-*nitroindole*, m.p. 140°, from indole-3-carboxylic acid by nitration and decarboxylation (*R. Majima* and *M. Kotake*, Ber., 1930, **63,** 2237); 4-, 5-, 6- and 7-*nitroindoles*, m.p. 205–206°, 141–142°, 144–145° and 95–96°, respectively (*S. M. Parmerter, A. G. Cook* and *W. B. Dixon*, J. Amer. chem. Soc., 1958, **80,** 4621); 2,3-**dimethyl**-4-, -5-, -6- and -7-**nitroindoles,** m.p. 130°, 188°, 141° and 164°, respectively (*Plant et al.*, J. chem. Soc., 1933, 955; 1940, 283; *H. Bauer* and *E. Strauss*, Ber., 1932, **65,** 308); 2-*methyl*-3,4-*dinitroindole*, m.p. 284–285° (decomp.), 2-*methyl*-3,5-*dinitroindole*, m.p. 285–287°, 2-*methyl*-3,6-*dinitroindole*, m.p. 305° (decomp.), 2-**methyl**-3,4,6-**trinitroindole,** m.p. 254–256° (decomp.); 1,2-**dimethyl-4-nitroindole,** m.p. 94–97°; *1,2-dimethyl-6-nitroindole*, m.p. 106–107°; 1,2-*dimethyl*-3,5-*dinitroindole*, m.p. 201–203°; 1,2-*dimethyl*-3,6-*dinitroindole*, m.p. 299°; 1,2-*dimethyl*-3,4,6-*trinitroindole*, m.p. 242–244°; 1,2-*dimethyl*-3,5,6-*trinitroindole*, m.p. 267–269° (decomp.) (*Noland, Smith* and *Rush, loc. cit.*).

1-**Methyl**-3-**nitroso**-2-**phenylindole,** m.p. 144° (*Campbell* and *Cooper, loc. cit.*).

Aminoindoles are relatively unstable compounds. Indole-2-carboxylic acid yields by the Curtius reaction 2-indolylcarbamate which cannot be hydrolysed to 2-aminoindole. Hydrogenation of the benzyl carbamate or cyclisation of *o*-aminobenzyl cyanide with sodium ethoxide gives an unstable base, which affords indole on reduction with sodium and alcohol, and is probably better formulated as 2-amino-3*H*-indole, a cyclic amidine. 3-Aminoindoles which are also unstable have been obtained by the reduction of oximino-3*H*-indoles, or 1-alkyl-3-nitrosoindoles (*Huang-Hsinmin* and *Mann*, J. chem. Soc., 1949, 2903). With nitrous acid these yield deep yellow, photosensitive, weakly basic diazo compounds, which are comparatively stable towards boiling dilute acids; they are usually formulated as 3-diazo-3*H*-indoles (*V. Castellana* and *A. d'Angelo*, Gazz., 1906, **36,** II, 56):

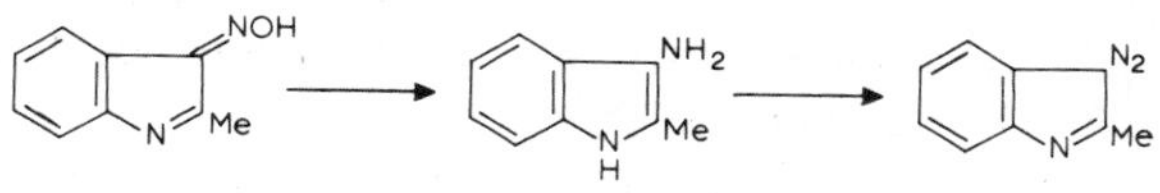

NOH, N, Me → NH$_2$, N, H, Me → N$_2$, N, Me

With indoles nitrosation proceeds readily but the nitroso compounds exists almost entirely in the oxime form and there is no further reaction. To obtain 3-**diazoindoles** it is necessary to follow the conventional method of nitrosation, reduction and diazotisation. 2-Phenyl-3-diazoindole couples with 2-naphthol in chloroform (*H. P. Patel* and *J. M. Tedder*, J. chem. Soc., 1963, 4593). The Stephen reduction of some *o*-nitrobenzyl cyanides with anhydrous stannous chloride yields 2-aminoindoles (*Snyder et al.*, J. Amer. chem. Soc., 1958, **80,** 4622):

$$\text{R-C}_6\text{H}_3(\text{NO}_2)\text{-CH(CN)CO}_2\text{Et} \xrightarrow[\text{Et}_2\text{O - HCl}]{\text{SnCl}_2} \text{2-amino-3-(CO}_2\text{Et)indole (R)}$$

R = CO_2Et or NO_2 → R = CO_2Et or NH_2

Benzenediazonium salts couple, with 2-substituted indoles, giving 3-**benzeneazoindoles** (*J. H. Binks* and *J. H. Ridd*, J. chem. Soc., 1957, 2398) reducible to 3-aminoindoles (*P. Wagner*, Ann., 1887, **242,** 383). Aminoindoles having the amino group in the benzene ring are obtainable by reducing the appropriate nitro derivative.

2-**Aminoindole,** m.p. 165°, *acetyl* deriv., m.p. 167°, *bisethoxycarbamoyl* deriv., m.p. 93° (*H. Rinderknecht et al.*, J. org. Chem., 1953, **18,** 971; *J. Kebrle* and *K. Hoffmann*, Helv., 1956, **39,** 116); 3-*aminoindole*, m.p. 117° (decomp.), *acetyl* deriv., m.p. 162–163° (decomp.) (*W. Madelung*, Ann., 1914, **405,** 92); 4-*aminoindole*, m.p. 108° (*Plieninger*, *loc. cit.*); 5-*aminoindole*, m.p. 127–129°, from 2-(5-amino-2-methoxyphenyl)ethylamine following demethylation with hydrobromic acid and oxidation with silver oxide (*J. Harley-Mason* and *A. H. Jackson*, J. chem. Soc., 1954, 1158); 6-*aminoindole*, m.p. 66–67°, *acetyl* deriv., m.p. 170–171° (*R. K. Brown* and *N. A. Nelson*, J. Amer. chem. Soc., 1954, **76,** 5149; *Plieninger, loc. cit.*); 3-**amino-2-methylindole,** m.p. 112–113°, obtained from the benzeneazo compound (*H. Erdtman*, Ber., 1936, **69,** 2482); 3-**amino-2-phenylindole,** m.p. 180° (*F. Angelico* and *S. Capuano*, Gazz., 1937, **67,** 633, 710); 3-*amino-2-(4-methoxyphenyl)indole*, m.p. 150–151° (*Patel* and *Tedder*, *loc. cit.*).

(vi) Hydroxyindoles

Indoles hydroxylated in the pyrrole ring, except perhaps indoxyl, 3-hydroxyindole, are tautomeric substances better represented as oxoindolines; they are discussed therefore in the section dealing with these compounds (pp. 448 *et seq.*).

Hydroxyindoles having the substituent in the benzene nucleus are sensitive and difficult to prepare, although their ethers are accessible by syntheses (*i*) (p. 398) and (*iv*) (p. 402); hydrogenolysis of the benzyl ether has been most serviceable (*A. Stoll et al.*, Helv., 1955, **38,** 1452).

When dihydroxyphenylethylamines in which the hydroxyl groups are *ortho* or *para* to one another are oxidised, the quinones first produced readily cyclise to hydroxyindoles, or to dihydroxyindolines, further oxidised to indolinequinones which are isomerised to dihydroxyindoles (*H. S. Raper*, Biochem. J., 1927, **21,** 89; *W. Dulière* and *Raper*, *ibid.*, 1930, **24,** 239; *Harley-Mason*, Chem. and Ind., 1952, 173; *J. D. Bu'Lock* and *Harley-Mason*, J. chem. Soc., 1951, 2248):

Adrenaline (XXII) is oxidised by catechol oxidase or by ferricyanide to the red quinone *adrenochrome* (XXIII), decomp. 115–120°, which gives on hydrogenation 5,6-dihydroxy- and 3,5,6-trihydroxy-1-methylindole (*D. E. Green* and *D. Richter*, Biochem. J., 1937, **51,** 596; *Harley-Mason*, J. chem. Soc., 1950, 1276; *Bu'Lock* and *Harley-Mason*, *ibid*., 1951, 712). Oxidation with iodic acid gives 2-iodoadrenochrome (XXIV) (*Richter* and *H. Blaschko*, *ibid*., 1937, 601):

(XXIV) (XXII) (XXIII)

Adenenochrome forms a 1,4- (or more accurately a 9,7- or 3a,7-) addition compound with sodium bisulphite giving a compound having the following formula (*R. Marchelli*, *W. S. Powell* and *R. A. Heacock*, Chem. and Ind., 1971, 1021):

4-, 5-, 6-**Hydroxyindoles,** m.p. 97–99°, 107–108° and 124–126°; *picrates*, m.p. 159–160°, 145° and 137°; *benzyl ethers*, m.p. 72–74°, 103–105° and 118°, respectively (*Stoll et al., loc cit.*). 7-*Hydroxyindole, methyl ether*, b.p. 157°/17 mm (*K. G. Blaikie* and *W.*

H. Perkin, J. chem. Soc., 1924, **125,** 327). 5-**Ethoxy**-1,3-**dimethylindole,** *physostigmal ethyl ether,* m.p. 86° (*E. Stedman, ibid.,* 1924, **125,** 1373).

5,6-**Dihydroxyindole,** m.p. 140°, *dimethyl ether,* m.p. 154–155° (*A. Robertson et al., ibid.,* 1948, 2223; *C. F. Huebner et al.,* J. Amer. chem. Soc., 1953, **75,** 5887).

Melanins are black, insoluble, polymeric pigments occurring widely in the animal kingdom. The melanins of hair, skin, melanoma tumours, and *Sepia* ink are all derived biologically by oxidation of tyrosine or 3,4-dihydroxyphenylalanine (dopa). Melanin can also be prepared from the latter *in vitro* by autoxidation or by oxidation in the presence of tyrosinase. Typical melanins are produced from 1-, 2-, 4- and 7-methyl-5,6-dihydroxyindoles, but not from the 3-methyl- and 4,7-dimethyl- derivatives (*R. I. T. Cromartie* and *Harley-Mason,* Chem. and Ind., 1953, 972). Dihydroxyphenylalanine, tyrosine, and tyramine have been oxidised quantitatively *in vitro* into melanins by polyphenol oxidase (*D. Kertesz,* Pubbl. staz. zool. Napoli, 1957, **29,** 33). About one mole of carbon dioxide is produced per two indole units oxidised, and experiments with dihydroxyphenylethylamine having labelled carbon atoms show that this comes mainly from nuclear carbons, 3, 4 and probably 5 (*G. A. Swann* and *D. Wright,* J. chem. Soc., 1954, 381; 1956, 1549). It is suggested that the sequence leading to melanin is:

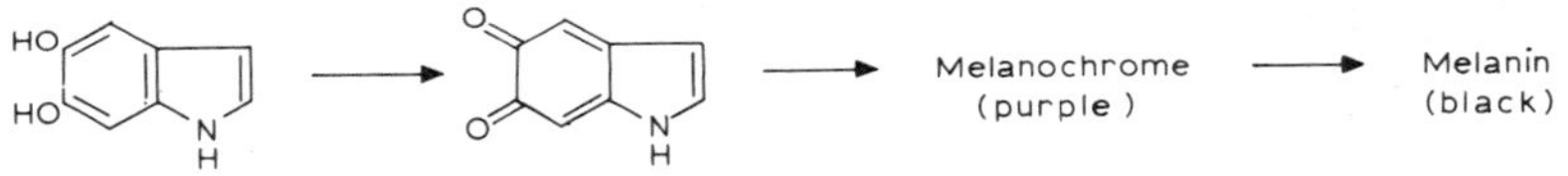

Melanochrome (purple) → Melanin (black)

and it is generally agreed that the biosynthesis of melanin involves oxidative polymerisation of 5,6-dioxoindole. Both thermodynamic and kinetic considerations suggest polymerisation with intermediate dimers of the type (*B. Pullman,* Biochim. biophys. Acta, 1963, **66,** 164):

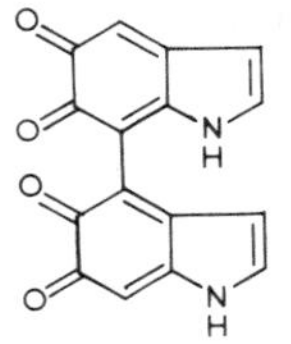

It would appear that animal melanins degraded by oxidation and alkali fusion, are indoles, and plant melanins catechols, except that *Daldinia concentrica* also contains 1,8-dihydroxynaphthalene (*E. Fattorusso, M. Piattelli* and *R. A. Nicolaus,* Rend. Accad. Sci. Fis. Mat., 1965, **32,** 57).

Ultraviolet spectroscopic studies of the formation *in vitro* of melanin from 2,4,5-trihydroxyphenylethylamine and from 2,4,5-trihydroxyphenylalanine showed that these compounds were unlikely intermediates in the autoxidation conversion of 3,4-dihydroxyphenylethylamine and 3,4-dihydroxyphenylalanine, respectively, into melanin (*Swann,* Ann. N.Y. Acad. Sci., 1963, **100,** 1005).

Electron spin resonance studies on squid, and synthetic melanins produced by autoxidation or by enzymatic oxidation of a number of biphenols showed that the e.s.r. signals of the natural and synthetic melanin were strikingly similar with respect to line width, line shape, and *g*-value. It is concluded that the unpaired electrons observed are associated with trapped free-radicals in the polymer, that the biosynthesis of melanin may involve a free-radical mechanism, and that melanin is a highly irregular three-dimensional polymer (*M. S. Blois Jr., A. B. Zahlan* and *J. E. Maling*, Biophys. J., 1964, **4**, 471). Specimens of 3,4-dihydroxyphenylalanine (dopa) labelled with tritium singly in the α-, β-, 2-, 5- and 6-positions and triply in the 2,5,6-positions have been mixed with (±) [α-^{14}C]dopa, and each converted into melanin by oxidation with oxygen in the presence of tyrosinase and catalase. The tritium loss during melanin formation was measured and it was concluded that melanin produced enzymatically from dopa had a highly irregular structure and differed significantly from that produced by autoxidation or by an intact organism (*G. W. Kirby* and *L. Ogunkoya*, Chem. Comm., 1965, 546). Similar results have been obtained with α- and β-deuterated dopa (*Swann*, Rend. Accad. Sci. Fis. Mat. (Napoli), 1964, Serie 4, 31), and the deuterium content of the precursor and melanins measured. [*carboxyl*-^{14}C]-3,4-Dihydroxyphenylalanine has also been converted into melanin by autoxidation and enzymatically, and the specific activities of the precursors, the resulting melanins and evolved carbon dioxide measured. It is concluded that dopa melanin contains dopachrome units, that more indole-5,6-quinone units are linked at the 2- than at the 3-position and that in the enzymatically formed melanin, more units are linked at both the 2- and 3-positions than in autoxidation melanin (*N. C. Robson* and *Swann*, Int. Pigment Cell Conf., 6th, Sofia, 1965/1966, 155). For electron paramagnetic resonance, ultraviolet and infrared absorption and electron spin resonance studies of melanin see *G. Tollin* and *C. Steelink*, Biochim. biophys. Acta, 1966, **122**, 377; *Zahlan, Maling* and *Blois*, Photochem. Photobiol., 1966, **5**, 269; *Blois*, J. Invest. Dermatol., 1966, **47**, 162.

(vii) Metal and metalloid derivatives

Although mercuration of indole with aqueous mercuric acetate is reported to give 2,3-**diacetoxymercuri-indole** (*Q. Mingoia*, Gazz., 1930, **60**, 509; *L. K. Ramachandran* and *B. Witkop*, Biochemistry, 1964, **3**, 1603), experiments with [2-^{2}H]- and [2-^{3}H]-labelled indole indicate that the 1,3- rather than the 2,3-derivative is formed (*G. W. Kirby* and *S. W. Shan*, Chem. Comm., 1965, 381).

With phosphorus trichloride indolylmagnesium bromide (p. 408) gives 1- and 3-**triindolylphosphines,** m.p. 223–225° and 195–196°, respectively; phosphoryl chloride yields *triindolylphosphine oxide*, m.p. 138–140°, and *diindolylphosphinic acid*, m.p. 190° (*Mingoia*, Gazz., 1930, **60**, 144; 1932, **62**, 333).

Arsenic acid arsonates indoles giving **indole**-3-**arsonic acid** only (*E. Funakubo*, C.A., 1928, **22**, 1775).

An ethereal solution of indolylmagnesium bromide when boiled with potassium tetrafluoroborate gives **potassium tetra(1-indolyl)boron** (*V. A. Sazonova* and *V. I. Karpov*, Zh. obshchei Khim., 1963, **33,** 3313):

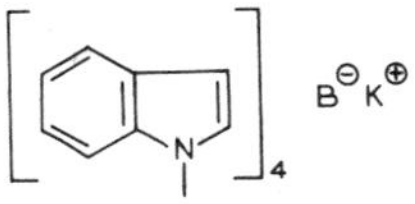

(*viii*) *Indolylcarbinols, aminoalkylindoles and related compounds*

(*1*) *Carbinols.* 2-**Indolylcarbinol,** m.p. 75–77°, is obtained by the reduction of ethyl indole-2-carboxylate with lithium tetrahydridoaluminate (*W. I. Taylor*, Helv., 1950, **33,** 164); similar reduction of ethyl indole-3-carboxylate or indole-3-carbaldehyde affords skatole, but with sodium tetrahydridoborate 3-**indolylcarbinol,** m.p. 101°, is formed. It results also in good yield from the action of alkali on gramine methiodide (see below) and easily forms di-3-indolylmethane with loss of formaldehyde (*E. Leete* and *L. Marion*, Canad. J. Chem., 1953, **31,** 775):

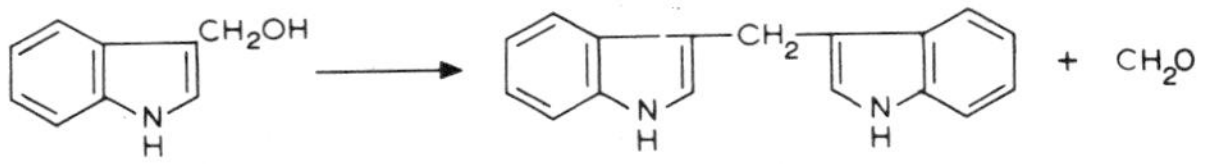

Tryptophol (3-*β*-**hydroxyethylindole**), m.p. 59° (*benzoyl deriv.*, m.p. 76°), which occurs in wine and is produced from tryptophan by yeast in presence of sucrose, has been synthesised from indolylmagnesium bromide and ethylene oxide (*B. Oddo* and *F. Cambieri*, Gazz., 1939, **69,** 19), by the Bouveault–Blanc reduction of ethyl 3-indolylacetate (*R. W. Jackson*, J. biol. Chem., 1930, **88,** 659) and by reduction of 3-indolylacetic acid with lithium tetrahydridoaluminate (*H. R. Snyder* and *F. J. Pilgrim*, J. Amer. chem. Soc., 1948, **70,** 3770), but a more convenient method of preparation is by reduction of ethyl 3-indolylglyoxylate, obtained by boiling 3-indolylglyoxyl chloride in ethanol with triethylamine, with lithium tetrahydridoaluminate in tetrahydrofuran (*T. Nogrady* and *T. W. Doyle*, Canad. J. Chem., 1964, **42,** 485):

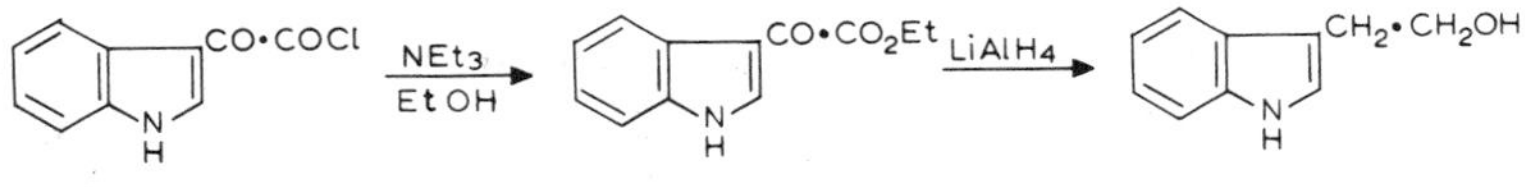

(*2*) *Aminoalkylindoles.* Hydrogenation of indole-3-aldoxime acetate over platinum oxide gives unstable 3-**aminomethylindole,** which is converted by nitrous acid into quinoline (*N. I. Putochin* and *N. P. Dawydowa*, Chem. Ztbl., 1933, I, 3714). 3-Aminomethyl- and 3-methylaminomethyl-indole are constituents of barley (*S. H. Mudd*, Nature, 1961, **189,** 489). The dimethyl derivative, **gramine, 3-dimethylaminomethylindole,** m.p. 134° (*methiodide,* m.p. 168–169°; *picrate,* m.p. 141.5°), is an important intermediate in synthetic work. It occurs in the germ of Swedish barley and in the reed *Arundo donax* (*H. von Euler et al.*, Ber., 1936, **69,** 743). First synthesised by treating indolylmagnesium iodide with dimethylaminoacetonitrile (*T. Wieland* and *C. Y. Hsing,* Ann., 1936, **526,** 188), it is produced in excellent yield by a Mannich reaction between indole, formaldehyde and dimethylamine (*H. Kühn* and *O. Stein,* Ber., 1937, **70,** 567). Gramine, which may be regarded as a vinylogous methylenediamine, is a very reactive indolylmethylating agent, liberating dimethylamine. Thus with sodium bisulphite it gives sodium 3-indolylmethanesulphonate $C_8H_6NCH_2SO_3Na$, a heteroauxin antagonist (*Wieland et al.*, Ann., 1949, **561,** 47), and takes part in malonic ester syntheses (*Snyder et al.*, J. Amer. chem. Soc., 1944, **66,** 200):

$$C_8H_6NCH_2NMe_2 \quad + \quad H_2C(CO_2Et)_2 \quad \xrightarrow{Na} \quad C_8H_6NCH_2{\cdot}CH(CO_2Et)_2$$

It also condenses with nitroalkanes, giving products reducible to tryptamines (*Snyder* and *L. Katz, ibid.*, 1947, **69,** 3140):

$$C_8H_6NCH_2NMe_2 \quad + \quad Me_2CHNO_2 \quad \rightarrow \quad C_8H_6NCH_2{\cdot}CMe_2NO_2 \quad \rightarrow \quad C_8H_6NCH_2{\cdot}CMe_2NH_2$$

Gramine metho-salts are still more reactive, and the conditions for their preparation are critical (*A. P. Gray, ibid.*, 1953, **74,** 1252); they yield 3-indolylacetonitrile with alkali cyanides (*K. T. Potts* and *R. Robinson,* J. chem. Soc., 1955, 2675), and react with alkylated diethyl malonate, cyanoacetate, and acetoacetate and many of their derivatives (*Snyder et al., loc. cit.;* see also tryptophan, p. 436). Similar alkylations can be effected with 1-methylgramine and its methiodide (*Snyder* and *E. L. Eliel,* J. Amer. chem. Soc., 1949, **71,** 663), but much less easily with the position isomeride, **2-dimethylaminomethylindole, isogramine,** b.p. 143–145°/6 mm (*Snyder* and *P. L. Cook, ibid.*, 1956, **78,** 969).

Indoles combine easily at the 3- or 2-position with Schiff bases, giving products which are decomposed by strong acids (*M. Passerini* and *T. Bonciani,* Gazz., 1933, **63,** 138):

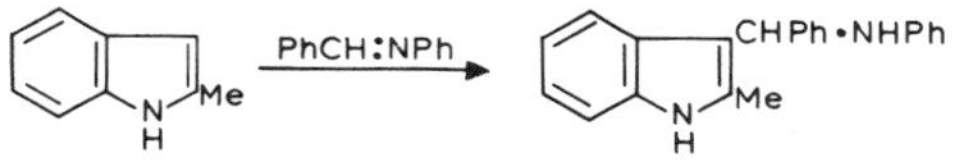

Tryptamine [**3-(2′-aminoethyl)indole**], m.p. 120° (*picrate,* decomp. 242–243°; *benzoyl* deriv., m.p. 141–142°), the product of decarboxylation of tryptophan, is important on account of its physiological properties and those of its hydroxylated derivatives; it is also produced by drastic degradation of strychnine. Tryptamine has been synthesised by the Fischer method (*A. J. Ewins*, J. chem. Soc., 1911, **99,** 270):

$PhNH{\cdot}NH_2$ + $(EtO)_2CH{\cdot}CH_2{\cdot}CH_2{\cdot}CH_2NH_2$ ⟶ Tryptamine ($CH_2{\cdot}CH_2NH_2$)

and more conveniently from gramine, either *via* 3-indolylacetonitrile (*J. Thesing* and *F. Schülde*, Ber., 1952, **85,** 324) or *via* ethyl 3-indolylmethylnitromalonate, $C_8H_6N{\cdot}CH_2{\cdot}C(NO_2)(CO_2Et)_2$ (*D. A. Lyttle* and *D. I. Weisblat*, J. Amer. chem. Soc., 1955, **77,** 5747).

Tryptamine may also be obtained by the reduction of the product formed by the nitroethylation of indole (*W. R. Noland* and *P. J. Hartman*, *ibid.*, 1954, **76,** 3227); indolylmagnesium bromide gives an increased yield:

Indole —($CH_2{:}CHNO_2$, C_6H_6)→ 3-($CH_2{\cdot}CH_2NO_2$)indole —(H_2, Pt)→ 3-($CH_2{\cdot}CH_2NH_2$)indole

Indole —(EtMgI)→ N-MgI indole —($CH_2{:}CHNO_2$, Et_2O)→ 3-($CH_2{\cdot}CH_2NO_2$)indole

Isotryptamine, [**2-(2′-aminoethyl)indole**], may be obtained by the following route (*W. Schindler*, Helv., 1957, **40,** 2156):

Indole-2-CO_2H —(*(1)* $SOCl_2$, C_6H_6, N_2; *(2)* Me_2NH, C_6H_6)→ 2-$CONMe_2$ —($LiAlH_4$, THF)→ 2-CH_2NMe_2 (m.p. 60-61°) —(EtAc, MeI)→ 2-$CH_2\overset{\oplus}{N}Me_3I^{\ominus}$ (m.p. 161-162°) —(N_2, KCN, MeOH)→ 2-$CH_2{\cdot}CN$ (m.p. 96-98°) —(H_2 Raney Ni, EtOH sat. NH_3, 80-190°, 120 atm; or $LiAlH_4$, poor yield)→ 2-$CH_2{\cdot}CH_2NH_2$ (m.p. 100-101°)

but a better method of preparation is from indole-2-carbaldehyde (p. 432).

The vasoconstrictor **serotonine** *(enteramine)* is 5-*hydroxytryptamine*, m.p.

214° (*picrate*, m.p. 193–194° decomp.). It has been synthesised from 5-benzyloxyindole, either through the corresponding 3-acetonitrile, or by production of the 3-glyoxyloyl derivative with oxalyl chloride, followed by reduction of the derived dibenzylamide (*M. E. Speeter et al.*, J. Amer. chem. Soc., 1951, **73,** 5514; 1954, **76,** 6208), and also from 2,5-dimethoxybenzaldehyde (*J. Harley-Mason* and *A. H. Jackson*, J. chem. Soc., 1954, 1165).

The poisonous secretion of several species of toad affords the basic constituents **bufotenine** and **bufotenidine;** the former also occurs in the fungus *Amanita mappa* (*Wieland et al.*, Ann., 1953, **581,** 10) and (0.94%) in the shrub *Piptadenia peregrina* (*V. L. Stromberg*, J. Amer. chem. Soc., 1954, **76,** 1707):

Bufotenine

Bufotenidine

Bufotenine contains a dimethylamino group; has two active hydrogen atoms, one phenolic, and gives indole colour reactions. Its methyl ether was synthesised by treating 5-methoxyindolylmagnesium bromide with chloroacetonitrile and conversion of the resulting $MeOC_8H_5N{\cdot}CH_2{\cdot}CN$ *via* the ester into carbinol and thence through the bromide into the amine. This was finally demethylated to bufotenine (*H. Wieland et al.*, Ann., 1934, **513,** 1; *T. Hoshino* and *K. Shimodaira*, Bull. chem. Soc., Japan, 1936, **11,** 221). For more convenient syntheses, see *Speeter et al. (loc. cit.)*; *A. Stoll et al.*, Helv., 1955, **38,** 1452. The associated compound **bufothionine** is the acid sulphuric ester of a phenol, $C_{12}H_{14}ON_2$, which can be hydrogenated to bufotenine (*H.* and *T. Wieland*, Ann., 1937, **528,** 234).

Wolff–Kishner reduction of 3,4-dihydro-β-carboline (XXV), prepared by treatment of *N*(b)-thioformyltryptamine with acid, gives 2-methyltryptamine (*I. Fleming* and *Harley-Mason*, J. chem. Soc., C, 1966, 425):

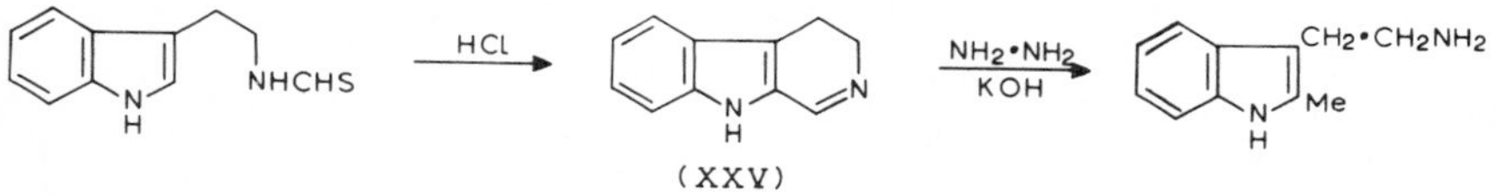

(XXV)

(ix) Aldehydes and ketones

Indole-2-carbaldehyde may be obtained in rather poor yield by oxidation of 2-hydroxymethylindole with potassium permanganate, but the yield is substantially increased by using activated manganese dioxide in ether.

Indole-2-carbaldehyde condenses readily with nitromethane to yield 2-(2′-nitrovinyl)-indole, which with lithium tetrahydridoaluminate affords 2-(2′-aminoethyl)indole (iso-tryptamine) (*Harley-Mason* and *E. H. Pauri*, J. chem. Soc., 1963, 2565):

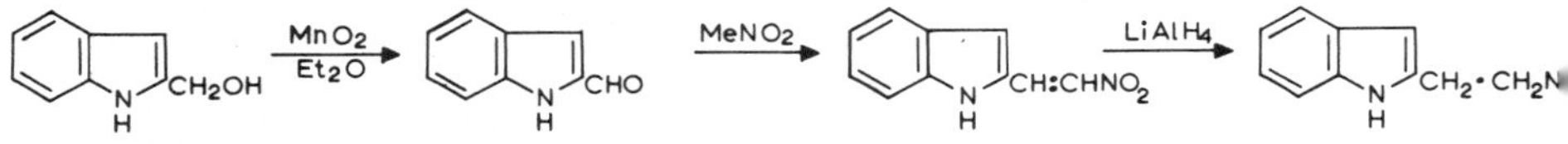

Indolecarbaldehydes are prepared by treating indole with *N*-dimethylformamide and phosphorus oxychloride (*G. F. Smith*, *ibid*., 1954, 3842), from magnesylindoles and ethyl formate (*R. Majima* and *M. Kotake*, Ber., 1922, **55**, 3859), and by the Reimer–Tiemann (*W. J. Boyd* and *W. Robson*, Biochem. J., 1935, **29**, 555) or the Gatterman synthesis (*H. Fischer* and *K. Pistor*, Ber., 1923, **56**, 2313), in each case with preferential substitution in position 3.

Indole-3-carbaldehyde is best prepared by the first method; in the Reimer–Tiemann method it is obtained along with 3-chloroquinoline and related reactions suggest that two mechanistic paths are followed (*C. W. Rees* and *C. E. Smithen*, J. chem. Soc., 1964, 928), but it is likely that the aldehyde is obtained without the formation of the intermediate shown on p. 415:

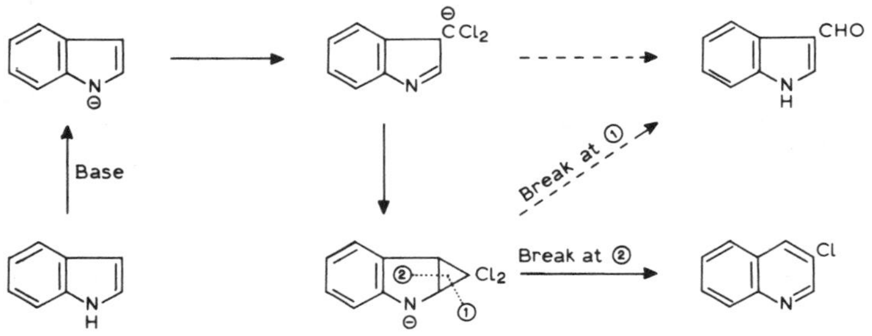

Indole-3-carbaldehyde, an important starting point for synthetic work, can also be produced by treating gramine with phenylhydroxylamine, oxidising the resulting hydroxylamine derivative to a nitrone, and hydrolysing it (*Thesing*, Ber., 1954, **87**, 507); and, very simply, from potassio-indole and carbon monoxide (135–150°/480 atm) (*F. T. Tyson* and *J. T. Shaw*, J. Amer. chem. Soc., 1952, **74**, 2273); it is formed when tryptophan is oxidised with ferric chloride.

Indole-3-carbaldehyde possesses only some of the properties of an aromatic aldehyde: it does not give a cyanohydrin (*D. E. Ames et al.*, J. chem. Soc., 1956, 1984) and this lack of reactivity is consistent with the view that it may be regarded as the vinylogue of an amide *(Thesing, loc: cit.)*; i.r. absorption spectroscopy confirms that structure XXVI is an important contributor to the resonance hybrid which represents the aldehyde (*D. G. O'Sullivan* and *P. W. Sadler*, J. chem. Soc., 1959, 877):

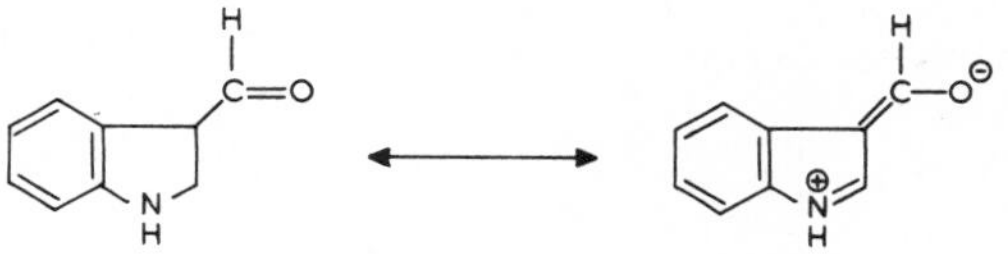

(XXVI)

under normal conditions it fails to undergo Cannizzaro's reaction or the benzoin reaction; it is methylated at nitrogen with dimethyl sulphate and aqueous alkali (*H. Wieland et al.*, Ann., 1934, **513,** 1). Heated with mineral acid, the aldehydes yield pigments of the diindolylmethene series (p. 468) with loss of formic acid.

Indigo is quickly decomposed by alkali at 150° into anthranilic acid and 3-**hydroxyindole**-2-**carbaldehyde** (indoxylaldehyde):

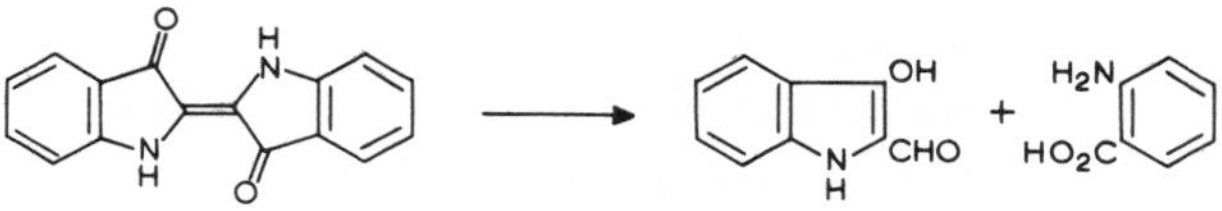

Indigo

In the same way Thioindigo Scarlet R (C.I., 1225) yields the isomeric 2-hydroxyindole-3-carbaldehyde and thiosalicylic acid (*P. Friedlander* and *S. Kielbasinski*, Ber., 1911, **44,** 3098).

Ketones are produced by treating magnesylindoles with acid chlorides (*B. Oddo* and *L. Sessa*, Gazz., 1911, **41,** I, 234), or by the Hoesch (*R. Seka*, Ber., 1923, **56,** 2058) or the Friedel–Crafts synthesis (*W. Borsche* and *H. Groth*, Ann., 1941, **549,** 238), in each case with preferential substitution in the 3-position. Free indole, with acetic anhydride gives mainly 1,3-diacetylindole (*J. E. Saxton*, J. chem. Soc., 1952, 3592).

Indole-2-carbaldehyde, m.p. 140–142°, *thiosemicarbazone*, m.p. 229° (decomp.), 2,4-*dinitrophenylhydrazone*, m.p. 315–320° (decomp.) (*Harley-Mason* and *Pauri, loc. cit.; W. I. Taylor*, Helv., 1950, **33,** 164); **indole-3-carbaldehyde,** m.p. 195°, *oxime*, m.p. 197–200°, *semicarbazone*, m.p. 265–270° (decomp.), *phenylhydrazone*, m.p. 198°, 2,4-*dinitrophenylhydrazone*, m.p. 300° *(Smith, loc. cit.; Thesing, loc. cit.)*; 1-**methylindole-3-carbaldehyde,** m.p. 65° (*Wieland et al., loc. cit.)*; 2-**methylindole-3-carbaldehyde,** m.p. 202–203°, *oxime*, m.p. 156–157°, *semicarbazone*, m.p. 224° (decomp.), *phenylhydrazone*, m.p. 201° (*G. Barger* and *Ewins*, Biochem. J., 1917, **11,** 59); 5-**methylindole-3-carbaldehyde,** m.p. 151° (*W. Robson*, J. biol. Chem., 1924, **62,** 507); 7-**methylindole-3-carbaldehyde,** m.p.

202° (*W. J. Boyd* and *Robson*, Biochem. J., 1935, **29,** 555). 2-**Hydroxyindole**-3-**carbaldehyde** *(oxindolealdehyde)*, m.p. 213°, 1-*methyl* deriv., m.p. 192° (from methyloxindole, *cf.* p. 453); 3-**hydroxyindole**-2-**carbaldehyde** *(indoxylaldehyde)*, decomp. 160°, *acetyl* deriv., m.p. 145°; 3-**indolylacetaldehyde,** *dimedone* deriv., m.p. 148–152° (*E. R. H. Jones et al.*, J. chem. Soc., 1952, 3172).

2-**Acetyl**-1-**methylindole,** m.p. 72° (*O. Diels* and *A. Köllisch*, Ber., 1911, **44,** 263); 2-**acetyl**-3-**methylindole,** m.p. 147° (*B. Oddo*, Gazz., 1913, **43,** II, 190); 3-**acetylindole,** m.p. 194° (*J. W. Baker*, J. chem. Soc., 1964, 461); 3-**chloroacetylindole,** m.p. 214° (*R. Majima*, Ber., 1923, **55,** 3865); 3-**acetyl**-2-**methylindole,** m.p. 196°; 3-**benzoyl**-2-**methylindole,** m.p. 181°. Magnesylindoles with phthalic anhydride yield *o*-carboxybenzoylindoles (*Oddo* and *C. Toffoli*, Gazz., 1930, **60,** 3). 3-**Indolylacetone,** m.p. 115 117.5° *(Jones et al., loc. cit.)*; 2-**acetyl**-3,3-**dimethyl**-3*H*-**indole,** m.p. 130°, obtained from the 2-aldoxime (p. 443) *via* the nitrile (*G. Plancher* and *D. Giumelli*, Chem. Ztbl., 1910, I, 451).

(x) Carboxylic acids

Indoles may be carb(eth)oxylated, preferentially in position 3, by treating the magnesyl derivatives with carbon dioxide at a high temperature or with ethyl chloroformate (p. 412), and also by heating the free indole with sodium in carbon dioxide. The Fischer synthesis has been used most frequently, notably in combination with the Japp–Klingemann reaction, to produce indole-2-carboxylic acids (p. 401), but is least effective for *N*-unsubstituted 3-carboxylic acids, since the initial hydrazone of a *β*-oxo-acid is liable to yield a pyrazolone.

Indole-2-carboxylic acids are accessible by synthetic method *iv* (p. 402). Indole-2,3-dicarboxylic ester is obtained by pyrolysis of the primary adduct of hydrazobenzene with dimethyl acetylenedicarboxylate (*O. Diels* and *J. Reese*, Ann., 1934, **511,** 168) and related monomethyl esters by treating with aqueous methanolic hydrogen chloride the orange oil, one of the products formed by the reaction between arylhydroxylamines and dimethyl acetylenedicarboxylate (*E. H. Huntress*, *T. E. Lesslie* and *W. H. Hearon*, J. Amer. chem. Soc., 1956, **78,** 419).

The acids lose carbon dioxide easily, the 3-compounds on simply boiling with water. Indole-2-carboxylic acid yields a dilactam analogous to pyrocoll (p. 362). Indole-3-carboxylic acid occurs in cauliflower and savory cabbage (*Z. Procházka*, Coll. Czech. chem. Comm., 1959, **24,** 1368).

Indoxylic acid (3-**hydroxyindole**-2-**carboxylic acid**), is produced by potash fusion of phenylglycine-*o*-carboxylic acid (*D. Vorländer* and *J. von Pfeiffer*, Ber., 1919, **52,** 325), and its ester by treating diethyl phenylglycine-*o*-carboxylate with sodium ethoxide (*Vorländer*, *ibid.*, 1902, **35,** 1683), or by adding to powdered sodium and dimethyl phenylglycine-*o*-carboxylate in benzene one drop of anhydrous methanol (*A. Robertson*, J. chem. Soc., 1927, 1937); and by heating diethyl anilinomalonate above 250° (*R. Blank*, Ber., 1898, **31,** 1812):

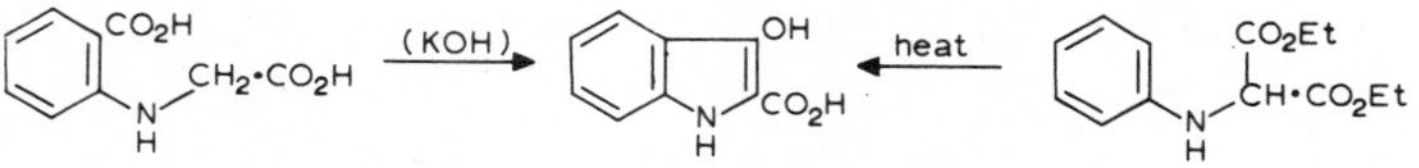

Indoxylic acid

or by reducing ethyl *o*-nitrophenylpropiolate or the derived isatogenic ester (see p. 443) with ammonium sulphide (*A. von Baeyer*, Ber., 1881, **14,** 1741).

Indoxylic acid is decarboxylated by boiling with water, and is easily oxidised to indigo. Hydrolysis of 1-acetylindoxylic acid with boiling dilute sodium carbonate solution gives about 20% of indigo and 50% of *N*-acetylanthranilic acid (*G. Spencer*, J. Soc. chem. Ind., 1931, **50,** 63T):

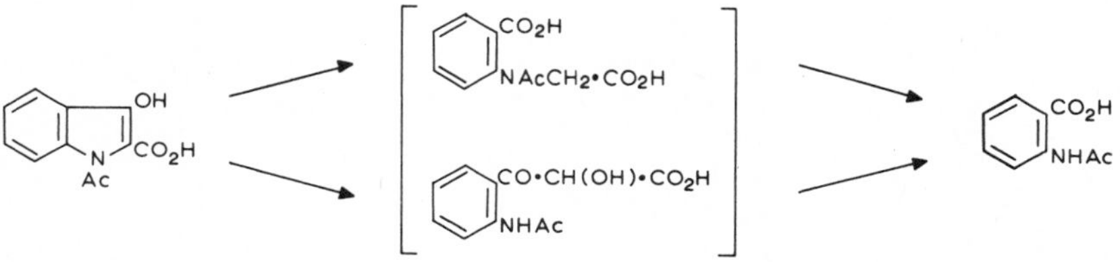

Aldehydes and other reactive carbonyl compounds condense with indoxylic acid with elimination of carbon dioxide and formation of the same indogenides as are obtained from indoxyl, and nitrosobenzene similarly yields isatin-2-anil (p. 462) (*R. Pummerer*, Ber., 1911, **44,** 338). Ethyl indoxylate is easily oxidised to ethyl indoxanthate (p. 458) (*von Baeyer, loc. cit.*); it is soluble in sodium carbonate solution and yields by direct ethylation ethyl 3-ethoxyindole-2-carboxylate.

Indole-3-acetic acid (heteroauxin) is a plant growth hormone; it is formed from tryptophan by irradiation (*A. Berthelot* and *G. Amoureux*, Compt. rend., 1938, **206,** 699), occurs in wine, and is easily decarboxylated to skatole. *J. B. Koepfli et al.* (J. biol. Chem., 1938, **122,** 763) discuss the constitutional specificity of auxin activity, which is shared by many homocyclically and heterocyclically substituted aliphatic acids. Heteroauxin can be synthesised *via* the ester, from indole and ethyl diazoacetate (*R. W. Jackson* and *R. H. Manske*, Canad. J. Res., 1935, **13B,** 170); and *via* the nitrile, from indolylmagnesium bromide and chloroacetonitrile (*R. Majima* and *T. Hoshino*, Ber., 1925, **58,** 2042), or from gramine (p. 429). It is obtained in good yield in one operation by treating the crude condensation product from indole, formaldehyde, and methylaniline with sodium cyanide and sodium hydroxide (*J. Thesing et al.*, *ibid.*, 1955, **88,** 1295), or by heating indole and potassium glycolate at 250° (*H. E. Johnson* and *D. G. Crosby*, J. org. Chem., 1963, **28,** 1246):

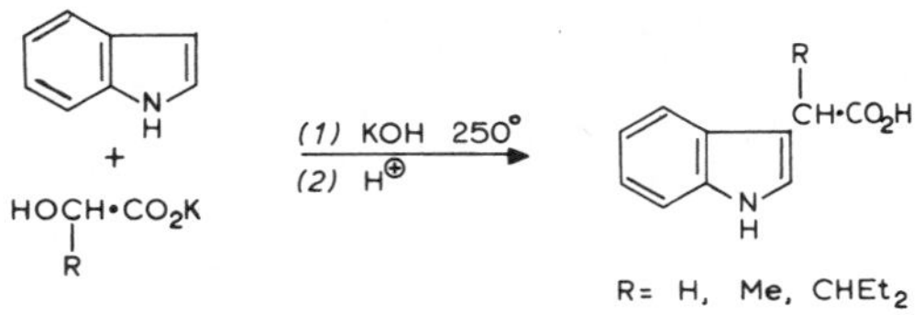

+ HOCH·CO$_2$K (R) —(1) KOH 250°; (2) H$^{\oplus}$→ CH(R)·CO$_2$H

R= H, Me, CHEt$_2$

The cyanohydrins of acetaldehyde and acetone give with indole in presence of alkali α-3-**indolylpropionic** and α-3-**indolylisobutyric acids**; the latter is obtained also (30% yield) from indole, acetone, chloroform and sodium hydroxide (*H. Erdtman* and *Å. Jönsson*, Acta Chem. Scand., 1954, **8,** 119). Substituted 2-methylindole-3-acetic acids are made from the arylhydrazones of laevulic acid (*F. J. Stevens* and *D. H. Higginbotham*, J. Amer. chem. Soc., 1954, **76,** 2206; *M. W. Bullock* and *J. J. Hand*, *ibid.*, 1956, **78,** 5852).

1-(4-**Chlorobenzoyl**)-5-**methoxy**-2-**methylindole**-3-**acetic acid** is a powerful anti-inflammatory and antipyretic agent (*T. Y. Shen et al.*, *ibid.*, 1963, **85,** 488).

The auxin 3-**indolylacetonitrile** is probably derived in plants from 3-indolylpyruvic acid oxime (*A. Ahmad* and *I. D. Spencer*, Canad. J. Chem., 1960, **38,** 1625).

3-Indolylacet-(β-phenylethyl)amide on treatment with polyphosphoric acid is converted in compounds XXVII and XXVIII, both being formed by opening of the indole ring (*Thesing* and *F. H. Funk*, Ber., 1958, **91,** 1546):

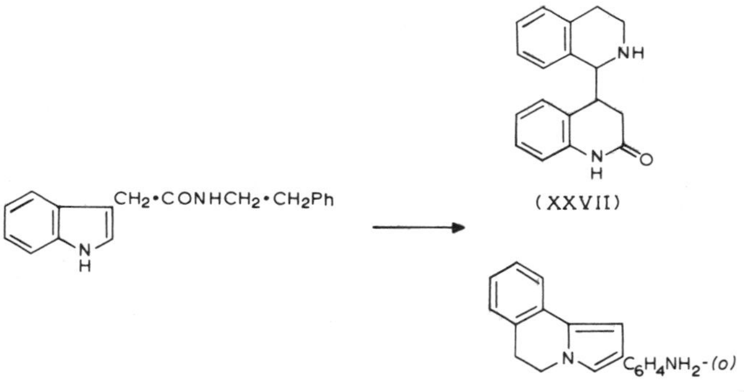

CH$_2$·CONHCH$_2$·CH$_2$Ph →

(XXVII)

C$_6$H$_4$NH$_2$-(*o*)

(XXVIII)

Tryptophan, L-α-amino-β-3-indolylpropionic acid, an indispensable amino acid not synthesised in the mammal, can be isolated by pancreatic digestion of casein and purified through the mercury complex (*G. J. Cox* and *H. King*, Org. Synth., 1930, **10,** 100). It is destroyed with formation of "humin" in acid hydrolysis of protein, but survives in the racemic form when baryta is used. DL-Tryptophan (resolution: *C. P. Berg*, J. biol. Chem., 1933, **100,** 79) can be prepared from indole-3-carbaldehyde by the hippuric acid or the hydantion method (*B. A. Hems et al.*, J. chem. Soc., 1944, 629), or, better, by the action of ethyl acetamidomalonate on gramine (*H. R. Snyder et al.*, J. Amer. chem. Soc., 1945, **67,** 38) or gramine methosulphate (*N. F. Albertson et al.*, *ibid.*, p. 59):

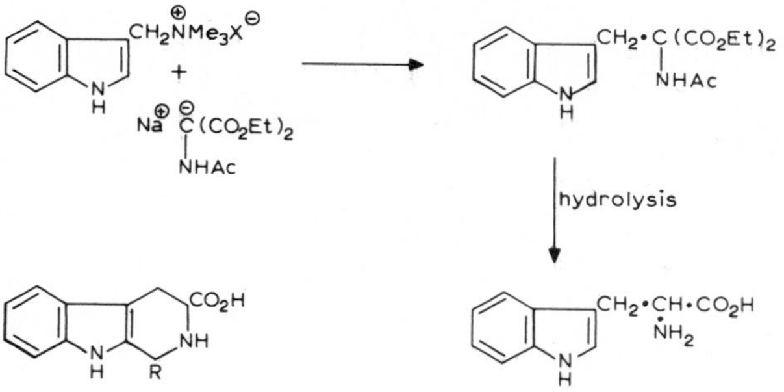

$CH_2\overset{\oplus}{N}Me_3X^{\ominus}$ + $Na^{\oplus}\ \overset{\ominus}{C}(CO_2Et)_2$ NHAc → $CH_2{\cdot}C(CO_2Et)_2$ NHAc

hydrolysis

$CH_2{\cdot}CH{\cdot}CO_2H$ NH_2

(XXIX)

Tryptophan

Tryptophan gives the Ehrlich test, but is better estimated (*J. L. D. Shaw* and *W. D. Macfarlane*, J. biol. Chem., 1940, **132,** 387) by the Adamkiewicz–Hopkins reaction, a violet colour with glyoxylic acid in concentrated sulphuric acid. The reaction, in which glyoxylic acid may be replaced by other aldehydes, depends on ring-closure to a tetrahydrocarbolinecarboxylic acid (XXIX) (*W. Robson et al.*, J. chem. Soc., 1941, 153).

Tryptophan can be synthesised, at least in lower organisms, *via* indole, from anthranilic acid:

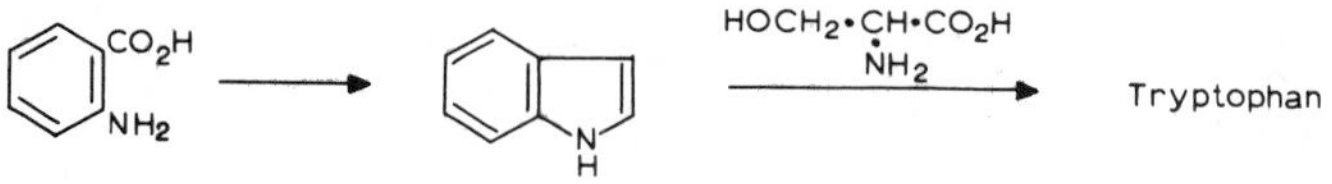

CO_2H NH_2 → (indole) $\xrightarrow{HOCH_2{\cdot}CH(NH_2){\cdot}CO_2H}$ Tryptophan

it thus belongs to the same stereochemical series as most natural amino acids. By bacterial degradation it has afforded indolyl-acetic, -propionic, and -lactic acids, tryptamine and tryptophol, skatole, and indole. The acid is broken down in the animal body to formylkynurenine and then to kynurenine, which can (*a*) complete the metabolic cycle by degradation to anthranilic acid, (*b*) suffer further oxidation and ring-closure to kynurenic acid, or (*c*) undergo nuclear hydroxylation, followed by ring-opening and re-synthesis, giving the essential nutrient, nicotinic acid; alternatively, hydroxykynurenine yields hydroxykynurenic acid (xanthurenic acid) (*A. Neuberger*, Ann. Reports, 1944, **41,** 237; *E. Kodicek*, *ibid.*, 1951, **48,** 282):

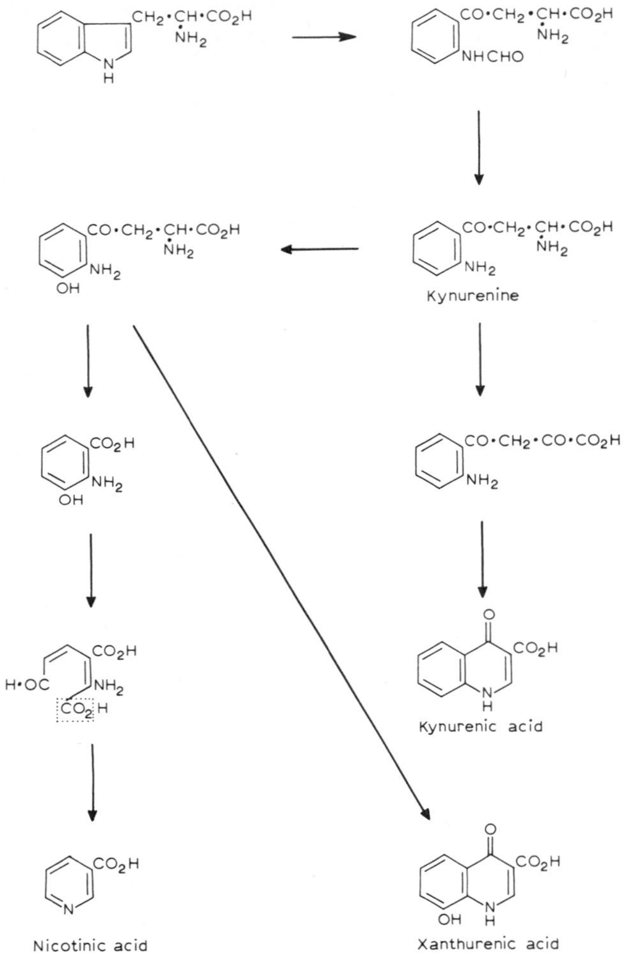

Kynurenine

Kynurenic acid

Nicotinic acid

Xanthurenic acid

Urinary excretion of kynurenine or of xanthurenic acid by dogs is associated with a diet including tryptophan but deficient in pyridoxine (*S. Lepkovsky et al.*, J. biol. Chem., 1945, **160,** 155).

Phalloidine, a toxic crystalline peptide from the fungus *Amanita phalloides*, yields on hydrolysis 2-**hydroxytryptophan** (XXX), along with alanine, cysteine, and the "unnatural" L-hydroxyproline (*H. Wieland* and *B. Witkop*, Ann., 1940, **543,** 171). Hydroxytryptophan has been synthesised and resolved (*T. Miwa et al.*, J. chem. Soc. Japan, 1953, **74,** 113):

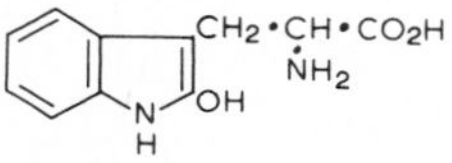

(XXX) 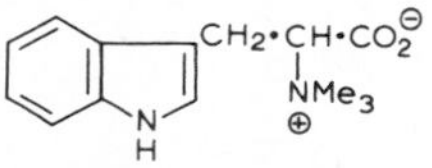 Hypaphorine

Abrine, present in the seeds of *Abrus precatorius* (*T. Hoshino*, Ann., 1935, **520,** 31), is tryptophan monomethylated on the side-chain nitrogen atom, and like tryptophan it yields on further methylation the ester iodide of the betaine **hypaphorine,** one of the alkaloids of *Erythrina* spp. (*K. Folkers* and *F. Koniuszy*, J. Amer. chem. Soc., 1939, **61,** 1232).

The biological interest of tryptophan has instigated the preparation of substitution products: 5- and 7-hydroxytryptophan (*A. Ek* and *Witkop*, *ibid*., 1954, **76,** 5579); 2-, 4-, 5-, 6-, and 7-methyltryptophan (*H. N. Rydon*, J. chem. Soc., 1948, 705).

Melting point and optical data on indolecarboxylic acids and their derivatives are collected together in Table 2.

(d) 3H-*Indoles (indolenines)*

Alkylation of indoles leads to successive replacement of the hydrogen atoms in the pyrrole ring to fully alkylated 3*H*-indoles and their quaternary derivatives. Thus indole, 1-, 2- and 3-methylindole, 2,3-dimethylindole and 1,2,3-trimethylindole all yield tetramethyl-3*H*-indolium (tetramethylindoleninium) iodide on methylation:

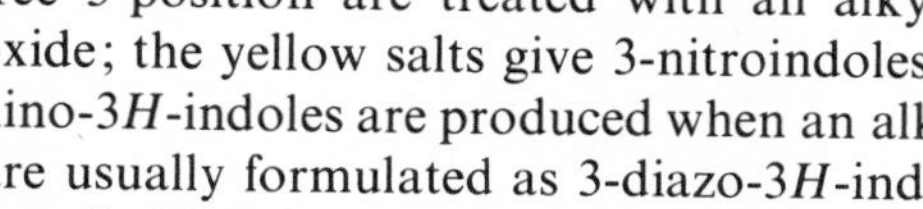

As shown on p. 422 salts of 3*H*-indole-3-nitronic acids are obtained when indoles with a free 3-position are treated with an alkyl nitrate and an alkali metal alkoxide; the yellow salts give 3-nitroindoles on acidification. Similarly, 3-oximino-3*H*-indoles are produced when an alkyl nitrite is used. 3-Diazoindoles are usually formulated as 3-diazo-3*H*-indoles (p. 423).

3*H*-Indoles are well-defined bases in which the carbon-nitrogen double bond is reactive and may add acetic anhydride, ammonia and amines (*H. Leuchs et al*., Ann., 1928, **461,** 27) as shown for 3,3-dimethyl-3*H*-indole:

TABLE 2
INDOLECARBOXYLIC ACIDS AND DERIVATIVES

Substituents	*m.p.°*	*Ref.*	*Substituents*	*m.p.°*	*Ref.*
1-CO_2H	108	1	1,3-$(CO_2Et)_2$	102–103	19
1-$CONH_2$	198–199	2	2,3-$(CO_2Me)_2$	114	18, 20
3-Me-1-CO_2H	162.5	3	1-Me-2,3-$(CO_2H)_2$	dec. 218	21
2-CO_2H	203	4	3-OH-2-CO_2H (*Indoxylic acid*)	dec. 122–123	
2-CO_2Me	151–152		3-OH-2-CO_2Et	116	
2-CN	101	5	3-OEt-2-CO_2H	dec. 160	
1-Me-2-CO_2H	212	4, 6	*Ethyl isatogenate*	115	22
1-OH-2-CO_2H	dec.* 160		2,6-$(CO_2H)_2$	310	17
3-Me-2-CO_2H	164–167	7, 8	2,6-$(CO_2Et)_2$	132	
3-Me-2-CO_2Me	148.	9	5,6-$(OH)_2$-2-CO_2H	dec. 234	23
5-Me-2-CO_2H	229–230	10	3-$CH_2 \cdot CO_2H$ (*Heteroauxin*)	168	24
6-Me-2-CO_2H	217	11	3-CH_2-CO_2Et	42–43	
7-Me-2-CO_2H	170	10	1-Me-3-$CH_2 \cdot CO_2H$	128	25
3-CO_2H	dec. 220–224	5	5-MeO-3- $CH_2 \cdot CO_2H$	dec. 146	26
3-CO_2Me	147–148		3-$CH(OH) \cdot CO_2Me$	82.5	27
3-CO_2NH_2	200		3-$CO \cdot CO_2H$	dec. 216	27
3-CN	178		3-$CH(NH_2) \cdot CO_2H$	dec. 221	27
1-Me-3-CO_2H	dec. 212	12	3-$CH_2 \cdot CH_2 \cdot CO_2H$	134	28
2-Me-3-CO_2H	190	10, 13	3-$CH:CH \cdot CO_2H$	195	29
2-Me-3-CO_2Me	165		3-$CH_2 \cdot CH \cdot CO_2H$ | NH_2 (*Tryptophan*) $[\alpha]_D$	L dec. 289 $-32.1°$, H_2O DL dec. 265	
4-Me-3-CO_2H	189	14			
5-Me-3-CO_2H	202	10	3-$CH_2 \cdot CH \cdot CONH_2$ | NH_2	L 167–170	30
7-Me-3-CO_2H	228		3-$CH_2 \cdot CH \cdot CO_2H$ | NHAc	L 189–190 DL 207–208	
4-CO_2H	213–214	15	3-$CH_2 \cdot CH \cdot CO_2H$ | NHCONHPh	L 166	
5-CO_2H	208–209	16	3-$CH_2 \cdot CH \cdot CO_2H$ | NHMe	L dec. 295	

* dec. = decomposition point.

TABLE 2 *(continued)*

Substituents	*m.p.°*	*Ref.*	*Substituents*	*m.p.°*	*Ref.*
6-CO_2H	243–244	17	(*Abrine*) $[\alpha]_D$	+44.4° HCl	
7-CO_2H	198–199	18	3-$CH_2 \cdot CH(N^{\oplus}Me_3) \cdot CO_2^{\ominus}$	L dec. 236–237	
			(*Hypaphorine*) $[\alpha]_D$	+113.1° H_2O	
			2-$CH_2 \cdot CH(NH_2) \cdot CO_2H$ (*Isotryptophan*)	DL dec. 220	31

References

1 *B. Oddo* and *L. Sessa*, Gazz., 1911, **41,** [i], 234.
2 *J. D. Dutcher* and *A. K. Kjaer*, J. Amer. chem. Soc., 1951, **73,** 4139.
3 *E. O. von Lippmann*, Ber., 1924, **57,** 257.
4 *J. R. Johnson et al.*, J. Amer. chem. Soc., 1945, **67,** 423.
5 *F. P. Doyle et al.*, J. chem. Soc., 1956, 2853.
6 *E. Fischer* and *O. Hess*, Ber., 1884, **17,** 559.
7 *Oddo*, Gazz., 1912, **42,** [i], 370.
8 *W. O. Kermack, W. H. Perkin* and *R. Robinson*, J. chem. Soc., 1921, **119,** 1634.
9 *J. A. Elvidge* and *F. S. Spring*, *ibid.*, 1949, S135.
10 *O. Kruber*, Ber., 1926, **59,** 2752.
11 *A. Reissert*, *ibid.*, 1897, **30,** 1030.
12 *W. B. Whalley*, J. chem. Soc., 1954, 1651.
13 *J. Houben* and *W. Fischer*, Ber., 1931, **64,** 2640.
14 *Kruber*, *ibid.*, 1929, **62,** 2877.
15 *F. C. Uhle*, J. Amer. chem. Soc., 1949, **71,** 761.
16 *H. Singer* and *W. Shive*, J. org. Chem., 1955, **20,** 1458.
17 *Kermack*, J. chem. Soc., 1924, **125,** 2285.
18 *Singer* and *Shive*, J. Amer. chem. Soc., 1955, **77,** 5700.
19 *R. Majima* and *M. Kotake*, Ber., 1930, **63,** 2237.
20 *E. H. Huntress, T. E. Lesslie* and *W. H. Hearon*, J. Amer. chem. Soc., 1956, **78,** 419.
21 *G. Reif*, Ber., 1909, **42,** 3036.
22 *P. Pfeiffer*, Ann., 1916, **411,** 151.
23 *A. Robertson et al.*, J. chem. Soc., 1949, 2061.
24 *H. E. Johnson* and *D. G. Crosby*, J. org. Chem., 1963, **28,** 1246.
25 *F. E. King* and *P. l'Écuyer*, J. chem. Soc., 1934, 1901.
26 *S. P. Findlay* and *G. Dougherty*, J. org. Chem., 1948, **13,** 560.
27 *J. W. Baker*, J. chem. Soc., 1940, 458.
28 *R. H. F. Manske* and *Robinson*, *ibid.*, 1927, 240.
29 *J. S. Moffatt*, *ibid.*, 1957, 1442.
30 *L. C. Bauguess* and *C. P. Berg*, J. biol. Chem., 1934, **106,** 615.
31 *H. L. Snyder* and *P. L. Cook*, J. Amer. chem. Soc., 1956, **78,** 969.

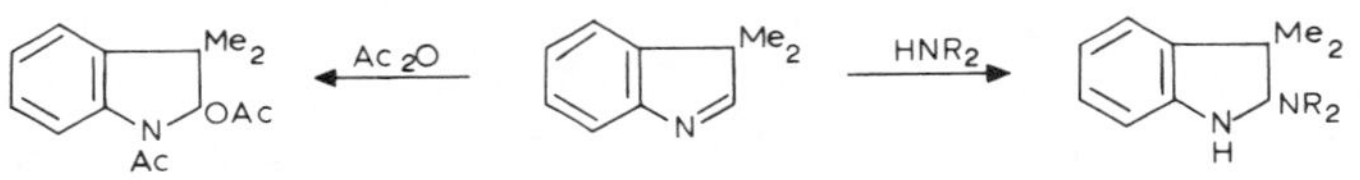

2-Halogeno- and 2-alkyl-3*H*-indoles show similar reactivity to the corresponding pyridines.

Tetramethyl-3*H*-indolium iodide undergoes the transformations outlined below; compounds XXXI–XXXVI have been synthesised:

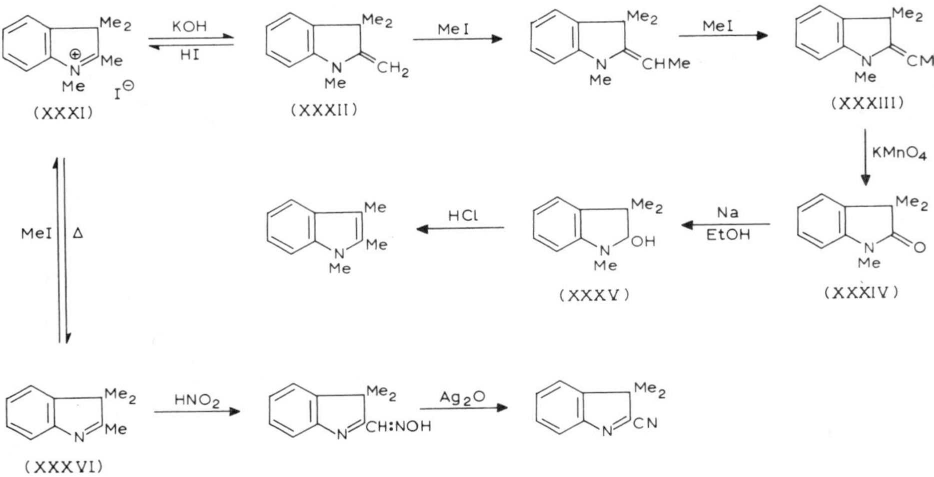

The salt XXXI possesses a reactive methyl group, as shown by its reaction with ethyl orthoformate to give the red dye Astraphloxine FF.

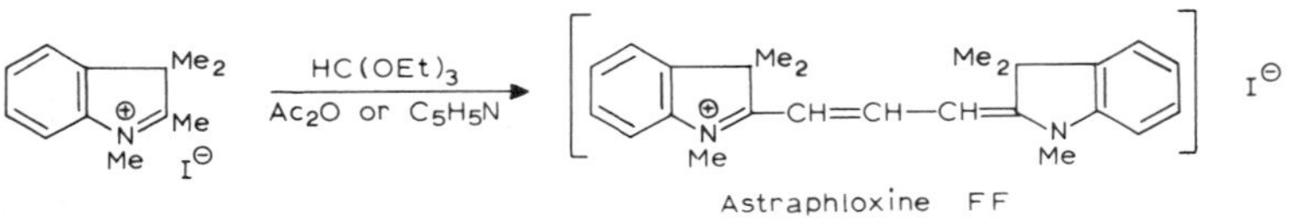

This reactivity makes the compound an important intermediate in the preparation of polymethine colouring matters (see also trimethylmethylene-3*H*-indole [p. 447] and the subsequent chapter on cyanine dyes).

Unstable 3,3-**dimethyl**-3*H*-**indole** forms a *trimer*, m.p. 216° (*K. Brunner*, Monatsh., 1895, **16,** 849). 2,3,3-**Trimethyl**-3*H*-**indole** (XXXVI), m.p. 229° (*P. Grammaticakis*, Compt. rend., 1940, **210,** 569), *methiodide* (XXXI), decomp. 253° (*Brunner*, Ber., 1898, **31,** 612). With chloroform and alkali 2,3-dimethylindole gives 2,3-*dimethyl*-3-*dichloromethyl*-3H-*indole* (p. 416), m.p. 73–74°.

3-**Oximino**-3*H*-**indole,** decomp. 170° (*W. Madelung*, Ann., 1914, **405,** 58), 2-*methyl*-3-*oximino*-3H-*indole*, decomp. 198° (*M. Spica* and *F. Angelico*, Gazz., 1899, **29,** II, 49), 3-*oximino-2-phenyl*-3H-*indole*, decomp. 259°, O-*ethyl* deriv., m.p. 43°) (*T. Ajello*, *ibid*., 1939, **69,** 646).

Salicylaldehyde and 2-methylindole give 2-**methyl-3-salicylidene**-3*H*-**indole,** m.p. 185° (*G. O. Burr* and *R. A. Gortner*, J. Amer. chem. Soc., 1924, **46,** 1224).

3,3-**Dimethyl**-3*H*-**indole-2-carbaldoxime** m.p. 156°, and the derived 2-*cyano*-3,3-*dimethyl*-3H-*indole*, b.p. 150–151°/30 mm (*G. Plancher* and *D. Bettinelli*, Gazz., 1899, **29,** I, 106).

3-**Diazo-2-methyl**-3*H*-**indole,** m.p. 94° (*V. Castellano* and *A. d'Angelo*, Chem. Ztbl., 1905, II, 899); 3-*diazo-2-phenyl*-3H-*indole*, m.p. 107–108°; 3-*diazo-5-methoxy-2-(4-methoxyphenyl)*-3H-*indole*, m.p. 136–138°; 3-*diazo-2-phenyl-6,7-benzo*-3H-*indole*, m.p. 166–167° (*H. P. Patel* and *J. M. Tedder*, J. chem. Soc., 1963, 4593); 3-**benzeneazoindole,** m.p. 133–134° (*Madelung*, Ber., 1924, **57,** 241); for azo-indoles (*Huang-Hsinmin* and *F. G. Mann*, J. chem. Soc., 1949, 2903).

(i) *Oxo*-3H-*indoles (indolones)* and isatogens*

3-Oxo-3*H*-indole (indolone) is a dehydrogenated indoxyl (3-oxoindoline), and the deep scarlet 3-oxo-2-phenyl-3*H*-indole is obtained by the following route (*L. Kalb* and *J. Bayer*, Ber., 1912, **45,** 2150):

NH_2 PbO$_2$ → NH → HCL → O → Ph, OH

On treatment with alcoholic sodium hydroxide the oxo-3*H*-indole undergoes a benzilic type rearrangement and the C:N double bond will add ammonia, methanol, and sodium bisulphite.

Isatogens are the *N*-oxides of 3-oxo-3*H*-indoles, which may be isolated as intermediates in the reduction of isatogens to indoxyls. *o*-Nitrophenylacetylenes are isomerised in sulphuric acid or by irradiation in pyridine to isatogens (3-oxo-3*H*-indole *N*-oxides, 3-oxo-3*H*-indole 1-oxides):

C:C·CO_2Et, NO_2 → O, $N^{\oplus}$, CO_2Et, $O^{\ominus}$

The related *o*-nitrostilbene dichloride, *o*-nitrostilbene, and *o*-nitrotolane all give on irradiation the same 2-arylisatogen (*P. Pfeiffer*, Ann., 1916, **411,** 72; *J. S. Splitter* and *M. Calvin*, J. org. Chem., 1955, **20,** 1086):

* This older nomenclature is no longer acceptable.

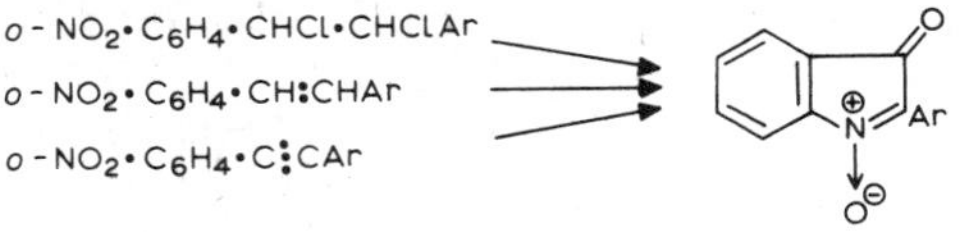

o-$NO_2 \cdot C_6H_4 \cdot CHCl \cdot CHClAr$
o-$NO_2 \cdot C_6H_4 \cdot CH{:}CHAr$
o-$NO_2 \cdot C_6H_4 \cdot C{⋮}CAr$

Isatogens are also obtained by the sequence:

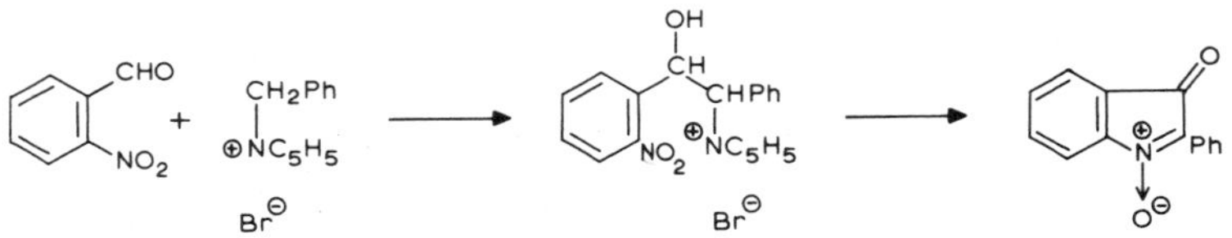

and by the action of alkaline reagents on substituted 1-vinylpyridinium salts (*F. Kröhnke et al.*, Ber., 1951, **84,** 932; 1952, **85,** 376; 1953, **86,** 1500).

They are yellow to red solids which liberate iodine from potassium iodide and yield, in addition to the normal oxime, an isomer formulated for example as XXXVII. They combine with alcohols, acetic anhydride and phenylmagnesium bromide, giving, for example, XXXVIII, XXIX and XL, respectively and can be isomerised by acids to isoisatogens XLI having an oxaziridine ring (*P. Ruggli et al.*, Helv., 1939, **22,** 140, 411):

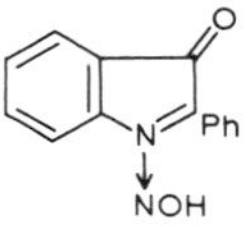

(XXXVII) 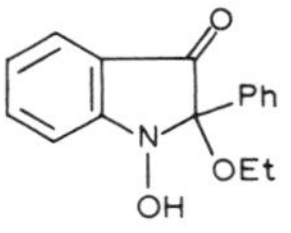 (XXXVIII) 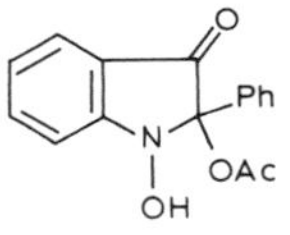 (XXXIX) 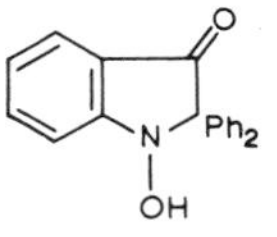 (XL) 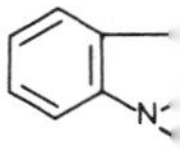 (XLI)

They undergo ring expansion and nitrogen insertion reactions with tetracyanoethylene or trichloroacetonitrile, ammonia, and acetylenes to give quinazolines, cinnolinols and quinolines (*W. E. Noland* and *D. A. Jones*, J. org. Chem., 1962, **27,** 341; *Noland* and *R. F. Modler*, J. Amer. chem. Soc., 1964, **86,** 2086):

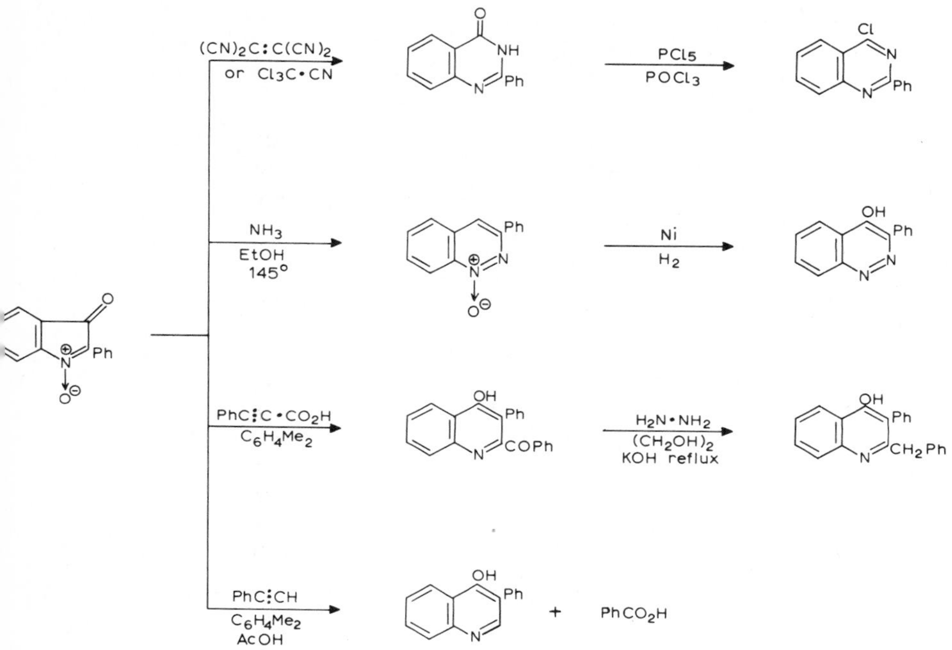

On hydrolysis by alkali, 2-ethoxycarbonylisatogen (the ethyl ester of isatogenic acid, the tautomeric 3*H*-indole form of indoxylic acid, see p. 434) affords isatin:

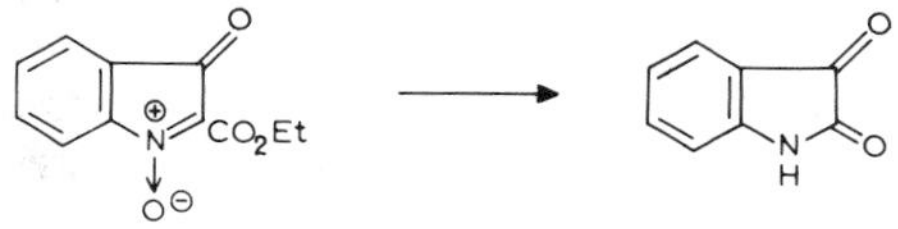

3-**Oxo**-2-**phenyl**-3*H*-**indole,** m.p. 102°, *imine*, m.p. 114.5°, *oxime*, m.p. 259°; 2-**phenylisatogen,** m.p. 186° (*Ruggli et al.*, Helv., 1937, **20,** 250), 3-*oxime*, decomp. 240°, 1-*oxime*, m.p. 167°; 2-*phenylisoisatogen*, m.p. 94°; 2-**chloro**-3-**oxo**-3*H*-**indole** *(isatin chloride)*, decomp. 180° (*A. von Baeyer*, Ber., 1879, **12,** 456).

2. Hydroindoles

(a) Indolines (dihydroindoles)

(i) Synthesis

(*1*) Indolines (2,3-dihydroindoles) are obtained by reducing indoles, best electrolytically (*J. von Braun* and *W. Sobecki*, Ber., 1911, **44,** 2158) or catalytically over copper chromite (*H. Adkins* and *H. L. Coonradt*, J. Amer. chem. Soc., 1941, **63,** 1563); 3*H*-indoles can also be reduced to indolines. Indole and 2,3-dimethylindole may be reduced to indoline and 2,3-dimethylindoline, respectively, using zinc dust and phosphoric acid (*L. J. Dolby* and *G. W. Gribble*, J. Heterocyclic Chem., 1966, **3,** 124). Indoline is produced also by treatment of oxindole (2-oxoindoline) with phosphorus pentasulphide followed by electrolytic reduction (*S. Sugasawa et al.*, J. chem. Soc., Japan, 1938, **58,** 137).

(*2*) Indoline is formed by dehydration of *o*-amino-*β*-phenylethyl alcohol (*G. M. Bennett* and *M. M. Hafez*, J. chem. Soc., 1941, 287, 652).

(*3*) 2-Ethylindoline has been obtained along with 2-methyltetrahydroquinoline by the reaction between *o*-*n*-butylnitrobenzene and triethyl phosphite, which involves the reductive cyclisation of a nitro-group (*R. J. Sundberg*, J. Amer. chem. Soc., 1966, **88,** 3781; Tetrahedron Letters, 1966, 477; *G. Smolinsky* and *B. I. Feuer*, J. org. Chem., 1966, **31,** 3882):

$(CH_2)_3Me$, NO_2 — $P(OEt)_3$ → N–H, Et + N–H, Me

(*4*) Intramolecular amination with dialkyl-*N*-chloramines and reducing metal salts also affords indolines (*F. Minisci* and *R. Galli*, Tetrahedron Letters, 1966, 2531):

HN, Me — NaOCl → ClN, Me — $FeSO_4$, H_2SO_4, $-5°$ → N, Me

(*5*) The nonoxidative photocyclisation of *N*-aryl enamines, prepared from *N*-alkylanilines and ketones, provides a useful synthesis of indolines. The cyclisation is not limited to the above enamines, for irradiation of α-(*N*-methylanilino)styrene gives 1-methyl-2-phenylindoline in a 66% yield (*O. L. Chapman* and *G. L. Eian*, J. Amer. chem. Soc., 1968, **90,** 5329):

CH_2, Ph, N, Me — $h\nu$ → N, Me, Ph

(ii) General properties and reactions

Indoline is a weak base showing all the properties of an alkylaniline: 2-methyl-1-nitrosoindoline undergoes normal Fischer–Hepp rearrangement to the green 5-nitro-compound. Unlike indole, to which it reverts on distillation over silver sulphate, it is not resinified by acids.

Phosphorus pentachloride opens the ring of 1-benzoylindolines, giving *o*-β-chloroethylbenzanilides (*von Braun* and *Sobecki*, *loc. cit.*), while Emde fission of dimethylindolinium salts goes in all three possible directions (*von Braun* and *L. Neuman*, Ber., 1916, **49,** 1283):

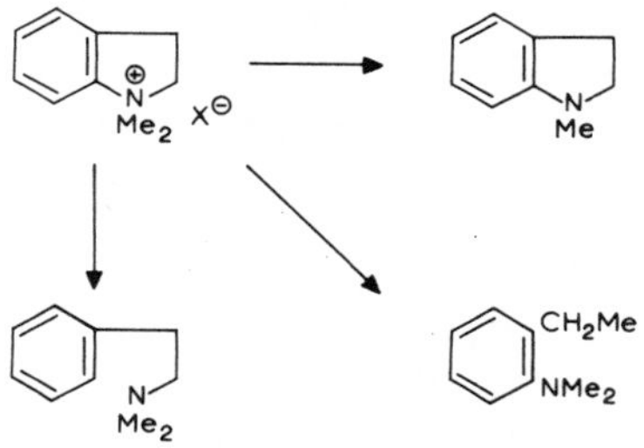

trans-1-Benzoyl-2,3-dihydroxy-2,3-dimethylindoline (I) with aqueous alkali gives 2-methyl-2-phenylindoxyl and 3-hydroxy-3-methyl-2-phenyl-3*H*-indole, whereas, the *cis*-isomer (II) affords 2,2-dimethylindoxyl. *cis*- and *trans*-Indolines are readily distinguished by their u.v. spectra (*J. W. Kershaw* and *A. Taylor*, J. chem. Soc., 1964, 4320):

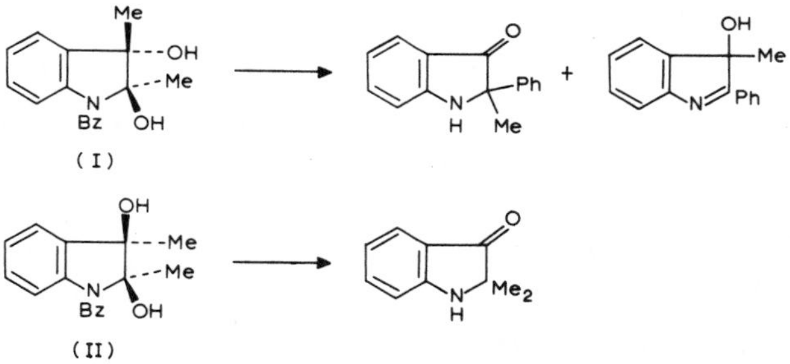

(I)

(II)

In the n.m.r. spectra of *cis*- and *trans*-2,3-dimethylindolines the coupling constant between the 2- and 3-hydrogen atoms depends strongly on the nature of the *N*-substituent (*F. A. L. Anet* and *J. M. Muchowski*, Chem. and Ind., 1963, 81).

In 1,3,3-trimethyl-2-methyleneindoline, Fischer's base (XXXII, p. 442; *O. Munn et al.*, Ber., 1939, **72,** 2107), and its analogues, the methylene group is very reactive, undergoing direct acetylation and reaction with phenyl isocyanate (*A. Ferratini*, Gazz., 1894, **24,** II, 190; *M. Coenen*, Ber., 1947, **80,** 546), methylation and diazo-coupling (*W. König* and *J. Müller*, *ibid.*, 1924, **57,** 144).

2-Hydroxy-1,3,3-trimethylindoline (XXXV, p. 442), also prepared by

Fischer synthesis from isobutyraldehyde methylphenylhydrazone, behaves as the pseudo-base of the 1,3,3-trimethyl-3*H*-indolium salts; energetic treatment with acids causes rearrangement to 1,2,3-trimethylindole (*K. Brunner*, Monatsh., 1896, **17,** 253; *G. Ciamician* and *A. Piccinini*, Ber., 1896, **23,** 2465). For 5,6-dihydroxyindolines and related quinones see p. 424.

(iii) Individual indolines

Indoline, b.p. 228–230°, d_4^{20} 1.069, n_D^{20} 1.5923; *acetate*, m.p. 105°, *benzoyl*, m.p. 118°, *picrate*, m.p. 174° (*P. A. S. Smith* and *Tung-Yin Yu*, J. Amer. chem. Soc., 1952, **74,** 1096; *R. Stollé et al.*, J. pr. Chem., 1930, **128,** 1). 1-**Methylindoline,** b.p. 100–102°/14 mm, *picrate*, m.p. 165°; 2-*methylindoline* (±), b.p. 228–229°, *picrate*, m.p. 151°, N-*acetyl* deriv. 55–56° (*O. Carrasco*, Gazz., 1908, **38,** 301); 3-*methylindoline*, b.p. 231–232°/744 mm, *picrate*, 149–150° (*M. Wenzing*, Ann., 1887, **239,** 239); 4-*methylindoline*, b.p. 245°, *picrate*, m.p. 188°; 7-*methylindoline*, b.p. 240–243°, *picrate*, m.p. 186° (*O. Kruber*, Ber., 1926, **59,** 2752; 1929, **62,** 2877). 3-**Ethylindoline,** b.p. 109–110°/7 mm. 2-**Phenylindoline,** m.p. 46° (*H. Adkins* and *H. L. Coonradt*, J. Amer. chem. Soc., 1941, **63,** 1563); 7-*phenylindoline*, m.p. 68–70° (*K. Wiesner et al.*, *ibid.*, 1955, **77,** 675). 1,2-**Dimethylindoline,** b.p. 227–229° (*Adkins* and *Coonradt*, *loc. cit.*). 2-**Methyl-5-nitroindoline,** m.p. 82° (*E. Bamberger* and *H. Sternitzki*, Ber., 1893, **26,** 1291); 2-*methyl-6-nitroindoline*, m.p. 50°; 3-*methyl-6-nitroindoline*, m.p. 75° (*J. v. Braun*, *A. Grabowski* and *M. Rawicz*, Ber., 1913, **46,** 3169). 4-**Methylindoline-3-carboxylic acid,** m.p. 223°; 7-*methylindoline-3-carboxylic acid*, m.p. 237° (*Kruber*, *loc. cit.*; Ber., 1929, **62,** 2877).

(b) Oxoindolines (indolinones) and hydroxyindoles

Derivatives of 2- and 3-oxoindoline, some of which may react as hydroxyindoles (indoxyl, p. 454), are closely related to indigo. Isatin, obtained by the oxidation of indigo with nitric or chromic acids and first described by *Laurent* and by *Erdman* in 1841, on treatment with sodium amalgam affords dioxindole, further reduced in acid solution to oxindole (2-oxoindoline; indolin-2-one):

Isatin → Dioxindole → Oxindole

(i) 2-Oxoindolines (oxindoles)

(*1*) *Synthesis.* (*a*) Oxindoles may be prepared by heating acylphenylhydrazines with lime at 190–200°, a modification of Fischer's synthesis (*Brunner*, Monatsh., 1897, **18,** 95, 531); the lime may be replaced by sodamide with advantage (*J. Staněk* and *D. Ribář*, Chem. Listy, 1946, **40,** 173):

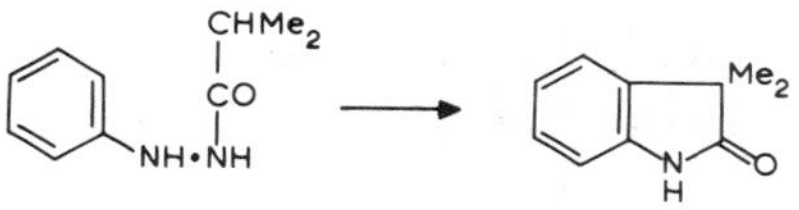

(*b*) by treating chloroacetyl derivatives of primary or secondary arylamines with aluminium chloride (*R. Stollé*, J. pr. Chem., 1930, [ii], **128,** 1; *A. Abramovitch* and *D. H. Hey*, J. chem. Soc., 1954, 1697):

(*c*) by hydrolysis of sodium indole-2-sulphonates; (*d*) by cyclisation of the amino acids obtained by the reduction of *o*-nitrophenylacetic acids (*P. W. Neber*, Ber., 1922, **55,** 835; *G. Hahn* and *H. J. Schulz*, *ibid.*, 1939, **72,** 1308):

(*e*) Reductive hydrolysis of the 2-disulphide derivatives of tryptamine (III) and *O*-benzylserotonin (IV) yields oxindoles (*K. Treter et al.*, J. Amer. chem. Soc., 1958, **80,** 983):

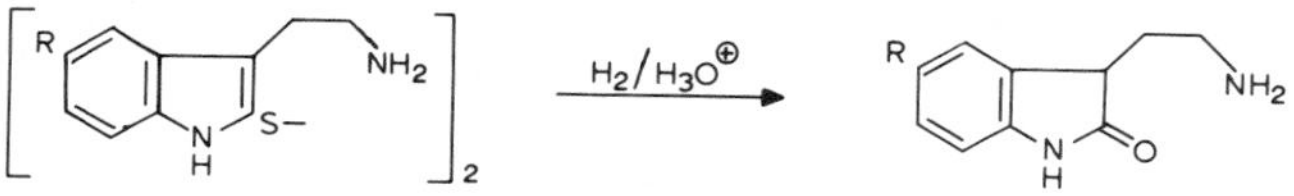

(III) R = H
(IV) R = $OCH_2 \cdot C_6H_5$

(*f*) *N*-Acyl-*N*-methyl-*o*-chloroanilines on treatment with potassamide in liquid ammonia afford oxindoles (*J. F. Bunnett et al.*, J. org. Chem., 1963, **28,** 1):

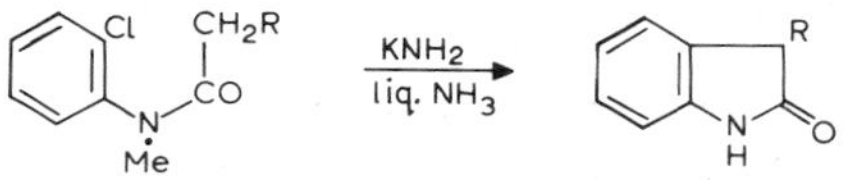

(*g*) Oxindole is prepared by Friedel–Crafts ring closure of chloroacetanilide (*S. Sugasawa*, *I. Satoda* and *J. Yamagasawa*, J. pharmac. Soc., Japan, 1938, **58,** 29). (*h*) A much more satisfactory synthesis giving a high yield is afforded by the decomposition of isatin 3-hydrazone with hot ethanolic sodium ethoxide; oxindoles substituted in the benzene ring can be obtained by this method (*A. H. Jackson*, Chem. and Ind., 1965, 1652):

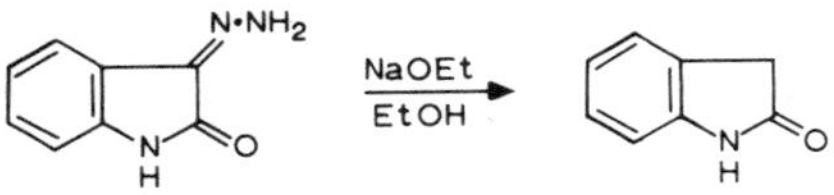

N·NH$_2$ — NaOEt / EtOH →

(*i*) Desulphurisation of isatin ethenedithioacetal in benzene or ethanol with Raney nickel affords oxindole, but with W2 Raney nickel in various alcohols for longer periods gives 3-alkyloxindoles (*E. Wenkert* and *N. V. Bringi*, J. Amer. chem. Soc., 1958, **80,** 5575):

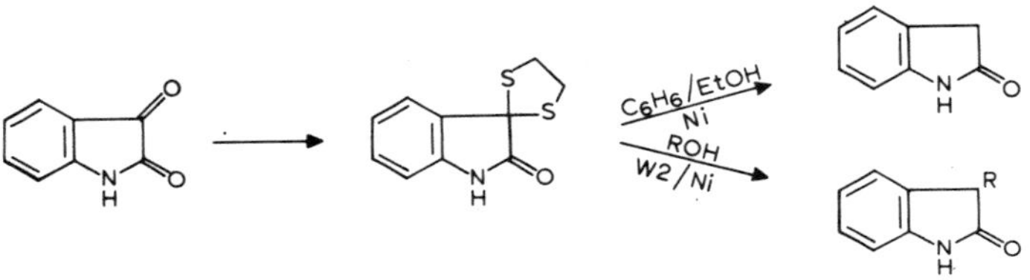

C_6H_6/EtOH, Ni; ROH, W2/Ni

(*j*) The reaction of α-chloro-α,α-diphenylacetanilide with sodium hydride in an inert medium gave a mixture of three compounds, 2,2-diphenylindoxyl, 3,3-diphenyloxindole and 1,3-diphenyloxindole (*J. C. Sheehan* and *J. W. Frankenfield*, *ibid*., 1961, **83,** 4792):

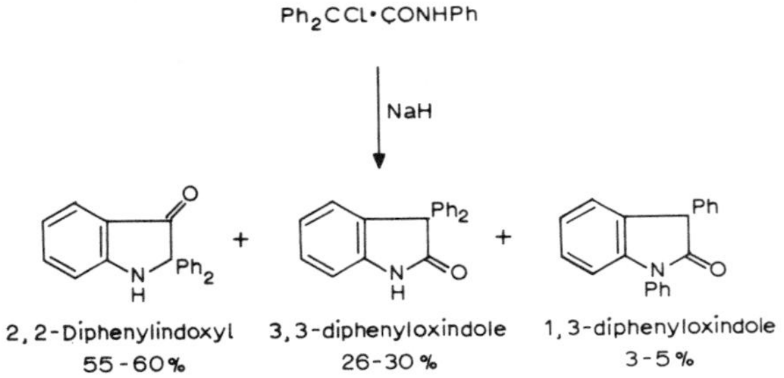

$Ph_2CCl\cdot CONHPh$

NaH

2,2-Diphenylindoxyl 55-60% + 3,3-diphenyloxindole 26-30% + 1,3-diphenyloxindole 3-5%

(*k*) The decomposition of 3-bromo-oxindole-3-acetic acid (p. 454) in aqueous ethanol gives 3-methyleneoxindole (*R. L. Hinman* and *C. P. Bauman*, J. org. Chem., 1964, **29,** 2431):

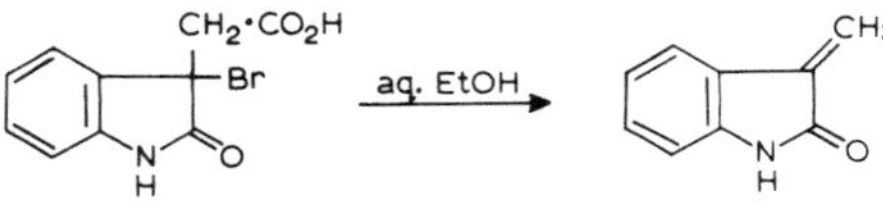

$CH_2\cdot CO_2H$, Br — aq. EtOH → CH_2

(*l*) Treatment of the diesters of *N*-benzoyldiphenylaminedicarboxylic acids (V) and (VI) with excess sodium methoxide gives oxindoles, but V with a deficient amount of sodium methoxide on acidification yields an indole, while under similar conditions base VI affords the enol VII (*J. W. Schulenberg*, J. Amer. chem. Soc., 1968, **90,** 7008):

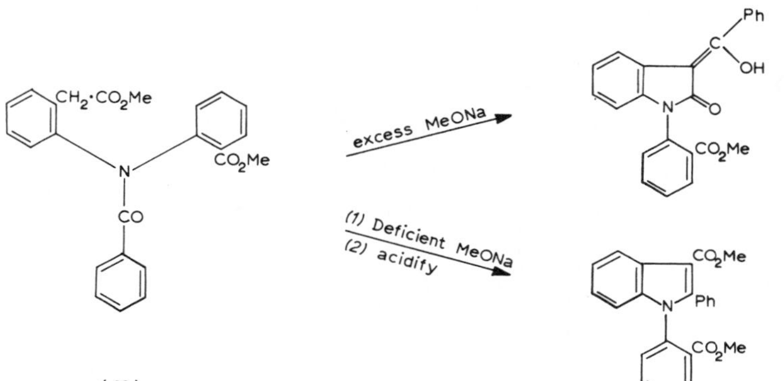

(V)

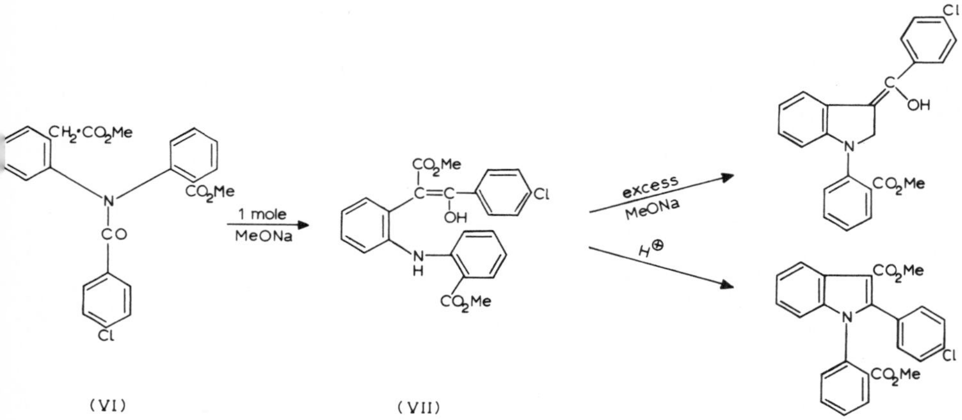

(VI)

(VII)

(*2*) *Properties and reactions.* **Oxindole** (2-oxoindoline), is feebly basic and still more feebly acidic, and behaves in general as an oxoindoline, being directly alkylated and acylated on nitrogen, but with phosphorus pentachloride it affords 2,3-dichloroindole. The ultraviolet absorption supports the lactam formulation (*P. Ramart-Lucas* and *D. Biquard*, Bull. Soc. chim. Fr., 1935, [v], **2,** 1383).

Oxindoles react at the methylene group with aldehydes and some ketones, giving isoindogenides (VIII) (*A. Wahl* and *P. Bagard*, *ibid.*, 1909,

[iv], **5,** 1033) or derivatives of di(oxindolyl)methane (*Stollé*, J. pr. Chem., 1932, [ii], **135,** 345), and with esters, giving 3-acyl-2-hydroxyindoles (IX) (*P. L. Julian et al.*, J. Amer. chem. Soc., 1934, **56,** 1797; 1935, **57,** 2026; *L. Horner*, Ann., 1941, **548,** 117).

Nitrosation affords isatin-3-oxime, yielding on reduction 3-amino-oxindole, which can be oxidised to isatin.

N-Substituted oxindoles are readily alkylated at $C_{(3)}$; it is often difficult to avoid dialkylation (*Wenkert et al.*, J. Amer. chem. Soc., 1956, **78,** 797). 3-Phenyloxindole is alkylated at $C_{(3)}$ and acylated at N, but 1-substituted 3-phenylindoles though they alkylate at $C_{(3)}$, afford *O*-acyl derivatives of 2-hydroxy-3-phenylindole (*J. M. Bruce* and *F. K. Sutcliffe*, J. chem. Soc., 1957, 4789).

Oxindoles are reduced by lithium tetrahydridoaluminate to indoles, with some indoline (*Julian* and *H. C. Printy*, J. Amer. chem. Soc., 1949, **71,** 3206). 3-Acyloxindole oximes on catalytic hydrogenation over palladium–charcoal give 3-(1′-aminoalkylidene)oxindoles (X) and 2-alkylindoles. Treatment of the oximes with potassium acetate leads to indoles (*Wenkert*, *B. S. Bernstein* and *H. H. Udelhofen*, *ibid.*, 1958, **80,** 4899):

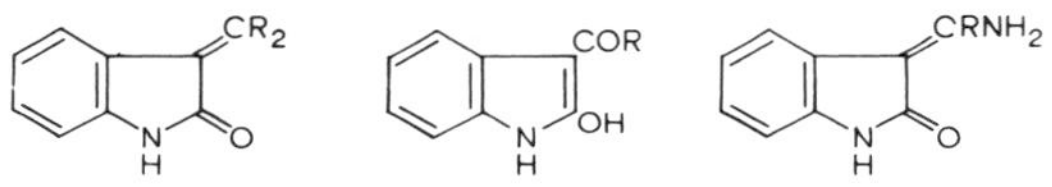

CR_2 N H O (VIII) COR OH N H (IX) $CRNH_2$ N H O (X)

Oxindoles are halogenated in positions 3, 5, and 7, the first being favoured by non-polar solvents (*W. C. Sumpter et al.*, *ibid.*, 1945, **67,** 1656). Heating oxindoles with phosphorus pentasulphide and treating the product with Raney nickel yields indoles (*H. Plieninger* and *G. Werst*, Angew. Chem., 1958, **70,** 272).

Oxindole is ethylated with triethyloxonium tetrafluoroborate to give 2-ethoxyindole, which on vacuum-sublimation is converted into 2-ethoxy-3*H*-indole. In carbon tetrachloride solution the indole and the 3*H*-indole exist in equilibrium (*J. Harley-Mason* and *T. J. Leeney*, Proc. chem. Soc., 1964, 369).

1-**Hydroxyoxindoles** have been prepared by reduction of *o*-nitrophenylacetic acids, with zinc and sulphuric acid, and with dimethyl sulphate they give 1-methoxyoxindoles. Treatment of the resultant methoxyoxindole with aqueous hydrogen bromide affords 5-bromo-oxindole also obtained by the reaction of hydrogen bromide on 1-hydroxyoxindole (*W. B. Wright Jr.* and *K. H. Collins*, J. Amer. chem. Soc., 1956, **78,** 221):

3,3-**Diaryloxindoles** are obtained by condensing isatin with phenol, dimethylaniline or toluene in the presence of sulphuric acid (*S. Inagaki* and *T. Nakato*, J. pharm. Soc., Japan, 1933, **53,** 686, 719):

Dehydration of *N*-methylbenzilanilide (from phenylmagnesium bromide and *N*-methylphenylglyoxoylanilide) gives 3,3-diphenyl-1-methyloxindole (*R. F. Reeves* and *H. G. Lindwall*, J. Amer. chem. Soc., 1942, **64,** 1086).

The i.r., u.v., and n.m.r. spectra of a number of oxindoles are reported (*Hinman* and *Bauman*, loc. cit.).

Oxindole, 2-**oxoindoline,** m.p. 127° (*Stollé*, J. pr. Chem., 1930 [ii], **128,** 1; *F. J. Dicarlo*, J. Amer. chem. Soc., 1944, **66,** 1420); 1-*methyloxindole*, m.p. 89° (*Stollé, loc. cit.*); 1-*ethyloxindole*, m.p. 97° (*Wenkert* and *Bringi, loc. cit.*); 1-*phenyloxindole*, m.p. 121° (*Stollé*, Ber., 1914, **47,** 2120); 1-**acetyloxindole,** m.p. 127°; 1-*benzoyloxindole*, m.p. 65–67° (*G. Palazzo* and *V. Rosnati*, Gazz., 1953, **83,** 211); 1-**hydroxyoxindole,** m.p. 199–201° (*Dicarlo, loc. cit.*); 1-*acetoxyindole*, m.p. 101° (*Sumpter et al., loc. cit.*).

4-**Hydroxyoxindole,** m.p. 267°; 6-*hydroxyoxindole*, m.p. 243°; 4-**methoxyoxindole,** m.p. 196°; 5-*methoxyoxindole*, m.p. 153°; 6-*methoxyoxindole*, m.p. 158°; 7-*methoxyoxindole*, m.p. 146° (*C. F. Koelsch*, J. Amer. chem. Soc., 1944, **66,** 2019; *T. Wieland* and *O. Unger*, Ber., 1963, **96,** 253); 5,6-*dimethoxyoxindole*, m.p. 204–205° (*Hahn* and *Schulz, loc. cit.*).

5-**Nitro-oxindole,** m.p. 240–242° (*Sumpter et al., loc. cit.*); 3-**amino-oxindole,** not isolated in free state (*R. E. Schachat et al.*, J. org. Chem., 1951, **16,** 1349). 3,3-**Dibromo-oxindole,** m.p. 165° (*Sumpter et al., loc. cit.*).

3-**Methyloxindole,** m.p. 123–124° (*H. Behringer* and *H. Weissauer*, Ber., 1952, **85,** 743); 3,3-**dimethyloxindole,** m.p. 152–153° (*Brunner, loc. cit.*); 1,3-*dimethyloxindole*, m.p. 55°; 1,3,3-*trimethyloxindole*, m.p. 50° (*Julian et al.*, J. Amer. chem. Soc., 1934, **56,** 1797); 3-*ethyl*-1-*methyloxindole*, b.p. 103–107°/0.5 mm (*M. W. Rutenberg* and *E. C. Horning*, Org. Synth., 1950, **30,** 62).

3-**Bromo-3-methyloxindole,** m.p. 143° (decomp.); 3-**bromo-oxindole-3-acetic acid,** m.p. 151–152° (decomp.); 3-**bromo-oxindole-3-butyric acid,** m.p. 140–141° (decomp.) (*Hinman* and *Bauman*, J. org. Chem., 1964, **29,** 1206).

Oxindole-3-acetic acid, m.p. 147°; *oxindole-3-propionic acid,* m.p. 174°; *oxindole-3-butyric acid,* m.p. 170–171°; *oxindole-3-caproic acid,* m.p. 160° (*Julian et al.*, J. Amer. chem. Soc., 1953, **75,** 5301, 5305; *Hinman* and *Bauman, loc. cit.*).

3,3-**Di(4-hydroxyphenyl)oxindole (phenolisatin),** m.p. 260–261°; 3-**benzylidineoxindole,** m.p. 176°.

On diazotisation 3-(3-amino-5-methoxycarbonyl-2-pyridylmethyl)-1-methyloxindole cyclises to the tetracyclic compound XI which contains the lysergic acid skeleton (*Plieninger, M. S. von Wittenau* and *B. Kiefer*, Ber., 1958, **91,** 2095; Angew. Chem., 1958, **70,** 168):

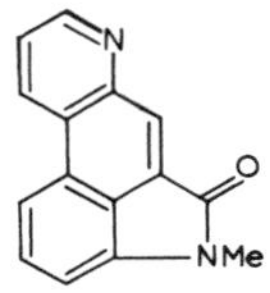

(XI)

(ii) 3-Oxoindoline; indoxyl

Indoxyl is found in *Indigofera* spp. as the β-glucoside, "plant indican", m.p. 176–178°, $[\alpha]^{19°}_{5461}$ −77–75° in water, which is hydrolysed by the enzyme indimulsin to indoxyl and D-glucose (*A. G. Perkin et al.*, J. chem. Soc., 1907, **91,** 1715; 1909, **95,** 793, 824). It has been synthesised by *A. Robertson* (*ibid.*, 1927, 1937).

Indoxyl is an unstable yellow, amphoteric solid, m.p. 85°, which resinifies, more so in presence of concentrated hydrochloric acid, and is easily oxidised by air, especially in alkaline solution, to indigo, together with some indirubin produced by condensation of indoxyl with isatin; under other conditions isatin can be the main product of oxidation. Apart from being soluble in alkali indoxyl gives a red ferric chloride colour followed by oxidation to indigo, distils on heating with zinc dust to afford indole, also obtained by reduction with sodium amalgam (*D. Vorlander* and *O. Apelt*, Ber., 1904, **37,** 1134), and reacts at position 2 with benzenediazonium chloride. Indoxyl with acetic anhydride gives the *N*-mono and the *O,N*-di-acetyl derivatives, but in the presence of sodium hydroxide *O*-monoacetylindoxyl (3-acetoxyindole) is obtained. These reactions suggest that indoxyl is 3-hydroxyindole. The diacetyl derivative is produced by boiling phenylglycine-*o*-carboxylic acid with acetic anhydride and sodium acetate (*Bayer and Co.*, G.P. 113,240/1898; Chem. Ztbl., 1900, II, 615). Methylation with dimethyl sulphate and aqueous alkali gives 3-methoxyindole, but with methyl iodide and solid sodium hydroxide, 3-oxo-2,2-dimethylindoline is obtained (*A. Étienne*, Bull. Soc. chim. Fr., 1948, 651):

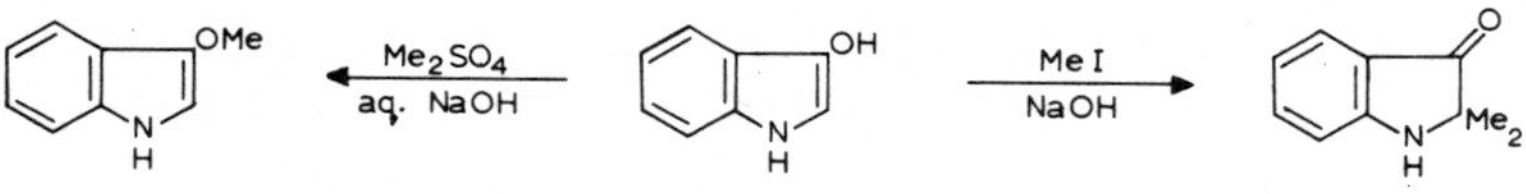

Indoxyl also reacts at position 2 with nitrous acid to give isatin-α-oxime and *O*-methyl- and *O*-acetyl-indoxyl are attacked at the same position (*Étienne, loc. cit.*); nitrosobenzene, and aldehydes, and some ketones in the presence of acetic acid (*A. von Baeyer*, Ber., 1883, **16**, 2188) yield coloured, halochromic "indogenides", simple prototypes of indigoid dyes:

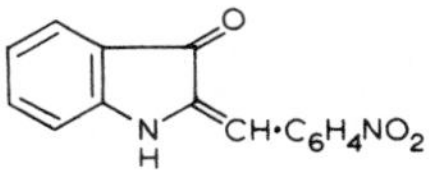

The last reaction has been used to support the 3-indolinone structure XII, but it is equally possible that the reaction of the aldehyde takes place directly at the activated 2-position of XIII and XIV. As yet no decisive physical evidence is available on the actual state of the indoxyl molecule (*S. J. Holt et al.*, J. chem. Soc., 1958, 1217):

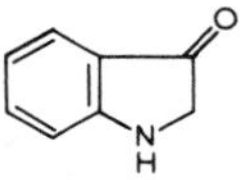

(XII) 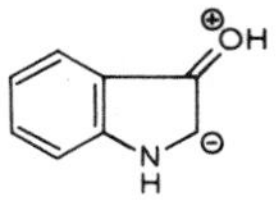 (XIII) 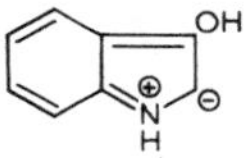 (XIV)

Indoxyl may be manufactured by alkaline fusion of phenylglycine (*Deutsche Gold- und Silber-Scheideanstalt*, G.P. 137,955/1901; Chem. Ztbl., 1903, I, 110) and can be prepared also from *N*-methylanthranilic acid by alkali fusion and by boiling indoxylic acid (p. 434) with water (*Vorlander* and *B. Drescher*, Ber., 1901, **34**, 1854).

The acid sulphate ester of indoxyl, "urine indican" (XV) is normally present in the urine of mammals, apparently as a metabolite of indole. It is hydrolysed by acids, oxidised by ferric chloride to indigo, and has been synthesised from *N*-acetylindoxyl by treatment with chlorosulphonic acid, followed by hydrolysis (*E. Schwenck* and *A. Jolles*, Biochem. Z., 1915, **69**, 467):

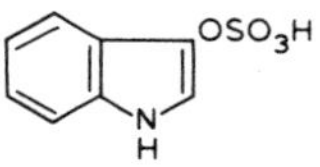

(XV)

Indoxyl derivatives (*Étienne, loc. cit.*): O-*methyl*, m.p. 70°; O-*acetyl*-N-*methyl*, m.p. 57° (*L. Ettinger* and *P. Friedländer*, Ber., 1912, **45**, 2074); O-*acetyl*, m.p. 126°; N-*acetyl*, m.p. 136°; O,N-*diacetyl*, m.p. 82°; 3-*benzylidene*, m.p. 175–176° (*Baeyer, loc. cit.*).

(iii) 3-Hydroxy-2-oxoindolines; 3-hydroxyoxindoles; dioxindoles

(*1*) *Preparation.* Dioxindole (3-hydroxyoxindole, 3-hydroxy-2-oxo-2,3-dihydroindole, 3-hydroxy-2-oxoindoline), is prepared (*a*) by reducing isatin with sodium dithionite (*L. Kalb*, Ber., 1911, **44,** 1455), and (*b*) rapidly from *o*-aminomandelic acid (*o*-amino-α-hydroxyphenylacetic acid) in acid solution (*Neber, loc. cit.; Hahn* and *Schulz, loc. cit.*).

Dioxindole exists substantially as 3-hydroxy-2-oxoindoline (*cf.* p. 448), since reduction of optically active *o*-nitromandelic acid with ferrous hydroxide yields active dioxindole; this is quickly racemised by alkali no doubt through tautomeric change into 2,3-dihydroxyindole (*A. McKenzie* and *P. A. Stewart*, J. chem. Soc., 1935, 104). Reduction of the acid with zinc dust and ammonia gives active 1-hydroxydioxindole (trioxindole), a strongly acid substance which yields isatin with acetic anhydride (*G. Heller*, Ber., 1909, **42,** 470).

(*c*) 3-Substituted dioxindoles are obtained by reacting isatins with Grignard reagents (*M. Kohn*, Monatsh., 1910, **31,** 747; *Kohn* and *A. Ostersetzer*, *ibid.*, 1911, **32,** 905; *M. E. Baumgarten* and *P. L. Creger*, J. Amer. chem. Soc., 1960, **82,** 4634; *B. Mills* and *K. Schofield*, J. chem. Soc., 1961, 5558) or aryl-lithium reagents (*J. M. Bruce*, *ibid.*, 1959, 2366).

(*d*) The treatment of 3-diazo-oxindoles in an alkanol with perchloric acid affords 3-alkoxyoxindoles (*Creger*, J. org. Chem., 1965, **30,** 3610).

(*e*) Depending on the amount of 2 *N* sodium hydroxide solution with which it is reacted 3,3-dichloro-2,4-dioxo-1,2,3,4-tetrahydroquinoline (XVI) gives, in addition to 3-chloro-4-hydroxycarbostyril (XVII), a dehydrogenation product of dioxindole and dioxindole or isatin, respectively (*E. Ziegler*, *T. Kappe* and *R. Salvador*, Monatsh., 1963, **94,** 453):

2 *N* NaOH

or

(XVI)

(XVII)

(*f*) 3-Methyl- and 3-phenyl-dioxindole (XX) are formed by the autoxidation of the related indole (*H. H. Wasserman* and *M. B. Floyd*, Tetrahedron Letters, 1963, 2009; *F. Ying-Hsiueh Chen* and *E. Leete*, *ibid.*, p. 2013). 1-Acetyl-2-hydroxy-3-oxo-2-phenylindoline (XVIII) exists partly as the acyclic tautomer XIX in solution and rapidly hydrolyses and rearranges in dilute aqueous alkali to give 3-phenyldioxindole (*C. W. Rees* and *C. R. Sabet*, J. chem. Soc., 1965, 870):

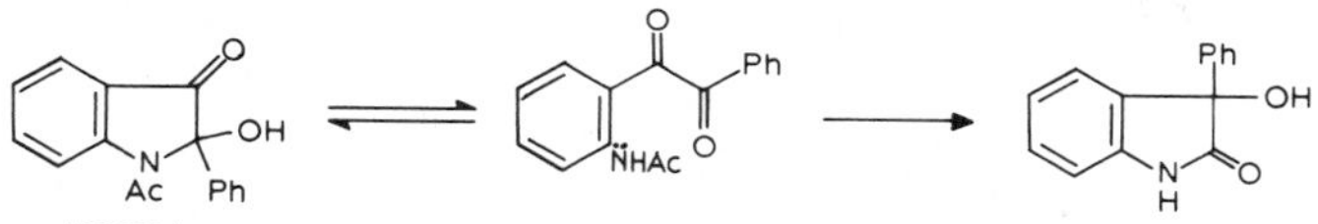

(XVIII) (XIX) (XX)

(*g*) 3-Substituted 1-aminodioxindoles may be obtained by the following route (*Baumgarten* and *Creger*, *loc. cit.*):

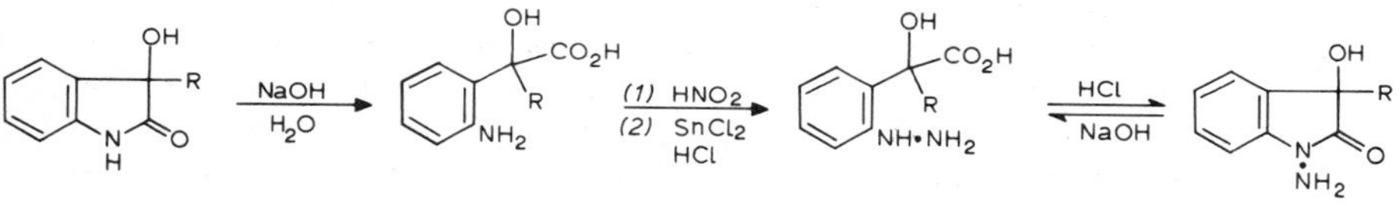

NaOH / H_2O; (1) HNO_2 (2) $SnCl_2$ HCl; HCl / NaOH

For the synthesis of some further dioxindole derivatives see *G. Hallmann*, Ber., 1962, **95,** 1138 and for the i.r., u.v., and n.m.r. spectra of a variety of dioxindoles, *R. L. Hinman* and *C. P. Bauman*, J. org. Chem., 1964, **29,** 2431.

(*2*) *Properties.* Dioxindole is weakly amphoteric. In aqueous solution it oxidises readily to isatide and when warmed in glycerol water is eliminated with the formation of isoindigotin:

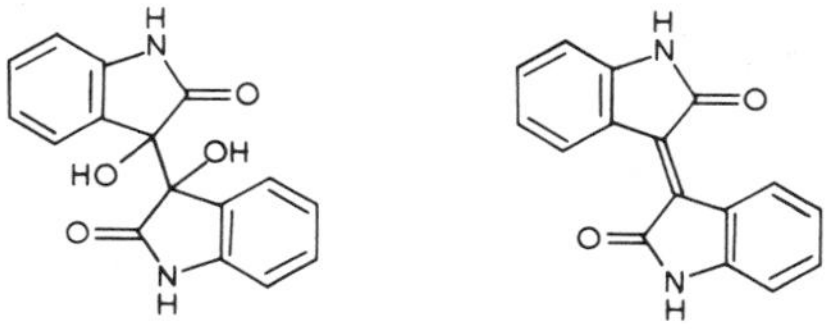

Isatide Isoindigotin

Dioxindole is readily oxidised to isatin and is reduced to oxindole. Reduction of dioxindole in ether with lithium tetrahydridoaluminate affords a mixture of compounds including 3-hydroxyindoline and indole (*E. Giovannini* and *T. Lorenz*, Helv., 1957, **40,** 2287):

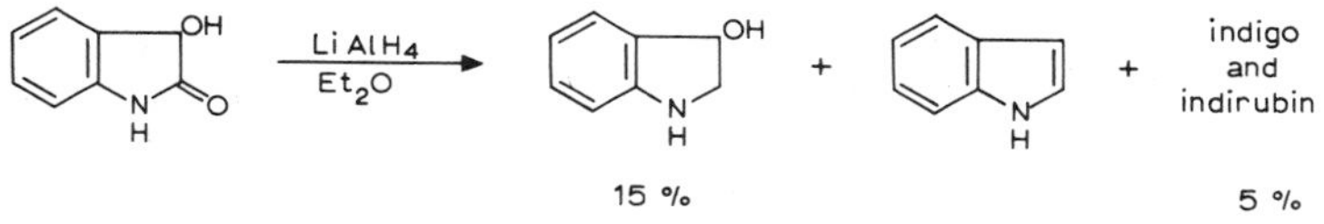

$LiAlH_4$ / Et_2O → 15 % + + indigo and indirubin 5 %

3-Phenyldioxindole on oxidation in alkali with hydrogen peroxide (*S. Inagaki*, J. pharm. Soc., Japan, 1939, **59,** 5) or better with potassium ferricyanide (*Mills* and *Schofield*, *loc. cit.*) gives 2-aminobenzophenone:

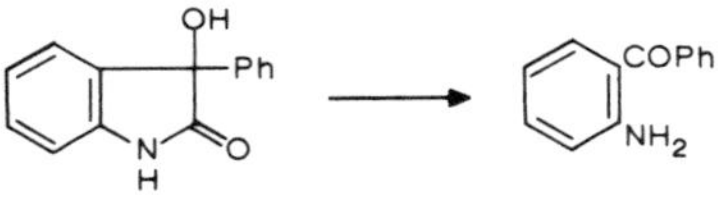

Dioxindole, 3-hydroxy-2-oxoindoline, m.p. 167–168° (+ or −) (*McKenzie* and *Stewart*, *loc. cit.*), *diacetyl* deriv., m.p. 88–89° (*E. J. Alford* and *Schofield*, J. chem. Soc., 1952, 2102). 1-*Methyldioxindole*, m.p. 147° (*J. Martinet*, Ann. Chim., 1919, **11,** 78); 1-*hydroxydioxindole*, m.p. 159° (decomp.) (+ or −), 172° (±). 3-*Methyldioxindole*, m.p. 161–162° (*Hinman* and *Bauman*, J. org. Chem., 1964, **29,** 2435).

Ethyl 2-hydroxy-3-oxoindoline-2-carboxylate (indoxanthic ester), m.p. 107°, is produced by oxidising ethyl indoxylate with ferric chloride (*A. von Baeyer*, Ber., 1882, **15,** 775; *L. Kalb*, *ibid.*, 1911, **44,** 1455). It is a yellow solid which yields a nitrosamine and reverts to ethyl indoxylate on reduction. Acids convert it into **ethyl dioxindole-3-carboxylate** perhaps by benzilic change of an open-chain modification:

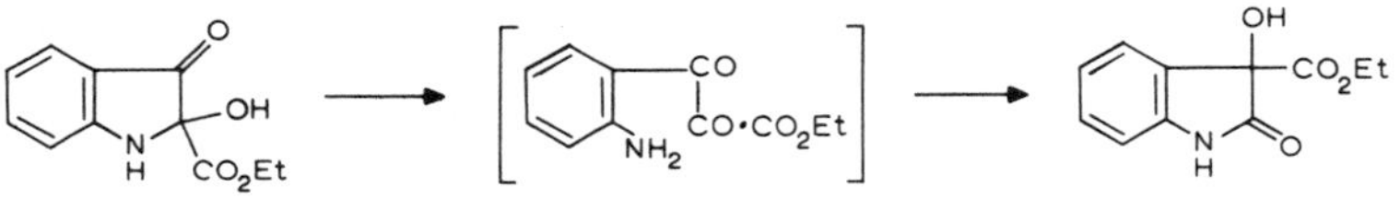

Derivatives of ethyl dioxindole carboxylate, *e.g.*, *ethyl 3-hydroxy-5-methyl-2-oxoindoline-3-carboxylate*, m.p. 130°, are obtained by warming diethyl mesoxalate with primary or secondary amines (preferably *para*-substituted) in acetic acid (*J. Martinet*, Ann. Chim., 1919, [ix], **11,** 15, 85):

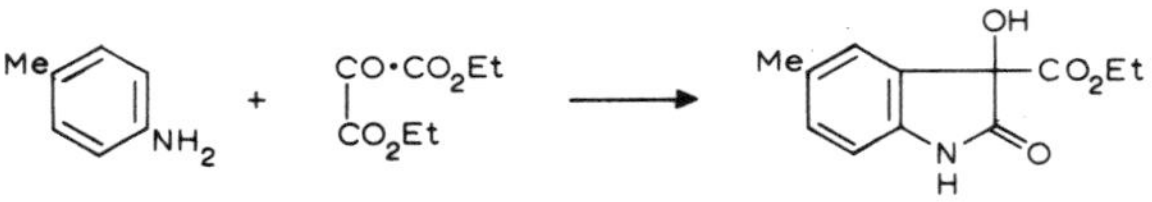

(iv) 2,3-Dioxoindolines; isatins

(*1*) *Synthesis.* (*a*) The structure of isatin (XXI), of major importance in elucidating that of indigotin, was established by its synthesis from *o*-nitrobenzoyl cyanide *via* isatinic acid (*L. Claisen* and *J. Shadwell*, Ber., 1879, **12,** 350):

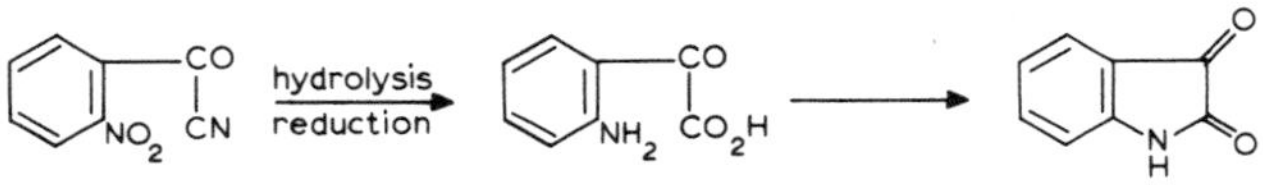

Isatinic acid

(XXI)

(*b*) Isatin was also obtained from oxindole (and more usefully *o*-nitrophenylpropionic acid) by treatment with alkali in presence of air (*A. von Baeyer*, *ibid.*, 1880, **13,** 2254). (*c*) Isatin and many of its analogues can be synthesised by reacting together aniline, trichloroacetaldehyde and hydroxylamine, treating the product with sulphuric acid and hydrolysing the resulting compound with dilute acid (*T. Sandmeyer*, Helv., 1919, **2,** 234):

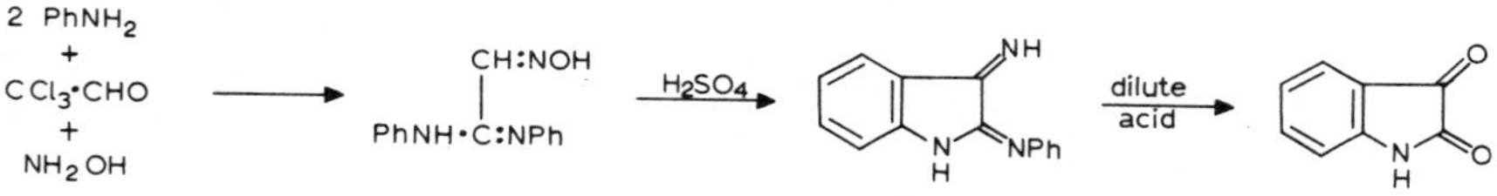

2 $PhNH_2$ + $CCl_3{\cdot}CHO$ + NH_2OH ⟶ PhNH·C(:NPh)·CH:NOH —H_2SO_4⟶ [2-phenylimino-3-imino-indoline, NH, NPh, N–H] —dilute acid⟶ [isatin]

(*d*) By treating oxamoyl chlorides from secondary amines with aluminium chloride (*R. Stollé*, J. pr. Chem., 1930, [ii], **128,** 1; *A. Abramovitch* and *D. H. Hey*, J. chem. Soc., 1954, 1697):

[PhN(R)·CO·COCl] ⟶ [N-R isatin]

(*e*) From dioxindolecarboxylic esters by simultaneous hydrolysis, decarboxylation and oxidation in alkali (*Martinet, loc. cit.*). (*f*) By the following Sandmeyer α-isatinanilide process (*Geigy and Co.*, J. Soc. chem. Ind., 1900, **19,** 657; *J. Bonnefoi* and *Martinet*, Compt. rend., 1921, **172,** 220):

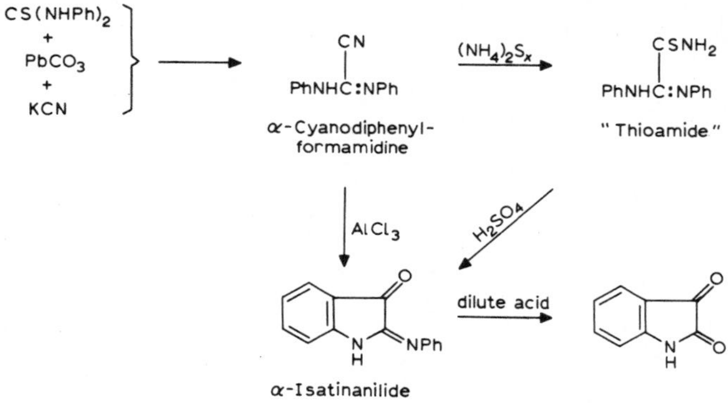

$CS(NHPh)_2$ + $PbCO_3$ + KCN ⟶ PhNHC(:NPh)CN (α-Cyanodiphenylformamidine) —$(NH_4)_2S_x$⟶ PhNHC(:NPh)$CSNH_2$ ("Thioamide")

α-Cyanodiphenylformamidine —$AlCl_3$⟶ α-Isatinanilide; "Thioamide" —H_2SO_4⟶ α-Isatinanilide

α-Isatinanilide —dilute acid⟶ [isatin]

(*g*) From the amino acid obtained by the reduction of *o*-nitrophenylglyoxylic acid (*P. W. Neber*, Ber., 1922, **55,** 835; *G. Hahn* and *H. J. Schulz*, *ibid.*, 1939, **72,** 1308):

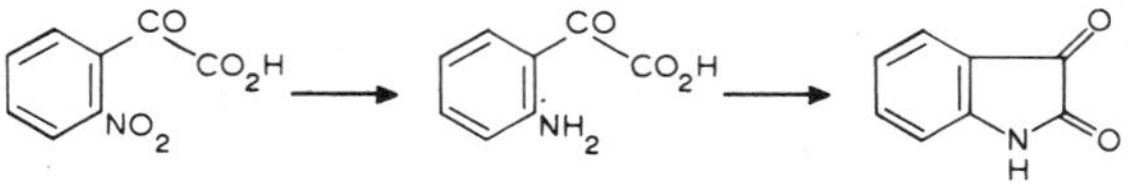

(2) *Properties and reactions.* **Isatin** is a red, weakly acidic substance whose deep-violet solutions in sodium hydroxide turn yellow on standing through ring-opening to sodium isatinate (*o*-aminobenzoylformate):

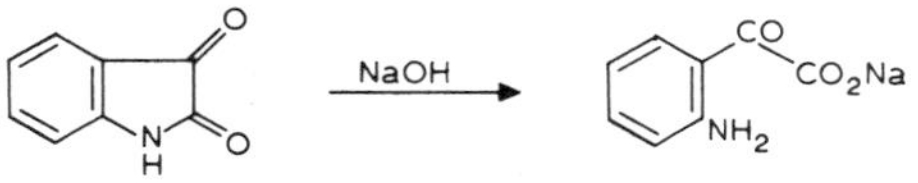

NaOH

Isatin is tautomeric and in accordance with the lactam formulation XXI, ordinary methods of acylation or alkylation yield 1-acyl- or 1-alkyl-isatins. The alternative "lactim" formula XXII is indicated by the alkylation of the silver salt of isatin to 2-alkoxy-3-oxo-3*H*-indoles (2-alkylisatins), and by the action of phosphorus pentachloride on isatin, giving 2-chloro-3-oxo-3*H*-indole (isatin chloride) (XXIII), which condenses with indoxyl or thioindoxyl to 2,2′-bisindole indigo (indigotin) and 2′-indole-2-thionaphthene indigo respectively:

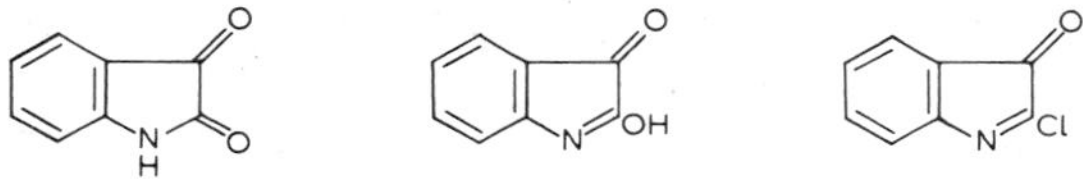

(XXI) (XXII) (XXIII)

Attempts to establish the structure of isatin by spectroscopy are indecisive (*R. G. Ault et al.*, J. chem. Soc., 1935, 1653) and X-ray diffraction suggests that the lactam structure is dominant (*G. H. Goldschmidt* and *F. J. Llewellyn*, Acta Cryst., 1950, **3**, 294).

(*a*) *Oxidation* of isatin gives anthranilic acid, isatoic anhydride or anhydro-isatin-α-anthranilide (*P. Friedlander* and *N. Roschdestwensky*, Ber., 1915, **48**, 1841):

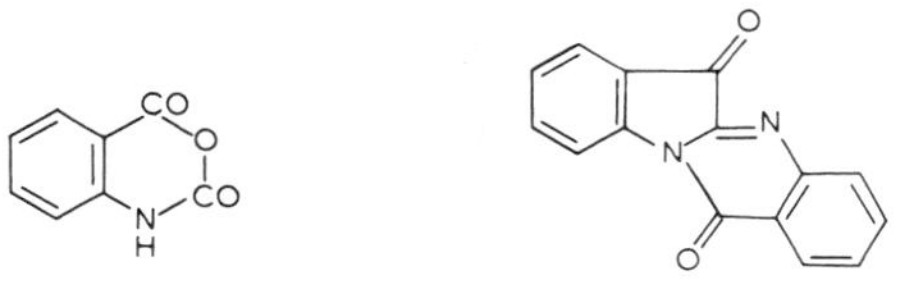

Isatoic anhydride Anhydroisatin-α-anthranilide

(*b*) Isatin yields by *reduction* a variety of bimolecular products; hydrogen sulphide in alcohol gives disulphisatide, which in the presence of pyridine or cuprous chloride affords oxindole and isoindigotin (see p. 451), whilst in the presence of alkali isoindigotin and isatin are obtained:

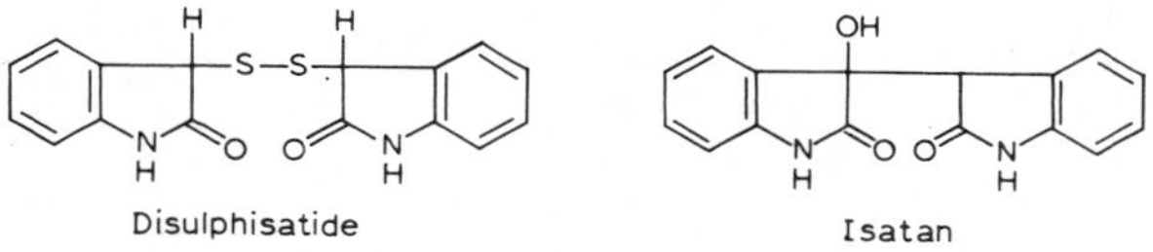

Disulphisatide Isatan

With a limited amount of ammonium sulphide isatin is reduced to the pinacol isatide (p. 457) (*Stollé* and *M. Merkle*, J. pr. Chem., 1934, [ii], **139,** 329; *E. D. Bergmann*, J. Amer. chem. Soc., 1955, **77,** 1549), which easily disproportionates to isatin and dioxindole, and may be prepared by condensation of these substances in presence of piperidine. Isatin condenses with oxindole in the same way yielding isatan, also obtained from isatin or disulphisatide with excess of ammonium sulphide, which gives isoindigo when heated (*C. W. Hansen*, Ann. Chim., 1924, **1,** 94). A useful technique for the reduction of isatins involves their conversion to 3-diazo-oxindoles by oxidation of the hydrazone with mercuric oxide or by reaction of the *p*-toluenesulphonylhydrazone with aqueous base. Treatment of the 3-diazo-oxindole in methanol with a suitable Lewis acid affords 3-methoxy-oxindoles (*P. L. Creger*, J. org. Chem., 1965, **30,** 3610):

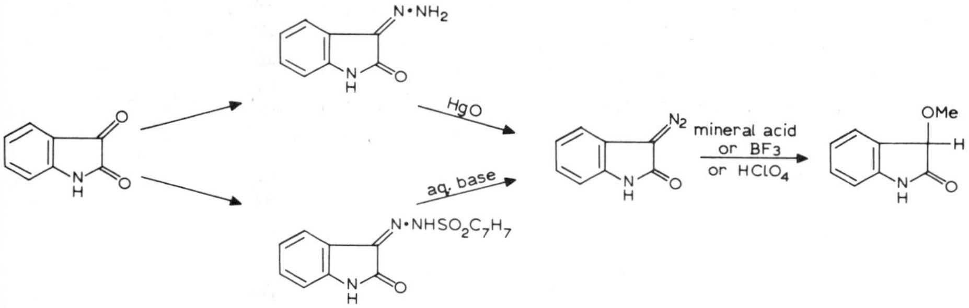

(*c*) *Carbonyl group reactivity*. The 3-carbonyl group of isatin shows normal reactivity: it forms an oxime and a hydrazone; reacts with amines and with ammonia yields isatin-3-imine; combines with hydrogen cyanide and bisulphites; condenses with a wide variety of "reactive" methylene compounds (*H. G. Lindwall et al.*, J. Amer. chem. Soc., 1932, **54,** 4739; 1933, **55,** 325; 1935, **57,** 735; 1936, **58,** 1236). Thienylmagnesium iodide, like other Grignard reagents, attacks the same position, giving 3-(2′-thienyl)dioxindole (XXIV), an intermediate in the indophenin reaction (*W. Steinkopf* and *W. Hanske*, Ann., 1939, **541,** 238). Phenylmagnesium bromide attacks both carbonyl groups in 1-methylisatin, giving a mixture of the indoxyl and the oxindole apparently produced by alternative pinacolic rearrangements of the glycol XXV (*B. Witkop* and *A. Ek*, J. Amer. chem. Soc., 1961, **73,** 5664):

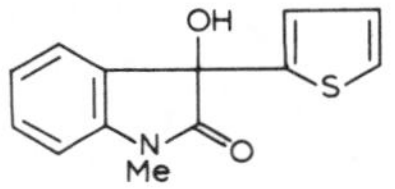

(XXIV)

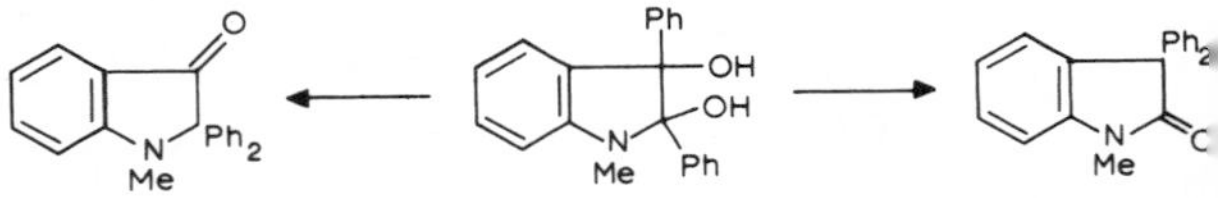

(XXV)

The Reformatsky reaction with a 1-alkylisatin yields an ester converted on hydrolysis into a quinolone (*F. J. Myers* and *Lindwall*, *ibid.*, 1938, **60**, 644):

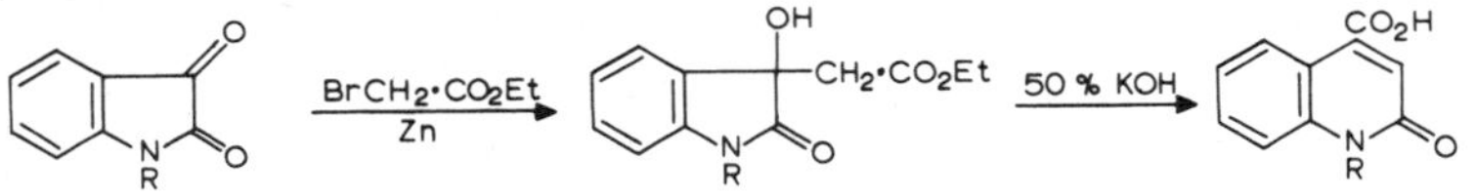

$$\xrightarrow[\text{Zn}]{\text{BrCH}_2\cdot\text{CO}_2\text{Et}} \qquad \xrightarrow{50\ \%\ \text{KOH}}$$

Another example of such a ring transformation is the Pfitzinger quinoline synthesis.

(*3*) *Isatin derivatives.* Adding bromide and chlorine successively to isatin in dilute hydrochloric acid at <17° affords 5-**bromoisatin** (B.I.O.S. 986). Dibromination in concentrated sulphuric acid (FIAT 1313 II; *I. Weinberg*, J. Amer. chem. Soc., 1931, **53,** 318) gives 5,7-**dibromoisatin,** whilst 5,7-**dichloroisatin** is obtained by chlorination in glacial acetic acid solution.

Isatin-2-**anil** (2-phenyliminoisatin) exists in two readily interconvertible modifications (*R. Pummerer*, Ber., 1911, **44,** 338, 810) each of which resembles the corresponding stable *N*-methyl derivative (*R. K. Callow* and *E. Hope*, J. chem. Soc., 1929, 1191):

orange

violet

Isatin anil gives with active methylene compounds deeply coloured indogenides (see p. 455), with hydrogen sulphide 2-thioisatin, and with phenylhydrazine **isatin**-2-**phenylhydrazone,** also obtained from benzenediazonium chloride and indoxyl. The phenylhydrazone can be prepared from phenylhydrazine and *O*-methylisatin, which reacts similarly with aniline to form the anil and with hydroxylamine to form **isatin**-2-**oxime** (*G. Heller*, Ber., 1907, **40,** 1291; 1916, **49,** 2757), also formed from nitrous acid and *O*-ethylindoxylic acid, or by treatment of thiocarbamoylformanilide oxime with sulphuric acid (*H. Wieland* and *E. Gmelin*, *ibid.*, 1908, **41,** 3512). In warm alkaline solution the oxime yields dihydroxyquinazoline by Beckmann change:

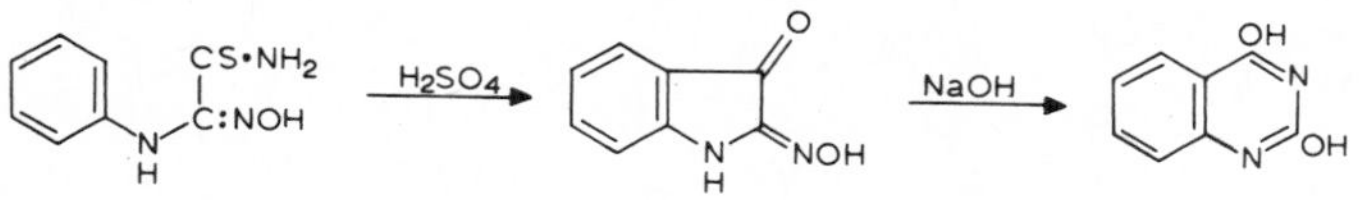

The reactivity of the free carbonyl group in isatin-2-anil and -2-phenylhydrazone is shown by the production of **isatin dianil** and **isatin osazone.** Isatin reacts at both positions with *o*-phenylenediamine giving quinoxaline (*E. Schunck* and *L. Marchlewski, ibid.*, 1895, **28,** 2527).

2-Methylisatin (isatin *O*-methyl ether) on aerial oxidation gives methylisatoid (*C. W. Bird*, Tetrahedron, 1963, **19,** 901), and the purple crystalline salt XXVI is obtained from the reaction between isatin, acetic anhydride and pyridine (*J. A. Ballantine, A. W. Johnson* and *A. S. Katner*, J. chem. Soc., 1964, 3323):

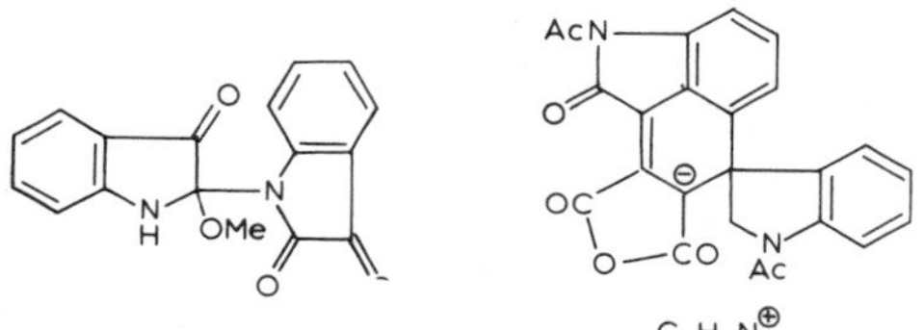

Methylisatoid (XXVI)

Isatin blue, derived from isatin and pipecolic acid, is formulated as the hybrid XXVII (*Johnson* and *D. J. McCaldin, ibid.*, 1957, 3470):

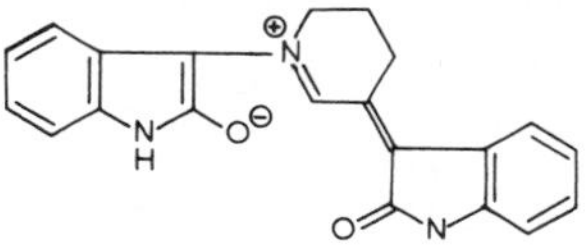

(XXVII)

1-**Hydroxyisatin,** obtained from hydroxyoxindole, or better by the action of sulphuric acid on *o*-nitrobenzoyldiazomethane, is an acidic substance readily isomerised to anthranil-3-carboxylic (anthroxanic) acid (*F. Arndt et al.*, Ber., 1927, **60,** 1364). When diazomethane labelled with ^{14}C is used in the preparation this becomes C-2 in the isatin (*E. C. Taylor* and *D. R. Eckroth*, Tetrahedron, 1964, **20,** 2059):

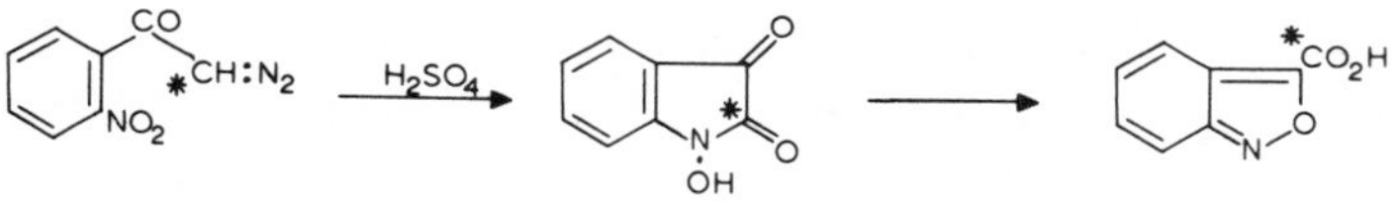

N-**Methylisatin-3-thiosemicarbazone** appears to be an extremely valuable prophylactic agent against smallpox and is the first synthetic compound found to possess this type of anti-viral activity (*D. J. Bauer et al.*, Lancet, 1963, [ii], 494).

Isatin derivatives are assembled in Table 3.

TABLE 3

DERIVATIVES OF ISATIN

Derivative	*m.p.* °	*Ref.*	*Derivative*	*m.p.* °	*Ref.*
Isatin	203.5	1	Ozazone	183 (dec.)	
2-Methyl			2-Oxime	200 (dec.)	
(*O*-methyl ether)	102	2	3-Oxime	222 (dec.)	8
1-Methyl	134	2	3-Diazo-oxindole	161 (dec.)	9
1-Acetyl	141				
1-Hydroxy	200–201 (dec.)		5-Nitro	254–255	10
1-Carboxymethyl	206–207	3	5-Methoxy	202	11
3-Imine	175–176	4	5-Methyl	187	1
2-Anil	126		5-Bromo	255	12
3-Anil	226	5	5-Chloro	247 (dec.)	13
Dianil	212	6	5,7-Dibromo	248–250	
2-Phenylhydrazone	242 (dec.)	6	5,7-Dichloro	223 and 313	14
3-Phenylhydrazone	211	7			

References
1 *C. S. Marvel* and *G. S. Hiers*, Org. Synth., 1925, **5**, 71.
2 *R. G. Ault et al.*, J. chem. Soc., 1935, 1653.
3 *A. D. Ainley* and *R. Robinson*, *ibid.*, 1934, 1508.
4 *A. Reissert* and *H. Happmann*, Ber., 1924, **57**, 972.
5 *E. Knoevenagel*, J. pr. Chem., 1914, [ii], **89**, 46.
6 *G. Heller*, Ber., 1907, **40**, 1291.
7 *A. Krause*, *ibid.*, 1890, **23**, 3617.
8 *W. Borsche* and *W. Sander*, *ibid.*, 1914, **47**, 2815.
9 *T. Curtius* and *H. Lang*, J. pr. Chem., 1891, [ii], **44**, 544; *P. L. Creger*, J. org. Chem., 1965, **30**, 3610.
10 *W. C. Sumpter* and *W. F. Jones*, J. Amer. chem. Soc., 1943, **65**, 1802.
11 *E. Ferber* and *G. Schmolke*, J. pr. Chem., 1940, [ii], **155**, 234.
12 *Heller*, Ber., 1920, **53**, 1545.
13 *Idem*, *ibid.*, 1910, **43**, 2895; *Buu-Hoï*, Rec. trav. chim., 1954, **73**, 197.
14 *Heller*, Ber., 1922, **55**, 2692.

(v) Indolines with side-chain carbonyl groups

An indoline derivative with the carbonyl group in the side-chain is 1,3,3,-**trimethyl-2-indolinylideneacetaldehyde,** m.p. 117–118°, "Fischer's aldehyde", a source of polymethine dyes, obtainable from tetramethyl-3*H*-indolium bromide:

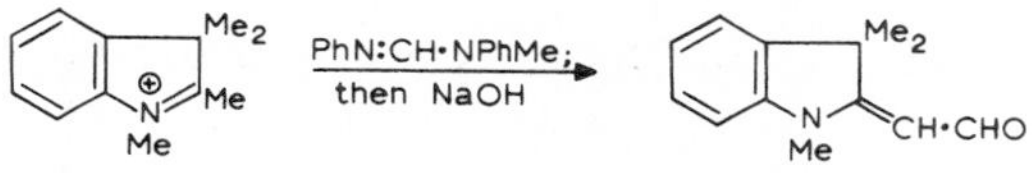

The aldehyde can also be obtained by direct formylation of 1,3,3-trimethyl-2-methyleneindoline with methylformanilide and phosphorus oxychloride (*I. G. Farbenind.*, B.P. 486,780/1936). The same base gives with diphenylformamidine the anil of Fischer's aldehyde, another valuable intermediate (*H. A. Piggott* and *E. H. Rodd*, B.P. 344,409/1929). Polymethine dyes will be included in a subsequent chapter on cyanine dyes.

(c) Tetra-, hexa-, and octa-hydroindoles

Indole unlike pyrrole may be reduced by metals in liquid ammonia, in the presence of alcohol or ammonium chloride; with a large excess of lithium and methanol in ammonia a product is obtained consisting of equal amounts of 4,7-**dihydroindole** and 4,5,6,7-**tetrahydroindole** (*S. O'Brien* and *D. C. C. Smith*, J. chem. Soc., 1960, 4609):

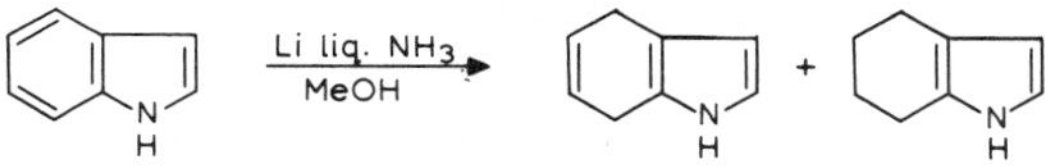

Derivatives of 4,5,6,7-tetrahydroindole are obtained by the application of general pyrrole syntheses to cyclohexanone (*A. Treibs* and *D. Dinelli*, Ann., 1935, **517,** 152; *Treibs, ibid.*, 1936, **524,** 285). Thus chlorocyclohexanone and ethyl oxaloacetate with ammonia yield the half ester XXVIII, and hence 4,5,6,7-**tetrahydroindole,** m.p. 55°, easily resinified, and resembling pyrrole homologues:

(XXVIII)

The combination of oximinocyclohexanone and ethyl acetoacetate, and cyclohexanone and ethyl oximinoacetoacetate yield by the Knorr synthesis isomeric ethyl methyltetrahydroindolecarboxylates and hence 2-*methyl*-4,5,6,7-*tetrahydroindole*, b.p. 105–107°/10 mm, and 3-*methyl*-4,5,6,7-*tetrahydroindole*, m.p. 58°. 1-*Phenyl*-4,5,6,7-*tetrahydroisatin*, m.p. 183–184° (*L. Horwitz*, J. Amer. chem. Soc., 1953, **75,** 4060).

2,3,4,5,6,7-**Hexahydroindole,** b.p. 80°/19 mm, has been prepared as follows (*R. Belleau*, *ibid.*, 1953, **75,** 5765):

CO_2Et, K, O + $BrCH_2$–CH_2–$N(CO)_2C_6H_4$ → CO_2Et, CH_2–CH_2–$N(CO)_2C_6H_4$, O —acid→ N H

A *hexahydroindole*, b.p. 93–95°/0.15 mm, is obtained by the elimination of pyrrolidine from the octahydroindole formed by the cyclo-addition of 1-ethoxycarbonylaziridine to the enamine of pyrrolidine and cyclohexanone (*J. E. Dolphini* and *J. D. Simpson*, *ibid.*, 1965, **87,** 4381):

N + NCO_2Et → N CO_2Et N —p-$MeC_6H_4SO_3H$→ N

The catalytic hydrogenation of indole over Raney nickel at 220°/250 atm. affords *cis*-**octahydroindole,** b.p. 73–76°/18 mm (*picrate*, m.p. 137°) (*H. Adkins* and *H. L. Coonradt*, *ibid.*, 1941, **63,** 1563); under similar conditions with Raney nickel in methanol *cis*-1-**methyloctahydroindole,** b.p. 179° and a little *trans*-isomer are obtained. The stereoisomeric 2-ethoxyethylcyclohexylamines, prepared from 2-ethoxyethylcyclohexanone, are cyclised to *cis*- and *trans*-**octahydroindole,** b.p. 72–73°/19 mm (*picrate*, m.p. 147°) (*F. E. King et al.*, J. chem. Soc., 1953, 250).

trans-Octahydroindole is best synthesised by the following route starting from 2-allylcyclohexanone, the oxime of which, when reduced using sodium and alcohol, affords the *trans*-base XXIX as the sole product (*H. Booth* and *King*, *ibid.*, 1958, 2688):

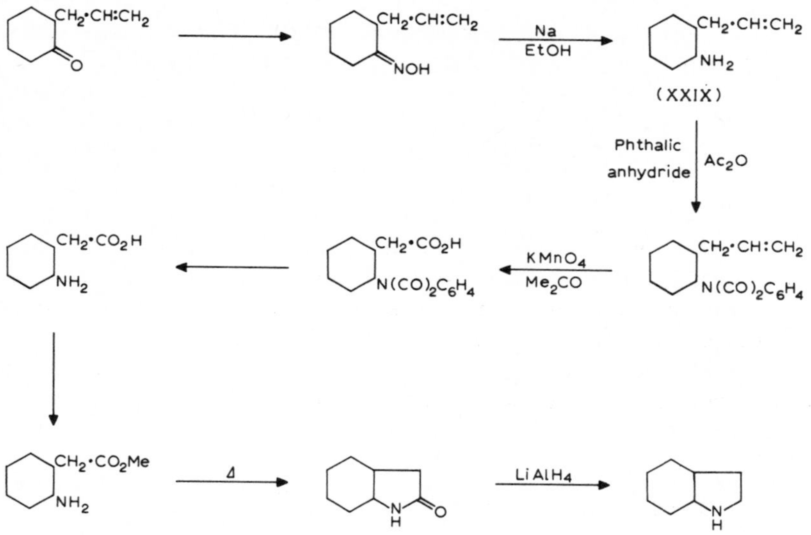

Na / EtOH
(XXIX)
Phthalic anhydride / Ac_2O
$KMnO_4$ / Me_2CO
Δ
$LiAlH_4$

1,5-Dimethyl-4-oxohexahydroindole (XXX) on treatment with formic acid is reduced with the formation of 1,5-dimethyl-4-oxo-octahydroindole (XXXI), which in acid solution undergoes fission, recyclisation and further reduction to the 1,7-dimethyloctahydroindole (XXXII) (*O. Červinka*, Chem. and Ind., 1959, 1129):

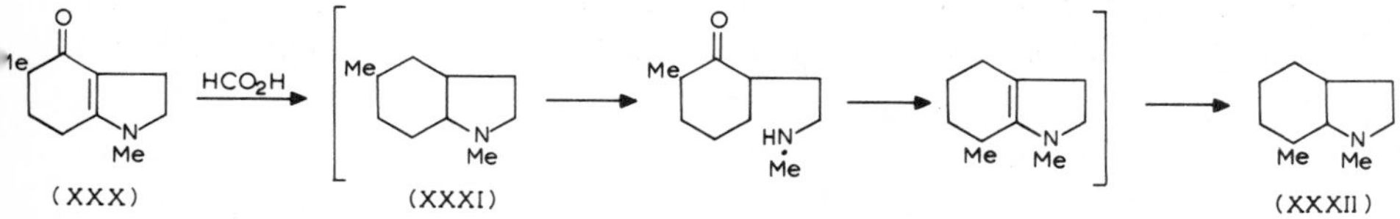

HCO_2H
(XXX) (XXXI) (XXXII)

(d) Compounds having more than one indole nucleus

The most important diindolyl derivatives are the colouring matters, indigotin, indirubin, and isoindigo, to be described in a later volume, with the bimolecular reduction products of isatin (p. 458).

Unsubstituted 2,2′-**diindolyl,** m.p. 302–303°, is obtained by ring-closure of oxal-*o*-toluidine (*W. Madelung*, Ann., 1914, **405,** 58); and the isomeric 3,3′-**diindolyl,** m.p. 286–287°, by heating indole with sulphur at 120° (*B. Oddo* and *L. Raffa*, Gazz., 1939, **69,** 562).

Indoles condense with aldehydes and ketones, giving derivatives of 3,3′-diindolylmethane (*M. Scholtz*, Ber., 1913, **46,** 1082, 2138):

2 (2-methylindole) + RCOR′ ⟶ R R′ C (2-methyl-3-indolyl)₂

Alternatively the carbonyl compound may be treated with an indolylmagnesium halide (*R. Majima*, Ber., 1922, **55,** 3865). The simplest example, 3,3′-**diindolylmethane,** m.p. 168°, is prepared from indole and formaldehyde in aqueous solution (*J. Thesing, ibid.*, 1954, **87,** 692).

When R or R′ is hydrogen, the colourless products are easily oxidised to coloured diindolomethenes, the analogues of the numerous dipyrromethenes, the mono-acid salts of which are red dyes:

(3-indolyl)–CHR–(3-indolyl) ⟶ (3-indolyl)–CR=(3-indolenine) ← 2 C_8H_7N + RCOCl

The same products, "rosindoles", are obtained by treating indoles with benzoyl chloride in presence of zinc chloride (*E. Fischer* and *P. Wagner*, *ibid.*, 1887, **20,** 815) or with formic acid *(Scholtz, loc. cit.)*. If R is an alkyl radical, the yellow diindolomethene may be less stable than the isomeric colourless diindolylalkene (*A. Kiange* and *F. G. Mann*, J. chem. Soc., 1953, 594):

(2-phenyl-3-indolyl)–CMe=(2-phenyl-3-indolenine)   (2-phenyl-3-indolyl)–C(=CH₂)–(2-phenyl-3-indolyl)

Diindolomethenes can combine, reversibly, with a third molecule of indole, giving a triindolylmethane, encountered as a by-product in the synthesis of the methene from indole and formic acid (*M. Passerini*, Gazz., 1938, **68,** 480).

Skatole (3-methylindole) gives, less readily, analogous methanes and methenes by substitution in the 2-position (*H. von Dobeneck* and *H. Prietzel*, Z. physiol. Chem., 1955, **299,** 214).

2,2-**Di-(3′-indolyl)propane,** m.p. 163–165°; 3,3′-**diindolylphenylmethane,** m.p. 149–152° *(Majima, loc. cit.)*; **di(2-methyl-3-indolyl)methane,** m.p. 230–231° (*R. von Walther* and

J. Clemen, J. pr. Chem., 1900, [ii], **61,** 256); 1,1-**bis(methylindolyl)ethane,** m.p. 191°; **bis(methylindolyl)phenylmethane,** 250–251° (*E. Fischer*, Ann., 1887, **242,** 372); 2,2-**bis-(methylindolyl)propane,** m.p. 197° (*Scholtz, loc. cit.*).

Di-3-indolomethene, m.p. 283–285° (*W. König*, J. pr. Chem., 1911, [ii], **84,** 194; *B. Oddo* and *G. Sanna*, Gazz., 1924, **54,** 684), 2,2-*dimethyl* deriv., m.p. 230° (*K. Kunz* and *W. Stühlinger*, Ber., 1925, **58,** 1860); **phenyldi-3-indolomethene hydrochloride,** m.p. 245–248° (*Majima, loc. cit.*), 2,2-*dimethyl* deriv. (free base), m.p. 270° (*Fischer* and *Wagner, loc. cit.*).

(i) Polymethine dyes and related compounds derived from bisindoles

Trimethylmethyleneindoline yields, by a curious reaction with nitrous acid, the methine dye XXXIII (*R. Kuhn et al.*, Ber., 1930, **63,** 3176):

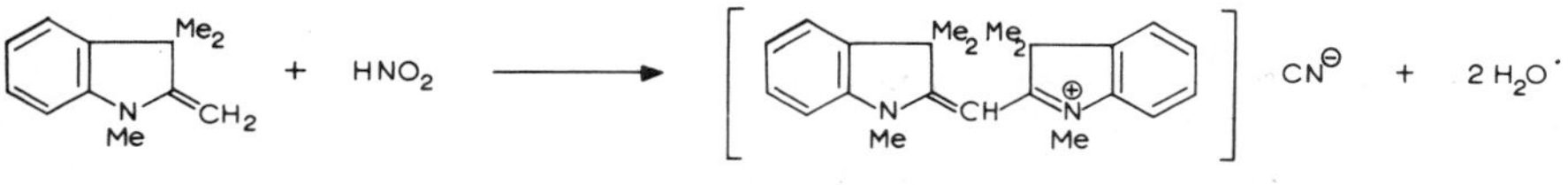

(XXXIII)

1,2-Bis-(3′-oxo-2′-indolinylidene)ethane (XXXIV) is obtained from indoxyl and glyoxal; it yields an orange-yellow vat giving blue-violet dyeings (*P. Friedländer* and *F. Risse*, Ber., 1914, **47,** 1919):

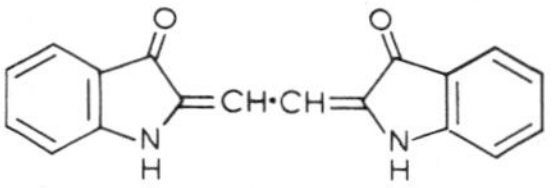

(XXXIV)

Indoles react with dihydroresorcinol giving trimethines in which the trimethine link is part of a ring:

The same products are obtained from cyclic monoketones due to atmospheric oxidation, and analogues having an open-chain trimethine system XXXV are obtained from acraldehyde and its derivatives (*A. Treibs* and *E. Herrmann*, Ann., 1954, **589,** 207; 1955, **592,** 1):

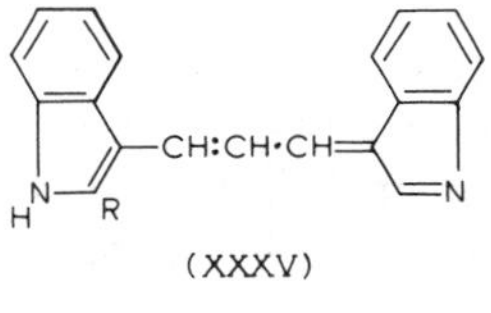

(XXXV)

By condensation with ethyl orthoformate or with *N,N'*-diphenylformamidine, trimethylmethyleneindoline yields the brilliant pink trimethine dye Astraphloxine FF (p. 442).

With pyridine and cyanogen bromide (giving the elements of glutacondialdehyde) 2-methylindoles yield deeply coloured pentamethines *e.g.* XXXVI (*W. König* and *R. Schreckenback*, J. pr. Chem., 1913, [ii], **87,** 241):

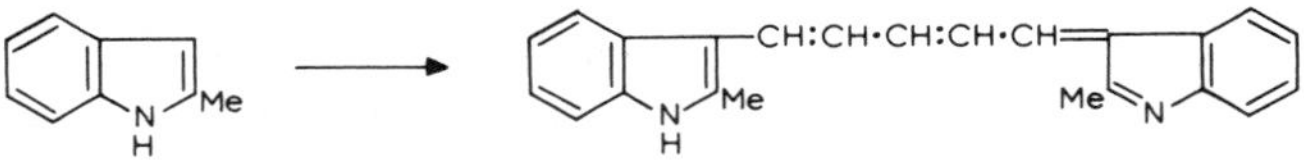

CH:CH·CH:CH·CH=

Me Me Me N N H H

(XXXVI)

(See also the subsequent chapter on cyanine dyes).

3. Isoindoles and isoindolines

The quinonoid isoindole (3,4-benzopyrrole) (I) is in principle tautomeric with the benzenoid 3*H*-isoindole (2-azaindene) (II) form and according to molecular-orbital calculations the more important charged resonance structures are III and IV (*H. C. Longuet-Higgins* and *C. A. Coulson*, Trans. Faraday Soc., 1947, **43,** 87):

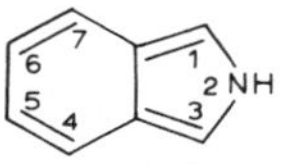

(I) 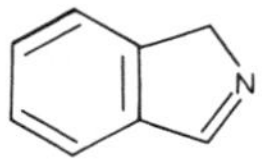 (II) 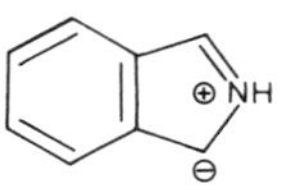 (III) 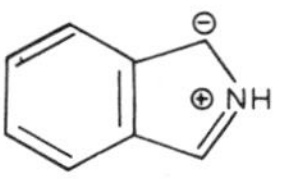 (IV)

Isoindoles are more difficult to produce than indoles, the first uncontested examples having been described by *G. Wittig* in 1951. Chemically, they appear to be less stable, although approximate quantum-mechanical calculations have indicated that indole and isoindole should differ little in energy (*M. J. S. Dewar*, *ibid.*, 1946, **42,** 764). Isoindolines (dihydroisoindoles), on the other hand, are stable and readily accessible; phthalimidines and phthalimides are formally oxoisoindolines.

(a) Isoindoles

Isoindole has been prepared in solution by treating compound V with potassium *tert*-butoxide in dimethyl sulphoxide, and trapped by the formation of adducts with maleic anhydride and *N*-phenylmaleimide (*R. Kreher* and *J. Seubert*, Z. Naturforsch., 1965, **20b,** 75):

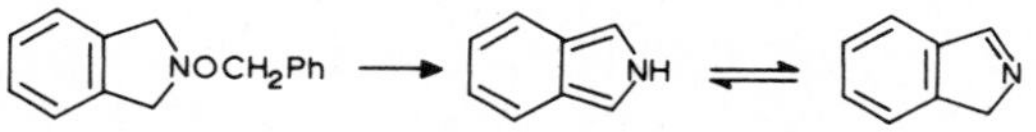

(V)

2-**Methylisoindole,** m.p. 90–91°, is produced from 2-methylisoindoline by treating 2,2-dimethylisoindolinium bromide with phenyl-lithium, or by oxidation *via* the *N*-oxide (*G. Wittig et al.*, Ann., 1951, **572,** 1; *J. Thesing, W. Schäfer* and *D. Melchior*, *ibid.*, 1964, **671,** 119):

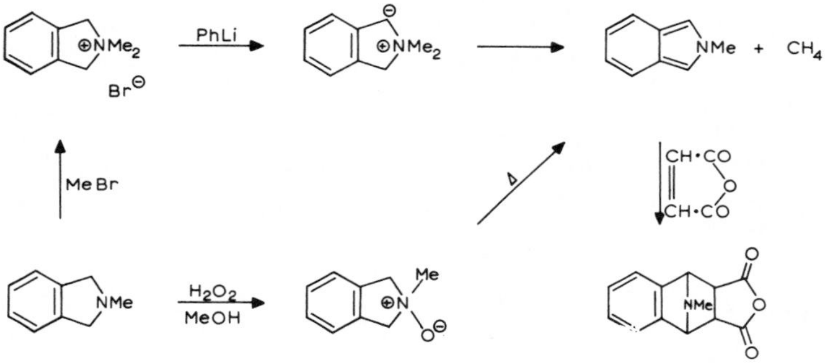

It resinifies slowly, but quickly in the presence of hydrochloric or even picric acid and forms an adduct with maleic anhydride.

N-Methylphthalimidine (1-oxo-2-methylisoindoline) with organolithium compounds gives 1-substituted-2-methylisoindoles (*W. Theilacker et al.*, *ibid.*, 1953, **584,** 87; 1955, **597,** 95; *Wittig* and *H. Streib*, *ibid.*, 1953, **584,** 1):

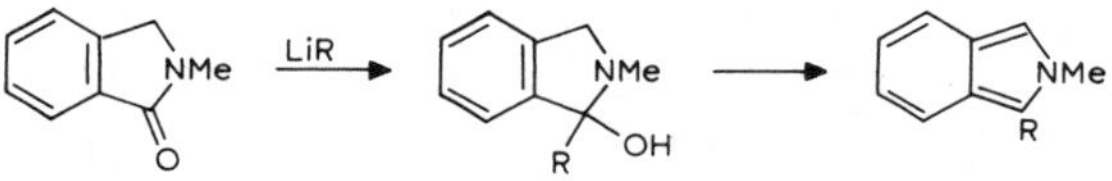

1-**Phenylisoindole,** m.p. 90–100° (decomp.), may be obtained from *N*-(2-benzoylbenzyl)phthalimide and hydrazine in refluxing ethanol. Spectral studies indicate that tautomers VI and VII are in equilibrium, and in deuterochloroform about 9% of VII is present, but possible tautomer VIII could not be detected (*D. F. Veber* and *W. Lwowski*, J. Amer. chem. Soc., 1963, **85,** 646):

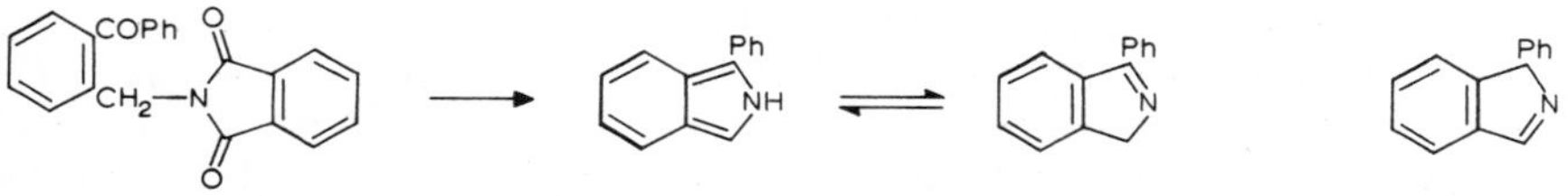

(VI) (VII) (VIII)

In solution 1-phenylisoindole exists predominantly as such, but with 1-(4-methoxyphenyl)isoindole, the composition of the equilibrium mixture varies with solvent (*idem, ibid.*, 1964, **86,** 4152). 1-Phenylisoindole reacts with maleic anhydride, condenses with Ehrlich reagent, is reduced with zinc dust in acid, and resinifies with acids and on standing in air.

2-Arylisoindoles may be obtained from 7,8-dihalogenobicyclo[4.2.0]octa-1,3,5-trienes and 2-arylisoindoline *N*-oxides (*Krehner* and *Seubert*, Tetrahedron Letters, 1966, 3015; Angew. Chem., intern. Edn., 1966, **5,** 967):

Br Br 100° ArNH2 Me2SO NAr NAr NEt3 Ar ⊕N O⊖ Ac2O −10° Ar ⊕N O C O Me CH3CO2⊖

By means of the Leuckart reaction *o*-dibenzoylbenzene on treatment with ammonium formate or methylammonium formate at 145°, is converted to 1,3-**diphenylisoindole,** m.p. 148–150° (decomp.), or 2-**methyl**-1,3-**diphenylisoindole,** m.p. 149–150°, respectively (*J. C. Emmett, Veber* and *Lwowski*, Chem. Comm., 1965, 272). Diels–Alder addition of butadiene to dibenzoylacetylene affords 1,2-dibenzoylcyclohexa-1,4-diene, which condenses with 40% aqueous methylamine to give 2-methyl-1,3-diphenylisoindole (*M. E. Mann* and *J. D. White, ibid.*, 1969, 420):

+ C·COPh ||| C·COPh MeC6H5 150-160° COPh COPh H2O MeNH2 Ph NMe Ph

1,3-Diphenylisoindole yields stable Diels–Alder adducts with acetylenedicarboxylic esters and benzyne, but the one with maleic anhydride readily reverts to the starting materials (*Emmett* and *Lwowski*, Tetrahedron, 1966, **22,** 1011). Reduction of 1,3-diphenylisoindole with zinc and acetic acid affords *cis*-1,3-**diphenylisoindoline,** m.p. 108–109° (*D. R. Boyd* and *D. E. Ladhams*, J. chem. Soc., 1928, 2089), (*sulphonamide*, m.p. 238–239°). Treatment of the sulphonamide with potassium *tert*-butoxide in dimethyl sulphoxide–benzene yields 1,3-diphenylisoindole:

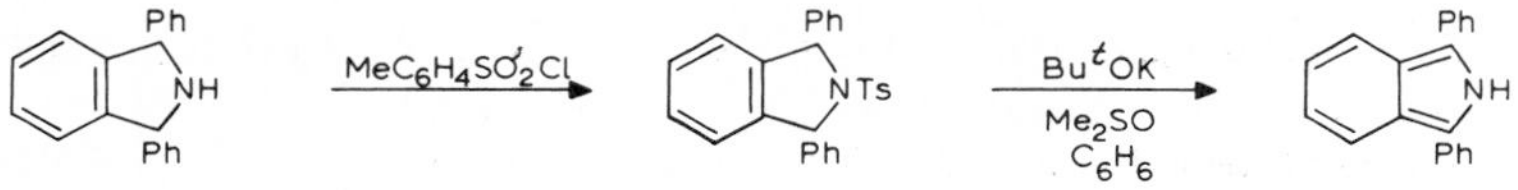

Acid-catalysed self-condensation of 2,5-dimethylpyrrole affords a product (p. 336), 1,3,4,7-tetramethylisoindole, which is predominantly in the 3*H*-isoindole form (*C. O. Bender* and *R. Bonnett*, Chem. Comm., 1966, 198):

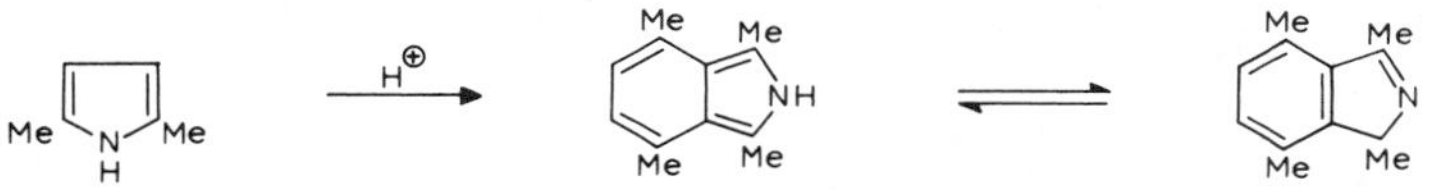

1-Alkyl-substituted 2,3-dihydro-5-phenyl-1*H*-1,4-benzo[*f*]diazepin-2-ones (IX) in dimethylformamide with one equivalent of sodium hydride readily undergo rearrangement to give extremely stable *N*-alkyl-3-phenylisoindole-1-carboxamides (X). In this case it is possible to isolate both tautomers, for treatment of X in dimethylformamide with sodium carbonate affords the 3*H*-isoindole (XI), which with dilute acid or strong aqueous base is reconverted to the thermodynamically more stable isoindole (*R. I. Fryer*, *J. V. Earley* and *L. H. Sternbach*, J. Amer. chem. Soc., 1966, **88**, 3173):

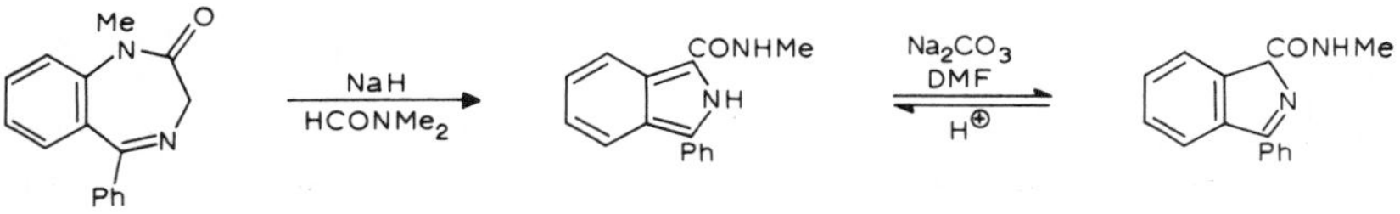

(IX) (X) (XI)

When 6-chlorophenyl-1,2,3,4,5,6-hexahydrobenzo[*g*]-2,5-diazocine (XII) is treated with potassium amide in liquid ammonia, simultaneous ring contraction and dehydrochlorination occur to give 2-(β-aminoethyl)-1-phenylisoindole (XIII). This is unusual in that the leaving group is on an aromatic ring (*M. Winn* and *H. E. Zaugg*, J. org. Chem., 1969, **34**, 249):

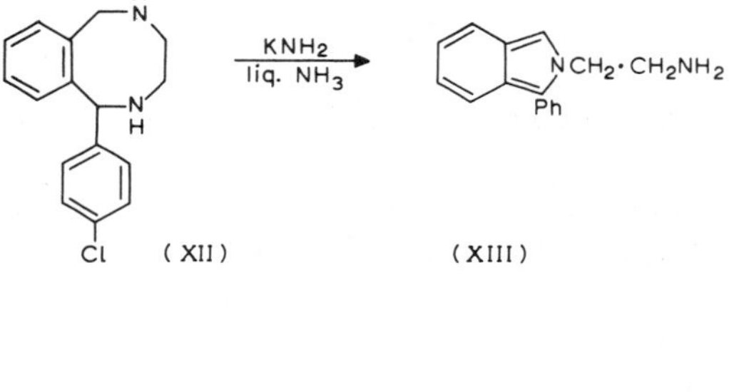

(XII) (XIII)

1-**Benzyl-2-methylisoindole,** m.p. 89–91°; 1-*ethyl*- and 1-*butyl*-analogues are oils (*Theilacker et al., loc. cit.*).

1,2,3-**Triphenylisoindole,** m.p. 234–235°, originally obtained by the reaction of phenyl-lithium with 2,3-diphenylisoindolone (*Theilacker* and *W. Schmidt*, Ann., 1957, **605,** 43), may be prepared in good yield from 1,3-diphenylisobenzofuran and aniline sulphoxide in the presence of boron trifluoride (*M. P. Cava* and *R. H. Schlessinger*, J. org. Chem., 1963, **28,** 2464):

Ph NPh O —PhLi→ Ph NPh Ph ←BF3— Ph O Ph + PhN:SO

It forms a peroxide which with water affords *o*-benzoylbenzophenone and aniline (*Theilacker et al., loc. cit.*):

Ph NPh Ph —CS_2 hν, O_2→ Ph O–O NPh Ph —H_2O→ COPh COPh + $PhNH_2$

Thermal equilibrium has been established at 300° between 1,2,3-triphenylisoindole and 1,3,3-triphenyl-3*H*-isoindole and this favours the latter (*Theilacker et al.*, Ann., 1964, **673,** 96):

Ph NPh Ph ⇌ Ph N Ph_2

2-**Methyl**-1,3-**diphenylisoindole,** m.p. 149.5°, is fairly resistant to light and air, but can be autoxidised to *o*-dibenzoylbenzene; it dissolves in acids without resinification.

Phthalonitrile reacts with organometallic compounds (including diethyl sodiomalonate and the like), giving strongly basic derivatives of 1-amino-3*H*-isoindole (*R. P. Linstead et al.*, J. chem. Soc., 1939, 1809; 1940, 1076, 1079):

CN CN —MeLi→ Me N NH_2

(*b*) *Isoindolines*

Isoindoline (1,3-**dihydroisoindole**), b.p. 213°, *picrate* m.p. 196–197° (decomp.), first obtained by reduction of 1-chlorophthalazine with ring-contraction and loss of nitrogen, and by dry distillation of *o*-xylylenediamine hydrochloride, can be prepared by electrolytic reduction of phthalimide (*E. Späth* and *F. Breusch*, Monatsh., 1928, **50,** 354; *E. W. Cook* and *W. G. France*, J. phys. Chem., 1932, **36,** 2383) and by treating *N-p*-toluenesulphonyldihydroisoindole, obtained from 2-bromomethylbenzyl bromide and *p*-toluenesulphonamide, with hydrogen bromide (*J. Bornstein et al.*, J. org. Chem., 1957, **22,** 1255; Org. Synth., 1967, **47,** 110; 1968, **48,** 65):

$C_6H_4(CH_2Br)_2$ + $H_2NSO_2C_6H_4Me$ —NaOEt→ N-$SO_2C_6H_4Me$-isoindoline —HBr→ isoindoline (NH)

Isoindolines are obtained in good yield from *o*-xylylene bromide and primary or secondary amines, least smoothly with *ortho*-substituted anilines (*M. Schlotz*, Ber., 1898, **31,** 414, 1707; *J. von Braun*, Ber., 1910, **43,** 1353; *Wittig et al.*, Ann., 1951, **572,** 1; *A. H. Sommers*, J. Amer. chem. Soc., 1956, **78,** 2439).

Reaction of isoindoline formed from ammonia with a second molecule of halide gives a quaternary spiro-compound XIV, yielding with phenyl-lithium 2-o-*methylbenzylisoindole*, m.p. 80°, which like the simpler compounds is scarcely basic and does not combine with methyl iodide:

NR ←NH_2R— $C_6H_4(CH_2Br)_2$ —NHR_2→ ⊕NR_2 $Br^{\ominus}$

$C_6H_4(CH_2Br)_2$ —NH_3→ NH —$C_6H_4(CH_2Br)_2$→ spiro-N⊕ $Br^{\ominus}$ (XIV) —PhLi→ N—CH_2—(o-Me-phenyl)

An isoindoline and not the expected tetrahydroisoquinoline is obtained by the reaction between 2-chloromethylphenacyl chloride and aniline (*I. G. Hinton, F. G. Mann* and *A. Vanterpool*, J. chem. Soc., 1959, 610):

$C_6H_4(CO{\cdot}CH_2Cl)(CH_2Cl)$ + 2 $PhNH_2$ —K_2CO_3, EtOH→ NPh, NPh, Me

and 2,5-dimethylpyrrole when boiled with tin and hydrochloric acid yields 1,3,4,7-tetramethylisoindoline, probably *via* a Diels–Alder type of addition followed by reduction and aromatisation (*Bonnett* and *White*, Proc. chem. Soc., 1961, 119):

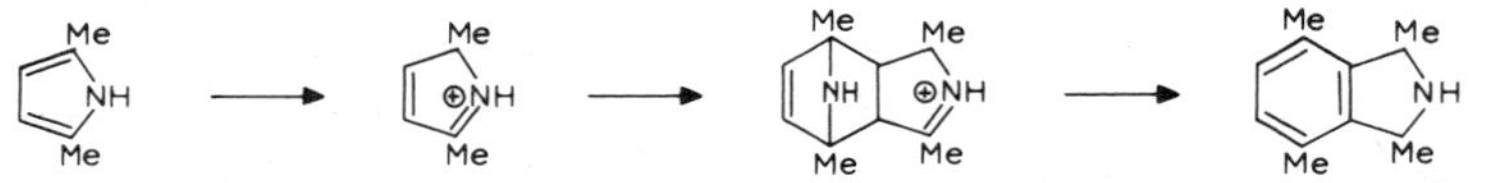

6-Phenyl-1,2,3,4,5,6-hexahydrobenzo[*g*]-2,5-diazocine (XV) undergoes ring contraction in liquid ammonia containing potassium amide to give 2-(β-aminoethyl)-1-phenylisoindoline (XVI) (see also p. 473) (*Winn* and *Zaugg*, *loc. cit.*):

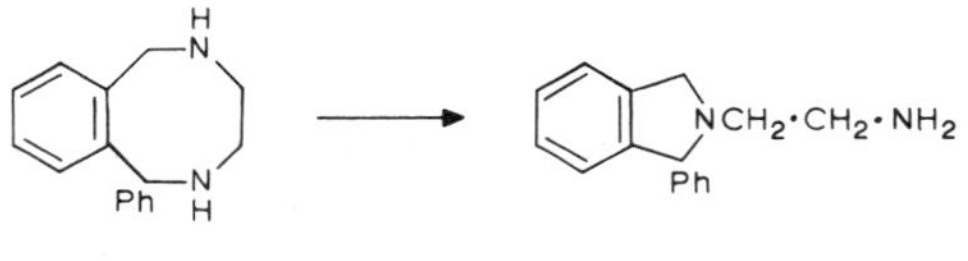

(XV) (XVI)

Isoindoline possesses the properties of a typical secondary amine. 2-Alkylisoindoline *N*-oxides on pyrolysis afford 2-alkylisoindoles (*Thesing*, *Schäfer* and *Melchior*, *loc. cit.*).

1-**Methylisoindoline,** b.p. 213° (*Linstead* and *E. G. Noble*, J. chem. Soc., 1937, 933); 2-*methylisoindoline*, b.p. 81–82°/13 mm (*von Braun* and *Z. Köhler*, Ber., 1918, **51,** 100); 2-*ethylisoindoline*, b.p. 219–220° (*Scholtz*, *ibid.*, 1898, **31,** 1700); 2-*allylisoindoline*, b.p. 125°/17 mm; 2-*carbamoylisoindoline*, m.p. 183° (*von Braun*, *ibid.*, 1910, **43,** 1353). 2-**Benzylisoindoline,** m.p. 41°; 2-**phenylisoindoline,** m.p. 172° (*Scholtz*, *ibid.*, 1898, **31,** 414, 629). 2-**Acetylisoindoline,** m.p. 77° (*E. Späth* and *F. Breusch*, Monatsh., 1928, **50,** 354); 2-*benzoylisoindoline*, m.p. 100°; 2-*aminoisoindoline* (oil) (*K. Fränkel*, Ber., 1900, **33,** 2808); 2-*nitrosoisoindoline*, m.p. 96–97° (*S. Gabriel* and *A. Neumann*, *ibid.*, 1893, **26,** 521).

Whereas 2,2-*dimethylisoindolinium bromide*, m.p. 242°, gives with phenyl-lithium 2-methylisoindole, degradation under Hofmann conditions affords 2-methylisoindoline and *o*-dimethylaminomethylbenzyl alcohol, $Me_2NCH_2{\cdot}C_6H_4{\cdot}CH_2OH$ (*von Braun* and *Köhler*, *loc. cit.*).

Phthalimide is formally 1,3-**dioxoisoindoline;** similarly, *iminophthalimidine*, m.p. 203°, obtained by heating *o*-cyanobenzamide above its melting point (*Linstead et al.*, J. chem. Soc., 1934, 1017), is 1-**iminoisoindoline** and *di-iminophthalimide*, m.p. 196° (decomp.), is 1,3-**di-iminoisoindoline,** obtained, for example, by heating phthalonitrile with hot methanolic ammonia (*J. A. Eldridge* and *Linstead*, *ibid.*, 1952, 5000). All these are important intermediates in the synthesis of phthalocyanines.

Phthalonitrile is converted by sodium sulphide into 1-**iminoisoindoline**-3-**thione** (XVII), an orange-yellow powder which gives *S*-alkyl derivatives also convertible into phthalocyanines (*H. France et al.*, B.P. 731,257/1955; *M. E. Baguley* and *Eldridge*, J. chem. Soc., 1957, 709):

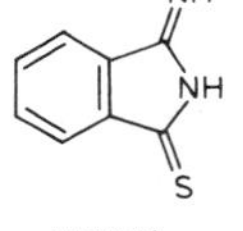

(XVII)

(c) Other hydroisoindoles

(i) Tetrahydroisoindoles

Isoindoles hydrogenated in the benzene nucleus are the unstable products of normal pyrrole synthesis from 1,2-diacylcyclo-hexenes or -hexanes and ammonium acetate (*G. O. Schenk*, Ber., 1947, **80**, 226):

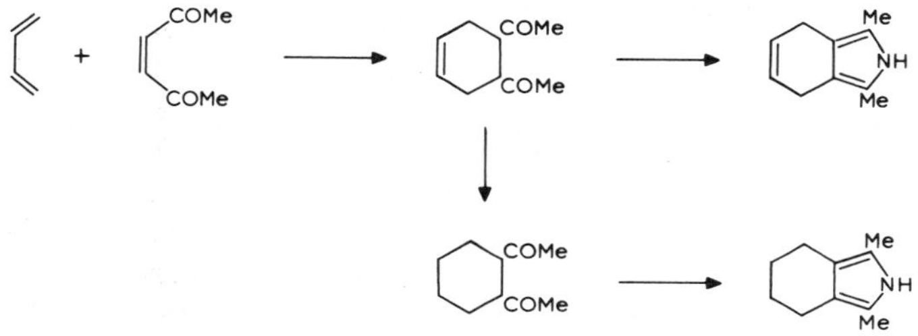

(ii) Hexa- and octa-hydroisoindoles

Reduction of *cis*-tetrahydrophthalimide with lithium tetrahydridoaluminate gives *cis*-1,3,3a,4,7,7a-**hexahydroisoindole,** b.p. 76–80°/15 mm, n_D^{20} 1.5110 (*L. M. Rice* and *C. H. Grogan*, J. org. Chem., 1955, **20,** 1687), 2-*methyl* derivative, prepared similarly, b.p. 60–63°/11 mm (*Rice et al.*, J. Amer. chem. Soc., 1955, **77,** 616).

Hexahydroisoindole is hydrogenated to *cis*-**octahydroisoindole,** *picrate*, m.p. 162–163°. Reduction of isoindoline over platinum oxide gave an *octahydroisoindole*, b.p. 188–189°, n_D^{20} 1.4917, *picrate*, m.p. 154–155° (*A. Dunet et al.*, Bull. Soc. chim. Fr., 1956, 906).

With ammonia or primary amines 2,3-bis(bromomethyl)naphthalene gives 1,3-dihydronaphtho[2,3-*c*]pyrroles (5,6-benzoisoindolines) (*W. Wenner*, U.S.P. 2,740,795/1956; C.A., 1957, **51,** 485).

4. Other bicyclic pyrrole systems

(a) Spirocyclic systems

These include the spirobispyrrolidinium ion (I) and the related tetracyclic spirobisindolinium ion (II) (*cf.* p. 475)

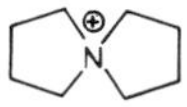

(I) 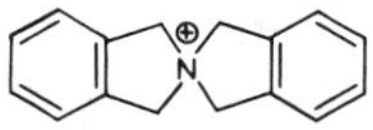 (II)

Catalytic reduction of γ-phenoxybutyronitrile gives the secondary amine di-(4-phenoxybutyl)amine, which by loss of phenol affords a pyrrolidine derivative converted by successive treatment with hydrogen bromide and alkali into **spirobispyrrolidinium bromide** (III), m.p. 256–258° (*J. von Braun et al.*, Ber., 1922, **55,** 2059; 1924, **57,** 185; Ann., 1925, **445,** 247), which with ammonia at 170–180° yields 1-δ-aminobutylpyrrolidine (IV), thought at first to be 1,6-diazacyclodecane:

(III) (IV)

Spirobisindolinium bromide (II) reacts in a similar way with ammonia. With phenyl-lithium it gives 2-*o*-methylbenzylisoindole (p. 475) and also by rearrangement the isoindoloisoquinoline (V) (*G. Wittig et al.*, Ann., 1951, **572,** 1):

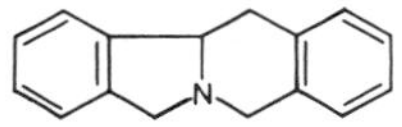

(V)

Salts of three active forms of the spirobis-3,4-dimethylpyrrolidinium ion (VI) have been prepared, as well as one inactive *meso* modification which has neither a plane nor a centre of symmetry and is probably the first recorded instance of its kind (*G. E. McCasland* and *S. Proskow*, J. Amer. chem. Soc., 1955, **77,** 4688):

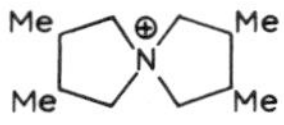

(VI)

(b) *Cycloalkenopyrroles*

2-Methyl-4,5-cyclopentenopyrrole-3-carboxylic ester (VII) has been obtained in low yield by a modified Knorr synthesis (*A. Treibs*, Ann., 1936, **524,** 285) and the simple, fully reduced *trans*-**cyclopentanopyrrolidine** (VIII), b.p. 167° (*picrate*, m.p. 87–90°) from ethyl cyclopentanonecarboxylate (*V. Prelog* and *S. Szpilfogel*, Helv., 1945, **28,** 178):

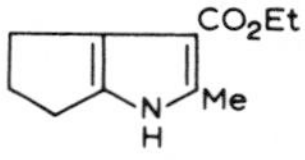

(VII)

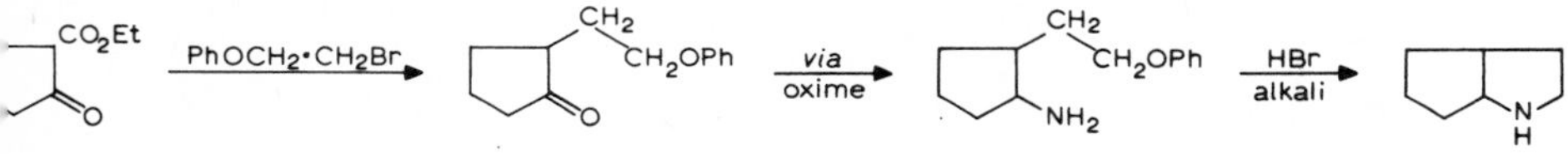

(VIII)

F. E. King et al. (Chem. and Ind., 1956, 466) have obtained the *cis-isomer*, b.p. 161° (*picrate*, m.p. 101°):

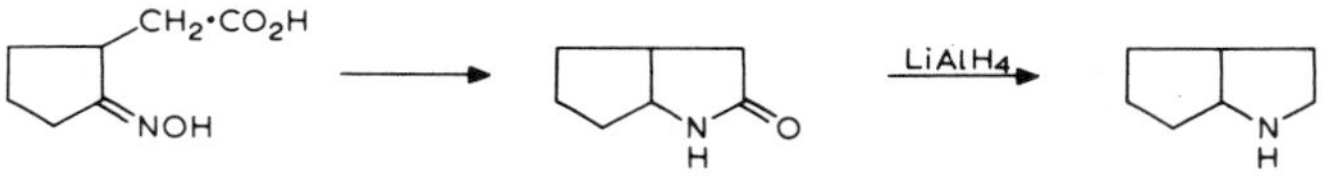

Compounds with additional fused benzene rings are often more readily accessible; *e.g.* 3-acetyl-2-methyl-1,8-dihydroindeno[2,1-*b*]pyrrole (IX) (*G. Bondietti* and *F. Lions*, J. Proc. roy. Soc. N.S.W., 1933, **66,** 477); 5,6-dihydroindeno[2,1-*b*]indole (X) (*J. W. Armit* and *R. Robinson*, J. chem. Soc., 1922, **121,** 827); 5,10-dihydroindeno[1,2-*b*]indole (XI) (*H. Leuchs* and *G. Kowalski*, Ber., 1925, **58,** 2822) and compound XII (*A. C. Sircar* and *M. D. R. Gopalan*, J. Indian chem. Soc., 1932, **9,** 297).

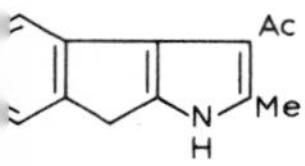

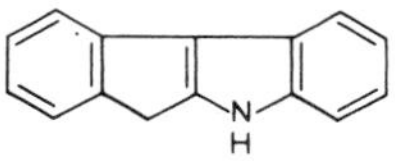

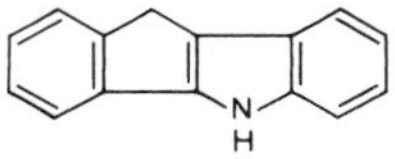

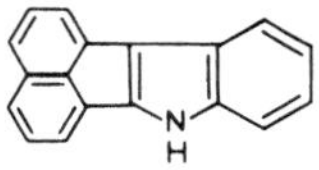

(IX) (X) (XI) (XII)

2,3-Cycloheptanopyrrolidine was prepared in the same way as VIII above, and separated by chromatography of the benzoyl derivatives into the *cis-* and *trans*-stereoisomers (*Prelog* and *U. Geyer*, Helv., 1945, **28,** 576). Neither form could be dehydrogenated to 1-**aza-azulene,** which *T. Nozoe et al.* (Chem. and Ind., 1954, 1357) obtained as a red oil (*picrate*, m.p. 197–198° decomp.) by another route from 2-aminotropone:

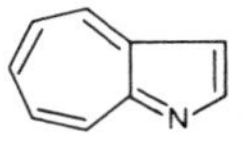

Aza-azulene

(c) Fused systems of two heterocyclic rings

A 2,3,2′,3′-pyrrolopyrrole skeleton is present in the alkaloid physostigmine. The dibenzo derivative (XIII, X = NH; "diindole") of 2,3,3′,2′-pyrrolopyrrole is described by *G. Heller* (Ber., 1917, **50,** 1202); the related indolocoumarone (XIII, X = O) (*S. R. Cawley* and *S. G. P. Plant*, J. chem. Soc., 1938, 1214) and benzo[*b*]thieno[3,2-*b*]indole (indolothionaphthene) (XIII, X = S) (*E. W. McClelland* and *J. L. D'Silva*, *ibid.*, 1932, 227) were obtained

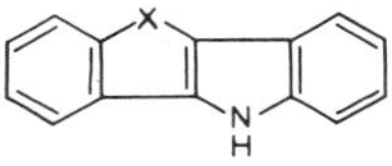

(XIII)

by Fischer synthesis from the phenylhydrazones of coumaranone and thioindoxyl, respectively. *N. P. Buu-Hoï et al.* (J. org. Chem., 1949, **14,** 802) describe several thiophenoindoles.

A more important case is that in which two pyrrolidine rings have the nitrogen and one carbon atom in common; this pyrrolizidine system (XIV) is present in the alkaloids of *Senecio* and *Heliotropium* species.

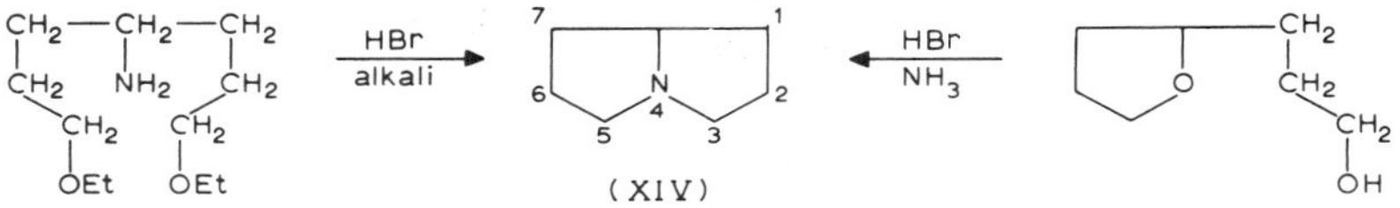

(XV) —HBr, alkali→ (XIV) Pyrrolizidine ←HBr, NH_3— (XVI)

Pyrrolizidine (*N. K. Kochetkov* and *A. M. Likhosherstov*, Adv. Heterocyclic Chem., 1965, **5,** 315), b.p. 148° (*picrate*, m.p. 257°) has been obtained from XV (*Prelog* and *S.Heimbach*, Ber., 1939, **72,** 1101) or from XVI (*R. Seiwerth*, Arkiv. Kem., 1951, **23,** 77), and best from γ-oxopimelic acid (*F. Michael* and *W. Flitsch*, Ber., 1955, **88,** 509):

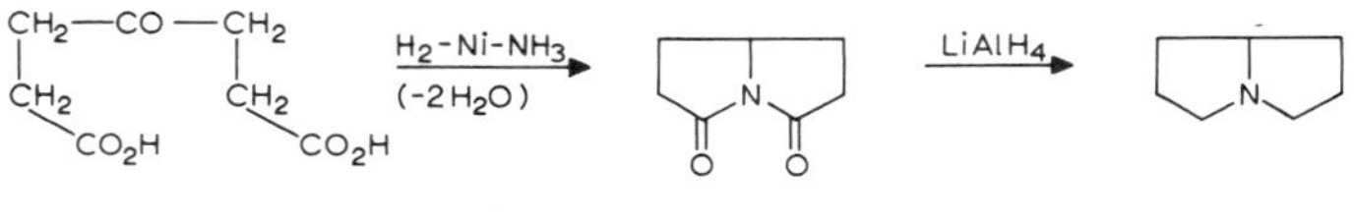

CH_2—CO—CH_2 (CH_2CO_2H)$_2$ —H_2-Ni-NH_3 (−2H_2O)→ dioxopyrrolizidine —$LiAlH_4$→ pyrrolizidine

The following synthesis devised by *N. J. Leonard et al.* (J. Amer. chem. Soc., 1947, **69,** 690; 1948, **70,** 2504) is of general application:

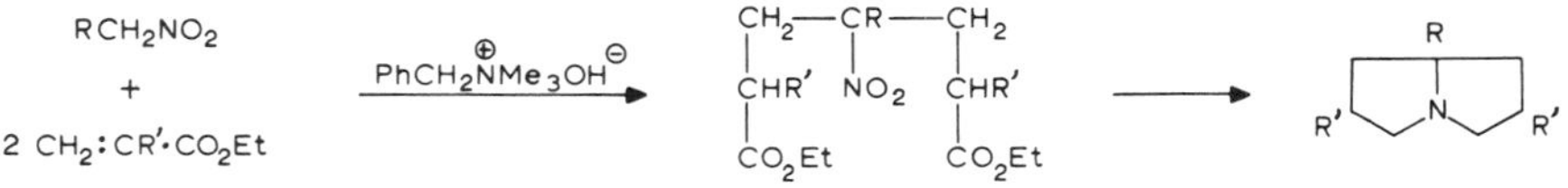

RCH_2NO_2 + 2 CH_2:CR′·CO_2Et —$PhCH_2\overset{\oplus}{N}Me_3OH^{\ominus}$→ CH_2(CHR′CO_2Et)–CR(NO_2)–CH_2(CHR′CO_2Et) ⟶ R, R′-substituted pyrrolizidine

The Michael addition, which can be made in two stages with different ester components, is followed by hydrogenation, first over platinum oxide, then over copper chromite at high temperature and pressure.

The appropriate analogue of XV gives inactive 1-*methylpyrrolizidine* (*picrate*, m.p. 234–236°); a (−)-rotatory form is heliotridan from *Heliotropium* alkaloids (*Prelog* and *E. Zalán*, Helv., 1944, **27**, 531). A methyl derivative of XVI affords 3-*methylpyrrolizidine*, b.p. 158–159° (*B. O. Majhofer* and *Seiwerth*, Monatsh., 1952, **83**, 1298). For stereoisomeric 1-hydroxymethylpyrrolizidines and pyrrolizidine-1-carboxylic acids, see *A. S. Labenskii et al.* (Doklady Akad. Nauk, S.S.S.R., 1953, **88**, 467).

A Dieckmann ring-closure on pyrrolidine-1,2-diacetic acid leads to XVII; and the ketone XVIII with one unreduced pyrrole ring, is produced by an intramolecular Hoesch reaction from β-1-pyrrylpropionitrile (*G. R. Clemo* and *T. A. Melrose*, J. chem. Soc., 1942, 424).

(XVII)  (XVIII)

The lithium salt of the 4-azapentalenyl anion (XIX) may be obtained from 3*H*-pyrrolizine and *n*-butyl-lithium; it is quenched with deuterium oxide to give 3-deuterio-3*H*-pyrrolizine and it reacts with benzophenone to yield a fulvene derivative which is probably the result of attack at position 3 (*W. H. Okamura* and *T. J. Katz*, Tetrahedron, 1967, **23**, 2941).
The 4-azapentalenyl anion is of interest because it is isoelectronic with the pentalenyl dianion and the above results suggest that the resonance form (XIX) makes an important contribution to its structure.

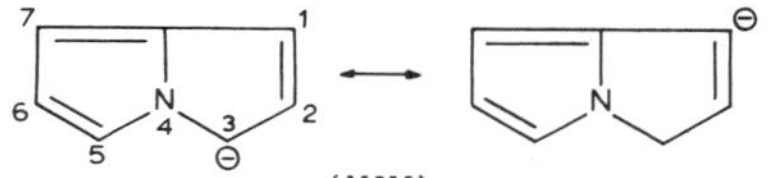

(XIX)

J. G. Jackson and *J. Kenner* suggested (J. chem. Soc., 1928, 573) that in a fused pyrrolizidine system of the type XX all three rings must be coplanar and thus a planar configuration be imposed on the nitrogen valencies. Since however the rigorous planarity of five-membered rings is no longer presumed, derivatives of XX have been synthesised, substituted so as to permit of optical activity in case of non-planarity, but instability of alkaloidal salts of the carboxylic acids has prevented a decision being reached. The compound XXI was obtained from *N*-*o*-aminophenylcarbazole by diazotisation and heating (*H. G. Dunlop* and *S. H. Tucker*, *ibid.*, 1939, 1945):

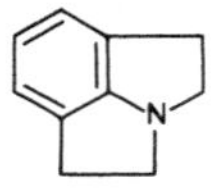

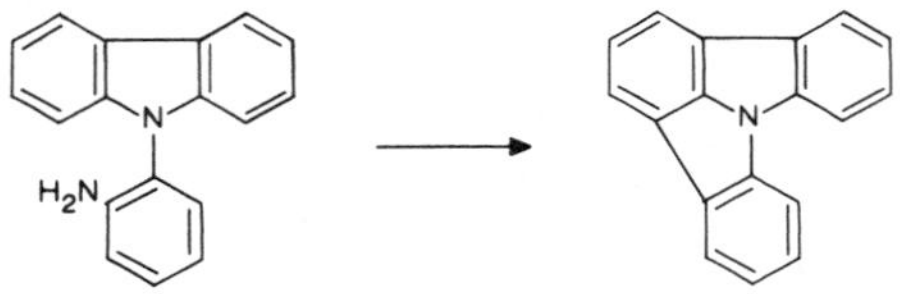
H2N
N

(XX) (XXI)

In XXII two pyrrolidine rings have the nitrogen and two carbon atoms in common. Prepared from 4-piperidylcarbinol, this 1-**azabicyclo**[1.2.2]**heptane** has m.p. 78–79°, *picrate*, m.p. 285° (decomp.) (*G. R. Clemo* and *Prelog*, *ibid.*, 1938, 400). For related compounds see *R. Lukeš* and *M. Ferles*, Coll. Czech. chem. Comm., 1955, **20,** 1227.

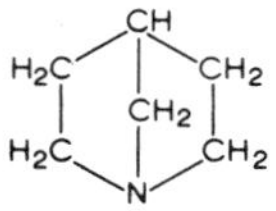
CH
H2C
CH2
CH2
H2C
CH2
N

(XXII)

5. Benzoindoles

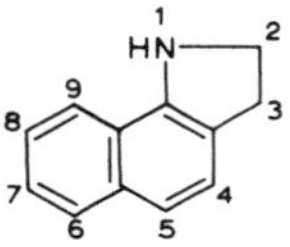

1*H*-Benz[*g*]indole
(6,7-Benzoindole)

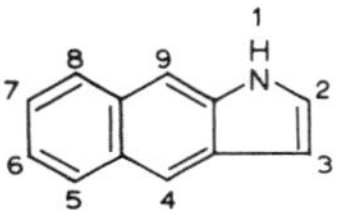

1*H*-Benz[*f*]indole
(5,6-Benzoindole)

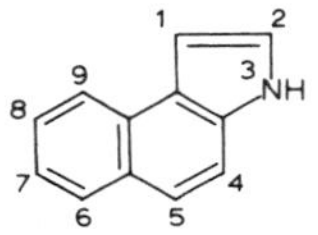

3*H*-Benz[*e*]indole
(4,5-Benzoindole)

Apart from carbazole (p. 486) benzoindoles have been little studied. 3*H*-**Benz**[*e*]**indole,** m.p. 39–40° (*O. Hinsberg* and *A. Simcoff*, Ber., 1898, **31,** 250), and 1*H*-**benz**[*g*]**indole,** m.p. 174° (*R. Pschorr* and *E. Kuhtz*, *ibid.*, 1905, **38,** 217) are readily accessible owing to the reactivity of position 2 and position 1 in 1- and 2-naphthylhydrazones. They may be obtained by the oxidation, using sodium periodate in alkali, of the related 3-hydroxypiperidine prepared by reaction between epichlorohydrin and the appropriate naphthylamine (*F. C. Pennington*, *M. Jellinet* and *R. D. Thurn*, J. org. Chem., 1959, **24,** 565; *J. R. Merchant* and *S. S. Salgar*, J. Indian chem. Soc., 1963, **40,** 23):

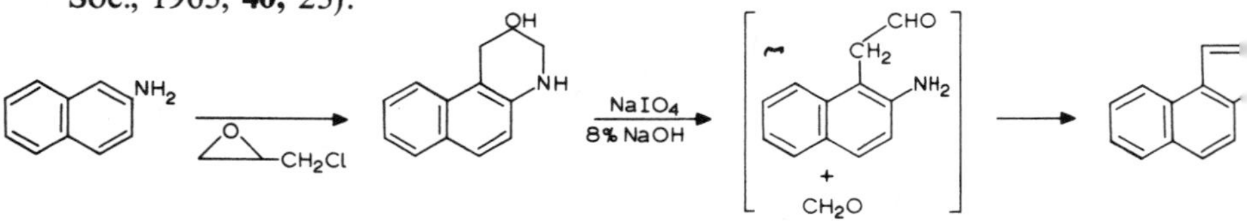
NH2
CH2Cl
OH
NH
NaIO4
8% NaOH
CHO
CH2
NH2
+
CH2O

1*H*-**Benz[*f*]indole,** m.p. 68–70° (*I. G. Farbenindustrie*, G.P. 516,675/1928). Linear benzoindoles are obtained only with difficulty (*O. Süs et al.*, Ann., 1953, **583,** 150) and known syntheses involve blocking of position 1 in 2-naphthylhydrazones (*E. A. Goldsmith* and *H. G. Lindwall*, J. org. Chem., 1953, **18,** 507).

2-**Phenyl**-1*H*-**benz[*g*]indole,** m.p. 166–167°, may be prepared by the cyclisation of the hydrazone obtained from 1-naphthylhydrazine and acetophenone; nitrosation, reduction and diazotisation affords 3-*diazo-2-phenyl-*1H-*benz*[g]*indole*, m.p. 166–167°.

2-**Phenyl**-3*H*-**benz[*e*]indole,** m.p. 133–134°, 1-*diazo-2-phenyl-*3H-*benz*[e]*indole*, m.p. 142–143° (*H. P. Patel* and *J. M. Tedder*, J. chem. Soc., 1963, 4593):

NH·NH₂ → (PhCOMe, AcOH) → hydrazone (HN–N=C(Ph)Me) → (170°, ZnCl₂) → 2-phenyl-1H-benz[g]indole → 3-diazo derivative

Derivatives of 1*H*-[*g*]- and 3*H*-[*e*]-benzindoles have been obtained by the Fischer and the Hinsberg indole syntheses and the respective *benzoisatins*, m.p. 255° and 252° (*Hinsberg*, Ber., 1888, **21,** 114) have been prepared by the mesoxalic ester method (p. 458) (*J. Martinet*, Ann. Chim., 1919, **11,** 39) and from naphthyloxamic acid chlorides (*I.G. Farbenindustrie*, G.P. 448,946/1925). For 1*H*-Benz[*f*]indole derivatives see *A. Étienne* and *A. Staehelin*, Compt. rend., 1950, **230,** 1960; *W. Dethloff* and *K. Schreiber*, Ber., 1950, **83,** 157; for 4,5- and 5,6-benzoindoles see *Goldsmith* and *Lindwall, loc. cit.*

4,5-Dihydro-3*H*-benz[*e*]indolecarboxylic acids can be obtained *via* the reaction of dimethyl acetylenedicarboxylate with 2-amino-1-tetralone hydrochloride (*U. K. Pandit* and *H. O. Huisman*, Rec. Trav. chim., 1964, **84,** 50):

2-amino-1-tetralone·HCl (NH$_2$HCl) → ($(\vdots C\cdot CO_2Me)_2$, MeOH, 0°) → dimethyl ester (MeO$_2$C, CO$_2$Me, NH), m.p. 172–174° → (hydrolysis, 20°) → half ester (MeO$_2$C, CO$_2$H, NH), m.p. 187–189° → (decarboxylation) → MeO$_2$C ester, m.p. 146–147°

dimethyl ester → (hydrolysis, boil) → CO$_2$H acid → (CH$_2$N$_2$) → CO$_2$Me ester, m.p. 167–169°

4,5-Dihydro-1*H*-benz[*g*]indole, m.p. 118° (*T. Severin* and *B. Brueck*, Ber., 1965, **98,** 3847).

1*H*-**Benz[*f*]indole-3-acetic acid,** m.p. 210°, may be prepared by the hydrolysis of *N*-acetyl-5,6-benzoindol-3-ylacetonitrile, obtained by the condensation of *N*-acetyl-5,6-benzoindoxyl with cyanoacetic acid using phenol and xylene as solvents and ammonium acetate as catalyst. Reduction of the acetonitrile derivatives with lithium tetrahydridoaluminate affords (2-(1*H*-**benz[*f*]indol-3-yl)ethylamine,** *hydrochloride*, m.p. 230° (*S. Y. Ambekar* and *S. Siddappa*, J. chem. Soc., C, 1966, 477):

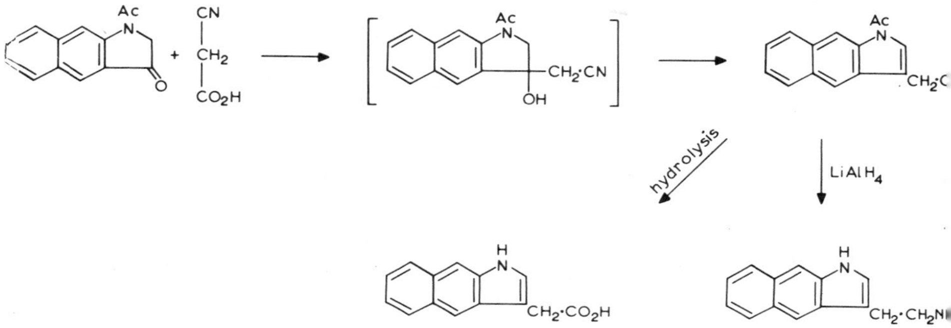

Other compounds containing a five-membered ring with a nitrogen atom fused to a naphthalene ring or to one benzene and a reduced six-membered ring are the naphthastyrils and the 3,4-trimethyleneindoles.

A benz[*cd*]indole derivative (3,4-**trimethyleneindole**) (I), m.p. 59°, is formed by reducing naphthastyril with sodium and butyl alcohol (*W. A. Jacobs* and *R. G. Gould*, J. biol. Chem., 1937, **120,** 141), and by the ring-closure of *N*-formyl-5,6,7,8-tetrahydro-1-naphthylamine (*F. C. Uhle et al.*, J. Amer. chem. Soc., 1955, **77,** 3334):

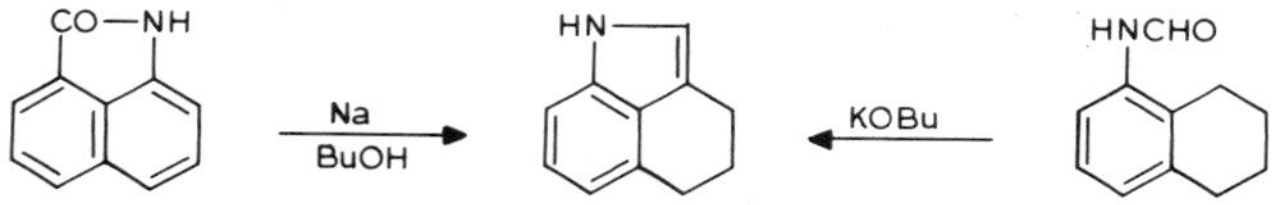

(I)

4-Cyanomethylindole with diethyl oxalate and sodium ethoxide in pyridine yields the dipolar benzindole derivative (II), but neither 3,4-biscyanomethylindole nor the related diacid could be converted into a tricyclic compound (*H. Plieninger* and *K. Suhr*, Ber., 1957, **90,** 1980):

(II)

Under Mannich conditions 4-acetonyl-1-acetyl-3-indolylacetaldehyde yields III (*Plieninger* and *G. Werst*, *ibid.*, 1956, **89,** 2783).

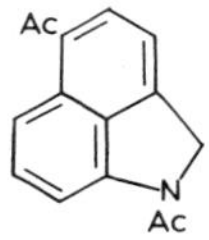

(III)

The direct cyclisation of 3-(2′-carboxyethyl)-1,2-dimethyl-5-methoxyindole, followed by remethylation of the intermediate 6-hydroxy derivative affords 1,3,4,5-tetrahydro-6-methoxy-1,2-dimethyl-5-oxobenz[*cd*]indole (IV) (*F. G. Mann* and *A. J. Tetlow*, J. chem. Soc., 1957, 3352):

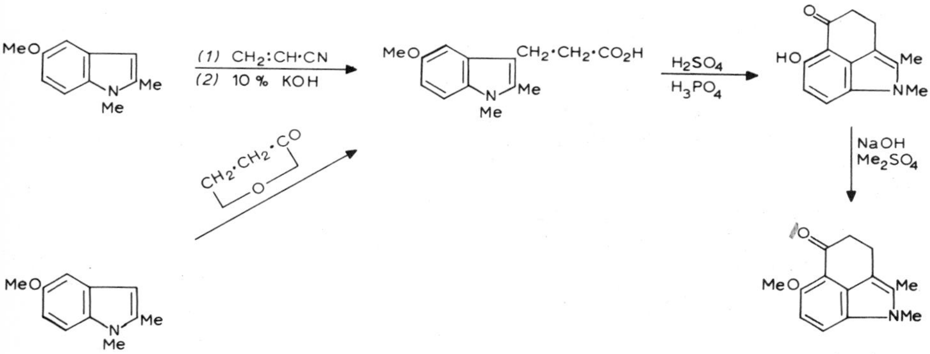

(1) $CH_2{:}CH{\cdot}CN$ (2) 10 % KOH

$CH_2{\cdot}CH_2{\cdot}CO_2H$

H_2SO_4 / H_3PO_4

NaOH / Me_2SO_4

(IV)

The ring-closure of 8-amino-1,2,3,4-tetrahydro-1-naphthoic acid affords 1,2,2a,3,4,5-hexahydro-2-oxobenz[*cd*]indole (*C. A. Grob*, *H. Kappeler* and *W. Meier*, Helv., 1961, **44,** 1517):

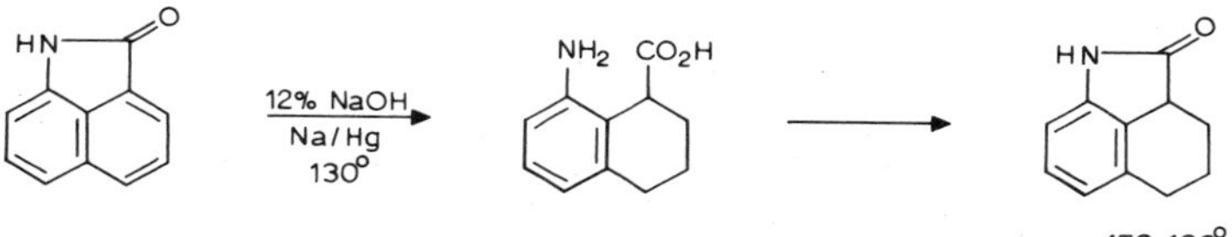

12% NaOH / Na/Hg / 130°

m.p. 159-160°

The addition of methyl propiolate to 1,2,2a,3,4,5-hexahydro-2,4-dioxobenz[*cd*]indole (V) yielded compound VI (*Kappeler* and *E. Renk*, *ibid.*, 1961, **44,** 1541):

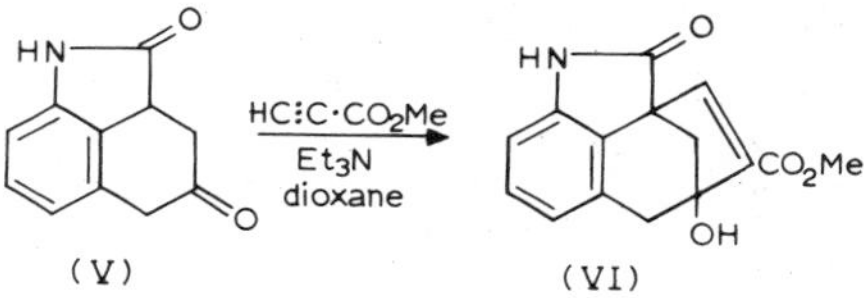

$HC{:}C{\cdot}CO_2Me$ / Et_3N / dioxane

(V) (VI)

The ring cleavage of clavine alkaloids *e.g.*, agroclavine (VII, R = H) and elymoclavine (VII, R = OH) gives benz[*cd*]indoline derivatives (*J. P. Dickinson, J. Harley-Mason* and *J. H. New*, J. chem. Soc., 1964, 1858):

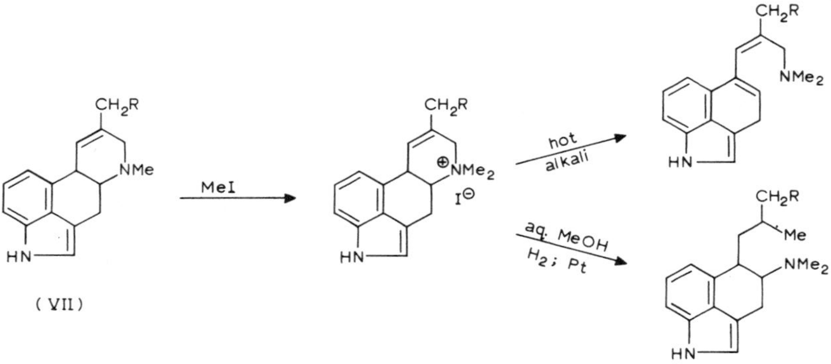

Compounds related to naphthastyril possessing a fully aromatic naphthalene ring are often interconvertible with those having an aromatic indole system and hydroxyl derivatives offer additional examples (*Grob et al.*, Helv., 1952, **35**, 2095; 1953, **36**, 839, 847):

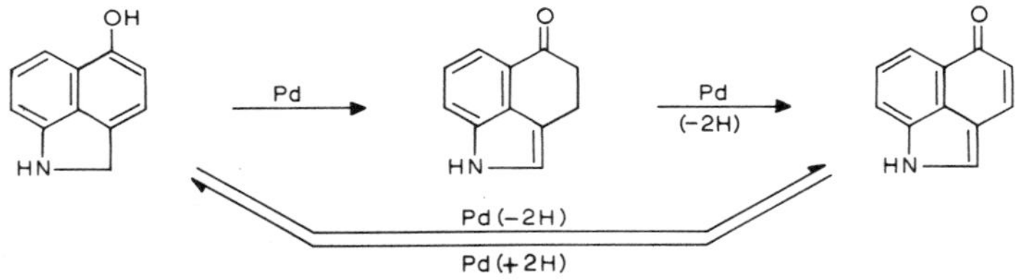

6. Compounds containing the carbazole nucleus

(a) Carbazole and its derivatives*

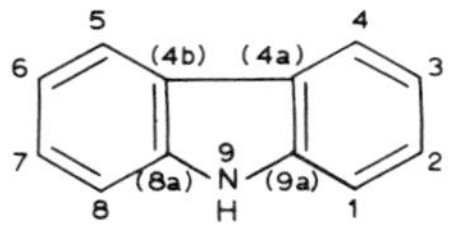

Carbazole was discovered in coal tar (*C. Graebe* and *C. Glaser*, Ber., 1872, **5**, 12), and its structure is inferred from its degradation to derivatives of

* For general reviews see *G. Cohn*, "Die Carbazolgruppe", Thieme, Leipzig, 1919; *N. Campbell* and *B. M. Barclay*, Chem. Reviews, 1947, **40**, 359. The numbering in the carbazole skeleton is as shown and is therefore an exception to the usual systematic method for numbering heterocyclic compounds.

indole and of biphenyl, and from numerous syntheses. It is the parent of several types of important dyestuffs, is encountered in the degradation of the *Strychnos* alkaloids, and is formed in the zinc dust distillation of humic acids (*M. V. Cheshire et al.*, Tetrahedron, 1967, **23,** 1669). The alkaloid murrayamine is a derivative of carbazole.

(i) Synthetic methods

(*1*) The Fischer indole synthesis applied to arylhydrazones of cyclohexanones affords a wide variety of tetrahydrocarbazoles (*W. Borsch et al.*, Ann., 1908, **359,** 49; *W. H. Perkin* and *S. G. P. Plant*, J. chem. Soc., 1921, **119,** 1825), which can be dehydrogenated to carbazole by heating with chloranil (*B. M. Barclay* and *N. Campbell*, *ibid.*, 1945, 530) or catalytically (*H. Adkins* and *L. G. Lunsted*, J. Amer. chem. Soc., 1949, **71,** 2964):

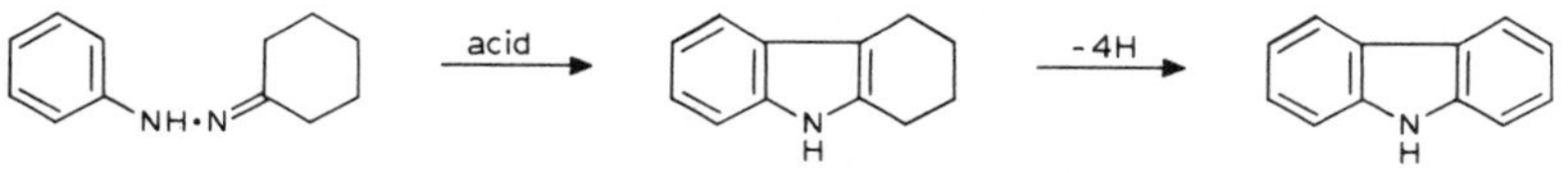

NH•N acid N H -4H N H

The mechanism of the cyclisation of cyclohexanone phenylhydrazones to give carbazoles has been investigated by using phenylhydrazine and aniline containing a heavy nitrogen atom, and shown to be intramolecular and similar to the benzidine rearrangement in early stages (*B. A. Geller* and *L. K. Skrunts*, Zhur. obshchei Khim., 1964, **34,** 661). Hydrazones obtained from *meta*-substituted phenylhydrazines give mixtures of 5- and 7-substituted tetrahydrocarbazoles (*Barclay* and *Campbell. loc. cit.*) and that from mesitylhydrazine on boiling with acetic acid affords 6,7,8-trimethyl-1,2,3,4-tetrahydrocarbazole, involving a novel 1,4-migration of a methyl group (*R. B. Carlin* and *M. S. Moores*, J. Amer. chem. Soc., 1959, **81,** 1259; 1962, **84,** 4107):

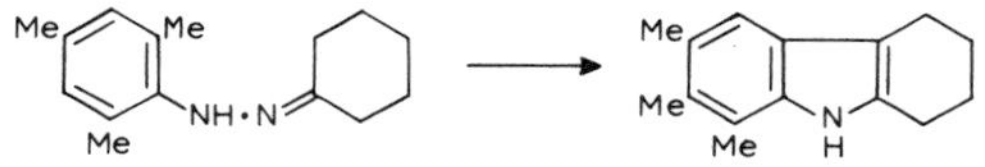

Me Me Me NH•N → Me Me Me N H

Depending on the conditions under which 2-(2-cyanoethyl)cyclohexanone phenylhydrazone is cyclised using zinc chloride either tetrahydrocarbazole or a mixture of tetrahydrocarbazole and 1-(2-cyanoethyl)-1,2,3,4-tetrahydrocarbazole (I) is obtained. If cyclisation is effected using 20% sulphuric acid then 4a-(2-cyanoethyl)-1,2,3,4-tetrahydro-4a*H*-carbazole (Ia) is also formed besides I (*A. N. Kost*, *L. G. Yudin* and *Yu-Chu Chiu*, Zhur. obshchei Khim., 1964, **34,** 3444):

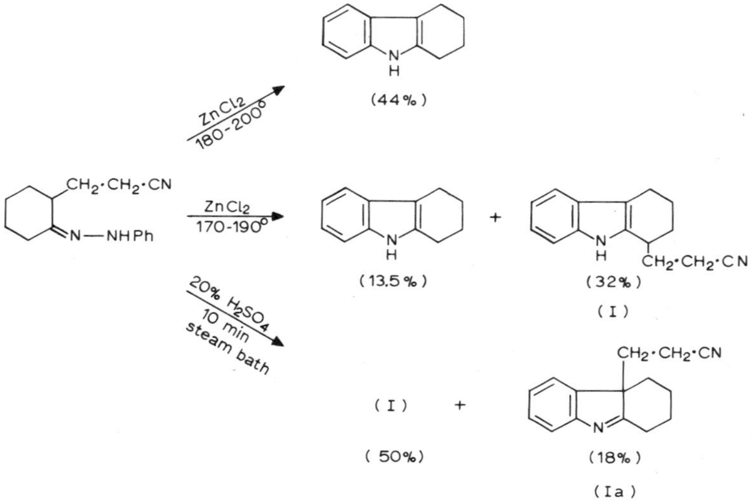

CH2·CH2·CN
N—NHPh
ZnCl2
180-200°
ZnCl2
170-190°
20% H2SO4
10 min
steam bath
(44%)
(13.5%)
+
CH2·CH2·CN
(32%)
(I)
CH2·CH2·CN
(I)
+
(50%)
(18%)
(Ia)

The phenylhydrazone of 2-cyclohexanonepropionic acid when treated with 20% sulphuric acid affords 1-*oxo*-10*b*-*aza*-1,2,3,3*a*,4,5,6,10*b*-*octahydrobenz*[a]*acenaphthylene* (II), m.p. 126°, and the aqueous solution after treating with ammonium hydroxide gives 1,2,3,4-*tetrahydro*-4*a*H-*carbazol*-4*a*-*ylpropionic acid* (III), m.p. 226°, on acidification. Hydrolysis of II with potassium hydroxide solution yields 3-(1,2,3,4-*tetrahydrocarbazol*-1-*yl*)*propionic acid* (IV), m.p. 123°, also obtained similarly from I:

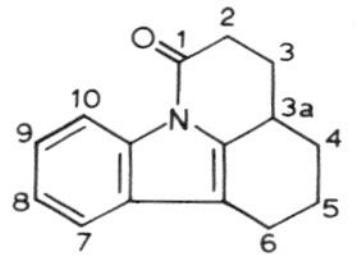

(II)

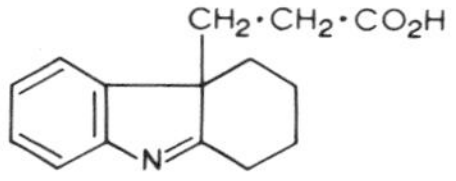

CH2·CH2·CO2H

(III)

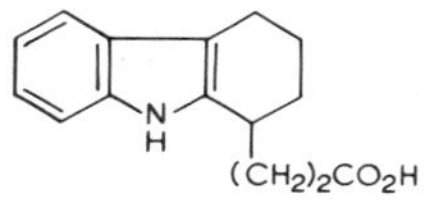

(CH2)2CO2H

(IV)

In a useful modification of the Fischer indole synthesis a hydroxymethylene- or a carboxy-cyclohexanone affords by the Japp–Klingemann method a cyclohexanedione monophenylhydrazone and thence a 1-oxo-1,2,3,4-tetrahydrocarbazole (*S. Coffey*, Rec. Trav. chim., 1923, **42,** 531; *A. Kent* and *D. McNeil*, J. chem. Soc., 1938, 8):

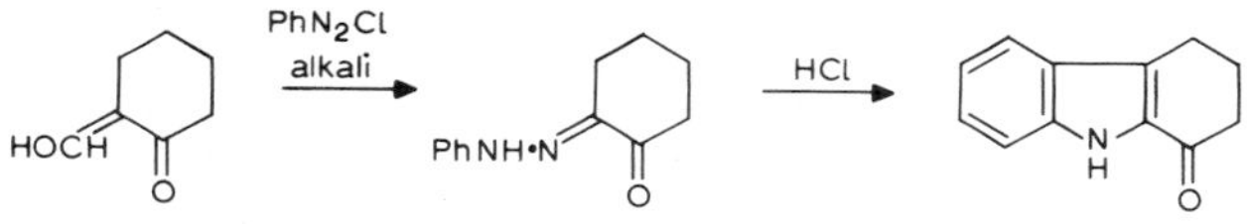

PhN2Cl
alkali
HOCH
PhNH·N
HCl

The monophenylhydrazone of cyclohexane-1,3-dione similarly gives 4-oxo-1,2,3,4-tetrahydrocarbazole (*G. R. Clemo* and *D. Felton*, *ibid*., 1951, 700). Methyl, dimethyl, chloro and carbomethoxy derivatives of 1,2,3,4-tetrahydrocarbazole are readily obtained by treating 2-chlorocyclohexanone with excess of the appropriate aniline (*Campbell* and *E. B. McCall*, *ibid.*, 1950, 2870):

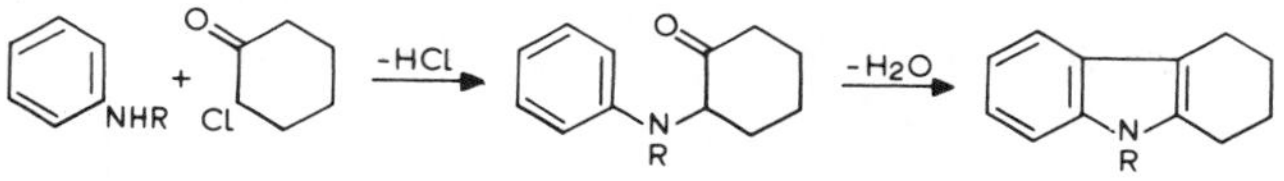

The reaction of benzenonium tetrafluoroborate in cyclohexanone results in the formation of tetrahydrocarbazole, with loss of aniline, instead of ammonia as in the typical Fischer reaction (*A. N. Nesmeyanov* and *R. V. Golovnya*, Doklady Akad. Nauk S.S.S.R., 1960, **133,** 1337).

(*2*) Another method of general application (Graebe–Ullmann) depends on the pyrolysis of 1-arylbenzotriazoles obtained from 2-aminodiphenylamines (*F. Ullmann*, Ann., 1904, **332,** 82; *Plant et al.*, J. chem. Soc., 1951, 110); this method is not readily applicable to nitro-substituted compounds but by a related method 2-aminodiarylamines afford *N*-substituted carbazoles (*D. H. Hey* and *R. D. Mulley*, *ibid.*, 1952, 2276):

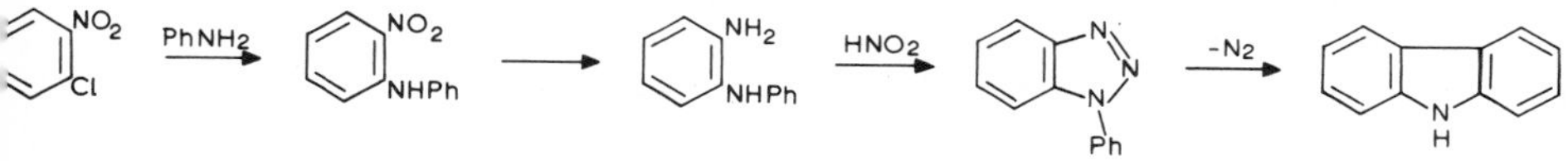

(*3*) 2-Nitrosobiphenyl and related compounds, *e.g.* 2′- and 4′-halogeno-2-nitrobiphenyls, with triethyl phosphite or triphenylphosphine are readily deoxygenated and cyclised in high yield, possibly *via* the intermediate azene, to the corresponding carbazole (*J. I. G. Cadogan et al.*, *ibid.*, 1963, 42; 1965, 4831):

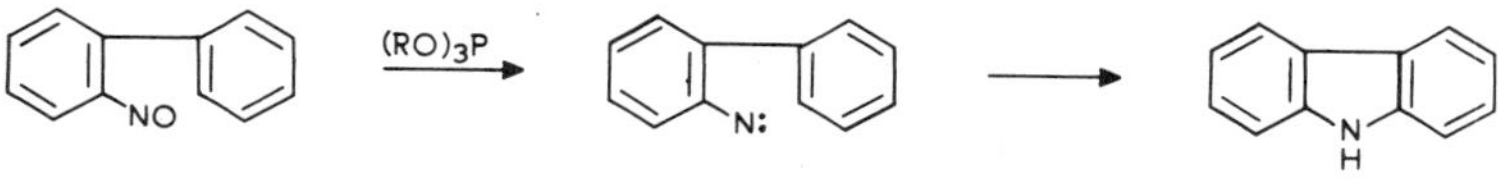

alternatively it is possible that these reductions may proceed without the intermediacy of a discrete nitrene, thus:

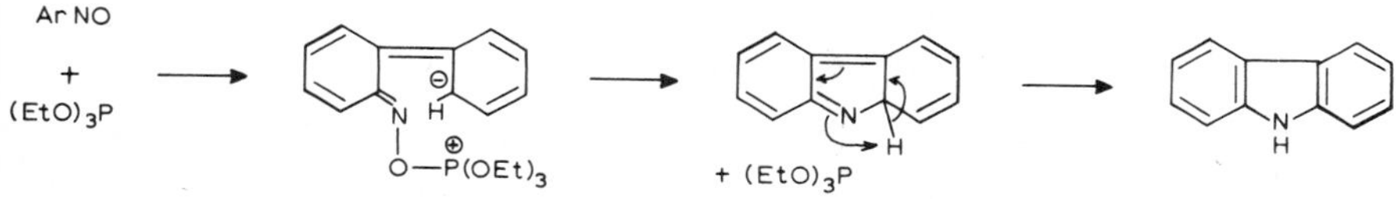

Diphenylnitrosoamine reacts with 2-aminodiphenylamine to give carbazole (*H. Sieper*, Ber., 1967, **100,** 1646). A small amount of carbazole along with other compounds is formed by the decomposition of nitrobenzene at 600° (*E. K. Fields* and *S. Meyerson*, J. Amer. chem. Soc., 1967, **89,** 3224). It is produced by pyrolysis of diphenylamine or aniline, and better by heating thiodiphenylamine (phenothiazine) with copper. The pyrolysis of 2-nitrobiphenyl with ferrous oxalate (FeC_2O_4) gives mainly carbazole and some 2-aminobiphenyl, but on heating alone only tar and starting material is obtained (*R. H. Smith* and *H. Suschitzky*, Tetrahedron, 1961, **16,** 80):

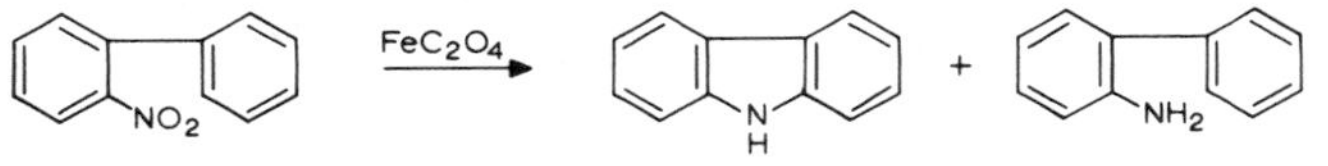

FeC_2O_4 NO_2 N H   + NH_2

Carbazole can be obtained in acceptable yield from 2,2′-diaminobiphenyl (*E. Täuber* and *R. Löwenhertz*, Ber., 1891, **24,** 1033; *H. Leditschke*, *ibid.*, 1953, **86,** 522), and from 2-aminobiphenyl (*G. T. Morgan* and *L. P. Walls*, J. Soc. chem. Ind., 1938, **57,** 358):

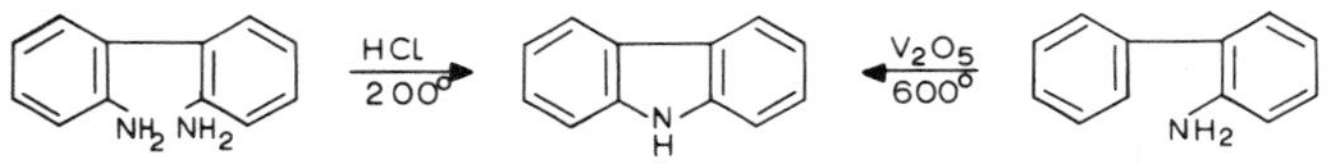

NH_2 NH_2 HCl 200°   N H V_2O_5 600° NH_2

Carbazoles can be a main product of thermal or photochemical decomposition of 2-azidobiphenyls (*P. Smith* and *P. Brown*, J. Amer. chem. Soc., 1951, **73,** 2435). Treated with sulphuric acid in acetic acid, *N,N*-diphenylhydroxylamine gives carbazole (*H. Wieland* and *C. Müller*, Ber., 1913, **46,** 3306).

Carbazole is formed when a solution of diphenylamine in deoxygenated methanol, isopropanol, or hexane is irradiated with u.v. light at ordinary temperatures (*E. J. Bowen* and *J. H. D. Eland*, Proc. chem. Soc., 1963, 202). For example, the photocyclisation of 4-methoxy-4′-methyldiphenylamine gives the alkaloid glycozoline (V) (*W. Carruthers*, Chem. Comm., 1966, 272; *D. P. Chakraborty*, Tetrahedron Letters, 1966, 661):

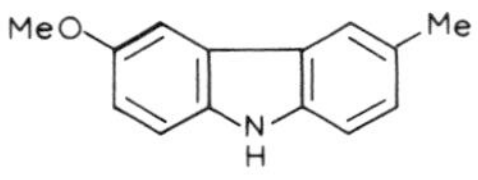

MeO   Me   N H

(V)

Triphenylamine and methyldiphenylamine on illumination in hexane exposed to air yield *N*-phenyl- and *N*-methyl-carbazole (*K. H. Grellman, G. M. Sherman* and *H. Linschitz*, J. Amer. chem. Soc., 1963, **85,** 1881).

(ii) General properties and reactions

Carbazole and its simple derivatives are very stable to heat, acids, and alkali, but sensitive to oxidation which may yield indefinite, coloured products. However, *oxidation* of carbazole with dichromate or permanganate in acetic acid affords 3,3′-bicarbazyl (VI), while 9,9′-bicarbazyl (VII) is formed by oxidation with permanganate in acetone; the formation of the two different isomers depends on the type of solvent rather than the type of oxidant (*M. Kuroki*, Uki Gosei Kagaku Kyokai Shi, 1965, **23,** 447):

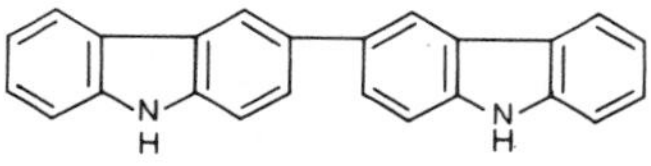

(VI 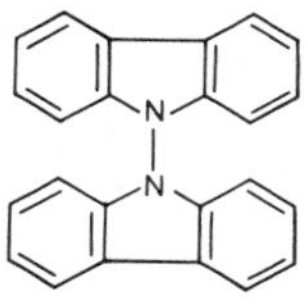) (VII)

The reduction of carbazole with lithium in *n*-propylamine gives a 90% yield of 1,2,3,4-tetrahydrocarbazole, which is resistant to further reduction by this reagent. Under similar conditions 9-methylcarbazole affords 9-methyl-1,2,3,4,4a,9a-hexahydrocarbazole (*H. Dressler* and *M. E. Baum*, J. org. Chem., 1961, **26,** 102). When carbazole is reduced, either by sodium–ammonia–ethanol or by sodium–ammonia–ammonium chloride, 1,4-dihydrocarbazole (VIII) is formed, whereas under the former conditions 9-methylcarbazole affords 1,4,5,8-tetrahydro-9-methylcarbazole (IX), but under the latter conditions proceeds only to 1,4-dihydro-9-methylcarbazole (*S. O'Brien* and *D. C. C. Smith*, J. chem. Soc., 1960, 4609):

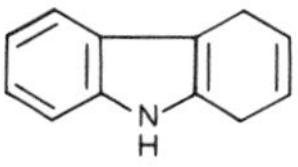

(VIII 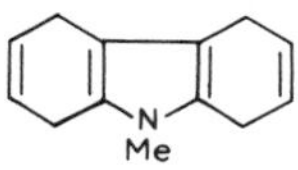) (IX)

Hydrogenation of carbazole over Raney nickel at 260° and 100 atmospheres yields octahydrocarbazole, together with a small amount of tetrahydrocarbazole (*M. Yamada*, Koru. Taru, 1960, **12,** 668; C.A., 1964, **61,** 11969c).

(iii) Carbazole

Carbazole, colourless, slightly fluorescent plates, m.p. 246°, b.p. 354°, *picrate*, m.p. 185°, *trinitrobenzene adduct*, m.p. 203°, occurs in anthracene oil of coal tar. It is separated by formation of its potassium derivative, or its insolubility in naphtha and ready solubility in pyridine bases (*J. M. Clark*, Ind. Eng. Chem., 1919, **11,** 204): residual anthracene is preferentially sulphonated by cold concentrated sulphuric acid. Separation from anthracene may be effected by the use of 2-furfuraldehyde (*S. Ashida* and *Y. Shimizu*, Shigen Gijutsu Shikensho Hokoku, 1962, **56,** 1; C.A., 1964, **61,** 6975d).

Carbazole is insoluble in water and in dilute acids and alkali; its basic properties, much feebler than those of diphenylamine, are shown by the unstable colourless perchlorate (*K. A. Hofmann et al.*, Ber., 1910, **43**, 178). Carbazole gives a red pine-shaving reaction and an intense blue colour with isatin and sulphuric acid.

(iv) N-*substituted carbazoles*

9-**Alkylcarbazoles** can be prepared by the action of alkyl halides or sulphates on carbazole in acetone or alcohol in the presence of alkali (*T. S. Stevens* and *S. H. Tucker*, J. chem. Soc., 1923, **123**, 2140). The substitution of powdered sodium or potassium hydroxide for concentrated aqueous alkali solutions increases the yields and reduces the reaction times (*V. P. Lopatinskii, E. E. Sirotkina* and *M. M. Anosova*, Tr. Tomskogo Gos. Univ., Ser. Khim., 1964, **170**, 49; C.A., 1965, **63**, 565d). Some 9-alkylcarbazoles have been made by ring-synthesis 2 (p. 489). 9-Methylcarbazole-1-carboxylic, acid, obtained by treating carbazole-1-carboxylic acid with sodamide and methyl iodide in liquid ammonia, decarboxylates with soda lime to give 9-methylcarbazole (*W. E. Noland* and *G. J. Meisters*, J. org. Chem., 1960, **25**, 2060). 9-Alkylcarbazoles and several of their substitution derivatives on heating with sulphur undergo smooth dealkylation (*N. P. Buu-Hoï* and *G. Saint-Ruf*, J. chem. Soc., C, 1966, 924).

9-**Phenylcarbazole** is obtained by the reaction of benzyne, generated by the decomposition of 4,5-benzothiadiazole 1,1-dioxide in tetrahydrofuran, with nitrosobenzene (*G. W. Steinhoff* and *M. C. Henry*, J. org. Chem., 1964, **29**, 2808).

9-**Vinylcarbazole** is prepared industrially from carbazole and acetylene under pressure with an alkaline catalyst, but for laboratory purposes the dehydration of 9-(2-hydroxyethyl)carbazole with 10% potassium hydroxide at 250–270° affords a convenient method (*Lopatinskii et al.*, Metody Polucheniyakkim. Reaktivov i Preparatov, 1964, **11**, 37; *H. Otsuki et al.*, Japan Pat. 174,356/1946). 9-(2-Hydroxyethyl)carbazole may be obtained from carbazole, ethylene oxide and powdered potassium hydroxide in ethyl methyl ketone (*Lopatinskii et al.*, Tr. Tomskogo Gos. Univ. Ser. Khim., 1964, **170**, 29; Metody Polycheniyakhim. Reaktivov i Preparatov, 1964, **11**, 94). Removal of hydrogen chloride from 9-(2-chloroethyl)carbazole also gives 9-vinylcarbazole.

9-Vinylcarbazole polymerises easily to form plastics which are somewhat brittle but resistant to heat. A cationic mechanism has been suggested for its spontaneous polymerisation in the presence of a dilute aqueous solution of perchloric acid (*O. F. Solomon, I.Z. Ciuta* and *N. Cobianu*, J. Polymer Sci., 1964, **2**, Pt.B, 311). In the electrolytically induced polymerisation of 9-vinylcarbazole catalysed by mercuric cyanide in acetonitrile a *dimer*, m.p. 196°, is formed besides the polymeric products. Its proposed structure is supported by mass spectral and n.m.r. data (*J. W. Breitenbach, O. F. Olaj* and *F. Wehrmann*, Monatsh., 1964, **95**, 1007):

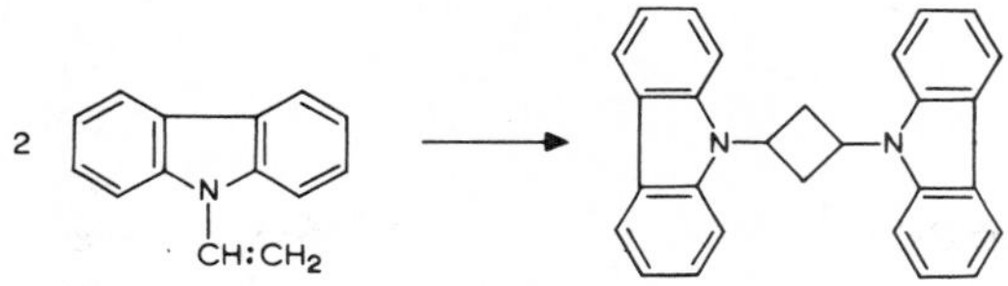

2 CH:CH₂ →

Treatment of 9-vinylcarbazole with a high concentration of ferric nitrate in methanol–water affords almost immediately a dimer, shown by mass spectral and n.m.r. analysis to be *trans*-1,2-dicarbazylcyclobutane (*S. McKinley, J. V. Crawford* and *Chi-Hua Wang*, J. org. Chem., 1966, **31**, 1963), previously reported as a by-product from its reaction with chloranil or tetranitromethane (*L. P. Ellinger, J. Fenney* and *A. Ledwith*, Monatsh., 1965, **96**, 131):

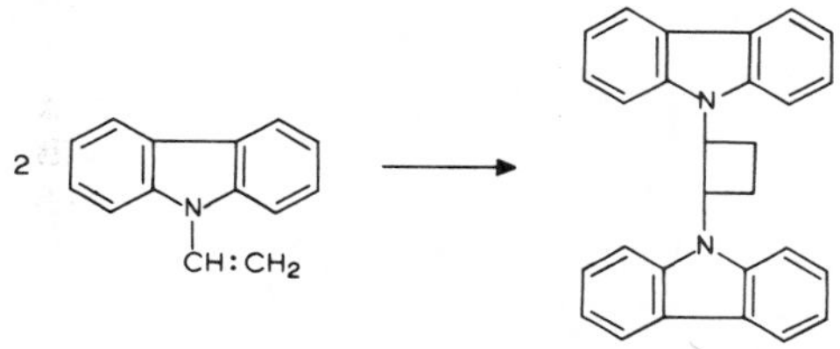

2 CH:CH₂ →

Carbazylmagnesium halides with acyl halides give 9-**acylcarbazoles** (*B. Oddo*, Gazz., 1911, **41**, I, 255); the potassium salt affords similarly both 9-alkyl and 9-acyl derivatives.

Acetylation of carbazole with acetic anhydride proceeds best (91–93% yield) when catalysed by 66.74 millimoles of sulphuric acid per mole of carbazole at 45°; also formed are 3-acetyl- and 3,9-diacetyl-carbazole (*E. Funakubo et al.*, Koru Taru, 1963, **15**, 402). The rates and products of chlorination of 9-acetylcarbazole in acetic acid have been examined and the results establish that the influence of the resonance of the heterocyclic ring on reactivity, though significant is small in magnitude (*P. B. D. de la Mare, O. M. H. el Dusouqui* and *E. A. Johnson*, J. chem. Soc., B, 1966, 521).

9-**Carbazolylacetic acids** may be obtained by treating carbazole, its halogeno and nitro derivatives with chloroacetic acid in the presence of potassium hydroxide in dimethylformamide (*Lopatinskii* and *I. P. Zherebtsov*, U.S.S.R. Pat. 183,210/1965; C.A., 1966, **65**, P 18562e).

Carbazole with formaldehyde in aqueous dioxane in the presence of small amounts of sulphuric acid affords 9-(**hydroxymethyl**)**carbazole** and 9,9′-**dicarbazolylmethane,** but in the presence of large amounts of sulphuric acid resinous products are obtained (*S. Nakade* and *M. Imoko*, Kogyo Kagaku Zasshi, Japan, 1966, **69**, 100).

(v) C-Alkylcarbazoles

Alkylcarbazoles have been prepared by the Graebe–Ullmann synthesis, then by the Fischer synthesis of tetrahydrocarbazoles, followed by dehydrogenation, and more recently *via* the Bischler reaction between 2-chlorocyclohexanone and arylamines in a high boiling solvent (Cellosolve) in the presence of sodium carbonate and a small amount of quinoline or pyridine (*E. Campaigne* and *R. D. Lake*, J. org. Chem., 1959, **24,** 478):

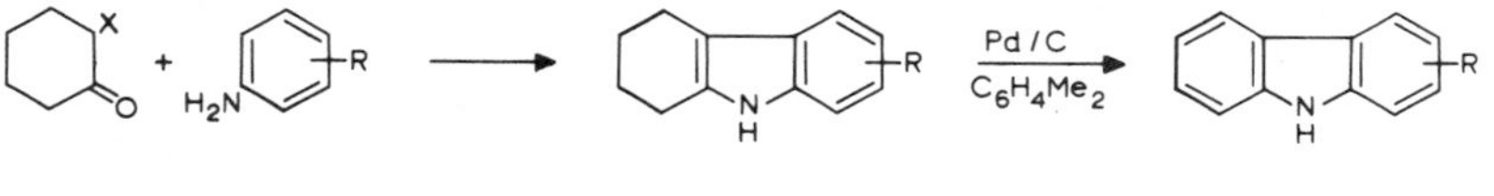

Pd / C
$C_6H_4Me_2$

Carbazole reacts with olefins in the presence of aluminium amides to give *peri*-alkyl derivatives *e.g.* 1-ethylcarbazole (*R. Stroh* and *W. Hahn*, Ann., 1959, **623,** 176).

Irradiation of 2-methyldiphenylamine in light petroleum through quartz with light from a mercury-vapour lamp gives 1-methylcarbazole. Similarly 2,2′-dimethyldiphenylamine affords 1,8-dimethylcarbazole, and polymethylcarbazoles which are sometimes troublesome to prepare by other routes may be obtained by this method (*W. Carruthers*, Chem. Comm., 1966, 272).

2-Methylindole reacts with methyl vinyl ketone in the presence of hydroquinone at 280° to give 2-methylcarbazole (*J. Szmuszkovicz*, J. Amer. chem. Soc., 1957, **79,** 2819):

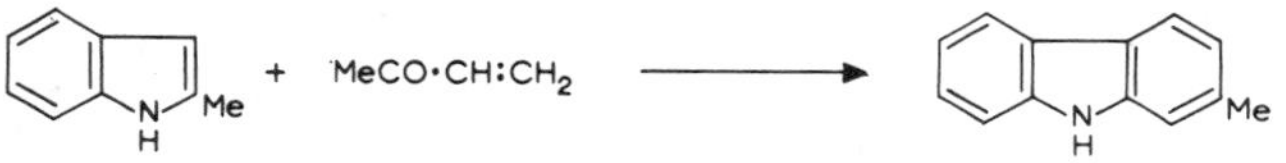

+ MeCO·CH:CH₂

1,4-Dimethylcarbazole is obtained when the product formed by passing dry hydrogen chloride into a solution of indole and hexane-2,5-dione in ethanol, is gently heated with semicarbazide hydrochloride and sodium acetate in aqueous ethanol (*P. A. Cranwell* and *J. E. Saxton*, J. chem. Soc., 1962, 3482). Formylation of 1,4-dimethylcarbazole yields 1,4-dimethylcarbazole-3-carbaldehyde which by Wolff–Kishner reduction affords 1,3,4-trimethylcarbazole. The formylation also gives a small yield of 1,4-dimethylcarbazole-3,6-dicarbaldehyde and hence by reduction 1,3,4,6-tetramethylcarbazole.

Some *C*-alkyl derivatives of carbazole are assembled in Table 4.

TABLE 4

C-ALKYLCARBAZOLES

Substituent	*m.p.* °	*Ref.*	*Substituent*	*m.p.* °	*Ref.*
1-Methyl	120.5–121	1, 2	1,4-Dimethyl	97–98	7, 8
2-Methyl	261–262	1, 3	1,8-Dimethyl	175–176	9
3-Methyl	206.5–207.5	1, 4	2,3-Dimethyl	251–253	10
4-Methyl	129.5–130	1, 5	2,4-Dimethyl	124–126	11
2-Ethyl	225	6	2,6-Dimethyl	224	12
3-Ethyl	142	6	3,6-Dimethyl	219	4, 12
			1,3,4-Trimethyl	146–146.5	7
			1,4,6-Trimethyl	127–128	7
			1,3,4,6-Tetramethyl	190–191.5	7

References
1 *E. Campaigne* and *R. D. Lake*, J. org. Chem., 1959, **24,** 483.
2 *F. Ullmann*, Ann., 1904, **332,** 82.
3 *N. Campbell* and *B. Barclay*, J. chem. Soc., 1945, 530.
4 *S. H. Oakeshott* and *S. G. P. Plant*, *ibid.*, 1926, 1210.
5 *K. H. Pausacker* and *R. Robinson*, *ibid.*, 1947, 1557.
6 *Plant* and *S. B. C. Williams*, *ibid.*, 1934, 1142.
7 *P. A. Cranwell* and *J. E. Saxton*, *ibid.*, 1962, 3482.
8 Sir *R. Robinson* and *Saxton*, *ibid.*, 1952, 976.
9 *C. S. Barnes*, *Pausacker* and *C. I. Schubert*, *ibid.*, 1949, 1381.
10 *Campbell* and *E. B. McCall*, *ibid.*, 1950, 2870.
11 *J. von Braun* and *W. Haensel*, Ber., 1926, **59,** 2006.
12 *G. T. Morgan* and *L. P. Walls*, J. chem. Soc., 1930, 1502.

(vi) Halogenocarbazoles

Monohalogenocarbazoles of known orientation have been prepared either by direct ring-synthesis or by synthesis of halogenotetrahydrocarbazoles followed by dehydrogenation (*B. M. Barclay* and *N. Campbell*, J. chem. Soc., 1945, 530). Direct halogenation by chlorine or sulphuryl chloride (*V. P. Lopatinskii* and *I. P. Zherebtsov*, Metody Polucheniya Khim. Reaktivov i Preparatov, 1964, **11,** 102), by bromine or potassium bromide–bromate, by treatment with *N*-bromosuccinimide in carbon tetrachloride in the presence of benzoyl peroxide (*R. O. Matevosyan*, *I. Ya. Postovskiĭ* and *A. K. Chirkov*, Zhur. obshcheĭ Khim., 1960, **30,** 3186), and by potassium iodide–iodate, takes place in position 3 with rapid further attack in 6. Carbazolediazonium salts afford 3-halogenocarbazoles free from higher substitution products (*S. H. Tucker*, J. chem. Soc., 1924, **125,** 1144). Similarly 1,6- and 3,6-dichlorocarbazole may be prepared (*H. M. Grotta*, *C. J. Riggle* and *A. E. Bearse*, J. org. Chem., 1964, **29,** 2474). As would be expected, the halogen atoms are firmly held.

Cyclohexanone 2,6-dibromophenylhydrazone, on heating with zinc chloride, gives, after dehydrogenation of the products with chloroanil, a monobromocarbazole of unknown orientation (*R. B. Carlin* and *G. W. Larson*, J. Amer. chem. Soc., 1957, **79**, 934).

3-Bromo- and 3-chloro-carbazole with sodium nitrite and acetic acid yield the 9-nitroso derivatives, reduced to the 9-amino compounds with zinc and acetic acid (*Matevosyan, Postovskiĭ* and *Chirkov, loc. cit.*).

Nitrosation of 1-bromocarbazole leads to a rearranged (Fischer–Hepp rearrangement) and oxidised product, 1-bromo-3-nitrocarbazole rather than to 1-bromo-9-nitrosocarbazole (*N. L. Drake et al.*, J. org. Chem., 1962, **27**, 1026):

HONO → NO_2

Halogenocarbazoles can be obtained from 2′- and 4′-halogeno-2-nitrobiphenyls (see p. 489).

The melting points of representative halogenocarbazoles are given in Table 5.

TABLE 5

HALOGENOCARBAZOLES

Substituent	*m.p. (°C)*	*Ref.*	*Substituent*	*m.p. (°C)*	*Ref.*
1-Chloro	110 (125)	1	4-Bromo	104–105	1
2-Chloro	244	1, 2	3-Iodo	192–194	4, 5
3-Chloro	201.5	1, 2	1,4-Dichloro	84–85	1
4-Chloro	96	1, 3	2,7-Dichloro	204	6
1-Bromo	111–112	1	3,6-Dichloro	202–203	7
2-Bromo	250–251	1	1,3-Dibromo	106–107	8
3-Bromo	199	1, 4	3,6-Dibromo	212–213	9, 10
			2,7-Diiodo	265–266	11
			3,6-Diiodo	202–204	5

References

1 *B. M. Barclay* and *N. Campbell*, J. chem. Soc., 1945, 530.
2 *F. Ullmann* and *D. Kogan*, Ann., 1904, **332**, 97.
3 *R. C. G. Moggridge* and *S. G. P. Plant*, J. chem. Soc., 1937, 1125.
4 *S. H. Tucker*, *ibid.*, 1924, **125**, 1144.
5 *Tucker*, *ibid.*, 1926, 546.
6 *H. Leditschke*, Ber., 1953, **86**, 522.
7 *G. Mazzara* and *M. Lamberti-Zanardi*, Gazz., 1896, **26**, II, 236.
8 *P. A. S. Smith* and *B. B. Brown*, J. Amer. chem. Soc., 1951, **73**, 2, 435.
9 *H. Lindemann* and *F. Mühlhaus*, Ber., 1925, **58**, 2371.
10 *J. McLintock* and *Tucker*, J. chem. Soc., 1927, 1216.
11 *D. Ponte*, Chem. Ztbl., 1934, II, 3116.

(vii) Nitro-, nitroso- and amino-carbazoles

When "nitrous fumes" containing some nitric acid are passed into a suspension of carbazole in acetic acid, 3-*nitro*-9-*nitrosocarbazole*, m.p. 166°

(decomp.), is formed, which gives 3-**nitrocarbazole** when boiled with alcoholic potash (*H. Lindemann*, Ber., 1924, **57,** 555). The following scheme represents an improved procedure for the preparation of 3-nitrocarbazole (*R. K. Eĭkhmann, V. O. Lukashevich* and *E. A. Silaeva*, Org. Chem. Ind., U.S.S.R., 1939, **6,** 93):

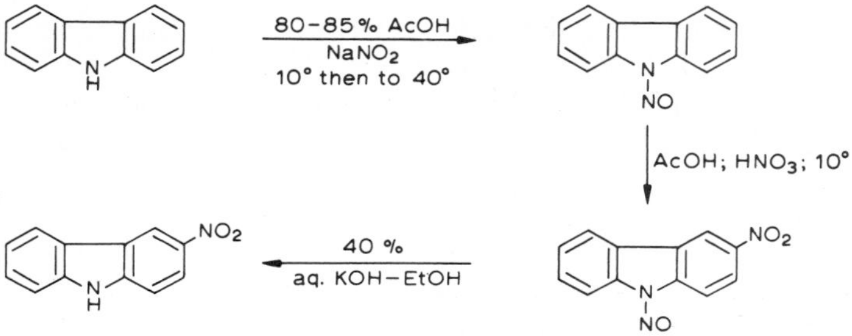

80–85% AcOH, $NaNO_2$, 10° then to 40°; AcOH; HNO_3; 10°; 40 % aq. KOH–EtOH

Treatment of 3-nitrocarbazole with acetic acid-nitric acid mixture at 70° yields 3,6-**dinitrocarbazole.** Further nitration yields 1,3,6,8-tetranitrocarbazole. Dinitration of carbazole leads to a mixture of 1,6- and 3,6-dinitrocarbazole, which on treatment with alcoholic potassium hydroxide affords a red solid and a red solution. The former on warming with dilute hydrochloric acid gives the 1,6-derivative and the latter the 3,6-derivative on acidification (*Grotta, Riggle* and *Bearse, loc. cit.; Grotta*, U.S.P. 3,144,462; 3,258,468; 3,291,787/1963). Mononitration also gives a small amount of 1-*nitrocarbazole* and 9-*p*-toluenesulphonylcarbazole affords 1-nitro-9-*p*-toluenesulphonylcarbazole (*D. H. Peacock et al.*, J. chem. Soc., 1942, 509). By the Fischer–Hepp rearrangement 9-nitrosocarbazole can be converted in low yield to 3-nitrocarbazole which could involve the formation and oxidation of an intermediate *C*-nitroso compound (*H. Wieland* and *H. Lecher*, Ann., 1912, **392,** 127, 156; *N. L. Drake et al.*, J. org. Chem., 1962, **27,** 1026):

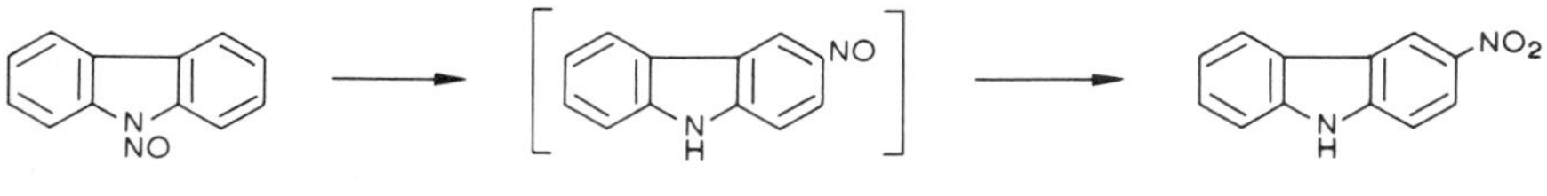

9-Nitrosocarbazole slowly decomposes in pyridine (*M. Matrka* and *F. Navratil*, Chem. Prumysl, 1962, **12,** 309). 9-Nitroso-3-nitrocarbazole is a mild nitrosating reagent converting *N*-methylaniline to *N*-nitrosomethylaniline in benzene solution, and reacting with aziridine under similar conditions to give ethylene and nitrous oxide, probably by the decomposition of *N*-nitrosoaziridine (*C. L. Bumgardner, K. S. McCallum* and *J. P. Freeman*, J. Amer. chem. Soc., 1961, **83,** 4417):

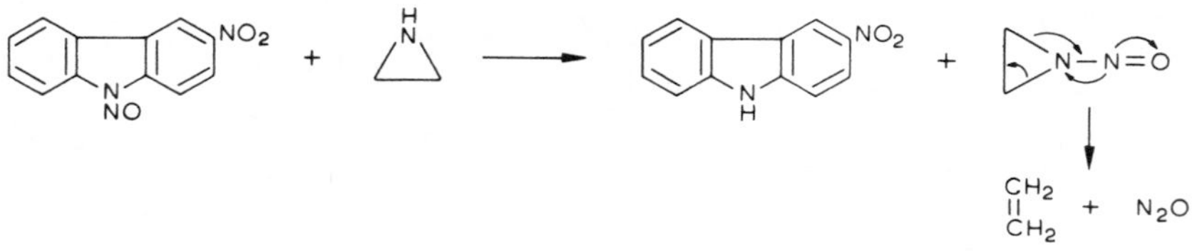

3-Vinylindole and nitroethylene react in benzene solution at room temperature during 6 days to form largely polymers and 2% of 1-nitrocarbazole (*W. E. Noland* and *R. J. Sundberg*, J. org. Chem., 1963, **28,** 884):

The isomeric 2- and 4-nitrocarbazoles are obtained by dehydrogenation of the corresponding tetrahydrocarbazoles (*B. M. Barclay* and *N. Campbell*, J. chem. Soc., 1945, 530). 2-**Nitrocarbazole** is best prepared by the thermal decomposition of 2-azido-4′-nitrobiphenyl (*G. D. Mendenhall* and *P. A. S. Smith*, Org. Synth., 1966, **46,** 85):

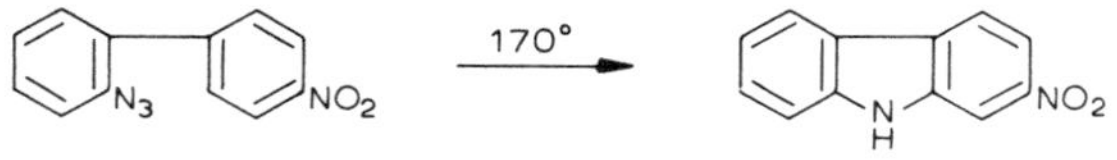

It may also be obtained by the deamination of 3-amino-2-nitrocarbazole (*G. Anderson* and *Campbell*, J. chem. Soc., 1950, 2904). All the nitrocarbazoles show their enhanced acidity by giving red solutions in alcoholic potash.

Melting points of nitrocarbazoles are given in Table 6.

TABLE 6

NITROCARBAZOLES

Substituent	*m.p. (°C)*	*Ref.*	*Substituent*	*m.p. (°C)*	*Ref.*
1-Nitro	187	1, 2	1,6-Dinitro	344–346	4
2-Nitro	165–166	3	3,6-Dinitro	386–387	4
3-Nitro	214 (205)	1, 3	1,3,6,8-Tetranitro	296	5
4-Nitro	182–183	3			

References
1 *H. Lindemann*, Ber., 1924, **57,** 555.
2 *R. W. G. Preston et al.*, J. chem. Soc., 1942, 500.
3 *B. M. Barclay* and *N. Campbell*, *ibid.*, 1945, 530.
4 *H. M. Grotta, C. J. Riggle* and *A. E. Bearse*, J. org. Chem., 1964, **29,** 2474.
5 *D. B. Murphy et al.*, J. Amer. chem. Soc., 1953, **75,** 4289.

Aminocarbazoles, photosensitive bases which darken rapidly in air, have been obtained by reducing the nitro compounds and also by ring-syntheses which establish the orientation. 3-Aminocarbazole is formed when 3-nitrocarbazole, sodium sulphite and water is autoclaved at 125–130°. Similarly 3,6-dinitrocarbazole is converted to 3,6-diaminocarbazole (*Eĭkhman, Lukashevich* and *Silaeva, loc. cit.*). 2,7- and 3,6-Diaminocarbazoles have been synthesised from benzidine (*E. Täuber*, Ber., 1890, **23,** 3267; 1892, **25,** 131):

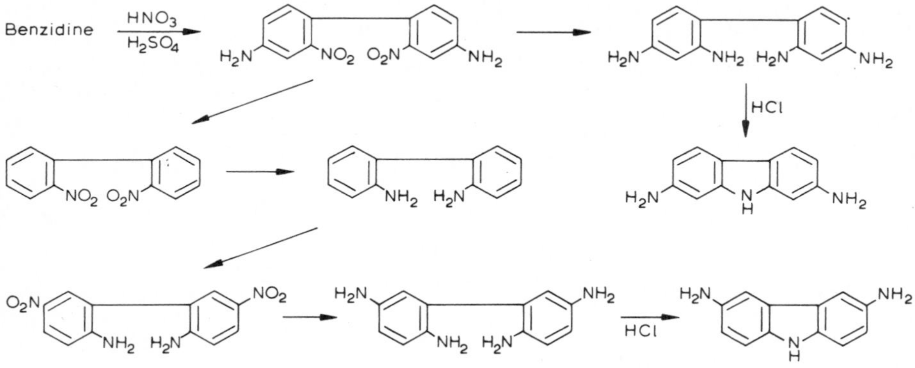
Benzidine
HNO3
H2SO4
H2N
NO2
O2N
NH2
HCl
NH2
N
H
HCl

When 1-morpholinocyclohexene, 2,4-dinitrofluorobenzene and dimethylformamide are heated together at 100°, 2-(2,4-dinitrophenyl)-cyclohexanone is obtained following hydrolysis of the reaction product, and may be converted to 2-aminocarbazole (*K. S. Murthy, R. Srinivasan* and *K. Venkataraman*, J. Sci. Ind. Res., India, 1962, **21B,** 290):

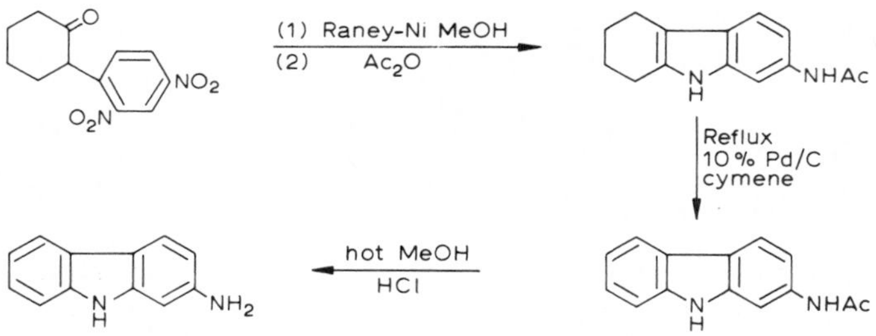
O
NO2
O2N
(1) Raney-Ni MeOH
(2) Ac2O
NHAc
N
H
Reflux
10 % Pd/C
cymene
hot MeOH
HCl
NH2

2-Aminocarbazole on diazotisation and coupling with 2-naphthol affords *2-carbazole-azo-2′-naphthol*, deep red needles, m.p. 273°, from benzene. **Carbazole-3-diazonium salts** are bright yellow, very stable towards heat, and so photosensitive that they have been made the basis of photoprinting processes (*O. Ruff* and *V. Stein*, Ber., 1901, **34,** 1668; *G. T. Morgan* and *H. N. Read*, J. chem. Soc., 1922, **121,** 2709).

Ammonia converts the diazonium ion reversibly into the unstable, orange-red *carbazole-3-diazoimine*, which explodes at 95°, and couples with phenols and amines.

4-Aminocarbazole has been obtained by irradiating 2,2′-diazidobiphenyl (*J. H. Boyer* and *G. J. Mikol*, Chem. Comm., 1969, 734).

The products obtained by the catalytic hydrogenation of the Schiff bases formed from cyclohexanone and aminocarbazoles, are useful as non-volatile antioxidants in rubber compounding (*C. S. Sheppard* and *M. H. Wilt*, U.S.P. 3,041,348/1962).

Amino derivatives of carbazole are assembled in Table 7.

TABLE 7

AMINOCARBAZOLES

Substituent	*m.p. (°C)*	*Ref.*	*Substituent*	*m.p. (°C)*	*Ref.*
1-Amino	195	1	2,7-Diamino	260	7, 8
2-Amino	239	2, 3	3,6-Diamino	>290	9
3-Amino	254	4, 5	2-Amino-9-methyl	135–136.5	3
4-Amino	183–185 (188–192)	2	3-Amino-9-methyl	173	3
3-Azido	176–177 (decomp.)	6			

References

1 *N. Campbell* and *J. A. R. MacLean*, J. chem. Soc., 1942, 504.
2 *G. Anderson* and *Campbell*, *ibid.*, 1950, 2904.
3 *E. Sawicki*, J. Amer. chem. Soc., 1954, **76**, 664.
4 *C. G. Schwalbe* and *S. Wolff*, J. chem. Soc., 1911, **99**, 103.
5 *T. C. Whitner*, J. Amer. chem. Soc., 1924, **46**, 2326.
6 *G. T. Morgan* and *H. N. Read*, J. chem. Soc., 1922, **121**, 2709.
7 *B. L. Manjunath*, J. Indian chem. Soc., 1927, **4**, 271.
8 *P. A. S. Smith* and *B. B. Brown*, J. Amer. chem. Soc., 1951, **73**, 2435.
9 *P. Ziersch*, Ber., 1909, **42**, 3798.

Hydron Blue R (CI. Vat Blue 43). Carbazole and *p*-nitrosophenol in presence of sulphuric acid give the indophenol (X), also produced by oxidising a mixture of carbazole and *p*-aminophenol; it is easily reduced to "carbazolylaminophenol" (XI). When heated with sodium polysulphide, either X or XI gives first a product soluble in sodium sulphide which behaves as an ordinary sulphur dye, and then an insoluble material, Hydron Blue R, which is dyed from a hydrosulphite vat. This important cotton colour gives dyeings faster than indigo towards light, washing, and hypochlorite:

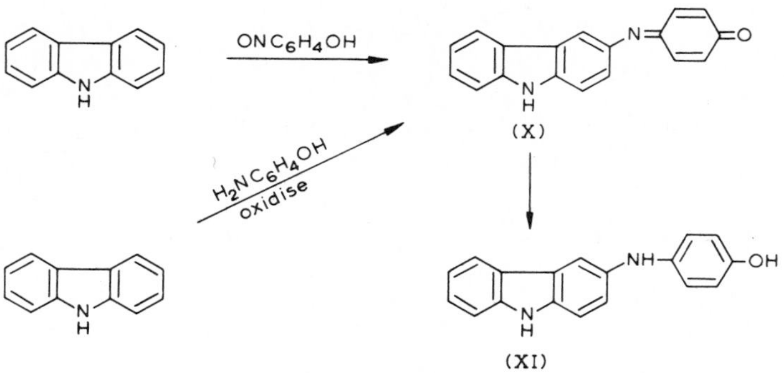

In the same way, *N*-ethylcarbazole yields the ethyl derivative **Hydron Blue G** (CI. Vat Blue 42). The basic skeleton of these dyes is indicated (*A. von Weinberg*, Ber., 1930, **63A,** 117; *E. Bernasconi*, Helv., 1932, **15,** 287) by the preparation of a product apparently identical with Hydron Blue R by treating with sodium tetrasulphide the indolothiazine XIa, synthesised as shown:

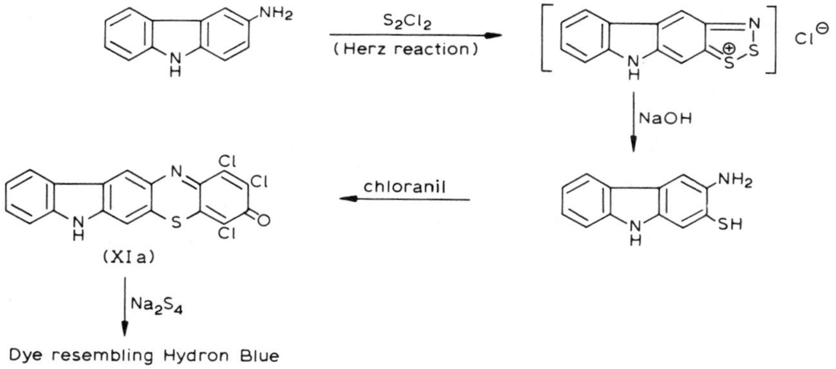

Bernasconi suggests that Hydron Blue R in its after-oxidised state is a polymer in which the fundamental unit has the structure:

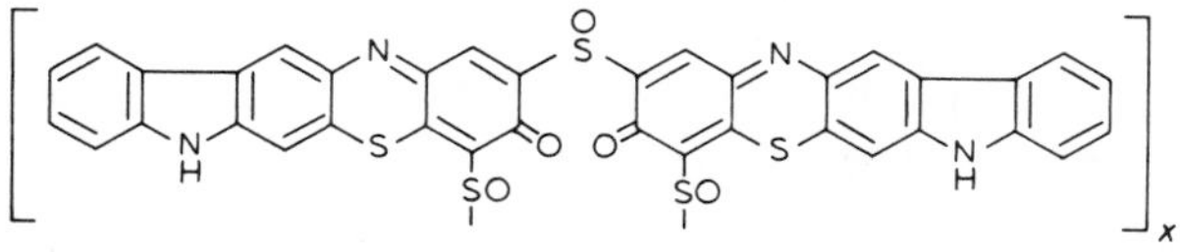

For a more detailed discussion on the constitution of Hydron Blue see *K. Venkataraman*, "The Chemistry of Synthetic Dyes", Vol. II, Academic Press, New York, 1952.

(viii) Hydroxycarbazoles

These are phenols, sensitive to oxidation, of which the 2-isomer is an important dye intermediate (p. 509). 1-, 2-, 3-, and 4-**Hydroxycarbazoles,** m.p. 164°, 276°, 255–256° and 169–170°, respectively, are obtained, the first three from the corresponding amines by the diazo-reaction, and the last by dehydrogenating 4-oxo-1,2,3,4-tetrahydrocarbazole (*O. Ruff* and *V. Stein*, Ber., 1901, **34,** 1668; *S. H. Tucker*, J. chem. Soc., 1924, **125,** 1144; *J. A. Cummins* and *M. L. Tomlinson*, *ibid.*, 1955, 3475).

The so-called "12-hydroxy-1,2,3,4,-tetrahydroisocarbazole" (*C. S. Barnes, K. H. Pausacker* and *C. I. Schubert*, J. chem. Soc., 1949, 1381) has proved to be identical with 6-hydroxy-1,2,3,4-tetrahydrocarbazole. Conversion of the latter to the methoxy derivative followed by dehydrogenation and demethylation affords 3-hydroxycarbazole (*A. H. Milne* and *Tomlinson*, *ibid.*, 1952, 2789):

HO–(tetrahydrocarbazole) —(Me_2CO, NaOH, Me_2SO_4)→ MeO–(tetrahydrocarbazole) —((1) Pd/C, CO_2, 280–300°; (2) aq. HBr)→ HO–(carbazole)

2-Hydroxycarbazole may be produced also by controlled alkali fusion of carbazole-2,3,6,8-tetrasulphonic acid, followed by desulphonation of the resultant hydroxytrisulphonic acid (*A. Schmelzer et al.*, U.S.P. 1,807,682/1931 and 1,983,301/1934; *P. P. Karpukhin* and *O. I. Levchenko*, Khim. Prom., Nauk Tekhn. Zb., 1963, 18):

Carbazole —(60% oleum, 15 h 15°)→ carbazole tetrasulphonic acid (HO_3S, SO_3H, SO_3H, SO_3H) (90%) ——→ Na Salt

—(50% NaOH, 30 h 250°, 30 atm.)→ hydroxycarbazole trisulphonic acid (HO_3S, SO_3H, OH, SO_3H) (96%) —(H_2SO_4, H_2O, 15 h 180°)→ 2-hydroxycarbazole (OH)

Other routes have been patented.

3-Hydroxycarbazole is obtained by heating sodium carbazole-3-sulphonate with potassium hydroxide at 290–350°.

Similarly 3,6-**dihydroxycarbazole,** m.p. >300°, is obtained from potassium carbazole-3,6-disulphonate. 3-*Acetoxy-carbazole*, m.p. 203–205°; 3,6-*diacetoxycarbazole*, m.p. 212–214° (*L. A. Stepanova* and *V. I. Shishkina*, Izv. Vyssh. Ucheb. Zaved, Khim. Khim. Technol., 1967, **10,** 298; C.A., 1967, **67,** 73475b).

The alkaloid **glycozoline,** m.p. 181–182°, from *Glycosmispentaphylla* is 3-methoxy-6-methylcarbazole (*D. P. Chakraborty*, Tetrahedron Letters, 1966, 661) and it has been synthesised by the photocyclisation of 4-methoxy-4′-methyldiphenylamine in petroleum (*W. Carruthers*, Chem. Comm., 1966, 272):

MeO, Me, HBr, HO, Me, Tosyl derivative, N H, Glycozoline, m.p. 228–300°, Zn dust, Raney Ni EtOH, hν, MeO, Me, N H, Me, N H, m.p. 204–205°

2- and 3-Methoxycarbazoles on treatment with lithium and ethanol in liquid ammonia yield the corresponding 1,4-dihydro derivatives, which on cleavage with acid are converted into 2-*oxo*-1,2,3,4-*tetrahydrocarbazole*, m.p. 131–133°, and 3-*oxo*-1,2,3,4-*tetrahydrocarbazole*, m.p. 156°, respectively (*H. J. Teuber* and *D. Cornelius*, Ann., 1964, **671,** 127):

Li EtOH liq. NH_3, OMe, HCl/Et_2O, OMe, O

(ix) Sulphonic acids and related compounds

Direct thiocyanation of carbazole gives 3-**carbazolyl thiocyanate,** m.p. 112°, also obtained from 3-aminocarbazole. On reduction it yields **carbazole**-3-**thiol,** m.p. 199–202°, which can be oxidised to **carbazole-3-sulphonic acid** (*K. G. Mizuch*, J. gen. Chem., U.S.S.R., 1940, **10,** 844). Sulphonation of carbazole with warm sulphuric acid gives the same sulphonic acid, and then di-, tri- and tetra-sulphonic acids. Carbazolesulphonic acids are in general poorly characterised, largely owing to the common failure of the usual methods of replacing the sulpho group by -OH or -CN for orientation; the tetrasulphonic acid obtained by treatment with oleum below 70° is regarded as the 2,3,6,8-isomeride *(Schmelzer, loc. cit.)*. **Carbazole**-2-**sulphinic acid** has been prepared from 2-aminocarbazole (*I.G. Farbenind.*, B.P. 303,520/1928).

(x) Organometallic derivatives

Successive treatment of 9-ethylcarbazole with butyl-lithium and carbon dioxide gives 9-ethylcarbazole-1-carboxylic acid, whereas direct mercuration takes place in positions 3 and 6.

3-**Acetoxymercuri-9-ethylcarbazole,** m.p. 156°, is converted by hydrazine into *bis*-9-*ethylcarbazolylmercury*, m.p. 217° (*H. Gilman* and *R. H. Kirby*, J. org. Chem., 1936,

1, 146). A bromine atom in position 3 can be replaced by lithium by the action of butyl-lithium (*Gilman* and *S. M. Spatz*, J. Amer. chem. Soc., 1941, **63,** 1553).

Arsenic has been introduced into the carbazole nucleus by the Bart reaction (*H. Burton* and *C. S. Gibson*, J. chem. Soc., 1927, 2386; *S. M. Scherlin* and *A. J. Berlin*, C.A., 1936, **30,** 1055); 9-**methyl**-3-**carbazolyldichloroarsine,** m.p. 121–122°. Reaction of the appropriate lithio-9-ethylcarbazole with triphenylchlorosilane gave 9-**ethyl**-1-**carbazolyltriphenylsilane,** m.p. 197–198°, and 9-**ethyl**-3-**carbazolyltriphenylsilane,** m.p. 218–220° (*R. H. Meen* and *Gilman*, J. org. Chem., 1955, **20,** 73).

(xi) Alcohols, aldehydes and ketones derived from carbazole

Carbazolylcarbinols, which would be expected to show the instability of aminobenzyl alcohols, have been little studied. Condensation of carbazole with benzophenone in presence of phosphorus oxychloride, followed by hydrolysis, gives 3-**carbazolyldiphenylcarbinol,** m.p. 172°, which forms with acids violet salts that dye wool (*S. Dutt*, J. chem. Soc., 1924, **125,** 802).

Carbazolecarboxylic acids afford, *via* their *p*-toluenesulphonohydrazides, the rather unreactive **carbazole**-1-, -2-, -3-, and -4-**carbaldehydes,** m.p. 143°, 155°, 153–154° and 163–164° (*S. G. P. Plant et al.*, *ibid.*, 1957, 2210; 1958, 1843). Aldehydes are obtained in good yield from 9-alkylcarbazoles with *N*-methylformanilide and phosphorus oxychloride; 9-*methylcarbazole*-3-*carbaldehyde*, m.p. 74°, so prepared gives by Kishner–Wolff reduction 3,9-dimethylcarbazole (*N. P. Buu-Hoï* and *N. Hoán*, J. Amer. chem. Soc., 1951, **73,** 98); also 1,4-dimethylcarbazole yields 1,4-**dimethylcarbazole**-3-**carbaldehyde,** m.p. 215–216°, *diphenylhydrazone*, m.p. 326° (decomp.), which on reduction affords 1,3,4-trimethylcarbazole (*P. A. Cranwell* and *J. E. Saxton*, Chem. and Ind., 1962, 45). **Carbazole**-1-**carbaldehyde** may be obtained in the following way starting from carbazole and a diethyl monoalkylmalonate (*M. Harfenist*, J. org. Chem., 1962, **27,** 4326):

N H
+ $RCH(CO_2Et)_2$
N O R O
(1) Br_2 CCl_4
(2) R'_2NH
Et_2O
N O R O NR'_2
$LiAlH_4$
Et_2O
N HO R OH NR'_2
$OH^{\ominus}$
N H CHO

The alkaloid **murrayanine** has been isolated from the petroleum extract of the mature stem-bark of *Murrayakoenigii* Spreng (Fam. *Rutaceae*) and shown to be 1-methoxycarbazole-3-carbaldehyde (*D. P. Chakraborty, B. K. Barman* and *P. K. Bose*, Tetrahedron, 1965, **21,** 681). It has been synthesised by the following route (*J. D. Crum* and *P. W. Sprague*, Chem. Comm., 1966, 417):

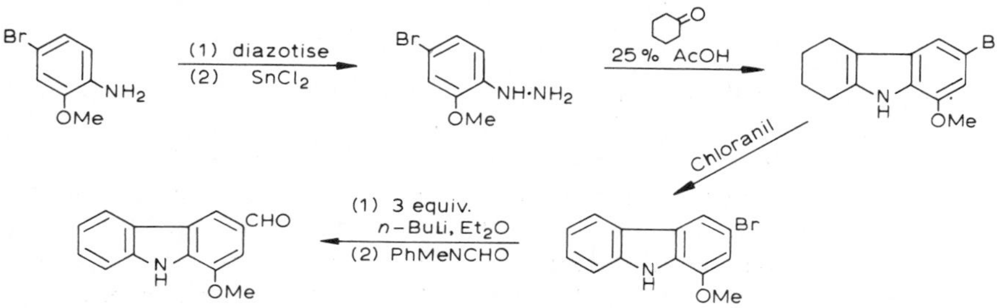

A few ketones have been prepared, and thus orientated by the Graebe–Ullmann or the Borsche synthesis, but the chief source is the Friedel–Crafts reaction. Applied to carbazole itself, this gives the 3,6-diacyl derivative, the 3-monoacyl compound being produced in only small yield under any conditions and best prepared by heating the 9-acylcarbazole with aluminium chloride. An acyl group may enter the 2- and/or the 9-positions depending upon the nature of the substituted H of the imino group (*H. Hannig* and *B. Schobless*, Pharmazie, 1963, **18,** 456; C.A., 1964, **60,** 5435h). Generally, acetylation of carbazole with acetyl halides or acetic anhydride gives both monoacetyl- and diacetyl-carbazole (*Y. Nagai* and *C. C. Huang*, Bull. chem. Soc., Japan, 1965, **38,** 951).

Acetylation of carbazole with acetic anhydride in the presence of zinc chloride at 80° affords 9-acetylcarbazole (*V. P. Lopatinskii, E. E. Sirotkina* and *M. M. Anosova*, Izv. Tomskogo Politekhn. Inst., 1961, III, 36):

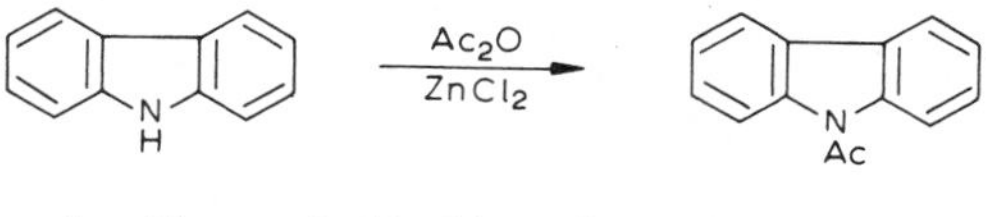

and 9-methylcarbazole with acetyl chloride and aluminium chloride in benzene yields 9-methyl-3,6-diacetylcarbazole (*idem, ibid.*, p. 40).

2-Acylcarbazoles may be obtained by treating 9-(bromoacetyl)carbazole in carbon disulphide in the presence of aluminium chloride with an aliphatic acid halide, and refluxing the product with alcoholic potassium hydroxide (*Hannig* and *Schobless, loc. cit.*); here the acylated nitrogen atom no longer directs substitution and the system behaves like biphenyl (*Plant et al.*, J. chem. Soc., 1935, 741; *E. Meitzner*, J. Amer. chem. Soc., 1935, **57,** 2327). The effects of basic factors on the yields of 3,6-diacetyl-9-alkylcarbazoles obtained by the acetylation of 9-alkylcarbazoles have been studied (*Lopatinskii et al.*, Izv. Tomskoga Politekhn. Inst., 1965, **136,** 18). With phthalic anhydride and aluminium chloride, carbazoles give 3,6-di(2-carboxybenzoyl) derivatives (*D. R. Mitchell* and *Plant*, J. chem. Soc., 1936, 1295), which are intermediates in the production of "anthraquinonecarbazoles" (*R. Scholl* and *W. Neovius*, Ber., 1911, **44,** 1249; *W. Bradley et al.*, J. chem. Soc., 1953, 1085; 1954, 2311; 1955, 3399; 1957, 819).

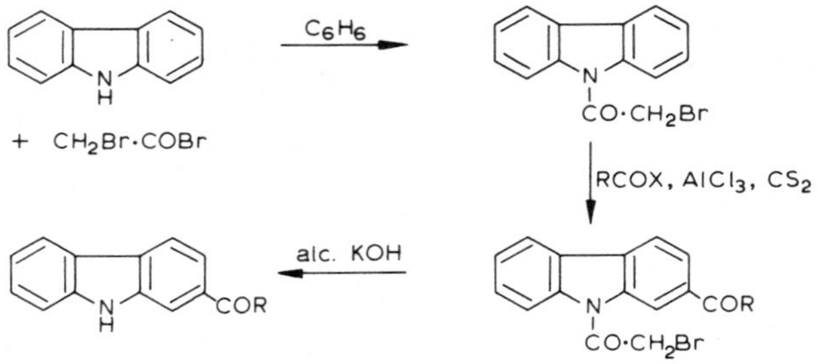

1-**Acetylcarbazole,** m.p. 136°; 2-*acetylcarbazole*, m.p. 227° (*W. Borsche* and *M. Feise*, Ber., 1907, **40,** 378); 3-*acetylcarbazole*, m.p. 167°; 9-*acetylcarbazole*, m.p. 74°; 3,6-*diacetylcarbazole*, m.p. 232°; 2-*propionylcarbazole*, m.p. 221°; 2-*butyrylcarbazole*, m.p. 218°; 2-*valerylcarbazole*, m.p. 213°; 2-*caproylcarbazole*, m.p. 215° (*Hannig* and *Schobless*, *loc. cit.*); 2-**benzoylcarbazole,** m.p. 163° (*Plant et al.*, and *Meitzner*, *loc. cit.*); 3-*benzoylcarbazole*, m.p. 206°; 3,6-*dibenzoylcarbazole*, m.p. 258° (*Plant* and *M. L. Tomlinson*, J. chem. Soc., 1932, 2188); 2,6-**diacylcarbazoles** (*D. G. Brooke* and *Plant*, *ibid.*, 1956, 2212); 9-*methyl*-3,6-*diacetylcarbazole*, m.p. 197–198° (*Lopatinskii*, *Sirotkina* and *Anosova*, *loc. cit.*).

(xii) Carbazolecarboxylic acids

These have been prepared by ring-synthesis, by potash fusion of carbazolyl methyl ketones, and by hydrolysis of carbazolyl trichloromethyl ketones (*H. G. Dunlop* and *S. H. Tucker*, J. chem. Soc., 1939, 1945), *e.g.* for carbazole-3-carboxylic acid:

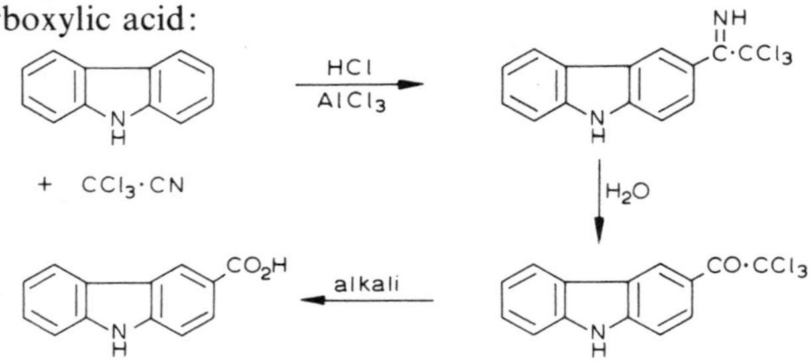

Carbonation of potassium- or magnesyl-carbazole yields at room temperature a salt of the unstable **carbazole-9-carboxylic acid,** which passes above 250° into that of **carbazole-1-carboxylic acid,** also obtained by treating carbazole successively with butyl-lithium and carbon dioxide (*H. Gilman* and *R. B. Kirby*, J. org. Chem., 1936, **1,** 146). It is also reported that when the adduct of carbazole and potassium hydroxide is autoclaved at 250–260°, followed by carbonation with gaseous carbon dioxide at atmospheric pressure, **carbazole-3-carboxylic acid** is formed (*Y. L. Slominskii*, Tr. Stud. Nauchn. Obshchestu, Khar Kousk. Politekhn. Inst., 1962, **5,** 155; C.A., 1964, **61,** 4300h).

Nitrosation of carbazole-3-carboxylic acid with sodium nitrite in glacial acetic acid affords 9-nitrosocarbazole-3-carboxylic acid, which on subsequent heating with hydrochloric and acetic acid rearranges to 6-*nitrosocarbazole-3-carboxylic acid*, m.p. 296–307°, *ethyl ester*, m.p. 190–196°.

Carbazole-1-carboxylic acid and its esters may be obtained by the following route (*M. Julia* and *J. Lenzi*, Bull. Soc. Chim., Fr., 1962, 2262):

CO_2Et, Br, O — $PhNH_2$ → O, CO_2Et, NH Ph — $ZnCl_2$ or boiling $PhNH_2$ → N H, CO_2Et — 5 % Pd/C → N H, CO_2Et — sap. → N H, CO_2H — CH_2N_2 → N H, CO_2Me

2-Methylcarbazole-4-carboxylic acid is prepared by a similar method. Carbazole-1-carboxylic acid is obtained when 1,9-(dichloromalonyl)carbazole is warmed with 2 *N* sodium hydroxide for a time at 50° and then acidified with acetic acid (*E. Ziegler* and *Th. Knappe*, Monatsh., 1963, **94,** 698).

Dimethyl 1-**methylcarbazole**-2,3-**dicarboxylate** is formed when the product of 3-indolylacetic acid and acetic anhydride in boron trifluoride etherate is heated with dimethyl acetylenedicarboxylate (*H. Plieniger*, *W. Mueller* and *K. Weinerth*, Ber., 1964, **97,** 667):

CH_2CO_2H, N H — Ac_2O, Et_2OBF_3 → O, O, N H, Me — $(:C\cdot CO_2Me)_2$, 5 min 130° → CO_2Me, CO_2Me, N H, Me

Several carbazoledicarboxylic acids are prepared in quantitative yields by oxidising the corresponding diacetylcarbazoles (*Y. Nagai* and *C. C. Huang*,

Bull. chem. Soc., Japan, 1965, **38,** 951). Some 9-alkylcarbazole-3,6-dicarboxylic acids have been prepared by hypobromite oxidation of the corresponding 9-alkyl-3,6-diacetylcarbazoles. The acid chlorides of these acids with allyl alcohol form diallyl esters, which polymerise in the presence of benzoyl and *tert*-butyl peroxide and copolymerise with methyl acrylate but not with styrene in the presence of benzoyl peroxide (*N. D. Negodyaev* and *Z. V. Pushkareva*, Khim. Geterotsikl. Soedin., Akad. Nauk, Latv. S.S.R., 1966, 46). The acid chlorides have also been converted to the diphenyl esters of 9-alkylcarbazole-3,6-dicarboxylic acids (*Negodyaev*, *G. D. Melnikov* and *Pushkareva*, Khim. Geterotsikl. Soedin., 1966, 634).

Esters of 5,6,7,8-tetrahydrocarbazole-1-, -2-, -3-, and -4-carboxylic acids can be dehydrogenated (*P. H. Carter*, *S. G. P. Plant* and *M. Tomlinson*, J. chem. Soc., 1957, 2210).

Kinetic studies made on the thermal degradation of ethyl carbazole-9-carboxylate and the rate of evolution of carbon dioxide at 293–318° show that the decomposition is first-order to at least 50% completion (*E. Dyer* and *G. C. Wright*, J. Amer. chem. Soc., 1959, **81,** 2138):

$$\text{Carbazole-}N\text{-}CO_2Et \xrightarrow{\Delta} \text{Carbazole (N-H)} + CO_2 + CH_2{:}CH_2$$

9-Carbazolylacetyl chloride when treated with aluminium chloride in benzene at 5–7°, followed by a mixture of ice and hydrochloric acid affords carbazole, a carbazole–formaldehyde condensate and carbon monoxide. The decarbonylation of the acid chloride is thought to proceed *via* the 9-carbazolylmethyl cation (*B. I. Stepanov* and *V. F. Traven*, Khim. Geterotsikl. Soedin., 1967, 369).

Melting points of some carbazolecarboxylic acids and their esters are collected in Table 8.

TABLE 8

CARBAZOLECARBOXYLIC ACIDS

Substituent	*m.p. (°C)*	*m.p. (°C) methyl ester*	*m.p. (°C) ethyl ester*	*Ref.*
1-carboxylic acid	271–273	135–137	106–107	1, 2
2-carboxylic acid	320–322		184	3
3-carboxylic acid	276–278		165	1, 4
4-carboxylic acid	244–245	96–97		5
1-methyl-4-carboxylic acid	232–234		49.5–51	2
3,6-dicarboxylic acid	360		206–207	4

References
1 *B. M. Barclay* and *N. Campbell*, J. chem. Soc., 1945, 530.
2 *M. Julia* and *J. Lenzi*, Bull. Soc. chim. Fr., 1962, 2262.
3 *S. G. P. Plant* and *S. B. C. Williams*, J. chem. Soc., 1934, 1142.
4 *H. G. Dunlop* and *S. H. Tucker*, *ibid.*, 1939, 1945.
5 *Plant et al.*, *ibid.*, 1957, 2210.

Arylamides of 2-*hydroxycarbazole*-3-*carboxylic acid*, m.p. 273–274° (*I.G. Farbenind.*, B.P. 343,164/1929; 347,113/1930), *e.g.* Naphthol AS LB, are used as azoic dye components (Vol. III C, p. 202), giving brown shades. The hydroxy acid is readily prepared from 2-hydroxycarbazole by the Kolbe synthesis (*E. Wolthuis* and *A. M. Shafer*, U.S.P. 2,453,105/1948).

(b) Hydrocarbazoles

(i) Tetrahydrocarbazoles

Reduction of carbazole with sodium and amyl alcohol gives 1,2,3,4-**tetrahydrocarbazole,** m.p. 119°, also obtained by controlled hydrogenation over copper chromite (*H. Adkins* and *H. L. Coonradt*, J. Amer. chem. Soc., 1941, **63,** 1563). A respectable number of 1,2,3,4-tetrahydrocarbazoles, with or without a methyl group in the 9-position have been obtained by the Borsche synthesis from arylhydrazones of cyclohexanones (*C. U. Rogers* and *R. B. Carson*, Org. Synth., 1950, **30,** 90). 1,2,3,4-Tetrahydrocarbazole is also produced in high yield by heating 2-phenylcyclohexanoneoxime (*A. Löffler* and *D. Ginsburg*, Nature, 1953, **172,** 820). 1,2,3,4-Tetrahydrocarbazole or the 8-piperidino derivative are precipitated within minutes on the addition of cyclohexanone to a previously warmed mixture of 3-phenyl- or 3-(2′-piperidinophenyl)-sydnone, hydrochloric acid and aqueous ethanol (*D. P. Ainsworth* and *H. Suschitzky*, J. chem. Soc., C, 1967, 315):

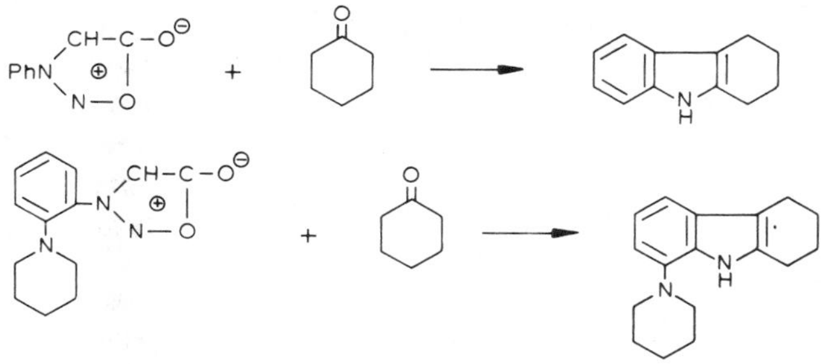

1-Pyrrolidinocyclohexene reacts exothermically with quinone dibenzenesulphonimide and the tetrahydrocarbazole XII is obtained when methanol and aqueous hydrochloric acid are added to the reaction mixture (*M. E. Kuehne*, J. Amer. chem. Soc., 1962, **84,** 837):

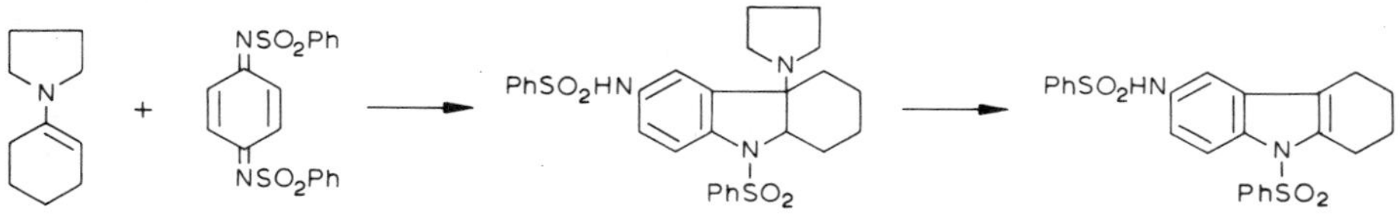

(XII)

Tetrahydrocarbazole is colourless, stable, insoluble in dilute acids, and is dehydrogenated only under rather energetic conditions; it forms a *picrate*, m.p. 147°. It reacts with chloroform and alkali, *e.g.* alcoholic sodium ethoxide to give 4a-*dichloromethyl*-1,2,3,4-*tetrahydro*-4a*H*-*carbazole*, m.p. 160–162° (*B. Robinson*, Tetrahedron Letters, 1962, 139):

4a*H*-Carbazoles are normal by-products in the Borsche synthesis with arylhydrazones of 2-alkylcyclohexanones, and *N*-methylated derivatives are obtained by direct methylation of tetrahydrocarbazoles (*G. Plancher* and *O. Carrasco*, Chem. Ztbl., 1904, II, 342):

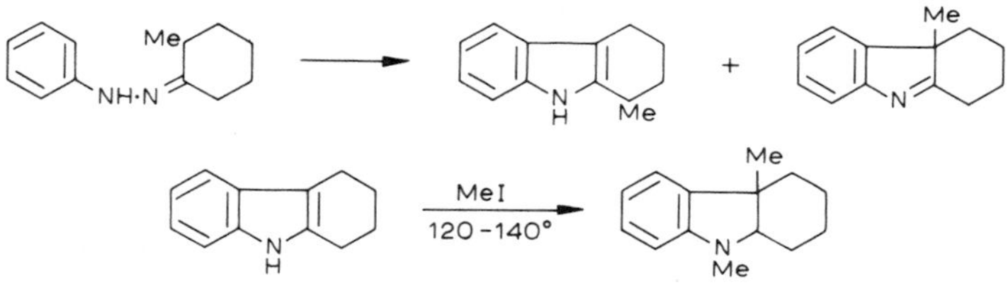

Tetrahydrocarbazole is alkylated in position 9 by reacting with an alkyl halide in liquid ammonia in the presence of either sodium or sodamide (*M. Nakazaki* and *S. Isoe*, Nippon Kagaku Zasshi, 1955, **76,** 1159; *Nakazaki*, Bull. chem. Soc. Japan, 1961, **34,** 334):

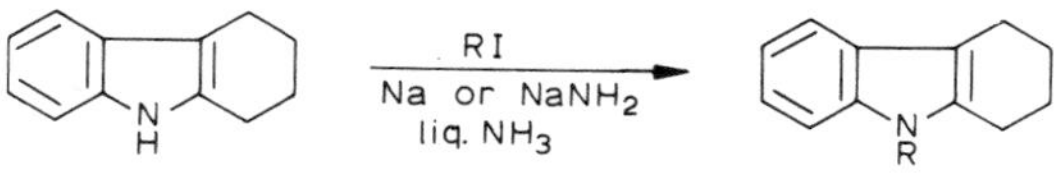

9-Acyl derivatives of tetrahydrocarbazole are prepared from the appropriate magnesium derivative and an ethyl ester, *e.g.*, ethyl acetate or benzoate; this method also affords some diacylated products (*Nakazaki*, *loc. cit.*).

With *N*-bromosuccinimide in the presence of pyridine, tetrahydrocarbazole yields the pyridinium bromide XIII, which condenses with acetoacetic ester and malonic ester in the presence of potassium carbonate. Diethyl malonate gives diethyl tetrahydrocarbazole-1-malonate and its hydrolysis affords tetrahydrocarbazole-1-acetic acid (*H. Sakakibara* and *T. Kobayashi*, Tetrahedron, 1966, **22,** 2475):

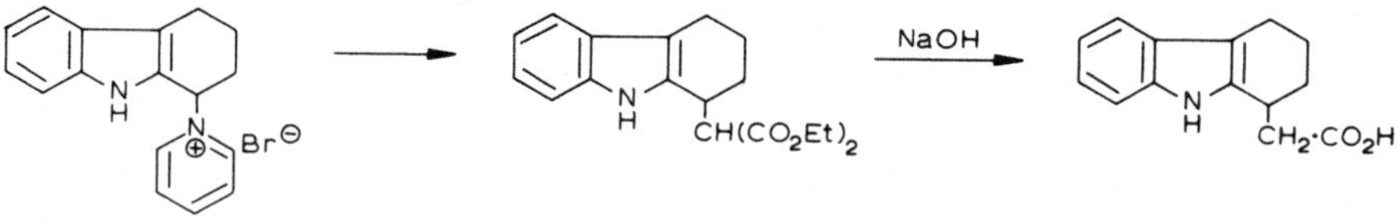

(XIII)

While nitration of tetrahydrocarbazole and its *N*-methyl derivative takes place in position 6, the *N*-acyl derivatives are substituted in position 7. In the latter case these products are accompanied by substances in which the 4a–9a double bond has been attacked; thus benzoyltetrahydrocarbazole gives XIV by addition of nitric acid, while acetyltetrahydrocarbazole is hydroxylated to XV. The heterocyclic ring of XIV is broken by alkali, with final reclosure to a quinoline derivative. By the action of acetic anhydride XV undergoes a pinacolic rearrangement to the spiro-compound XVI, while treatment with alkali gives the isomeric spiro-compound XVII; the structures assigned to XVI and XVII are based on rational syntheses (*S. G. P. Plant et al.*, J. chem. Soc., 1923, **123,** 676; 1950, 2127; 1951, 3475; *B. Witkop*, J. Amer. chem. Soc., 1950, **72,** 614). Similar additive compounds are encountered in the halogenation of acyltetrahydrocarbazoles (*Plant* and *M. L. Tomlinson*, J. chem. Soc., 1931, 3324; 1933, 298):

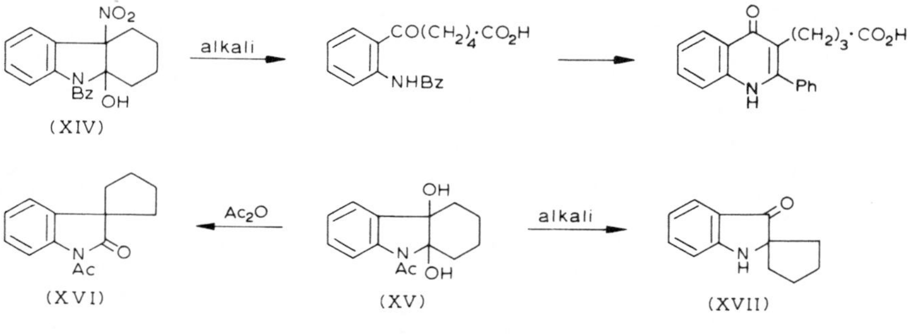

(XIV)

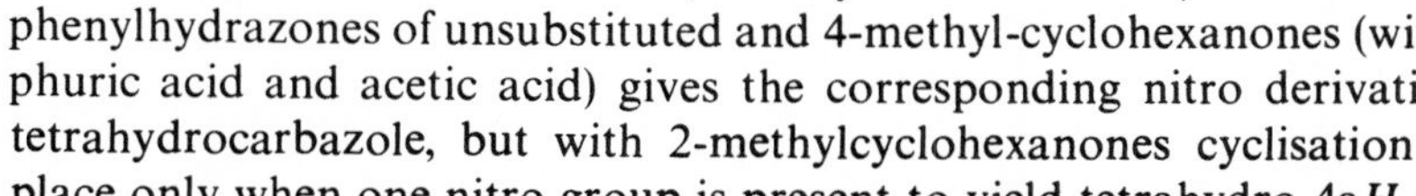

(XVI) (XV) (XVII)

The cyclisation of mononitro- (with hydrochloric acid) and of dinitrophenylhydrazones of unsubstituted and 4-methyl-cyclohexanones (with sulphuric acid and acetic acid) gives the corresponding nitro derivatives of tetrahydrocarbazole, but with 2-methylcyclohexanones cyclisation takes place only when one nitro group is present to yield tetrahydro-4a*H*-carbazoles (*D. S. Deohra* and *S. S. Joshi*, J. org. Chem., 1961, **26,** 3527):

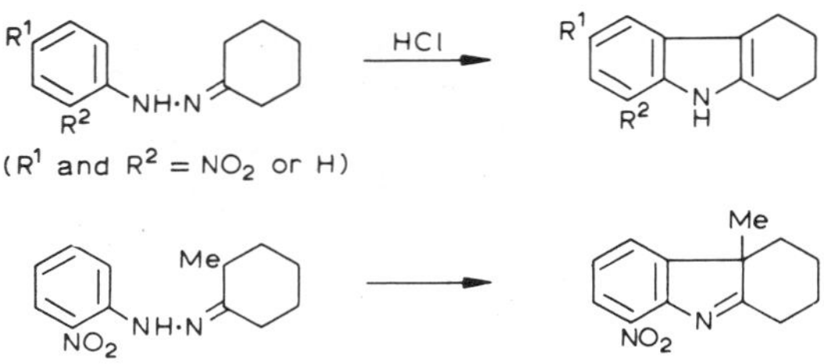

(R^1 and R^2 = NO_2 or H)

When polyphosphoric acid containing phosphorus pentoxide (83%) is used as a catalyst higher temperatures are required and lower yields are obtained (*A. R. Fasca*, Anales. Asoc. Quim. Arg., 1962, **50,** 162; C.A., 1963, 127446).

6-, 7-, and 8-Hydroxytetrahydrocarbazoles, formaldehyde and piperidine or dimethylamine in alcohol or acetic acid give Mannich adducts when kept at 20° or when refluxed in an atmosphere of nitrogen (*S. A. Monti, W. O. Johnson* and *D. H. White*, Tetrahedron Letters, 1966, 4459):

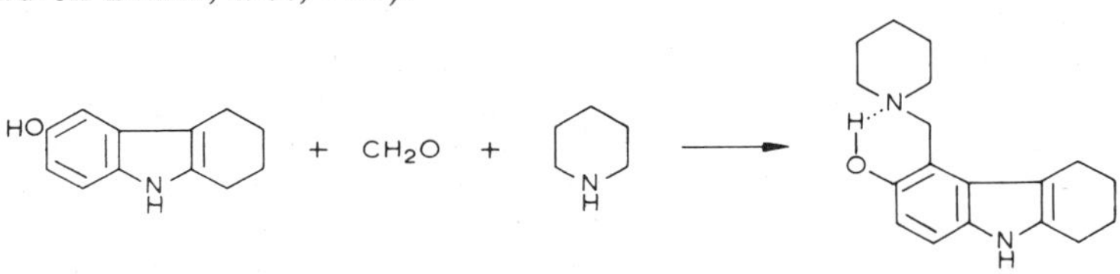

Aminomethylation takes place when tetrahydrocarbazole, 37% formaldehyde and an amine are boiled in ethanol, and stable hydrochlorides are formed by passing dry hydrogen chloride through an ethereal solution of the product; allylamine affords *compound* XVIII, m.p. 148–150° (*W. E. Hahn, H. Zawadzka* and *R. Bartnik*, Lodz. Tow. Nauk, Wyez III Acta Chim., 1966, II, 76; C.A., **66,** 75869m).

R	*m.p.(°C)*	*HCl m.p. (°C)*
N-Pyrrolidinyl	56.5 – 57	150 – 151
Morpholino	68.5 – 70.5	144 – 146 (decomp.)
Piperidino	49.5 – 50	152 – 154 (decomp.)

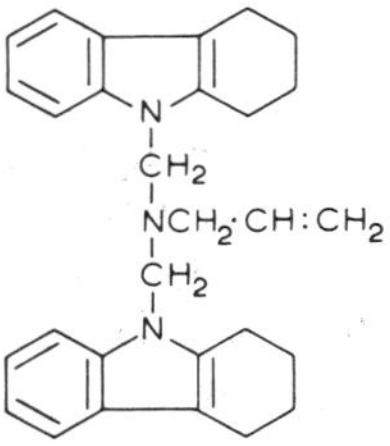

(XVIII)

When a ligroin solution of tetrahydrocarbazole is exposed to air, or a solution in ethyl acetate treated with oxygen in presence of platinum, 4a-hydroperoxy-1,2,3,4-tetrahydro-4a*H*-carbazole (XIX) is formed, which is weakly amphoteric and can be reduced catalytically to 4a-hydroxy-1,2,3,4-tetrahydroxy-4a*H*-carbazole (XX). Reduction with ferrous sulphate in sulphuric acid convert XIX by rearrangement into XXIII, also obtainable from XX. Reaction of tetrahydrocarbazolylmagnesium iodide and hydrogen peroxide also yields XX. With sodium hydroxide XIX gives cyclopentenoquinolone (XXI). Methyl-lithium reacts normally with XXIII giving a carbinol which undergoes pinacolic rearrangement to 4a-methyl-1,2,3,4-tetrahydro-4a*H*-carbazole (XXII), so regenerating the carbazole system (*R. J. S. Beer et al.*, J. chem. Soc., 1950, 2118, 3283; *Witkop* and *J. B. Patrick*, J. Amer. chem. Soc., 1950, **72**, 633; 1953, **75**, 2572; Experientia, 1950, **6**, 183).

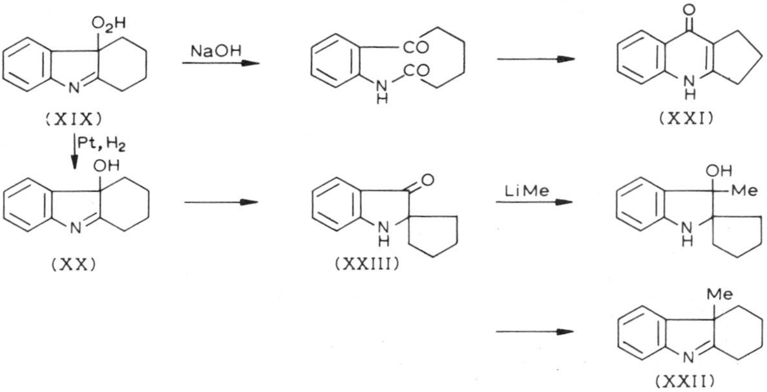

(XIX) —NaOH→ ... → (XXI)
(XIX) —Pt, H_2→ (XX) → (XXIII) —LiMe→ ... → (XXII)

On treating the residue obtained by taking to dryness 1-oxo-1,2,3,4-tetrahydrocarbazole, dioxane, ethanol and sodium, with ethyl chloroacetate in dimethylformamide, *ethyl* 1-*oxo*-1,2,3,4-*tetrahydrocarbazole*-9-*acetate*, m.p. 120–121°, is obtained in 90% yield. Hydrolysis with alcoholic sodium hydroxide gives 1-**oxo**-1,2,3,4-**tetrahydrocarbazole**-9-**acetic acid,** m.p. 212–213° (*V.I. Shvedov, L. B. Altukhova* and *A. N. Grinev*, Khim.-Farm. Zh., 1967, **1,** 9).

Oxidation of 5,8-**dihydroxytetracarbazole,** m.p. 225° (decomp.) gives red 1,2,3,4-**tetrahydrocarbazole**-5,8-**quinone,** m.p. 298–310° (decomp.) (*A. Blackwell* and *R. H. Thomson*, J. chem. Soc., 1954, 3916).

Some derivatives of 1,2,3,4-tetrahydrocarbazole are listed in Table 9.

TABLE 9

DERIVATIVES OF 1,2,3,4-TETRAHYDROCARBAZOLE

Substituent	*m.p. (°C)*	*Ref.*	*Substituent*	*m.p. (°C)*	*Ref.*
9-Methyl	50–52	1, 2	8,9-Dimethyl	151–152	8
9-Vinyl	b.p. 210/14 mm	3	6-Bromo	153	9
9-Phenyl	86	5	8-Chloro		10
9-Acetyl	77	1	5-Nitro	153–154	11
9-Chloroacetyl	124.5	6	6-Nitro	174	1, 9
9-Benzoyl	85	1	7-Nitro	168–169	11
9-Ethoxycarbonyl	65	1	8-Nitro	148–149	11
8-Methyl	97–99	7	6-Amino	152	1
8-Phenyl	90–91	4	7-Amino	103–104	12, 13
7,8-Dimethyl	84–86	7	7-Amino-1-methyl	113–114	13
5,8-Dimethyl	88–89	7	7-Methoxycarbonyl		13
6,8-Dimethyl	92–94	7	5-Methoxy	130	14
6,9-Dimethyl	89–90	8	6-Methoxy	94–95	9
			7-Methoxy	148	14
			8-Methoxy	Syrup	14
			1-Oxo	169	15
			4-Oxo	223	16

References
1 *W. H. Perkin* and *S. G. P. Plant*, J. chem. Soc., 1921, **119**, 1825; 1923, **123**, 676.
2 *M. Nakazaki* and *S. Isoe*, Nippon Kagaku Zasshi, 1955, **76**, 1159.
3 *G. R. Clemo* and *Perkin*, J. chem. Soc., 1924, **125**, 1804.
4 *A. N. Kost, I. P. Rudakova* and *A. P. Yakubov*, Zhur. org. Khim., 1965, **1**, 124.
5 *W. H. Linnell* and *Perkin*, J. chem. Soc., 1924, **125**, 2451.
6 *V. Ettel* and *J. Myska*, C.A., 1956, **50**, 5619.
7 *G. Anderson* and *N. Campbell*, J. chem. Soc., 1950, 2855.
8 *K. H. Bloss* and *C. E. Timberlake*, J. org. Chem., 1963, **28**, 267.
9 *W. Borsche et al.*, Ann., 1908, **359**, 49.
10 *F. P. Robinson* and *R. K. Brown*, Can. J. chem., 1964, **42**, 194.
11 *A. R. Frasca*, Anales Asoc. Quim. Arg., 1962, **50**, 162.
12 *S. G. P. Plant*, J. chem. Soc., 1936, 899.
13 *M. E. Kuehne*, J. Amer. chem. Soc., 1962, **84**, 837.
14 *H. T. Openshaw et al.*, J. chem. Soc., 1957, 1115.
15 *S. Coffey*, Rec. Trav. chim., 1923, **42**, 531; *A. Kent* and *D. McNeil*, J. chem. Soc., 1938, 8.
16 *Clemo* and *D. Felton*, *ibid.*, 1951, 700.

(ii) More fully reduced carbazoles

Hexahydrocarbazole (1,2,3,4,4a,9a-hexahydrocarbazole) is obtained by reducing tetrahydrocarbazole with tin and hydrochloric acid (*W. H. Perkin et al.*, J. chem. Soc., 1927, 2676) or electrolytically (*Perkin* and *S. G. P. Plant*, *ibid.*, 1924, **125**, 1503).

It exists in two modifications; that produced in greater quantity, m.p. 99° (*acetyl* deriv., m.p. 98°; *benzoyl* deriv., m.p. 106°) is regarded as *cis*-hexahydrocarbazole, and the other, m.p. 127° (*acetyl* deriv., m.p. 113°; *benzoyl* deriv., m.p. 133°) as the somewhat strained *trans* form.

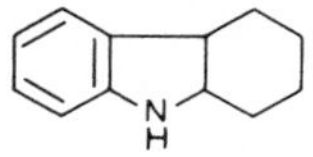

Reduction of 1,2,3,4-tetrahydrocarbazole using Raney nickel and copper chromite catalysts at 120 atm. pressure affords the hexahydro derivative in yields not exceeding 20% (*H. Booth, F. E. King* and *J. Parrick, ibid.*, 1958, 2302). Quantitative yields of the following hexahydrocarbazoles are obtained when the tetrahydrocarbazoles are hydrogenated in ethanol-fluoroboric acid solution over platinum oxide at 1 atm. and room temperature:

R^1	R^2	*m.p. (°C)*
H	H	95–96
Me	H	30–31
H	Me	54–56
Bu^t	H	64–65
H	Bu^t	75–77

All the 3-substituted hexahydrocarbazoles prepared are sharply-melting compounds giving one peak on a vapour-phase chromatograph. This suggests that only one of the possible stereoisomers is formed (*A. Smith* and *J. H. P. Utley*, Chem. Comm., 1965, 427).

1,2,3,4-Tetrahydrocarbazoles may be reduced to the hexahydro derivative by zinc dust and phosphoric acid (p. 446).

9-Methyl-1,2,3,4,4a,9a-hexahydrocarbazole is prepared by reduction of 9-methyl-1,2,3,4-tetrahydrocarbazole with tin-hydrochloric acid (*J. von Braun* and *H. Ritter*, Ber., 1922, **55,** 3795). When 9-methyl-1,2,3,4-tetrahydrocarbazoles containing one or more methyl groups in the benzene ring are reduced by metal–acids, or sodium and ethanol, the corresponding hexahydrocarbazoles are obtained, but the yields fail to exceed 13%. Using Adam's platinum oxide and glacial acetic acid, 9-methyl- and the (5–8),9-dimethyl-1,2,3,4-tetrahydrocarbazoles give the related hexahydro derivative in yields of 85% and better (*K. H. Bloss* and *C. E. Timberlake*, J. org. Chem., 1963, **28,** 267).

Hexahydrocarbazole behaves normally as a secondary aromatic amine, substitution, except nitration in strong sulphuric acid, takes place in position 6 both of the parent substance and of its 9-acyl derivatives. The 9-methylhexahydrocarbazoles are basic liquids, virtually insoluble in water or aqueous alkali, but reversibly soluble in 2 *N* mineral acids. The *cis* benzoyl compound, treated with phosphorus pentachloride, followed by hydrolysis and dehydrogenation, suffers ring-fission with ultimate production of *o*-benzamidobiphenyl (*von Braun*, Ber., 1910, **43,** 2879):

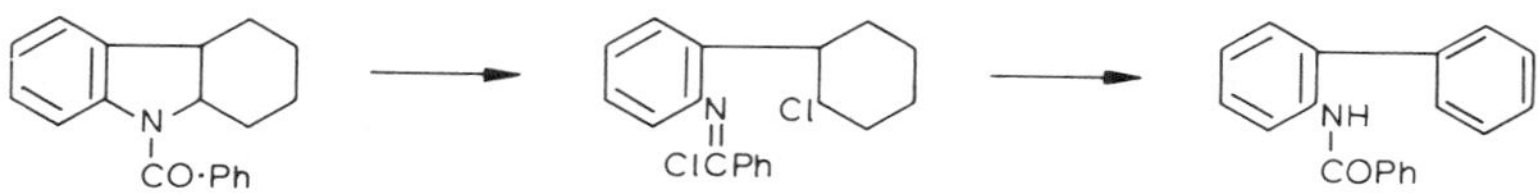

When the phenylhydrazone of 2-ethylcyclohexanone is boiled in freshly distilled formic acid, carbon dioxide is evolved to give 4a-ethyl-1,2,3,4,4a,9a-hexahydro-9*H*-carbazole-9-carbaldehyde (XXIV) and the 4a-ethyltetrahydro-4a*H*-carbazole (XXV), *picrate*, m.p. 147–149°. The aldehyde is hydrolysed with hydrochloric acid to afford 4a-*ethyl*-1,2,3,4,4a,9a-*hexahydro-9H-carbazole hydrochloride*, m.p. 193–195°, and acetic anhydride converts the free base to the 9-*acetyl* derivative, m.p. 68.5–69.5° (*Y. Ban et al.*, Chem. Pharm. Bull., Tokyo, 1967, **15,** 531):

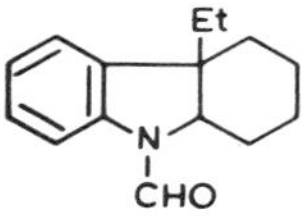

(XXIV) 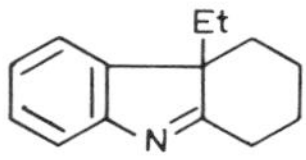 (XXV) 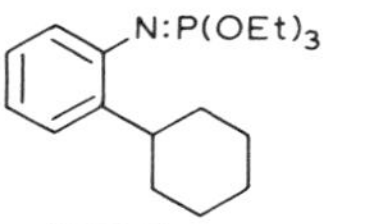 (XXVI)

Treatment of *o*-nitrophenylcyclohexane with boiling triethyl phosphite gives triethyl phosphate, triethyl *o*-cyclohexylphenylphosphorimidate (XXVI), *cis*- and *trans*-1,2,3,4,4a,9a-hexahydrocarbazole and some unidentified matter indicating the intervention of an electron-deficient nitrogen atom during deoxygenation (*R. J. Sundberg*, Tetrahedron Letters, 1966, 477).

1,2,3,4,5,6,7,8-**Octahydrocarbazole,** b.p. 161.5–162.5/13 mm, m.p. 96°, may be prepared by heating cyclohexanone azine and zinc chloride at 220–230°, or by the Piloty synthesis when the cyclisation is effected by hydrochloric acid (*B. Robinson*, Tetrahedron, 1964, **20,** 215). Boiling the azine with acetyl chloride in dioxane affords 9-*acetyloctahydrocarbazole*, m.p. 71° (*A. N. Kost* and *I. I. Grandberg*, Zhur. obshcheĭ Khim., 1956, **26,** 565), and with phosphorus tribromide in dioxane a mixture of bis-2-bromoethyl ether and octahydrocarbazole is formed. Phosphorus trichloride does not react under these conditions (*Kost*, *Grandberg* and *E. B. Eureinova*, *ibid*., 1958, **28,** 512). Octahydrocarbazole is obtained on boiling 2,2′-dioxobicyclohexyl and ammonium acetate in acetic acid (*T. Masqamune*, *M. Ohno* and *M. Koshi*, J. Fac. Sci., Hokkaido Univ., 1957, III, **5,** 37; C.A., **52,** 14579a).

Hydrogenation of 9-methylcarbazole over nickel gives the feebly basic 9-**methyl-1,2,3,4,5,6,7,8-octahydrocarbazole,** m.p. 94° (*von Braun et al.*, Ber., 1922, **55,** 3792; 1925, **58,** 2156), also formed by a Paal-Knorr synthesis from 2,2′-dioxobicyclohexyl (*Plant*, J. chem. Soc., 1930, 1595):

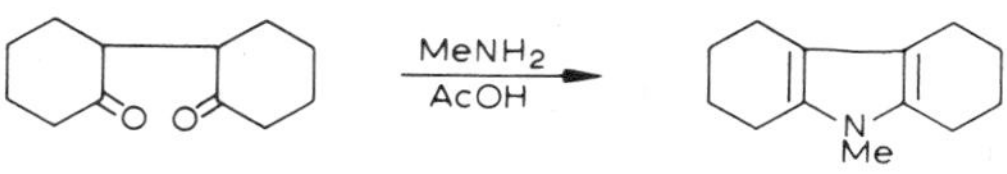

$\xrightarrow[AcOH]{MeNH_2}$

Methylation of 1,2,3,4,5,6,7,8-octahydrocarbazole gives a product which is different from the above compound and spectral data indicates that it is 8a-**methyl-1,2,3,4,5,6,7,8-octahydro-**8a*H*-**carbazole** (XXVII), b.p. 148–150°/16 mm, *picrate*, m.p. 147°, showing that methylation occurs on carbon atom 8a rather than on nitrogen (*Robinson, loc. cit.*):

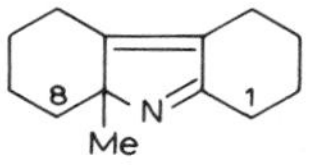

Further reduction of 9-methyloctahydrocarbazole by tin and hydrochloric acid gives a 9-**methyldecahydrocarbazole,** b.p. 138–139°/12 mm, for which the formula XXVIII is suggested by its degradation to 2-cyclohexenylcyclohexanone:

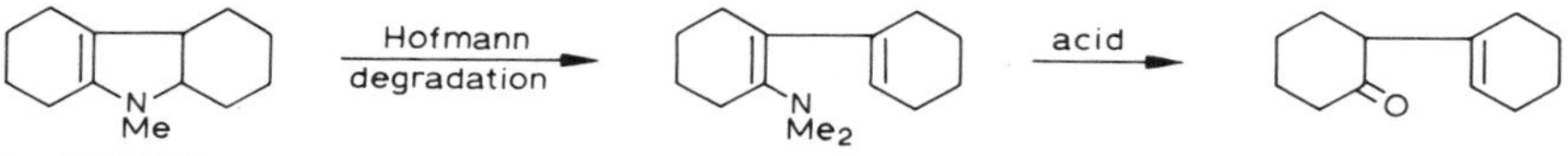

Cyclohexanoneazine, with hydrogen chloride at 180°, gives a basic product, formulated as 1,2,3,6,7,8,8a,9a-**octahydrocarbazole,** m.p. 102°, (*Perkin* and *Plant, loc. cit.*; *Plant, loc. cit.*; *E. Benary,* Ber., 1934, **67**, 708):

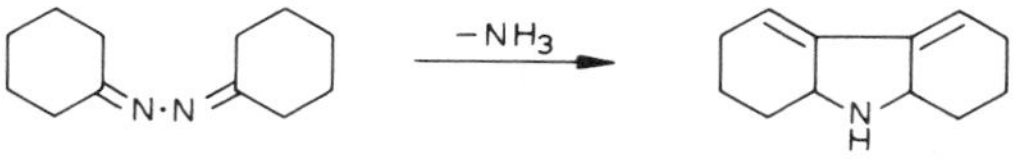

By hydrogenating carbazole over nickel or copper chromite, *tetra-*, *hexa-*, and *dodeca-hydrocarbazole* are obtained, the last as a solid, m.p. 74°, accompanied by a probably stereoisomeric liquid form (*H. Adkins* and *H. L. Coonradt*, J. Amer. chem. Soc., 1941, **63,** 1563).

(c) Compounds having more than one singly linked carbazole nucleus

Oxidation of carbazole with permanganate in acetone affords 9,9′-**bicarbazolyl,** m.p. 220°, and an *isomer*, m.p. 265° (*J. McLintock* and *S. H. Tucker*, J. chem. Soc., 1927, 1214). With chromic acid, carbazole yields 3,3′-*bicarbazolyl* (*diacetyl* deriv., m.p. 247–249°), identified by synthesis. The following have been obtained also: 1,1′-*bicarbazolyl*, m.p. 205–207°; 3,9′-*bicarbazolyl*, m.p. 208–210° (*M. C. Nelmes* and *Tucker*, *ibid.*, 1933, 1523); 1,9′-*bicarbazolyl*, m.p. 191° (*A. C. Geale, J. M. C. Linnell* and *M. L. Tomlinson*, *ibid.*,

1956, 1124); the preparations of a number of derivatives are described.

When carbazole is heated with oxalic acid, a deep blue substance is obtained — **"Carbazole Blue".** This appears to be tris-3-carbazolylmethyl formate; it yields on hydrolysis the colourless **tris-3-carbazolylcarbinol** (XXIX), m.p. 117–119°, and by reduction **tris-3-carbazolylmethane,** m.p. 217–218° (*M. Copisarow, ibid.*, 1920, **117,** 1542):

(XXIX)

(d) Carbazoles with additional fused rings

(i) Benzocarbazoles

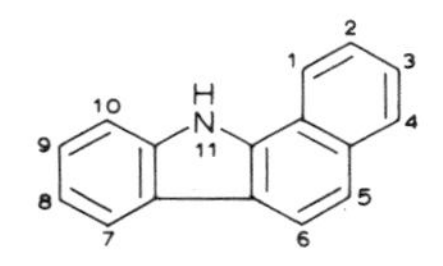

11*H*-Benzo[*a*]carbazole
(1,2-Benzocarbazole)

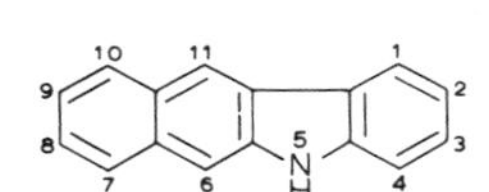

5*H*-Benzo[*b*]carbazole
(2,3-Benzocarbazole)

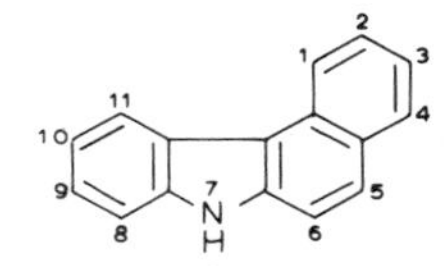

7*H*-Benzo[*c*]carbazole
(3,4-Benzocarbazole)

The three possible monobenzocarbazoles, and various dibenzocarbazoles are accessible by the Graebe–Ullman carbazole synthesis (p. 489); and by Borsche's synthesis of tetrahydrocarbazoles (p. 509), especially its variant the Bucherer method in which a naphthol or naphthylamine is boiled with an arylhydrazine and sodium bisulphite in aqueous solution; α-naphthol or α-naphthylamine with phenylhydrazine yields the phenylhydrazone of 1-tetralone-3-sulphonic acid, which under the influence of acids is converted partially to 11*H*-benzo[*a*]carbazole and partially to a diamino compound; similarly β-naphthol or β-naphthylamine affords 7*H*-benzo[*c*]carbazole and 5,6-dihydro-7*H*-benzo[*c*]carbazole-5-sulphonic acid forms as an intermediate (*A. Rieche* and *H. Seeboth*, Ann., 1960, **638,** 81).

2-Naphthol reacts with phenyl [2-^{15}N]hydrazine in the presence of sulphur dioxide to give ammonia and benzo[*c*]carbazole, which retains some of the ^{15}N. Phenylnaphthylhydrazine is probably formed during the reaction and is converted into the carbazole mainly through an *o,o'*-benzidine rearrangement and through an *o*-semidine rearrangement. The intermediate 1-(2-aminophenyl)-2-[^{15}N]naphthylamine produced in the main reaction eliminates ^{15}N exclusively on decomposition under the same reaction conditions (*P. F. Holt* and *C. J. McNae*, J. chem. Soc., 1964, 1759).

Reduction of 1-nitro-2-(2-nitrophenyl)naphthalene and 2-nitro-1-(2-nitrophenyl)naphthalene with zinc and acetic acid gives 11*H*-benzo[*a*]carbazole and 7*H*-benzo[*c*]carbazole, respectively (*J. F. Corbett* and *Holt*, *ibid.*, 1960, 3646). 11*H*-Benzo[*c*]carbazole is prepared in high yield from both 2-(2-azidophenyl)naphthalene and 1-azido-2-phenylnaphthalene, and 7*H*-benzo[*c*]carbazole from 1-(2-azidophenyl)naphthalene (*P. A. S. Smith*, *J. M. Clegg* and *J. H. Hall*, J. org. Chem., 1958, **23**, 524).

The cyclodehydrogenation of phenyl-2-naphthylamine and phenyl-1-naphthylamine at 500° under the influence of a platinum–magnesium oxide catalyst produces 5*H*-benzo[*b*]carbazole and 11*H*-benzo[*a*]carbazole, respectively (*H. M. Grotta*, *C. J. Riggle* and *A. E. Bearse*, *ibid.*, 1961, **26**, 1509). The former compound is also obtained by pyrolysis of phenyl-2-naphthylamine, but phenyl-1-naphthylamine yields a different product (*C. Graebe* and *W. Knecht*, Ann., 1880, **202**, 1):

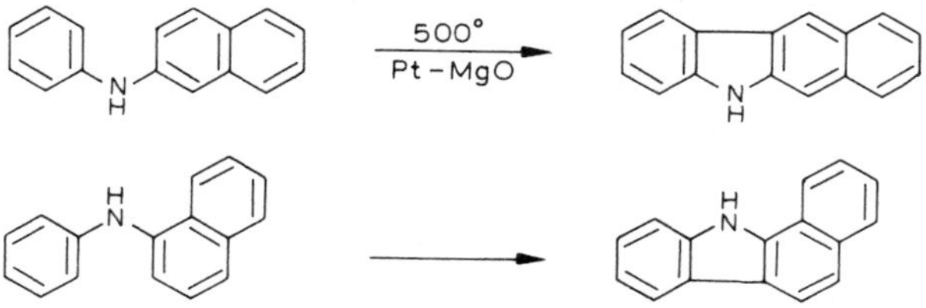

500°
Pt – MgO

Oxidation of 2-amino-1-(2-aminophenyl)naphthalene with sodium perborate in acetic acid gives mainly a cinnoline and a trace of 7*H*-benzo[*c*]carbazole; under similar conditions 1-amino-2-(2-aminophenyl)naphthalene yields ~5% of 11*H*-benzo[*a*]carbazole (*Corbett* and *Holt*, J. chem. Soc., 1961, 3695).

cis-*β*-Decalone phenylhydrazone undergoes Fischer indolisation with hydrogen chloride to yield a mixture of linear (XXX) and angular (XXXI) octahydrobenzocarbazoles, which on dehydrogenation with chloranil are converted to 5*H*-benzo[*b*]- and 7*H*-benzo[*c*]-carbazoles, respectively (*N. P. Buu-Hoï*, *P. Jacquignon* and *D. Lavit*, *ibid.*, 1956, 2593):

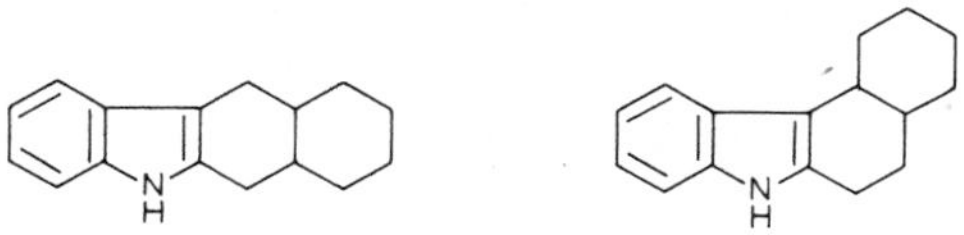

(XXX) (XXXI)

2-Methylindole reacts with 2-hydroxymethylenecyclohexanones in boiling acetic acid to form 7,8,9,10-tetrahydro-5*H*-benzo[*b*]carbazoles (*W. E. Noland* and *J. E. Johnson*, Tetrahedron Letters, 1962, 589):

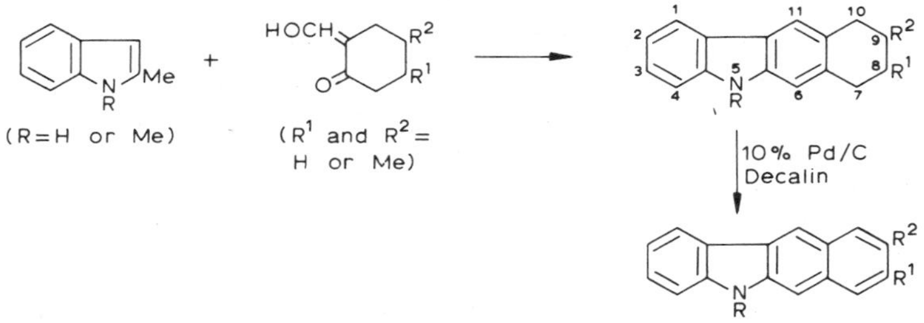

(R=H or Me)

(R^1 and R^2 = H or Me)

10% Pd/C Decalin

Although the Friedel–Crafts reaction between acid anhydrides and 9-alkylcarbazoles was generally believed to lead to disubstitution, it has been shown that 9-ethylcarbazole readily undergoes monosuccinacylation, the product being converted by reduction into γ-(9-ethylcarbazol-3-yl)butyric acid, which undergoes cyclisation with polyphosphoric acid to 5-ethyl-7,8,9,10-tetrahydro-7-oxo-5*H*-benzo[*b*]carbazole (XXXII). The ketone on Wolff–Kishner reduction followed by dehydrogenation over palladised charcoal, affords 5-ethyl-5*H*-benzo[*b*]carbazole in excellent yield (*Buu-Hoï* and *G. Saint-Ruf*, J. chem. Soc., 1965, 5464). XXXII following reaction with methylmagnesium iodide may be converted into 5-ethyl-7-methyl-5*H*-benzo[*b*]carbazole (XXXIII). Indolisation of the phenylhydrazone and dehydrogenation of the product with palladium–charcoal gives 12-ethyl-12,14-dihydrocarbazole[2,3-*a*]carbazole (XXXIV):

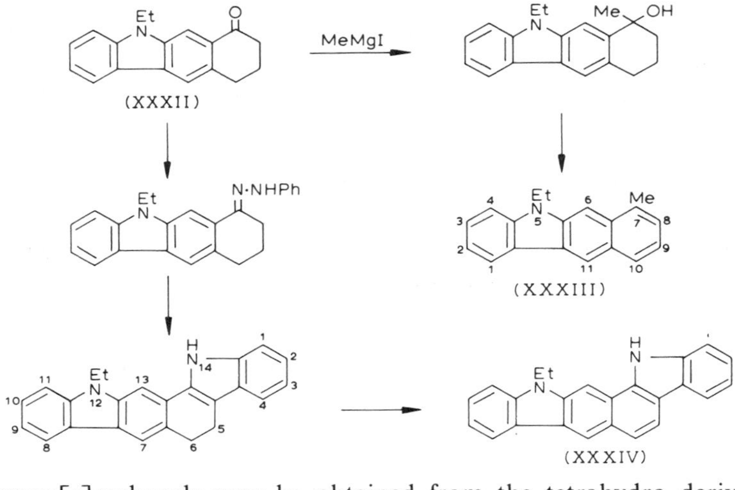

(XXXII) MeMgI (XXXIII) (XXXIV)

7*H*-benzo[*c*]carbazole may be obtained from the tetrahydro derivative, prepared by the Bischler reaction between 2-chlorocyclohexanone and β-naphthylamine in boiling cellosolve in the presence of sodium carbonate and quinoline (*E. E. Campaigne* and *R. D. Lake*, J. org. Chem., 1959, **24,** 478):

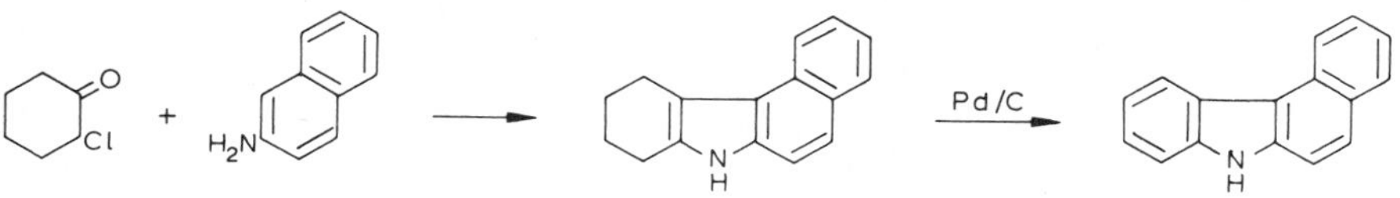

Pd/C

Methyl and dimethyl 7*H*-benzo[*c*]carbazoles are prepared by the dehydrogenation of the respective 8,9,10,11-tetrahydro-7*H*-benzo[*c*]carbazoles obtained by a modified Fischer–Borsche reaction. The product of the reaction between 3-methylcyclohexanone and *β*-naphthylhydrazine is 9-methyl-8,9,10,11-tetrahydro-7*H*-benzo[*c*]carbazole (XXXV) and not the 11-methyl isomer (*Campaigne et al., ibid.*, 1959, **24,** 487):

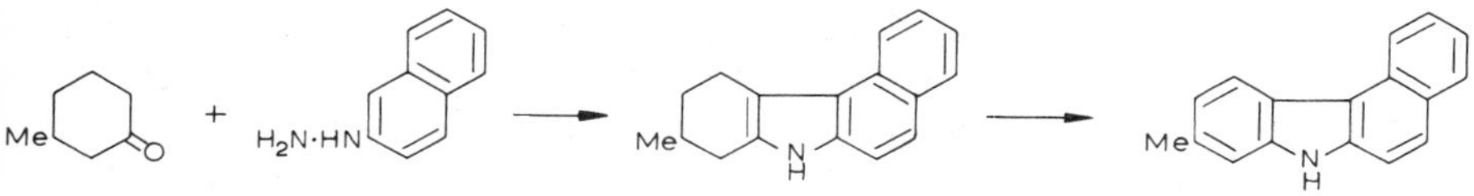

(XXXV)

The synthesis of several 6-methyl-11*H*-benzo[*a*]carbazoles (XXXVI) and 6-methyl-13*H*-dibenzo[*a,i*]- and 13-methyl-7*H*-dibenzo[*a,g*]-carbazoles (XXXVII) and (XXXVIII) is possible by indolisation of the appropriate 3-methyl-1-tetralone arylhydrazone, followed by dehydrogenation with palladised charcoal of the resulting 5,6-dihydro-compounds (*Buu-Hoï, M. Mangane* and *Jacquignon*, J. chem. Soc., C, 1967, 662):

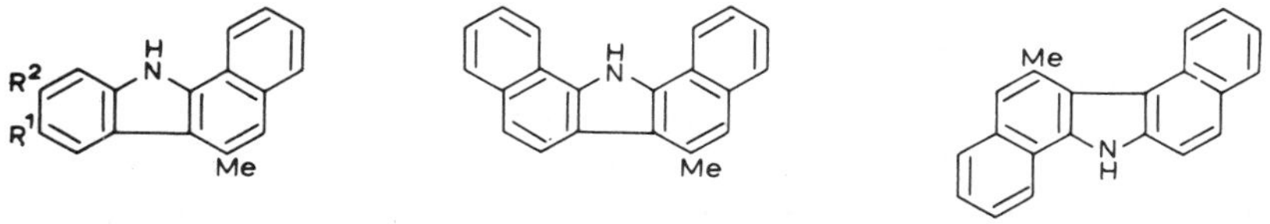

(XXXVI) (XXXVII) (XXXVIII)

β-Tetralone and phenylhydrazine when boiled in acetic acid afford 5,6-dihydro-7*H*-benzo[*c*]carbazole, which dehydrogenates to give the corresponding benzocarbazole. If the dihydro derivative is heated with aluminium chloride at 220–240° rearrangement takes place to yield 11*H*-benzo[*a*]carbazole (*M. Nakazaki*, Bull. chem. Soc., Japan, 1960, **33,** 461). 7*H*-Benzo[*c*]carbazole is isomerised by aluminium chloride to 11*H*-benzo[*a*]carbazole (*M. Zander* and *W. H. Franke*, Ber., 1967, **100,** 2649), and 7*H*-dibenzo[*c,g*]carbazole (XXXIX) into 13*H*-dibenzo[*a,g*]-carbazole (XL) (*idem*, Angew. Chem., 1964, **76,** 922):

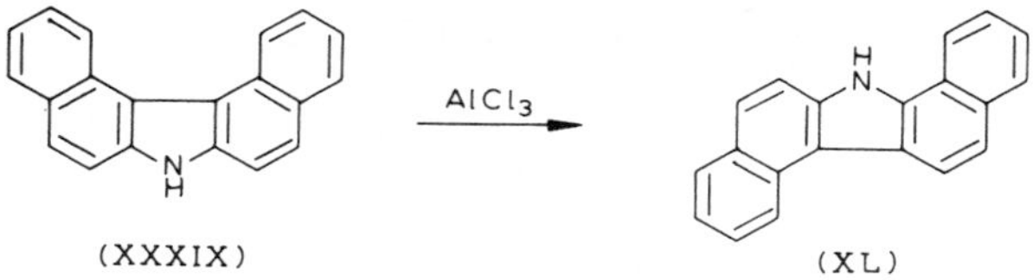

$AlCl_3$

(XXXIX) (XL)

5,7-, 5,8-, and 6,7-dimethyl-1-tetralones are susceptible to the Fischer reaction, thus giving a series of polymethyl-11*H*-benzo[*a*]carbazoles (*Buu-Hoï et al.*, J. chem. Soc., 1963, 2274).

1,4-Dimethylcarbazole condenses with hexane-2,5-dione to give 1,4,7,10-tetramethyl-5*H*-benzo[*b*]carbazole (XLI). Under similar conditions 1,4,6-trimethylcarbazole did not give a derivative of 5*H*-benzo[*b*]-carbazole with hexane-2,5-dione, but 1,3,4-trimethylcarbazole yielded 1,2,4,7,10-pentamethyl-5*H*-benzo[*b*]carbazole. This novel route to benzocarbazoles illustrates the ability of hexane-2,5-dione to react with reactive vicinal aromatic positions with the resulting formation of a new aromatic ring (*P. A. Cranwell* and *J. E. Saxton*, J. chem. Soc., 1962, 3482):

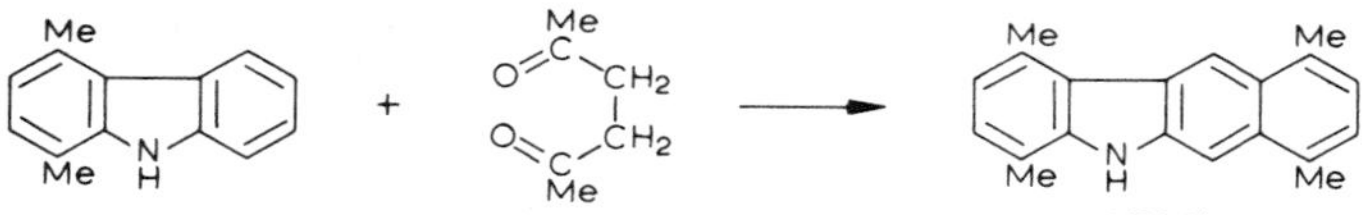

(XLI)

1-Benzylpyrrole reacts with 2-bromofluorobenzene and magnesium in dry tetrahydrofuran and on chromatographing the product on alumina, gives 11-benzyl-1,2-dihydro-11*H*-benzo[*a*]carbazole, which on dehydrogenation over palladium–carbon yields 11-benzyl-11*H*-benzo[*a*]carbazole, also obtained by reacting 11*H*-benzo[*a*]carbazole in ether with butyl-lithium and then with benzyl bromide (*G. Wittig* and *B. Reichel*, Ber., 1963, **96,** 2851):

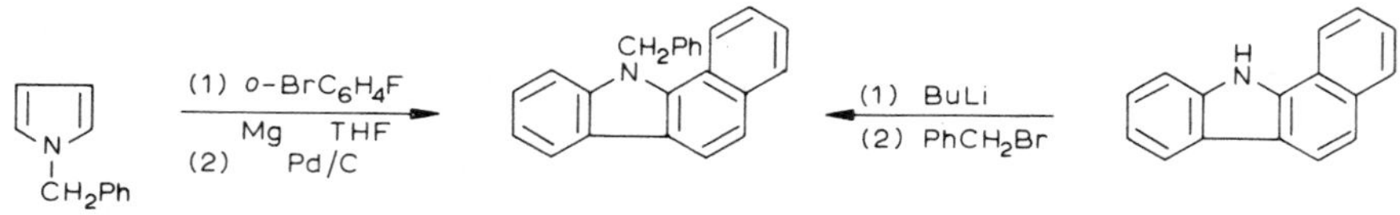

(1) o-BrC_6H_4F, Mg, THF; (2) Pd/C

(1) BuLi; (2) $PhCH_2Br$

Benzo[*a*]- and benzo[*c*]-carbazole form addition compounds with tetrachlorophthalic anhydride. A number of π-complexes are formed between hexachloronaphthalic anhydride, the anhydride, amide and esters of 3,5-dinitrobenzoic acid, the nitro derivatives of naphthalene-2,3-dicarboxylic acid, the bromo derivatives and dianhydride of naphthalene-1,4,5,8-tetracarboxylic acid and benzocarbazoles (*Buu-Hoï et al.*, Bull. Soc. chim. Fr., 1962, 312; 1963, 2543; 1964, 2517; Rec. Trav. chim., 1963, **82,** 370). Acetylation of linear 5*H*-benzo[*b*]carbazole in methylene chloride in the presence of aluminium chloride yields 5-acetyl-5*H*-benzo[*b*]carbazole, which on Wolff–Kishner reduction furnishes an ethyl-5*H*-benzo[*b*]carbazole (*P. Mabille* and *Buu-Hoï*, J. org. Chem., 1960, **25,** 1937).

The nitration of 5*H*-benzo[*b*]carbazole in acetic acid affords a *dinitro*-compound, m.p. >310°, but the 5-acetyl derivative is mononitrated to give a mixture of two isomers probably 5-acetyl-6-nitro-5*H*-benzocarbazole and 5-acetyl-2(?)-nitro-5*H*-benzocarbazole. Deacetylation of the probable 6-nitro derivative, followed by reduction and oxidation yields 5*H*-benzo[*b*]carbazole-6,11-quinone. Bromo derivatives are obtained on treating the 5-acetyl compound with *N*-bromosuccinimide; in acetic acid a 5-acetyl-tribromobenzo[*b*]carbazole is obtained, whilst in carbon tetrachloride a monobromo-derivative is formed (*H. Hoellinger*, *Buu-Hoï* and *Mabille*, J. chem. Soc., 1968, 2209).

A number of dyes have been produced from benzocarbazole derivatives, for example fast azo dyes for cotton have been prepared by coupling tetrazotised diamines with compounds XLII and XLIII; a bluish black dye is obtained by tetrazotising *p*-phenylenediamine and coupling with XLII (*K. Harada*, *H. Mitoguchi* and *Y. Nakatsukasa*, C.A., 1967, **69,** 3693p).

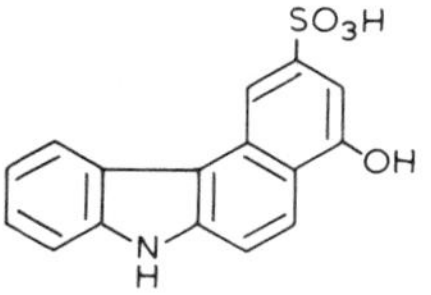

SO_3H OH N H (XLII)

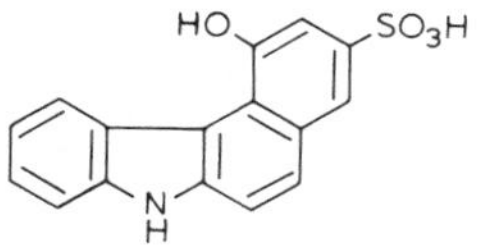

HO SO_3H N H (XLIII)

11*H*-**Benzo**[*a*]**carbazole,** m.p. 228°, *tetrachlorophthalic anhydride adduct*, m.p. 206–207°; 5*H*-**benzo**[*b*]**carbazole,** m.p. 337°; 7*H*-**benzo**[*c*]**carbazole,** m.p. 135°, *tetrachlorophthalic anhydride adduct*, m.p. 202°; 9*H*-**dibenzo**[*a,c*]**carbazole,** m.p. 188–189°; 13*H*-**dibenzo**[*a,g*]**carbazole,** m.p. 237–238°; 13*H*-**dibenzo**[*a,i*]**carbazole,** m.p. 221°; 7*H*-**dibenzo**[*c,g*]**carbazole,** m.p. 158°.

Some derivatives of benzo- and dibenzo-carbazoles are assembled in Table 10.

TABLE 10

DERIVATIVES OF BENZO- AND DIBENZO-CARBAZOLES

Substituent	*m.p. (°C)*	*Ref.*
6-Methyl-11*H*-[*a*]	187	1
picrate	185	
π-complex with tetrachlorophthalic anhydride	202	
6,8,9-Trimethyl-11*H*-[*a*]	195	1
picrate	205	
8-Methoxy-6-methyl-11*H*-[*a*]	174	1
picrate	177	
2-Isopropyl-11*H*-[*a*]	150	2
picrate	216	
8,9-Dimethyl-11*H*-[*a*]	295	2
picrate	213	
6-Methyl-13*H*-[*a,i*]	194	1
dipicrate	198	
13-Methyl-7*H*-[*a,g*]	163	1
picrate	217	
Dinitro-5*H*-[*b*]	>310	3
6-Nitro-5*H*-[*b*]	253	3
5-Acetyl-6-nitro-5*H*-[*b*]	248	3
6-Amino-5*H*-[*b*]	204	3
11-Benzyl-11*H*-[*a*]	147–148	4
11*H*-[*a*]-1-carboxylic acid	230–231	5

References
1 *N. P. Buu-Hoï, M. Mangane* and *P. Jacquignon*, J. chem. Soc., C, 1967, 662.
2 *Buu-Hoï et al., ibid.*, 1963, 2274.
3 *H. Hoellinger, Buu-Hoï* and *P. Mabille, ibid.*, 1968, 2209.
4 *G. Wittig* and *B. Reichel*, Ber., 1963, **96**, 2851.
5 *N. S. Dokunikhin* and *G. I. Bystritskii*, Zhur. obshcheĭ Khim., 1963, **33**, 974.

(ii) More complicated fused ring carbazoles

More complicated compounds containing the carbazole nucleus include 1,12-iminobenz[*a*]anthracene (XLIV), obtained by reducing 1-nitrobenz-[*a*]anthracene (*G. M. Badger* and *A. R. M. Gibb*, J. chem. Soc., 1949, 799); and the indolocarbazoles. Of the latter, the linear compound **indolo**[3,2-*b*]**carbazole** (XLV) is obtained along with other products when cyclohexane-1,4-dione bisphenylhydrazone is treated with sulphuric acid (*B. Robinson*, J. chem. Soc., 1963, 3097; *J. Harley-Mason* and *E. D. Pavri*, *ibid.*, 1963, 2504), **indolo**[3,2-*c*]**carbazole** (XLVI) is obtained by a double Borsche synthesis (p. 509) followed by dehydrogenation (*M. L. Tomlinson*, *ibid.*, 1951, 809) and **indolo**[1,2-*a*]**carbazole** (XLVII) from the phenylhydrazone of 1-oxotetrahydrocarbazole in a similar way (*B. D. Tilak et al.*, Chem. and Ind., 1957, 363):

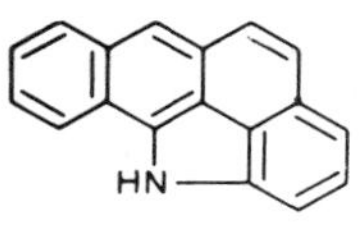

(XLIV) 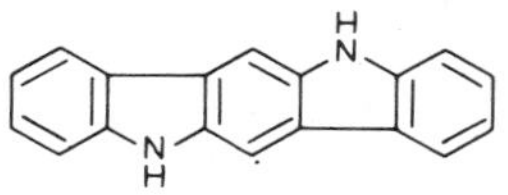 (XLV) 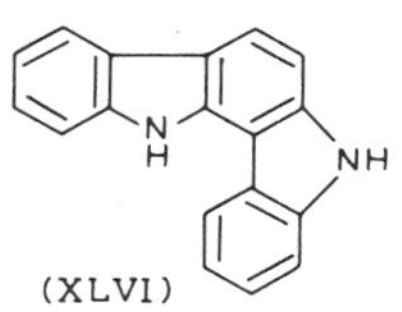 (XLVI)

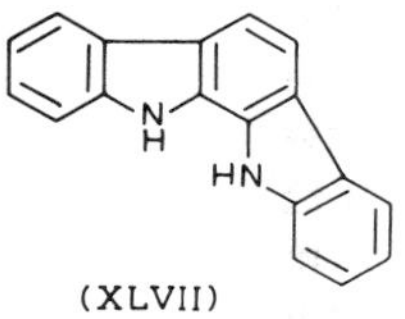

(XLVII) 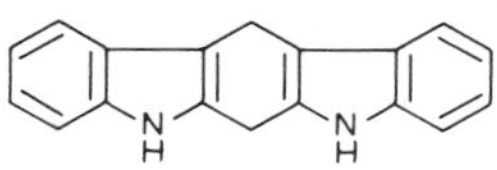 (XLVIII)

Cyclohexadienobis[1,2-*b*:5,4-*b'*]indole (XLVIII), formed by condensation of indole with formaldehyde (*H. von Dobeneck* and *I. Maas*, Ber., 1954, **87,** 455; *cf. M. L. Swindells* and *Tomlinson*, J. chem. Soc., 1956, 1135), is not strictly speaking a carbazole but it is structurally not far removed from the compounds under discussion.

Compound IL is formed when 1,2-phthaloylanthraquinone is warmed with benzylamine (*R. Scholl et al.*, Ber., 1936, **69,** 2428):

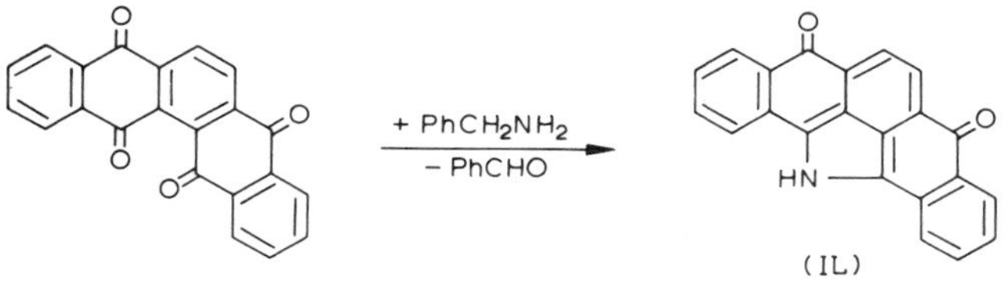

$+ PhCH_2NH_2$ / $- PhCHO$

(IL)

Phthaloylcarbazoles; "Anthraquinone-carbazoles". By the Friedel–Crafts reaction with phthalic anhydride, carbazole yields 3,6-bis(*o*-carboxybenzoyl)carbazole, which is cyclised to the very stable golden-yellow 2,3-6,7-**diphthaloylcarbazole** (L) (*Scholl* and *W. Neovius*, *ibid.*, 1911, **44,** 1249); this dyes cotton yellow from a dark brown dithionite vat:

(L)

The *N*-ethyl derivative is the vat dyestuff Hydron Yellow G (*Cassella and Co.*, G.P. 261,495/1910).

A series of phthaloylcarbazole vat colouring matters can be obtained by oxidative cyclisation of 1-arylaminoanthraquinones (*F. Bayer and Co.*, G.P. 288,824/1914; Chem. Ztbl., 1916, I, 85), including dianthraquinonylamines (*W. Bradley et al.*, J. chem. Soc., 1953, 1085; 1954, 2311; 1955, 3309; 1957, 819), *e.g.*:

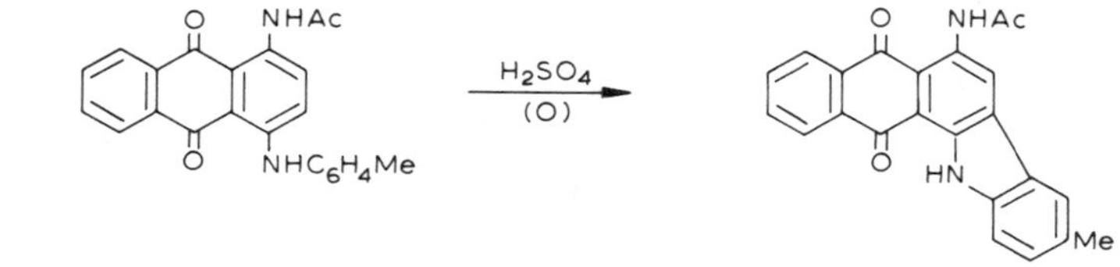

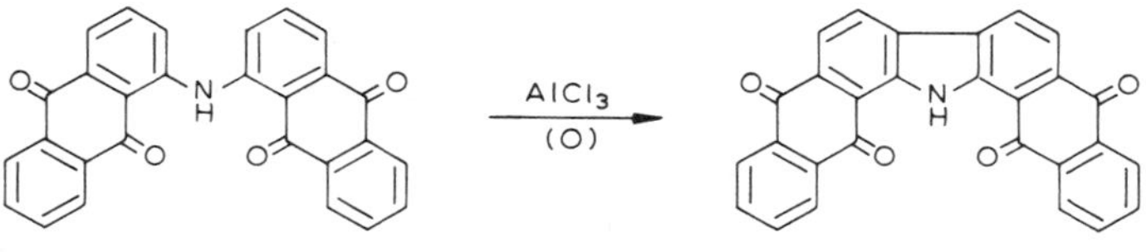

F. Ullman and *E. Illgen* (Ber., 1914, **47,** 380) prepared the simpler 1,2-**phthaloylcarbazole** (LI) by the Graebe–Ullman synthesis; the brick-red material dyes cotton orange from a red vat:

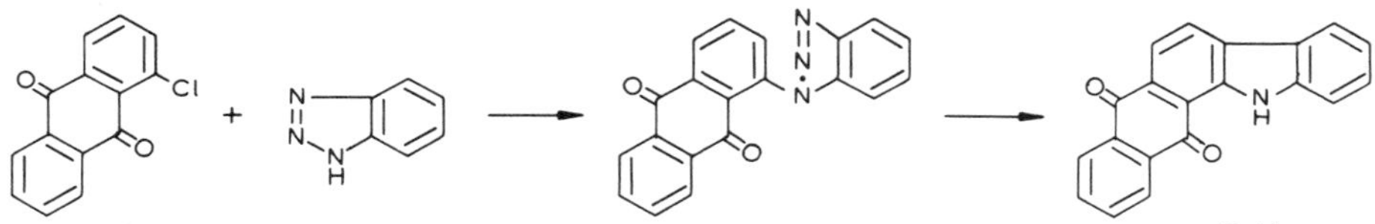

(LI)

Anthracene[2,3-*b*]**carbazole** (LII), m.p. 400°, and **anthracene**[2,3-*c*]**carbazole** (LIII), m.p. 286°, have been obtained from carbazole and 3-methyl-2-naphthoic acid (*M. Sander* and *W. H. Franke, ibid.*, 1964, **97,** 212):

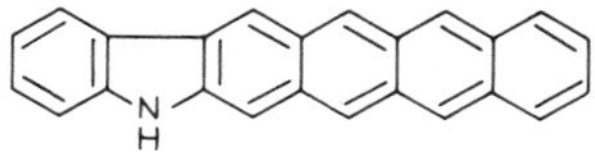

(LII) 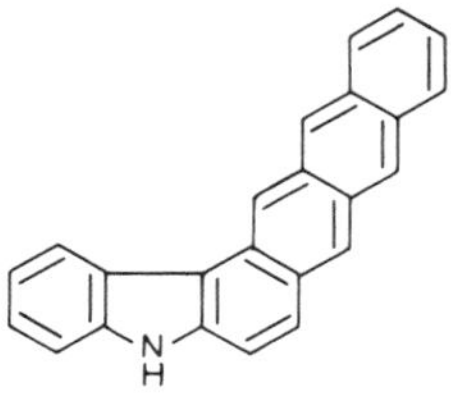 (LIII)

7. Other tricyclic pyrrole systems

The 2,3-**polymethyleneindoles** (I, $n = 1$ to 4; m.p. 108°, 119°, 144°, and 72–74°) are obtained by Borsche synthesis in the same way as tetrahydrocarbazole. Their autoxidation and the reaction of their *N*-acyl derivatives with nitric acid and halogens have been compared with the corresponding behaviour in the tetrahydrocarbazole series (*S. G. P. Plant et al.*, J. chem. Soc.,

1923, 123, 3242; 1928, 2586; 1929, 2934; 1933, 298; *B. Witkop et al.*, Experientia, 1950, **6,** 461).

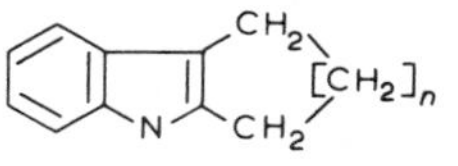

(I)

The compound (I, $n = 3$) has been prepared from cyclohexanone using pinacolic ring expansion (*V. F. Martynov*, C.A., 1955, **49,** 3124) as follows:

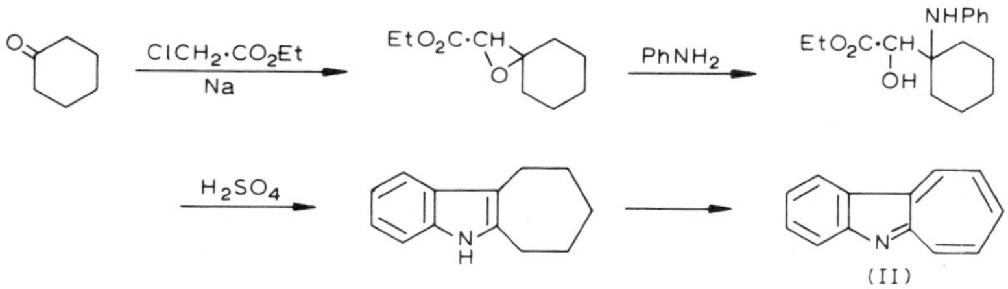

$ClCH_2 \cdot CO_2Et$ / Na; $EtO_2C \cdot CH$; $PhNH_2$; NHPh; $EtO_2C \cdot CH$; OH; H_2SO_4

(II)

It has been dehydrogenated to the permanganate-coloured **benzo[*b*]-1-aza-azulene** (1-**aza**-2,3-**benzazulene**) (II), m.p. 143° (*A. G. Anderson* and *J. Tazuma*, J. Amer. chem. Soc., 1952, **74,** 3455; *W. Treibs et al.*, Ann., 1953, **581,** 54). A convenient synthesis of benzo[*b*]-1-aza-azulenes (III) involves hydride abstraction from *o*-tropylanilines. On spectral evidence structure III is preferred to the alternative *o*-benzoquinone tropimidimine formulation (IV) (*K. Takase, T. Asao* and *N. Hirata*, Bull. chem. Soc., Japan, 1968, **41,** 3027).

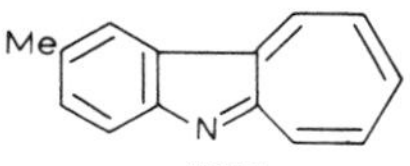

(III) 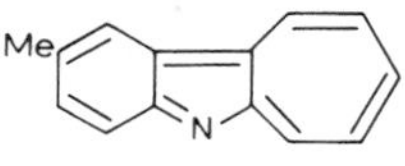 (IV)

Tricyclic pyrrole systems containing one carbon atom common to two rings have been obtained (see p. 400).

2-(2-Indolyl)ethyl *p*-toluenesulphonate reacts with ethyl cyanoacetate in the presence of sodium ethoxide to give the cyclohexanonoindole V (*T. Sakan et al.*, Tetrahedron Letters, 1968, 4925):

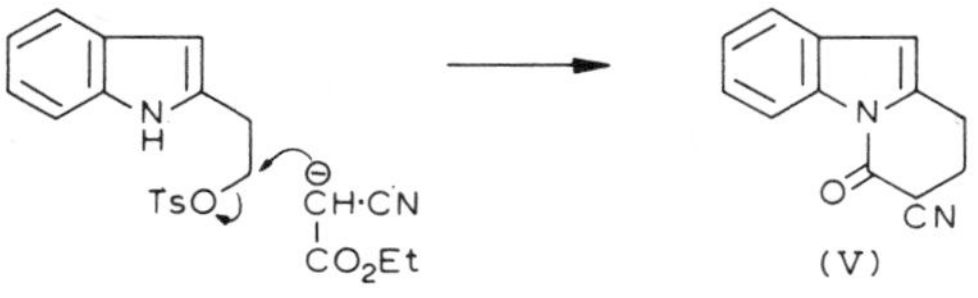

TsO; $\overset{\ominus}{C}H \cdot CN$; CO_2Et; CN

(V)

On the other hand, the reaction of 2-(5- or 7-methoxy-2-indolyl)ethyl *p*-sulphonatotoluene under the same conditions provides a ring enlargement reaction to give VI:

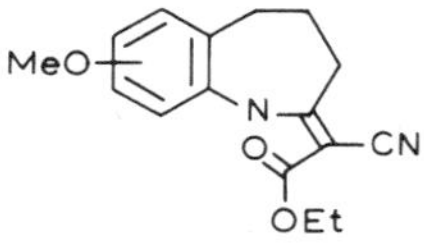

MeO, N, CN, O, OEt, (VI)

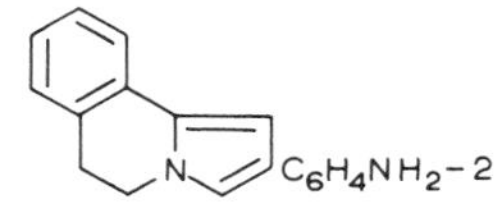

$C_6H_4NH_2$-2, N, (VII)

Heteroauxin phenethylamide on treatment with polyphosphoric acid is converted into 2-(2′-aminophenyl)benzo[*g*]-5,6-dihydropyrrocoline (VII), by opening of the indole ring (*J. Thesing* and *F. H. Funk*, Ber., 1958, **91,** 1546).

The symmetrical *ammonium salt* VIII, m.p. > 350°, is prepared by the routes shown (*F. Šorm* and *J. Beránek*, Coll. Czech. chem. Comm., 1954, **19,** 298):

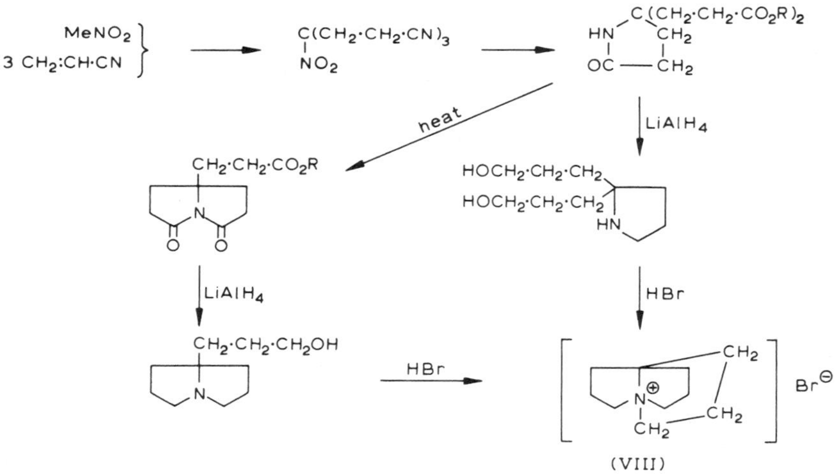

$MeNO_2$ + 3 CH_2:CH·CN → $C(CH_2·CH_2·CN)_3$ / NO_2 → HN–C($CH_2·CH_2·CO_2R)_2$–CH_2–CH_2–OC

heat; $CH_2·CH_2·CO_2R$; N; O; O

LiAlH$_4$; $CH_2·CH_2·CH_2OH$; N

LiAlH$_4$; $HOCH_2·CH_2·CH_2$; $HOCH_2·CH_2·CH_2$; HN

HBr; HBr; [N$^{⊕}$, CH_2, CH_2, CH_2] $Br^{⊖}$

(VIII)

Reduction of the corresponding phthalimide analogue with lithium tetrahydridoaluminate gives the base, 4,7-*endomethylene*-1,3,3a,4,7,7a-*hexahydro-2-methylisoindole* (IX), b.p. 70–74°/10 mm (*L. M. Rice et al.*, J. Amer. chem. Soc., 1955, **77,** 616):

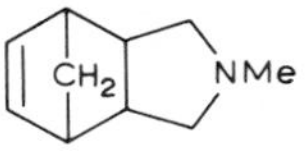

(IX)

2-Ethoxycarbonyl-4,5-dihydropyrrolo[3,2,1-*hi*]indole when heated with Pd/C in nitrobenzene gives 2-ethoxycarbonylpyrrolo[3,2,1-*hi*]indole, which on heating with cuprous oxide and quinoline yields **pyrrolo**[3,2,1-*hi*]**indole** (X), m.p. 33–36° (*W. W. Paudler* and *H. G. Shin*, J. heterocycl. Chem., 1969, **6,** 415):

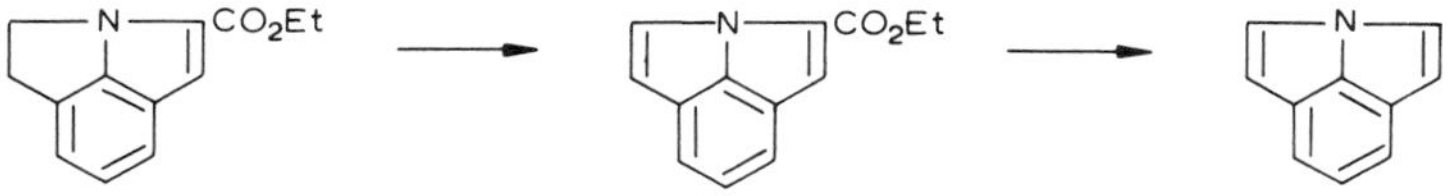

(X)

The antibacterial agent 7-**methoxymitrosene,** m.p. 206–207°, has been assigned structure XI (*G. R. Allen, J. F. Poletto* and *M. J. Weiss*, J. Amer. chem. Soc., 1964, **86,** 3877):

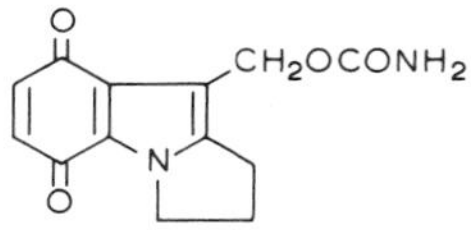

(XI)

Chapter 6

Other Five-Membered Ring Compounds with One Hetero Atom in the Ring from Groups, 3, 4 and 5

R. LIVINGSTONE

1. Phosphorus compounds

(a) Mononuclear compounds

(i) Phospholes

Pentaphenylphosphole (pentaphenylphosphacyclopentadiene) (I), m.p. 256–257°, is prepared from dilithiotetraphenylbutadiene and phenylphosphonous dichloride:

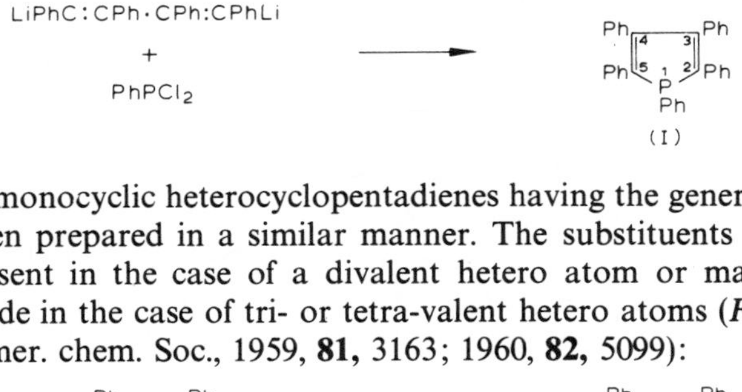
LiPhC:CPh·CPh:CPhLi
+
PhPCl2
Ph
Ph
Ph
Ph
P
Ph
4
3
5
1
2

(I)

A series of monocyclic heterocyclopentadienes having the general structure II have been prepared in a similar manner. The substituents R^1 and R^2 may be absent in the case of a divalent hetero atom or may be alkyl, aryl or halide in the case of tri- or tetra-valent hetero atoms (*F. C. Leavitt et al.*, J. Amer. chem. Soc., 1959, **81**, 3163; 1960, **82**, 5099):

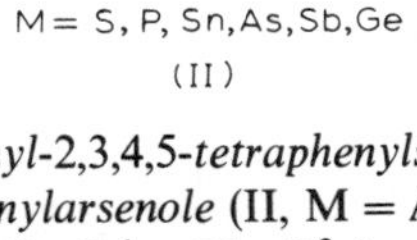
Ph
Ph
Ph
Ph
M
R1
R2
M = S, P, Sn, As, Sb, Ge

(II)

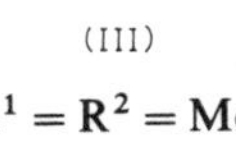
Ph
Ph
Ph
Ph
P
Ph
O

(III)

e.g. 1,1-*Dimethyl-2,3,4,5-tetraphenylstannole* (II, M = Sn, $R^1 = R^2$ = Me), m.p. 192–193°; *pentaphenylarsenole* (II, M = As, R^1 = Ph, R^2–), m.p. 215–216°; *pentaphenylstibiole* (II, M = Sb, R^1 = Ph, R^2–), m.p. 160°; *hexaphenylstannole* (II, M = Sn, $R^1 = R^2$ = Ph), m.p. 173–174°.

The reaction between dilithiotetraphenylbutadiene and phenyldichlorophosphine oxide affords **pentaphenylphosphole oxide** (III), m.p. 292–293°. Pentaphenylphosphole may be obtained also by reacting hexacarbonylbis-(diphenylacetylene)diiron [$Fe_2(CO)_6(PhC{:}CPh)_2$] with phenyldichlorophosphine at 140° (*E. H. Braye* and *W. Hübel*, Chem. and Ind., 1959, 1250), and by treating 1,4-diiodotetraphenylbutadiene with disodium phenylphosphide, $PhPNa_2$ (*Braye*, *Hübel* and *I. Caplier*, J. Amer. chem. Soc., 1961, **83**, 4406).

Pentaphenylphosphole is oxidised slowly in solution by the action of air to pentaphenylphosphole oxide. 1-Alkyl substituted phospholes take up oxygen quite readily; in the case of the reaction between dilithiotetraphenylbutadiene and methyldiiodophosphine only 1-*methyltetraphenylphosphole oxide*, m.p. 240–241°, has been isolated. Pentaphenylphosphole also reacts easily with sulphur and selenium yielding pentaphenylphosphole sulphide or selenide, respectively. This shows that the phosphorus atom in phospholes possesses a reactive lone electron pair. The pentaphenyl derivative also undergoes Diels–Alder reactions and reacts with iron carbonyls, thus resembling a great variety of non-aromatic conjugated dienes.

1,2,5-**Triphenylphosphole** (IV), m.p. 187–189°, is obtained by heating 1,4-diphenyl-1,3-butadiene and phenylphosphonous dichloride at 214–217° for ten hours. It forms an *oxide*, m.p. 237–239°, *sulphide*, m.p. 215–216.5°, and *selenide*, m.p. 205.5–206.5°, although the sulphide appears to dimerise, gives only the 1,1-dibromide, forms a methiodide, and reacts with methyl diazoacetate and with diazomethane, affording a variety of products depending on the conditions. It appears both from physical and chemical evidence that the heterocyclic ring in 1,2,5-triphenylphosphole has little or no aromatic character (*I. G. M. Campbell et al.*, J. chem. Soc., 1965, 2184):

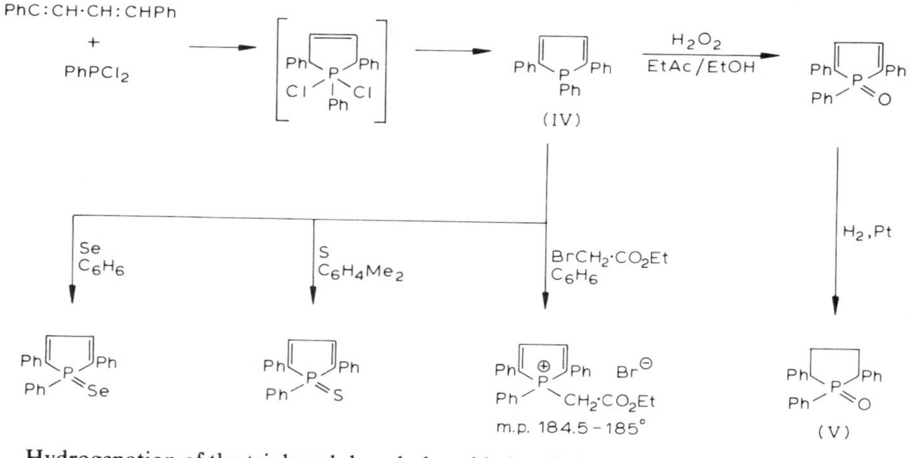

Hydrogenation of the triphenylphosphole oxide in ethyl acetate–acetic acid over platinum gives 1,2,5-**triphenylphosphacyclopentane** 1-**oxide** (V), m.p. 205–207°. The phosphole cannot be hydrogenated, presumably because the tervalent phosphorus poisons the catalyst.

1,2,5-Triphenylphosphole and dimethyl acetylenedicarboxylate give a stable yellow adduct, the tricycloallylidenephosphorane (VI) (*N. E. Waite* and *J. C. Tebby*, J. chem. Soc., C, 1970, 386):

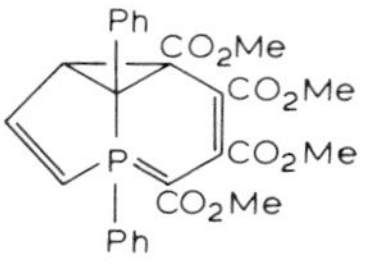

(VI)

When 1,2,3-triphenylphosphole and pentaphenylphosphole are treated with potassium in dimethoxyethane and in tetrahydrofuran they give three distinct interconvertible e.s.r. signals, which are associated with the phenyl group. The radicals obtained are not phosphole anions (*C. Thomson* and *D. Kilcast*, Angew. Chem., intern. Edn., 1970, **9,** 310).

1,3-Butadiynes on treatment with bis(hydroxymethyl)phenylphosphine, $PhP(CH_2OH)_2$, or with triphenylphosphine in the presence of phenyl-lithium give 1-**phenylphospholes** (VII); 2,4-hexadiyne with the latter two reagents affords 1-*phenyl-2,5-dimethylphosphole* (VII, R = Me), b.p. 66–69°/0.2 mm; 1-*phenyl-2,5-bis*(p-*tolyl*)*phosphole* (VII, $R = p\text{-}MeC_6H_4\text{-}$, from $p\text{-}MeC_6H_4C{:}C{\cdot}C{:}C{\cdot}C_6H_4\text{-}p$ Me), m.p. 194–196° (*G. Maerkl* and *R. Potthast*, *ibid.*, 1967, **6,** 86):

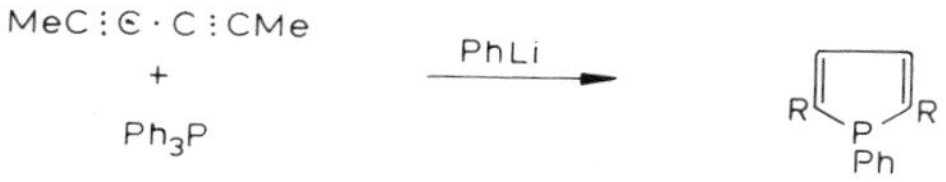

(VII)

The synthesis of the very reactive 1-**ethoxyphosphole** 1-**oxide** (VIII) has been reported. It dimerises too readily to allow its isolation, but it has been identified by its u.v. spectrum and by trapping it as a Diels–Alder adduct IX with cyclopentadiene (*D. A. Usher* and *F. H. Westerheimer*, J. Amer. chem. Soc., 1964, **86,** 4732):

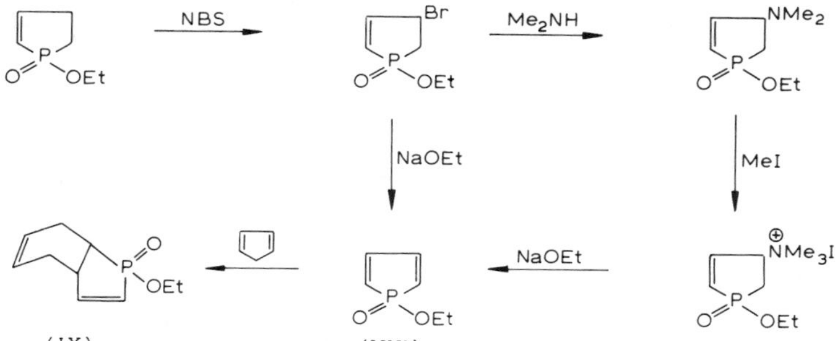

(IX) (VIII)

Dicyanoacetylene in acetonitrile reacts with triphenylphosphine to give **tetracyano-**1,1,1-**triphenylphosphole** (X), m.p. 237–239° (decomp.) (*G. S. Reddy* and *C. D. Weis*, J. org. Chem., 1963, **28,** 1822):

NC CN
NC CN
P
Ph_3
(X)

(ii) Phospholenes and phospholanes

The treatment of 1,3-dienes with phosphorus trichloride or tribromide in the presence of copper stearate or aromatic polynitro compounds as polymerisation inhibitors gives 1,1,1-trihalogenophospholenes (1,1,1-trihalogeno-1-phosphacyclopentenes) (*U. Hasserodt, K. Hunger* and *F. Korte*, Tetrahedron, 1963, **19,** 1563).

Phosphorus tribromide reacts more energetically than does the trichloride; the latter tends to yield much tar (*B. A. Arbuzov* and *A. O. Vizel*, Doklady Akad. Nauk, S.S.S.R., 1964, **158,** 1105). The position of the double bond as indicated by n.m.r. spectra shows that it depends not only on the method of synthesis but also on the nature of the diene and on the degree of symmetry of the arrangement of substituents on the ring (*Arbuzov et al., ibid.*, 1964, **159,** 1062). The spin–spin coupling constants of the olefinic protons with ^{31}P as well as the n.m.r. fine structure are characteristic of the position of the double bond in phospholenes (*H. Weitkamp* and *Korte*, Z. anal. Chem., 1964, **204,** 245).

The 1,4-cycloaddition of butadiene to phenyldibromophosphine gives 1-**phenyl**-3-**phospholene** 1-**oxide** (XI), which is reduced at room temperature with phenylsilane, $PhSiH_3$, to 1-**phenyl**-3-**phospholene** (XII) (*benzylphosphonium bromide*, m.p. 233–234°). Compound XI with bromine in chloroform yields 3,4-**dibromo**-1-**phenylphospholane** 1-**oxide** (XIII), m.p. 104–106°, which on treatment with potassium *tert*-butoxide in dimethylformamide gives Diels–Alder *adduct* XIV, m.p. 234–237°, rather than the expected 1-phenylphosphole 1-oxide. The dibromo derivative XIII on reduction with phenylsilane affords 3,4-*dibromo*-1-*phenylphospholane* (XV) (*benzylphosphonium bromide*, m.p. 156–157°), which on treatment with *tert*-butyl-lithium gives 1-phenyl-3-phospholene and with 1,5-diazobicyclo[5.4.0]undec-5-ene it yields 1-**phenylphosphole** (XVI), b.p. 64–65°/0.4 mm. The latter with hydrogen peroxide in acetone gives the tricyclo *oxide* XIV and with sulphur in boiling benzene the corresponding *sulphide* XVII, m.p. 183–184° (*Maerkl* and *Potthast*, Tetrahedron Letters, 1968, 1755):

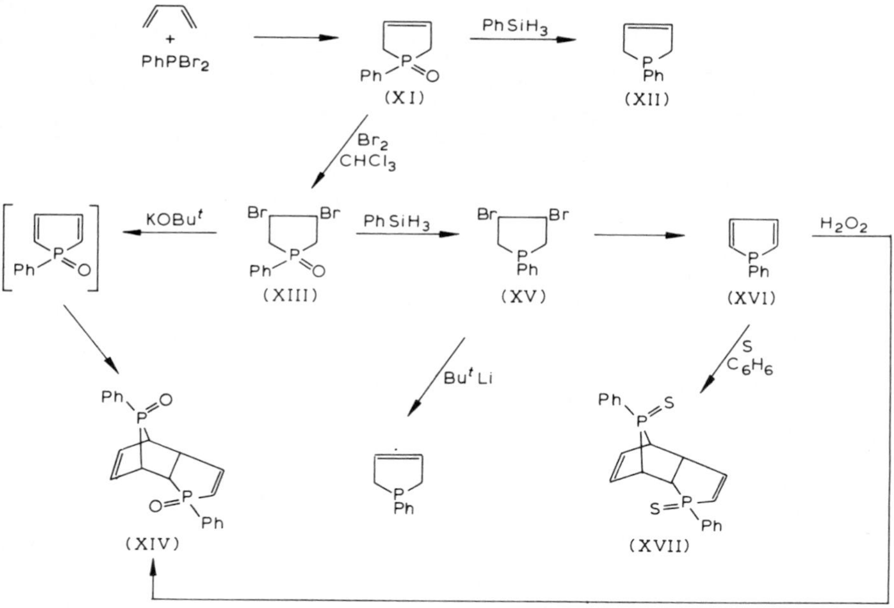

Heating butadiene with butyl dichlorophosphite, $BuOPCl_2$, in the presence of copper stearate in an autoclave at 120° gives 1-**chloro**-3-**phospholene** 1-**oxide** (XVIII), m.p. 53–54°. Similar reaction of butadiene, with a mixture of butyl dichlorophosphite and phosphorus trichloride yields a product, whose i.r. spectrum indicated the presence of 2- and 3-phospholene ring isomers (*Arbuzov et al.*, Izvest. Akad. Nauk, S.S.S.R., Ser. Khim., 1967, 672). Repeated fractional distillation of the mixture affords pure 1-**chloro**-2-**phospholene** 1-**oxide** (XIX), b.p. 84–86°/0.04 mm (*Arbuzov* and *Vizel*, *ibid.*, 1969, 460):

(XVIII) (XIX)

Butadiene and methyldichlorophosphine in cyclohexane containing copper stearate yield in three months 1,1-dichloro-1-methyl-3-phospholene, reduced with magnesium in tetrahydrofuran to give 1-**methyl**-3-**phospholene,** b.p. 114–115° (*benzylphosphonium bromide*, m.p. 114–115°) (*L. D. Quin et al.*, Tetrahedron Letters, 1964, 3689). 1,2-*Dimethyl*-3-*phospholene*, b.p. 130–134° (*Quin*, *J. P. Gratz* and *R. E. Montgomery*, *ibid.*, 1965, 2187). 1,1,1-**Tribromo**-3-**phospholene,** m.p. 125–127°; 1,1,1-*tribromo*-3-*methyl*-3-*phospholene*, m.p. 114–117°; 1,1,1-*tribromo*-3,4-*dimethyl*-3-*phospholene*, m.p. 151–153° (*Arbuzov* and *Vizel*, Doklady Akad. Nauk, S.S.S.R., 1964, **158,** 1105).

1-Alkoxy-2-phospholene 1-oxides (XX) are obtained from alcohols and 1,1,1-trichloro-3-phospholenes, and 1-alkoxy-3-phospholene 1-oxides (XXI) from alcohols and

1,1,1-tribromo-3-phospholenes (*Hasserodt, Hunger* and *Korte*, Tetrahedron, 1964, **20,** 1593):

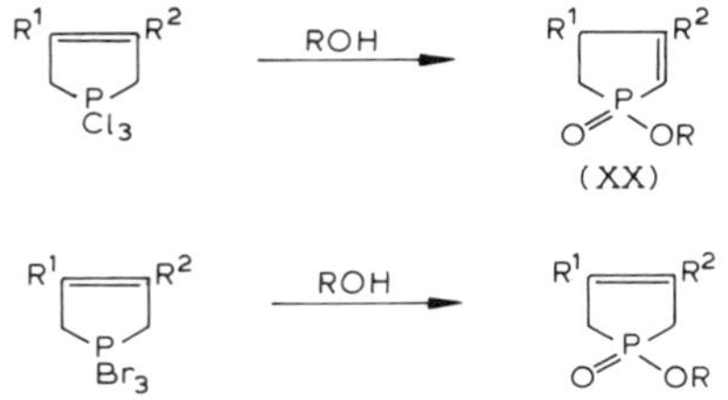

1,1,1-Trichloro-3-phospholenes on treatment with sulphur dioxide give a mixture of 1-chlorophospholene 1-oxides. Pure isomers are prepared by reacting the corresponding alkoxy compounds with carbonyl chloride in carbon tetrachloride at 0°. Since the 1,1,1-trichloro-3-phospholenes contain a labile double bond, subject to prototropic isomerisation, 1,1,1-trichloro-3,4-dimethyl-3-phospholene on hydrolysis yields 1-*hydroxy*-3,4-*dimethyl*-2-*phospholene* 1-*oxide*, m.p. 65–70°, and on treatment with two moles of acetic anhydride it affords acetyl chloride and 1-*chloro*-3,4-*dimethyl*-2-*phospholene* 1-*oxide*, b.p. 94–96°/0.05 mm, m.p. 78–82°. 1,1,1-Tribromo-3-phospholenes not being prone to isomerisation on hydrolysis, 1,1,1-tribromo-3,4-dimethyl-3-phospholene gives 1-*hydroxy*-3,4-*dimethyl*-3-*phospholene* 1-*oxide*, m.p. 122–123°, and affords acetyl bromide and 1-*bromo*-3,4-*dimethyl*-3-*phospholene* 1-*oxide*, m.p. 85–87°. 2,3-Dimethyl-1,3-butadiene reacts with ethyl dichlorophosphite to form 1-chloro-3,4-dimethyl-3-phospholene 1-oxide with the loss of ethyl chloride (*Arbuzov et al.*, Doklady Akad. Nauk, S.S.S.R., 1964, **159,** 582):

Some phospholene 1-oxides are listed in Table 1.

3,4-**Dimethyl**-2-**phospholene,** b.p. 146–148° is obtained by heating 1-hydroxy-3,4-dimethyl-2-phospholene 1-oxide and diphenylsilane at 150–190°; 3-**methyl**-2-**phospholene,** b.p. 80°; 2-**phospholene,** b.p. 75°, is prepared from 1-hydroxy-2-phospholene 1-oxide and phenylsilane (*H. Fritsch et al.*, Ber., 1965, **98,** 1681).

The Arbuzov rearrangement reaction of diene hydrocarbons and ethyl dichlorophosphinyl sulphide results in the formation of 1-chlorophospholene 1-sulphides:

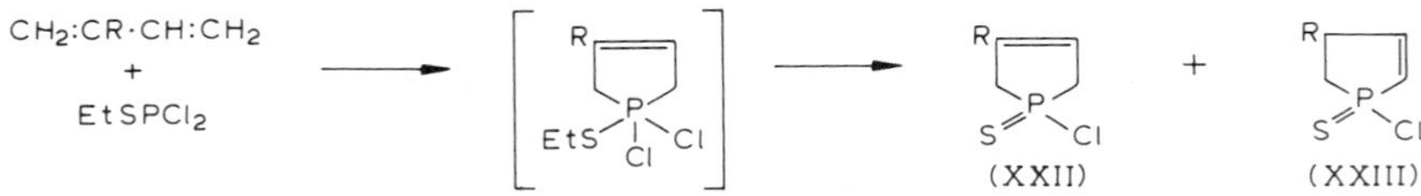

TABLE 1

PHOSPHOLENE 1-OXIDES

Substituents				*m.p. (°C)*	*Ref.*
1	*2*	*3*	*4*		
2-Phospholene 1-oxides					
Cl				b.p. 83–85/0.25 mm	1
OH		Me		116–117	2
3-Phospholene 1-oxides					
Cl				53	1
Cl		Me	Me	74 (83–86)	1
Br				46–48	3
Br		Me		b.p. 123–125°/0.02 mm	3
MeO				b.p. 55–60°/0.06 mm	1
EtO				b.p. 64–66°/0.2 mm	1
MeO		Me		b.p. 117–119°/10 mm	2
EtO		Me		b.p. 118–119°/9 mm	2
PrO		Me		b.p. 123–124°/8 mm	2
MeO		Me	Me	b.p. 86–90°/0.09 mm	1
EtO		Me	Me	42–43	3
PrO		Me	Me	b.p. 135–136°/9 mm	3
PhO				60–62	4
PhO	Me			b.p. 162–164°/3 mm	4
PhO		Me		56–58	4
Ph				75–77	4

References
1 *U. Hasserodt, K. Hunger* and *F. Korte*, Tetrahedron, 1964, **20,** 1593.
2 *A. O. Vizel et al.*, Doklady Akad. Nauk, S.S.S.R., 1965, **160,** 826.
3 *B. A. Arbuzov* and *Vizel*, *ibid.*, 1964., **158,** 1105.
4 *Arbuzov* and *L. A. Shapshinskaya*, Izvest. Akad. Nauk, S.S.S.R., Otdel. khim. Nauk, 1962, 65; C.A., 1962, **57,** 13791b.

Reaction with isoprene gives only 1-**chloro-4-methyl-2-phospholene** 1-**sulphide** (XXIII, R = Me), b.p. 78–80°/0.5 mm, m.p. 43°; butadiene yields a mixture of 1-*chloro*-3-*phospholene* 1-*sulphide* (XXII, R = H), b.p. 51–53°/0.5 mm, and 1-chloro-2-phospholene (XXIII, R = H). Hydrolysis gives the 1-hydroxy analogues, but the chlorine atom cannot be replaced by a thiol grouping by heating with RSH. Treatment with sodium ethanethiolate affords 1-*ethylthio*-3-*phospholene* 1-*sulphide*, b.p. 98–101°/0.5 mm, m.p. 32° (*N. A. Razumova, L. S. Kovalev* and *L. I. Zubtsova*, Zhur. obshcheĭ Khim., 1967, **37,** 1919).

The reaction of pyrocatechol phosphorochloridite with a diene in a closed vessel during 3–4 days gives a crystalline adduct, identified as 1-halogeno-1,1-(*o*-phenylenedioxy)-3-phospholene (XXIV). The normally expected Arbuzov rearrangement is prevented by the strength of the bond between O and the aromatic ring. Similar adducts have been obtained using pyrocatechol phosphorobromidite and phosphorofluoridite (*Razumova et al.*, *ibid.*, 1963, **33**, 3858; 1969, **39**, 176). The melting points of some adducts are given in Table 2.

(XXIV)

TABLE 2
1-HALOGENO-1,1-(*o*-PHENYLENEDIOXY)-3-PHOSPHOLENES (XXIV)

	Substituents				*Substituents*			
	R	*R'*	*X*	*m.p. (°C)*	*R*	*R'*	*X*	*m.p. (°C)*
(a)	H	H	Cl	86–88	Me	H	Br	62–65
(b)	Me	H	Cl	98–100	Me	Me	Br	63–65
	H	Me	Cl	96–98	H	H	F	82–83
	H	H	Br	67–68	Me	H	F	88–89

Hydrolysis of (a) at room temperature gives XXIV (R = R' = H; X = OH), m.p. 125–126°, and heating with ethanol affords XXIV (R = R' = H; X = OEt), m.p. 129–130°; (b) on boiling with water gives pyrocatechol.

The volatile liquid 1-dimethyliminophospholane–boron trihydride complex (XXV) on heating forms **phospholane (phosphacyclopentane)** (XXVI), m.p. −88°, b.p. est. 105° (*A. R. Burg* and *P. J. Slota*, J. Amer. chem. Soc., 1960, **82**, 2148):

$$Me_2NPC_4H_8{\cdot}BH_3 \longrightarrow$$

(XXV) (XXVI)

Tetrafluoroethene, iodine, and phosphorus on heating several hours at 150–250° under pressure give **octafluoro-1-iodophospholane** (*C. G. Krespan*, U.S.P. 2,931,803/1960).

(*b*) *Polynuclear, fused ring compounds*

(*i*) *Dibenzophospholes*

5-**Phenyldibenzophosphole** 5-**oxide,** *9-phenyl-9-phosphafluorene 9-oxide* (XXVII), m.p. 167–168°, has been obtained by cyclisation of 2-biphenylylphenylphosphinic acid, and reduction of the oxide yields 5-**phenyldibenzophosphole** (XXVIII), m.p. 90–92° (*Campbell* and *J. K. Way*, J. chem. Soc., 1961, 2133):

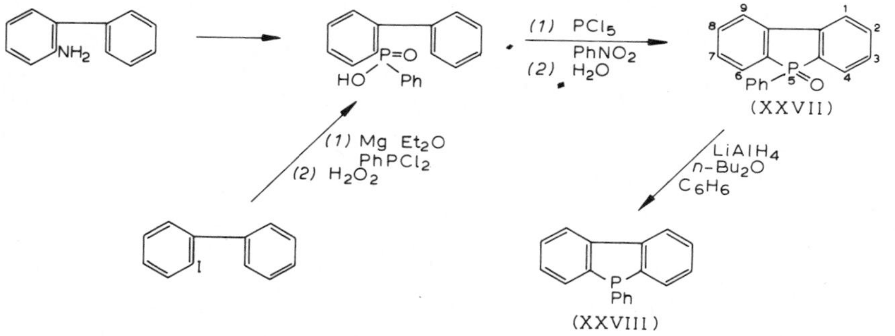

5-Phenyldibenzophosphole-3-carboxylic acid has been prepared and resolved.

5-Phenyldecahydrodibenzophosphole 5-oxide, *9-phenyldecahydro-9-phosphafluorene 9-oxide* (XXIX), b.p. 200–220°/0.14 mm, is prepared from 1,1′-bicyclohexenyl and phenyldichlorophosphine (*Campbell et al., ibid.*, 1965, 2186):

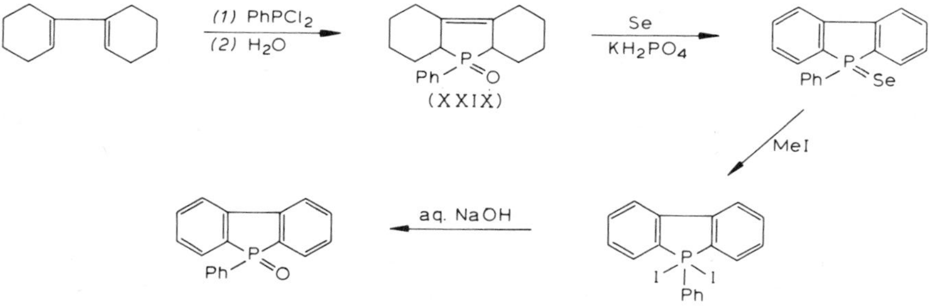

Dehydrogenation of compound XXIX using selenium affords *5-phenyldibenzophosphole 5-selenide*, m.p. 162–164°, which may be converted to 5-phenyldibenzophosphole 5-oxide.

2. Silicon compounds

(a) Mononuclear compounds

(i) Silacyclopentadienes; siloles

Silacyclopentadiene (I), b.p. 60–62°, n_D^{25} 1.4265, is obtained by reacting 1,1-*dichlorosilacyclopentadiene* (II), b.p. 128–130°, with lithium tetrahydridoaluminate in dry "diglyme" at 0° (*R. A. Benkeser, R. F. Grossman* and *G. M. Stanton*, J. Amer. chem. Soc., 1962, **84,** 4723):

Treatment of 1,1-dichlorosilacyclopentane (III) with two equivalents of sulphuryl chloride and catalytic amounts of benzoyl peroxide yields a monochlorination product IV and a dichlorination product V. Thermal dehydrochlorination of the latter at 550°, under nitrogen, affords 1,1-dichlorosilacyclopentadiene (II). A vapour phase chromatogram of IV indicates that it is a mixture of both the 2- and 3-chloro derivatives in an approximate ratio of 1:9; dehydrochlorination of mixture IV with quinoline affords a mixture of 1,1-dichlorosila-2-cyclopentene (VI) and 1,1-dichlorosila-3-cyclopentene (VII):

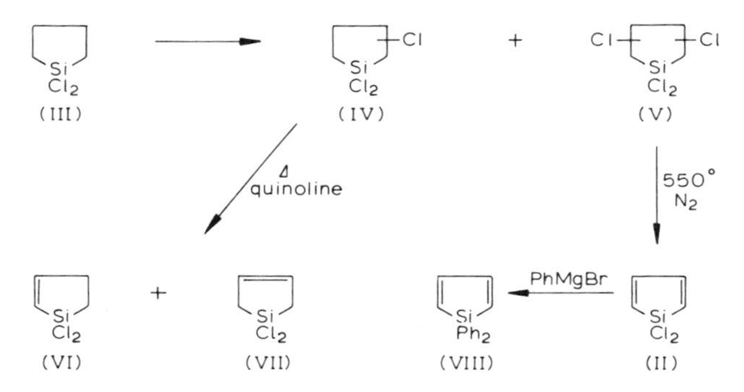

The far-i.r. spectra of sila-2-cyclopentene, sila-3-cyclopentene, and sila-3-cyclopentene-1,1-d_2 have been recorded (*J. Laane*, J. chem. Phys., 1969, **50,** 776; 1970, **52,** 358).

1,1-Dichlorosilacyclopentadiene reacts with phenylmagnesium bromide to form 1,1-diphenylsilacyclopentadiene (VIII), and silacyclopentadiene (I) with hydrogen over platinum gives **silacyclopentane** *(cyclotetramethylenesilane)* (IX), b.p. 70–72°, also obtained by the reduction of 1,1-dichlorosilacyclopentane with lithium tetrahydridoaluminate:

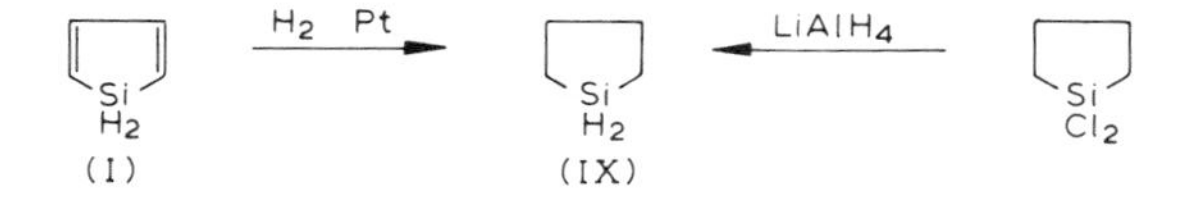

Hexaphenylsilacyclopentadiene (hexaphenylsilole) (X), m.p. 190–191°, is prepared from 1,4-dilithio-1,2,3,4-tetraphenylbutadiene and diphenyldichlorosilane (*E. H. Braye, W. Hübel* and *I. Caplier*, J. Amer. chem. Soc., 1961, **83,** 4406):

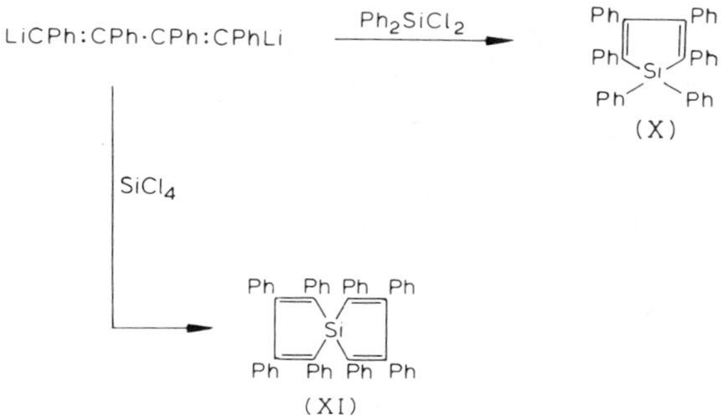

A very small amount of **octaphenyl-**1,1′-**spirobisilole** (XI), m.p. 288–295° (decomp.), is obtained from 1,4-dilithio-1,2,3,4-tetraphenylbutadiene and silicon tetrachloride. The dilithio compound on treatment with trichlorosilane yields 1-chloro-2,3,4,5-tetraphenylsilacyclopentadiene, which on reduction with lithium tetrahydridoaluminate affords 2,3,4,5-tetraphenylsilacyclopentadiene. 1,2,5-Triphenyl-, 1-chloro-2,5-diphenyl, and 2,5-diphenyl-silacyclopentadiene may be obtained in a similar way (*V. Hagen* and *K. Ruehlmann*, Z. Chem., 1968, **8,** 262).

1,1-Dimethyl-2,5-diphenylsilacyclopentane prepared from styrene, dichlorodimethylsilane and lithium is converted to 1,1-dimethyl-2,5-diphenylsilacyclopentadiene by the following route; some of the reactions of the silacyclopentadiene are also indicated (*W. H. Atwell* and *D. R. Weyenberg*, J. org. Chem., 1967, **32,** 885):

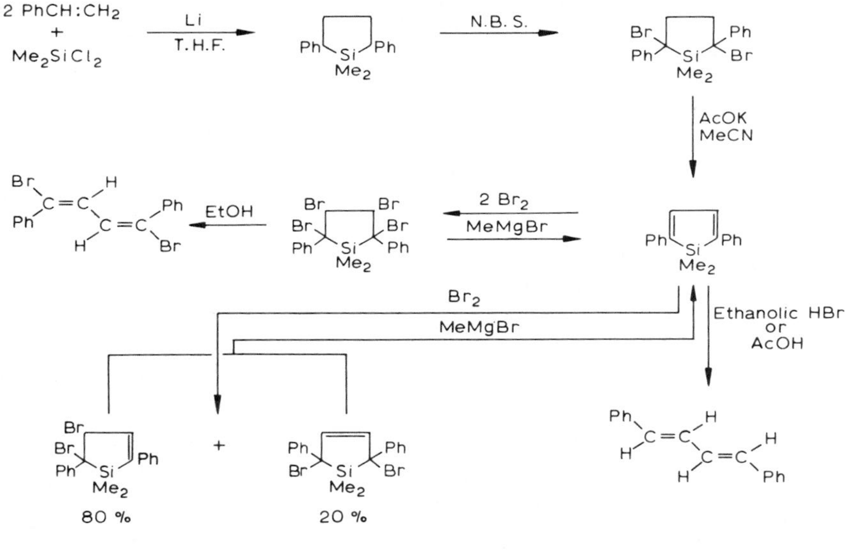

The reduction of 1,1-dimethyl-2,5-diphenylsilacyclopentadiene by sodium or potassium to the *radical anion* XII has been detected by e.s.r. spectra, which also show absorption due to the *dianion* XIII. Solution of the dianion mixed with the parent organosilane produces the radical anion (*E. G. Janzen, J. B. Pickett* and *Atwell*, J. organometal. Chem., 1967, **10,** 6). Similar observations are made with 1,1-dimethyl-2,3,4,5-tetraphenylsilacyclopentadiene:

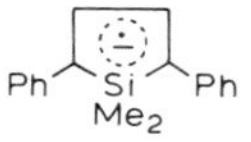

Decaphenyl-1,1-**bis**(1-**silacyclopentadiene**), m.p. 237–238°, is obtained by reacting sodium pentaphenylsilacyclopentadienide with dimethyl sulphate or 1-chloropentaphenylsilacyclopentadiene or by treating the latter compound with magnesium. It does not form radicals on irradiation or on boiling in xylene in the presence of oxygen (*Hagen* and *Ruehlmann*, Z. Chem., 1968, **8,** 114).

(ii) Silacyclopentenes

2,3-Dimethylbutadiene reacts with dimethylsilylene generated from either the pyrolysis of *sym*-dimethoxytetramethyldisilane (XIV) or the 7-silanorbornadiene (XV) to give 1,1,3,4-**tetramethylsila**-3-**cyclopentene** (XVI) (*Atwell* and *Weyenberg*, J. Amer. chem. Soc., 1968, **90,** 3438), b.p. 72°/68 mm (XIV and dimethylbutadiene heated at 400°):

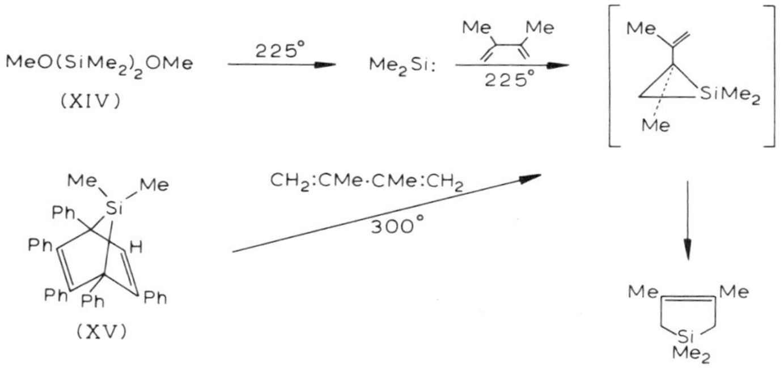

1,1,4-**Trimethylsila**-3-**cyclopentene,** b.p. 123–124°; 1-*methoxy*-1,3,4-*trimethylsila*-3-*cyclopentene*, b.p. 158–159°. A mixture of 1-chloro-1-methylsila-3-cyclopentene and 1-chloro-1-methylsila-2-cyclopentene is obtained by heating trichlorotrimethyldisilane, $Me_3Si_2Cl_3$, or a mixture of this and tetrachlorodimethyldisilane, $Me_2Si_2Cl_4$, with butadiene at 550°. 1,1-Dichlorosila-3-cyclopentene is prepared by heating a mixture of hexachlorodisilane and butadiene at 500°. Trifluorotrimethyldisilane with butadiene at 550° yields a mixture of 1-fluoro-1-methylsila-3-cyclopentene and 1-fluoro-1-

methylsila-2-cyclopentene (*Atwell*, Ger. Offen. 1,921,833/1969; C.A., 1970, **72,** 31611r).

1,3,3,5-Tetramethyl-6,6-dichloro-3-silabicyclo[3.1.0]hexane (XVII), formed by allowing 1,1,3,4-tetramethylsila-3-cyclopentene to react with phenyl(bromodichloromethyl)mercury in benzene at 80°, is unstable at room temperature (*D. Seyferth et al.*, J. Amer. chem. Soc., 1970, **92,** 657):

(XVII)

(iii) Silacyclopentanes

Silacyclopentane has already been mentioned (p. 540).

1,1-**Dichlorosilacyclopentane (cyclotetramethylenedichlorosilane)** (III, p. 540), b.p. 143°, is prepared by treating silicon tetrachloride with the Grignard reagent obtained from 1,4-dibromobutane (*R. West*, *ibid.*, 1954, **76,** 6012).

1,1-**Dialkylsilacyclopentan**-3-**ols** (XX and XXI) are synthesised by converting the sila-3-cyclopentene (XVIII, R = an alkyl group, R^1 and R^2 = H and alkyl groups) into the epoxide **XIX**, which is then treated with lithium tetrahydridoaluminate (*G. Manuel*, *P. Mazerolles* and *J. C. Florence*, Compt. rend., Ser. C, 1969, **269,** 1553; C.A., 1970, **72,** 79151t):

p-$NO_2C_6H_4{\cdot}CO_3H$ and

(XVIII) (XIX) (XX) (XXI)

Compound XVII may be reduced with tetrahydridoborate and the products treated with hydrogen peroxide in sodium hydroxide solution to give alcohols XX and XXI.

(b) Fused ring compounds

The reaction of diphenylacetylene with *n*-butyl-lithium yields a dilithium intermediate XXII, which on subsequent reaction with diphenyldichlorosilane in boiling tetrahydrofuran affords 3-*n*-**butyl**-1,1,2-**triphenyl**-1-**silaindene** (XXIII), m.p. 84–84.5° (*M. D. Rausch* and *L. P. Klemann*, J. Amer. chem. Soc., 1967, **89,** 5732):

PhC⋮CPh + 2 n-BuLi → (XXII) $\xrightarrow{Ph_2SiCl_2}$ (XXIII)

(XXII) (XXIII)

1,1-**Diphenyl**-1-**silaindan,** 2,3-**benzo**-1,1-**diphenylsilacyclopentane,** m.p. 62–63°, is prepared by converting 1-chloro-2-(2-bromoethyl)-benzene to the corresponding Grignard reagent, reacting it with chlorodiphenylsilane to obtain compound XXIV, and adding a toluene solution of XXIV to molten sodium in hot toluene (*H. Gilman* and *O. L. Marrs*, Chem. and Ind., 1961, 208):

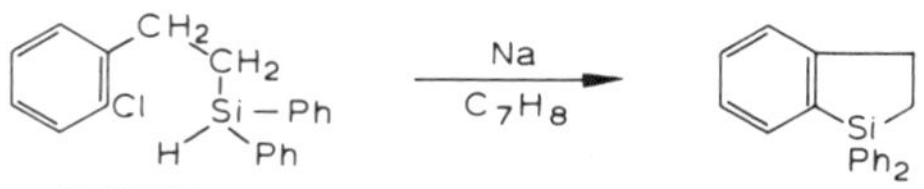

$\xrightarrow[C_7H_8]{Na}$

(XXIV)

1,1-Dimethyl-1-silaindan, 1,1-dimethyl-2,3-benzosilacyclopentane (*C. Eaborn, D. R. M. Walton* and *M. Chan*, J. organometal. Chem., 1967, **9,** 251).

3,4-**Benzo**-1,1-**dichlorosilacyclopentane** (XXV), b.p. 82–84°/4 mm, n_D^{20} 1.5505, d^{20} 1.2517, is prepared by boiling benzylchloromethyldichlorosilane with aluminium chloride in benzene. XXV on treatment with lithium tetrahydridoaluminate gives 3,4-**benzosilacyclopentane,** b.p. 190.5–192.5°, n_D^{20} 1.5590, d^{20} 0.9865; with methylmagnesium bromide it affords 3,4-*benzo*-1,1-*dimethylsilacyclopentane*, b.p. 209–211°; and on heating with acetic anhydride it yields 1,1-*diacetoxy*-3,4-*benzosilacyclopentane*, b.p. 143–145°/5 mm, m.p. 38°. 1,1-*Dibutyl*-3,4-*benzosilacyclopentane*, b.p. 159–160°/5 mm (*N. S. Nametkin et al.*, Doklady Akad. Nauk, S.S.S.R., 1964, **154,** 383):

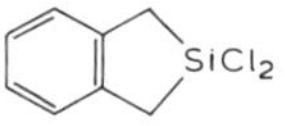

(XXV)

The i.r. and u.v. spectra of 3,4-benzosilacyclopentane and its derivatives have been recorded (*V. D. Oppengeim* and *E. Sh. Finkel'shtein*, Tr. Komis. po Spektroskopii, Akad. Nauk, S.S.S.R., 1964, **3,** 99; C.A., 1966, **64,** 10597g).

2,2′-Dilithiobiphenyl reacts with appropriately disubstituted dichlorosilanes to yield **silafluorenes, dibenzosiloles** (XXVI):

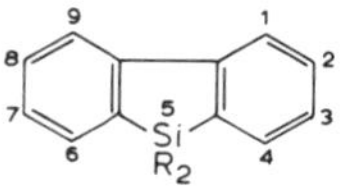

(XXVI)

Methyl-, *n*-dodecyl-, and phenyl-trichlorosilane all react with the dilithio compound to give excellent yields of 5-chloro-5-methyl-, 5-chloro-5-*n*-dodecyl-, and 5-chloro-5-phenyl-dibenzosilole, respectively, in addition to small amounts of 5,5′-spirobi-(dibenzosilole) (XXVII) (*Gilman* and *R. D. Gorsich*, J. Amer. chem. Soc., 1958, **80,** 1883):

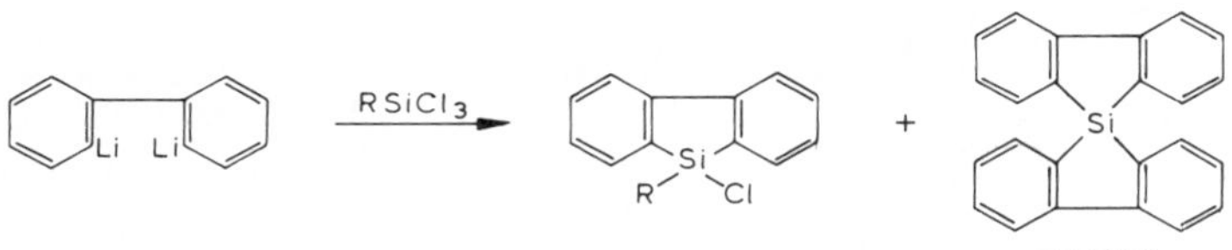

RSiCl3 → (XXVII)

5-Chloro-5-methyldibenzosilole (**XXVIII**) when treated with an equimolar quantity of 2,2′-dilithiobiphenyl in ether yields almost equimolar amounts of **XXVII** and 5,5-dimethyldibenzosilole (*idem, ibid.*, p. 3243):

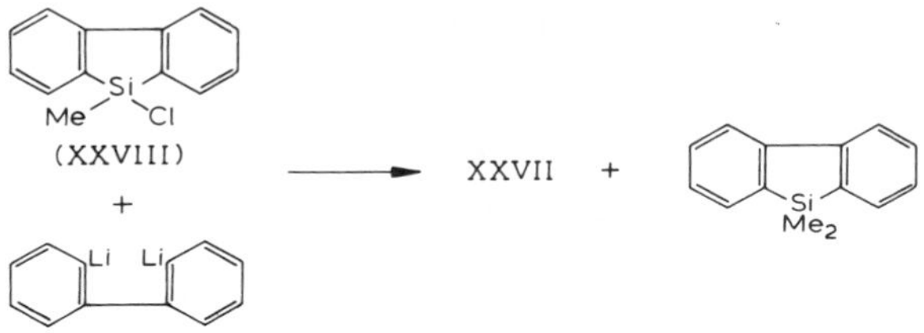

(XXVIII) + → XXVII +

3. Germanium compounds

(a) Five-membered mononuclear compounds

Germacyclopentane (I), b.p. 91–92°, n_D^{20} 1.4838, d^{20} 1.2261, is prepared by treating dichlorogermacyclopentane (II) with lithium tetrahydridoaluminate. The reaction also yielded the spiro compound 5-**germaspiro**[4.4]**nonane** (III), b.p. 188–189°, n_D^{20} 1.5118, d^{20} 1.1837. The dichloro derivative II is obtained by reacting the Grignard reagent from 1,4-dibromobutane with germanium tetrachloride (*P. Mazerolles*, Bull. Soc. chim. Fr., 1962, 1907):

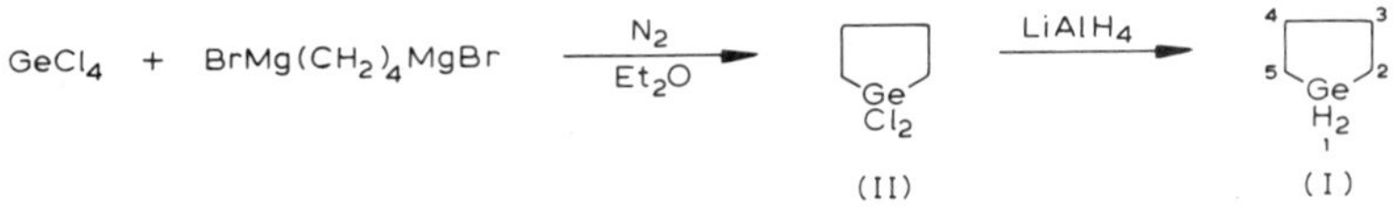

$$GeCl_4 + BrMg(CH_2)_4MgBr \xrightarrow[Et_2O]{N_2} \text{(II)} \xrightarrow{LiAlH_4} \text{(I)}$$

(II) (I)

(III)

Germacyclopentane on shaking with oxygen is rapidly converted into the **germacyclopentane 1-oxide,** m.p. 166°, also obtained by shaking an ethereal solution of dibromogermacyclopentane with the theoretical amount of aqueous sodium hydroxide. Heating germacyclopentane with *sec*-butyl iodide in the dark under nitrogen affords **diiodogermacyclopentane,** b.p. 107°/5 mm, n_D^{20} 1.6770, d^{20} 2.500. *Dibromogermacyclopentane,* b.p. 97°/15 mm, n_D^{20} 1.5720, may be obtained in a similar manner. Boiling diiodogermacyclopentane and silver chloride in heptane affords *dichlorogermacyclopentane,* b.p. 94°/64 mm, n_D^{20} 1.5101. Germacyclopentane when cooled to 0° under nitrogen and treated with iodine yields *iodogermacyclopentane*, b.p. 107°/48 mm, n_D^{20} 1.5956. The reaction between dibromogermacyclopentane and ethylmagnesium bromide gives 1,1-**diethylgermacyclopentane,** b.p. 173–174°/750 mm, n_D^{20} 1.4725; 1,1-**diphenylgermacyclopentane,** m.p. 33°, b.p. 115°/0.2 mm, n_D^{20} 1.5971; 1-**ethyl**-1-**phenylgermacyclopentane,** b.p. 140°/20 mm, n_D^{20} 1.5441. Treatment of the latter compound with bromine results in the cleavage of the Ph–Ge bond to give 1-bromo-1-ethylgermacyclopentane. It does not react with iodine in the cold, 1-ethyl-1-iodogermacyclopentane may be obtained by reacting the bromo derivative with sodium hydroxide followed by hydriodic acid (*Mazerolles* and *J. Dubac*, Compt. rend., 1963, **257,** 1103).

1,1-Diphenylgermacyclopentane reacts with bromine in ethyl bromide to give 1-*bromo*-1-*phenylgermacyclopentane*, b.p. 116°/0.9 mm, n_D^{20} 1.5878, which on boiling with lithium tetrahydridoaluminate is converted to 1-**phenylgermacyclopentane** (IV), b.p. 127°/30 mm, n_D^{20} 1.5601 *(Mazerolles, loc. cit.):*

Ge Ph2 —(Br2, EtBr)→ Ph–Ge–Br —($LiAlH_4$, Et_2O)→ Ph–Ge–H (IV)

1-*Butylgermacyclopentane*, b.p. 179°/745 mm, n_D^{20} 1.4756, is prepared similarly.

The germacyclopentane ring is stable to the action of acids and bases, except that sulphuric acid opens the ring. Strong organic acids cleave the allyl–Ge bond; dichloroacetic acid when mixed with 1-allyl-phenylgermacyclopentane produces 1-*dichloroacetoxy*-1-*phenylgermacyclopentane*, b.p. 123°/0.28 mm, n_D^{20} 1.5505.

Dichlorocarbene insertion into germanium–carbon bond of 1,1-diethylgermacyclobutane results in ring expansion and the formation of 2,2-*dichloro*-1,1-*diethylgermacyclopentane*, n_D^{25} 1.5061 (*D. Seyferth et al.*, J. organometal. Chem., 1969, **16,** 503):

$GeEt_2$ (cyclobutane) + $PhHgCCl_2Br$ → Cl_2 / Ge Et_2 (cyclopentane) + PhHgBr

The reaction of germanium diodide with conjugated olefins leads to 1,1-diiodogerma-3-cyclopentenes, (V), which are moderately stable to substitution, alkylation and reduction, but not to hydrolysis or halogenation. The compounds react with freshly prepared silver chloride in boiling heptane to give 1,1-dichlorogerma-3-cyclopentenes (VI) (*Mazerolles* and *G. Manuel*, Bull. Soc. chim. Fr., 1966, 327):

Me, Me diene + GeI_2 → Me–Ge(I_2)–Me (V) —(AgCl, C_7H_{16})→ Me–Ge(Cl_2)–Me (VI)

(V) —$LiAlH_4$→ Me–Ge(H_2)–Me

(V) —MeMgI→ Me–Ge(Me_2)–Me

Mixing butadiene and germanium hydrogen trichloride at $-60°$, followed by warming to room temperature and heating at 60° gives germanium tetrachloride and a mixture of 1,1-dichlorogerma-3-cyclopentene (70%) and 2-butenylgermanium trichloride, $Cl_3GeCH_2CH{:}CHMe$. Treatment of the mixture with methylmagnesium chloride produces 1,1-dimethylgerma-3-cyclopentene and 2-butenyl-trimethylgermanium, $Me_3GeCH_2CH{:}CHMe$. The above reaction probably occurs as a result of an equilibrium existing between $HGeCl_3$ and $GeCl_2 + HCl$ (*V. F. Mironov* and *T. K. Gar*, Izvest. Akad. Nauk, S.S.S.R., Ser. Khim., 1966, 482; *O. M. Nefedov et al.*, Doklady Akad. Nauk, S.S.S.R., 1964, **154,** 1389). Germanium hydrogen trichloride and 1,4-dioxane react with elimination of hydrogen chloride and formation of a complex, $C_4H_8O_2{\cdot}GeCl_2$, which reacts with butadiene to yield 1,1-dichlorogerma-3-cyclopentene and $(CH_2CH{:}CHCH_2GeCl_2)_n$ (*S. P. Kolesnikov, V. I. Shiryaev* and *Nefedov*, Izvest. Akad. Nauk, S.S.S.R., Ser. Khim., 1966, 584).

Similarly germanium hydrogen tribromide and butadiene yield two products, but germanium dibromide and butadiene in acetone give 1,1-dibromogerma-3-cyclopentene and a powdery polymer (*Gar* and *Mironov*, *ibid.*, 1965, 755, 855).

The Raman and i.r. spectra of 1,1-dichloro- and 1,1-dimethylgerma-3-cyclopentene (*L. A. Leites, Gar* and *Mironov*, Doklady Akad. Nauk, S.S.S.R., 1964, **158,** 400), the m.s. of some substituted germacyclopentanes and pentenes (*A. M. Duffield et al.*, J. organometal. Chem., 1968, **12,** 123), and the p.m.r. spectra of some germacyclopentanes (*A. S. Khachaturov* and *Nefedov*, Khim. Geterotsikl. Soedin., 1969, 927) are reported.

Germa-3-cyclopentenes (VII, R = alkyl; R^1 and R^2 = H or alkyl) on treatment with *p*-nitroperbenzoic acid give an epoxide, which with lithium tetrahydridoaluminate yields the germacyclopentanols (VIII and IX), also obtained on reacting the products from the hydroboration of VII, with hydrogen peroxide in sodium hydroxide solution (*Manuel, Mazerolles* and *J. C. Florence*, Compt. rend., Ser. C, 1969, **269,** 1553):

(VII) → → (VIII) and (IX)

1,1-Dialkylgermacyclopentan-3-ols (3-hydroxygermacyclopentanes) are ring-opened on treatment with phosphoryl chloride or bromide, *p*-toluenesulphonyl chloride (in the presence of sodium ethoxide), and acids, *e.g.*, hydrogen halides and acetic acid to give 3-alkenylgermanium compounds, of the type $Et_2GeXCH_2{\cdot}CHRCR'{:}CH_2$ (X = Cl, Br, OEt, OAc and O_2CCHCl_2; R and R′ = H or Me) (*Mazerolles* and *Manuel*, Compt. rend., Ser. C, 1968, **267,** 1158).

Some germa-3-cyclopentenes are listed in Table 3.

(b) Polycyclic compounds with germanium in a five-membered ring

1,1-Diethylgerma-3-cyclopentene with hexachlorocyclopentadiene gives the *endo*

TABLE 3
GERMA-3-CYCLOPENTENES

Substituents	*b.p. (°C/mm)*	n_D^{20}	*Ref.*
3-Me	117	1.5212	1
1,1-DiMe	121	1.4723	2
3,4-DiMe	142	1.502	1
1,1-DiCl, 3-Me	98/34	1.5128	1
1,1-DiBr, 3-Me	125/25	1.5825	1
1,1-DiI, 3-Me		1.651	1
1,1-DiEt, 3-Me	89/24	1.4805	1
1,1-DiCl, 3,4-DiMe	120/26	1.5178	1
1,1-DiI, 3,4-DiMe	m.p. 32–34°		1
1,1,3,4-TetraMe	71/27	1.4799	1
1,1-DiEt, 3,4-DiMe	102/20	1.4855	1
1,1-DiBu, 3,4-DiMe		1.4820	1
1,1-DiPh, 3,4-DiMe	m.p. 40–41°		1

References:
1 *P. Mazerolles* and *G. Manuel*, Bull. Soc. chim. Fr., 1966, 327.
2 *V. F. Mironov* and *T. K. Gar*, Izvest. Akad. Nauk, S.S.S.R., Ser. Khim., 1966, 482.

adduct X, and with ethyl diazoacetate it yields ethyl 3,3-diethyl-3-germabicyclo[3.1.0]-hexane-6-carboxylate (XI) (*Mazerolles, Manuel* and *F. Thoumas*, Compt. rend., Ser. C, 1968, **267,** 619):

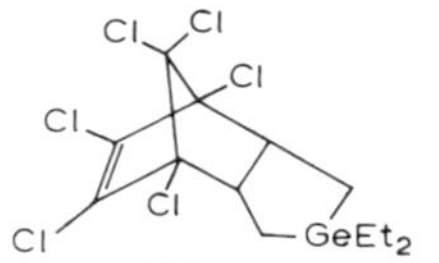

(X) 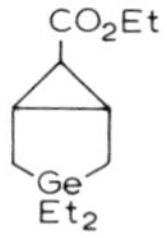 (XI)

The reaction of dichlorocarbene with 1,1-diethylgerma-3-cyclopentenes gives the expected 3,3-diethyl-6,6-dichloro-3-germabicyclo[3.1.0]hexanes (XII), but these are not very stable thermally. Only 3,3-diethyl-6,6-dichloro-3-germabicyclo[3.1.0]hexane itself is isolated as a pure substance. Methyl substituents at $C_{(3)}$ and $C_{(4)}$ of the germacyclopentene ring tend to decrease the stability of the dichlorocarbene adduct. Decomposition of these dichloro compounds gives either a 1,1-diethyl-4-chloro-1-germacyclohexa-2,4-diene (XIII) or a diethylchlorogermyl-substituted pentadiene (*e.g.*, $Et_2ClGeCH_2CH{:}CCl{\cdot}CH{:}CH_2$). The 3,3-diethyl-6,6-difluoro-3-germacyclo[3.1.0]-hexane structure is much more stable than the analogous chloro system (*Seyferth et al.*, J. Amer. chem. Soc., 1970, **92,** 657):

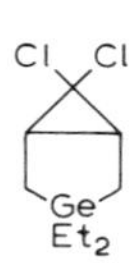

(XII) 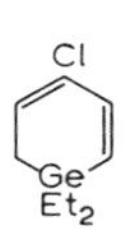 (XIII)

5,5-**Diphenyldibenzogermole** (XIV), m.p. 152–153°, is prepared by heating a mixture of diphenylgermanium dichloride and 2,2′-dilithiobiphenyl in ether (*I. M. Gverdtsiteli et al.*, Soobshch. Akad. Nauk Gruz. S.S.R., 40, 1965, 333; C.A., 1966, **64,** 11239a):

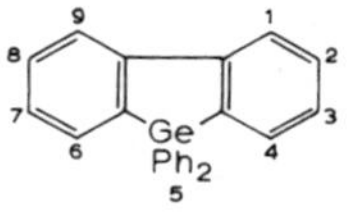

(XIV)

4. Compounds containing aluminium, boron, antimony and tin as the hetero atom

The interaction between triphenylaluminium and diphenylacetylene at 200° leads to the formation of benzene and 1,2,3-**triphenylbenzaluminole** (I), m.p. 285–288°. The reaction proceeds more smoothly in phenyl ether solution (*J. J. Eisch* and *W. C. Kaska*, J. Amer. chem. Soc., 1962, **84,** 1501):

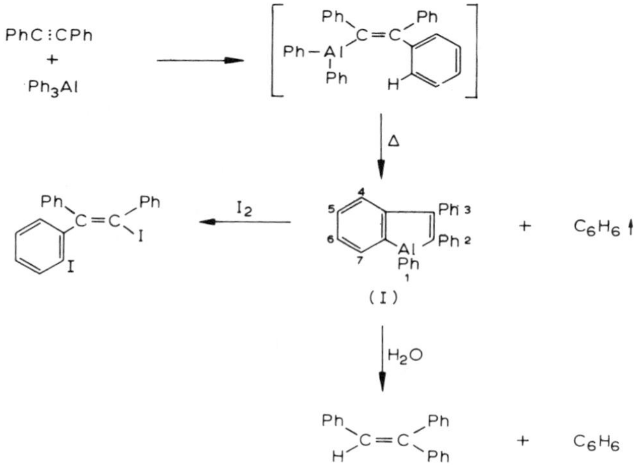

o-Biphenylyldiphenylaluminium prepared from *o*-biphenylyl-lithium and diphenylaluminium chloride, on heating gives 5-**phenyldibenzoaluminole** (II), m.p. 225–230°:

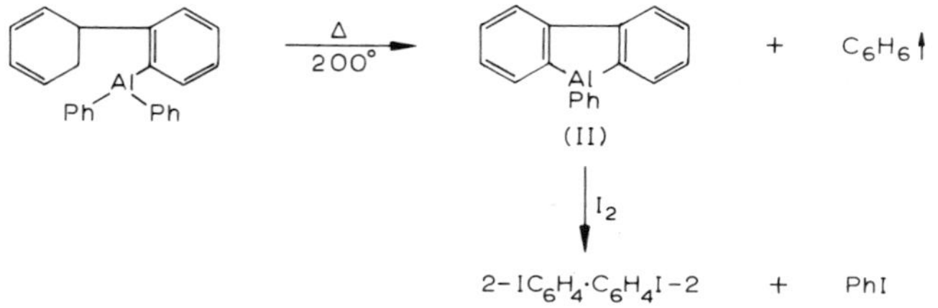

2-Biphenylyldialkylboranes split out alkane at 180–200° to yield 5-alkyldibenzoboroles (III) (9-alkylborafluorenes):

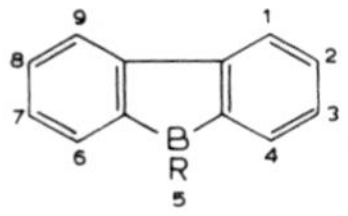

(III)

5-**Ethyldibenzoborole,** b.p. 112°/0.3 mm, m.p. 16°; 5-*phenyldibenzoborole*, m.p. 118° (*R. Koester* and *G. Benedikt*, Angew. Chem., 1963, **75,** 419); 5-**phenyldibenzobismole,** m.p. 167–168°; 5-**phenyldibenzostibole** (crude), m.p. 101–102° (*G. Wittig* and *D. Hellwinkel*, Ber., 1964, **97,** 789); 5-**methyldibenzostannole,** m.p. 123–124° (*F. Johnson*, U.S.P. 3,234,239/1966; C.A., 1966, **64,** 11251c).

Appendix

IUPAC Commission on the Nomenclature of Organic Chemistry

B. Fundamental heterocyclic systems

Rule B–1. Extension of Hantzsch–Widman system

1.1—Monocyclic compounds containing one or more hetero-atoms in a three- to ten-membered ring are named by combining the appropriate prefix or prefixes from Table I (eliding "a" where necessary) with a stem from Table II. The state of hydrogenation is indicated either in the stem, as shown in Table II, or by the prefixes "dihydro-", "tetrahydro-", *etc.*, according to Rule **B–1.2.**

TABLE I

Element	*Valence*	*Prefix*	*Element*	*Valence*	*Prefix*
Oxygen	II	Oxa	Antimony	III	Stiba[a]
Sulfur	II	Thia	Bismuth	III	Bisma
Selenium	II	Selena	Silicon	IV	Sila
Tellurium	II	Tellura	Germanium	IV	Germa
Nitrogen	III	Aza	Tin	IV	Stanna
Phosphorus	III	Phospha[a]	Lead	IV	Plumba
Arsenic	III	Arsa[a]	Mercury	II	Mercura

[a] When immediately followed by "-in" or "-ine", "phospha-" should be replaced by "phosphor-", "arsa-" should be replaced by "arsen-" and "stiba-" should be replaced by "antimon-".

TABLE II

Number of members in the ring	*Rings containing nitrogen*		*Rings containing no nitrogen*	
	Unsaturation[a]	*Saturation*	*Unsaturation*[a]	*Saturation*
3	-irine	-iridine	-irene	-irane[e]
4	-ete	-etidine	-ete	-etane
5	-ole	-olidine	-ole	-olane
6	-ine[b]	c	-in[b]	-ane[d]
7	-epine	c	-epin	-epane
8	-ocine	c	-ocin	-ocane
9	-onine	c	-onin	-onane
10	-ecine	c	-ecin	-ecane

Footnotes to Table II

[a] Corresponding to the maximum number of non-cumulative double bonds, the hetero elements having the normal valences shown in Table I.

[b] For phosphorus, arsenic, antimony, see the special provisions of Table I.

[c] Expressed by prefixing "perhydro" to the name of the corresponding unsaturated compound.

[d] Not applicable to silicon, germanium, tin and lead. In this case, "perhydro-" is prefixed to the name of the corresponding unsaturated compound.

[e] The syllables denoting the size of rings containing 3, 4 or 7–10 members are derived as follows: "ir" from t*ri*, "et" from t*et*ra, "ep" from h*ep*ta, "oc" from *oc*ta, "on" from n*on*a, and "ec" from d*ec*a.

Examples:

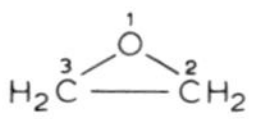

Oxirane

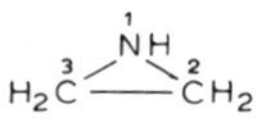

Aziridine

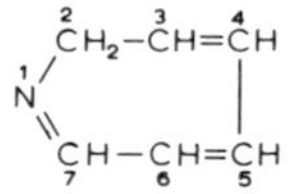

2*H*-Azepine

1.2—Heterocyclic systems whose unsaturation is less than the one corresponding to the maximum number of non-cumulative double bonds are named by using the prefixes "dihydro-", "tetrahydro-", *etc.*

In the case of 4- and 5-membered rings, a special termination is used for the structures containing one double bond, when there can be more than one non-cumulative double bond.

Number of members of the partly saturated rings	*Rings containing nitrogen*	*Rings containing no nitrogen*
4	-etine	-etene
5	-oline	-olene

Examples:

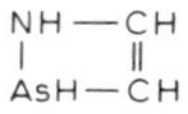

Azarsetine

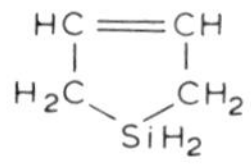

Silolene

1.3—Multiplicity of the same hetero atom is indicated by a prefix "di-", "tri-", *etc.*, placed before the appropriate "a" term (Table I).

Example:

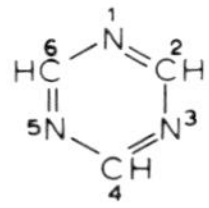

1,3,5-Triazine

1.4—If two or more kinds of "a" terms occur in the same name, their order of citation is by descending group number of the Periodic Table and increasing atomic number in the group as illustrated by the sequence in Table I.

Examples:

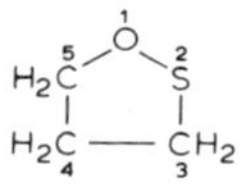

1,2-Oxathiolane

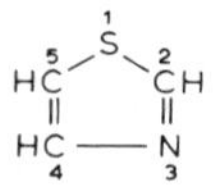

1,3-Thiazole

1.51—The position of a single hetero atom determines the numbering in a monocyclic compound.

Example:

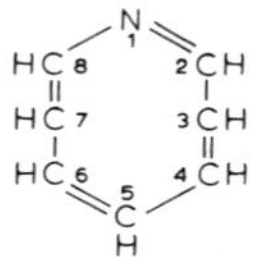

Azocine

1.52—When the same hetero atom occurs more than once in a ring, the numbering is chosen to give the lowest locants to the hetero atoms.

Example:

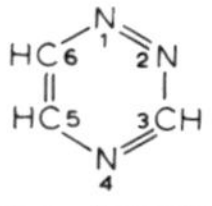

1,2,4-Triazine

1.53—When hetero-atoms of different kinds are present, the locant 1 is given to a hetero atom which is as high as possible in Table I. The numbering is then chosen to give the lowest locants to the hetero atoms.

Examples:

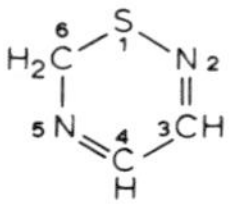

6*H*-1,2,5-Thiadiazine
(not: 2,1,4-Thiadiazine)
(not: 1,3,6-Thiadiazine)

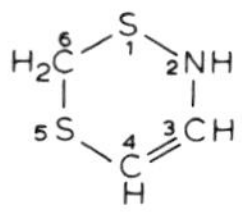

2*H*,6*H*-1,5,2-Dithiazine
(not: 1,3,4-Dithiazine)
(not: 1,3,6-Dithiazine)
(not: 1,5,4-Dithiazine)

The numbering must begin with the sulfur atom. This condition eliminates 2,1,4-thiadiazine. Then the nitrogen atoms receive the lowest possible locant, which eliminates 1,3,6-thiadiazine.

The numbering has to begin with a sulfur atom. The choice of this atom is determined by the set of locants which can be attributed to the remaining hetero atoms of any kind.
As the set 1,2,5 is lower than 1,3,4 or 1,3,6 or 1,5,4 in the usual sense, the name is 1,5,2-dithiazine.

Rule B–2. Trivial and semi-trivial names

2.11—The following trivial and semi-trivial names constitute a partial list of such names which are retained for the compound and as a basis of fusion names. The names of the radicals shown are formed according to Rule **B–5.**

	Parent compound	Radical name
(1)	Thiophene	Thienyl (2- shown)
(2)	Benzo [*b*] thiophene (replacing thianaphthene)	Benzo [*b*] thienyl (2- shown)
(3)	Naphtho [2,3-*b*] thiophene (replacing thiophanthrene)	Naphtho [2,3-*b*] thienyl (2- shown)
(4)	Thianthrene	Thianthrenyl (2- shown)
(5)	Furan	Furyl (3- shown)
(6)	Pyran (2*H*- shown)	Pyranyl (2*H*-Pyran-3-yl shown)
(7)	Isobenzofuran	Isobenzofuranyl (1- shown)
(8)	Chromene (2*H*- shown)	Chromenyl (2*H*-Chromen-3-yl shown)

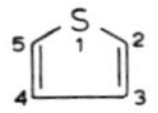

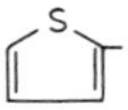

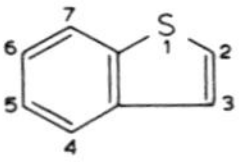

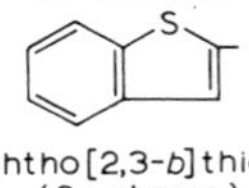

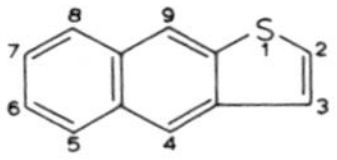

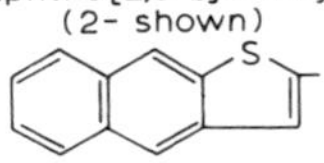

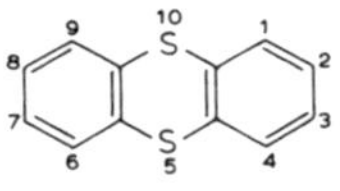

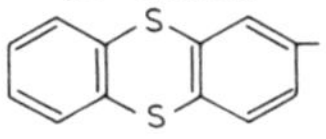

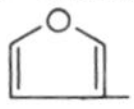

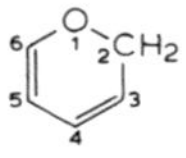

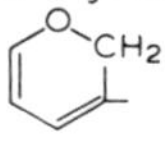

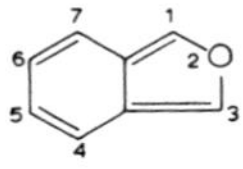

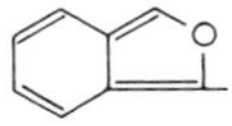

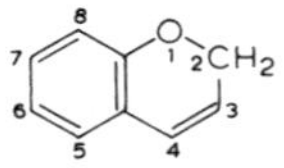

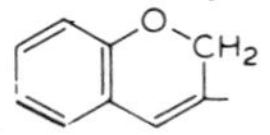

	Parent compound	Radical name
(9)	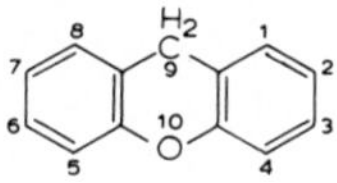 Xanthene*	Xanthenyl* (2- shown) 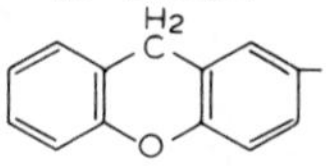
(10)	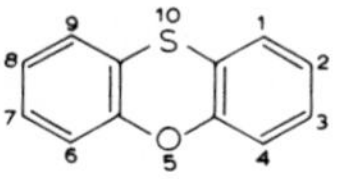 Phenoxathiin	Phenoxathiinyl (2- shown) 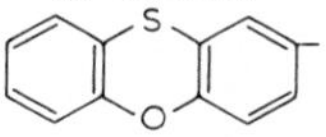
(11)	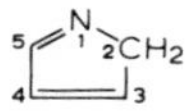 2*H*-Pyrrole	2*H*-Pyrrolyl (2*H*-Pyrrol-3-yl shown) 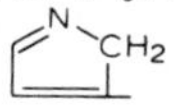
(12)	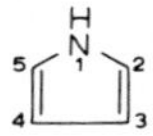 Pyrrole	Pyrrolyl (3- shown) 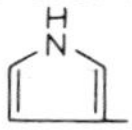
(13)	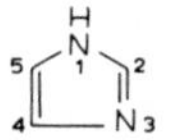 Imidazole	Imidazolyl (2- shown) 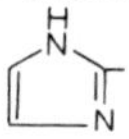
(14)	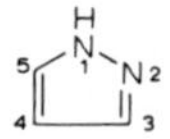 Pyrazole	Pyrazolyl (1- shown) 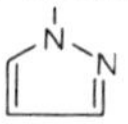
(15)	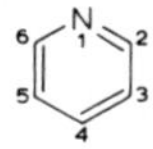 Pyridine	Pyridyl (3- shown) 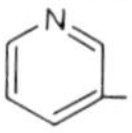
(16)	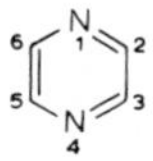 Pyrazine	Pyrazinyl 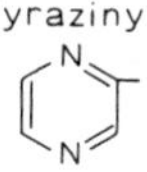

* Denotes exceptions to systematic numbering.

	Parent compound	Radical name
(17)	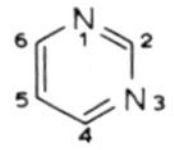 Pyrimidine	Pyrimidinyl (2- shown) 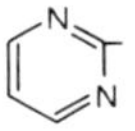
(18)	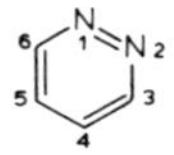 Pyridazine	Pyridazinyl (3- shown) 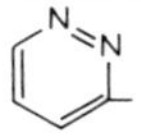
(19)	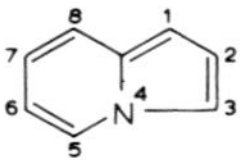 Indolizine	Indolizinyl (2- shown) 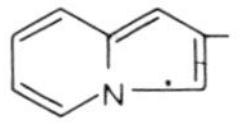
(20)	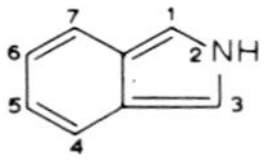 Isoindole	Isoindolyl (2- shown) 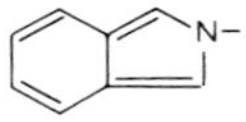
(21)	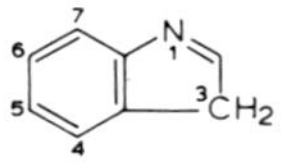 3*H*-Indole	3*H*-Indolyl (3*H*-Indol-2-yl shown) 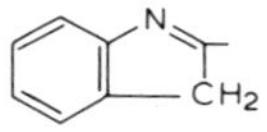
(22)	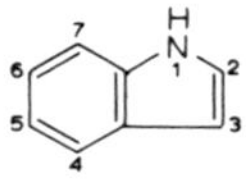 Indole	Indolyl (1- shown) 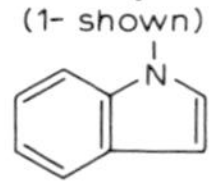
(23)	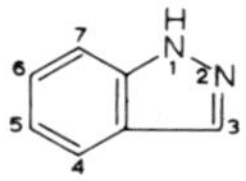 1*H*-Indazole	Indazolyl (1*H*-Indazol-3-yl shown) 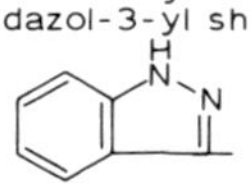
(24)	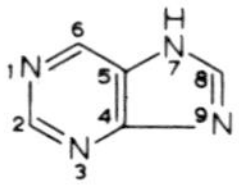 Purine*	Purinyl* (8- shown) 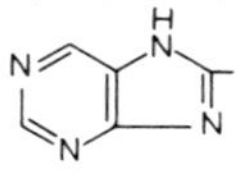

* Denotes exceptions to systematic numbering.

	Parent compound		Radical name
(25)	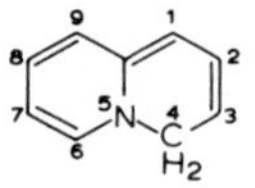	4*H*-Quinolizine	4*H*-Quinolizinyl (4*H*-Quinolizin-2-yl shown) 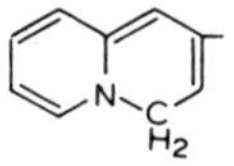
(26)	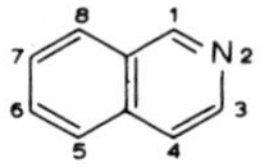	Isoquinoline	Isoquinolyl (3- shown) 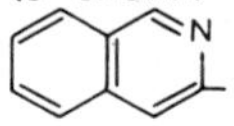
(27)	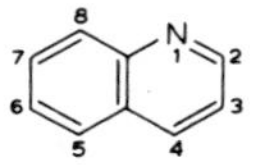	Quinoline	Quinolyl (2- shown) 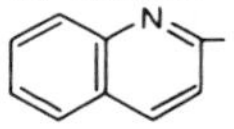
(28)	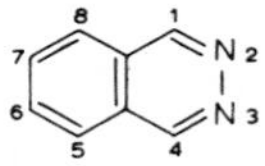	Phthalazine	Phthalazinyl (1- shown) 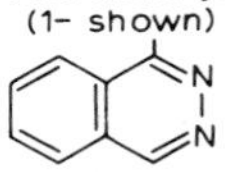
(29)	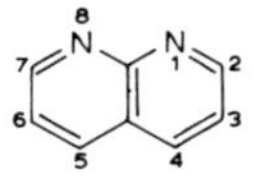	Naphthyridine (1,8- shown)	Naphthyridinyl (1,8-Naphthyridin-2-yl shown) 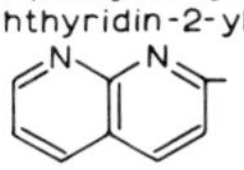
(30)	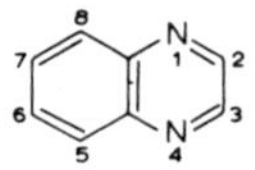	Quinoxaline	Quinoxalinyl (2- shown) 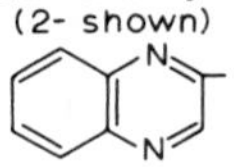
(31)	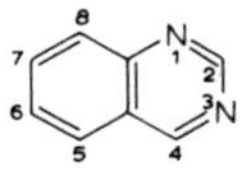	Quinazoline	Quinazolinyl (2- shown) 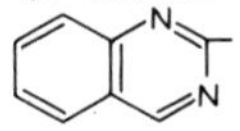
(32)	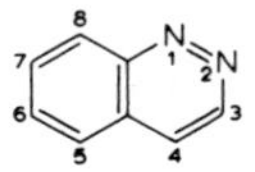	Cinnoline	Cinnolinyl (3- shown) 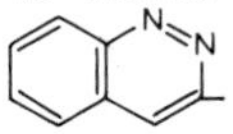

	Parent compound		Radical name
(33)	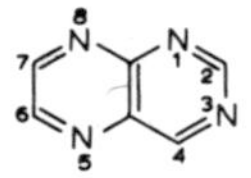	Pteridine	Pteridinyl (2- shown) 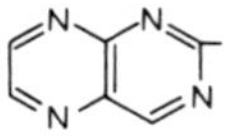
(34)	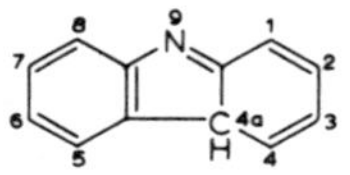	4a*H*-Carbazole*	4a*H*-Carbazolyl* (4a*H*-Carbazol-2-yl shown) 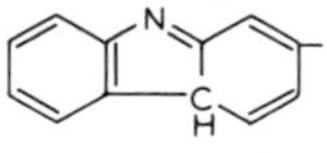
(35)	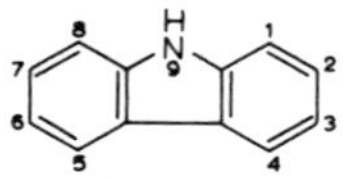	Carbazole*	Carbazolyl* (2- shown) 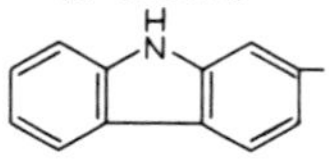
(36)	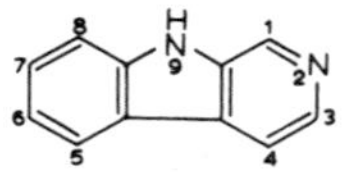	*β*-Carboline	*β*-Carbolinyl (*β*-Carbolin-3-yl shown) 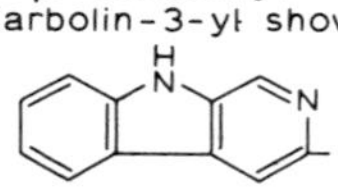
(37)	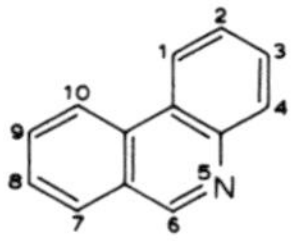	Phenanthridine	Phenanthridinyl (3- shown) 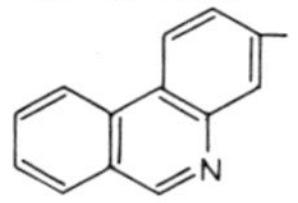
(38)	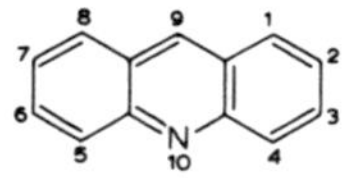	Acridine*	Acridinyl* (2- shown) 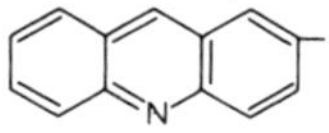
(39)	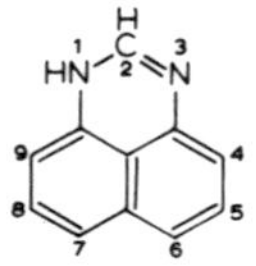	Perimidine	Perimidinyl (2- shown) 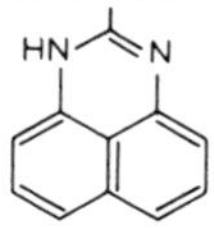

* Denotes exceptions to systematic numbering.

	Parent compound	Radical name
(40)	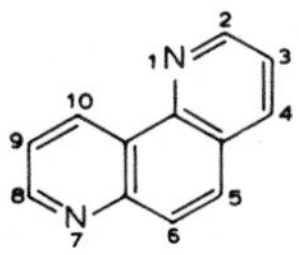 Phenanthroline (1,7- shown)	Phenanthrolinyl (1,7-Phenanthrolin-3-yl shown) 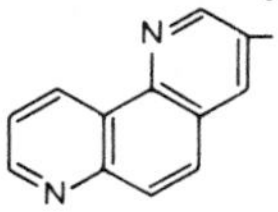
(41)	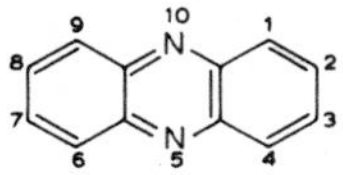 Phenazine	Phenazinyl (1- shown) 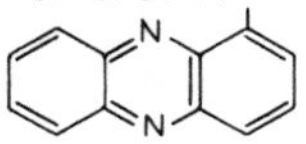
(42)	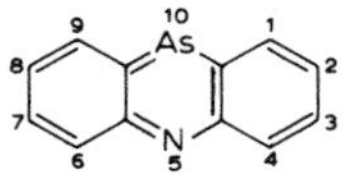 Phenarsazine	Phenarsazinyl (2- shown) 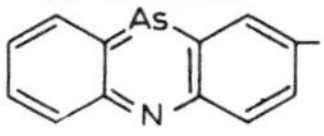
(43)	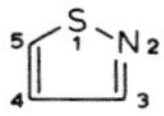 Isothiazole	Isothiazolyl (3- shown) 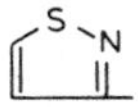
(44)	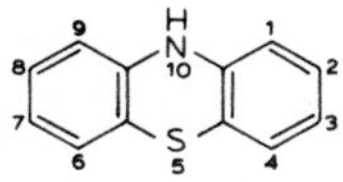 Phenothiazine	Phenothiazinyl (2- shown) 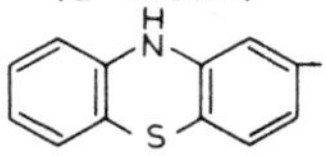
(45)	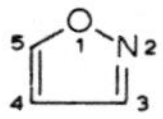 Isoxazole	Isoxazolyl (3- shown) 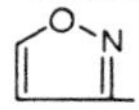
(46)	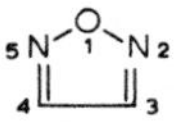 Furazan	Furazanyl (3- shown) 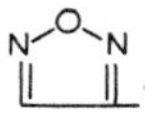
(47)	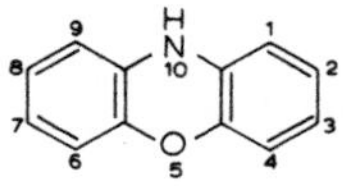 Phenoxazine	Phenoxazinyl (2- shown) 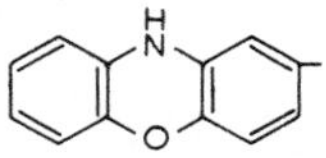

B–2.12—The following trivial and semi-trivial names are retained but are not recommended for use in fusion names. The names of the radicals shown are formed according to Rule **B–5.**

	Parent compound	Radical name
(1)	Isochroman	Isochromanyl (3- shown)
(2)	Chroman	Chromanyl (7- shown)
(3)	Pyrrolidine	Pyrrolidinyl (2- shown)
(4)	Pyrroline (2- shown*)	Pyrrolinyl (2-Pyrrolin-3-yl* shown)
(5)	Imidazolidine	Imidazolidinyl (2- shown)
(6)	Imidazoline (2- shown*)	Imidazolinyl (2-Imidazolin-4-yl* shown)
(7)	Pyrazolidine	Pyrazolidinyl (2- shown)

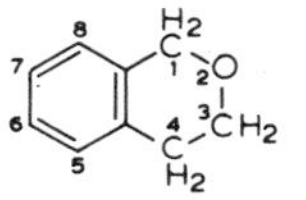

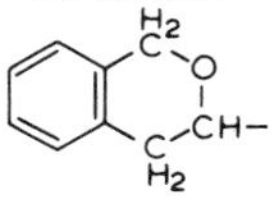

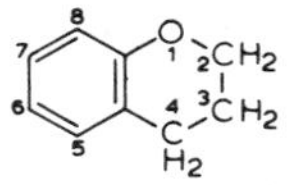

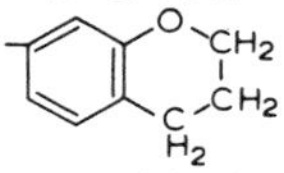

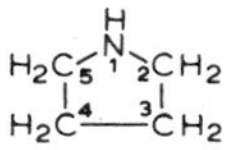

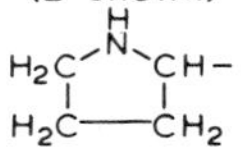

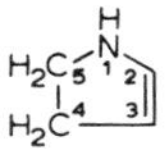

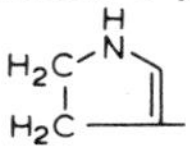

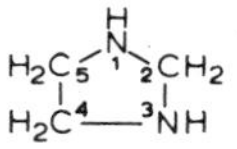

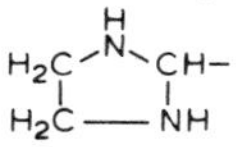

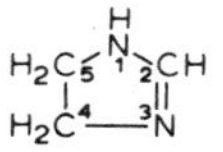

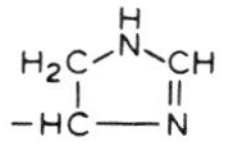

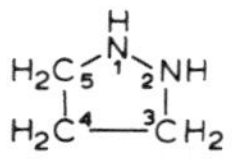

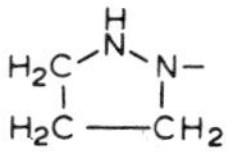

*The "2-" denotes the position of the double bond.

Parent compound — Radical name

(8) H; H_2C_5 N 1 2NH; 4 3 — Pyrazoline (3- shown*) — Pyrazolinyl (3-Pyrazolin-2-yl* shown): H; H_2C N N–

(9) H; H_2C_6 N 1 2CH_2; H_2C_5 4 3CH_2; C H_2 — Piperidine — Piperidyl[a] (2- shown): H; H_2C N CH–; H_2C C CH_2; H_2

(10) H; H_2C_6 N 1 2CH_2; H_2C_5 4 3CH_2; N H — Piperazine — Piperazinyl (1- shown): N; H_2C CH_2; H_2C N CH_2; H

(11) 7 6 5 4; H N 1 2CH_2; 3CH_2 — Indoline — Indolinyl (1- shown): N CH_2; CH_2

(12) 7 6 5 4; H_2 C 1; 2NH; 3 C H_2 — Isoindoline — Isoindolinyl (1- shown): H C; NH; C H_2

(13) N; H_2C_6 1 7CH_2 2CH_2; H_2C_5 8CH_2 3CH_2; C H 4 — Quinuclidine — Quinuclidinyl (2- shown): N; H_2C CH_2 CH–; H_2C CH_2 CH_2; C H

(14) O; H_2C_6 1 2CH_2; H_2C_5 4 5CH_2; N H — Morpholine — Morpholinyl[b] (3- shown): O; H_2C CH_2; H_2C N CH–; H

* The "3-" denotes the position of the double bond.

[a] For 1-Piperidyl use piperidino.

[b] For 4-Morpholinyl use morpholino.

Rule B–3. Fused heterocyclic systems

3.1—"*Ortho*-fused" and "*ortho*- and *peri*-fused" ring compounds containing hetero atoms are named according to the fusion principle described in Rule **A–21** for hydrocarbons. The components are named according to Rules **A–21, B 1** and **B–2.** The base component should be a heterocycle. If there is a choice, the base component should be, by order of preference:

(*a*) A nitrogen-containing component.

Example:

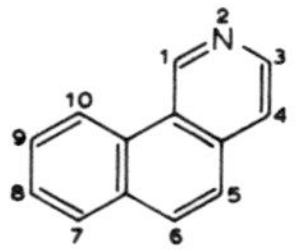

Benzo[*h*]isoquinoline
not Pyrido[3,4-*a*]naphthalene

(*b*) A component containing a hetero atom (other than nitrogen) as high as possible in Table I.

Example:

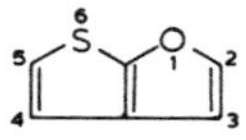

Thieno[2,3-*b*]furan
not Furoi[2,3-*b*]thiophene

(*c*) A component containing the greatest number of rings.

Example:

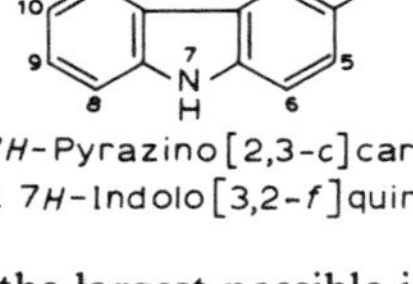

7*H*-Pyrazino[2,3-*c*]carbazole
not 7*H*-Indolo[3,2-*f*]quinoxaline

(*d*) A component containing the largest possible individual ring

Example:

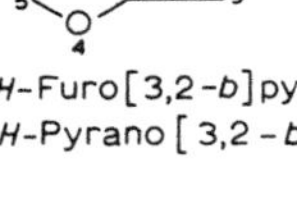

2*H*-Furo[3,2-*b*]pyran
not 2*H*-Pyrano[3,2-*b*]furan

(*e*) A component containing the greatest number of hetero atoms of any kind.

Example:

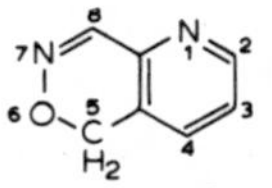

5*H*-Pyrido[2,3-*d*]-*o*-oxazine
not *o*-Oxazino[4,5-*b*]pyridine

(*f*) A component containing the greatest variety of hetero atoms.

Examples:

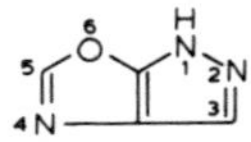

1*H*-Pyrazolo[4,3-*d*]oxazole
not 1*H*-Oxazolo[5,4-*c*]pyrazole

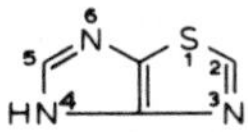

4*H*-Imidazo[4,5-*d*]thiazole
not 4*H*-Thiazolo[4,5-*d*]imidazole

(*g*) A component containing the greatest number of hetero atoms first listed in Table I.

Example:

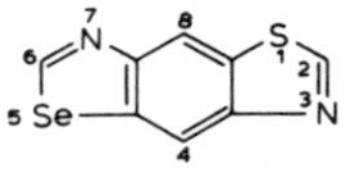

Selenazolo[5,4-*f*]benzothiazole*
not Thiazolo[5,4-*f*]benzoselenazole

(*h*) If there is a choice between components of the size containing the same number and kind of hetero atoms choose as the base component that one with the lower numbers for the hetero atoms before fusion.

Example:

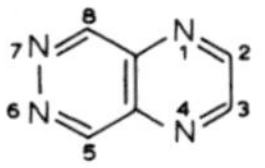

Pyrazino[2,3-*d*]pyridazine

3.2—If a position of fusion is occupied by a hetero atom, the names of the component rings to be fused are so chosen as both to contain the hetero atom.

* In this example the hetero atom first listed in Table I is sulfur and the greatest number of sulfur atoms in a ring is one.

Example:

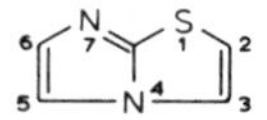

Imidazo[2,1-*b*]thiazole

3.3—The following contracted fusion prefixes may be used: furo, imidazo, isoquino, pyrido, pyrimido, quino and thieno.

Examples:

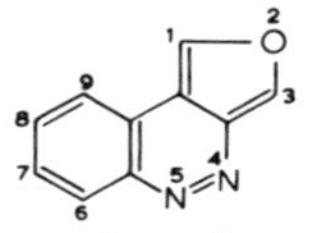

Furo[3,4-*c*]cinnoline

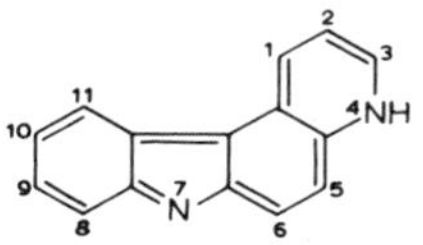

4*H*-Pyrido[2,3-*c*]carbazole

3.4—In peripheral numbering of the complete fused systems, the ring system is oriented and numbered according to the principles of Rule **A–22.** When there is a choice of orientations, it is made in the following sequence in order to:

(*a*) Give low numbers to hetero atoms, thus:

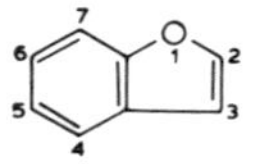

Benzo[*b*]furan

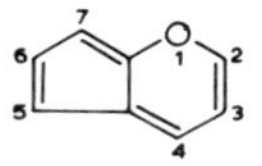

Cyclopenta[*b*]pyran

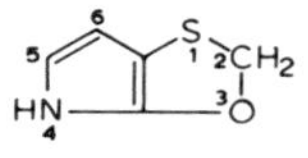

4*H*-1,3-Oxathiolo[5,4-*b*]pyrrole
(N.B. 1,3,4 lower than 1,3,6)

(*b*) Give low numbers to hetero atoms in order of Table I, thus:

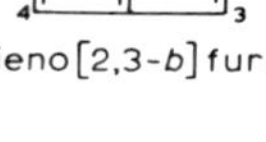

Thieno[2,3-*b*]furan

(*c*) Allow carbon atoms common to two or more rings to follow the lowest possible numbers (see Rules **A–22.2** and **A–22.3**). [A hetero atom common to two rings is numbered according to Rule **B–3.4**(*e*)], thus:

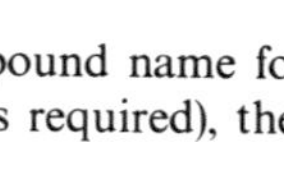

Imidazo[1,2-*b*]-[1,2,4]-triazine   not   or

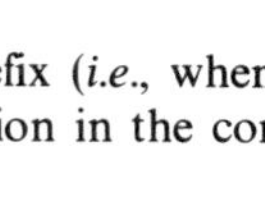

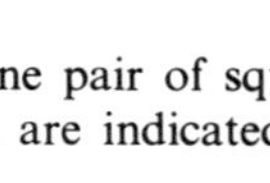

In a compound name for a fusion prefix (*i.e.*, when more than one pair of square brackets is required), the points of fusion in the compound prefix are indicated by the use of unprimed and primed numbers, the unprimed numbers being assigned to the ring attached directly to the base component, thus:

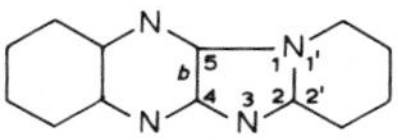

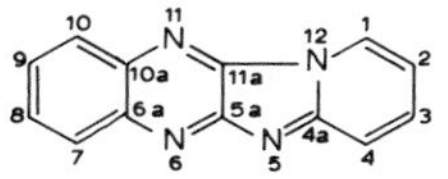

Pyrido[1',2':1,2]imidazo-[4,5-*b*]quinoxaline

not

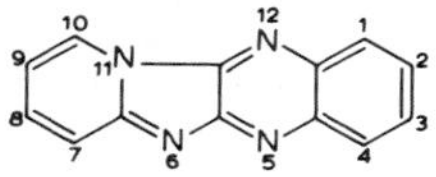

or

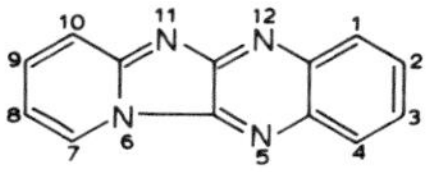

or

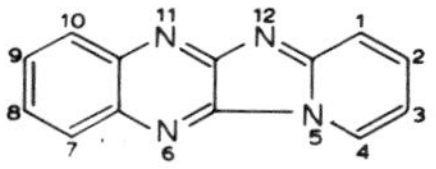

(*d*) Give hydrogen atoms lowest numbers possible, thus:

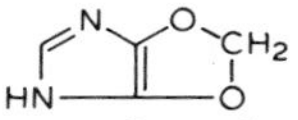

4*H*-1,3-Dioxolo[4,5-*d*]imidazole

(*e*) The ring is numbered as for hydrocarbons but numbers are given to all hetero atoms even when common to two or more rings. Interior hetero atoms are numbered last following the shortest path from the highest previous number.

Rule B–4. "a" Nomenclature

4.1—Names of heterocyclic compounds may also be formed by prefixing "a" terms (see Table I of Rule **B–1.1**) to the name of the corresponding homocyclic compound. The letter *a* should not be elided. There are two methods of applying this principle:

.1(*a*)—*Stelzner Method:* In this method, the "a" term name relates to that of the hydrocarbon with the same distribution of bonds in the rings. Thus, I is not so related to benzene but to 1,4-cyclohexadiene, and II is not so related to naphthalene but to 1,4-dihydronaphthalene.

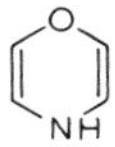

I 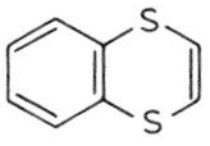 II

.1(*b*)—*Chemical Abstracts Method:* If the corresponding homocyclic compound is partially or completely hydrogenated and if this state of hydrogenation is denoted in its name without the use of hydro prefixes, as indan and cyclohexane, the procedure is the same as in (*a*). In other cases, positions in the skeleton of the corresponding homocyclic compound which are occupied by hetero atoms are denoted by the "a" terms, and the parent heterocyclic compound is considered to be that which contains the maximum mumber of conjugated or isolated* double bonds; hydrogen is added, as necessary, as hydro-prefixes and/or as *H* to the "a" name thus obtained.

Examples:

Stelzner method	Chemical Abstracts method
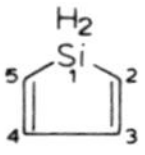 Sila-2,4-cyclopentadiene	Sila-2,4-cyclopentadiene
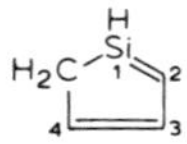 Sila-1,3-cyclopentadiene	Sila-1,3-cyclopentadiene
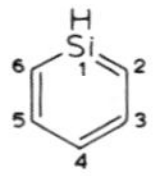 Silabenzene	Silabenzene
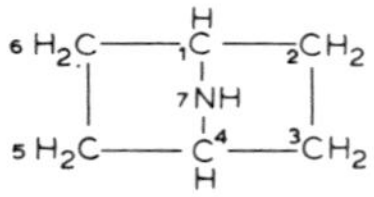 7-Azabicyclo[2.2.1]heptane	7-Azabicyclo[2.2.1]heptane
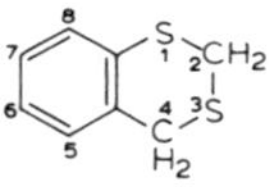 1,3-Dithia-1,2,3,4-tetrahydro-naphthalene	4*H*-1,3-Dithianaphthalene
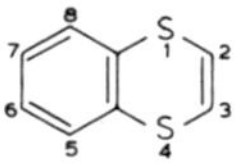 1,4-Dithia-1,4-dihydronaphth-alene	1,4-Dithianaphthalene
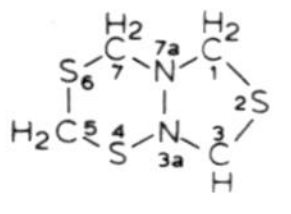 2,4,6-Trithia-3a,7a-diazaper-hydroindene	2,4,6-Trithia-3a,7a-diazaindene

* Isolated double bonds are those which are neither conjugated nor cumulative as in

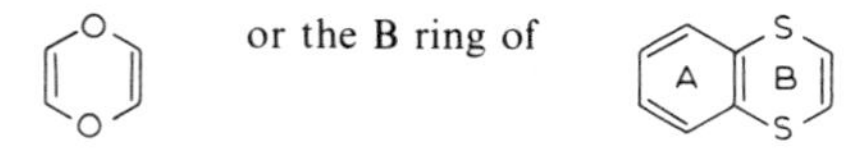

	Stelzner method	Chemical Abstracts method
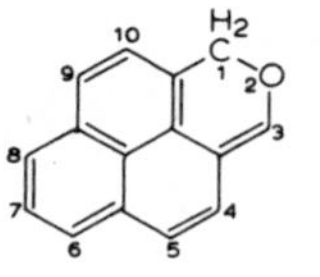	2-Oxa-1,2-dihydropyrene	1*H*-2-Oxapyrene
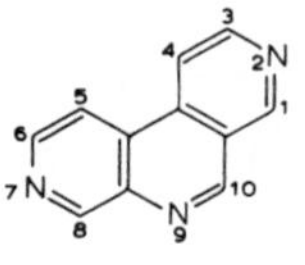	2,7,9-Triazaphenanthrene	2,7,9-Triazaphenanthrene

4.2—In fusion names, the "a" terms precede the completed name of the parent hydrocarbon. If two or more kinds of "a" terms occur in the same name, the procedure described in Rule **B–1.4** applies. Prefixes denoting ordinary substitution precede the "a" terms.

Example:

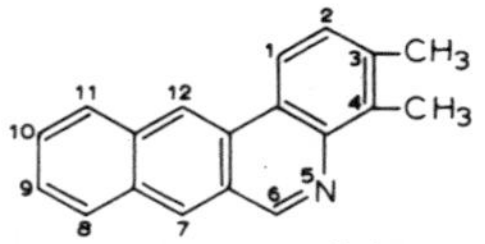

3,4-Dimethyl-5-azabenz[*a*]anthracene

Rule B–5. Radicals

5.11—Univalent radicals derived from heterocyclic compounds by removal of hydrogen from a ring are in principle named by adding "yl" to the names of the parent compounds (with elision of final "e" if present).

Examples:

Indolyl	from	indole
Pyrrolinyl	from	pyrroline
Triazolyl	from	triazole
Triazinyl	from	triazine

(For further examples see Rule **B–2.11**.)

The following exceptions are retained: furyl, pyridyl, piperidyl, quinolyl, isoquinolyl and thienyl (from thiophene) (see also Rule **B–2.12**).

As exceptions, the names "piperidino" and "morpholino" are preferred to "1-piperidyl" and "4-morpholinyl".

5.12—Bivalent radicals derived from univalent heterocyclic radicals whose names end in "-yl" by removal of one hydrogen atom from the atom with the free valence are named by adding "-idene" to the name of the corresponding univalent radical.

Example:

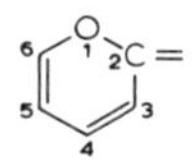

2-Pyranylidene

5.13—Multivalent radicals derived from heterocyclic compounds by removal of two or more hydrogen atoms from different atoms in the ring are named by adding "-diyl", "-triyl", *etc.*, to the name of the ring system.

Example:

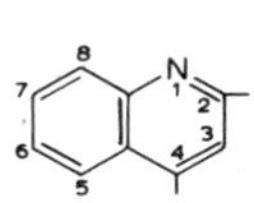

2,4-Quinolinediyl

5.21—The use of "a" terms (Rule **B-4**) does not affect the formation of radical names. Such names are strictly analogous to those of the hydrocarbon analogs except that the "a" terms establish enumeration in whole or in part.

Examples:

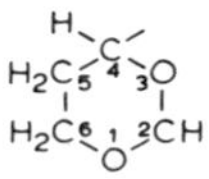

1,3-Dioxa-4-cyclohexyl

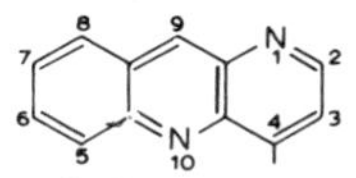

1,10-Diaza-4-anthryl

Rule B-6. Cationic hetero-atoms

6.1—According to the "a" nomenclature, heterocyclic compounds containing cationic hetero-atoms are named in conformity with the preceding rules by replacing "oxa-", "thia-", "aza-", *etc.*, by "oxonia-", "thionia-", "azonia-", *etc.*, the anion being designated in the usual way.

Examples:

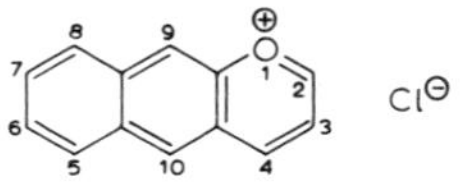

$Cl^{\ominus}$ 1-Oxoniaanthracene chloride

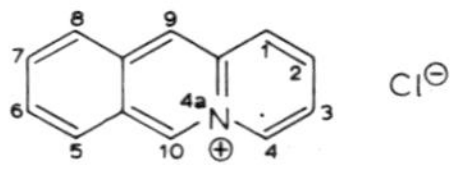

$Cl^{\ominus}$ 4a-Azoniaanthracene chloride

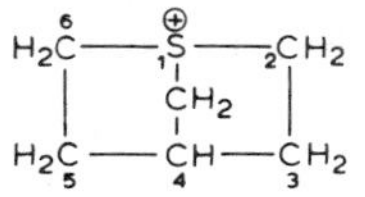

$Cl^{\ominus}$ 1-Thioniabicyclo[2.2.1]heptane chloride

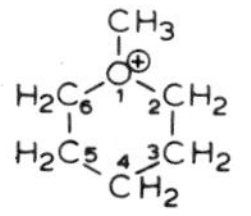

$Cl^{\ominus}$ 1-Methyl-1-oxoniacyclohexane chloride

Heterocyclic Spiro Compounds

Rule B–10 (Alternate to Rule B–11)

10.1—Heterocyclic spiro compounds containing single-ring units only may be named by prefixing "a" terms (see Table I, Rule **B–1.1**) to the names of the spiro hydrocarbons formed according to Rules **A–41.1, A–41.2, A–41.3** and **A–41.6.** The numbering of the spiro hydrocarbon is retained and the hetero-atoms in the order of Table I are given as low numbers as are consistent with the fixed numbering of the ring. When there is a choice, hetero-atoms are given lower numbers than double bonds.

Examples:

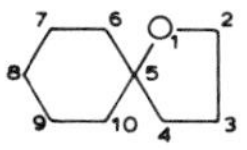

1-Oxaspiro[4.5]decane

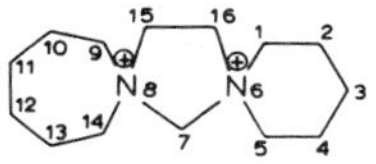

$2\ Cl^{\ominus}$

6,8-Diazoniadispiro[5.1.6.2]hexadecane dichloride

10.2—If at least one component of a mono- or poly-spiro compound is a fused polycyclic system, the spiro compound is named according to Rule **A–41.4** or **A–41.7,** giving the spiro atom as low a number as possible consistent with the fixed numberings of the component systems.

Examples:

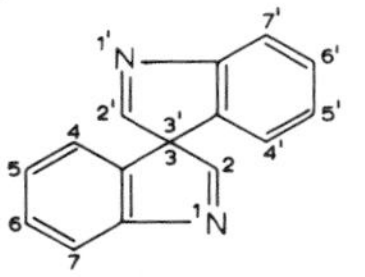

3,3'-Spirobi(3*H*-indole)

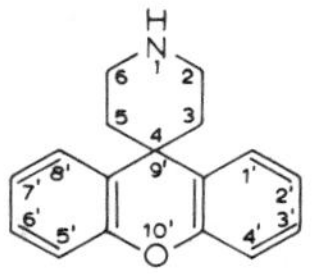

Spiro[piperidine-4,9'-xanthene]

Rule B–11 (Alternate to B–10)

11.1—Heterocyclic spiro compounds are named according to Rule **A–42,** the following criteria being applied where necessary: (*a*) spiro atoms have numbers as low as consistent with the numbering of the individual component systems; (*b*) heterocyclic components have priority over homocyclic components of the same size; (*c*) priority of heterocyclic components is decided according to Rule **B–3.** Parentheses are used where necessary for clarity in complex expressions.

Examples:

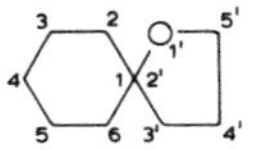

Cyclohexanespiro-2-(tetrahydrofuran)

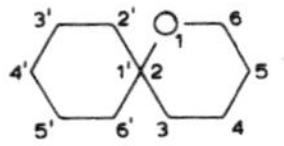

Tetrahydropyran-2-spirocyclohexane

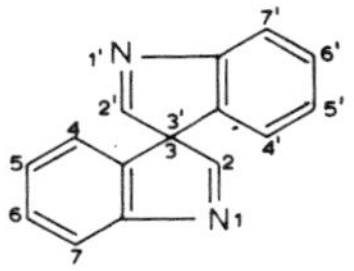

3,3'-Spirobi-(3*H*-indole)

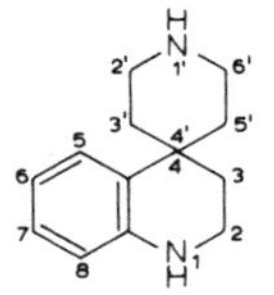

1,2,3,4-Tetrahydroquinoline-4-spiro-4'-piperidine

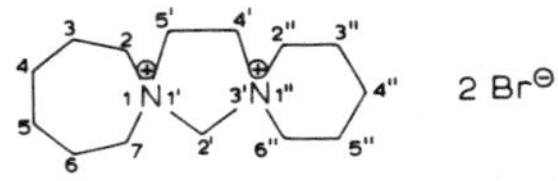

Hexahydroazepinium-1-spiro-1'-imidazolidine-3'-spiro-1''-piperidinium dibromide

C–0.14. Seniority of ring systems*

Rule C–14.1

14.11—Seniority of ring systems is decided by applying the following criteria, successively in the order given, until a decision is reached:

(*a*) All heterocycles are senior to all carbocycles.

Example:

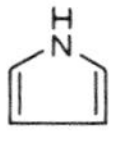

senior to

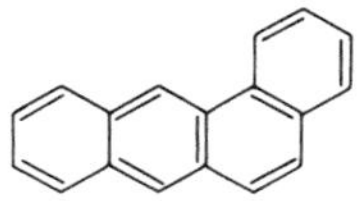

(*b*) For heterocycles the criteria based on the nature and position of the heteroatoms set out in Rule **B–3.1** (*a*)–(*h*)**.

Examples:

See Rule **B–3.1**, such as:

senior to

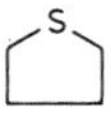

(*c*) Largest number of rings:

Example:

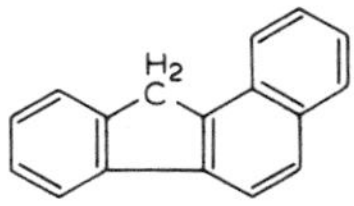

senior to

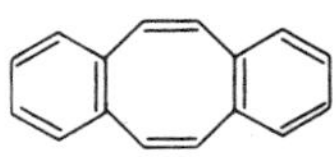

* Compare Rule **C–12.5.**

** In Rule **B–3.1** (*b*) "other than nitrogen" means "in the absence of nitrogen".

(*d*) Largest individual ring at first point of difference.

Examples:

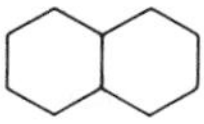

senior to

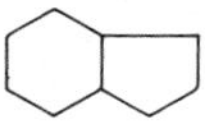

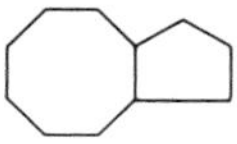

senior to

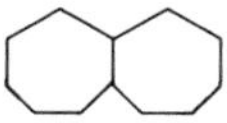

(*e*) Largest number of atoms in common among rings.

Examples:

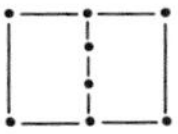

senior to 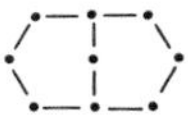 senior to

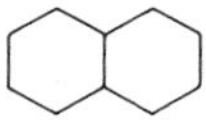

senior to  senior to

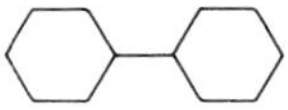

Note: Rings joined by a link (single or double) are included in this choice only when identical and named by the bi-, ter-, quater, *etc.*, system (see Rule **A–54.1**).

(*f*) Lowest letters* (*a*, *b*, *etc.*, see Rule **A–21.5**) in the expression for ring junctions.

Example:

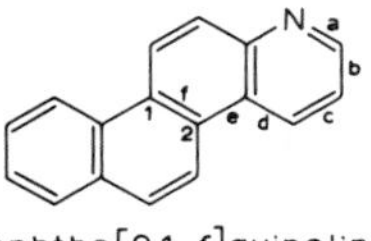

Naphtho[2,1-*f*]quinoline senior to Naphtho[1,2-*g*]quinoline

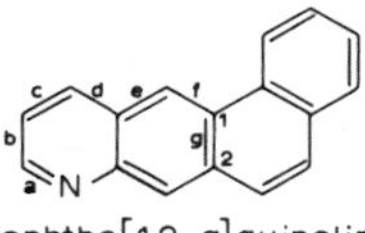

(*g*) Lowest numbers at the first point of difference in the expression for ring junctions. (See Rules: **A–21.5** for *ortho*-fusion and *ortho–peri*-fusion; **A–32** for tricyclo, *etc.*, systems; **A–41, A–42, B–10,** and **B–11** for spirans; **A–52** for assemblies of identical units.)

*Lowest means *a* before *b* before *c*, *etc.*

Examples:

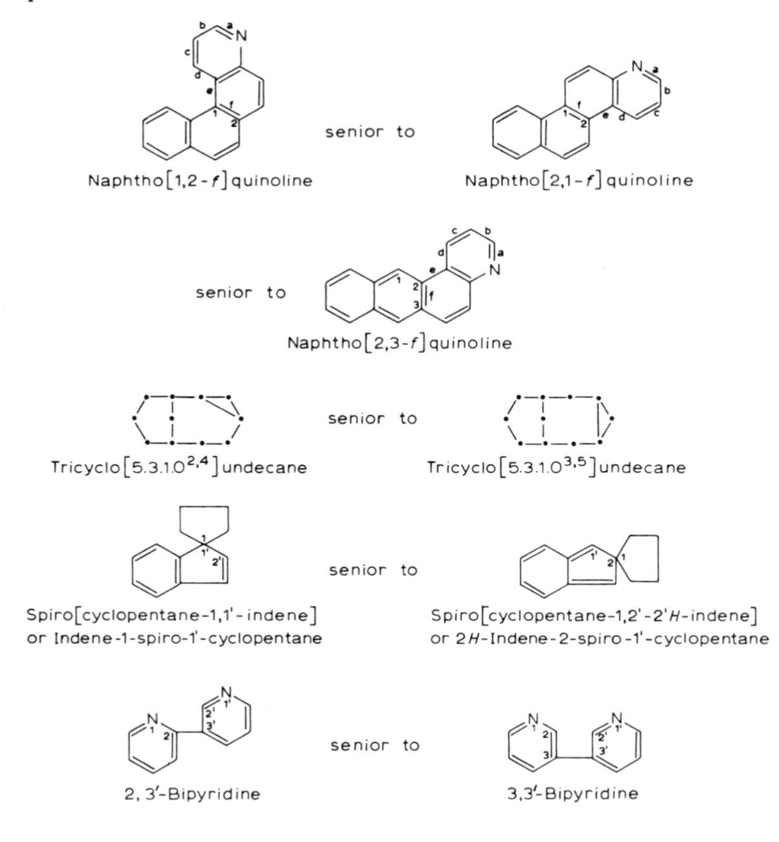

(*h*) Lowest state of hydrogenation.

Example:

senior to senior to

(*i*) Lowest locant for indicated hydrogen.

Example:

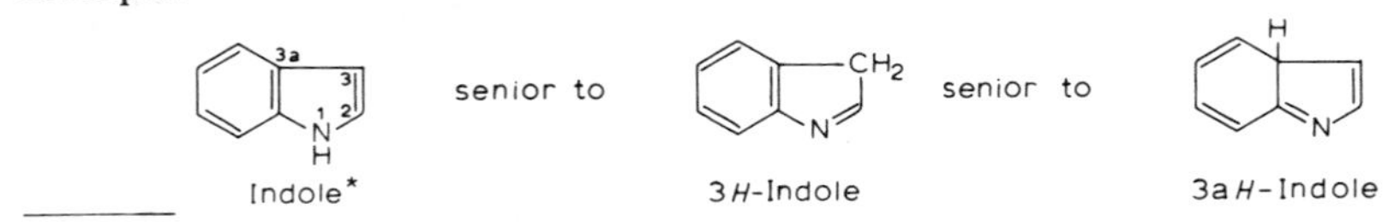

* 1*H*- is understood to be present.

(*j*) Lowest locant for point of attachment (if a radical).

Example:

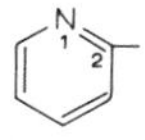

2-Pyridyl senior to 3-Pyridyl

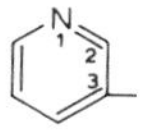

(*k*) Lowest locant for an attached group expressed as suffix.

Example:

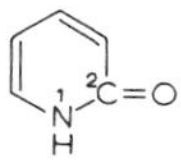

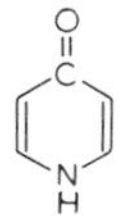

2(1*H*)-Pyridone senior to 4(1*H*)-Pyridone

(*l*) Lowest locant for substituents named as prefixes, hydro-prefixes, -ene, and -yne, all considered together in one series in ascending numerical order independently of their nature.

Examples:

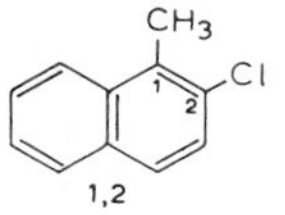

1,2 2-Chloro-1-methyl senior to 2,3 2-Chloro-3-methyl

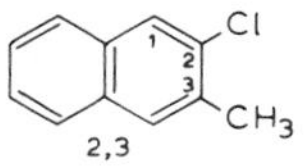

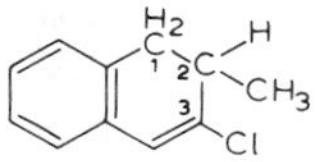

1,2,2,3 3-Chloro-1,2-dihydro-2-methyl senior to 2,2,2,3 2-Chloro-2,3-dihydro-3-methyl

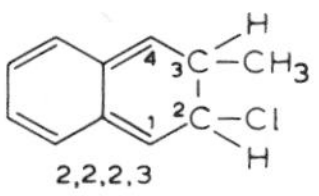

(*m*) Lowest locant for that substituent named as prefix which is cited first in the name, either (*i*) in order of complexity, or (*ii*) in alphabetical order (see Subsection **C–0.16**).

Example:

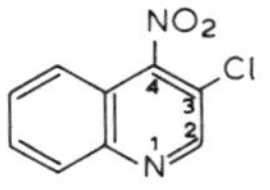

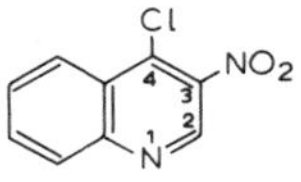

3-Chloro-4-nitro	senior to	4-Chloro-3-nitro
(*i*) 1-Methyl-2-ethyl	senior to	2-Methyl-1-ethyl
or (*ii*) 2-Ethyl-1-methyl	junior to	1-Ethyl-2-methyl

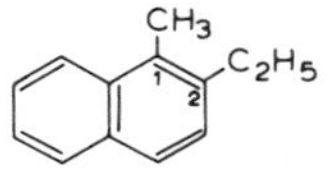

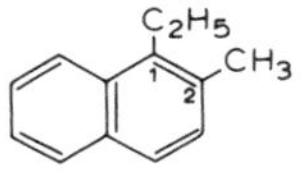

Note: Hydro prefixes, if treated as detachable (see Rule **C–16.11**), are considered along with prefixes for substituents when this criterion is applied.

GUIDE TO THE INDEX

This index is constructed in a similar manner to the volume indexes of the first edition of the Chemistry of Carbon Compounds. However, to make the index easier to use, more descriptive entries have been made for the commonly occurring individual, and groups of chemicals.

The indexes cover primarily the chemical compounds mentioned in the text, and also include reactions and techniques, where named, and some sources of chemical compounds such as plant and animal species, oils, etc.

Chemical compounds have been indexed alphabetically under the names used by authors, editing being restricted to ensuring uniformity of entries under the same heading. In view of the alternative nomenclature that can often be used, a limited amount of cross-referencing has been done where it is considered to be helpful, but attention is particularly drawn to Convention 2 below.

For this and the succeeding volumes, the indexing conventions listed below have been adopted.

1. Alphabetisation

(a) The following prefixes have not been counted for alphabetising:

n-	*o-*	*as-*	*meso-*	D	*C-*
sec-	*m-*	*sym-*	*cis-*	DL	*O-*
tert-	*p-*	*gem-*	*trans-*	L	*N-*
	vic-				*S-*
		lin-			*Bz-*
					Py-

Some prefixes and numbering have been omitted in the index, where they do not usefully contribute to the reference.

(b) The following prefixes have been alphabetised:

Allo	Epi	Neo
Anti	Hetero	Nor
Cyclo	Homo	Pseudo
	Iso	

(c) A letter by letter alphabetical sequence is followed for entries, firstly for the main entry, followed by the descriptive entry. The only exception to this sequence is the placing of plural entries in front of the corresponding individual entries to prevent these being overlooked by a strict alphabetical sequence which could lead to a considerable separation of plural from individual entries. Thus "butanes" will come before *n*-butane, "butenes" before 1-butene, and 2-butene, etc.

2. *Cross references*

In view of the many alternative trivial and systematic names for chemical compounds, the indexes should be searched under any alternative names which may be indicated in the main body of the text. Only a limited amount of cross-referencing has been carried out, where it is considered that it would be helpful to the user.

3. *Esters*

In the case of lower alcohols esters are indexed only under the acid, *e.g.* propionic methyl ester, not methyl propionate. Ethyl is normally omitted *e.g.* acetic ester.

4. *Derivatives*

Simple derivatives are not normally indexed if they follow in the same short section of the text.

5. *Collective and plural entries*

In place of "— derivatives" or "— compounds" the plural entry has normally been used. Plural entries have occasionally been used where compounds of the same name but differing numbering appear in the same section of the text.

6. *Main entries*

The main entry of the more common individual compounds is indicated by heavy type. Where entries relate to sections of three pages or more, the page number is followed by "ff".

7. *Italicised entries*

Italicised entries other than names of species refer to the Appendix (IUPAC Commission on the Nomenclature of Organic Chemistry).

INDEX

Abrine, **439**, 441
Abrus precatorius, 439
Acenaphthylene, 312
Acetals, cyclic, 48
Acetaldehyde, 115, 203
—, reaction with furfural, 140
—, — with thiophene-2-carbaldehyde, 250
Acetaldehyde cyanohydrin, 436
Acetamide, 47
Acetamidoacetaldehyde, 351
2-Acetamidoacetophenone, 422
3-Acetamido-1-acetylpyrrole, 351
2-Acetamidobenzothiophene-3-carboxylic ester, 275
5-Acetamido-3-bromobenzothiophene, 276
3-Acetamidodibenzofuran, 201
3-Acetamido-2,4-dimethylpyrrole, 351
Acetamidomalonic ester, 436
3-Acetamidopyrrole, **351**
Acetamidothiophenes, 243, 246
Acetanilide, 77
Acetic acid, tri-substituted, 120
Acetic anhydride, reaction with isatogens, 444
Acetic ester, 140
Acetic thieno[3,4-*b*]thiophene-2-carboxylic anhydride, 297
Acetoacetic acid, 205
Acetoacetic ester, 85, 118, 137, 178, 354, 429, 465, 511
Acetoacetic synthesis, 84
5-Acetoacetyl-2-furoic esters, 120
Acetone, 31, 51, 100, 111, 129, 140, 146, 159, 206, 209
—, reaction with fumaronitrile, 46
—, — with furan, 95
Acetone cyanohydrin, 436
Acetonedicarboxylic acid, reaction with phenols, 166
Acetonedicarboxylic ester, 85
Acetone phenylhydrazone, 415
Acetonylacetoacetic ester, 118
Acetonylacetone, 220, 237, 289, 313, 314, 350, 404
4-Acetonyl-1-acetyl-3-indolylacetaldehyde, 485
Acetophenone, 4, 37, 99, 129, 146, 162, 237, 271, 314, 483
Acetophenone 2,6-dimethylphenylhydrazone, 414
4-Acetoxybenzofuran-6-carboxylic ester, 156
4-Acetoxybenzofuran-6-carboxylic methyl ester, 165
3-Acetoxybenzothiophene, 267, 282, 283
3-Acetoxycarbazole, 502
5-Acetoxy-2,5-dihydro-2-oxofuran, 107
2-Acetoxy-2,5-diphenyl-2,3-dihydrofuran-3-one, 99
2-Acetoxyfuran, **107**
1-Acetoxyindole, 453
3-Acetoxyindole, 454
3-Acetoxymercuri-9-ethylcarbazole, 503
3-Acetoxymercurifuran, 97
2-Acetoxy-5-nitro-2,5-dihydrofuran, 104
3-Acetoxytetrahydropyran, 133
Acetylacetone, 268
Acetylalkylthiophenes, 253
4-Acetyl-2-amino-3,5-dimethylpyrrole, 351
3-Acetylamino-2-phenylbenzofuran, 156
2-Acetylaminothiophene, 243
Acetylangelicalactone, 107
N-Acetylanthranilic acid, 435
1-Acetylaziridine, 28
5-Acetyl-5*H*-benzo[*b*]carbazole, 523
2-Acetylbenzofuran, 159, 160, 176
3-Acetylbenzofuran, **160**
N-Acetyl-5,6-benzoindol-3-ylacetonitrile, 484
N-Acetyl-5,6-benzoindoxyl, 484
2-Acetylbenzo[*b*]selenophene, 322
Acetylbenzothiophenes, 277, 279
Acetylcarbazoles, 493, 505, **506**

Acetyl chloride, 9, 278, 377, 415
—, reaction with thietane, 54
Acetyldibenzofuran, 201
2-Acetyl-2,3-dihydrobenzofuran, 176
5-Acetyl-2,4-dihydroxybenzoic acid, 162
1-Acetyl-3,3-dimethylazetidine, 385
7-Acetyl-2,3-dimethylbenzofuran, 161
3-Acetyl-2,5-dimethylfuran, 116
1-Acetyl-2,3-dimethylindole, 422
2-Acetyl-3,3-dimethyl-3*H*-indole, 434
2-Acetyl-3,5-dimethylpyrrole, 351
3-Acetyl-2,4-dimethylpyrrole, 351, 382
3-Acetyl-2,5-dimethylthiophene, 251
3-Acetyl-2,5-diphenylpyrrole, 349
Acetylenes, 444
—, conversion to furans, 88
—, from halogenofurans, 131
—, oxidation with peracetic acid, 5
—, reaction with aminoketones, 331
Acetylene, 37, 217, 360, 384
—, benzofuran synthesis, 144
—, conversion to pyrrole, 337
—, — to thiophene, 229
—, reaction with aniline, 411
—, — with benzenethiol, 271
—, — with carbazole, 492
—, — with formaldehyde, 125
—, — with germanium diiodide, 71
—, — with indole, 412
—, — with pyrrole, 339, 341
—, — with selenium, 315, 325
—, — with sulphides, 229
—, — with sulphur, 296
Acetylenedicarboxylic esters, 113, 272, 472, 483
Acetylenedicarboxylic ester, 93, 253, 333, 340, 341, 360, 364, 372
Acetylenedicarboxylic methyl ester, 88, 99, 137, 165, 193, 199, 269, 287, 434, 507, 533
Acetylene oxides, 5
—, reaction with amines, 353
Acetylene sulphides, 14
Acetylenic amino-alcohols, 334
Acetylenic α-epoxides, 314
Acetylenylthiophenes, 237
5-Acetyl-2-ethylfuran, 116
2-Acetyl-3-ethylindole, 409
5-Acetyl-2-ethylthiophene, 245
2-Acetylfuran, 116
3-Acetylfuran, 115
2-Acetylfuran enamines, 116
α-Acetylglutaric acid cyanohydrin, 364
3-Acetyl-1-hydroxy-2,5-dimethylpyrrole, 352
5-Acetyl-6-hydroxy-2-isopropylcoumaran, 162
3-Acetyl-5-hydroxy-2-methylnaphtho-[1,2-*b*]furan, 207
3-Acetyl-1-hydroxy-2-methyl-5-phenylpyrrole, 352
1-Acetyl-2-hydroxy-3-oxo-2-phenylindoline, 456
N-Acetylimidazole, 341
N-Acetylindoles, 408
1-Acetylindole, 413
3-Acetylindole, 434
Acetylindoxyl, 454, 455
1-Acetylindoxylic acid, 435
1-Acetylisatin, 464
2-Acetylisoindoline, 476
Acetylmethylbenzothiophenes, 279
5-Acetyl-2-methyl-2,3-dihydrobenzofuran, 176
3-Acetyl-2-methyl-1,8-dihydroindeno-[2,1-*b*]pyrrole, 479
2-Acetyl-1-methylindole, 434
2-Acetyl-3-methylindole, 434
3-Acetyl-2-methylindole, 434
O-Acetyl-*N*-methylindoxyl, 455
3-Acetyl-2-methylnaphtho[2,1-*b*]furan, 208
2-Acetylnaphtho[2,1-*b*]thiophene, 309
3-Acetylnaphtho[2,3-*b*]thiophene, 310
N-Acetylneuraminic acid, 363
Acetyl nitrate, 337, 348
5-Acetyl-6(?2)-nitro-5*H*-benzocarbazole, 523
5-Acetyl-3-nitro-2-hydroxythiophene, 245
9-Acetyloctahydrocarbazole, 516
1-Acetyloxindole, 453
3-Acetyl-2-oxo-2,3-dihydrobenzofuran, 170
4-Acetyl-5-oxo-1,3-diphenyl-2-pyrroline, 378
o-Acetylphenylselenoacetic acid, 322
1-Acetylpyrrole, **341**
2-Acetylpyrrole, 352
C-Acetylpyrroles, 344
3-Acetylselenophene, **317**

Acetyltetrahydrocarbazole, 511, 514
1-Acetyl-3-thiocyanoindole, 421
Acetylthioglycollyl chloride, 257
2-Acetylthioindoxyl, 283
3-Acetylthio-oxindole, 281
2-Acetylthiophene, 250, **251**, 252, 254
3-Acetylthiophene, 247, **251**
Acetyl-*o*-toluidine, 402, 415
5-Acetyltribromobenzo[*b*]carbazole, 523
Acid anhydrides, reaction with pyrroles, 356
—, — with thiophenes, 250
Acid chlorides, reaction with thiophenes, 250
Acraldehyde, 469
Acridine, 558
Acrolein, 383
Acrylic ester, 364, 380, 382, 384, 387
Acrylic methyl ester, 508
Acrylonitrile, 20, 186, 364, 384
—, reaction with indole, 412
—, — with pyrrole, 339
Acylacetates, 375
5-Acyl-2-alkyl furans, 116
Acylamidomalonic esters, 354
N-Acylamines, conversion to pyrroles, 340
o-Acylaminoketones, 408
γ-Acylaminovaleraldehydes, 383
2-Acyl-3-arylaziridines, 17
1-Acylaziridines, 25
Acylbenzofurans, 142, 146, 160, 166
2-Acylcarbazoles, 505
9-Acylcarbazoles, **493**, 505
Acyl chlorides, 115, 316, 338
N-Acyl-*o*-chloro-*N*-methylanilines, 449
Acyl cyanides, 166
2-Acyl-2,3-dihydrobenzofuran-3-ones, 176
Acylfurans, 97, 116
Acylglycines, 249
3-Acyl-2-hydroxyindoles, 452
1-Acyl-2-hydroxypyrrolidines, 383
1-Acylindoles, 412
1-Acylisatins, 460
Acylisothiocyanates, 222
Acylmethylene succinates, 375
C-Acyl-*N*-methylpyrroles, 341
8-Acyl-1-naphthols, 213
2-Acyloxaziridines, 41
3-Acyloxindole oximes, 452
2-Acyl-3-oxo-2,3-dihydrobenzofurans, 171
3-Acyloxybenzofurans, 171
o-Acyloxy-ω-chloroacetophenones, 176
β-Acyloxynitro compounds, 332
2-Acyloxytetrahydrofurans, 123
Acylphenylhydrazines, 448
2-Acylpyrroles, 339, 357
N-Acylpyrroles, isomerisation, 356
4-Acyl-2-pyrrolethiolcarboxylates, 357
Acylpyruvates, 357
Acyltetrahydrocarbazoles, 510, 511
Acylthiiranes, 9
2-Acylthiomethylthiiranes, 9
3-Acylthiopropenes, 9
Acyl-*o*-toluidines, 402
Adamkiewicz–Hopkins reaction, 437
Adrenaline, 425
Adrenochrome, 425
Aetioporphyrin, 353
Agroclavine, 486
Alanine, 27, 438
β-Alanine esters, 387
Alcohols, from furans, 92
—, from tetrahydrofurans, 131
—, reaction with isatogens, 444
—, — with oxetanes, 47, 48
—, — with thiiranes, 9
Aldehydes, arylhydrazones, 398
—, conversion to oxaziridines, 40
—, — to pyrroles, 333
—, from oxaziridines, 40
—, reaction with aziridinium salts, 21
—, — with α-halogenoketones, 3
—, — with indoles, 410, 468
—, thiophenes, 249
—, $\alpha\beta$-unsaturated, 130
Aldehyde–urea condensation products, 78
Aldoses, conversion to furans, 87
—, reaction with 1,3-dicarbonyl compounds, 87
Aliphatic isocyanides, 44
Alkaloids, 329, 330
—, ergot, 389
—, indole derivatives, 397
—, morphine group, 194
Alkanesulphenyl 2,4,6-trinitrobenzenesulphonates, 12
Alkanesulphonic acids, 186
Alkanoic acids, 230

Alkanoyl 2,2′-bithienyls, 262
Alkenes, 128
—, reaction with iodine azide, 35
N-Alkenylanilines, 414
3-Alkenylgermanium compounds, 547
α-Alkoxyacetophenones, 48
β-Alkoxyamino compounds, 38
Alkoxy-3-halogenodiazirines, 44
5-Alkoxyisoxazoles, 35
N-(3-Alkoxy-3-methylbutyl)-*N*-methyl benzylamine perchlorate, 66
Alkoxyoxetanes, 48
3-Alkoxyoxindoles, 456
2-Alkoxy-3-oxo-3*H*-indoles, 460
1-Alkoxy-2-phospholene 1-oxides, 535
1-Alkoxy-3-phospholene 1-oxides, 535
Alkoxysalicylaldehydes, 156
2-Alkoxytetrahydrofurans, 123
Alkoxythiophenes, 220
Alkylamidines, 44
N-Alkylamines, conversion to pyrroles, 340
Alkylaminoacetylbenzofurans, 161
1-Alkylamino-3-chloropropanes, 66
3-Alkylaminopropanols, 63
N-Alkylanilines, 446
Alkylarylamines, 412, 420
2-Alkyl-3-aryloxaziridines, 41
1-Alkylazetidin-2-ones, 67
α-Alkylazidoethylenes, 33
Alkylbenzofurans, 148, 150
—, properties, 151
2-Alkylbenzofurans, 150, 160
3-Alkylbenzofurans, 148
Alkylbenzothiophenes, 272
Alkyl benzyl ethers, 26
Alkyl 2,2′-bithienyls, 262
9-Alkylborafluorenes, 550
9-Alkylcarbazoles, **492**, 504, 506, 520
C-Alkylcarbazoles, 494, 495
9-Alkylcarbazole-3,6-dicarboxylic acids, 508
Alkylchlorothiophenes, 257
1-Alkyl-2-cyanoaziridines, 30
2-Alkylcyclohexanones, 510
1-Alkyldiaziridines, 44
N-Alkyldiaziridines, 42
5-Alkyldibenzoboroles, 550
Alkyl dichloromethyl ethers, 249
1-Alkyl-3,4-dihydroxypyrrolidines, 383
2-Alkyl-3,3-diphenyloxaziridines, 42
Alkylene-1,3-dihydroisobenzofurans, 189
Alkylfurans, 91, 97
—, properties, 101
—, reactions, 98
Alkyl furoylacetates, 136
Alkyl halides, 225, 233, 338
Alkyl-3-halogenodiazirines, 44
Alkylhydrazines, 44
Alkyl hydroperoxides, 188
β-Alkylhydroxylamines, 40
Alkylidenebisaziridines, 24
Alkylidenebisindoles, 410
Alkylidenebis(3-oxo-2,3-dihydrobenzofurans), 171
2-Alkylidene-1-methylpyrrolidines, 369
Alkylindoles, 414
1-Alkylindoles, 412
2-Alkylindoles, 402, 422, 452
3-Alkylindoles, 415, 416, 420
C-Alkylindoles, physical data, 418
2-Alkyl-3*H*-indoles, 442
Alkyl iodides, 82
1-Alkylisatins, 460, 462
2-Alkylisatins, 460
Alkyl isocyanates, 70
2-Alkylisoindoles, 476
2-Alkylisoindoline *N*-oxides, 476
Alkyl isothiocyanates, 222
Alkylisoureas, 44
Alkyl-lithium compounds, reaction with β-propiolactone, 50
C-Alkyl-*N*-methylpyrroles, 341
1-Alkyl-2-methyl-2-pyrrolines, 412
2-Alkyl-1-methyl-2-pyrrolines, 369
Alkyl nitrites, 439
1-Alkyl-3-nitrosoindoles, 423
3-Alkyloxindoles, 450
2-Alkyl-3-oxo-2,3-dihydrobenzofurans, 148
N-Alkyl-3-phenylisoindole-1-carboxamides, 473
1-Alkylphospholes, 532
Alkylpyrroles, 339
—, picrates, 346
—, thermal rearrangement, 342
2-Alkylpyrroles, 336, 344
C-Alkylpyrroles, 335, 341, 343, 346
—, physical constants, 345
N-Alkylpyrroles, 333, 335
C-Alkylpyrrolidines, 382

Alkylselenophenes, 317
Alkylsulphones, 235
9-Alkyltetrahydrocarbazole, 510
1-Alkyltetrahydroindoles, 412
Alkyl thienyl ketones, 251
Alkyl thienyl sulphides, 246
Alkylthiophenes, 233, 236, 257
—, oxidation, 235
2-Alkylthiophenes, 251
3-Alkylthiophenes, 236
S-Alkylthiophenium salts, 233
Alkylthiothiophenes, 220
Alkyltricarbonylpyrrolylmanganese, 344
Alkyl vinyl ketones, 50
Allenes, conversion to β-lactams, 68
Allene mono-oxides, 4
Allenic esters, 34
Allohydroxyproline, 390
Alloxan, 228
Allyl acetate, 132
Allyl alcohols, 167
Allylamine, 512
Allyl bromide, 340
2-Allylcyclohexanone, 466
4-Allyldibenzofuran, 200
Allyl halides, 52, 167
1-Allylindole, 413
2-Allylisoindoline, 476
Allylphenol, 175
o-Allylphenols, 148, 149, 166, 167, 176
o-Allylphenyl acetate, 175
1-Allylphenylgermacyclopentane, 546
Allyl phenyl selenide, 321
1-Allylpyrrole, 340
2-Allylpyrrole, 345
o-Allylselenophenol, 321
2-Allylthiophene, 237
Aluminium, heteroatoms, in compounds, 549
Amanita mappa, 431
Amanita phalloides, 390, 438
Amides, conversion to β-lactams, 68
Amides, α-substituted, 32
Amines, 316
—, from oxaziridines, 40
—, reaction with formaldehyde, 43
—, — with ketones, 42
—, — with malonyl chloride, 70
—, — with oxetane, 47
—, — with thiiranes, 9
Amines, *(continued)*
—, unsaturated, oxidation, 19
Aminoacetaldehydes, 357
2-Aminoacetylthiophene, 251
Amino-acids, 357, 437
—, aliphatic, 230
—, cyclisation, 449
—, esters, 377
—, lactams, 384
Amino-alcohols, acetylenic, 334
Aminoaldehydes, conversion to pyrroles, 331
α-Aminoaldehydes, 356
3-Aminoalkyl hydrogen sulphates, 62
3-(1-Aminoalkylidene)oxindoles, 452
Aminoalkylindoles, 428
1-Amino-2-(2-aminophenyl)naphthalene, 519
2-Amino-1-(2-aminophenyl)naphthalene, 519
o-Aminoarenethiols, 267
1,12-Aminobenz[*a*]anthracene, 524
Aminobenzocarbazoles, 524
Aminobenzofurans, 146, 154
2-Aminobenzofuran, 155
5-Aminobenzofuran, **155**
2-Aminobenzophenone, 458
Aminobenzothiophenes, **275,** 276
2-Aminobenzothiophene, 281
Aminobenzoxazolin-2-one, 19
o-Aminobenzoylformic acid, sodium salt, 460
o-Aminobenzyl cyanide, 423
3-Amino-1-benzyl-2,5-diphenylpyrrole, 350
2-Aminobiphenyl, 302, 490
o-Aminobiphenyl ethers, 195
2-Aminobiphenyl 2,4-xylyl ether, 195
γ-Amino-α-bromobutyric acid, 67
1-δ-Aminobutylpyrrolidine, 478
Aminocarbazoles, 496, **499,** 500
2-Aminocarbazole, 499, 503
3-Aminocarbazole, 499
4-Aminocarbazole, 500
o-Aminocarbonyl compounds, 404
2-Amino-4-chloroacetophenone, 323
o-Amino-ω-chlorostyrene, 265, 397
β-Aminocrotonic ester, 164, 222, 376, 403
2-Amino-3-cyano-4,5-dimethylfuran, 106
2-Aminocyclohexanol, 27
2-Aminodiarylamines, 489

Aminodibenzofurans, 201
2-Aminodibenzoselenophene, 324
3-Aminodibenzoselenophene, 324
Aminodibenzothiophenes, 305
3-Aminodibenzothiophene, 302
3-Amino-2,3-dihydrobenzofuran, 170
3-Amino-2-(2,5-dihydroxyphenyl)crotonic ester, 164
3-Amino-2,5-dimethylfuran, **107**
1-Amino-2,5-dimethylpyrrole, 332, 350
3-Amino-2,4-dimethylpyrrole, 351
1-Amino-2,5-dimethylpyrrole-3,4-dicarboxylic ester, 332
2-Amino-5,5-dimethyl-1-pyrroline, 384
2-Amino-5,5-dimethyl-1-pyrroline 1-oxide, 384
1-Aminodioxindoles, 457
2-Aminodiphenylamines, 489
1-Amino-2,5-diphenylpyrrole, 350
3-Amino-2,5-diphenylpyrrole, 350
2-Aminodiphenyl sulphide, 300
Amino esters, 32
α-Amino-esters, 387
β-Amino-esters, reaction with Grignard reagents, 67
Aminoethanol, 22
3-Amino-5-ethoxycarbonyl-2,4-dimethylpyrrole, 351
β-Aminoethylalkyl sulphide, 23
β-Aminoethylaryl ethers, 23
β-Aminoethylaryl sulphide, 23
Aminoethylation, 22
o-Aminoethylbenzene, 280
2-(2-Aminoethyl)indole, **430,** 432
3-(2-Aminoethyl)indole, **429**
2-(β-Aminoethyl)-1-phenylisoindole, 473
2-(β-Aminoethyl)-1-phenylisoindoline, 476
2-Aminoethylpyrrole, 384
2-Aminoethylpyrrolidine, 384
2-Aminoethyltetrahydrofuran, 135
Aminofurans, 90, 105, 106
—, diazo reactions, 106
5-Amino-2-furoic ester, 105
3-Amino-1-furylbutane, 106
3-Amino-1-furylpentane, 106
Aminohydrins, 23
o-Amino-α-hydroxyphenylacetic acid, 456
Aminoindoles, 413, 421, **423,** 424
2-Amino-3*H*-indole, 423
L-α-Amino-β-3-indolylpropionic acid, 436
1-Amino-3*H*-isoindole, 474
2-Aminoisoindoline, 476
Aminoketones, conversion to pyrroles, 331
—, reaction with acetylenes, 331
α-Aminoketones, 18, 333, 350
γ-Aminoketones, conversion to pyrrolines, 369
Aminomalonic esters, 67
o-Aminomandelic acid, 456
2-Amino-4-methoxyacetophenone, 323
3-(3-Amino-5-methoxycarbonyl-2-pyridylmethyl)-1-methyloxindole, 454
2-(5-Amino-2-methoxyphenyl)ethylamine, 424
3-Amino-2-(4-methoxyphenyl)indole, 424
2-Aminomethylbenzothiophene, 278
Aminomethylcarbazoles, 500
2-Aminomethylfuran, 111
3-Amino-2-methylfuran, **107**
5-Aminomethylfurfuryl alcohol, 112
3-Aminomethylindole, 424, 428
2-Aminomethylpyrrole, **354**
7-Amino-1-methyltetrahydrocarbazole, 514
2-Aminomethylthiolane, 260
2-Aminomethylthiophene, 248
Aminonitrenes, 19
3-Amino-2-nitrocarbazole, 498
3-Amino-oxindole, 452, 453
o-Aminophenols, 116
p-Aminophenol, reaction with carbazole, 500
2-(2′-Aminophenyl)benzo[*g*]-5,6-dihydropyrrocoline, 528
6-Amino-2-phenylbenzofuran, 152, 156
N-*o*-Aminophenylcarbazole, 481
o-Amino-β-phenylethyl alcohol, 168, 446
1-Amino-3-phenylindole, 413
3-Amino-2-phenylindole, 424
1-(2-Aminophenyl)naphthylamine, 518
5-Aminophthalide, 189
N-Aminophthalimide, 350
β-Aminopropionic acids, conversion to β-lactones, 49
β-Aminopropionic esters, 67
1-Aminopyrroles, 332, **350**
3-Aminopyrroles, 349
2-Aminopyrrole-1-carboxylic ester, 360
Aminotetrahydrocarbazoles, 514
Aminotetrahydrofurans, 132
8-Aminotetrahydro-1-naphthoic acid, 485

2-Amino-1-tetralone hydrochloride, 483
3-Aminotetramethyltetrahydrofuran, 135
3-Aminothietane 1,1-dioxides, 57
Aminothiophenes, 241
2-Aminothiophene, **243**
3-Aminothiophene, 240, **243**
3-Amino-2,4,5-triphenylpyrrole, 350, 351
2-Aminotropone, 479
6-Amino-2,4-xylyl phenyl ether, 195
Ammonium mucate, 333, 337
Ammonium rhamnonate, 333
Ammonium saccharate, 333
tert-Amyl alcohol, 272
3-*tert*-Amylbenzothiophene, 272
Amyl nitrite, 348, 349, 351, 413
Angelica lactone, **107**
Anhydroisatin-α-anthranilide, 460
2,5-Anhydro-D-mannose, 87
2,5-Anhydro-sugars, 87
Aniline, 65, 77, 175, 293, 355, 375, 462, 474, 475, 487
—, conversion to isatin, 459
—, pyrolysis, 490
—, reaction with acetylene, 411
—, — with 1,3-dibromopropane, 62
—, — with furfural, 113
Aniline acetate, 114, 159
Aniline hydrochloride, 401
Aniline sulphoxide, 474
1-Anilinocyclohexanecarboxylic acid, 377
Anilinomalonic ester, 434
5-Anilinomethyl-1-phenyl-3-phenylimino-2*H*-pyrroline, 113
1-Anilino-3-phenylindole, 413
Animal melanins, 426
o-Anisoyldiazomethane, 171
3-*p*-Anisoyl-4,6-dimethoxybenzofuran, 143
Annulene polyoxides, 95
Anthracene, adduct with thietene dioxide, 57
—, condensed rings with furan, 214
Anthracenecarbazoles, **526**
Anthracene oil, 491
Anthrafurans, 214
Anthranil-3-carboxylic acid, 463
Anthranilic acid, 267, 322, 397, 408, 433, 437, 460
Anthraquinonecarbazoles, 506, 525
Anthroxanic acid, 463
Antimony, heteroatoms in compounds, 549
Anti-phenylalanine properties, 254
Antirrhinum majus, 173
Antituberculosis agents, 5
Anti-viral activity, 464
Arbuzov rearrangement, 536, 538
Arenediazohydroxides, 234
Arenethiols, 272
Areusin, **173**
Argentic fluoride, 79
Arndt–Eistert reaction, 166, 280
N-Aroylaziridines, 24
Aroyl-3-oxo-2,3-dihydrobenzofurans, 174
2,2′-Arsenothiophene, 232
Artemisia fragrans, 212
Arundo donax, 429
Arylamidines, 44
Arylamines, chloroacetyl derivatives, 449
—, conversion to indoles, 401
—, — to pyrroles, 340
—, reaction with furfural, 113
Arylamine salts, 267
1-Arylaminoanthraquinones, 525
α-Arylamino-arylacetonitriles, 333
Arylaminoketones, 402
—, conversion to indoles, 401
1-Arylamino-3-phenoxypropanes, 63
N-Arylazetidines, 63
α-Arylazetidines, 63
α-Arylazidoethenes, 33
Arylaziridines, 17
Arylbenzofurans, 148, 149, 150
—, properties, 151
2-Arylbenzofuran-3-carboxylic acids, 149
Arylbenzothiophenes, 272
1-Arylbenzotriazoles, 489
Arylbromoacetic esters, 148
Aryl chloroglyoxylates, 175
1-Aryl-1-cyanocyclopropanes, 368
Aryldiazonium cations, 400
Aryldibenzothiophenes, 303
1-Aryl-3,4-dihydroxypyrrolidines, 383
Aryldiindolylmethanes, 411
N-Aryl enamines, 446
Arylfurans, 97, 99
—, properties, 101
Aryl-3-halogenodiazirines, 44
Arylhydrazines, conversion to indoles, 399
Arylhydrazones, cyclisation, 398, 399

Arylhydroxylamines, 340, 434
5-Aryl-2-hydroxythiophenes, 244
Arylidene-bis-indoles, 410
Arylidene-3*H*-indoles, 410
Arylidenepyruvic acids, 387
N-Arylimines, 70
3-Arylimino-3*H*-indoles, 410
Arylindoles, 414
1-Arylindoles, 412
2-Arylindoles, 417
3-Arylindoles, 415
C-Arylindoles, physical data, 418
2-Arylisatogens, 443
2-Arylisoindoles, 472
2-Arylisoindoline *N*-oxides, 472
Aryl isothiocyanates, 222
Arylisoureas, 44
o-Aryloximes, 144
Aryloxyacetals, 142
Aryloxyacetones, 142
Aryloxyacetophenones, 149, 171
α-Aryloxydeoxybenzoins, 149
Aryl phenacyl sulphides, 269
1-Arylpyrroles, 391
C-Arylpyrroles, 341, 342, 343, 346
—, physical constants, 345
N-Arylpyrroles, 333, 339
N-Arylpyrroletetracarboxylic esters, 340
C-Arylpyrrolidines, 382
Arylpyrrolyl sulphides, **353**
Arylsulphenyl 2,4,6-trinitrobenzenesulphonates, 273
Arylsulphones, 235
Arylthiiranes, 9
Arylthioacetaldehyde dialkylacetals, 268
Arylthioacetaldehyde diethylacetals, 306
Arylthioacetic acids, 282
Arylthioalkyl ketones, 269
2-Arylthioindoxyls, 283
Arylthiophenes, 233, 234, 236
Aryl vinyl ketones, 50
Arynes, 93, 186, 225
Ascaridic acid, **136**
Ascaridole, 136, 193
Aspergillus niger, 293
Astraphloxine FF, 442, 470
Atractylis ovata, 213
Atractylodes japonica, 213
Atractylon, **213**
Aucuba japonica, 111
Aucubin, **111**
Aurones, 173, 174
Aurone epoxides, 174
Auxin activity, 280, 435
1-Aza-azulene, **479**
1-Aza-2,3-benzazulene, 527
1-Azabicyclo[1.1.0]butane, 20
1-Azabicyclo[1.2.2]heptane, 482
7-Azabicyclo[2.2.1]heptane, 556
7-Azabicyclo[4.1.0]heptane, 27
7-Azabicyclo[4.2.0]octanes, 65
Azaferrocenes, 344
2-Azaindene, 470
7-Azanorbornadiene-1,2-dicarboxylic ester, 340
4-Azapentalenyl anions, 481
3-Azaquadricyclanes, 340
Azarsetine, 552
2H-Azepine, 552
4*H*-Azepine, 362
Azete, 1
Azetidines, 20, 61
Azetidine, **63**
Azetidine-2-carboxylic acid, 67
Azetidine-2,4-diones, 70
Azetidine-3-sulphonic acid, 63
Azetidinium perchlorates, 66
Azetidinium salts, 61
Azetidin-3-ols, 66
Azetidinones, 62, 67
—, 3,3-disubstituted, 67
—, *N*-unsubstituted, 63
Azetidin-2-one, **67**
—, structural data, 2
Azetines, 1, 63
Azides, 17
—, conversion to azirines, 34
Azidoacetyl chloride, 69
2-Azidobiphenyls, 490
3-Azidocarbazole, 500
Azidoformic ester, 19, 224
2-Azido-4′-nitrobiphenyl, 498
1-Azido-2-phenylbut-2-ene, 37
3-Azido-2-phenylbut-1-ene, 37
1-Azido-1-phenylethylene, 34
Azidophenylnaphthalenes, 519
3-Azido-2-phenylprop-1-ene, 20
Aziridines, 1, 15, 35
—, fission, 25
—, from olefins, 17

Aziridines, *(continued)*
—, highly strained bridge, 19
—, properties, 27
—, reactions, 20
—, reaction with cyanogen bromide, 22
—, ring expansion, 24
—, stereospecific synthesis, 18
Aziridine, 15, 17, 23, 24, **27**, 497, *552*
—, halogenation, 21
—, polymerisation, 22
—, structural data, 2
Aziridine-2-carboxanilides, 20
Aziridine carboxylic acids, 30
Aziridinium derivatives, 16
Aziridinium ions, 22, 23
Aziridinium perchlorates, 21, 39
Aziridinium rings, expansion, 21
Aziridinium salts, 21, 38
Aziridinium vinylcarbamate, 24
Aziridinones, 31, 32, 44
—, reactivity, 32
Azirines, 1, 33, 34, 37
—, properties, 35
—, reactions, 35
—, ring expansion, 36
Azirinium salts, 36
Azobenzene, 76, 77
Azocine, *553*
Azo compounds, 15, 400
—, from pyrroles, 349
—, reaction with ketene, 76
Azodicarboxylic ester, 80, 413
Azo dyes, 523
Azo-indoles, 443
Azomethines, 15
—, oxidation, 40
—, reaction with ketenes, 66
Azomethine ylides, 25
4a-Azonia-anthracene chloride, *568*
3-Azoniaspiro [2.5]octane perchlorate, 39
2,3′-Azopyrroles, **350**
3,3′-Azopyrroles, **349**
Azosulphones, reaction with ketene *N*,*N*-acetals, 80
Azoxy-*tert*-butane, 45

Bacteriochlorophyll, 393
Bakelite, 114
Bamford–Stevens reaction, 255
Barbituric acid, 357
Barium furoate, 83
Bart reaction, 504
Beckmann rearrangement, 162, 243, 253, 287
—, cyclohexanone oxime, 204
Belladonna, 375
Benzalaniline, 69
Benzaldehyde, 3, 48, 106, 146, 160, 170, 171, 244, 251, 290, 378, 415
—, reaction with indoles, 410
Benzaldehyde acetal, 26
o-Benzamidobiphenyl, 515
α-Benzamidoisobutyryl chloride, 387
γ-Benzamidopropyl methyl ketone, 371
Benzazocines, 381
Benzene, in coal tar, 219
—, reaction with furoic ester, 119
—, — with nitrobenzene, 197
—, — with oxetane, 47
—, — with sulphur dioxide, 301
Benzeneazoformic ester, 76
3-Benzeneazoindoles, 424, 443
Benzenediazohydroxide, 171
Benzenediazonium chloride, 454, 462
Benzenediazonium salts, 424
Benzene rings, condensed system with furan nucleus, 207
—, with more than one furan nucleus, 215
Benzeneselenol, 320
Benzenesulphonamide, 78
Benzenethiol, 271
Benzenonium tetrafluoroborate, 489
Benzhydrylidene-2*H*-pyrrole, 363
Benzidine, 302, 499
o-Benzidine conversion, 399
Benzidine rearrangement, 487, 518
Benzidine sulphone, 302
Benzil, 176, 197, 228
Benzilic acid, 197
Benzilic type rearrangement, 443
1*H*-Benz [*f*]indole, 483
1*H*-Benz [*f*]indole-3-acetic acid, 484
2-(1*H*-Benz [*f*]indol-3-yl)ethylamine, 484
Benzo [*b*]-1-aza-azulene, 527
Benzocarbazoles, 518, **523**, 524
5*H*-Benzo [*b*]carbazole, 519, 522
7*H*-Benzo [*c*]carbazole, 518, 519, 520, 521
11*H*-Benzo [*a*]carbazole, 518, 519, 521, 522
Benzocarbazole dyes, 523
5*H*-Benzo [*b*]carbazole-6,11-quinone, 523

4,5-Benzocoumarone, 207
5,6-Benzocoumarone, 208
6,7-Benzocoumarone, 207
Benzocyclobutadiene, 186
1,2-Benzocyclobutadiene quinone, 291
Benzo[3,4]cyclobuta[1,2-*c*]thiophene, 291
Benzo[3,4]cyclobuta[1,2-*c*]thiophene 2,2-dioxide, 291
Benzocyclobutene, 291, 292
3,4-Benzo-1,1-dichlorosilacyclopentane, 544
Benzo[1,2-*b*:4,3-*b'*]difuran, 215
Benzo[1,2-*b*:5,4-*b'*]difuran, 216
3,4-Benzo-1,1-dimethylsilacyclopentane, 544
2,3-Benzo-1,1-diphenylsilacyclopentane, 544
Benzodithiophenes, **299**
Benzofurans, 83, 141
—, from dihydrobenzofurans, 168
—, hydrogenation, 145, 166
—, methylated, 207
—, naturally occurring, 157
—, reactions, 145
—, 2-substituted, 144
—, 3-substituted, 171
—, synthetic methods, 141
—, with side chain amino groups, 156
Benzofuran, 141, **147**, 148, 159
—, acylation, 150
—, from phenol, 144
—, irradiation, 146
—, nitration, 154
—, pyrolysis, 147
—, reaction with bromine, 169
—, — with hydrogen sulphide, 147
Benzo[b]furan, 564
Benzo[*c*]furan, **180**
3,4-Benzofuran, **180**
Benzofuranacetic acids, **166**
Benzofuran-3-acetic-2-carboxylic acids, 166
Benzofuran-2-carbaldehyde, 155, **159**
Benzofuran-3-carbaldehyde, 159
2-Benzofurancarbonyldiazomethane, 161
Benzofurancarboxylic acids, 163
Benzofuran-2-carboxylic acid, 141, **165**, 176
Benzofuran-3-carboxylic acid, 164, **166**
Benzofuran-3-carboxylic ester, 159
Benzofuran-2,3-dicarboxylic acid, 165, **166**
Benzofuran-4,5-dicarboxylic methyl ester, 165
Benzofuran-3(2*H*)-one, 170, 171
Benzofuran picrate, 148
Benzofuran-2-propionic acid, **166**
2-Benzofurfurylamines, 165
β-3-Benzofurylalanine, 166
Benzofurylalkanoic acids, 166
2-Benzofurylcarbamic ester, 155
2-Benzofurylcarbinol, **158**
3-Benzofurylcarbinol, **159**
2-Benzofuryldiphenylcarbinol, 159
1-(2-Benzofuryl)ethanol, 160
α-2-Benzofurylethylamine, 160
Benzofurylmethanols, 158, 159
2-Benzofurylmethylcarbinol, 160
Benzoic acid, 40
Benzoic ester, 278
Benzoin, 59, 88, 99, 142, 150, 197, 208, 214, 216, 401
Benzoin condensation, 90, 139, 249
Benzoindoles, 482, 483
Benzoin oxime, 413
Benzoin reaction, 433
Benzoin yellow, 214
Benzoisatins, 483
5,6-Benzoisoindolines, 477
Benzo[h]isoquinoline, 562
Benzo[*d*]naphtho[1,2-*d'*]benzo dithiophene, 299
Benzo[*b*]naphtho[1,2-*d*]furan, 215
Benzo[*b*]naphtho[2,1-*d*]furan, 215
Benzo[*b*]naphtho[2,3-*d*]furan, 214
Benzo[*b*]naphtho[1,2-*d*]thiophene, 311
Benzo[*b*]naphtho[2,1-*d*]thiophene, 311, 312
Benzo[*b*]naphtho[2,3-*d*]thiophene, 311
Benzophenone, 48, 92, 94, 129, 146, 224, 481, 504
Benzophenone anyl, 77, 80
Benzophenone chloroacetylhydrazone, 78
2,3-Benzopyrroles, 397, 398
3,4-Benzopyrrole, 470
2,3-Benzopyrrolines, 398
o-Benzoquinodimethane, 289
Benzoquinones, reaction with enamines, 156
Benzoquinone, 109, 143, 156, 164, 166, 235, 308, 313, 403, 416
p-Benzoquinone dibenzoylimide, 403
o-Benzoquinone dimethide, 289
Benzo[*b*]selenophene, **320**

Benzo [*b*]selenophene, *(continued)*
—, derivatives, 320, 324
2,3-Benzoselenophenes, 320
Benzo [*b*]selenophene-3-carbaldehyde, 321, 322
Benzo [*b*]selenophene-2-carboxylic acid, 322
Benzoselenopheno [2,3-*b*]benzoselenophene, 326
3,4-Benzosilacyclopentane, 544
Benzotellurophenes, 327
4,5-Benzothiadiazole 1,1-dioxide, 492
Benzo-1,3-thiaza-2-thionium chlorides, 267
Benzothienobenzothiophenes, 298
Benzo [*b*]thieno [3,2-*b*]indole, 480
3-Benzothienylacetic acid, 283
β-3-Benzothienylalanine, 280
β-3-Benzothienylethyl alcohol, 278
β-2-Benzothienylethylamine, 278
Benzothienyl-lithium, 272, 273, 277, 278, 281
3-Benzothienylmagnesium bromide, 279
2-Benzothienyl-sodium, 277, 279
2-Benzothienyltriphenylsilicane, 273
Benzothiepins, 56
3,4-Benzothiocoumarin, 303
6,7-Benzothioindoxyl, 306
Benzothiolenes, **280**
Benzothiophenes, 268, 269
—, reduced, 280
Benzothiophene, 147, 230, 305
—, acetylation, 279
—, acylation, 270
—, bromination, 270, 273
—, chloromethylation, 277
—, isopropylation, 271
—, mercuration, 270
—, nitration, 270, 274
—, oxidation, 270
—, reactions, 270
—, reduction, 270, 280
—, sulphonation, 277
Benzo [*b*]thiophenes, 265, 312
—, synthesis, 266
Benzo [*b*]thiophene, 219, 265, **271**, *554*
Benzo [*c*]thiophene, 288, **289**
2,3-Benzothiophene, 265
3,4-Benzothiophenes, 288
Benzothiophene-2-acetic acid, **280**
Benzothiophene-3-acetic acid, 280
Benzothiophene-4-acetic acid, 280
Benzothiophene alcohols, 277
Benzothiophene aldehydes, 277
Benzothiophenecarbaldehydes, 279
Benzothiophene-2-carbaldehyde, 278
Benzothiophene-3-carbaldehyde, 277
Benzothiophene-2-carboxylamide, 278
Benzothiophenecarboxylic acids, **279**, 280
Benzo [*c*]thiophene-1-carboxylic acid, **290**
Benzothiophene-2-carboxylic acid, 266, 267, 277
Benzothiophene-3-carboxylic acid, 279
Benzo [*c*]thiophene-5-carboxylic acid, 290
Benzothiophene-2,3-dicarboxylic acid, 279
Benzothiophene-4,5-dicarboxylic acid anhydride, 269
Benzothiophene 1,1-dioxide, **271**, 272, 273, 275, 281, 306
Benzothiophene ketones, 277
Benzothiophenequinone, 280
Benzothiophene-2,3-quinone, **284**
Benzothiophene-4,7-quinone, **276**
Benzothiophenesulphonic acids, **277**
Benzothiophene-2-thiol, **277**
Benzothiopheno [2,3]cycloheptene, 313
Benzothiopheno [2,3-*b*]cyclopentenone,313
N-Benzoxazolinonylaziridines, 30
Benzoylacetone, 268
1-Benzoylalkylpyrroles, 344
1-Benzoylazetidine, 63, 64
1-Benzoylaziridine, 28
2-Benzoylbenzofuran, 159
o-Benzoylbenzoic esters, 189
2-Benzoylbenzophenones, 184, 474
3-Benzoylbenzothiophene, 279
N-(2-Benzoylbenzyl)phthalimide, 471
3-Benzoyl-1-*tert*-butyl-2-phenylazetidines, 64
2-Benzoyl-1-*tert*-butyl-3-phenylaziridine, 25
Benzoylcarbazoles, **506**
Benzoyl chloride, 37, 413, 468
N-Benzoylchloroaziridines, 37
2-Benzoyl-1-cyclohexyl-3-phenylaziridine, 25
2-Benzoyldibenzofuran, 201
1-Benzoyl-2,3-dihydroxy-2,3-dimethylindoline, 447
N-Benzoyldiphenylaminedicarboxylic acids, 451

3-Benzoyl-2,5-diphenylpyrrole, 349
cis-Benzoylethene, 99
2-Benzoylfuran, 116, 145
2-Benzoyl-3-hydroxybenzothiophene, 268
3-Benzoyl-5-hydroxy-2-methylnaphtho-[1,2-*b*]furan, 207
1-Benzoylindole, 413, 419
1-Benzoylindolines, 447
2-Benzoylisoindoline, 476
3-Benzoyl-2-methylindole, 434
3-Benzoyl-2-methylnaphtho[2,1-*b*]furan, 208
Benzoyl nitrate, 241
1-Benzoyloxindole, 453
Benzoyl peroxide, 495
2-Benzoyl-2-phenyl-3-diphenylmethylene-β-sultone, 60
2-Benzoyl-3-phenylindole, 409
Benzoyl(phenyl)isobenzofuran, 182
2-Benzoyl-3-phenyloxirane, 3
β-Benzoylpropionic acid, 237
β-Benzoylpropionitrile, 371
Benzoylpyrroles, 341
1-Benzoylpyrrolidine, 382
1-Benzoyl-3-pyrroline, 375
Benzoyltetrahydrocarbazole, 511, 514
2-Benzoylthioindoxyl, 268
2-Benzoylthiophene, 251
3-Benzoylthiophene, **251**
1-Benzoyl-1,2,2-triphenylcyclopropane, 124
Benzylamine, 31, 525
—, *N*-methylene derivative, 81
β-Benzylaminocrotonic ester, 332
N-Benzylaniline, 26
1-Benzylaziridine, 28
2-Benzyl-1-azirine, 35
11-Benzyl-11*H*-benzo[*a*]carbazole, 522
2-Benzylbenzofuran, 151
Benzyl bromide, 522
N-Benzyl-*N*-*tert*-butylhydroxylamine, 41
Benzyl carbamate, 423
Benzyl chloride, 100, 303
Benzylchloromethyldichlorosilane, 544
Benzyl cyanide, 36
11-Benzyl-1,2-dihydro-11*H*-benzo[*a*]-carbazole, 522
1-Benzyl-2,3-dioxopyrrolidine, 387
1-Benzyl-2,5-diphenylpyrrole, 348
2-Benzylfuran, 101
α-Benzyl-β-2-furylacetylene, 100
Benzylideneaniline, 26, 68, 108
N-Benzylidene-*tert*-butylamine, 41
Benzylidenecrotonolactone, 109
1-Benzylidene-1,3-dihydroisobenzofuran, 190
2-Benzylidene-2,3-dihydrobenzo[*b*]selenophene, 323
2-Benzylidene-2,3-dihydroselenonaphthen-one, 323
4-Benzylidene-2-methyl-5-oxo-2-thiolene, 244
3-Benzylideneoxindole, 454
Benzylidene-2-oxo-2,3-dihydrobenzofurans, 170
2-Benzylidene-3-oxo-2,3-dihydrobenzofuran, 173
2-Benzylidene-2-oxo-2,3-dihydrobenzofuran, 171
3-Benzylidene-2-oxo-2,3-dihydrobenzofuran, 170
Benzylidene-3*H*-pyrroles, 378
1-Benzylindole, 413
3-Benzylindole, 418
2-Benzylisoindoline, 476
Benzyl ketoximes, 18
Benzylmagnesium chloride, reaction with β-propiolactone, 50
1-Benzyl-2-methylisoindole, 474
1-Benzyl-3-nitro-2,5-diphenylpyrrole, 350
α-Benzyloxyacetophenone, 48
o-Benzyloxybenzaldehydes, 167
5-Benzyloxyindole, 431
2-Benzyl-3-phenylindole, 409
1-Benzylpyrrole, 340, 522
1-Benzylpyrrolidine, 382
O-Benzylserotonin, 449
N-Benzyltetramethylpyrrole, 344
2-Benzylthiophene, 237, 251
1-Benzyl-1,2,2-trimethylazetidinium perchlorates, 66
Benzyltriphenoxyphosphonium bromide, 130
Benzyne, 186, 344, 472, 492
—, reaction with indole, 410
Betonica officinalis, 390
Betonicine, **390**
1,1′-Biaziridine, 21
2,2′-Bibenzothienyl, 276
Bicarbazolyls, **517**
3,3-Bicarbazyl, 490

9,9-Bicarbazyl, 490
Bicyclic pyrrole systems, 477
Bicyclic systems, with a furan ring, 191
Bicyclodiaziridine valence bond tautomer, 78
1,1′-Bicyclohexenyl, 539
Bicyclohexyl, 203
Bicyclo-oxin, 94
2,2′-Bifuran, 136, **137**
2,3′-Bifuran, 138
3,3′-Bifuran, **137**
2,2′-Bifuran-5-carbaldehyde, 137, 138
2,2′-Bifuran-5,5′-dicarboxylic methyl ester, 138
β-5-(2,2′-Bifuranyl)acrylic acid, 138
2,2′-Bifuryls, 136
Bile pigments, 329, 375, 393
Bilirubic acids, 395
Biotin, 133, 219
β-Biotin, 258, 260
Biphenols, 427
Biphenyls, 487
—, *o*-substituted, 194
Biphenyl, 328, 506
—, reaction with sulphur, 300
Biphenylene, 304
Biphenyl ethers, 194
2-Biphenylmagnesium iodide, 164
Biphenyl-2-sulphonic acid, 304
2-Biphenylyl-2,2′-biphenylyleneselenonium iodide, 325
2-Biphenylyldialkylboranes, 550
o-Biphenylyldiphenylaluminium, 549
o-Biphenylyl-lithium, 549
2-Biphenylylphenylphosphinic acid, 538
Biphthalyls, 187
Bipyridines, nomenclature, 572
1,1-Bipyrrolyls, 332, **391**
2,2′-Bipyrrolyls, **392**
2,5-Bis(acetoxymethyl)furan, 98
3,4-Bisacetoxymethyltetrahydrothiophene, 233
Bis(acetylamino)dibenzoselenophenes, 325
Bis(1-acetylethylidene)-1,3-dithietane, 74
Bis-β-aminoethyl sulphide, 23
Bisbenzofuranquinone, 163
Bis(benzofurylidene)acetone, 159
Bis(2-biphenyl) disulphide, 302
Bis(2,2′-biphenylylene)selenium, 325
Bis(bromobenzofuran)mercury, 198
Bis-2-bromoethyl ether, 516
2,3-Bis(bromomethyl)naphthalene, 477
1,1-Bis(4-bromophenyl)ethene, 326
Bis(*o*-bromotetrafluorophenyl) sulphide, 304
2,3-Bis(*tert*-butylimino)azetidines, 70
3,6-Bis(*o*-carboxybenzoyl)carbazole, 525
5,5′-Bis(2-carboxyethyl)-2,2′-difurylmethanes, 138
Bis(1-carboxy-2-β-naphthylvinyl) disulphide, 307
Bischler indole synthesis, 401
Bischler reaction, 370, 494, 520
Bis[3-chloro-2-(3-chromenyl)] ether, 147
3,3-Bis(chloromethyl)oxetane, 47, 75
2,5-Bischloromethylselenophene, 317
1,1-Bis(4-chlorophenyl)ethene, 326
3,4-Biscyanomethylindole, 484
1,4-Bis(diazomethyl)benzene, 138
Bisdiazonium salts, 195
1,1′-Bis(1,3-dihydroisobenzofuranyls), 187, 188
2,2′-Bis(1,4-diiodobutyl), 261
1,4-Bis(ω-dimethoxyethylthio)benzene, 299
1,4-Bis(dimethylamino)-1,3-butadiene, 340
Bis(dimethylamino)methane, 243
Bis(4,4′-dimethyl-2,2′-biphenylylene) selenium, 325
Bis-2-diphenyl diselenide, 324
2,2′-Biselenienyl, **320**
2,5-Bis(ethoxycarbonylamino)thiolane, 260
2,4-Bisethoxycarbonyl-3,5-dimethylpyrrole, 370
1,3-Bis(ethoxycarbonyl)-4-methoxy-3-pyrroline, 361
Bis-9-ethylcarbazolylmercury, 503
2,5-Bisethylene-imino-*p*-benzoquinol, 15
Bis(ethylsulphonyl)methane, 144
1,2-Bis(2-furyl)ethene, 139
1,2-Bis(2-furyl)tetraphenylethane, 139
Bis(3-halogenopropyl) disulphide, 55
1,2-Bishydroxymethylcyclohexanes, 294
1,2-Bishydroxymethylcyclopentanes, 191
Bis(hydroxymethyl)phenylphosphine, 533
2,5-Bis(hydroxymethyl)pyrrole, 354
2,5-Bishydroxymethylthiolane, 259
Bisindoles, 469
2,2-Bisindole indigo, 460
Bis(iodomethyl)dimethylsilane, 71
1,4-Bismethanesulphonyloxybutane, 130

1,2-Bis(methoxycarbonyl)cyclo-octatetraene, 192
Bismethoxycarbonyldibenzofurans, 201
3,3-Bismethoxycarbonyl-1,1-dimethylsiletane, 71
2,2-Bis(6-methoxy-3-methyl-2-benzofuryl)propane, 160
1,2-Bis(*o*-methoxyphenyl)naphtho[2,1-*b*]furan, 208
1,2-Bis(*p*-methoxyphenyl)naphtho[2,1-*b*]furan, 208
2,2-Bis(3-methyl-2-benzofuryl)propane, 160
1,1-Bis(methylindolyl)ethane, 469
2,5-Bis[2-(3-methylindolyl)]hydroquinone, 416
Bis(methylindolyl)phenylmethane, 469
2,2-Bis(methylindolyl)propane, 469
1,1-Bis(4-methylphenyl)ethene, 326
Bis(β-2-naphthylethyl) disulphide, 307
1,2-Bis(3′-oxo-2′-indolinylidene)ethane, 469
Bis(2-phenyl-3-benzofuranyl)methane, 158
Bispyrrobenzoquinones, 362
Bis-selenophene, 320
3,3′-Bistellurolane, 327
Bis(tetrahydrofurfuryl)amine, 136
Bis(1,2,3,3-tetramethyl-3*H*-indolium) sulphate, 411
2,2′-Bisthiobenzoic acid, 283
ββ-Bis(trifluoromethyl)acrylic ester, 52
Bis(trifluoromethyl)-1,2-dithietene, **75**
2,4-Bis[2,2,2-trifluoro-1-(trifluoromethyl)ethylidene]1,3-dithietane, 73
3,3′-Bi(tetrahydrothiophene), 261
2,2′-Bithienyl, 228, 230, 231, **260**, 261, 262 263, 264
2,3′-Bithienyl, 231, **261**, 263
3,3′-Bithienyl, **260,** 261
2,2′-Bithienyl-5-carbaldehyde, 262
2,2′-Bithienyl-5-carboxylic acid, 262
2,2′-Bithienyl-5-carboxylic methyl ester, 262
2(5,2′-Bithienyl)vinyl ketones, 262
3,3′-Bithiolane, **261**
Bithiophene, **260**
Biurets, trisubstituted, 79
Bone oil, 329
Boron, hetero atoms in compounds, 549
Boron trifluoride dietherate, 171
Borsche synthesis, 509, 510, 524, 526
Bouveault–Blanc reduction, 111, 383, 428
Bractein, **174**
α-Brazan, **215**
β-Brazan, **214,** 215
Brazilin, 214
Bromoacetals, 297, 320
Bromoacetic acid, 352
Bromoacetic ester, 143, 216, 283, 322
2-Bromoacetalbenzofuran, 161
Bromoacetyl bromide, 106
9-(Bromoacetyl)carbazole, 505
2-Bromoacetyldibenzofuran, 201
2-Bromoacetylthiophene, 251
β-Bromo-acid chlorides, 46
1-Bromoaziridine, 21
Bromobenzene, 260, 410
Bromobenzofurans, 152, 153
3-Bromobenzofuran, 146, 164
Bromobenzothiophenes, **273,** 274
Bromobenzothiophene 1,1-dioxides, 273
5-Bromo-2,2′-bithienyl, 262
4-Bromobutylmethanesulphonate, 130
2-Bromo-*N*-*tert*-butyl-2-methylpropionamide, 31
Bromocarbazoles, 496
α-Bromocarbonyl compounds, 222
γ-Bromocrotonic methyl ester, 378
α-Bromodeoxybenzoin, 150
Bromodibenzofurans, 197, 200
2-Bromodibenzofuran, 196
Bromodibenzothiophenes, 305
1-Bromodibenzothiophene, 301
2-Bromodibenzothiophene, 303, 304
7-Bromo-2,3-dimethylbenzofuran, 153
4-Bromodimethylhexahydronaphtho-[2,1-*b*]furan, 212
1-Bromo-3,4-dimethyl-3-phospholene 1-oxide, 536
3-Bromo-2,7-dinitrobenzothiophene-5-diazo-4-oxide, 276
1-Bromo-1,3-diphenyl-1,3-dihydrobenzo-[*c*]thiophene 2,2-dioxide, 292
α-Bromoethanesulphonyl chloride, 14
β-Bromoethylamine, 15, 16
1-Bromo-1-ethylgermacyclopentane, 546
2-Bromofluorobenzene, 410, 522
Bromofuran, 102, 104
Bromofurancarboxylic acids, 122
3-Bromofuran-2-carboxylic acid, 117

3-Bromofurfural, 114
5-Bromofurfural, 109
5-Bromo-2-furoic methyl ester, 107
Bromoindoles, **420**
5-Bromoisatin, **462**, 464
3-Bromo-2-lithiobenzofuran, 198
Bromomalonic methyl ester, 150
Bromomethylbenzothiophenes, 278
2-Bromomethylbenzyl bromide, 475
2-Bromomethyl-4,7-di-*tert*-butyl-2,3-dihydrobenzofuran, 167
2-Bromomethyl-2,3-dihydrobenzofuran, 175
2-Bromomethylfuran, **110**
6-Bromo-1-methylindole-2,3,4-tricarboxylic ester, 419
3-Bromo-3-methyloxindole, 454
Bromomethyl phenyl ketone, 3
2-Bromo-2-methyl-*N*-*n*-propylpropionamide, 32
Bromomethylpyrroles, 355, 393, 394
3-Bromomethylselenophene, 316
2-Bromomethyltetrahydrofuran, 135
2-Bromo-2-methylthiirane 1,1-dioxide, 14
2-Bromomethylthiophene, 248
3-Bromomethylthiophene, 248
4-Bromo-2-methylthiophene, 261
Bromonaphtho[2,1-*b*]thiophene, 309
3-Bromo-2-nitrobenzofuran, 154
3-Bromo-2-nitrobenzothiophene, 274
1-Bromo-3-nitrocarbazole, 496
5-Bromo-2-nitrofuran, 103, 105
1-Bromo-9-nitrosocarbazole, 496
5-Bromo-4-nitrothiophene-2-carboxylic acid, 242
3-Bromo-oxindoles, 416, 420
5-Bromo-oxindole, 452
3-Bromo-oxindole-3-acetic acid, 450, 454
3-Bromo-oxindole-3-butyric acid, 454
2-Bromo-3-oxo-2,3-dihydrobenzofuran, 172
2-Bromophenylacetic ester, 150
3-Bromo-2-phenylbenzofuran, 152
5-Bromo-2-phenylbenzofuran, 152, 153
1-Bromo-1-phenylgermacyclopentane, 546
4-Bromo-3-phenylpyrrole-5-carboxylic acid, 347
Bromophenylthiophenes, 239
α-Bromopropionic ester, 212
5-Bromo-2-*n*-propylbenzofuran, 152, 153
γ-Bromopropylmalonic ester, 388
5-Bromopyromucic acid, 105
5-Bromopyrrole-2-carboxylic acids, 376
Bromopyrrolecarboxylic esters, 347
Bromoselenophenes, 316, 318
N-Bromosuccinimide, 29, 71, 110, 199, 235, 239, 248, 287, 290, 322, 330, 416, 420, 495
5-Bromotetrachloro-1,3-dihydroisobenzofuran, 190
5-Bromotetrafluoro-1,3-dihydroisobenzofuran, 190
6-Bromotetrahydrocarbazole, 514
1-Bromotetraphenylpyrrole, 347
5′-Bromo-2,2′-thienyl-5-carboxylic acid, 262
Bromothiophenes, 239, 240, 241
2-Bromothiophene, 231, 232, 239, 243, 247
3-Bromothiophene, 231
5-Bromothiophene-2-carbaldehyde, 249
5-Bromothiophene-2-carboxylic ester, 242
3-Bromothiophene-4-thiol, 247
Bufotenidine, **431**
Bufotenine, **431**
Bufothionine, **431**
Butadienes, 339
—, reaction with 1,2-diacylethenes, 182
—, — with dibenzoylacetylene, 183
Butadiene, 269, 272 306, 374, 472, 534, 535, 537, 542, 547
—, conversion to pyrrole, 337
—, reaction with 2,3-dichlorotetrahydrofuran, 131
—, — with furfural, 113
—, — with sulphur dioxide, 255
Butadienecarboxylic esters, 360
1,3-Butadiynes, 533
Butane, 227, 229, 259
Butanetetracarboxylic acid, 138, 260
Butanethiol, 227, 259
Butane-1,2-4-triol, 132
Butan-1-ol, 130
Butenes, 10, 259
But-2-ene, 12
But-2-ene sulphoxide, 12
Butenolides, 107, 125
3-Butenyldibenzofuran, 200
2-Butenylgermanium trichloride, 547
2-(Butenyl)thiophene, 269
2-Butenyltrimethylgermanium, 547

2-*tert*-Butoxy-2-methyl-*N*-*n*-propylpropionamide, 32
5-*tert*-Butoxy-2-selenienyl-lithium, 319
2-*tert*-Butoxythiophene, 244
tert-Butyl acetoacetate, 361
tert-Butylacetylene, 168
tert-Butyl alcohol, 31
N-*tert*-Butylamides, 32
Butylamines, *N*-halogenated, 380
Butylamine, 375
2-*tert*-Butylamino-2-methyl-1-propanol, 33
3-*tert*-Butylamino-2-phenylindoles, 70
3-*tert*-Butylaminopropionhydrazide, 31
1-*tert*-Butylaziridine-2-carboxylic acid, 68
N-*tert*-Butylaziridine-2-carboxylic acid, 24
1-*tert*-Butylaziridine-2-carboxylic ester, 31
2-Butyl-1-azirine, 35
tert-Butylazomethine, 43
tert-Butyl-1,4-benzoquinones, 150, 167
3-*tert*-Butylbenzothiophene, 272, 273
Butyl bromide, 260
tert-Butyl-2-*tert*-butylaminopropionate, 32
tert-Butyl carbazate, 350
1-*tert*-Butyl-3-chloroazetidine, 65
1-*tert*-Butyl-2-chloroazetidin-2-one, 24
1-*tert*-Butyl-3-chloroazetidin-2-one, 68
1-*tert*-Butyl-2-chloromethylaziridine, 65
N-*tert*-Butyl-*N*-chlorophenylacetamide, 31
Butyl dichlorophosphite, 535
3-*tert*-Butyl-2,3-dihydrobenzofuran, 168
1-*n*-Butyl-3,4-dihydroxypyrrolidines, 383
1-*tert*-Butyl-3,3-dimethylaziridinone, 31, 32
5-*tert*-Butyl-2,2-dimethyl-1,4-dioxaspiro-[2.2]pentane, 4
3-*tert*-Butyl-2,5-dimethylthiophene, 235
1-*tert*-Butyldiphenylpyrroles, 64
2-*tert*-Burylfuran, 101
1-Butylgermacyclopentane, 546
tert-Butyl hydroperoxide, 174
tert-Butyl *N*-hydroxycarbamate, 154
tert-Butyl hypochlorite, 43
2-*tert*-Butylindole, 418
3-*tert*-Butylindole, 416
tert-Butyl isocyanide, 31, 70
2-*tert*-Butyl-3-isopropyloxaziridine, 40
Butyl-lithium, 55, 97, 103, 114, 117, 146, 153, 231, 240, 244, 247, 261, 271, 297, 301, 303, 317, 325, 359, 360, 381, 481, 503, 504, 507, 522
N-*tert*-Butylmethylacrylamide, 31
5-*tert*-Butyl-2-methyl-2,3-dihydrofuran, 169
1-Butyl-2-methylisoindole, 474
tert-Butyl 2-methyl-2-*n*-propionaminopropionate, 32
o-*n*-Butylnitrobenzene, 446
tert-Butyl perbenzoate, 132
tert-Butyl peroxide, 128
1-*tert*-Butyl-3-phenylaziridinone, 31
2-*tert*-Butyl-3-phenyloxaziridine, 41
Butylpyrroles, 339
1-*n*-Butylpyrrole, 340
1-*n*-Butylpyrrolidine, 382
Butylthiophenes, 233, 237
—, isomerisation, 235
5-*tert*-Butylthiophene-2-carbaldehyde, 249
3-*n*-Butyl-1,1,2-triphenyl-1-silaindene, 543
Butynedial tetraethyl acetal, 137
Butynediol, 125, 337
Butyrolactone, 130, 384
Butyrolactone phenylhydrazone, 399
Butyrylcarbazole, 506
n-Butyryl chloride, 251

Cafestol, 214
Cannabinoid, 72
Cannizzaro reaction, 110, 112, 119, 158, 247, 249, 277, 433
Cantharidin, 193
Caproic acid, labelled, 252
2-Caproylcarbazole, 506
Carbamates, 27
Carbamic ester, 359
1-Carbamoylazetidine, 63
1-Carbamoylaziridine, 28
Carbamoyl chloride, 253, 362
2-Carbamoylisoindoline, 476
Carbazoles, fused ring, 524
—, *N*-substituted, 489, 492
—, reduced, 514
—, with additional fused rings, 518
Carbazole, **486**, 489, 490, **491**, *558*
—, acetylation, 493, 505
—, alcohols, 504
—, aldehydes, 504
—, hydrogenation, 517
—, ketones, 504
—, organometallic derivatives, 503
—, oxidation, 490, 517
—, reactions, 490

Carbazole, *(continued)*
—, reaction with acetylene, 492
—, — with chloroacetic ester, 493
—, — with ethylene oxide, 492
—, — with formaldehyde, 493
—, — with *p*-nitrosophenol, 500
—, — with olefins, 494
—, — with oxalic acid, 518
—, — with phthalic anhydride, 525
—, reduction, 491
—, sulphonation, 503
—, thiocyanation, 503
4*aH*-Carbazoles, 510, *558*
2-Carbazoleazo-2′-naphthol, 499
Carbazole Blue, **518**
Carbazolecarbaldehydes, **504**
Carbazolecarboxylic acids, 492, 504, 506, **507**, 508
Carbazole-3-diazoimine, 500
Carbazolediazonium salts, 495, **499**
Carbazoledicarboxylic acids, 507
Carbazole-3,6-disulphonic acid, 502
Carbazole dyestuffs, 487
Carbazole–formaldehyde condensate, 508
Carbazole-2-sulphinic acid, 503
Carbazole-3-sulphonate, 502
Carbazolesulphonic acids, **502**
Carbazole-2,3,6,8-tetrasulphonic acid, 502
Carbazole-3-thiol, **503**
9-Carbazolylacetic acids, **493**
9-Carbazolylacetyl chloride, 508
Carbazolylaminophenol, 500
Carbazolylcarbinols, 504
3-Carbazolyldiphenylcarbinol, 504
Carbazolylmethanols, 504
Carbazolyl methyl ketones, 506
3-Carbazolylthiocyanate, **503**
Carbazolyl trichloromethyl ketones, 506
Carbazylmagnesium halides, 493
Carbene, 43
Carbinolamines, 188
Carbodiimides, 44
Carbohydrates, conversion to furans, 86
β-Carboline, *558*
Carbonates, cyclic, 6
Carbon–carbon fission, aziridines, 25
Carbon chain, lengthening, 131
Carbon dioxide, reaction with aziridine, 24
Carbon disulphide, reaction with aziridines, 24
1,1′-Carbonylbisaziridine, 28
Carbonylbispyrrole, 341, 393
Carbonyl compounds, 46
—, conversion to oxiranes, 4
—, reaction with aziridine, 23
—, — with ketenes, 49
—, — with vinyl ethers, 48
Carbonyl group reactivity, isatins, 461
Carbonyl sulphide, reaction with oxiranes, 6
Carbopyrotritaric acid, 118
Carboxanilides, 20
o-Carboxybenzenesulphonylacetic acid, 286
2-Carboxybenzofurans, 142
o-Carboxybenzoylindoles, 434
Carboxycyclohexanones, 488
Carboxydibenzofurans, 201
3-(2-Carboxyethyl)-1,2-dimethyl-5-methoxyindole, 485
α-Carboxylamides, 333
Carboxylic acids, furans, 117
Carboxylic acid methyl esters, hydrolysis rates, 165
2-Carboxymethoxy-4,6,4′-trimethoxybenzil, 143
1-Carboxymethylisatin, 464
o-Carboxyphenoxyacetic ester, 166
2′-Carboxyphenyl-2-methylpyrrole-3-carboxylic acid, 391
o-Carboxyphenyl thienyl sulphide, 247
o-Carboxyphenylthioglycollic acid, 267
5-(3′-Carboxypropyl)-2,3-dihydrobenzo[*b*]thiophene, 310
2-Carboxytetrahydrofuran, 135
Carlina acaulis, 100
Carlina oxide, **100**
Casein, 436
Catalase, 427
Catechol, 145
Catechol carbonate, 252
Catechol oxidase, 425
Cernuoside, **173**
Chalkones, 173
Chemical Abstracts Nomenclature, *566*
Chitin, 343
Chloral, 382
Chloramine-T, 41, 284
Chloranil, 163, 225, 487, 493, 519
Chlorins, 393

Chloroacetaldehyde, 85, 115, 287, 288, 332
Chloroacetanilide, 31, 449
Chloroacetic acid, 267
Chloroacetic ester, 4, 115, 513
α-Chloroacetoacetic ester, 163, 177, 216
Chloroacetonitrile, 172, 431, 435
Chloroacetophenone, 207
α-Chloroacetoxy-3,4-dihydroxybenzene,172
Chloroacetyl chloride, 173, 376
Chloroacetyldiazomethane, 50
2-Chloroacetylfuran, 116
3-Chloroacetylindole, 434
9-Chloroacetyltetrahydrocarbazole, 514
2-(2-Chloroallyl)cyclopentanones, 177
β-Chloroallyl ketones, 84, 177
α-Chloroamides, 20
N-Chloroamines, cyclisation, 380
Chloroanil, 496
7-Chloroazabicyclo [4.1.0]heptane, 30
Chloroaziridines, 27, 28, 29, 30
1-Chloroaziridine, 21
5-Chloro-1,2-benzisothiazole-3-carboxamide, 285
3-Chlorobenzo [*f*]coumarin, 207
Chlorobenzofurans, 152, 153
3-Chlorobenzofuran-2-carbaldehyde, 159
5-Chlorobenzofuran-2-carboxylates, 163
Chlorobenzothiophenes, **273**
3-Chlorobenzothiophene 1,1-dioxide, 282
1-(4-Chlorobenzoyl)-5-methoxy-2-methylindole-3-acetic acid, 436
5-Chloro-2,2′-bithienyl, 262
5′-Chloro-2,2′-bithienyl-5-carboxylic acid, 262
1-Chloro-2-(2-bromoethyl)benzene, 544
3-Chlorobutan-2-one, 159, 161
2-Chloro-*N*-butylpropionamide, 32
γ-Chlorobutyronitrile, 368
Chlorocarbazoles, 496
Chlorocarbene, 28, 338
5-Chloro-2-chloromethylthiane, 259
1-Chloro-4-(3-chlorotetrahydrofur-2-yl)-but-2-ene, 132
3-Chloro-4-(3-chlorotetrahydrofur-2-yl)-but-1-ene, 132
3-Chlorocoumarin, 163
Chlorocyclohexanone, 178, 203, 204, 465, 489, 494, 520
1-Chlorocyclopentane-2,3-dione, 249
Chlorodibenzofurans, 200
3-Chlorodibenzofuran, 196
Chlorodibenzothiophenes, 305
3-Chloro-4,5-dihydrofuran, 131
7-Chloro-4,6-dimethoxy-2,3-dihydrobenzofuran-3-one, 175
7-Chloro-2,3-dimethylbenzofuran, 152
1-Chloro-3,4-dimethyl-2-phospholene 1-oxide, 536
1-Chloro-3,4-dimethyl-3-phospholene 1-oxide, 536
3-Chloro-2,4-dimethylquinoline, 416
Chloro-2,4-dinitrobenzene, 154, 421
3-Chloro-1,2-dioxocyclopentane, 224
5-Chloro-2,3-dioxo-2,3-dihydrobenzothiophene, 285
α-Chloro-αα-diphenylacetanilide, 33, 450
Chlorodiphenylacetyl chloride, 76
5-Chloro-2,3-diphenyl-6-nitrobenzofuran, 155
1-Chloro-2,5-diphenylsilacyclopentadiene, 541
Chlorodiphenylsilane, 544
Chlorodithienylarsine, 232
5-Chloro-5-*n*-dodecyldibenzosilole, 544
2-Chloroethanol, 3
β-Chloroethylamines, 16, 21, 22, 39, 158
o-β-Chloroethylbenzanilides, 447
9-(2-Chloroethyl)carbazole, 492
3-β-Chloroethylcyclohex-1-ene, 288
2-(2′-Chloroethyl)-1-methylpyrrolidine, 381
2-α-Chloroethylthiophene, 237
Chloroform, 442
Chloroformic ester, 338, 360, 434
Chloroformic methyl ester, reaction with pyrrole, 359
2-Chlorofuran, 102, 104
Chlorofurancarboxylic acids, 122
5-Chlorofurfural, 109
3-Chloro-4-hydroxycarbostyril, 456
Chloroindoles, 418, **420**
5-Chloroisatin, 464
α-Chloroketones, 85
Chloromercurifurans, 97
Chloromercuri-indoles, 419
2-(Chloromercurimethyl)-2,3-dihydrobenzofuran, 176
2-Chloromercurithiophene, 233, 247
N-Chloromethylamine, 40, 42
1-Chloro-2-methylaziridine, 16, 29
2-Chloromethylbenzofuran, 158

2-Chloromethylbenzothiophene, 278
3-Chloromethylbenzothiophene, 277, 278, 280
2-Chloromethyl-5-(3′-chloropropen-1-yl)-thiophene, 235
5-Chloro-5-methyldibenzosilole, 544, 545
2-Chloromethyl-2,3-dihydrobenzofuran, 175
2-Chloro-2-methyl-1,3-diphenylaziridine, 28
2-Chloromethylfuran, **109**, 110
3-Chloromethylfuran, **110**
5-Chloromethylfuran-2-carboxylic methyl ester, 119
5-Chloromethylfurfural, 95, 115
3-Chloro-2-methylindole, 419
3-Chloromethyl-2-methylbenzofuran, 158
Chloromethylmethylbenzothiophenes, 278
2-Chloromethyl-5-methylfuran, 110
2-Chloromethylphenacyl chloride, 475
4-Chloro-3-methylphenoxyacetyl chlorides, 171
3-Chloromethyl-2-phenylbenzofuran, 158
1-Chloro-4-methyl-2-phospholene 1-sulphide, 537
2-Chloromethylselenophene, 317
1-Chloro-1-methylsilacyclopentenes, 542
2-Chloromethyltetrahydrofuran, 135
3-Chloro-2-methyltetrahydrofuran, 135
Chloromethylthiirane, 9, 13, 53
2-Chloromethylthiophene, 248
3-Chloronaphtho[2,3-*b*]furan, 211
o-Chloronitrobenzenes, 403
1-Chloro-1-nitrosocyclohexane, 360
6-Chloro-3-oxo-2,3-dihydrobenzo[*b*]-selenophene, 323
2-Chloro-3-oxo-3*H*-indole, 445, 460
2-Chloropenta-2,4-diene-1-al, 124
1-Chloropentaphenylsilacyclopentadiene, 542
o-Chlorophenol, 152
Chloro-2-phenylbenzofurans, 152, 153
2-(4-Chlorophenyl)-1-cyano-4,4-diphenyl-diazetidin-3-one, 77
p-Chlorophenyl diazocyanide, 77
5-Chloro-5-phenyldibenzosilole, 544
6-Chlorophenylhexahydrobenzo[*g*]-2,5-diazocine, 473
3-Chloro-4-phenylpyrrole, 347
1-Chloro-2-phospholene, 537
1-Chlorophospholene 1-oxides, 535, 536
1-Chlorophospholene 1-sulphides, 536, 537
1-Chlorophthalazine, 474
Chlorophyll, 329, 342, 363, 393
3-Chloropropanol, 46
2-(3-Chloro-1-propenyl)-5-chloromethyl-thiophene, 248
3-Chloropropyl acetate, 46
β-Chloropropylamine, 158
γ-Chloropropylamine, 64
2-Chloropropyl *S*-thioacetate, 9
3-Chloropyridine, 337, 355
2-Chloropyrrole, **346**
Chloropyrrolecarboxylic esters, 347
N-Chloropyrrolidine, 369
2-Chloropyrroline, 371
3-Chloroquinoline, 432
2-Chloroselenophenes, 316, 318
N-Chlorosuccinimide, 29, 419
Chlorosulphonic acid, 455
2-Chlorosulphonyldibenzofuran, 201
Chlorosulphonyl isocyanate, 68, 69
1-Chlorosulphonyl-3-methyl-4-phenyl-azetidin-2-one, 68
5-Chlorotetrafluoro-1,3-dihydroisobenzo-furan, 190
8-Chlorotetrahydrocarbazole, 514
2-Chlorotetrahydrofuran, 131
3-Chlorotetrahydrofuran, 135
1-Chlorotetraphenylsilacyclopentadiene, 541
1,2-Chlorothiocyanates, 7
1,2-Chlorothiols, 6
3-Chlorothiol acetates, 52
3-Chlorothiol esters, 52
Chlorothiophenes, 238, 241, 257
2-Chlorothiophene, 231
5-Chlorothiophene-2-carbaldehyde, 249
2-Chlorovinyl aldehydes, 221
Chromans, 167, 560
Chromene, 554
Chrysene, 311
1,4-Cineole, 193
Cinnamic acid, 197
Cinnamic ester, 245
2-Cinnamoylbenzofuran, 160
2-Cinnamoylfuran, 116
Cinnamoylglycine, 249
Cinnamoylpyrroles, 344
2-Cinnamoylthiophene, 251

Cinnamylidenearylamines, 387
Cinnolines, 519, *557*
Cinnolinols, 444
Citral, 72
Citric acid, 296
Civet, 415
Claisen condensation, 140
Clavine alkaloids, 486
Clemmensen reduction, 380
Coal tar, 156, 194, 207, 214, 219, 265, 300, 306, 311, 329, 397, 411, 414, 486, 491
Coal-tar naphthalene, 271
Convallaria majalis, 67
Coproporphyrin ester, 395
Coreopsis grandiflora, 174
Cosmos sulphureus, 174
Coumalinic methyl ester, 85
Coumarans, 141, 145, 166
Coumaran, **168**
Coumarandione, 165
Coumaranones, 141, 146, 156, 480
Coumarilamide, 155
Coumarilic acid, 141, 142, 147, 158, **165**, 176
—, esters, 163, 164, 165
Coumariloyl cyanide, 159
Coumarins, 141, 163
Coumarin-4-acetic acids, 166
Coumarone, 141
Coumarone resins, 141
Courbonia virgata, 390
Cresol, 150
p-Cresol, oxidation, 206
Croton gubouga, 390
Crotyl 2,3-dichloro-5-methylphenyl ether, 168
o-Crotylthiobenzoic acid, 281
Cryptopyrrole, 343, 345, 378
Cryptopyrrolecarboxylic acid, 353, 367
Cryptopyrrolidine, 382
Cuprous acetylides, 144
Cuprous phenylacetylide, 152
Cuprous *n*-propylacetylide, 152
Curcuma zedoaria, 210
Curtis reaction, 243, 250
Curtius reaction, 170, 423
Curzerenone, 208, 210
Cuscohygrine, 386
Cyanine dyes, 442, 465
Cyanoacetic acid, 484
Cyanoacetic esters, 70, 429
Cyanoacetic ester 396, 403, 527
Cyanoaziridines, 30
o-Cyanobenzamide, 476
2-Cyanobenzofuran, 159
3-Cyanobenzofuran, 146, 160, **166**
5-Cyano-2,2′-bithienyl, 262
Cyanodibenzofurans, 201
2-Cyano-3,3-dimethyl-3*H*-indole, 443
2-(2-Cyanoethyl)cyclohexanone phenylhydrazone, 487
1-*β*-Cyanoethylindole, 412, 413
1-(2-Cyanoethyl)tetrahydrocarbazole, 487
4*a*-(2-Cyanoethyl)-1-tetrahydro-4*aH*-carbazole, 487
Cyanofuran, 103, 105, 138
Cyanogen, 79
Cyanogen bromide, 22, 470
Cyanogen chloride, 253
Cyanomaclurin, **169**
2-Cyano-3-methylbenzofuran, 165
2-Cyano-5-methylfuran, 110
4-Cyanomethylindole, 484
2-Cyanomethyl-5-methoxythiophene, 244
2-Cyanomethyl-5-methylfuran, 110
4-Cyano-1-methyl-3-oxopyrrolidine, 386
2-Cyano-1-methylpyrrole, 349
2-Cyanomethyl-5-(2′,4′, 6′-trihydroxyphenyl)thiophene, 244
2-Cyanopyrroles, 349
3-Cyanopyrrole, 351
2-Cyanopyrrolidines, 371
2-Cyanothiophene, 242, **253**
3-Cyanothiophene, 242, 251, **253**
Cyanuryl chloride, 355
Cyclic hydrocarbons, from thiophenes, 227
2,3-Cycloalkanotetrahydrothiophenes, 294
Cycloalkanothiophenes, 294
Cycloalkenopyrroles, 478
Cycloalkyldibenzothiophenes, 303
Cyclobutanes, 93
Cyclobutene episulphides, 8
2,3-Cycloheptanopyrrolidine, 479
Cycloheptanothiophene, 295
Cycloheptene, 17
2,3-Cycloheptenofuran, 192
Cyclohexadiene, 217
Cyclohexadienobis[1,2-*b*:5,4-*b*′] indole, 525

Cyclohexane-1,2-dicarboxylic anhydride, 293
Cyclohexane-1,3-diones, 177, 204
Cyclohexane-1,4-dione bisphenylhydrazone, 524
Cyclohexanedione phenylhydrazone, 488
Cyclohexane spiro-5-(2′-methyl-4′-oxo-1′-phenyl-2′-pyrroline), 377
Cyclohexanespiro-2-(tetrahydrofuran), 570
Cyclohexanols, 116
Cyclohexanones, 116, 286, 303
—, reaction with phenols, 203
Cyclohexanone, 48, 176, 217, 288, 306, 399, 465, 466, 489, 500, 509, 527
—, reaction with sulphuric acid, 204
Cyclohexanone arylhydrazones, 487, 509
Cyclohexanone azine, 516, 517
Cyclohexanone 2,6-dibromophenylhydrazone, 495
Cyclohexanone oxime, Beckmann rearrangement, 204
Cyclohexanone phenylhydrazones, 487
2-Cyclohexanonepropionic acid, 488
Cyclohexanone pyrrolidine enamine, 57
Cyclohexanonoindole, 527
Cyclohexenes, 180
Cyclohexene, 18, 26, 190
Cyclohexene-1,2-dicarboxylic anhydride, 293
Cyclohexene episulphide, 13
Cyclohexeneimine, 27
Cyclohexene oxide, 179
Cyclohexene sulphide, 11
Cyclohexenylcyclohexanone, 306, 517
(1-Cyclohexenyl)dibenzothiophenes, 303
o-(1-Cyclohexenyl)phenol, 203
Cyclohexylamine, 44
—, *N*-methylene derivative, 81
Cyclohexylcyclohexanol, 203
1-Cyclohexyl-3-ethyldiaziridine, 44
o-Cyclohexylphenylphosphorimidic ester, 516
Cyclo-octa[*c*]furan, 192
Cyclo-octanes, 13
Cyclo-octene, 13, 17, 18
Cyclo-octene methylepisulphonium 2,4,6-trinitrobenzenesulphonate, 13
Cyclo-octene sulphide, 13
Cyclopentadienes, 334
Cyclopentadiene, 217, 223
Cyclopentadiene, *(continued)*
—, ultraviolet absorption, 230
Cyclopentadithiophene, 299
Cyclopentanedicarboxylic acids, 295
Cyclopentane-1,2-dicarboxylic acid, 191
Cyclopentanonecarboxylic ester, 478
Cyclopentanopyrrolidine, **478**
3,4-Cyclopentanotetrahydrothiophene, 295
Cyclopentano[*b*]thiophene, 295
Cyclopenta[b]pyran, 564
Cyclopentene, 17
Cyclopentenoquinolone, 513
Cyclopent-2-enylacetoxime, 370
Cyclopentyne, 186
Cyclopropanes, 3, 124
Cyclopropanecarbaldehyde, 124
Cyclopropenes, 93, 94
Cyclopropenecarbaldehyde, 93
Cyclopropenethiocarbaldehyde, 234
Cyclopropylimines, 370
Cyclopropyl ketone oximes, 379
Cyclopropyl methyl ketone, 134
Cyclotetramethylenedichlorosilane, 543
Cyclotetramethylenesilane, 540
p-Cymene, 415
Cysteine, 260, 438

Daedalea junisperina, 250
Daldinia concentrica, 426
Darzens condensation, 3
Decaborane, 230
Decahydrodibenzothiophene 5,5-dioxide, 306
β-Decalone phenylhydrazone, 519
Decaphenyl-1,1-bis(1-silacyclopentadiene), 542
Decarbousnic acid, 206
Dehydrobenzofuran, **198**
Dehydromucic acid, 117, 121
3,4-Dehydroproline, 389
3,4-Dehydropyrrolidine-2-carboxylic acid, 389
Dehydrotremetone, **169**
Dendrolasin, **100**
Dendrolasius fuliginosus lao, 100
3-Deoxy-D-erythrohexosone, 87
Desaurins, 73
Deuterioacetic acid, 236, 263
5-Deuterio-2-pentadeuteriophenylthiophene, 234

Deuteriopyrroles, 338
3-Deuterio-3*H*-pyrrolizine, 481
Deuterioselenophenes, 315
Deuterium chloride, 230
Deuterium oxide, 274, 315
2,8-Diacetodibenzofuran, 199
Diacetosuccinic ester, 107, 118, 332, 351
Diacetoxyacetophenone, 172
1,1-Diacetoxy-3,4-benzosilacyclopentane, 544
3,6-Diacetoxycarbazole, 502
2,5-Diacetoxy-2,5-dihydrofuran, 107, 126
2,3-Diacetoxymercuri-indole, 427
3,4-Di(acetoxymethyl)tetrahydrofuran, 134
Diacetyl, 140
3,6-Diacetyl-9-alkylcarbazoles, 506, 508
Diacetylcarbazoles, 505, 507
3,6-Diacetylcarbazole, 506
3,9-Diacetylcarbazole, 493
2,8-Diacetyldibenzofuran, 201
Diacetylenes, conversion to pyrroles, 333, 337
—, — to thiophenes, 222
—, reaction with amines, 333
—, — with hydrogen sulphide, 222
—, — with sodium telluride, 326
1,3-Diacetylindole, 433
O,N-Diacetylindoxyl, 455
O,O-Diacetylporphyrilic methyl ester, 204
2,6-Diacylcarbazoles, 506
1,2-Diacylcyclohexanes, 477
1,2-Diacylcyclohexenes, 182, 477
Dialdehydes, 84
(2′,2′-Dialkoxyethylthio)thiophenes, 296
Dialkylaminoacetylbenzofurans, 161
α-Dialkylaminoketones, 66
3-Dialkylaminomethylbenzothiophenes, 278
3-Dialkylaminomethylfurans, 110
Dialkyl(2-biphenyl)boranes, 550
2,2-Dialkyl-3-bromopropan-1-ols, 46
Dialkyl-*N*-chloramines, 446
1,2-Dialkyldiaziridines, 42
1,3-Dialkyl-1,3-dihydroisobenzofurans, 181
1,1-Dialkylgermacyclopentan-3-ols, 547
2,2-Dialkylindolin-3-ones, 409
Dialkylpyrroles, 336, 342, 344, 346
Dialkyl selenides, 319
1,1-Dialkylsilacyclopentan-3-ols, 543
4,4-Dialkyl-1,2-thiazetidin-3-one 1,1-dioxides, 82
Dialkylthiophenes, 253
Diamines, 316, 374
—, pyrolysis, 379
2,2′-Diaminobiphenyl, 195, 300, 490
Diaminocarbazoles, 499, 500
3,7-Diaminodibenzoselenophene, 325
2,5-Diamino-3,4-dicyanothiophene, 333
1,3-Diaminopropane, 62
3,4-Diaminotetrahydrofuran, 133
3,4-Diaminothiolanes, 260
1,3-Di-*p*-anisylisobenzofuran, 187
Dianthraquinonylamines, 525
o-Diaroylbenzenes, 185
Diaroylethenes, 85
2,3-Diarylaziridines, 25
2,3-Diarylbenzothiophenes, 273
Diaryldiazomethanes, 7
1,2-Diarylethenes, reaction with butadienes, 182
Diarylisobenzofurans, 182, 185
Diarylketoximes, 7
2,3-Diaryloxaziridines, 41
3,3-Diaryloxindoles, 453
1,10-Diaza-4-anthryl, 568
1,6-Diazacyclodecane, 478
2,6-Diazaspiro[3.3]heptane, 70
Diazetidines, 76, 78
1,2-Diazetidinones, 76
Diazetines, 79, 80
2,2′-Diazidobiphenyl, 500
Diaziridines, 42
Diaziridineimine, 43
Diaziridinones, 44
Diazirines, 42
Diazirine, **43**
—, structural data, 2
Diazoacetic ester, 88, 94, 224, 435, 548
—, reaction with pyrroles, 363
Diazoacetic methyl ester, 532
ω-Diazoacetophenone, 108
Diazoalkanes, 11
Diazobenzenesulphonic acid, 343
Diazobenzil, 60
1,5-Diazobicyclo[5,4,0]undec-5-ene, 534
3-Diazo-2-butanone, 74
Diazo-coupling, 337
2-Diazo-3,5-diphenylpyrrole, 350
3-Diazoindoles, **423,** 439
3-Diazo-3*H*-indoles, 423
Diazoketones, 107

Diazoketones, *(continued)*
—, reaction with pyrroles, 356
α-Diazoketones, 76
Diazomethane, 11, 37, 38, 43, 259, 281, 284, 341, 532
—, photochemical decomposition, 129
—, reaction with furan, 94
3-Diazo-5-methoxy-2-(4-methoxyphenyl)-3*H*-indole, 443
3-Diazo-2-methyl-3*H*-indole, 443
Diazoniadispirohexadecane, 569
2-Diazo-4-nitro-3,5-diphenylpyrrole, 350
Diazonium salts, 99, 130, 273, 281, 284
—, from pyrroles, 349
Diazonium tetrafluoroborate, 189
3-Diazo-oxindoles, 456, 461, 464
2-Diazo-2-phenylacetophenone, 74
1-Diazo-2-phenyl-3*H*-benz[*e*]indole, 483
3-Diazo-2-phenyl-1*H*-benz[*g*]indole, 483
3-Diazo-2-phenyl-6,7-benzo-3*H*-indole, 443
3-Diazo-2-phenyl-3*H*-indole, 443
3-Diazo-2-pyridone, 361
3-Diazo-4-pyridone, 361
2-Diazopyrroles, **350**
3-Diazopyrroles, **349**
Diazo reactions, furans, 106
3-Diazothianaphthen-2-one, 285
1,3-Dibenzenesulphonyl-1,3-diazetidine, 78
Dibenzocarbazoles, 518, **523,** 524
7*H*-Dibenzo[*c,g*]carbazole, 521
13*H*-Dibenzo[*a,g*]carbazole, 521
Dibenzofurans, natural products, 204
—, physical properties, 200
—, reduced, 203
—, substituted, 197
Dibenzofuran, 83, **194,** 195, **197**, 198, 304
—, chloromethylation, 197
—, from phenol, 196
Dibenzofuran 2-aldehyde, 197
Dibenzofuran 2-carbaldehyde, 201
Dibenzofuran-2-carboxylic acid, 196
Dibenzofuran-4-carboxylic acid, 197
Dibenzofuran-2,8-dicarboxylic acid, 199
Dibenzofuran-1,2-dicarboxylic methyl ester, 199
Dibenzophospholes, 538
Dibenzoselenophene, **324,** 325
Dibenzoselenophene 5-oxide, 324
Dibenzoselenophene *Se*-oxide, 324
Dibenzosiloles, 544
Dibenzotellurophene, 328
Dibenzothiophenes, 300
—, reduced, 305
—, substituted, 305
Dibenzothiophene, 300, **302**
—, monophenyl derivatives, 301
—, nitration, 302
—, phenylation, 302, 303
—, reactivity, 303
—, synthesis, 300
Dibenzothiophenecarboxylic acids, 304
Dibenzothiophene-4-carboxylic acid, 303
Dibenzothiophene 5,5-dioxide, 302, 304
Dibenzothiophene 5-oxide, 302
Dibenzoylacetylene, 183, 472
o-Dibenzoylbenzene, 184, 186, 472, 474
3,6-Dibenzoylcarbazole, 506
1,2-Dibenzoylcyclohexa-1,4-dienes, 183, 472
1,2-Dibenzoylcyclohexane, 190
1,2-Dibenzoyl-1-cyclohexene, 190
1,2-Dibenzoyl-4,5-dimethylcyclohexa-1,4-diene, 290
2,3-Dibenzoyl-6,7-dimethyl-1,4-endoxo-1,4-diphenyl-1,4-dihydronaphthalene, 184
2,3-Dibenzoyl-2,3-diphenylthiirane, 11, 12
Dibenzoylethylene, 217
2,3-Dibenzoylquinol, 183
Dibenzoylquinol diacetate, 183
Dibenzoyl peroxide, 231, 248
Dibenzoylstilbene, 12, 235, 335
o,o'-Dibenzoyl tetraphenylethene, 186
Dibenzyl disulphide, 273
Dibenzyl sulphide, 273
Dibenzyl sulphone, 12
Diborane, 357, 395
1,3-Dibromoacetone dimethyl acetal, 59
αα'-Dibromoadipic acid, 260, 390
αα'-Dibromoadiponitrile, 340
2,3-Dibromobenzofuran, 152
Dibromobenzoselenopheno[2,3-*b*]benzoselenophene, 326
2,3-Dibromobenzothiophene, 273
1,4-Dibromobiacetyl, 133
1,4-Dibromobutane, 130, 258, 543, 545
1,4-Dibromobutane-2,3-diol, 383
Dibromocarbazoles, 496
2,8-Dibromodibenzofuran, 200
2,8-Dibromodibenzothiophene, 305
4,4'-Dibromodibutyl ether, 130

2,3-Dibromo-2,3-dihydrobenzofuran, 169
1,2-Dibromo-2,3-dihydro-1*H*-naphtho-[2,1-*b*]thiopyran, 309
3,4-Dibromo-3,4-dihydro-2*H*-naphtho-[1,2-*b*]thiopyran, 309
4,4′-Dibromo-5,5′-dimethyl-3,3′-bithienyl-2,2′-dicarboxylic acid, 261
1,3-Dibromo-1,3-diphenyl-1,3-dihydrobenzo[*c*]thiophene 2,2-dioxide, 292
2,2′-Dibromo-3,3′-dithienyl, 262
2′,5-Dibromo-2,3′-dithienyl, 262
5,5′-Dibromo-2,2′-dithienyl, 262
1,2-Dibromoethyl acetate, 362
2,3-Dibromofuran, 114, 117
2,5-Dibromofuran, 104
Dibromofurancarboxylic acids, 122
Dibromogermacyclopentane, 545, 546
1,1-Dibromogerma-3-cyclopentene, 547
1,4-Dibromo-2-hydroxybutane, 245
5,7-Dibromoisatin, **462,** 464
2,5-Dibromonaphtho[2,1-*b*]thiophene, 309
2,8-Dibromononan-5-one, 192
3,3-Dibromo-oxindole, 453
2,2-Dibromo-3-oxo-2,3-dihydrobenzofurans, 172
2,4-Dibromophenol, 152
3,6-Dibromo-2-phenylbenzofuran, 152
3,4-Dibromo-1-phenylphospholane, 534
3,4-Dibromo-1-phenylphospholane 1-oxide, 534
2,3-Dibromo-4-phenylpyrrole, 347
2,5-Dibromo-3-phenylthiophene, 239
1,3-Dibromopropane, reaction with aniline, 62
—, — with sodium sulphide, 53
$\alpha\beta$-Dibromopropionitrile, 30
4-(2,3-Dibromopropyl)-2,6-di-*tert*-butyl-4-methylcyclohexa-2,5-dienone, 167
1,3′-Dibromopropylmalonic ester, 388
Dibromopyrrolecarboxylic esters, 347
Dibromoselenophenes, 318
1,1-Dibromotellurophene, 326
3,4-Dibromothiolane, 256
Dibromothiophenes, 241
2,5-Dibromothiophene, 224, 231, 239, 245
3,4-Dibromothiophene-2,5-dicarbaldehyde, 250
Dibromotriacetonamine, 389
$\alpha\alpha'$-Dibromo-*o*-xylene, 291
1,3-Di-*tert*-butylallene oxide, 4
1,3-Di-*tert*-butylaziridinone, 32
1,1-Dibutyl-3,4-benzosilacyclopentane, 544
N,*N*′-Di-*tert*-butyl-*N*-chlorourea, 44
2,3-Di-*tert*-butylcyclopropanone, 4
Di-*tert*-butyldiaziridinone, 44
2,5-Di-*tert*-butylfuran, 94
N,*N*′-Di-*tert*-butyl-*N*″-methylguanidine, 43
Di-*n*-butyloxirene, 5
2,5-Di-*tert*-butylthiophene, 235
9,9′-Dicarbazolylmethane, **493**
1,2-Dicarbazylcyclobutane, 493
1,2-Dicarbonyl compounds, 143 228, 331
—, conversion to furans, 84
—, enol-ethers, 89
1,3-Dicarbonyl compounds, reaction with aldoses, 87
—, — with diglycollic esters, 87
1,4-Dicarbonyl compounds, 91, 331
—, conversion to thiophenes, 220
Dicarbonylcyclopentadienyliron iodide, 344
Dicarbonyl(thiophene)iron, 231
Dicarboxylic acids, benzofurans, 165
2,5-Dicarboxytetrahydrofuran, 135
Dichloramine, 43
1-Dichloroacetoxy-1-phenylgermacyclopentane, 546
Dichloroaziridines, 27, 28
Dichlorobenzofurans, 153
3,5-Dichlorobenzofuran-2-carbaldehyde, 159
Dichlorobenzoselenopheno[2,3-*b*]benzo selenophene, 326
2,3-Dichlorobenzothiophene, 273
1,4-Dichlorobutane, 258
1,4-Dichlorobut-2-ene, 374
Dichlorocarbazoles, 495, 496
Dichlorocarbene, 27, 32, 124, 147, 416, 546, 548
3,6-Dichlorocoumarins, 163
2,2-Dichloro-1,1-diethylgermacyclopentane, 546
3,3-Dichloro-2,3-dihydrobenzofuran, 169
Dichloro-1,3-dihydroisobenzofuran-4,7-quinone, 189
2,3-Dichloro-5,6-dihydro-2*H*-pyran, 124
Dichlorodihydroxybutane, 260
5,6-Dichloro-4,7-dihydroxy-1,3-dihydroisobenzofuran, 189, 190
Dichlorodimethylsilane, 71, 541

3,3-Dichloro-2,4-dioxo-1,2,3,4-tetrahydroquinoline, 456
4,5-Dichloro-3,6-endoxotetrahydrophthalic anhydride, 103
1,2-Dichloroethane, 233
1,2-Dichloroethyl ether, 85, 332
2,5-Dichlorofuran, 102
3,4-Dichlorofuran, 102, 104
Dichlorofurancarboxylic acids, 122
Dichlorogermacyclopentane, 545
1,1-Dichlorogerma-3-cyclopentenes, 546, 547
2,3-Dichloroindole, 420, 451
5,7-Dichloroisatin, **462,** 464
Dichloromaleimide, 346
1,9-(Dichloromalonyl)carbazole, 507
3-Dichloromethyl-2,3-dimethyl-3*H*-indole, 416
2-Dichloromethyl-2,5-dimethyl-2*H*-pyrrole, 347
1,2-Dichloromethylenedioxybenzene, 252
2,4-Dichloro-3-methyl-6-(α-methylallyl)-phenol, 168
2,4-Dichloro-5-methyl-6-(α-methylallyl)-phenol, 168
1,1-Dichloro-1-methyl-3-phospholene, 535
Dichloromethylpyrroles, 355
4*a*-Dichloromethyltetrahydro-4*aH*-carbazole, 510
Dichloropyrrole, **346, 347**
Dichloropyrrole-carboxylic esters, 347
2,5-Dichloroselenophene, 318
1,1-Dichlorosilacyclopentadiene, 539, 540
1,1-Dichlorosilacyclopentane, 540, 543
1,1-Dichlorosilacyclopentenes, 540, 542
αα-Dichlorosulphones, 14
2,3-Dichlorotetrahydrofuran, 131, 135
2,5-Dichlorotetrahydrofuran, 131, 359
Dichlorothienylarsine, 232
2,5-Dichloro-3-thiolene 1-oxide, 257
Dichlorothiophenes, 238, 241, 257
2,4-Dichlorothiophene-2-carbaldehyde, 250
5,7-Dichloro-2,3,4-trimethyl-2,3-dihydrobenzofurans, 168
2,2-Dichloro-1,3,3-triphenyl-aziridine, 29
Dicyanoacetylene, 534
2,3-Dicyano-4,4′-dimethyloxetane, 46
3,4-Dicyanofuran, 106
Dicyanomethylene-1,3-dithietane, 73
Dicyclohexanofuran, 203
Dicyclohexenyl, 204, 306
Dideuterioselenophenes, 315
Dieckmann ring-closure, 166, 176, 258, 259, 286, 386, 481
Diels–Alder reaction, 57, 93, 107, 137, 186, 193, 225, 336, 340, 472, 532
1,3-Dienes, 19, 128, 167
—, reaction with phosphorus halides, 534
Diene addition, 235
Dienecarboxylic esters, 334
Diene synthesis, 120
—, pyrroles, 364
5,6-Diethoxybenzothiophene, 276
3,3-Diethoxythietane 1,1-dioxide, 54
Diethylaluminium hydride, 212
Diethylamine, conversion to pyrrole, 337
1-Diethylaminepropyne, 58
3-Diethylaminobenzothiophene 1,1-dioxide, 282
1,1-Diethylaziridinium perchlorate, 39
2,3-Diethyl-1-azirine, 35
1,1-Diethyl-4-chloro-1-germacyclohexa-2,4-diene, 548
3,3-Diethyl-6,6-dichloro-3-germabicyclo-[3.1.0]hexanes, 548
3,3-Diethyl-6,6-difluoro-3-germacyclo-[3.1.0]hexane, 548
2,5-Diethylfuran, 98
3,3-Diethyl-3-germabicyclo[3.1.0]hexane-6-carboxylic ester, 548
1,1-Diethylgermacyclobutane, 546
1,1-Diethylgermacyclopentane, 546
1,1-Diethylgerma-3-cyclopentenes, 547, 548
2,3-Diethyl-3-hydroperoxy-3*H*-indole, 409
2,3-Diethylindole, 409
Difluoramine, 26, 43
Di-2-furfurylamine, **111**
Difurfuryl disulphide, 111
Difurfuryl ether, 110
1,2-Di-(2-furyl)ethene, 215
Di-2-furyl ketone, 138
Di-2-furylmethane, 138
Digenea simplex, 390
Diglycollic esters, 87
1,4-Dihalides, 258
—, from tetrahydrofuran, 130
—, reaction with amines, 379
4,5-Dihalogeno alcohols, 128
1,3-Dihalogenoalkanes, 62
—, reaction with alkali sulphides, 52

7,8-Dihalogenobicyclo [4.2.0]octa-1,3,5-trienes, 472
2,3-Dihalogeno-2,3-dihydrobenzofurans, 145
4,5-Dihalogeno-3,6-endoxohydrophthalic anhydride, 102
1,3-Dihalogenopropanes, 47
Dihalogenosulphoacetic acids, 74
3,4-Dihalogenothiochromans, 277
Dihydroanthracenes, 293
Dihydroazetes, 1
4,5-Dihydro-1*H*-benz [*g*]indole, 484
4,5-Dihydro-3*H*-benz [*e*]indolecarboxylic acids, 483
5,6-Dihydro-7*H*-benzo [*c*]carbazole, 521
5,6-Dihydro-7*H*-benzo [*c*]carbazole-5-sulphonic acid, 518
Dihydrobenzofurans, 141, 142, 145, 166, 167, 169
2,3-Dihydrobenzofuran, 145, **168**
2,3-Dihydrobenzofuran-3-acetate-2-carboxylic diethyl ester, 176
2,3-Dihydrobenzofuran-2-carboxylic acid, 170,176
2,3-Dihydrobenzofuran-2,3-dione, 171
2,3-Dihydrobenzofuran-3-ones, 173, 175
2,3-Dihydrobenzofur-3-one-2-carboxylic ester, 176
2-(2,3-Dihydrobenzofuryl)carbamic ester, 170
Dihydrobenzothiophenes, 280
1,3-Dihydrobenzo [*c*]thiophene, 289, 291
2,3-Dihydrobenzothiophene, 270, **280**
1,3-Dihydrobenzo [*c*]thiophene 2,2-dioxide, 289, 291
2,3-Dihydrobenzothiophene 1,1-dioxides, 280, 281
3a,7a-Dihydrobenzo [*b*]thiophene 1,1-dioxide, 226
1,3-Dihydrobenzo [*c*]thiophene 2-oxide, 289, 292
1,4-Dihydrocarbazole, 491
3,4-Dihydro-β-carboline, 431
Dihydroconiferyl alcohol, 161
Dihydrodiazetes, 79, 80
1,4-Dihydrodibenzothiophene, 305
2,3-Dihydro-2,3-diphenyleneindole, 410
Dihydrodithiin, 75
Dihydrofurans, 121, 123, 126, 375
—, physical properties, 127
2,3-Dihydrofuran, **123,** 124, 125
2,5-Dihydrofuran, **125,** 128, 189, 217
4,5-Dihydrofuran-3-carboxylic ester, 121
Dihydrofuranolones, 109
Dihydrofuran-2-ones, 125
4,5-Dihydro-2-furoic acid, 125
Dihydrogriseofulvin, 174
2,5-Dihydrohexachlorofuran, 102
2,3-Dihydro-3-hydroxybenzofuran, 167
5,6-Dihydroindeno [2,1-*b*]indole, 479
5,10-Dihydroindeno [1,2-*b*]indole, 479
Dihydroindoles, 445
4,7-Dihydroindole, 405, **465**
1,3-Dihydroisobenzofurans, 180, 181, 188, **189**
4,5-Dihydroisobenzofurans, 190
4,7-Dihydroisobenzofurans, 182, 190
1,3-Dihydroisobenzofuran-4-carboxylic acid, 190
Dihydroisoindoles, 470
1,3-Dihydroisoindole, **474**
3,4-Dihydroisoquinoline, 42
1,4-Dihydro-9-methylcarbazole, 491
Dihydromuconic dinitrile, 351
1,4-Dihydronaphthalene-1,4-endo-oxide, 184
1,2-Dihydronaphtho [2,1-*b*]furans, 210
4,5-Dihydronaphtho [1,2-*b*]furan-2-carboxylic ester, 210
1,3-Dihydronaphtho [2,3-*c*]pyrroles, 477
Dihydronaphtho [1,2-*b*]thiophene, 307
Dihydronaphtho [2,3-*b*]thiophene, 310
Dihydro-1,3-oxazines, 64
2,3-Dihydro-5-phenyl-1*H*-1,4-benzo [*f*]-diazepin-2-ones, 473
2,3-Dihydro-7-phenylbenzo [*b*]thiophene, 302
Dihydroporphyrins, 393
Dihydropyran, 133, 134
Dihydropyridazines, 332
Dihydropyrroles, 329, 331
Dihydroresorcinol, 469
Dihydroselenophene dioxides, 314
4,7-Dihydrotetraphenylbenzo [*c*]furan, 291
Dihydrothianaphthalene 1,1-dioxide, 226
5,12-Dihydrothienonaphthacene, 310
1,3-Dihydrothieno [3,4-*c*]thiophene, 298
4,6-Dihydrothieno [3,4-*b*]thiophene, 297
Dihydrothiophenes, 221
2,3-Dihydrothiophene, **255**

2,5-Dihydrothiophene, **255**
2,3-Dihydrothiophene 1,1-dioxide, 269
Dihydroxybenzenes, 216
3,6-Dihydroxybenzo[2,1-*b*:3,4-*b*′]dithiophene, 311
1,7-Dihydroxybenzofuran, 204
6,7-Dihydroxybenzofuran, 157
Dihydroxybenzothiophenes, **276**
2-(3′,4′-Dihydroxybenzylidene)-4,6-dihydroxy-3-oxo-2,3-dihydrobenzofuran, 173
2,2′-Dihydroxybiphenyls, 194, 195, 196, 197
2,3-Dihydroxy-α-bromoacetophenone, 172
3,6-Dihydroxycarbazole, **502**
3,4-Dihydroxyfuran-2,5-dicarboxylic acid, 121
Dihydroxyfuranotropolone, 192
Dihydroxyindoles, 424
2,3-Dihydroxyindole, 456
5,6-Dihydroxyindole, 426
Dihydroxyindolines, 424
2,3-Dihydroxyindolines, 408
5,6-Dihydroxyindolines, 448
γγ-Dihydroxyketone dianhydrides, 191
1,7-Dihydroxy-3-methyldibenzofuran, 205
5,6-Dihydroxy-1-methylindole, 425
1-(3′,4′-Dihydroxy-4-methylpentyl)cyclohexanol, 191
1,8-Dihydroxynaphthalene, 426
Dihydroxy-3-oxo-2,3-dihydrobenzofurans, 172, 173
2-(2,5-Dihydroxyphenyl)acetoacetic ester, 164
Dihydroxyphenylalanine, 426, 427
Dihydroxyphenylethylamines, 424, 426
3,3-Di(4-hydroxyphenyl)oxindole, **454**
3,4-Dihydroxy-1-phenylpyrrolidines, 383
Dihydroxypyrroles, 352
Dihydroxyquinazoline, 462
5,8-Dihydroxytetracarbazole, 513
3,4-Dihydroxytetrahydrofuran, **132**
3,4-Dihydroxytetrahydrothiophene, 227
3,3′-Dihydroxytetraphenyl-1,1′-bis-1,3-dihydroisobenzofuranyl, 184
3,4-Dihydroxythiolane, 260
2,4-Dihydroxythiophene, **245**
3,4-Dihydroxythiophene, **245**
4,7-Dihydroxy-2,3,6-trimethylbenzofuran, 157
1,3-Diiminoisoindoline, 476
Diiminophthalimide, 476
2,5-Diiminopyrrolidine, 387
Diindole, 480
Diindolomethenes, 411, 468, 469
2,2′-Diindolyl, **467**
3,3′-Diindolyl, **467**
Diindolylalkenes, 468
2,2-Diindolyl disulphides, 421
3,3-Diindolyl disulphide, 420, 421
3,3′-Diindolylmethane, 428, **468**
Diindolylmethene pigments, 433
3,3′-Diindolylphenylmethane, 468
Diindolylphosphinic acid, 427
2,2-Di-(3′-indolyl)propane, 468
Diindolyl sulphide, 421
Diindolyl sulphone, 421
Diindolyl sulphoxide, 421
1,6-Diiodo-3,4-bisiodomethylhexane, 320, 327
5,5′-Diiodo-2,2′-bithienyl, 262
3,3′-Diiodo-2,2′-bithienyl-5-carboxylic acid, 262
Diiodocarbazoles, 496
2,8-Diiododibenzofuran, 199, 200
Diiododimethylpyrroles, 347
1,1-Diiodo-2,3-diphenylgermirene, 71
2,5-Diiodofuran, 104
Diiodogermacyclopentane, 545
1,1-Diiodogerma-3-cyclopentenes, 546
1,1-Diiodogermirene, 71
Diiodomethane, 73
Diiodoselenophene, 318
1,2-Diiodotetrafluorobenzene, 325
Diiodotetramethylpyrrole, 347
1,4-Diiodotetraphenylbutadiene, 315, 327, 532
Diisobutylaluminium hydride, 86, 178, 212
Diisopentylcadmium, 117
Diketene, 387
1,2-Diketones, 357
—, conversion to thiophenes, 220
1,3-Diketones, 268
—, conversion to indoles, 403
1,4-Diketones, 85, 177
—, conversion to thiophenes, 220
—, dehydration, 84
—, reaction with hydrazine, 391
1,5-Diketones, 116
2,2′-Dilithiobiphenyl, 325, 544, 545, 549

1,4-Dilithiotetraphenylbutadiene, 315, 327, 531, 532, 540, 541
2,3-Dilithiothiophene, 240
Dimercaptoalkanols, 6
Dimesitoylmesitylethane, 84
2,5-Dimethoxybenzaldehyde, 431
4,6-Dimethoxybenzofuran, 157
Dimethoxybenzothiophenes, 276
2,5-Dimethoxy-2,5-dihydrofurans, 91, 126
2,2-Dimethoxyethyl 1-naphthyl sulphide, 306
3,4-Dimethoxyfuran, 94
5,6-Dimethoxy-1-methylbenzo[*c*]thiophene, 290
5,6-Dimethoxyoxindole, 453
3,4-Dimethoxyphenol, 163
2-(3,4-Dimethoxyphenyl)-5-(3-hydroxypropyl)-7-methoxybenzofuran, 162
Dimethoxytetramethyldisilane, 542
Dimethylamine, 59, 429
Dimethylaminoacetonitrile, 429
p-Dimethylaminobenzaldehyde, 281, 363, 410, 411
2-*p*-Dimethylaminobenzylidene-2*H*-pyrrole-5-carboxylic acid, 363
1-Dimethylaminobutan-3-one, 383
2-Dimethylaminocarbonylpyrroles, 356, 395
3-Dimethylaminodibenzofuran, 201
Dimethylaminoethene, 381
Dimethylaminoethyl 2-thianyl ketone, 251
6-Dimethylamino-3-methylbenzofuran, 155
o-Dimethylaminomethylbenzyl alcohol, 476
2-Dimethylaminomethyl-3-dimethylamino-4-phenylthietane 1,1-dioxide, 58
2-Dimethylaminomethylindole, **429**
3-Dimethylaminomethylindole, **429**
3-Dimethylaminomethyl-4-methylenethiolane, 259
5-Dimethylaminopent-2-yne, 381
p-Dimethylaminophenyl-2-pyrrolyl-2-(2*H*-pyrrolylidene)methanedicarboxylic acid, 363
2-Dimethylaminothiophene, 243
Dimethylaniline, 112, 453
3,4-Dimethyl-5-azabenz[*a*]*anthracene, 567*
Dimethylaziridines, 28
2,2-Dimethylaziridine, 20
2,3-Dimethylaziridine, 25, 26
Dimethylbenzocarbazoles, 524
Dimethyl-7*H*-benzo[*c*]carbazole, 521
1,3-Dimethylbenzocyclobutene, 292
3,5-Dimethylbenzo[1,2-*b*:5,4-*b*′]difuran, 216
3,8-Dimethylbenzo[1,2-*b*:3,4-*b*′]difuran, 216
3,6-Dimethylbenzo[1,2-*b*:6,5-*b*′]difurandicarboxylic ester, 216
Dimethylbenzofurans, 151, 197
2,3-Dimethylbenzofuran, 150, 161
2,3-Dimethylbenzofuran-7-carbaldehyde, 159
3,5-Dimethylbenzofuran-2-carboxylic methyl ester, 163
5,6-Dimethylbenzofuran-2-carboxylic methyl ester, 166
Dimethylbenzoselenopheno[2,3-*b*]benzoselenophene, 326
1,1-Dimethyl-2,3-benzosilacyclopentane, 544
Dimethylbenzothiophenes, **272**
2,3-Dimethylbenzothiophene, 274
3,5-Dimethylbenzothiophene, 267
Dimethylbetaines, 390
5,5′-Dimethyl-2,2′-bifuran, 137, 138
5,5′-Dimethyl-2,2′-bifuran-4,4′-dicarboxylic acid, 138
5,5′-Dimethyl-2,2′-bifuran-4,4′-dicarboxylic ester, 137
4,4′-Dimethyl-2,2′-bifuran-5,5′-dicarboxylic methyl ester, 138
2,2′-Dimethyl-1,1-bipyrrolyl-3,3′-dicarboxylic acid, 391
Dimethylbis(2,5-dimethylpyrrole-3-yl)-methane, 343, 344
1,3-Dimethyl-4,6-bis(1′-pyrrolyl-3′-carboxy)benzene, 391
5,5′-Dimethyl-2,2′-bithienyl, 262
5,5′-Dimethyl-3,3′-bithienyl, 261
5,5′-Dimethyl-2,2′-bithienyl-3-carbaldehyde, 262
5,5′-Dimethyl-3,3′-bithienyl-2,2′-dicarboxylic acid, 261
Dimethylbutadiene, 237, 536, 542
Dimethylcadmium, 251
Dimethylcarbazoles, 495
1,4-Dimethylcarbazole, 494, 504, 522
1,8-Dimethylcarbazole, 494
3,9-Dimethylcarbazole, 504

1,4-Dimethylcarbazole-3-carbaldehyde, 494 504
1,4-Dimethylcarbazole-3,6-dicarbaldehyde, 494
1,3-Dimethyl-6*H*-cyclohepta[*c*]thiophen-6-one, 295
2,3-Dimethyl-4,6-diacetylbenzofuran, 161
3a,6a-Dimethyl-2,3:5,6-dibenzo-3a,6a-dihydrofuro[3,2-*b*]furan, 215
2,4-Dimethyldibenzofuran, 195, 200
5,5-Dimethyldibenzosilole, 545
2,8-Dimethyldibenzothiophene, 305
2,3-Dimethyl-3-dichloromethyl-3*H*-indole, 442
1,3-Dimethyl-1,3-dihydrobenzo[*c*]thiophene, 292
2,2-Dimethyl-2,3-dihydrofuran, 125, 169
2,2-Dimethyl-2,5-dihydrofuran, 128
2,5-Dimethyl-2,5-dihydrofuran, 128
Dimethyl-1,3-dihydroisobenzofurans, 190
1,3-Dimethyl-4,7-dihydroisobenzofuran, 190
3,9-Dimethyl-2,3-dihydronaphtho[2,3-*b*]-furan, 210
2,6-Dimethyl-4,5-dihydro-3*H*-naphtho-[1,8-*bc*]furan, 213
5,5′-Dimethyl-4,4′-di(hydroxymethyl)-2,2′-bifuran, 138
3,3-Dimethyldiindolyl 2,2-disulphide, 421
2,5-Dimethyl-2,5-dimethoxy-2,5-dihydrofuran, 98
1,2-Dimethyldinitroindoles, 423
2,2′-Dimethyldiphenylamine, 494
5,6-Dimethyl-1,3-diphenylbenzo[*c*]-thiophene, 290
5,6-Dimethyl-1,3-diphenyl-4,7-dihydroisobenzofuran, 190
5,6-Dimethyl-1,3-diphenylisobenzofuran, 184, 187
4,7-Dimethyl-1,3-diphenylisobenzofuran, 187
5,5-Dimethyl-2,4-diphenyl-1-pyrroline 1-oxide, 373
1,1-Dimethyl-2,5-diphenylsilacyclopentadiene, 541
1,1-Dimethyl-2,5-diphenylsilacyclopentane, 541
1,1-Dimethyl-2,3-diphenylsilirene, 71
3a,5a-Dimethyldodecahydronaphtho-[2,1-*b*]furan, 212
3,4-Dimethylenetetrahydrofuran, **134**
3,4-Dimethylenethiolane, 233, 259
Dimethylformamide, 111, 250, 262, 316, 355, 392, 432, 499
Dimethylfurans, 98, 101
2,4-Dimethylfuran, 94
2,5-Dimethylfuran, 94, 96, 98, 99
2,4-Dimethylfuran-3-carbaldehyde, 115
Dimethylfurancarboxylic acids, 122
2,4-Dimethylfuran-3-carboxylic acid, 86
Dimethylfurfurylamine, 111
1,1-Dimethylgerma-3-cyclopentene, 547
6,6-Dimethylhexahydrobenzofuran-2,4-dione, 178
Dimethylhexahydronaphtho[2,1-*b*]furan, 212
2,5-Dimethyl-3-hexyne-2,5-diol, 50
2,3-Dimethyl-4-hydroxybenzofuran, 158
2,3-Dimethyl-6-hydroxybenzofuran, 158
1-Dimethyliminophospholane boron trihydride complex, 538
Dimethylindoles, physical data, 418
1,3-Dimethylindole, 407, 415
2,3-Dimethylindole, 409, 416, 439, 446
2,4-Dimethylindole, 336
3,3-Dimethyl-3*H*-indole, 439, 442
3,3-Dimethyl-3*H*-indole-2-carbaldoxime, 443
1,2-Dimethylindoline, **448**
2,3-Dimethylindoline, 446, 447
Dimethylindolinium salts, 447
1,1-Dimethylindolium perchlorate, 417
Di((2-methyl-3-indolyl)methane, 468
Di-(3-methyl-3-indolyl)phenylmethane, 415
2,2-Dimethylindoxyl, 447
2,2-Dimethylisoindolium bromide, 471, 476
Dimethylketene, 51
$\alpha\beta$-Dimethyllaevulic acid, 237
Dimethylmaleic anhydride, 92, 224
3,3-Dimethyl-1-methylene-1,3-dihydroisobenzofuran, 190
Dimethyl-5-methyleneoctahydronaphtho-[2,3-*b*]furan, 213
1,1-Dimethyl-2-methylenepyrrolidinium hydroxide, 381
3,5-Dimethylnaphtho[2,3-*b*]furan, 209
2,2-Dimethyl-2*H*-naphtho[1,8-*bc*]furan, 213
2,3-Dimethyl-6-nitrobenzothiophene, 274

4,7-Dimethyl-6-nitro-2,3-diphenylbenzofuran, 155
Dimethylnitroindoles, 423
4,4-Dimethyl-5-nitropentan-2-one, 371
2,5-Dimethyl-3-nitrosopyrrole, 349
2,6-Dimethylocta-2,6-diene, 256
1,7-Dimethyloctahydroindole, 467
Dimethyloctahydronaphtho [2,3-*b*]furan, 212
Dimethyloxetanes, 47
2,2-Dimethyloxetane, 45
3,3-Dimethyloxetane, 47
Dimethyloxetone, 192
Dimethyloxindoles, 453
2,5-Dimethyl-3-oxo-2,3-dihydrobenzofuran, 172
1,5-Dimethyl-4-oxohexahydroindole, 467
1,5-Dimethyl-4-oxo-octahydroindole, 467
1,5-Dimethyl-2-oxopyrrolidine, 384
5,5-Dimethyl-2-oxopyrrolidine, 384
1,2-Dimethyl-5-oxo-2-pyrroline, 378
3,4-Dimethyl-2-oxo-3-pyrroline, 376
5,5-Dimethyl-3-oxopyrroline 1-oxide, 378
3,4-Dimethylpent-3-en-2-one, 52
2,3-Dimethylphenol, 116
3,4-Dimethylphenoxyacetyl chlorides, 171
2,2-Dimethyl-3-phenylazirine, 36
3,3′-Dimethyl-2-phenyl-1-azirine, 34
3,3-Dimethyl-1-phenyl-1,3-dihydrobenzo[*c*]thiophene, 292
4,7-Dimethyl-2-phenylindole, 414
2,3-Dimethyl-3-phenyl-3*H*-indole, 414
3,3-Dimethyl-2-phenyl-3*H*-indole, 414
2,5-Dimethyl-1-phenylphosphole, 533
Dimethyl-2-phenyl-2-propoxyaziridine, 34
2,5-Dimethyl-1-phenylpyrrole, 345
2,3-Dimethyl-1-(*N*-phenylthiocarbamoyl)aziridine, 26
Dimethylphospholene, 535, 536
Dimethylphthalide, 190
2,5-Dimethyl-1-phthalimidopyrrole, 350
2,2-Dimethyl-5-propylpyrrolidine, 380
4,6-Dimethyl-α-pyrone-5-carboxylic acid, 86
Dimethylpyrroles, 345
1,2-Dimethylpyrrole, 342
2,5-Dimethylpyrrole, 343, 344, 374, 473, 475
3,5-Dimethylpyrrole-2-carboxylic acids, 394
2,4-Dimethylpyrrole-3-carboxylic esters, 347
2,5-Dimethylpyrrole-1-carboxylic ester, 360
1,2-Dimethylpyrrolidine, 373
3,3-Dimethylpyrrolidine, 382
3,4-Dimethylpyrrolidine, 382
1,1-Dimethylpyrrolidinium bromide, 381
Dimethylpyrrolidinium iodide, 382
5,5-Dimethylpyrrolid-2-one, 373
Dimethyl-3-pyrrolines, 375
1,1-Dimethyl-2-pyrroline, 373
1,2-Dimethyl-2-pyrroline, 369
2,5-Dimethyl-1-pyrroline, 374
2,5-Dimethyl-3-pyrroline, 344, 374
5,5-Dimethylpyrroline, 373
5,5-Dimethyl-1-pyrroline 1-oxide, 66, 373, 377, 380
Dimethyl-(2-pyrrolylmethyl)amine, 354
3-(3′,5′-Dimethylpyrrolyl-2)propenal, 359
3,3-Dimethylselenetane, 61
2,5-Dimethylselenophene, 313, 314
1,1-Dimethyl-1-silaindan, 544
1,1-Dimethylsiletane, 71
Dimethylsilylene, 542
αα-Dimethylsuccinic acid, 118
Dimethyl sulphate, 301
Dimethylsulphonium methylide, 36, 89, 234, 404
Dimethylsulphoxonium methylide, 167
2,7-Dimethyltetrahydrobenzofuran, 177
1,3-Dimethyltetrahydrobenzo [*c*]thiophene, 293
Dimethyltetrahydrocarbazoles, 514, 515
Dimethyltetrahydrofuran, 135
Dimethyltetrahydronaphtho [2,1-*b*]furan, 211, 212
Dimethyl-1-tetralones, 522
1,1-Dimethyltetraphenylsilacyclopentadiene, 542
1,1-Dimethyltetraphenylstannole, 531
4,6-Dimethylthieno [3,4-*b*]thiophene, 297
Dimethyl-2-thienylcarbinol, 247
Dimethylthietanes, 53
4,4-Dimethylthietan-2-one, 59
2,2-Dimethylthiirane, 13
2,3-Dimethylthiiranes, 10
Dimethylthiomaleic anhydride, 228
Dimethylthiophenes, 236, 237
2,5-Dimethylthiophene, 220, 235, 250, 252, 289, 293, 359

3,4-Dimethylthiophene, 233, 259
—, oxidation, 228
2,5-Dimethylthiophene-3-thiol, 297
4,4′-Dimethyltolane, 273
1,2-Dimethyl-3,5,6-trinitroindole, 423
2,5-Dimethyl-2-vinyl-2,3-dihydrofuran, 125
Dinaphthofurans, **215**
Dinaphthols, 215
1,3-Di-α-naphthylisobenzofuran, 187
2,4-Dinitrobenzenesulphenyl chloride, 246
Dinitrobenzocarbazoles, 524
5,7-Dinitrobenzofurans, 154
3,5-Dinitrobenzoic acid, 523
Dinitrobenzothiophenes, 275
3,4-Dinitrobenzothiophene, 274
5,5′-Dinitro-2,2′-bifuran, 138
Dinitrocarbazoles, 497, 498
3,6-Dinitrocarbazole, 499
2,4-Dinitrochlorobenzene, 421
3,8-Dinitrodibenzofuran, 201
2,7-Dinitrodibenzoselenophene, 325
3,7-Dinitrodibenzoselenophene, 325
2,8-Dinitrodibenzothiophene, 305
5,5′-Dinitro-2,2′-difuryl sulphide, 105
3,5-Dinitro-2,2′-dithienyl, 262
5,5′-Dinitro-2,2′-dithienyl, 262
2,4-Dinitrofluorobenzene, 499
2,5-Dinitrofuran, **104,** 105
3,5-Dinitro-2-hydroxythiophene, 245
3,6-Dinitro-2-phenylbenzofuran, 154, 155
2-(2,4-Dinitrophenyl)cyclohexanone, 499
Dinitrophenylhydrazine, 114
O-(2,4-Dinitrophenyl)hydroxylamine, 154
2,4-Dinitrophenyl-3-indolyl sulphide, 421
O-(2,4-Dinitrophenyl)oximes, 154
Dinitropyrroles, **348**
Dinitrothiophenes, 224
2,4-Dinitrothiophene, 242
2,5-Dinitrothiophene, 242
Diols, 116, 372
1,2-Diols, cyclic carbonates, 6
1,3-Diols, cyclic carbonates, 53
2,5-Diones, 415
1,3-Dioxa-4-cyclohexyl, 568
1,4-Dioxane, 547
Dioxane, dibromide, 102
Dioxane–sulphur trioxide complex, 74
2,6-Dioxaspiro[3.3]heptane, 48
Dioxetanes, 72
Dioxindole, 397, 426, 448, 456, 457, **458,** 461
Dioxindolecarboxylic esters, 458, 459
Di(oxindolyl)methane, 452
2,2′-Dioxobicyclohexyl, 516
5,5′-Dioxo-2,2′-bipyrrolidinyl, 385
2,3-Dioxo-2,3-dihydrobenzofuran, 165, 175
2,3-Dioxo-2,3-dihydrobenzo[*b*]selenophene, 322
2,3-Dioxo-2,3-dihydrobenzothiophene, 280, 281, 284
2,3-Dioxo-2,3-dihydrobenzothiophene 1,1-dioxide, 284
2,3-Dioxo-2,3-dihydrobenzothiophene-2-oxime, 282
2,3-Dioxo-2,3-dihydrobenzothiophene 3-oxime, 285
2,3-Dioxo-2,3-dihydrobenzothiophene 3-*p*-tosylhydrazone, 285
Dioxodihydroisobenzofurans, 189
1,2-Dioxo-1,2-dihydronaphtho[2,1-*b*]-furan, 211
2,5-Dioxo-2,5-dihydroselenophene, 319
4,7-Dioxo-1,3-diphenyl-4,7-dihydroisobenzofuran, 183
3,4-Dioxo-1-ethylpyrrolidine-2,5-dicarboxylic ester, 388
2,3-Dioxoindolines, 458
1,3-Dioxoisoindoline, **476**
4H-1,3-Dioxolo[4,5-d]imidazole, 565
2,3-Dioxomethyl-2,3-dihydrobenzothiophenes, 285
2,3-Dioxo-octahydrobenzofuran, 180
1,2-Dioxo-1-phenyl-2-(2′-thienyl)ethane, 251
Dioxopiperazines, 362, 389
Dioxopyrrolidines, 108, 352, **387**
Dioxopyrrolines, 378
Dioxotetrahydrofurans, **133**
1,3-Diperoxy-1,3-dihydroisobenzofuran, 188
Diphenacyl, 237
Di-(4-phenoxybutyl)amine, 477
Diphenylacetaldehyde, 77
Diphenylacetic acid, 77
Diphenylacetylene, 15, 71, 549
—, reaction with *n*-butyl-lithium, 543
Diphenylaluminium chloride, 549
Diphenylamine, 490
—, pyrolysis, 489

1,2-Diphenyl-2-azetin-4-one, 70
2,3-Diphenylaziridines, 28
1,2-Diphenylbenzocyclobutene, 292
2,3-Diphenylbenzofurans, 149, 151, 154
Diphenyl-2-benzofurylmethanol, 165
1,3-Diphenylbenzo[*c*]thiophene, **290**
1,4-Diphenyl-1,3-butadiene, 532
1,4-Diphenyl-1,4-butanedione, 350
Diphenylcarbodiimide, 79
Diphenylcyclopropanes, 55, 56
1,2-Diphenylcyclopropanesulphinic acid, 56
1,3-Diphenylcyclopropanesulphinic acid, 55
1,2-Diphenylcyclopropanethiol, 56
1,2-Diphenylcyclopropene-3,3-dicarboxylic acid, 108
Diphenylcyclopropenone, 15, 25
1,3-Diphenyl-1,3-diazetidine-2,4-dione, 79
Diphenyl diazetidin-2-ones, 78
Diphenyldiazomethane, 11, 33, 292
Diphenyldibenzofurans, 198, 200
5,5-Diphenyldibenzogermole, 549
Diphenyldichlorosilane, 540, 543
2,3-Diphenyl-5,6-dihydrobenzofuran, 177
1,3-Diphenyl-1,3-dihydrobenzo[*c*]-thiophene 2,2-dioxide, 292
Diphenyl-1,3-dihydroisobenzofurans, 190
1,3-Diphenyl-4,5-dihydroisobenzofuran, 190
1,3-Diphenyl-4,7-dihydroisobenzofuran, 190
2,5-Diphenyl-1,4-dithiin, 223
$\beta\beta'$-Diphenyldivinyl sulphide, 8
Diphenylene oxide, 83, **194**
Diphenylene selenides, 324
Diphenylene sulphide, 300
2,2-Diphenylethanol, 77
1,1-Diphenylethene, 81, 326, 413
Diphenyl ether, 196
1,1-Diphenylethylene glycol, 77
N,N'-Diphenylformamidine, 465, 470
Diphenylfurans, 101
2,4-Diphenylfuran, **99**
2,5-Diphenylfuran, 91, 99
3,4-Diphenylfuran, 131
1,1-Diphenylgermacyclopentane, 546
Diphenylgermanium dichloride, 549
Diphenylglycolaldehyde phenylhydrazone, 413
Diphenyl hydrogen phosphate, 167
2,5-Diphenyl-3-hydroxyfuran, 108
N,N-Diphenylhydroxylamine, 490
2,2-Diphenyl-4-imidazolidinone, 78
Diphenylindenone, 182
Diphenylindenone epoxide, 182
2,3-Diphenylindole, 418
2,2-Diphenylindoxyl, 33, 450
1,3-Diphenylisobenzofuran, 59, 181, 182, 183, 184, 185, 186, 187, 474
1,3-Diphenylisobenzofuran peroxide, 186
1,3-Diphenylisobenzofuran-4,7-quinone, 183
1,3-Diphenylisoindole, **472**
1,3-Diphenylisoindoline, **472**
2,3-Diphenylisoindolone, 474
Diphenylketene, 49, 60, 76, 77, 80, 108
2-Diphenylmethyldiazetidin-3-one, 78
N-Diphenylmethylenemethylamine, 41
2-Diphenylmethylnaphtho[2,1-*b*]furan, 208
2-Diphenylmethylpyrrole, 345
1,2-Diphenylnaphtho[2,1-*b*]furan, 208
2,3-Diphenyl-6-nitrobenzofuran, 155
Diphenylnitrosoamine, 489
3,5-Diphenyl-1,2-oxathiolane 2-oxide, 55
2,5-Diphenyloxazole, 25
2,3-Diphenyloxetan-3-ols, 48
Diphenyl oxide, 102
1,3-Diphenyloxindole, 450
3,3-Diphenyloxindole, 33, 450
Diphenyloxirene, 5
4,5-Diphenyl-2-oxo-2,5-dihydrofuran, 108
1,3-Diphenyl-5-oxo-2-pyrroline, 378
1,3-Diphenyl-4-(phenylimino)-1,3-diazetidin-2-one, 79
Diphenyl-(3-phenylpropyl)silane, 71
1,4-Diphenylphthalazine *N*-oxide, 183
Diphenylpyrroles, 345
2,4-Diphenylpyrrole, 37, 350
2,5-Diphenylpyrrole-3-diazonium chloride, 349
2,5-Diphenylpyrrolidine, 382
2,4-Diphenylselenophene, 314
1,1-Diphenylsilacyclopentadiene, 540
2,5-Diphenylsilacyclopentadiene, 541
1,1-Diphenyl-1-silaindan, 544
Diphenylsilane, 536
Diphenyl sulphide, 300
Diphenyl sulphone, 301
Diphenyl sulphoxide, 300

5,5″-Diphenyl-2,2′:5′,2″-terthienyl, 264
2,3-Diphenyltetrahydrobenzofuran, 177
Diphenyltetrahydrofuran, **131**, 135
1,3-Diphenyltetrahydroisobenzofuran, 190
2,4-Diphenyltetrahydrothiophene, 56
2,3-Diphenyl-5-thiabicyclo[2.1.0]pentane, 8
Diphenyl-2-thienylmethanol, 247
2,4-Diphenylthietane, **55,** 56
Diphenylthietane 1,1-dioxides, 55
2,4-Diphenylthietane 1-oxides, 56
2,3-Diphenylthiirane, 13
Diphenylthiirane 1,1-dioxide, 11, 12
2,3-Diphenylthiirene 1,1-dioxide, 14, 15
Diphenylthiophenes, 237, 271
3,4-Diphenylthiophene, 226
2,5-Diphenyl-1-*p*-tolylpyrrole, 348
Diphthaloylcarbazole, **525**
Dipotassio-1,1-dimercapto-2,2-dicyanoethene, 73
Dipotassium tetrapyrroleferrate, 339
Di-*n*-propyloxirene, 5
2,2′-Dipyrrolidinyl derivatives, 373
1,4-Di(2′-pyrrolyl)butane, 335
Dipyrrolyl disulphides, **353**
2,2′-Dipyrrolyl ketones, 357, 395
Dipyrrolylmethanes, 353, 393, 394, 395
Dipyrrolylmethanols, **395**
Dipyrrolyl peroxides, 376
Di-2-pyrrolylphthalide, 357
Dipyrrolylsuccinic acid, 364
Dipyrrolyl sulphides, **353**
2,2′-Dipyrromethane, **393**
Dipyrromethenes, 353, 355, 394, 395, 468
—, vinylogues, 359
o,o′-Diselenodibenzoic acid, 323
Disodiocatechol, 216
Disodium phenylphosphide, 532
Disulphene, **74**
Disulphisatide, 460, 461
Di-2-thenylamine, 248
N,N′-Di-(2-thenyl)-1,3-diazacyclobutane, 248
1,4-Dithia-1,4-dihydronaphthalene, 566
1,4-Dithianaphthalene, 566
4H-1,3-Dithianaphthalene, 566
1,4-Dithiane, 9
2,6-Dithiaspiro[3.3]heptane, 60
1,3-Dithiatetrahydronaphthalene, 566
Dithiazines, 553
4,7-Di(2-thienyl)benzo[*b*]thiophene, 225
1,2-Di(2-thienyl)ethene, 299
2,2′-Dithienyl ketone, 264
2,3′-Dithienyl ketone, 264
3,3′-Dithienyl ketone, 264
Dithienylmercury, 232, 233
2,2′-Dithienylmethane, 264
Di-2-thienylmethanol, 247
2,2′-Dithienyl sulphide, **247**
Dithienyl sulphone, **247**
Dithietanes, 73
1,3-Dithietane tetroxide, 74
Dithietenes, 75
1,2-Dithietene, 75
1,2-Dithiocyanates, 7
2,5-Dithiocyanopyrrole, 353
2,5-Dithiocyanothiophene, **247**
Dithiolane, 75
1,3-Di-*p*-tolyl-1,3-diazetidine, 78
1,3-Di-*p*-tolylisobenzofuran, 187
2,5-Di(triphenylmethyl)-3-pyrroline, 336
Divinyl sulphone, 259
Dodecacarbonyltriiron, 263
Dodecahydrocarbazole, 517
n-Dodecyltrichlorosilane, 544
Döbner's cinchoninic acid synthesis, 387
Dopa, 426, 427
Dopachrome units, 427
Duroquinone, 146
(*o*-Duroylphenyl)phenylmethanol, 183
1-Duryl-3-phenylisobenzofuran, 183

Egonol, **161,** 162
Ehrlich colour reactions, 411, 437
Ehrlich reagent, 363, 472
Elliptone, 166
Elsholtzia cristata, 117
Elsholtzic acid, 117, 122
Elsholtzione, **117**
Elymoclavine, 486
Enamines, 116, 156, 412, 446
—, reaction with sulphonyl chlorides, 57
4,7-Endomethylenehexahydro-2-methylisoindole, 528
Endoxohydrophthalic acids, 102
Enteramine, 430
Eosin, 335
Epichlorohydrin, 111, 388, 482
Episulphonium salts, 12
α-Epoxides, acetylenic, 314

α-Epoxides, *(continued)*
—, vinylacetylenic, 314
1,4-Epoxycyclohexa-2,5-diene-2,3-dicarboxylic dimethyl ester, 193
3,6-Epoxycyclohexene, 96
3,6-Epoxyhexahydrobenzoic acid, 120
3,6-Epoxyhexahydrophthalic acid, 120
3,4-Epoxy-3-methylbutanal diethylacetal, 98
2,3-Epoxypropanols, 5, 49
3,6-Epoxy-3,4,5,6-tetrahydrophthalic dimethyl ester, 120
3,4-Epoxytetramethyltetrahydrofuran, 192
Eremophila latrobei, 117
Ergot alkaloids, 389
Erivanin, **212**
Erythrina spp., 439
Erythritol, 87
Esters, reaction with aziridine, 20
—, αβ-unsaturated, 369, 372
Ethanolamine, 384
Ethers, unsaturated, 128
Ethoxalyl chloride, 254
Ethoxyacetylene, 52
2-Ethoxy-1-benzenesulphonylpyrrolidine, 383
3-(α-Ethoxybenzyl)indoles, 410
2-Ethoxy-4-bistrifluoromethyloxet-2-ene, 52
1-Ethoxycarbonylamino-3,3-diphenyloxindole, 77
1-Ethoxycarbonylaziridine, 28, 466
Ethoxycarbonylcarbene, 88, 210
Ethoxycarbonyldiazomethane, 210
2-Ethoxycarbonyl-4,5-dihydropyrroloindole, 529
5-Ethoxycarbonyl-2,4-dimethyl-3-dimethylaminopyrrole, 351
3-Ethoxycarbonyl-2-ethylbenzofuran, 146
2-Ethoxycarbonylisatogen, 445
2-Ethoxycarbonyl-4-isopropylidene-3,3-dimethyl-1,2-thiazetidine 1,1-dioxide, 81
Ethoxycarbonylmethyl-β-cyanoethylmethylamine, 386
Ethoxycarbonylnitrene, 224, 359
1-Ethoxycarbonyl-3-oxopyrrolidine, 386
2-Ethoxycarbonylpyrroloindole, 529
Ethoxycarbonylsulphamoyl chloride, 81
6-Ethoxycarbonyltetrafluorophenoxyacetic ester, 153
1-Ethoxycarbonyltetrahydrobenzo[c]-thiophene, 290
9-Ethoxycarbonyltetrahydrocarbazole, 514
2-Ethoxycarbonyltetrahydrofuran, 135
2-Ethoxycarbonyl-1,4,4-triphenyl-1,2-diazetidinone, 76
5-Ethoxy-1,3-dimethylindole, 426
2-Ethoxy-3,4-dimethylpyrrole, 376
4-Ethoxy-2-ethoxycarbonyl-1-hydroxybicyclo[4,3,0]-3-thianonane, 290
2-Ethoxyethylcyclohexanone, 466
2-Ethoxyethylcyclohexylamines, 466
2-Ethoxy-5-hydroxybenzofuran, 143, 156
2-Ethoxyindole, 452
2-Ethoxy-3*H*-indole, 452
3-Ethoxyindole-2-carboxylic ester, 435
6-Ethoxy-2-isopropylidene-2,2-dimethyl-2,3-dihydro-1,4,5-oxathiazine 4,4-dioxide, 81
Ethoxymagnesium malonic ester, 317
2-Ethoxymethylenecyclohexanone, 289
1-Ethoxyphosphole 1-oxide, 533
6-Ethoxythioindoxyl-2-carbaldehyde, 283
Ethylaziridines, 28
Ethylbenzene, 116, 270, 321
o-Ethylbenzeneselenol, 321
2-Ethylbenzenethiol, 270, 271
Ethyl-5*H*-benzo[*b*]carbazole, 520, 523
3-Ethylbenzofuran, 151
2-Ethylbenzofuran-2-acetic acid, 166
2-Ethylbenzofuran-3-butyric acid, 166
2-Ethylbenzofuran-3-carbaldehyde, 161
2-Ethylbenzofuran-3-propionic acid, 166
2-Ethyl-3-benzofurfurylamine, 161
Ethylbenzothiophenes, **272**
Ethylcarbazoles, 495
1-Ethylcarbazole, 494
9-Ethylcarbazole, 503, 520
N-Ethylcarbazole, 501
9-Ethylcarbazole-1-carboxylic acid, 503
γ-(9-Ethylcarbazol-3-yl)butyric acid, 520
9-Ethylcarbazolyltriphenylsilanes, 504
2-Ethyl-3-cyanobenzofuran, 146
2-Ethylcyclohexanone phenylhydrazone, 516
5-Ethyldibenzoborole, 550
2-Ethyldibenzofuran, 200
Ethyldichlorophosphinyl sulphide, 536

Ethyl dichlorophosphite, 536
2-Ethyl-2,3-dihydrobenzothiophene, 281
12-Ethyl-12,14-dihydrocarbazole [2,3-*a*]-carbazole, 520
3-Ethyl-2,4-dimethyl-5-oxo-2-pyrroline, 378
4-Ethyl-2,3-dimethyl-5-oxo-2-pyrroline, 378
Ethyldimethylpyrroles, 345
3-Ethyl-2,4-dimethylpyrrolidine, 382
Ethylene, benzofuran synthesis, 144
—, reaction with benzenethiol, 271
—, — with furan, 96, 98
Ethylenediamine hydrochloride, 15
Ethylenediaminetetra-acetic acid, 213
Ethylene dibromide, reaction with phenol, 168
Ethyleneimines, 1, 15
Ethylene oxides, 3, 353
Ethylene oxide, 1, **3**, 240, 247, 428, 492
Ethylene sulphides, 1, 5
Ethylene sulphoxide, 11
2-Ethylfuran, 97, 98, 101
Ethyl furfuryl sulphide, 111
Ethyl 3-furyl ketone, 116
Ethylfurylmethanol, 97
Ethyl glycinate, 31
4*a*-Ethylhexahydro-9*H*-carbazole-9-carbaldehyde, 516
4*a*-Ethylhexahydro-9*H*-carbazole hydrochloride, 516
2-Ethylidene-1-methylpyrrolidine, 382
Ethyliminodiacetic ester, 388
Ethylindoles, physical data, 418
1-Ethylindole, 413
2-Ethylindoline, 446
3-Ethylindoline, **448**
O-Ethylindoxylic acid, 462
1-Ethyl-1-iodogermacyclopentane, 546
2-Ethylisoindoline, 476
5-Ethyl-2-isopropyl-5-methyl-2-thiolene, 257
Ethylmagnesium iodide, 55, 231, 339, 546
Ethylmethylmaleimide, 343, 364
5-Ethyl-7-methyl-5*H*-benzo [*b*]carbazole, 520
1-Ethyl-2-methylisoindole, 474
3-Ethyl-1-methyloxindole, 453
Ethylmethylpyrroles, 345
7-Ethylnaphtho [2,1-*b*]furan, 208
2-Ethyloxetane, 47
1-Ethyloxindole, 453
2-Ethylphenol, 144
3-Ethylphenol, 116
1-Ethyl-1-phenylgermacyclopentane, 546
2-Ethyl-5-propylpyrrolidine, 106
Ethylpyrroles, 345
1-Ethylpyrrole, 340, 341
1-Ethylpyrrolidine, **382**
2-Ethylpyrrolidine, 382
1-Ethylpyrrolidine 1-oxide, 371, 383
1-Ethyl-2-pyrrolidone, 384
2-Ethylpyrroline, 371
o-Ethylstyrene, 292
2-Ethylsulphonylbenzofuran, 144
4*a*-Ethyltetrahydro-4*aH*-carbazole, 516
1-Ethyltetrahydrofuran, 135
5-Ethyltetrahydro-7-oxo-5*H*-benzo [*b*]-carbazole, 520
2-Ethylthiirane, 13
Ethylthioethyne, 111
2-Ethylthiolane, 251
Ethylthiophenes, 237
2-Ethylthiophene, 251
2-Ethylthiophene-5-sulphonic acid, 245
1-Ethylthio-3-phospholene 1-sulphide, 537
3-Ethyl-2,4,5-trimethylpyrrole, 345
2-Ethynylfuran, 101
1-Ethynyl-1-hydroxy-8a-methyl-6-oxo-octahydronaphthalenes, 213
Euparin, **162**
Eupatorium purpureum, 162
Eupatorium urticaefolium, 169
Euryops spp., 213
Euryopsol, **213**
Evodia hortensis, 177
Evodone, **177**
Evonymus atropurpureus, 121

Feist–Benary synthesis, 85, 117, 204
Fischer cyclisation, 154
Fischer–Hepp rearrangement, 446, 496, 497
Fischer indole synthesis, 143, 398, 401, 414, 434, 448, 480, 483, 487, 488, 494
Fischer reaction, 521, 522
Fischer's aldehyde, 464, 465
Fish poisons, 141
Five-membered ring compounds, 531

Five-membered ring compounds, *(continued)*
—, with one hetero atom from Group V, 397
—, with one hetero atom from Group VI, 219
—, with one hetero nitrogen atom, 329
—, with one hetero oxygen atom, 83
—, with one selenium atom, 313
—, with one tellurium atom, 326
Flavanols, 149, 173, 174
Flavones, 173
Fluorene, 299
Fluorobenzene, 273
2-Fluorobenzothiophene, 273
Fluorodibenzofurans, 200
Fluoro-2,4-dinitrobenzene, 499
Fluorofurans, 103
Fluoroindoles, **420**
1-Fluoro-1-methylsila-2-cyclopentene, 542
1-Fluoro-1-methylsila-3-cyclopentene, 542
2-Fluoro-5-methylthiophene, 240
Fluoro-olefins, 316
—, reaction with sulphur trioxide, 74
5-Fluorophthalide, 189
5-Fluorotetrachloro-1,3-dihydroisobenzofuran, 189
Fluorothiophenes, 240
Formaldehyde, 81, 110, 247, 277, 278
—, reaction with allyl acetate, 132
—, — with carbazole, 493
—, — with chloramine, 42
—, — with cyclohexene, 190
—, — with furfural, 114
—, — with indole, 525
—, — with primary amines, 42
—, — with pyrrole, 353, 354
—, — with *p*-toluidine, 78
Formamido compounds, 105
Formic esters, 378, 432
—, conversion to pyrrolecarbaldehydes, 355
β-Formylacrylic acid, 113
3-Formylbenzofurans, 146
2-Formylbenzoyl chloride, 181
o-Formylbenzyl alcohol, 189
4-Formyl-1,3-butadiene-1-carboxylic ester, 94
α-Formyl-*β*-chloro-olefins, 334
1-Formylindole, 413
Formylkynurenine, 437
3-Formyl-2-methylpyrrole-5-carboxylic acid, 356
5-Formyl-2-methylpyrrole-3-carboxylic ester, 354
2-Formylnaphtho[2,1-*b*]thiophene, 309
3-Formyl-2-naphthyloxyacetic acid, 208
γ-Formyl-*α*-oxobutenoic acid, 113
β-Formylpropionic acid, 107
1-Formylpyrrolidine, 382
N-Formyltetrahydro-1-naphthylamine, 484
Formyl-*o*-toluidine, 411
3-Formyl-1,2,5-trimethylpyrrole, 354
Four-membered rings, containing one nitrogen atom, 61
—, — silicon and phosphorus, 71
—, with one nitrogen and one oxygen atom, 80
—, with one oxygen atom, 45
—, with one selenium atom, 61
—, with one sulphur atom, 52
—, with two hetero-atoms, 72
Friedel–Crafts acylation, 160, 297
Friedel–Crafts alkylation, 97, 233
Friedel–Crafts reaction, 47, 115, 145, 197, 211, 246, 254, 287, 304, 337, 356, 433, 505, 520, 525
Fries rearrangement, 170, 173
Fructose, conversion to furans, 87
Fulvenes, 481
Fumaronitrile, reaction with acetone, 46
Fungicides, 189
3-Furaldehyde, **115**
Furans, addition reactions, 91, 94, 126
—, aldehydes, 112
—, amines, 105
—, aminomethylation, 112
—, bicyclic systems, 191
—, catalytic reduction, 129
—, diazo reactions, 106
—, 2,3-disubstituted, 88
—, formylation, 112
—, from acetylenes, 88
—, from carbohydrates, 86
—, halogenation, 91
—, in essential oils, 83
—, mercuration, 96
—, metallic derivatives, 96, 97
—, naturally occurring, 100
—, nitration, 91

Furans, *(continued)*
—, oxidation, 91, 120
—, ozonolysis, 91
—, photochemistry, 93
—, properties, 89
—, reactions, 89
—, reaction with acyl chlorides, 115
—, — with ketones, 14
—, — with mineral acids, 91
—, ring-fission, 96
—, 3-substituted, 97
—, sulphonation, 103
—, with two or more unfused nuclei, 136
Furan, **96**, 131, 186, 224, *554*
—, acetylation, 115
—, adducts, 94
—, alkylation, 98
—, chlorination, 102
—, chloromercuri derivatives, 96
—, conversion to pyrrole, 337
—, diene addition, 92
—, halogenation, 102
—, nitration, 103, 105
—, polyene derivatives, 140
—, reaction with acetone, 95
—, — with diazomethane, 94
—, — with ethylene, 98
—, — with hydrogen selenide, 315
—, — with maleic anhydride, 92
—, substitution products, 84
—, substitution reactions, 96
—, ultraviolet absorption, 230
Furanaldehyde, 112
Furanamines, 334, 342
Furan arsenicals, 97
Furan-2-carbaldehyde, 83, 86, **112**
Furan-3-carbaldehyde, **115**
Furan-5-carbaldehyde-2-carboxylic methyl ester, 119
Furan-1-carbonyl chloride, 117
Furancarboxylic acids, 84, 117, 122
Furan-2-carboxylic acid, 83, 87, **119,** 122
Furan-3-carboxylic acid, **121,** 122
Furancarboxylic acid azides, 105
Furancarboxylic esters, 96
Furandicarboxylic acids, 122
Furan-2,3-dicarboxylic acid, 117
Furan-2,4-dicarboxylic acid, 85
Furan-2,5-dicarboxylic acid, 87, **121,** 147
Furan-3,4-dicarboxylic acid, 93, 107, **120**
Furan-2,5-dimethyl-3,4-dicarboxylic acid, 122
Furan-2,5-disulphonic acid, **103**
Furan group, 83
Furanoquinone, 109
Furan organo-metallic compounds, 117
Furanosesquiterpenes, 213
Furanose sugars, 132
Furan ozonides, 91
Furan rings, condensed system with benzene rings, 207
—, condensed with anthracene, 214
—, — with phenanthrene, 214
—, in polycyclic systems, 207
Furansulphonic acids, **103**
Furantetracarboxylic acid, **120,** 122
Furantricarboxylic acids, **120,** 122
Furazan, 559
Furfural, 83, 86, 90, 96, 97, 110, 111, **112,** 115, 119, 121, 127, 159
—, autoxidation, 113
—, benzoin condensation, 139
—, conversion to furan, 96
—, — to pyrrole, 337
—, determination, 114
—, hydrogenation, 136
—, oxidation, 114
—, reaction with acetaldehyde, 140
—, — with aniline, 113
—, — with arylamines, 113
—, — with butadiene, 113
—, — with formaldehyde, 114
Furfural acetals, 113
Furfuraldazine, 139
2-Furfuraldehyde, 491
Furfuraldoxime, 111
Furfural oximes, 114,
Furfuramide, 112, 114
Furfuranethiol, 111
Furfuryl acetate, 110
β-2-Furfurylacraldehyde, 129
Furfuryl alcohol, 96, 109, 110, 130
—, oxidation, 114
Furfurylamine, **111,** 112
Furfurylamine mucate, 393
Furfuryl bromide, 97
Furfuryl chloride, 109
Furfurylidenealkyl ketones, 342
Furfurylidene-amines, 111
Furfurylidenenitromethane, 114

Furfuryl methyl ether, 110
N-Furfurylpyrrole, 393
Furil, **139**
Furil dioxime, 139
Furilic acid, **139**
Furo[3,4-c]cinnoline, *564*
2-Furodimethylamide, 115
Furoflavones, 162
2-Furoic acid, 83, 96, 97, 103, **119**, 122, 136
—, halogenation, 102
3-Furoic acid, 115, 117, **121**, 122
Furoic esters, 97, 111, 120
Furoin, 112, 113, **139**
2-Furonitrile, 115
2H-Furo[3,2-b]pyran, 562
Furopyrroles, 106
Furotetralones, 207
Furo[2,3-b]thiophene, *562*
2-Furoylacetic ester, 121
Furoyl azide, 107
Furoyl chloride, 115, 119
Furoylglycine, 119
2-Furylacetaldehyde, 115
2-Furylacetic acid, **121**
2-Furylacetic ester, 111
2-Furylacetylene, 100
Furylacraldehyde, 113, 115
Furylacrylic acid, 112, 121
β-2-Furylalanine, 121
3-Furylamines, 106
N-2-Furylbenzamide, 107
N-2-Furylcarbamic methyl ester, 107
Furyl carbinols, 96, **110**, 111
Furyl-3,4-dicarbamate, 107
2-Furyldimethylcarbinol, 111
3-Furyldimethylcarbinol, 111
2-Furyldiphenylcarbinol, 111
β-2-Furylethanol, 111
2-Furylglycollic acid, 121
2-Furylglyoxal hydrate, 116
2-Furylglyoxylic acid, 121
2-Furylideneacetone, 116
2-Furyl isocyanate, 107
2-Furyl ketones, 115
3-Furyl ketones, 115
3-Furyl-lithium, 97, 103, 111, 137
Furylmagnesium iodide, 97, 137, 138
Furylmercuric chloride, 102
Furyl metallic compounds, 97
4-(2-Furyl)-3-methoxycarbonylbut-3-enoic acid, 156, 165
Furylmethylcarbinol, 97, 111
2-Furylnitroethene, 114
Furylpentadienal, 113
2-Furylphenylmethanol, 111
β-2-Furylpropionic acid, 121
1-(2-Furyl)-2-(2-thienyl)ethene, 299
2-Furyl 2-thienyl ketone, 264
δ-2-Furylvaleric acid, 192
Fused ring compounds, 397
—, polynuclear, 538
Fused oil, 138

Gallic acid, 214
Gattermann reaction, 112, 137, 145, 159, 197, 285, 337, 342, 355, 432
Gelatine, 389
Germacyclopentane, **545**, 547
Germacyclopentane 1-oxide, 545
Germacyclopentanols, 547
Germa-3-cyclopentanes, 547
Germanium, in organo-ring compounds, 71
Germanium compounds, 545
Germanium diiodide, 71
Germanium tetrachloride, 545
5-Germaspiro[4.4]nonane, 545
Glucosamine, 354, 360
Glucose, 86, 454
D-Glucuronamide, degradation, 355
Glutacondialdehyde, 470
Glutamic acid, conversion to pyrrole, 337
Glutaric acid, *o*-ethyl ester chloride, 295
Glyceraldehyde, 87
Glycerol *α*-monoethers, 48
Glycidol, 5
Glycines, 334
Glycine ester, 348
Glycols, dehydration, 189
—, from phthalic esters, 181
—, from phthalides, 181
1,4-Glycols, 128
—, from tetrahydrofuran, 130
—, reaction with amines, 379
Glycolic acid, 435
Glycosmispentaphylla, 502
Glycozoline, 490, **502**
Glyoxal, 99, 228, 469
—, conversion to pyrroles, 340
—, from furan, 96

Glyoxal bisulphite, 412, 420
Glyoxylic acid, 166, 437
Gmelina leichhardtii, 193
Gmelinol, **193**
Gomberg reaction, 99, 251
Graebe–Ullmann synthesis, 494, 518, 526
Gramine, **429,** 430, 432, 435, 436
Gramine methiodide, 428
Gramine metho-salts, 429
Grignard reagents, 42, 46, 47, 48, 67, 96, 97, 103, 111, 115, 146, 148, 171, 186, 368, 375
—, indoles, 408
—, reaction with ketoximes, 18
Grignard synthesis, 353
Griseofulvic acid, 174
Griseofulvin, **174,** 175

Haematic acid, 364
Haematomma coccineum, 204
Haematomma porphyrium, 204
Haemin, 329, 342, 363, 393
Haemoglobin, 393
Haemopyrrole, 343, 345, 364, 378
Haemopyrrolecarboxylic acid, 367
ω-Halogenoacetophenones, 149
α-Halogeno acids, 67
β-Halogeno acids, reaction with basic reagents, 49
1,4-Halogeno-alcohols, 128
—, from tetrahydrofuran, 130
—, reaction with alkalis, 45
3-Halogenoalkanethiols, 52
1-Halogeno-2-alkylaziridines, 30
Halogenoalkylthiophenes, 248
1,4-Halogenoamines, 379
Halogenoanilines, 387
Halogenoaziridines, 27
Halogenobenzofurans, 145, 146, 152
—, physical properties, 153
Halogenobenzothiophenes, 273, 274
γ-Halogenobutyrohydroxamic acids, 385
Halogenocarbazoles, 495, 496
5-Halogeno-2,5-dihydro-2-oxofuran, 107
Halogeno-2,2-dimethylaziridines, 29
β-Halogenoethers, 131, 146
N-β-Halogenoethylsulphonamides, 16
Halogenofurans, 96, 102, 104
2-Halogeno-2′-hydroxybiphenyls, 194
Halogeno-indoles, 418, 419, 420
2-Halogeno-3*H*-indoles, 442
α-Halogenoketones, condensation with aldehydes, 3
—, conversion to indoles, 401
—, — to pyrroles, 332
α-Halogenoketoximes, 361
2-Halogenomethyl-2,3-dihydrobenzofuran, 176
p-Halogenonitrobenzenes, 144
Halogeno-2-nitrobiphenyls, 489, 496
o-Halogenophenols, 144
1-Halogeno-1,1(*o*-phenylenedioxy)-3-phospholenes, 537, 538
Halogenopropylamines, 61, 64
Halogenopyrroles, 346
Halogenotetrahydrocarbazoles, 495
Halogenotetrahydrofurans, 131
Halogenotetramethylaziridines, 29
Halogenothiophenes, 238, 241
Hantzsch–Widman nomenclature system, 551
Helichrysum bracteatum, 174
Heliotridan, 481
Heliotropium alkaloids, 481
Heliotropium sp., 480
Heptafluorodibenzothiophene, 305
Heptaheterohelicene, 299
Herz reaction, 267, 283
Heteroauxin, 397, **435,** 440
Heteroauxin phenethylamide, 528
Heterocyclic compounds, 3-membered, 1
—, 4-membered, 1
—, nomenclature, 1
Heterocyclic spiro compounds, nomenclature, 569
Heterocyclic systems, nomenclature, 551
Heterocyclopentadienes, 531
Hexacarbonylbis(diphenylacetylene)-diiron, 532
Hexachlorobuta-1,3-diene, 239, 327
Hexachlorocyclopentadiene, 547
Hexachloro-2,5-dihydrothiophene, 257
Hexachlorodisilane, 542
Hexachloronaphthalic anhydride, 523
Hexachlorotellurophene, 327
Hexachlorotetrahydroselenophene, 318
2,4-Hexadiyne, 533
Hexafluoroacetone, 52
Hexafluorobenzene, 198
Hexafluoro-2-butyne, 75

Hexafluoro-Dewar benzene, 336
Hexafluoro-1,4-diiodobutane, 241, 259
Hexaheterohelicenes, 299
Hexahydroazepinium-1-spiro-1′-imidazolidine-3′-spiro-1″-piperidinium dibromide, 570
Hexahydrobenzofurans, 178, **179**
Hexahydrobenzofuran-2,4-dione, 178
Hexahydrobenzofuran-2-one, 179
Hexahydrobenzothiophenes, 287, 293
Hexahydrocarbazoles, **514**, 515, 516, 517
Hexahydrodibenzofuran, **203**
Hexahydro-2,4-dioxobenz[*cd*]indole, 485
Hexahydroindoles, 382, 465, **466**
Hexahydroisobenzofurans, 190, 191
Hexahydroisoindoles, 477
Hexahydro-2-oxobenz[*cd*]indole, 485
Hexahydrophthalic esters, 191
Hexahydrophthaloyl chloride, 190
Hexamethoxybiphenyl, 195
Hexamethyl-2,2′-bipyrrolyl-4,4′-dicarboxylic acid, 392
Hexamethylenetetramine, 278, 322
Hexamethylphosphoramide, 235
Hexamine, 250, 251
Hexane-2,5-dione, 494, 522
Hexaphenylethane, 139, 264
Hexaphenylisobenzofuran, 187
Hexaphenylsilacyclopentadiene, **540**
Hexaphenylsilole, **540**
Hexaphenylstannole, 531
Hexoses, conversion to furans, 87
1-*n*-Hexyl-2-methylpyrrole, 343
Hinsberg synthesis, 245, 253, 254, 314, 483
Hippuric acid, 436
Hoesch reaction, 172, 173, 244, 337, 342, 433, 481
Hofmann reaction, 64, 212, 243, 380, 381
Huang–Minlon reduction, 150, 251
Humic acids, 487
Humin, 436
Hydantoin, 436
3-Hydrazinodibenzofuran, 201
Hydrazobenzene, 434
Hydrobenzofurans, 166
Hydrobenzo[*c*]thiophenes, 291
Hydrocarbazoles, 509
Hydrocinnamanilide, 28
Hydrofurfuramide, 112, 114
Hydroindoles, 445
Hydroisobenzofurans, 188
Hydrolysis rates, carboxylic acid methyl esters, 165
Hydron Blue G, **501**
Hydron Blue R, **500,** 501
Hydron Yellow G, 525
3-Hydroperoxy-3*H*-indole, 409
4*a*-Hydroperoxytetrahydro-4*aH*-carbazole, 513
Hydrophthalide, 189
Hydropyrroles, 389
Hydroquinone, 164
Hydroselenophenes, 319
Hydrotellurophenes, 327
Hydrothiophenes, monocyclic, 219
3-Hydroxyacetals, 48
γ-Hydroxyacetoacetic acid lactone, 133
o-Hydroxyacetophenones, 161, 167, 215
2-Hydroxyacetylbenzofuran, 161
2-Hydroxyacetylfuran, 117
β-Hydroxy acids, reaction with orthoesters, 49
3-Hydroxyaldehydes, 48
γ-Hydroxyaldehydes, 123, 132
β-Hydroxyamino compounds, 38
Hydroxyamino-ketones, 402
3-Hydroxyanthra[1,2-*b*]thiophene, 311
3-Hydroxyanthra[2,3-*b*]thiophene, 311
1-Hydroxyanthra[2,1-*b*]thiophene-6,11-quinone, 311
3-Hydroxy-1-arylpyridinium salts, 115
Hydroxybenzofurans, 146, 156, 157
—, physical properties, 159
6-Hydroxybenzofuran-5-carbaldehyde, 216
4-Hydroxybenzofuran-6-carboxylic acid, 156, 157
7-Hydroxybenzo[*b*]furan-2-carboxylic acid, 120
3-Hydroxybenzofuran-2-carboxylic ester, 156, 166
Hydroxybenzoic acids, 127
2′-Hydroxybenzoin, 152
3-Hydroxybenzo[*b*]selenophene, 320, 322
3-Hydroxybenzo[*b*]tellurophene, 327
Hydroxybenzothiophenes, **276**
3-Hydroxybenzothiophene, 267, 268, 270
2-Hydroxybenzothiophene-2-carboxylic acid, 267
3-Hydroxybenzothiophene-2-carboxylic 1,1-dioxide, 286

δ-*o*-Hydroxybenzoylvaleric acid, 203
3-Hydroxybiphenyl-1,2-dicarboxylic esters, 99
4-Hydroxybutyraldehyde, 130
Hydroxycarbazoles, **502**
2-Hydroxycarbazole, 509
2-Hydroxycarbazole-3-carboxylic acid, 509
o-Hydroxycarbonyl compounds, ring-closure, 142
2′-Hydroxychalcones, 149
2-Hydroxy-α-chloroacetophenones, 170, 172
2-Hydroxycyclohexanone, 179
Hydroxydibenzofurans, 195, 200
Hydroxydibenzothiophenes, 305
Hydroxydibenzothiophenecarboxylic acids, 304
3-Hydroxy-2,3-dihydrobenzofuran, 170
5-Hydroxy-2,3-dihydrobenzofurans, 169
6-Hydroxy-2,3-dihydrobenzofuran, 170
3-Hydroxy-2,3-dihydrobenzothiophene, 283
Hydroxy-1,2-dihydroisobenzofurans, 181
Hydroxy-1,3-dihydroisobenzofuran, 190
3-Hydroxy-2,3-dihydronaphtho[2,1-*b*]-furan, 211
2-(2-Hydroxy-4,6-dimethoxybenzoyl)-6-methoxy-3-oxo-2,3-dihydrobenzofuran, 173
Hydroxy-2,3-dimethylbenzofuran, 157
1-Hydroxy-3,3-dimethyl-1,3-dihydroisobenzofurans, 187
2-Hydroxy-4,4-dimethyl-6-oxocyclohex-1-enylacetic acid, 178
1-Hydroxy-3,4-dimethyl-2-phospholene 1-oxide, 536
1-Hydroxy-3,4-dimethyl-3-phospholene 1-oxide, 536
1-Hydroxy-2,5-dimethylpyrrole, 341, 351
2-Hydroxy-3,5-dinitrothiophene, 245
1-Hydroxydioxindole, 456, 458
Hydroxy-2,3-diphenylbenzofuran, 157
1-Hydroxy-1,3-diphenyl-1,3-dihydroisobenzofuran, 184
Hydroxy-α-epoxides, 314
β-Hydroxyethylamines, 17
9-(2-Hydroxyethyl)carbazole, 492
2-(2-Hydroxyethyl)cyclohexanone, 179
2-α-Hydroxyethyl-2,3-dihydrobenzofuran, 176
3-β-Hydroxyethylindole, **428**
1-Hydroxyethyl-2-oxopyrrolidine, 384
2-α-Hydroxyethyltetrahydrofuran, 135
3-Hydroxyflavanones, 170
Hydroxyfurans, 89, 90, 107, 109
3-Hydroxygermacyclopentanes, 547
Hydroxyhydroquinone, 172
4-Hydroxyhygric acid, 390
Hydroxyimino-β-diketones, 357
3-Hydroxyimino-2,5-dimethyl-2*H*-pyrrole *N*-oxide, 351
3-Hydroxyimino-2,5-dimethyl-1-pyrroline *N*-oxide, 341
Hydroxyindoles, 402, 403, 424, **425,** 448
1-Hydroxyindole, 413
3-Hydroxyindole, 454, 457
3-Hydroxy-3*H*-indoles, 408
2-Hydroxyindole-3-carbaldehyde, 433, 434
3-Hydroxyindole-2-carbaldehyde, 433, 434
3-Hydroxyindole-2-carboxylic acid, 411, 434
1-Hydroxyisatin, **463,** 464
4-Hydroxy-3-isopropenyl-2,3-dihydrobenzofuran-5-carboxylic acid, 176
6-Hydroxy-2-isopropylbenzofuran, 162
4-Hydroxy-3-isopropylbenzofuran-5-carboxylic acid, 166
Hydroxyketones, from phthalic esters, 181
—, from phthalides, 181
α-Hydroxyketones, 125
β-Hydroxyketones, 180
γ-Hydroxyketones, 123
Hydroxykynurenic acid, 437
Hydroxykynurenine, 437
Hydroxylamine-*O*-sulphonic acid, 41, 374
2-(6-Hydroxy-2-methoxy-3,4-methylenedioxyphenyl)benzofuran, 162
2-Hydroxymethylbenzofuran, 158
3-Hydroxymethylbenzofuran, 159
6-Hydroxy-3-methylbenzofuran, 157
5-Hydroxy-2-methylbenzofuran-3-carboxylic ester, 164
3-Hydroxy-2-methyl-1,4-benzopyrone, 160
2-Hydroxymethylbenzothiophenes, 277
3-Hydroxymethylbenzothiophenes, 277
5-Hydroxy-6-methylbenzothiophene-4,7-quinone, 276
9-(Hydroxymethyl)carbazole, **493**
3-Hydroxymethylcyclohexanol, 192
2-Hydroxymethyldibenzofuran, 175, 201

Hydroxymethyleneacetone, 115
Hydroxymethylene cyclohexanones, 488, 519
Hydroxymethylenesuccinates, 375
2-Hydroxymethylfuran, 110
3-Hydroxy-2-methylfuran, 109
Hydroxymethylfurfural, 87, 114, 116
Hydroxymethylfurfuraldehyde, 114
5-Hydroxymethyl-2-furoic acid, 121
2-Hydroxymethylindole, 431
2-Hydroxymethyl-3-methylindole, 409
2-Hydroxy-3-methyl-1,4-naphthoquinone, 277
2-Hydroxymethylnaphthothiophenes, 309
2-Hydroxymethyloxiranes, 5, 49
3-Hydroxy-5-methyl-2-oxoindole-3-carboxylic ester, 458
3-Hydroxy-3-methyl-2-phenyl-3*H*-indole, 447
5-Hydroxy-2-methylpyridine, 112
Hydroxymethylpyrroles, 353
1-Hydroxymethylpyrrole, 341
2-Hydroxymethylpyrrole, **354**
2-Hydroxymethylpyrrolidine, 384
3-Hydroxy-1-methylpyrrolidine, 386
1-Hydroxymethylpyrrolizidines, 481
2-Hydroxymethyltetrahydrofuran, 135
3-Hydroxy-3-methyltetrahydrothiophene-2-carboxylic ester, 221
5-Hydroxy-6-methyl-4,7-thianaphthenequinone, 277
2-Hydroxymethylthiophene, **247**
4-Hydroxy-2-methylthiophene-3-carboxylic ester, 255
3-Hydroxymethyl-1,2,5-trimethylpyrrole, 354
Hydroxynaphthaldehydes, 209
Hydronaphthofurans, 207
8-Hydroxy-1-naphthoic acid lactone, 213
1-Hydroxynaphtho[2,1-*b*]thiophene, 311
3-Hydroxynaphtho[1,2-*b*]thiophene, 311
β-(2-Hydroxynaphthyl)ethanol, 210
5-Hydroxy-4-nitrobenzothiophene, 275
3-Hydroxy-4,5-oxidophenanthrene, 214
Hydroxyoxindoles, **453,** 463
1-Hydroxyoxindoles, **452**
3-Hydroxyoxindoles, 456, **458**
2-Hydroxy-3-oxindole-2-carboxylic ester, 458
2-Hydroxy-6-oxocyclohex-1-enylacetic acid, 178
Hydroxy-3-oxo-2,3-dihydrobenzofurans, 172, 173
3-Hydroxy-2-oxo-2,3-dihydrobenzofuran, 170
3-Hydroxy-2-oxo-2,3-dihydroindole, 456
3-Hydroxy-2-oxoindolines, 456
3-Hydroxy-2-oxo-octahydrobenzofuran, 179
1-Hydroxy-2-oxopyrrolidines, 385
2-Hydroxy-5-oxo-3-pyrroline, 335
5-Hydroxy-2-oxo-3-pyrroline, 376
4-Hydroxy-2-oxo-3-thiolene-3-carboxylic ester, 257
o-Hydroxyphenylacetic acid, 141
o-Hydroxyphenylacetic acid lactone, 170
o-Hydroxyphenylacetylene, 146
o-Hydroxyphenylallene, 158
(2-Hydroxyphenyl)-β-enamino ketones, 146
o-Hydroxy-β-phenylethyl alcohols, 166
o-Hydroxy-β-phenylethyl bromide, 166
2-Hydroxyphenylglyoxal cyclic hemiacetal, 173
o-Hydroxyphenylglyoxylic acid, 175
1-Hydroxy-2-phenylindole, 413
2-Hydroxy-3-phenylindole, 452
5-Hydroxy-2-phenylnaphtho-[1,2-*b*]furan 3-carboxylic ester, 207
o-Hydroxyphenyl styryl ketones, 173
3-Hydroxy-5-phenylthiophene, 245
2-Hydroxyphenyl 2,4-xylyl ether, 195
1-Hydroxy-2-phospholene 1-oxide, 536
3-Hydroxypiperidines, 369, 482
1-Hydroxypiperidin-2-one, 369
Hydroxyproline, 330, 360, 388, **390,** 438
γ-Hydroxypropylamines, 64
5-Hydroxy-2-propylbenzofuran-3-carboxylic ester, 166
3-Hydroxypyridines, 116, 126
3-Hydroxypyridinium salts, 113
Hydroxypyrroles, 375
1-Hydroxypyrroles, 332, **351,** 352
3-Hydroxypyrroles, 377, 378
C-Hydroxypyrroles, 352
Hydroxypyrrolidines, **383**
2-Hydroxy-3-pyrrolin-5-one, 335
Hydroxysalicylaldehydes, 156
3-Hydroxystachydrines, 390

β-Hydroxysulphonic acids, 74
3-Hydroxytelluranaphthene, 327
Hydroxytetrahydrocarbazoles, 502, 512
Hydroxytetrahydrofurans, 123, 132, 135
12-Hydroxytetrahydroisocarbazole, 502
4*a*-Hydroxytetrahydroxy-4*aH*-carbazole, 513
3-Hydroxy-1-thiacoumarin, 284
3-Hydroxythietane, 9
3-Hydroxy-1-thiochromone, 284
2-Hydroxythiolane, **259**
Hydroxythiophenes, 220, 221, 222, 243
2-Hydroxythiophene, **244**
3-Hydroxythiophene, **245**
3-Hydroxythiophene-2-carboxylic acid, 245
Hydroxythiophenotropolone, 295
Hydroxytremetone, **169**
2-Hydroxy-4,6,4′-trimethoxybenzil, 143
Hydroxytrimethylbenzofurans, 157
9-Hydroxy-2,5,9-trimethyl-6,7-benzomorphan methiodide, 212
1-Hydroxy-1,3,3-trimethyl-1,3-dihydroisobenzofuran, 187
2-Hydroxy-1,3,3-trimethylindoline, 447
5-Hydroxytryptamine, 430
Hydroxytryptophans, 439
2-Hydroxytryptophan, **438**
6-Hydroxy-2,4-xylyl phenyl ether, 195
Hygric acid, 388, **389**
Hygric ester, 383
Hygrine, 386, 389
Hygrinol, 384
Hypaphorine, **439,** 441

Imidazole, 555
Imidazolidine, 560
Imidazoline, 560
Imidazolinium rings, 39
Imidazolinium salts, 36
Imidazo[2,1-b]thiazole, 564
4H-Imidazo[4,5-d]thiazole, 563
Imidazo[1,2-b]-[1,2,4]-triazine, 564
Imines, 69
—, conversion to oxaziridines, 40
Iminocoumarans, 146
Iminodihydrobenzofurans, 146
Iminodihydrofurans, 90
Imino groups, aminoethylation, 23
1-Iminoisoindoline, 476
1-Iminoisoindoline-3-thione, 476
Iminophthalimidine, 476
5-Iminopyrrolidine, 385
Iminotriphenyl-3*H*-pyrrole, 351
Indano[1,2-*b*]aziridine, 30
1H-Indazole, 556
Indene, 147, 148, 417
Indene-spiro-cyclopentanes nomenclature, 572
Indigo, 329, 397, 408, 411, 433, 435, 448, 454, 455, 500
Indigofera spp., 454
Indigoid dyestuffs, 281, 282, 455
Indigoid type pigments, 244
Indigotin, 460, 467
Indimulsin, 454
Indirubin, 454, 467
Indogenides, 455, 462
Indoles, 334, 336, 341, 397, 451, 487
—, addition reactions, 405
—, aldehydes, 431
—, alkylation, 411, 439
—, alkyl-substituted, 414
—, autoxidation, 408
—, 2-carbonyl substituted, 401
—, carboxylic acids, 434
—, characterisation, 411
—, dimerisation, 416
—, from acyltoluidines, 402
—, from *o*-aminocarbonyl compounds, 404
—, from arylhydrazones, 399
—, from isocyanides, 404
—, from ketones, 404
—, from oxindoles, 452
—, from pyrroles, 404
—, from δ-sultams, 405
—, Grignard reagents, 408
—, ketones, 431
—, metal compounds, 408
—, metal and metalloid derivatives, 427
—, nitration, 422
—, nitrosation, 423
—, *nomenclature, 572*
—, oxidation, 408
—, ozonolysis, 408
—, polymerisation, 405
—, reactions, 405
—, reaction with aldehydes and ketones, 410, 468
—, — with benzoyl chloride, 468
—, — with carbonyl compounds, 410

Indoles, reaction *(continued)*
—, — with nitroso compounds, 410
—, — with sulphuryl chloride, 419
—, reduction, 405, 445
—, substituent positions, 416
—, 2-substituted, 400
—, *N*-substituted, 413
—, substitution reactions, 407
—, sulphur compounds, 420
—, synthesis, 398
—, — from benzoquinones, 403
—, — from nitro compounds, 402
3*H*-Indoles, 397, 400, 410, 439, 446
—, 3,3-disubstituted, 407
Indole, 36, 147, **411**, 415, 418, 437, 454, 457, 494, *556*
—, *N*-derivatives, 412
—, dimer, 405, 406, **411**
—, maleic anhydride adduct, 405
—, mercuration, 427
—, oxidation to indigo, 411
—, reaction with acetylene, 412
—, — with acrylonitrile, 412
—, — with benzyne, 410
—, — with formaldehyde, 468, 525
—, — with formaldehyde and dimethylamine, 429
—, — with sulphur, 467
—, — with thiourea, 420
—, reduction to hydroindoles, 465
—, sulphonation, 420
—, thiocyanation, 421
—, trimer, 405, **411**
3*H*-Indole, 398, *556*
Indole-3-acetic acid, 420, **435**
Indole-3-aldoxime acetate, 428
Indole-2-alkanoic acids, 420
Indole-3-arsonic acid, 427
Indole-2-carbaldehyde, 430, 431, 432, **433**
Indole-3-carbaldehyde, 428, 432, 433, 436
Indolecarboxylic acids, physical data, 440
Indole-1-carboxylic acid, 413
Indole-2-carboxylic acid, 402, 411, 423, 434
Indole-3-carboxylic acid, 423, 434
Indole-2-carboxylic ester, 428
Indole-3-carboxylic ester, 428
Indole-2,3-dicarboxylic ester, 434
Indolenines, 398, 439
3*H*-Indole-3-nitronic acids, 439
Indole-5,6-quinone, 427
Indolesulphonic acids, **420**, 449
3-Indolethiol, **420**
2-Indole-2-thionaphthene indigo, 460
Indolines, 405, 445, 447
—, alkyl-substituted, 414
—, with side-chain carbonyl groups, 464
Indoline, 446, **448**, *561*
Indoline-2,3-diol, 422
Indolinequinones, 424
Indolinol, 400
Indolinones, 448
3*H*-Indolium salts, 417
Indolizine, *556*
Indolocarbazoles, 524
Indolocoumarone, 480
Indolones, 443
7H-Indolo[3,2-f]quinoxaline, *562*
Indolothiazines, 501
Indolothionaphthene, 480
3-Indolylacetaldehyde, 434
3-Indolylacetic acid, 428, 437, 507
3-Indolylacetone, 434
3-Indolylacetonitrile, 429, 430, 436
3-Indolylacet-(β-phenylethyl)amide, 436
β-3-Indolylalanine, 397
2-Indolylcarbamate, 423
2-(2-Indolyl)ethyl *p*-toluenesulphonate, 527
3-Indolylglyoxyl chloride, 428
3-Indolylglyoxylic ester, 428
3-Indolyl hydroperoxides, 409
α-3-Indolylisobutyric acid, 436
S-(3-Indolyl) isothiuronium iodide, 420
3-Indolyl ketone, 412
Indolyl-lactic acid, 437
Indolylmagnesium bromide, 427, 428, 430, 435
Indolylmagnesium halides, 408, **412**, 468
Indolylmagnesium iodide, 429
3-Indolylmethanesulphonic acid, 429
Indolylmethanols, 428
3-Indolylmethylnitromalonic ester, 430
Indolylpropionic acid, 436, 437
3-Indolylpyruvic acid oxime, 436
α-Indophenins, 228
β-Indophenins, 228
Indophenin reaction, 219, 228, 230, 315, 461
Indophenols, 500
Indoxanthic ester, 435, 458
Indoxyl, 424, 435, 443, **454**, 455, 460, 461, 462, 469

Indoxyl, *(continued)*
—, acid sulphate ester, 455
Indoxyl aldehyde, 433, 434
Indoxyl derivatives, **455**
Indoxylic acid, 411, **434,** 435, 440, 445, 455
Indoxylic ester, 458
Iodine azide, 35
2-Iodoadrenochrome, 425
β-Iodoazides, 18
Iodobenzene, 232, 234
Iodobenzothiophenes, **273**
4-Iodobenzothiophene, 303
5-Iodo-2,2′-bithienyl, 262, 264
5′-Iodo-2,2′-bithienyl-5-carboxylic acid, 262
5′-Iodo-2,2′-bithienyl-5-carboxylic methyl ester, 262
3-Iodocarbazole, 496
Iododibenzofurans, 200
2-Iododibenzofuran, **199**
2-Iododibenzoselenophene, 325
Iododibenzothiophenes, 305
Iodofurans, 97
2-Iodofuran, 102, 103
3-Iodofuran, 102, 103
Iodogermacyclopentane, 546
3-Iodoindole, **420**
Iodomethyl styryl ketone, 386
2-Iodo-8-nitrodibenzoselenophene, 325
5-Iodo-2-nitrofuran, 105
Iodoselenophenes, 315, 316, 318
2-Iodoselenophene, 320
Iodosobenzene, 256
β-Iodo sulphides, 10
β-Iodosulphonium iodide, 10
Iodothiophenes, 240, 241
2-Iodothiophene, 231, 232, 247, 259, 260, 261, 264
2-Iodothiophene-5-carboxylic acid, 232
Iodotrimethylpyrroles, 347
2-Iodo-3,4,5-trimethylthiophene, 237
Ipomeamarone, **117,** 138
Ipomeanine, **117**
Iron carbonyls, 532
Iron pentacarbonyl, 231
Isatan, 461
Isatide, 457, 461
Isatins, 378, 458
—, physical data, 464
Isatin, 175, 228, 445, 448, 452, 453, 454, 456, 457, 458, 459, **460,** 461, 463, 492
—, derivatives, 462
Isatin anils, 464
Isatin 2-anil, 282, 435, **462,** 463
Isatin 3-anil, 281
α-Isatin anilide, 459
Isatin blue, **463**
Isatin chloride, 282, 445, 460
Isatin dianil, **463,** 464
Isatin ethenedithioacetal, 450
Isatin 3-hydrazone, 449
Isatinic acid, 458
Isatinimine, 461, 464
Isatin-*O*-methyl ether, 463
Isatin osazone, **463,** 464
Isatinoximes, 464
Isatin-2-oxime, 455, **462**
Isatin-3-oxime, 452
Isatinphenylhydrazones, 462, 464
Isatogens, **443,** 444
Isatogenic acid, 445
Isatogenic ester, 435, 440
Isatoic anhydride, 460
Isoamyl nitrate, 348
2-Isoazolines, 19
Isobenzofurans, 57, 180, 181, **187**
—, aryl-substituted, 186
—, reactions, 185
—, substituted, 184
Isobenzofuran, 554
Isobenzofurylium perchlorate, 187
Isobenzofurylium salts, 187
Isobenzofurylium tetrachloroferrates, 187
Isobenzothiophenes, 288
Isobutene, 272
Isobutenyl methyl ketone, 160
Isobutylene, 98, 237, 361
—, reaction with thiophene, 233
N-Isobutylidene-*tert*-butylamine, 40
Isobutyl 3-methyl-2-furyl ketone, 117
Isobutyraldehyde methylphenylhydrazone, 448
Isobutyrophenone dimethylhydrazone methiodide, 34
Isobutyrophenone phenylhydrazone, 414
Isochroman, 560
Isocyanides, conversion to indoles, 404
—, reaction with ketones, 410
—, — with pyrroles, 355

Isodehydracetic acid, 86
Isodiazomethane, 43
Isogmelinol, **193**
Isogramine, **429**
Isohemipinic acid, 161
3-Isohexenylfuran, 100
Isohexylsuccinic acid, 100
Isoindigo, 461, 467
Isoindigotin, 457, 460
Isoindogenides, 451
Isoindoles, 470
Isoindole, 180, **470,** *556*
3*H*-Isoindole, 470
Isoindolines, 470, 474, 475
Isoindoline, **474,** *561*
—, reduction, 477
Isoindoloisoquinoline, 478
Isoisatogens, 444
Isomannide, 193
Iso-oxazolium salts, 378
Isoprene, 537
Isopropenyl 2-thienyl ketone, 295
Isopropylbenzothiophenes, 272, 273
Isopropyl chloride, 114
7-Isopropyl-2,4-dimethyl-2,3-dihydrofuran, 169
Isopropylfurans, 101
4-Isopropylfurfural, 114
2-Isopropylidene-1-oxaspiro[4.5]decane, 191
1-Isopropylindole, 413
3-Isopropylindole, 416, 418
Isopropylmagnesium bromide, 244
5-Isopropyl-2-methyl-2-thiolene, 256
Isopropylthiophenes, 237
Isopulegone, 177
Isoquinoline, 30, *557*
Isothiachromanone enamine, 290
Isothiazole, *559*
Isotryptamine, **430,** 432
Isotryptophan, 441
Isovaleric acid, 117
Isoxazoles, 160
Isoxazole, *559*
Isoxazole oximes, 350
Isoxazolidines, 372

Japanese egonoki fruits, 161
Japp–Klingemann reaction, 400, 401, 434, 488
Jasmine oil, 411
Junipol, **250**

Kainic acid, **390**
Karanjin, 157, 166
Ketenes, 69, 70
—, reaction with azomethines, 67
—, — with carbonyl compounds, 49
—, — with diazoketones, 107
Ketene, 43
—, reaction with azo compounds, 76
Ketene *N*,*N*-acetals, reaction with azosulphones, 80
Ketene diethyl acetal, 54, 60, 143, 156
Keten dimer, 89
Keten phenylimine, 34
Ketoaldehydes, 84
Ketones, arylhydrazones, 398
—, conversion to dithietanes, 73
—, — to indoles, 404
—, — to oxaziridines, 40
—, — to pyrroles, 334
—, from furans, 92
—, from oxaziridines, 40
—, from thiophenes, 227, 250
—, furans, 115
—, reaction with aziridine, 20
—, — with aziridinium salts, 21, 38
—, — with furans, 140
—, — with α-halogenoketones, 3
—, — with indoles, 468
—, — with isocyanides, 410
—, — with primary amines and chloramine, 42
—, thiophenes, 249
—, $\alpha\beta$-unsaturated, 130, 361, 369, 415
γ-Ketonic acids, reaction with phosphorus pentasulphide, 243
β-Ketonic esters, 85
—, conversion to pyrroles, 332
Ketopimelic acid, 112
Ketoximes, reaction with Grignard reagents, 18
Ketoxime sulphonyl esters, 33
Kischner–Wolff reduction, 97, 127, 341, 342, 393
Knoevenagel condensation, 316
Knorr synthesis, 331, 361, 465, 478
—, pyrroles, 330, 370
Kolbe synthesis, 509

Kynurenic acid, 437
Kynurenine, 437, 438

Labelled acids, from thiophenecarboxylic acids, 252
α-Lactams, 31, 32, 33
β-Lactams, 63, 67, 69
—, from allenes, 68
—, from amides, 68
—, stereochemistry, 69
δ-Lactams, 69
Lactols, 132
Lactones, 107
β-Lactones, 32, 48, 49
γ-Lactones, αβ-unsaturated, 86
Laevulic acid, 100, 110, 236, 384
—, arylhydrazones, 436
Laevulic aldehyde, 373
3-Laevuloylfuran, 117
Lapachol, 211
Leptosin, **174**
Leuckart reaction, 248, 472
Lewis acids, 461
Lichen products, 194
Lignin, 136
Lindestrene, 209
o-Lithioanisole, 198
1-Lithioaziridine, 21
Lithiodibenzofurans, 197
4-Lithiodibenzofuran, 203
Lithio-9-ethylcarbazole, 504
3-Lithiothiophene, 240
Lithium telluride, 327
Lithium tetrahydridoaluminate, 18, 19, 25, 32, 41, 44, 47, 48, 53, 62, 70, 72, 77, 130, 148, 158, 159, 161, 165, 188, 189, 198, 247, 249, 251, 277, 280, 301, 341, 353, 354, 372, 379, 380, 384, 386, 428, 432, 457, 477, 528, 539, 540, 541, 543, 544, 545, 547
Lithium tetrahydridoborate, 12
Lysergic acid, 454

Madelung indole synthesis, 402
Magnesylcarbazole, 507
Magnesylindoles, 413, 421, 432, 433, 434
Malachite green, 112, 247
Malealdehyde, 126
Maleic acids, 102
—, from furans, 91
Maleic acid, 364
Maleic anhydride, 93, 102, 103, 109, 113, 235, 236, 269, 272, 286, 289, 310, 313, 472
—, adduct with indole, 405
—, adducts with isoindole, 470
—, hydrogenation, 130
—, reaction with furan, 92
Maleic ester, 128
Maleimide, 92, 335, 338, 376
Maleimidoximes, 335
Malonic acid, 137, 254, 316
Malonic ester, 179, 270, 388, 403, 511
—, syntheses, 429
Malonic methyl ester, 71
Malonimides, 70
Malonodinitrile, 350
Malonyl chloride, 70
Mandelic acid, 121
Mannich adducts, 512
Mannich reaction, 248, 251, 337, 343, 354, 407, 429
Mannitol, 193
Maritimein, **174**
Markownikow's rule of addition, 167
Melanins, 362, **426**
—, biosynthesis, 426, 427
Menthofuran, **177**
o-Mercaptobenzaldehyde, 270
o-Mercaptobenzoic acid, 267, 270
2-Mercaptobenzothiophene, 277
2-Mercaptobiphenyl-2′-carboxylic acid, 303
Mercaptocinnamic acid, 266, 271, 276, 279
2-Mercaptocyclohexanone, 287
3-Mercaptocyclohexanone, 287
3-Mercaptocyclohexanone ethylene acetal, 288
Mercaptoethylation, thiiranes, 8
β-Mercaptoisovaleric acid lactone, 59
2-Mercaptomethylthiiranes, 6
2-Mercaptomethylthiophene, 248
o-Mercaptophenylacetic acid lactone, 281
o-Mercaptophenylglyoxylic acid, 284
1-Mercaptopropan-2-one, 60
Mercuric cyanate, 233
Mercurithiophenes, 232
Mesitylhydrazine, 487
Mesityl oxide, 130, 407
Mesoxalic ester, 458

Metacycloprodigiosin, 396
Methanesulphenyl chloride, 421
o-Methanesulphonylbenzoic acid, 282
o-Methanesulphonylbenzonitrile, 282
Methanesulphonyl bromide, 130
Methanesulphonyl chloride, 54, 60, 74
Methine colouring matters, 285
Methoxybenzofurans, **156,** 157
4-Methoxybenzofuran-5-carboxylic ester, 162
Methoxybenzothiophenes, 276
2-Methoxybenzothiophene, 274
3-Methoxybenzothiophene, 282
3-Methoxybenzothiophene-2-carboxylic acid, 286
o-Methoxybenzoyldiazomethane, 171
Methoxycarbazoles, 503
1-Methoxycarbazole-3-carbaldehyde, 505
Methoxycarbonyldibenzofurans, 201
2-Methoxycarbonyl-2-phenyl-1-azirine, 35
7-Methoxycarbonyltetrahydrocarbazole, 514
Methoxydibenzofurans, 201
3-Methoxydibenzofuran-1,2-dicarboxylic methyl ester, 199
2-Methoxydibenzothiophenes, 305
5-Methoxy-2,6-dimethylbenzofuran, 150
2-Methoxy-3,5-dinitrothiophene, 242
Methoxyethynyl prop-1-enyl ketone, 175
2-Methoxyfuran, **107**
3-Methoxyfuran, 103, 109
3-Methoxyindole, 454
(Methoxy-2-indolyl)ethyl *p*-sulphonatotoluenes, 528
5-Methoxyindolylmagnesium bromide, 431
5-Methoxyisatin, 464
α-Methoxylaevulic acid, 96
6-Methoxy-3-methylbenzofuran, 160
4-(6-Methoxy-3-methyl-2-benzofuryl)-4-methyl-3-pentanone, 160
3-Methoxy-6-methylcarbazole, 502
5-Methoxy-1-methyl-4,7-dihydroindole, 405
4-Methoxy-4′-methyldiphenylamine, 490, 503
α-Methoxymethylene ketones, 88
Methoxymethylenetriphenylphosphorane, 159
5-Methoxy-1-methylindole, 405
5-Methoxy-1-methylindoline, 405
7-Methoxymitrosene, **529**
5-Methoxynaphtho[1,2-*b*]furan, 209
8-Methoxy-1,4-naphthoquinone-2-carboxylic ester, 94
1-Methoxyoxindoles, 452
3-Methoxyoxindoles, 461
6-Methoxy-3-oxo-2,3-dihydrobenzofuran, 173
6-Methoxy-3-oxo-2,3-dihydrobenzo[*b*]-selenophene, 323
6-Methoxy-3-oxo-2(2,4,6-trimethoxybenzoyl)-2,3-dihydrobenzofuran, 173
2-Methoxyphenylacetic acid, 158
o-Methoxy-β-phenylethyl alcohol, 168
1-(4-Methoxyphenyl)isoindole, 472
2-(4-Methoxyphenyl)naphtho[1,2-*b*]thiophene, 307
2-(4-Methoxyphenyl)naphtho[2,1-*b*]thiophene, 307
1-(*p*-Methoxyphenyl)-2-phenylnaphtho-[2,1-*b*]furan, 208
Methoxypyrroles, 376
2-Methoxy-1-pyrroline, 392
5-Methoxytetrahydrobenzofurans, 145, 178
Methoxytetrahydrocarbazoles, 514
Methoxytoluquinone, 174
Methoxytrifluorobenzofurans, 153
Methoxytrimethylbenzofurans, 157
1-Methoxy-1,3,4-trimethylsila-3-cyclopentene, 542
2-(β-Methoxyvinyl)benzofuran, 159, 199
Methronic acid, 118
1-Methyl-2-alkenylpyrrolidines, 374
2-Methylallyl chloride, 98
4-Methylallyldibenzofuran, 200
Methylamine, 259, 269, 390
Methylamine hydrochloride, 155
N-Methylaminoacetic ester, 334
3-Methylaminodibenzofuran, 201
3-Methylaminomethyl-indole, 428
2-Methylaminothiophene, 243
Methylaniline, 435, 497
α-(*N*-Methylanilino)styrene, 446
N-Methylanthranilic acid, 455
Methylaziridines, 27, 28
N-Methylbenzilanilide, 453
Methylbenzocarbazoles, 524
Methyl-7*H*-benzo[*c*]carbazole, 521
6-Methyl-11*H*-benzo[*a*]carbazoles, 521
Methylbenzofurans, 148, 150, 151

2-Methylbenzofuran, 149, 158
3-Methylbenzofuran, 160, 167
2-Methylbenzofuran-3-acetic acid, 158
5-Methylbenzofuran-3-acetic-2-carboxylic acid, 166
3-Methylbenzofuran-2-carbaldehyde, 150, 159
3-Methylbenzofuran-2-carboxylic acid, 163, 165
3-Methylbenzofuran-2-carboxylic ester, 165
Methyl 2-benzofuryl carbamate, 155
3-(Methyl-2-benzofuryl)diphenylmethanol, 165
2-Methylbenzo[*b*]selenophene, 321
3-Methylbenzo[*b*]selenophene, 321
3-Methylbenzo[*b*]selenophene-2-carbaldehyde, 322
3-Methylbenzo[*b*]selenophene-2-carboxylic acid, 322
Methylbenzothiophenes, **272**, 274
3-Methylbenzothiophene, 278, 281
Methylbenzothiophene-2-carbaldehyde, 274, 279
Methylbenzothiophenecarboxylic acids, 280
2-*o*-Methylbenzylisoindole, 475, 478
5-Methyl-2,2′-bifuran, 137, 138
3-Methyl-2,2′-bifuran-3′-carboxylic acid, 138
5-Methyl-2,2′-bifuran-3-carboxylic acid, 138
5-Methyl-2,2′-bithienyl, 262
5′-Methyl-2,2′-bithienyl-5-carbaldehyde, 262
2-Methylbutadiene, 69
2-Methyl-2-butene, 46
3-Methyl-3-buten-1-ylamines, 66
Methylcarbazoles, 495
1-Methylcarbazole, 494
2-Methylcarbazole, 494
9-Methylcarbazole, 491, 492, 505, 516
N-Methylcarbazole, 490
9-Methylcarbazole-3-carbaldehyde, 504
2-Methylcarbazole-4-carboxylic acid, 507
9-Methylcarbazole-1-carboxylic acid, 492
1-Methylcarbazole-2,3-dicarboxylic methyl ester, 507
9-Methyl-3-carbazolyldichloroarsine, 504
3-Methylcoumarilonitrile, 165
3-Methylcoumariloyl chloride, 159
Methyl cyanide, 92
1-Methylcyclohexane-3,5-dione, 177
2-Methylcyclohexanones, 511
3-Methylcyclohexanone, 521
4-Methylcyclohexanones, 511
2-Methyl-2,3-cyclopentano-4,5-cyclohex-5-eno-2,3,4,5-tetrahydrofuran, 204
4-Methyl-4-cyclopentenone, 125
2-Methyl-4,5-cyclopentenopyrrole-3-carboxylic ester, 478
9-Methyldecahydrocarbazole, 517
9-Methyl-2-decalone, 212
9-Methyl-3,6-diacetylcarbazole, 505, 506
6-Methyl-13*H*-dibenzo[*a,i*]carbazole, 521
13-Methyl-7*H*-dibenzo[*a,g*]carbazole, 521
Methyldibenzofurans, 200
2-Methyldibenzofuran, 196
4-Methyldibenzofuran, 197
5-Methyldibenzostannole, 550
Methyldibenzothiophenes, 305
1-Methyldibenzothiophene, 301
Methyldichlorophosphine, 535
2-Methyl-2,3-dihydrobenzofuran, 167, 169, 176
5-Methyl-2,3-dihydrobenzofuran-2,3-dione, 175
5-Methyl-2,3-dihydrobenzofuran-3-one, 175
2-Methyl-2,3-dihydrobenzo[*b*]selenophene, 321
3-Methyl-2,3-dihydrobenzothiophene, 281
2-Methyl-4,5-dihydrofuran, 125, 134
1-Methyl-1,2-dihydronaphtho[2,1-*b*]furan, 210
2-Methyl-2,3-dihydronaphtho[2,3-*b*]furan, 210
3-Methyl-2,3-dihydronaphtho[1,2-*b*]furan, 210
Methyl-2,3-dihydropyrans, 134
Methyl-5,6-dihydroxyindoles, 426
Methyldiiodophosphine, 532
2-Methyl-5,7-dinitrobenzofuran, 154
2-Methyldinitroindoles, 423
1-Methyldioxindole, 458
3-Methyldioxindole, 456, 458
Methyl-2,3-dioxo-2,3-dihydrobenzofurans, 175
Methyldiphenylamine, 490, 494
Methyl-2,3-diphenylbenzofurans, 150, 151

2-Methyl-1,3-diphenylisoindole, 472, 474
2-Methyl-3,3-diphenyloxaziridine, 41
4-Methyl-2,5-diphenyloxazole, 37
1-Methyl-3,3-diphenyloxindole, 453
1-Methyl-2,5-diphenylpyrrole, 348
Methylene bromide, 4
Methylene chloride, 415
Methylenecyclohexane, 4
Methylenediamine, 429
Methylenediammonium sulphate, 43
2-Methylene-2,5-dihydrofuran, 127
Methylene-1,3-dihydroisobenzofurans, 181
Methyleneindolines, 400
Methylenemalonic ester, 81
3-Methyleneoxindole, 450
2-Methylene-4-phenylthietene 1,1-dioxide, 59
Methylenequinones, 109
2-Methylenetetrahydrofuran, 125, **134,** 135
2-Methylenethietene dioxide, 58
Methylformanilide, 249, 279, 287, 465, 504
—, vinylogues, 359
Methylfurans, 98, 101
2-Methylfuran, 94, 97, 99, 127, 224
3-Methylfuran, 94, 98, 101
4-Methylfuran-2-acetic-3-carboxylic acid, 121
3-Methylfuran-2-acetic-4,5-dicarboxylic acid, 205
2-Methylfuran-3-carbaldehyde, 115
Methylfurancarboxylic acids, 122
3-Methylfuran-2-carboxylic acid, 117
5-Methylfuran-2-carboxylic methyl ester, 119
2-Methylfuran-3,5-dicarboxylic acid, 87
5-Methylfurfural, 86, 114
Methylfurfuraldehyde, 114
Methylfurfuryl alcohols, 134
Methyl 3-furoate, 111
5-Methyl-2-furoic acid, 104
Methylglyoxal, 99
1-Methylgramine, 429
2-Methyl-2-hepten-6-one, 256
2-Methylhexahydrobenzofuran, 145, 179
3-Methylhexahydrobenzofuran-2-one, 178
9-Methylhexahydrocarbazoles, 491, 515
1-Methylhexahydronaphtho[2,1-*b*]furan, 212
N-Methylhydroxylamine, 285
O-Methylhydroxylamine, 197
N-Methylhydroxylamine-*O*-sulphonic acid, 40
N-Methyl-2-hydroxymethylpyrrole, 354
1-Methyl-2-hydroxymethylpyrrolidine, 383
N-Methylhydroxyproline, **390**
1-Methyl-3-hydroxypyrrolidine, 383
Methylindoles, 439
—, physical data, 418
1-Methylindole, 408, 412, **413**
2-Methylindole, 36, 402, **415,** 416, 419, 443, 470, 494, 519
3-Methylindole, **415,** 419, 468
2-Methylindole-3-acetic acids, 436
Methylindole-3-carbaldehydes, 433
Methylindolecarboxylic acids, physical data, 440
1-Methylindole-2,3,4-tricarboxylic ester, 419
Methylindolines, **448**
Methylindoline-3-carboxylic acids, **448**
3-(2-Methylindolyl) phosphonic dichloride, 419
O-Methylindoxyl, 455
Methylisatins, 464
1-Methylisatin, 461
2-Methylisatin, 463
5-Methylisatin, 464
O-Methylisatin, 462
N-Methylisatin 3-thiosemicarbazone, 464
Methylisatoid, 463
2-Methylisoindole, **471,** 476
Methylisoindolines, **476**
2-Methylisoindoline, 471
2-Methyl-5-isopropyltetrahydrofuran-2,5-dicarboxylic acid, 136
Methylketole, 415
β-Methyllaevulic acid, 236
Methyl-lithium, 77, 212, 303, 415, 513
Methylmagnesium bromide, 544
Methylmagnesium chloride, 547
Methylmagnesium iodide, 56, 111, 160, 187, 209, 212, 213, 251, 316
Methylmaleimidopropionic acid, 364
9*b*-Methyl-8-methoxyhexahydronaphtho-[2,1-*b*]furan, 212
4-Methyl-4-(3′-methyl-2′-benzofuryl)-2-pentanone, 160
1-Methyl-3-(1′-methyl-2′-pyrrolidinyl)-2-pyrroline, 373
1-Methyl-2-methylthio-2-pyrroline, 374

Methylnaphthalene, 399
2-Methylnaphthofurans, 209
2-Methylnaphtho [1,2-*b*]furan, 207
2-Methylnaphtho [2,1-*b*]furan, 208
2-Methylnaphtho [2,3-*b*]furan, 208
3-Methylnaphtho [2,3-*b*]furan, 209
3-Methyl-2-naphthoic acid, 526
1-Methylnaphtho [2,1-*b*]thiophene, 308
Methyl 2-naphthyl ether, 211
2-Methyl-3-nitrobenzothiophene, 274
3-Methyl-2-nitrobenzothiophene, 275
7-Methyl-3-nitrobenzothiophene, 275
3-Methyl-3-nitro-*n*-butyl cyanide, 384
5-Methyl-6-nitro-2,3-diphenylbenzofuran, 155
7-Methyl-6-nitro-2,3-diphenylbenzofuran, 155
2-Methyl-3-nitroindole, 422, 423
Methylnitroindolines, **448**
3-Methyl-2-nitromethylbenzothiophene, 274
3-Methylnitrosoaminothiolane 1,1-dioxide, 259
2-Methyl-3-nitrosoindoles, 422
2-Methyl-1-nitrosoindoline, 446
1-Methyl-3-nitroso-2-phenylindole, 423
5-Methyl-3-nitroso-2-phenylpyrrole, 349
2-Methyl-2-nitrosopropane, 44
2-Methyloctahydrobenzofuran, 179
9-Methyloctahydrocarbazole, 516, 517
8*a*-Methyloctahydro-8*aH*-carbazole, 517
Methyloctahydroindole, 466
2-Methyloxetane, 47
2-Methyl-3-oximino-3*H*-indole, 443
Methyloxindole, 415, 434, 453
5-Methyl-6-oxocyclopenteno [*b*]thiophene, 295
Methyl-2-oxo-2,3-dihydrobenzofurans, 170
Methyl-3-oxo-2,3-dihydrobenzofurans, 172
3-Methyl-3-oxo-2,3-dihydrobenzo [*b*]selenophene, 323
5-Methyl-2-oxodihydrofurans, 107
5-Methyl-2-oxo-2,3-dihydrothiophene, 244
5-Methyl-2-oxomethylpyrrolidine, 384
1-Methyl-1-oxoniacyclohexane chloride, 569
2-Methyl-4-oxo-1-phenyl-2-pyrrolines, 377
1-Methyl-3-oxopyrrolidine, 386
1-Methyl-3-oxopyrrolidine-4-carboxamide, 386
2-Methyl-4-oxo-2-pyrroline-3-carboxylic ester, 378
2-Methyl-5-oxo-2-pyrroline-3-carboxylic ester, 378
3-Methyl-4-oxotetrahydrobenzofurans, 177
2-Methyl-5-oxo-2-thiolene, 244
2-Methyl-5-oxo-3-thiolene, 244
Methylpentoses, 114
—, conversion to furans, 86
2-Methyl-3-phenyl-1-azirine, 36
3-Methyl-2-phenyl-1-azirine, 35, 37
5-Methyl-3-phenylbenzo [*b*]selenophene-2-carbaldehyde, 322
1-Methyl-3-phenyl-1,3-dihydrobenzo [*c*]-thiophene, 292
3-Methyl-2-phenyl-2,3-dihydrofurans, 124
4-Methyl-2-phenylfuran, 101
N-Methylphenylglyoxoylanilide, 453
1-Methyl-2-phenylindole, 418
1-Methyl-2-phenylindoline, 446
2-Methyl-2-phenylindoxyl, 447
1-Methyl-2-phenylnaphtho[2,1-*b*]furan, 208
2-Methyl-1-phenylnaphtho [2,1-*b*]furan, 208
2-Methyl-3-phenylnaphtho [1,2-*b*]furan, 207
3-Methyl-2-phenylnaphtho [1,2-*b*]furan, 207
1-Methyl-2-phenylpyrrolidine, 382
2-Methyl-4-phenylpyrroline, 371
Methylphloracetophenone, 206
1-Methyl-3-phospholene, 535
3-Methyl-2-phospholene, 536
N-Methylphthalimidine, 471
N-Methyl-L-proline, **389**
Methyl propenyl ketone, 133, 134
2-Methyl-5-propylpyrrolidine, 106
Methylpyrroles, 342, 345
1-Methylpyrrole, 333, **340**, 341, 348, 355
2-Methylpyrrole, 336, 353
3-Methylpyrrole, 336
C-Methylpyrroles, 341, 346
N-Methylpyrroles, 336, 339, 346, 360, 364, 374
Methylpyrrolecarbaldehydes, physical data, 358
1-Methylpyrrole-2-carbaldehyde, 355
Methylpyrrolecarboxylic acids, physical data, 365, 366
N-Methylpyrrole-2-carboxylic acid, 360
1-Methylpyrrolidine, 370, 373, 381, **382**
2-Methylpyrrolidine, 382
3-Methylpyrrolidine, 382

Methylpyrrolidine-2-carboxylic ester, 383
1-Methylpyrrolidine-2,5-dicarboxylic acid, 390
1-Methyl-2-pyrrolidone, 374, 384
5-Methyl-2-pyrrolidone, 384
1-Methyl-3-pyrroline, 375
2-Methylpyrroline, **371**
4-Methylpyrroline, 371
1-Methylpyrrolizidine, 481
3-Methylpyrrolizidine, 481
4-Methylquinoline, 415
Methylsalicylaldehyde cyanohydrins, 170
2-Methyl-3-salicylidene-3*H*-indole, 443
o-Methylselenoacetophenone, 322
Methylselenophenes, 316, 319
Methylstyrene, 68
Methylsuccinic acid, 174, 236
1-Methylsulphonylaziridine, 16
9-Methyltetrahydro-7*H*-benzo [*c*]carbazole, 521
2-Methyltetrahydrobenzofuran, 177
3-Methyltetrahydrobenzofuran, 178
2-Methyltetrahydrobenzofuran-3-carboxylic acid, 178
2-Methyltetrahydrobenzothiophene, 287
5-Methyltetrahydro-2,3′-bifuran, 138
Methyltetrahydrocarbazoles, 513, 514, 515
Methyltetrahydrofurans, 129, 135
Methyltetrahydroindoles, 466
Methyltetrahydroindolecarboxylic esters, 466
2-Methyltetrahydroquinoline, 446
2-Methyl-5-tetrahydroxybutylpyrrole-3-carboxylic ester, 354
3-Methyl-1-tetralone arylhydrazones, 521
1-Methyltetraphenylphosphole oxide, 532
N-Methyl-2-thenylamine, 248
Methylthianaphthenequinone, 285
2-Methylthieno[3,2-*b*]thiophene, 297
Methyl thienyl ketoxime, 243
Methyl-2-thienylmethanol, 247
Methyl 3-thienylmethanol, 237, 247
5-Methyl-5′-(2-thienylmethyl)-2,2′-bithienyl, 264
Methyl 2-thienyl selenide, 231
Methyl 2-thienyl sulphone, 247
2-Methylthietane, 53
Methylthiiranes, 6
2-Methylthiirane, 13
Methylthiirene 1,1-dioxide, 14
2-Methyl-5-thiocyanopyrrole, 353
Methylthiomaleic anhydride, 228
Methylthiophenes, 235, 236, 239
2-Methylthiophene, 230, 233, 242, 244, 252
—, halogenation, 238
3-Methylthiophene, 231, 248, 250, 252
—, oxidation, 228
Methylthiophene-2-carbaldehydes, 249
3-Methylthiophene-2-carbaldehyde, 250
3-Methylthiophene-5-carbaldehyde, 250
3-Methylthiophene-2-carboxylic ester, 221
Methylthiophenium hexafluorophosphate, 226
2-Methylthio-3-phenylindole, 421
2-Methylthiopyrrole, **352**
2-Methylthiopyrrole-5-carbaldehyde, 352
5-Methylthiothiophene-2-carboxylic ester, 223
Methyl-*o*-toluidine, 411
Methyltrichlorosilane, 544
2-Methyl-3,4,6-trinitroindole, 423
2-Methyltryptamine, 431
Methyltryptophans, 439
Methyl vinyl ketone, 221, 412, 494
Michael addition, 59, 118, 176, 396, 481
Michael-type adducts, 94
Micro-organisms, growth inhibition, 254
Mills–Nixon effect, 303
Morphenol, 214
3-Morpholinodithioacrylic methyl ester, 223
Morpholine, 561
1-Morpholinocyclohexene, 499
3-Morpholinothietene 1,1-dioxide, 60
Mucates, conversion to pyrroles, 333
Mucic acid, 87, 117, 119, 223, 252
Mucoproteins, 360
Murrayakoenigii Spreng, 505
Murrayamine, 487, **505**
Myosmine, 368

Naphthalene, 271
—, in coal tar, 219
Naphthalene-2,3-dicarboxylic acid, 289, 523
Naphthalenetetracarboxylic acid, 523
β-Naphthaquinonesulphonic acid, 412
Naphthastyrils, 484, 486
Naphtho[1,2-*b*]furan, **207**
Naphtho[2,1-*b*]furan, **207,** 208, 209

Naphtho[2,3-*b*]furan, **208**, 209
Naphtho[2,1-*b*]furan-2-carbaldehyde, 208
Naphtho[2,1-*b*]furan-2-carboxylic acid, 207
α-Naphthoic acid, 119
Naphthols, 209, 211
1-Naphthol, 207, 518
2-Naphthol, 106, 208, 210, 349, 350, 423, 499, 518
Naphthol AS LB, 509
Naphthoquinolines, nomenclature, 572
Naphtho[1,2-g]quinoline, 571
Naphtho[2,1-f]quinoline, 571
1,4-Naphthoquinone, 313
Naphtho[1,8-*bc*]thiolane, **311**
Naphthothiophenes, 309
Naphtho[1,2-*b*]thiophene, **306**, 307
Naphtho[2,1-*b*]thiophene, **308**
Naphtho[2,3-*b*]thiophene, **310**, *554*
Naphtho[1,2-*b*]thiophene-2-carboxylic acid, 307
Naphtho[2,1-*b*]thiophene-2-carboxylic acid, 309
Naphthothiophene dioxide, 308, 309
Naphtho[2,1-*b*]thiophene-6,9-quinone, 308
Naphtho[1,8-*bc*]thiopyran, 306
Naphthoxyacetic acids, ring closure, 210
2-Naphthoxyacetone, 207
2-Naphthoxyacetophenone, 208
Naphthoxy ketones, 209
Naphthylamines, 482
1-Naphthylamine, 518
2-Naphthylamine, 518, 520
1-Naphthylhydrazine, 483
β-Naphthylhydrazine, 521
Naphthylhydrazones, 482, 483
Naphthyloxamic acid chlorides, 483
1-Naphthyl phenacyl sulphide, 307
Naphthylphenanthrene, 309
2-Naphthylthioacetaldehyde diethyl acetal, 308
Naphthyridine, 557
Neber rearrangement, 18, 34
Nef reaction, 86
Nenitzescu indole synthesis, 403
Neopentyl glycol, 45
N-Neopentylidene-*tert*-butylamine, 32
Ngaione, **117**, 138
Nicotine, 389
Nicotinic acid, 437
Ninhydrin, 67
Nitrenium ion, 30
Nitriles, reaction with aziridinium salts, 21, 39
Nitrile oxides, 19
Nitrile ylides, 374
Nitroalkanes, 44, 429
β-Nitroamines, 332
Nitroanilines, 387
1-Nitroazetidine, 63, 64
1-Nitrobenz[*a*]anthracene, 524
Nitrobenzene, 234
—, pyrolysis, 489
—, reaction with benzene, 197
Nitrobenzocarbazoles, 524
Nitrobenzofurans, 154, 168
2-Nitrobenzofuran, **154**, 155
5-Nitrobenzofuran, 155
6-Nitrobenzofuran, 155
7-Nitrobenzofuran, 155
2-Nitrobenzofuran-3-ol, 154
Nitrobenzothiophenes, **274**
3-Nitrobenzothiophene, 276
2-Nitrobenzothiophene-3-carboxylic acid, 274
5-Nitrobenzothiophene-2-carboxylic acid, 274
6-Nitrobenzothiophene 1,1-dioxide, 275
o-Nitrobenzoyl cyanide, 458
o-Nitrobenzoyldiazomethane, 463
m-Nitrobenzoylproline, 389
2-Nitrobenzyl cyanides, 402, 403, 423
2-Nitrobiphenyl, pyrolysis, 490
3-Nitro-2,2′-bithienyl, 262
5-Nitro-2,2′-bithienyl, 262
Nitrocarbazoles, 496, **498**
1-Nitrocarbazole, **497**
3-Nitrocarbazole, **497**, 499
γ-Nitrocarbonyl compounds, 372
Nitro compounds, conversion to indoles, 402
Nitrodibenzofurans, 196, 201
2-Nitrodibenzoselenophene, 324
3-Nitrodibenzoselenophene, 324
2-Nitrodibenzotellurophene, 328
4-Nitrodibenzotellurophene, 328
Nitrodibenzothiophenes, 305
3-Nitrodibenzothiophene 5,5-dioxide, 302
Nitroethenes, 407, 498
Nitrofurans, 103

2-Nitrofuran, **103**
3-Nitrofuran, **104**
5-Nitrofuran-3-carbaldehyde, 115
Nitrofurancarboxylic acids, 122
5-Nitrofurfural, 114, 251
Nitroindoles, **421**, **423**, 439
3-Nitroindole, 422
5-Nitroisatin, 464
Nitromalondialdehyde, 348
o-Nitromandelic acid, 456
Nitromethane, 155, 249, 432
5-Nitro-2-methylbenzofuran, 154
2-Nitronaphtho[2,1-*b*]thiophene, 309
Nitrones, 66, 371
—, conversion to oxaziridines, 40
1-Nitro-2-(2-nitrophenyl)naphthalene, 519
3-Nitro-9-nitrosocarbazole, 496
Nitronium tetrafluoroborate, 104, 168
Nitro-olefins, 86
—, conversion to pyrroles, 332
5-Nitro-oxindole, 453
Nitroparaffins, 369
p-Nitroperbenzoic acid, 547
o-Nitrophenylacetic acids, 449, 452
o-Nitrophenylacetylenes, 443
Nitro-2-phenylbenzofurans, 154, 155
o-Nitrophenylcyclohexane, 516
o-Nitrophenylglyoxylic acid, 459
o-Nitrophenylpropiolic ester, 435
o-Nitrophenylpropionic acid, 459
o-Nitrophenylpyruvic acid, 402
2-(3′-Nitrophenyl)thienyl-3-carboxylic acid, 236
(*p*-Nitrophenylthio)acetaldehyde dimethyl acetal, 275
2-Nitropropane, 384
2-Nitropyrrole, **348**
3-Nitropyrrole, **348**
Nitropyrrolecarbaldehydes, 349
4-Nitropyrrole-2-carboxylic acid, 348
Nitroselenophenes, 318
2-Nitroselenophene, 316
3-Nitroselenophene, 316
4-Nitroselenophene-2-carboxylic acid, 316
N-Nitrosoacetanilide, 234, 251
Nitrosoalkane, dimers, 41
1-Nitrosoazetidine, 63, 64
N-Nitrosoaziridine, 497
Nitrosobenzene, 80, 81, 360, 435, 455, 492
2-Nitrosobiphenyl, 489
Nitrosocarbazoles, 496
9-Nitrosocarbazole, 497
Nitrosocarbazole-3-carboxylic acids, 507
Nitroso compounds, 63, 334
—, reaction with indoles, 410
Nitrosodimethylaniline, 119, 171
1-Nitroso-2,3-dimethylindole, 413
Nitrosoindoles, **422**
1-Nitrosoindole, 413
2-Nitrosoisoindoline, 476
N-Nitrosomethylaniline, 497
9-Nitroso-3-nitrocarbazole, 497
p-Nitrosophenol, reaction with carbazole, 500
Nitrosopyrroles, 349
1-Nitrosopyrrolidine, 382
2-Nitrostilbenes, 417, 443
o-Nitrostilbene dichloride, 443
Nitrosyl chloride, reaction with aziridine, 21
Nitrotetrahydrocarbazoles, 514
β-Nitro-2-thienylethene, 249
Nitrothiophenes, 241, 243
2-Nitrothiophene, 224, **241**, 242
3-Nitrothiophene, **242**
o-Nitrotolane, 443
1-Nitro-9-*p*-toluenesulphonylcarbazole, 497
3-Nitro-2,4,5-triphenylpyrrole, 348
2-(2′-Nitrovinyl)benzofuran, 155
2-(2-Nitrovinyl)indole, 432
Nomenclature, *Chemical Abstracts method*, *566*
—, *heterocyclic systems*, *551*
—, *seniority of ring systems*, *570*
—, *Stelzner method*, *565*
Norbornadiene, 185

Octa-arylbipyrrolyls, 391
Octachlorothiolane, 259
Octafluorodibenzofuran, 200
Octafluorodibenzoselenophene, 325
Octafluorodibenzotellurophene, 328
Octafluorodibenzothiophene, 303, 305
Octafluorodibenzothiophene 5,5-dioxide, 198
Octafluoro-1-iodophospholane, 538
Octahydrobenzocarbazoles, 519
Octahydrobenzofurans, 145, **179**
Octahydrobenzothiophenes, 287, 288

Octahydrobenzo[*c*]thiophenes, 293, 294
Octahydrobenzothiophene 1,1-dioxide, 288
Octahydro-2,2′-bifuran, 138
Octahydro-3,3′-bifuran, 138
Octahydro-3,3′-biselenienyl, 320
Octahydro-3,3′-biselenophene, 320
Octahydro-3,3′-bitellurophene, 327
Octahydrocarbazole, 491, **516**, 517
Octahydrodibenzofuran, **203**
Octahydrodibenzothiophene, 306
Octahydroindoles, 465, **466**
Octahydroisobenzofurans, 191
Octahydroisoindoles, 477
Octamethyltetraoxaquaterene, 95
Octanones, 116
Octaphenyl-1,1′-spirobisilole, 541
tert-Octylazomethine, 43
Oil, jasmine, 411
—, orange blossom, 411
—, peppermint, 177
Olefins, alkylation of furans, 129
—, conversion to aziridines, 17, 19, 36
—, from thiiranes, 9, 12
—, nitrosochlorination, 17
—, reaction with aziridine, 21
Olefinic alcohols, 128
Olivetol, 72
Opsopyrrole, 343, 345
Opsopyrrolecarboxylic acid, 363, 367
Orange blossom oil, 411
Organocadmium compounds, reaction with β-propiolactone, 50
Organolithium compounds, 10, 47, 93, 186, 381
Organomagnesium compounds, reaction with β-propiolactone, 50
Organometallic compounds, reaction with phthalic esters, 181
Orthoformic ester, 98, 171, 172, 249, 279, 442, 470
2-Oxabicyclo[3.2.1]octane, 192
3-Oxabicyclo[3.3.0]octanes, 191
Oxadiaziridine, 45
2-Oxa-1,2-dihydropyrene, 567
Oxalbis(phenylimidyl)dichloride, 211
Oxalic acid, 114, 179, 228
—, reaction with carbazole, 518
Oxalic esters, 378, 387, 484
Oxalic ester, 221, 351, 387, 388
Oxalis cernua, 173
Oxaloacetic ester, 465
Oxal-*o*-toluidine, 467
Oxalyl chloride, 211, 431
Oxamoyl chlorides, 459
1H-2-Oxapyrene, 567
2-Oxaspiranes, 47
1-Oxaspiro[4,5]decane, 569
2-Oxaspiro[3.2]hexane, 47
1-Oxaspiro[3.5]nonan-3-one, 50
2,6-Oxathiaspiro[3.3]heptane, 75
Oxathietanes, 74
1,2-Oxathiolane, 553
4H-1,3-Oxathiolo[5,4-b]pyrrole, 564
Oxazetidines, 60, 80
1,2-Oxazines, 360
o-Oxazino[4,5-b]pyridine, 563
Oxaziranes, 40
Oxaziridines, 40, 373
—, hydrolysis, 40
Oxaziridine rings, 41, 444
Oxazolidines, 23
Oxazolidinium rings, 38
Oxazolines, 24, 25, 26
Oxazolinium salts, 36
1H-Oxazolo[5,4-c]pyrazole, 563
Oxetanes, 1, 45
—, disubstituted, 45
—, substituted, 46
Oxetane, 2, **46**
Oxetanols, 5, 48, 49
Oxetanones, 2, 49, 50
Oxetan-3-one 2,4-dinitrophenylhydrazone, 50
Oxetes, 1, 52
Oxetenes, 1, 52
Oxetones, 191
4-Oxidocyclohexane, bridged system, 193
Oximes, 15
Oxime tosylates, 18
Oximinoacetoacetic ester, 465
Oximinocyclohexanone, 465
Oximino-β-diketones, conversion to pyrroles, 331
Oximino-3*H*-indoles, 422, 423, 439, 443
Oximinoketones, 251, 357
—, conversion to pyrroles, 331
2-Oximino-3-oxo-2,3-dihydrobenzofuran, 154
3-Oximino-2-phenyl-3*H*-indole, 443

3-Oximino-2-phenyl-3*H*-indole 1-oxide, 413
Oximinotetrahydrothiophenes, 243
Oxindigotin, 154
Oxindoles, 403, 412, 419, 420, 448, 450, 451
Oxindole, 397, 418, 446, 449, 450, **451**, **453**, 457, 459, 460, 461
Oxindole-3-acetic acid, **454**
Oxindolealdehyde, 434
Oxindole-3-butyric acid, 454
Oxindole-3-caproic acid, 454
Oxindole-3-propionic acid, 454
Oxindolesulphurous esters, 420
Oxiranes, 1, 3, 25, 45, 53
—, reaction with carbonyl sulphide, 6
—, — with thioamides, 5
Oxirane, **3**, *552*
—, structural data, 2
3,4-Oxiranotetrahydrofuran, 132
Oxirenes, 1, 5
α-Oxo-acids, 180
1-Oxo-10*b*-aza-octahydrobenz[*a*]acenaphthylene, 488
Oxoazetidines, 67, 80
2-Oxocyclohexaneglyoxylic ester, 179
2-Oxocyclohexylglyoxylic acid, 180
2-Oxocyclohexyl phenyl sulphide, 305
2-Oxo-3-diazo-2,3-dihydrobenzothiophene, 285
Oxodihydrobenzofurans, 146, 156
2-Oxo-2,3-dihydrobenzofuran, 155, 160, 170
3-Oxo-2,3-dihydrobenzofuran, 148, 153, 159, 169, 170, 171, 216
3-Oxo-2,3-dihydrobenzofuran oxime, 170
3-Oxo-2,3-dihydrobenzo[*b*]selenophene, 323
Oxodihydrobenzothiophenes, 276, 285
1-Oxo-1,3-dihydrobenzo[*c*]thiophene, 293
2-Oxo-2,3-dihydrobenzothiophene, **281**
3-Oxo-2,3-dihydrobenzothiophene, 267, **281**
3-Oxo-2,3-dihydrobenzothiophene-2-carbaldehyde, 285
2-Oxo-2,3-dihydrobenzothiophene-3-carbaldehyde, 285
Oxodihydrofurans, 90, 107, 108
2-Oxo-2,3-dihydrofurans, 107
2-Oxo-2,5-dihydrofurans, 108
Oxodihydroisobenzofurans, 189
1-Oxo-1,2-dihydronaphtho[2,1-*b*]furan, 210
3-Oxo-2,3-dihydronaphtho[1,2-*b*]furan, 210
3-Oxo-2,3-dihydronaphtho[2,3-*b*]furan, 209, 211
3-Oxo-2,3-dihydronaphtho[2,3-*b*]furan-2-carboxylic ester, 211
2-Oxodihydroselenophenes, 319
2-Cxodihydrothiophenes, 243, 245
2-Oxo-2,5-dihydrothiophene, 244
3-Oxo-2,2-dimethylindoline, 454
4-Oxo-1,3-dimethyltetrahydrobenzo[*c*]-thiophene, 293
4-Oxo-1,3-dioxane, 49
2-Oxo-4,5-diphenyl-2,5-dihydrofuran, 108
3-Oxo-1,5-diphenylpyrrolidine, 386
5-Oxo-1,3-diphenyl-2-pyrroline, 378
β-Oxo-esters, 86, 361, 362, 377
—, conversion to pyrroles, 332
1-Oxohexahydrobenzo[*c*]thiophene, 293
4-Oxohexahydrobenzothiophene, 288
2-Oxohexahydronaphthalene, 213
Oxo-3*H*-indoles, 443
3-Oxo-3*H*-indole *N*-oxides, 443
Oxoindolines, 424, 448
2-Oxoindoline, 446, **451**, **453**
3-Oxoindoline, 443, **454**
Oxoisoindolines, 470
3-Oxo-2-methyl-2,3-dihydrofuran, 109
1-Oxo-2-methylisoindoline, 471
1-Oxoniaanthracene chloride, 568
Oxonium salts, 13, 130
2-Oxo-octahydrobenzofurans, 179
1-Oxo-octahydrodibenzofuran, 204
3-Oxo-2-phenyl-3*H*-indole, 443, 445
γ-Oxopimelic acid, 480
3-Oxopiperidines, 380
Oxopyrrolidines, 384
2-Oxopyrrolidine, **384**, 385, 392
3-Oxopyrrolidine, 386
3-Oxopyrrolines, 377
4-Oxopyrrolines, 376, 377, 378
5-Oxopyrrolines, 375, 376, 378
5-Oxorhamnolactone, 109
Oxotetrahydrobenzothiophenes, **286**, 287
Oxotetrahydrocarbazoles, 488, 503, 514
1-Oxotetrahydrocarbazole, 513, 524
4-Oxotetrahydrocarbazole, 502

1-Oxotetrahydrocarbazole-9-acetic acid, 513
Oxotetrahydrofurans, 133
3-Oxotetrahydrofuran, 135, 137
4-Oxotetrahydro-3*H*-indole, 414
3-Oxothiolane, **259**, 261
3-Oxothiolane 1,1-dioxide, 259
2-Oxo-3-thiolene, 244
5-Oxothiolenes, 243
2-Oxo-3,5,5-trimethylpyrrolidine, 385
2-Oxo-3,3,5-triphenyl-2,3-dihydrofuran, 108
2-Oxo-3-vinylpyrrolidine, 384
Oxybiotin, 107

Paal–Knorr synthesis, 117, 350, 351, 516
—, furans, 84
—, pyrroles, 331
Penicillin, 61
Penicillium griseofulvum, 174
Pentacarbonylmanganese, 344
Pentacarbonylmanganese bromide, 344
Pentachloro-1,3-dihydroisobenzofuran, 190
Pentachloropyrrole, 337
Pentachloro-3*H*-pyrrole, **346**
Pentadiyn-3-ones, 377
Pentaerythritol chlorides, 45
Pentafluoro-1,3-dihydroisobenzofuran, 189, 190
Pentafluorothiophenolate anions, 303
Pentalenyldianions, 481
Pentamethines, 470
Pentamethyl-5*H*-benzo[*b*]carbazole, 522
2,2-Pentamethylene-1,1-tetramethyleneaziridinium perchlorate, 38
Pentamethylpyrrole, 345
Pentanediols, 130
Pentaphenylarsenole, 531
Pentaphenylazetidin-4-one, 80
Pentaphenylphosphacyclopentadiene, **531**
Pentaphenylphosphole, **531**, 532, 533
Pentaphenylphosphole oxide, **532**
Pentaphenylphosphole sulphide, 532
Pentaphenylpyrrole, 345
Pentaphenylstibiole, 531
2-Pentene-2,3,5-tricarboxylic acid 2,3-anhydride, 364
4-Pentenyldibenzofuran, 200
Pentosans, conversion to furfural, 112
Pentoses, conversion to furans, 86
—, — to furfural, 112
—, determination, 114
Pentyl alcohol, 310
Pentylsodium, 125, 134
4-Pentynol, 134
Peppermint oil, 177
Perbenzoic acid, 4, 302
Perchloryl fluoride, 273
Perfluorodibenzofuran, 198
Perfluoro-2-methyl-1-oxa-2-azacyclobutane, 81
Perfluoro-2-methyl-1,2-oxazetidine, 81
Perfluorotetrahydrofuran, 135
Perfluorothiolane, 259
Perfluorothiophene, 241
Perhydroazepines, 381
Perhydrodibenzofuran, **203**
Perhydro-1,5-dimethylazonine-2,6-dione, 382
Perilla citriodora oil, 100
Perilla ketone, **117**
Perillen, **100**
Perimidine, 558
Perkin coumarilic acid synthesis, 166
Perkin reaction, 112, 121, 254, 316
Peroxyacetic acid, 226
Peroxy acids, reaction with azomethines, 40
(1*S*)-Peroxycamphoric acid, 41
Peroxytrifluoroacetic acid, 64
Perphthalic acid, 98
Petroleum, 258, 300
Petroleum oils, 265
Pfitzinger quinoline synthesis, 462
Phalloidine, **390**, 438
Phenacylanilines, 401
Phenacyl bromides, 222
7-Phenacylindolines, 407
Phenanthrafurans, 214
Phenanthraquinone, 228
Phenanthrene, condensed rings with furan, 214
Phenanthridine, 558
Phenanthroline, 559
Phenarsazine, 559
Phenazine, 559
Phenazones, 195
Phenols, 116, 150
—, complexes with pyrroles, 346

Phenols, *(continued)*
—, dihydric, 142, 173
—, oxidation, 194
—, polyhydric, 172
—, reaction with acetonedicarboxylic acid, 166
—, — with cyclohexanones, 203
—, — with imines, 23
—, — with oxetanes, 47, 48
—, — with phenylpyruvic acid, 170
Phenol, 148, 194, 453
—, chlorination, 249
—, conversion to benzofuran, 144
—, — to dibenzofuran, 197
—, pyrolysis, 196
—, reaction with 1,2-dibromoethane, 168
—, — with *tert*-butylacetylene, 168
Phenolisatin, **454**
Phenothiazine, 490, *559*
Phenoxathiin, 555
Phenoxazine, 559
Phenoxides, 53
γ-Phenoxybutyronitrile, 477
2-Phenoxydibenzothiophene, 303
Phenylacetamide, 388
Phenylacetic acid, 100, 158
Phenylalanine, 148
3-Phenyl-1-azabicyclo[1.1.0]butane, 20, 36
1-Phenylazetidine, 62, 63
1-Phenylazetidin-2-ones, 69
Phenyl azide, 17
1-Phenylaziridine, 28
2-Phenyl-1-azirine, 34, 35, 36, 37
3-Phenyl-1-azirine, 36
2-Phenyl-1*H*-benz[*g*]indole, 483
2-Phenyl-3*H*-benz[*e*]indole, 483
1-Phenyl-2,3-benzoazetidine, 65
Phenylbenzofurans, 151
2-Phenylbenzofuran, 147, 152, 154, 158
3-Phenylbenzo[*b*]selenophene-2-carbaldehyde, 322
Phenylbenzothiophenes, 269, **273**, 298
Phenylbenzoylcarbene, 60
Phenylbenzoyleneisobenzofuran, 214
1-Phenyl-2,5-bis(*p*-tolyl)phosphole, 533
Phenyl(bromodichloromethyl)mercury, 543
3-Phenyl-2-butanone phenylhydrazone, 414
4-Phenylbut-3-enylamine, 370
1-(*N*-Phenylcarbamoyl)azetidine, 63
1-(*N*-Phenylcarbamoyl)-2,3-dimethylaziridine, 26
9-Phenylcarbazole, **492**
N-Phenylcarbazole, 490
Phenylcarbene, 26
Phenyl chloroacetate, 170
2-Phenylcyclohexanoneoxime, 509
5-Phenyldecahydrodibenzophosphole 5-oxide, 539
9-Phenyldecahydro-9-phosphafluorene 9-oxide, 539
2-Phenyl-3-diazoindole, 423
Phenyldiazomethane, 11
5-Phenyldibenzoaluminole, 549
5-Phenyldibenzobismole, 550
5-Phenyldibenzoborole, 550
Phenyldibenzofuran, 198, 200
5-Phenyldibenzophosphole, 538
5-Phenyldibenzophosphole-3-carboxylic acid, 539
5-Phenyldibenzophosphole 5-oxide, 538, 539
5-Phenyldibenzophosphole 5-selenide, 539
5-Phenyldibenzostibole, 550
Phenyldibenzothiophenes, 305
Phenyldibromophosphine, 534
Phenyldichlorophosphine, 532, 539
Phenyldichlorophosphine oxide, 531
1-Phenyl-1,3-dihydrobenzo[*c*]thiophene, 292
3-Phenyl-3,4-dihydrobenzotriazine, 65
2-Phenyl-2,3-dihydrofurans, 124
1-Phenyl-1,3-dihydroisobenzofuran, 189
2-Phenyldihydro-1,3-thiazine, 64
Phenyldi-3-indolomethene hydrochloride, 469
3-Phenyldioxindole, 409, 456, 457, 458
Phenyldipyrrolylmethane-*o*-carboxylic acid, 357
o-Phenylenediamine, 463
p-Phenylenediamine, 523
2-(β-Phenylethyl)-1-azirine, 35
2-Phenylfurans, 101
3-Phenylfurans, **99**, 100
1-Phenylgermacyclopentane, 546
N-Phenylglycines, αα-dialkyl substituted, 377
Phenylglycine, 455
Phenylglycine-*o*-carboxylic acid, 434, 454
Phenylglyoxal, 91

Phenylglyoxalic acid, 228
6-Phenylhexahydrobenzo[*g*]-2,5-diazocine, 476
2-Phenylhexahydrobenzofuran, 179
N'-Phenylhydrazine-*N*-carboxylic ester, 76
Phenylhydrazones, conversion to indoles, 400
O-Phenylhydroxylamine, 144
2-Phenyliminoisatin, 462
Phenylindoles, 410
—, physical data, 418
1-Phenylindole, **413**
2-Phenylindole, 401, **417**
3-Phenylindole, 421
Phenylindolines, **448**
2-Phenylisatogen, 445
Phenyl isocyanate, 33, 79, 447
1-Phenylisoindole, **471,** 472
2-Phenylisoindoline, 93, 476
2-Phenylisoisatogen, 445
Phenyl-lithium, 303, 471, 474, 475, 476, 478, 533
—, reaction with thietane, 55
Phenylmagnesium bromide, 109, 128, 165, 181, 251, 273, 444, 453, 461, 540
N-Phenylmaleimide, 65, 298
—, adducts with isoindole, 470
Phenylmethanesulphonyl chlorides, 58
Phenylmethanethiol, 236
Phenylmethylglycidic ester, 4
1-Phenylnaphtho[2,1-*b*]furan, 208
2-Phenylnaphtho[1,2-*b*]furan, 207
2-Phenylnaphtho[2,1-*b*]furan, 208
3-Phenylnaphtho[1,2-*b*]furan, 207
Phenylnaphtho[2,1-*b*]thiophene, 308
2-Phenylnaphtho[1,2-*b*]thiophene, 307
Phenyl-1-naphthylamine, 519
Phenyl-2-naphthylamine, 519
Phenylnaphthylhydrazine, 518
4-Phenyl-1,2-oxathietane dioxide, 74
2-Phenyloxetane, 47
O-Phenyloximes, 143
1-Phenyloxindole, 453
3-Phenyloxindole, 452
o-Phenylphenol, 196
5-Phenyl-3-phenylamino-2,5-dihydrofuran, 108
4-Phenyl-3-phenylimino-1-azabicyclo-[2.1.0]pentane, 34
9-Phenyl-9-phosphafluorene 9-oxide, 538
Phenyl phosphate, 194
1-Phenylphosphole, **533,** *534*
1-Phenyl-3-phospholene, 534
1-Phenyl-3-phospholene 1-oxide, 534
1-Phenylphosphole 1-oxide, 534
Phenylphosphonous dichloride, 531, 532
γ-Phenylpropyl alcohol, 47
2-γ-Phenylpropylfuran, 100
1-Phenylpyrrole, 340, 341, 359
2-Phenylpyrrole, 336, 345
Phenylpyrrolecarboxylic acids, physical data, 365, 366
1-Phenylpyrrolidine, **382**
2-Phenylpyrrolidine, 382
3-Phenylpyrrolidine, 382
2-Phenyl-7*a*-(*N*-pyrrolidino)perhydrobenzofuran, 179
1-Phenylpyrroline, 371, 375
2-Phenylpyrroline, 371, 375
Phenylpyruvic acid, 170
Phenylselenophenes, 319
Phenylsilane, 534, 536
Phenylsodium, 300
Phenylsuccinic acid, 237
3-Phenylsydnone, 509
Phenyltetrahydrocarbazoles, 514
1-Phenyltetrahydroisatin, 466
Phenyl 2-thienyl sulphone, 247
Phenyl 2-thienyl sulphoxide, 247
Phenylthioacetic acid, 282
Phenylthioglycollic acid, 282
Phenylthioglycollic-2-carboxylic acid, 282, 286
Phenylthiophenes, 230, 234, 237, 239
2-Phenylthiophene, 231, 232
3-Phenylthiophenes, 236
3-Phenylthiothiophenes, 236
Phenyltrichlorosilane, 544
Philodienes, 185
Phloroglucinol, 114, 216
Phosphacyclopentane, **538**
Phosphetane, four-membered, 72
Phospholane, 534, **538**
Phospholes, 531
Phospholenes, 534, **536**
Phospholene 1-oxides, physical data, 537
Phosphorus, in organo-ring compounds, 71
Phosphorus compounds, five-membered rings, 531

Photochemical alkylation, furans, 129
Photochemical decarbonylation, 93
Photochemistry, 2-alkylfurans, 98
Phthalaldehyde, 181, 188
Phthalan, 180
Phthalazine, *557*
Phthalic acid, 92
Phthalic anhydride, 180, 188, 312, 434, 506
—, reaction with carbazole, 525
—, — with pyrrole, 357
—, reduction by hydrogen sulphide, 293
Phthalic esters, 189
—, reaction with organo-metallic compounds, 181
Phthalic ester, 187, 357
Phthalide, 180, 181, 188, 189, 204
2-Phthalidylidene-2*H*-pyrroles, 344, 357
Phthalimide, 470, 474, 476
Phtalimidines, 470
Phthalimidomalonic ester, 388
Phthalocyanines, 329, 476
Phthalonitrile, 474, 476
1,2-Phthaloylanthraquinone, 525
2,3-Phthaloylbenzofuran, 215
Phthaloylcarbazoles, 525
1,2-Phthaloylcarbazole, **526**
Phthaloylcarbazole vat colouring matters, 525
Phthaloyl chloride, 188, 357
Phthalylene tetrachlorides, 188
Phthalylium salts, 187
Phthiocol, 277
Phyllopyrrole, 343, 345
Phyllopyrrolecarboxylic acid, 367
Physostigmine, 480
Picric acid, 64, 336
Piloty synthesis, 516
n-Pimelic acid, 112
Pinacols, 356
Pinacol rearrangements, 461
Pine-shaving reaction, 411, 492
Pine wood tar, 96
Pipecolic acid, 463
Piperazine, 15, 22, *561*
Piperazinium salts, 16
Piperidines, 380
Piperidine, 103, 144, 236, 512, *561*
3-(2′-Piperidinophenyl)-sydnone, 509
4-Piperidylcarbinol, 482
Piperonylic acid, 161
Piptadenia peregrina, 431
Plant Indican, 454
Plant melanins, 426
Polyalkylpyrroles, 339
Polyamines, 374
Polybenzofuran, 148
Polycyclic compounds, 185
—, germanium, 547
Polycyclic furans, 217
Polycyclic systems, with a furan ring, 207
Polycyclic thiahydrocarbons, 313
Polyenes, from furan, 140
Polyene aldehydes, 140
Polyene salts, 113
Polygonatum multiflorum, 67
Polyhalogenothiophenes, 241
Polymers, from pyrroles, 336
Polymethine colouring matters, 359, 442, 464, 465, 469
Polymethyl-11*H*-benzo[*a*]carbazoles, 522
Polymethylcarbazoles, 494
2,3-Polymethyleneindoles, **526**
Polymethylsilicone, 71
Polynuclear fused ring compounds, 538
Polyoxobipyrrolidylidenes, 393
Polyoxopyrrolidines, 387
Polyphenoloxidase, 426
Polythienyls, 264, 265
Polyvinylcarbazole, 492
Polyvinylpyrrolidone, 384
Pongamoa glabra, 162
Pongamol, **162**
Porphin, 393
Porphyrilic acid, **204**
Porphyrins, 140, 355, 393, 394
Potassio-indole, 432
Potassiopyrrole, **338,** 339, 340, 341, 359
Potassium *tert*-butoxide, 402, 409, 411
Potassium tetrafluoroborate, 320
Potassium tetrafluorocobaltate, 240
Potassium tetra(1-indolyl)boron, 428
Potassium tetra(2-selenienyl)borate, 320
Potassium thiocyanate, 6
—, reaction with oxiranes, 5
Potassium trifluorocobaltate, 240
Potassium xanthate, 265
Prodigiosin, C-25 analogue, 396
Proline, 61, 330, 371, 388, **389**
—, dimethylbetaine, 389

Propane-1,3-diols, 50
Propane-1-thiol, 57
Propan-2-ol, 34
β-Propiolactone, 49
—, reaction with organometal compounds, 50
Propiolic ester, 485
Propionaldehyde, 48
Propionic acids, 3-substituted, 50
Propionoacetic ester, 166
2-Propionylcarbazole, 506
2-Propionylnaphtho [2,1-*b*]thiophene, 309
Propiophenone, 146
n-Propylamine, 44
2-*n*-Propylbenzofuran, 151
N-*n*-Propylcyclohexylamine, 44
Propylene, 237
Propylene sulphide, 6, 11
2-Propylfuran, 97
3-*n*-Propylindole, 416
2-Propynylcyclohexanone, 287
Proteins, 330
—, acid hydrolysis, 436
—, hydrolysis, 389
—, indole derivatives, 397
Protein hydrolysates, 390
Protoanemonin, 109
1-Proto-1-pyridinium sulphonate, 337
Pseudoindole, 398
Pteridine, *558*
Pulegone, 177
*Purine**, *556*
Putrescine, 380
PVP, 384
Pyrans, 141
Pyran, *554*
2H-Pyrano [*3,2-b*]*furan*, *562*
2-Pyranylidene, *568*
Pyrazine, *555*
7H-Pyrazino [*2,3-c*]*carbazole*, *562*
Pyrazino [*2,3-d*]*pyridazine*, *563*
Pyrazoles, 160, 255
Pyrazole, *555*
Pyrazolidine, *560*
Pyrazoline, *561*
Pyrazolones, 434
1H-Pyrazolo [*4,3-d*]*oxazole*, *563*
Pyridazines, 91
Pyridazine, *556*
Pyridines, 347
Pyridine, 85, 170, 338, 354, 470, *555*
4H-Pyrido [*2,3-c*]*carbazole*, *564*
Pyrido [*1′,2′*: *1,2*]*imidazo-* [*4,5-b*]*quinoxaline*, *565*
Pyrido [*3,4-a*]*naphthalene*, *562*
Pyridones, 377, *573*
5H-Pyrido [*2,3-d*]*-o-oxazine*, *563*
Pyridoxine, 116, 126, 438
2,3′-Pyridyl-1-pyrroline, 368
Pyrimidine, *556*
Pyrocatechol phosphorobromidite, 538
Pyrocatechol phosphorochloridite, 537
Pyrocoll, 362, 434
Pyrocurzerenone, 210
Pyroglutamic acid, 389
Pyroglutamic ester, 384
Pyromucic acid, 83, 117, **119**
α-Pyrone-5-carboxylic ester, 85
Pyrotritaric acid, 118, 122
Pyrousnic acid, 205
Pyrroles, 85, 126, 131, 329
—, 3-acylated, 357
—, aldehydes, 355
—, alkylated, 339
—, *C*-alkylation, 342
—, *C*-amines, 351
—, conversion to indoles, 404
—, diazonium salts, 349
—, diene synthesis, 364
—, 2,5-disubstituted, 343
—, from aminoketones, 331
—, from diacetylenes, 333
—, from dienecarboxylic esters, 334
—, from furanamines, 334
—, from oximinoketones, 331
—, from δ-sultams, 334
—, from tetracyanoethane, 333
—, from αβ-unsaturated ketones, 334
—, ketones, 355
—, Knorr synthesis, 330
—, molecular complexes, 346
—, oxidation, 335
—, Paal–Knorr synthesis, 331
—, picrates, 436
—, picrolonates, 346
—, polymerisation, 335
—, reaction with acid anhydrides, 356
—, — with diazoacetic ester, 363
—, — with diazoketones, 356
—, — with formaldehyde, 394

Pyrroles, reaction *(continued)*
—, — with hydrogen peroxide, 376
—, — with isocyanides, 355
—, reduction, 379
—, — to pyrrolines, 374
—, side-chain carboxylic acids, 363
—, styphnates, 346
—, 1-substituted, 338
—, *N*-substituted, 331, 340
—, sulphur derivatives, 352
—, tautomerism, 348
—, tricyclic systems, 526
—, 1,2,3-trisubstituted, 333
—, unstable complexes, 346
—, *N*-unsubstituted, 371
Pyrrole, 329, **337, 338,** 392, *555*
—, carbonation, 360
—, chlorination, 346
—, *N*-derivatives, 338
—, formylation, 355
—, from ammonium compounds, 333
—, halogenation, 337, 346
—, hydrogenation, 382
—, nitration, 348
—, photo-oxidation, 376
—, potassium salt, **338**
—, protonation, 335
—, reactions, 335
—, reaction with acetylene, 339, 341
—, — with acrylonitrile, 339
—, — with amines and formaldehyde, 354
—, — with chloroformic ester, 359
—, — with dimethylformamide, 355
—, — with formaldehyde, 341, 353
—, — with phthalic anhydride, 357
—, reduction, 336
—, structure, 330
—, sulphonation, 353
—, synthesis, 330
—, thiocyanation, 352, 353
—, ultraviolet absorption, 230
2H-Pyrrole, 555
α-Pyrroleacetic esters, **363**
Pyrrole-2-aldoxime, 354
Pyrrole-black, 338
Pyrrolecarbaldehydes, 355, 395
—, physical data, 358
Pyrrole-2-carbaldehyde, 337, 349, 354, **355,** 356, 357, 360
Pyrrolecarboxylic acids, 359, 394
Pyrrolecarboxylic acids, *(continued)*
—, esters, 346, 353, 363
—, physical data, 365, 366
Pyrrole-1-carboxylic acid, **359**
Pyrrole-2-carboxylic acid, 339, **360,** 361, 363
Pyrrole-3-carboxylic acids, **361**
Pyrrole-1-carboxylic ester, 225
Pyrrole-2-carboxylic methyl ester, 347
Pyrrole-3,4-dicarboxylic ester, 360
Pyrrole homologues, characterisation, 344
Pyrrole Mannich bases, 342
Pyrrolenines, 329
Pyrrolenephthalides, 344, 357
Pyrrolephthalein, 357
Pyrrole pigments, 393
Pyrrole-polycarboxylic acids, **362**
Pyrrolepolyenecarbaldehydes, 359
Pyrrole-red, 338
Pyrrole rings, linked through methylene or methine groups, 393
Pyrrole-2-sulphonic acid, 337, **353**
Pyrrole systems, bicyclic, 477
Pyrrolethiols, 333
Pyrrole-1,3,4-tricarboxylic acid, 360
Pyrrole-2,3,5-tricarboxylic acid, 362
Pyrrolidines, 329, 369, 371, 372, 374, 379
—, dehydrogenation, 334
—, reactions, 380
—, substituted, 383
Pyrrolidine, 130, 336, 375, **382,** 466, *560*
—, conversion to pyrrole, 337
—, reaction with cyanogen, 381
Pyrrolidine acetate, 369
Pyrrolidine-1-carbaldehyde, 382
Pyrrolidinecarboxylic acids, 388
Pyrrolidine-2-carboxylic acid, **389**
Pyrrolidine-1,2-diacetic acid, 481
Pyrrolidine-1,5-disulphonic acid, 336
Pyrrolidine enamine, 212
1-Pyrrolidinocyclohexene, 179, 509
2,2′-Pyrrolidinylpyrrole, 392
2,3′-Pyrrolidinylpyrrole, 371
Pyrrolidinyl radicals, 330
Pyrrolidones, 369, 370, 384
—, reduction, 379
2-Pyrrolidone, 371, **384**
1-*N*-Pyrrolidylcyclohexene, 38
Pyrrolines, 329, 368
—, synthesis, 369

1-Pyrrolines, 368, 370, 371
2-Pyrrolines, 373, 374
3-Pyrrolines, 24, 374
—, reduction, 379
1-Pyrroline, 369, **370**, 392, *560*
3-Pyrroline, 374, **375**
1-Pyrroline 1-oxides, 39, **371**, 372, 373
3-Pyrroline-2-carboxylic acid, 375
2-Pyrrolin-4-ols, 333
Pyrrolinones, 352, 375
1-Pyrrolino[2,3-*b*]oxolane, 384
Pyrrolinylbipyrrolyl, 392
1,2-(1′-Pyrrolinyl)pyrrole, 392
2,2′-(1′-Pyrrolinyl)pyrrole, 392
Pyrrolinyl radicals, 330
Pyrrolizidine, **480**
Pyrrolizidine-1-carboxylic acids, 481
Pyrrolizidine system, 480, 481
3*H*-Pyrrolizine, 356, 481
Pyrrolo[3,2,1-*hi*]indole, 529
Pyrrolones, 352, 375
Pyrrolopyrrole system, 480
2-Pyrrolylacetamide, 354
2-Pyrrolylacetonitrile, 354
Pyrrolylalkanoic acids, 363
Pyrrolylalkanols, 353
Pyrrolylcarbinols, 353
o-(2-Pyrrolylcarbonyl)benzoic acid, 357
β-2-Pyrrolylethanols, 353
β-2-Pyrrolylethylamine, **354**
Pyrrolyl ketones, 356, 357
—, physical data, 358
Pyrrolyl-lithium, 360
Pyrrolylmagnesium bromide, 357, 360
Pyrrolylmagnesium halides, 339, 342, 353, 356, 394
Pyrrolylmagnesium iodide, 355
Pyrrolylmaleic acid, 364
Pyrrolylmethanols, 353
Pyrrolylmethylamines, 354
β-Pyrrolylpropionic acids, **363**, 364
β-*N*-Pyrrolylpropionitrile, 339
5-(2-Pyrrolyl)-2-pyrrolidinone, 338
Pyrrolyl radicals, 330
Pyrrolylsuccinic acid, 364
2-Pyrrolylthioacetic acid, 352
Pyrromethanes, 357
Pyrromethenes, 395, 396
Pyrryl ketones, 347
β-1-Pyrrylpropionitrile, 481

Pyruvic acid, 108, 117, 118, 360, 387, 399
Pyruvic ester, 221
Pyrylium salts, 89

Quadricyclane, 80
Quaterthienyl, **264**
Quinazolines, 444
Quinazoline, 557
Quinols, 167, 216
Quinolines, 416, 444
Quinoline, 124, 137, 232, 321, 428, *557*
2,4-Quinolinediyl, 568
4H-Quinolizine, 557
Quinolones, 400, 462
Quinones, 169, 214, 424
—, reaction with β-dicarbonyl compounds, 143
—, — with indoles, 407
Quinone dibenzenesulphonimide, 509
Quinoxaline, 463, *557*
Quinuclidine, 561

Radicals, nomenclature in heterocyclic compounds, 567
Red oil, 329
Reformatsky reaction, 254, 283, 322, 462
Reformatsky reagents, 68
Reimer–Tiemann reaction, 337, 355, 432
Resorcinol, 216, 349
Resorcylaldehyde 4-benzyl ether, 170
Rhodanilic acid, 389
Ring-closure, *o*-hydroxycarbonyl compounds, 142
Ring expansion reactions, aziridines, 24
Ring systems, seniority, 570
Rosenmund reaction, 112, 115, 159
Rosindoles, 468
Rotenone, 157, 176

Saccharates, conversion to pyrroles, 333
Saccharic acid, 147
Salicylaldehyde, 144, 145, 148, 150, 152, 159, 167, 443
Salicylaldehyde cyanohydrin, 170
Salicylic acid, derivatives, 145
Salvia milthiorrhizae, 214
Sandmeyer reaction, 152
Schiff bases, 17, 27, 28, 69, 106, 161, 377, 500
—, oxidation, 40

Schiff bases, *(continued)*
—, reaction with indoles, 429
Schotten–Baumann method, 341
Selenonaphthenes, 320
Selenonaphthenequinone, 322
Selenonaphtheno [2,3-*b*]selenanaphthene, 326
Selenanthrene, 324
*Selenazolo [5,4-f]benzothiazole**, *563*
Selenetane, **61**
β-(2-Selenienyl)acrylic acid, 316
2-Selenienylcarbinol, 316
Selenienyl ketones, 317
3-Selenienyl-lithium, 316, 317
2-Selenienylmagnesium iodide, 317, 320
β-2-Selenienylpropionic acid, 318
Selenochroman, 321
Selenochromene, 321
Selenochromylium perchlorates, 322
o-Selenocyanatoacetophenone, 322
Selenoflavanone, 323
Selenoindigo, 323
Selenolanes, 316
Selenoloselenophenes, **325**
Selenophenes, 313, 314, **315**, 317, **325**
—, acylation, 316
—, derivatives, 318
—, formylation, 316
—, halogenation, 316
—, mercuration, 315
—, metalloid derivatives, 320
—, nitration, 316
—, sulphonation, 316
—, with two fused rings, 325
Selenophenecarbaldehydes, **316**, 317, 318
Selenophene-2-carbaldehyde-5-sulphonic acid, 316
Selenophene-3-carbonitrile, 316
Selenophenecarboxylic acids, **317**, 318, 319
Selenophene-2,5-dicarboxylic acid, **317**
Selenophenesulphonic acids, 318
Selenophenol, 320
Selenosalicylic acid, 322
o-Semidine rearrangement, 518
Semi-trivial names, heterocyclic compounds, 554
Senecio sp., 480
Sepia ink, 426
Serotonine, **430**
Sesquiterpene derivatives, 212
Shale oil, 219, 265
Shihunine, **190**
Silabenzene, *566*
Silacyclopentadiene, **539**, 540, *566*
Silacyclopentane, **540**, 543
Silacyclopentenes, 540, 542, 543
Silafluorenes, 544
7-Silanorbornadiene, 542
Silicon, in organo-ring compounds, 71
Silicon compounds, 539
Silicon tetrachloride, 322
Silolene, *552*
Siloles, 539
Silver 2,4,6-trinitrobenzenesulphonate, 13
Skatole, 411, 412, 415, 416, 428, 435, 437, 468
Sodamide, 271, 373, 381, 402, 410, 415, 448
Sodiobenzothiophene, 271
Sodiocyanoacetic ester, 9
Sodioindole, 412, 413
Sodiomalonic ester, 179, 257
—, reaction with thiophosgene, 73
Sodio-oxaloacetic ester, 120
Sodiopyrrole, 348
Sodium *n*-butoxide, 79
Sodium isatinate, 460
Sodium pentaphenylsilacyclopentadienide, 542
Sodium tetrahydridoborate, 78, 148, 283, 354
Sodium tetraphenylpyrrole, 347
Sommelet reaction, 316
Sonn–Müller process, 112, 115
5,5′-Spirobi(dibenzosilole), 544
3,3′-Spirobi(3H-indole), *569*, *570*
Spirobis-3,4-dimethylpyrrolidinium ions, 478
Spirobisindolinium bromide, 478
Spirobisindolinium ions, 477
Spirobispyrrolidinium bromide, **478**
Spirobispyrrolidinium ions, 477
3,3′-Spirobithietane, 60
Spiro compounds, 4, *569*
Spirocyclic systems, pyrroles, 477
Spiro [cyclopentane-indenes], *nomenclature*, *572*
Spiro dioxide structure, 4
Spiro-oxadiazole, 281
Spiro [piperidine-4,9′-xanthene], *569*
Stachydrine, **389**, 390

Stachydrinol, 384
Stachys tuberifera, 389
Staphylococcus, 105
Starch, conversion to furfural, 112
Stearic acid, labelled, 252
Stelzner method, nomenclature, 565
Stephen reduction, 423
Stilbene, 12, 15, 237
Stilbene episulphone, 12
Streptomyces spp., 396
Strychnine, 430
Strychnos alkaloids, 487
Styrax spp., 161
Styrax officinalis, 162
Styrene, 74, 271, 320, 417, 508, 541
—, reaction with aziridine, 21
Styrene sulphide, 11
2-Styrylfuran, 208
2-Styrylthiophene, 237, 308
Succinaldehyde, 260, 336
Succindialdehyde, 131, 147, 332
Succinic acid, 100, 237, 364
—, conversion to thiophene, 229
—, from furans, 91
—, from tetrahydrofuran, 130
—, from thiolane, 259
—, methyl ester-chloride, 286
Succinic anhydride, 118, 223, 293, 305
Succinic ester, 145
Succinimides, 379, 387
—, reduction, 384
Succinimide, 338, 375
—, conversion to pyrrole, 337
Sucrose, conversion to methylfurfural, 114
Sugar, conversion to furfural, 112
Sulphenes, reaction with diazoalkanes, 11
Sulphene dimer, **74**
Sulphenyl halides, 14
Sulphinic esters, unsaturated cyclic, 58
Sulphobenzeneazopyrroles, 350
Sulphones, 54, 236
—, from thiiranes, 11
—, reaction with aziridine, 20
Sulphonium salts, 13
Sulphonium salt formation, 9
N-Sulphonylcarbamic ester, 81
N-Sulphonylimines, 80
Sulphoxides, from thiiranes, 11
Sulphurein, **174**
δ-Sultams, 334
δ-Sultams, *(continued)*
—, conversion to indoles, 405
β-Sultones, 74

Tagetes erecta, 264
Tanshinones, 214
Tartaric acid, 118
Taurine, 23
Tellurium tetrachloride, 327, 328
Tellurolane, **327**
Tellurophenes, 326
—, 2,5-disubstituted, 326
Tellurophene, **326**
Terpenes, synthesis, 212
Terpyrrolyl, **392**
2,2′:4′,2″-Terthienyl, 225
2,2′:5,2″-Terthienyl, 263, **264**
Tetra-acetoxymercurithiophene, 240
Tetra-acyloxysilanes, 316
Tetra-alkylpyrroles, 336
Tetra-arylpyrroles, 391
Tetra-arylthiiranes, 7
Tetrabromobutane, 340
Tetrabromofuran, 104
Tetrabromoneopentane, 60, 70
Tetrabromopyrrole, 346
Tetrabromoselenophene, 318
Tetrabromotetrahydroselenophene, 318
Tetrabromothiophenes, 239, 241, 253
Tetrachlorobenzene-2-diazo-1-oxide, 199
Tetrachlorobutane-1,4-diol, 102
Tetrachlorocyclopentadiene, 295
Tetrachlorodibenzofuran, **199**, 200
Tetrachloro-1,3-dihydrobenzo[*c*]thiophene, 293
Tetrachloro-1,3-dihydroisobenzofuran, 188
Tetrachlorodimethyldisilane, 542
Tetrachloro-4,7-endoxohexahydroisobenzofuran, 189
Tetrachloro-5-fluoro-1,3-dihydroisobenzofuran, 190
Tetrachlorofuran, 102, 189
Tetrachloro-5-iodo-1,3-dihydroisobenzofuran, 190
Tetra(chloromercuri)-thiophene, 230
Tetrachlorophthalic anhydride, 523
Tetrachloropyrrole, **346**
Tetrachloroselenophene, 318
Tetrachlorotellurophene, 327
Tetrachlorotetrahydroselenophene, 318

Tetrachlorothiolane, **258**
Tetrachlorothiophene, 238, 239, 241
Tetracyano-1,4-dithiin, 253
Tetracyanoethane, 333
Tetracyanoethene, 99, 192, 350, 444
Tetracyanofuran, 106
Tetracyanothiophene, 253
Tetracyano-1,1,1-triphenylphosphole, 534
Tetracyclic systems, 214
Tetracyclone, 198
Tetradeuterioselenophenes, 315
Tetradeuterothiophene, 230
Tetrafluorobenzofuran, 153
Tetrafluorobenzyne, 225, 303
Tetrafluoro-Δ'-1,2-diazetine, 79
Tetrafluorodibenzofuran, **198,** 200
Tetrafluoro-1,3-dihydrobenzo[*c*]thiophenes, 293
Tetrafluoro-3,4-dihydro-1,2-diazete, 79
Tetrafluoro-1,2-dithietane, 73
Tetrafluoroethene, 81, 538
Tetrafluorofuran, 103, 104
Tetrafluoro-5-iodo-1,3-dihydroisobenzofuran, 190
1,2,3,4-Tetrafluoronaphthalene, 225
Tetrafluorophenoxyacetic acid, 153
Tetrafluorothiophene, 240
Tetrafurylgermanium, 97
Tetrafuryl-lead, 97
Tetrahalogeno-*o*-benzoquinones, 403
Tetrahydrobenzaldehyde, 180
Tetrahydro-5*H*-benzo[*b*]carbazoles, 519
Tetrahydro-7*H*-benzo[*c*]carbazoles, 521
Tetrahydrobenzofurans, 176, 177, 178
Tetrahydrobenzothiophenes, **286,** 287, 293
Tetrahydrobenzothiophene-3-carboxylic methyl ester, 287
Tetrahydrocarbazoles, 487, 494, 498, **509,** 514, 515, 518
Tetrahydrocarbazole, 488, 489, 491, 510, 512, 513, 517, 526
Tetrahydro-4*aH*-carbazoles, 511
Tetrahydrocarbazole-1-acetic acid, 511
Tetrahydrocarbazolecarboxylic acids, 508
Tetrahydrocarbazole-1-malonic ester, 511
Tetrahydrocarbazole-5,8-quinone, 513
Tetrahydrocarbazolylmagnesium iodide, 513
Tetrahydrocarbazolylpropionic acids, 488
Tetrahydrocarbolinecarboxylic acids, 437
Tetrahydrocyclopentano[*c*]thiophene, 295
Tetrahydrodibenzofuran, **203**
Tetrahydrodibenzothiophene, 305
Tetrahydrodibenzothiophene 5,5-dioxide, 306
Tetrahydroeuparin, 162
Tetrahydrofurans, 91, 126, 128
—, physical properties, 127
—, polymerisation, 129
—, reaction with amines, 379
—, side-chain substituted, 133
—, substituted, 135
Tetrahydrofuran, 83, 103, 123, **130**
—, alkylation, 129
—, chlorination, 131
—, photodimerisation, 138
Tetrahydrofuran-2-carbaldehyde, 135
Tetrahydrofuran-2-carboxamide, 135
Tetrahydrofuran-2,5-dicarboxylic acid, **136**
Tetrahydrofuran-3-ones, 123
Tetrahydrofuran peroxide, 130
Tetrahydrofuran rings, bridged, 193
Tetrahydrofurfuramide, 136
Tetrahydrofurfuryl acetate, 133
Tetrahydrofurfuryl alcohol, 123, **133,** 134 136
Tetrahydrofurfurylamine, **136**
Tetrahydrofurfuryl bromide, 134
Tetrahydrofurfuryl esters, 133
Tetrahydrofuroic acid, **136**
Tetrahydrofuroic methyl ester, 123
Tetrahydrofuronitrile, 123
Tetrahydro-2-imino-1,3-thiazines, 64
Tetrahydroindoles, 405, 465
Tetrahydroisobenzofurans, 190
Tetrahydroisoindoles, 477
Tetrahydroisoquinoline, 475
Tetrahydro-6-methoxy-1,2-dimethyl-5-oxobenz[*cd*]indole, 485
Tetrahydro-9-methylcarbazole, 491
Tetrahydronaphtho[2,1-*b*]thiophene, 308
Tetrahydro-1-naphthyl hydrogen peroxide, 244
Tetrahydro-4-oxonaphtho[2,1-*b*]thiophene, 308
Tetrahydrophthalic anhydride, 191
—, reduction with hydrogen sulphide, 294
Tetrahydrophthalimide, 477
Tetrahydroporphyrins, 393
Tetrahydropyrans, 116

Tetrahydropyran-2-spirocyclohexane, 570
Tetrahydropyromucic acid, **136**
Tetrahydropyrroles, 329, 379
Tetrahydroquinoline, 62
Tetrahydroquinoline, 62
1,2,3,4-Tetrahydroquinoline-4-spiro-4-piperidine, 570
Tetrahydroselenophenes, 316, 319
Tetrahydroselenophene-2,5-dicarboxylic acids, 319
Tetrahydrotellurophene, **327**
Tetrahydrothiophene, 219, 221, 227, 257, **258**
Tetrahydrothiophthalide, 293
Tetrahydroxybiphenyl, 195
Tetrahydroxydibenzofuran, 195
Tetraiodofuran, 104
Tetraiodopyrrole, 346, 347
Tetraiodoselenophene, 318
Tetraiodothiophene, 240
β-Tetralone, 521
1-Tetralone-3-sulphonic acid, 518
Tetramethyllallene, 81
Tetramethylazetidinium perchlorates, 66
Tetramethylaziridinium perchlorate, 39
Tetramethyl-5*H*-benzo[*b*]carbazole, 522
Tetramethylbenzo[*c*]thiophene, 289
Tetramethylcarbazoles, 494, 495
Tetramethyl-6,6-dichloro-3-silabicyclo-[3.1.0]hexane, 543
Tetramethyl-2,5-dihydrofuran, 128
Tetramethyldodecahydronaphtho[2,1-*b*]-furan, 212
Tetramethylene dibromide, 319, 327, 380
Tetramethylene glycols, 128, 258
—, reaction with amines, 379
Tetramethylethene, 75
Tetramethylfuran, 98, 101
Tetramethyl-3-hydroxyoxetane-3-carboxylic acid, 50
Tetramethylindoleninium iodide, 439
Tetramethyl-3*H*-indolium bromide, 464
Tetramethyl-3*H*-indolium iodide, 439
Tetramethylisoindole, 473
Tetramethyl isoindoline, 475
Tetramethyloxetan-3-one, 50, 51
Tetramethyloxet-2-ene, 52
Tetramethyloxirane, 51
Tetramethylpyrrole, 342, 345, 346
Tetramethylpyrrolidine-3-carboxylic acid, 388
Tetramethyl-3-pyrroline, 375
Tetramethylsila-3-cyclopentene, **542**, **543**
Tetramethyltetrahydrofuran, 135
Tetramethylthiophene, 237
Tetranitro-2,2′-bithienyl, 262
Tetranitrocarbazole, 497, 498
Tetranitromethane, 493
Tetraphenylacetone, autoxidation, 51
Tetraphenylbenzodifurans, 216
Tetraphenylbenzo[*c*]thiophene, **291**
Tetraphenylbutane, 227, 235
Tetraphenylcyclopentadienone, 184, 235
Tetraphenyldiazetidin-3-one, 77
Tetraphenyldibenzofuran, 198
Tetraphenyldihydrobenzo[*b*]thiophene 1,1-dioxide, 227
Tetraphenyldihydrofuran, 124, 128
Tetraphenylfuran, **99**, 101
Tetraphenylisobenzofuran, 187
Tetraphenyloxetan-3-one, 51
Tetraphenylpyrrole, 335
Tetraphenylselenophene, **315**
Tetraphenylsilacyclopentadiene, 541
Tetraphenyltellurophene, 327
Tetraphenylthiirane, 13
Tetraphenylthiirane 1,1-dioxide, 11
Tetraphenylthiophene, 227, 235, 237, 312
Tetra-*N*-pyrrolylsilicon, 339
Tetra(2-selenienyl)boric acid, 320
Tetronic acid, 133
Thelephoric acid, **163**
2-Thenal-5-oxazolones, 249
Thenil, **249**
Thenilic acid, 249
2-Thenoic acid, **252**
3-Thenoic acid, **252**
2,2′-Thenoin, **249**
3,3′-Thenoin, **250**
β-2-Thenoylpropionic acid, 286
Thenoyltrifluoroacetone, 251
Thenyl alcohol, **247**
2-Thenylamine, **248**
2-Thenyl bromide, **248**
2-Thenyl chloride, **248**, 249, 254, 264
Thenyldimethylamine, 248
2-Thenylhydroxylamine, 248
Thenylmalonic azide, 250
2-(1-Thia-alkyl)thiophenes, 246

2-Thiabicyclo [2.2.1]heptane, 223
7-Thiabicyclo [4.1.0]heptane, 13
2-Thiabicyclo [3.1.0]-3-hexene-6-carboxylic ester, 224
2-Thiabicyclo [3.3.1]nonane, 288
Thiadiazines, nomenclature, 553
2-Thiahydrindan, 294
Thiahydrocarbons, polycyclic, 313
Thianaphthene, 265, *554*
Thianaphthenequinone, 281, 284
Thianaphthenequinone *S*-dioxide, 284
β-3-Thianaphthenylalanine, 280
Thianthrene, 301, *554*
Thianthrene 5-oxide, 301
Thianthrenetetroxide, 328
Thiazetidines, 81
1,3-Thiazole, 553
Thiazolines, 26
Thiazolo [5,4-f]benzoselenazole, 563
4H-Thiazolo [4,5-d]imidazole, 563
Thieno [3,2-*e*]benzobisbenzothiophene, 299
Thieno [3,2-*e*]benzofuran, 299
Thieno [2,3-b]furan, 562, 564
Thienothiophenes, **296,** 297, 298
Thieno [3,4-*b*]thiophene-2-carboxylic acid, 297
Thienotropylium perchlorate, 296
2-Thienylacetaldehyde, **250**
2-Thienylacetic acid, **254**
3-Thienylacetic acid, **254**
Thienylacetonitrile, 249
Thienylacetylenes, 234
β-2-Thienylacraldehyde, 250
2-Thienylacrylic acid, 254
2-Thienylalanine, 254
3-Thienylalanine, 254
ω(3-Thienyl)alkanoyl chlorides, 296
Thienylarsenic acids, 232
2-Thienylarsine oxide, 232
2-Thienylboronic acid, 240, 244
2-(2′-Thienyl)but-2-ene, 261
γ-(2-Thienyl)butyric ester, 300
2-Thienylcarbamate, 243
2-Thienylcarbinol, **247,** 248
2-Thienylcopper, **232,** 261
2-Thienylcyanoacetic acids, 254
3-(2′-Thienyl)dioxindole, 461
2-Thienylethanols, 247
β-2-Thienylethylamine, 248
2-Thienyl-fatty acids, 254
2-Thienylglycollic acid, **254**
2-Thienylglyoxal, 251
2-Thienylglyoxylic acid, **254**
Thienyl Grignard compounds, 252
β-Thienyl-β-hydroxypropionic acids, 254
2-Thienyl ketoximes, 253
2-Thienyl-lithium, **231,** 240, 244, 246, 261, 264
3-Thienyl-lithium, **231,** 251, 261
Thienylmagnesium bromide, 244, 245, 247, 252, 253, 260
2-Thienylmagnesium chloride, 239
Thienylmagnesium halides, **231,** 246, 249
Thienylmagnesium iodide, 231, 232, 247, 261, 264, 461
2-Thienylmalonic acid, 254
Thienylmercuric salts, 229
2-Thienylmethanethiol, 248
2-Thienylmethanol, **247,** 248
Thienylnitroethene, 249
2-Thienylpropiolic acid, **254**
β-Thienylpropionic acid, 249
2-Thienylpyruvic acid, **254**
2-Thienylsodium, **231,** 247, 252
3-Thienylthioacetic acid, 247
2-Thienyl thiocyanate, 247
2-Thienyl thio-ethers, 246
β-2-Thienylvinyl bromide, 237
Thietanes, 1, 52
—, complexes with mercuric chloride, 54
—, polymerisation, 54
—, reactions, 53
Thietane, **53**
—, oxidation, 55
—, reaction with acetyl chloride, 54
—, — with halogens, 55
—, structural data, 2
Thietane dioxides, 53, 54
Thietane 1,1-dioxide, 53, **56,** 57
Thietane 1-oxides, 54
Thietan-3-ol, 57, 59
Thietan-3-ol 1,1-dioxide, 60
Thietan-2-ones, 59
Thietan-3-one, 59
Thietan-3-one 1,1-dioxide, **60**
Thietes, 1, 56
Thietenes, 56
—, aliphatic, 57
Thietene, 1, 57
Thietene 1,1-dioxides, 57, 58

Thiiranes, 1, 5, 6, 7, **12,** 53
—, desulphurisation, 10
—, mercaptoethylation, 8
—, oxidation, 11
—, reactions, 8
—, physical properties, 13
—, structural data, 2
—, unsymmetrical, 9
Thiirane 1,1-dioxides, 11, 12
Thiirane 1-oxides, 11, 12, 75
Thiiranium ion intermediates, 14
Thiiranium salts, 10, 12
Thiirenes, 1, 14
Thiirene 1,1-dioxides, 14
1-Thioacylazetidines, 64
Thioamides, 60
—, reaction with oxiranes, 5
Thiobarbituric acid, 114
Thiobenzaldehyde, 235
Thiobenzilic acid, 290
(α-Thiobenzoyl)-acetophenone, 222
1-Thiobenzoylazetidines, 64
1-Thiocarbamoylazetidines, 64
Thiocarbamoylformanilide oxime, 462
Thiocarbonyl compounds, reaction with aziridine, 23
Thiocyanates, reaction with cyclic carbonates, 53
Thiocyanogen, 247
—, reaction with pyrrole, 352
3-Thiocyanoindole, 421
Thiocyanopyrroles, **352,** 353
2-Thiocyanothiophene, **247**
Thiodiglycollic ester, 220
Thiodiphenylamine, 490
Thioethers from thietane, 55
Thioformaldehyde, trimeric, 81
N(b)-Thioformyltryptamine, 431
Thiofurfural, 139
Thiofuroic acid, 119
Thioglycollic acid, 267
Thioglycollic ester, 221, 289
Thioindigo, 282, 286
Thioindigo 1,1-dioxide, 284
Thioindigoid dyes, 265, 267
Thioindigoid pigments, 285
Thioindigo Scarlet R, 433
Thioindigotin di-imine, 276
Thioindigo vat dyes, 265
Thioindoxyl, 267, 268, 271, 273, 276, 280, **281,** 282, 284, 285, 322, 460, 480
—, polycyclic analogues, 311
—, substituted, 283
Thioindoxyl-2-carbaldehyde, **285**
Thioindoxylcarboxylic acid, 271, 282, **286**
Thioindoxyl 1,1-dioxide, **282**
2-Thioisatin, 462
Thioketones, reduction, 7
Thiols, from thietane, 55
—, reaction with imines, 23
—, — with thiiranes, 9
Thiolacetic acid, 52
β-Thiolactams, 69
β-Thiolactones, 59
Thiolanes, 257
—, 3,4-disubstituted, 260
Thiolane, 230, **258**
—, oxidation, 259
Thiolane-2-carboxylic acid, **260**
Thiolane-1,1-dioxide, 258
Thiolane 1-oxide, 253, 259
Thiolan-3-one, 255
Thiolenes, 3,4-disubstituted, 256
2-Thiolene, **255,** 256
3-Thiolene, **255,** 256
Thiolene 1,1-dioxides, **255, 256**
Thiolene 1-oxides, **256**
Thiomaleic anhydride, 228
Thionaphthene, 219, 265
Thionessal, 235
1-Thioniabicyclo[2.2.1]heptane chloride, 300, *569*
Thionylarylamines, 339
Thio-oxindole, 244, 275, 276, 280, **281,** 285
Thiophanthrene, *554*
Thiophanthrenequinones, 312
Thiophenes, aromaticity, 236
—, 3,4-disubstituted, 220
—, from aliphatic hydrocarbons, 224
—, from α-bromocarbonyl compounds, 222
—, from diacetylenes, 222
—, from 1,4-dicarbonyl compounds, 220
—, from 1,4-diketones and hydrogen sulphide, 220
—, indophenin reaction, 228
—, mercury derivatives, 232
—, metal derivatives, 231
—, monocyclic, 219

Thiophenes, *(continued)*
—, phosphorus compounds, 232
—, reactions, 224
—, reaction with acid chlorides, 250
—, reduced, 219
—, systems of two fused rings, 296
—, tetra-substituted, 222
—, tri- and poly-cyclic ring systems, 306
—, with two or more unfused rings, 260
Thiophene, **220,** 221, **229,** 230, 237, 240, 243, 249, 251, 252, 259, 264, 359, *554*
—, acetylation, 115, 250
—, acylation, 251
—, addition reactions, 224
—, bromination, 238, 239
—, chlorination, 238, 257, 259
—, chloromethylation, 248
—, conversion to dibenzothiophene, 300
—, from butane, 247
—, from succinic anhydride, 223
—, halogenation, 238
—, hydrogenation, 257
—, hydrosulphurisation, 227
—, *S*-labelled, 224
—, mercuration, 233
—, metallation, 231
—, nitration, 241
—, oxidation, 228
—, — by hydrogen peroxide, 226
—, ozonisation, 228
—, polymerisation, 225
—, pyrolysis, 230, 260
—, reaction with arynes, 225
—, — with carbamoyl chloride, 253
—, — with formaldehyde, 248
—, — with hypochlorous acid, 257
—, — with isobutylene, 233
—, — with phosgene, 252
—, reduction, 227
—, — with sodium, 225
—, structure, 229
—, sulphonation, 245
—, ultraviolet absorption, 230
2-[^{13}C]Thiophene, 224
3-[^{13}C]Thiophene, 224
[^{35}S]-Thiophene, 230
Thiophene alcohols, 247
Thiophene-3-aldoxime, 253
Thiophenecarbaldehydes, 249, 254
—, thiosemicarbazones, 250
Thiophene-2-carbaldehyde, 224, 242, 247, 248, **249,** 254, 264
Thiophene-3-carbaldehyde, 242, **250**
Thiophene-2-carbaldimine, 249
Thiophene-2-carbaldoxime, 260
Thiophene-2-carboxamides, 253
Thiophenecarboxylic acids, 235, 252
—, methyl esters, 254
Thiophene-2-carboxylic acid, 223, 231, 234, 235, 243, **252,** 260
Thiophene-3-carboxylic acid, 235, 251, **252**
Thiophene-2-[^{14}C]carboxylic acid, 252
Thiophene-2-carboxylic ester, 264
Thiophene-2-carboxylic methyl ester, 297
Thiophenedicarboxylic acids, **253**
Thiophene-2,5-dicarboxylic acid, 232
Thiophene 1,1-dioxides, 226, 235
Thiophene 2,2′-disulphide, **246**
Thiophene-2,4-disulphonic acid, 246
Thiophene 3,4-disulphonic anhydride, 245
Thiophene-3,4-dithiol, **247**
Thiophene 1-oxide, 227
Thiophene sesquioxide, 226
Thiophene-2-sulphinic acid, **246**
Thiophene sulphonic acids, 245
Thiophene-3-sulphonic acid, 246, 247
Thiophenesulphonyl chlorides, 242, **246**
Thiophenetetracarboxylic acid, **253**
Thiophene-2-thiol, **246**
Thiophene-3-thiol, **247**, 296
Thiophenetricarboxylic acids, **253**
Thiophenoindoles, 480
Thiophenols, 270, 272
Thiophosgene, 223
—, reaction with sodiomalonic ester, 73
1-Thiophthalide, 293
2-Thiophthalide, 293
Thiophthenes, 296
Thiophyne, 225, 230
Thiopyrylium iodide, 249
Thiosalicylic acid, 247, 268, 433
Thiosalicylic methyl ester, 269
Thiotetronic acid, 245, 257
2-Thiothiazolidones, 24
Thiourea, 248
—, reaction with dihalogenoalkanes, 52
—, — with oxiranes, 5, 6
Thiouronium salts, 6, 52
Thioxanthone, 146

Three-membered rings, containing one nitrogen atom, 15
—, — one oxygen atom, 3
—, — one sulphur atom, 5
—, — silicon and germanium, 71
—, with more than one hetero-atom, 40
Thymoquinone, 150
Tin, heteroatoms in compounds, 549
Toad poisons, 397
Tolane, 273
Toluene, 453
Toluenesulphonamide, 62, 380
p-Toluenesulphonic acid, 64, 213, 244
1-*p*-Toluenesulphonylazetidine, 63
1-*p*-Toluenesulphonylaziridine, 28
9-*p*-Toluenesulphonylcarbazole, 497
p-Toluenesulphonyl chloride, 5
N-*p*-Toluenesulphonyldihydroisoindole, 475
Toluene-*p*-sulphonylglycine, 361
5-(*p*-Toluenesulphonylimino)dibenzoselenophene, 325
1-Toluenesulphonylpyrrolidine, 380
o-Toluic acid, 204
p-Toluidine, reaction with formaldehyde, 78
o-Tolylmagnesium bromide, 164
1-*p*-Tolylsulphonylaziridine, 16
1-*p*-Tolylsulphonylpyrrole, 340
3-Tosyloxyoxetane, 5
Toxol, **169**
Tremetone, **169**
Trialkyl borates, 240
Trialkylpyrroles, 336, 344
Triarylmethane-*o*-carboxylic acids, 182
2,7,9-Triazaphenanthrene, 567
1,2,4-Triazine, 553
1,3,5-Triazine, 552
1,2,3-Triazolines, 17, 18
1,1,1-Tribromo-3,4-dimethyl-3-phospholene, 535, 536
1,1,1-Tribromo-3-methyl-3-phospholene, 535
1,1,1-Tribromo-3-phospholene, 535, 536
Tribromoselenophene, 316, 318, 319
2,3,5-Tribromothiophene, 239, 240
3,4,5-Tribromothiophene-2-carboxylic acid, 253
Trichloroacetaldehyde, conversion to isatin, 459
Trichloroacetonitrile, 444
Trichloro-1,3-dihydroisobenzofuran, 190
1,1,1-Trichloro-3,4-dimethyl-3-phospholene, 536
Trichlorofuran, 102, 104
Trichloro-7-hydroxy-1,3-dihydroisobenzofuran, 189
1,1,1-Trichloro-3-phospholenes, 535, 536
Trichloropyrrole, **346**
Trichlorosilane, 541
Trichlorotetrahydrofurans, **131**, 135
Trichlorothiophenes, 238
Trichlorotrimethyldisilane, 542
Tricyclic furans, 217
Tricyclic pyrrole systems, 526
Tricyclic systems, with a furan ring, 207
Tricycloallylidenephosphorane, 533
Tricyclohexyloxyaluminium, 310
Tricyclo-oxepane, 94
Tricycloundecanes, nomenclature, 572
Triethoxonium tetrafluoroborate, 63
Triethylamine, 14, 54, 69, 81
Triethyl borate, 240
Triethyloxonium tetrafluoroborate, 452
Triethylphosphine, 79
Triethyl phosphite, 417, 489, 516
Trifluoroacetic acid, 154
Trifluorodibenzofuran, 200
Trifluoronitrosomethane, 81
Trifluorotrimethyldisilane, 542
1,1,1-Trihalogeno-1-phosphacyclopentenes, 534
1,1,1-Trihalogenophospholenes, 534
Trihydroxybiphenyl, 195
1,2,4-Trihydroxybutane, 99
2,4,5-Trihydroxy-α-chloroacetophenone, 172
Trihydroxyfuranoeremophilane, 213
2,4,6-Trihydroxymethylacetophenone, 205
3,5,6-Trihydroxy-1-methylindole, 425
2,4,5-Trihydroxyphenylalanine, 426
2,4,5-Trihydroxyphenylethylamine, 426
Trihydroxytrimethyloctahydronaphtho[2,3-*b*]furan, 213
Triindolylmethanes, 468
Triindolylphosphines, 427
Triindolylphosphine oxide, 427
Triiodoselenophene, 318
1,2,4-Triketones, 109
Trimethines, 469

1,3,5-Trimethoxybenzene, 173
5,7,4′-Trimethoxyisoflavone, 143
2,4,5-Trimethoxyphenylacetic acid, 158
Trimethylamine, 74, 188
Trimethylammonium iodide, 20, 34
Trimethylazetidines, 63
Trimethylbenzocarbazoles, 524
Trimethylbenzofurans, 151
Trimethylbenzotrifurantricarboxylic ester, 216
Trimethylcarbazoles, 495, 522
1,3,4-Trimethylcarbazole, 494, 504
2,2,6-Trimethyl-2,3-dihydrobenzofuran,168
2,4,6-Trimethyl-2,3-dihydrofuran, 169
1,5,8-Trimethyl-6,7-dihydronaphtho-[2,1-*b*]furan, 210
Trimethylene bromide, 61, 388
3,4-Trimethyleneindoles, 484
Trimethyleneimines, 1, 61
Trimethylene oxides, 1, 45
Trimethylene selenide, **61**
Trimethylene sulphides, 52
Trimethylene sulphide, 1
Trimethylfurans, 101
2,5,5-Trimethyl-2,3-hexadiene, 4
Trimethylindoles, physical data, 418
1,2,3-Trimethylindole, 418, 439, 448
2,3-Trimethyl-3*H*-indole, 442
1,3,3-Trimethyl-2-indolinylidene-acetaldehyde, 464
1,3,3-Trimethyl-3*H*-indolium salts, 448
1,3,3-Trimethylisobenzofurylium perchlorate, 187
Trimethylmethylene-3*H*-indole, 442
Trimethylmethyleneindoline, 447, 465, 469, 470
1,5,8-Trimethylnaphtho[2,1-*b*]furan, 208
2,3,9-Trimethylnaphtho[2,3-*b*]furan, 208
2,2,4-Trimethyl-6-oxa-1-azabicyclo[3.1.0]-hexane, 385
2,2,5-Trimethyl-6-oxa-1-azabicyclo[3.1.0]-hexane, 385
3,3,5-Trimethyl-6-oxa-1-azabicyclo[3.1.0]-hexane, 385
1,3,3-Trimethyloxindole, 453
2,2,5-Trimethyl-3-oxo-2,3-dihydrobenzofuran, 172
Trimethyloxonium tetrafluoroborate, 225
Trimethyloxonium 2,4,6-trinitrobenzenesulphonate, 13
2,4,4-Trimethyl-2-pentene, 72
Trimethylpyrroles, 345
2,3,5-Trimethylpyrrole, 342
1,2,2-Trimethylpyrrolidine, 370, 374, 382
1,3,4-Trimethylpyrrolidine, 374
1,3,4-Trimethylpyrroline, 374
2,4,4-Trimethyl-1-pyrroline, 385
2,4,4-Trimethyl-1-pyrroline 1-oxide, 371
4,5,5-Trimethyl-1-pyrroline 1-oxide, 66, 371
1,1,4-Trimethylsila-3-cyclopentene, 542
6,7,8-Trimethyltetrahydrocarbazole, 487
Trimethyltetrahydronaphtho[2,3-*b*]furan, 211
Trimethylthiophenes, 237
2,4,6-Trinitrobenzene diazonium salts, 360
3,5,3′-Trinitro-2,2′-dithienyl, 262
3,5,5′-Trinitro-2,2-dithienyl, 262
Trioxindole, 456
2,4,5-Trioxo-3-phenylpyrrolidine, 388
Triphenoxyphosphine dibromide, 130
Triphenylaluminium, 549
Triphenylamine, 490
1,2,3-Triphenylaziridine, 26
1,2,3-Triphenylbenzaluminole, 549
Triphenyl-4-benzoylthietan-3-one 1,1-dioxide, 60
Triphenylchlorosilane, 504
Triphenylcyclopropenium bromide, 234
Triphenyldiazetidin-3-one, 77
Triphenyldiazetidin-3-one-1-carboxylic ester, 76
Triphenyldibenzofuran, 198
1,1,3-Triphenyl-1,3-dihydrobenzo[*c*]-thiophene 2,2-dioxide, 292
Triphenyl-1,3-dihydroisobenzofuran, 190
1,2,3-Triphenylisoindole, **474**
1,3,3-Triphenyl-3*H*-isoindole, 474
Triphenylmethyl perchlorate, 264
2-Triphenylmethylpyrrole, 345
Triphenylmethyl radicals, 336
1,2,3-Triphenyloctahydroisoindole, 26
Triphenyloxazetidin-3-ones, 80
1,2,5-Triphenylphosphacyclopentane 1-oxide, 532
Triphenylphosphate, 130
Triphenylphosphine, 12, 489, 533, 534
Triethyl phosphite, 446
1,2,5-Triphenylphosphole, **532,** 533
Triphenylphosphole oxide, 532

Triphenylpyrrole, 345, 348
1,2,5-Triphenylsilacyclopentadiene, 541
1,1,2-Triphenylsiletane, 71
2,3,4-Triphenylthiophene 234, 237
Triphenylvinylphosphonium bromide, 356
Triptycene, 310
Tripyrrole, **335**, 336
Tripyrrole hydrochloride, 338
Tripyrroline, 370
Tris-3-carbazolylmethane, 518
Tris-3-carbazolylmethanol, 518
Tris-3-carbazolylmethylformate, 518
2,4,6-Trithia-3a,7a-diazaindene, 566
2,4,6-Trithia-3a,7a-diazaperhydroindene, 566
Tri-2-thienylarsine, 232
Tri-2-thienylmethane, 264
Tri-2-thienylmethanol, 264
Tritium, 230
Trivial names, heterocyclic compounds, 554
o-Tropylanilines, 527
Tsung Huan-tsao, 190
Tubaic acid, **176**
Turicine, **390**
Tryptamine, **429,** 430, 437
Tryptamine 2-disulphides, 449
Tryptophan, 397, 415, 428, 429, 432, 435, **436,** 437, 438, 439, 440
Tryptophan antagonist, 280
Tryptophol, **428,** 437
Tyramine, 426
Tyrosinase, 426, 427
Tyrosine, 426

Ullmann reaction, 136
Undecaheterohelicene, 299
Undecylprodigiosin, 396
Ureas, 27, 63
Urine indican, 455
Uronic acids, determination, 114
Usnea barbata, 205
Usneol, 205
Usnetic acid, 205, 206
Usnetol, 205
Usnic acid, 157, 194, **205,** 206
Uvic acid, 118

Valeric acid, labelled, 252
γ-Valerolactone, 192
2-Valerylcarbazole, 506
Vat dyes, 265, 397
Vat Red I, 267
Vinhaticoic acid, 214
Vinylacetylenic α-epoxides, 314
Vinylacetylenic oxides, reaction with amines, 353
Vinyl azides, 34, 35
2-Vinylaziridines, 24
2-Vinylbenzofuran, 151
3-Vinylbenzo[*b*]thiophene, 313
9-Vinylcarbazole, **492,** 493
1-Vinylcycloalkenes, 294
4-Vinyldibenzofuran, 200, 203
Vinyl ethers, reaction with carbonyl compounds, 48
2-Vinylfuran, 93, 101, 165
Vinylhydrazines, conversion to indoles, 400
1-Vinylindole, 412, 413
3-Vinylindole, 498
1-Vinylnaphthalene, 272
o-Vinylphenols, 168
1-Vinylpyridinium salts, 444
1-Vinylpyrrole, 341
2-Vinylpyrrole, 345
N-Vinylpyrrole, 339
1-Vinyl-2-pyrrolidone, 384
9-Vinyltetrahydrocarbazole, 514
Vinylthiophenes, 237
2-Vinylthiophene, 235, 240, 248, 269, 286, 308
3-Vinylthiophene dibromide, 237
Vinyltriphenylphosphonium bromide, 125, 126, 179
Vitamin B_{12}, 329
Vitamin E, 169

Walden inversion, 27
Willgerodt–Kindler reaction, 166
Willgerodt reaction, 254
Wilsmeier–Haack reaction, 414
Wittig reaction, 159, 291
Wolff–Kishner reduction, 233, 262, 281, 431, 494, 520, 523
Wolf rearrangement, 60, 74
Wood tar, 97
Wurtz–Fittig synthesis, 233

Xanthene, 555*
Xanthone, 146

Xanthurenic acid, 437, 438
Xanthydrol, 411
o-Xylene, 116
—, oxidation, 188
o-Xylylene bromide, 475
o-Xylylenediamine, 474
o-Xylylene dibromide, 188
o-Xylylene glycol, 188
o-Xylylene oxide, 180

Yeast, 162

Zerewitinoff reactions, 376